MECHANICS of MATERIALS

MECHANICS of MATERIALS

Daryl L. Logan

University of Wisconsin—Platteville

HarperCollins*CollegePublishers*

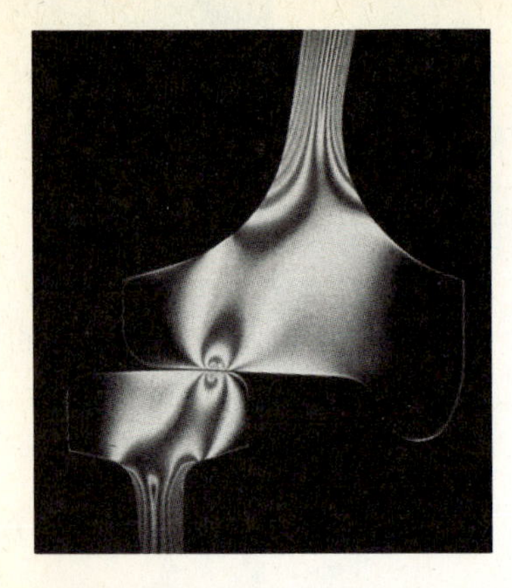

About the cover: The image depicted on the cover is of a photoelastic fringe pattern of the contact area between a railroad rail and wheel. The colorful patterns are directly related to the stresses and strains on the loaded objects and are observable when two-dimensional plane stress models, fabricated from photoelastic material, are subjected to simulated scale loads and illuminated by a transmission polariscope. (Cover image courtesy of the Photolastic Division of Measurements Group, Inc., Raleigh, NC, USA.)

To my family, Diane, Katherine, Daryl Jr. and Paul

Sponsoring Editor: Don Childress
Project Editor: Kristin Syverson/Elizabeth Fresen
Art Direction: Julie Anderson
Text Design: Lucy Lesiak Design
Cover Design: Matthew J. Doherty
Photo Research: Nina Page
Production: Beth Maglione
Compositor: Ruttle Graphics
Printer and Binder: R.R. Donnelley & Sons Company
Cover Printer: Lehigh Press Lithographers

Mechanics of Materials

Library of Congress Cataloging-in-Publication Data

Logan, Daryl L.
Mechanics of materials / Daryl L. Logan
p. cm.
Includes bibliographical references and index.
ISBN 0-06-044108-9
1. Strength of materials. I. Title.
TA405.L57 1991
620.1'12—dc20 90-20828
CIP

94 9 8 7 6 5 4 3 2

Contents

Preface xi

1 INTERNAL FORCES 1

1.1 Introduction 1
1.2 Systems of Units and Conversions 2
1.3 Internal Forces 6
1.4 General Steps in Problem Solving 21
1.5 Summary of Important Definitions and Equations 22

Problems 22

2 STRESSES, STRAINS, AND DEFORMATIONS IN AXIALLY LOADED MEMBERS 28

2.1 Introduction 28
2.2 Stresses Due to Axial Loading 30
2.3 Stresses on an Inclined Plane Under Axial Loading 37
2.4 Strain Due to Axial Loading 42
2.5 Stress-Strain Diagram and Mechanical Properties of Materials 44
2.6 Deflections Due to Axial Loading 61
2.7 Statically Indeterminate Problems 70
2.8 Temperature Effects 83
2.9 Allowable Stresses and Factor of Safety 91
2.10 Stress Concentrations in Axial Loading 96
2.11 Summary of Important Definitions and Equations 102

References 104
Problems 105
Computer Problems 122

3 DIRECT SHEAR AND BEARING STRESSES 124

3.1 Introduction 124
3.2 Direct Shear Stress and Analysis of Welds 127
3.3 Shear Strain and Deformation Due to Direct Shear Forces 143
3.4 Bearing Stress 146
3.5 Summary of Important Definitions and Equations 153

References 153
Problems 154
Computer Problems 163

4 TORSION 165

4.1 Introduction 165
4.2 Angle of Twist—Shear Strain of Circular Shaft 165
4.3 Shear Stress Due to Torsion in Circular Shaft 168
4.4 Other Stresses Due to Applying Pure Torque 174
4.5 Angle of Twist—Torque Relationship in Circular Cross-Sections 176
4.6 Statically Indeterminate Shafts 183
4.7 Power Transmission 188
4.8 Nonlinear Stress Distributions 190
4.9 Torsion in Noncircular Cross-Sections 194
4.10 Torsion of Thin-Walled Closed Tubes 200
4.11 Stress Concentrations in Torsional Loading 206
4.12 Summary of Important Definitions and Equations 209

References 210
Problems 211
Computer Problems 219

5 BEAM STRESSES 221

5.1 Introduction 221
5.2 Basic Beam Equations 223
5.3 Shear Force and Bending Moment Diagrams 226
5.4 Beam Bending Stress 241
5.5 Beam Shear Stress 256
5.6 Design of Prismatic Beams for Strength 273
5.7 Moving Loads on Beams 279
5.8 Composite Beams (Beams of More than One Material) 286
5.9 Stress Concentrations in Bending Members 292
5.10 Unsymmetric Bending 296
5.11 Principal Moments of Inertia 304
5.12 Shear Flow in Beams with Thin Sections and Shear Center 311
5.13 Summary of Important Definitions and Equations 324

References 326
Problems 326
Computer Problems 344

6 BEAM DEFLECTIONS 346

6.1 Introduction 346
6.2 Derivation of Beam Deflection Equation 348
6.3 Integration Method for Deflection 352
6.4 Beam Deflections by Superposition 365
6.5 Moment-Area Method for Deflection 371
6.6 Use of Discontinuity Functions 385
6.7 Statically Indeterminate Beams 404
6.8 Deflections in Unsymmetric Bending Members 412
6.9 Summary of Important Definitions and Equations 415

References 417
Problems 417
Computer Problems 424

7 STRESSES UNDER COMBINED LOADINGS, PRESSURE VESSELS 426

7.1 Introduction 426
7.2 Superposition of Normal Stresses 427
7.3 Superposition of Shear Stresses 434
7.4 Stresses in Thin-Walled Pressure Vessels 444
7.5 Code Design of Pressure Vessels 454
7.6 Summary of Important Definitions and Equations 457

References 458
Problems 458
Computer Problems 469

8 ANALYSIS OF PLANE STRESS AND STRAIN 471

8.1 Introduction 471
8.2 Plane Stress and Equations for the Transformation of Plane Stress 472
8.3 Principal Stresses 478
8.4 Maximum Shear Stress 484
8.5 Mohr's Circle for Plane Stress Analysis 492
8.6 Absolute Maximum Shear Stress 505
8.7 Failure Theories 513
8.8 Plane Strain, the Equations for the Transformation of Plane Strain, and Mohr's Circle for Plane Strain 521
8.9 Strain Measurements Using Rosettes 535
8.10 Plane Stress-Strain Equations 541
8.11 Three-Dimensional Stress-Strain Equations and Dilatation 548
8.12 Summary of Important Definitions and Equations 550

References 551
Problems 552
Computer Problems 559

9 COLUMN BUCKLING 561

9.1 Introduction 561
9.2 Euler's Column Formula for Pinned Ends 563
9.3 Euler's Column Formula for Other End Conditions 568
9.4 Inelastic Column Behavior/Tangent Modulus 575
9.5 Inelastic Behavior/Johnson's Formula 578
9.6 Code Design Formulas for Columns Under Centric Load 584
9.7 Summary of Important Definitions and Equations 591

References 591
Problems 592
Computer Problems 598

10 ENERGY METHODS 600

10.1 Introduction 600
10.2 External Work and Strain Energy 601
10.3 Elastic Strain Energy Due to Axial Force 607
10.4 Elastic Strain Energy Due to Shear Stress 608
10.5 Elastic Strain Energy Due to Torsion 610
10.6 Elastic Strain Energy Due to Bending 611
10.7 Elastic Strain Energy Due to Transverse Shear Stress in Beam 613
10.8 Strain Energy For a General State of Stress 615
10.9 Principle of Work-Energy 618
10.10 Virtual Work Method For Deflections 621
10.11 Method of Virtual Work For Trusses 623
10.12 Method of Virtual Work For Beams and Frames 628
10.13 Castigliano's Theorem 635
10.14 Analysis of Statically Indeterminate Structures Using Castigliano's Theorem 654
10.15 Impact Problems Using Energy Methods 662
10.16 Summary of Important Definitions and Equations 675

References 677
Problems 677
Computer Problems 691

11 FATIGUE 693

11.1 Introduction 693

11.2 The S-N Diagram 694

11.3 High Cycle Fatigue Under Completely Reversed Loading 697

11.4 Factors Affecting Endurance Limit of Members Subjected to Completely Reversed Loading 699

11.5 Influence of Stress Concentration on Fatigue Under Completely Reversed Fatigue Loading 703

11.6 Fatigue Strength Under Fluctuating Stresses 706

11.7 Summary of Important Definitions and Equations 710

References 711

Problems 711

Computer Problems 714

Appendix A Properties of Plane Areas 715

Appendix B Typical Properties of Selected Engineering Materials 719

Appendix C Properties of Structural Steel Shapes, Aluminum, and Timber 722

Appendix D Beam Diagrams and Formulas 751

Appendix E Numbering Systems for Some Metals 763

Index 767

Answers to Selected Problems 778

Preface

The purpose of this book is to provide a clear, logical approach to mechanics of materials (or strength of materials, as it has traditionally been called) that can be understood by sophomore- and junior-level college students who have had a basic course in statics and in calculus. The book is written primarily as a basic learning tool for undergraduate students who are interested in the theoretical aspects of mechanics of materials and want to apply mechanics of materials to solve practical physical problems.

General theoretical principles are presented for each topic, followed by applications of these principles. The book proceeds from basic to advanced topics and is suitable for use in a two-course sequence.

The book begins with a review of systems of units and a review of the important concept of determining internal reactions, which is essential to successful study of mechanics of materials. In subsequent chapters the component element method of teaching is used to provide a logical sequence for introducing concepts. For instance, the basic concepts of stress, strain, and deformation for the axially loaded member are introduced. Next stress, strain, and deformation for members loaded in direct shear and bearing are treated. Then the stress, strain, and deformation in torsionally loaded members are considered. This is followed by bending and shear stresses in beams and immediately by beam deflections. Finally, the individual kinds of loadings are combined, and stresses under combined loading are considered. This combined loading problem is followed by the development of stress transformation equations leading to principal stresses, maximum shear stress, and the concept of Mohr's circle to obtain the principal stresses. Next, column buckling is considered. An extensive chapter on energy methods in structural mechanics, including the analysis of structures subjected to impact loading, is included. Finally, the topic of fatigue loading is treated.

Special or advanced topics are included throughout the text in the chapters where they logically belong. For instance, the topic of stress concentrations appears in each of the chapters on axial loading, torsional loading, and bending. Torsion of noncircular cross sections appears in Chapter 4. Unsymmetric bending, composite beams, moving loads, and shear flow and shear center occur in Chapter 5. The moment area method and discontinuity functions for beam deflections occur in Chapter 6. Failure theories, used to predict failure in both ductile and brittle materials, are presented in Chapter 8. Nonlinear behavior in torsional members occurs in Chapter 4 and in buckling problems in Chapter 9. Other topics often included in an advanced course are energy methods and impact loading (Chapter 10) and fatigue loading (Chapter 11).

Appendices cover the following topics: (1) properties of plane areas; (2) typical properties of engineering materials; (3) properties of structural steel shapes, aluminum shapes, and timber sizes; (4) an extensive table of beam diagrams and formulas for beam slopes and deflections; and (5) a description of the numbering system used to identify many metals. Answers to numerous problems are provided in the back of the book. The book also includes "steps in solution" sections to guide students in formulating proper procedures for solving problems, and summary sections at the end of each chapter to provide a central quick reference of important concepts and equations.

More than 200 solved example problems appear throughout the text. These examples illustrate the theoretical concepts. Nearly 1000 end-of-chapter problems in an equal mix of S.I. and U.S. customary units are provided to reinforce concepts. Computer problems are included at the end of each chapter to promote the use of computers and programming in mechanics of materials.

Following is an outline of suggested topics for a first course (approximately 40 lectures of 50 minutes each) using this textbook.

Topic	*Number of Lectures*
Chapter 1	2
Chapter 2	8
Chapter 3	3
Exam 1	1
Chapter 4, Sections 4.1–4.7	5
Chapter 5, Sections 5.1–5.6	5
Exam 2	1
Chapter 6, Sections 6.1–6.4	2
Chapter 7, Sections 7.1–7.4	3
Chapter 8, Sections 8.1–8.5	5
Exam 3	1
Chapter 9, Sections 9.1–9.5	4
Final Exam	

I use this outline in a one-quarter course. There are enough additional topics for the book to be used in a second-quarter course in advanced mechanics of materials.

I express my deepest appreciation to the staff at HarperCollins Publishers, especially J. Donald Childress, Kristin Syverson, Julie Anderson, Janet Tilden, and Elizabeth Fresen for their assistance in producing the book. I appreciate the initial reviews of the manuscript by Edward C. Ting, Purdue University; Terry L. Kohutek, Texas A & M University; Kyung Kim, Brown University; Blaine I. Leidy, University of Pittsburgh; Tim Kennedy, Oregon State University; James Wang, Georgia Tech University; and Philip Perdikaris, Case Western Reserve University. Many of their recommendations were included in the text.

I thank the many students who used the notes that developed into this text. Special thanks to Chili Bao, Quishi Chao, Joachim Vedder, Dennis Wagner, and Jiechi Xu for proofreading, checking, and solving problems in the text. Thanks also to Paula Duggins and my wife Diane for skillful typing of portions of the manuscript and to Steve White for helping me with the photos used in the book.

Finally, a very special thank you to my family, Diane, Kathy, Daryl Jr., and Paul, for their many sacrifices during the preparation of this book.

Daryl L. Logan

MECHANICS of MATERIALS

1 Internal Forces

1.1 INTRODUCTION

Mechanics of materials, or *strength of materials* as it traditionally was called in the past, deals with the determination of the relationship between the loads and resulting internal forces and deformations of solid bodies. Solid bodies include basic members, such as axially loaded rods, shafts in torsion, beams, columns, and pressure vessels, as well as machines and other load-resisting structures composed of the basic members. Specifically, mechanics of materials refers to the formulation and use of analytical methods for determining (1) *strength,* (2) *stiffness* (deflections or deformations), and (3) *stability* (or buckling) of load-resisting members making up machines and structures.

For instance, the analysis and design of a simple jib crane (Figure 1.1) requires the determination of the size and shape of (1) an axially loaded member, (2) a column, (3) a beam, and (4) the sizing of the connectors, such as brackets, bolts, and welds, to resist a lifted load. The propeller shaft of a ship (Figure 1.2) or a shaft in a gearbox must be of sufficient size to resist the required torque and stiff enough so as not to twist excessively. A box-shaped beam (Figure 1.3) must be of proper size to resist the heavy loads transmitted to it by a trolley car moving across it and stiff enough not to deflect excessively. The many beams and columns of a building (Figure 1.4) must be of proper size to resist the various loadings transmitted to them such as from wind, snow, or earthquake. The walls of a pressurized vessel or tank must be of proper thickness to contain the internal pressure (Figure 1.5).

The purpose of this chapter is to review some of the basic concepts from statics that are necessary prerequisites for solving problems in mechanics of materials.

We first review the two common systems of units used in solving problems in mechanics of materials. Conversions between the two systems of units is also included. We then describe how to obtain the internal forces (resultants) needed to resist the effects of the externally applied forces. These resultants are necessary to analyze any load resisting members making up machines and structures. Finally, we present a brief summary of the steps frequently used to solve problems in mechanics of materials.

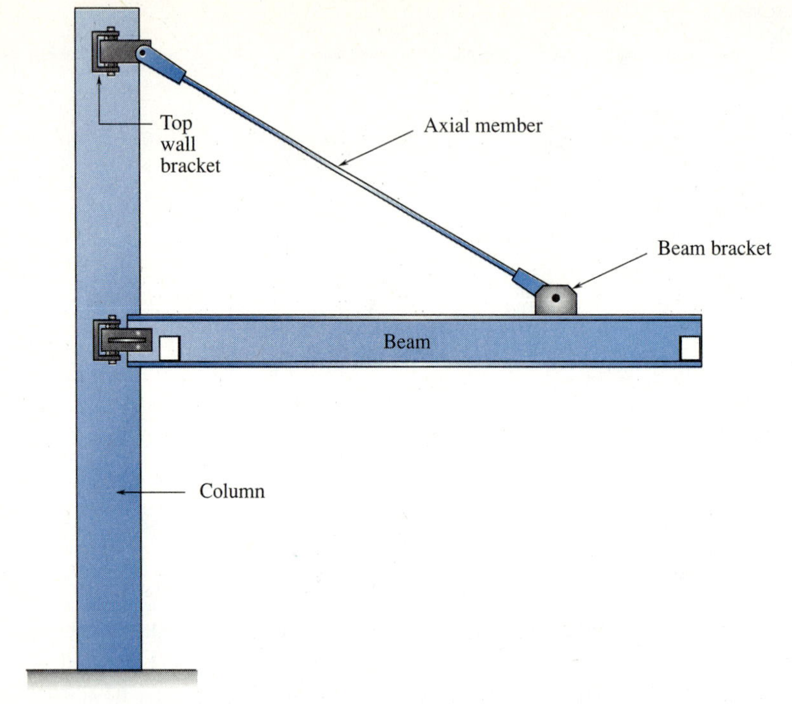

Figure 1.1 Jib crane (a crane with brackets that allow beam to rotate through approximately 180°).

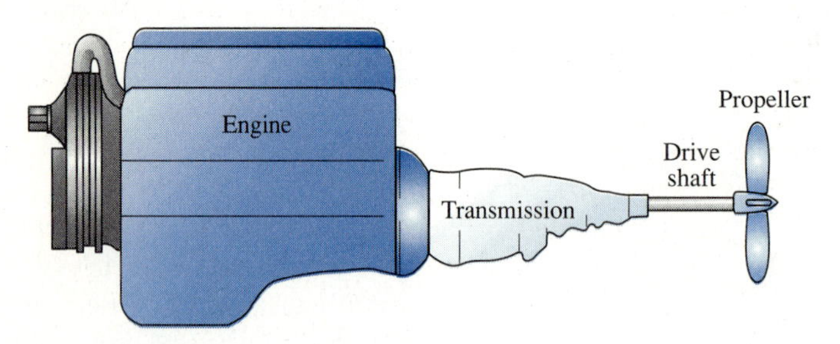

Figure 1.2 Drive train of inboard motor.

1.2 SYSTEMS OF UNITS AND CONVERSIONS

There are two systems of units in common use in engineering. These are the International System of Units (Système International d'Unités), referred to as the SI system, and the U.S. Customary System (USCS) or British system. The International System is based on the primary or base units of *mass, length,* and *time*. This system is called an *absolute system* of units because quantities measured using the primary units of mass, length, and time are the same at any location where the measurements are made. The USCS is based on the

Figure 1.3 Box-shaped beam.

Figure 1.4 Building frame.

Figure 1.5 Pressurized vessel.

primary units of force, length, and time. The unit force is defined as the weight of a standard mass and since weight depends on the gravitational attraction, which varies with location, the USCS is called a *gravitational system*.

SI Units

In SI, the fundamental unit of mass m is the *kilogram,* kg, and the unit of force F is the *Newton,* N. The unit of length is the *meter,* m, and the unit of time is the *second,* s. Using Newton's second law, $F = ma$ (where a denotes acceleration), we derive the unit of force as

$$1 \text{ newton} = (1 \text{ kilogram})(1 \text{ meter/second}^2)$$

or

$$1 \text{ N} = 1 \text{ kg} \cdot \text{m/s}^2 \tag{1.1}$$

In SI, we are often presented with the mass of an object and need to convert it to weight (a force) to use the weight of an object, for instance, in writing equilibrium equations. The *weight* of an object is the force exerted on it by gravity. This force of gravity, and hence weight, varies depending on the location of the object with respect to the earth. The mass does not change with location and hence is called an *invariant property*.

The mass, m, and weight, W, are related thus

$$W = mg \tag{1.2}$$

where W is the weight in Newtons, m is the mass in kilograms, and g is the *acceleration of gravity* which varies with elevation and latitude on the earth. Standard gravity on the surface of the earth is taken as

$$g = 9.81 \text{ m/s}^2 \tag{1.3}$$

Hence, if a mass of an object is, say, 10 kg, using Eqs. (1.2) and (1.3), the weight is given by

$$W = (10 \text{ kg})(9.81 \text{ m/s}^2) = 98.1 \text{ kg} \cdot \text{m/s}^2 = 98.1 \text{ N}$$

A list of the SI symbols and units of numerous quantities is provided inside the front cover of the text.

Often *prefixes* are attached to SI units to form multiples and submultiples of based units. Using the prefixes listed inside the front cover of the text we have, for example, 1 GN (giganewton) = 10^9 N, 1 MN (meganewton) = 10^6 N, and 1 kN (kilonewton) = 10^3 N. The use of prefixes avoids the use of cumbersomely large or small numbers. As examples, a force of 30,000 N = 30 kN, and a bar diameter of 0.025 m = 25 mm, where here m stands for meter and mm for millimeter.

U.S. Customary Units

In the USCS the fundamental unit of mass is the *slug,* the unit of force is the *pound* (lb), and the unit of length is the *foot* (ft). Time is again expressed in units of *seconds* (s). The mass is derived from the equation $m = F/a$ as

$$1 \text{ slug} = \frac{1 \text{ lb}}{1 \text{ ft/s}^2}$$

or

$$1 \text{ slug} = 1 \text{ lb} \cdot \text{s}^2/\text{ft} \tag{1.4}$$

In the USCS, the weight of an object is normally given and can then be used directly in writing the equations of equilibrium. If, however, we desire the mass of the object, we solve Eq. (1.2) for mass, m, as

$$m = \frac{W}{g} \tag{1.5}$$

where standard gravity on the surface of the earth is now taken as

$$g = 32.2 \text{ ft/s}^2 \tag{1.6}$$

Hence, if an object weighs, say, 32.2 lb, then using Eqs. 1.5 and 1.6, its mass is

$$m = \frac{W}{g} = \frac{32.2 \text{ lb}}{32.2 \text{ ft/s}^2} = 1 \text{ lb} \cdot \text{s}^2/\text{ft}$$

or by Eq. 1.4 the mass is

$$m = 1 \text{ slug}$$

A list of the USCS symbols and units of numerous quantities is provided inside the front cover of the text.

To facilitate manipulations involving large forces acting on machine parts and structures, the unit *kip* (standing for kilopound) is used in the USCS units. The kip is defined as equal to 1000 lb. For example 20,000 lb = 20 kips.

Table 1.1 summarizes the primary units used for each system of units.

TABLE 1.1 PRIMARY UNITS FOR SI AND USCS

Unit	SI	USCS
	Unit Name and Symbol	
Force	newton (N)	pound (lb)
Mass	kilogram (kg)	slug
Length	meter (m)	foot (ft)
Time	second (s)	second (s)

Conversion of Units

It is very useful to be able to do engineering work in either system of units. Although a great emphasis is now being placed on the use of the SI system, particularly in the engineering sciences, mechanics of materials is a design-oriented subject. Design courses rely on data of mechanical properties of materials, and on commercially available standard sizes of members, such as structural tubing, angles, bolts, steel beams, and timber. These data and standard sizes are still mainly given in the U.S. Customary system of units. Hence, to become a successful engineer it is necessary to become well acquainted with both SI and USCS units.

Therefore, engineers need to develop a physical understanding of the quantities they are working with in both systems of units. For instance, the magnitude of a force, a length, or a stress should be understood in both systems of units.

Conversion factors are listed inside the front cover of the text to help convert quantities given in either system of units to the other. For instance, from the table: 1 ft = 0.305 m, 1 lb = 4.45 N, and 1 ft^2 = 0.0929 m^2. Example 1.1 further illustrates how to use the table.

EXAMPLE 1.1

Convert the following from USCS units to SI units:

(a) length (L) = 20 ft,

(b) cross-sectional area (A) = 2 in.^{2}

(c) force, (F) = 10,000 lb,

(d) moment (M) = 5000 lb · ft

(e) stress (σ) = 10,000 psi.

Solution

Using the table inside the front cover, we have

(a) $L = 20 \text{ ft} \left(\dfrac{0.305 \text{ m}}{1 \text{ ft}}\right) = 6.1 \text{ m}$

(b) $A = 2 \text{ in.}^2 \left(\dfrac{645 \text{ mm}^2}{1 \text{ in.}^2}\right) = 1290 \text{ mm}^2$

(c) $F = 10{,}000 \text{ lb} \left(\dfrac{4.45 \text{ N}}{1 \text{ lb}}\right) = 44{,}500 \text{ N} = 44.5 \text{ kN}$

(d) $M = 5000 \text{ lb} \cdot \text{ft} \left(\dfrac{1.36 \text{ N} \cdot \text{m}}{1 \text{ lb} \cdot \text{ft}}\right) = 6800 \text{ N} \cdot \text{m} = 6.8 \text{ kN} \cdot \text{m}$

(e) $\sigma = 10{,}000 \text{ psi} \left(\dfrac{6890 \text{ N/m}^2}{1 \text{ psi}}\right) = 68{,}900{,}000 \dfrac{\text{N}}{\text{m}^2} \text{ (or Pa)} = 68.9 \dfrac{\text{MN}}{\text{m}^2} = 68.9 \text{ MPa}$

1.3 INTERNAL FORCES

To determine the load carrying capacity of a machine part or a structural member, we must be able to obtain the internal forces needed to balance or resist the effects of the externally applied forces. *External forces* are those forces acting on the body from other

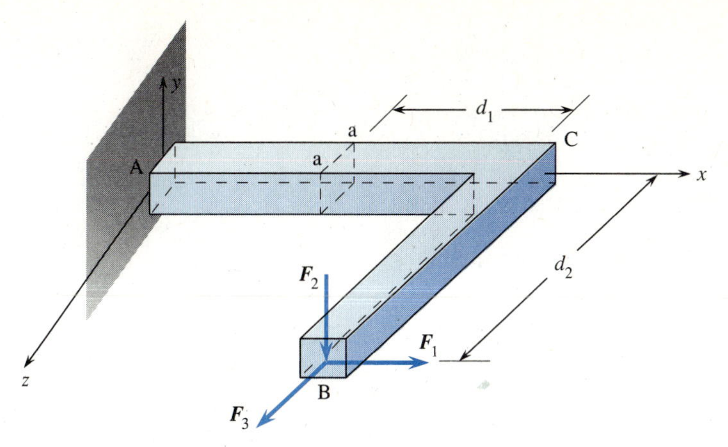

Figure 1.6 Fixture subjected to loads.

bodies, from the reactive forces caused by the supports, and from the weight of the body. The internal forces result in stresses and deformations in the member. In particular, the stresses are obtained as formulas expressed in terms of the internal forces. These stresses can then be compared to the allowable stresses for a particular material such as steel, aluminum, or wood. The development and use of these stress formulas in the analysis and design of load-carrying machines and structures is a major objective of mechanics of materials.

To describe the physical effects caused by the different kinds of internal forces, we consider the fixture shown in Figure 1.6. The fixture is in static equilibrium under the loads F_1, F_2, and F_3 shown. We want to obtain the internal forces (and moments) at section a-a located at distance d_1 from point C. Also d_2 represents the length of the bent portion BC.

We draw a free-body diagram of the right section of the fixture from section a-a to the end B as shown in Figure 1.7(a). That is, we pass a plane perpendicular to the x-axis through the member at a-a. The forces are shown by the single-headed arrows and the moments by the double-headed arrows.

For clarity, we use the double-headed arrow notation to denote the moments. That is, we use the usual right-hand rule in which the thumb of the right hand, pointing in the direction of the double arrow, is used to obtain the axis that the moment is attempting to rotate the body about, while the fingers of the right hand indicate this motion (Figure 1.7(b)). Now since the whole fixture is in equilibrium, any part of it must also be in equilibrium. It is the internal forces and moments at the cut section a-a that balance the externally applied forces. Hence, for equilibrium, we must have the vector equations $\Sigma \boldsymbol{F} = 0$ and $\Sigma \boldsymbol{M} = 0$ on a cut section, where $\Sigma \boldsymbol{F}$ includes all external and internal forces acting on the cut and $\Sigma \boldsymbol{M}$ includes moments from the external and internal forces acting on the cut section.

By statics (or the equations of equilibrium), on summing forces in the x, y, and z directions and moments about the x-, y- and z-axes at a-a in Figure 1.7(a), we obtain in

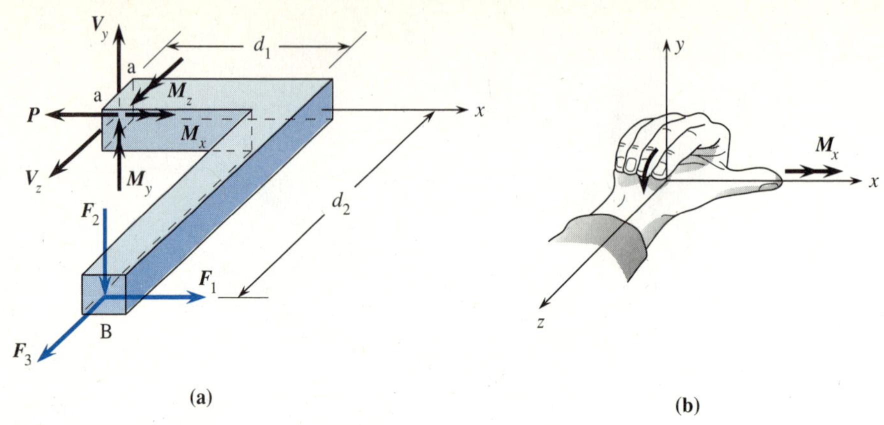

Figure 1.7 (a) Free-body diagram of right section of fixture, and (b) illustration of right-hand rule for moment about x-axis.

scalar form (taking forces positive and double-headed moment arrows positive in positive coordinate directions)

$$\Sigma F_x = 0: \quad F_1 - P = 0 \qquad P = F_1$$

$$\Sigma F_y = 0: \quad -F_2 + V_y = 0 \qquad V_y = F_2$$

$$\Sigma F_z = 0: \quad F_3 + V_z = 0 \qquad V_z = -F_3$$

$$\Sigma M_x = 0: \quad M_x + F_2 d_2 = 0 \qquad M_x = -F_2 d_2$$

$$\Sigma M_y = 0: \quad M_y + F_1 d_2 - F_3 d_1 = 0 \qquad M_y = F_3 d_1 - F_1 d_2$$

$$\Sigma M_z = 0: \quad M_z - F_2 d_1 = 0 \qquad M_z = F_2 d_1$$

Internal forces and moments are classified according to their physical effects caused on the member. These physical effects are (1) an axial force P, which tends to pull (stretch, elongate) or push (compress, shorten) the member in the x direction; (2) a shear force V, which tends to cause sliding of one part of a member with respect to an adjacent part; (3) a torsional moment (torque) M_x or T, which tends to twist the member about the x-axis; and (4) bending moments M_y and M_z, which tend to bend or flex the member about the y- and z-axes, respectively. These physical effects are clearly illustrated in Figure 1.8.

At section *a-a* in Figure 1.7(a), we have the following physical interpretation of the internal forces and moments: (1) an internal axial force P which is directed away from the cut cross-section to resist external force F_1, and so tends to stretch the member at *a-a* or pull fibers of material of Figure 1.7(a) to the left. (2) a shear force V_y directed upward to resist the downward-directed external force F_2, which tends to cause sliding (shearing) in the y direction (up and down) of the left portion from A to *a-a* past the right portion (shown in Figure 1.7(a)) at cross-section *a-a*, and V_z resisting the external force F_3 and tending to shear at section *a-a* in the z direction. (3) a torsional moment M_x resisting twisting about the x-axis due to the external force F_2. (4) bending moments M_y and M_z tending to bend the member about the y- and z-axes, respectively.

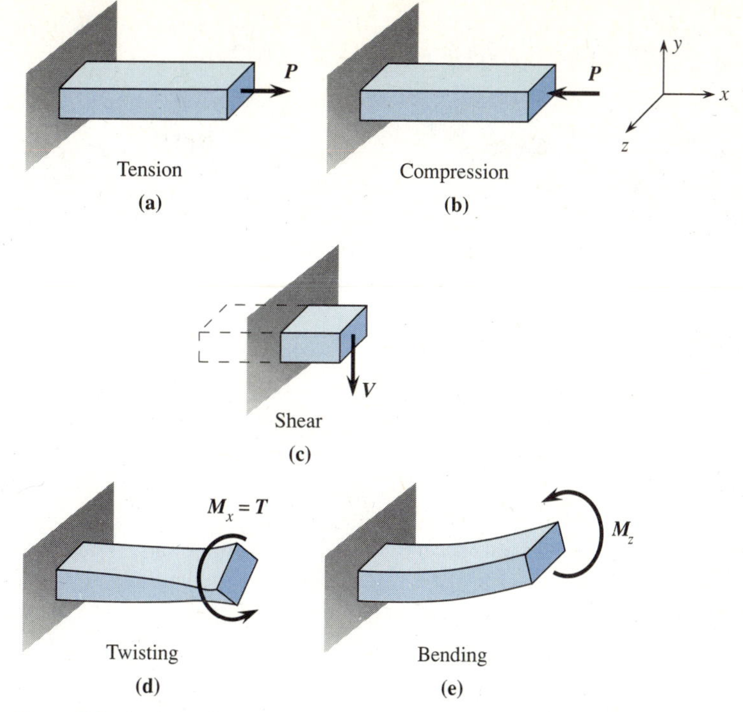

Figure 1.8 Physical effects caused by forces and moments acting on a member.

STEPS IN SOLUTION

The following steps are generally used to obtain the internal reactions in a load-carrying member.

1. Draw the *free-body diagram* of the whole structure or of the member. For a simply supported member this is a necessary step. For a cantilevered member this step can be avoided unless the reactions at the fixed end are desired.
2. Replace any distributed loading by a statically equivalent concentrated load.
3. Write the *equilibrium equations* (or equations of statics) to obtain the reactions. For cantilevered members this step can be avoided as illustrated in Examples 1.4 and 1.5.
4. Now place any original distributed loading back on the member and pass a plane through the section where the internal reactions are desired.
5. Draw a *free-body diagram* of the section showing the previously obtained reaction(s) associated with the section and the unknown internal reactions. These internal reactions may be one or more of the following type: a normal force, shear forces, a torsional moment (or torque), and bending moments.
6. Replace any distributed loading acting on the section by its statically equivalent load.
7. Write the *equilibrium equations* and solve for the internal reactions.

Steps 2, 4, and 6 are not relevant if distributed loading is not present on the member.

Examples 1.2 through 1.9 illustrate the steps used in the solution to obtain internal reactions in various load-carrying members.

EXAMPLE 1.2

A connecting rod (Figure 1.9) is subjected to the axial forces shown. Determine the internal reactions at sections **(a)** a-a, **(b)** b-b, and **(c)** c-c.

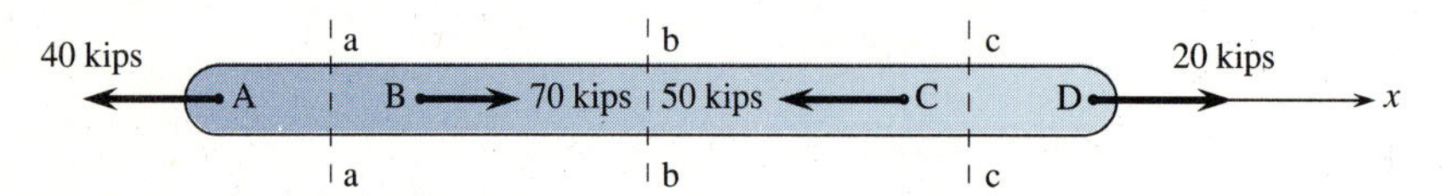

Figure 1.9 Connecting rod subjected to axial forces.

Solution

(a) Using the method of sections, we pass a plane through the rod at a-a perpendicular to the length of it and draw a free-body diagram of the section showing the internal axial force P acting on the cross-section.

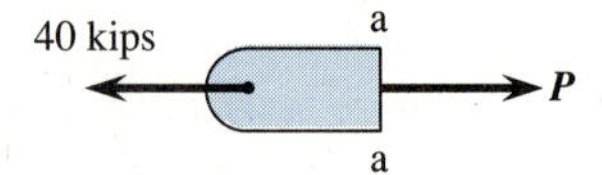

By statics, we write

$$\Sigma F_x = 0: \quad -40 \text{ kips} + P = 0 \qquad P = 40 \text{ kips (to the right)}$$

(b) Similarly, passing a plane through section b-b, the free-body diagram is

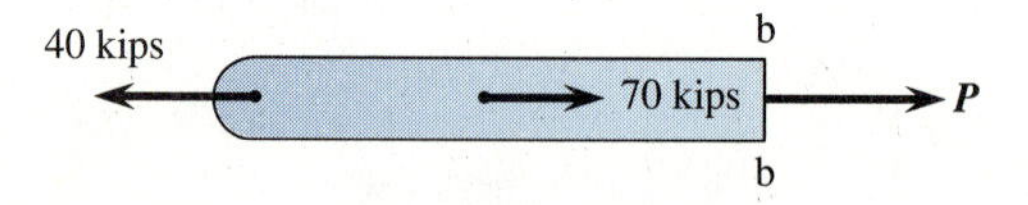

By statics, we have

$$\Sigma F_x = 0: \quad -40 \text{ kips} + 70 \text{ kips} + P = 0 \qquad P = -30 \text{ kips (or to the left)}$$

The negative sign means the force P at section b-b is in the opposite direction from the one assumed in the free-body diagram.

(c) Similarly, passing a plane through section c-c, we obtain the free-body diagram shown below.

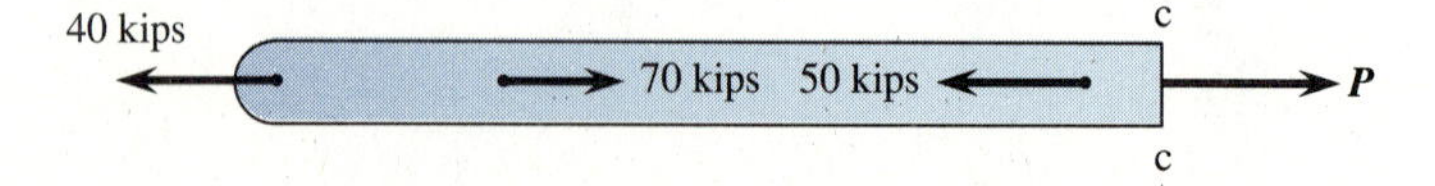

By statics, we obtain

$$\Sigma F_x = 0: \quad -40 \text{ kips} + 70 \text{ kips} - 50 \text{ kips} + P = 0 \qquad P = 20 \text{ kips (to the right)}$$

NOTE: The internal force on a plane perpendicular to the axis of the rod is an axial force only. The axial force is positive and pulling away on sections a-a and c-c. The axial force is negative at section b-b, and hence is really in the opposite direction from that assumed and is pushing back into the cross-section so as to compress the rod at b-b.

EXAMPLE 1.3

The transmission shaft has four gears attached to it that transmit the torques shown in Figure 1.10. Determine the internal torques at sections **(a)** a-a, **(b)** b-b, and **(c)** c-c.

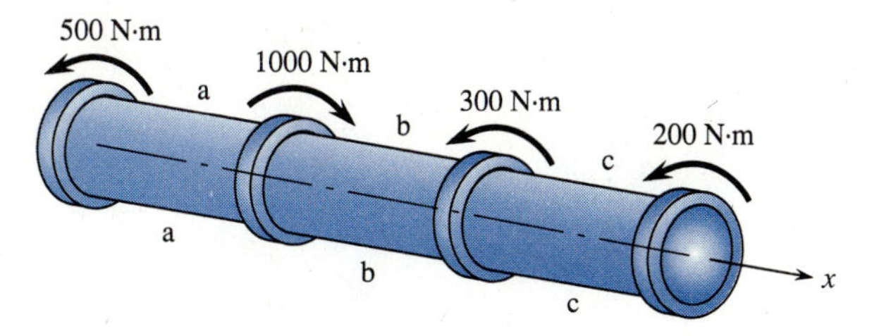

Figure 1.10 Transmission shaft subjected to torques.

Solution

(a) We cut the shaft at section a-a with the internal torque T as shown by the curved arrow or equivalently by the double-headed arrow.

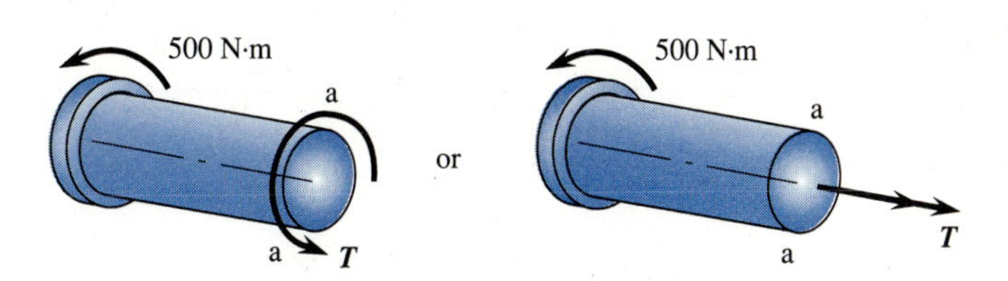

By statics (summing torques about the x-axis), we obtain

$$\Sigma M_x = 0: \qquad 500 \text{ N} \cdot \text{m} + T = 0 \qquad T = -500 \text{ N} \cdot \text{m}$$

The negative sign on T indicates the internal torque is in the opposite direction from that shown on the free-body diagram.

(b) We cut the shaft at section b-b and use the left portion.

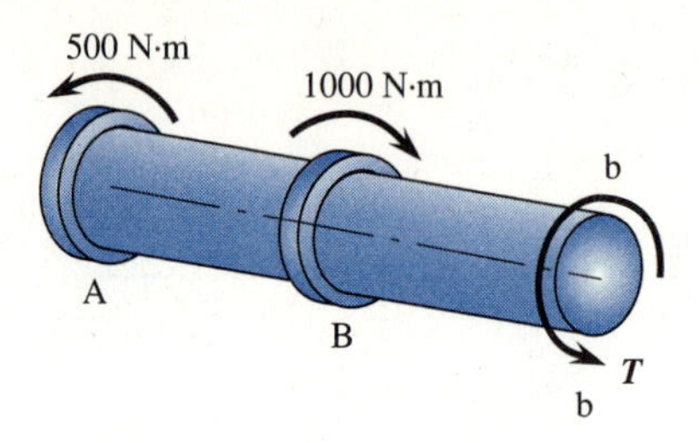

By statics, we obtain

$$\Sigma M_x = 0: \quad 500 \text{ N} \cdot \text{m} - 1000 \text{ N} \cdot \text{m} + T = 0 \qquad T = 500 \text{ N} \cdot \text{m}$$

(c) Cut the shaft at section c-c and use the left portion.

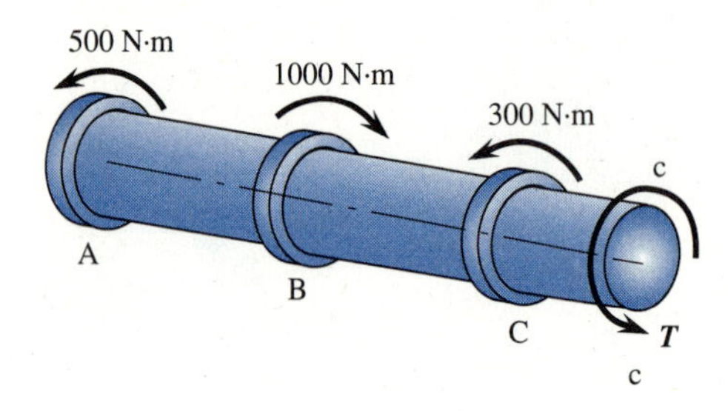

By statics, we obtain

$$\Sigma M_x = 0: \quad 500 - 1000 + 300 + T = 0 \qquad T = 200 \text{ N} \cdot \text{m}$$

EXAMPLE 1.4

For the cantilever beam shown in Figure 1.11, determine the internal reactions at **(a)** section a-a and **(b)** section b-b.

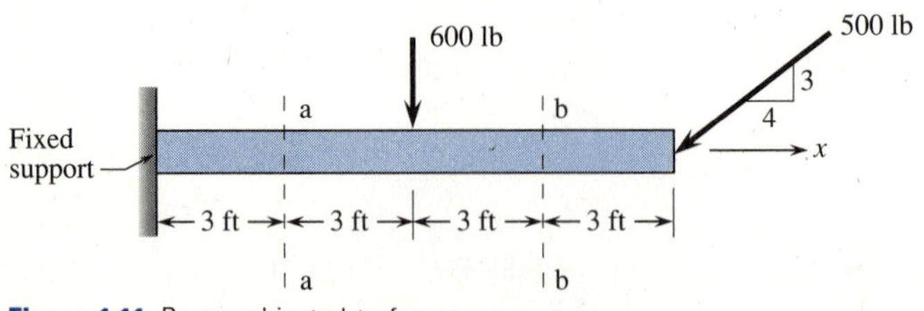

Figure 1.11 Beam subjected to forces.

Solution

(a) We first cut the beam at a-a, and draw a free-body diagram of the portion to the right of a-a showing P, V, and M at a-a.

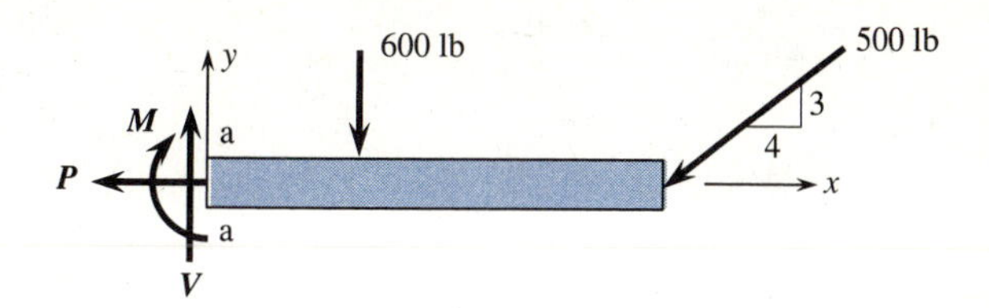

By statics, we have

$$\Sigma F_x = 0: \quad -P - 500\left(\frac{4}{5}\right) = 0$$

$$P = -400 \text{ lb (or to the right)}$$

$$\Sigma F_y = 0: \quad V - 600 - 500\left(\frac{3}{5}\right) = 0$$

$$V = 900 \text{ lb (upward)}$$

$$\Sigma M_z = 0: \quad -M - 600\,(3\text{ ft}) - 500\left(\frac{3}{5}\right)(9\text{ ft}) = 0$$

$$M = -4500 \text{ lb} \cdot \text{ft (or counter-clockwise, ccw)}$$

The negative signs on the answers for P and M indicate that P and M are in the opposite directions from those shown on the free-body diagram.

(b) Now cut the beam at section b-b and draw the free-body diagram of the portion to the right of b-b.

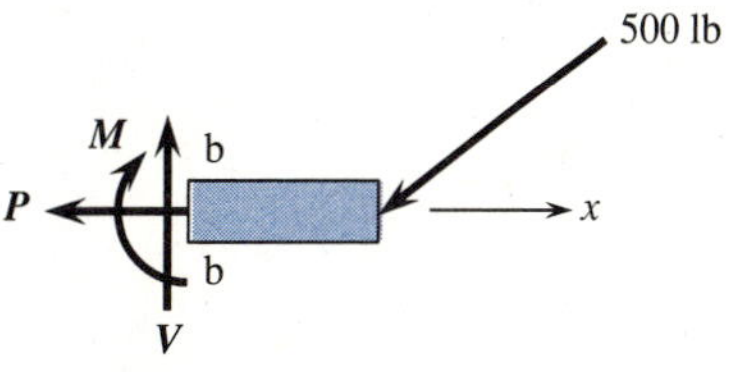

By using statics, we have

$$\Sigma F_x = 0: \quad -P - 500\left(\frac{4}{5}\right) = 0 \qquad P = -400 \text{ lb}$$

$$\Sigma F_y = 0: \quad V - 500\left(\frac{3}{5}\right) = 0 \qquad V = 300 \text{ lb}$$

$$\Sigma M_z = 0: \quad -M - 500\left(\frac{3}{5}\right)(3\text{ ft}) = 0 \qquad M = -900 \text{ lb} \cdot \text{ft}$$

Remember the physical interpretation of the internal reactions is (1) P is an axial force, (2) V is a shear force, and (3) M is a bending moment about the z-axis. For the loading applied to the beam of Figure 1.11, P, V, and M are necessary for equilibrium.

EXAMPLE 1.5

For the bent pipe shown in Figure 1.12, determine the internal reactions at section a-a.

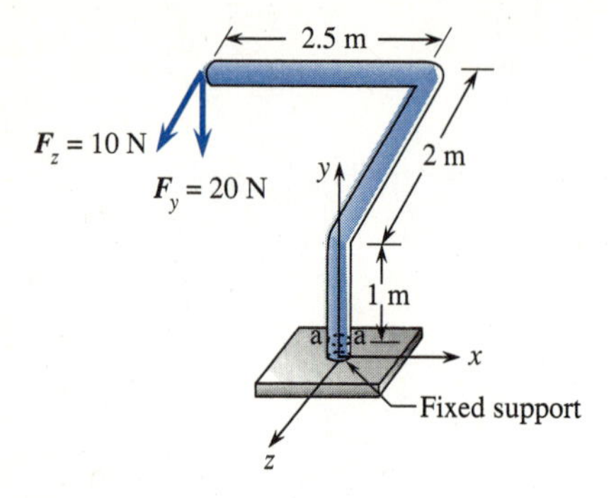

Figure 1.12 Bent pipe subjected to forces.

Solution

A free-body diagram of a section through a-a is shown.

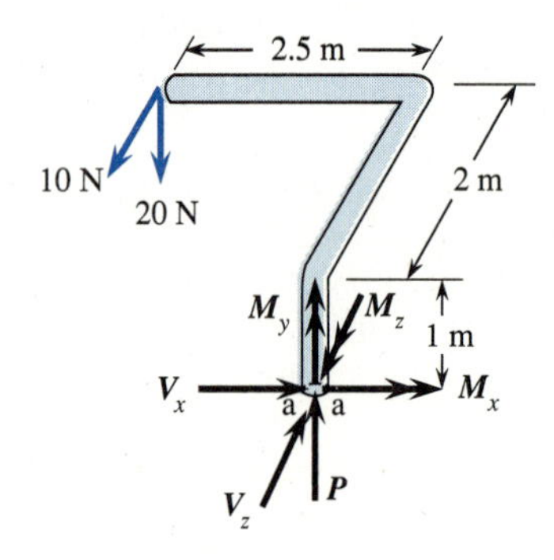

By statics, we write

$$\Sigma F_x = 0: \quad V_x = 0$$

$$\Sigma F_y = 0: \quad P - 20\text{ N} = 0$$
$$P = 20\text{ N (as shown)}$$

$$\Sigma F_z = 0: \quad -V_z + 10\text{ N} = 0$$
$$V_z = 10\text{ N (as shown)}$$

$$\Sigma M_x = 0: \quad M_x - (20\text{ N})(2\text{ m}) + (10\text{ N})(1\text{ m}) = 0$$
$$M_x = 30\text{ N}\cdot\text{m (as shown)}$$

$$\Sigma M_y = 0: \quad M_y + (10\text{ N})(2.5\text{ m}) = 0$$
$$M_y = -25\text{ N}\cdot\text{m (opposite direction shown)}$$

$$\Sigma M_z = 0: \quad M_z + (20\text{ N})(2.5\text{ m}) = 0$$
$$M_z = -50\text{ N}\cdot\text{m (opposite direction shown)}$$

NOTE: At section a-a, P is an axial force, V_x and V_z are shear forces, M_x and M_z are bending moments, and M_y is a twisting or torsional moment.

EXAMPLE 1.6

For the jib crane shown in Figure 1.13, determine the internal reactions at section a-a (just to the left of the 2000-lb force).

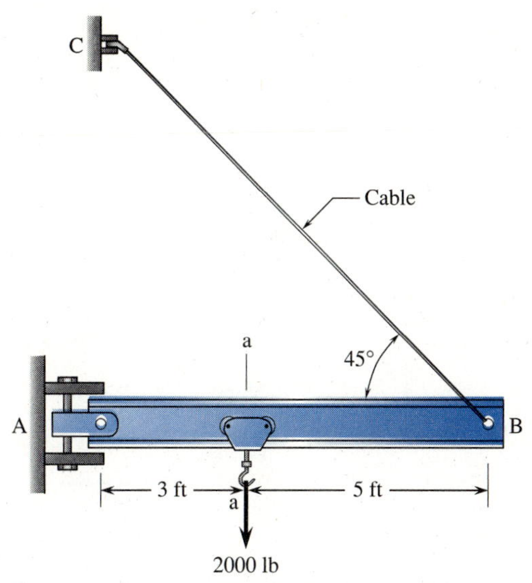

Figure 1.13 Beam subjected to concentrated force.

Solution

We first draw the free-body diagram of member AB. This is accomplished by replacing (1) the pin at A with reactions A_x and A_y and (2) the cable support at B with reaction F_{BC}. Also the 2000-lb force is assumed to act as a concentrated force 3 ft from end A as the wheels of the trolley are close together.

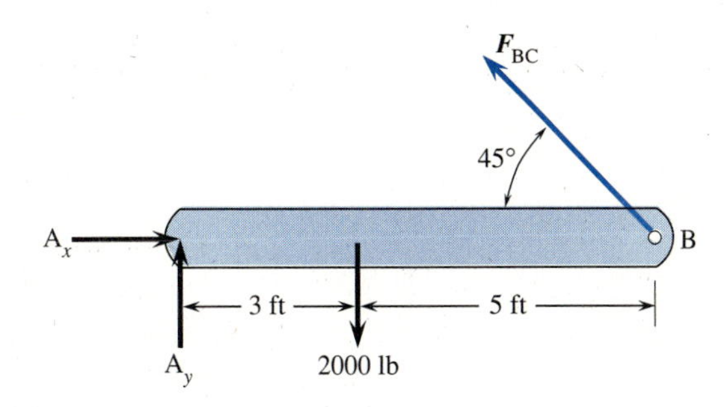

By summing moments at A, we write

$$+)\ \Sigma M_A = 0: \quad -(3\text{ ft})(2000\text{ lb}) + (8\text{ ft})(F_{BC}\sin 45°) = 0 \qquad F_{BC} = 1060\text{ lb}$$

$$\Sigma F_x = 0: \quad A_x - F_{BC}\cos 45° = 0 \qquad A_x = 750\text{ lb}$$

$$\Sigma F_y = 0: \quad A_y - 2000 + F_{BC}\sin 45° = 0 \qquad A_y = 1250\text{ lb}$$

Now we draw a free-body diagram of a section of the member to the left of a-a.

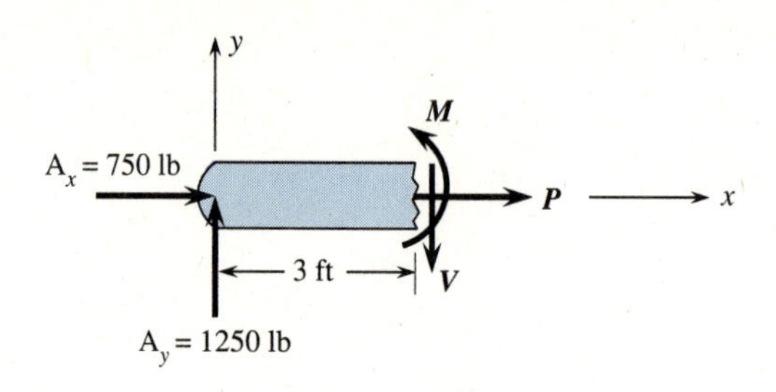

Using statics, we write

$$\Sigma F_x = 0: \quad 750 + P = 0 \qquad P = -750 \text{ lb (directed to left)}$$

$$\Sigma F_y = 0: \quad 1250 - V = 0 \qquad V = 1250 \text{ lb (down)}$$

$$\Sigma M_{\text{a-a}} = 0: \quad -(3 \text{ ft})(1250 \text{ lb}) + M = 0 \qquad M = 3750 \text{ lb} \cdot \text{ft (counter-clockwise)}$$

EXAMPLE 1.7

For the simply supported beam subjected to the uniformly distributed load shown in Figure 1.14, determine the internal reactions at section a-a.

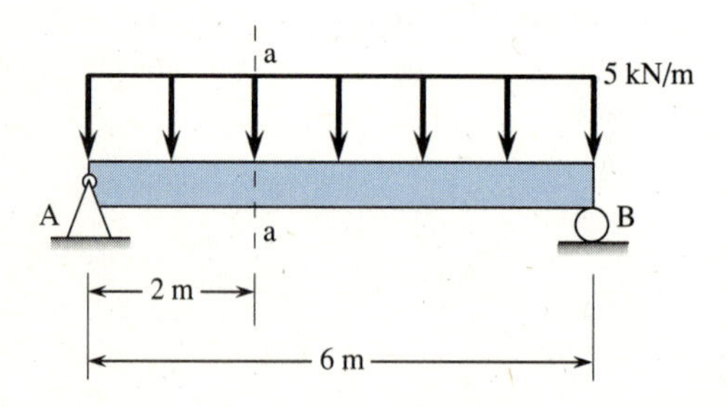

Figure 1.14 Beam subjected to uniformly distributed load.

Solution

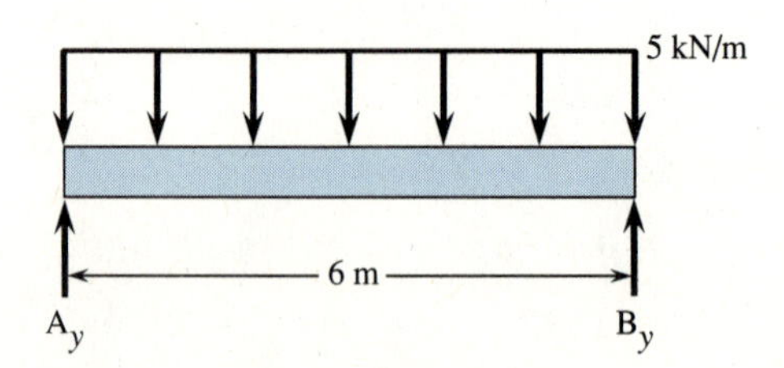

Replace the distributed load by a statically equivalent concentrated force. The distributed load is rectangular. Therefore, we have F = base × height (the area under the load distribution)

$$F = (6\text{ m})(5\text{ kN/m}) = 30\text{ kN}$$

Force F is located at the center of force distribution, 3 m from the left end.

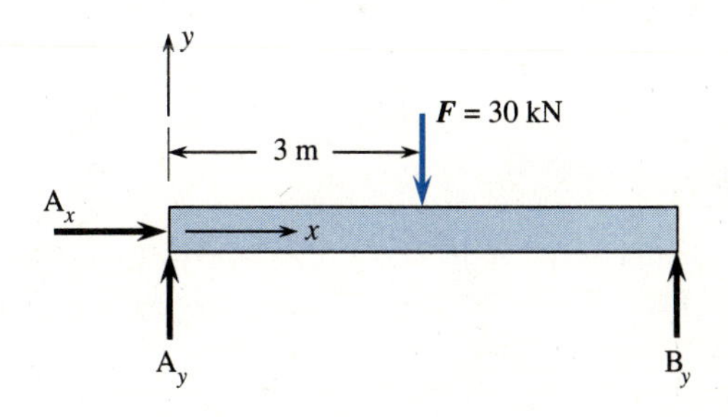

Write equilibrium equations as follows:

$$\Sigma F_x = 0: \quad A_x = 0$$

$$\Sigma M_A = 0: \quad -(3\text{ m})(30\text{ kN}) + (6\text{ m})\,B_y = 0 \qquad B_y = 15\text{ kN}$$

$$\Sigma F_y = 0: \quad A_y + B_y = 30 \qquad A_y = 30 - B_y = 15\text{ kN}$$

Place the original distributed loading back on the beam and then cut the section at a-a.

Draw the free-body diagram of the left (or right) section of the beam. Here we draw the left section.

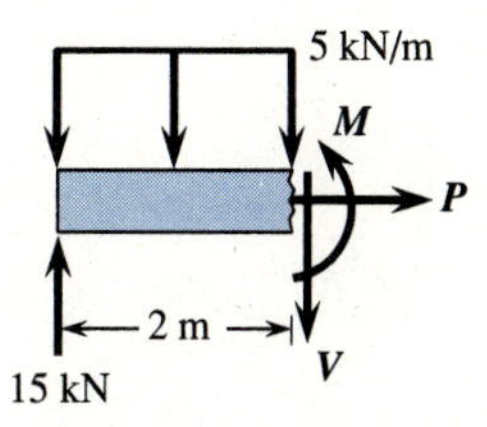

Replace the uniformly distributed load by a statically equivalent concentrated load

$$F = (2\text{ m})(5\text{ kN/m}) = 10\text{ kN}$$

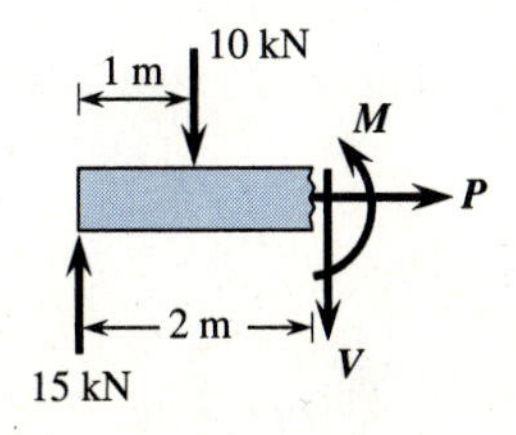

Write the equilibrium equations as follows:

$$\Sigma M_{a\text{-}a} = 0: \quad -(2\text{ m})(15\text{ kN}) + (1\text{ m})(10\text{ kN}) + M = 0$$
$$M = 20\text{ kN}\cdot\text{m}$$

$$\Sigma F_y = 0: \quad 15\text{ kN} - 10\text{ kN} - V = 0$$
$$V = 5\text{ kN}$$

$$\Sigma F_x = 0: \quad P = 0$$

EXAMPLE 1.8

For the member BC supported by a bar AB and pin at C and subjected to the linearly varying line load shown in Figure 1.15, determine the internal reactions at section a-a.

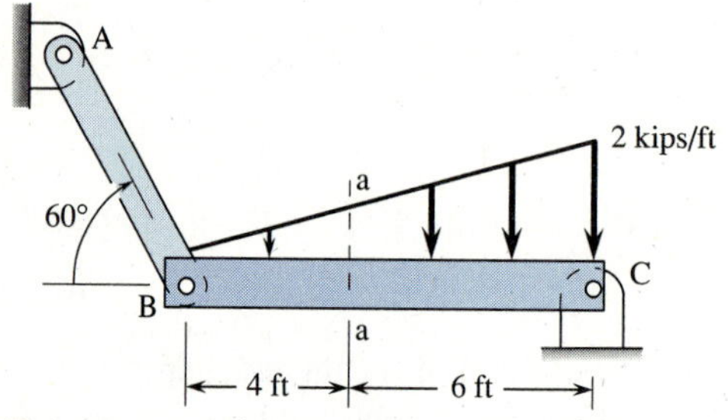

Figure 1.15 Member subjected to linearly varying line load.

Solution

Bar AB is a two-force member and we have a pin at C; therefore, we have the free-body diagram shown below.

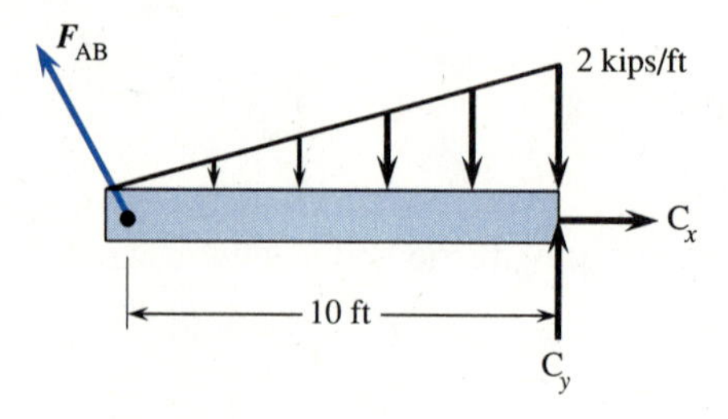

Replace the distributed load by a statically equivalent concentrated force, F. For a linearly varying line load, F becomes

$$F = \frac{1}{2}\text{ bh} = \frac{1}{2}(10\text{ ft})\left(2\,\frac{\text{kips}}{\text{ft}}\right) = 10\text{ kips}$$

and is placed 1/3 the distance the load acts over from the highest value of the load. This distance is $1/3 \times 10\text{ ft} = 3.33\text{ ft}$ from point C.

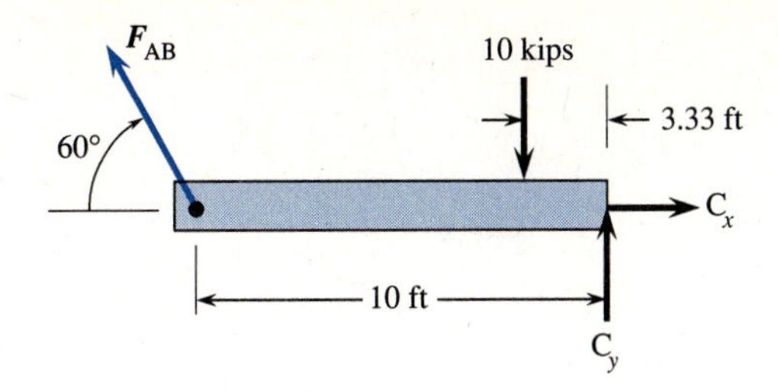

Write equilibrium equations as follows:

$$\Sigma M_C = 0: \quad -(10 \text{ ft}) F_{AB} \sin 60° + (3.33 \text{ ft})(10 \text{ kips}) = 0$$
$$F_{AB} = 3.85 \text{ kips}$$

$$\Sigma F_x = 0: \quad -F_{AB} \cos 60° + C_x = 0$$
$$C_x = 1.93 \text{ kips}$$

$$\Sigma F_y = 0: \quad F_{AB} \sin 60° - 10 + C_y = 0$$
$$C_y = 6.67 \text{ kips}$$

Place the original distributed loading back on the beam and cut the section at a-a, and draw the free-body diagram of the left section of the member.

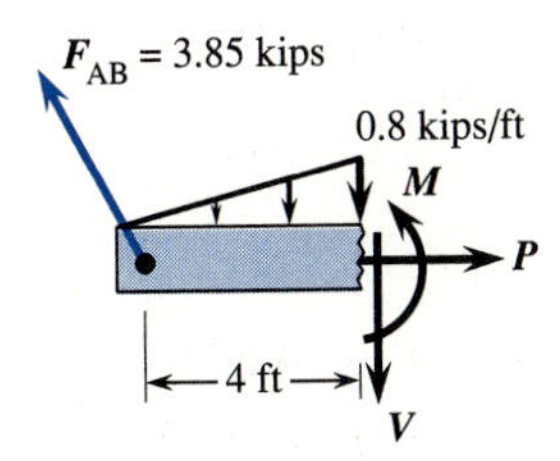

Based on similar triangles, the distributed load at 4 ft has a value of

$$\frac{4 \text{ ft}}{10 \text{ ft}} (2 \text{ kips/ft}) = 0.8 \text{ kips/ft}$$

Replace the linearly varying load by its statically equivalent concentrated load as follows:

$$F = \frac{1}{2} \text{bh} = \frac{1}{2} (4 \text{ ft})(0.8 \text{ kips/ft}) = 1.6 \text{ kips}$$

at $\frac{1}{3}$ (4 ft) = 1.33 ft from the right end of the section.

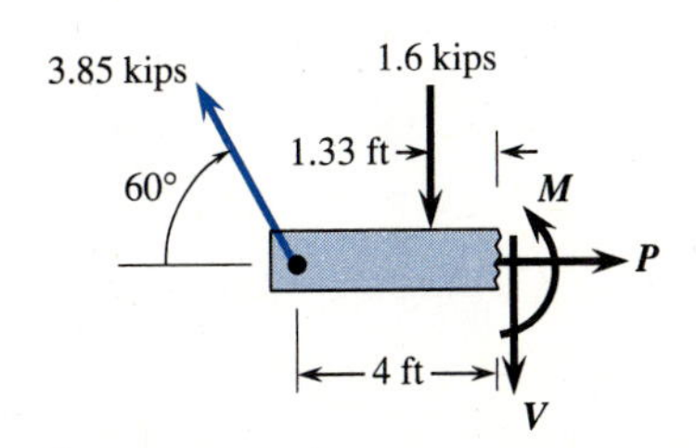

Write the equilibrium equations as follows:

$$\Sigma M_{\text{a-a}} = 0: \quad -(4\text{ ft})(3.85\sin 60^\circ)\text{ kips} + (1.33\text{ ft})(1.6\text{ kips}) + M = 0$$
$$M = 11.20\text{ kip}\cdot\text{ft}$$

$$\Sigma F_x = 0: \quad -(3.85\cos 60^\circ) + P = 0$$
$$P = 1.93\text{ kips}$$

$$\Sigma F_y = 0: \quad 3.85\sin 60^\circ - 1.6 - V = 0$$
$$V = 1.73\text{ kips}$$

EXAMPLE 1.9

For the simply supported member shown in Figure 1.16, determine the internal reactions at section a-a.

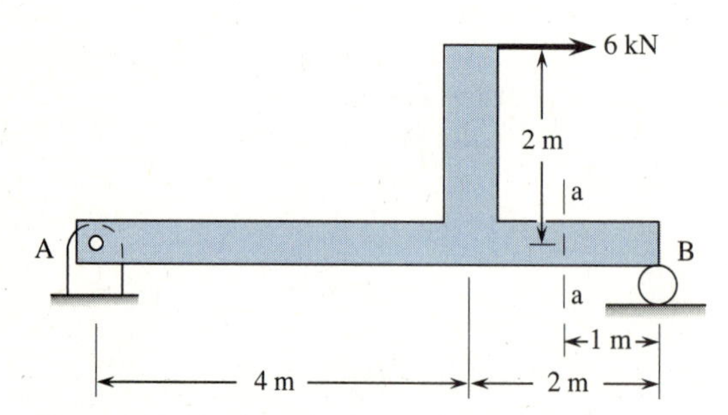

Figure 1.16 Member subjected to horizontal force.

Solution

Draw the free-body diagram of member AB.

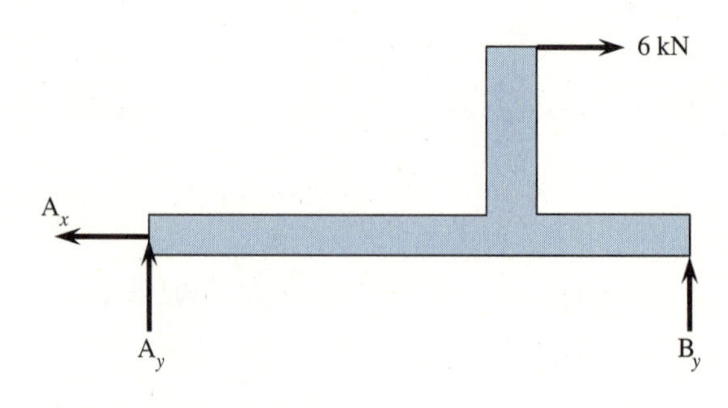

Write the equilibrium equations as follows:

$$\Sigma F_x = 0: \quad -A_x + 6\text{ kN} = 0 \qquad A_x = 6\text{ kN}$$

$$\Sigma M_A = 0: \quad (6\text{ m})\,B_y - (2\text{ m})(6\text{ kN}) = 0 \qquad B_y = 2\text{ kN}$$

$$\Sigma F_y = 0: \quad A_y + 2\text{ kN} = 0 \qquad A_y = -2\text{ kN (downward)}$$

Draw a free-body diagram of the section to the right of a-a.

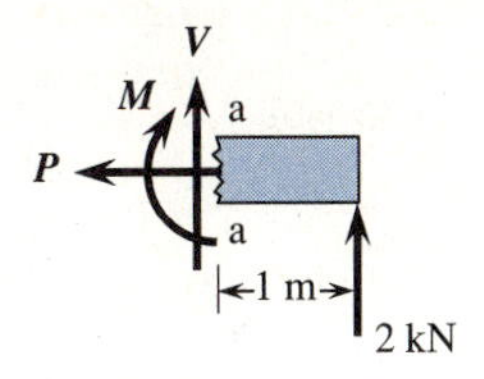

Write the equilibrium equations as follows:

$$\Sigma F_x = 0: \quad P = 0$$

$$\Sigma F_y = 0: \quad V = -2 \text{ kN}$$

$$\Sigma M_{a\text{-}a} = 0: \quad (1 \text{ m})(2 \text{ kN}) - M = 0 \qquad M = 2 \text{ kN} \cdot \text{m}$$

1.4 GENERAL STEPS IN PROBLEM SOLVING

Most of the problems you solve in mechanics of materials follow similar steps. We now present a brief summary of these steps and use them throughout this text.

1. A **free-body diagram** of the member or part to be analyzed is constructed. This means that the member is isolated from its supports and/or connections and all the **applied forces** and **reactive forces** (those forces provided by the supports and/or connections to other members) acting on it are shown.

2. **Statics** is used. This means that the **equations** of **statics** or **equilibrium,** that is, $\Sigma \boldsymbol{F} = 0$ and $\Sigma \boldsymbol{M} = 0$, are used to determine the reactions for statically determinate problems. For statically indeterminate problems, additional equations from **compatibility** of **deformations**, are needed.

3. A **section** or **plane** (normally perpendicular to the axis of the member) is passed through the member at the point where the stress is desired, and the portion of the member to one side of the section or the other is completely removed.

4. The **internal reactions** at the cut section necessary to keep the portion of the member in equilibrium are determined. These internal reactions may be an *axial force,* a *shear force,* a *torsional moment* (or torque), and a *bending moment*. For statically determinate problems, these internal reactions are determined by writing equations of statics for the free-body diagram of the portion of the member under consideration. For statically indeterminate problems, both equations of statics and compatibility of deformations are needed.

5a. For an **analysis** problem (one in which the size and shape of the member are known) **stresses** at the cut section are determined using the formulas from mechanics of materials, which are expressed in terms of the internal reactions. The formulas to obtain stresses due to axial force, shear force, torque, and bending moment are developed throughout the following chapters of the text.

Or 5b. For a **design** problem, a **member** of adequate size and shape to resist the stresses is selected for a given material whose allowable stresses are known.

6. The **deformation** of the member is determined using formulas from mechanics of materials. This deformation is compared to allowable deformations.

1.5 SUMMARY OF IMPORTANT DEFINITIONS AND EQUATIONS

1. Relationship between force and mass
1 newton (N) = 1 $kg \cdot m/s^2$
1 slug = 1 $lb \cdot s^2/ft$

2. Weight and mass
$W = mg$
$g = 9.81\ m/s^2$ or
$g = 32.2\ ft/s^2$

3. Equilibrium equations
$\Sigma F_x = 0, \Sigma F_y = 0, \Sigma F_z = 0$
$\Sigma M_x = 0, \Sigma M_y = 0, \Sigma M_z = 0$

4. Internal reactions
P = axial force
V = shear force
M = bending moment
T = torsional moment or torque

PROBLEMS

Section 1.2

1.1 Express the units in the U.S. Customary System for the following quantities:
(a) Length **(f)** Moment of inertia (area)
(b) Area
(c) Force **(g)** Stress
(d) Mass **(h)** Power
(e) Weight density **(i)** Temperature
(j) Moment

1.2 Express the units in the SI system of units for the quantities listed in Problem 1.1.

1.3 A truck carries 2000 kg of gravel. What is the weight of the gravel in newtons?

1.4 What is the weight in newtons of an object that has a mass of **(a)** 8 kg, **(b)** 4 g, and **(c)** 760 Mg?

1.5 A four-wheeled truck having a total mass of 5000 kg is sitting on a bridge. If 60% of the weight is on the rear wheels and 40% is on the front wheels, compute the force exerted on the bridge at each wheel.

1.6 Measure the length, width, and thickness of this book in millimeters.

1.7 Express the weight determined in Problem 1.3 in pounds.

1.8 Express the forces determined in Problem 1.5 in pounds.

1.9 Wood has a mass density of 1.20 $slug/ft^3$. What is its density expressed in SI units?

1.10 Using the table on the inside front cover of your text, determine the mass in kilograms, and weight in newtons of a 200-pound man.

1.11 Represent each of the following as a number between 0.1 and 1000 using an appropriate prefix: **(a)** 55,320 kN, **(b)** $468(10^5)$mm, and **(c)** 0.001 Mg.

1.12 Determine the number of cubic millimeters contained in 1 cubic inch.

1.13 The pascal (Pa) is actually a very small unit of pressure. To show this, convert 1 Pa = 1 N/m^2 to lb/in^2. Atmospheric pressure at sea level is 14.7 lb/in^2. How many pascals is this?

1.14 Convert **(a)** 40 $lb \cdot ft$ to $N \cdot m$, **(b)** 900 lb/ft^3 to kN/m^3, and **(c)** 5000 $lb/in.^2$ to kPa.

1.15 An aluminum ingot has a width of 600 mm, a thickness of 300 mm, and a length of 1200 mm. If the mass density of aluminum is 2710 kg/m^3, determine the weight of the ingot in pounds.

1.16 A steel disk has a diameter of 500 mm and a thickness of 35 mm. If the mass density of steel is 7850 kg/m^3, determine the weight of the disk in pounds.

1.17 If an object has a mass of 80 slugs, determine its mass in kilograms.

1.18 A pressure vessel contains a gas at 1500 psi. Express the pressure in pascals.

1.19 A structural steel has an allowable stress of 21,600 psi. Express this in pascals.

1.20 An electric motor shaft rotates at 1850 rpm. Express the rotational speed in radians per second.

1.21 Express an area of 2.0 in.^{2} in the units of square millimeters.

1.22 An allowable deformation of a certain beam is 0.090 in. Express the deformation in millimeters.

1.23 A base for a building column measures 16.0 in. × 16.0 in. on a side and 10.0 in. high. Compute the cross-sectional area in both square inches and square millimeters. Compute the volume in **(a)** cubic inches, **(b)** cubic feet, **(c)** cubic millimeters, and **(d)** cubic meters.

1.24 Compute the area of a rod having a diameter of 0.25 in. in square inches. Then convert the result to square millimeters.

Section 1.3

1.25 The forces act on the shaft shown. Determine the internal axial force at points A, B, and C.

Figure P1.25

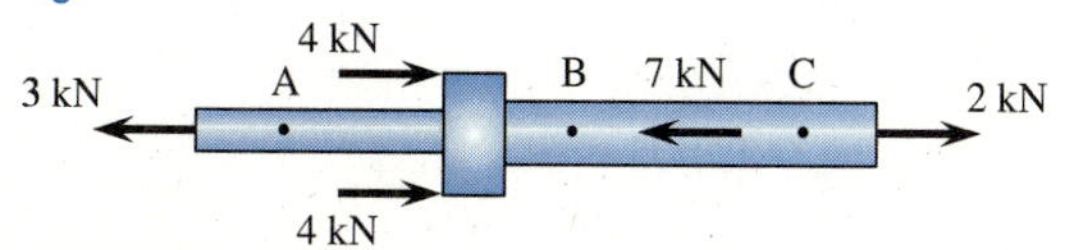

1.26 The rod is subjected to the forces shown. Determine the internal axial force at points A, B, and C.

Figure P1.26

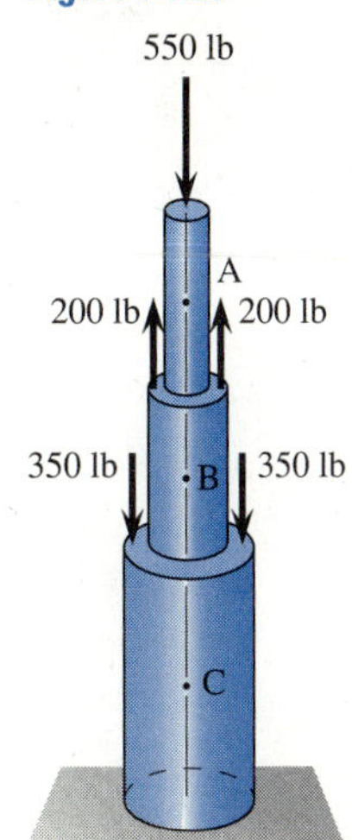

1.27 The shaft is supported by the two smooth bearings A and B. The four pulleys attached to the shaft are used to transmit power to adjacent machinery. If the torques applied to the pulleys are as shown, determine the internal torques at points C, D, and E.

Figure P1.27

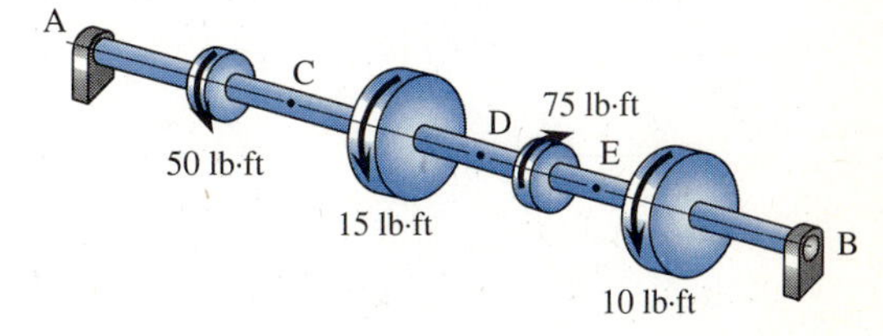

1.28 The shaft is supported by smooth bearings at A and B and subjected to the torques shown. Determine the internal torques at points C, D, and E.

Figure P1.28

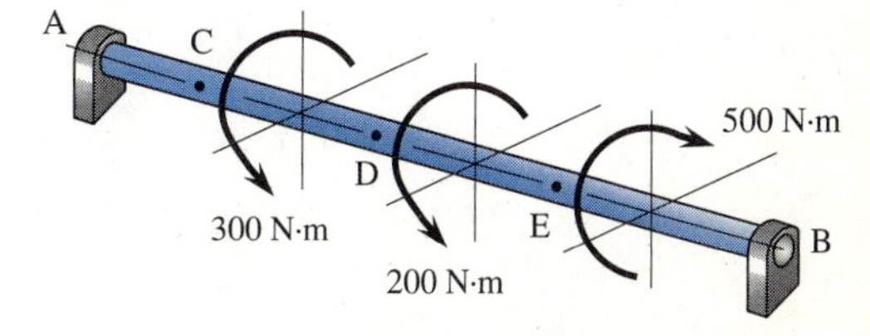

1.29 Three torques act on the shaft as shown. Determine the internal torques at points A, B, C, and D.

Figure P1.29

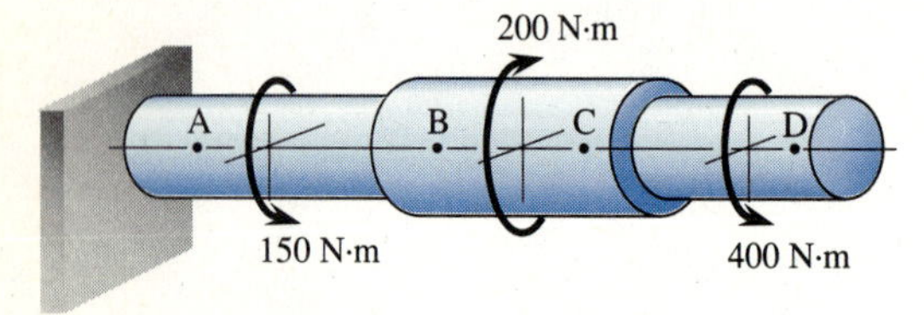

1.30 The beam-column is fixed to the floor and supports the loads shown. Determine the internal axial force, shear force, and moment at points A and B due to this loading.

Figure P1.30

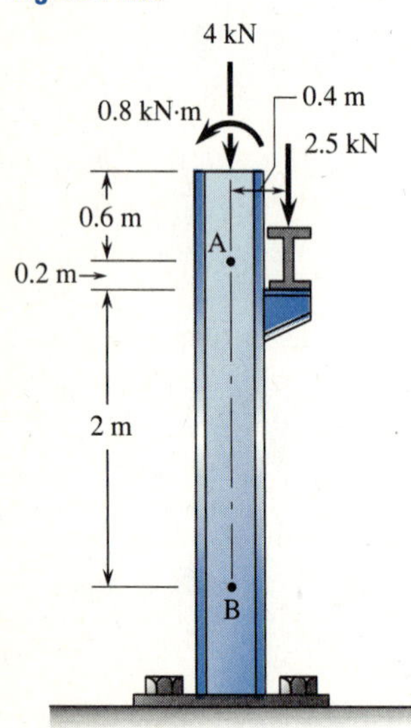

1.31–1.34 Determine the internal reactions at sections a-a and b-b for the beams shown in terms of the physical components.

Figure P1.31

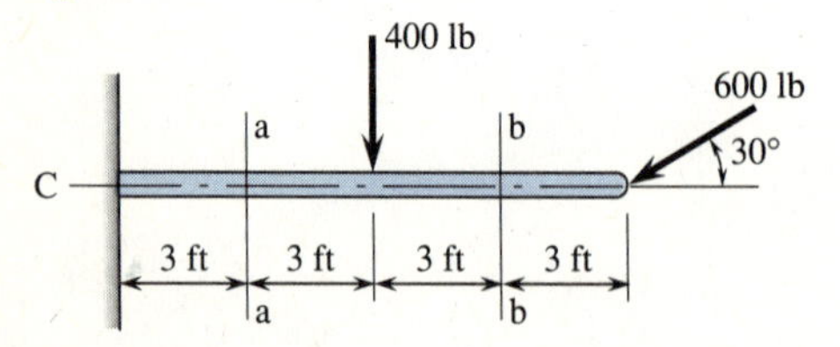

Figure P1.32

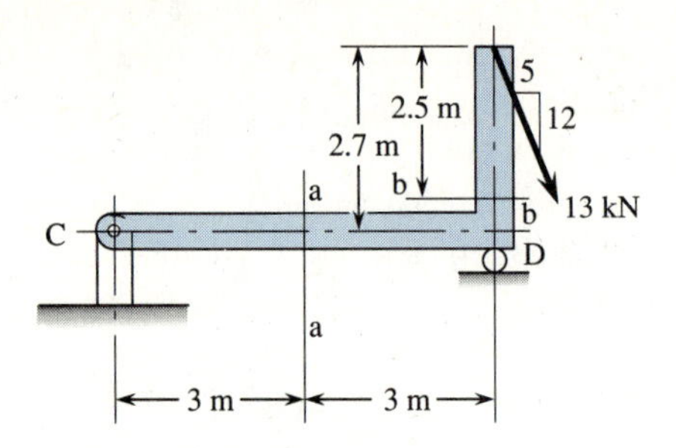

Figure P1.33

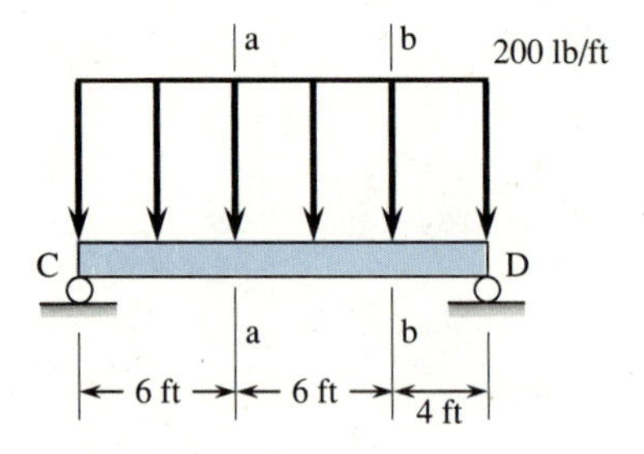

Figure P1.34

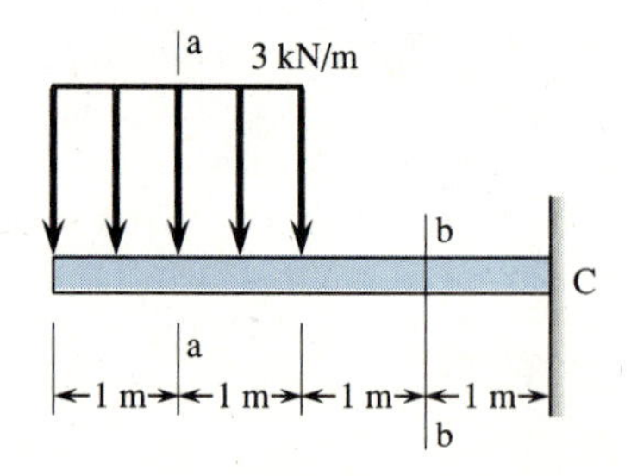

1.35 Determine the axial force, shear force, and moment at point C.

Figure P1.35

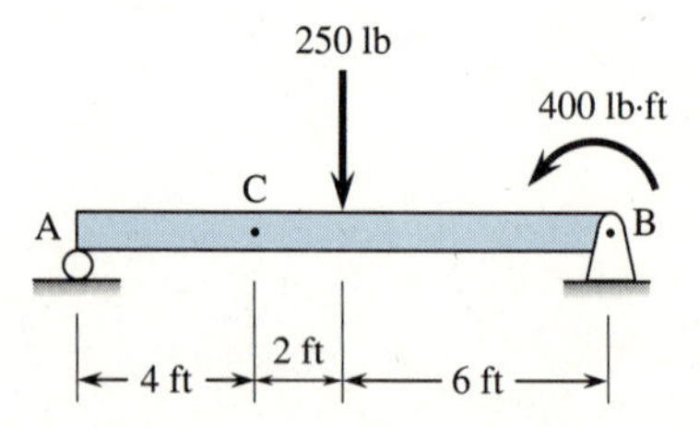

1.36 Determine the axial force, shear force, and moment at point A if the clamp exerts a compressive force of $F = 95$ lb on board B. Force F acts along the centroidal axis of the screw.

Figure P1.36

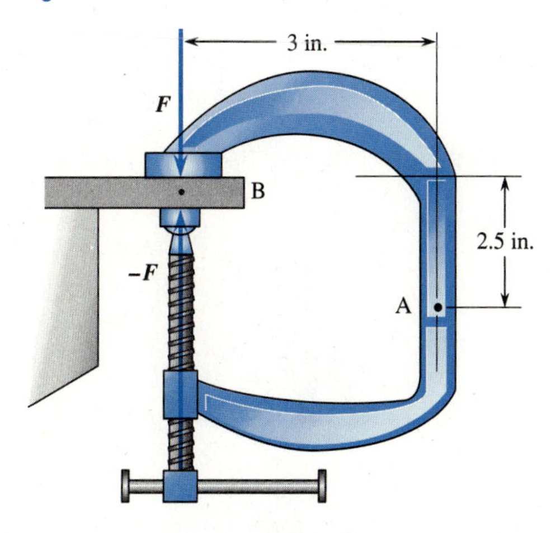

1.37 Determine the internal axial force, shear force, and moment at point D.

Figure P1.37

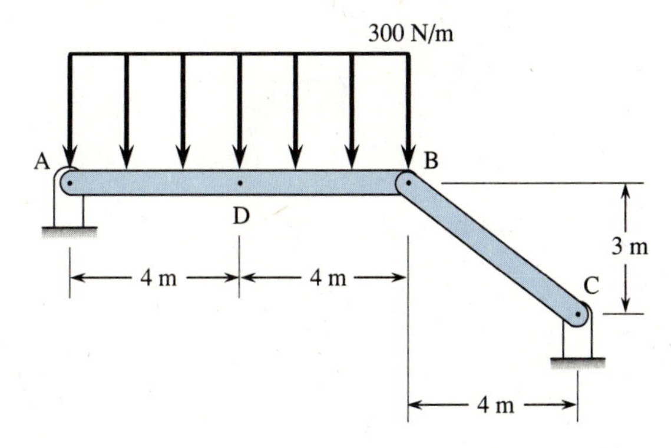

1.38–1.41 For the fixture shown, determine the component internal reactions at the indicated locations A and B.

Figure P1.38

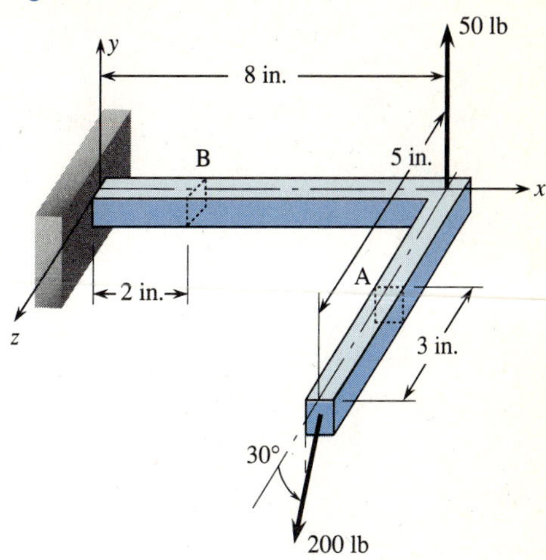

Figure P1.39

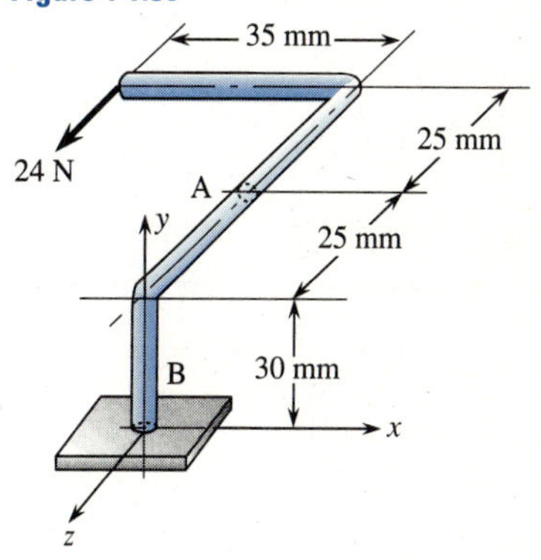

Figure P1.40

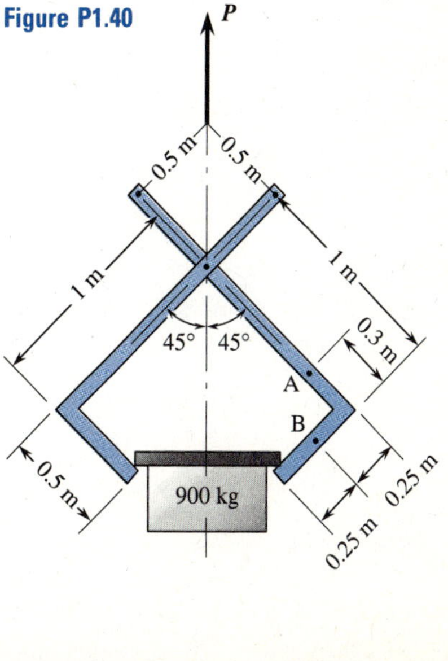

Figure P1.41

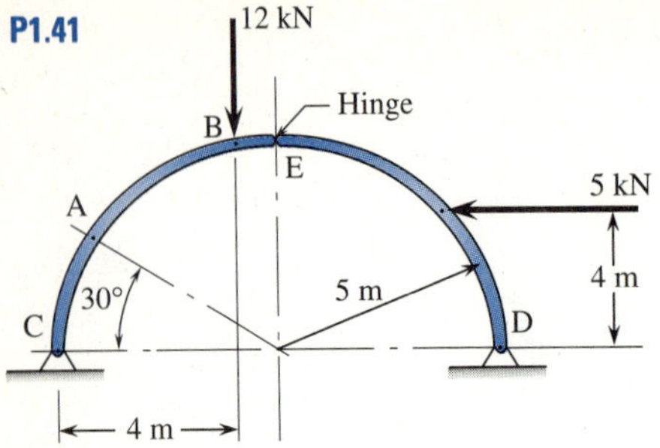

1.42 A vertical force of 80 N is applied to the handle of the pipe wrench as shown. Determine the x, y, and z components of internal force and moment at point A.

Figure P1.42

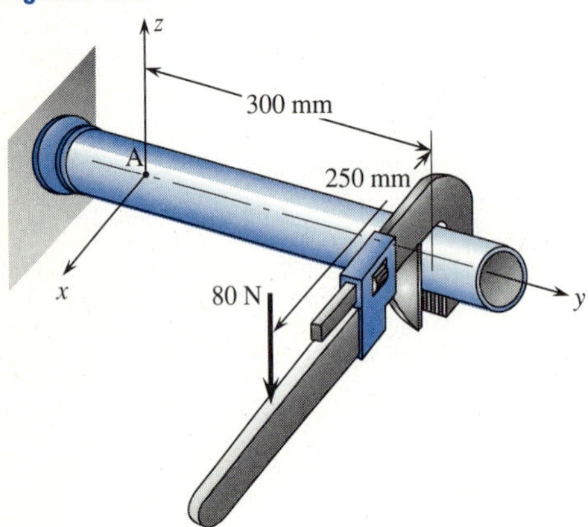

1.43 Determine the x, y, and z components of internal force and moment at point C in the pipe assembly. Neglect the weight of the pipe.

Figure P1.43

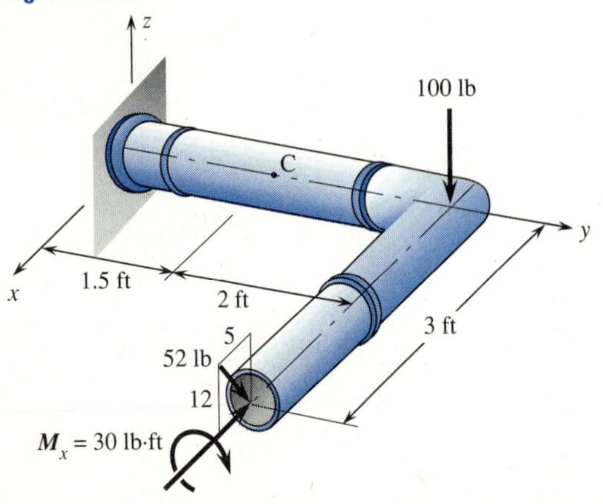

1.44 Determine the x, y, and z components of the internal axial force, shear force, and moment at point B of the rod.

Figure P1.44

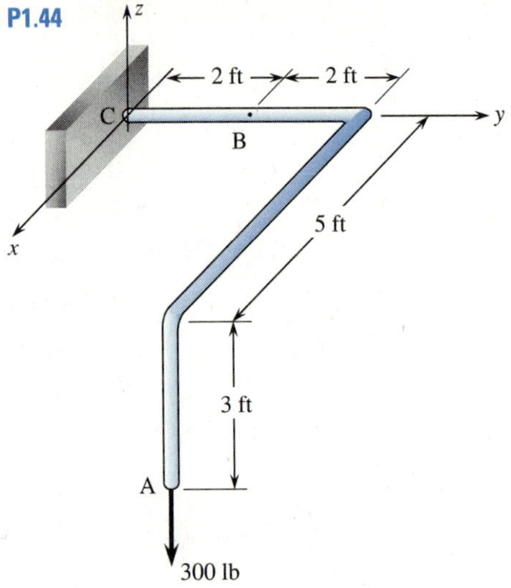

1.45–1.51 Find the internal reactions at the indicated sections for the structures shown.

Figure P1.45

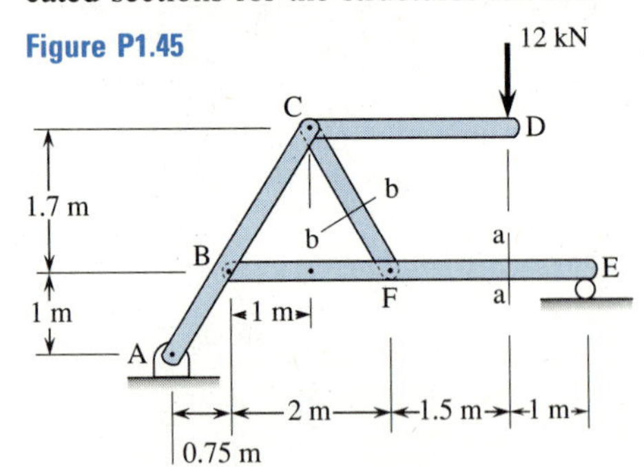

Load frame

Figure P1.46

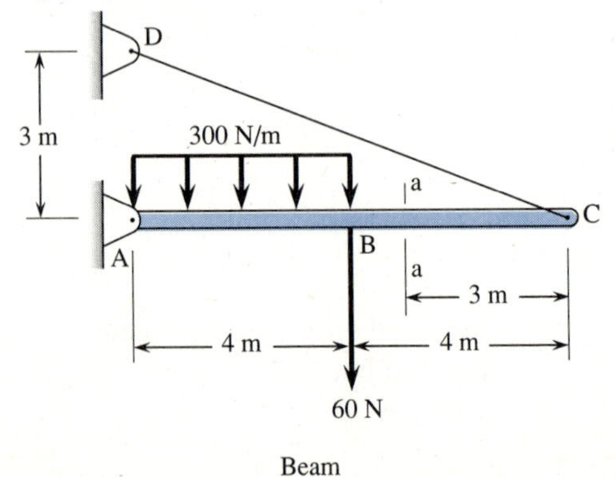

Beam

Figure P1.47

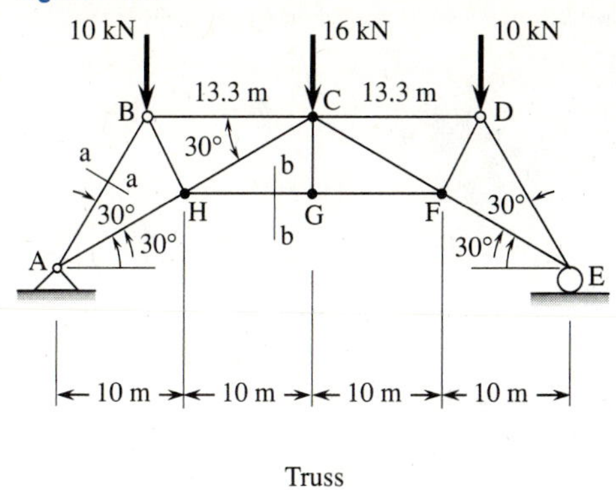

Truss

Figure P1.48

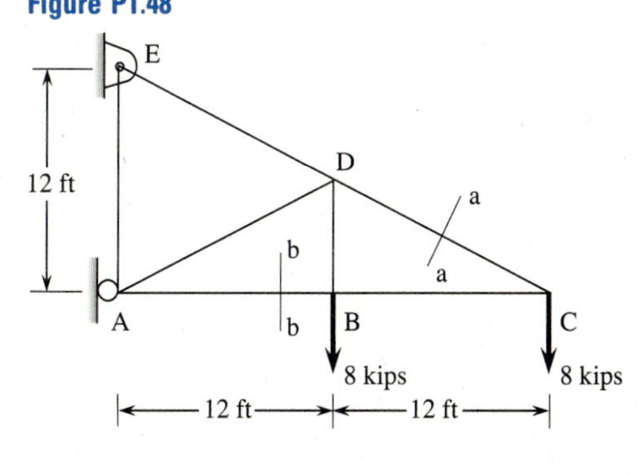

Truss

Figure P1.49

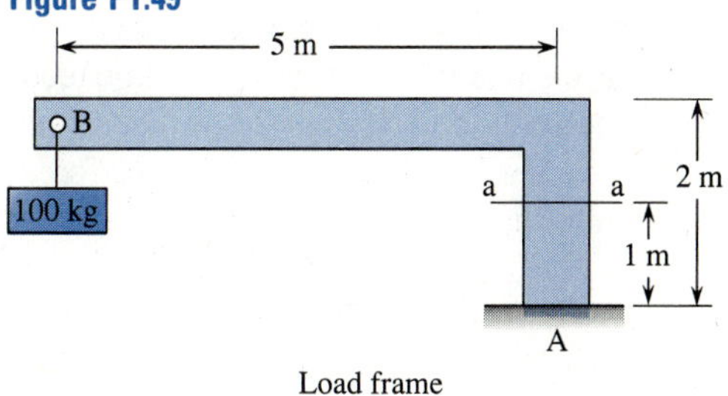

Load frame

Figure P1.50

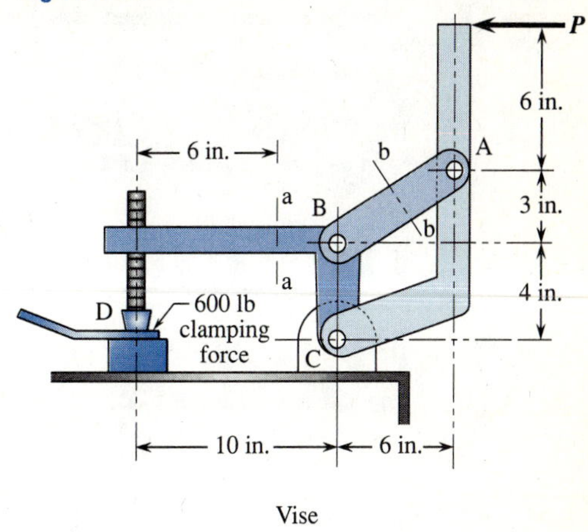

Vise

Figure P1.51

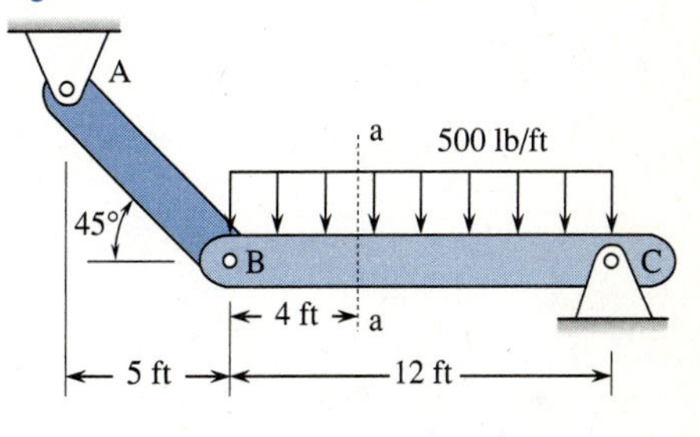

Load frame

Stresses, Strains, and Deformations in Axially Loaded Members

2.1 INTRODUCTION

In this chapter, we consider the concept of stresses and strains developed in axially loaded members. These members are generally the simplest to analyze and hence are considered first. They play important roles as load-carrying members in framed structures and machines; specific examples include trusses, landing gears, and hydraulic cranes as shown in Figure 2.1. These members are two-force members in that they are pin-ended members that transmit force only at those pins (one force at each end of the member, hence only two forces act on these members). In the truss, all members are two-force members. In the landing gear and the hydraulic crane, the member labeled BC is a two-force member.

We then consider mechanical properties of materials. Mechanical properties are used in this chapter to calculate deformations and to compare actual stresses to allowable ones. We then learn how to calculate axial displacement. Next, we treat statically indeterminate problems. Then we consider temperature effects. We describe the concept of allowable stress and considerations in selecting a factor of safety (FS). Finally, we consider stress concentrations in axially loaded members.

Figure 2.1 Examples of axially loaded members
(a) Truss
(b) Landing gear
(c) Hydraulic crane.

(a)

(b)

(c)

2.2 STRESSES DUE TO AXIAL LOADING

To evaluate the strength of structural members, the concept of stress is necessary. Stress is the quantity used to measure the strength of a member. Determining the internal reactions, as in Section 1.3, is a major step toward determining the stress in a member. However, the actual strength of a member depends on its size and shape and the material it is made of. For instance, we know a 1/2-in.-diameter steel rod is stronger than a 1/4-in.-diameter steel rod. This is because the 1/2-in.-diameter rod has a bigger cross-sectional area.

We also know that a 1-in.^{2} steel hanger can carry more load than a 1-square-inch aluminum hanger since the same size steel is a stronger material than aluminum. However, all structural members do not have the same size and shape and loading conditions (or resulting internal reactions). To compare strengths of different structural members requires the concept of stress, which incorporates the size, shape, and internal reactions.

Stress is defined as the internal force acting per unit of area. To determine stresses due to axial loading, consider the homogeneous rod subjected to equal but opposite tensile forces so as to maintain equilibrium of the rod as shown in Figure 2.2(a). The forces F are assumed to act through the centroid of the cross-sectional area A perpendicular to the longitudinal axis of the rod. To obtain an expression for the normal stress, we first determine the internal forces P at any cross-section within the length of the rod by passing a plane (say B-B) through the cross-section and perpendicular to the axis of the rod. Then since the whole rod is in equilibrium, any parts of it, equivalently represented in the two-dimensional Figures 2.2(b) and (c), are in equilibrium. Force equilibrium in the axial direction results in

$$P = F$$

The force P acting perpendicular to the cross-sectional area A actually represents the resultant of an infinite number of smaller forces acting over the cross-section. Because the load F is applied axially and through the centroid, and the material is homogeneous, these smaller forces are assumed to be uniformly distributed over and directed perpendicular to the cross-sectional area of the rod, as shown in Figures 2.2(d) and (e). Then, by the definition of *normal stress* [denoted by the Greek letter *sigma* (σ)], we have

$$\sigma = \frac{F}{A} = \frac{P}{A} \quad \left(\frac{\text{force}}{\text{area}}\right) \qquad \left(\frac{\text{lb}}{\text{in.}^2}\right) \text{ or } \left(\frac{\text{N}}{\text{m}^2}\right) \tag{2.1}$$

In the U.S. Customary System of units forces are expressed in units of pounds (lb) and area is given in units of inches squared (in.^{2}). Then, by Eq. (2.1), stress has units of pounds per square inch (lb/in.^{2}) (abbreviated psi), or kilopounds (kips) per square inch (ksi), where 1 kilopound is equal to 1000 pounds. In the Système International d'Unités (SI system) of units, force is in units of newtons and area is given in units of meters squared (m^2). Then, stress has units of newtons per square meter (N/m^2) or *pascals* (Pa), where 1 N/m^2 = 1 Pa. In practice, using the SI system the N/m^2 (Pa) is a small quantity. Hence, stresses are normally expressed in multiples of this unit, such as kilopascals (kPa) or megapascals (MPa), where 1 kPa = 10^3 Pa and 1 MPa = 10^6 Pa. The table of conversion factors provided on the inside front cover shows that 1 psi = 6890 Pa or 6.89 kPa.

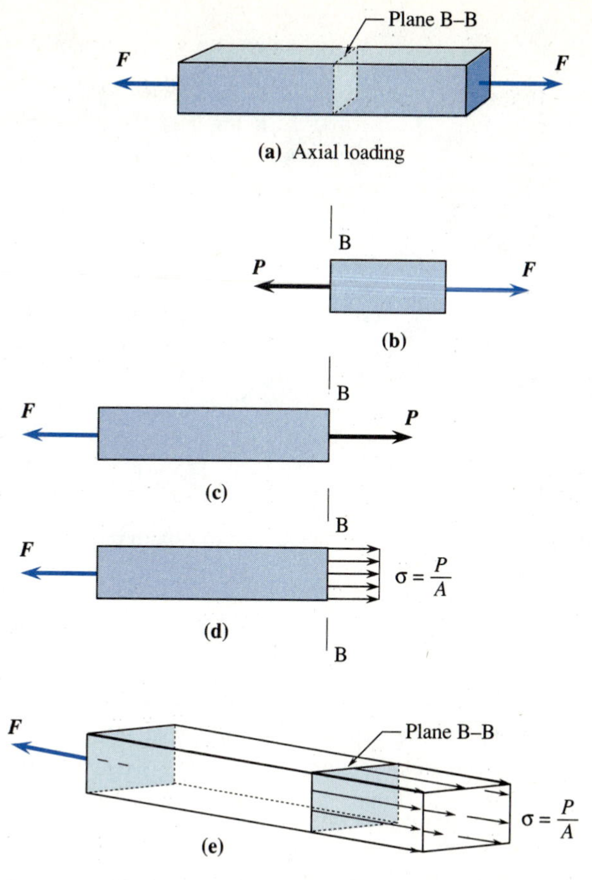

Figure 2.2 Rod subjected to tensile forces and resulting normal stress.

If the internal force P acts away or pulls on the cross-section, the resulting stress is a *tensile stress;* if the force acts into or pushes on the cross-section, the resulting stress is a *compressive stress*. It is then customary to define tensile stress as positive and compressive stress as negative.

In summary, Eq. (2.1) is valid for the following assumptions:

1. A uniform cross-section (prismatic) member
2. A centric (centroidal) axial load
3. A material that is homogeneous

For slightly tapered members, Eq. (2.1) is reasonably accurate. For compression members, we assume buckling does not occur. As a rule of thumb, a member whose smallest cross-sectional dimension is greater than or equal to one-tenth its length will probably not buckle and can be considered as a short member. In practice, then, short compression members must be used. The problem of buckling instability is dealt with in Chapter 9.

STEPS IN SOLUTION

The following steps are generally taken to obtain the normal or axial stress in an axially loaded member.

1. Draw a free-body diagram of the member showing the axial forces resisted by the member. Sometimes these forces are given as a starting point. However, for axially loaded two-force members in structures and machines these forces must be determined through proper statics.
2. Pass a plane through the member normal to the longitudinal axis of the member where you want to determine the internal force P and then the normal stress σ.
3. Write the force equilibrium equation for the member to determine the internal force P.
4. Use Eq. (2.1) to obtain the normal stress, that is,

$$\sigma = \frac{P}{A}$$

Examples 2.1 through 2.4 illustrate the steps used to obtain the normal or axial stress in axially loaded members.

EXAMPLE 2.1

A connecting rod is subjected to the axial forces applied to the centroid of the cross-section shown in Figure 2.3. The cross-sectional area is 2 in.^{2} Determine the normal stress **(a)** 10 in. and **(b)** 50 in. from the left end.

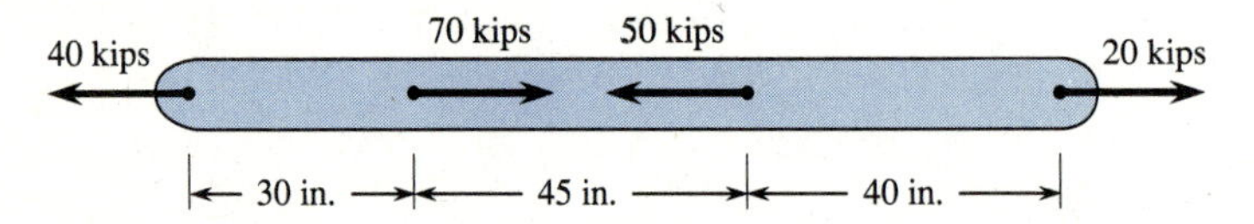

Figure 2.3 Connecting rod subjected to axial forces.

Solution

(a) Making a cut through the cross-section, we obtain the free-body diagram of the 10-in. segment as

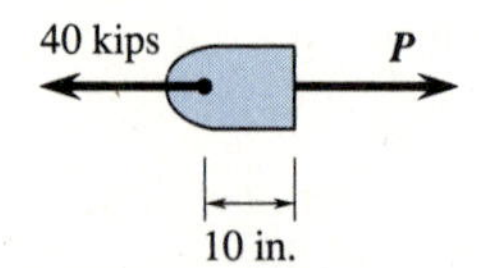

By force equilibrium, the internal normal force is

$$\Sigma F_x = 0: \quad P = 40 \text{ kips (to the right)}$$

The normal stress is obtained using Eq. (2.1) as

$$\sigma = \frac{P}{A} = \frac{40 \text{ kips}}{2 \text{ in.}^2} = 20 \text{ ksi (tensile) } (T)$$

(b) Cutting through the rod 50 in. from the left end, the free-body diagram is

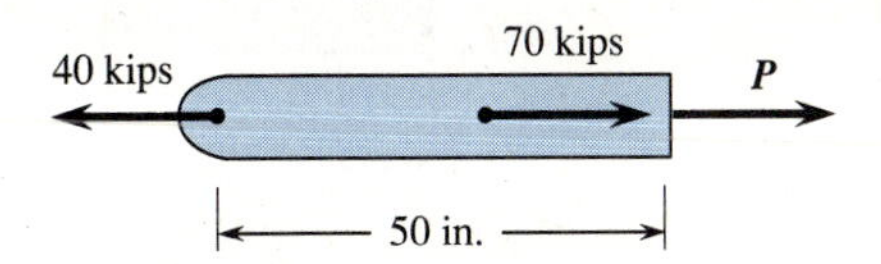

By statics, we obtain

$$\Sigma F_x = 0: \quad -40 + 70 + P = 0$$

$$P = -30 \text{ kips (or } P = 30 \text{ kips to the left)}$$

$$\sigma = \frac{P}{A} = \frac{-30 \text{ kips}}{2 \text{ in.}^2} = -15 \text{ ksi (compressive) (C)}$$

EXAMPLE 2.2

For the beam supported by the two-force members A and B shown in Figure 2.4, the axial stress in members A and B have been experimentally determined to be $\sigma_A = 20{,}000$ psi (tensile) and $\sigma_B = 1500$ psi (compressive). The cross-sectional areas of members A and B are 0.50 in.^{2} and 5 in.^{2}, respectively. Determine the applied force F.

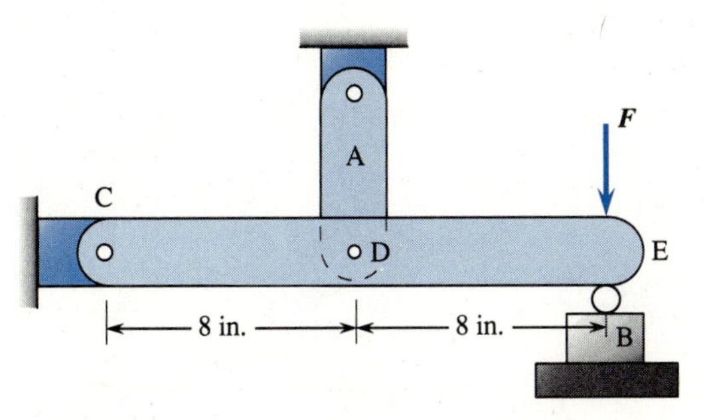

Figure 2.4 Beam supported by two-force members.

Solution

Using Eq. (2.1), we determine the axial forces in members A and B as

$$F_A = \sigma_A A_A = (20{,}000 \text{ psi})(0.5 \text{ in.}^2) = 10{,}000 \text{ lb } (T)$$

$$F_B = \sigma_B A_B = (1500 \text{ psi})(5 \text{ in.}^2) = 7500 \text{ lb (C)}$$

Next draw the free-body diagram of member CE (shown below).

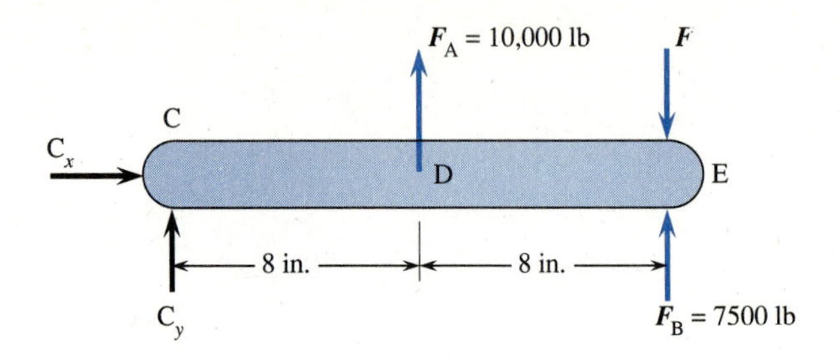

Here F_A pulls up on the horizontal member CE while F_B pushes up on CE: By statics we obtain

$$+\circlearrowleft \; \Sigma M_c = 0: \; (10{,}000 \text{ lb})(8 \text{ in.}) - F\,(16 \text{ in.}) + (7500 \text{ lb})(16 \text{ in.}) = 0$$

$$F = 12{,}500 \text{ lb}$$

EXAMPLE 2.3

For the truss subjected to the wind loading shown in Figure 2.5, determine the axial stress in member AD. The cross-sectional area of AD is a rectangle with dimensions of 50 mm and 100 mm. The truss is one of a series of trusses spaced 6 m apart.

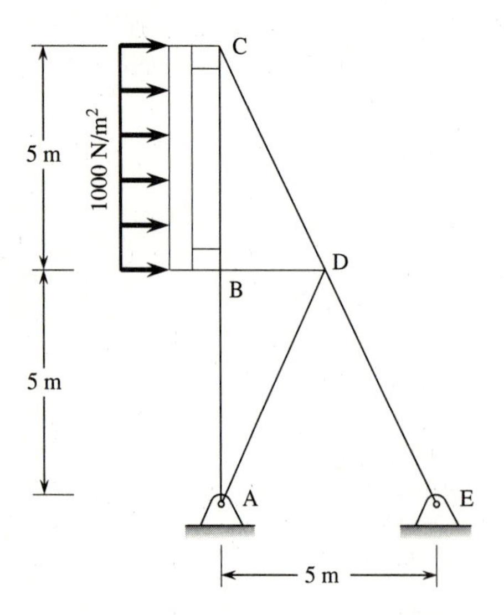

Figure 2.5 Truss subjected to wind loading.

Solution

The total force F acting on the sign supported by the truss is

$$F = (5.0 \text{ m} \times 6.0 \text{ m})(1000 \text{ N/m}^2) = 30 \text{ kN}$$

The force transmitted to joints B and C is

$$F_B = F_C = \frac{F}{4} = \frac{30\text{ kN}}{4} = 7.5\text{ kN}$$

The free-body diagram of the whole truss with the concentrated wind loads at B and C is

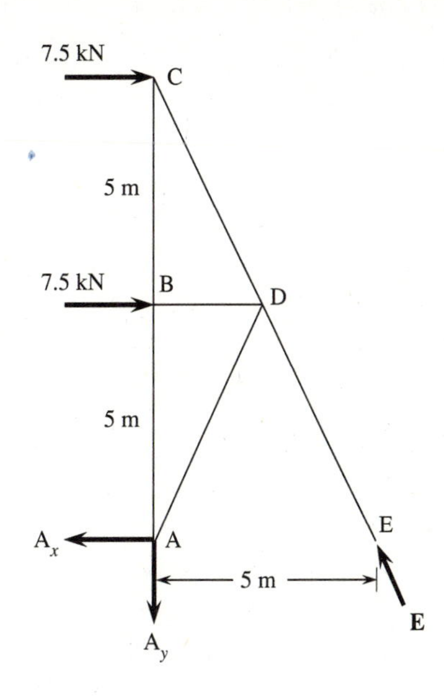

By statics, we obtain

$$\Sigma M_E = 0:\quad A_y(5.0\text{ m}) - (7.5\text{ kN})(5.0\text{ m}) - (7.5\text{ kN})(10.0\text{ m}) = 0$$

$$A_y = 22.5\text{ kN}$$

$$\Sigma F_y = 0:\quad \frac{5}{5.6}E - A_y = 0$$

$$E = \frac{5.6}{5}(22.5) = 25.2\text{ kN}$$

$$\Sigma F_x = 0:\quad -A_x - \frac{2.5}{5.6}E + 7.5 + 7.5 = 0$$

$$A_x = 3.75\text{ kN}$$

A free-body diagram of joint A is

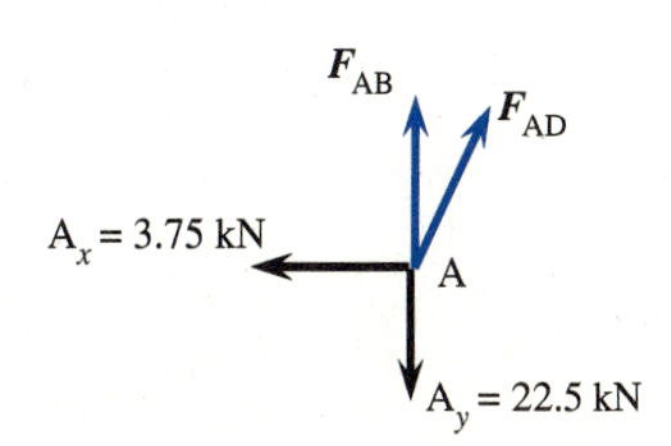

By statics, we obtain

$$\Sigma F_x = 0: \quad \frac{2.5}{5.6} F_{AD} - 3.75 = 0 \qquad F_{AD} = 8.4 \text{ kN } (T)$$

Using Eq. (2.1), the axial stress in member AD is

$$\sigma_{AD} = \frac{F_{AD}}{A_{AD}} = \frac{8.4 \text{ kN}}{0.05 \text{ m} \times 0.1 \text{ m}}$$
$$= 1680 \frac{\text{kN}}{\text{m}^2} = 1.68 \text{ MPa } (T)$$

EXAMPLE 2.4

The landing gear shown in Figure 2.6(a) is subjected to a force of 30 kN at the point of contact with the ground. Determine the axial stress in the two-force member BC. Member BC has a cross-sectional area of 1500 mm^2.

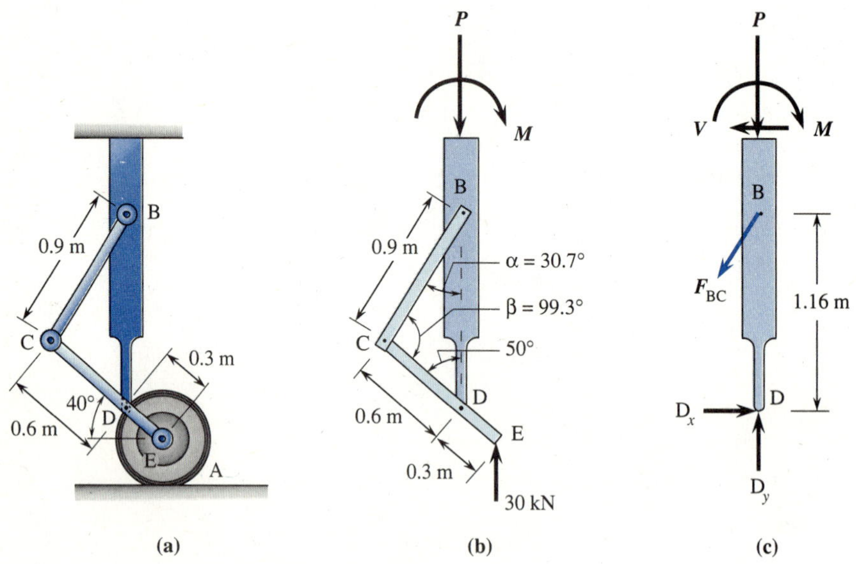

Figure 2.6 (a) Landing gear subjected to force. (b) Free-body diagram of landing gear and (c) Free-body diagram of member BD.

Solution

First, draw the free-body diagrams of the landing gear (Figure 2.6(b)) and member BD (Figure 2.6(c)).

Using the free-body diagram of the landing gear (Figure 2.6(b)) and statics, we have

$$\Sigma F_y = 0: \quad -P + 30 \text{ kN} = 0$$

$$P = 30 \text{ kN}$$

$$+\circlearrowleft \Sigma M_E = 0: \quad -M + P\,(0.3 \sin 50°) = 0$$

$$M = 30(0.3 \sin 50°)$$

$$M = 6.89 \text{ kN} \cdot \text{m}$$

From the free-body diagram of member BD (Figure 2.6(c)) and statics, we write

$$+\circlearrowleft \Sigma M_D = 0: \quad (F_{BC} \sin 30.7°)(1.16 \text{ m}) - 6.89 \text{ kN} \cdot \text{m} = 0$$

$$F_{BC} = 11.6 \text{ kN (tensile)}$$

Using Eq. (2.1), we obtain the normal stress in member BC as

$$\sigma_{BC} = \frac{F_{BC}}{A_{BC}} = \frac{11.6 \text{ kN}}{1500 \text{ mm}^2 \left(10^{-3} \dfrac{\text{m}}{\text{mm}}\right)^2} = 7.75 \times 10^3 \frac{\text{kN}}{\text{m}^2} = 7.75 \text{ MPa } (T)$$

2.3 STRESSES ON AN INCLINED PLANE UNDER AXIAL LOADING

In the preceding section we considered axial forces acting on two-force members and determined the normal stresses in the members by considering only planes perpendicular to the axis of the members. However, axial forces cause both normal and shear stresses on planes that are inclined to the axis of the member.

Consider the two-force member subjected to axial forces F in Figure 2.7(a). Now pass a plane inclined at an angle θ to the normal plane as shown in Figure 2.7(b). The free-body diagram of the portion of the member to the left of the inclined plane is shown in Figure 2.7(c). Force equilibrium in the axial direction requires that internal force $P = F$ on the inclined plane.

Now resolving P into normal N and tangential V force components to the section as shown in Figure 2.7(d), we obtain

$$N = P \cos \theta \qquad V = P \sin \theta \tag{2.2}$$

Using Eq. (2.1), the average normal stress σ is

$$\sigma = \frac{N}{A_\theta} \tag{2.3a}$$

The average shear stress, denoted by the Greek letter τ (*tau*), is obtained by dividing the shear force V in the plane of the section by the area A as

$$\tau = \frac{V}{A_\theta} \tag{2.3b}$$

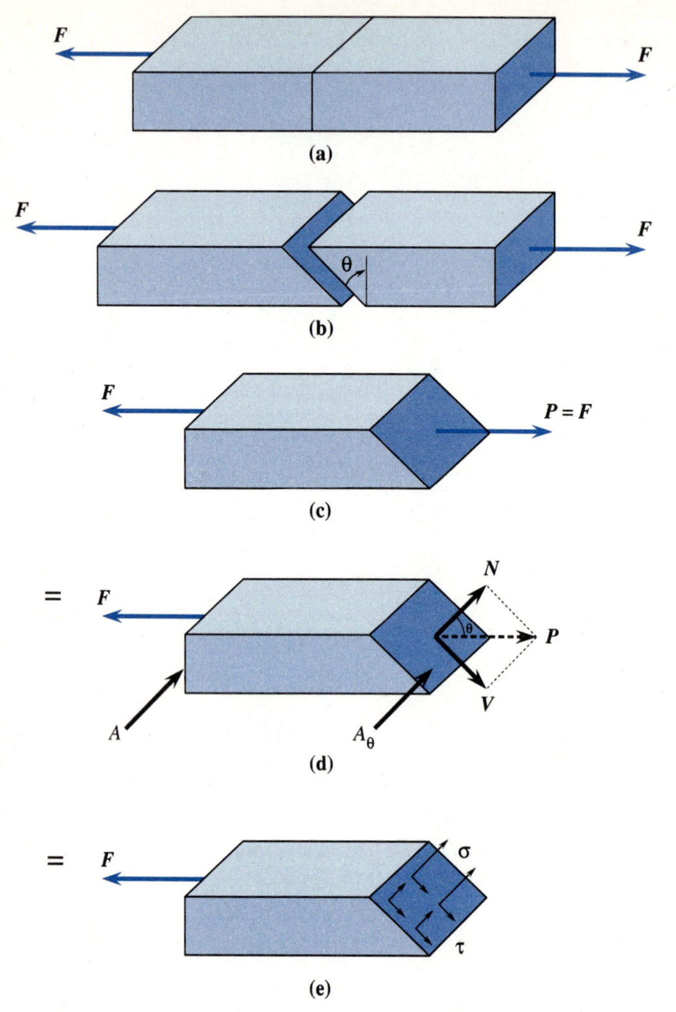

Figure 2.7 Two-force member subjected to axial forces.

where A_θ is the cross-sectional area of the inclined plane over which N and V act. Area A_θ is related to the cross-sectional area A perpendicular to the axis of the member; this relationship may be found by using trigonometry to be

$$A_\theta = \frac{A}{\cos\theta} \tag{2.4}$$

The stresses in Eq. (2.3) are actually average stresses on the area A_θ. Substituting Eq. (2.2) for N and V and Eq. (2.4) for A_θ into Eq. (2.3), we obtain

$$\sigma = \frac{P}{A}\cos^2\theta \qquad \tau = \frac{P}{A}\cos\theta\sin\theta \tag{2.5}$$

Based on Eqs. (2.5), we observe that the normal stress is a maximum when $\theta = 0°$, that is, this stress occurs on a plane perpendicular to the axis of the member and is given by

$$\sigma_{\max} = \frac{P}{A} \tag{2.6}$$

The second of Eqs. (2.5) yields $\tau = 0$ when $\theta = 0°$ and $\theta = 90°$, and yields the maximum shear stress when $\theta = 45°$ and is given by

$$\tau_{\max} = \frac{P}{A} \cos 45° \sin 45° = \frac{P}{2A}$$

or using Eq. (2.6), we have

$$\tau_{\max} = \frac{\sigma_{\max}}{2} \tag{2.7}$$

Equation (2.7) shows that the maximum shear stress in an axially loaded member is $\sigma_{\max}/2$ and occurs at $\theta = 45°$.

Equations (2.6) and (2.7) can be used to explain the manner in which brittle and ductile materials fail when subjected to axial loading. Brittle materials generally fail on planes where the maximum normal stress occurs. Hence, they fail on planes where $\theta = 0°$ or 180° as shown in Figure 2.8(a). Ductile materials fail on planes where the maximum shear stress occurs. These planes are identified with $\theta = 45°$ or 135° as shown in the ductile failure of Figure 2.8(b). We will further discuss the behavior of brittle and ductile materials in Section 2.5.

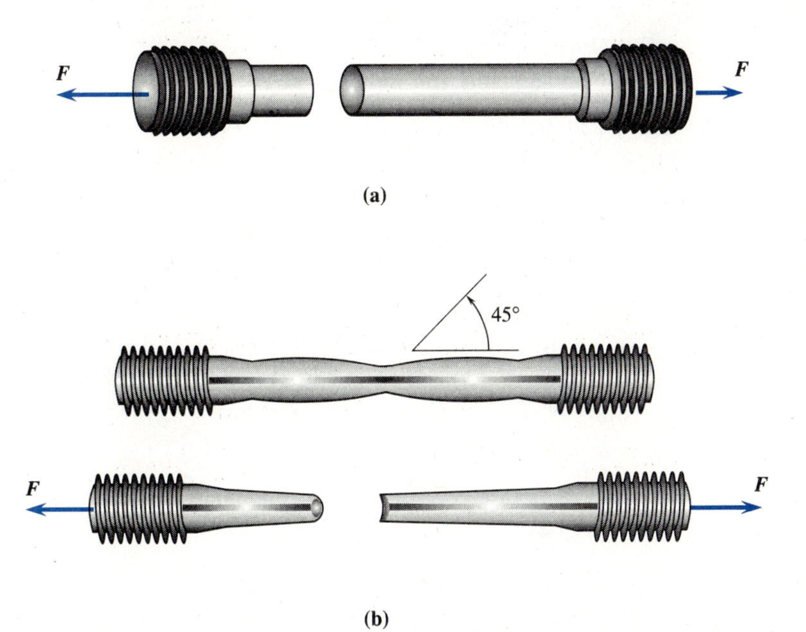

Figure 2.8 Typical (a) brittle and (b) ductile material failures in axially loaded members.

STEPS IN SOLUTION

The following steps, identified by methods 1 and 2 below, are generally taken to obtain the normal and shear stresses on an inclined plane of an axially loaded member.

METHOD 1

1. Draw a free-body diagram of the member showing the axial forces acting on the member.
2. Pass the desired inclined plane through the member and draw a free-body diagram of the portion of the member showing the axial force $P = F$ on the inclined plane.
3. Resolve P into a force component N normal to the inclined plane and a force component V tangential to, or in the plane of, the inclined plane.
4. Determine the area A_θ of the inclined plane.
5. Use Eqs. (2.3) to obtain the average normal stress σ and the average shear stress τ acting on the inclined plane.

METHOD 2

1. Use steps 1 and 2 from method 1 above to obtain P.
2. Obtain the angle θ between the axial force P and the normal force N to the inclined plane where positive θ is considered to be counterclockwise from P to N and negative θ is clockwise from P to N.
3. Use Eqs. (2.5) (with the given cross-sectional area A normal to the axis of the member, along with the determined $P = F$ and θ) to obtain the average normal stress σ and the average shear stress τ acting on the inclined plane.

Using Method 2, one has the advantage of having only to properly determine θ, as A and P (or F) are either given or obtained very easily. Resolving P into force components and obtaining A_θ are then not necessary.

Examples 2.5 and 2.6 illustrate the steps used to obtain normal and shear stresses on an inclined plane.

EXAMPLE 2.5

Two wooden members of uniform cross-sectional areas 1500 mm^2 are connected together by a glued scarf connection shown in Figure 2.9. The axial tensile load to be carried by the member is $F = 150$ kN. Determine the average normal and shear stresses on the glued connection.

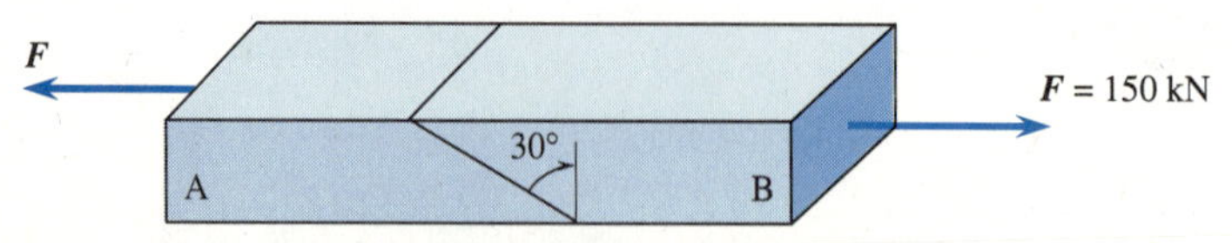

Figure 2.9 Wooden members connected by glued scarf joint.

Solution

Draw a free-body diagram of the left piece of the splice as shown below.

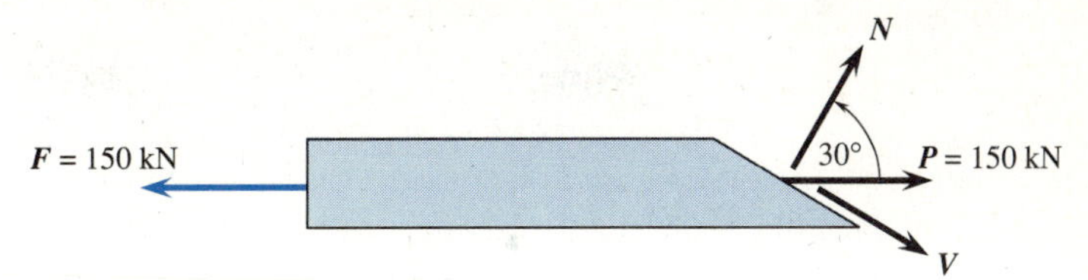

Using Eqs. (2.2), the forces normal to, and in the plane of, the area are

$$N = 150 \cos 30° = 130 \text{ kN}$$

$$V = 150 \sin 30° = 75 \text{ kN}$$

Using Eq. (2.4), we obtain A_θ as

$$A_\theta = \frac{1500 \text{ mm}^2}{\cos 30°} = 1732 \text{ mm}^2 = 1732 \times 10^{-6} \text{ m}^2$$

Using Eqs. (2.3), the normal and shear stresses on the glued connection are

$$\sigma = \frac{130 \text{ kN}}{0.001732 \text{ m}^2} = 75{,}100 \frac{\text{kN}}{\text{m}^2} = 75.1 \text{ MPa}$$

$$\tau = \frac{75 \text{ kN}}{0.001732 \text{ m}^2} = 43{,}300 \frac{\text{kN}}{\text{m}^2} = 43.3 \text{ MPa}$$

We could have obtained these same stresses by directly using Eqs. (2.5) with $\theta = 30°$ (the angle between forces $F = P$ and N).

EXAMPLE 2.6

The rear wheel of a 4000-lb automobile has been removed and the brake housing is resting on a 2.5 in. × 2.5-in. block as shown in Figure 2.10. Determine the shear stress parallel to the wood grain and the normal stress perpendicular to the wood grain. Assume the rear wheels support 40% of the weight of the automobile.

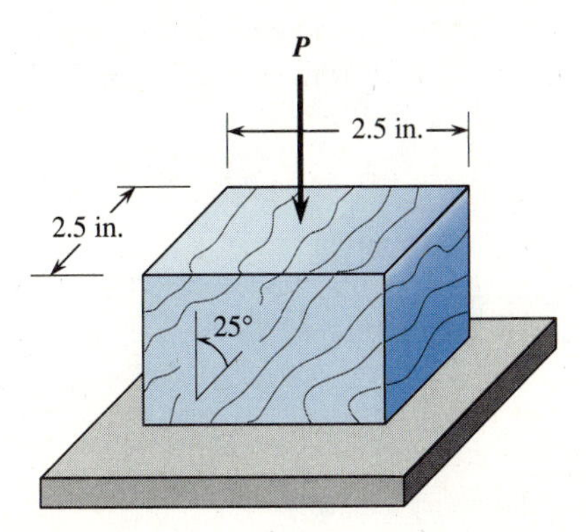

Figure 2.10 Wooden block subjected to compressive force.

Solution

We first determine the vertical force P acting on the block (the force transferred from one of the rear wheels) as

$$P = \frac{(4000 \text{ lb})(0.4)}{2} = 800 \text{ lb}$$

Using Eqs. (2.2), (2.4), and (2.3) in that order, we obtain the normal force, shear force, A_θ, and then the normal and shear stresses as

$$N = 800 \cos 65° = 338 \text{ lb}$$

$$V = 800 \sin 65° = 725 \text{ lb}$$

$$A_\theta = \frac{2.5 \text{ in.} \times 2.5 \text{ in.}}{\cos 65°} = 14.8 \text{ in.}^2$$

$$\sigma = \frac{338 \text{ lb}}{14.8 \text{ in.}^2} = 22.8 \text{ psi (compressive)}$$

$$\tau = \frac{725 \text{ lb}}{14.8 \text{ in.}^2} = 49.0 \text{ psi}$$

2.4 STRAIN DUE TO AXIAL LOADING

In addition to the importance of stresses in engineering design, we must also be concerned with deformations caused by the loads applied to a structure. It is important to avoid excessive deformations that may prevent a structure from serving its intended purpose. For instance, excessive deformations in a shaft may cause the shaft to bind up in its bearing and hence not operate properly. Excessive deformations of a member may result in unacceptable cracks in supporting walls or unacceptable clearances between members. For many structures it is not possible to determine the forces in the members by only using statics. For the analysis of statically indeterminate structures, which is initially considered in Section 2.7, deflection considerations will be necessary.

We now consider the bar, of length L, and of uniform cross-sectional area A subjected to applied tensile load F at the right end and held from displacing at the left end (Figure 2.11(a)). The total change in length of the bar is given by δ as shown in Figure 2.11(b).

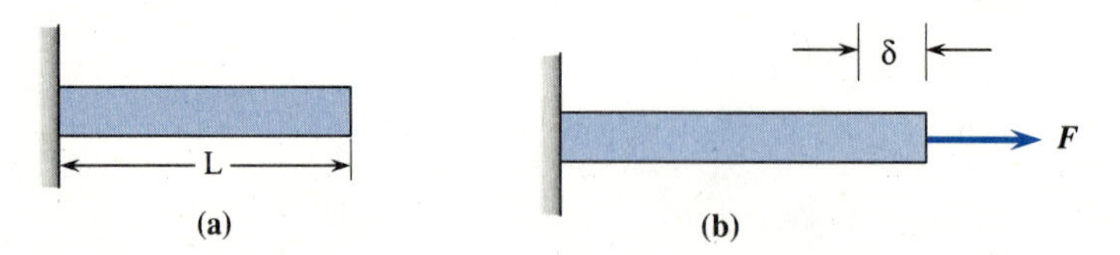

Figure 2.11 (a) Bar, and
(b) bar subjected to axial force and showing deformation.

This deformation is the sum resultant of the stretching throughout the whole length of the bar. We then define the normal strain by the Greek letter ϵ (*epsilon*) as the change in length of the original bar length divided by the original bar length and given as

$$\epsilon = \frac{\delta}{L} \tag{2.8}$$

If the bar is in tension, the strain is a *tensile strain* and the bar elongates. If the bar is in compression, the strain is a *compressive strain* and the bar shortens. Because both δ and L are expressed in the same units, normal strain ϵ is a dimensionless quantity. Hence, strain is a pure number with no units. Example 2.7 illustrates the calculation of normal strain.

EXAMPLE 2.7

The bar of length L = 50 in. elongates 0.01 in. when subjected to axial tensile forces shown in Figure 2.12. Determine the normal strain ϵ in the bar.

Figure 2.12 Bar subjected to axial tensile forces.

Solution

Using Eq. (2.8), we obtain

$$\epsilon = \frac{\delta}{L} = \frac{0.01\text{ in.}}{50\text{ in.}} = 200 \times 10^{-6}$$

In practice, it is customary to associate the original units with the strain and express the answer in Example 2.7 as

$$\epsilon = 200 \times 10^{-6}\text{ in./in.}$$

or

$$\epsilon = 200\ \mu\text{in./in.}$$

where the Greek letter μ (*mu*) stands for micro ($\mu = 10^{-6}$).

In the general case of a member with variable cross-sectional area the normal stress will vary along the length of the member and we must define strain by considering a small element of undeformed length Δx (Figure 2.13(a)) which then undergoes a deformation $\Delta\delta$ (Figure 2.13(b)) when subjected to load F. Then by our definition of normal strain

$$\epsilon = \frac{\Delta\delta}{\Delta x} \tag{2.9}$$

The normal strain at a point B in the member is then given by taking the limit of Eq. (2.9) as Δx approaches zero

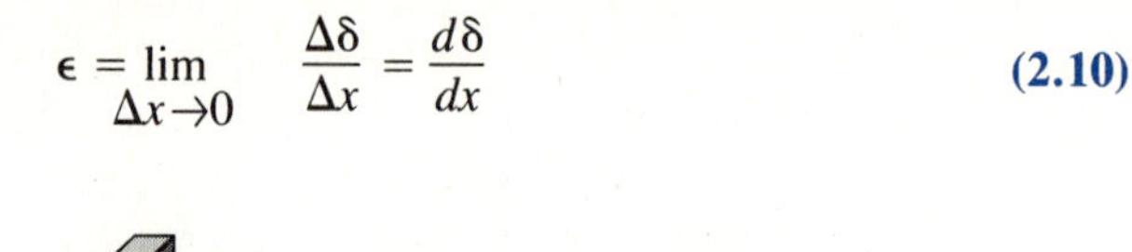

$$\epsilon = \lim_{\Delta x \to 0} \frac{\Delta\delta}{\Delta x} = \frac{d\delta}{dx} \tag{2.10}$$

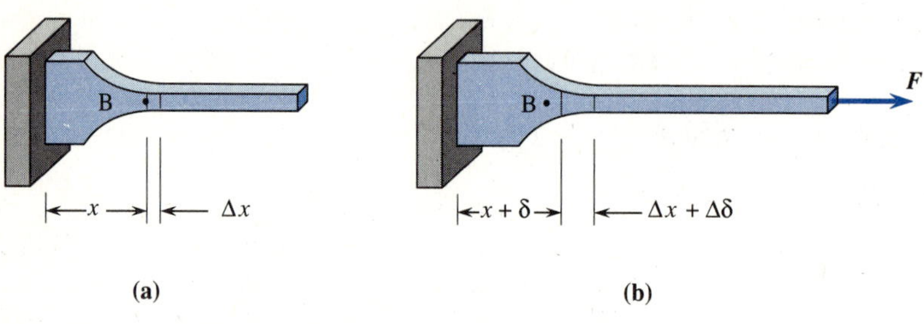

Figure 2.13 (a) Member with variable cross section, and (b) same member subjected to axial force.

2.5 STRESS-STRAIN DIAGRAM AND MECHANICAL PROPERTIES OF MATERIALS

In this section we present the important basic concepts involved with stress-strain diagrams and address the most common mechanical properties needed in subsequent concepts presented in the text. For a more complete description of the recommended standard procedure to obtain a stress-strain diagram consult the American Society for Testing materials (ASTM) [1]. For additional information on mechanical properties consult references [2] and [3].

We now describe some of the most important mechanical properties of materials. Some of these properties are shown to be needed to determine displacements in Section 2.6, to analyze statically indeterminate problems in Section 2.7, and to compare actual stress to the allowable stress in Section 2.9. These properties are then quite necessary in this chapter. We will see that they are found through a simple pulling of a bar and then measuring the axial stress and corresponding strain.

Having defined normal stress and strain for axially loaded members in Sections 2.1 and 2.4, we now describe their relationship in this section. Axial stress and strain are related through the *stress-strain diagram*. To obtain the stress-strain diagram a standard specimen of material (Figure 2.14) is subjected to a *tensile test* by applying tensile forces

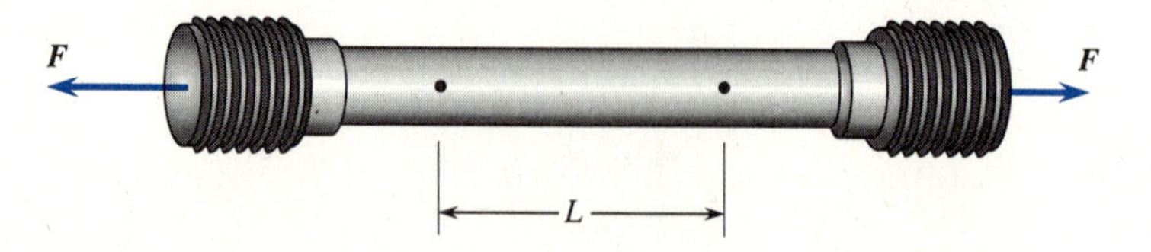

Figure 2.14 Standard tensile specimen.

to each end of the specimen. The standard size specimen is described in the ASTM specification for standards for materials and testing [1].

The specimen is enlarged at its ends so that failure will occur in the central uniform cross-sectional area portion. It is placed in a testing machine (Figure 2.15) which is capable of applying load through the centroid of the cross-section. A dial gauge is then attached to the specimen to measure the elongation during loading. The ASTM standard gauge length is 2.0 in. This is the original length L whose change is then recorded during loading. Sometimes a second gauge is used to measure the change in diameter. The ASTM standard specimen diameter is 0.5 in. The load F is applied either continuously or in increments and is recorded from the loading gauge reading, along with the elongation from the dial gauge reading. If the load is slowly increased at a deformation of about 0.1 in./min (or a strain rate of about 0.001 in./(in./sec)), the test is called a *static test*. If the load is applied rapidly, then a *dynamic test* is performed. Recommended loading rates are provided in the ASTM standard for static and dynamic testing. Materials are generally loaded gradually (static test) and at room temperatures. Figure 2.29(b) shows the influence of strain rate on the σ-ϵ curve of steel, and Figure 2.29(a) shows the influence of strain rate on tensile properties. We will describe the influence of strain rate subsequently under Other Influences on Mechanical Properties.

Figure 2.15 Typical testing machine used for tensile testing.

From each pair of readings, the stress is computed by dividing F by the cross-sectional area A and the strain by dividing the change in length δ by the original gauge length L. If the original cross-sectional area is used to obtain the stresses, the resulting stress is called the *engineering stress* or *nominal stress*. If the actual cross-sectional area is used to obtain the stress at each load, the stress is called the *true stress*. If the original gauge length is used to obtain the strains, the resulting strain is the *nominal strain*. If the actual distance between gauge marks is used to obtain the strain, we call this the *true strain*.

Each pair of stress-strain data is then plotted to obtain the stress versus strain curve. It is conventional to plot strain on the abscissa (horizontal axis) and stress on the ordinate (vertical axis). The plot is then called a *stress-strain diagram*. The engineering stress-strain diagram is most frequently used. From it a number of mechanical properties used for analysis of load bearing structures are determined.

Typical engineering stress-strain diagrams of a low-carbon mild steel, a low-alloy, high-strength steel, and a tool steel are shown in Figure 2.16(a). *Mild steel* or *structural steel* (steel with less than 0.25% carbon) is commonly used in the construction of buildings, bridges, or cranes. *Low-alloy, high-strength steels* are used for grader blades, mining equipment, construction equipment, railroad cars, offshore oil and gas lines, ship construction, and for other critical structural applications such as pressure vessels and nuclear applications. *Tool steel* is used for shock and wear, such as for pneumatic tooling parts, chisels, and punches. These diagrams were obtained by slowly applying a load (static test) at room temperature to the specimen. Enlarged diagrams of some typical steels are shown in Figure 2.16(b).

The diagram of the mild steel shows an initial linear portion up to point A and a nonlinear portion thereafter. From the diagram, a number of mechanical properties of the material are described. Some of these properties, even though they do exist, are difficult to measure and of limited practical use. Others are very important in engineering analysis, such as for predicting yielding of materials, deformation, and buckling.

1. ***Proportional limit.*** The *proportional limit* (σ_p) represents the maximum value of the stress at which the stress is proportional to the strain. This is shown as point A in Figures 2.16(a) and (b). A typical value of σ_p is approximately 30 ksi (200 MPa) for low carbon steel. The proportional limit is a difficult property to obtain experimentally and is rarely used in engineering analysis.

2. ***Elastic limit.*** The *elastic limit* (σ_e) represents the maximum value of stress at which the material will regain its original dimensions if the load causing this stress is removed. For any stress level below σ_e, the material will return to its original undeformed state. However, for stresses beyond σ_e, the material does not return to its original dimensions and some permanent deformation results. Point B in Figures 2.16(a) and (b) represents the elastic limit. This value is difficult to determine experimentally and is therefore difficult to exactly locate on the stress-strain curve, although it is normally somewhat higher than the proportional limit. For practical purposes, the elastic and proportional limits are often taken to be equal.

If the load is such that the stress is beyond the elastic limit and then the load is removed, the material follows the stress-strain line DE shown in Figure 2.16(c). This line is parallel to the linear-elastic straight line OA. When the load is totally removed as represented by point E, a *residual* or *permanent strain* remains in the material. The strain EG

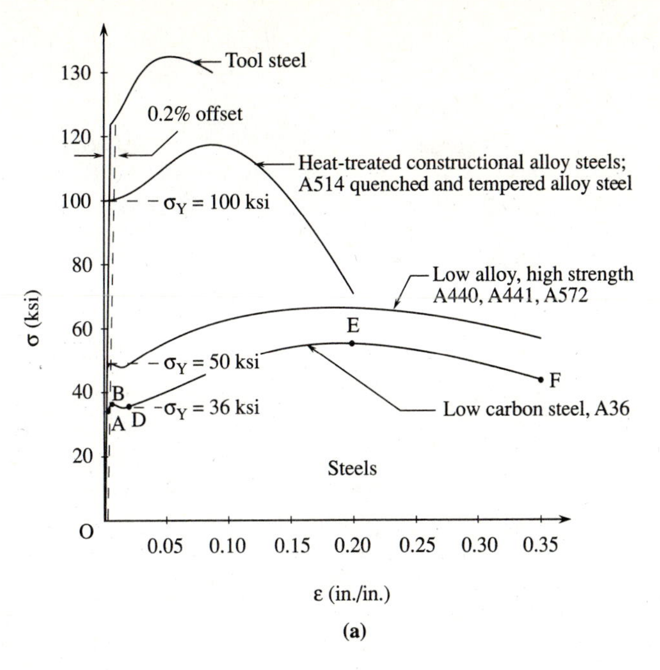

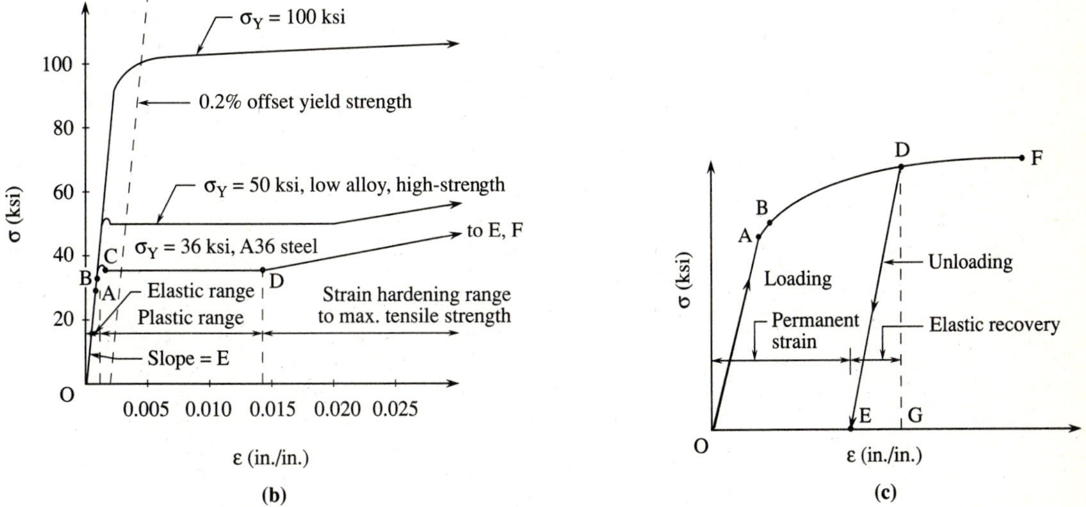

Figure 2.16 (a) Typical stress-strain diagrams of some steels, (b) enlarged diagrams of some typical steels, and (c) path taken during unloading.

has been recovered as it is elastic strain, while the strain OE remains in the bar as permanent strain. The strains beyond the elastic limit are also called *plastic strains*.

3. ***Modulus of Elasticity.*** The *modulus of elasticity* is the proportionality constant relating stress to strain in the linear portion of the stress-strain diagram. The modulus of elasticity E (sometimes called Young's modulus) is then the slope of the straight line OA

on the diagram in Figure 2.16(b). The modulus is then a measure of the stiffness of a material. We then relate stress to strain in the segment OA of the diagram by

$$\sigma = E\epsilon \tag{2.11}$$

Equation (2.11) is also known as *Hooke's law,* named after Robert Hooke who established the linear relationship between applied load F and elongation δ in 1676. This equation relates stress to strain for the uniaxial tension or compression state of stress. For two- and three-dimensional states of stress, more elaborate relationships exist among the stresses and strains and are described in Chapter 8. From Eq. (2.11), we observe the units on E must be those of stress, psi or ksi in the USCS, and pascals (Pa) (Newton's per square meter, N/m^2) (or multiples of Pa) in the SI system. Typical approximate values of E are: for steel $E = 29 \times 10^6$ psi (200 GPa), and for aluminum alloys $E = 10 \times 10^6$ psi (70 GPa). The modulus of elasticity is an important mechanical property in engineering analysis, therefore a table of values of E for numerous materials is provided in Appendix B.

4. ***Yield stress.*** The *yield stress* (σ_Y) is the value of the stress at which the material continues to deform without additional applied load. This region from C to D in Figure 2.16(b) shows considerable elongation taking place at constant stress. Region CD is called the *perfectly plastic* region. When a material undergoes inelastic strains (beyond those up to the elastic limit), the material is behaving in a plastic manner. Some materials, such as aluminum alloys (Figure 2.17) and steels with high yield strengths (see Figure 2.16(b)), do not have a well-defined yield stress point. In this case the yield stress is defined by an

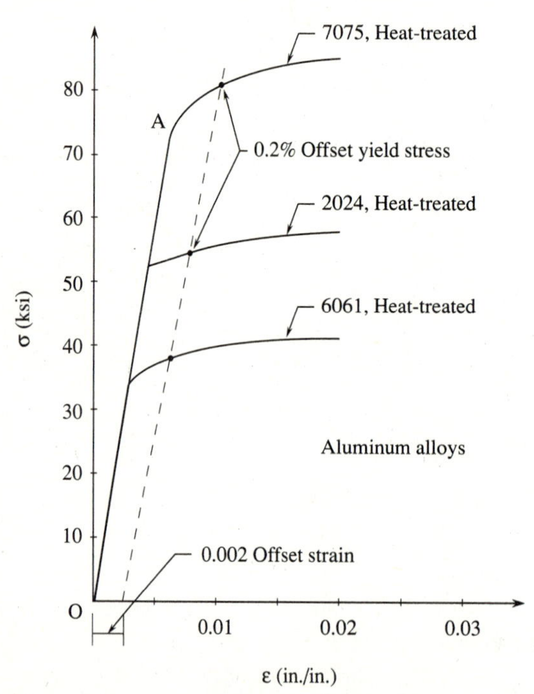

Figure 2.17 Typical stress-strain diagrams of some aluminum alloys.

offset. Using the offset method, a line is drawn on the stress-strain diagram parallel to the linear portion OA of the diagram but is offset by a standard amount of strain normally at $\epsilon = 0.002$ (or called the 0.2% offset) as shown in Figures 2.16(b) and 2.17. The aluminum alloy 2024 is often used in aircraft and race car structures and truck wheels, while aluminum alloy 6061 is often used for truck and marine parts, corrosion resistant structures, pipelines and race cars, and aluminum alloy 7075 is used for aircraft structural components.

For many metals the yield stress is just slightly higher than the proportional limit. Typical yield stresses for structural steel range from 30 to 100 ksi (200 to 700 MPa) and for aluminum alloys from 5 to 70 ksi (35 to 500 MPa). The yield stresses of numerous materials are listed in Appendix B as this property is frequently used in engineering analysis.

5. ***Ultimate stress.*** After point D for some steels, the stress must be increased to keep elongating the material until the ultimate stress is reached.This is due to strain hardening taking place (Figure 2.16(b)). During strain hardening, the material undergoes changes in its atomic and crystalline structure. These changes produce increased resistance of the material to additional deformation. The ultimate stress, σ_u, is the value of the highest stress point on the stress-strain diagram. This is point E in Figure 2.16(a). This point corresponds to the maximum load that the material can resist. For structural steel ultimate stresses range from 50 to 120 ksi (100 to 550 MPa).

6. ***Fracture stress.*** The fracture or rupture stress, σ_f, is the stress at which the material fractures or breaks into separate parts. This is point F in Figure 2.16(a). Between points E and F on the diagram in Figure 2.16(a) there is actually a decrease in the load resisted by the material.

7. ***Poisson's ratio.*** When a uniform cross-section bar is subject to a tensile load, the axial stretching is accompanied by a contraction in the lateral dimension. For circular cross-sections, this lateral dimension is the diameter d of the bar. For rectangular bars, the lateral dimensions are the width b and height h of the bar (Figure 2.18(b)). Figure 2.18(a) shows the lateral contraction for a bar in tension.

The lateral contraction d (or b or h) is often measured with a dial gauge so that the lateral strain, ϵ_l, can be determined from

$$\epsilon_l = \frac{\text{lateral contraction}}{\text{original lateral dimension}} = \frac{\Delta d}{d} = \frac{\Delta b}{b} = \frac{\Delta h}{h}$$

The lateral strain is proportional to the axial strain, ϵ_a, in the linear elastic range (OA in Figure 2.16(a) or (b)) of a material, provided the material is homogeneous and isotropic.

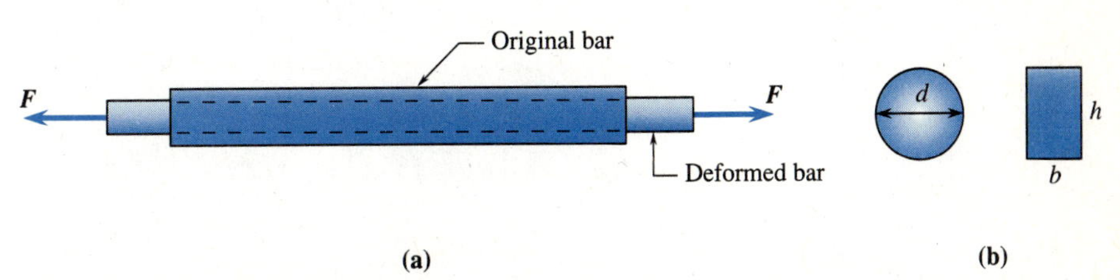

Figure 2.18 (a) Lateral contraction for a bar in tension with (b) cross-sectional dimensions.

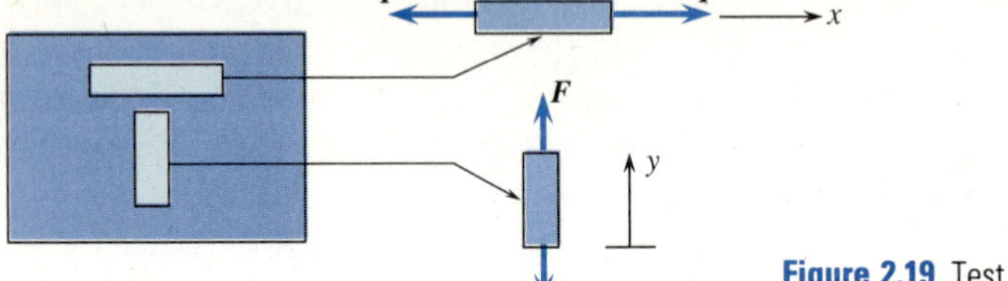

Figure 2.19 Test specimens taken from typical isotropic material.

A *homogeneous* material has the same composition throughout and hence the same mechanical properties at every position in the material. An *isotropic* material has the same mechanical properties in all directions. For instance, cutting out two test specimens, one in the x-direction and one in the y-direction (Figure 2.19) and then performing a tension test on each, results in the same value of E as measured from the stress-strain diagrams of each. Most metals consist of a large number of crystals randomly oriented and are considered to be isotropic. An *anisotropic* material has mechanical properties that change with orientation. For instance, E measured from a specimen cut in the x-direction will be different than E measured from a specimen cut in the y-direction. Fiberglass with directional fibers (Figure 2.20(a)) and wood (Figure 2.20(b)) both exhibit anisotropic material behavior.

Poisson's ratio, denoted by the Greek letter ν (nu), is defined to be the absolute value of the lateral strain, ϵ_l, divided by the axial strain, ϵ_a, or the negative of the lateral strain divided by the axial strain and is given by

$$\nu = \left| \frac{\text{lateral strain}}{\text{axial strain}} \right| = \left| \frac{\epsilon_l}{\epsilon_a} \right| = \frac{-\epsilon_l}{\epsilon_a} \tag{2.12}$$

Typical values of Poisson's ratio are: for steel $\nu = 0.30$, for aluminum alloys $\nu = 0.33$, and for concrete $\nu = 0.1 - 0.2$. A theoretical upper limit for Poisson's ratio is 0.5. This can be determined by considering the volume change that occurs in a bar during a tensile or compressive loading. The volume must increase under a tensile loading and hence the limiting value of $\nu = 0.5$. A value of ν greater than 0.5 would result in a volume decrease which is not physically possible. This topic is discussed further in Section 8.12.

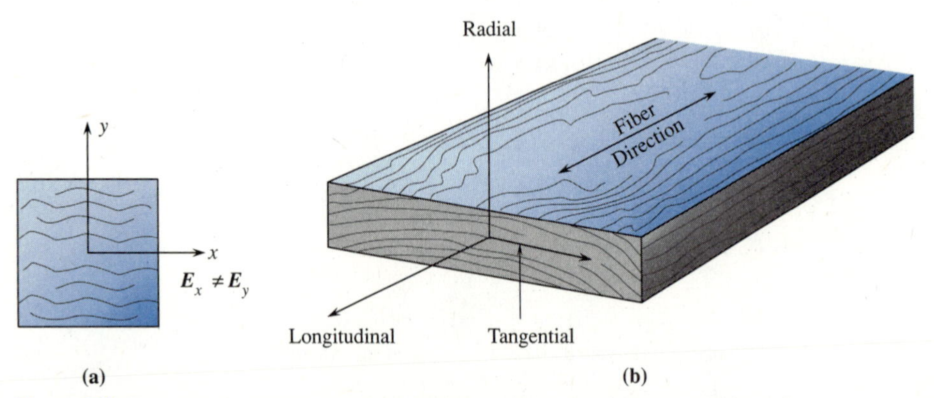

Figure 2.20 Typical anisotropic material (a) fiberglass with directional fibers and (b) wood.

EXAMPLE 2.8

During a tensile test on a high-strength steel rod, the original diameter of $d = 0.5$ in. decreases by 0.000105 in. and the original gauge length of 2 in. increases by 0.00141 in. when subjected to a tensile load $F = 4000$ lb. Calculate the modulus of elasticity E and Poisson's ratio ν.

Solution

Using Eq. (2.1), the axial stress is

$$\sigma = \frac{F}{A} = \frac{4000 \text{ lb}}{\pi(0.25 \text{ in.})^2} = 20{,}400 \text{ psi}$$

Using Eq. (2.8), the axial strain is

$$\epsilon_a = \frac{\delta}{L} = \frac{0.00141 \text{ in.}}{2 \text{ in.}} = 0.000705$$

and the lateral strain is

$$\epsilon_l = \frac{\Delta d}{d} = \frac{-0.000105 \text{ in.}}{0.5 \text{ in.}} = -0.000211$$

Since E is the slope of the linear portion of the stress-strain diagram and $\sigma = 20{,}400$ psi is less than the proportional limit stress of steel ($\sigma_p = 30{,}000$ psi is a typical proportional limit stress of a high-strength steel) then

$$E = \frac{\sigma}{\epsilon_a} = \frac{20{,}400 \text{ psi}}{0.000705} = 28.9 \times 10^6 \text{ psi (200 GPa)}$$

Using Eq. (2.12), Poisson's ratio is

$$\nu = \left| \frac{\epsilon_l}{\epsilon_a} \right| = \left| \frac{-0.000211}{0.000705} \right| = 0.299$$

8. ***Ductility.*** *Ductility* is a measure of a material's ability to deform in the plastic range. Materials that can undergo large strains before failure are called *ductile*. If the region from B to F in Figure 2.16(a) is long, a material is ductile. Mild steel, aluminum, and many other metals are examples of ductile materials. Ductile materials, when subjected to tensile loading, have a large reduction in area just before failure. This phenomenon is called *necking* as seen in Figure 2.21.

A material's ductility is measured by its elongation and reduction in area. These characteristics can be quantified by defining (1) the *percent elongation* (e_l) as

$$\text{Percent elongation} = e_l = \frac{L_f - L}{L} \times 100 \tag{2.13}$$

Figure 2.21 Ductile material showing necking just before failure.

where L_f is the length between the gauge marks at fracture and L is the original gauge length before loading; and (2) the *percent reduction in area* (r_a) defined as

$$\text{Percent reduction in area} = r_a = \frac{A - A_f}{A} \times 100 \tag{2.14}$$

where A_f is the final cross-sectional area at the fractured section and A is the original cross-sectional area. Great care must be taken to obtain L_f and A_f experimentally. Percent elongation based on L = 2-in. gauge length, is approximately 5–40% for steel and from 1% to 60% for aluminum, depending on their chemical composition and treatment.

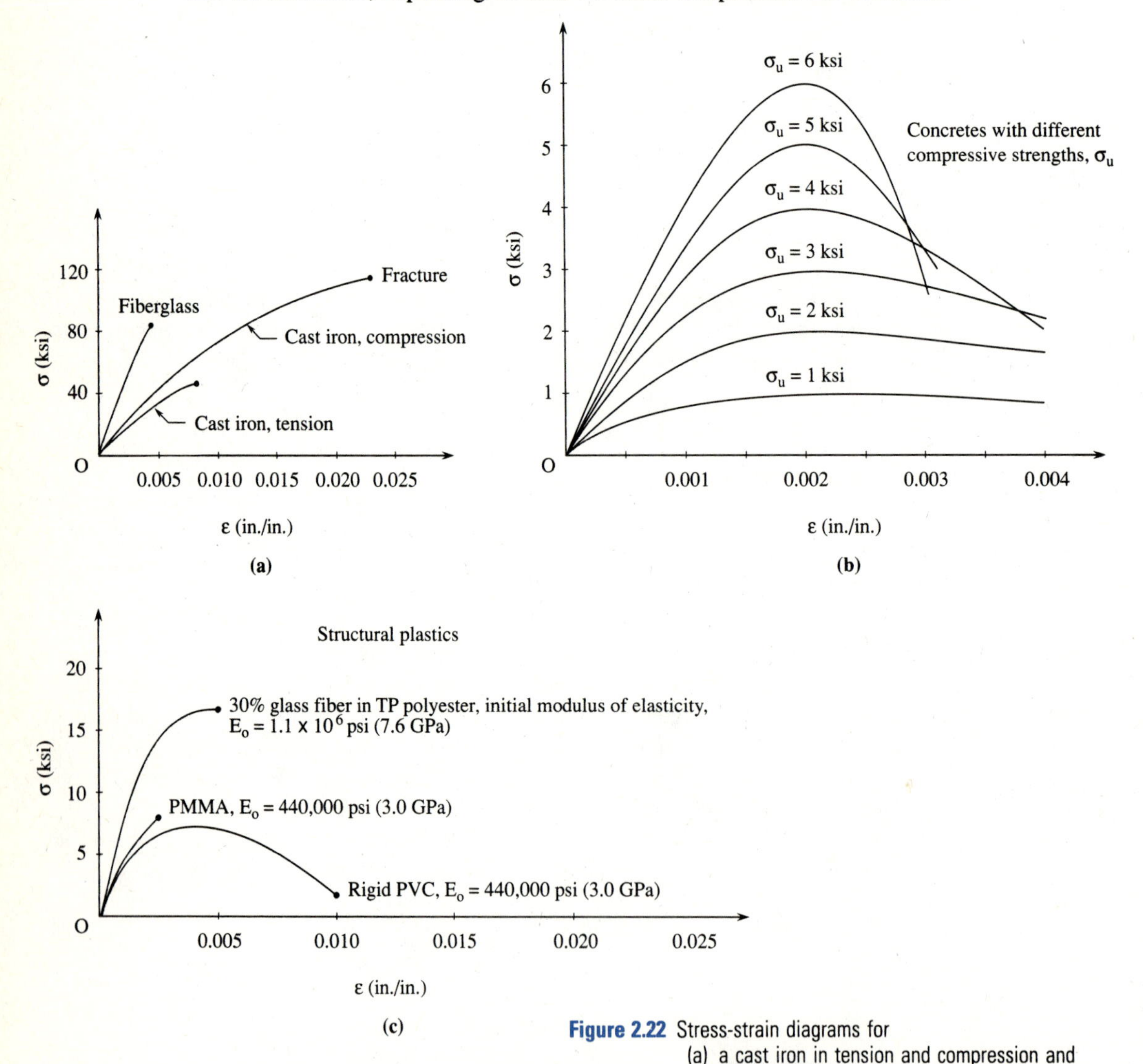

Figure 2.22 Stress-strain diagrams for
(a) a cast iron in tension and compression and for fiberglass in tension,
(b) for concretes of various compressive strengths, and
(c) for several structural plastics.

9. ***Brittle.*** Materials that fracture at low strains and with little plastic or permanent strain are classified as *brittle*. That is, brittle materials generally fail with small total strains and sometimes linear stress-strain behavior (Figure 2.22(a)), and with no definite yield strength. Brittle materials are often defined as ones with less than 5% elongation. High-carbon steel, cast iron, concrete, glass, and ceramics are examples of brittle materials. A typical tensile stress-strain diagram for the brittle materials cast iron, fiberglass, polymethyl methacrylate (PMMA) or acrylic, rigid polyvinyl chloride (PVC), and the thermal plastic (TP) polyester with 30% glass fibers is shown in Figure 2.22.

Brittle materials are often much stronger in compression than in tension. For instance, cast iron may reach an ultimate stress of 100 ksi (700 MPa) in compression but only 40 ksi (280 MPa) in tension (see Figure 2.22(a)). Similarly, concrete (see Figure 2.22(b)) may have ultimate stresses of 3000–10,000 psi (20–70 MPa) in compression but only approximately 300–1000 psi (2–7 MPa) in tension.

10. ***Energy absorption capacity.*** The capacity of a material to absorb energy is most important when considering the response of a machine or structure subjected to static, dynamic, and impact loads. In Chapter 10, Section 10.15, we describe the energy method used to analyze structures under impact loads. Here we find it convenient to briefly describe two useful energy quantities often used to measure the energy absorption capacity of materials.

a. The *modulus of resilience* (MR) represents the amount of energy absorbed per unit volume of material when the material is stressed to the proportional limit. It is obtained by calculating the area under the stress-strain diagram up to the proportional limit as shown in Figure 2.23.

Therefore, MR is

$$\text{MR} = \frac{\sigma_p^2}{2E} \quad \text{(psi) or (Pa)} \tag{2.15}$$

Because the proportional limit is difficult to obtain experimentally, the modulus of resilience is also an approximate property. For a structural steel with σ_p approximately equal to 30,000 psi and E of 30×10^6 psi, the modulus of resilience is approximately given by

$$\text{MR} = \frac{30{,}000^2}{2 \times 30 \times 10^6} = 15 \text{ psi (103 kPa)}$$

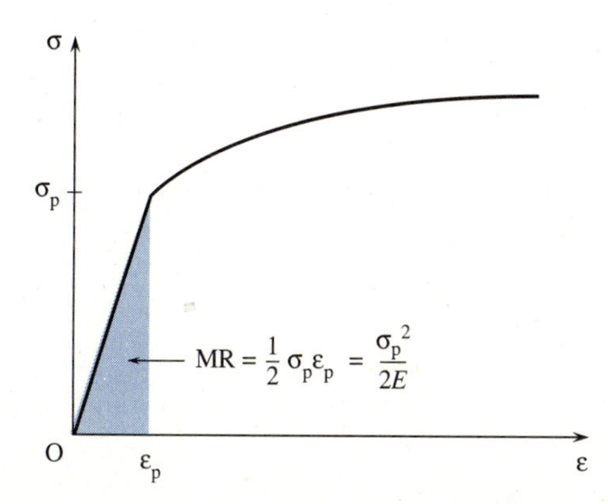

Figure 2.23 Typical stress-strain curve showing modulus of resilience, MR.

b. The *modulus of toughness* (MT) represents the amount of energy absorbed per unit volume of material when the material is stressed to fracture. It is obtained by calculating the area under the entire stress-strain diagram as shown in Figure 2.24.

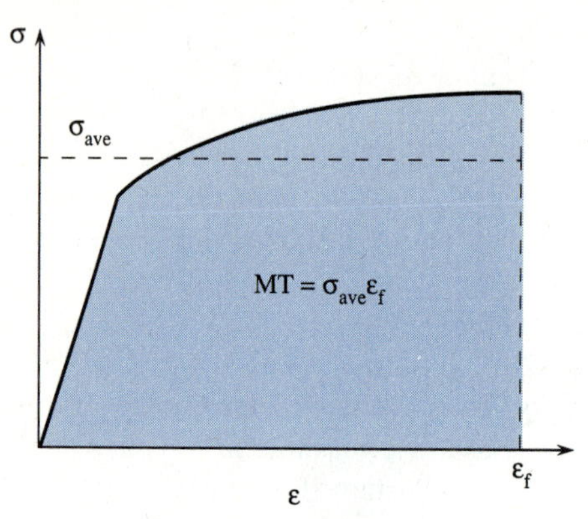

Figure 2.24 Typical stress-strain diagram showing modulus of toughness, MT.

We sometimes calculate the modulus of toughness by estimating an average stress σ_{ave} as shown in Figure 2.24, then MT is approximately given by

$$MT = \sigma_{ave}\,\epsilon_f \tag{2.16}$$

EXAMPLE 2.9

A steel rod 0.127 in. in diameter, with a gauge length of 4 in., is subjected to a gradually increasing tensile load. The load versus deformation table obtained from the test is given below. Construct the engineering stress-strain diagram and determine the following: (1) the modulus of elasticity, E, (2) the 0.2% yield stress, σ_Y, and (3) the ultimate stress, σ_u.

Load, F, (lb)	Deformation, δ, (in.)
250	0.0025
500	0.0050
750	0.0075
850	0.0095
950	0.0115
1050	0.0171
1100	0.0212
1150	0.0305
1200	0.0356
1150	0.0410
1100	0.0461 (fracture)

Solution

The cross-sectional area of the rod is

$$A = \frac{\pi}{4}(0.127 \text{ in.})^2 = 0.0127 \text{ in.}^2$$

Stresses and strains are calculated for each pair of load-deformation data. For example, at $F = 750$ lb and $\delta = 0.0075$ in.

$$\sigma = \frac{750 \text{ lb}}{0.0127 \text{ in.}^2} = 59{,}055 \text{ psi}$$

$$\epsilon = \frac{0.0075 \text{ in.}}{4 \text{ in.}} = 0.00188$$

Other pairs of stress-strain data can be developed. The final stress-strain diagram is then constructed as shown.

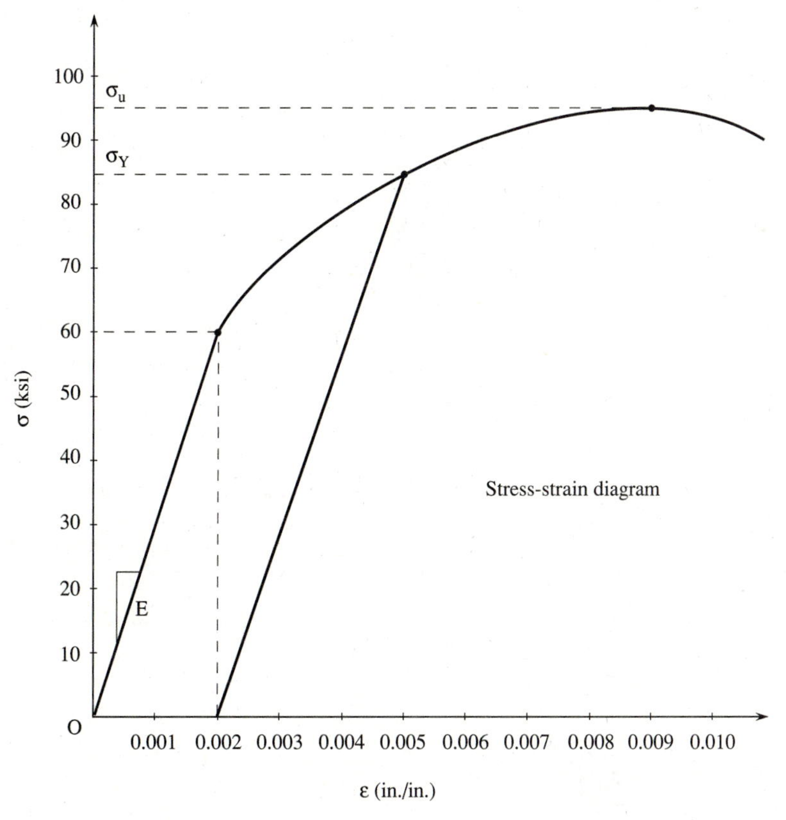

1. $E = \dfrac{\Delta\sigma}{\Delta\epsilon} = 30.0 \times 10^6$ psi (the slope of the straight-line portion of the stress-strain diagram)
2. $\sigma_Y = 85{,}000$ psi (based on the 0.2% = 0.002 offset strain on the diagram)
3. $\sigma_u = 94{,}500$ psi (the largest stress on the diagram)

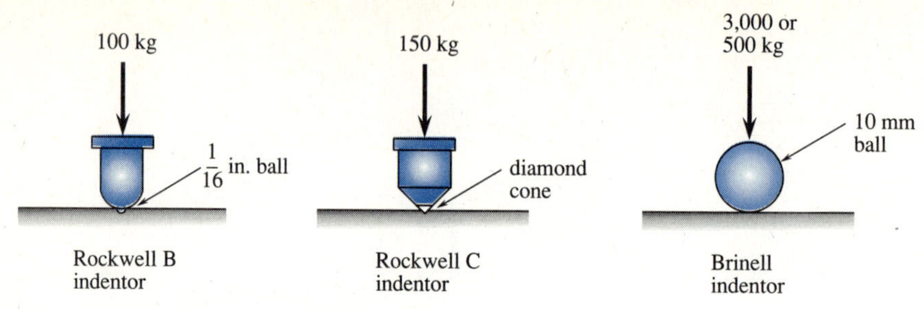

Figure 2.25 Indentors used in hardness testing.

11. ***Hardness.*** The resistance of a material to penetration by a hardened ball or other form of indentor is *hardness*. The two most common methods for measuring hardness are the *Rockwell* and *Brinell* test methods. Indentors used for these tests are shown in Figure 2.25. The hardness numbers from these tests can be related to the ultimate strength of the material.

The *Rockwell hardness test* is an easy and quick test with good reproductibility. The hardness is read directly from the simple testing machine dial. Rockwell scales are designated A, B, C, etc. The indenting tools are numbered 1, 2, or 3. The applied load is either 60, 100, or 150 kg. For instance, the common Rockwell B scale uses a 100-kg load and a No. 2 indentor ball with a 1/16 in. diameter. The Rockwell C scale uses a 150-kg load with a diamond cone indentor, which is called a No. 1 indentor. Hardness numbers obtained are relative to the material being tested. For a given material a table of hardness numbers corresponds to an approximate ultimate tensile strength. The hardness of metals can usually be increased either by cold working or by heat treating, or by both methods. (Cold working and heat treating are discussed subsequently in this section.) Table 2.1 shows Rockwell and Brinell hardness numbers for corresponding ultimate tensile strengths for carbon and alloy steels (not for cast iron). For more complete tables, see American Society of Metals handbook [19].

The *Brinell hardness test* is also a common method used to obtain the hardness of a material and then to relate it to the tensile strength. In this test, the indenting tool is a ball and the hardness number is determined as a number equal to the applied load divided by the spherical surface area of the indentation. The units of Brinell hardness are those of stress. This test takes more time to get the results than the Rockwell test, as the Brinell hardness number is computed from the data and not read directly as in the Rockwell test. The primary advantage of the Brinell hardness number is that it is directly related to the ultimate strength of the material tested. Both kinds of hardness tests are nondestructive and therefore can be used in the field to determine the ultimate strength of a material whose ultimate strength is not known for some reason, such as sometimes occurs in older existing structures.

For steels and cast iron, the relationship between the minimum ultimate tensile strength and the Brinell hardness number H_B is recognized by ASTM as

Steel:

$$\sigma_u = 0.450\, H_B \quad \text{in ksi}$$

$$\sigma_u = 3.10\, H_B \quad \text{in MPa} \tag{2.17}$$

TABLE 2.1 HARDNESS VERSUS APPROXIMATE TENSILE STRENGTH OF STEEL

Rockwell Hardness		Brinell Hardness		
C Scale	B Scale	(Steel Ball)	σ_u (ksi)	Common Steel*
—	79	136	70	1018 hot-rolled
2	86	160	81	1018 cold-rolled
6	89	171	85	
8	90.3	177	88	
12	93.4	190	93	
14	94.9	197	97	
16	96.2	206	100	
22	100.2	235	112	E4340 steel—normalized
34	—	318	150	Machined 4140, heat-treated
46	—	442	208	4340
54	—	—	256	Torsion bars for high performance

*An explanation of the numbers used here is given in Appendix E.

Cast iron:

$$\sigma_u = 0.230\, H_B - 12.5 \quad \text{in ksi}$$

$$\sigma_u = 1.58\, H_B - 86 \quad \text{in MPa} \tag{2.18}$$

From Eq. (2.18), we observe that to obtain a gray cast iron of 25 ksi tensile strength, we need

$$H_B = \frac{\sigma_u + 12.5}{0.23} = \frac{25 + 12.5}{0.23} = 163$$

12. ***Other influences on mechanical properties.*** We emphasize that many of the material's properties change due to such factors as temperature changes in the material, addition of alloys, type of heat treatment, and manufacturing process used. We briefly explain some of these additional factors in the following. For more details on how these factors affect material properties, consult a materials science reference such as [2].

Cold Working

Cold working is a process of stressing and deforming a material in the plastic region of the σ-ϵ diagram by working it by rolling, drawing or pressing, etc. at normal temperature. The effect of cold working on the stress-strain diagram of low-carbon (mild) steel is shown in Figure 2.26. From Figure 2.26(b) we observe that yield strength, ultimate strength, and hardness all increase, while ductility decreases with increased cold working. Cold rolling is commonly used for production of wide sheets, while cold drawing is often used for cold finished bars.

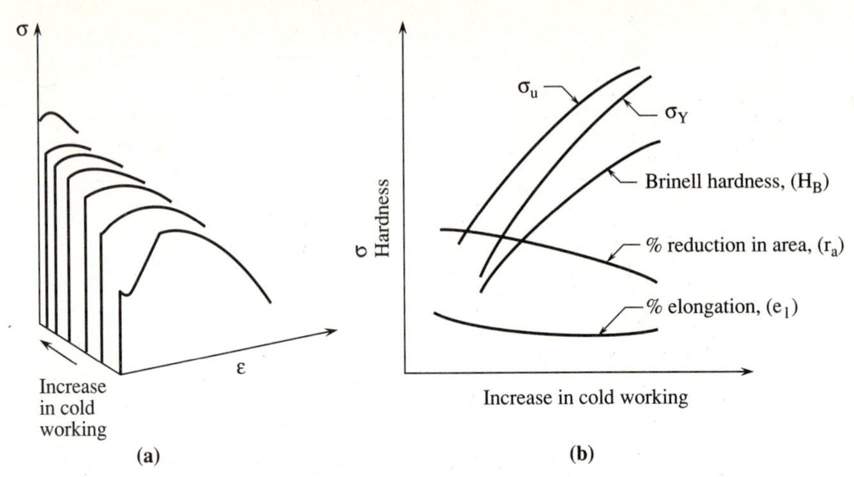

Figure 2.26 (a) Effect of cold working on σ-ϵ diagram of mild steel and (b) effect of cold working on mechanical properties.

Hot Working

Hot working is a process of deforming a material by such means as hot rolling, forging, hot extrusion, and hot pressing while the material is heated sufficiently to make it plastic and easily worked. *Hot rolling* is often used to form a bar of material of a specific shape and size. Figure 2.27(a) shows a typical hot-rolled bar, while Figure 2.27(b) shows common shapes produced by hot rolling. Figure 2.27(c) shows the hot rolling process of a flat plate into a tube. The edges of the plate are rolled together to form the tube. These butted edges are then welded together. Seamless tubing is manufactured by piercing a solid, round, heated rod with a piercing roll (Figure 2.27(d)). The difference in σ-ϵ diagrams for a cold-drawn bar and a hot-rolled bar is shown in Figure 2.28.

Extrusion is a process in which a metal billet (or blank) is heated and then forced through a restricted orifice or a die to form extruded tubing, pipe, bar, or various shapes (Figure 2.27(e)). This process is generally reserved for metals with lower melting points, such as aluminum, copper, magnesium, lead, tin, and zinc.

Forging is the hot working of a metal by hammering, pressing, or with a forging machine. The forging process results in a refined grain structure which results in increased strength and ductility of the metal.

Heat treating effects the properties of steel. We briefly describe the common heat treating operations of annealing, quenching, tempering, and case hardening here. For more details, again consult a materials science text [2].

Annealing is used to soften a material and make it more ductile and to relieve residual stresses due to hot or cold working. Annealing is a heating operation in which the temperature of the steel is raised about 100°F above the critical temperature based on the equilibrium diagram [2]. It is held at this temperature sufficiently long for carbon to become dissolved and diffused through the material.

Quenching indicates the manner in which a heated steel is cooled. Steel is cooled in air, water, or oil. *Oil quenching* is relatively slow, but prevents quenching cracks caused by rapid cooling of the steel.

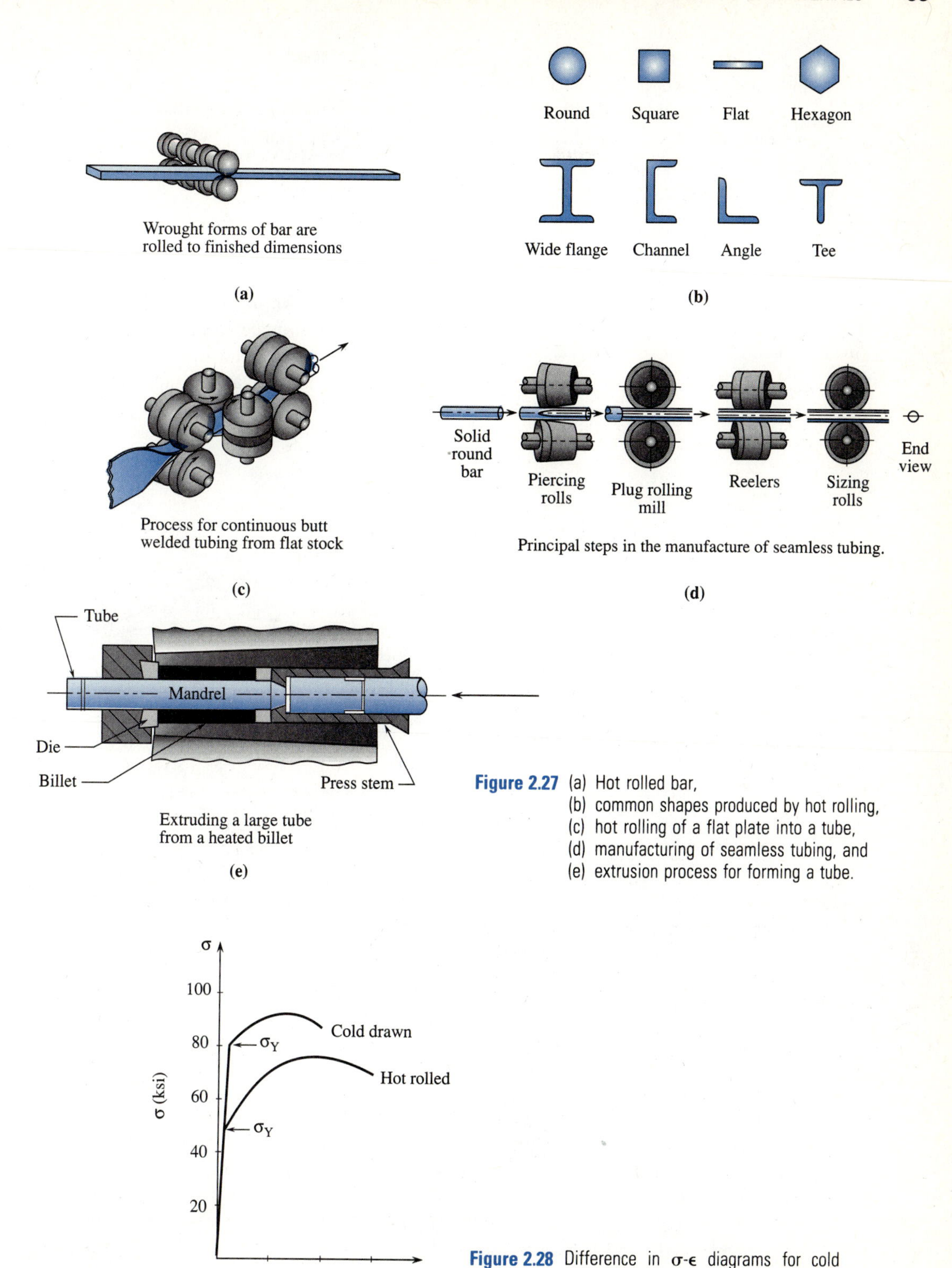

Figure 2.27 (a) Hot rolled bar,
(b) common shapes produced by hot rolling,
(c) hot rolling of a flat plate into a tube,
(d) manufacturing of seamless tubing, and
(e) extrusion process for forming a tube.

Figure 2.28 Difference in σ-ϵ diagrams for cold drawn and hot rolled bars of steel.

Tempering is the reheating of steel after being quenched to relieve the steel of stress and soften or temper it.

Case hardening is used to produce a hard outer surface while maintaining a ductile and tough inner core. This harder shell is obtained by increasing the carbon content at the surface of low-carbon steel.

Rate of Loading

The *rate of loading* effects the stress-strain diagram as shown in Figure 2.29. When the strain rate is increased, as under dynamic loading, the ultimate strength of a material increases while the yield strength appears to approach the ultimate strength. The strain rate corresponding to 0.001 in./(in. · s) is normally used for obtaining the usual stress-strain diagrams. This rate is considered to be similar to a static load condition.

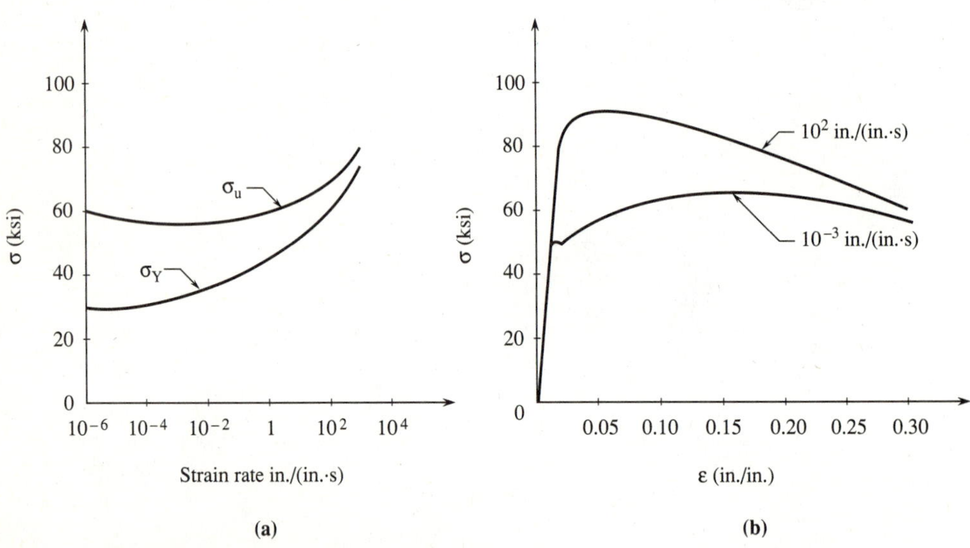

Figure 2.29 (a) Effect of strain rate on tensile properties and (b) effect of strain rate on σ-ϵ diagram.

Temperature of Material

The *temperature of a ductile material* effects the mechanical properties as shown in Figure 2.30(a) for structural (mild) steel. The ultimate tensile strength changes only slightly until reaching a critical temperature (this temperature depends on the material). At this temperature, the tensile strength decreases rapidly. The yield strength tends to decrease slowly as the temperature is increased until some critical temperature is reached. Then the yield strength drops off rapidly. The modulus of elasticity gradually decreases as the temperature increases. The ductility increases with temperature increase. For very low tempera-

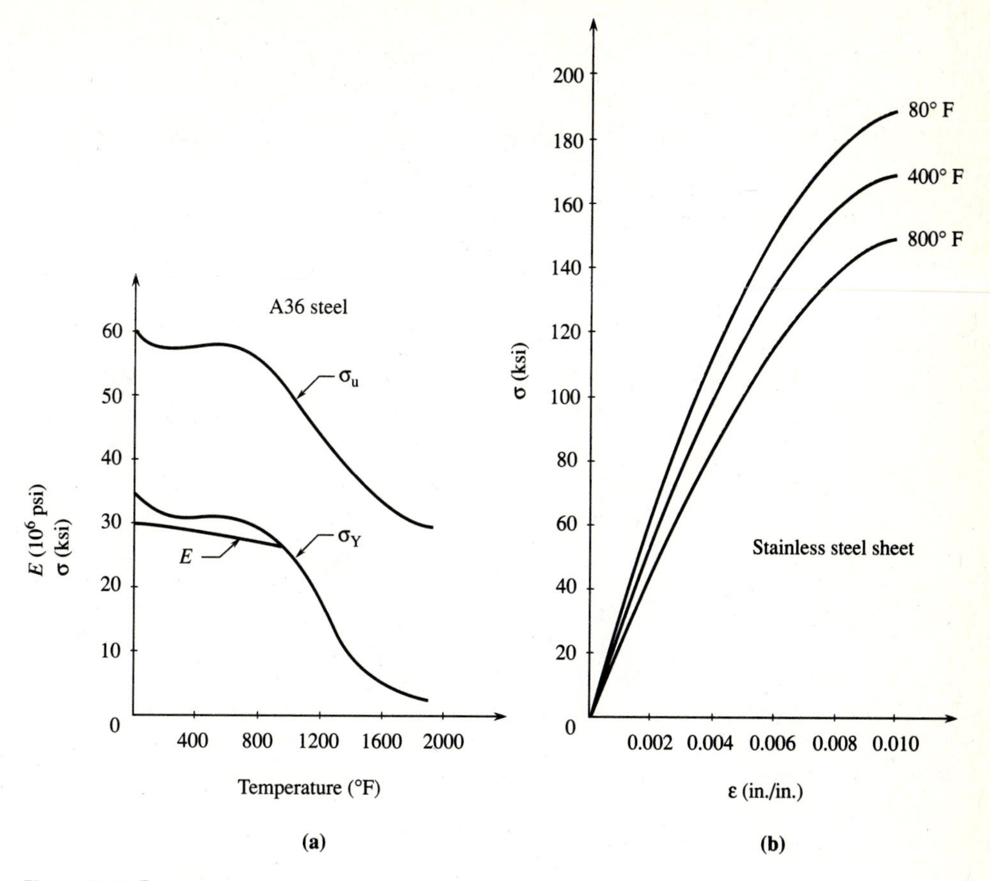

Figure 2.30 Temperature effect on mechanical properties of
(a) mild steel and
(b) stainless steel.

tures, many materials behave in a brittle fashion. For materials with no well-defined yield stress, such as stainless steel (Figure 2.30(b)), the ultimate strength decreases and the modulus of elasticity decreases with temperature increases.

2.6 DEFLECTIONS DUE TO AXIAL LOADING

To determine deflections in members subjected to axial loading, consider the bar of length L, cross-sectional area A, and modulus of elasticity E subjected to the axial force P shown in Figure 2.31. We assume the loading P results in axial stress $\sigma = P/A$ which does not exceed the proportional limit stress of the material. Hence, Hooke's law, Eq. (2.11), relating stress to strain is used along with Eq (2.1) to obtain

$$\epsilon = \frac{\sigma}{E} = \frac{P}{AE} \tag{2.19}$$

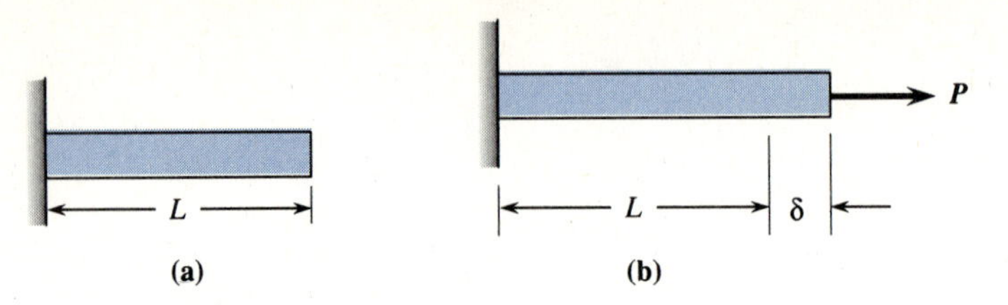

Figure 2.31 (a) Undeformed bar and
(b) bar subjected to axial force showing deformation.

Now using Eq. (2.8) for ϵ, we obtain

$$\epsilon = \frac{\delta}{L} = \frac{P}{AE} \tag{2.20}$$

Equation (2.20) is now rewritten as

$$\delta = \frac{PL}{AE} \quad \text{(in.) or (mm)} \tag{2.21}$$

Equation (2.21) can be directly applied only if the bar has a constant E, uniform cross-sectional area A, and constant internal force P over the whole length L. If the bar is subjected to a series of concentrated forces applied to different locations on the bar (Figure 2.32) or if segments of the bar have different materials (different E's) or different uniform cross-sectional areas (Figure 2.33) then we must apply Eq. (2.21) to each segment with the different E's, A's, or internal axial forces P's as follows:

$$\delta = \sum_{i=1}^{N} \frac{P_i L_i}{A_i E_i} \quad \text{(in.) or (mm)} \tag{2.22}$$

where N = the total number of parts the equation is separated into based on the abrupt changes in P, A, or E and i is a dummy variable from 1 to N. For instance, consider the

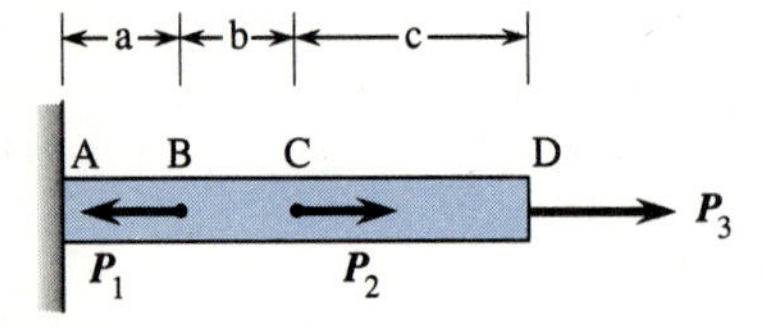

Figure 2.32 Bar with intermediate axial loads.

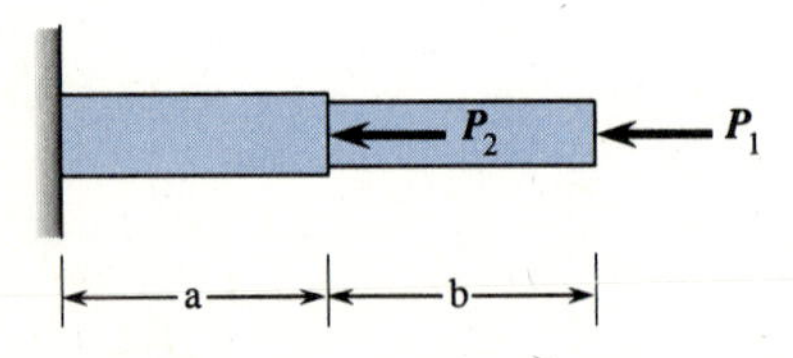

Figure 2.33 Bar with changes in cross section and intermediate loads.

uniform cross-sectional bar loaded as shown in Figure 2.32. From the free-body diagrams obtained by cutting sections of the bar between CD, BC, and AB as shown, we obtain the internal forces used in Eq. (2.22) as

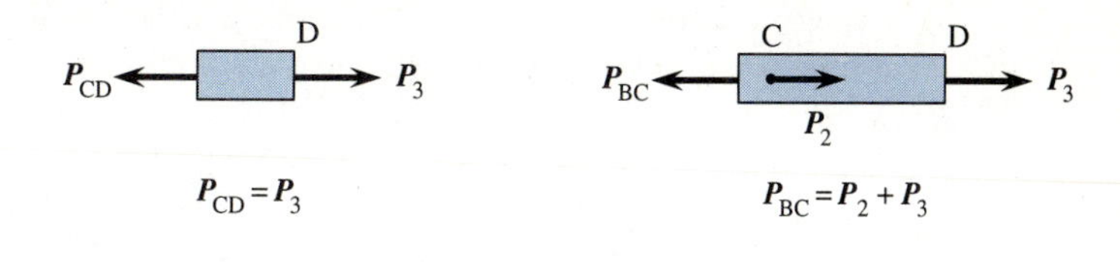

$$P_{AB} = P_2 + P_3 - P_1$$

Equation (2.22) then yields the axial deformation of the entire rod as

$$\delta = \frac{P_{CD}c}{AE} + \frac{P_{BC}b}{AE} + \frac{P_{AB}a}{AE}$$

For bars with variable cross-sectional areas (that is, $A = A(x)$, meaning A a function of x) (Figure 2.34(a)) and/or variable modulus of elasticity and/or if the loading P is a function of x (Figure 2.34(b)), we must express the deformation of an element length dx as $d\delta$ where

$$d\delta = \epsilon dx = \frac{P(x)}{A(x)E(x)} dx \qquad \textbf{(2.23)}$$

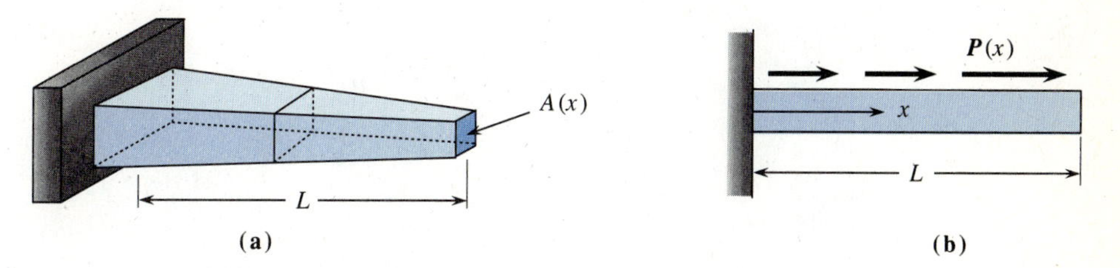

Figure 2.34 (a) Bar with variable cross sectional area and (b) bar subjected to loading that varies with x.

The total deformation of the bar is then obtained by integrating Eq. (2.23) over the length L of the bar as

$$\delta = \int_0^L \frac{P(x)}{A(x)\,E(x)} dx \quad \text{(in.) or (mm)} \qquad \textbf{(2.24)}$$

STEPS IN SOLUTION

The following steps are generally taken to obtain the axial deflection in axially loaded members.

1. Draw a free-body diagram of the member.
2. Pass a plane through the member normal to the axis of the member and determine the internal axial force P acting over the member length L.
3. **(a)** For uniform cross-sectional area A and modulus of elasticity E over the length L, use Eq. (2.21) to determine the total axial deflection of the member.
   (b) If a number of intermediate concentrated loads, or abrupt changes in cross-sectional area A or in the modulus of elasticity E occur, use Eq. (2.22) to determine the total deflection of the member.
   (c) If the internal force, cross-sectional area, or modulus of elasticity varies with axial coordinate x, that is, if $P = P(x)$ or $A = A(x)$, or $E = E(x)$, then use Eq. (2.24) to determine the total axial deflection of the member.

Examples 2.10 through 2.14 illustrate the steps used to obtain axial deflections for members included in steps 3(a), (b), and (c).

EXAMPLE 2.10

For the column subjected to the compressive end load $P = 50$ kips with uniform cross-sectional area $A = 3.0$ in.^{2} and made of a steel with $E = 29 \times 10^6$ psi (Figure 2.35(a)), determine the axial shortening of the upper end. The initial length of the column is 10 in.

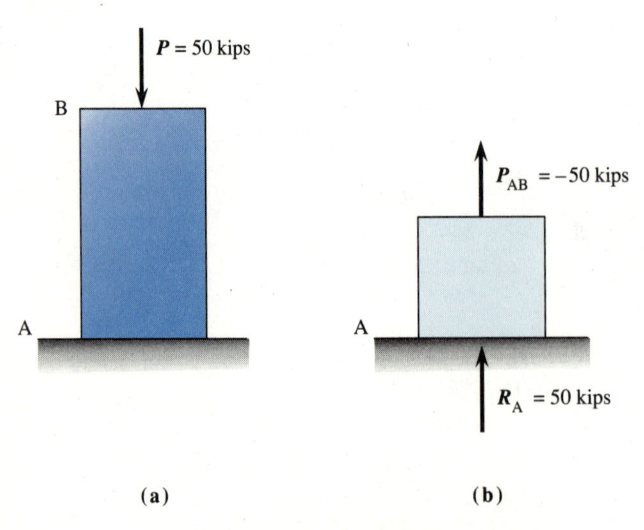

Figure 2.35 (a) Column subjected to compressive force, and (b) free-body diagram of lower part of column.

Solution

The free-body diagram is shown in Figure 2.35(b) where R_A is the reaction at A. By statics, the internal force in the column is

$$P_{AB} = -50 \text{ kips}$$

This force is used in Eq. (2.21) to obtain the deformation as

$$\delta = \frac{P_{AB} L_{AB}}{A_{AB} E} = \frac{(-50 \text{ kips})(10 \text{ in.})}{(3.0 \text{ in.}^2)(29 \times 10^3 \text{ ksi})} = -5.75 \times 10^{-3} \text{ in.}$$

EXAMPLE 2.11

A two-story building has columns AB and BC supporting loads shown (Figure 2.36(a)). The cross-sectional areas of AB and BC are 20 in.^{2} and 10 in.^{2}, respectively. Determine the shortening of each column and the total deformation of point C. Let $E = 30 \times 10^6$ psi for each column.

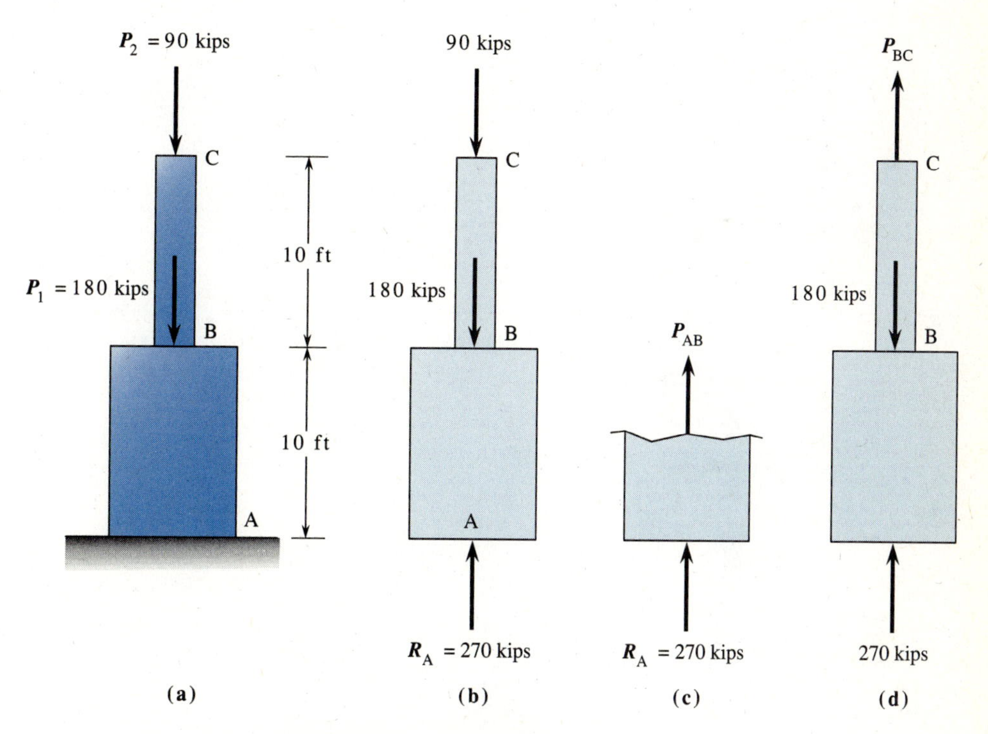

Figure 2.36 (a) Two-story building column supporting loads, (b) free-body diagram showing support reaction, (c) free-body diagram showing internal axial force in section AB, and (d) free-body diagram showing internal axial force in section BC.

Solution

We must first obtain the internal force P_{AB}. The reaction at A is $R_A = 90 + 180 = 270$ kips. Therefore passing a plane through AB (see Figure 2.36(c))

$$P_{AB} = -R_A = -270 \text{ kips}$$

Using Eq. (2.21) applied to column AB, we obtain

$$\delta_{AB} = \frac{(-270 \text{ kips})(10 \text{ ft} \times 12 \text{ in./ft})}{(20 \text{ in.}^2)(30 \times 10^3 \text{ ksi})} = -54 \times 10^{-3} \text{ in.}$$

as the shortening of column AB. Passing a plane through BC (see Figure 2.36(d)), we obtain

$$P_{BC} = -90 \text{ kips}$$

Using Eq. (2.21), we then obtain the shortening of column BC as

$$\delta_{BC} = \frac{(-90 \text{ kips})(10 \text{ ft.} \times 12 \text{ in./ft})}{(10 \text{ in.}^2)(30 \times 10^3 \text{ ksi})} = -36 \times 10^{-3} \text{ in.}$$

The total deformation of point C is then

$$\delta = \delta_{AB} + \delta_{BC} = -90 \times 10^{-3} \text{ in.}$$

which is equivalent to that obtained using Eq. (2.22).

EXAMPLE 2.12

The bar (Figure 2.37) has weight density γ, uniform cross-sectional area A, length L, and modulus of elasticity E. Determine the total elongation due to the bar's own weight.

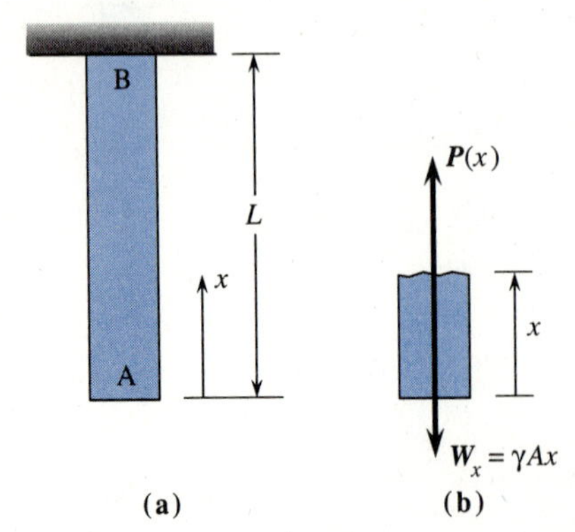

Figure 2.37 (a) Bar subjected to own weight and (b) free-body diagram of lower section of bar.

Solution

The bar's weight W_x varies with axial coordinate x as

$$W_x = \gamma V(x)$$

where $V(x)$ is the volume of the bar at location x and $V(x) = Ax$ (volume varies linearly with x). Therefore, from the free-body diagram of the lower section and by statics, we obtain

$$\Sigma F_y = 0: \quad W_x = \gamma A x = P(x)$$

Since the weight is the axial load $P(x)$, we must apply Eq. (2.24) to obtain

$$\delta = \int_0^L \frac{P(x)}{AE}\, dx = \int_0^L \frac{\gamma A x}{AE}\, dx = \frac{\gamma L^2}{2E}$$

Expressing γ as $\gamma = W/(AL)$ (where W is the total weight of the bar), we obtain the total deformation of the lower end of the bar as

$$\delta = \left(\frac{W}{AL}\right)\frac{L^2}{2E} = \frac{WL}{2AE}$$

EXAMPLE 2.13

For the rigid beam supported by the bar BC (Figure 2.38), determine the elongation of bar BC. Let $E = 200$ GPa and the diameter of the bar be 12.5 mm.

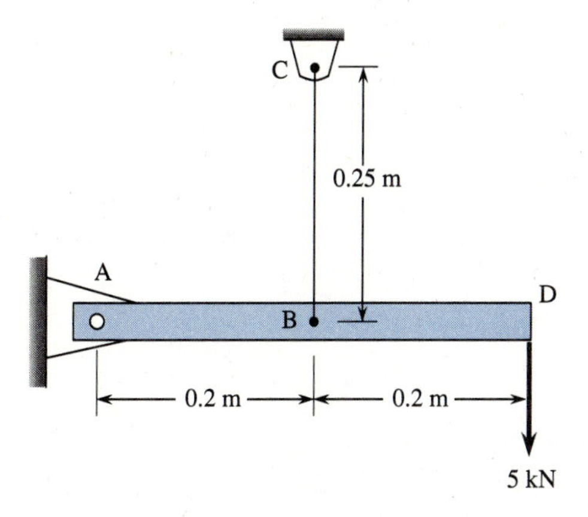

Figure 2.38 Rigid beam supported by bar.

Solution

First we draw the free-body diagram of the beam to determine the force in bar BC.

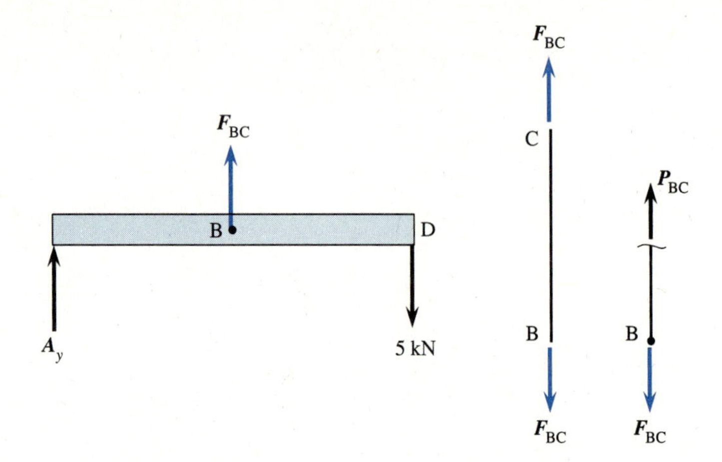

By statics, we obtain

$$+\circlearrowleft \Sigma M_A = 0: \quad (0.2 \text{ m})\, F_{BC} - (0.4 \text{ m})(5 \text{ kN}) = 0$$

$$F_{BC} = 10 \text{ kN } (T)$$

The internal force in BC is then $P_{BC} = F_{BC}$.

Applying Eq. (2.21), we obtain

$$\delta_{BC} = \frac{F_{BC}L_{BC}}{A_{BC}E_{BC}} = \frac{(10 \text{ kN})(0.25 \text{ m})}{\frac{\pi}{4}(12.5 \times 10^{-3} \text{ m})^2(200 \times 10^6 \text{ kN/m}^2)}$$

$$= 1.02 \times 10^{-4} \text{ m} = 0.102 \text{ mm}$$

EXAMPLE 2.14

The catwalk is supported by the bar hangers (each 20 ft long) shown in Figure 2.39. The bars have cross-sectional areas of 5 in.^{2} and $E = 30 \times 10^6$ psi. The design dead load (self-weight) of the catwalk is 25 psf and the live load is 100 psf. **(a)** Determine the axial stress in each bar and the elongation of each bar. **(b)** If hanger DH has failed, determine the axial stress in the remaining three bars and their elongation.

Solution

First determine the total force F due to dead load (F_{DL}) plus live load (F_{LL}) on the catwalk:

$$F_{DL} = 25\,(40 \text{ ft} \times 80 \text{ ft}) = 80{,}000 \text{ lb}$$
$$F_{LL} = 100\,(40 \text{ ft} \times 80 \text{ ft}) = 320{,}000 \text{ lb}$$
$$F = 80{,}000 + 320{,}000 = 400{,}000 \text{ lb}$$

Then draw a free-body diagram of the catwalk showing all bar forces and the force F at the center of gravity of the slab (Figure 2.39(b)).

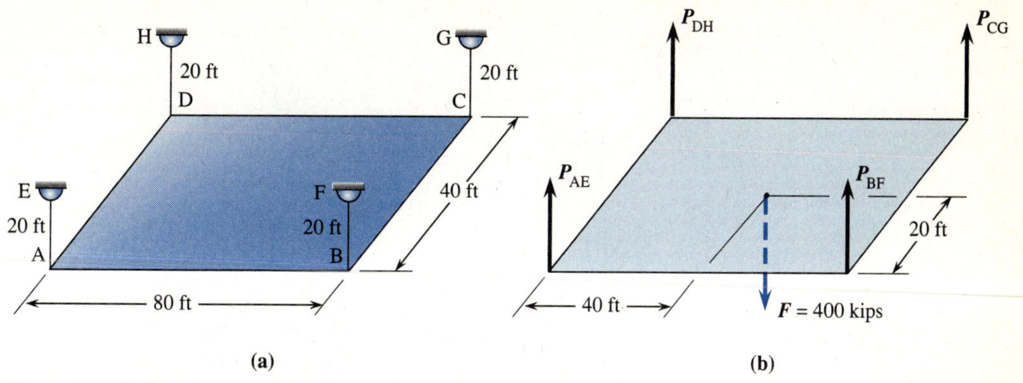

Figure 2.39 (a) Catwalk supported by bar hangers and (b) free-body diagram of catwalk.

(a) Assuming each bar to support 1/4 of the total force, from $\Sigma F_y = 0$, we have

$$P_{AE} = P_{BF} = P_{CG} = P_{DH} = 1/4(F) = 1/4(400{,}000) = 100{,}000 \text{ lb} = 100 \text{ kips}$$

Using Eq. (2.1), the axial stress in each bar is

$$\sigma = \frac{P}{A} = \frac{100 \text{ kips}}{5 \text{ in.}^2} = 20 \text{ ksi} \quad (T)$$

Using Eq. (2.21), the elongation of each bar is

$$\delta = \frac{PL}{AE} = \frac{(100 \text{ kips})(20 \text{ ft} \times 12 \text{ in./ft})}{(5 \text{ in.}^2)(30 \times 10^3 \text{ ksi})} = 0.160 \text{ in.}$$

(b) The free-body diagram of the catwalk with bar DH removed is shown.

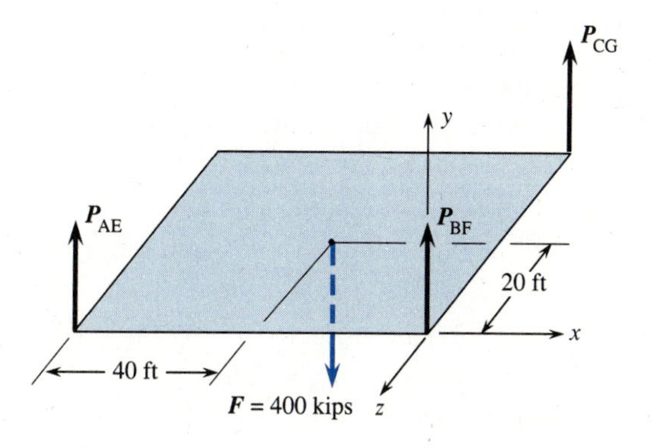

From statics, we obtain

$$\Sigma M_x = 0: \quad P_{CG}\,(40 \text{ ft}) - 400\,(20 \text{ ft}) = 0$$
$$P_{CG} = 200 \text{ kips}$$

$$\Sigma M_z = 0: \quad -P_{AE}\,(80 \text{ ft}) + 400\,(40 \text{ ft}) = 0$$
$$P_{AE} = 200 \text{ kips}$$

$$\Sigma F_y = 0: \quad P_{BF} = 0$$

Using Eq. (2.1), the stress in bar hangers AE and CG is

$$\sigma = \frac{P}{A} = \frac{200 \text{ kips}}{5 \text{ in.}^2} = 40 \text{ ksi}$$

Using Eq. (2.21), the elongation of bars AE and CG is

$$\delta = \frac{PL}{AE} = \frac{(200 \text{ kips})(20 \text{ ft} \times 12 \text{ in./ft})}{(5 \text{ in.}^2)(30 \times 10^3 \text{ ksi})} = 0.320 \text{ in.}$$

The stress in hanger BF is now 0 since the bar BF does not resist any load.

2.7 STATICALLY INDETERMINATE PROBLEMS

In Section 2.6 we applied the equations of equilibrium to determine the internal forces acting in axially loaded members and then used Eqs. (2.21) or (2.22) to determine the axial deformation of the member. When the equations of equilibrium are sufficient to determine the internal forces the problems are called *statically determinate*.

In this section we consider problems in which the equations of equilibrium (or statics) alone do not provide enough equations to determine the internal forces. These problems are called statically indeterminate and require additional relationships involving the geometry of the deformed structure, along with the load-deformation relation such as Eq. (2.21). This consideration of the geometry of deformation is often called *structural compatibility*. Compatibility then means that the overall deformation of a structure must follow that forced on it by the support conditions, and the deformations of connecting parts must be such that the parts remain together as a unit.

In general, the degree of static indeterminacy is the additional number of unknowns (these unknowns include internal reactions and support reactions) greater than the total number of relevant equilibrium equations. For instance, statically indeterminate to one degree means there is one unknown internal force or support reaction that cannot be determined by use of equilibrium equations alone. One additional equation (called a compatibility equation), obtained through consideration of the geometry of the deformed structure, is needed.

STEPS IN SOLUTION

The following steps are generally taken to obtain the solution to a structure that is statically indeterminate to one degree. Here we consider the extra unknown to involve axially loaded members.

1. Draw a free-body diagram of the structure.
2. Write the relevant equilibrium equations.
3. Consider the geometry of the deformed structure to write a compatibility equation. This equation will involve deformations of one or more axially loaded members.
4. Because the compatibility equation involves deformations of one or more axially loaded members, use Eq. (2.21) (or Eq. (2.22) if necessary) to express the compatibility equation in terms of internal axial forces.
5. This additional equation, along with the equilibrium equations, yields a sufficient number of equations to solve for both the internal and support reactions.

Examples 2.15 through 2.17 illustrate the steps used to obtain the solution of statically indeterminate structures involving axially loaded members.

EXAMPLE 2.15

A rod of uniform cross-sectional area $A = 3$ in.^{2} is fixed to rigid end supports at A and C (Figure 2.40). A load of 9 kips is then applied to the rod at B. Determine the axial stress in sections AB and BC.

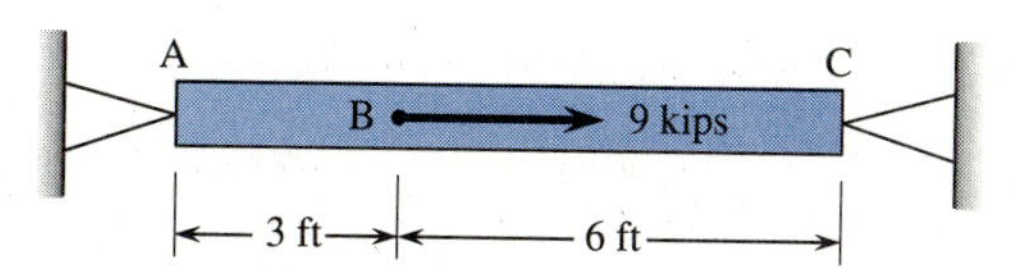

Figure 2.40 Rod with rigid end supports subjected to load.

Solution

The free-body diagram of the rod is

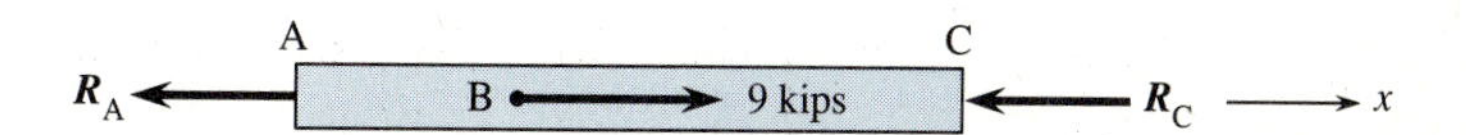

From force equilibrium, we obtain

$$\Sigma F_x = 0: \quad -R_A + 9 - R_C = 0 \tag{a}$$

We cannot solve for R_A or R_C from this single equation of statics. All other equations of statics are trivially satisfied and hence do not yield additional relations between R_A and R_C.

Therefore, we consider the geometry of deformation and obtain a compatibility equation by observing that the total deformation of the bar must be zero. Therefore

$$\delta_{AC} = \delta_{AB} + \delta_{BC} = 0 \tag{b}$$

where δ_{AB} and δ_{BC} are the deformations in segments AB and BC, respectively.

Making use of Eq. (2.21), we relate the deformations to the internal axial force in each segment as

$$\delta_{AB} = \frac{P_{AB}\ L_{AB}}{AE} \qquad \delta_{BC} = \frac{P_{BC}\ L_{BC}}{AE} \tag{c}$$

From free-body diagrams of sections passing through AB and BC, we have

A R_A $P_{AB} = R_A$ P_{BC} C R_C

$$P_{AB} = R_A \qquad P_{BC} = -R_C \tag{d}$$

where $P_{AB} = R_A$ is positive since it acts away on the cross-section while $P_{BC} = -R_C$ is negative since it actually pushes into the cross-section. Using Eq. (d) in (c), and this result in (b), we obtain

$$\delta_{AB} + \delta_{BC} = \frac{R_A(3 \text{ ft} \times 12 \text{ in./ft})}{AE} + \frac{(-R_C)(6 \text{ ft} \times 12 \text{ in./ft})}{AE} = 0 \quad \text{(e)}$$

Simplifying the above Eq. (e) yields

$$R_C = 1/2\, R_A \quad \text{(f)}$$

Equations (a) and (f) can now be solved to obtain R_A and R_C. Substituting Eq. (f) into (a), we obtain

$$-R_A + 9 - 1/2\, R_A = 0$$

$$R_A = 6 \text{ kips}$$

and from Eq. (f), we have

$$R_C = 3 \text{ kips}$$

The stresses in each segment are then

$$\sigma_{AB} = \frac{P_{AB}}{A} = \frac{R_A}{A} = \frac{6 \text{ kips}}{3 \text{ in.}^2} = 2 \text{ ksi } (T)$$

$$\sigma_{BC} = \frac{P_{BC}}{A} = \frac{-R_C}{A} = \frac{-3 \text{ kips}}{3 \text{ in.}^2} = -1 \text{ ksi } (C)$$

EXAMPLE 2.16

A steel cylinder is filled with concrete (Figure 2.41) to be used as a column, often called a "Lally-column" in a building. The compressive load to be supported by the column is $P = 100$ kN. Determine the axial stress induced in the steel cylinder and in the concrete. Also determine the axial shortening of the column. Let E of steel = 200 GPa and E of concrete = 24 GPa.

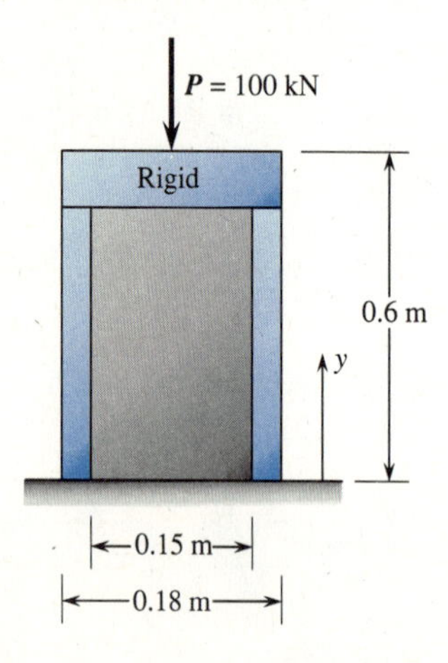

Figure 2.41 Column made of steel cylinder with concrete core.

Solution

A free-body diagram of the rigid plate yields

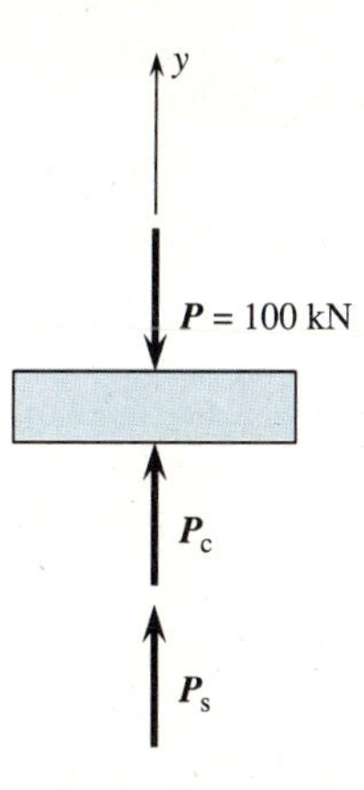

where P_C and P_S are the forces from the concrete and steel acting on the rigid plate, respectively. Writing an equilibrium equation, we obtain

$$\Sigma F_y = 0: \quad P_c + P_s - 100 = 0 \tag{a}$$

We now have one equation with two unknown forces P_C and P_S. A compatibility equation results in the second equation as follows. The deflection of the concrete must equal the deflection of the steel due to the centroidally applied force of 100 kN. We then write

$$\delta_C = \delta_S \tag{b}$$

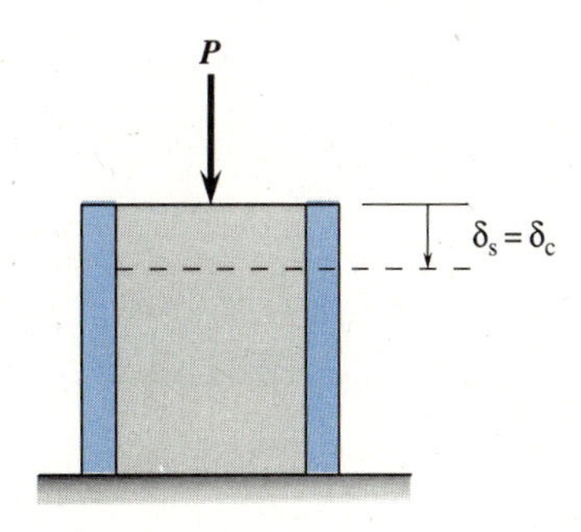

Using Eq. (2.21) and Eq. (b), we obtain

$$\delta_C = \frac{P_C\,L_C}{A_C\,E_C} = \delta_S = \frac{P_S\,L_S}{A_S\,E_S} \tag{c}$$

We can now simultaneously solve Eqs. (a) and (c). From (c), we obtain

$$P_C = \frac{P_S\,L_S}{A_S\,E_S}\left(\frac{A_C\,E_C}{L_C}\right) \tag{d}$$

Now

$$L_S = L_C = 0.6 \text{ m}, \ E_S = 200 \text{ GPa}, \ E_C = 24 \text{ GPa}$$

$$A_C = \frac{\pi}{4} d^2 = \frac{\pi}{4} (0.15 \text{ m})^2 = 0.0177 \text{ m}^2$$

$$A_S = \frac{\pi}{4} (d_0^2 - d_i^2) = \frac{\pi}{4} (0.18^2 - 0.15^2) = 0.0077 \text{ m}^2$$

Using Eq. (d), we obtain

$$P_C = P_S \frac{(0.0177 \text{ m}^2)(24 \times 10^6 \text{ kN/m}^2)}{(0.0077 \text{ m}^2)(200 \times 10^6 \text{ kN/m}^2)} = 0.276 P_S \tag{e}$$

Substituting Eq. (e) into (a), we have

$$P_S + 0.276 P_S - 100 = 0$$

$$P_S = 78.4 \text{ kN} \tag{f}$$

Substituting the result for P_S into Eq. (e) yields

$$P_C = 0.276 (78.4) = 21.6 \text{ kN}$$

The axial stresses are then given by

$$\sigma_S = -\frac{P_S}{A_S} = -\frac{78.4 \text{ kN}}{0.0077 \text{m}^2} = -10200 \frac{\text{kN}}{\text{m}^2} = -10.2 \frac{\text{MN}}{\text{m}^2} = -10.2 \text{ MPa (C)}$$

$$\sigma_C = -\frac{P_C}{A_C} = -\frac{21.6 \text{ kN}}{0.0177 \text{ m}^2} = -1220 \frac{\text{kN}}{\text{m}^2} = -1.22 \frac{\text{MN}}{\text{m}^2} = -1.22 \text{ MPa (C)}$$

where P_S and P_C are pushing on the steel and concrete so as to cause compressive stresses. Using Eq. (c), the axial shortening of the column is

$$\delta_S = \delta_C = \frac{P_C L_C}{A_C E_C} = \frac{(-21.6 \text{ kN})(0.6 \text{ m})}{(0.0177 \text{ m}^2)(24 \times 10^6 \text{ kN/m}^2)} = -0.0305 \times 10^{-3} \text{ m}$$

$$= -0.0305 \text{ mm}$$

It should be noted that on removal of the concrete the stress in the steel-only column would be $\sigma_S = -100 \text{ kN}/(0.0077) = -13.0$ MPa.

EXAMPLE 2.17

A rigid beam is pin-supported at A and supported by two wire hangers at B and D (Figure 2.42(a)). A hoisting crane is located at C and lifts 4000 lb. Determine the axial stress and deformation in the wire hangers. The wire hangers are 0.10 in. in diameter with $E = 30 \times 10^6$ psi.

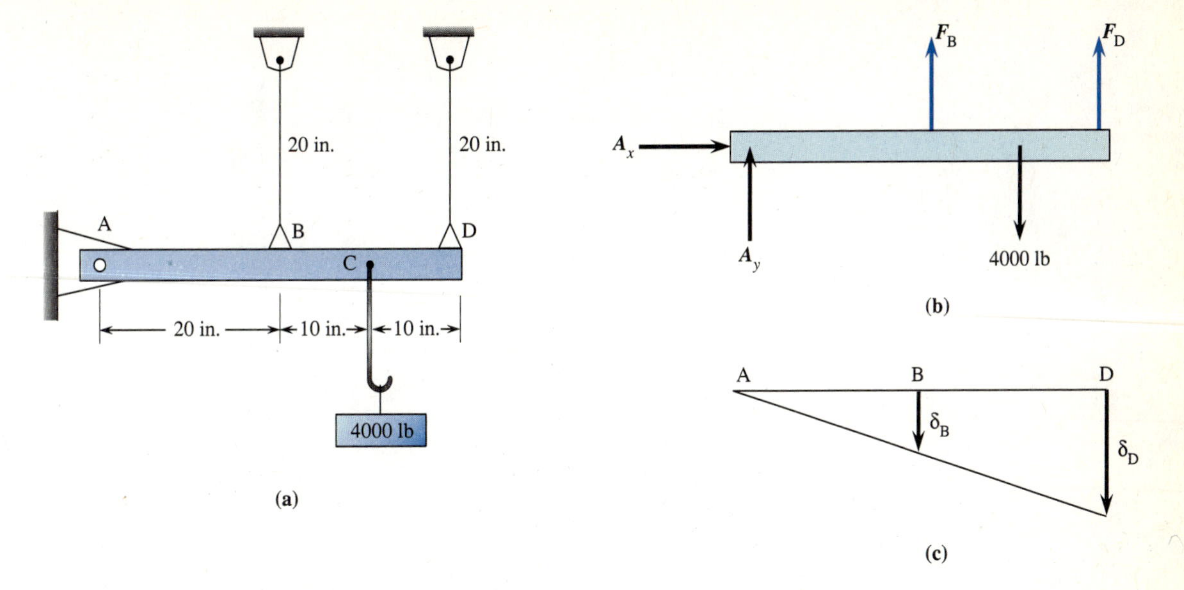

Figure 2.42 (a) Rigid beam supported by hangers,
(b) free-body diagram of beam, and
(c) deformed position of beam showing displacements of hangers.

Solution

From the free-body diagram of member AD (Figure 2.42(b))

$$+\circlearrowleft\ \Sigma M_A = 0: \quad F_B(20 \text{ in.}) - (4000 \text{ lb})(30 \text{ in.}) + F_D(40 \text{ in.}) = 0 \tag{a}$$

$$\Sigma F_y = 0: \quad A_y + F_B + F_D - 4000 = 0 \tag{b}$$

$$\Sigma F_x = 0: \quad A_x = 0$$

Equations (a) and (b) result in three unknowns.

Compatibility of deformation yields a relationship between the deformation at B and that at D. By considering member AD as rigid, similar triangles (Figure 2.42(c)) yield

$$\delta_B = \frac{1}{2}\delta_D \tag{c}$$

Using Eq. (2.21) in (c), and realizing that δ_B and δ_D represent the axial deformations in the wires at B and D, we obtain

$$\delta_B = \frac{F_B L_B}{A_B E_B} = \frac{1}{2}\delta_D = \frac{1}{2}\left(\frac{F_D L_D}{A_D E_D}\right) \tag{d}$$

Since $L_B = L_D$ and $E_B = E_D$ and $A_B = A_D$, Eq. (d) yields

$$F_B = \frac{1}{2}F_D \tag{e}$$

Using Eq. (e) in (a), we obtain

$$\left(\frac{1}{2}F_D\right)(20 \text{ in.}) - (4000 \text{ lb})(30 \text{ in.}) + F_D(40 \text{ in.}) = 0$$

$$F_D = 2400 \text{ lb}$$

Equation (e) then yields

$$F_B = \frac{1}{2}F_D = 1200 \text{ lb}$$

Using Eq. (2.1), the stresses are

$$\sigma_B = \frac{F_B}{A_B} = \frac{1200 \text{ lb}}{0.1 \text{ in.}^2} = 12{,}000 \text{ psi} \quad \text{(tensile)}$$

$$\sigma_D = \frac{F_D}{A_D} = \frac{2400 \text{ lb}}{0.1 \text{ in.}^2} = 24{,}000 \text{ psi} \quad \text{(tensile)}$$

Using Eq. (2.4), the deformations are

$$\delta_B = \frac{F_B L_B}{A_B E_B} = \frac{(1200 \text{ lb})(20 \text{ in.})}{(0.1 \text{ in.}^2)(30 \times 10^6 \text{ psi})} = 0.008 \text{ in.}$$

$$\delta_D = 2\delta_B = 0.016 \text{ in.}$$

Flexibility Method of Superposition

Another method commonly used to solve statically indeterminate problems involves superposition. The specific method described is called the *flexibility method.*

The method involves the introduction of redundant reactions which are included in the solution of the problem by considering separately the deformations caused by the applied loads and by the redundant reactions, and by adding together, or superimposing, the resulting displacement expressions, which then allow the determination of the redundants.

To describe the method, we consider the statically indeterminate bar shown in Figure 2.43(a). The bar is attached to supports at each end and is subjected to an applied axial force P.

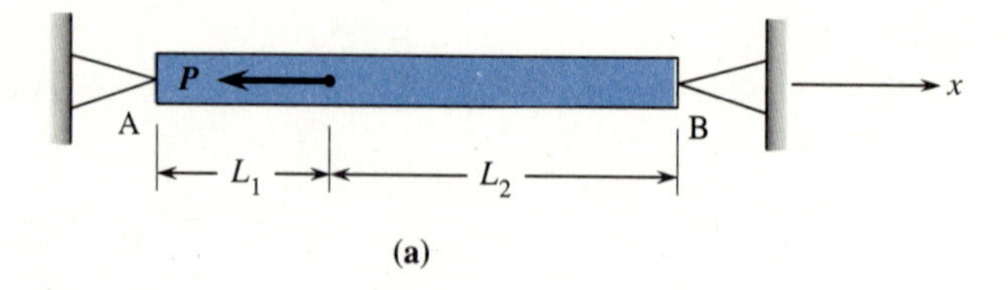

(a)

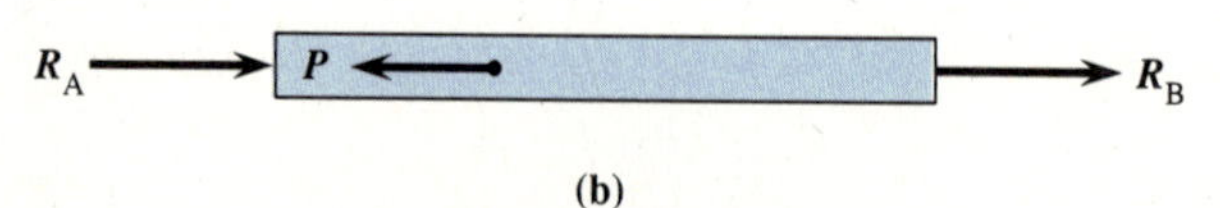

(b)

Figure 2.43 (a) Statically indeterminate bar and (b) free-body diagram of bar.

Reactions R_A and R_B (Figure 2.43(b)) develop due to the applied load. These reactions cannot be determined by statics alone because only one independent equation of static equilibrium exists for the bar. Upon summing forces along the x-axis of the bar, we obtain

$$R_A + R_B = P \tag{a}$$

Another equation must be obtained from the deflection equation for the bar using superposition of solutions due to the applied load and due to a redundant. To begin the analysis, we consider the reaction at B as the redundant. This force is redundant in that it is considered to be the extra force that cannot be determined by statics alone. We remove the support at B and consider R_B as an unknown load acting on the structure (Figure 2.44(a)). We then treat the primary structure subjected to the applied load (Figure 2.44(b)) separately from the structure subjected to the redundant force as shown in Figure 2.44(c).

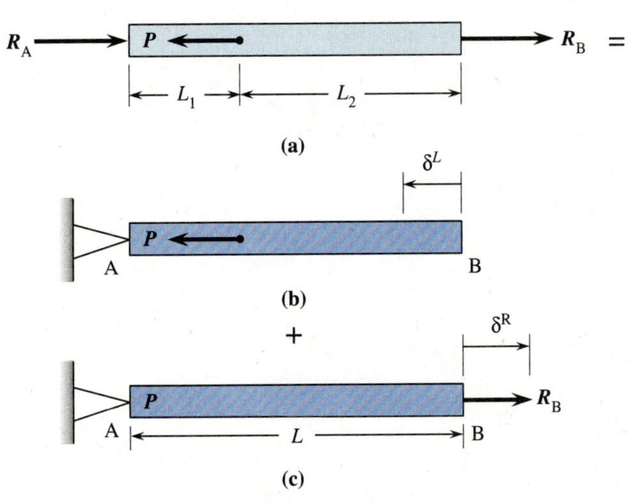

(a) Free-body diagram of bar,
(b) primary structure subjected to actual loads, and
(c) structure subjected to redundant force.

The solution is obtained by considering the deformation of point B due to the load using Eq. (2.21) as

$$\delta^L = -\frac{PL_1}{AE} \tag{b}$$

and the deformation of point B due to the redundant as

$$\delta^R = \frac{R_B L}{AE} \tag{c}$$

The total deformation of the bar due to both P and R_B must be zero and yields

$$\delta^L + \delta^R = 0 \tag{d}$$

Substituting Eqs. (b) and (c) into (d) yields

$$\frac{-PL_1}{AE} + \frac{R_B L}{AE} = 0$$

Solving for R_B, we obtain

$$R_B = \frac{PL_1}{L} \tag{e}$$

We can now substitute Eq. (e) into Eq. (a) to obtain reaction R_A as

$$R_A = P - R_B = P - \frac{PL_1}{L} = P\left(\frac{L - L_1}{L}\right) = \frac{PL_2}{L} \tag{f}$$

STEPS IN SOLUTION

In summary, the following steps are used in the flexibility method of superposition for structures that are statically indeterminate to one degree.

1. Select one of the unknown reactions as a redundant and release the structure from that redundant's support making the structure statically determinate.
2. Consider the primary structure subjected to the actual loads separate from the structure subjected to the redundant load.
3. Calculate the displacements caused by the actual load and then by the redundant (such as Eqs. (b) and (c), above).
4. Combine the displacement expressions from step 3 into a single equation of *compatibility of displacements* based on known geometry of the deformation of the structure (such as Eq. (d) above).
5. Solve the compatibility equation for the redundant (such as Eq. (e) above).
6. Substitute the now known redundant into the equilibrium equation (such as Eq. (a) above) to determine the other reaction.

This method can be applied to structures with more than one redundant by introducing as many redundants as necessary to make the primary structure statically determinate and then treating the primary structure subjected to the actual applied loads separately from each redundant load case.

Examples of the above procedure for analyzing statically indeterminate structures with one redundant now follow.

EXAMPLE 2.18

A steel bar of two different cross-sectional areas $A_{AB} = 3$ in.^{2} and $A_{BC} = 1$ in.^{2} is fixed to rigid end supports at A and C (Figure 2.45). A load of 20 kips is then supported by the bar at B. Determine the axial stress in sections AB and BC.

Solution

We will use the flexibility method to solve this problem. The free-body diagram is drawn with reactions R_A and R_C (Figure 2.46(a)).

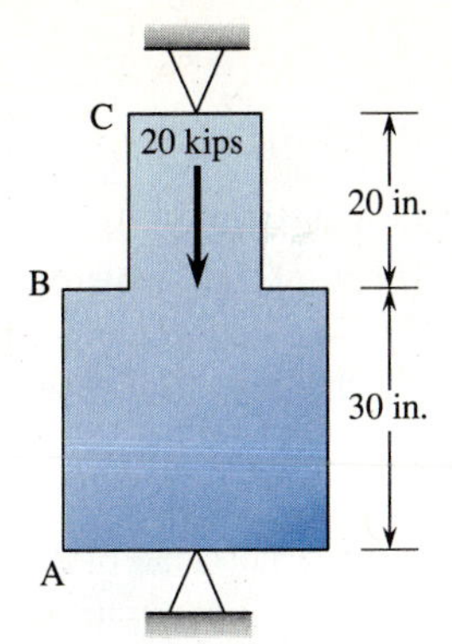

Figure 2.45 Bar with different cross sectional areas.

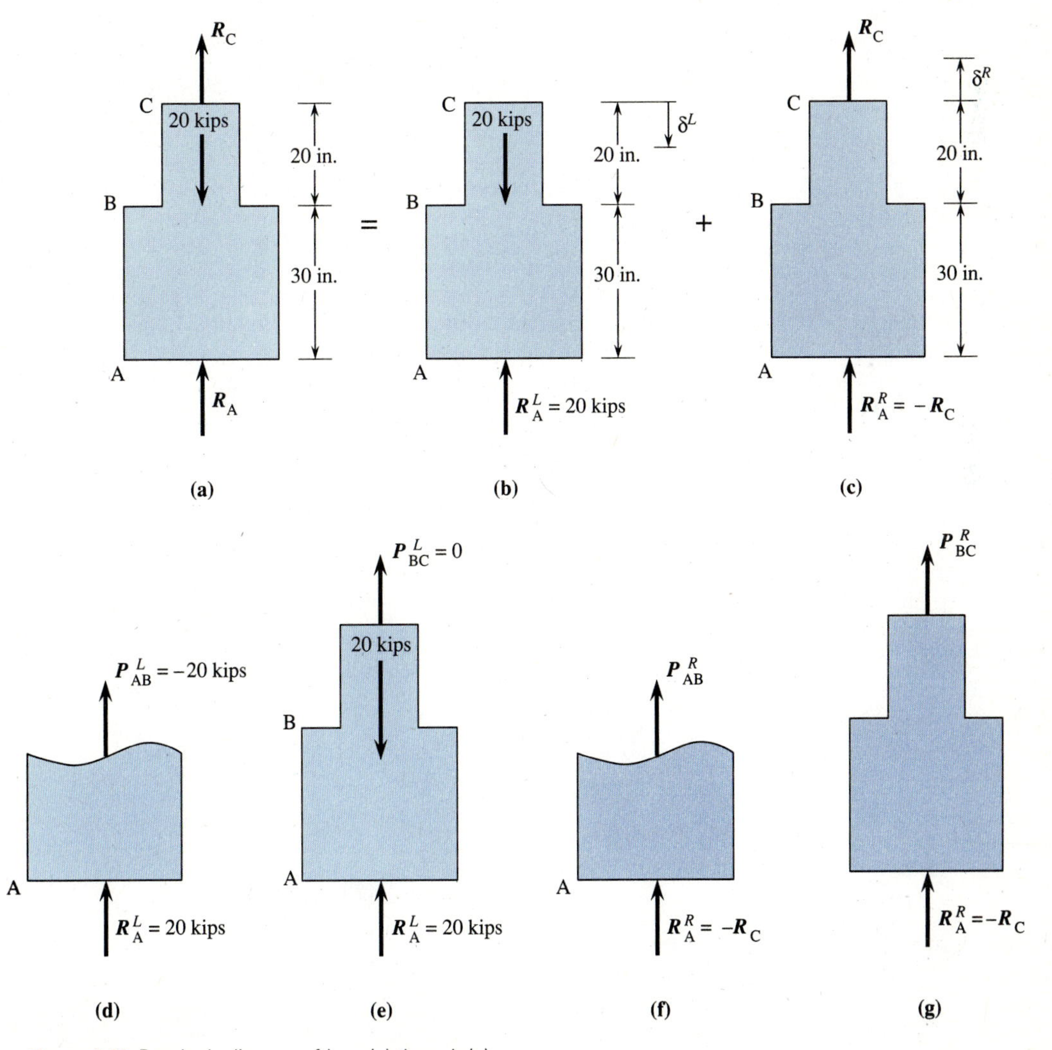

Figure 2.46 Free-body diagram of bars (a) through (g).

Using the force equilibrium equation, we obtain

$$\Sigma F_y = 0: \quad R_A + R_C = 20 \tag{a}$$

There are two unknown reactions in Eq. (a) and we have applied the only useful equation of equilibrium. We release the bar from its upper support C. By superposition we treat the actual loading and the redundant load R_C separately as shown in Figure 2.46. Applying the compatibility equation with the total deformation of the bar equal to zero, we obtain

$$\delta^L + \delta^R = 0 \tag{b}$$

where the δ^L is the total deflection of the bar due to the actual load of 20 kips and δ^R is the deflection due to the redundant R_C. Cutting into section AB in Figure 2.46(b), we obtain the internal axial force P^L_{AB} due to the applied load as shown in Figure 2.46(d).

$$P^L_{AB} = -20 \text{ kips} \tag{c}$$

Cutting section BC in Figure 2.46(b), we obtain P^L_{BC} as shown in Figure 2.46(e).

$$P^L_{BC} = 0 \tag{d}$$

Using Eq. (2.22), the deflection due to the actual load is

$$\delta^L = \frac{P^L_{AB} L_{AB}}{A_{AB} E_{AB}} + \frac{P^L_{BC} L_{BC}}{A_{BC} E_{BC}} = \frac{(-20 \text{ kips})(30 \text{ in.})}{(3 \text{ in.}^2)(30 \times 10^3 \text{ ksi})} = -0.00667 \text{ in.} \tag{e}$$

Similarly, cutting into section AB in Figure 2.46(c), we obtain the internal axial force P^R_{AB} due to the redundant R_C as shown in Figure 2.46(f).

$$P^R_{AB} = -R^R_A = R_C \tag{f}$$

and then cutting section BC (Figure 2.46(g)), we obtain

$$P^R_{BC} = -R^R_A = R_C \tag{g}$$

Using Eq. (2.22), the deflection due to the redundant force R_C is written similar to Eq. (e) as

$$\delta^R = \frac{R_C (30 \text{ in.})}{(3 \text{ in.}^2)(30 \times 10^3 \text{ ksi})} + \frac{R_C (20 \text{ in.})}{(1 \text{ in.}^2)(30 \times 10^3 \text{ ksi})}$$

$$\delta^R = \frac{30 R_C}{30 \times 10^3} = 0.001 R_C \tag{h}$$

Using Eqs. (e) and (h) in (b), we obtain

$$-0.00667 + 0.001 R_C = 0$$

$$R_C = 6.67 \text{ kips} \tag{i}$$

Substituting Eq. (i) into (a) yields

$$R_A = 20 - 6.67 = 13.33 \text{ kips} \tag{j}$$

The axial stresses are

$$\sigma_{AB} = \frac{P_{AB}}{A_{AB}} = \frac{-R_A}{A_{AB}} = \frac{-13.33 \text{ kips}}{3 \text{ in.}^2} = -4.44 \text{ ksi } (-30.6 \text{ MPa}) \quad \text{(compressive) } (C)$$

$$\sigma_{BC} = \frac{P_{BC}}{A_{BC}} = \frac{6.67 \text{ kips}}{1 \text{ in.}^2} = 6.67 \text{ ksi } (46.0 \text{ MPa}) \quad \text{(tensile) } (T)$$

EXAMPLE 2.19

A brass bolt of 10 mm diameter is fitted inside an aluminum tube of 18 mm outside diameter and 3 mm wall thickness (Figure 2.47). A nut is placed on the end of the bolt and is just bearing against the plate B. If the nut is then tightened one-half turn and the bolt is single-threaded with a 2-mm pitch (the pitch is the distance a bolt moves in a single revolution turn), determine the normal stress (a) in the bolt and (b) in the tube. Let $E = 105$ GPa for the brass bolt and $E = 70$ GPa for the aluminum tube.

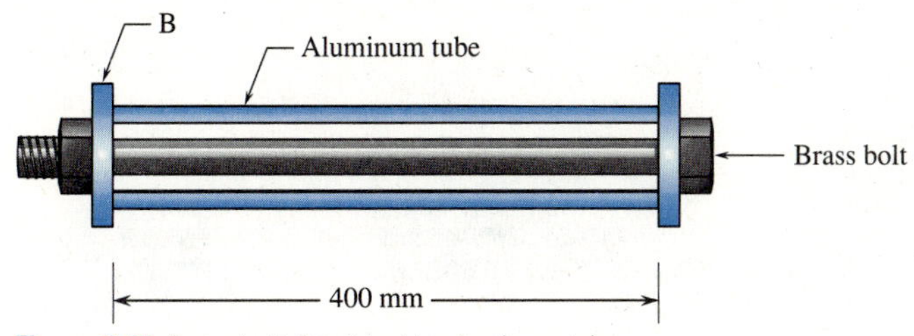

Figure 2.47 Brass bolt fitted inside aluminum tube.

Solution

First the free-body diagrams of the plate B and the tube and bolt are drawn.

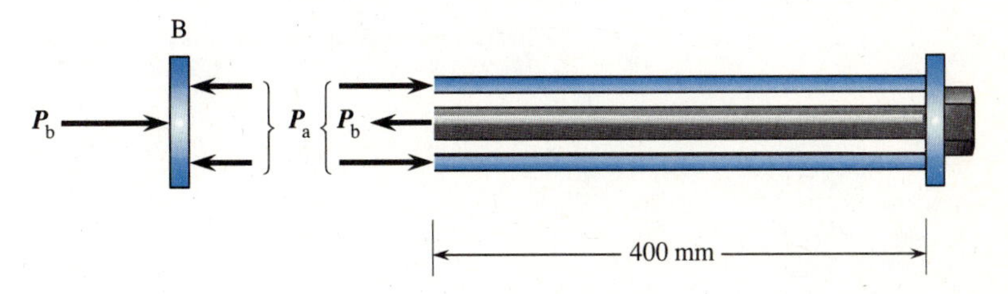

By statics

$$\Sigma F = 0: \quad P_a - P_b = 0 \tag{a}$$

The deformation due to tightening the nut is equal to the amount turned (1/2) times the pitch (2 mm) times the number of threads (1) and is determined as

$$\delta = \frac{1}{2}(2 \text{ mm}) = 1 \text{ mm} \tag{b}$$

This deformation is equivalent to the shortening of the aluminum tube plus the elongation of the brass bolt that occurs during the tightening (as the nut moves to the right relative to the bolt head). By superposition we apply force P_a to the right to the aluminum tube and force $P_b = R_b$ (the redundant) back to the left to the bolt, and use Eq. (2.21) to obtain the deformations in the aluminum and brass as

$$\delta_a = \frac{P_a L}{A_a E_a} = \frac{P_a L}{(141 \times 10^{-6}\ \text{m}^2)(70 \times 10^9\ \text{N/m}^2)} = 101 \times 10^{-9} P_a L \qquad \text{(c)}$$

and

$$\delta_b = \frac{R_b L}{A_b E_b} = \frac{R_b L}{(78.5 \times 10^{-6}\ \text{m}^2)(105 \times 10^9\ \text{N/m}^2)} = 121 \times 10^{-9} R_b L \qquad \text{(d)}$$

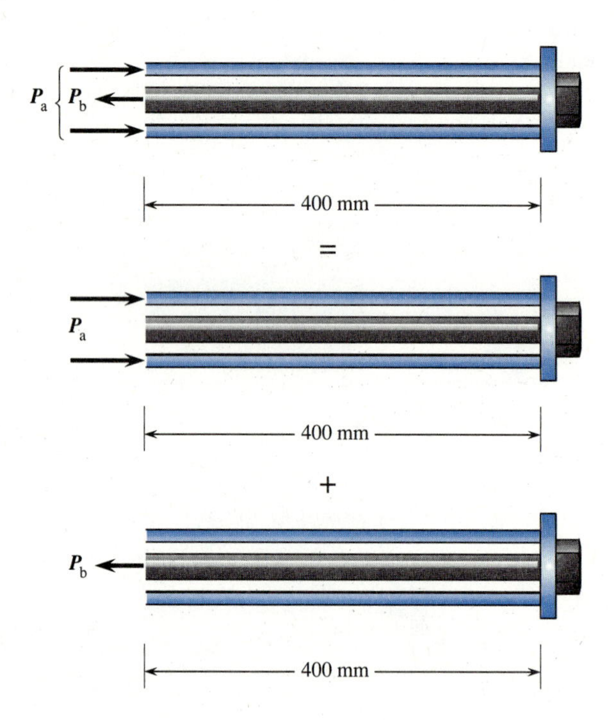

where

$$A_a = \frac{\pi}{4}(0.018\ \text{m}^2 - 0.012\ \text{m}^2) = 141 \times 10^{-6}\ \text{m}^2$$

$$A_b = \frac{\pi}{4}(0.01\ \text{m})^2 = 78.5 \times 10^{-6}\ \text{m}^2$$

By superposition, then,

$$\delta = \delta_a + \delta_b \qquad \text{(e)}$$

Or using Eqs. (b), (c), and (d) in Eq. (e), we obtain

$$1.0 \times 10^{-3}\ \text{m} = 101 \times 10^{-9} P_a L + 121 \times 10^{-9} R_b L \qquad \text{(f)}$$

Using Eq. (a) in (f) (with $R_b = P_b$), we obtain

$$P_a = \frac{1.0 \times 10^{-3}}{222 \times 10^{-9}\,(0.4\text{ m})} = 11{,}250\text{ N} = 11.25\text{ kN}$$

From Eq. (a), we have

$$P_b = P_a = 11.25\text{ kN}$$

The stresses in the aluminum tube and brass bolt are

$$\sigma_a = \frac{-P_a}{A_a} = \frac{-11.25\text{ kN}}{141 \times 10^{-6}\text{ m}^2} = -79{,}800\,\frac{\text{kN}}{\text{m}^2} = -79.8\text{ MPa} \quad \text{(compressive)}$$

$$\sigma_b = \frac{P_b}{A_b} = \frac{11.25\text{ kN}}{78.5 \times 10^{-6}\text{ m}^2} = 143{,}000\,\frac{\text{kN}}{\text{m}^2} = 143\text{ MPa} \quad \text{(tensile)}$$

The results indicate that the aluminum tube is in compression while the brass bolt is in tension.

2.8 TEMPERATURE EFFECTS

Temperature changes in a structure can result in large stresses if not considered properly in design. In bridges, improper constraint of beams and slabs can result in large compressive stresses and resulting buckling failures due to temperature changes (Figure 2.48(a)). In statically indeterminate trusses, members subjected to large temperature changes can result in stresses induced in members of the truss. Similarly, machine parts constrained from expanding or contracting may have large stresses induced in them due to temperature changes. Composite members made of two or more different materials may experience large stresses due to temperature change if they are not thermally compatible; that is, if the materials have large differences in their coefficients of thermal expansion, stresses may be induced even under free expansion (Figure 2.48(b)).

When a member undergoes a temperature change the member attempts to change dimensions. For an unconstrained member AB (Figure 2.49) undergoing uniform change in temperature, the change in the length L is given by

$$\delta_T = \alpha(\Delta T)L \tag{2.25}$$

where α is called the *coefficient of thermal expansion* and ΔT is the change in temperature. The coefficient α is a mechanical property of the material having units of 1/°F (where °F is degrees Fahrenheit) in the USCS of units or 1/°C (where °C is degrees Celsius) in the SI system. In Eq. (2.25), δ_T is considered to be positive when expansion occurs and negative when contraction occurs. Typical values of α are: for structural steel $\alpha = 6.5 \times 10^{-6}$/°F ($12 \times 10^{-6}$/°C) and for aluminum alloys $\alpha = 13 \times 10^{-6}$/°F ($23 \times 10^{-6}$/°C). A more extensive list of coefficients of thermal expansion for various common materials is given in Appendix B and inside of the back cover.

Based on the definition of normal strain, given by Eq. (2.8), we can determine the strain due to a uniform temperature change. For the bar subjected to a uniform tempera-

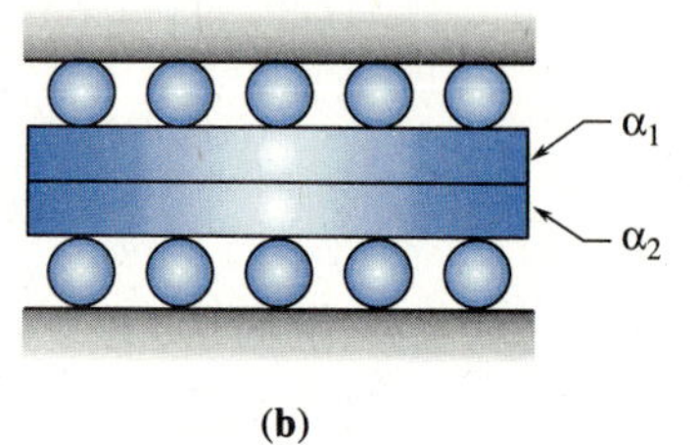

(a) (b)

Figure 2.48 (a) Member buckled by stress induced by temperature increase, and (b) composite member of two different materials.

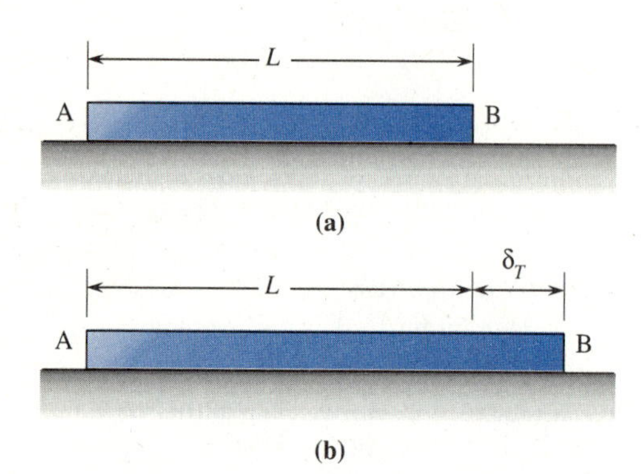

Figure 2.49 (a) Unconstrained member, and (b) same member subjected to uniform temperature increase.

ture change ΔT (Figure 2.49), the strain is the change in a dimension due to a temperature change divided by the original dimension. Considering the axial direction, we then have

$$\epsilon_T = \alpha(\Delta T) \tag{2.26}$$

Since the bar in Figure 2.49 is free to expand, that is, it is not constrained by other members or supports, the bar will not have any stress in it. In general, for statically determinate structures, a uniform temperature change in one or more members does not result in stress in any of the members. That is, the structure will be stress-free. For statically indeterminate structures, a uniform temperature change in one or more members of the structure usually results in stress σ_T in one or more members. We can have strain due to temperature change ϵ_T without stress due to temperature change, and we can have σ_T without any actual change in member lengths or without strains. Statically indeterminate structures subjected to temperature change can be solved using the flexibility method of superposition.

STEPS IN SOLUTION

The following steps are generally used in the flexibility method of superposition for statically indeterminate structures involving axially loaded members in which one or more members is subjected to a uniform temperature change.

1. Select one of the unknown reactions as the redundant and release the structure from that redundant's support, making the structure determinate.
2. Consider the primary structure subjected to the temperature change separate from the structure subjected to the redundant force.
3. Calculate the displacements caused by the temperature change and then by the redundant.
4. Combine the displacement expressions from step 3 into a single compatibility equation based on the geometry of the deformed structure.
5. Because the compatibility equation involves deformations of one or more axially loaded members, use Eqs. (2.21) and (2.25) to express the compatibility equation in terms of the redundant axial force and in terms of the temperature change.

Examples 2.20 through 2.23 illustrate the flexibility method for solving statically indeterminate structures subjected to temperature change.

EXAMPLE 2.20

The bar AB (Figure 2.50) of length L is fixed between supports A and B and then subjected to a temperature rise of ΔT. Determine the stress in the bar. Let the bar have cross sectional area A, coefficient of thermal expansion α, and modulus of elasticity E.

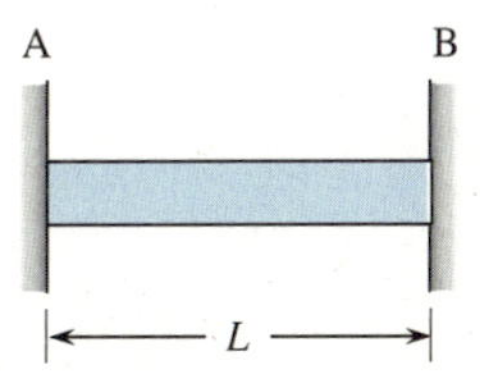

Figure 2.50 Bar with fixed ends subjected to temperature rise.

Solution

The problem is statically indeterminate since we have two reactions and only one useful equation of equilibrium. We use the method of superposition to solve for the stress. First, we release the bar from support B (Figure 2.51(a)) and allow it to elongate freely due to the temperature increase ΔT (Figure 2.51(b)). Using Eq. (2.25), the elongation is

$$\delta_T = \alpha(\Delta T)L \qquad \text{(a)}$$

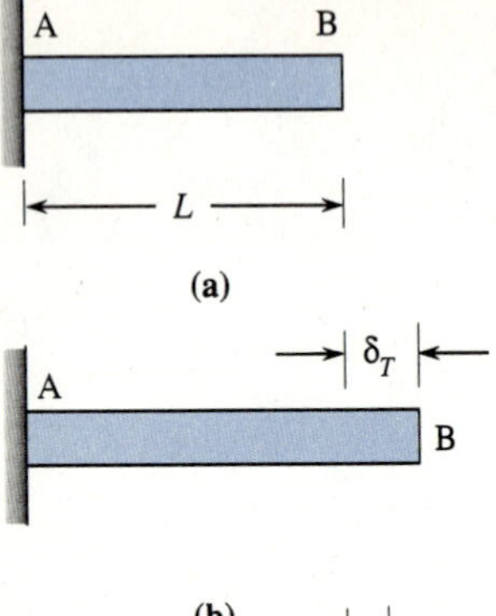

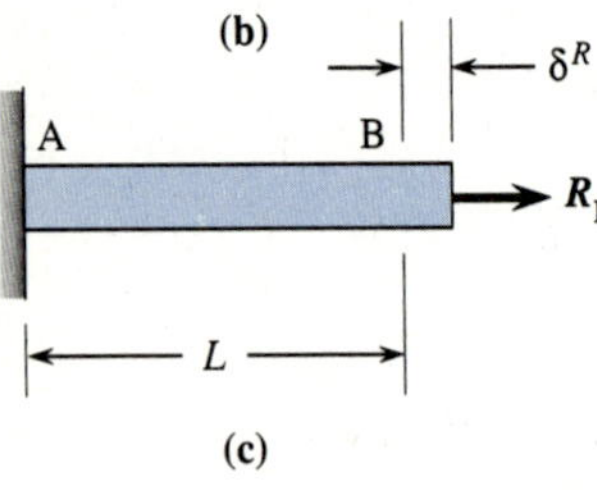

Figure 2.51 (a) Bar released from support B, (b) bar allowed to elongate freely due to temperature increase, and (c) bar with redundant reaction force.

We then apply the redundant reaction force R_B (Figure 2.51(c)), and using Eq. (2.21), calculate the deformation due to R_B as

$$\delta^R = \frac{R_B L}{AE} \qquad \textbf{(b)}$$

Applying the compatibility equation, that says the total deformation of the bar is zero, we superimpose Eqs. (a) and (b) to obtain

$$\delta = \delta_T + \delta^R = \alpha(\Delta T)L + \frac{R_B L}{AE} = 0$$

Solving for R_B, we have

$$R_B = -AE\,\alpha(\Delta T)$$

The negative sign indicates that R_B is to the left (to force end B back to its original position).

The stress in the bar is

$$\sigma_T = \frac{R_B}{A} = -E\alpha(\Delta T) \quad (C) \qquad \textbf{(2.27)}$$

Hence, we observe that the statically indeterminate constant cross-sectional area bar has no strain in it because it does not change length but does have a stress due to the temperature rise.

EXAMPLE 2.21

The bar (Figure 2.52) is made up of two different cross-sectional areas of steel. It is rigidly supported at ends A and C and then subjected to a temperature rise of 75°F. The cross-sectional area of AB is 2 in.^{2} and that of BC is 3 in.^{2} The bar has $E = 30 \times 10^6$ psi and $\alpha = 6.5 \times 10^{-6}/°\text{F}$. Determine **(a)** the axial force in the bar, **(b)** the maximum axial stress in the bar, and **(c)** the displacement of point B.

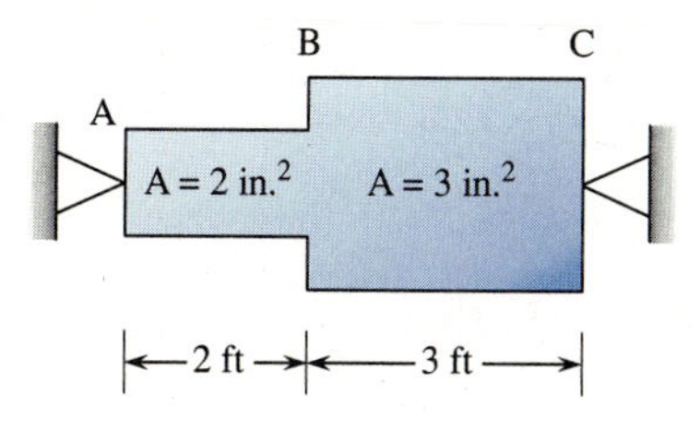

Figure 2.52 Bar with rigid end supports subjected to temperature rise.

Solution

We use the superposition method. Therefore release end C (Figure 2.53(a)) and allow the bar to elongate freely (Figure 2.53(b)). Then calculate the deformation due to temperature increase using Eq. (2.25) as

$$\delta_T = \alpha(\Delta T)L = (6.5 \times 10^{-6})(75°\text{F})(5 \text{ ft}) = 0.00244 \text{ ft}$$

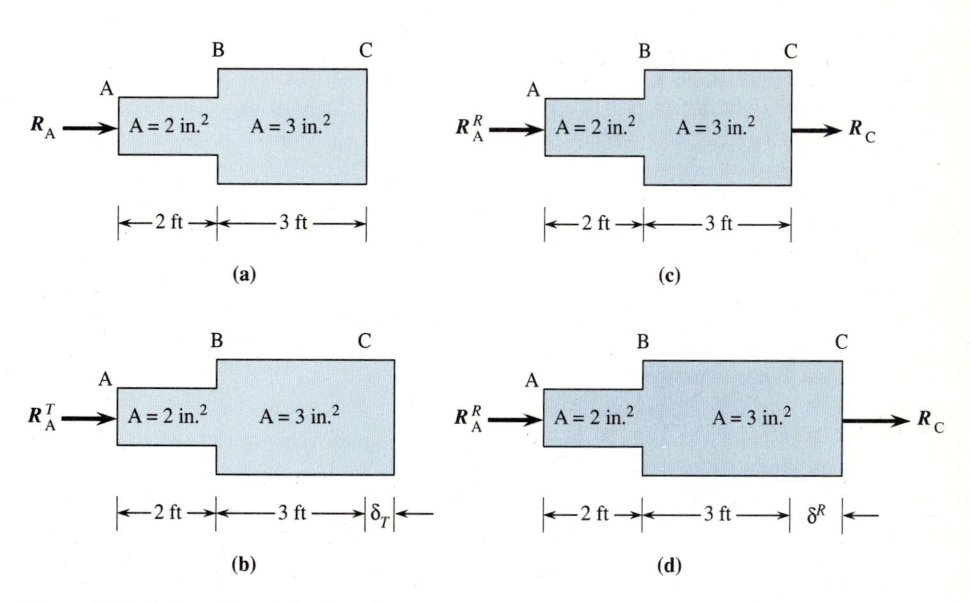

Figure 2.53 (a) Bar with end C released,
(b) bar elongating freely due to temperature increase,
(c) bar subjected to redundant reaction at C, and
(d) bar deformation due to redundant reaction at C.

Apply the redundant reaction force R_C (Figure 2.53(c)), and use Eq. (2.22) to calculate the deformation (Figure 2.53(d)) due to R_C as

$$\delta^R = \frac{R_C L_{AB}}{A_{AB}E} + \frac{R_C L_{BC}}{A_{BC}E} = \frac{R_C}{E}\left(\frac{3\text{ ft}}{3\text{ in.}^2} + \frac{2\text{ ft}}{2\text{ in.}^2}\right) = \frac{2R_C}{E}\frac{\text{ft.}}{\text{in.}^2}$$

where the internal force in sections AB and BC is R_C in Figure 2.53 (d). Using the compatibility equation, with the total deformation of the bar equal to zero, we have

$$\delta = \delta_T + \delta^R = 0 = 0.00244\text{ ft} + \frac{2R_C}{30 \times 10^6}$$

$$R_C = -36{,}600\text{ lb} = -36.6\text{ kips}$$

The negative sign indicates that R_C is opposite in direction to that initially assumed. Then by statics, we obtain

$$\Sigma F_x = 0: \quad R_A + R_C = 0, \qquad R_A = 36.6\text{ kips}$$

(a) The axial force in the bar is

$$P_{AB} = P_{BC} = -R_A = -36.6\text{ kips} \quad (C)$$

(b) The axial stress in each cross-section is

$$\sigma_{AB} = \frac{P_{AB}}{A_{AB}} = \frac{-36.6\text{ kips}}{2\text{ in.}^2} = -18.3\text{ ksi (130 MPa)} \quad \text{(compressive)}$$

$$\sigma_{BC} = \frac{P_{BC}}{A_{BC}} = \frac{-36.6\text{ kips}}{3\text{ in.}^2} = -12.2\text{ ksi (85 MPa)} \quad \text{(compressive)}$$

$$\sigma_{max} = \sigma_{AB} = -18.3\text{ ksi} \quad \text{(compressive)}$$

(c) The displacement of point B is due to both temperature change and loading and is given by

$$\delta_B = \delta_{BT} + \delta_B^R$$

$$= \alpha(\Delta T)\, L_{AB} + \frac{P_{AB} L_{AB}}{A_{AB} E}$$

$$= (6.5 \times 10^{-6}/°\text{F})(75°\text{F})(24\text{ in.}) + \frac{(-36.6\text{ kips})(24\text{ in.})}{(2\text{ in.}^2)(30 \times 10^3\text{ ksi})}$$

$$= 0.0117\text{ in.} - 0.0146\text{ in.} = -0.0029\text{ in.} \ (-0.074\text{ mm})$$

The negative sign indicates that point B moves to the left. Therefore, even though the total deformation of the bar is zero, for bars with nonuniform cross-sectional areas, such as that in Figure 2.52, the deformation of an interior point such as B is not zero. If the bar of Figure 2.52 was of uniform cross-section then the displacement of point B would have been zero. You can show this by choosing one of the cross-sectional areas of Example 2.21 for the whole bar and reworking the problem.

EXAMPLE 2.22

A structural steel beam from a bridge structure is 30 m long (Figure 2.54). A 25-mm gap is allowed for expansion at the right end. The beam was placed in position at a temperature of 10°C. Determine (a) the highest temperature the beam can reach before it just bears against the adjacent abutment, and (b) the stress in the beam if the temperature increases by 5°C over the final temperature in part (a). Let $E = 210$ GPa and $\alpha = 12 \times 10^{-6}/°C$.

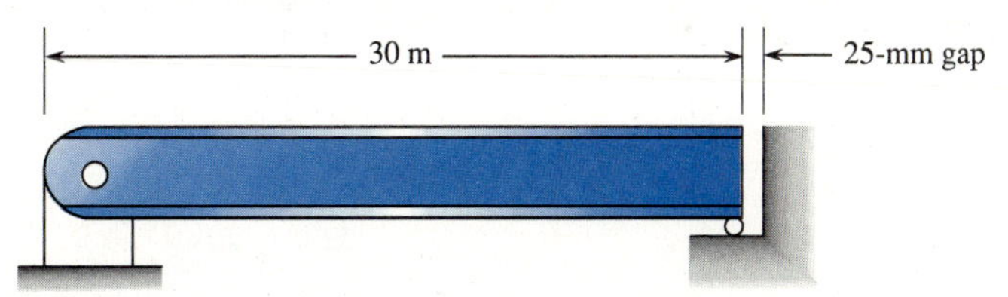

Figure 2.54 Beam subjected to uniform temperature rise.

Solution

(a) By Eq. (2.25), the temperature change needed to close the 25-mm gap is

$$\Delta T = \frac{\delta_T}{\alpha L} = \frac{0.025 \text{ m}}{(12 \times 10^{-6}/°\text{C})(30 \text{ m})} = 69.4°\text{C}$$

(b) If ΔT is now 5°C when the beam bears against the abutment, then by Eq. (2.26), the strain due to temperature change is

$$\epsilon_T = \alpha(\Delta T) = (12 \times 10^{-6}/°\text{C})(5°\text{C}) = 60 \times 10^{-6}$$

and the stress due to temperature change, given by Eq. (2.27), is

$$\sigma_T = -E\epsilon_T = -\left(210 \times 10^6 \frac{\text{kN}}{\text{m}^2}\right)(60 \times 10^{-6}) = -12{,}600 \frac{\text{kN}}{\text{m}^2} \quad (-1830 \text{ psi})$$
(compressive)

EXAMPLE 2.23

A rigid beam is pin supported at the left end and supported by vertical bars of steel at points B and D (Figure 2.55). The right end bar DE is then subjected to a temperature change of 100°F. Determine the normal stresses in the vertical bars. Let $E = 30 \times 10^6$ psi, $\alpha = 7 \times 10^{-6}/°F$, and $A = 1$ in.^{2} for each bar.

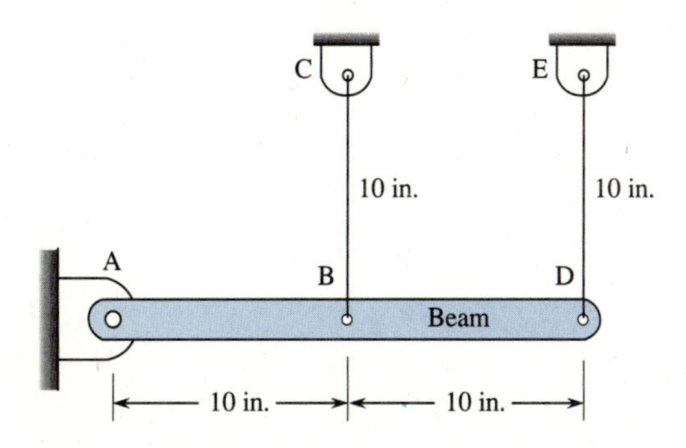

Figure 2.55 Rigid beam structure.

Solution

The structure is statically indeterminate to one degree. That is, one vertical bar could be removed and equilibrium of the beam still maintained. The free-body diagram of the beam is shown below.

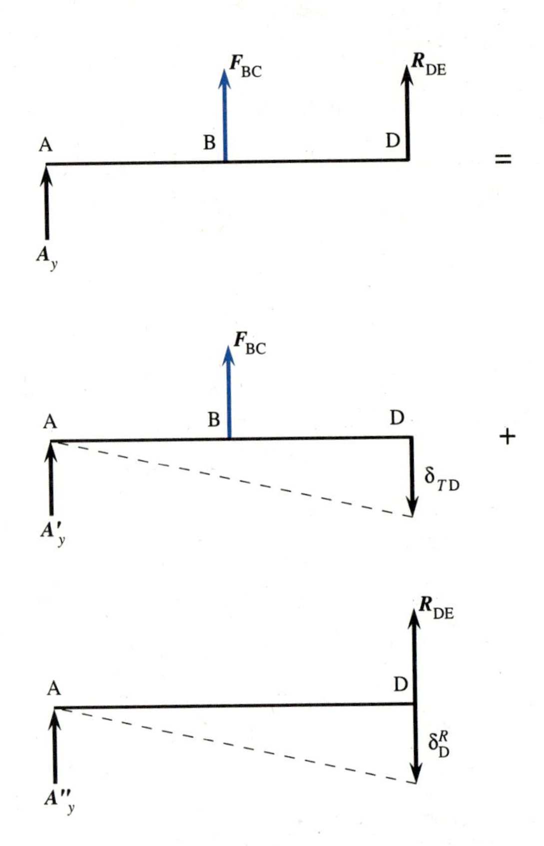

We will call R_{DE} the redundant. By statics, we obtain one equation as

$$\Sigma M_A = 0: \quad (10 \text{ in.})\, F_{BC} + (20 \text{ in.})\, R_{DE} = 0 \tag{a}$$

We need another equation to obtain the forces in the bars. This equation comes from compatibility as

$$\delta_D = 2\delta_B \tag{b}$$

The deflection at D in bar DE is given by that due to temperature change plus that due to the redundant R_{DE}

$$\delta_D = \delta_{TD} + \delta_D^R \tag{c}$$

Using Eqs. (2.25) and (2.21), Eq. (c) becomes

$$\delta_D = \alpha(\Delta T)L + \frac{R_{DE}\, L}{AE} \tag{d}$$

Also by Eq. (2.21), the deflection at B in bar BE is

$$\delta_B = \frac{F_{BC} L}{AE} \tag{e}$$

Substituting Eqs. (d) and (e) into (b), we obtain

$$\alpha(\Delta T)L + \frac{R_{DE} L}{AE} = 2\frac{F_{BC} L}{AE} \tag{f}$$

From Eq. (a), we have

$$F_{BC} = -2R_{DE} \tag{g}$$

Using Eq. (g) in Eq. (f), yields

$$\alpha(\Delta T) = \frac{-5R_{DE}}{AE}$$

or

$$R_{DE} = -\frac{AE\,\alpha(\Delta T)}{5} \tag{h}$$

Substituting the numerical quantities into Eq. (h) yields

$$R_{DE} = \frac{-(1)(30 \times 10^6)(7 \times 10^{-6})(100)}{5}$$

$$R_{DE} = -4200 \text{ lb} \quad \text{(compressive)}$$

Using Eq. (g)

$$F_{BC} = 8400 \text{ lb} \quad \text{(tensile)}$$

The stresses are then obtained by

$$\sigma_{DE} = \frac{R_{DE}}{A_{DE}} = \frac{-4200}{1} = -4200 \text{ psi} \quad \text{(compressive)}$$

$$\sigma_{BC} = \frac{F_{BC}}{A_{BC}} = \frac{8400}{1} = 8400 \text{ psi} \quad \text{(tensile)}$$

We interpret the signs on the stresses as follows. Bar DE was heated but not allowed to freely expand. That is, bar BC prevented the free expansion of bar DE, and this restraint resulted in a compressive stress in bar DE. Bar BC was stretched, due to the heating of bar DE, resulting in a tensile stress in bar BC.

2.9 ALLOWABLE STRESSES AND FACTOR OF SAFETY

We have now learned how to calculate stresses in a variety of axially loaded members. We now understand that stresses play a major role in the analysis of load-bearing structures and machines, and in the selection, sizing, and choice of materials in the design of new structures and machines.

To analyze and design structures against failure, we must know the load carrying capacity or the strength of the structure. The true strength of the structure must be great enough to resist the loads applied to it. The strength of a particular material must then be known in order to properly analyze or design a structure.

We have seen in Section 2.5 how the strength of a material is determined and also how to find a table in Appendix B, and inside the back cover of the text, of various material properties, such as the ultimate strength and yield strength, used in the analysis/design process.

For instance, knowing the ultimate strength of the material being used, our objective is to design a structure or machine so that the probability of reaching the ultimate load is very small. This smaller load is called the *design load, allowable load,* or *working load.* We use a concept called the *factor of safety,* (FS), in the design of any structure, where the factor of safety is defined as the ratio of the ultimate load P_u to the allowable load P_{allow} given by

$$FS = \frac{\text{ultimate load}}{\text{allowable load}} = \frac{P_u}{P_{allow}} \tag{2.28}$$

The FS must be greater than 1 to avoid failure. Usual values range from a value slightly above 1.0 to about 15.

For many applications, we want to avoid permanent deformations after loads are removed. That is, we want a linear-elastic relationship between the load and the stress caused by the load. For ductile materials, such as mild steel and aluminum, a FS against yielding of the material is often used. Then, we define the FS by

$$FS = \frac{\text{yield stress}}{\text{allowable stress}} = \frac{\sigma_Y}{\sigma_{allow}} \tag{2.29}$$

The *allowable stress* σ_{allow} is that largest stress deemed to be permissible in the material. Allowable stress is also called *working stress or design stress.* Allowable stresses for various materials are specified in numerous codes, which are listed at the end of this section.

A typical factor of safety in steel building design to protect against yielding of the steel is 1.67 for tension members under static loading (see American Institute of Steel Construction [AISC] Specification) [4]. For a typical mild structural steel, the yield stress is 36 ksi. Then, using Eq. (2.29), with FS = 1.67, and solving for the allowable stress, we obtain

$$\sigma_{allow} = \frac{\sigma_Y}{FS} = \frac{36 \text{ ksi}}{1.67} = 21.6 \text{ ksi}$$

For materials that do not have a clearly defined yield stress, such as wood, or brittle materials, such as cast iron, the ultimate stress is often used instead of the yield stress to define the factor of safety as

$$FS = \frac{\text{ultimate stress}}{\text{allowable stress}} = \frac{\sigma_u}{\sigma_{allow}} \tag{2.30}$$

Another type of safety factor used in the aircraft industry is called the *margin of safety* (MS) and is defined as the factor of safety minus 1. Therefore,

$$MS = FS - 1 \tag{2.31}$$

The choice of an appropriate factor of safety requires sound engineering judgment based on numerous considerations, such as the following:

1. The *type and number of loads* planned for the present structure and those anticipated for future use. The loads applied to a structure generally fall into one of a number of types based on how they vary with time. The following list provides the types of time-varying loads:

(a) *Static*—a slowly applied load assumed to be constant (or not varying with time) that acts for a short time, such as a snow load on a roof.

(b) *Sustained*—a load that acts for a long time, such as the dead load (self-weight) of a material.

(c) *Impact*—a rapidly applied load, such as when one body collides with another.

(d) *Repeated*—a cyclic or fluctuating load, such as occurs in machine parts and bridge structures, often causing fatigue of the material.

Larger factors of safety are generally required for impact and repeated loads.

The possibility of increased loads to be resisted by the structure at some future date should be considered. For instance, a bridge being considered for a rural road with light traffic, might with population growth require heavier loads and numerous repetitions where fatigue could be a major concern.

2. The *uncertainty in the material properties* from those expected. We must decide how confident we are in the strength and actual dimensions of the materials selected. For instance, concrete has much greater variability in strength than does steel and normally requires greater factors of safety than steel (actually called load factors in reinforced concrete design codes, such as those of the American Concrete Institute (ACI) [5]).

3. The *type or mode of failure* that is likely to occur. Here we mention four likely modes of failure to structures subjected to slowly applied or static loads.

(a) *Excessive stress* (of which two types occur).

(1) *Sudden fracture*. This failure mode usually occurs in a brittle material, such as cast iron, and is usually due to a tensile, compressive, or shear stress that exceeds the ultimate stress of the material.

(2) *General yielding*. This results in loss of structural stiffness and/or large deformations, and finally rupture and collapse of the structure. For ductile materials, such as structural steel or aluminum, this kind of failure is likely under static loading conditions.

(b) *Excessive deformations*. This type of failure may result in loss of service or function of the structure or machine even though rupture may not occur. For instance, if a beam in a building deflects too much, the walls may crack; or if a crankshaft deflects too much, a machine may not operate properly.

(c) *Buckling or sudden collapse*. This type of failure may occur suddenly and without warning when a small increase in load to a structure results in a sudden snap through motion. A member may fail in a buckling mode long before its ultimate or yield stress is reached. Long slender members subjected to compressive loads, such as columns of buildings, must be designed to prevent buckling. Column buckling is considered in Chapter 9. Generally, larger factors of safety are used for members susceptible to buckling than for members guided by strength or deformation requirements.

4. *The importance of a member* to the overall load resistance capability of the structure. A primary member, such as a beam or a column, is more important to the integrity of a structure than a secondary member, such as bracing.

5. *The capability of the theoretical/analytical method* in predicting the actual behavior. The assumption of uniform axial load acting at the centroid of a cross-section of a rod used as one of a number of hangers supporting, say a concrete catwalk, or the assumption of smooth pins connecting members of a truss together result in analytical methods that may neglect secondary bending effects in the actual structures.

6. The significance of *environmental factors*. For instance, the effects of corrosion of steel; or decay or termites to wood; or salt and freeze-thaw to concrete may be significant environmental concerns.

7. Finally, consideration of *weight* may be most important in design as in aircraft structures.

Along with sound engineering judgment, *design codes* and *specifications* are available to guide the engineer in choosing appropriate factors of safety and/or allowable loads and stress levels. These codes may be (1) general codes; (2) codes pertaining to a specific material, such as wood, concrete, steel, aluminum, or plastic; and (3) codes pertaining to a certain service or use provided by the structure.

A list including many organizations and some of the often used codes is provided below:

1. General codes
   (a) American National Standards Institute (ANSI) [6]
   (b) Basic Building Code (BBC) [7]
   (c) Uniform Building Code (UBC) [8]
   (d) Standard Building Code (SBC) [9]

2. Material and service
   (a) The Aluminum Association [10] (for design of aluminum structures)
   (b) American Institute of Steel Construction (AISC) [4] (for design of steel structures)
   (c) American Concrete Institute (ACI) [5] (for design of reinforced concrete structures)
   (d) American Institute of Timber Construction (AITC) [11] (for design of wood structures)
   (e) National Design Specification for Wood Construction (NDS) [12] (for design of wood structures)
   (f) American Association of State Highway and Transportation Officials (AASHTO) [13] (for design of highway bridges)
   (g) American Railway Engineers Association (AREA) [14] (for design of railroad bridges)
   (h) Crane Manufacturers Association of America (CMAA) [16] (for design of overhead cranes)
   (i) Structural Plastics Design Manual [15] (for design using structural plastics)
   (j) ASME Boiler and Pressure Vessel Code [17] (for design, fabrication, and inspection of boilers and pressure vessels, and nuclear power plant components during construction)

Examples 2.24 and 2.25 illustrate the concepts of allowable stress and factor of safety to determine the size of members.

EXAMPLE 2.24

Two steel hangers made of equal leg angles are used to suspend a 20,000-lb air conditioning unit. The steel is ASTM A36 with an allowable tensile stress of 21,600 psi. Determine the size of angle needed.

Solution

Each hanger is assumed to resist one-half the total load. Therefore the axial force in a hanger is 10,000 lb.

Using Eq. (2.1), we solve for the cross-sectional area as

$$A = \frac{P}{\sigma} = \frac{10{,}000 \text{ lb}}{21{,}600 \text{ psi}} = 0.463 \text{ in.}^2$$

Using Appendix C, the equal leg angle with cross-sectional area closest to the required is a 2 in. × 2 in. by 1/8 in.-thick angle with cross-sectional area of 0.484 in.^{2} This notation means the angle has legs of 2 in. length and of 1/8 in. thickness.

EXAMPLE 2.25

A mild steel pipe with a yield stress $\sigma_Y = 270$ MPa is subjected to a compressive load $P = 1500$ kN (Figure 2.56). Using a factor of safety against yielding of FS = 1.8, determine the diameter d of the pipe. The thickness $t = d/8$.

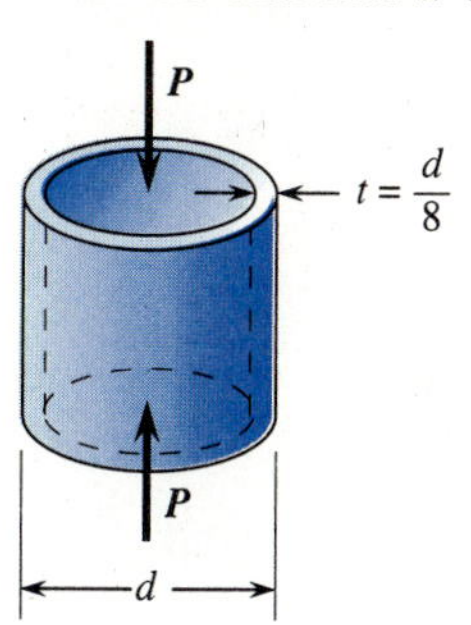

Figure 2.56 Steel pipe subjected to compressive load.

Solution

Using Eq. (2.29), the allowable stress is

$$\sigma_{\text{allow}} = \frac{\sigma_Y}{\text{FS}} = \frac{270 \text{ MPa}}{1.8} = 150 \text{ MPa}$$

Using Eq. (2.1), the axial stress is

$$\sigma = \frac{P}{A} \quad \text{or} \quad A = \frac{P}{\sigma} \tag{a}$$

where

$$A = \frac{\pi}{4}\left[d^2 - \left(d - 2\frac{d}{8}\right)^2\right] = \frac{7\pi\, d^2}{64} \qquad \text{(b)}$$

Equating Eqs. (a) and (b), we have

$$\frac{7\pi d^2}{64} = \frac{1500 \text{ kN}}{150 \times 10^3 \text{ kN/m}^2} = 10 \times 10^{-3} \text{ m}^2$$

$$d = 0.17 \text{ m} = 170 \text{ mm}$$

2.10 STRESS CONCENTRATIONS IN AXIAL LOADING

In the previous sections of this chapter, we have assumed the members to be straight or nearly so and to have a uniform cross-section. When this is true, the normal stress is given by Eq. (2.1) as $\sigma = P/A$. This normal stress is an average or nominal stress assumed to be acting uniformly over the cross-section.

In practice, it is sometimes difficult to design members without a change in the geometry due to, for instance, holes, grooves, or fillets. When these abrupt changes in geometry or discontinuities occur, sudden, often large, increases in stress result in the region near the discontinuity. These regions are then called areas of stress concentration. Two common examples where stress concentrations arise are the tensile members shown in Figure 2.57(a) with a hole and Figure 2.57(b) with grooves or notches. For the plate with the hole, we illustrate the stress concentration. The stress lines or trajectories in the plate must move closer together in the vicinity of the hole, that is across the plane B-B.

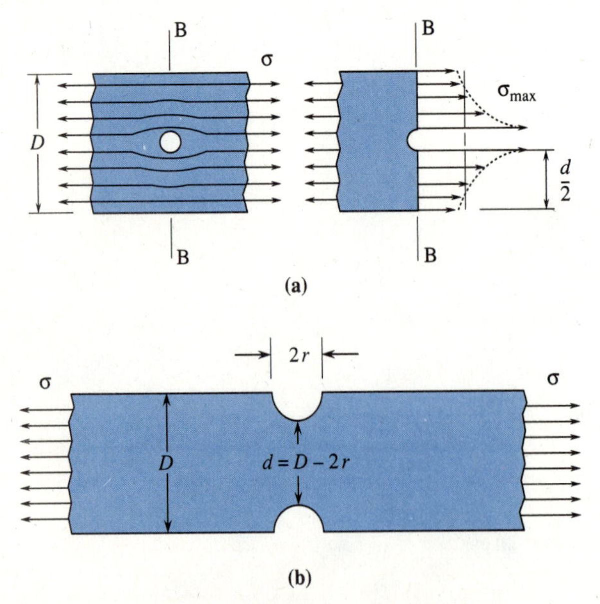

Figure 2.57 Tensile members with stress concentrations
(a) plate with hole and
(b) plate with notches.

The actual peak stress in the plate, σ_{max}, depends on the geometry of the concentration and the type of material. For the plate, the stress concentration depends on the width D of the plate, the width of the plate d at the hole and the radius r of the hole. We now define the stress concentration factor as

$$K = \frac{\sigma_{max}}{\sigma_{ave}} \tag{2.32}$$

Using Eq. (2.1), the average stress is given by

$$\sigma_{ave} = \frac{P}{A_{net}} \tag{2.33}$$

For the plate of thickness t with a hole, the net area is

$$\begin{aligned} A_{net} &= A_{total} - A_{hole} \\ &= t \times D - t \times 2r = t\,(D - 2r) = t \times d \end{aligned}$$

For the notched plate, the net area is

$$A_{net} = t \times d$$

There are numerous methods that have been used to obtain stress concentration factors for various geometries. Experimental methods, such as photoelasticity, brittle-coating methods, and electrical strain gauge methods have been used. For special geometries, the theory of elasticity has been used. More recently, the finite element method has been used when simple geometries do not exist. Extensive figures of stress concentration factors are available not only for axially loaded members, but also for shafts under torsion and beams in bending for a large number of geometries. A most notable source for stress concentration factors is R.E. Peterson's *Stress Concentration Factors* [18].

Generally, for linear-elastic material behavior, the stress concentration factor is primarily dependent on the geometry involved. For very abrupt changes in geometry, such as sharp radii or 90° reentrant corners, the stress concentration is higher.

Stress concentration factors for five common conditions, the axially loaded flat bar with a hole, grooves, or fillets, and the round bar with fillet and groove are shown in Figure 2.58. From Figure 2.58, we observe that the stress concentration is reduced as the ratio of r/d is increased.

Figure 2.59 illustrates the flow of stress and typical stress concentration factors that arise for different fillet radii. Again, sharp corners and small radii between changes in the size of the member should be avoided as these changes in geometry result in large stress concentrations. The influence of the stress concentrations is limited to regions near the geometric discontinuity. These localized effects disappear at a short distance equal to the width D of the bar from the fillets or notches. This is part of Saint-Venant's principle (after Barre de Saint-Venant, 1864). The principle also explains the validity of using $\sigma = P/A$ for the axial stress in a bar. This formula gives the stress quite accurately at distance D away from any concentrated load.

The concentration factors in Figure 2.58 have been developed for materials behaving linear-elastic. Therefore, since most brittle materials behave linear-elastic almost to rupture, these factors apply for all stresses to rupture. However, materials behaving in a ductile manner are able to deform and flow in the vicinity of the stress concentration and thus

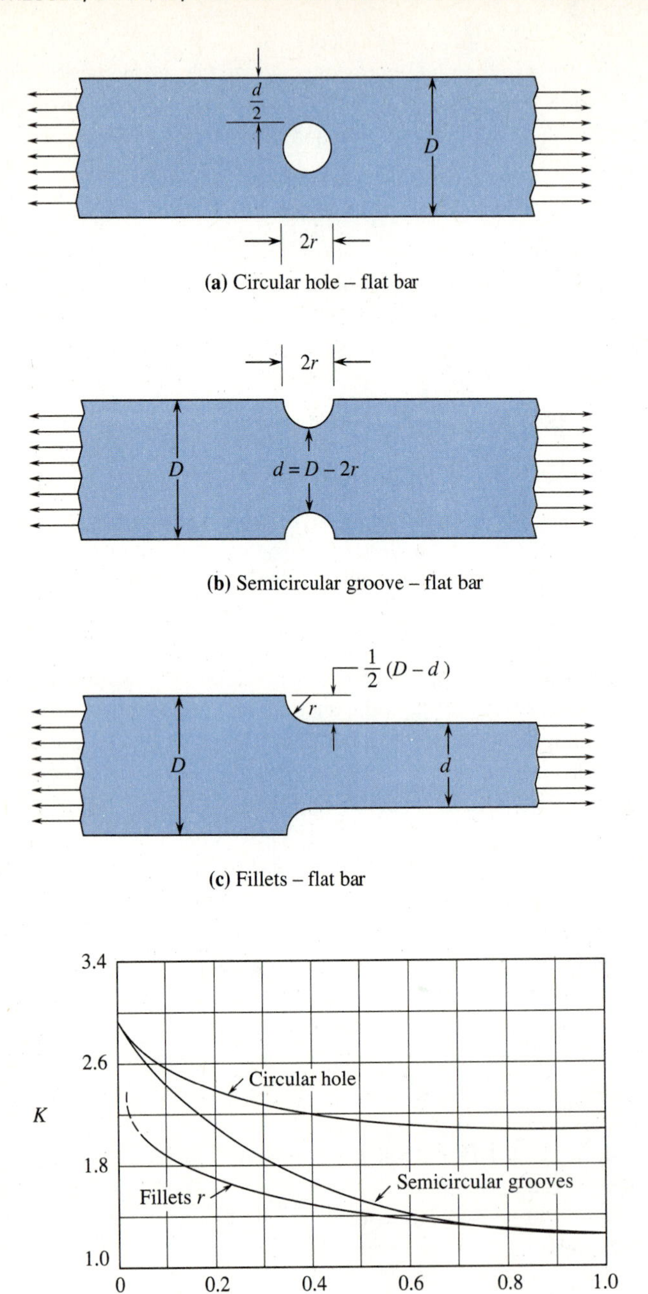

(d) Stress concentration factors for flat bars in **(a)**, **(b)**, and **(c)**

Figure 2.58 Stress concentration factors for five common axially loaded bars.

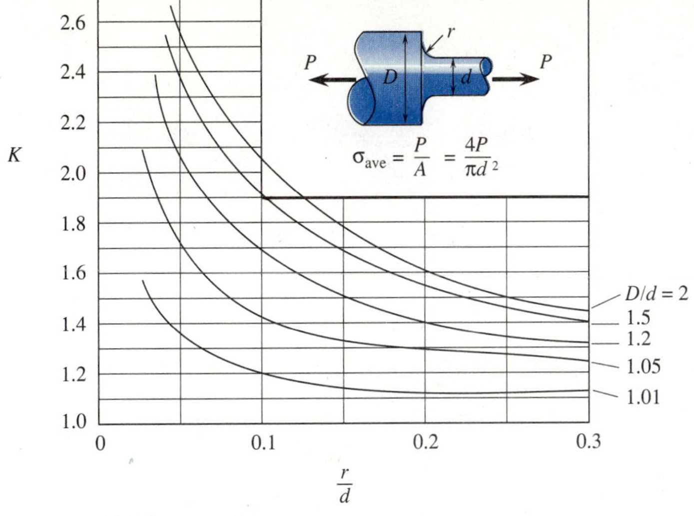

(e) Fillet – round bar

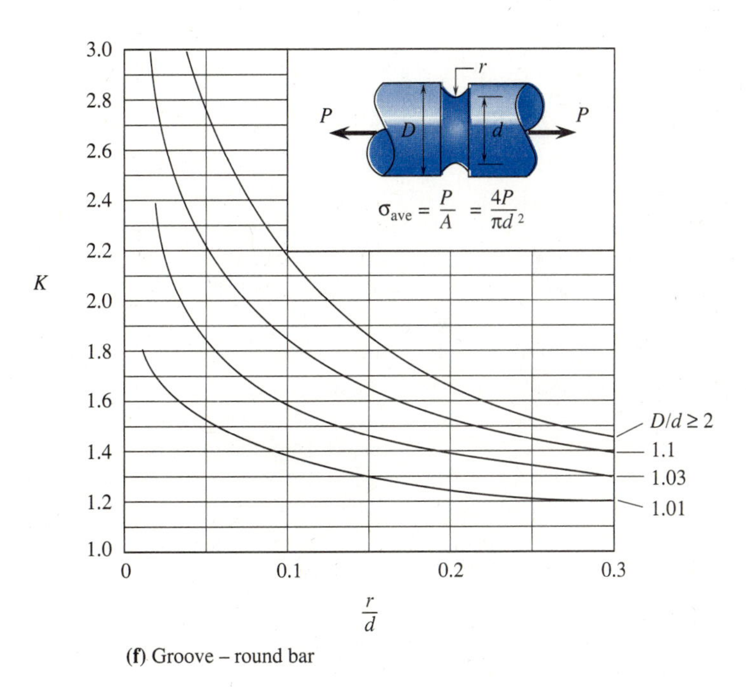

(f) Groove – round bar

they redistribute the stress in a way that the maximum stress is only slightly higher than the proportional limit stress. That is, if we multiply the average stress by the stress concentration factor and exceed a value greater than the proportional limit stress (or in practice exceed the yield stress), then the stress at the change in geometry has redistributed so

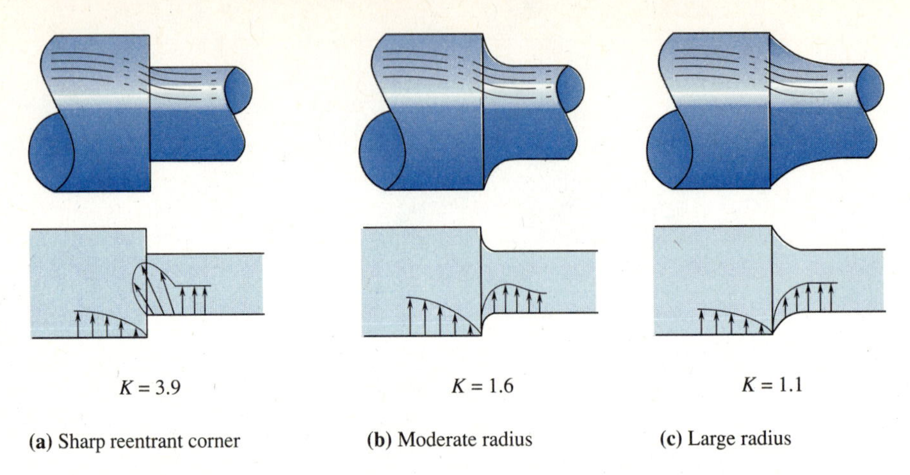

(a) Sharp reentrant corner (b) Moderate radius (c) Large radius

Figure 2.59 Flow of stress and typical stress concentration factors for different fillet radii.

that we assume the stress to be no greater than the yield stress of the material. The description of why this is so is shown in Figure 2.60 and is explained as follows.

Assume we have two flat tensile bars. Bar one (Figure 2.60(a)) is of uniform cross-section area A, while bar two (Figure 2.60(b)) is notched on each edge such that the cross sectional area at the notches (along section B-B) is also A. Each bar is made of a ductile material whose idealized stress-strain diagram is shown in Figure 2.60(c). The load F on the uniform bar (Figure 2.60(a)) can be increased to a value equal to the product of the area A times the yield stress σ_Y before failure by gross yielding occurs. This is illustrated in Figure 2.60(e). Assuming the geometry of the notched bar such that the stress concentration factor is $K = 2$, results in yielding beginning at one-half the load of that in the uniform bar. This is shown in Figure 2.60(d) and again as stage 1 in Figure 2.60(f). As the load is further increased, the stress distribution goes through typical stages shown as 2, 3, and finally 4. These curves show an increased region of local yielding, which begins at the root or edge of the notch (stage 1) and increases as the load is increased until gross yielding over the whole cross-section A begins at stage 4. You should notice that the load F_4 associated with curve 4 is identical to the unnotched load capacity shown in Figure 2.60 (e), that is, $F_4 = A\sigma_Y$. This increased loading can be done without significant elongation of the bar until yielding occurs in stage 4. The end result is that, for practical purposes, the notched bar will resist the same static load as the unnotched or uniform bar.

Finally, remember that for fatigue (see Chapter 11 for design under fatigue loading) and impact loading (see Section 10.15) of most engineering materials, stress concentrations must be considered, as even the normally ductile materials may start to behave in a brittle manner when subjected to fatigue and impact loading. For instance, recall the discussion on cold working in Section 2.5.

EXAMPLE 2.26

The machine part with fillets shown in Figure 2.61(a) is subjected to a static tensile force of 50 kips. Determine the maximum tensile stress in the member if it is made of (a) a mild steel or (b) an alloy steel whose partial stress-strain curves are shown in Figure 2.61(b).

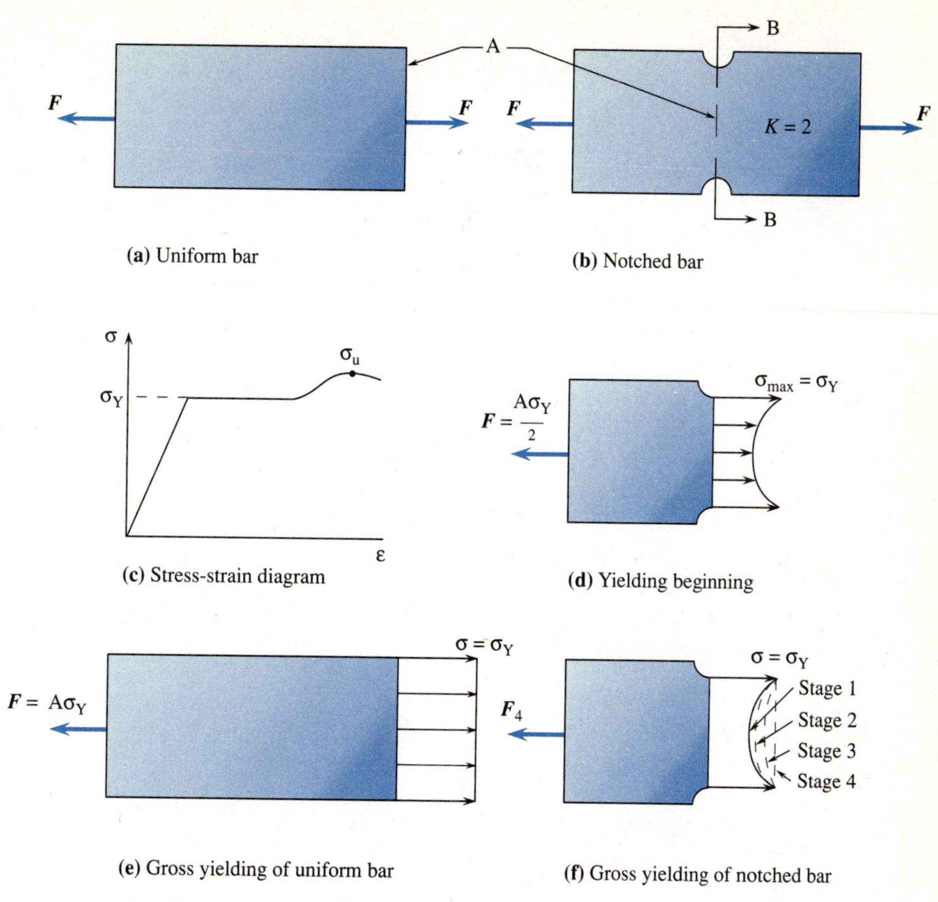

Figure 2.60 Uniform bar and notched bar used to illustrate stress concentration effects in ductile material behavior.

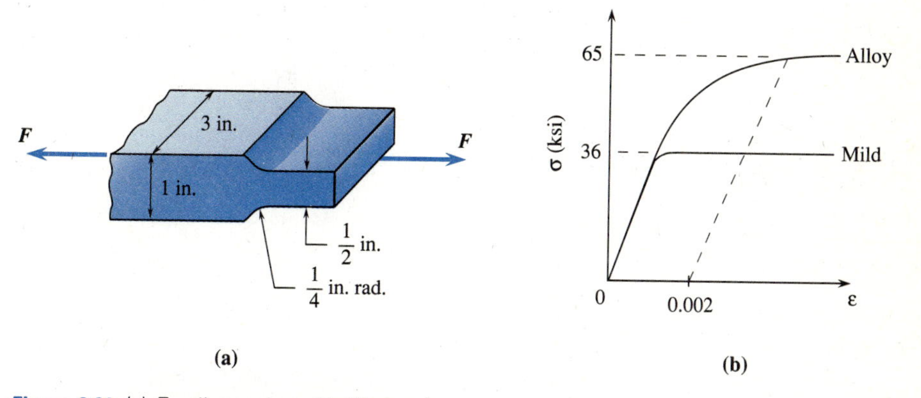

Figure 2.61 (a) Tensile member with fillets and (b) stress-strain diagrams of the materials.

Solution

From the geometry, Figure 2.61(a) and use of Figure 2.58(c), we have the ratio of r/d given by

$$\frac{r}{d} = \frac{0.25}{0.50} = 0.50$$

Therefore, using Figure 2.58(d), the value of the theoretical stress concentration factor is approximately

$$K = 1.4$$

The average tensile stress in the smaller section is

$$\sigma_{ave} = \frac{F}{A} = \frac{50 \text{ kips}}{3 \times 0.5}$$

$$= 33.33 \text{ ksi}$$

(a) For the mild steel, using Eq. (2.32), we then obtain the maximum stress as

$$\sigma_{max} = 1.4\ (33.33 \text{ ksi}) = 46.67 \text{ ksi}$$

This stress is above the yield stress of 36 ksi of the mild steel. Therefore, the maximum stress is the yield stress, that is,

$$\sigma_{max} = 36 \text{ ksi}$$

(b) For the alloy steel, using Eq. (2.32), we obtain

$$\sigma_{max} = K\sigma_{ave}$$

$$= 1.4\ (33.33 \text{ ksi}) = 46.67 \text{ ksi}$$

This is below the 0.2% yield stress and so is a valid answer.

2.11 SUMMARY OF IMPORTANT DEFINITIONS AND EQUATIONS

1. Stress = internal force acting per unit area in units of psi or N/m^2 (1 N/m^2 = 1 Pa) (1 psi = 6.895 kPa) (1 ksi = 6.895 MPa)

2. Normal stress in centric axial loading

$$\sigma = \frac{F}{A} = \frac{P}{A} \quad \text{(psi) or (MPa)} \tag{2.1}$$

(holds for centric load, uniform cross section, homogeneous material)

3. Average stresses on inclined plane of centric axial loaded member

$$\sigma = \frac{P}{A}\cos^2\theta \qquad \tau = \frac{P}{A}\cos\theta\sin\theta \tag{2.5}$$

$$\sigma_{max} = \frac{P}{A} \tag{2.6}$$

$$\tau_{max} = \frac{\sigma_{max}}{2} \tag{2.7}$$

4. Normal strain = change in original length, δ, divided by the origin length, L, in units of in./in. or mm/mm.

$$\epsilon = \frac{\delta}{L} \text{ (in./in.) or (mm/mm)} \tag{2.8}$$

5. Modulus of elasticity (E) = slope of linear portion of stress-strain diagram in units of psi or Pa.

6. Hooke's law

$$\sigma = E\epsilon \tag{2.11}$$

(valid for linear-elastic material)

7. Poisson's ratio

$$\nu = \left|\frac{\text{lateral strain}}{\text{axial strain}}\right| = \left|\frac{\epsilon_l}{\epsilon_a}\right| = -\frac{\epsilon_l}{\epsilon_a} \tag{2.12}$$

(valid only in linear-elastic range of material)

8. Modulus of resilience (MR)

$$\text{MR} = \frac{\sigma_p^2}{2E} \quad \text{(psi) or (Pa)} \tag{2.15}$$

(area under σ-ϵ diagram up to the proportional limit stress)

9. Modulus of toughness (MT)

$$\text{MT} = \sigma_{ave}\,\epsilon_f \quad \text{(psi) or (Pa)} \tag{2.16}$$

(area under the whole σ-ϵ diagram)

10. Deformation

$$\delta = \frac{PL}{AE}, \quad \delta = \sum_{i=1}^{N} \frac{P_i L_i}{A_i E_i} \quad \text{(in.) or (mm)} \qquad (2.21), (2.22)$$

$$\delta = \int_0^L \frac{P(x)}{A(x)\,E(x)}\,dx \tag{2.24}$$

11. Deformation due to temperature change (free expansion)

$$\delta_T = \alpha(\Delta T)L \quad (2.25)$$

12. Strain due to temperature change (free expansion or contraction of bar)

$$\epsilon_T = \alpha(\Delta T) \quad (2.26)$$

13. Stress due to temperature change (no expansion or contraction of bar)

$$\sigma_T = -E\,\alpha(\Delta T) \quad (2.27)$$

14. Factor of safety

$$\mathrm{FS} = \frac{P_u}{P_{allow}} \quad (2.28)$$

$$\mathrm{FS} = \frac{\sigma_Y}{\sigma_{allow}} \quad (2.29)$$

$$\mathrm{FS} = \frac{\sigma_u}{\sigma_{allow}} \quad (2.30)$$

15. Margin of safety

$$\mathrm{MS} = \mathrm{FS} - 1 \quad (2.31)$$

16. Stress concentration factor

$$K = \frac{\sigma_{max}}{\sigma_{ave}} \quad (2.32)$$

REFERENCES

1. American Society for Testing Materials (ASTM). *Annual Book of Standards*. 1916 Race St., Philadelphia, PA 19103.
2. Askeland D.R. *The Science and Engineering of Materials*. PWS Engineering, Boston, 1984.
3. Materials reference issue. *Machine Design*. Vol. 57 April 1985.
4. American Institute of Steel Construction, Inc. *Manual of Steel Construction,* 9th ed., One East Wacker Dr., Chicago, IL 60601, 1989.
5. American Concrete Institute, ACI Committee 318. *Building Code Requirements for Reinforced Concrete*. Latest Edition. Detroit.
6. American National Standards Institute (ANSI). *American National Standard Minimum Design Loads for Buildings and Other Structures, ANSI A58.1–1982,* 1430 Broadway, New York, NY 10018, 1982.
7. Building Officials and Code Administrators International, Inc. (BOCA). *Basic Building Code 1978,* 7th ed. 17926 South Halsted, Homewood, IL 60430, 1978.
8. International Conference of Building Officials. *Uniform Building Code*. Latest Edition. 5360 South Workman Mill Road, Whittier, CA 90601, 1979.
9. Southern Building Code Congress International, Inc. (SBCC). *Standard Building Code,* Latest Edition. 5200 Montclair Road, Birmingham, AL 35213.
10. The Aluminum Association. *Specifications for Aluminum Structures,* 9th edition. 900 19th St., N.W., Washington, D.C. 20006, 1988.
11. American Institute of Timber Construction. 333 West Hampden Avenue, Englewood, CO 80110. *Timber Construction Manual,* 3rd ed. John Wiley & Sons, New York, 1986.
12. National Forest Products Association. *National*

Design Specification for Wood Construction, 1986 ed. 1619 Massachusetts Avenue, N.W., Washington, D.C. 20036, 1986.

13. American Association of State Highway and Transportation Officials. *Standard Specifications for Highway Bridges,* 14th ed. 440 North Capital St., N.W., Washington, D.C., 20001, 1989.
14. American Railway Engineering Association. *Specifications for Steel Railway Bridges.* Chicago, Latest Ed.
15. Structural Plastics Design Manual. *ASCE Manuals and Reports on Engineering Practice* 63, American Society of Civil Engineers, 345 East 47th St., New York, NY 10017, 1984.
16. Crane Manufacturers Association of America, Inc. *Specifications for Electric Overhead Traveling Cranes.* CMAA Specification No. 70. 1326 Freeport Road, Pittsburgh, PA 15238, Rev. 1983.
17. American Society of Mechanical Engineers (ASME). *ASME Boiler and Pressure Vessel Code.* 345 East 47th St. N.W., New York, NY 10017, 1989.
18. Peterson RE. *Stress Concentration Factors.* John Wiley & Sons, New York, 1974.
19. American Society for Metals. *ASM Metals Reference Book—A Handbook of Data and Information.* Metals Park, OH, 1981.

PROBLEMS

Section 2.2

2.1 Compute the axial stress in a round bar subjected to a direct tensile force of 6400 N if the diameter of the bar is 10 mm.

2.2 A hollow pipe of inside diameter $d_1 = 4.0$ in. and outside diameter $d_2 = 4.25$ in. is compressed by an axial force $P = 60$ kips (Figure P2.2). Calculate the average compressive stress in the pipe.

Figure P2.2

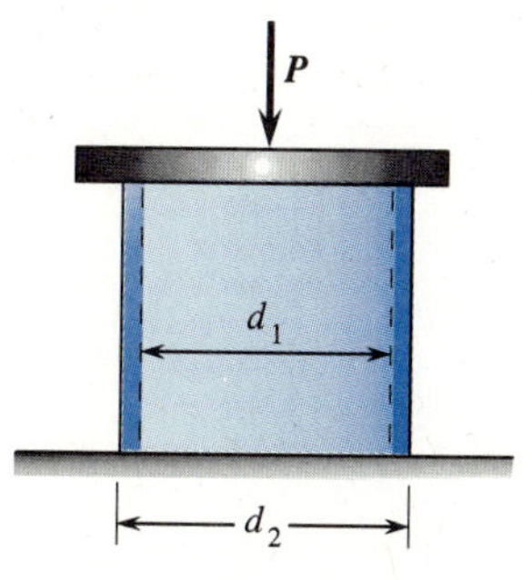

2.3 Compute the stress in a rectangular bar having cross-sectional dimensions of 10 mm × 30 mm if a direct tensile force of 40 kN is applied.

2.4 A link in a mechanism of an automobile hood is subjected to a tensile force of 50 lb. If the link cross-section is 0.25 in. × 0.50 in., compute the axial stress in the link.

2.5 A circular rod 3/8 in. in diameter supports an air conditioning unit weighing 2000 lb. Compute the axial stress in the rod.

2.6 A tension member in a wood truss is subjected to 5000 lb of force. A Douglas fir member, 1.5 in. × 3.5 in., construction grade, is used in the truss? (See Appendix B.) Determine the axial stress in the member.

2.7 A steel press has four tension members. Each member has a diameter of 15 mm. The largest load resisted by the press is to be 44 kN. Determine the axial stress in the tension members.

Figure P2.7

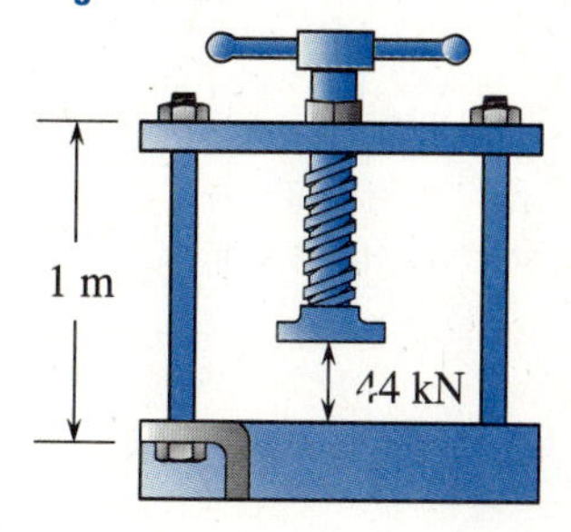

2.8 A guy wire for an antenna tower is to be steel, having an allowable stress of 22,000

psi. If the expected maximum load on the wire is 8000 lb, determine the required diameter of the wire.

2.9 A hopper having a mass of 1,000 kg is designed to hold a load of grain having a mass of 7000 kg. The hopper is to be suspended by four rectangular hangers, each carrying one-fourth of the load. Steel plate with a thickness of 8.0 mm and a width of 50 mm is used for the hangers. Determine the tensile stress in a hanger.

2.10 A concrete pedestal base is circular, 10 in. in diameter, and carries a direct compressive load of 80,000 lb. Compute the compressive stress in the concrete.

2.11 Three short, square, wood blocks, 3.5 in. on a side, support a machine weighing 30,000 lb. Compute the compressive stress in the blocks. Assume each block supports one-third of the total load.

2.12 A short link in a robotic mechanism carries an axial compressive load of 4000 N. If it has a square cross section, 8.0 mm on a side, compute the stress in the link.

2.13 A tension member in a truss is a steel angle with legs of 2.5 in. and a thickness of 1/4 in. Its cross-sectional area is 1.19 in.^{2} (see Appendix C). If the angle is A36 structural steel, how much tensile load can be applied if the allowable normal stress is 21,600 psi?

2.14 Refer to the composite member shown and assume the cross-sectional areas of the three parts of the member to be as follows: $A_{AB} =$ 3 in.^{2}, $A_{BD} = 2$ in.^{2}, and $A_{DE} = 4$ in.^{2} Determine the normal stress of largest magnitude in the composite member.

Figure P2.14

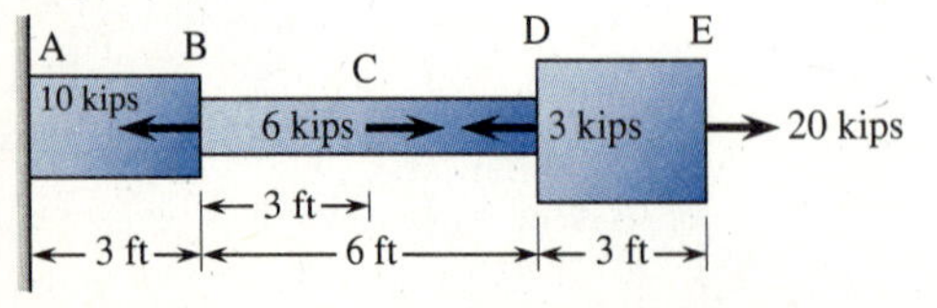

2.15 A composite member is shown in Figure P2.15. Let the cross-sectional areas of the three parts of the member be as follows: $A_{AB} = 5 \times 10^{-4}$ m^2, $A_{BC} = 3 \times 10^{-4}$ m^2, and $A_{CD} = 2 \times 10^{-4}$ m^2. If the normal stress of largest magnitude in the composite member is not to exceed 100 MPa, determine the magnitude of the load *P*.

Figure P2.15

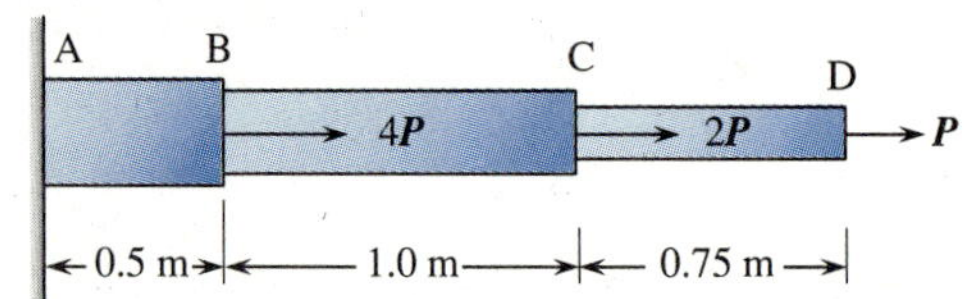

2.16 A composite member is shown in Figure P2.16. The cross-sectional areas of the three parts of the member are as follows: $A_{AB} =$ 2 in.^{2}, $A_{BC} =$ 3 in.^{2}, and $A_{CD} = 1.5$ in.^{2} Determine the normal stress of largest magnitude and indicate which part of the member this stress exists in.

Figure P2.16

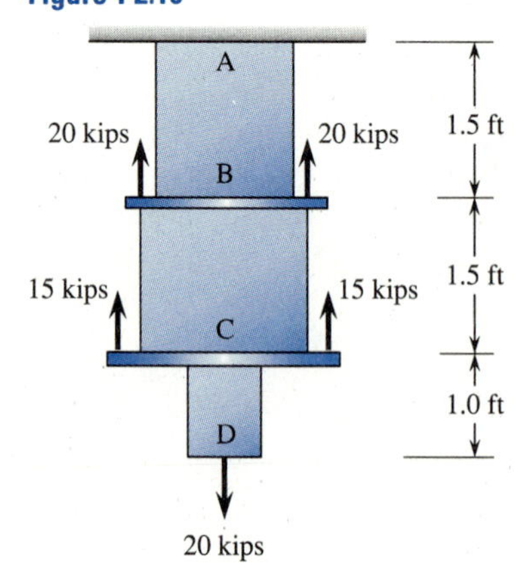

2.17 A crane is modeled by the truss shown. Determine the axial stress in members DE, DG, and HG under the load of the 3-ton tractor. Members DE, DG, and HG all have cross-sectional areas of 3.0 in.^{2}

Figure P2.17

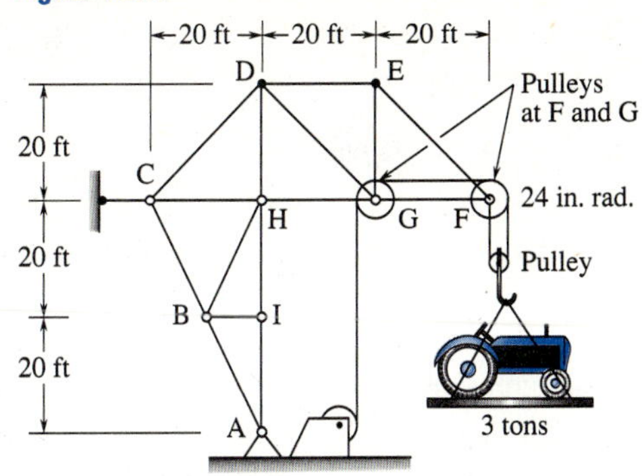

Section 2.3

2.18 The eye bar in Figure P2.18 is fusion welded across a 30° seam as shown. Find the average normal and shear stresses acting on the fused cross-section.

Figure P2.18

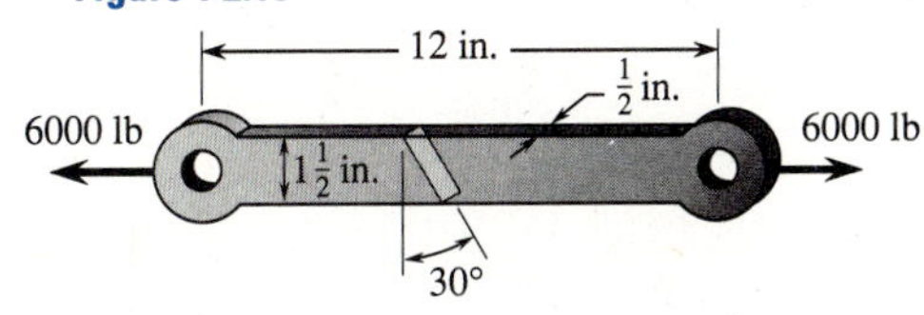

2.19 Two wooden members of uniform cross-section 3.5 × 5.5 in. are joined by the simple glued scarf splice shown. If the allowable shear stress for the glued splice is 80 psi, determine the largest axial load P that may be applied.

Figure P2.19

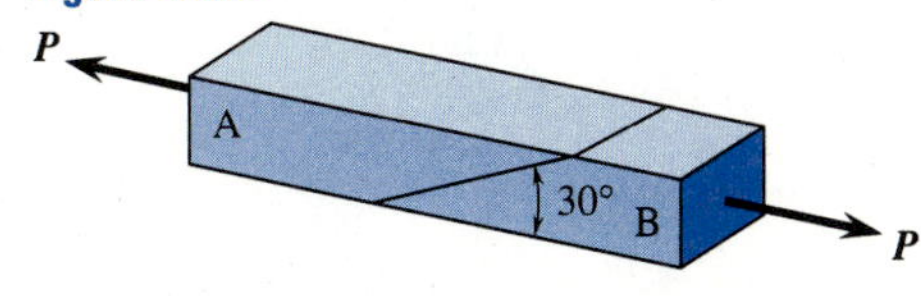

2.20 A steel pipe of 250 mm outside diameter is fabricated from a 6-mm-thick plate by groove (butt) welding along a helix which forms an angle of 20° with a plane perpendicular to the axis of the pipe. Knowing that an axial force P = 200 kN is applied to the pipe, determine the stresses normal and tangential to the weld.

Figure P2.20

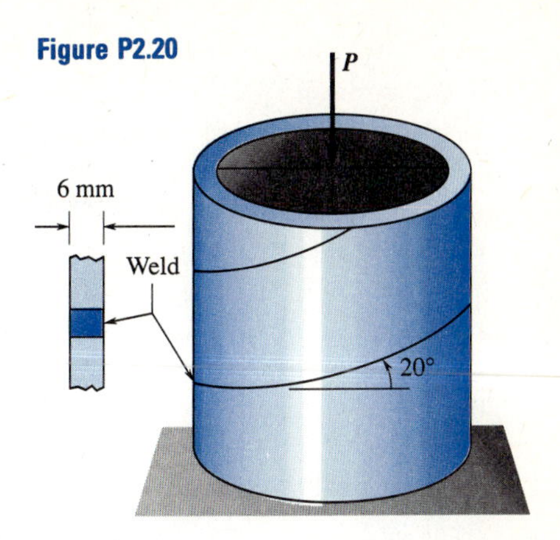

2.21 A grain bin is supported by four wooden blocks, such as to transmit P = 2000 lb to each block. The 4 × 4-in. block is positioned as shown. What is the shear stress parallel to the wood grain? What is the normal stress perpendicular to the wood grain?

Figure P2.21

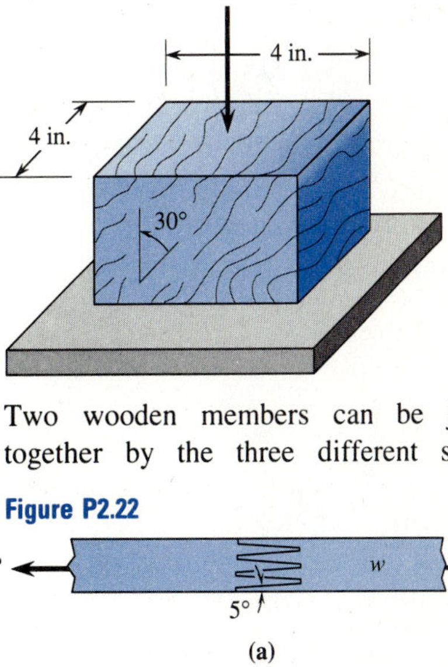

2.22 Two wooden members can be joined together by the three different splices

Figure P2.22

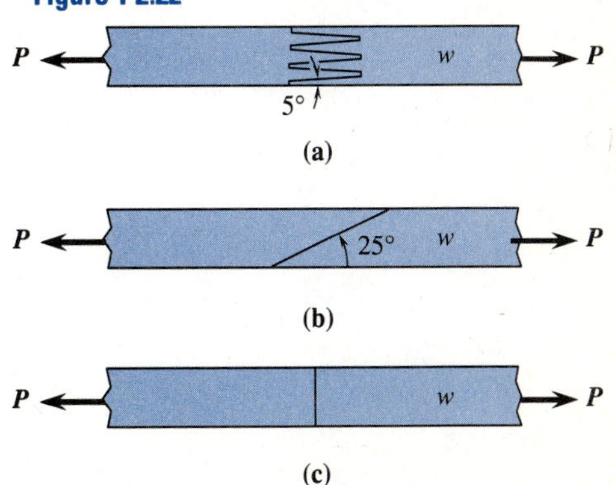

shown. Which of the three methods results in the strongest joint **(a)** if the glue is assumed to be twice as strong in shear as in tension or **(b)** the glue is assumed to be twice as strong in tension as in shear. In all three cases assume the member has width w and thickness t.

2.23 A bar of uniform cross-sectional area A is subjected to an axial tensile stress $\sigma = P/A$. The stresses on an inclined plane a–a are $\sigma_\theta = 70$ MPa and $\tau_\theta = -30$ MPa. Find the axial stress σ and the angle θ.

Figure P2.23

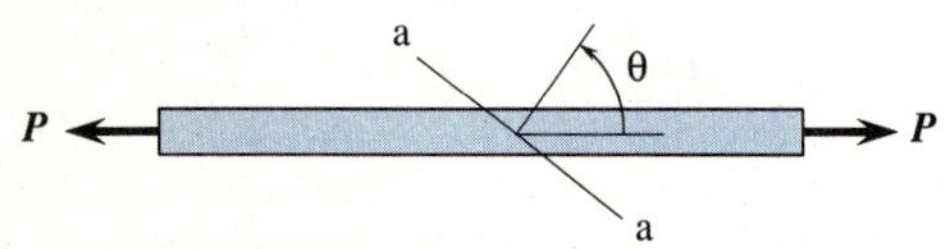

2.24 Two members are fused together as shown, determine the shear stress on the fused cross-section.

Figure P2.24

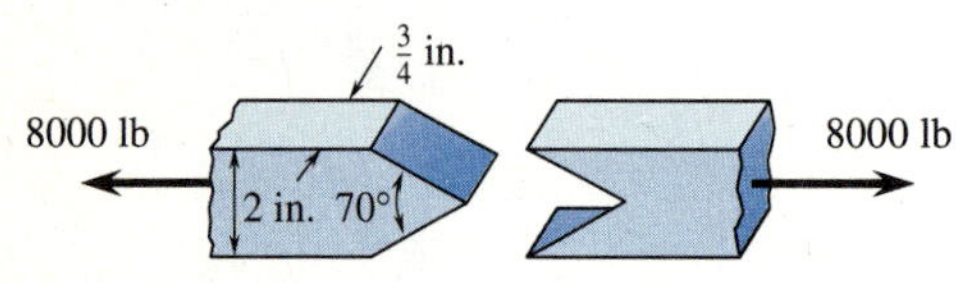

Sections 2.3 and 2.4

2.25 A steel rod BC of 1/2 in. diameter supports a crane beam subjected to a load of 500 lb as shown. Determine **(a)** the axial stress in the rod and **(b)** the maximum shear stress in the rod.

Figure P2.25

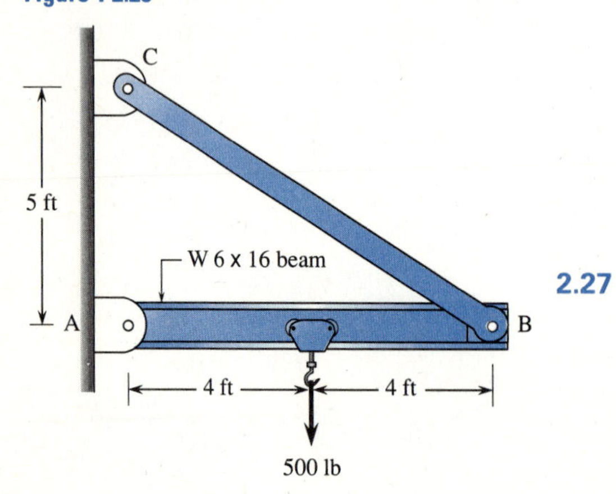

2.26 A strut and cable assembly ABC supports a vertical load $P = 20$ kN. The cable has an effective cross-sectional area of 100 mm^2, and the strut has an area of 250 mm^2. **(a)** Calculate the normal stresses in the cable and strut, and indicate whether they are tension or compression. **(b)** Calculate the maximum shear stress in the cable and strut. **(c)** If the cable elongates 1.5 mm, what is the normal strain? **(d)** If the strut shortens 0.60 mm, what is the normal strain?

Figure P2.26

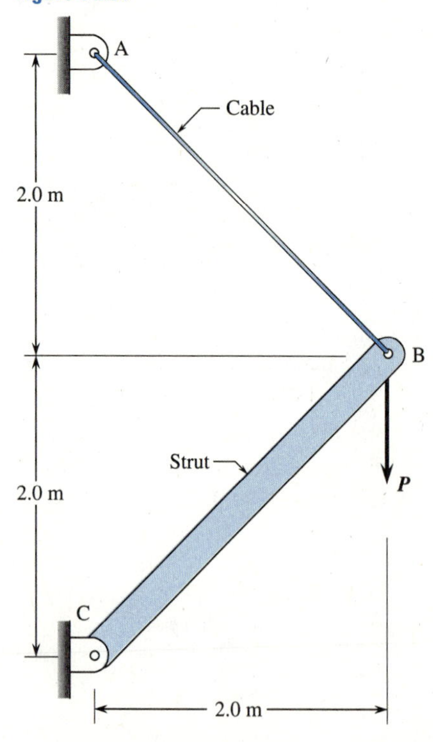

2.27 A steel bar with $E = 30 \times 10^6$ psi is 8 in. long. The bar is subjected to a compressive axial load which produces a shortening of 0.008 in. at the proportional limit for the material. Determine **(a)** the proportional limit stress **(b)** the modulus of resilience.

2.28 A compressive force of 800 kN is gradually applied to a bar whose uniform cross-section is 0.05 m × 0.05 m and whose length is 0.20 m. The 0.05 m dimension changed to 0.05008 m and the 0.20 m length changed to 0.1989 m. Determine **(a)** Poisson's ratio **(b)** the modulus of elasticity for the material.

2.29 A metal bar with a rectangular cross-section 0.02 m × 0.04 m and a gauge length of 0.40 m is subjected to a tensile load of 400 kN. The 0.02 m dimension changed to 0.01998 m and the 0.40 m gauge length changed to 0.402 m. If the behavior of the material is assumed to be within the range of proportionality between stress and strain, determine **(a)** Poisson's ratio **(b)** the final value of the 0.04 m dimension **(c)** the axial stress in the bar **(d)** the maximum shear stress in the bar **(e)** the modulus of elasticity.

2.30 Two bars, one of aluminum and one of steel, are subjected to tensile forces that produce normal stresses $\sigma = 22$ ksi in both bars. What are the lateral strains in the aluminum and steel bars, if $E = 10.6 \times 10^6$ psi and $\nu = 0.33$ for aluminum and $E = 29 \times 10^6$ psi and $\nu = 0.30$ for steel?

2.31 During testing of a cast iron in compression, the original diameter of 2 in. was increased by 0.00025 in. and the original length of 4 in. was decreased by 0.0025 in. under the action of a compressive load $P = 50{,}000$ lb. Calculate the modulus of elasticity E and Poisson's ratio ν.

2.32 A steel rod 0.50 in. in diameter and a gauge length of 4 in. is subjected to a gradually increasing tensile force. At the proportional limit, the value of the load was 12,000 lb, the gauge length measured 4.01 in., and the diameter measured 0.4997 in. Find **(a)** the modulus of elasticity **(b)** Poisson's ratio **(c)** the modulus of resilience.

2.33 A bar of aluminum is subjected to tensile forces that produce normal stress of 22 ksi in the bar. What is the lateral strain in the aluminum bar if $E = 10.6 \times 10^6$ psi and Poisson's ratio is 0.33 for the bar?

2.34 An elevator weighs 1000 lb and is supported by a 5/16-in.-diameter cable, 1500 ft long. When the elevator carries a 1500-lb load, the cable stretches 6 in. more. What is the modulus of elasticity of the cable?

2.35 A brass rod ($E = 103$ GPa, $\nu = 0.35$) has diameter of 25 mm before load. In order to maintain certain clearances, the diameter of the rod must not exceed 25.02 mm. What is the largest permissible compressive load P?

2.36 A compression member constructed from steel pipe ($E = 200$ GPa, $\nu = 0.30$) has an outside diameter of 80 mm and a cross-sectional area of 1600 mm^2. What axial force P will cause the outside diameter to increase by 0.0084 mm?

2.37 Determine the tensile force P required to produce an axial strain $\epsilon = 0.007$ in a copper bar ($E = 17 \times 10^6$ psi) of circular cross-section with diameter equal to 1 in.

2.38 A tension test was performed on a steel specimen whose original diameter was

Load (lb)	Deformation (in.)	Load (lb)	Deformation (in.)
0	0	12,300	0.0170
1,500	0.0004	12,200	0.0200
3,100	0.0010	12,000	0.0275
4,700	0.0016	13,000	0.0330
6,200	0.0022	15,000	0.0400
8,000	0.0026	16,000	0.0500
9,500	0.0032	17,500	0.0680
11,000	0.0035	19,000	0.1080
12,000	0.0041	19,600	0.1510
12,300	0.0051	20,100	0.2010
12,500	0.0071	20,100	0.2600
12,700	0.0100	18,700	0.3300
12,700	0.0131	17,200	0.4100
12,500	0.0150	16,400	0.4500 (fracture)

0.50 in. and whose gauge length was 2.0 in. Corresponding values of load and deformation are given in the following tabulation. Construct the engineering stress-strain curve for this material and determine the ultimate stress and the fracture stress.

2.39 Redraw the initial portion of the data given in Problem 2.38 using a larger scale and determine **(a)** the modulus of elasticity **(b)** the upper yield point **(c)** the lower yield point **(d)** the proportional limit.

2.40 If the final diameter of the specimen of Problem 2.38 was 0.355 in., determine **(a)** the percent reduction of area **(b)** the percent elongation.

2.41 Refer to the data in Problem 2.38 and determine **(a)** the modulus of resilience **(b)** the modulus of toughness.

2.42 A plain concrete compression specimen is tested to failure. This original specimen has a diameter of 6 in. and a length of 10 in. When subjected to compressive loads, the specimen fails suddenly by material breakdown, since the specimen is too short for instability to occur. Force (lb) and deformation (in.) data are as follows, respectively: (0.0, 0.0), (28,000, 0.0026), (56,000, 0.0040), (85,000, 0.0081), (99,100, 0.0100), (113,900, 0.0130), (127,000, 0.0160), (141,000, 0.0260), (137,000, 0.0320), (113,000, 0.0360). Carefully plot the stress-strain curve for this concrete specimen. What is **(a)** the ultimate stress of this concrete measured in psi and **(b)** the modulus of elasticity for a stress of 30% of the ultimate stress. (For concrete the modulus of elasticity is normally taken as that for a stress of 30% of ultimate.)

2.43 A tension test on a metallic specimen resulted in the following corresponding values of load and deformation. The diameter of the specimen is 0.0250 m and the gauge length is 0.1000 m. Construct the engineering stress-strain diagram and determine the ultimate stress and the fracture stress. Also determine the percent elongation and the percent reduction of area if the final diameter of the specimen is 0.01500 m.

Load (N)	Deformation (m)	Load (N)	Deformation (m)
0	0	17,700	29.52×10^{-5}
1,780	1.41×10^{-5}	18,600	36.60×10^{-5}
3,530	2.85×10^{-5}	19,500	46.26×10^{-5}
5,320	4.30×10^{-5}	20,400	57.55×10^{-5}
7,000	5.72×10^{-5}	20,880	74.92×10^{-5}
8,860	7.15×10^{-5}	20,850	90.20×10^{-5}
10,600	8.55×10^{-5}	20,400	125.00×10^{-5}
12,400	10.01×10^{-5}	19,000	150.00×10^{-5}
14,200	13.21×10^{-5}	16,800	170.00×10^{-5}
15,900	18.30×10^{-5}	13,100	194.00×10^{-5}
16,800	24.12×10^{-5}	(Fracture)	

2.44 Redraw the initial portion of the data given in Problem 2.43 using a larger scale and determine **(a)** the proportional limit **(b)** the 0.20% yield stress (i.e., the yield stress corresponding to a strain of 0.0020).

2.45 Refer to the data given in Problem 2.43 and determine **(a)** the modulus of toughness **(b)** the modulus of resilience.

2.46 Three different materials A, B, and C are tested in tension using standard test specimens having diameters of 0.500 in. and gauge lengths of 2.0 in. After the specimens are fractured, the distances between the gauge marks are found to be 2.12, 2.47, and 2.32 in., respectively. Also, the diameters are 0.482, 0.399, and 0.254 in., respectively, at the failure cross-sections. Determine the percent elongation and percent reduction in area of each specimen. Also, classify the materials as brittle or ductile.

2.47 A standard steel specimen of 1/2 in. diameter elongated 0.0043 in. in a 4-in. gauge length when it was subjected to a tensile force of 6000 lb. If the specimen was known to be in the elastic range, what is the elastic modulus of the steel?

Section 2.6

2.48 A steel rod 9 m long used in a control mechanism must transmit a tensile force of 6 kN without stretching more than 2.5 mm, or exceeding an allowable stress of 160 MN/m^2. What must the diameter of the rod be? Give the answer to the nearest millimeter. Let $E = 210$ GPa.

2.49 Determine the elongation of a strip of plastic, 0.80 mm thick × 12 mm wide by 375 mm long, if it is subjected to a load of 100 N and is made of **(a)** polystyrene or **(b)** melamine (see Appendix B).

2.50 A steel support must connect two 30,000-lb tensile loads separated by 200 in. The maximum allowable stress is 10,000 psi and the maximum allowable elongation is 0.02 in. What is the required cross-sectional area?

2.51 A 1-in.-diameter steel rod ($E = 30 \times 10^6$ psi) must carry a load in tension of 40,000 lb. as shown. If the initial length of the stressed portion of the rod is 22.0 in., what is its final length?

Figure P2.51

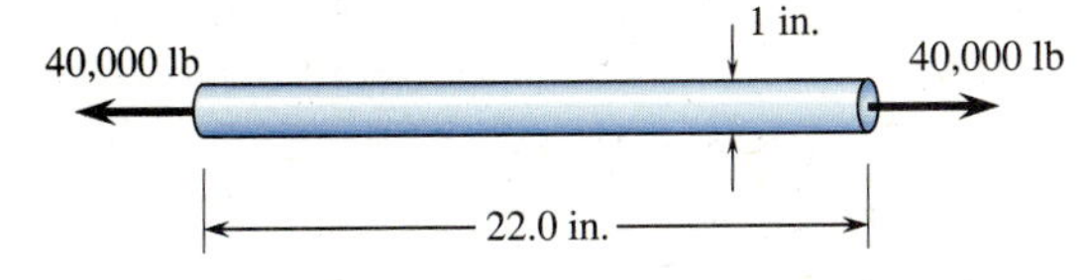

2.52 A 9-m-long round bar made of aluminum ($E = 70$ GPa) carries a tensile load of 600 kN. What is the minimum required diameter d of the bar if the maximum allowable elongation is 8 mm?

2.53 A steel bolt has a diameter of 10.0 mm in the unthreaded portion. Determine the elongation in a length of 200 mm if a force of 15.0 kN is applied. Let $E = 210$ GPa.

2.54 In an aircraft structure, a rod is designed to be 1.50 m long and have a square cross-section 9.0 mm on a side. Determine the amount of elongation which would occur if it is made of **(a)** titanium 6A1–4V and **(b)** stainless steel. The load is 6000 N. (See Appendix B.)

2.55 A tension member in a welded steel truss is 12.0 ft long and subjected to a force 20,000 lb. Choose an equal leg angle (see Appendix C) made of A36 steel which will limit the stress to 21 600 psi. Then compute the elongation in the angle due to the force. Use $E = 29.0 \times 10^6$ psi for structural steel.

2.56 A round steel bar having a cross-section of 0.8 $in.^2$ is attached at the top and is subjected to three axial forces, as shown in the figure. Find the deflection of the free end caused by these forces.

Figure P2.56

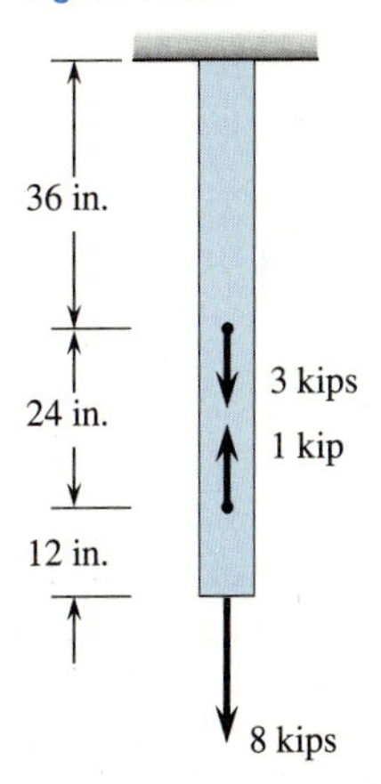

2.57 A link in a mechanism is subjected alternately to a tensile load of 500 lb and a compressive load of 60 lb. Compute the elongation and compression of the link if it is a rectangular steel bar 1/4 in. × 3/8 in. in cross-section and 8.00 in. long.

2.58 A two-story building has columns AB at the first floor and BC at the second floor. The columns are loaded as shown in the figure, with the roof load P_1 equal to 100 kips and the load P_2 applied at the second floor equal to 160 kips. The cross-sectional areas of the upper and lower columns are

7.06 in.^{2} and 19.41 in.^{2}, respectively, and each column has a length of 10 ft. Assuming that structural steel with $E = 29 \times 10^6$ psi is used, determine the shortening of each column due to the applied loads.

Figure P2.58

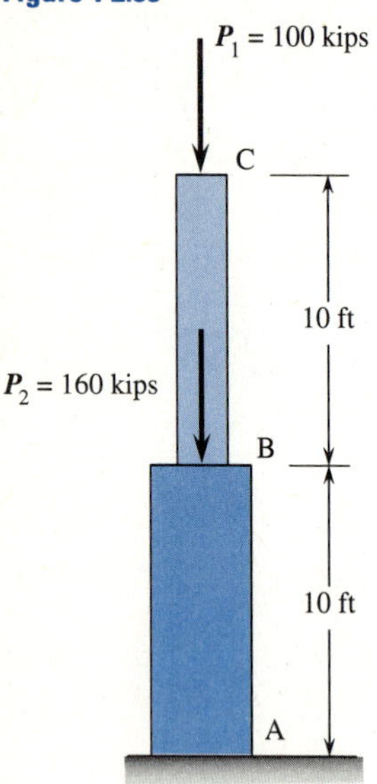

2.59 A concrete pedestal of circular cross-section has an upper part of diameter 0.6 m and height of 0.5 m and a lower part of diameter 1.0 m and height 1.0 m. It is subjected to loads $P_1 = 7$ MN and $P_2 = 18$ MN. Assuming $E = 25$ GPa, calculate the vertical deflection of the top of the pedestal at A.

Figure P2.59

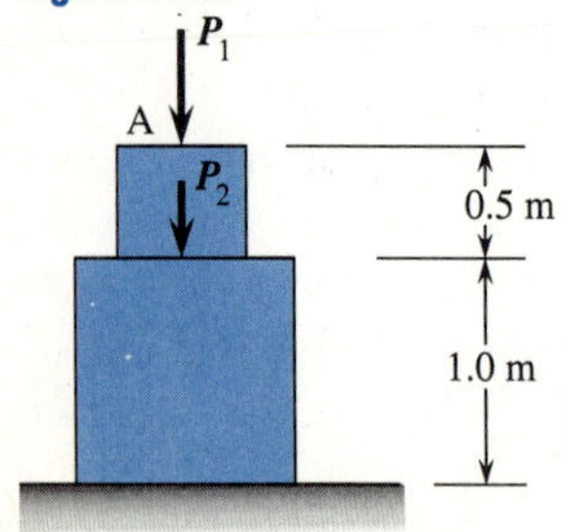

2.60 A uniform cross-section bar ABCD is subjected to the three loads shown. The bar is made of steel with a modulus of elasticity of $E = 200$ GPa and cross-sectional area A = 200 mm^2. Determine the deflection at the lower end of the bar due to the loads. Does the bar elongate or shorten?

Figure P2.60

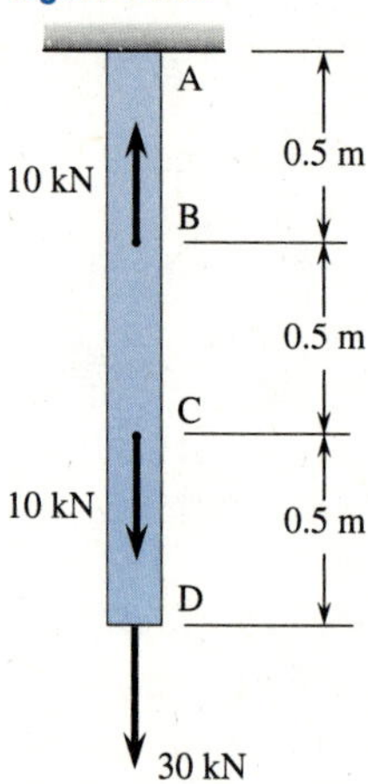

2.61 The steel bar AD has a length of 10 ft and a square cross-section 0.75 in. × 0.75 in. The bar is loaded by the forces shown. Let $E = 30 \times 10^6$ psi and determine the change in length of the bar due to the loads. Does the bar elongate or shorten?

Figure P2.61

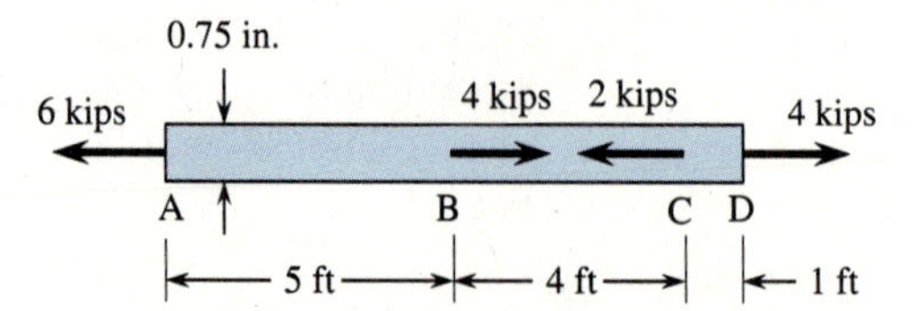

2.62 A steel pipe with $E = 200$ GPa is supported and loaded as shown. The cross-sectional area of the pipe is 200 mm^2. Determine the force P so that the lower end D of the bar does not move vertically when the loads are applied.

Figure P2.62

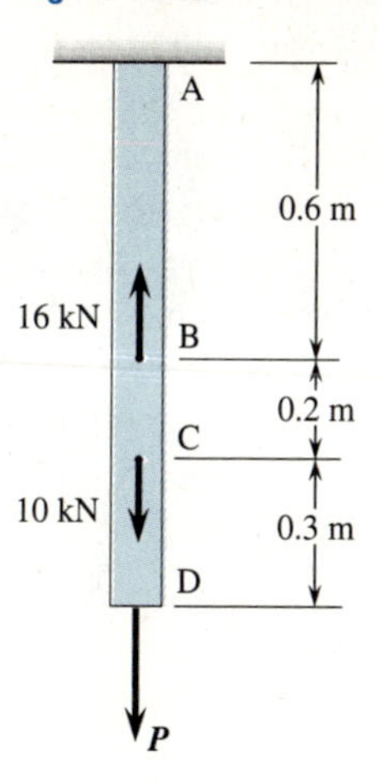

2.63 A steel bar 8 ft long has a circular cross-section of diameter $d_1 = 0.75$ in. over section AB and diameter $d_2 = 0.50$ in. over section BC. **(a)** Determine how much the bar elongates due to a tensile force of $P = 6000$ lb. **(b)** If the same volume of material is rolled into a bar of constant diameter d and same length of 8 ft, determine the elongation due to the same load P. Let $E = 30 \times 10^6$ psi.

Figure P2.63

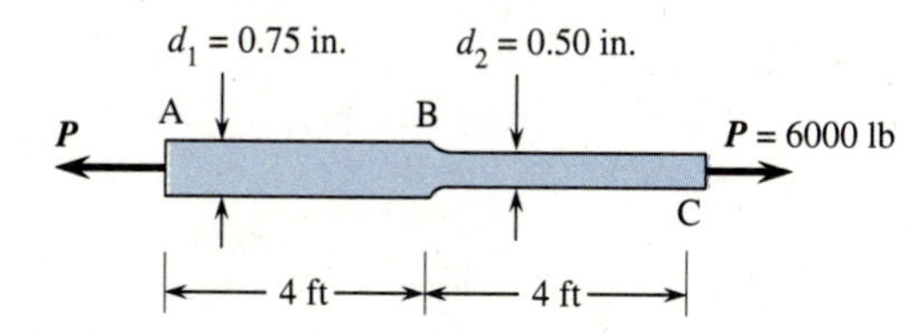

2.64 A long wire hangs vertically under its own weight. Determine its greatest length without yielding if it is made of: **(a)** steel having a yield stress of 36,000 psi, or **(b)** aluminum having a yield stress of 16,000 psi? (The specific weight of steel is 490 lb/ft^3 and that of aluminum is 170 lb/ft^3.)

2.65 A vertical rod of aluminum weighing 1.15 lb/ft is 2 in. × 2 in. square. What should its length be for the free end to elongate 0.30 in. under its own weight? Let $E = 10.0 \times 10^6$ psi.

2.66 For the truss shown, determine the total elongation of the member BC due to the load $P = 500$ kN. The member BC is made from steel with $E = 210$ GPa and cross-sectional area A = 60 mm^2.

Figure P2.66

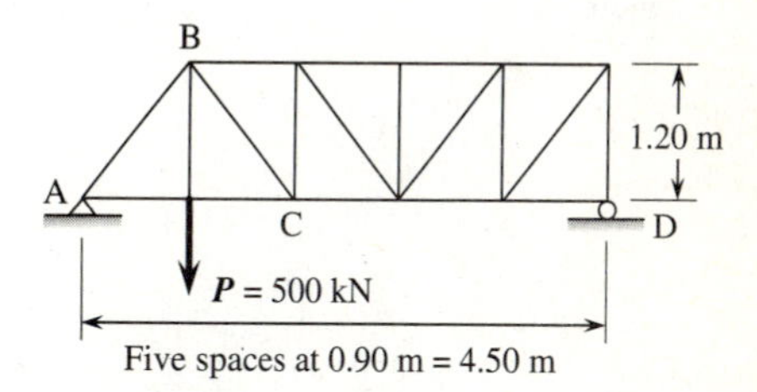

2.67 A concrete pier of square cross-section is 8 m high as shown. The sides taper uniformly from a width of 0.5 m at the top to 1.0 m at the bottom. Determine the shortening of the pier due to a compressive load of 2000 kN. (Neglect the self-weight of the pier.) Assume that the modulus of elasticity of the concrete is 20 GPa.

Figure P2.67

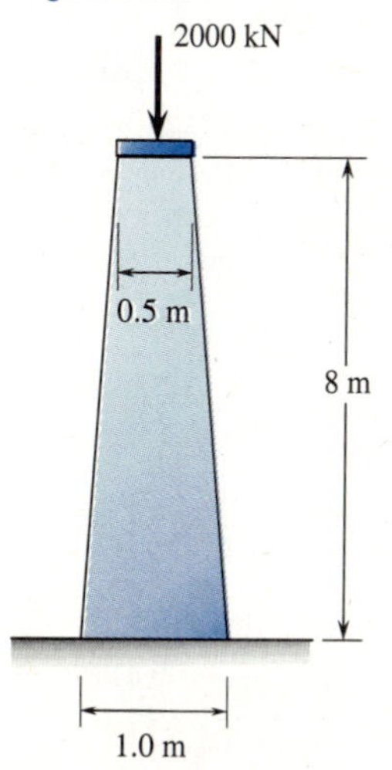

Section 2.7

2.68 A short post is made by welding steel plates into a square, as shown, and then filling the inside with concrete. The width $w = 200$ mm and thickness $t = 10$ mm. Determine the stress in the steel and concrete if E of steel is 200 GPa, E of concrete is 20 GPa, and the post supports an axial compressive force of 1.5 MN.

Figure P2.68

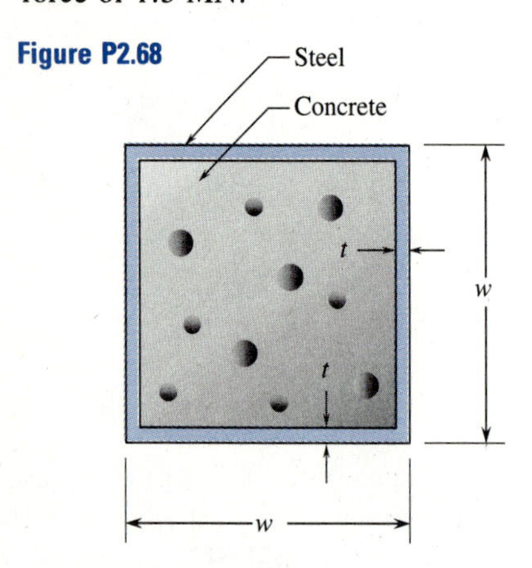

2.69 A short post is made by welding steel plates into a square as shown in Figure P2.68, and then filling it with concrete. The steel has an allowable stress of 21,600 psi and the concrete has an allowable stress of 2000 psi. If $w = 8.0$ in. and $t = 1/4$ in., determine the allowable axial load on the post. Let E of steel be 29×10^6 psi and E of concrete be 2.5×10^6 psi.

2.70 A steel bar AB having two different cross-sectional areas A_1 and A_2 is held between rigid supports and loaded at C by a force P as shown. Determine the reactions at A and B.

Figure P2.70

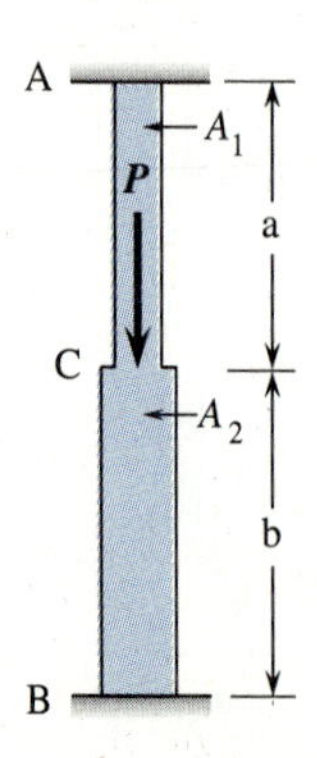

2.71 A square column of concrete reinforced with steel rods is compressed by an axial force P as shown. Determine the fraction of the load carried by the concrete if the total cross-sectional area of the steel bars is one-tenth that of the concrete and the modulus of elasticity of the steel is ten times that of the concrete.

Figure P2.71

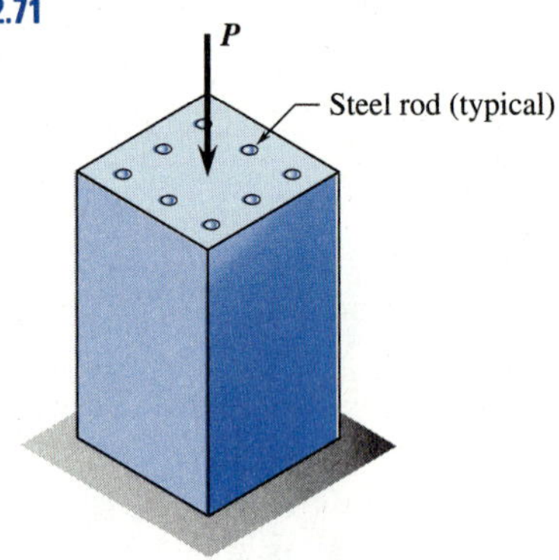

2.72 A short round column is formed of a 3 in. nominal diameter construction pipe with a wall thickness of 0.216 in. filled with concrete (see Appendix C). If the allowable stresses for the steel and concrete are 15,000 psi and 1500 psi, respectively, determine the maximum allowable load P. Assume E of steel is 29×10^6 psi and E of concrete is 3×10^6 psi.

Figure P2.72

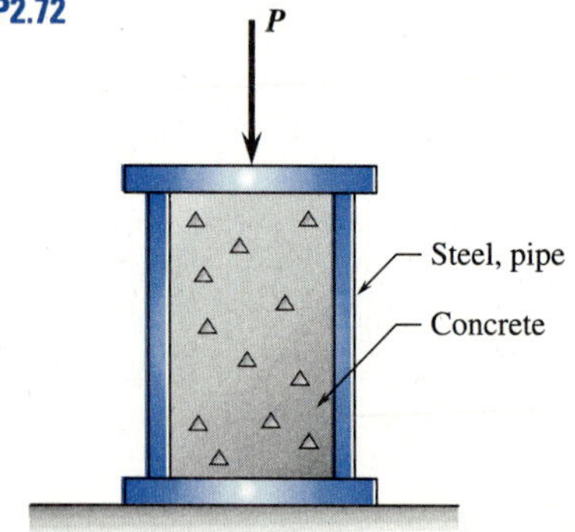

2.73 A square column is made of 20 mm thick plates welded together and then filled with concrete. The outside dimensions of the column are 300 mm × 300 mm. The steel casing has a modulus of elasticity of 205 GPa, and the concrete core has a modulus of elasticity of 20 GPa. Determine the maximum permissible load P on the column if the allowable stresses in the steel and concrete are 60 MPa and 6.0 MPa, respectively.

Figure P2.73

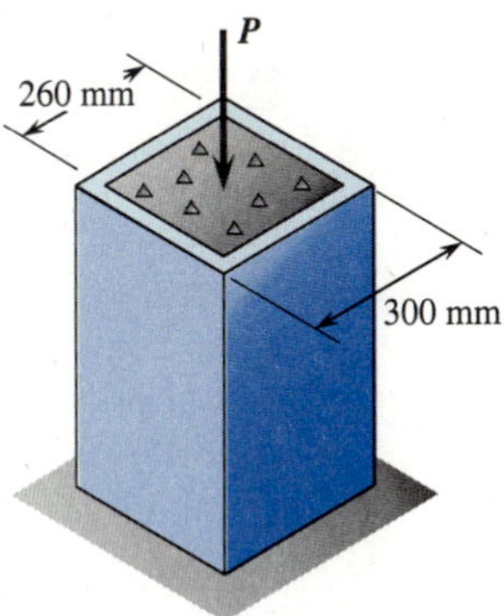

2.74 A rod ABCD has two different cross-sectional areas as shown. The rod is rigidly attached to immovable supports at the ends and is loaded by equal but oppositely directed forces $P = 20$ kN. The cross sectional areas of sections AB and CD are 300 mm^2 and that of section BC is 500 mm^2. Determine the axial stress in each section of the rod.

Figure P2.74

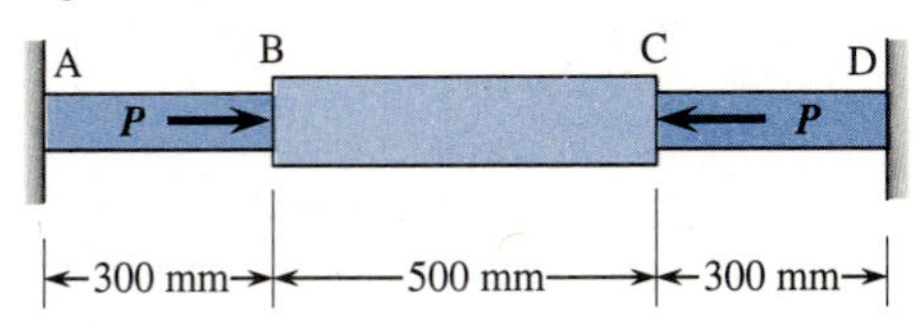

2.75 A round steel bar with a 9 $in.^2$ cross-section is rigidly attached to walls at each end. A 50,000-lb load is applied 3 in. from the left end. What is the reaction at the left end?

Figure P2.75

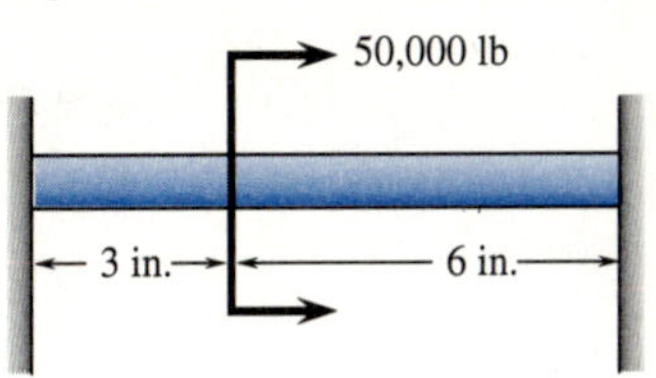

2.76 A steel cylinder is surrounded by a copper tube that is bonded to the steel cylinder. The diameter of the cylinder is 5 in. The outer diameter of the copper tube is 10 in. The composite must resist a uniform load of 100 kips applied axially. Determine the stress in the steel cylinder.

Figure P2.76

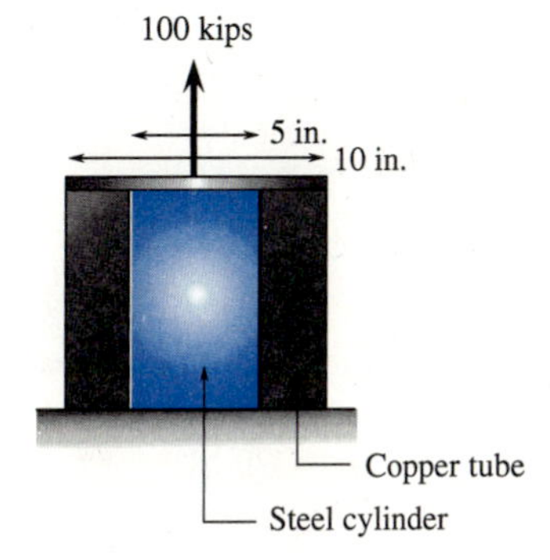

2.77 Bar AB is rigid and remains horizontal. It is supported by two steel rods on the outside, each with cross-sectional area of 0.2 $in.^2$ The central rod is copper with an area of 0.6 $in.^2$ All rods are 6 ft long. The modulus of elasticity of the steel is 30×10^6 psi and that of the copper is 17.5×10^6 psi. What is the force in the copper bar?

Figure P2.77

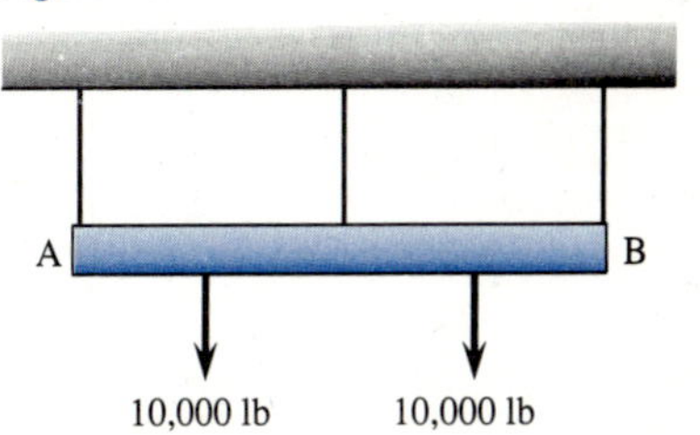

2.78 The structure shown consists of a rigid beam hinged at A and supported by a steel flexible cable at B (with modulus of elasticity of 200 GPa and cross-sectional area of 0.7×10^{-4} m^2) and a concrete support at C (with modulus of elasticity of 20 GPa and cross-sectional area of 5×10^{-3} m^2). A load of 100 kN is applied as shown. Determine the axial stresses induced in the steel cable and concrete support.

Figure P2.78

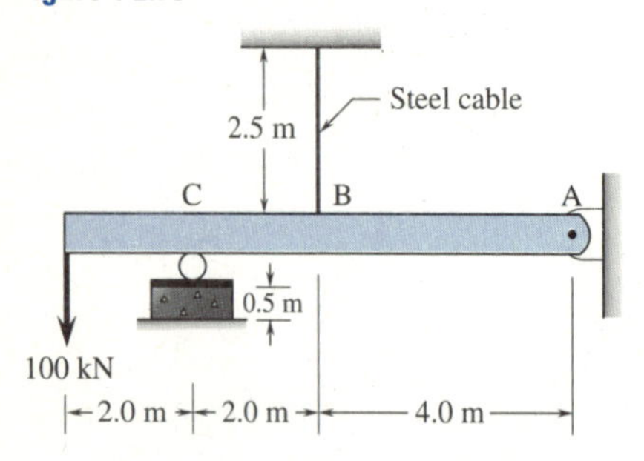

2.79 For the structure shown, determine the stress in bars BC and DE. Let $A = 3$ in.^{2}, $L = 12$ in., and $E = 29 \times 10^6$ psi for both bars. Assume horizontal member ABD is rigid.

Figure P2.79

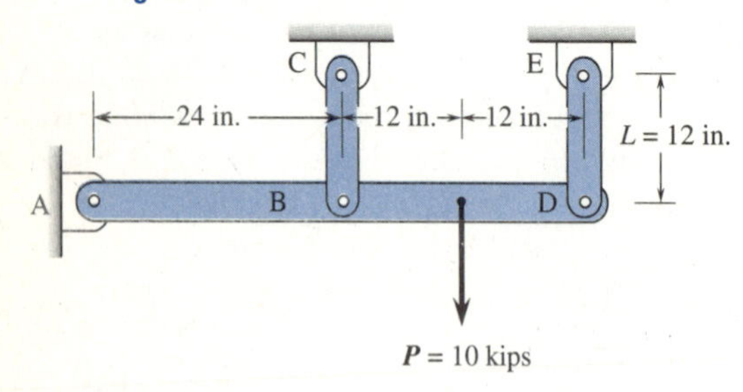

2.80 For the rigid horizontal member ABD supported by the pin at A, the bar BC, and the column at D, determine the stress in bar BC. Let $E = 205$ GPa and $A = 600$ mm^2 for both bar BC and column DE.

Figure P2.80

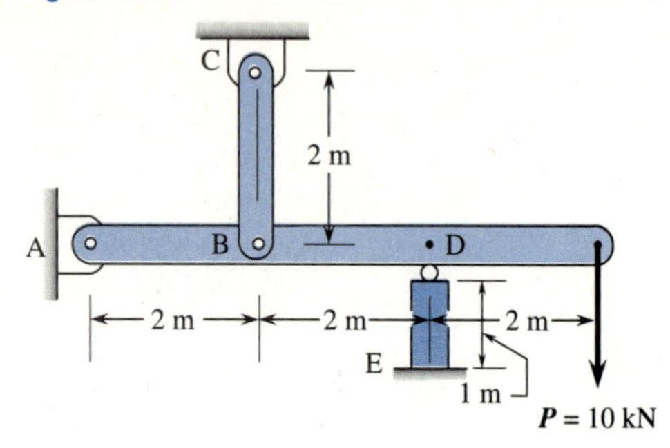

2.81 A copper tube and steel bolt are assembled as shown. Then the end nut is quarter-turned. The length of the bolt is 40 in., the cross-sectional area of the bolt is 1 in.^{2}, and the cross-sectional area of the tube is 1.5 in.^{2} The modulus of elasticity of the steel and the copper are 30×10^6 psi and 16×10^6 psi, respectively. The pitch of the threads is 1/8 in., where the pitch of the threads is the same as the distance advanced by the nut in one complete turn. Determine the stress in the tube and the bolt.

Figure P2.81

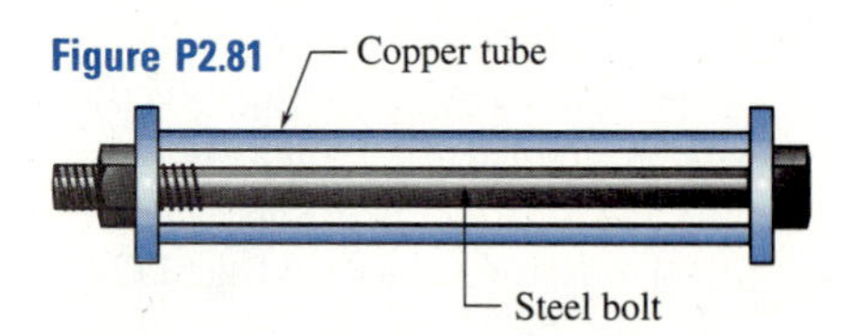

Section 2.8

2.82 A composite bar of steel and aluminum alloy are perfectly bonded together. The bar is then subjected to a uniform temperature increase of 50°F. Determine the normal stress in each material. Let $E_s = 30 \times 10^6$ psi, $E_{al} = 10 \times 10^6$ psi, $\alpha_s = 6.5 \times 10^{-6}$/°F, and $\alpha_{al} = 13.1 \times 10^{-6}$/°F. Both bars have cross-sectional areas of 1 in^2.

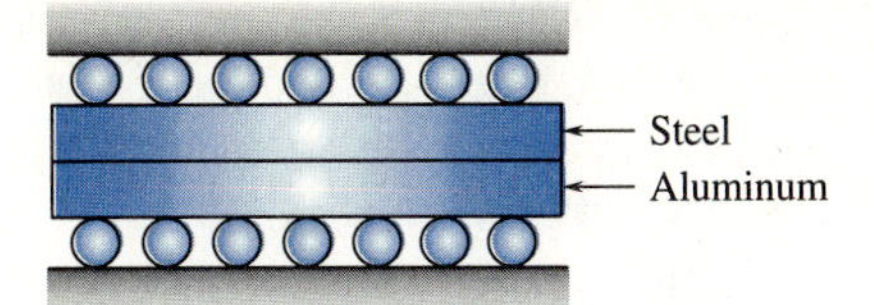

Figure P2.82

2.83 A concrete slab in a highway is 70 ft long. Determine the change in length of the slab if the temperature changes from −30°F to 115°F.

2.84 A steel rail for a railroad track is 10.0 m long. Determine the change in length of the rail if the temperature changes from −30°C to 45°C.

2.85 Determine the stress that would result in the rail of Problem 2.84 if it was completely restrained from expanding. Let $E = 210$ GPa, $\Delta T = 75°C$, and $\alpha = 12 \times 10^{-6}/°C$.

2.86 The pushrods that actuate the valves of an automobile engine are SAE 1090 steel with $\alpha = 11.7 \times 10^{-6}/°C$ and are 600 mm long and 10.0 mm in diameter. Determine the change in length of the rods if their temperature varies from −40°C to 120°C and the expansion is unconstrained.

2.87 If the pushrods described in Problem 2.86 were installed with zero clearance with other parts of the valve mechanism at 20°C, determine the following: **(a)** the clearance between parts at −30°C, and **(b)** the stress in the rod due to a temperature increase to 120°C. Assume that all mating parts are rigid.

2.88 A steel surveyor's tape is standardized at 68°F. It is used at 30°F to place two corner stakes exactly 80 ft apart. What will be the tape reading used to place the stakes? Use $\alpha = 6.5 \times 10^{-6}/°F$.

2.89 An aluminum rod is fastened to a nonyielding support at A. The free end can move 0.03 in. before touching a similar support at B. Determine the allowable temperature rise before the axial stress in the rod reaches 5000 psi.

Figure P2.89

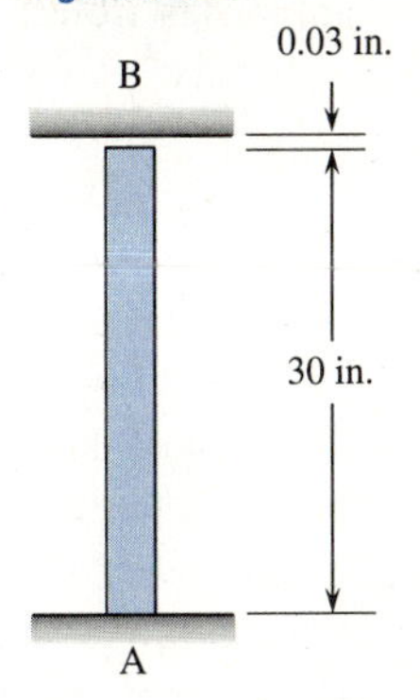

2.90 A carbon steel bar and an aluminum alloy bar are each secured to a rigid support and then fastened together at their free ends by a 1-in.-diameter pin as shown. Determine the maximum normal stress induced in the aluminum bar by a 60°F drop in temperature.

Figure P2.90

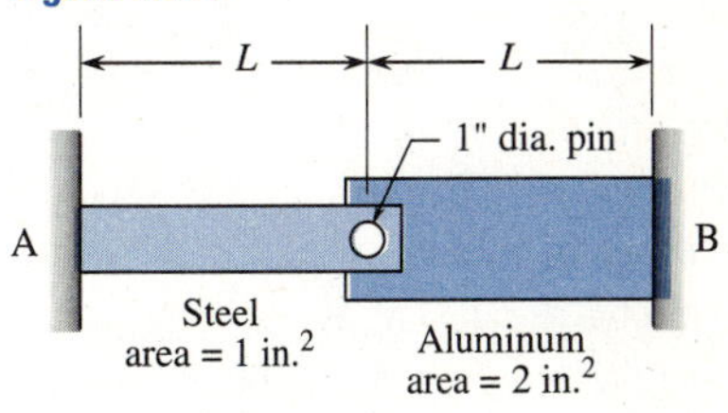

2.91 A 30-in.-long rod of steel with a 4 in.^{2} cross-section is fixed at both ends. The rod is then heated to 60°F above the neutral temperature. What is the axial stress in the rod? Use $E = 30 \times 10^6$ psi and $\alpha = 6.5 \times 10^{-6}$ in./(in. − °F).

2.92 What will be the elongation if one end of the rod described in Problem 2.91 is free to expand?

2.93 A bridge deck is made as one continuous concrete slab 125 ft long. Determine the required width of expansion joints at the ends of the bridge if no stress is to develop when the temperature varies from −25°F to 120°F. Assume the deck was built at 72.5°F.

2.94 For the bridge deck of Problem 2.93, assume that the deck is to just be in contact with the edge of the supporting abutment at a temperature of 100°F. If the deck is installed at 65°F, what should the gap be between the deck and its supports?

2.95 For the bridge deck of Problem 2.93, determine the stress developed in the slab if it was installed at 65°F with no expansion joints and the temperature increased to 110°F?

2.96 A heat exchanger is made by arranging several cold-rolled yellow brass tubes inside a cold-rolled stainless steel shell. Initially, when the temperature is 20°C, the tubes are 4.0 m long and the shell is 4.25 m long. Determine how much each material will elongate when the temperature is increased to 90°C.

2.97 In Alaska, a 50-ft section of steel pipe can vary in temperature from −50°F in the winter when empty to 150°F when carrying heated oil. Determine the change in the length of the pipe under these conditions.

2.98 A Copper Development Association 145 copper bar (see Appendix B) AB of length 1.5 m is placed in position at room temperature (20°C) with a gap of 0.10 mm between end A and a rigid wall. Determine the axial compressive stress in the bar if the temperature increases by 50°C.

Figure P2.98

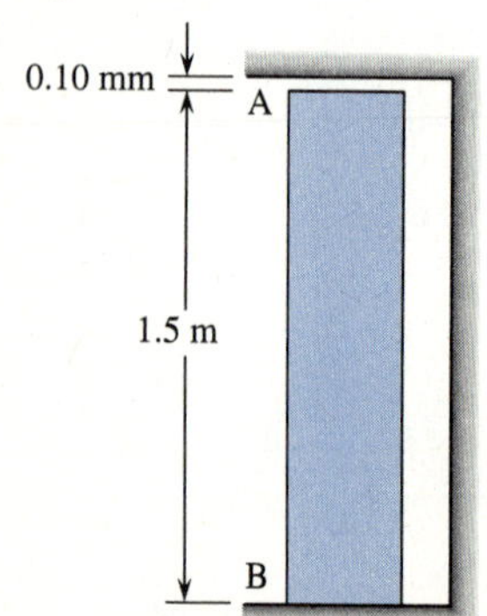

2.99 A steel bar AC with two different cross-sections is held between two rigid supports as shown. The cross-sectional areas of AB and BC are 1.0 in^2 and 3.0 in.^{2}, respectively. The modulus of elasticity E is 30×10^6 psi, and the coefficient of thermal expansion is 6.5×10^{-6}/°F. Determine the following quantities: **(a)** the axial force in the bar, **(b)** the maximum axial stress in the bar, and **(c)** the axial displacement of point B when the bar temperature increases by 50°F.

Figure P2.99

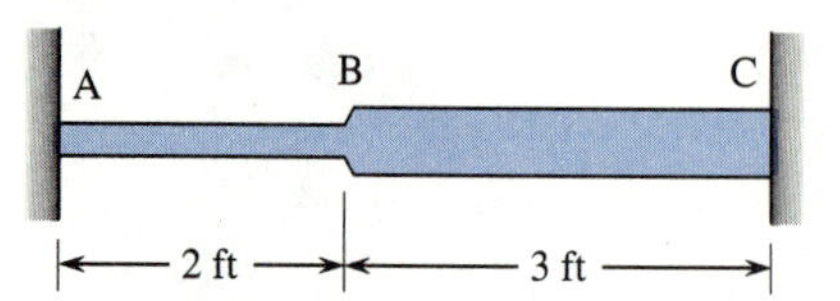

2.100 A horizontal rigid steel plate is supported by three posts of steel each having 250 mm × 250 mm square cross-section and length L = 2 m as shown. Before the load P is applied, the middle post is shorter than the two outer ones by an amount d = 0.5

Figure P2.100

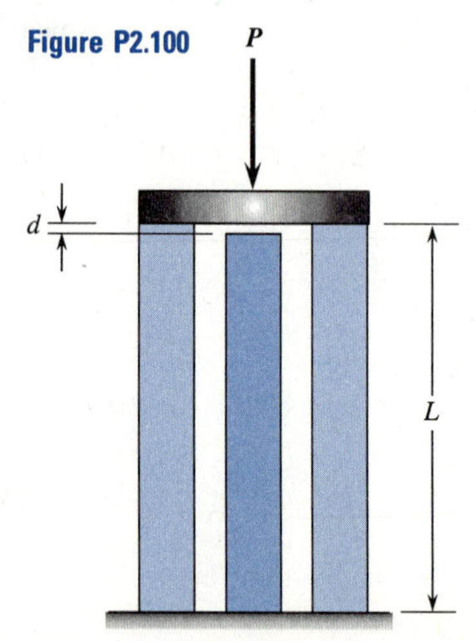

mm. Determine the maximum permissible load P if the modulus of elasticity of the steel is 200 GPa and the allowable stress in compression is 60 MPa.

2.101 A brass sleeve is fitted over a steel bolt and the nut is tightened until it is just snug. The bolt has a diameter of 20 mm, and the sleeve has inside and outside diameters of 21 mm and 35 mm, respectively. Determine the temperature rise ΔT that is required to yield a stress in the sleeve of 25 MPa compression. The brass is cold-rolled and the steel is ASTM-A242 (see Appendix B.)

Figure P2.101

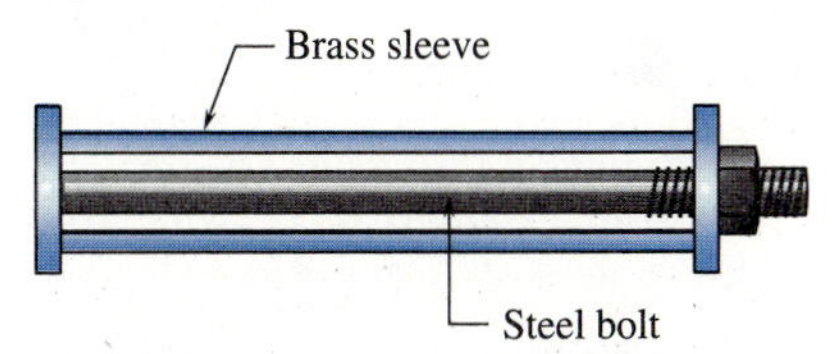

2.102 A steel bar fits between rigid supports at 70°F as shown. Determine the maximum normal and shear stresses in the bar if the temperature increases to 100°F. Assume the thermal expansion coefficient of steel to be 6.5×10^{-6}/°F and $E = 30 \times 10^6$ psi.

Figure P2.102

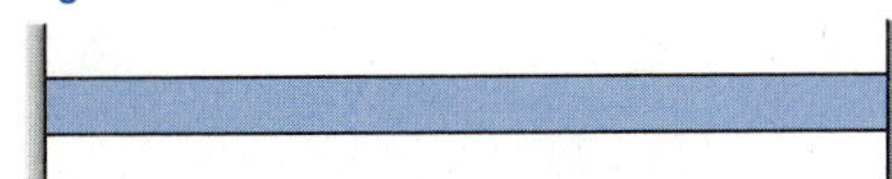

2.103 A bimetallic thermal control is made of a cold-rolled yellow brass bar and a magnesium alloy bar. The bars are arranged with a gap of 0.005 in. between them at 72°F. The brass bar has length 1.0 in. and cross-sectional area of 0.10 in.^{2}, while the magnesium bar has length 1.5 in. and cross-sectional area of 0.15 in.^{2} Determine the following: **(a)** the temperature at which the bars just come into contact and **(b)** the stress in the bars when the temperature increase is 250°F after the gap is just closed. (See Appendix B for material properties.)

Figure P2.103

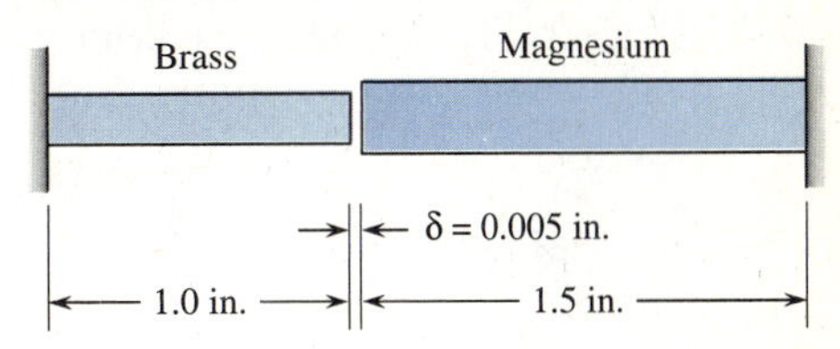

2.104 A copper wire of diameter 1/8 in. is stretched between fixed supports so that it is under a tension force of $T = 40$ lb. If the temperature of the wire is then dropped 40°F, what is the maximum tensile stress and shear stress in the wire? The coefficient of thermal expansion for the wire is 9.3×10^{-6}/°F and the modulus of elasticity is 16×10^6 psi.

Figure P2.104–105

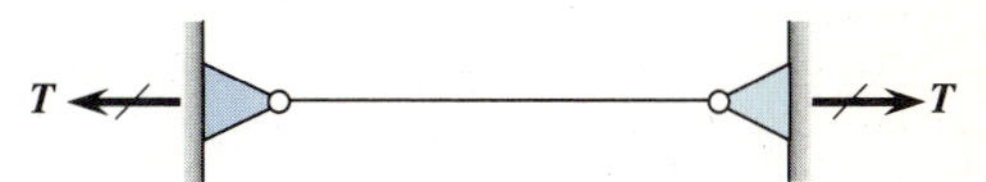

2.105 A steel wire of diameter 1/8 in. is stretched between fixed supports so that it is under an initial tension force of $T = 40$ lb. If the temperature of the wire drops 40°F, what is the tensile stress in the wire? Let $\alpha = 6.5 \times 10^{-6}$/°F and $E = 30 \times 10^6$ psi.

Section 2.9

2.106 Determine the required diameter of a steel member if the tensile design load is 7000 lb. Assume a factor of safety of 5 based on the ultimate strength of 60,000 psi.

2.107 For the structure shown, determine the cross-sectional area required for member BC based on a yield stress of 36,000 psi and a factor of safety of 2.0.

Figure P2.107

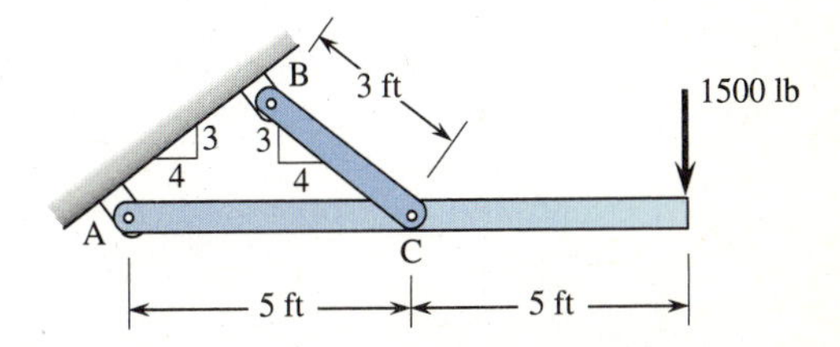

2.108 Knowing that the breaking strength of rod BD is 90 kN, determine the factor of safety with respect to rod failure for the load shown.

Figure P2.108

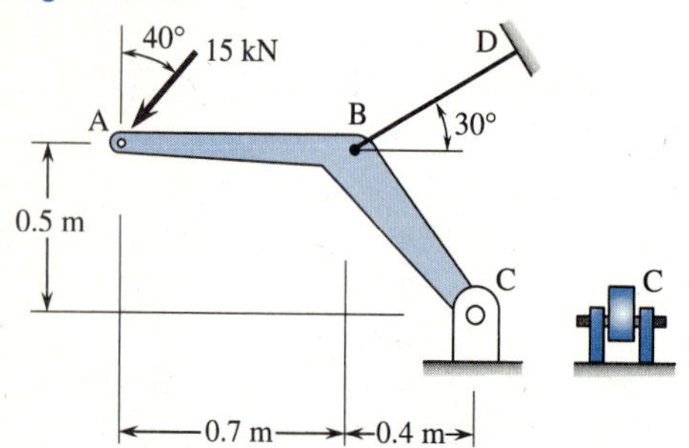

2.109 Link AB is made of a steel with an ultimate normal stress of 450 MPa. Determine the cross-sectional area of AB using a factor of safety of 3.0. Assume the link will not fail around the pins at A and B.

Figure P2.109

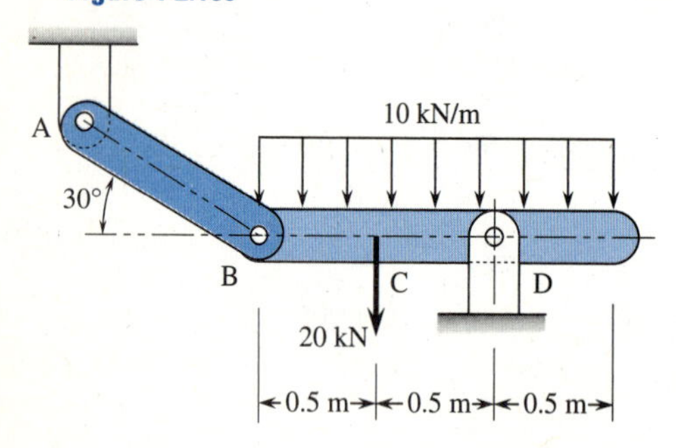

Section 2.10

In the problems in this section consider stress concentrations.

2.110 A hole of 20 mm diameter has been drilled through a long steel plate which is subjected to a centric axial load of $P = 20$ kN. The plate depth is 100 mm and its thickness is 15 mm. Determine the maximum normal stress at the hole.

Figure P2.110

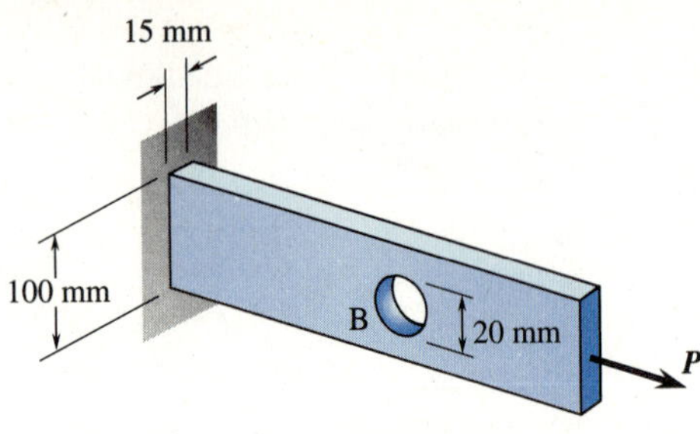

2.111 A hole of 1 in. diameter has been drilled through a steel plate that is subjected to a centric axial load of $P = 4000$ lb. The plate depth is 6 in. and its thickness is 0.25 in. Determine the maximum normal stress at the hole.

2.112 If the allowable stress is 24,000 psi for the plate in Problem 2.111, determine the allowable axial force P.

2.113 If the allowable stress is 100 MPa in the plate of Problem 2.110, determine the allowable axial force P.

2.114 A steel plate is machined as shown with the fillet radius of $r = 0.50$ in. The larger width is 4 in., while the smaller width is 2 in. The plate is subjected to a centric tensile force of $P = 8000$ lb. Determine the axial stress in the plate.

Figure P2.114

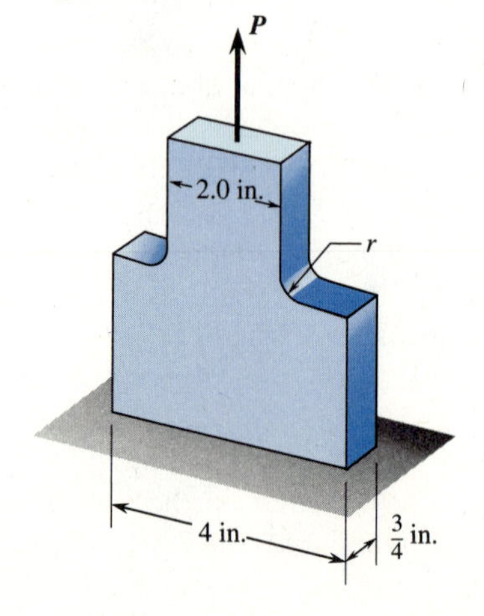

2.115 A steel plate shown with fillets of 10 mm is subjected to a centric axial load of 30 kN. The larger width is 100 mm and the smaller width is 80 mm. Determine the axial stress in the plate.

Figure P2.115

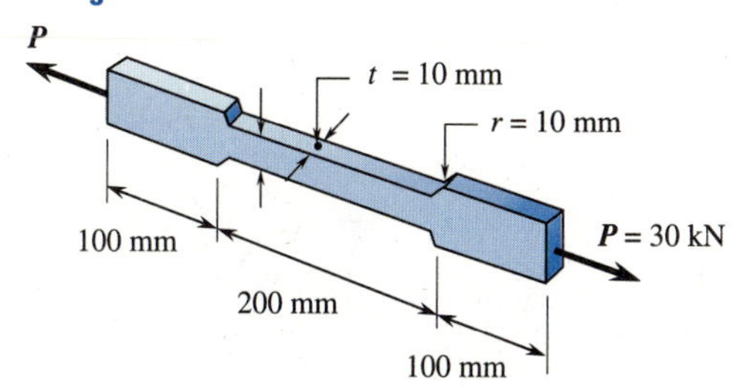

2.116 The fixture shown is made of a brittle material whose stress-strain diagram is linear to rupture at 18,000 psi. The fixture has a hole drilled through it and fillets as shown. Determine the allowable applied load P if a factor of safety of 2.0 is used.

Figure P2.116

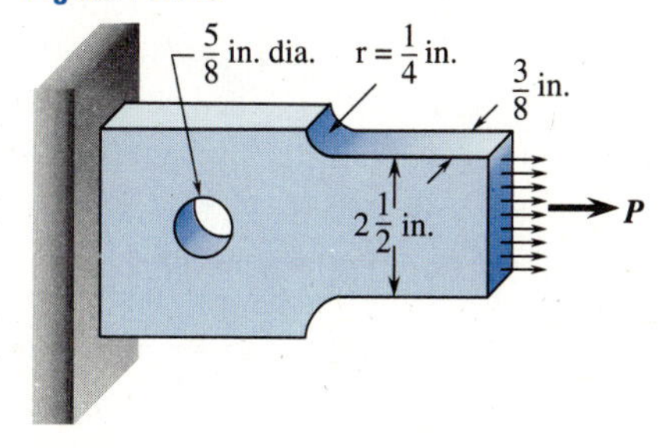

2.117 The fixture shown is made of a brittle material with a linear stress-strain diagram to rupture at 20,000 psi. The fixture thickness is 0.5 in. and the larger width is 2.5 in. Determine the narrow width d of the plate so that it can resist a centric load of 2000 lb. Use a factor of safety of 3.0 based on rupture mode of failure.

Figure P2.117

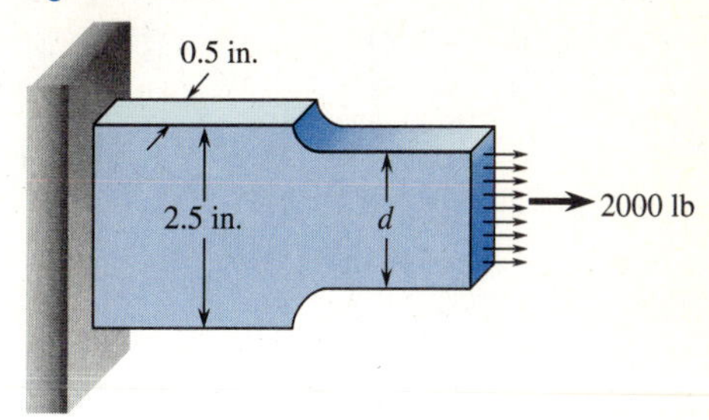

2.118 A brittle tensile plate 10 mm thick × 50 mm wide must transmit 30 kN without exceeding a tensile stress of 220 MPa in the material. Semicircular grooves must be machined into the edges at the center of the plate for clearance with another member. How deep can these grooves be made?

Figure P2.118

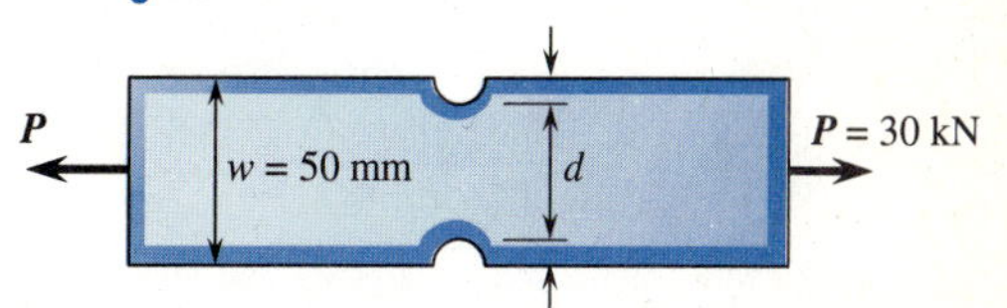

2.119 A fillet with a radius of 0.25 in. is used at the junction in a stepped round bar where the diameter is reduced from 4 in. to 3 in. Determine the maximum axial stress in the fillet when the bar is subjected to a centric axial force of 20 kips.

Figure P2.119

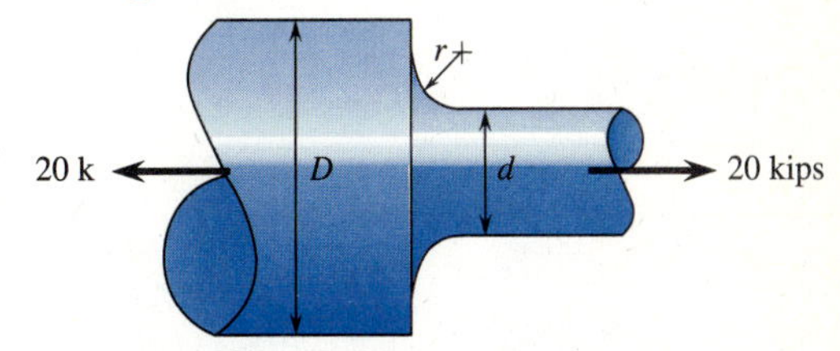

2.120 A fillet with radius of 10 mm is used at the junction in a stepped round bar where the diameter is reduced from 100 mm to 75

mm. Determine the maximum axial stress in the fillet when the bar is subjected to a centric axial force of 80 kN.

Figure P2.120

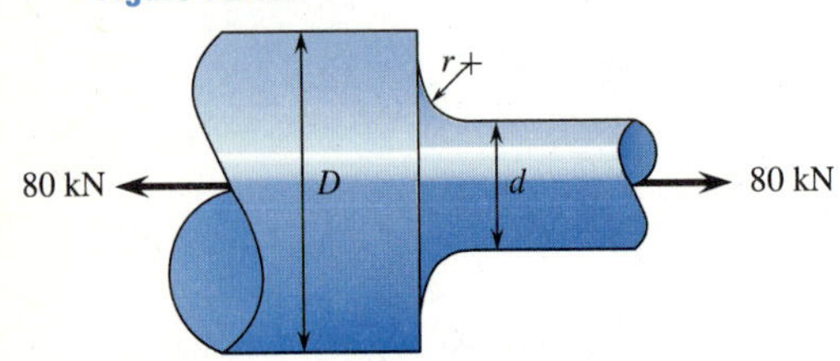

2.121 A semicircular groove with a 10-mm radius (r) is required in a 100-mm-diameter (D) round bar. If the maximum allowable axial stress is 200 MPa, determine the maximum axial force that can be resisted by the bar.

Figure P2.121

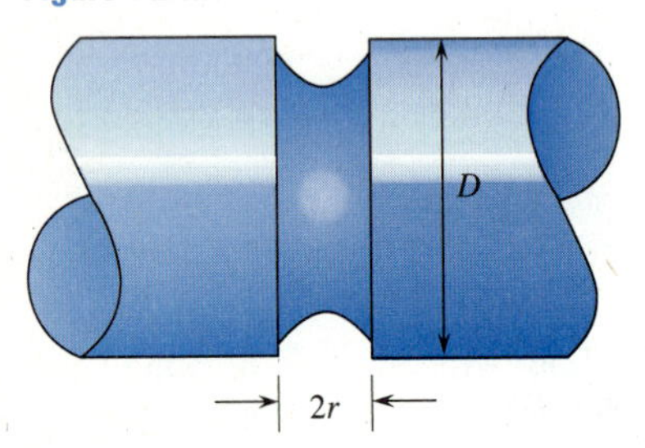

2.122 A semicircular groove of 0.25 in radius (r) is machined into a 5-in.-diameter (D) round steel bar. If the maximum allowable axial stress is 20 ksi, determine the maximum axial force that can be resisted by the bar.

Figure P2.122

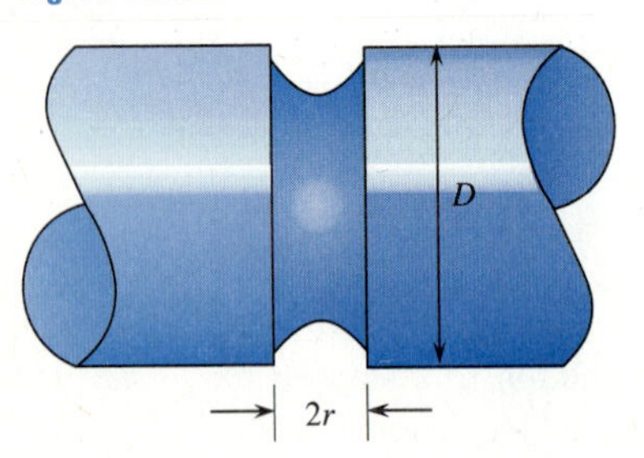

2.123 A stepped bar has a 5 in. diameter D for one-half its length and a 3 in. diameter d for the other half length. If the maximum allowable normal stress in the fillet between the two parts of the bar must be no greater than 20 ksi, determine the minimum radius r needed at the junction between the two parts when the axial force to be resisted is 60 kips.

Figure P2.123

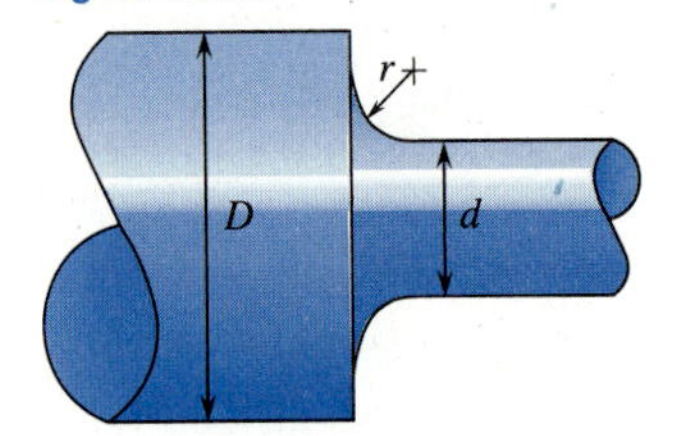

COMPUTER PROBLEMS

The following problems are suggested to be solved using a programmable calculator, microcomputer, or mainframe computer. It is suggested that you write a simple computer program in BASIC, FORTRAN, PASCAL, or some other language to solve these problems.

C2.1 An eye bar is fusion welded across a seam as shown. Determine the average normal and shear stresses acting on the fused cross-section, that is, **(a)** compute the normal stress σ for angles of θ for $-90° \leq \theta \leq 90°$, **(b)** compute the shear stress τ for angles of θ for $-90° \leq \theta \leq 90°$. Use increments of 5° for θ. Set the program up to input the load P and the cross-sectional area A. Check your program using the following data: $P = 1000$ lb and $A = 2$ in.^{2}

Figure C2.1

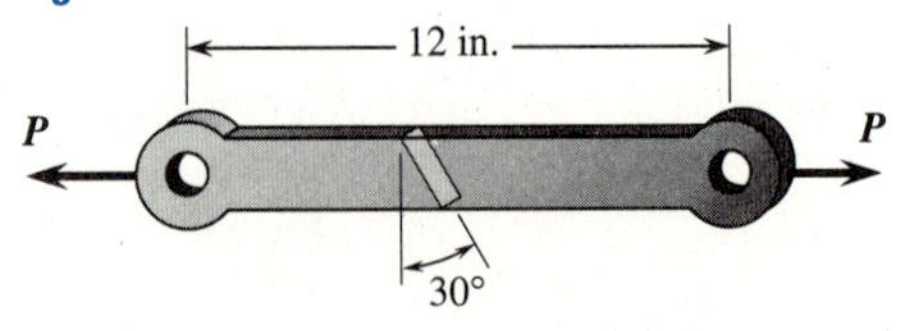

C2.2 A surveyor must correct a measured distance for a tape held at a tension different from the tension used to determine the true length of the tape. The following formula can be used to correct a measured length for variation in tension:

$$\delta = \frac{(T_a - T_s)\,L}{AE}$$

where

δ is the correction per tape length
T_a is the applied tension
T_s is the standard tension
L is the length of the tape
A is the cross sectional area of the tape
E is the modulus elasticity of the tape

The total correction is determined by multiplying the correction per tape length δ times the number of tape lengths T_l measured. The problem assumes that the applied tension T_a is the same during each measurement. Check your program using the following data: $T_a = 15$ lb, $T_s = 12$ lb, $L = 100$ ft, $A = 0.005$ in.^{2}, $E = 29 \times 10^6$ psi, and a total measured length of 800 ft.

C2.3 Write a program to obtain the stresses and deformation in a statically indeterminate (composite) axially loaded column. The input should include the cross-sectional area A of each material, the moduli of elasticity E of each material, the length L of the column, and the supported load P. Check your program using the following data: $A_{steel} = 0.0077$ m^2, $A_{concrete} = 0.0177$ m^2, $E_{steel} = 200$ GPa, $E_{concrete} = 24$ GPa, $L = 0.6$ m, and $P = 100$ kN.

Figure C2.3

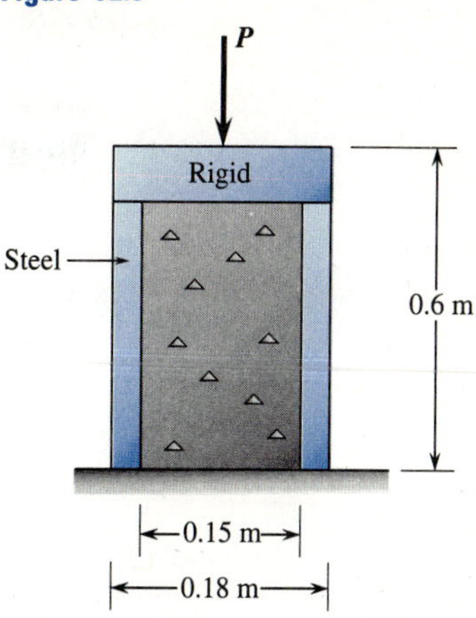

C2.4 Write a program to obtain the axial stress in a bolt fitted inside a tube and then tightened down by a nut as shown. The input should include the cross-sectional areas of the bolt and tube, the moduli of elasticity of the bolt and tube, the number of turns of the nut, the pitch and number of threads of the nut, and the length of the bolt tube assembly. Check your program using the following data: $A_{bolt} = 78.54 \times 10^{-6}$ m^2, $A_{tube} = 141.1 \times 10^{-6}$ m^2, $E_{bolt} = 105$ GPa, $E_{tube} = 70$ GPa, $L = 400$ mm, a nut with a 2-mm pitch, single-threaded, and turned one-half turn after bearing against the plate B shown.

Figure C2.4

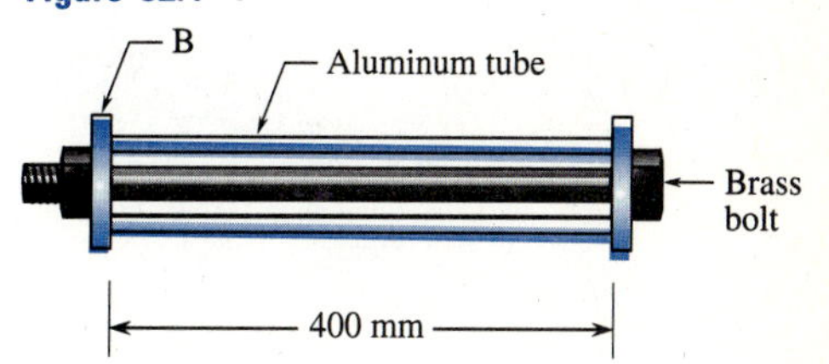

Direct Shear and Bearing Stresses

3.1 INTRODUCTION

In this chapter we consider direct shear and bearing stresses. These kinds of stresses occur most frequently in connectors used to assemble or join members together in structures and machines transmitting load.

Examples of direct shear stress are shown in Figures 3.1 through 3.3. In the bolted lap joint connection (Figure 3.1(a)), and glued lap joint connection (Figure 3.1(b)), the bolt and glue joint are subjected to direct shear stress along surfaces a-a.

Another example of direct shear occurs in the shear or drive pin shown in Figure 3.2. The boat propeller is attached to the drive shaft in a manner that the torque must be transferred from the drive shaft to the propeller through the drive pin. As the shaft rotates, it tends to cut the pin at the top and bottom of the shaft between the shaft and the propeller housing. Thus these two cross-sections labeled a-a in Figure 3.2 are subjected to direct shear.

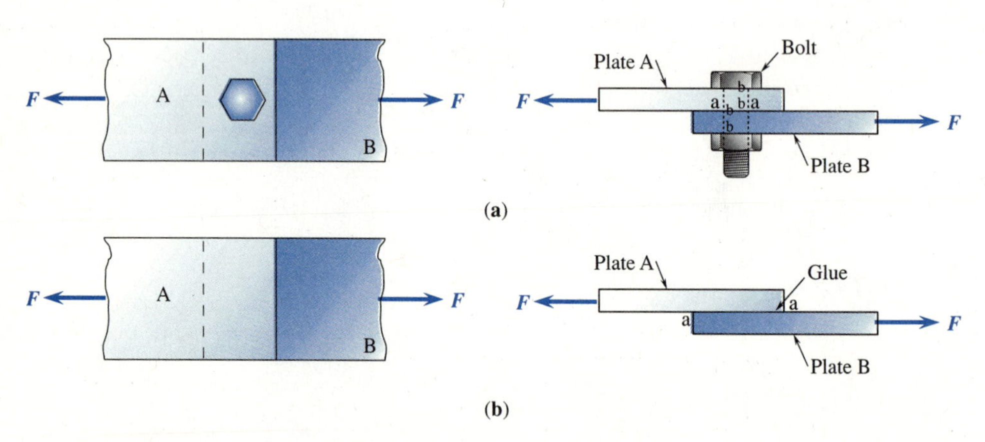

Figure 3.1 (a) Bolted lap joint connection and (b) glued lap joint connection.

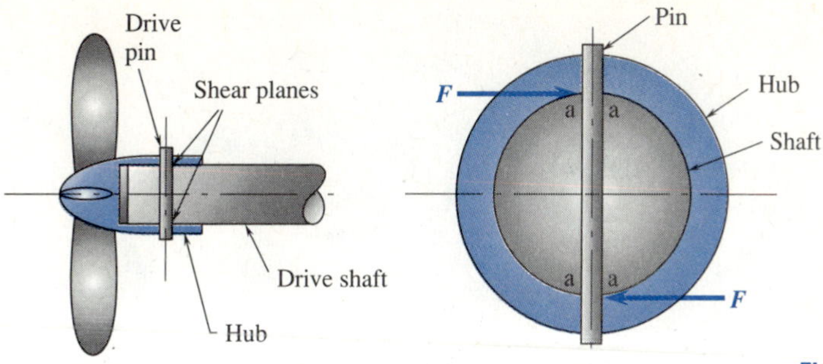

Boat propeller

Cross-section of hub and shaft at drive pin

Figure 3.2 Drive or shear pin used to connect boat propeller to drive shaft.

(**a**) Punch press punching plate

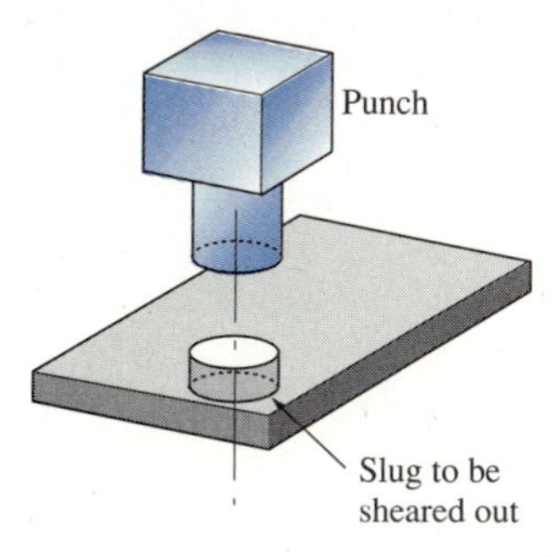

(**b**) Schematic of punching operation

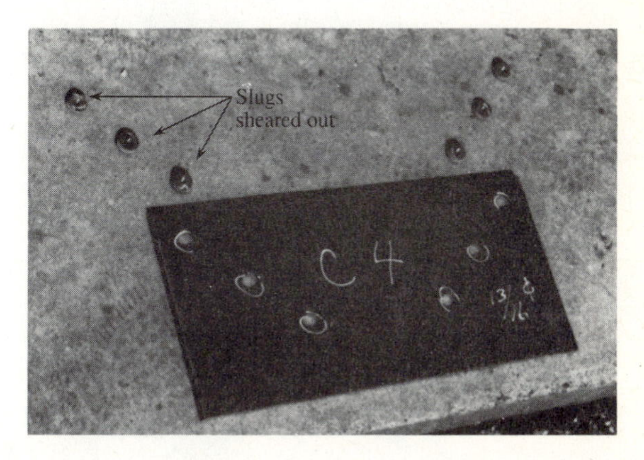

(**c**) Plate with six slugs sheared out

Figure 3.3 Punching operation illustrating shear in plate.

Figure 3.4 Concrete pedestal and base plate.

In a punching operation, in which a hole is punched in a metal plate, the action of punching shears the piece of metal creating the hole as shown in Figure 3.3. The *shear area* is the cylindrical surface of the metal piece that has been pushed out.

Examples of bearing stress include the vertical edge surface b-b between the upper plate and the bolt, and between the lower plate and the bolt in Figure 3.1(a), the stress between the base plate and the concrete pedestal of the column shown in Figure 3.4, and the stress between the washer and the wooden member shown in Figure 3.5.

In the subsequent sections of this chapter, we develop the equations to analyze direct shear and bearing stresses. We consider bolted, glued, and welded connections in load-carrying structures and machines as basic examples to illustrate direct shear and bearing stresses. We also consider shear strain and shear deformation due to direct shear forces.

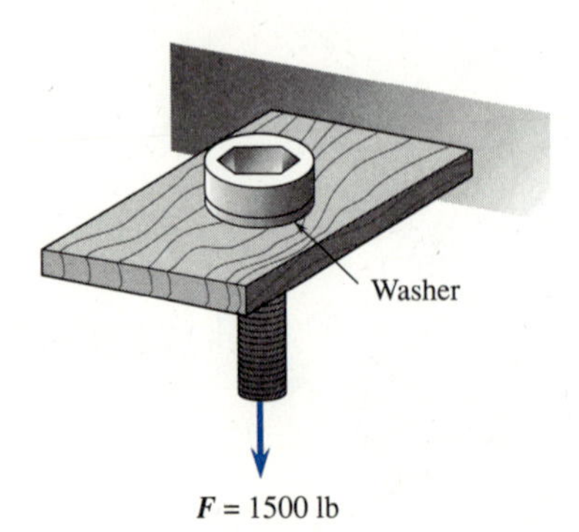

Figure 3.5 Wooden member with bolt and washer connector.

3.2 DIRECT SHEAR STRESS AND ANALYSIS OF WELDS

We now consider the two plates A and C, which are connected together by the bolt B in Figure 3.6. The plates are then subjected to tensile forces F, that result in shear stresses developing on the cross-section of the bolt along the interface a-a.

To obtain the shear stress on the cross-section a-a, we first draw a free-body diagram of the bolt as shown in Figure 3.7(a) with forces F transmitted to the bolt due to the bearing contact between the bolt and plates A and C.

We then pass a plane through the bolt at section a-a and draw a free-body diagram of the upper part of the bolt above plane a-a (Figure 3.7(b)). For equilibrium, the internal force V must act over the cross-section at a-a. By summing forces, we obtain

$$\Sigma F = 0: \quad F - V = 0 \qquad V = F$$

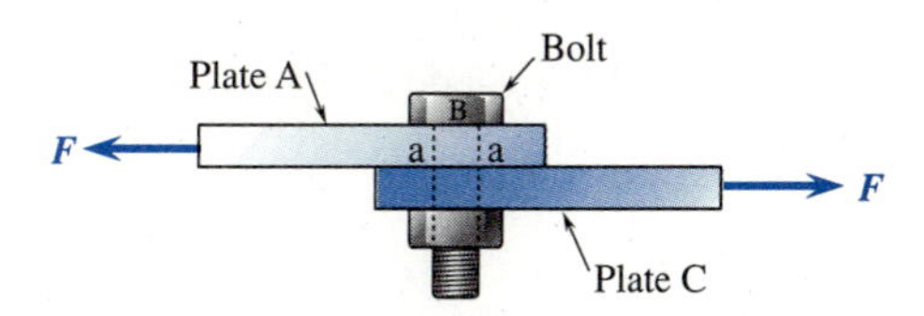

Figure 3.6 Bolt in single shear.

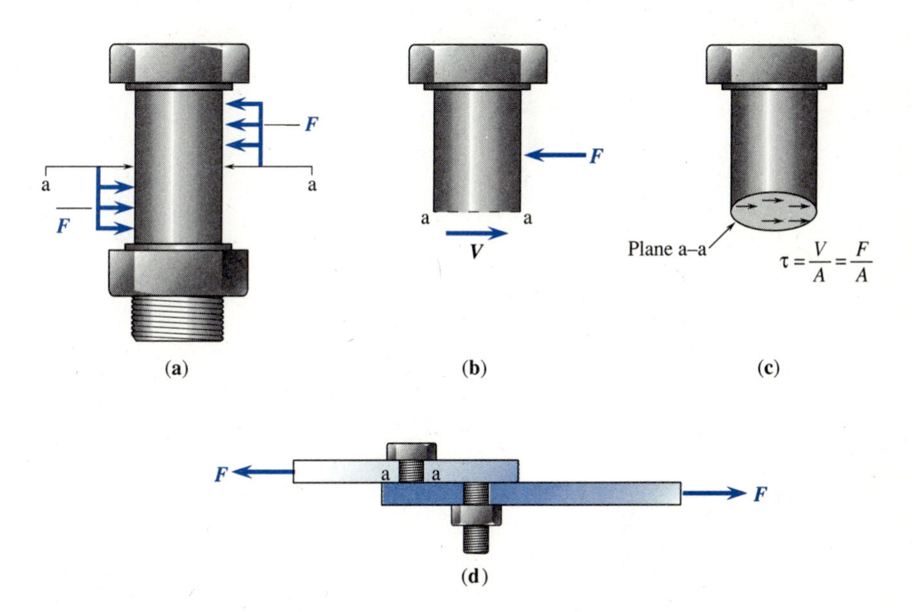

Figure 3.7 (a) Free-body diagram of bolt,
(b) free-body diagram of upper part of bolt, and
(c) average shear stress acting on cross-section
(d) Shear failure of bolt.

The shear stresses actually are distributed in a nonuniform manner over the bolt cross-section due to numerous factors sometimes beyond control of the engineer, such as poor alignment of holes, loose or unequal tightening, unexpected eccentric loading, and poor construction. However, it has been found through experience that reliable analysis can be based on the average shear stress. We then define the average or nominal shear stress τ_{ave} (Figure 3.7(c)) acting in the plane at a-a, where τ_{ave} is then obtained by dividing the internal shear force V by the area A over which V acts. Here A is the cross-sectional area of the bolt at a-a, since section a-a is the area being sheared.

$$\tau_{ave} = \frac{V}{A} = \frac{F}{A} \tag{3.1}$$

Because the bolt has one shear plane a-a, the bolt is in single shear. A typical shear failure of a bolt along plane a-a is shown in Figure 3.7(d).

For the connection shown in Figure 3.8, the bolt B has two shear planes, a-a and b-b. The bolt is then in double shear as shown in the free-body diagrams of the bolt and a section between a-a and b-b in Figures 3.9(a) and (b).

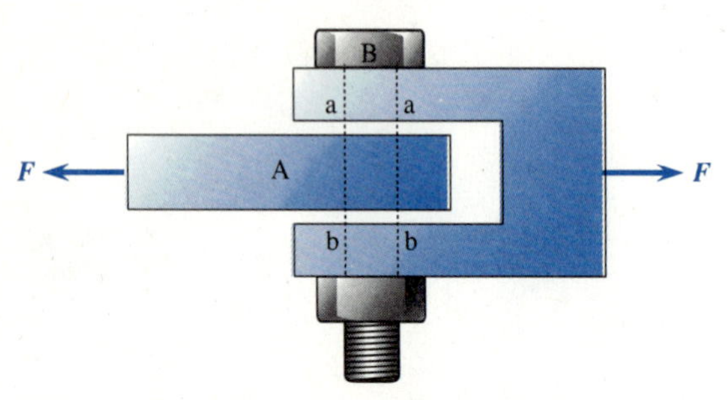

Figure 3.8 Bolted connection (bolt in double shear).

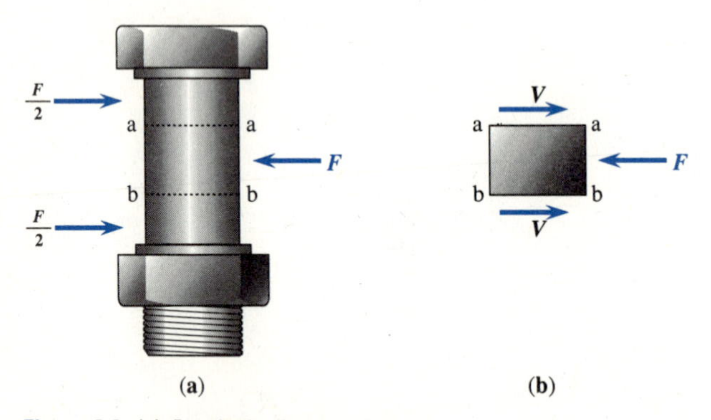

Figure 3.9 (a) Free-body diagram of bolt and (b) free-body diagram of section of bolt.

Using Figure 3.9(b) and statics, we obtain

$$\Sigma F = 0: \quad 2V - F = 0 \qquad V = \frac{F}{2}$$

and the average shear stress in the bolt is then

$$\tau_{ave} = \frac{V}{A} = \frac{F}{2A} \tag{3.2}$$

Comparing Eqs. (3.1) and (3.2), we observe that the average shear stress is one-half as large for a bolt in double shear as that in single shear under the same load F.

STEPS IN SOLUTION

The following steps are used to determine the average shear stress in members subjected to direct shear force.

1. Draw a free-body diagram of the member, such as a connector, showing the forces acting on it.
2. Pass a plane through the section of the member where the shearing is taking place and show the free-body diagram of the section.
3. Determine the shear force V acting on the shear plane and the area A that V acts over.
4. Obtain the average shear stress using $\tau_{ave} = V/A$.

Examples 3.1 through 3.7 illustrate how to determine the average shear stress in members subjected to direct shear forces.

EXAMPLE 3.1

Two steel plates are connected together by two bolts as shown in Figure 3.10. The bolts have diameters of 25 mm. The connection transmits 150 kN of load. Determine the average shear stress in the bolts.

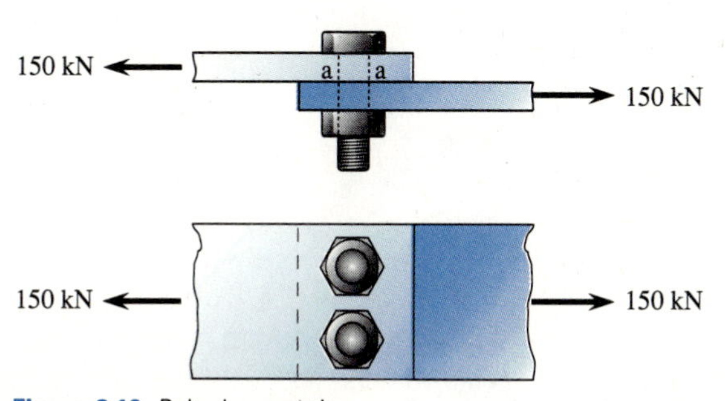

Figure 3.10 Bolted connection.

Solution

First draw a free-body diagram of one of the bolts. Forces of 150/2 kN are transmitted to the bolt from each plate as shown below, where we have assumed each bolt to resist one-half of the 150 kN force.

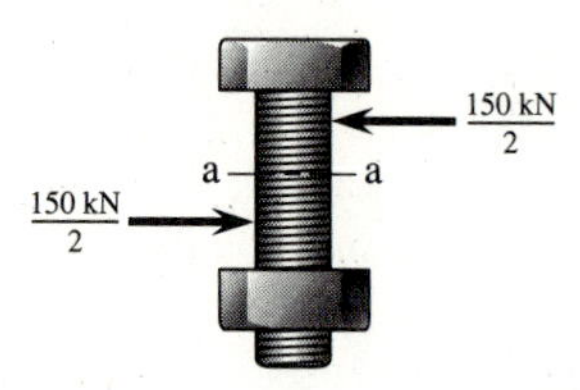

Then pass a plane through section a-a (at the intersection of the two plates where the shear stress occurs on the cross section of the bolt) and draw a free-body diagram of the upper part of the bolt.

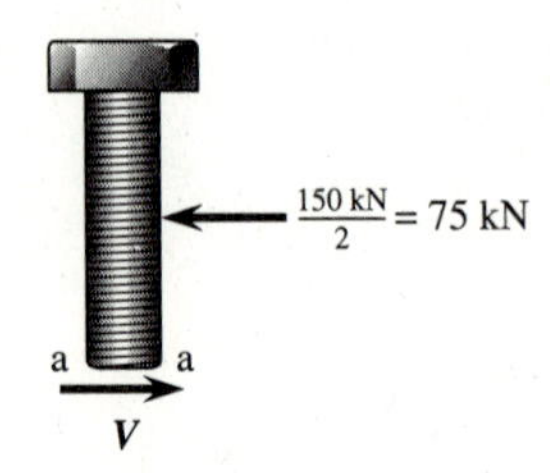

By statics, we obtain

$$V = 75 \text{ kN}$$

Using Eq. (3.1), the average shear stress is

$$\tau_{\text{ave}} = \frac{V}{A} = \frac{75 \text{ kN}}{\frac{\pi}{4}(0.025 \text{ m})^2} = \frac{75 \text{ kN}}{4.91 \times 10^{-4} \text{m}^2} = 153 \times 10^3 \frac{\text{kN}}{\text{m}^2}$$

$$= 153 \text{ MPa}$$

EXAMPLE 3.2

A beam AB is supported by a strut CD (Figure 3.11). A load of 6000 lb is applied to the beam at B. The strut consists of two members, one behind the other, and is connected to the beam by a bolt passing through each of the members at joint C. If the diameter of the bolt is 1.0 in., determine the average shear stress in the bolt.

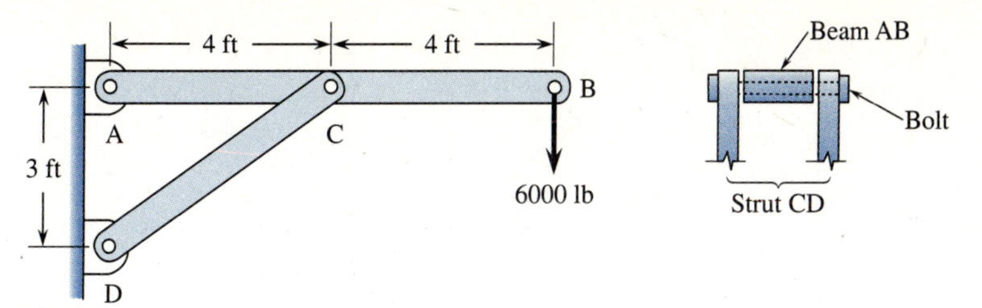

Figure 3.11 Beam supported by strut.

Solution

First draw a free-body diagram of the beam AB. (Remember member CD is a two-force member.)

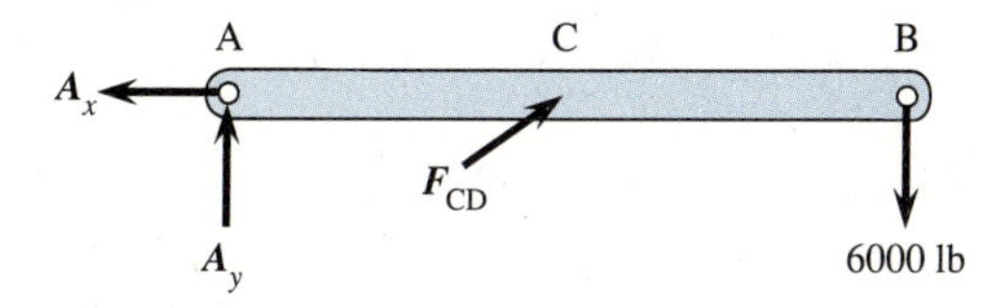

By statics, we find the force F_{CD} (representing the force in both struts together)

$$\Sigma M_A = 0: \quad \frac{3}{5}F_{CD}\,(4\text{ ft}) - (6000\text{ lb})(8\text{ ft}) = 0 \qquad F_{CD} = 20{,}000\text{ lb}$$

Each of the struts then resists a force of $F/2$ equal to 10,000 lb.

A free-body diagram of the bolt at C yields

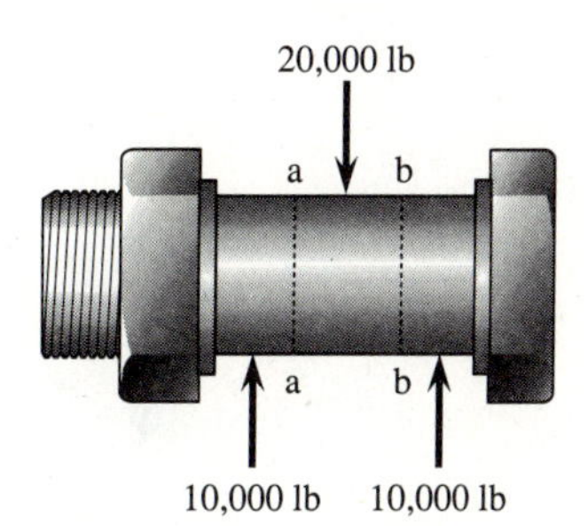

Cutting planes through the bolt at a-a and b-b, we draw a free-body diagram of the part of the bolt between sections a-a and b-b as shown.

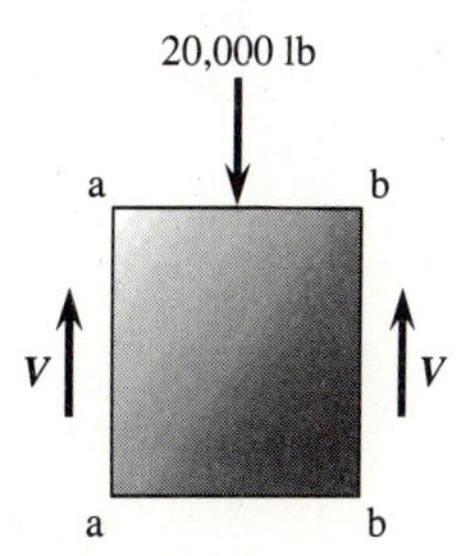

By statics, we have

$$2V - 20{,}000 = 0 \qquad V = 10{,}000 \text{ lb}$$

Using Eq. (3.2), the average shear stress in the bolt (at both a-a and b-b) is

$$\tau_{ave} = \frac{V}{A} = \frac{10{,}000 \text{ lb}}{\frac{\pi}{4}(1.00 \text{ in.})^2} = \frac{10{,}000 \text{ lb}}{0.785 \text{ in.}^2} = 12{,}700 \text{ psi}$$

In many codes (for instance, see AISC[3]), the allowable shear stress in a standard high strength steel bolt is greater than 12,700 psi. This bolt should then safely carry the 6000-lb load.

EXAMPLE 3.3

A punch with a diameter of $d = 20$ mm is used to punch a hole in an aluminum plate of thickness $t = 6$ mm (Figure 3.12). If the force P is equal to 50 kN, determine the shear stress in the plate.

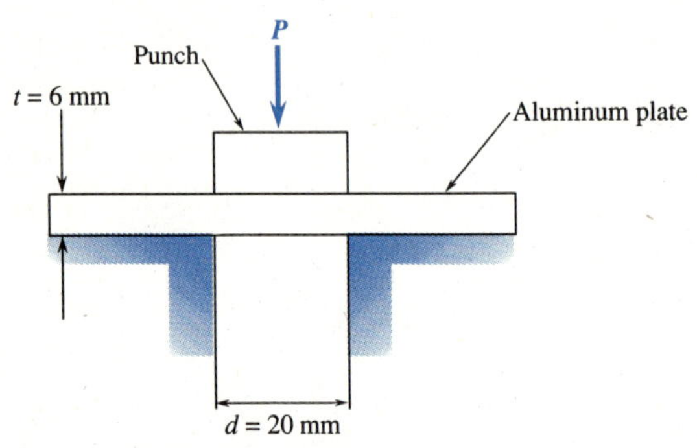

Figure 3.12 Punching operation.

Solution

Draw a free-body diagram of the portion of the plate below the punch.

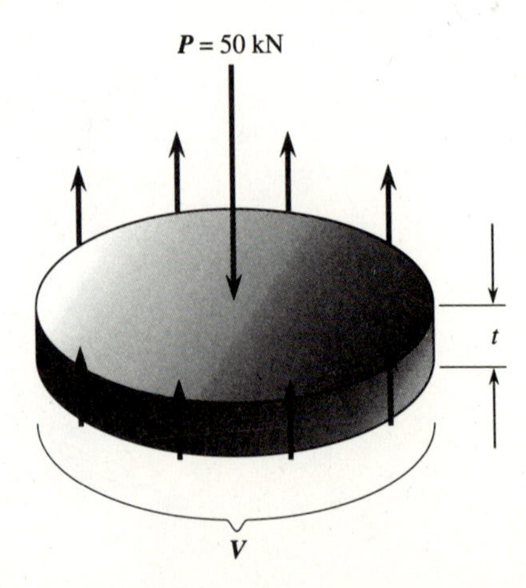

The total shear force V acts on the edge surface of the circular cylindrical area with diameter $d = 20$ mm and thickness $t = 6$ mm. The shear area is then $A = 2\pi rt = \pi dt$.

By statics, the shear force is

$$V = P = 50 \text{ kN}$$

By Eq. (3.1), the average shear stress is

$$\tau_{\text{ave}} = \frac{V}{A} = \frac{50 \text{ kN}}{\pi dt} = \frac{50 \text{ kN}}{\pi(0.02 \text{ m})(0.006 \text{ m})}$$
$$= 133 \times 10^3 \frac{\text{kN}}{\text{m}^2} = 133 \text{ MPa}$$

EXAMPLE 3.4

A rock crusher mechanism is shown in Figure 3.13. **(a)** Determine the largest vertical force F that can be applied so that the 1/2-in. diameter pin in single shear at B does not exceed a stress of 12,600 psi in shear. **(b)** If the pin fails at an ultimate stress of 40,000 psi, what is the factor of safety against shear failure in the pin at B?

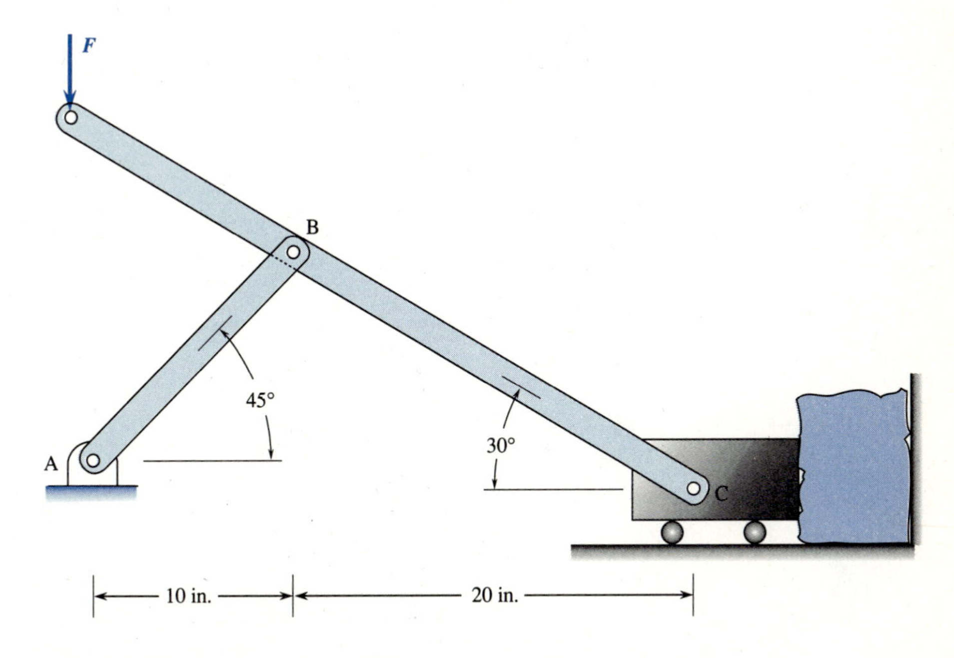

Figure 3.13 Rock crusher.

Solution

(a) Draw a free-body diagram of member BC. (Remember member AB is a two-force member.)

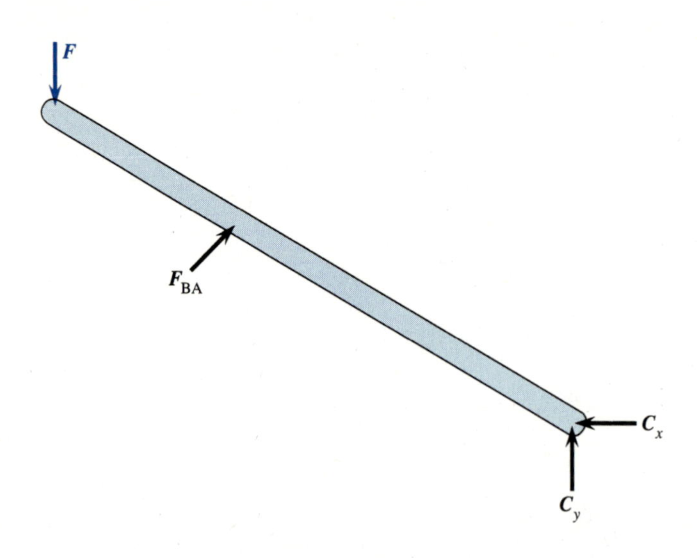

By statics, we write

$$\Sigma M_c = 0: \quad F(30 \text{ in.}) - \left(F_{BA}\frac{\sqrt{2}}{2}\right)(20 \text{ in.}) - \left(F_{BA}\frac{\sqrt{2}}{2}\right)(10 \text{ in.}) = 0$$

$$F = \frac{\sqrt{2}}{2}F_{BA}$$

The shear force in pin B equals the force F_{BA}. Using Eq. (3.1)

$$\tau_{ave} = \frac{F_{BA}}{A}$$

Solving for F_{BA}, we have

$$\begin{aligned} F_{BA} = \tau_{ave} A &= (12{,}600 \text{ psi})\frac{\pi}{4}\left(\frac{1}{2}\text{ in.}\right)^2 \\ &= (12{,}600)(0.196 \text{ in.}^2) \\ &= 2470 \text{ lb} \end{aligned}$$

$$F = \frac{\sqrt{2}}{2}F_{BA} = \frac{\sqrt{2}}{2}(2470 \text{ lb}) = 1750 \text{ lb}$$

(b) Replacing normal stresses by shear stresses in Eq. (2.30), the factor of safety is

$$\text{FS} = \frac{\tau_u}{\tau_{allow}} = \frac{40{,}000 \text{ psi}}{12{,}600 \text{ psi}} = 3.17$$

EXAMPLE 3.5

A steel bar 1 in. in diameter is loaded in double shear by the Clevis joint until failure (Figure 3.14). The ultimate load is determined to be 100 kips. If the allowable stress is based on a factor of safety of 4, what must be the diameter of a pin designed for an allowable load of 6 kips in single shear?

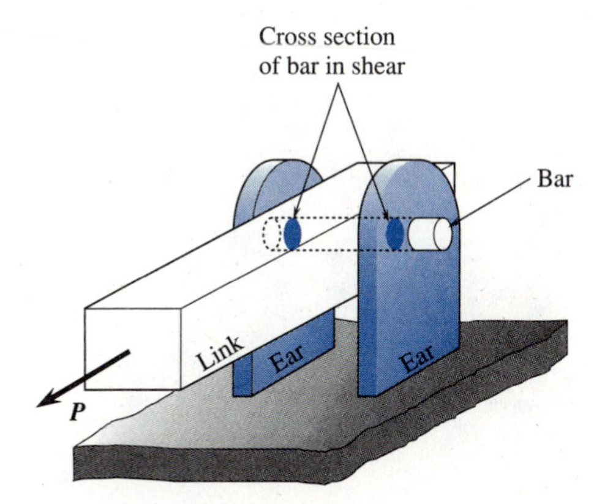

Figure 3.14 Clevis joint.

Solution

In double shear the ultimate load P_u is 100 kips. Therefore, P_u in single shear is

$$P_u = 50 \text{ kips}$$

The ultimate stress is

$$\tau_u = \frac{50 \text{ kips}}{\frac{\pi}{4}(1 \text{ in.})^2} = 63.7 \text{ ksi}$$

Using Eq. (2.30)

$$\tau_{allow} = \frac{\tau_u}{\text{FS}} = \frac{63.7}{4} = 15.9 \text{ ksi}$$

The pin diameter is now determined by

$$\tau_{allow} = 15.9 \text{ ksi} = \frac{6 \text{ kips}}{A}$$

$$A = 0.377 \text{ in.}^2$$

$$\frac{\pi}{4}d^2 = 0.377 \text{ in.}^2$$

$$d = 0.693 \text{ in. (use } \frac{3}{4} \text{ in. diameter)}$$

EXAMPLE 3.6

The hydraulic floor crane shown in Figure 3.15 is to be designed to lift 1/2 ton. Determine the size of the pin at B so that the shear stress does not exceed 10,000 psi. The pin is in double shear. Use a factor of safety of 3 against overloading.

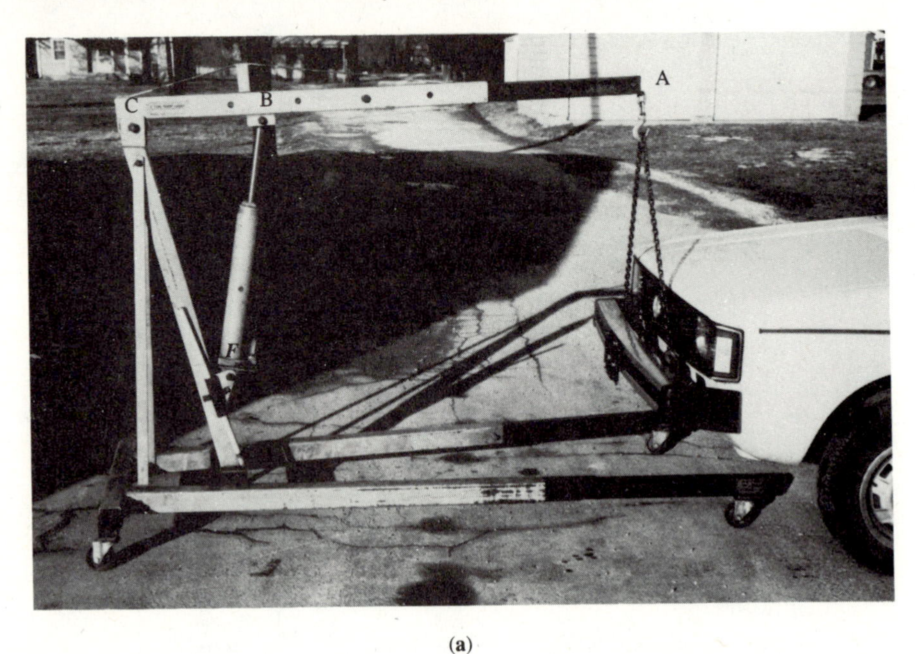

(a)

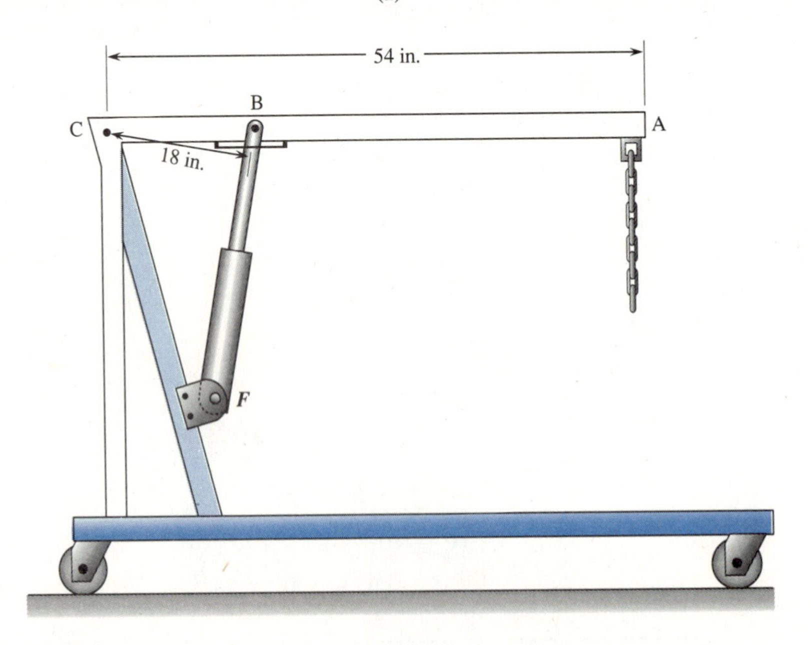

(b)

Figure 3.15 Hydraulic floor crane.

Solution

Using Eq. (2.28), with P_{allow} = 1000 lb (1/2 ton) and FS = 3, we solve for P_u as

$$P_u = (1000 \text{ lb})(3) = 3000 \text{ lb}$$

The force of 3000 lb is used to design the shear pin.

Draw a free-body diagram of member AC.

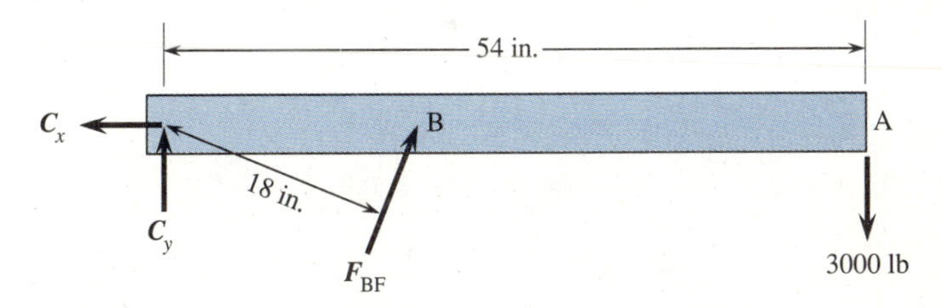

By statics, we have

$$\Sigma M_c = 0: \quad -(3000 \text{ lb})(54 \text{ in.}) + F_{BF}(18 \text{ in.}) = 0$$
$$F_{BF} = 9000 \text{ lb}$$

The pin is in double shear. Therefore, using Eq. (3.2), we have

$$\tau_{ave} = 10{,}000 \text{ psi} = \frac{F_{BF}}{2A} = \frac{9000 \text{ lb}}{2A}$$
$$A = 0.45 \text{ in.}^2$$
$$\frac{\pi}{4}d^2 = 0.45$$
$$d = 0.757 \text{ in. (use } \frac{7}{8} \text{ in. diameter)}$$

EXAMPLE 3.7

The pliers shown in Figure 3.16(a) is designed for a maximum gripping force of 120 N. Using a factor of safety of 2.85 against yielding, determine the diameter of the pin. The pin has a yield stress of 275 MPa.

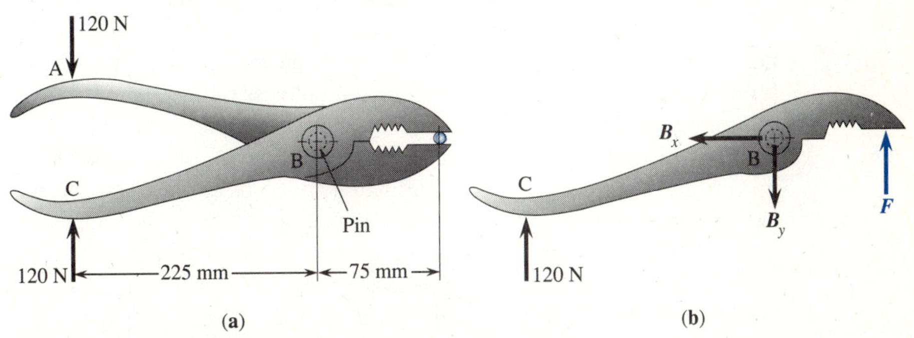

Figure 3.16 (a) Pliers subjected to forces and (b) free-body diagram of lower member BC.

Solution

By statics

$$\Sigma M_B = 0: \quad (120 \text{ N})(225 \text{ mm}) - F(75 \text{ mm}) = 0 \qquad F = 360 \text{ N}$$

$$\Sigma F_y = 0: \quad -B_y + 120 \text{ N} + 360 \text{ N} = 0 \qquad B_y = 480 \text{ N}$$

Using Eq. (2.29), the allowable shear stress is

$$\tau_{\text{allow}} = \frac{\tau_Y}{\text{FS}} = \frac{275 \text{ MPa}}{2.85} = 96.5 \text{ MPa}$$

The cross-sectional area of the pin is

$$A = \frac{480 \text{ N}}{96.5 \times 10^6 \dfrac{\text{N}}{\text{m}^2}} = 4.97 \times 10^{-6} \text{ m}^2$$

The diameter d of the pin is

$$\frac{\pi}{4} d^2 = 4.97 \times 10^{-6} \text{ m}^2 \qquad d = 2.52 \times 10^{-3} \text{ m} = 2.52 \text{ mm}$$

Analysis of Welded Joints

Many welded connections can be analyzed based on the concepts developed for direct shear stress in this chapter and for tensile stress from Chapter 2. Hence, we will introduce the basic concepts involved with welded connections as they are a very important type of connection used in engineering practice and many welds can be analyzed using the concepts presently developed. We emphasize that for a thorough treatment of welding types, processes, and design based formulas, consult references (1–5).

In general, welding implies the local heating and fusing of members together. Numerous methods of welding exist in engineering practice. The most common welding process uses heat from an electric current, although oxyacetylene torches are also used. For a thorough treatment of welding consult the *Welding Handbook* [1, 2]. Here we summarize some of the most important concepts involved with welding.

The theoretical strength of a weldment is difficult to achieve due to heating and cooling processes affecting the microstructure of the weld region, impurities and gases possibly being entrapped in the weld, and internal stresses developing in the final product. In critical welding applications, it is necessary to control the welding process carefully and to verify weld strength by performing destructive testing of samples and by inspecting all weldments of the finished product using one or more nondestructive testing methods. For our analysis of welded connections, we will assume that the weldments are homogeneous, free of internal stresses, and have uniform mechanical properties.

The types of welds are the groove or butt, fillet, slot, and plug welds as shown in Figure 3.17. Each type has certain advantages over the others for a specific application.

The groove and fillet welds are used in about 95% of all welded construction and hence will be considered here. The groove and fillet welds make up approximately 15%

and 80%, respectively, of all welded construction. Groove welds are principally used to connect structural members aligned in the same plane. A typical application is found in pressure vessels. Fillet welds are often used for general machine elements and structural applications (Figure 3.18).

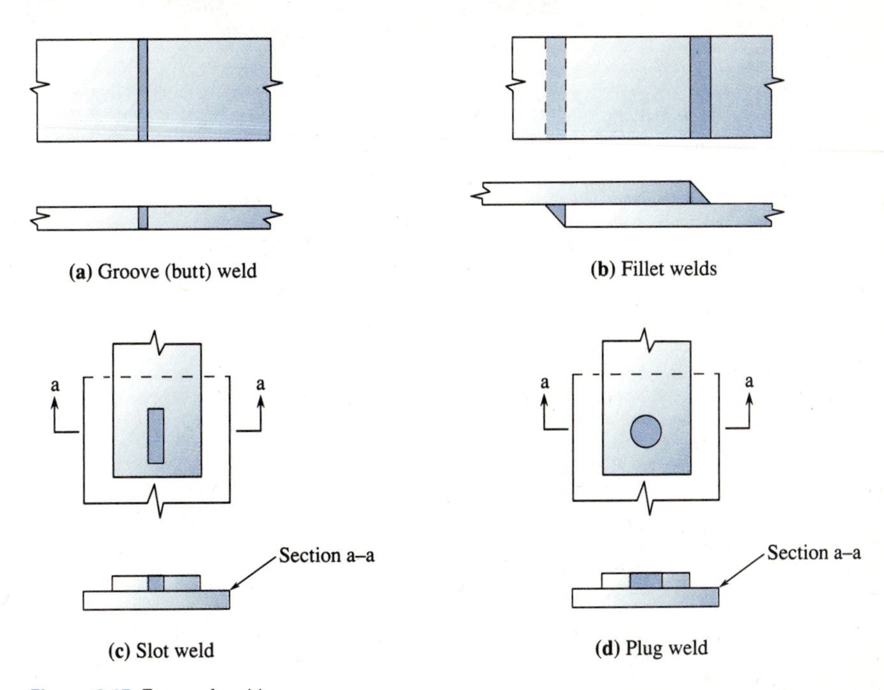

(a) Groove (butt) weld
(b) Fillet welds
(c) Slot weld
(d) Plug weld

Figure 3.17 Types of welds.

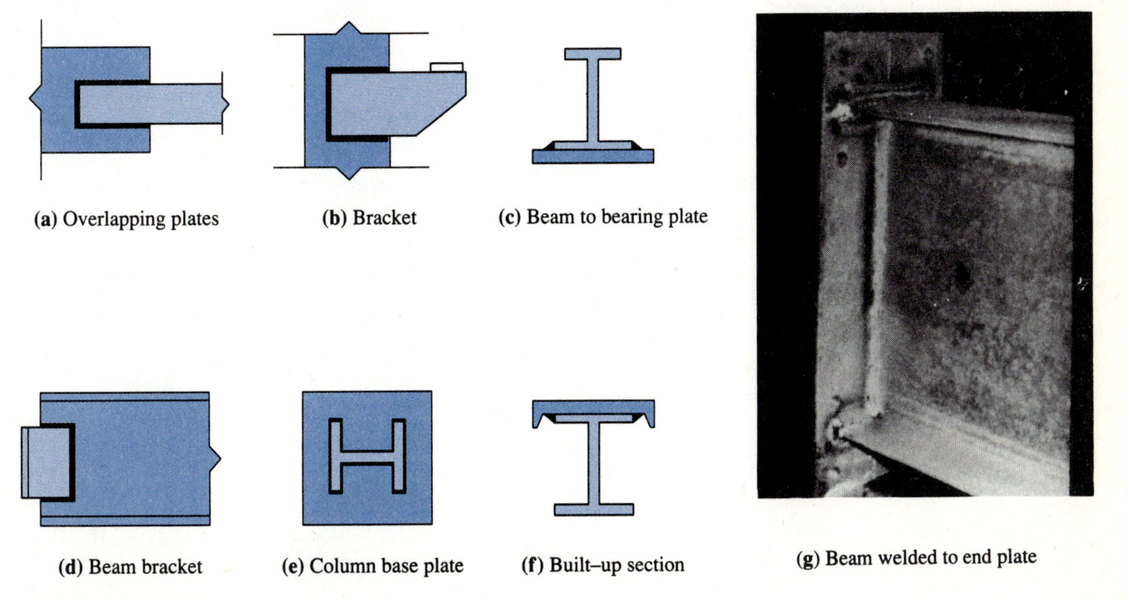

(a) Overlapping plates
(b) Bracket
(c) Beam to bearing plate
(d) Beam bracket
(e) Column base plate
(f) Built–up section
(g) Beam welded to end plate

Figure 3.18 Typical uses of fillet welds.

The groove weld is easy to analyze knowing the thickness t and length l of the weld, and the force F resisted by the weld (Figure 3.19).

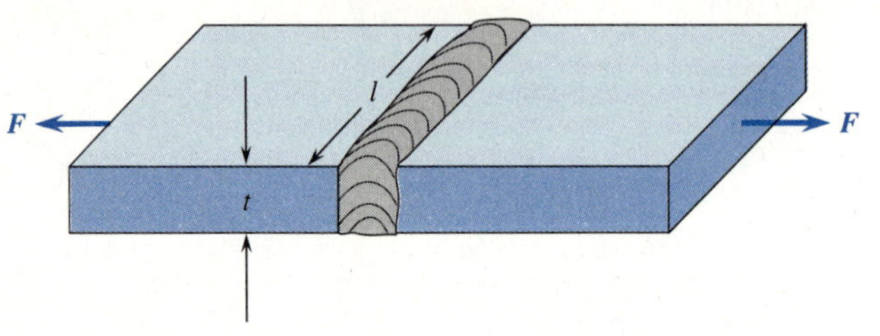

Figure 3.19 Groove weldment subjected to tensile force.

The average tensile or compressive stress is

$$\sigma = \frac{F}{tl} \tag{3.3}$$

The allowable tensile stress in a groove weld is the same as that of the connected material.

The fillet weld is the most often used weld due to its overall economy, ease of fabrication, and adaptability. Fillet welds come in 1/8 in. to 1/2 in. sizes in increments of 1/16 in. Fillet welds are slightly more complicated to analyze than groove welds. In the fillet weld (Figure 3.20), you might initially assume that area A is in tension and area B is in shear, and that these planes should be analyzed separately for the critical plane. However, the shear stress is taken on a plane between A and B where the weld metal is thinnest. Because shear strength is normally lower than tensile strength, the fillet weld strength is based on the allowable shear stress. It is then customary engineering practice to use the plane inclined to the plates as the critical plane. This plane is called the *throat area*. For fillet welds of equal legs, t, the shearing area per unit length of weld is taken as the throat and is $t \sin 45°$.

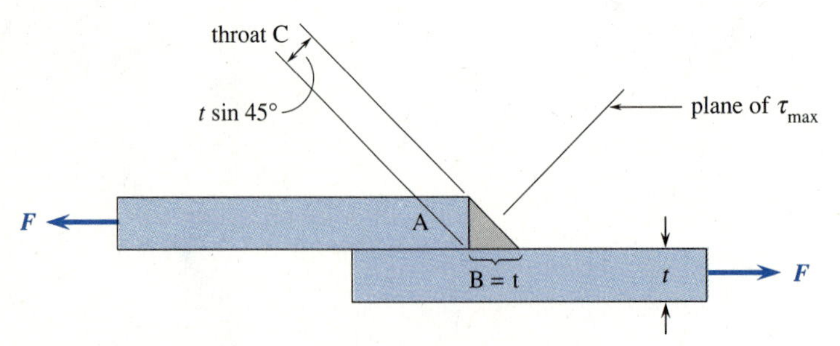

Figure 3.20 Fillet weld (t is called the *legs* of the weld).

This concept holds for welds parallel or transverse to the load F. The shear stress is then given by

$$\tau = \frac{F}{0.707\, tl} \tag{3.4}$$

The allowable shear stress for fillet welds is based on the electrode used. The electrode is given a code beginning with *E* followed by a two- or three-digit number indicating the tensile strength in ksi of the weld metal in the welding rod. For example, an *E* 60 electrode means one with a tensile strength of 60 ksi. The American Institute of Steel Construction (AISC) [3] takes the allowable shear stress on the throat area as 0.3 times the ultimate tensile strength of the weld metal, that is

$$\tau_{allow} = 0.3\,\sigma_u \tag{3.5}$$

However, the allowable shear stress in the connection also cannot exceed 0.4 times the yield stress of the base metal. That is

$$\tau_{allow} = 0.4\,\sigma_Y \tag{3.6}$$

For the numerous welding symbols and for complete tables of allowable stresses in weldments consult the American Institute of Steel Construction [3] and American Welding Society [4].

STEPS IN SOLUTION

The following steps are used to obtain the average shear stress in a fillet weld.

1. Draw a free-body diagram of the fillet weld showing the force F transmitted by the fillet weld. This force is also the shear force *V*.
2. Based on customary engineering practice, determine the critical shear area *A*. This area is taken as the throat thickness times the length of the weld *l*, where for equal legs the throat thickness is the leg width *t* times sin 45°. Therefore,

$$A = 0.707(tl)$$

3. Obtain the shear stress using Eq. (3.4) as

$$\tau = \frac{F}{0.707\,tl}$$

Examples 3.8 and 3.9 illustrate the steps outlined above to obtain the average shear stress in a fillet weld.

EXAMPLE 3.8

Two overlapping 1/4-in.-thick plates are connected together by the 45° welds shown on top and bottom (Figure 3.21). The connection transmits 20,000 lb. The E 60 electrode is used for the weldment. The plates are A36 steel. Therefore, their yield stress is 36,000 psi. Determine the length *l* of weld required on each plate.

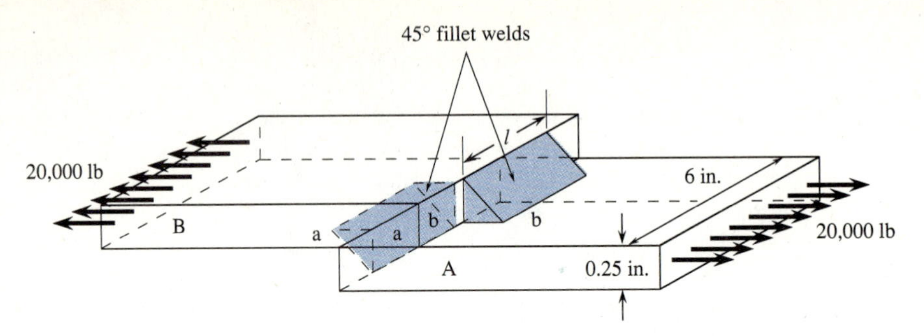

Figure 3.21 Welded connection.

Solution

Draw a free-body diagram of the lower plate A.

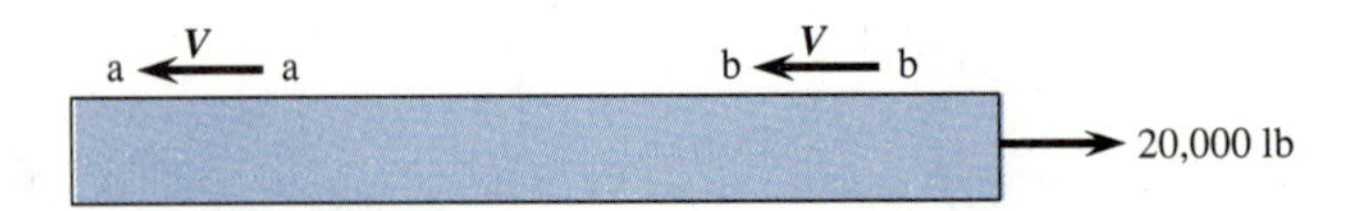

We assume the shear force V along faces a-a and b-b to be equal. By statics, we determine the shear force V

$$2V - 20{,}000 \text{ lb} = 0 \qquad V = 10{,}000 \text{ lb}$$

The shear area per unit length of weld is $t \sin 45°$ where $t = 1/4$ in. Therefore, the total shear area that V acts over is taken as $A = (1/4 \sin 45°)l$. Note that two shear areas exist to resist the 20,000-lb force. The allowable shear stress in the weld is the lesser of $0.3 \times 60{,}000$ psi = 18,000 psi or $0.4 \times 36{,}000$ psi = 14,400 psi. Using Eq. (3.4), and knowing the allowable shear stress acting on the upper weld, we solve for the length l of the weld as

$$\tau_{\text{ave}} = 14{,}400 \text{ psi} = \frac{10{,}000 \text{ lb}}{(0.25 \text{ in.})(\sin 45°)\, l}$$

$$l = 3.93 \text{ in.} \qquad \text{use } l = 4 \text{ in.}$$

Similarly, the length of the lower weld is also 4 in.

EXAMPLE 3.9

Connectors in the form of angles are often used to attach beams to columns as shown in Figure 3.22. The two angles are 1/4 in. fillet welded side, top, and bottom to the column. The allowable shear stress in the weld is 14 ksi. Determine the allowable load that can be transferred to the connections.

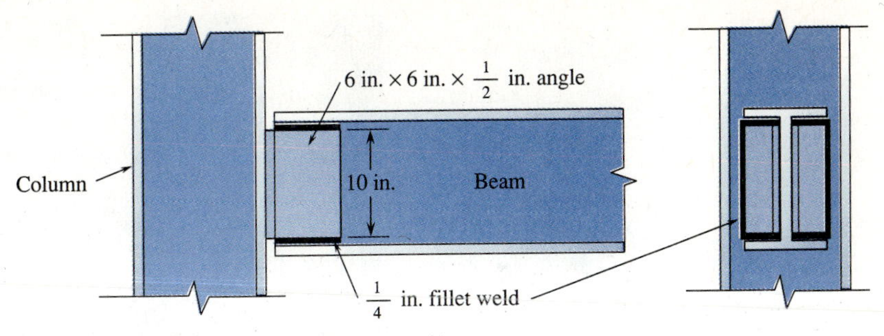

Figure 3.22 Beam to column welded connection.

Solution

The weld area on each angle is

$$
\begin{aligned}
A &= 0.707\, t \times L \\
L &= \text{the total weld length of one angle} \\
&= 6 \text{ in.} + 6 \text{ in.} + 10 \text{ in.} = 22 \text{ in.} \\
t &= 0.25 \text{ in. (size of fillet weld)} \\
A &= 0.707(0.25 \text{ in.})(22 \text{ in.}) \\
&= 3.89 \text{ in.}^2
\end{aligned}
$$

There are two angles, therefore

$$A_{\text{total}} = 2 \times 3.89 = 7.78 \text{ in.}^2$$

The allowable force is then obtained by solving Eq. (3.4) for F as

$$
\begin{aligned}
F &= 2(0.707\, t\, L)\, \tau \\
&= (A_{\text{total}})\, \tau \\
&= (7.78 \text{ in.}^2)(14 \text{ ksi}) \\
F &= 109 \text{ kips}
\end{aligned}
$$

3.3 SHEAR STRAIN AND DEFORMATION DUE TO DIRECT SHEAR FORCES

Direct shear deformation due to shear forces does not often occur in practice. In fact in considering fasteners, such as analyzed for shear stresses in Section 3.2, the shear stress occurs only in shear planes assumed to have no thickness. Hence shear deformation is a senseless quality to attempt to obtain in fasteners.

However, there are special cases where direct shear deformation occurs, such as in elastomeric bearing pads used in bridge structures (Figure 3.23). Also we want to relate shear stress to shear strain, therefore it is appropriate at this time to consider shear deformation and shear strain. Finally, in Chapter 4, we consider the very common and important practical case of direct shear deformation due to torsion.

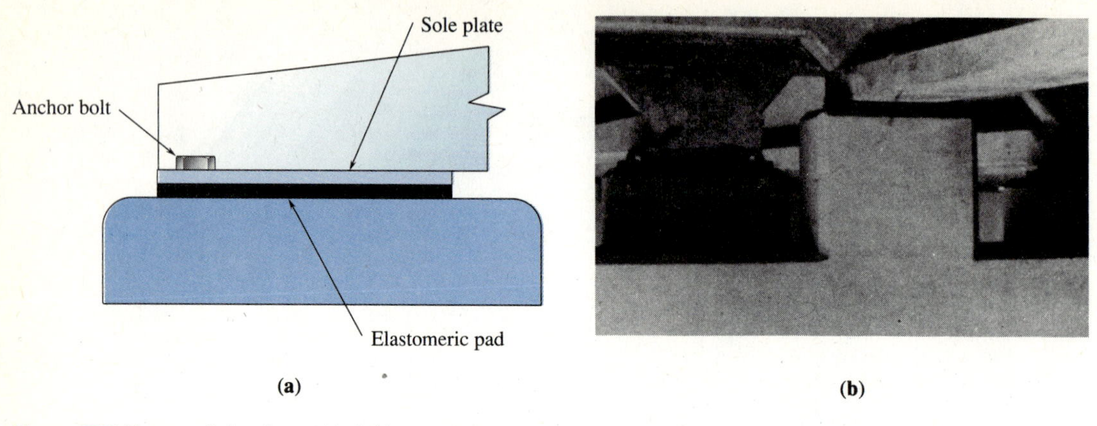

Figure 3.23 Elastomeric bearing pad in bridge structure.

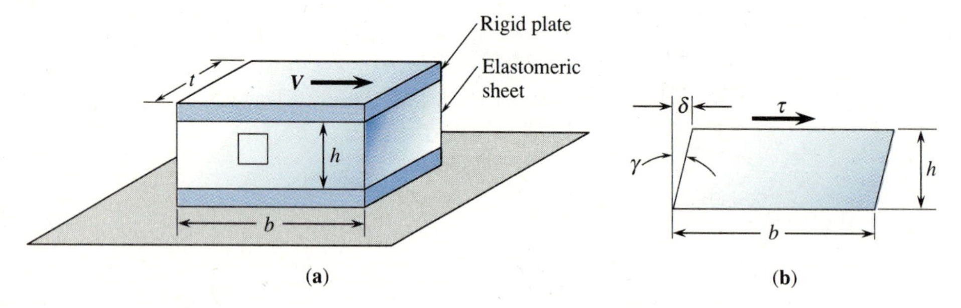

Figure 3.24 (a) Elastomeric sheet subjected to direct shear force and (b) resulting deformed sheet.

To visualize shear deformation, we consider an elastomeric sheet, such as rubber, bonded to two rigid horizontal plates shown in Figure 3.24(a).

When subjected to direct shear force V on the top surface, the elastomeric material deforms large enough to be seen. A square drawn on the side of the sheet deforms into a parallelogram. We already know from our definition of shear stress that the shear stress on the sheet is

$$\tau = \frac{V}{A} = \frac{V}{bt} \tag{3.7}$$

where $A = bt$ is the shear area.

The shear deformation of the whole sheet is shown as δ in Figure 3.24(b). The shear strain γ in Figure 3.24(b) is then defined as the change in right angle between two lines originally perpendicular to each other. This angle is normally expressed in radians.

Using the definition of the tangent of an angle, we express the shear strain as

$$\tan \gamma = \frac{\delta}{h} \tag{3.8}$$

For small shear strains, $\tan \gamma = \gamma$. Therefore, the shear strain becomes

$$\gamma = \frac{\delta}{h} \tag{3.9}$$

Equation (3.9) is valid in most engineering applications as shear strains are nearly always less than 0.105 radians (6°). For $\gamma = 0.105$ radians, Eqs. (3.8) and (3.9) are identical to three significant figures, that is, $\tan 0.105 = 0.105$.

In the linear-elastic range of a material behavior, the shear strain is related to the shear stress by Hooke's law in shear. This relationship is

$$\tau = G\gamma \tag{3.10}$$

where G (called the *shear modulus* or *modulus of rigidity*) is the slope of the linear portion of the τ-γ diagram. The shear modulus has the same units as shear stress, for instance psi, ksi, or MPa.

We can relate the shear force V to the shear deformation δ by using Eqs. (3.7) and (3.9) in (3.10) and solving for δ as

$$\delta = \frac{Vh}{GA} \tag{3.11}$$

Equation (3.11) is analogous to Eq. (2.21) which was used to obtain axial deformation due to axial loading. We merely replace shear force V with axial force P, shear modulus G with modulus of elasticity E, transverse length h with longitudinal length L, and understand that here A is a shear area while in Eq. (2.21) A was an area normal to axial force P.

Three of the important mechanical properties described in Chapters 2 and 3 are related to each other. The properties E, ν, and G that apply to isotropic materials behaving in their linear-elastic range, are related by

$$G = \frac{E}{2(1 + \nu)} \tag{3.12}$$

Equation (3.12) indicates that if, for instance, we determine the modulus of elasticity E and Poisson's ratio ν from a tension test, the shear modulus G can be determined using Eq. (3.12). We do not prove Eq. (3.12) here as it is probably easiest proven using a pure shear element, and the concept of Mohr's circle that is presented in Chapter 8.

EXAMPLE 3.10

An elastomeric bearing of size 5 in. × 5 in. × 1 in. high supporting a bridge beam is made of an elastomer bonded to two rigid horizontal plates (Figure 3.25). The lower plate is fixed while the upper one is subjected to a horizontal bridge force of 12 kips. The shear modulus of the elastomer is 100 ksi. Determine **(a)** the average shear stress in the bearing and **(b)** the horizontal (lateral) displacement.

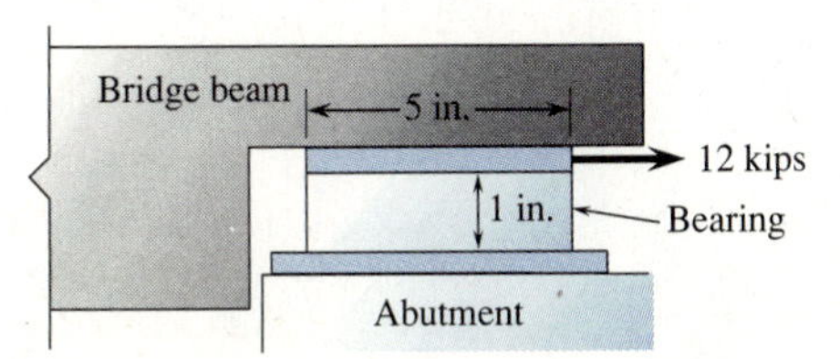

Figure 3.25 Bearing in shear.

(a) Using Eq. (3.7), the average shear stress in the bearing is

$$\tau_{ave} = \frac{V}{A} = \frac{12 \text{ k}}{5 \text{ in.} \times 5 \text{ in.}} = 0.48 \text{ ksi}$$

(b) Using Eq. (3.10), we obtain the shear strain as

$$\gamma = \frac{\tau}{G} = \frac{0.48 \text{ ksi}}{100 \text{ ksi}} = 0.0048 \text{ rad}$$

Using Eq. (3.9), the horizontal deformation is

$$\delta = \gamma h = (0.0048)(1 \text{ in.}) = 0.0048 \text{ in.}$$

3.4 BEARING STRESS

Bearing stresses develop in connectors such as bolts, pins or rivets due to the members being joined pressing or bearing against the connectors. In general, when one member pushes on another a bearing stress develops between the surfaces that press against each other.

For example, consider the two plates A and C connected together by the bolt B (Figure 3.26(a)), as was previously discussed in Section 3.2. The plate A transmits a force F against the bolt, while the bolt transmits the oppositely directed force F on the plate as shown in Figure 3.26(b). Similarly, plate C applies force F to the bolt. The force F acting on the bolt represents the resultant of a number of forces distributed over the half cylinder of diameter d of the bolt and length t equal to that of the thickness of the plate bearing against the bolt. The actual distribution of the resulting stress is difficult to obtain. In engineering practice, we define an average bearing stress σ_b as the force F divided by the projected area of the curved bolt surface onto the plate (Figure 3.26(c). This projected area is a rectangle of area $d \times t$, that is,

$$A_b = dt \tag{3.13}$$

The average bearing stress is then the force F divided by the area A_b that F is assumed to act over.

$$\sigma_b = \frac{F}{A_b} = \frac{F}{dt} \tag{3.14}$$

STEPS IN SOLUTION

The following steps are used to determine the average bearing stress in members subjected to direct bearing.

1. Draw a free-body diagram of the member showing the bearing forces, F.
2. Identify the bearing area A_b over which the force F acts.
3. Obtain the average bearing stress using Eq. (3.14) as

$$\sigma_b = \frac{F}{A_b}$$

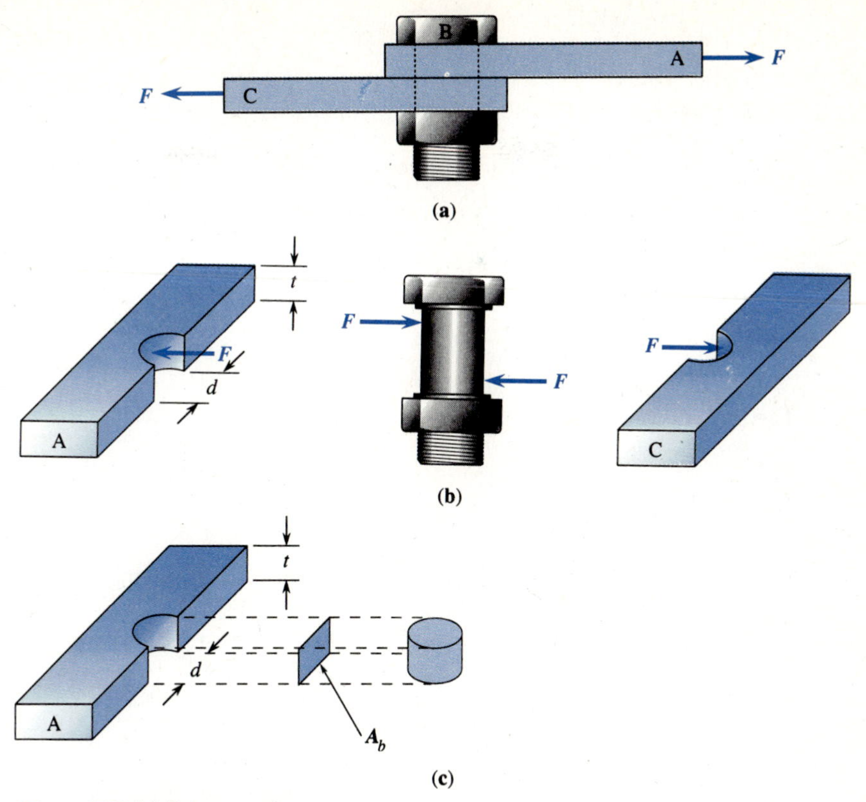

Figure 3.26 (a) Bolt connection,
(b) forces on plates and bolt, and
(c) bearing area A_b between plate and bolt.

Examples 3.11 through 3.15 illustrate the steps outlined above to obtain bearing stresses.

EXAMPLE 3.11

Two 6-mm-thick plates are connected together by using two 20-mm-diameter bolts (Figure 3.27). The plates are subjected to tensile forces of 50 kN. Determine the bearing stress between the bolts and the plates.

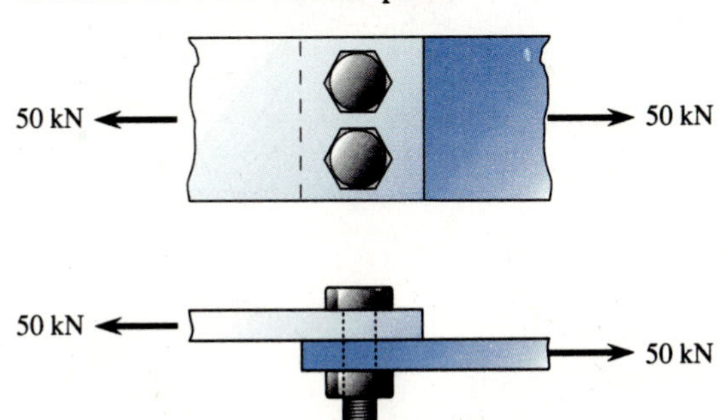

Figure 3.27 Plates connected by bolts.

Solution

A free-body diagram of a bolt with a force of 50 kN/2 = 25 kN acting on each side, as transmitted from the plates, is shown (assuming the two bolts carry equal force).

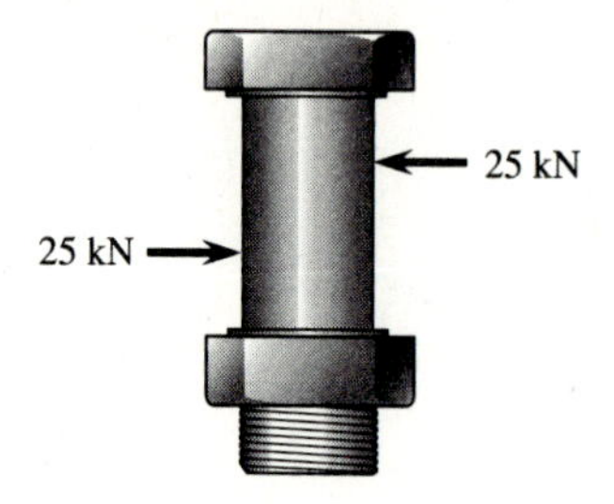

The bearing area is

$$A_b = dt = (0.02 \text{ m})(0.006 \text{ m}) = 1.2 \times 10^{-4} \text{ m}^2$$

Using Eq. (3.14), the bearing stress is

$$\sigma_b = \frac{F}{A_b} = \frac{25 \text{ kN}}{1.2 \times 10^{-4} \text{ m}^2} = 20.8 \times 10^4 \frac{\text{kN}}{\text{m}^2} = 208 \text{ MPa}$$

EXAMPLE 3.12

A 250 mm × 250 mm wooden post transmits a load of 50 kN to a concrete footing (Figure 3.28). **(a)** Determine the bearing stress in the concrete. **(b)** If the allowable bearing stress in the soil is 100 kPa, determine the size of the concrete footing needed.

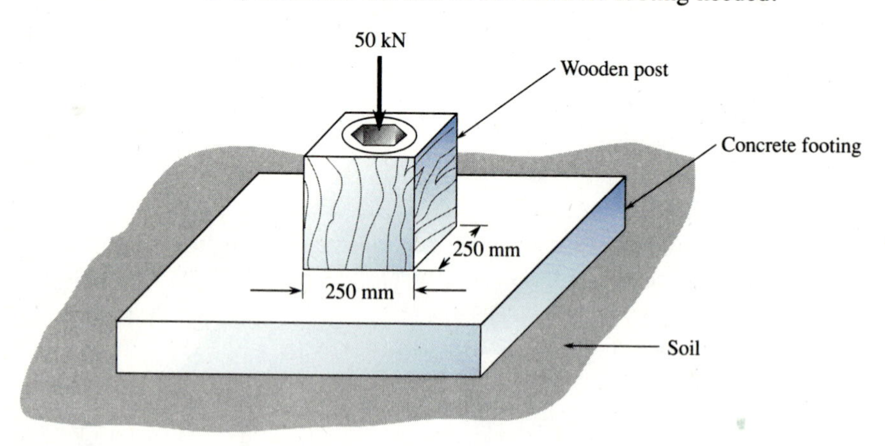

Figure 3.28 Post on concrete footing.

Solution

(a) The bearing force is 50 kN acting over the bearing area of 250 mm × 250 mm.

Using Eq. (3.14), the bearing stress is

$$\sigma_b = \frac{F}{A_b} = \frac{50 \text{ kN}}{0.25 \text{ m} \times 0.25 \text{ m}} = 800 \frac{\text{kN}}{\text{m}^2}$$

(b) Using Eq. (3.14), but now solving for the bearing area, we obtain

$$\sigma_b = 100\,\frac{\text{kN}}{\text{m}^2} = \frac{50\text{ kN}}{A_b}$$
$$A_b = 0.5\text{ m}^2$$

A 0.707 m × 0.707 m square concrete footing satisfies the allowable bearing stress in the soil. In practice, a 0.75 m × 0.75 m square footing would probably be used.

EXAMPLE 3.13

The 1/4-in.-diameter bolt is one of four holding up a wooden platform (Figure 3.29). Each bolt supports a load of 1500 lb. If the average bearing stress between the washer and the wooden plate is not to exceed 600 psi, determine the minimum diameter of the washer.

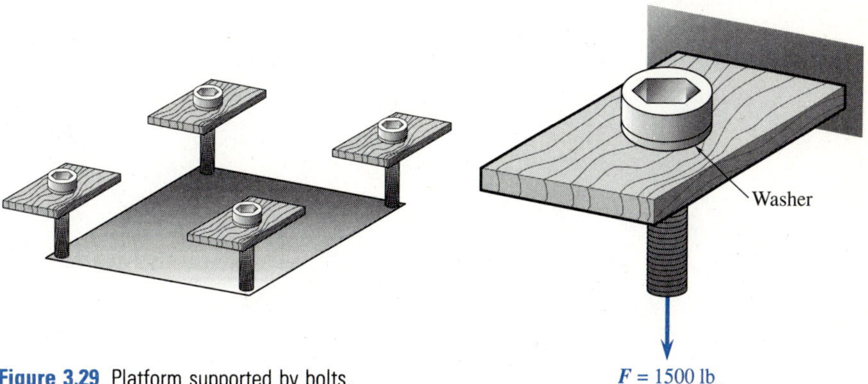

Figure 3.29 Platform supported by bolts.

Solution

The bearing area is equal to the area of the washer A_w, as shown in the free-body diagram of the washer.

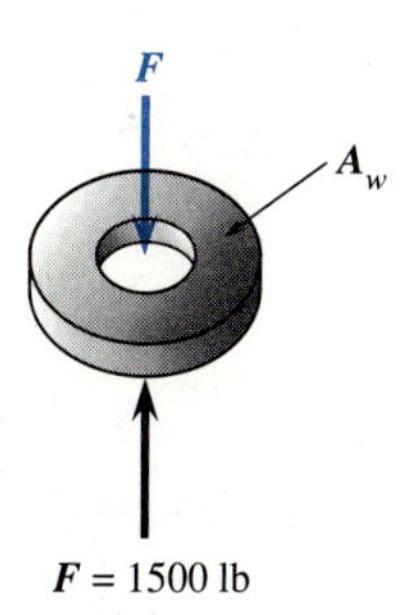

By statics, the force F representing the resultant of the many smaller forces acting on the cross-sectional area of the washer is

$$F = 1500\text{ lb}$$

Using Eq. (3.14), the diameter of the washer can be determined as

$$\sigma_b = 600 \text{ psi} = \frac{F}{A_b} = \frac{1500 \text{ lb}}{A_w} = \frac{1500 \text{ lb}}{\frac{\pi}{4} d_w^2 - \frac{\pi}{4}(0.25 \text{ in.})^2}$$

$$\frac{\pi}{4} d_w^2 = \frac{1500}{600} + 0.0491 \text{ in.}^2$$

$$d_w = 1.80 \text{ in.}$$

EXAMPLE 3.14

Based on the Crane Manufacturers Association of America (CMAA) code [6] for crane design, bolts supporting a crane cab must act in shear. Based on this requirement, a frame of a crane cab is hung using bolted connections from a beam above, as shown in Figure 3.30. The cab self-weight plus the weight of a person totals 2000 lb and must be resisted by the 3/4 in.-diameter-bolts in shear. Determine **(a)** the average shear stress in the bolts and **(b)** the bearing stress on the bolts.

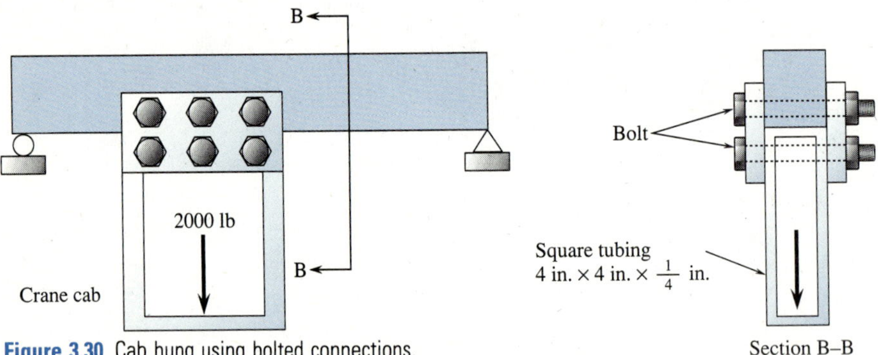

Figure 3.30 Cab hung using bolted connections.

Solution

(a) We begin with a free-body diagram of one of the three bolts as shown below (where we assume each bolt resists 1/3 of the total vertical force of 2000 lb).

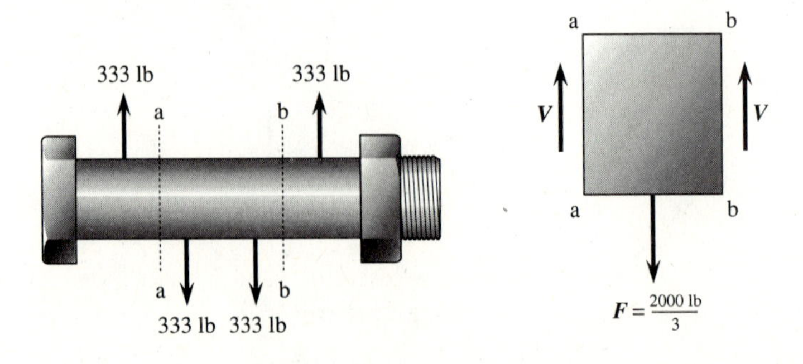

By statics

$$2V = F = \frac{2000}{3}\text{ lb} = 667\text{ lb} \qquad V = 333\text{ lb}$$

The bolt-cross sectional area is

$$A = \frac{\pi}{4}(0.75\text{ in.})^2 = 0.442\text{ in.}^2$$

Using Eq. (3.2), since the bolt is in double shear

$$\tau_{\text{ave}} = \frac{F}{2A} = \frac{667\text{ lb}}{2(0.442\text{ in.}^2)} = 755\text{ psi}$$

(b) The bearing area is given by Eq. (3.13) as

$$A_b = dt = \left(\frac{3}{4}\text{ in.}\right)\left(\frac{1}{4}\text{ in.}\right) = \frac{3}{16}\text{ in.}^2 = 0.188\text{ in.}^2$$

where the thickness of the tube has been used for t.

Using Eq. (3.14), the bearing stress is

$$\sigma_b = \frac{333\text{ lb}}{0.188\text{ in.}^2} = 1770\text{ psi}$$

NOTE: In practice, we must also consider the possibility of impact forces due to the crane hitting an end stop. This concept is beyond the scope of our discussion.

EXAMPLE 3.15

For the load frame shown in Figure 3.31, with pins B and D in single shear and pin C in double shear, determine **(a)** the average shear stress in each pin and **(b)** the bearing stresses in the support brackets at C and D.. The diameter of pins B and D is 5/16 in. and the diameter of pin C is 1/4 in.

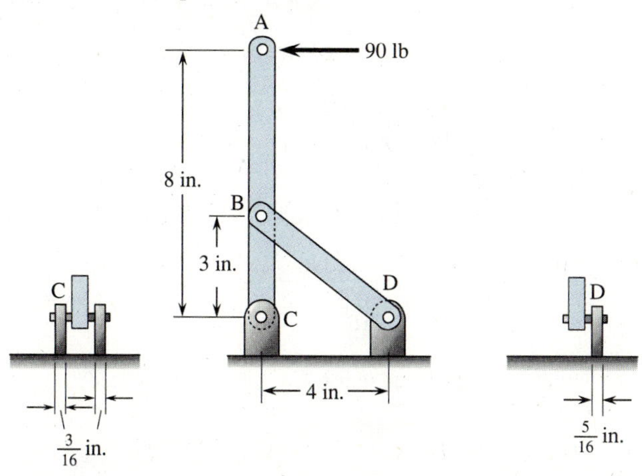

Figure 3.31 Load frame.

Solution

Shear Stresses in Pins

(a) Member BD is a two-force member. The free-body diagram of member ABC is

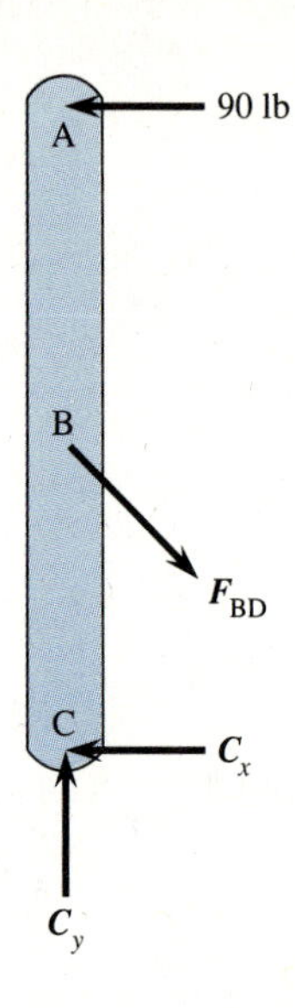

By statics

$$\Sigma M_c = 0: \quad -\left(F_{BD}\frac{4}{5}\right)(3 \text{ in.}) + 90(8 \text{ in.}) = 0 \qquad F_{BD} = 300 \text{ lb}$$

$$\Sigma F_x = 0: \quad -C_x - 90 + F_{BD}\left(\frac{4}{5}\right) = 0 \qquad C_x = 150 \text{ lb}$$

$$\Sigma F_y = 0: \quad C_y - F_{BD}\left(\frac{3}{5}\right) = 0 \qquad C_y = 180 \text{ lb}$$

$$C = \sqrt{C_x^2 + C_y^2} = \sqrt{150^2 + 180^2} = 234 \text{ lb}$$

Pin B is in single shear. Therefore, using Eq. (3.1), we have

$$\tau_{ave} = \frac{F_{BD}}{A_B} = \frac{300 \text{ lb}}{\frac{\pi}{4}\left(\frac{5}{16} \text{ in.}\right)^2} = 3910 \text{ psi}$$

Pin D is in single shear. Again using Eq. (3.1), we obtain

$$\tau_{ave} = \frac{F_{BD}}{A_D} = \frac{300 \text{ lb}}{\frac{\pi}{4}\left(\frac{5}{16} \text{ in.}\right)^2} = 3910 \text{ psi}$$

Pin C is in double shear. Therefore, using Eq. (3.2), we obtain

$$\tau_{ave} = \frac{C}{2A} = \frac{234 \text{ lb}}{2\frac{\pi}{4}\left(\frac{1}{4} \text{ in.}\right)^2} = 2380 \text{ psi}$$

Bearing Stresses in Support Brackets C and D

(b) *Bracket* D: Using Eq. (3.14), with the bearing area calculated using $t = 5/16$-in.-thick bracket and $d = 5/16$-in.-diameter pin, we obtain

$$\sigma_b = \frac{F_{BD}}{td} = \frac{300 \text{ lb}}{\frac{5}{16} \times \frac{5}{16}} = 3070 \text{ psi}$$

Bracket C: Using Eq. (3.14), with the bearing area now calculate using $t = 3/16$-in.-thick bracket and $d = 1/4$ in. diameter pin, we obtain

$$\sigma_b = \frac{C}{2td} = \frac{234 \text{ lb}}{2\left(\frac{3}{16} \times \frac{1}{4}\right)} = 2500 \text{ psi}$$

Notice that two bearing areas exist in bracket C.

3.5 SUMMARY OF IMPORTANT DEFINITIONS AND EQUATIONS

1. Average shear stress (single shear)

$$\tau_{ave} = \frac{F}{A} \quad (3.1)$$

2. Average shear stress (double shear)

$$\tau_{ave} = \frac{F}{2A} \quad (3.2)$$

3. Average normal stress in groove weld

$$\sigma = \frac{F}{tl} \quad (3.3)$$

4. Average shear stress in fillet weld

$$\tau = \frac{F}{0.707\, tl} \quad (3.4)$$

5. Shear strain

$$\gamma = \frac{\delta}{h} \quad (3.9)$$

6. Hooke's law in shear

$$\tau = G\gamma \quad (3.10)$$

7. G = shear modulus (the slope of the linear portion of the τ-γ diagram)

8. Shear deformation

$$\delta = \frac{Vh}{GA} \quad (3.11)$$

9. Relationship among E, ν, and G

$$G = \frac{E}{2(1 + \nu)} \quad (3.12)$$

10. Average bearing stress in bolt

$$\sigma_b = \frac{F}{A_b} = \frac{F}{dt} \quad (3.14)$$

REFERENCES

1. American Welding Society. *Welding Handbook,* Vol. 2, 6th ed. *Welding Processes: Gas, Arc, and Resistance*. New York, 1969.
2. American Welding Society. *Welding Handbook,* Vol. 1, 7th ed. *Fundamentals of Welding*. Miami, FL, 1976.
3. American Institute of Steel Construction, One East Wacker Drive, Chicago, 60601. *Manual of Steel Construction,* 9th ed., 1989.
4. American Welding Society. *Symbols for Welding and Nondestructive Testing,* AWS A2.4–76, Miami, FL, 1976.
5. Blodgett O.W. *Design of Welded Structures*. James F. Lincoln Arc Welding Foundation, 1966.
6. Crane Manufacturers Association of America. *Specifications for Electric Overhead Traveling Cranes, CMAA Specification No. 70*. (1326 Freeport Road, Pittsburgh, PA 15238.) 1983.

PROBLEMS

Section 3.2

3.1 Two members are connected by a bolt AB as shown in the figure. If the load $P = 30$ kN and the diameter of the bolt is 10 mm, determine the shear stress in the bolt.

Figure P3.1

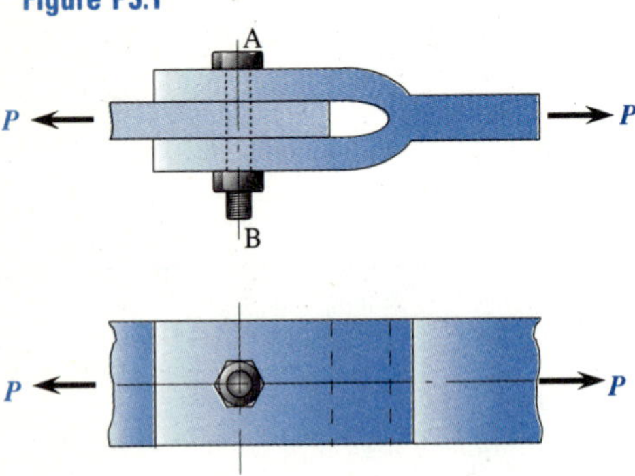

3.2 A block of wood is tested in direct shear using the standard test specimen shown in the figure. The load P produces shear in the specimen along the plane AB. The width of the specimen (perpendicular to the plane of the paper) is 2 in., and the height h of plane AB is 2 in. For a load $P = 500$ lb, what is the average shear stress τ_{ave} in the wood?

Figure P3.2

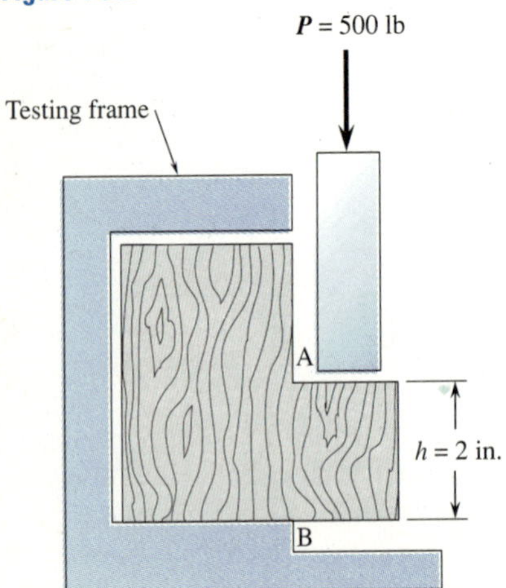

3.3 In a pair of pliers, the hinge pin is subjected to direct shear, as indicated in the figure. If the pin has a diameter of 4.0 mm and the force exerted at the handle, F, is 60 N, compute the stress in the pin.

Figure P3.3

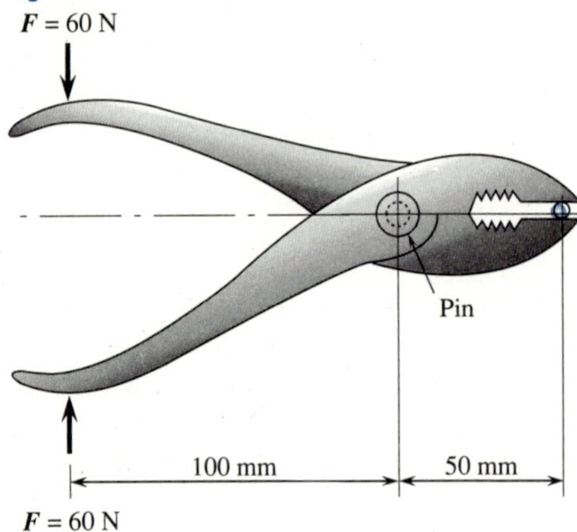

3.4 A hollow box beam ABC of length 24 ft is supported at A by a 1-in.-diameter pin that passes through the beam as shown in the figure. A roller connection at B supports the beam at a distance 6 ft from A. Calculate the average shear stress τ_{ave} in the pin if the load P equals 5000 lb.

Figure P3.4

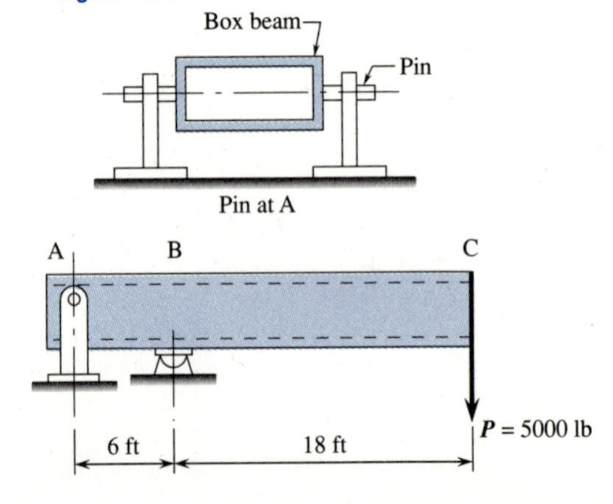

3.5 An angle bracket is attached to a column with two 10-mm-diameter bolts as shown in the figure. The bracket supports a load $P = 25$ kN. Calculate the average shear stress τ_{ave} in the bolts, disregarding friction between the bracket and the column.

Figure P3.5

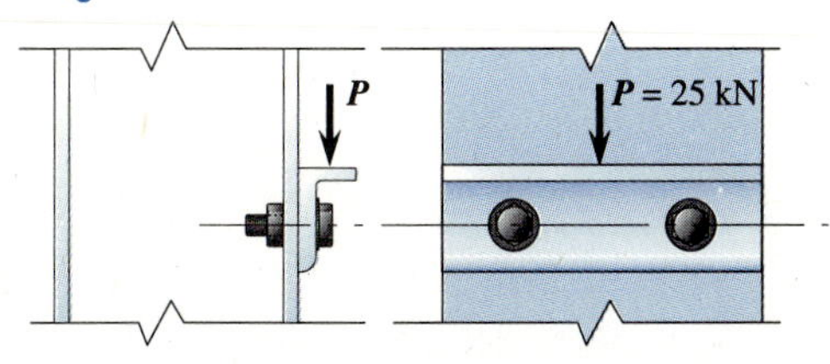

3.6 An automobile hood mechanism supports the 200 N force shown at D. The pin at C has a diameter of 4 mm. Determine the average shear stress in the pin at C. The pin at C is in single shear.

Figure P3.6

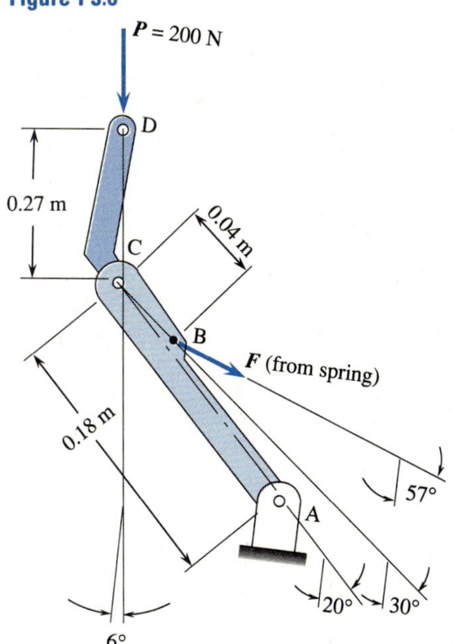

3.7 Three steel plates are connected by two rivets as shown in the figure. If the rivets have diameters of 10 mm and the ultimate shear stress in the rivets is 210 MPa, what force P is required to cause the rivets to fail in shear?

Figure P3.7

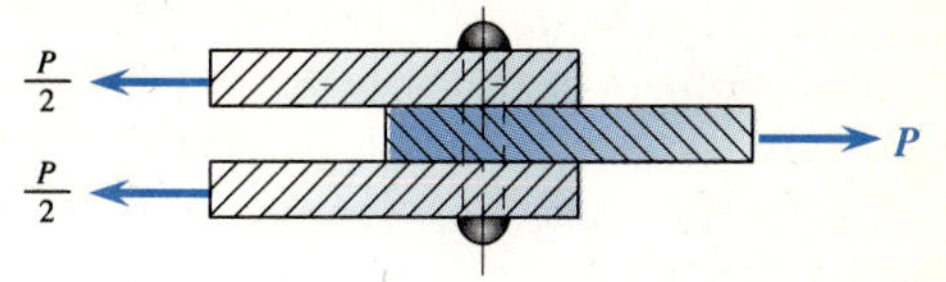

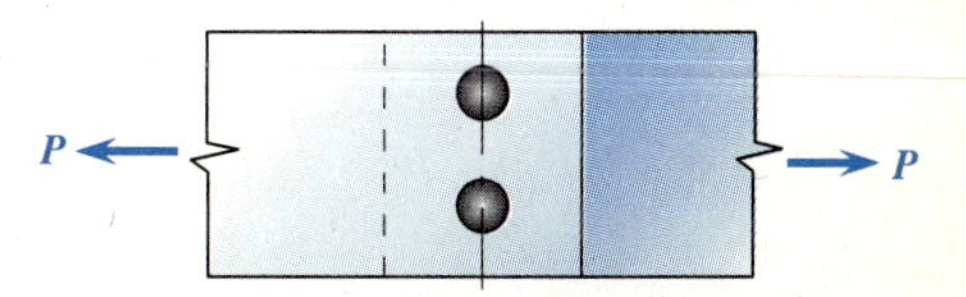

3.8 The caster C shown supports a load of 1000 lb. Determine the average shear stress in the axle. The axle diameter is 3/8 in.

Figure P3.8

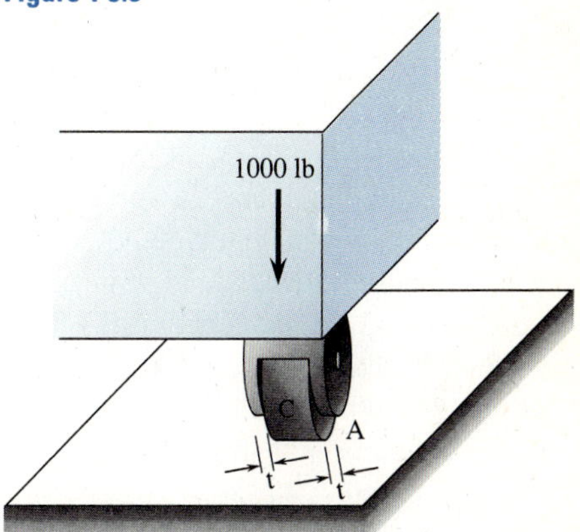

3.9 Two pieces of material of width 60 mm are interlocked as shown in the figure and are pulled by forces $P = 10$ kN. Determine the average shear stress in the material.

Figure P3.9

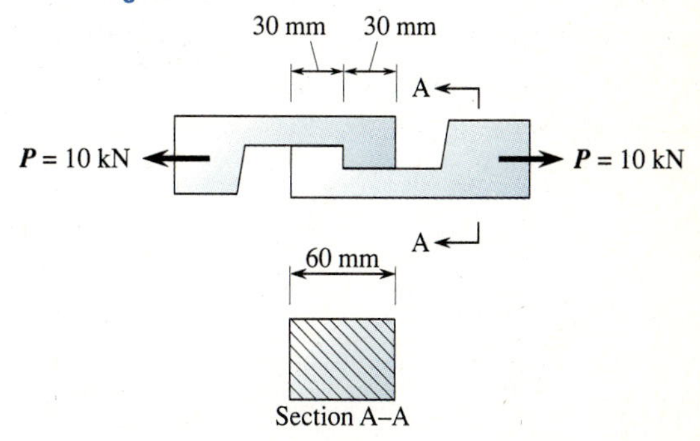

3.10 The key in Figure P3.10 has the dimensions $b = 5$ mm, $h = 8$ mm, and $L = 20$ mm. Determine the shear stress in the key when 100 N · m of torque is transferred from the 35-mm-diameter shaft to the hub.

Figure P3.10 & 11

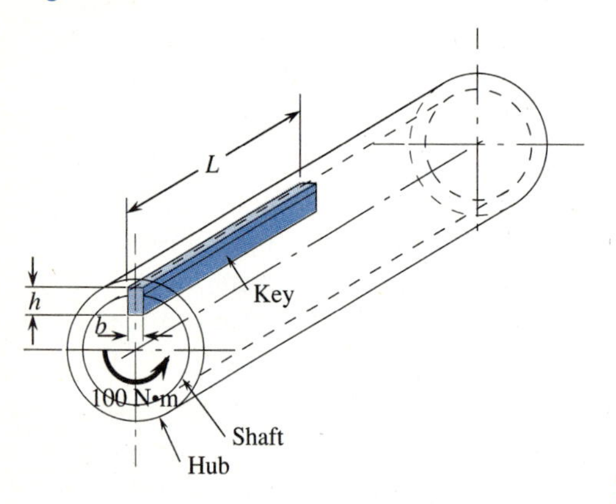

3.11 A key is used to connect a hub of a gear to a shaft, as shown in Figure P3.11 above. It has a rectangular cross-section with $b = 1/2$ in. and $h = 5/8$ in. The length is 2.25 in. Compute the shear stress in the key when it transmits 10,000 lb · in. of torque from the 2.0-in.-diameter shaft to the hub.

3.12 A small, hydraulic crane shown in Figure P3.12 carries a 5000-lb load. Determine the shear stress that occurs in the pin at B, which is in double shear. The pin diameter is 1 in.

Figure P3.12

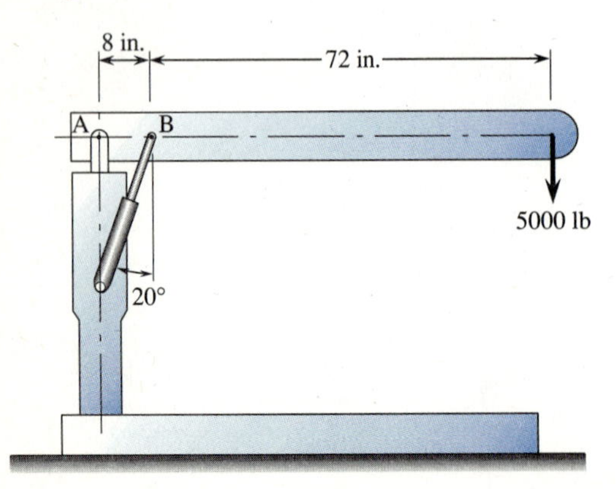

3.13 Determine the force required to punch a slug whose shape is shown in Figure P3.13 from an aluminum sheet 6.0 mm thick. The ultimate shear strength of the aluminum is 90 MPa.

Figure P3.13

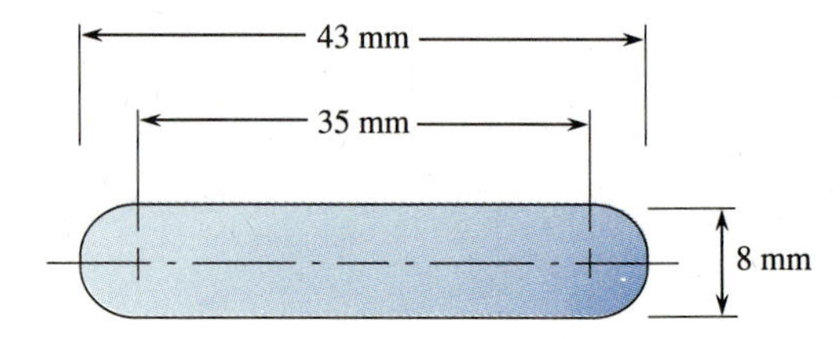

3.14 The two overlapped 1/4-in. plates are 45° fillet welded on top and bottom as shown. If the tensile force in the plates is 16,000 lb and the nominal shear stress in the welds cannot exceed 13,600 psi, how many total inches of weld are required?

Figure P3.14

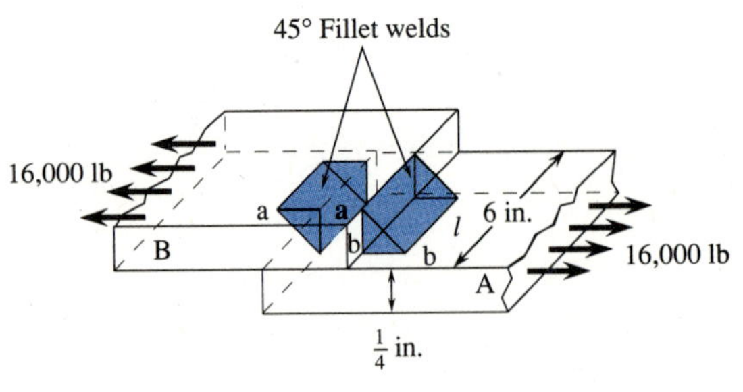

3.15 A notch is made in a piece of Douglas fir construction grade wood, as shown, in order to support a load of $F = 2000$ lb. Compute the shear stress in the wood. Is the notch safe based on an allowable shear stress of 85 psi?

Figure P3.15

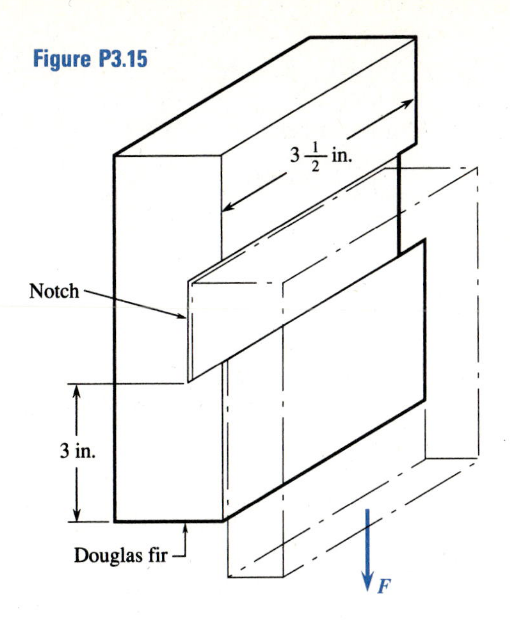

3.16 The ultimate shear strength of a 5/8-in.-thick steel plate is 42,000 psi. What force is necessary to punch a 3/4-in. diameter round hole in the plate?

3.17 The ultimate strength of a 15-mm-thick steel plate is 400 MPa. What force is necessary to punch a 15-mm-diameter round hole in the plate?

3.18 Two wood members are connected by the glued lap splice shown. A force of 20 kN is transmitted through the splice. Determine the average shear stress in the splice.

Figure P3.18

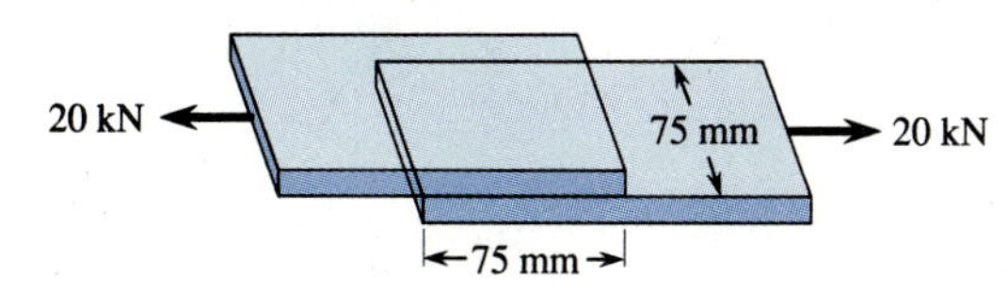

3.19 The two overlapped 6-mm plates are 45° fillet welded on top and bottom as shown. If the tensile force in the plates is 60 kN and the nominal shear stress in the welds cannot exceed 90 MPa, how many total millimeters of weld are required along one weld?

Figure P3.19

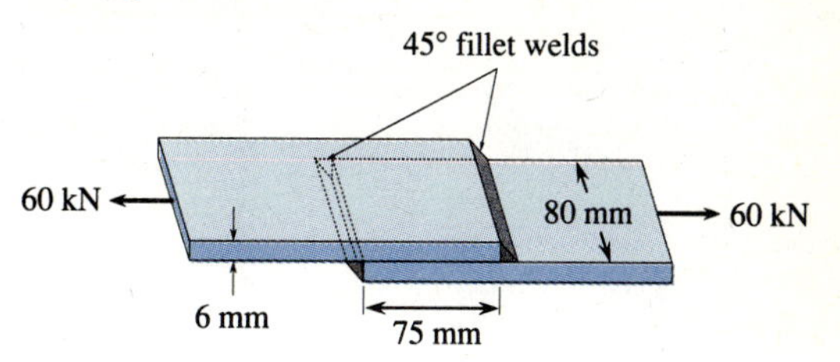

3.20 Two overlapped plates of 0.25 in. thickness are 45° fillet welded on top and bottom as shown. The length of the fillet welds is l = 5 in. If the allowable shear stress in the welds is 13,600 psi, determine the allowable tensile force in the plates.

Figure P3.20

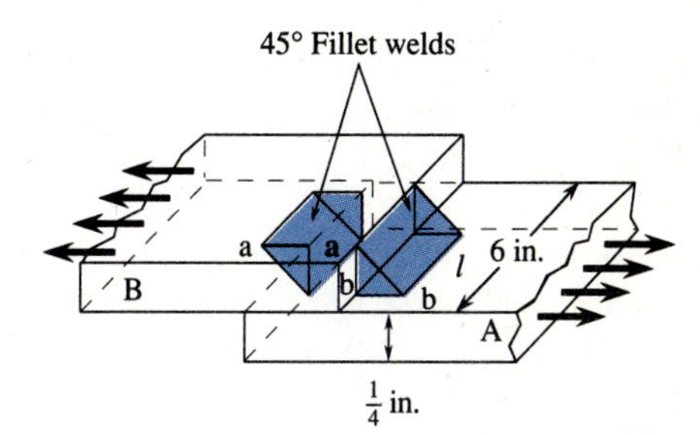

3.21 For the load frame of Problem 1.45, the pin of diameter 10 mm at C is in single shear. Determine the average shear stress in the pin.

3.22 For the beam of Problem 1.46, the pin at A is in single shear. The allowable shear stress in the pin is 75 MPa. Determine the required diameter of the pin.

3.23 For the vise of Problem 1.50, the pin at A is in single shear. The allowable shear stress in the pin is 12 ksi. Determine the required diameter of the pin.

3.24 For the crane shown, the pin at B is in single shear. The allowable shear stress in the pin is 75 MPa. Determine the required diameter of the pin.

Figure P3.24

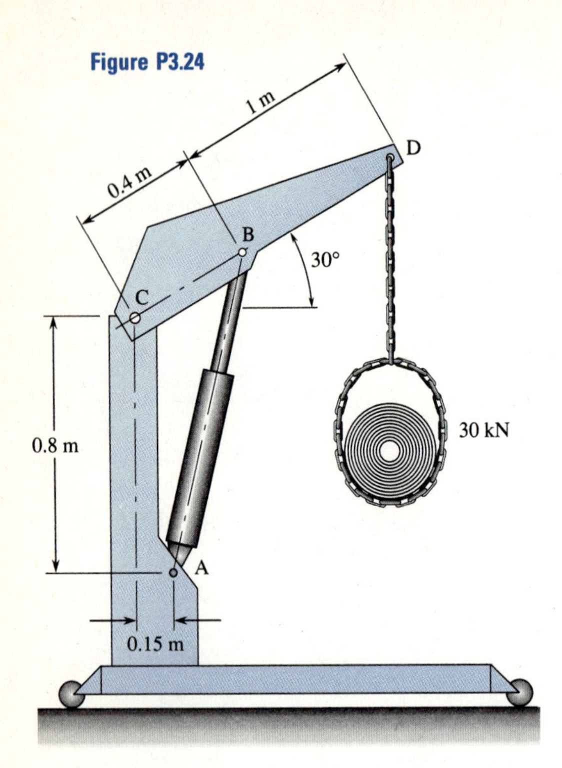

3.25 For the angle bracket shown in Figure P3.5, a 45° fillet weld 100 mm long and 10 mm thick across the top of the angle replaces the bolt. Determine the average shear stress in the weld.

3.26 A cleavis joint for transmitting a tensile force is shown. If the diameter of the rods being connected by the pin is *D*, determine the diameter *d* of the pin. Assume the allowable shear stress in the pin is one-half that of the maximum allowable tensile stress in the rods. This is a common assumption to make.

Figure P3.26

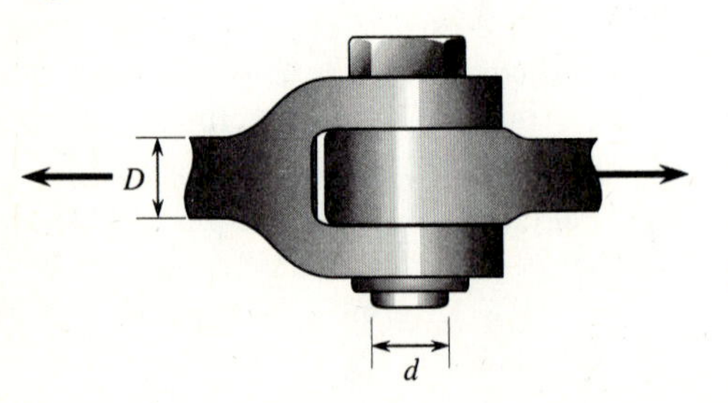

3.27 A torque *T* of 15 kN · m is transmitted between two flanged shafts by means of four 20-mm-diameter bolts. What is the average shear stress in each bolt if the diameter *d* of the bolt circle is 200 mm?

Figure P3.27

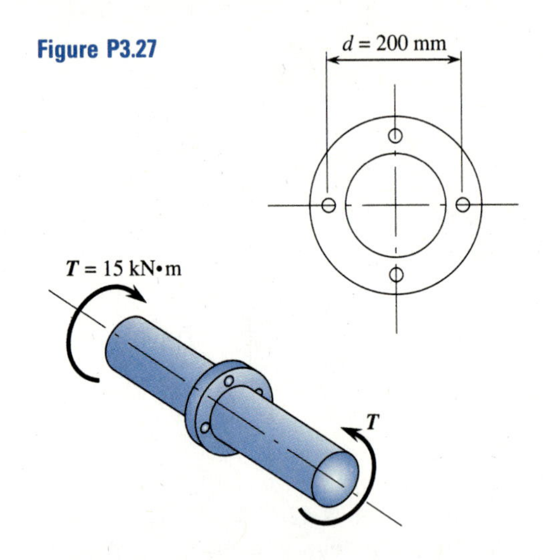

3.28 A key is used to connect a shaft to a hub as shown. The key has a rectangular cross-section of 0.5 in. by 1.00 in. and is 2.00 in. long. If the key fails when the average shear stress in it is 30 ksi what torque in the shaft will fail the key? The shaft diameter is 2 in.

Figure P3.28

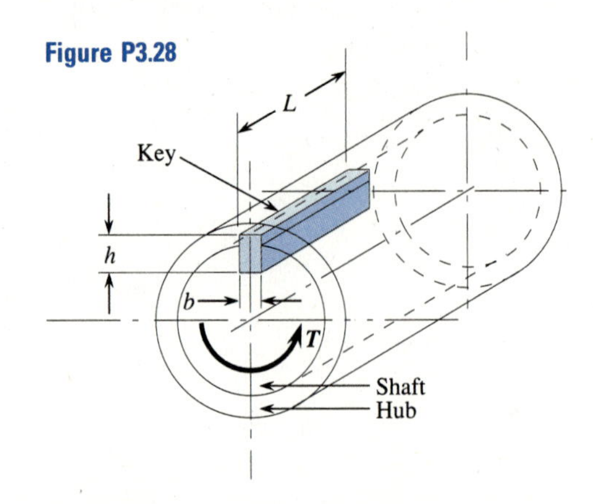

3.29 An angle bracket is attached to a column by 45° fillet welds as shown. The welds are 6 mm thick by 125 mm long. The bracket supports a vertical force of 20 kN. Calculate the average shear stress in the weld.

Figure P3.29

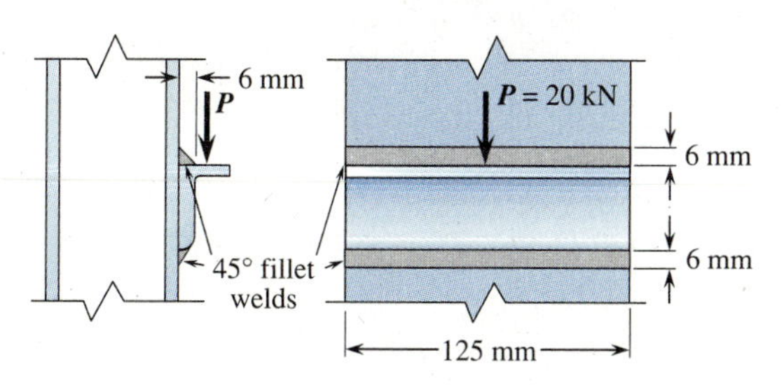

3.30 A water tank with contents weighs 200 kips and is supported by four legs. The load is transferred to the legs by 1/4-in.-thick fillet welds in shear. Determine the distance L up the vertical side of the tank that the pipe legs must extend to obtain sufficient length of weld. The allowable weld shear stress is 14 ksi.

Figure P3.30

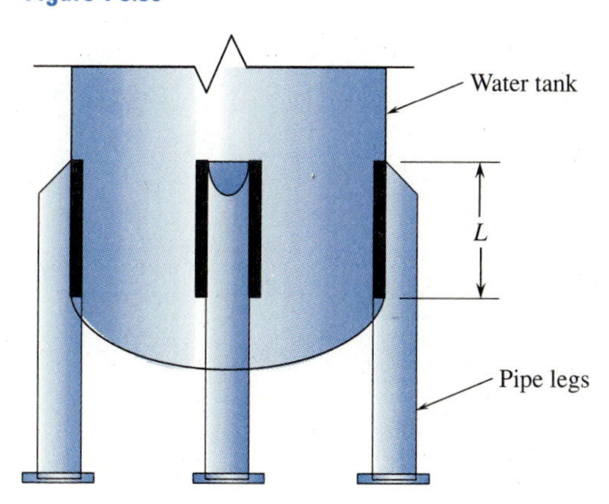

3.31 The log hoist lifts a log weighing 4000 lb. In the position shown, booms AF and EG are at right angles to each other and AF is perpendicular to AB. Determine the average shear stress in the pins at C and D. Both pins are 1 in. in diameter and in double shear.

Figure P3.31

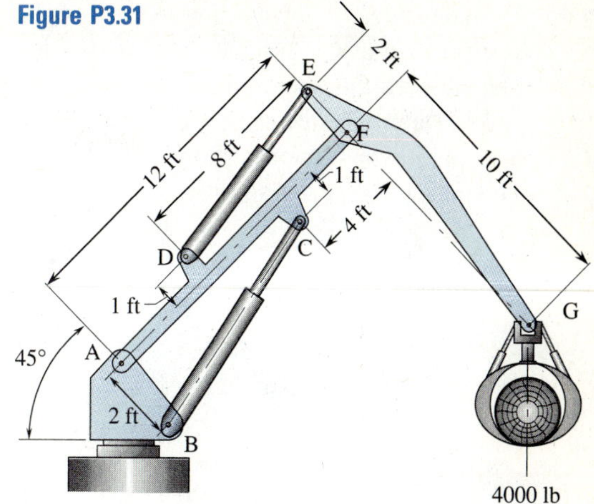

3.32 The movable floor crane in a mill lifts a spool of aluminum sheet weighing 800 lb. For the position shown, determine the shear stress in the pins at B and C. Both pins have diameters of 3/4 in. and are in double shear.

Figure P3.32

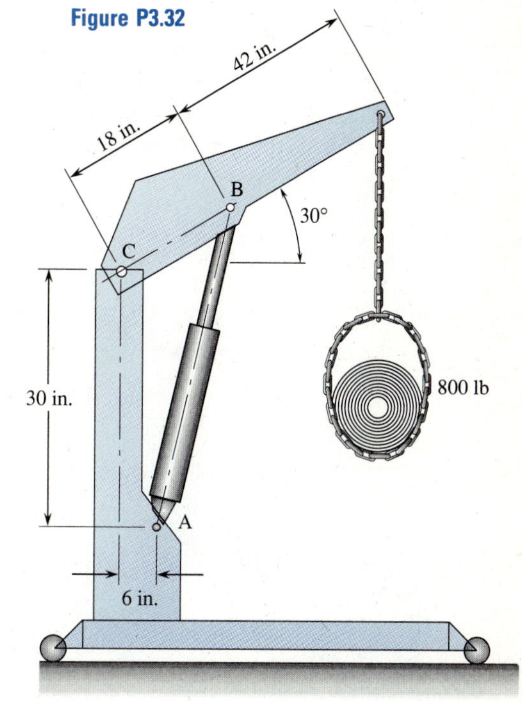

3.33 The step ladder shown supports a 1000 N vertical load at C. Assume there is no friction between the floor and the supports at D and E. Determine the shear stress in the pins at A and B whose diameters are both 15 mm. The pins are in single shear.

Figure P3.33

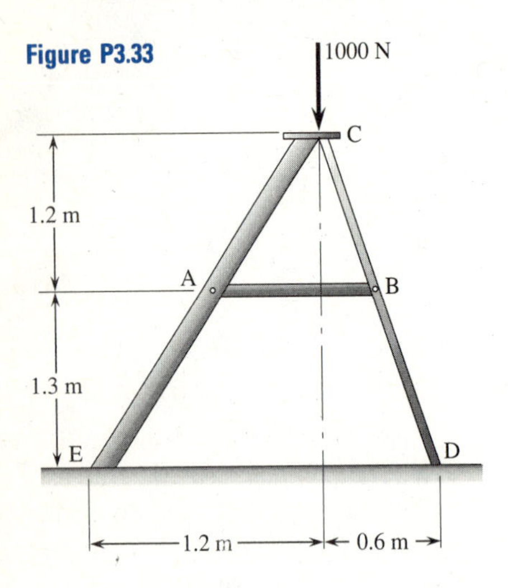

3.34 A support for a beam is made as shown. Determine the required thickness of the projecting ledge t if the maximum allowable shear stress is 12,000 psi. The load on the support is 25,000 lb.

Figure P3.34

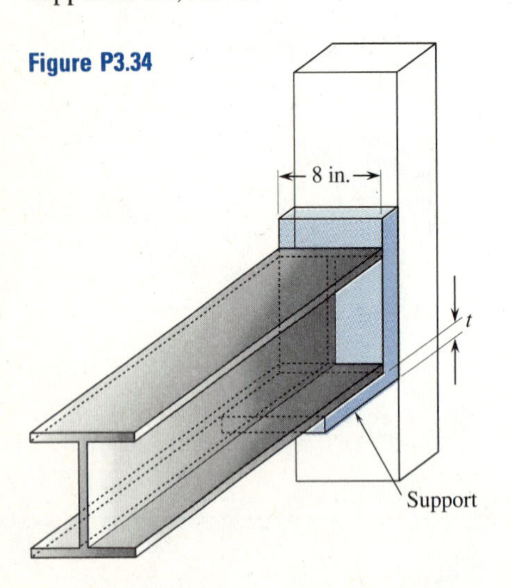

3.35 The machine shown is an overload protection device which releases the load when the shear pin S fails. The released position is also shown. Determine the maximum allowable tension T if the pin S will shear when the shear stress in it is 40 MPa. Also determine the shear stress in the hinge pin A just after pin S has failed. The diameter of pin S is 4 mm and that of pin A is 8 mm. The pins are in single shear. Note upper member BE rests on top of pin S.

Figure P3.35

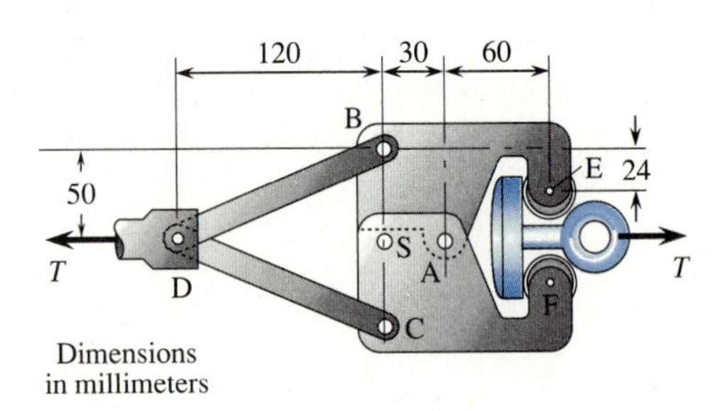

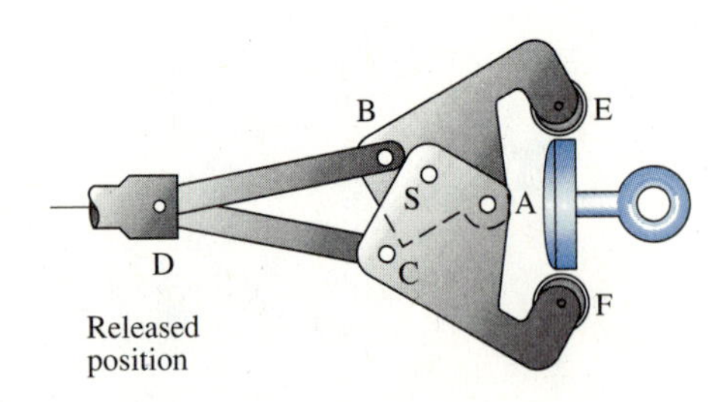

3.36 The crane shown is lifting 30 kN at point D. Determine the average stress in the bolt at A. The bolt is in double shear and has a diameter of 15 mm.

Figure P3.36

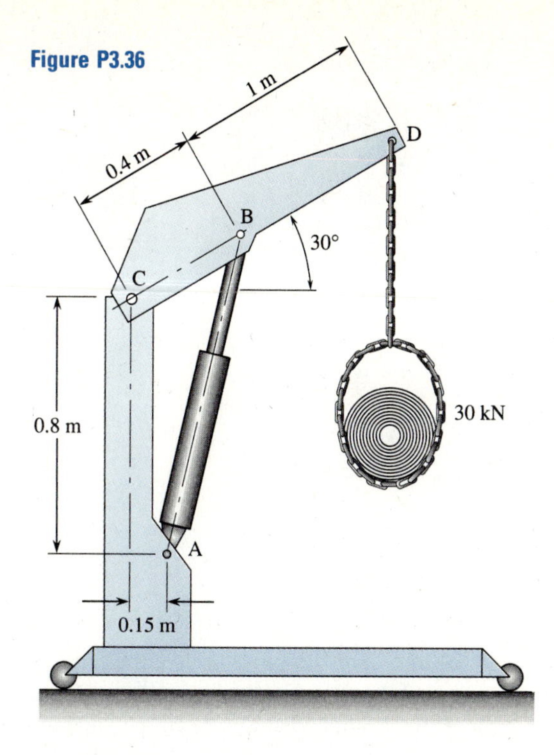

Section 3.3

3.37 A test for determining the shear modulus of rubber consists of bonding two identical pieces of rubber to three steel plates (assumed to be rigid) as shown in Figure P3.37. The center plate is then pulled with force P and the displacement δ in the direction of P is measured. Derive an expression for the shear modulus G in terms of the shear area A of the rubber, the rubber thickness h, and the measured values of P and δ.

Figure P3.37

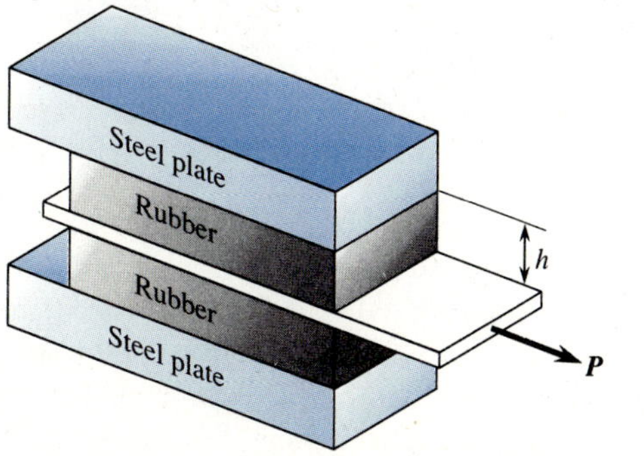

3.38 A plastic block is bonded to a fixed base and to a horizontal rigid plate which is subjected to a horizontal force of $P = 5000$ lb. The shear modulus of the plastic is $G = 50$ ksi. Determine the horizontal deflection of the top edge of the plate.

Figure P3.38

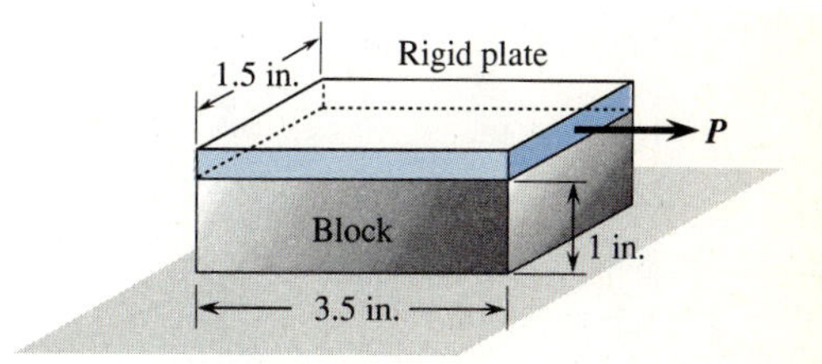

3.39 A vibration isolation support consists of two pieces of hard rubber pads bonded to a middle plate and to two outer rigid supports as shown in Figure P3.39. The rubber has an allowable shear stress of 1.5 MPa and $G = 20$ MPa. A vertical force of $P = 30$ kN acts at the center of the middle plate which causes a 2.5-mm vertical deflection of the middle plate. Determine the required minimum dimensions for the thickness h and length L of the rubber pads.

Figure P3.39

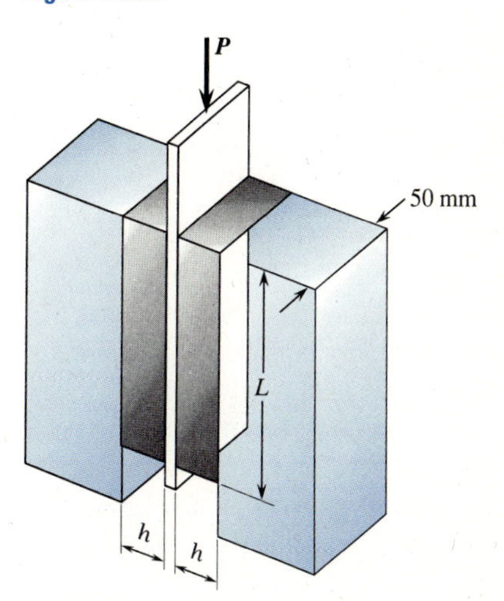

3.40 An elastomeric bearing pad of neoprene rubber supporting a bridge beam must resist a maximum lateral load of $P = 16{,}000$ lb. The lateral deflection must not be greater than $\delta = 0.1$ in. The shear modulus of the rubber is 50 ksi. Determine the maximum thickness h of the rubber. The bearing pad is 4 in. long × 4 in. deep.

Figure P3.40

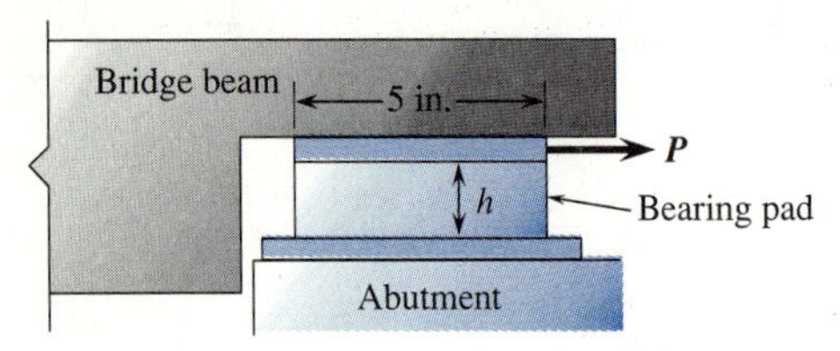

Section 3.4

3.41 Determine the bearing stress between the 2 in. × 2 in. square plate E and the workpiece F when a force of P equal to 40 lb is applied to the handle of the toggle clamp that holds the workpiece in place.

Figure P3.41

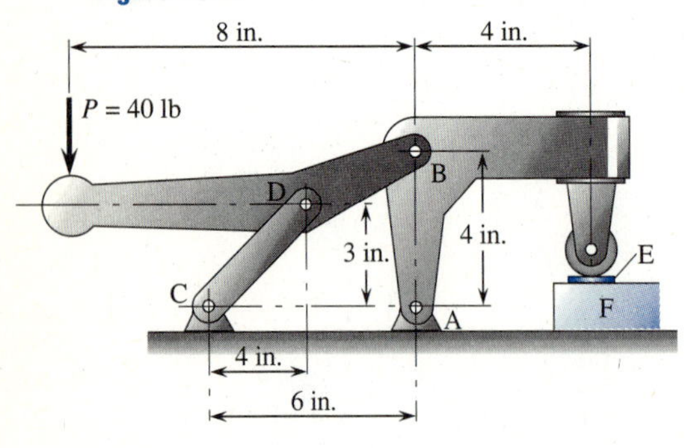

3.42 The vertical shaft shown is supported by a collar bearing and is loaded in compression. The maximum compressive stress in the shaft is 20,000 psi and the average bearing stress between the collar and the bearing is 6000 psi. Determine the diameter D of the collar.

Figure P3.42

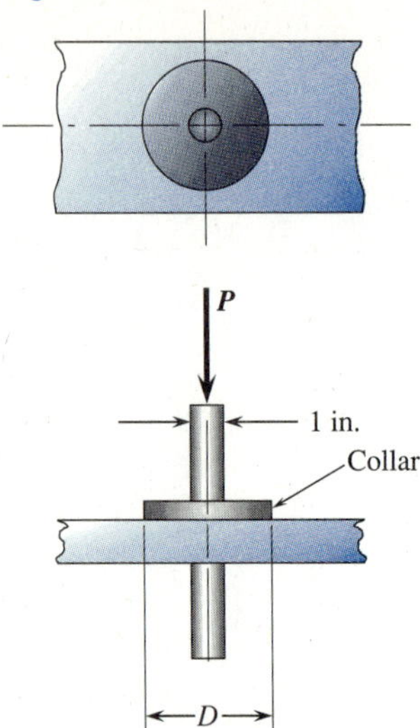

3.43 A 10-mm-diameter pin is used at the connection C of the foot pedal shown. When a force of 800 N is applied at D, determine **(a)** the average shear stress in the pin, **(b)** the average bearing stress in the pedal at C, and **(c)** the average bearing stress in each support bracket at C.

Figure P3.43

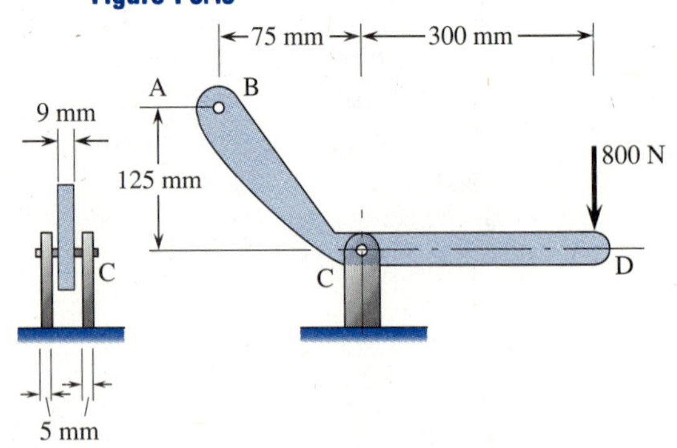

3.44 A wide flange section is used as a column supported on a square base plate and a concrete pedestal as shown. The wide flange is a W 10 × 100. (See Appendix C for dimensions of the W 10 × 100.) The allowable compressive stress in the concrete is 16 MPa and for the steel is 160 MPa. Determine the minimum required dimension d of the base plate if it is to support the maximum load P that the wide flange can carry.

Figure P3.44

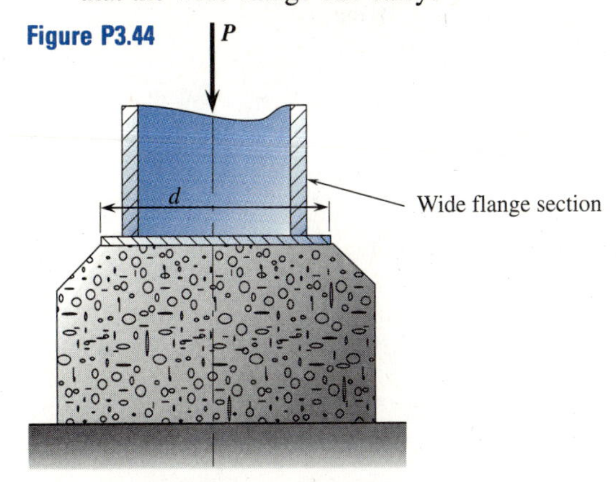

3.45 For the caster shown in Figure P3.8, how thick (t) should the support plates B be if the average bearing stress between them and the 3/8-in.-diameter axle is not to exceed 5000 psi?

3.46 In Problem 3.5, determine the bearing stress between one of the bolts and the angle bracket. Assume the angle bracket has legs of 2 in. × 2 in. and is 1/4 in. thick.

3.47 In Problem 3.7, determine the bearing stress in one of the rivets when $P = 66$ kN. Assume the plates all have thickness of 15 mm.

3.48 A 10 in. × 10 in. wood post transmits a load of 12,000 lb to a concrete footing. Determine **(a)** the bearing stress in the concrete and **(b)** the size of the concrete footing needed if the allowable bearing stress in the soil is 15 psi.

Figure P3.48

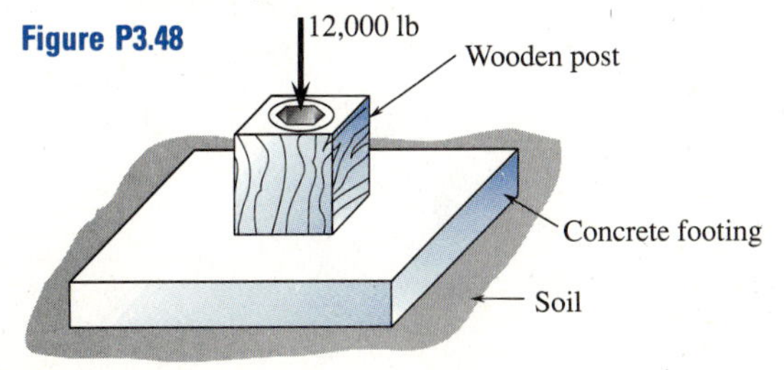

COMPUTER PROBLEMS

The following problems are suggested to be solved using a programmable calculator, microcomputer, or mainframe computer. It is suggested that you write a simple computer program in BASIC, FORTRAN, PASCAL, or some other language to solve these problems.

C3.1 Write a program to obtain the average shear stress in a key used to connect a shaft to a hub as shown. The key is rectangular. The input should include the cross-section dimensions and length of the key, the diameter of the shaft, and the torque to be resisted by the shaft. Check your program using the following data: Key cross-section of 0.5 in. × 1.00 in. and length (L) of 2.00 in., shaft diameter of 2 in., and torque (T) of 30 kip · in.

Figure C3.1

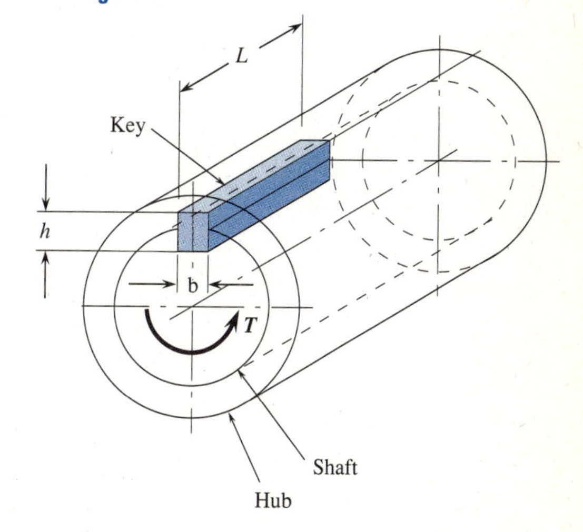

C3.2 Write a program to determine the shear modulus G of a material based on the test procedure shown. The input should include the pulling force P, the displacement δ, the shear area A, and the thickness h of the material. Check your program using the following data: $P = 500$ lb, $\delta = 0.001$ in., $A = 6$ in.^{2}, and $h = 1$ in.

Figure C3.2

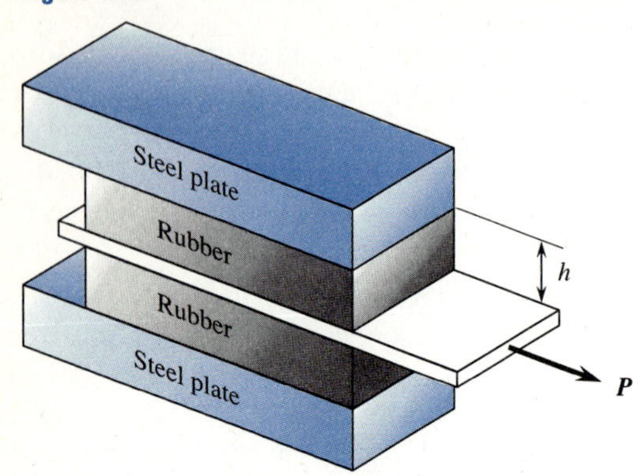

C3.3 Write a program to determine the size of a concrete footing needed if the allowable bearing stress in the supporting soil is given and the size of the column and load transmitted by the column to the footing are known. The input should include the size of the column, the load transmitted by the column, and the allowable bearing stress in the soil. Check your program using the following data: column size 10 in. × 10 in., load transmitted of 12,000 lb, and soil allowable stress of 15 psi.

Figure C3.3

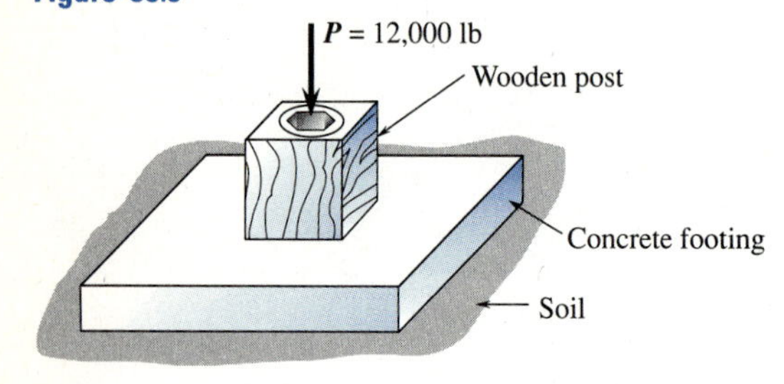

C3.4 Write a program to obtain the average shear stress in a block of wood tested in direct shear using the standard test specimen shown in the figure. The input should include the load P, the width of the specimen, and the height h of the specimen. Check your program using the following data: $P = 500$ lb, width of specimen = 2 in., and height of specimen = 2 in.

Figure C3.4

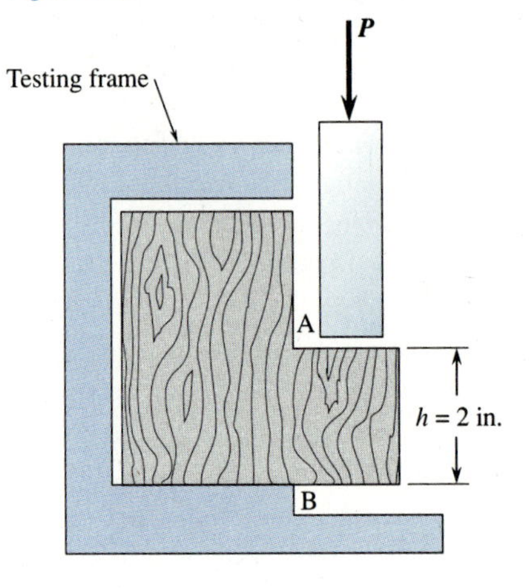

Torsion

4.1 INTRODUCTION

In this chapter, we consider the concept of stresses and strains developed in torsionally loaded members. Torsional members are frequently encountered in structures and machines. Examples include power transmission shafts, drive shafts, axles, ship propeller shafts, torsion bar suspensions, and tractor power takeoff shafts. Figure 4.1 shows some examples of torsionally loaded members. In the analysis of torsionally loaded members, we are primarily concerned with the torsional stress and the angle of twist of the shaft.

We first consider the circular cross-section shaft and describe how to determine stresses and angle of twist of the shaft. We also learn how to analyze statically indeterminate shafts. Next we consider power transmission in circular shafts. This is followed by nonlinear stress distributions in shafts. We then consider noncircular cross-section shafts, such as elliptical and rectangular, and thin-walled closed tubes subjected to torque. Finally, we consider stress concentrations in torsionally loaded members.

4.2 ANGLE OF TWIST—SHEAR STRAIN OF CIRCULAR SHAFT

In this section we consider the angle of twist taking place in a circular shaft when the shaft is subjected to pure torsion. We consider deformation before stress because a very useful relationship between angle of twist and shear strain results, and this relationship is directly used in developing the torsional stress formula in Section 4.3.

To develop the relationship between angle of twist and shear strain, we consider a circular cross-section shaft fixed at its left end and subjected to a torque T (represented by the curved arrow) at its right end (Figure 4.2(a)). The torque is directed through the longitudinal axis x of the shaft. The shaft is in equilibrium as the fixed end support resists the applied torque.

(a)

(b)

(c)

Figure 4.1 Examples of torsionally loaded members
(a) tug boat shaft and propeller
(b) tractor power takeoff
(c) truck drive shaft.

As we apply the torque, the shaft will twist through an angle ϕ at its free end. This angle ϕ is called the *angle of twist* [Figure 4.2(b)].

Two basic deformation characteristics occur when a circular shaft is subjected to torsional loading. Because the circular cross-section has its longitudinal axis as an axis of symmetry, (1) plane cross-sections of the bar rotate as rigid bodies about the longitudinal

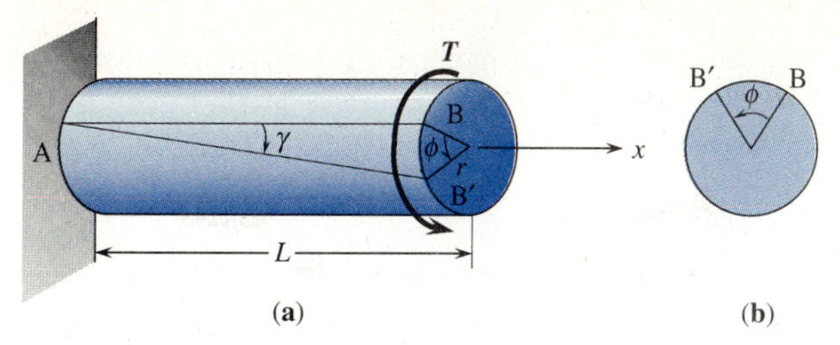

Figure 4.2 (a) Circular shaft subjected to torque, and
(b) cross-section showing angle of twist at free end.

axis and remain plane and circular during and after the torsional loading (whereas in all noncircular cross-sections, such as the square shaft shown in Figure 4.3(c), warping of the cross-section occurs [Figure 4.3(d)]); and (2) diameters of a cross-section remain straight and do not change length during or after the torsional loading.

To determine the shear strain-angle of twist relationship, we consider the circular shaft of length L with left end A fixed and right end B free to twist and subjected to an applied torque T (Figure 4.2(a)). The shaft free end twists through an angle ϕ with point B on the surface moving to B′. Therefore, the angle between lines AB and AB′ on the surface of the shaft is the shear strain γ.

Using the arc length formula in Figure 4.2(a) and assuming small angle γ, we have

$$BB' = L\gamma \tag{a}$$

Similarly using Figure 4.2(b), we observe

$$BB' = r\phi \tag{b}$$

Equating Equations (a) and (b) and solving for γ, we obtain the shear strain at the surface of the shaft as

$$\gamma_r = \frac{r\phi}{L} \tag{4.1}$$

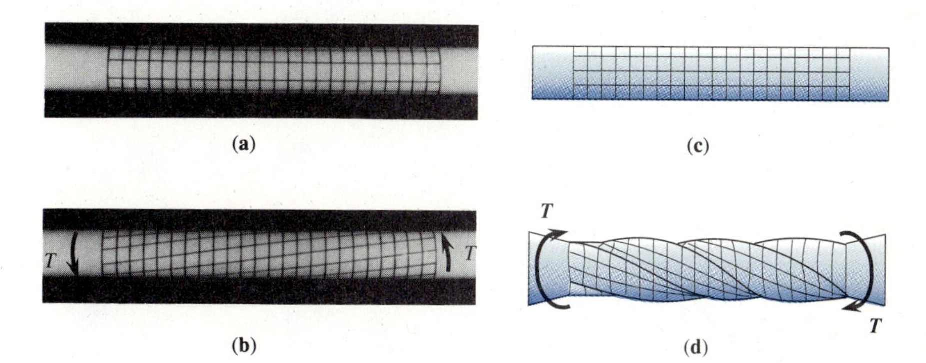

Figure 4.3 Circular shaft (a) before and
(b) after subjected to torque, and square shaft
(c) before and
(d) after subjected to torque.

where γ_r and ϕ are both expressed in radians. Similarly, the shear strain at an interior smaller concentric circle of radius ρ in the shaft is

$$\gamma = \frac{\rho\phi}{L} \tag{4.2}$$

Equation (4.2) indicates that the shear strain varies linearly with the radial distance from the center of the shaft.

Comparing Eqs. (4.1) and (4.2), we observe that γ_r is the maximum shear strain, that is, $\gamma_r = \gamma_{max}$ and we can relate the shear strains γ and γ_{max} by

$$\gamma = \gamma_{max}\frac{\rho}{r} \tag{4.3}$$

4.3 SHEAR STRESS DUE TO TORSION IN CIRCULAR SHAFT

In this section, we will develop the torsion formula relating the torque in a shaft to the shear stress. To obtain the formula, consider the shaft subjected to torque T shown in Figure 4.4(a). For equilibrium there is a reaction torque $T_A = T$.

We pass a plane perpendicular to the axis of the shaft through an arbitrary point B located along the shaft. Looking at the free-body diagram of section AB, small forces dV occur in the plane of the shaft and perpendicular to the radius of the shaft (Figure 4.4(b)). For torsional equilibrium the sum of all the internal torques at B must equal the torque T. Therefore, we write

$$\int \rho\, dV = T \tag{a}$$

where ρ is the radial coordinate locating the perpendicular distance from the axis of the shaft to force dV. Now

$$dV = \tau\, dA \tag{b}$$

where τ is the shear stress acting on the plane of the element of area dA. Using Eq. (b) in (a), we obtain the general *torque-shear stress relationship* given by

$$T = \int \rho(\tau\, dA) \tag{4.4}$$

Equation (4.4) will become useful if we know how the shear stress varies in the radial direction over the cross-section.

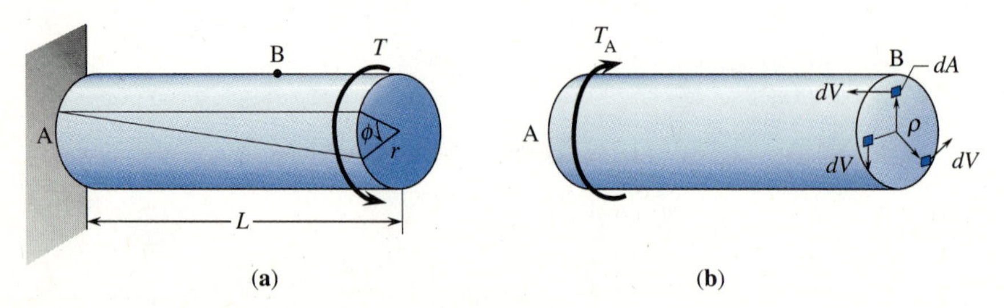

Figure 4.4 (a) Shaft subjected to torque and
(b) free-body diagram of section AB of shaft showing internal shear forces dV.

Linear-Elastic Material Behavior

Let us now consider a material that behaves in its linear-elastic range. From Hooke's law in shear (see Eq. 3.10), we relate the shear stress to the shear strain by

$$\tau = G\gamma \tag{4.5}$$

where G is the shear modulus. Using Eq. (4.3) for γ in (4.5), we have

$$\tau = G\gamma_{\max}\frac{\rho}{r} \tag{4.6}$$

Now since

$$\tau_{\max} = G\gamma_{\max} \tag{4.7}$$

for $\tau_{\max}$ in the linear-elastic range, we use Eq. (4.7) in (4.6) to express the shear stress as a function of ρ given by

$$\tau = \tau_{\max}\frac{\rho}{r} \tag{4.8}$$

To relate the shear stress to the torque, we substitute Eq. (4.8) into (4.4) as follows:

$$T = \int_A \rho\, \tau_{\max}\frac{\rho}{r}\, dA \tag{4.9}$$

A simple torsion test experiment to obtain G of a circular shaft is shown in Figure 4.5. A hanging load W is applied to the pulley connected to the shaft as shown. This load produces a torque $T = Wr$, where r is the radius of the pulley. Using Eq. (4.14), the shear stress is obtained for T. The angle of twist is read directly from the dial attached to the end of the pulley. Then using Eq. (4.1), we obtain the shear strain γ. This procedure can be repeated for increasing increments of weight (or torque). The sets of data of T versus ϕ can be converted to a shear stress versus shear strain plot. The slope of the linear portion of this curve is G.

Now we realize that $\tau_{\max}$ and r are not functions of position, but merely constants. We then simplify Eq. (4.9) to

$$T = \frac{\tau_{\max}}{r}\int_A \rho^2 dA \tag{4.10}$$

From statics, we recall that the polar moment of inertia is

$$J = \int_A \rho^2 dA \tag{4.11}$$

where for a solid circular cross-section J becomes

$$J = \frac{\pi r^4}{2} \qquad (\text{in.}^4 \text{ or m}^4) \tag{4.12}$$

while for a hollow circular cross-section J becomes

$$J = \frac{\pi}{2}(r_o^4 - r_i^4) \qquad (\text{in.}^4 \text{ or m}^4) \tag{4.13}$$

(a)

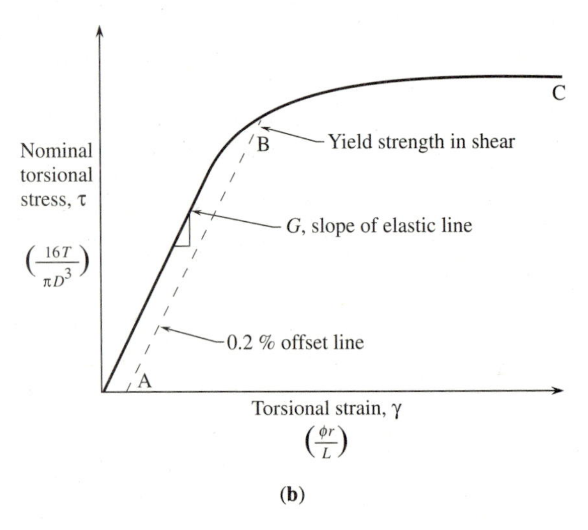

(b)

Figure 4.5 (a) Simple torsion experiment to obtain τ-γ curve and (b) resulting τ-γ diagram.

where r_o is the outer radius and r_i the inner radius of the hollow shaft. Using Eq. (4.11) in (4.10) and then solving for τ_{max}, we obtain

$$\tau_{max} = \frac{Tr}{J} \tag{4.14}$$

Equation (4.14) is called the *torsion formula*.

At any radial distance ρ from the shaft center line, the shear stress is obtained from Eqs. (4.8) and (4.14) as

$$\tau = \frac{T\rho}{J} \tag{4.15}$$

In summary, Eq. (4.14) is used to obtain the maximum shear stress that occurs at the radial distance $\rho = r$ from the shaft center line, while Eq. (4.15) is used to obtain the shear stress at any radial distance ρ from the shaft axis. The torsion formula, Eq. (4.14), applies only when the material behaves in the linear-elastic range. In practice we often use this formula up to $\tau_{max} \leq \tau_Y$, where τ_Y is the yield stress in shear.

STEPS IN SOLUTION

The following steps are used in determining the shear stress in circular cross-section members subjected to torsional loading.

1. Draw a free-body diagram of the section of the shaft under consideration. Then using equations of equilibrium, calculate the internal torque T on the section under consideration.
2. Calculate the polar moment of inertia J for the section of the torsional member under consideration using Eq. (4.12) or (4.13), depending on whether the shaft is a solid or hollow one.
3. Using Eq. (4.14), calculate the maximum shear stress on the section due to this torque. (Remember that this shear stress must be less than the yield stress in shear or else we use a nonlinear analysis [see Section 4.8]).
4. If the shear stress τ at any other radius, ρ, is desired, use Eq. (4.15).

EXAMPLE 4.1

The circular stepped shaft of Figure 4.6 is subjected to torques of $T_A = 5000$ lb · in. and $T_B = 20{,}000$ lb · in. as shown. Determine the maximum stress for both the steel and the brass section of the shaft. The shaft is fixed at the left end.

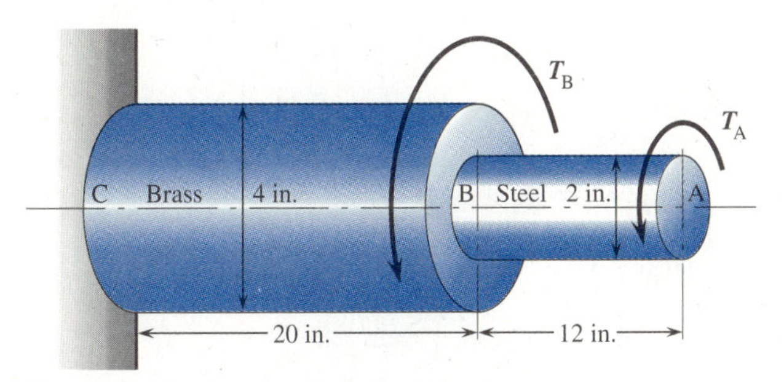

Figure 4.6 Circular stepped shaft subjected to torques.

Solution

Using the free-body diagram of a section of the shaft between A and B, we obtain the internal torque in the steel portion by summing the torques as

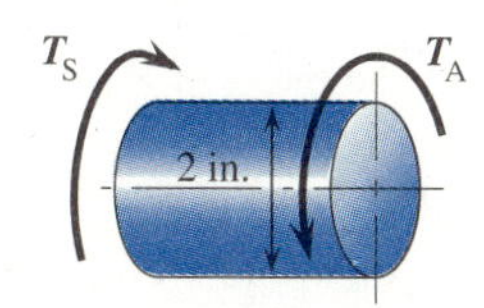

$$\Sigma T = 0: \quad T_S = T_A = 5000 \text{ lb} \cdot \text{in.}$$

For section AB of the shaft, J is obtained from Eq. (4.12) as

$$J_S = \frac{\pi r^4}{2} = \frac{\pi(1 \text{ in.})^4}{2} = 1.571 \text{ in.}^4$$

Therefore, using Eq. (4.14), we obtain the maximum shear stress in the steel shaft as

$$\tau_S = \frac{T_S r_S}{J_S} = \frac{(5000 \text{ lb} \cdot \text{in.})(1 \text{ in.})}{1.571 \text{ in.}^4}$$

$$\tau_S = 3180 \text{ psi}$$

We now pass a plane perpendicular to the axis of the shaft between B and C. The free-body diagram of section BC must resist the combined torque of T_A and T_B as shown. By summing the torques, we obtain

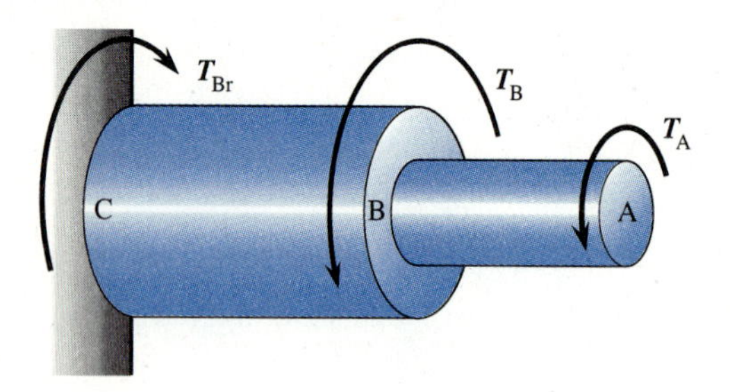

$$T_{Br} = T_A + T_B = 25{,}000 \text{ lb} \cdot \text{in.}$$

For section BC of the shaft, J is

$$J_{Br} = \frac{\pi r^4}{4} = \frac{\pi(2 \text{ in.})^4}{2} = 25.1 \text{ in.}^4$$

Again using Eq. (4.14), the maximum shear stress in the brass is

$$\tau_{Br} = \frac{T_{Br} r_{Br}}{J_{Br}} = \frac{(25{,}000 \text{ lb} \cdot \text{in.})(2 \text{ in.})}{25.1 \text{ in.}^4}$$

$$\tau_{Br} = 1990 \text{ psi}$$

EXAMPLE 4.2

Many circular shafts are hollow due to their method of fabrication. This is done to improve their strength to weight ratio. Figure 4.7 shows a hollow shaft subjected to torques $T_B = 4.0 \text{ kN} \cdot \text{m}$ and $T_C = 1.5 \text{ kN} \cdot \text{m}$. Determine the location and the magnitude of the largest shear stress in this shaft.

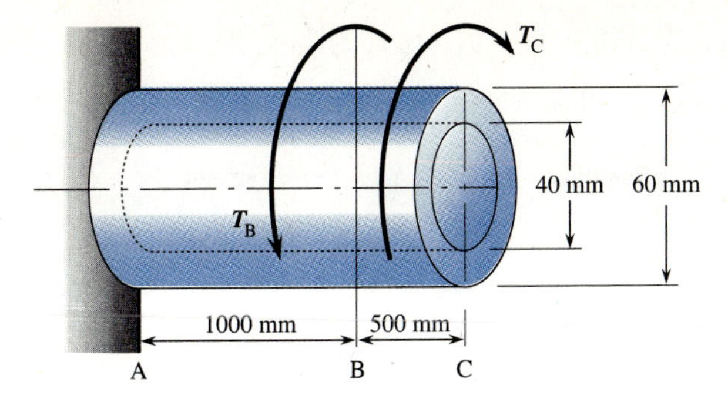

Figure 4.7 Hollow shaft subjected to torques.

Solution

From the free-body diagram of section AB of the shaft, we sum torques to obtain T_{AB} as

$$T_{AB} = T_B - T_C = 4.0 - 1.5 = 2.5 \text{ kN} \cdot \text{m}$$

Similarly, the torque in section BC is simply

$$T_{BC} = T_C = 1.5 \text{ kN} \cdot \text{m}$$

For a hollow circular shaft, we use Eq. (4.13) to obtain the polar moment of inertia as

$$J = \frac{\pi}{2}(r_{\text{outside}})^4 - \frac{\pi}{2}(r_{\text{inside}})^4$$

$$J = \frac{\pi}{2}[(30 \text{ mm})^4 - (20 \text{ mm})^4] = 1.021 \times 10^6 \text{ mm}^4$$

$$J = 1.021 \times 10^{-6} \text{ m}^4$$

Using Eq. (4.14), we obtain the maximum shear stress in section AB as

$$\tau_{AB} = \frac{(2.5 \text{ kN} \cdot \text{m})(0.03 \text{ m})}{1.021 \times 10^{-6} \text{ m}^4} = 73{,}500 \text{ kN/m}^2$$

$$\tau_{AB} = 73{,}500 \text{ kPa} = 73.5 \text{ MPa}$$

(Again, we note that when using metric units it is often preferable to express stress in units of megapascals [MPa].)

Similarly at section BC, we obtain

$$\tau_{BC} = \frac{(1.5 \text{ kN} \cdot \text{m})(0.03 \text{ m})}{1.021 \times 10^{-6} \text{ m}^4} = 44.1 \text{ MPa}$$

Comparing answers, the largest shear stress occurs in section AB.

4.4 OTHER STRESSES DUE TO APPLYING PURE TORQUE

We observed in Section 2.3 that shear stresses occurred in axially loaded members. Similarly, normal stresses occur in torsionally loaded members. Consider the circular shaft of radius r subjected to equal and opposite torques T in Figure 4.8.

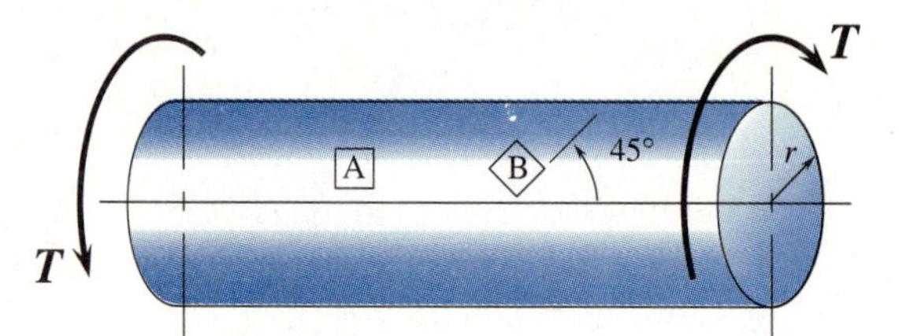

Figure 4.8 Shaft subjected to equal and opposite torques.

Now consider element A on the surface of the shaft of Figure 4.8 and shown in Figure 4.9.

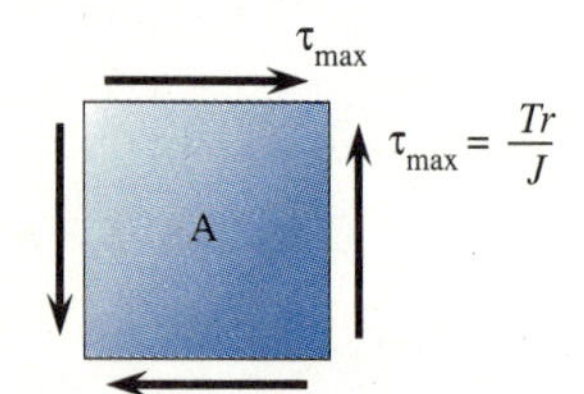

Figure 4.9 Element of material from shaft with stresses acting on it.

Figure 4.9 shows the stresses acting on the vertical and horizontal faces of the element. These are shear stresses obtained using the torsion formula (Eq. (4.14)). The stresses on the horizontal face must equal those on the vertical face to satisfy moment equilibrium.

Now consider element B whose faces are 45° to the axis of the shaft. To obtain the normal stress on a plane at 45° to the axis of the shaft, consider the triangular element shown in Figure 4.10(a). This is element A with a 45° plane passed through it. The vertical edge represents the vertical face. The stress on this face is given by the torsion formula. Multiplying the stresses by the areas they act on, we obtain Figure 4.10(b).

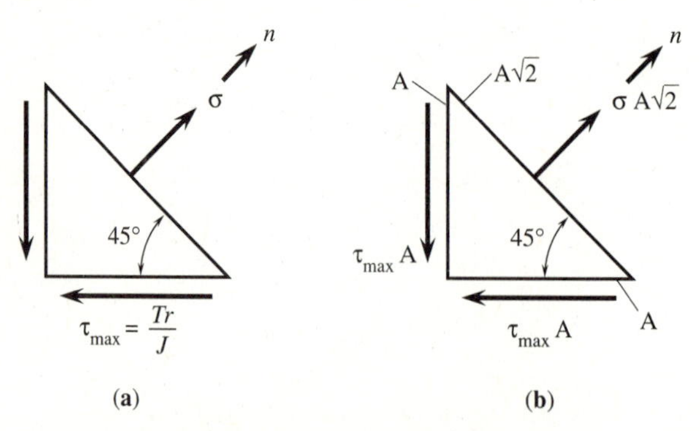

Figure 4.10 (a) Triangular element with stresses acting on it, and (b) same element with forces acting on it.

By summing forces in the n direction, we obtain

$$\Sigma F_n: \quad \sigma(A\sqrt{2}) = 2(\tau_{max}A)\cos 45° = \tau_{max}A\sqrt{2}$$

or

$$\sigma = \frac{\tau_{max}A\sqrt{2}}{A\sqrt{2}} = \tau_{max}$$

Since

$$\tau_{max} = \frac{Tr}{J}$$

we have

$$\sigma = \frac{Tr}{J} \tag{4.16}$$

The normal stress given by Eq. (4.16) can be shown to be the maximum tensile stress. There is a maximum tensile stress acting at 45° counterclockwise from the x-axis of magnitude given by Eq. (4.16) when a shaft is subjected to the pure torque T as applied in Figure 4.8. The largest compressive stress can be shown to act on a plane whose normal is at 45° clockwise from the x-axis for the torques applied in Figure 4.8. Equations (4.14) and (4.16) help to explain the failure of ductile and brittle materials due to torsion. *Ductile materials,* such as structural steel, generally fail in shear. Therefore, when subjected to torque they fail on planes of maximum shear stress that are at right angles to the shaft axis (Figure 4.11(a)).

Brittle materials, such as cast iron and chalk, normally fail in tension. Therefore, when subjected to torque they fail on planes of maximum tensile stress which are at 45° angles to the shaft axis (Figure 4.11(b)). You can observe this type of failure by twisting a piece of chalk until failure results.

(a)

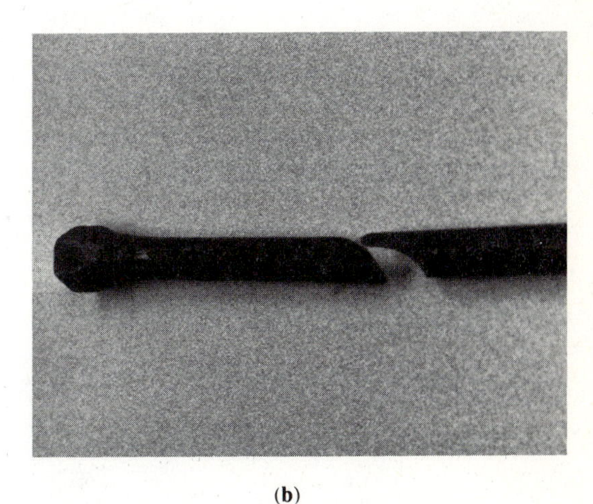

(b)

Figure 4.11 (a) Typical failure of ductile material shaft due to torsion and (b) typical failure of brittle material shaft due to torsion.

In general, we can show for the pure shear element that on a plane whose normal is at angle θ counterclockwise from the x-axis, the normal and shear stresses are

$$\sigma_\theta = \tau \sin 2\theta \quad \text{and} \quad \tau_\theta = \tau \cos 2\theta$$

These relations are similar to Eqs. (2.5) used for an axially loaded member.

EXAMPLE 4.3

A 1-in.-diameter solid circular steel shaft is subjected to a maximum torque $T = 2$ kip · in. Determine the largest tensile stress in the shaft.

Solution

First we determine J using Eq. (4.12) as

$$J = \frac{\pi}{2}(0.5)^4 = 0.0982 \text{ in.}^4$$

Then using Eq. (4.16), we determine the maximum tensile stress as

$$\sigma_{45^\circ} = \frac{Tr}{J} = \frac{(2 \text{ kip} \cdot \text{in.})(0.5 \text{ in.})}{0.0982 \text{ in.}^4}$$

$$\sigma_{45^\circ} = 10.19 \text{ ksi}$$

4.5 ANGLE OF TWIST—TORQUE RELATIONSHIP IN CIRCULAR CROSS-SECTIONS

Shafts, bars, and rods are used to transmit rotational power and torque. In addition to determining the stresses that act on shafts, it is necessary to be able to predict the angle of twist or angular deflection of these members. As an example, many automobile suspension systems use torsion rods and the design engineer must be able to predict in advance the angle of twist. In this section we derive an expression for the angle of twist of a circular shaft (solid or hollow) subjected to a torque about its center line.

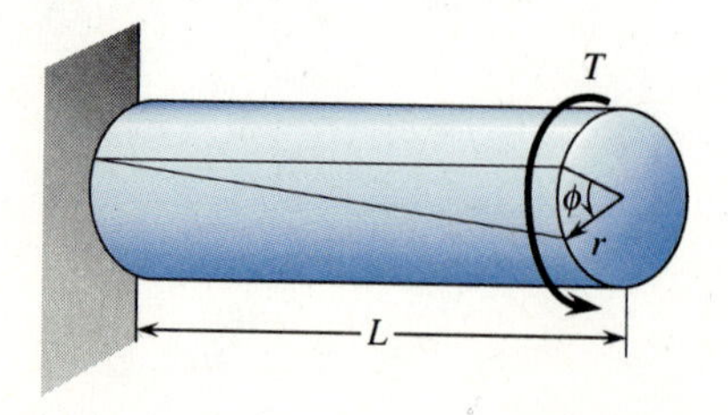

Figure 4.12 Angle of twist of circular shaft.

Figure 4.12 shows such a shaft of length L and radius r. Let the total angle of twist at the free end be ϕ as shown.

In the linear-elastic range of a material behavior the shear strain γ is related to the shear stress τ by Hooke's law in shear. This relationship was introduced as Eq. (3.10) and repeated as Eq. (4.5). Solving Eq. (4.5) for γ and using this result in Eq. (4.1), we obtain

$$\frac{\phi}{L} = \frac{\tau}{Gr} \tag{4.17}$$

Now using the shear stress relation, Eq. (4.14), in Eq. (4.17), we have

$$\frac{\phi}{L} = \frac{T}{GJ} \tag{4.18}$$

Equation (4.18) is the rate of change of angular twist over the length L. Solving Eq. (4.18) for the angle of twist, we have

$$\phi = \frac{TL}{GJ} \tag{4.19}$$

Equation (4.19) is analogous to Eq. (2.21), which is used to obtain the axial deflection for members subjected to axial loading. That is, ϕ and δ are displacements, ϕ rotational, δ longitudinal; T and F are torsional load and axial load, respectively; L is the same in either formula; G is analogous to E; and J is analogous to A. But Eq. (4.19) applies only to circular cross-sections, while Eq. (2.21) for axial deflection applies to any cross-sectional shape. Equation (4.19) is often called the *torque-twist formula*.

For the case of a stepped shaft (Figure 4.13), with different but constant Js in various portions of the shaft, or if the shaft has different constant Gs in various portions of the shaft, or if there are different concentrated torques on the shaft (Figure 4.13), then Eq. (4.19) must be used for individual portions of the shaft as

$$\phi = \sum_{i=1}^{N} \frac{T_i L_i}{G_i J_i} \tag{4.20}$$

where N is the number of times the formula is applied.

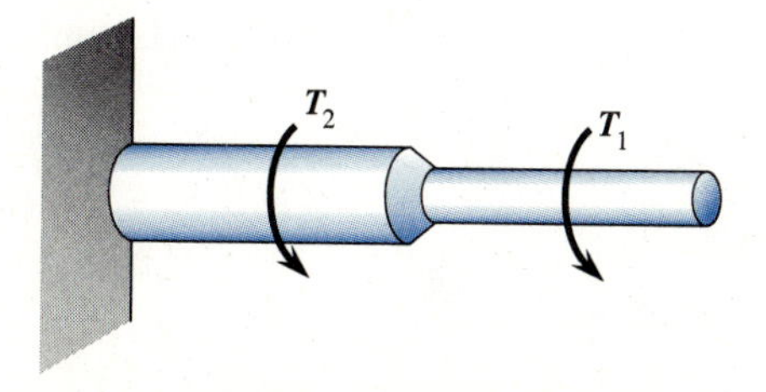

Figure 4.13 Stepped shaft with different concentrated torques.

In the general case of a shaft with variable cross-section, J is a function of x and Eq. (4.19) is applied to a small length of shaft dx. The angle of rotation of one face of the small length with respect to the other is $d\phi$ given by

$$d\phi = \frac{T dx}{GJ} \tag{4.21}$$

On integrating Eq. (4.21) in x from 0 to L, we obtain

$$\phi = \int_0^L \frac{T}{GJ} dx \tag{4.22}$$

STEPS IN SOLUTION

The following steps are used to obtain the angle of twist in circular cross-section shafts subjected to torque.

1. Cut as many sections and draw the free-body diagrams as necessary considering changes in cross-section, material Gs, and discontinuities in torque as described preceding Eq. (4.20), to determine the internal torques T.
2. Determine the polar moment of inertia J in the sections.
3. Use appropriate Eqs. (4.19), (4.20), or (4.22) to determine the angle of twist of the shaft.

EXAMPLE 4.4

Determine the angle of twist at the free end A of the solid steel circular shaft subjected to the torque shown. The shaft has a radius of 1 in. Let $G = 12 \times 10^3$ ksi.

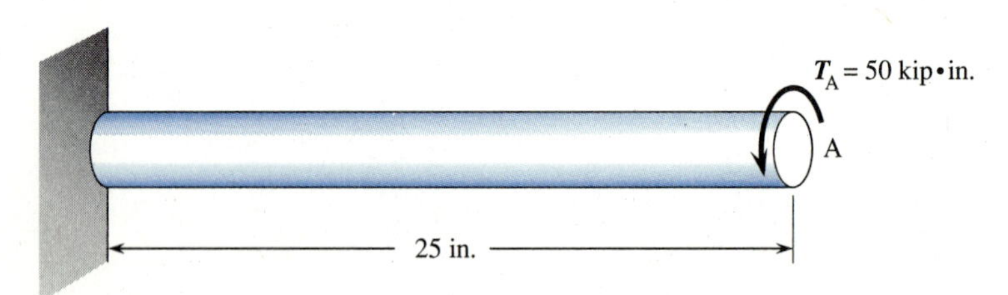

Figure 4.14 Shaft subjected to end torque.

Solution

Cut the member, draw a free-body diagram of the cut section, and determine the internal torque.

T

T_A = 50 kip•in.

$$\Sigma T = 0: \quad T = T_A = 50 \text{ kip} \cdot \text{in.}$$

Determine the polar moment of inertia using Eq. (4.12) as

$$J = \frac{\pi}{2} r^4 = \frac{\pi}{2} (1)^4 = 1.571 \text{ in.}^4$$

The torque is constant throughout the uniform shaft of length L. Therefore using Eq. (4.19), we obtain the total angle of twist as

$$\phi = \frac{TL}{GJ} = \frac{(50 \text{ kip} \cdot \text{in.})(25 \text{ in.})}{(12 \times 10^3 \text{ ksi})(1.571 \text{ in.}^4)}$$
$$= 0.0663 \text{ rad}$$

or

$$\phi = 0.0663 \text{ rad} \left(\frac{360°}{2\pi}\right) = 3.80° \quad \text{(counterclockwise looking in from the right end)}$$

EXAMPLE 4.5

For the stepped solid steel American Iron and Steel Institute (AISI) 1040 steel machine shaft subjected to the torques shown in Figure 4.15, determine the angle of twist of end D with respect to end A. Let $G = 70$ GPa. The diameters of sections AB and CD are 30 mm and the diameter of section BC is 50 mm.

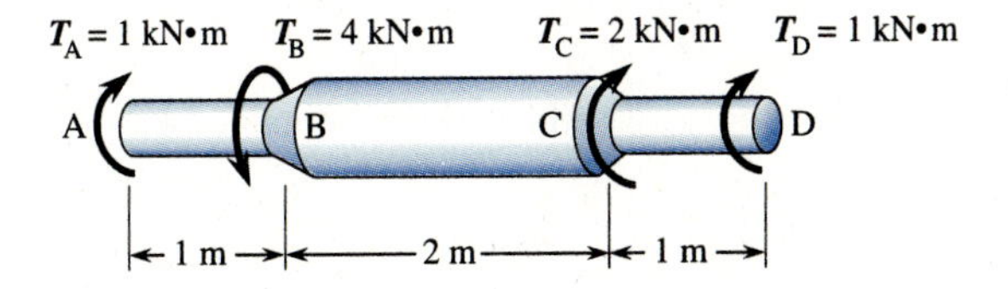

Figure 4.15 Stepped shaft subjected to torque.

Solution

First we cut sections one at a time through AB, BC, and CD to determine the internal torque in each section.

Section AB

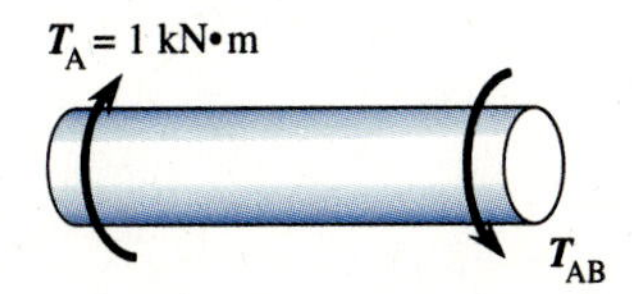

By statics

$$\Sigma T = 0: \quad T_{AB} = 1 \text{ kN} \cdot \text{m}$$

Section BC

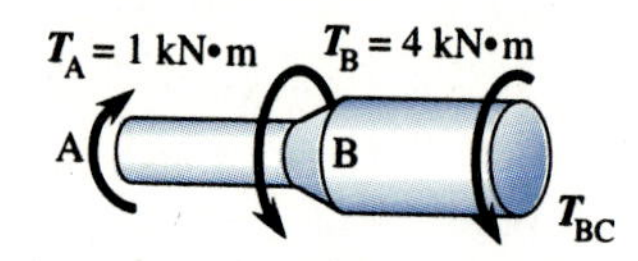

$$\Sigma T = 0: \quad T_{BC} = -3 \text{ kN} \cdot \text{m}$$

Section CD

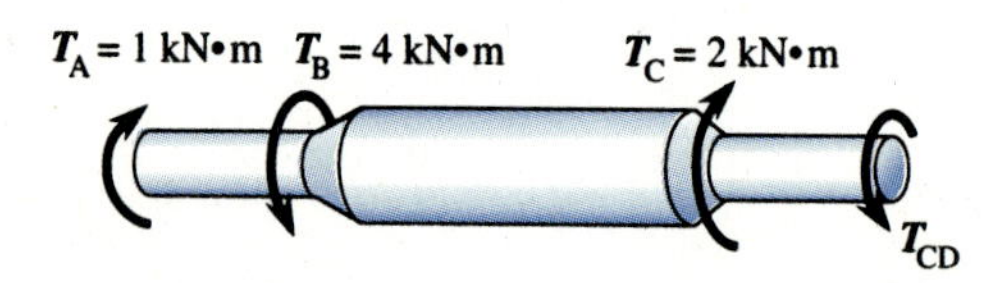

$$\Sigma T = 0: \quad T_{CD} = 1 + 2 - 4 = -1 \text{ kN} \cdot \text{m}$$

Next we calculate the polar moment of inertia for each section using Eq. (4.12) as follows:

$$J_{AB} = J_{CD} = \frac{\pi}{2} r^4 = \frac{\pi}{2}(15 \text{ mm})^4 = 79{,}500 \text{ mm}^4 = 0.0795 \times 10^{-6} \text{ m}^4$$

$$J_{BC} = \frac{\pi}{2}(25 \text{ mm})^4 = 614{,}000 \text{ mm}^4 = 0.614 \times 10^{-6} \text{ m}^4$$

The angle of twist of end D with respect to end A is then obtained using Eq. (4.20) as

$$\phi_{D/A} = \left(\frac{TL}{GJ}\right)_{B/A} + \left(\frac{TL}{GJ}\right)_{C/B} + \left(\frac{TL}{GJ}\right)_{D/C}$$

$$= \frac{1}{70 \times 10^6 \text{ kN/m}^2}\left[\frac{(1 \text{ kN} \cdot \text{m})(1 \text{ m})}{0.0795 \times 10^{-6} \text{ m}^4} + \frac{(-3 \text{ kN} \cdot \text{m})(2 \text{ m})}{0.614 \times 10^{-6} \text{ m}^4} + \frac{(-1 \text{ kN} \cdot \text{m})(1 \text{ m})}{0.0795 \times 10^{-6} \text{ m}}\right]$$

$$= \frac{1}{70}(12.58 - 9.77 - 12.58)$$

$$= -0.140 \text{ rad} = -8.02^\circ \quad \text{(or clockwise looking in from the right end)}$$

EXAMPLE 4.6

Determine the angle of twist at: **(a)** section D located 0.8 m from the fixed end of the hollow shaft made of 6061–T6 alloy aluminum shown in Figure 4.16; and **(b)** at the free end C.

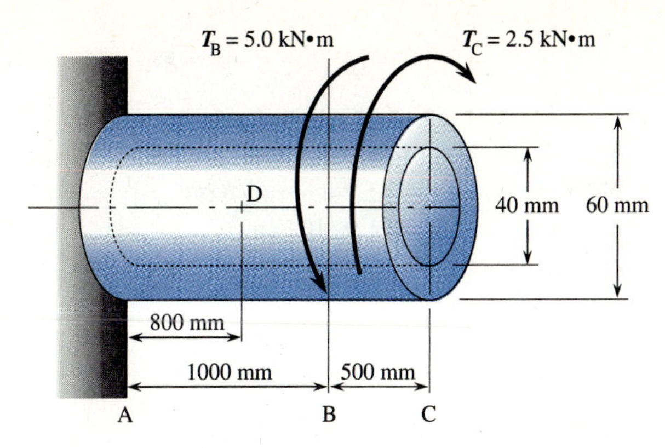

Figure 4.16 Hollow shaft subjected to torques.

Solution

(a) By statics, the torque at section D is

$$T_{AB} = T_D = T_B - T_C = (5.0 - 2.5)\ \text{kN} \cdot \text{m}$$

$$T_D = 2.5\ \text{kN} \cdot \text{m} \qquad \textbf{(a)}$$

The polar moment of inertia for this cross-section was found previously in Example 4.2 to be $J = 1.021 \times 10^{-6}\ \text{m}^4$.

From Eq. (4.19), the angle of twist at section D is therefore

$$\phi_D = \frac{T_D L}{GJ} \qquad \textbf{(b)}$$

Substituting $L = 0.8$ m, $G = 3.7 \times 10^6$ psi (from Appendix B), or $G = (3.7 \times 10^6\ \text{psi}) \times 6.895\ \text{kPa/psi} = 25.5 \times 10^6$ kPa, and the value for T_D given by Eq. (a) into Eq. (b), we have

$$\phi_D = \frac{(2.5\ \text{kN} \cdot \text{m})(0.8\ \text{m})}{(25.5 \times 10^6\ \text{kN/m}^2)(1.021 \times 10^{-6}\text{m}^4)}$$

$$\phi_D = 0.0768\ \text{rad} = 4.40^\circ$$ (counterclockwise looking in from the right end)

(b) To obtain the angle of twist of end C, we need to obtain the torque in section BC. By statics

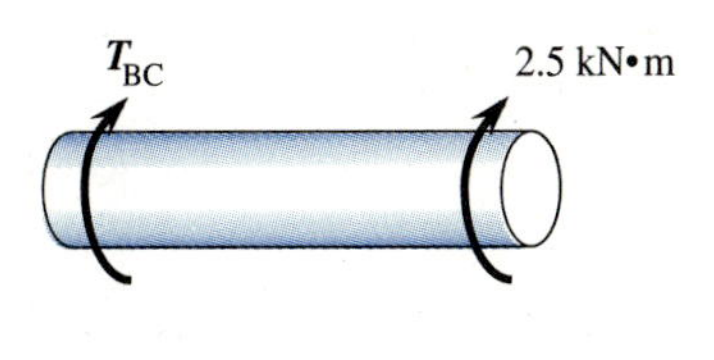

$$T_{BC} = -2.5\ \text{kN} \cdot \text{m}$$

Now using Eq. (4.20), we obtain the angle of twist at C as

$$\phi_C = \left(\frac{TL}{GJ}\right)_{B/A} + \left(\frac{TL}{GJ}\right)_{C/B}$$

$$= \frac{(2.5\ \text{kN}\cdot\text{m})(1\ \text{m}) + (-2.5\ \text{kN}\cdot\text{m})(0.5\ \text{m})}{(25.5 \times 10^6\ \text{kPa})(1.021 \times 10^{-6}\ \text{m}^4)}$$

$$= 0.0480\ \text{rad} = 2.75^\circ$$ (counterclockwise looking in from the right end)

EXAMPLE 4.7

An automotive engineer is designing a front suspension system and wants to use a torsion bar 3 ft long. If the material is to be alloy steel having a shear modulus of $G = 12 \times 10^6$ psi, determine the required diameter of the solid bar so that the resulting torsional spring constant (or torsional stiffness) of the bar will be 265 lb · in. per degree of angular rotation of the bar.

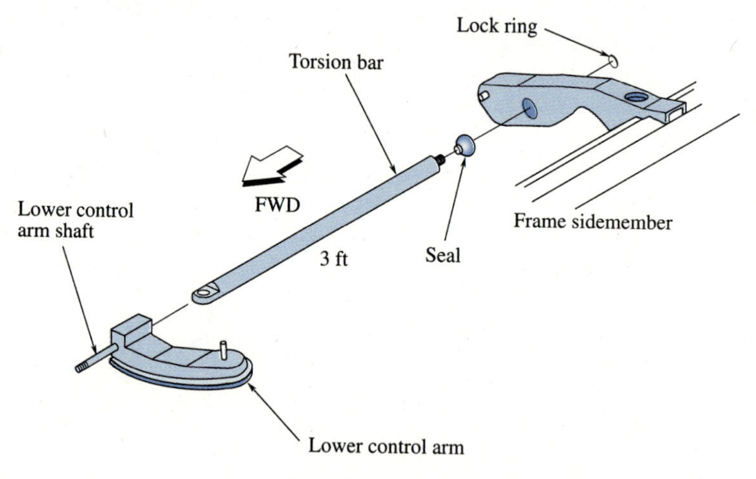

Typical torsion bar suspension

Solution

This given torsional stiffness will presumably give the automobile the desired stiffness in the front suspension, for instance, for optimal handling and is defined as the ratio of torque required on the end of the bar to angle of twist or

$$[\text{Torsional stiffness}] = K_T = \frac{T}{\phi}$$

From Eq. (4.19), the torsional stiffness is

$$K_T = \frac{T}{\phi} = \frac{GJ}{L} \tag{a}$$

Solving for J, we obtain

$$J = \frac{TL}{\phi G} = K_T \frac{L}{G}$$

Now $\dfrac{T}{\phi} = 265$ lb. · in./deg, therefore

$$J = \left(265 \, \frac{\text{lb.} \cdot \text{in.}}{\text{deg}}\right)\left(\frac{180}{\pi} \, \frac{\text{deg}}{\text{rad}}\right)\left(\frac{36 \text{ in.}}{12 \times 10^6 \text{ psi}}\right)$$

or

$$J = 0.0455 \text{ in.}^4 \tag{b}$$

Substituting Eq. (b) into Eq. (4.12) and solving for the radius r, we obtain

$$\frac{\pi}{2} r^4 = 0.0455 \text{ in.}^4$$

$$r = 0.413 \text{ in.}$$

The desired diameter of the torsion bar is then

$$d = 2r = 2(0.413 \text{ in.}) = 0.826 \text{ in.}$$

In practice, a $\frac{7}{8}$-in.-diameter bar would be used.

4.6 STATICALLY INDETERMINATE SHAFTS

We have shown in the previous sections of this chapter that to determine torsional stresses and/or angles of twist in a shaft it is necessary to determine the torque at any location in this shaft. In the torsional problems we have solved in the previous sections, we were able to determine the torque at any cross-section using only the equations of static equilibrium (specifically the sum of torques equal to zero). In many problems, however, the equations of statics alone are not sufficient to solve for the torques in the shaft. These kinds of problems are then called *statically indeterminate,* or *redundant,* and require that we utilize additional information concerning the geometry of deformation of the shaft. This geometry of deformation results in compatibility equations similar to those used in Section 2.7 when analyzing statically indeterminate axially loaded members.

STEPS IN SOLUTION

The following steps are used to solve statically indeterminate shaft problems.

1. Draw a free-body diagram of the shaft and write the torsional equilibrium equation. There will be more unknown torques than can be obtained from the torsional equilibrium equation.
2. Consider the geometry of deformation of the shaft. Often an equation based on the following geometry of deformation can be written.
   a. For a shaft fixed at both ends, the total angle of twist is zero.
   b. For a composite shaft, a shaft made of two or more different materials (say an inner shaft and an outer shell bonded together), the angle of twist for each shaft will be equal.
3. Then use the torque-angle of twist equation such as Eq. (4.19) to express the additional compatibility equation in terms of the unknown torques.
4. Use the compatibility and equilibrium equations to obtain the torques.

EXAMPLE 4.8

The stepped shaft shown is fixed at both ends and is subjected to a torque $T_B = 2000$ lb · ft at section B. If the material is hot-rolled AISI 1040 steel, determine the shear stress in both sections of the shaft and the angle of twist at section B.

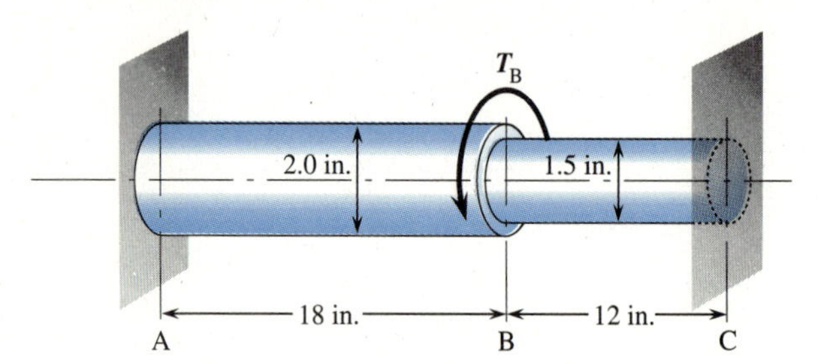

Figure 4.17 Shaft with both ends fixed.

Solution

Since this shaft is supported at both ends we cannot directly determine the torque in both sections from torsional equilibrium as we have two unknowns, the torques (T_A and T_C) and only one torsional equilibrium equation. The shaft is then statically indeterminate to the first degree. However, considering the geometry of deformation for the shaft, the additional equation needed to solve the problem is obtained.

First draw a free-body diagram of the shaft and express the torsional equation of equilibrium as

$$T_B = T_C + T_A \tag{a}$$

Next, considering the geometry of deformation, we observe that the total angle of twist is zero because the shaft is fixed at both ends. Therefore, we have the additional equation (called a compatibility equation) given by

$$\phi_{\text{total}} = 0 \tag{b}$$

or

$$\phi_{B/A} + \phi_{C/B} = 0 \tag{c}$$

Now using Eq. (4.19) relating internal torque to angle of twist we have

$$\phi_{B/A} = \frac{T_{BA} L_{BA}}{GJ_{BA}} \quad \text{and} \quad \phi_{C/B} = \frac{T_{CB} L_{CB}}{GJ_{CB}} \tag{d}$$

Substituting Eqs. (d) into Eq. (c), we have

$$\frac{T_{BA} L_{BA}}{J_{BA}} + \frac{T_{CB} L_{CB}}{J_{CB}} = 0 \tag{e}$$

The internal torques T_{BA} and T_{CB} can be expressed in terms of the reaction torques by drawing free-body diagrams of sections of the shaft as shown. The resulting equations are

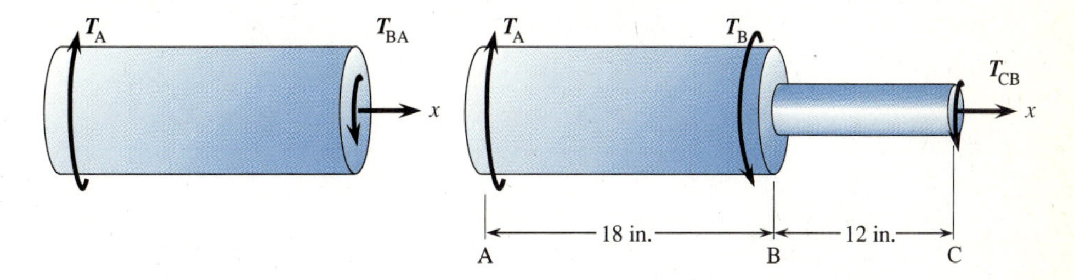

$$\begin{aligned} T_{BA} &= T_A \\ T_{CB} &= T_A - T_B = -T_C \end{aligned} \tag{f}$$

Substituting Eqs. (f) into Eq. (e), we obtain

$$\frac{T_A L_{BA}}{J_{BA}} + \frac{(T_A - T_B) L_{CB}}{J_{CB}} = 0 \tag{g}$$

Next using Eq. (4.12), we obtain the polar moments of inertia of each section of the shaft as

$$J_{AB} = \frac{\pi r^4}{2} = \frac{\pi (1 \text{ in.})^4}{2} = 1.571 \text{ in.}^4$$

$$J_{BC} = \frac{\pi \left(\frac{1.5 \text{ in.}}{2}\right)^4}{2} = 0.497 \text{ in.}^4$$

Therefore, substituting numerical values into Eq. (g) and solving for T_A, we have

$$\frac{T_A (18 \text{ in.})}{1.571 \text{ in.}^4} + \frac{(T_A - 24{,}000)(12 \text{ in.})}{0.497 \text{ in.}^4} = 0$$

or

$$T_A = 16{,}280 \text{ lb} \cdot \text{in.}$$

Using Eq. (a), we have

$$T_C = 24{,}000 - 16{,}280$$

$$T_C = 7720 \text{ lb} \cdot \text{in.}$$

From either of the two Eqs. (d), we can solve for the angle of twist at B relative to either one of the fixed ends. For instance,

$$\phi_{B/A} = \frac{T_A L_{BA}}{G J_{BA}} \tag{h}$$

Using G = 12×10^6 psi (from Appendix B) and other appropriate numeric values in Eq. (h), yields

$$\phi_{B/A} = \frac{(16{,}280 \text{ lb} \cdot \text{in.})(18 \text{ in.})}{(12 \times 10^6 \text{ psi})(1.571 \text{ in.}^4)} = 0.0155 \text{ rad} = 0.888°$$

Using Eq. (4.14), we solve for the maximum shear stress in each section of the shaft as

$$\tau_{AB} = \frac{T_A r_{AB}}{J_{AB}} = \frac{(16{,}280 \text{ lb} \cdot \text{in.})(1 \text{ in.})}{1.571 \text{ in.}^4} = 10{,}360 \text{ psi}$$

$$\tau_{BC} = \frac{T_C r_{BC}}{J_{BC}} = \frac{(7720 \text{ lb} \cdot \text{in.})\left(\frac{1.5}{2} \text{ in.}\right)}{0.497 \text{ in.}^4} = 11{,}650 \text{ psi}$$

EXAMPLE 4.9

A composite shaft is fabricated by attaching a rigid plate to a steel tube and a brass bar as shown in Figure 4.18. If this system is subjected to a torque of 20 kip · in., determine the maximum shear stress in the tube and in the bar and also determine the angular twist of both members.

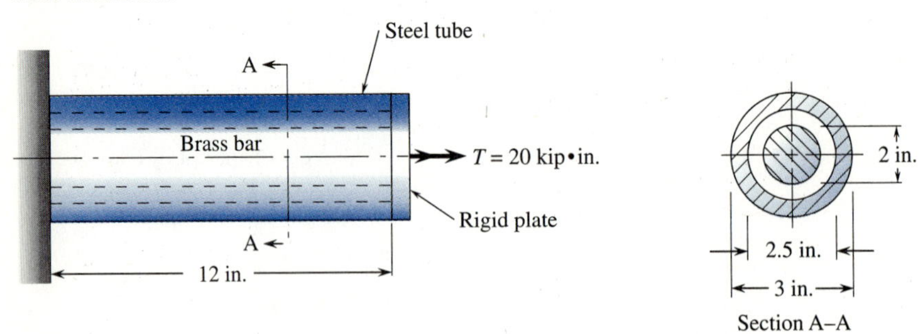

Figure 4.18 Composite shaft subjected to torque.

Solution

From statics we know that the applied torque is resisted by both the brass and steel shafts. Therefore

$$T = T_B + T_S$$

or

$$20 \text{ kip} \cdot \text{in.} = T_B + T_S \quad \textbf{(a)}$$

Equation (a) yields one equation with two unknown torques.

From the geometry of deformation, we observe that the bar and the tube will have the same angle of twist at the right end because they are both connected to the rigid plate. Therefore, we can express the compatibility equation as

$$\phi_B = \phi_S \quad \textbf{(b)}$$

Using Eq. (4.19), we express Eq. (b) as

$$\phi_B = \frac{T_B L_B}{G_B J_B} = \phi_S = \frac{T_S L_S}{G_S J_S} \quad \textbf{(c)}$$

Because the lengths of the two materials are the same, Eq. (c) becomes

$$\frac{T_B}{G_B J_B} = \frac{T_S}{G_S J_S} \quad \textbf{(d)}$$

Equations (a) and (d) yield two independent equations to solve for the two unknowns, T_B and T_S.

The polar moments of inertia for the solid brass bar and the steel tube are

$$J_B = \frac{\pi(1 \text{ in.})^4}{2} = 1.571 \text{ in.}^4$$

$$J_S = \frac{\pi[(1.5 \text{ in.})^4 - (1.25 \text{ in.})^4]}{2} = 4.117 \text{ in.}^4$$

The shear modulus of each material is obtained from Appendix B as $G_B = 5.6 \times 10^6$ psi and $G_S = 11.5 \times 10^6$ psi.

We now solve Eqs. (a) and (d) simultaneously to obtain

$$T_B = 3.13 \text{ kip} \cdot \text{in.}$$

and

$$T_S = 16.87 \text{ kip} \cdot \text{in.}$$

Using Eq. (4.14), we calculate the maximum torsional stresses as

$$\tau_B = \frac{(3.13 \text{ kip} \cdot \text{in.})(1 \text{ in.})}{1.571 \text{ in.}^4} = 1.99 \text{ ksi}$$

$$\tau_S = \frac{(16.87 \text{ kip} \cdot \text{in.})(1.5 \text{ in.})}{4.117 \text{ in.}^4} = 6.15 \text{ ksi}$$

The angle of twist at the free (right) end is obtained from Eq. (c) as

$$\phi = \frac{(3.13 \text{ kip} \cdot \text{in.})(12 \text{ in.})}{(5.6 \times 10^3 \text{ ksi})(1.571 \text{ in.}^4)}$$

$$\phi = 4.27 \times 10^{-3} \text{ rad} = 0.245°$$

4.7 POWER TRANSMISSION

In addition to transmitting static torque, shafts are typically used to transmit power, such as in drive shafts of automobiles, propeller shafts in boats, and gear trains. A shaft rotating with a constant angular speed ω and subjected to a torque will transmit power.

From basic mechanics power P is defined as the product of the torque T times the angular velocity ω of the shaft. That is

$$P = T\omega \tag{4.23}$$

If the torque is given in units of pound · foot and the angular velocity is expressed in radians per second, then in the U.S. Customary System or English system of units, the power would be in units of foot · pounds per second. If International System of Units (SI) units are used the torque is in newton · meters, the angular velocity is in radians per second, and the power is then in units of newton-meters per second or the watt (W). Oftentimes in engineering practice the angular speed is expressed in units of revolutions per second (or hertz), f. In this case, the power is given by

$$P = T(2\pi f) \tag{4.24}$$

The customary units for power are either horsepower (hp) in the English system of units or kilowatts (kW) in the SI system. To convert power into horsepower units, we often use the following relations:

$$1 \text{ hp} = 550 \text{ ft} \cdot \text{lb/s}$$

Also in SI, we have

$$1 \text{ kW} = 1000 \text{ N} \cdot \text{m/s}$$

To convert from the horsepower unit to the kilowatt, we have

$$1 \text{ hp} = 0.746 \text{ kW}$$

When English units are used, Eq. (4.24) can be expressed conveniently as

$$P = \frac{2\pi n T}{33{,}000} \tag{4.25}$$

where power P is expressed in horsepower; n, the angular speed of the shaft, in revolutions per minute (rpm); and T in lb · ft.

EXAMPLE 4.10

The propeller shaft on a small tugboat (similar to that shown in Figure 4.1(a)) transmits 240 hp at 2200 rpm (revolutions per minute). Determine the torque that the shaft must resist.

Solution

Using Eq. (4.24), we obtain the torque as

$$T = \frac{P}{2\pi f}$$

where

$$P = (240 \text{ hp})(550 \text{ ft}\cdot\text{lb/s}) = 132{,}000 \text{ ft}\cdot\text{lb/s}$$

$$f = (2200 \text{ rpm})(1 \text{ min}/60 \text{ s}) = 36.67 \text{ rev/s}$$

Using P and f in the above equation, the torque is

$$T = \frac{132{,}000 \text{ ft}\cdot\text{lb/s}}{\left[2\pi\left(\frac{\text{rad}}{\text{rev}}\right)\right]\left[36.67\left(\frac{\text{rev}}{\text{s}}\right)\right]} = 573 \text{ lb}\cdot\text{ft}$$

EXAMPLE 4.11

An electric motor is required that will transmit 200 N · m of torque at 1750 rpm. What is the minimum size motor in kilowatts of power that can be selected?

Solution

From Eq. (4.24), we have

$$P = T(2\pi f)$$

or

$$P = (200 \text{ N}\cdot\text{m})\left[2\pi\left(\frac{\text{rad}}{\text{rev}}\right)\right]\left[1750\left(\frac{\text{rev}}{\text{min}}\right)\right]\left(\frac{1 \text{ min}}{60 \text{ sec}}\right)$$

$$= 36{,}660 \frac{\text{N}\cdot\text{m}}{\text{s}} = 36{,}660 \text{ W}$$

$$= 36.66 \text{ kW}$$

EXAMPLE 4.12

A solid circular steel shaft is to be used to transmit 20 hp at 1200 rpm without exceeding an allowable shear stress of 10,000 psi. Determine the required diameter of the shaft to meet this requirement.

Solution

Because the power is given in horsepower units and the shaft speed in revolutions per minute, we can use Eq. (4.25) to obtain the torque as

$$T = \frac{P(33{,}000)}{2\pi n}$$

$$T = \frac{(20)(33{,}000)}{2\pi(1200)} = 87.5 \text{ lb}\cdot\text{ft} \tag{a}$$

Next we use Eq. (4.14) to determine the required value of J/r as

$$\frac{J}{r} = \frac{T}{\tau}$$

$$\frac{J}{r} = \frac{(87.5 \text{ lb} \cdot \text{ft})\left(12 \frac{\text{in.}}{\text{ft}}\right)}{10{,}000 \text{ psi}} = 0.105 \text{ in.}^3 \tag{b}$$

From Eq. (4.12), for a solid circular shaft

$$J = \frac{\pi r^4}{2} \tag{c}$$

Therefore, using Eq. (c) in (b), we obtain

$$\frac{J}{r} = \frac{\pi r^3}{2} = 0.105 \text{ in.}^3$$

Solving for r^3, we have

$$r^3 = \frac{2}{\pi}(0.105 \text{ in.}^3)$$

and

$$r = 0.406 \text{ in.}$$

The required diameter is then

$$d = 0.812 \text{ in.}$$

In practice, a $\frac{7}{8}$-in.-diameter shaft would probably be used.

4.8 NONLINEAR STRESS DISTRIBUTIONS

In the preceding discussions of shear stresses due to torsion, it was assumed that the shear stresses and shear strains were related in a linear fashion. In Figure 4.19 it is assumed that we were in the linearly elastic portion of the shear stress–strain curve (below the proportional limit stress τ_p).

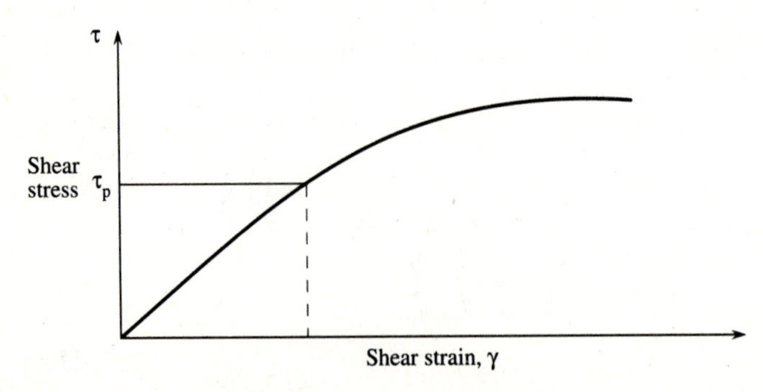

Figure 4.19 Nonlinear shear stress-shear strain diagram.

In situations where the shear stress levels exceed the shear proportional limit, τ_P, the stresses are no longer linearly related to the radial distance, nor to the shear strain, and Eqs. (4.14) and (4.15) are no longer valid. We will now look at some specific examples of nonlinear stress distributions. In each case the shear strains are assumed to be linearly related to radial distance, that is, Eq. (4.1) is still valid. However, the general form of the torque-shear stress equation given by Eq. (4.4) will have to be used when the material no longer behaves in the linear-elastic range.

Elastic-Plastic Stress Distribution

EXAMPLE 4.13

Determine the relationship between the applied torque, T, and the plastic or yield shear stress, τ_Y, for a solid circular bar with the assumed *elastic-plastic* (elastoplastic) stress distribution shown in Figure 4.20.

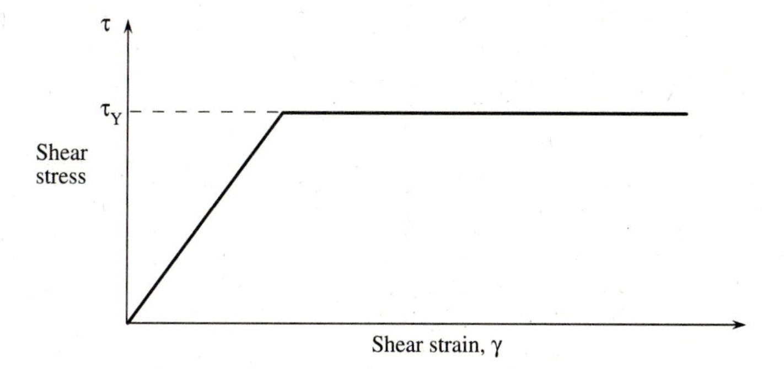

Figure 4.20 Elastic-plastic shear stress-shear strain distribution.

Solution

The above idealized state of stress will be distributed along the cross-section of a solid circular shaft as shown in Figure 4.21 where r is the radius of the shaft and r_p is the radial distance at which a pure plastic stress distribution begins.

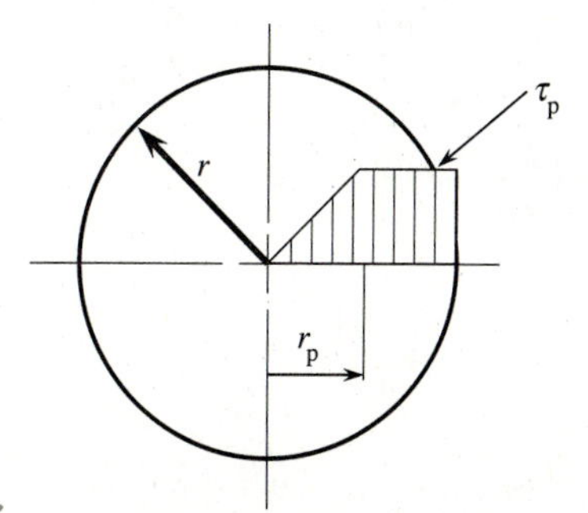

Figure 4.21 Elastic-plastic shear stress distribution.

As with all static stress problems, the external loading must be in equilibrium with the internal stresses. In this example, we can write

$$T = T_L + T_P \tag{a}$$

where T is the externally applied torque, T_L is the portion of the torque resisted by the linear stress distribution, and T_P is the portion of the torque resisted by the plastic stress distribution. From Eq. (4.14), we can write for the linear portion

$$\tau_P = \frac{T_L r_P}{J_P}$$

or

$$T_L = \frac{\tau_P J_P}{r_P} \tag{b}$$

where

$$J_P = \frac{\pi r_P^4}{2}$$

For the plastic portion from Eq. (4.4), we must have for equilibrium

$$T_P = \int_{r_P}^{r} \tau_P \, \rho \, dA \tag{c}$$

where ρ is any radius between r_P and r, τ_P is constant, and dA is the area of the annulus at radius ρ. Therefore,

$$dA = 2\pi\rho d\rho \tag{d}$$

Using Eq. (d) in (c), we obtain

$$T_P = \int_{r_P}^{r} \tau_P \, 2\pi\rho^2 \, d\rho \tag{e}$$

On explicit integration of Eq. (e), we have the torque resisted by the plastic stress distribution as

$$T_P = \frac{2\pi\tau_P}{3}(r^3 - r_P^3) \tag{4.26}$$

Finally, substituting Eqs. (b) and (4.26) into (a), we have the total torque resisted by the elastic-plastic shaft as

$$T = \tau_P \pi \left(\frac{2}{3} r^3 - \frac{1}{6} r_P^3\right) \tag{4.27}$$

Note that in the fully linear case ($r_P = r$), Eq. (4.27) reduces to

$$T = \frac{\pi \tau_P r^3}{2} \tag{f}$$

which is the same result as Eq. (4.14) when the explicit expression for J, given by Eq. (4.12), is substituted into Eq. (4.14).

Fully Plastic Stress Distribution

EXAMPLE 4.14

Determine the relationship between the applied torque, T, and the plastic shear stress, τ_P, for a solid circular bar with the assumed fully plastic stress distribution shown in Figure 4.22.

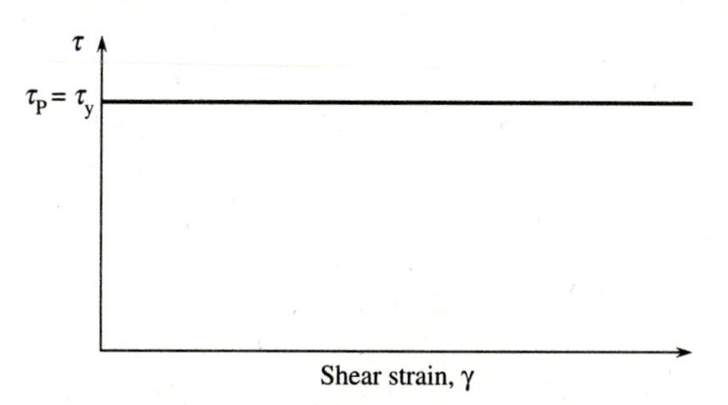

Figure 4.22 Fully plastic shear stress–shear strain distribution.

Solution

This idealized fully plastic state of stress will be distributed uniformly along the cross-section of the solid circular shaft as shown in Figure 4.23 where τ_P is the magnitude of the plastic stress.

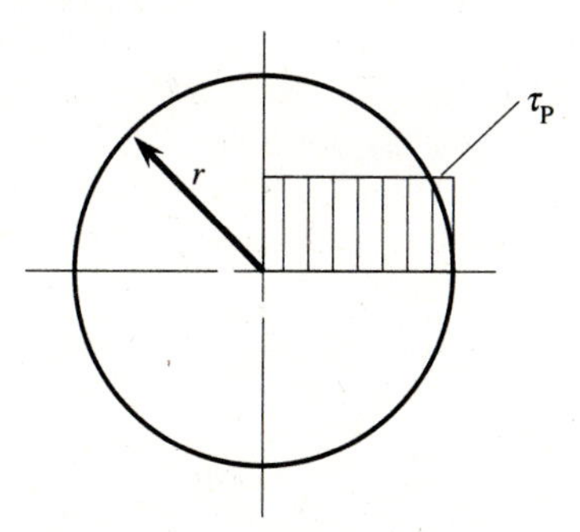

Figure 4.23 Idealized fully plastic shear stress distribution.

The solution is simply obtained by letting $r_P = 0$ in Eq. (4.27). The result is

$$T = \frac{2}{3}\pi\tau_P r^3 \qquad \textbf{(4.28(a))}$$

(Note that we could have used Eq. (4.4) with $\tau = \tau_P$ and integrated from $\rho = 0$ to $\rho = r$ to obtain the same result.)

This fully plastic condition is an idealized one, and can only approximately be realized as explained in the following. Recall from Eq. (4.2), that the shear strain is related to the angle of twist by

$$\gamma = \frac{\rho\phi}{L} \qquad \textbf{(4.2)}$$

Now let us denote by ρ_Y the radius of the elastic core when γ reaches the yield strain γ_Y. Also let ϕ_Y be the angle of twist at the initiation of yielding. Then by Eq. (4.2), we now have

$$\gamma_Y = \frac{\rho_Y \phi_Y}{L} \quad \textbf{(a)}$$

Therefore, for $\phi > \phi_Y$, we can express the ratios of radii to ratios of angles of twist at yielding by

$$\frac{\rho_Y}{\rho} = \frac{\phi_Y}{\phi} \quad \textbf{(b)}$$

Using Eq. (b) with $\rho = r$ in Eq. (4.27) (with $\rho_Y = r_P$), we obtain

$$T = \tau_P \pi \left(\frac{2}{3} r^3 - \frac{1}{6} r^3 \frac{\phi_Y^3}{\phi^3} \right)$$

$$= \tau_P \pi r^3 \left(\frac{2}{3} - \frac{1}{6} \frac{\phi_Y^3}{\phi^3} \right) \quad \textbf{(4.28(b))}$$

Therefore, for $\phi > \phi_Y$, as ϕ approaches infinity, T from Eq. (4.28(b)) approaches T_P given by Eq. (4.28(a)). But T does not actually reach T_P.

4.9 TORSION IN NONCIRCULAR CROSS-SECTIONS

Our discussion of torsion has so far been restricted to both solid and hollow circular cross-sections. Occasionally it is necessary to determine the maximum shear stress and angle of twist in noncircular cross-sections due to torsional loads. The development of these equations utilizes the theory of elasticity (see reference [1]) and is generally beyond the scope of this book.

We now present the equations used to obtain the maximum shear stress and angle of twist for various noncircular cross-sections.

Elliptical Cross-Section

Figure 4.24 shows an elliptical bar and its cross section subjected to an applied torque, T. The major and minor axes are a and b, respectively.

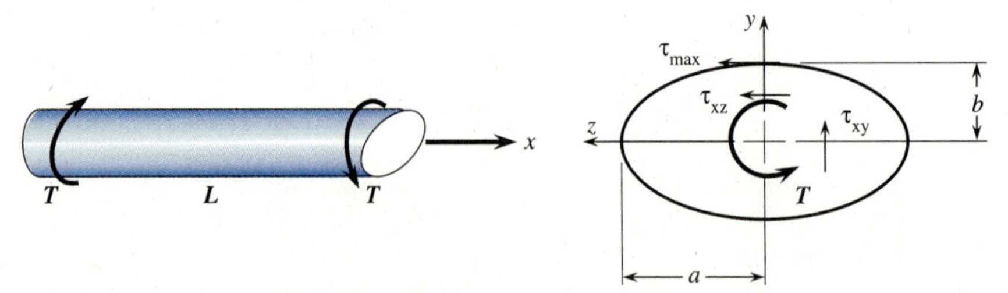

Figure 4.24 Elliptical bar and its cross section.

If the x-axis is the longitudinal axis of the bar, then the magnitudes of the shear stresses on the x-face are

$$\tau_{xz} = \frac{2Ty}{\pi ab^3} \qquad \tau_{xy} = \frac{2Tz}{\pi a^3 b} \tag{4.29}$$

The absolute value of the maximum shear stress occurs at $y = b$ (the edge of the minor axis) and is equal to

$$\tau_{max} = \frac{2T}{\pi ab^2} \tag{4.30}$$

Note that if $a = b$, Eq. (4.30) reduces to the well-known result for a solid circular cross-section subjected to torque.

The angle of twist per unit of length of the bar is given by

$$\theta = T\frac{a^2 + b^2}{\pi G a^3 b^3} \left(\frac{\text{radian}}{\text{length}}\right) \tag{4.31}$$

The total angle of twist over the whole length L is

$$\phi = \theta L = T\frac{(a^2 + b^2)L}{\pi G a^3 b^3} \tag{4.32}$$

Rectangular Cross-Sections

Figure 4.25 shows a rectangular bar subjected to a torque T.

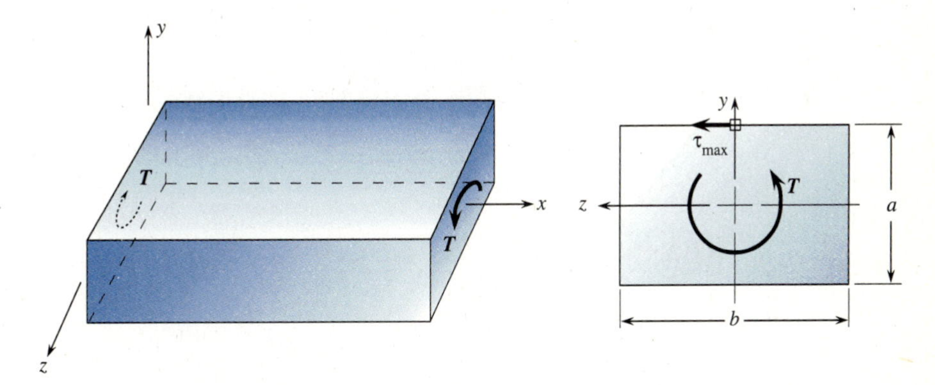

Figure 4.25 Rectangular bar and its cross section.

From the theory of elasticity, it has been shown that the maximum shear stress occurs at the middle of the longest side, that is, if $b \geq a$, then τ_{max} occurs at $z = 0$, $y = \pm a/2$, where

$$\tau_{max} = \frac{T}{k_2 a^2 b} \tag{4.33}$$

Also the shear stress at the corners is zero.

TABLE 4.1 CONSTANTS FOR TORSION OF A RECTANGULAR BAR

b/a ratio	k_1	k_2
1.0	0.1406	0.208
1.2	0.166	0.219
1.5	0.196	0.231
2.0	0.229	0.246
2.5	0.249	0.258
3	0.263	0.267
4	0.281	0.282
5	0.291	0.291
10	0.312	0.312
∞	0.333	0.333

The angle of twist per unit length of the bar is

$$\theta = \frac{T}{k_1 G a^3 b} \quad \left(\frac{\text{radian}}{\text{length}}\right) \tag{4.34}$$

The angle of twist is then

$$\phi = \theta L = \frac{TL}{k_1 G a^3 b} \tag{4.35}$$

In Eqs. (4.33) and (4.34), k_1 and k_2 are constants that depend on the geometry of the cross section. Table 4.1 gives values of k_1 and k_2 for various ratios of b/a.

Narrow Rectangular Cross-Section

Figure 4.26 shows the cross-section of a narrow bar subjected to an applied torque T.

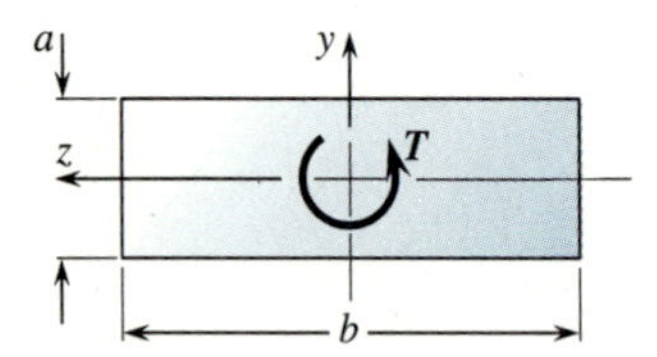

Figure 4.26 Narrow rectangular cross section.

The absolute value of the maximum shear stress on a narrow bar occurs in the middle of the long side and approaches (as $k_2 \rightarrow 0.333$)

$$\tau_{\max} = \frac{3T}{ba^2} \tag{4.36}$$

The corresponding value for the angle of twist per unit length is

$$\theta = \frac{3T}{ba^3 G} \quad \left(\frac{\text{radian}}{\text{length}}\right) \tag{4.37}$$

A narrow rectangular cross-section is defined as one that has the long length b at least 20 times the short length a in order for Eqs. (4.36) and (4.37) to be accurate. However, for b/a ratios as low as 10, the error is approximately only 6%. For smaller values of b/a, Table 4.1 should be used with Eqs. (4.33) and (4.34).

Equations (4.36) and (4.37) for narrow rectangular cross-sections can also be used for thin-walled bars having cross-sections such as those shown in Figure 4.27.

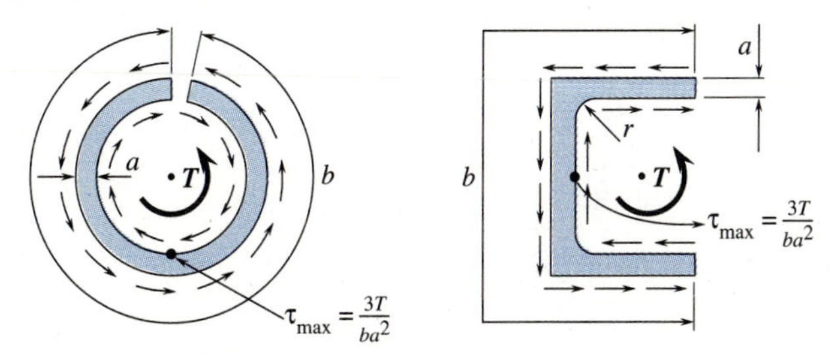

Figure 4.27 Equivalent narrow rectangular sections with shear stress distribution.

You should be cautioned that in the case of inside corner radii, such as in the channel section in Figure 4.27, there probably will exist stress concentration depending on the fillet radius, r. These areas of stress concentrations can significantly raise the nominal predicted values of stress and should be considered.

EXAMPLE 4.15

A 1-in.-square AISI 1040 steel power takeoff shaft transmits 50 hp at 500 rpm. Determine the maximum shear stress in the shaft and the angle of twist in a 3-ft section of shaft.

Solution

From Eq. (4.25), we solve for the torque, T, as

$$T = \frac{P(33{,}000)}{2\pi n}$$

$$T = \frac{(50)(33{,}000)}{2\pi(500)} = 525 \text{ lb} \cdot \text{ft}$$

$$= 6300 \text{ lb} \cdot \text{in} \qquad \text{(a)}$$

From Table 4.1 for $b/a = 1$ (square shaft), we obtain $k_1 = 0.1406$ and $k_2 = 0.208$. Using a value of the shear modulus for steel as $G = 11.5 \times 10^6$ psi (from Appendix B), Eqs. (4.33) and (4.35), and the torque from Eq. (a), yields the maximum shear stress and the angle of twist as

$$\tau_{max} = \frac{6300 \text{ lb} \cdot \text{in.}}{(0.208)(1 \text{ in.})^2(1 \text{ in.})} = 30{,}290 \text{ psi}$$

and

$$\phi = \frac{(6300 \text{ lb} \cdot \text{in.})(36 \text{ in.})}{(0.1406)(11.5 \times 10^6 \text{ psi})(1 \text{ in.})^3(1 \text{ in.})} = 0.140 \text{ rad} = 8.03°$$

EXAMPLE 4.16

An elliptical shaft 1 m long of AISI 1040 steel has a major axis of 12 mm and a minor axis of 6 mm. Determine the torsional stiffness in newton-meters per degree if the shear modulus for the steel is $G = 79$ GPa.

Solution

Using Eq. (4.31), we express the torsional stiffness as

$$\frac{T}{\theta} = \frac{\pi G a^3 b^3}{a^2 + b^2} \tag{a}$$

or substituting the numeric quantities into Eq. (a), we have

$$\frac{T}{\theta} = \frac{\pi(79 \times 10^9 \text{ N/m}^2)(0.012 \text{ m})^3 (0.006 \text{ m})^3}{(0.012 \text{ m})^2 + (0.006 \text{ m})^2}$$

$$\frac{T}{\theta} = 515 \frac{\text{N} \cdot \text{m}}{\text{rad/m}} = 8.99 \frac{\text{N} \cdot \text{m}}{\text{deg/m}} \tag{b}$$

This is the torsional stiffness per unit length, and since the length of the torsion rod is 1 m, dividing Eq. (b) by the length of the rod yields the total stiffness as

$$\frac{T}{\theta L} = \frac{T}{\phi}$$

Therefore, the torsional stiffness is

$$\left(\frac{T}{\phi}\right) = \left(8.99 \frac{\text{N} \cdot \text{m}}{\text{deg/m}}\right)\left(\frac{1}{1 \text{ m}}\right) = 8.99 \frac{\text{N} \cdot \text{m}}{\text{deg}}$$

EXAMPLE 4.17

A thin-walled tube shown in Figure 4.28(a) has a length L, a nominal radius r, a thickness t, and a shear modulus G. Determine the maximum shear stress, τ, due to torque T and the total torsional stiffness. Then compare these values for the closed tube with those for an identical tube which has been slit lengthwise as shown in Figure 4.28(b). Evaluate these results for a thin tube having a diameter of 10 in. and a thickness of 0.1 in.

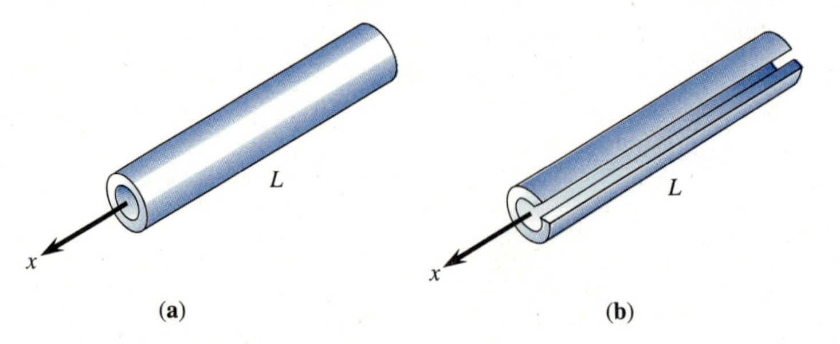

Figure 4.28 (a) Continuous and (b) slit thin-walled tubes.

Solution

In the case of the thin-walled tube, we recognize that all of the cross-sectional area, A, is concentrated at the nominal radius, r. Therefore, the polar moment of inertia will be

$$J = Ar^2 = (2\pi rt)r^2 = 2\pi r^3 t \quad \text{(See also Appendix A.)}$$

Using Eq. (4.14), the shear stress for the tube of Figure 4.28(a) is

$$(\tau)_a = \frac{Tr}{J} = \frac{Tr}{2\pi r^3 t} = \frac{T}{2\pi r^2 t}$$

Using Eq. (4.19), the total torsional stiffness will be

$$\left(\frac{T}{\phi}\right)_a = \frac{GJ}{L} = \frac{2\pi r^3 tG}{L}$$

For the slit tube of Figure 4.28(b), we can use Eq. (4.36) where $a = t$ and $b = 2\pi r$, to obtain the shear stress as

$$(\tau)_b = \frac{3T}{2\pi rt^2}$$

Using Eq. (4.37), with $\theta = \phi L$, the torsional stiffness for the slit tube becomes

$$\left(\frac{T}{\phi}\right)_b = \frac{2\pi rt^3 G}{3L}$$

Comparing maximum stresses for the closed versus the slit tube, we see that the ratio of the two stresses is

$$\frac{(\tau)_a}{(\tau)_b} = \frac{\dfrac{T}{2\pi r^2 t}}{\dfrac{3T}{2\pi rt^2}} = \frac{1}{3}\left(\frac{t}{r}\right)$$

and the ratio of the torsional stiffnesses is

$$\frac{\left(\dfrac{T}{\phi}\right)_a}{\left(\dfrac{T}{\phi}\right)_b} = \frac{\dfrac{2\pi r^3 tG}{L}}{\dfrac{2\pi rt^3 G}{3L}} = 3\left(\frac{r}{t}\right)^2$$

For the given 10-in. diameter tube with a 0.1-in. wall thickness, we observe that the ratio of the closed tube maximum shear stress to the slit tube maximum stress is

$$\frac{(\tau)_a}{(\tau)_b} = \frac{1}{3}\left(\frac{0.1}{\dfrac{10}{2}}\right) = 0.0067$$

or less than 1%. The ratio of the torsional stiffnesses is

$$\frac{\left(\dfrac{T}{\phi}\right)_a}{\left(\dfrac{T}{\phi}\right)_b} = 3\left(\frac{5}{0.1}\right)^2 = 7500$$

An appreciation for this last result can be easily experienced by cutting a cardboard tube lengthwise with scissors and noting the dramatic difference in stiffness from the uncut tube.

4.10 TORSION OF THIN-WALLED CLOSED TUBES

In previous sections of this chapter we determined shear stresses and angles of twist in solid and hollow circular cross-section members, and solid elliptical and rectangular members subjected to applied torques. In this section we will determine shear stresses and angles of twist in thin-walled closed members of arbitrary shape subjected to torque.

Torsion Formula

Consider the closed tube shown in Figure 4.29 subjected to a torque T. The walls are thin compared to the cross-sectional dimensions but they are not necessarily of constant thickness.

We now introduce the concept of shear flow, q, for closed thin-walled tubes. The shear flow is defined here as a shear force per unit length, acting on the edge of a thin sheet or wall. The shear flow also is obtained by multiplying the shear stress acting on the wall by the wall thickness t. That is

$$q = \tau t \tag{4.38}$$

In Figure 4.29, the shear flow q acts along the front edge of the tube due to the applied torque T. This shear flow is a constant throughout the member and around the edge. (This transfer of torque to the edge of the tube could be accomplished, for example, by means of an end plate.)

Now let the incremental moment or torque about point O due to the shear flow acting on the edge of element $dA(=tds)$ be dM_0 given by

$$dM_0 = dF(r) = (\tau dA)r = (\tau tds)r \tag{4.39}$$

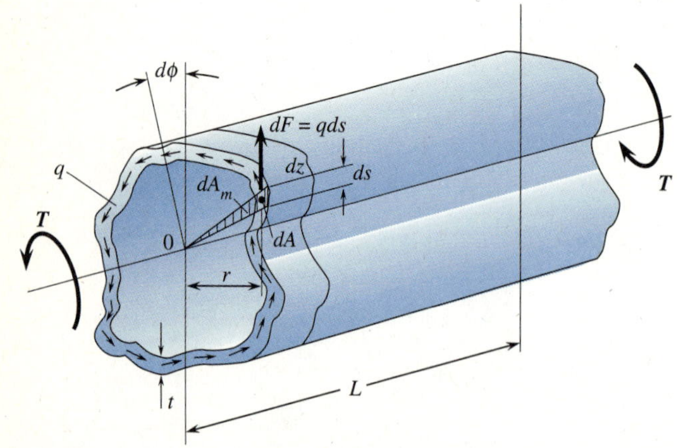

Figure 4.29 Closed thin-walled tube (single cell).

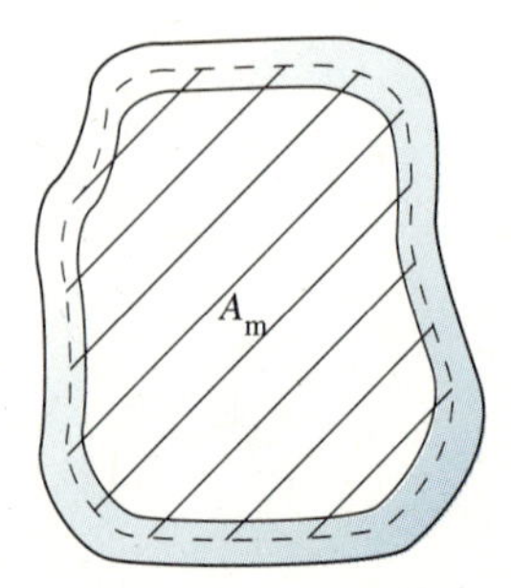

Figure 4.30 Enclosed area A_m.

or using Eq. (4.38) in (4.39), we have

$$dM_0 = (qds)r \tag{4.40}$$

where r is the moment arm about point O of the force qds in Figure 4.29. Now note that the incremental area dA_m of the shaded triangle shown in Figure 4.29 is

$$dA_m = 1/2\ rds \tag{4.41}$$

Therefore, using Eq. (4.41) in (4.40), we have

$$dM_0 = 2qdA_m \tag{4.42}$$

and therefore integrating Eq. (4.42) over the cross-section, we obtain the sum of the moments of all differential shearing forces acting on the wall section which is equal to the applied torque T. That is,

$$T = \oint dM_0 = \oint 2qdA_m \tag{4.43}$$

Since the shear flow q is constant around the edge of the tube, Eq. (4.43) becomes

$$T = 2qA_m \tag{4.43a}$$

or solving for q

$$q = \frac{T}{2A_m} \tag{4.44}$$

where A_m is the total enclosed area bounded by the center line of the wall-cross section of the tube shown in Figure 4.30.

Using Eq. (4.38) in (4.44), we express the shear stress at any location on the wall perimeter as

$$\tau = \frac{T}{2A_m t} \tag{4.45}$$

EXAMPLE 4.18

Determine the shear flow and the shear stress in a thin-walled structural cold-formed steel rectangular tube of 6 in. × 8 in. cross-section-loaded with a torque T = 500 lb · in. Due to fabrication error the thicknesses of the walls are t_1 = 1/8 in. and t_2 = 1/16 in. as shown in Figure 4.31.

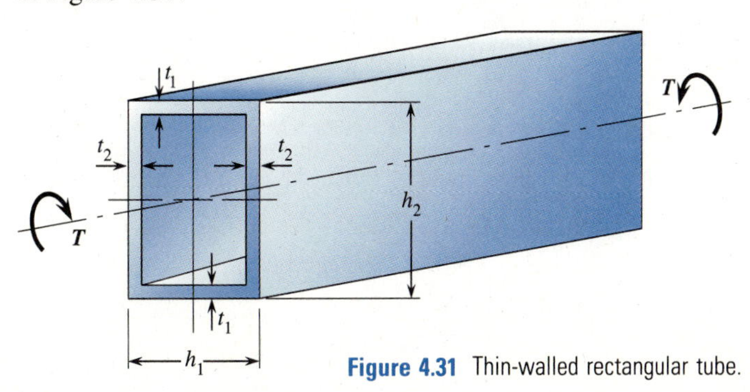

Figure 4.31 Thin-walled rectangular tube.

Solution

We first determine the area bounded by the center line of the tube as

$$A_m = \left[6 - \left(\frac{1}{16}\right)\right]\left[8 - \left(\frac{1}{8}\right)\right]$$
$$= 46.76 \text{ in.}^2$$

From Eq. (4.44), the constant shear flow around the edge of the tube is

$$q = \frac{T}{2A_m} = \frac{500 \text{ lb} \cdot \text{in.}}{2(46.76 \text{ in.}^2)} = 5.35 \text{ lb/in.}$$

Using Eq. (4.45), the shear stresses in the walls are

In the horizontal walls

$$\tau_h = \frac{T}{2A_m t_1} = \frac{q}{t_1} = \frac{5.35 \text{ lb/in.}}{(\frac{1}{16} \text{ in.})} = 85.6 \text{ psi}$$

In the vertical walls

$$\tau_v = \frac{T}{2A_m t_2} = \frac{q}{t_2} = \frac{5.35 \text{ lb/in.}}{(\frac{1}{8} \text{ in.})} = 42.8 \text{ psi}$$

Angle of Twist

To determine the angle of twist of a thin-walled tube subjected to torque T, first recall the angle of twist and shear strain for a circular cross-section as related by

$$\phi = \frac{\gamma L}{r} \tag{4.46}$$

where γ is the shear strain at a radial distance r. We now assume Eq. (4.46) applies, along with the shear stress-strain law ($\tau = G\gamma$), to the element of length ds in Figure 4.29. The angle that area dA_m rotates through is then

$$d\phi = \frac{\tau L}{Gr} \tag{4.47}$$

Using Eq. (4.45) for τ in Eq. (4.47), we have

$$d\phi = \frac{TL}{2A_m t G r} \tag{4.48}$$

We now assume that the average angle of twist of each dA_m section is the angle of twist of the whole section. Therefore

$$\phi = \frac{\oint d\phi \, dA_m}{A_m} \tag{4.49}$$

Substituting Eq. (4.48), along with Eq. (4.41) for dA_m, we obtain

$$\phi = \frac{\oint \frac{TL}{2A_m tGr}\left(\frac{rds}{2}\right)}{A_m} \tag{4.50}$$

Simplifying Eq. (4.50) yields

$$\phi = \frac{TL}{4A_m^2 G} \oint_0^{s_m} \frac{ds}{t} \tag{4.51}$$

where s_m is the mean circumference of the tube. Equation (4.51) is referred to as Bredt's formula after the German engineer, Rudolph Bredt.

Equation (4.51) gives the total angle of twist ϕ in radians, for a thin-walled closed tube subjected to a torque, T, where A_m is the enclosed area, G is the shear modulus, L is the length of the tube, t is the thickness, and s_m is the circumferential distance. The integral in Eq. (4.51) can often be evaluated discretely and Eq. (4.51) then takes the form

$$\phi = \frac{TL}{4A_m^2 G} \sum_{i=1}^{N} \frac{s_i}{t_i} \tag{4.52}$$

where s_i and t_i represent the length and corresponding thickness of a side of a thin-walled tube such as the rectangular tube in Figure 4.31 and N is the number of these separate lengths and thicknesses around the tube.

Recall that for a circular cross-section the angle of twist was

$$\phi = \frac{TL}{GJ} \tag{4.53}$$

If we equate Eqs. (4.51) and (4.53), we get

$$\frac{TL}{4A_m^2 G} \oint_0^{s_m} \frac{ds}{t} = \frac{TL}{GJ}$$

Solving this for $J = J_E$ gives

$$J_E = \frac{4A_m^2}{\oint_0^{s_m} \frac{ds}{t}} \tag{4.54}$$

This is the equivalent circular polar moment of inertia or J_E. If J_E is evaluated from Eq. (4.54) it can then be used in the circular relationship for angle of twist. That is, if J_E is evaluated for a thin-walled noncircular cross-section then the angle of twist can be found from

$$\phi = \frac{TL}{GJ_E} \tag{4.55}$$

EXAMPLE 4.19

Determine the angle of twist of the rectangular torque tube in Example 4.18. Let $G = 12 \times 10^6$ psi and $L = 200$ in.

Solution

Using Eq. (4.52), we first evaluate

$$\sum\frac{s}{t} = 2\frac{h_1}{t_1} + 2\frac{h_2}{t_2} = 2\frac{6}{(\frac{1}{8})} + 2\frac{8}{(\frac{1}{16})} = 352 \tag{a}$$

$$A_m = (h_1 - t_2)(h_2 - t_1)$$

$$= \left(6 - \frac{1}{16}\right)\left(8 - \frac{1}{8}\right) = 46.76 \text{ in.}^2 \tag{b}$$

The angle of twist is then

$$\phi = \frac{TL}{4A_m^2G}\left(2\frac{h_1}{t_1} + 2\frac{h_2}{t_2}\right) \tag{c}$$

Substituting numeric quantities into Eq. (c), we obtain

$$\phi = \frac{(500 \text{ lb}\cdot\text{in.})(200 \text{ in.})}{4(46.76 \text{ in.}^2)^2(12 \times 10^6 \text{ psi})}(352)$$

$$= 0.335 \times 10^{-3} \text{ rad}$$

EXAMPLE 4.20

A thin-walled torque tube is extruded in the form of a trapezoid and has the dimensions shown in Figure 4.32. If the material is 2024-T6 aluminum ($G = 28$ GPa), determine the maximum shear stress and the angle of twist for an applied torque of $T = 90$ N · m. All walls are 4 mm thick.

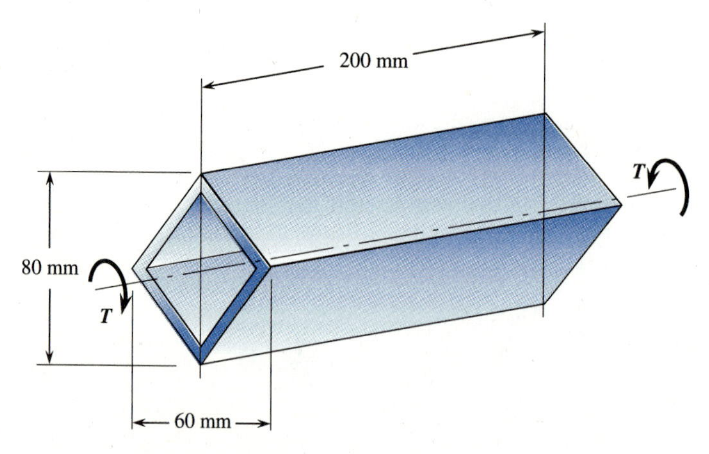

Figure 4.32 Trapezoidal torque tube.

Solution

The shear flow q will be constant around the edges of the tube. Therefore, from Eq. (4.44)

$$q = \frac{T}{2A_m}$$

$$q = \frac{90\ \text{N}\cdot\text{m}}{2[(0.060\ \text{m})(0.040\ \text{m})]} = 18{,}750\ \frac{\text{N}}{\text{m}}$$

Since the wall thickness is constant, the shear stress is also constant and is obtained from Eq. (4.38) as

$$\tau = \frac{q}{t} = \frac{18{,}750\ \text{N/m}}{0.004\ \text{m}} = 4.69\ \text{MPa}$$

The angle of twist is obtained from Eq. (4.52) as

$$\phi = \frac{TL}{4A_m^2 G} \sum \frac{s}{t}$$

$$\phi = \frac{(90\ \text{N}\cdot\text{m})(0.200\ \text{m})}{4[(0.060\ \text{m})(0.040\ \text{m})]^2\,(28 \times 10^9\ \text{N/m}^2)} \left(\frac{200\ \text{mm}}{4\ \text{mm}}\right)$$

$$\phi = 1.395 \times 10^{-3}\ \text{rad}$$

The previous equations for closed thin-walled tubes can be applied easily to torque tubes having multiple cells. Figure 4.33 shows an arbitrary cross-section of a multiple-celled torsional member subjected to a torque T.

In Figure 4.33, A_1, A_2, and A_3 represent the enclosed areas of cells 1, 2, and 3 and q_1, q_2, and q_3 represent the shear flows around the cells. Note that the net shear flow on webs separating two adjacent cells for example is $(q_1 - q_2)$, etc.

Now since we have three cells to distribute the applied torque, T, Eq. (4.43a) becomes

$$T = 2A_1q_1 + 2A_2q_2 + 2A_3q_3 \tag{4.56}$$

where $T_1 = 2A_1q_1$, etc., are the torques resisted by each cell.

In general, for n cells, this would become

$$T = 2A_1q_1 + 2A_2q_2 + \cdots + 2A_nq_n \tag{4.57}$$

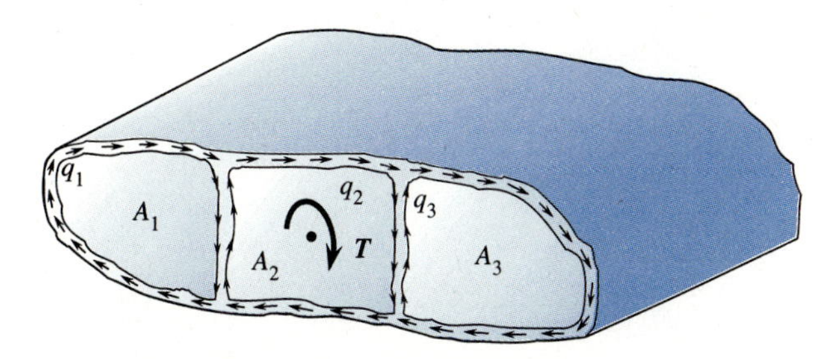

Figure 4.33 Closed thin-walled tube (multiple cell).

Since each cell of our three-cell closed tube will undergo the same angle of twist, we can write

$$\phi_1 = \phi_2 = \phi_3 = \phi \tag{4.58}$$

or in general for n cells

$$\phi_1 = \cdots = \phi_n = \phi \tag{4.58a}$$

Therefore from Eq. (4.51), we have

$$\phi = \frac{q_1 L}{2A_1 G} \oint_0^{s_1} \frac{ds}{t} \tag{4.59a}$$

$$\phi = \frac{q_2 L}{2A_2 G} \oint_0^{s_2} \frac{ds}{t} \tag{4.59b}$$

$$\phi = \frac{q_3 L}{2A_3 G} \oint_0^{s_3} \frac{ds}{t} \tag{4.59c}$$

where s_1, s_2, and s_3 represent the length around the edges of cells 1, 2, and 3, respectively, and T_1 has been replaced by $2A_1q_1$ in Eq. (4.59a), etc.

We have now generated four equations (Eqs. (4.56), (4.59a), (4.59b), and (4.59c)) to solve the four unknowns q_1, q_2, q_3, and ϕ. For a cross-section containing n cells, we would have similarly $(n + 1)$ unknowns and $(n + 1)$ equations.

4.11 STRESS CONCENTRATIONS IN TORSIONAL LOADING

In Section 2.10, we discussed stress concentrations in axially loaded members due to geometric changes such as holes, fillets, or notches in the members.

In this section, we discuss stress concentrations that occur in a circular shaft subjected to torsional loads when the shaft has changes in its geometry, such as fillets or grooves, as shown in Figure 4.34. In these cases, the largest torsional stresses occur near the discontinuities labeled (B) in Figure 4.34.

The shear stress in a uniform or slightly tapered shaft due to the torque can be determined using formulas developed in Section 4.3. For instance, in a shaft with a uniform cross-section, the shear stress on the cross-section at the outer radius of the shaft is given by Eq. (4.15) as

$$\tau = \frac{Tr}{J} \tag{4.60}$$

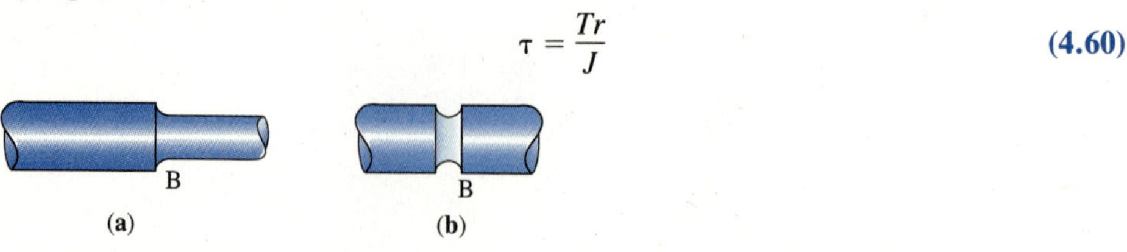

Figure 4.34 Torsion shaft with (a) fillets and (b) grooves.

Now if the shaft has an abrupt change in its diameter, stress concentrations will occur near the abrupt change. The stress can be reduced using, for instance, a fillet. The maximum shear stress is then

$$\tau_{max} = K\left(\frac{Tr}{J}\right) \tag{4.61}$$

where Tr/J is computed for the smaller diameter shaft and K is again the stress concentration factor. The concentration factor K depends only on the ratio of the diameters D/d of the shaft and the ratio of the radius r of the fillet to the diameter d of the smaller shaft. Figure 4.35(a) shows a plot of these K values for various D/d and r/d values for a fillet. Similarly, for a grooved shaft, the values of K depend on the ratio of D/d and r/d where now r is the radius of the groove (Figure 4.35(b)). Numerous plots of stress concentration factors for other geometric discontinuities are provided in Peterson's *Stress Concentration Factors* [2]. Again, as described in Section 2.10, Eq. (4.61) is valid only for the material remaining in the linear-elastic range, that is, for τ_{max} less than the proportional limit stress in shear. The reason for this is that the values of K obtained in Figure 4.35 were based on a linear-elastic relation between shear stress and shear strain.

For shafts with keyways (sometimes called Woodruff keyways) (Figure 4.36), which are often used to connect gears and pulleys to shafts, the American Society of Mechanical Engineers (ASME) allows a 25% reduction in the allowable stress and use of Eq. (4.60)

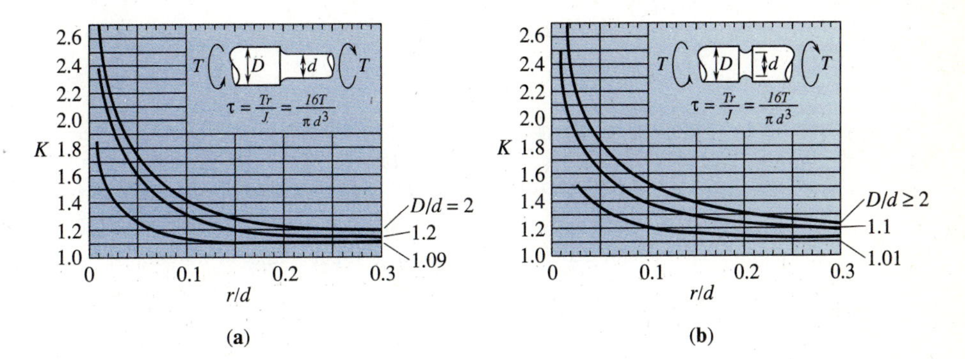

Figure 4.35 Stress concentrations for torsion shaft
(a) with fillet and
(b) with groove.

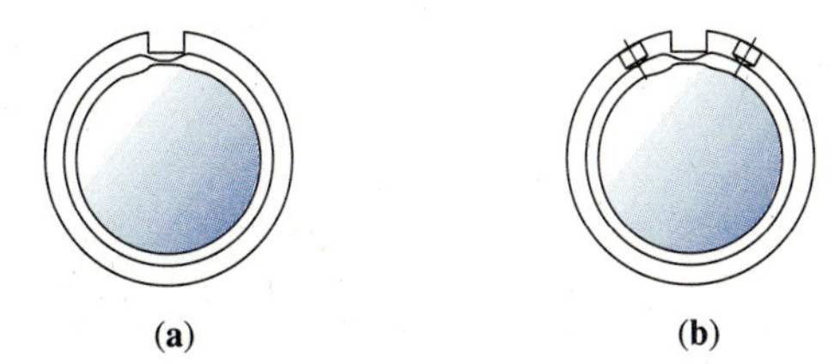

Figure 4.36 (a) Shaft with keyway and
(b) possible way to reduce the stress by drilling holes on each side of the keyway.

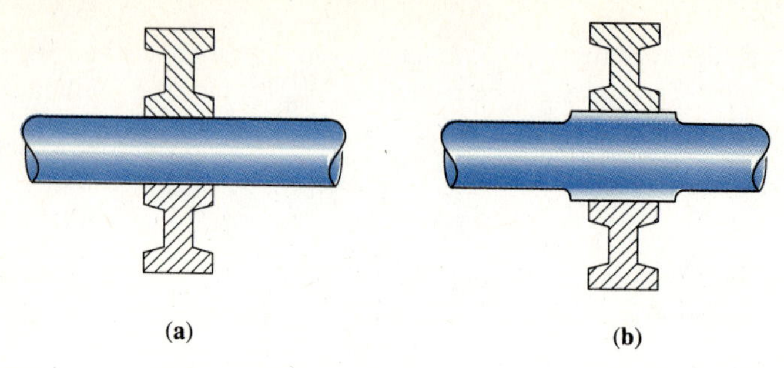

Figure 4.37 Shrink fitting pulley on shaft
(a) one way and
(b) preferred way.

for circular shafts subjected to torsion. The 25% reduction is assumed to account for the stress concentration. Figure 4.36(b) shows a possible way to reduce stress at the keyway. However, it is recommended to avoid keyways if possible as numerous failures, often due to fatigue loading (see Chapter 11) originate at keyways. For more on this type of failure, see [3]. One possible way to avoid the keyway and reduce stress concentrations is to press or shrink-fit the gear or pulley on the shaft as shown in Figure 4.37. Figure 4.37(b) shows the best method.

EXAMPLE 4.21

A stepped shaft transmits 20 hp as it rotates at 300 rpm. The grade of steel has an allowable shear stress of 8000 psi. The larger shaft has a 4-in. diameter and the smaller shaft a 3-in. diameter. The fillet radius at the juncture of the two shafts is 0.25 in. Use the stress concentration factor data to determine the shear stress in the smaller shaft.

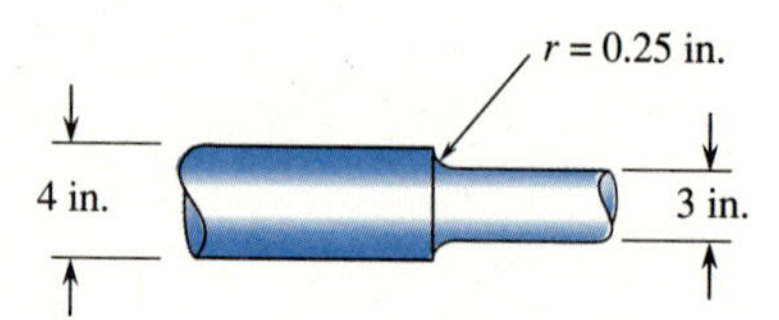

Figure 4.38 Stepped shaft for stress concentration analysis.

Solution

Using Eq. (4.25), we can solve for the torque in the shaft as

$$T = \frac{33{,}000P}{2\pi n} = \frac{33{,}000\ (20\ \text{hp})}{2\pi\ (300\ \text{rpm})} = 350\ \text{lb} \cdot \text{ft}$$

Using the geometry of the shaft, we determine the ratios r/d and D/d to obtain the stress concentration factor K as

$$\frac{r}{d} = \frac{0.25}{3} = 0.0833$$

$$\frac{D}{d} = \frac{4}{3} = 1.33$$

Using the r/d and D/d ratios in Figure 4.35(a), K is determined thus

$$K = 1.45$$

Using Eq. (4.61), the maximum shear stress in the smaller shaft is

$$\tau_{max} = K\left(\frac{Tr}{J}\right)$$

$$= 1.45\,\frac{(350 \times 12 \text{ in./ft})(1.5 \text{ in.})}{(\frac{\pi}{2})(1.5 \text{ in.})^4}$$

$$= 1150 \text{ psi}$$

Based on the allowable shear stress, the shaft is safe against yielding.

4.12 SUMMARY OF IMPORTANT DEFINITIONS AND EQUATIONS

1. Shear strain–angle of twist relationship

$$\gamma_r = \frac{r\phi}{L} \qquad \textbf{(4.1)}$$

2. Shear strain–maximum shear strain relationship

$$\gamma = \gamma_{max}\frac{\rho}{r} \qquad \textbf{(4.3)}$$

3. General torque-shear stress equation

$$T = \int \rho(\tau dA) \qquad \textbf{(4.4)}$$

4. Hooke's law in shear

$$\tau = G\gamma \qquad \textbf{(4.5)}$$

(valid for linear-elastic material range)

5. Polar moment of inertia

General form

$$J = \int_A \rho^2 dA \qquad \textbf{(4.11)}$$

For solid circular cross-section

$$J = \frac{\pi r^4}{2} \qquad \textbf{(4.12)}$$

For hollow circular cross-section

$$J = \frac{\pi}{2}(r_o^4 - r_i^4) \qquad \textbf{(4.13)}$$

6. Torsion formula

$$\tau = \frac{T\rho}{J} \qquad \textbf{(4.15)}$$

(valid for linear-elastic material range)

7. Largest normal stress in shaft subjected to torque

$$\sigma = \frac{Tr}{J} \qquad \textbf{(4.16)}$$

8. Angle of twist in circular shaft

$$\phi = \frac{TL}{GJ} \qquad \textbf{(4.19)}$$

$$\phi = \sum_{i=1}^{N} \frac{T_i L_i}{G_i J_i} \qquad \textbf{(4.20)}$$

$$\phi = \int_o^L \frac{T}{GJ}\,dx \qquad \textbf{(4.22)}$$

9. Power

$$P = T\omega \qquad \textbf{(4.23)}$$

$$P = T(2\pi f) \qquad \textbf{(4.24)}$$

$$P = \frac{2\pi n T}{33{,}000} \qquad \textbf{(4.25)}$$

Eq. (4.25) used for n in rpm
T in lb · ft
P in hp

1 hp = 550 ft · lb/s
1 hp = 0.746 kW

10. Elastic-plastic shaft

$$T = \tau_p \pi\left(\frac{2}{3} r^3 - \frac{1}{6} r_p^3\right) \qquad \textbf{(4.27)}$$

11. Ideally fully plastic shaft

$$T = \frac{2}{3}\pi\tau_p r^3 \qquad \textbf{(4.28a)}$$

12. Noncircular cross-sections

Elliptical

$$\tau_{max} = \frac{2T}{\pi a b^2} \qquad \textbf{(4.30)}$$

$$\phi = T\frac{(a^2 + b^2)L}{\pi G a^3 b^3} \qquad \textbf{(4.32)}$$

Rectangular

$$\tau_{max} = \frac{T}{k_2 a^2 b} \qquad \textbf{(4.33)}$$

$$\phi = \frac{TL}{k_1 G a^3 b} \qquad \textbf{(4.35)}$$

Narrow rectangular

$$\tau_{max} = \frac{3T}{ba^2} \qquad \textbf{(4.36)}$$

13. Thin-walled closed tubes

$$\tau = \frac{T}{2A_m t} \qquad \textbf{(4.45)}$$

$$\phi = \frac{TL}{4A_m^2 G}\sum_{i=1}^{N}\frac{s_i}{t_i} \qquad \textbf{(4.52)}$$

14. Maximum shear stress with stress concentrations

$$\tau_{max} = K\left(\frac{Tr}{J}\right) \qquad \textbf{(4.61)}$$

REFERENCES

1. Timoshenko SP, Goodier JN. *Theory of Elasticity*, 3rd ed. McGraw-Hill, New York, 1970.
2. Peterson RE. *Stress Concentration Factors*. John Wiley & Sons, New York, 1974.
3. Sachs NW. How to analyze shaft and bolt failures. *Power Transmission Design*, September, 43–47, 1988.

PROBLEMS

Section 4.3

4.1 A 2-in.-diameter solid circular steel shaft is subjected to a torque of 5000 lb · ft. Determine the maximum shear stress in the shaft.

4.2 If the shaft in Problem 4.1 has a 1-in.-diameter hole drilled through the center of its cross-section, determine the maximum shear stress. Also determine the shear stress at the inner surface.

4.3 A 2.5-in.-diameter steel shaft is 2 ft long. Its maximum shear stress is 10,000 psi. What is the applied torque?

4.4 A $\frac{3}{4}$-in.-diameter solid circular shaft is machined from a steel bar with an allowable shear stress of 20 ksi. Determine the allowable torque that can be resisted by the shaft.

4.5 A solid circular aluminum bar 50 mm in diameter is subjected to the torques shown. Determine the maximum shear stress **(a)** in section AB and **(b)** in section BC.

Figure P4.5

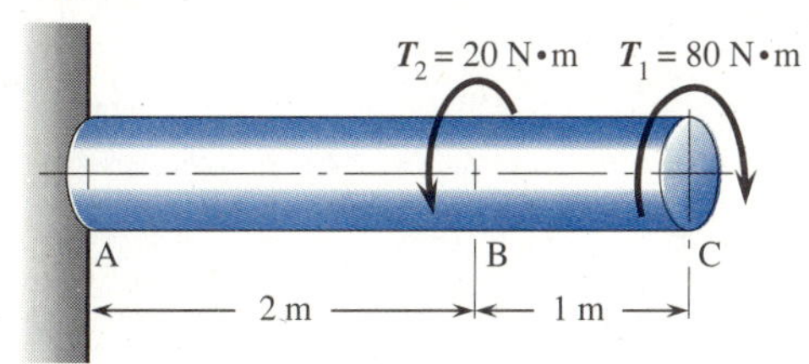

4.6 Determine the maximum shear stress in the solid stepped shaft shown in Figure P4.6 for an applied end torque of 350 lb · ft. Neglect any stress concentrations.

Figure P4.6

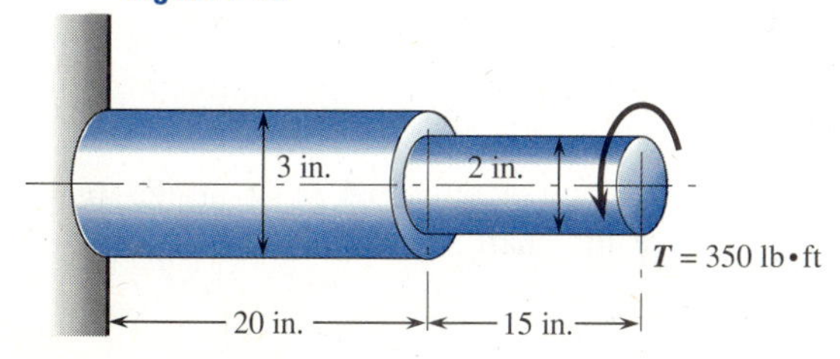

4.7 A 3-in.-diameter aluminum 6061-T6 shaft has a 2-in. hole bored 10 in. into its right end as shown in Figure P4.7. Determine the maximum shear stress in the shaft.

Figure P4.7

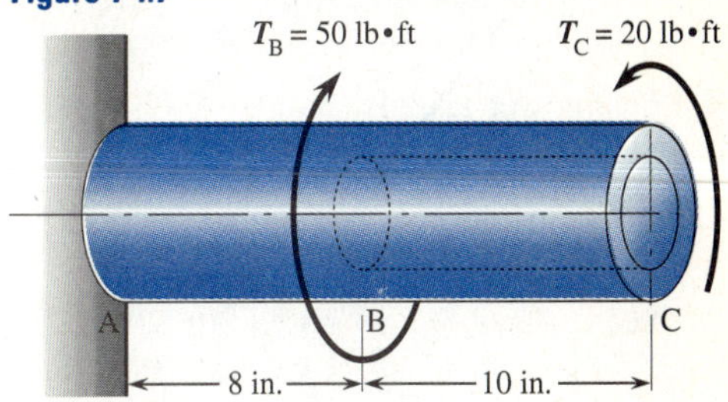

4.8 Determine the largest shear stress in the uniform solid circular shaft of radius 15 mm subjected to the torques shown in Figure P4.8.

Figure P4.8

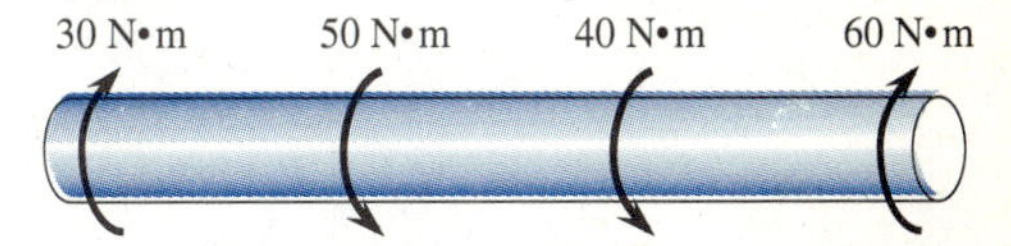

4.9 A tugboat solid shaft of AISI 1040 steel is subjected to a torque of 50 kip · ft. The shaft has a diameter of 10 in. Determine the maximum shear stress in the shaft.

4.10 A solid steel shaft of 25 mm diameter is used as a torsion bar to aid in lifting a garage door. If the allowable shear stress in the bar is 50 MPa, determine the maximum torque the bar can resist.

4.11 An ice auger made of a hollow steel shaft with inner diameter of 3/4 in. and outer diameter of 1 in. is subjected to a torque of $T = 100$ lb · in. Determine the maximum shear stress in the shaft.

Figure P4.11

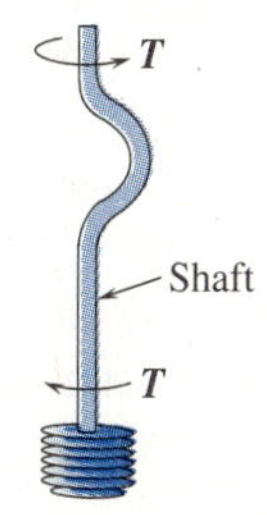

4.12 Compare the efficiency of a solid shaft of 0.75 in. diameter with a hollow one with 0.75 in. outer diameter and 0.5 in. inner diameter. Here we define efficiency to be the ratio of torsional strength (T/τ_{max}) to cross-sectional area of the shaft.

4.13 A power takeoff shaft of steel is required to resist a torque of 10 kip · in. and has an allowable shear stress of 12,000 psi. Determine the required diameter of the shaft.

4.14 A torsion bar in an automobile suspension system must resist a torque of 200 kip · in. The allowable shear stress is 7500 psi. Determine the required diameter.

4.15 A steel bolt must resist a torque of 2.0 kN · m. The allowable shear stress is 50 MPa. Determine the required diameter in millimeters.

Section 4.4

4.16 A 1-in.-diameter solid circular machine shaft is made of ASTM A-48 gray cast iron. Knowing that the allowable stresses are 10 ksi in tension, 50 ksi in compression, and 20 ksi in shear, determine the largest torque that may be applied.

4.17 A 1/2-in.-diameter plastic shaft is made of melamine with an allowable tensile stress of 3.0 ksi, an allowable compressive stress of 5.0 ksi, and an allowable shear stress of 3.5 ksi. Determine the maximum allowable torque.

4.18 A solid circular shaft with radius 1 in. resists a torque $T = 1.0$ kip · in. Determine the largest normal stress σ in the shaft.

4.19 Two pieces of steel bar are butt-welded together along a 45° plane as shown. Knowing that the allowable tensile stress in the weld is 14 ksi, determine the largest torque that can be applied.

Figure P4.19

4.20 A hollow pipe with inner diameter of 2 in. and outer diameter of 2.5 in. is made of ASTM A-48 gray cast iron. The allowable stresses in the pipe are 10 ksi in tension, 50 ksi in compression, and 20 ksi in shear. Determine the allowable torque that may be applied to the pipe.

4.21 A brittle cast iron rod with an allowable tensile stress of 70 MPa, an allowable compressive stress of 350 MPa, and an allowable shear stress of 140 MPa must resist a torque of 10 kN · m. Determine the diameter of the solid circular rod.

4.22 Two pieces of steel bar are butt-welded together along a 45° plane as shown in Figure P4.22. Knowing that the allowable tensile stress in the weld is 80 MPa, determine the largest torque T that can be applied.

Figure P4.22

4.23 A solid circular shaft of cast iron is to be designed for allowable stresses in tension, compression, and shear of 60 MPa, 150 MPa, and 80 MPa, respectively. What is the minimum required diameter of the shaft if it is subjected to a maximum torque of 600 N · m?

Section 4.5

4.24 Determine the torque T that is required to twist the end of the steel shaft of Figure P4.6 through 1°. Let $G = 11.5 \times 10^6$ psi.

4.25 Determine the angle of twist at the right end of the shaft of Problem 4.5. Let $G =$ 26 GPa.

4.26 Determine the angle of twist at the free end due to the applied torques in Figure P4.7.

4.27 Determine the angle of twist of B with respect to A for the stepped ASTM A36 steel shaft shown in Figure P4.27. Neglect stress concentrations.

Figure P4.27

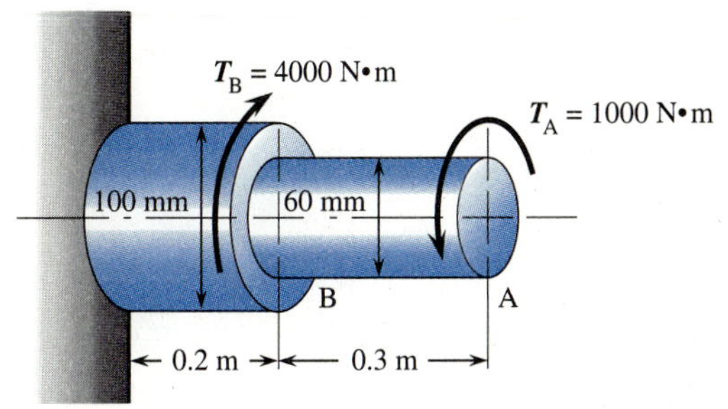

4.28 The angle of twist of a 2.5-m-long solid circular transmission shaft must not exceed 2° when a torque of 10 kN · m is applied to the shaft. Determine the required diameter of the shaft, if the shaft is made of steel with an allowable shear stress of 90 MPa and a modulus of rigidity of 79 GPa.

4.29 A solid circular steel rod BC is attached to a rigid lever AB and to the fixed support at C. The modulus of rigidity of the rod is $G = 11 \times 10^6$ psi. Determine the diameter of the rod so that the vertical deflection of point A will not exceed 1 in. and the maximum shear stress in the rod will not exceed 14 ksi. The load $P = 150$ lb.

Figure P4.29

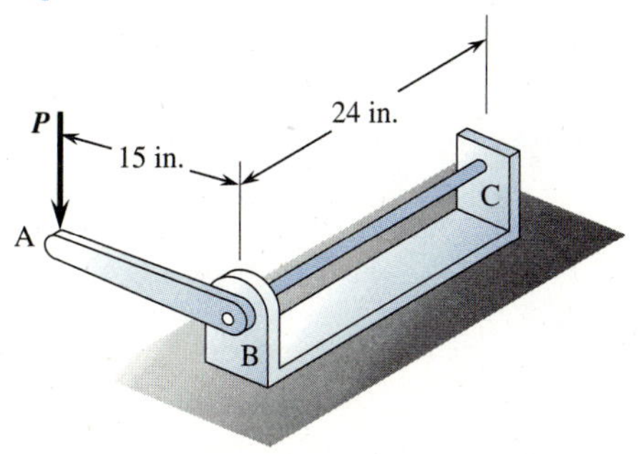

4.30 The 1.25-in.-diameter steel shaft carries an input torque of 500 lb · ft at pulley B and output torques of 300 lb · ft and 200 lb · ft at A and C, respectively. Determine the maximum shear stress in the shaft. Determine the angle of twist of end C with respect to end A. Let $G = 12 \times 10^6$ psi.

Figure P4.30

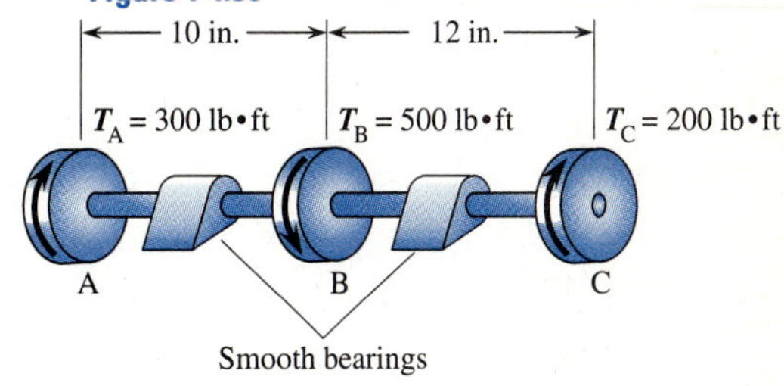

4.31 A bar in torsion with total length of 100 in. has a diameter of 2 in. over one-half its length and diameter of 1.5 in. over the other half. Determine the allowable torque T if the angle of twist is not to exceed 0.02 rad. Let $G = 12 \times 10^6$ psi.

4.32 For the stepped shaft shown in Figure P4.32, determine the angle of twist of point D with respect to A. Let $G = 79$ GPa.

Figure P4.32

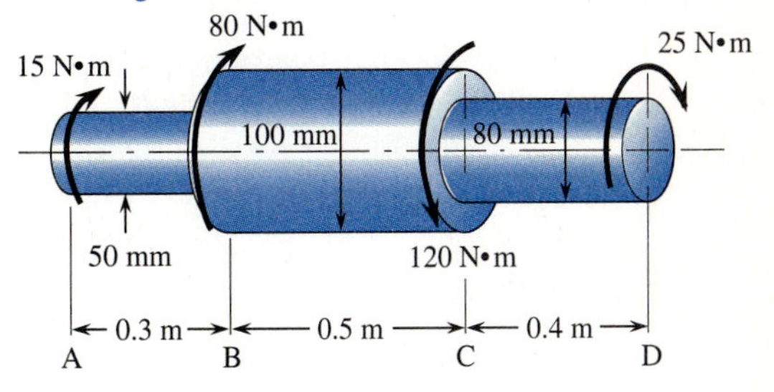

4.33 A solid circular shaft shown in Figure P4.33 tapers from a diameter of $2r_A$ to $2r_B$. Determine the angle of twist at B. Let the shear modulus be G.

Figure P4.33

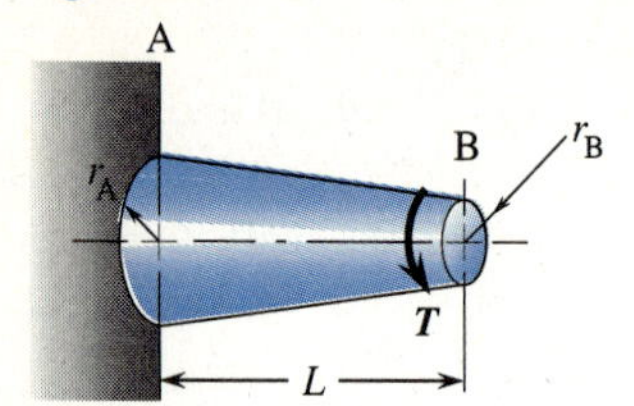

Section 4.6

4.34 The solid circular steel shaft shown in Figure P4.34 is fixed at both ends. Determine the angle of twist at point B and the maximum shear stress in the shaft.

Figure P4.34

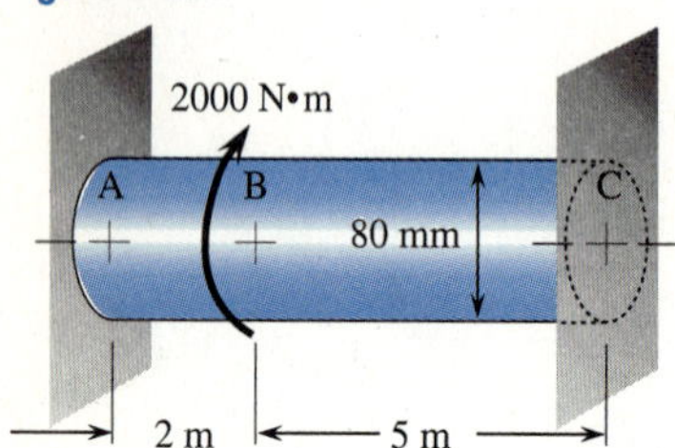

4.35 The two shafts are fixed at their left ends and are interconnected by spur gears at their right ends as shown in Figure P4.35. If the two shafts have different values for L, G, and J, and if the numbers of teeth on the gears are N_A and N_B, derive an expression for angle of twist of gear A due to the torque T. *Hint:* The ratio of the angles of twist of the shafts is inversely proportional to the ratio of the numbers of teeth; that is, $\phi_A/\phi_B = N_B/N_A$. Note also that $T = T_A + T_B$.

Figure P4.35

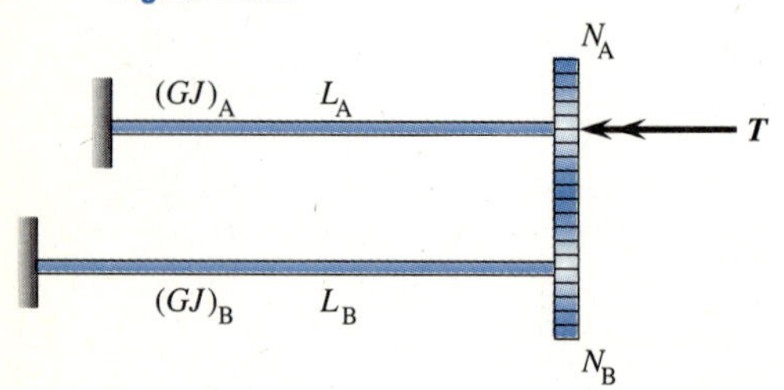

4.36 A solid steel shaft with a diameter of 2 in. and a hollow steel cylinder are both fastened to a rigid wall at one end and have their other ends welded to a rigid disk as shown in Figure P4.36. Determine **(a)** the angle of twist of the disk, **(b)** the maximum shear stress in the shaft, and **(c)** the maximum shear stress in the cylinder due to the applied torque $T = 1000$ lb · ft. Let $G = 11.5 \times 10^6$ psi.

Figure P4.36

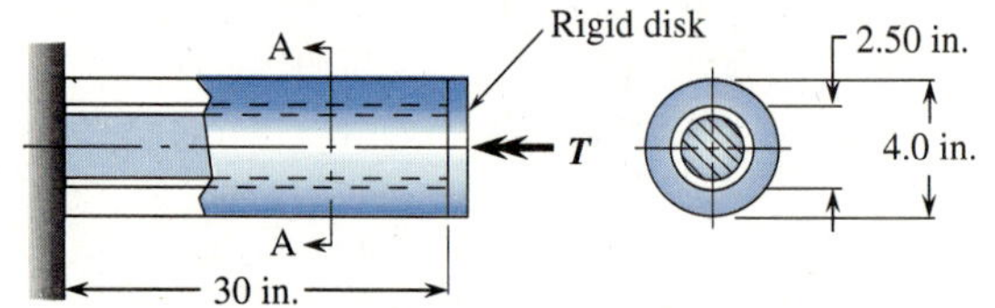

4.37 A composite shaft is fixed at the left end and subjected to a torque $T = 100$ lb · in. at the right end. Determine the amount of torque resisted by the outer shaft. Let the outer radius be 1.5 in. and the inner radius be 1 in. Also let the ratio of the outer material to inner material shear modulus be 1.57.

4.38 A composite shaft of aluminum and brass shown in Figure P4.38 is subjected to a torque T at the end. The allowable shear stress in the aluminum is 5000 psi and that in the brass is 8000 psi. Determine the maximum torque that can be applied. Let $G_{Al} = 3.7 \times 10^6$ psi and $G_{Br} = 5.6 \times 10^6$ psi.

Figure P4.38

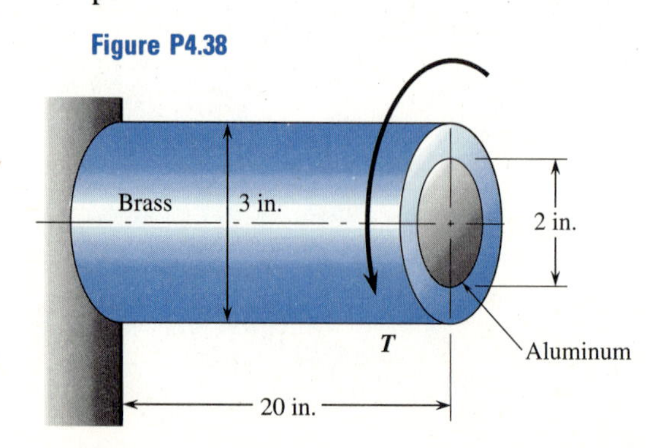

4.39 Two solid steel shafts are connected to a flange coupling at B and to the rigid supports at A and C. For the applied torque T shown, determine the maximum shear stress in **(a)** shaft AB and **(b)** shaft BC.

Figure P4.39

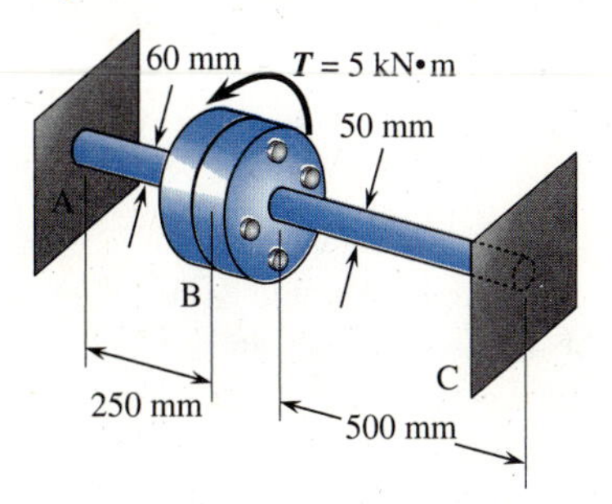

4.43 A motor delivers 200 hp at 200 rpm to a shaft at A as shown in Figure P4.43. The gears at B and C remove 125 hp and 75 hp, respectively. Determine the required diameter of the shaft if the allowable shear stress is 7500 psi and the angle of twist between the motor and the gear C is limited to 1.5°. Let $G = 11.5 \times 10^6$ psi.

Figure P4.43

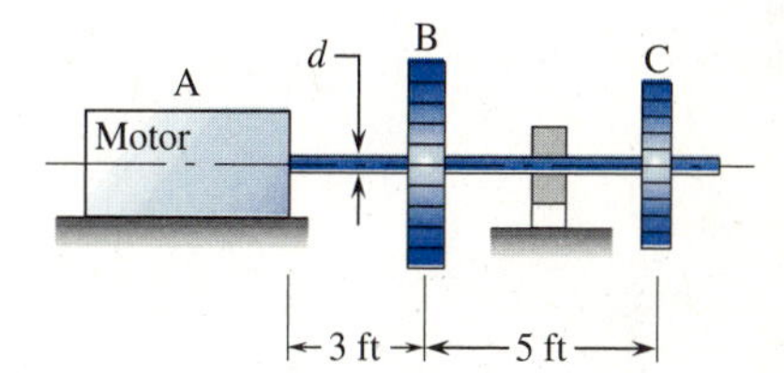

4.44 The propeller shaft of a certain boat is a hollow tube having outside diameter of 3 in. and inside diameter of 2 in. The shaft turns at 20 rpm. The maximum shear stress is limited to 5000 psi. How much horsepower can be transmitted by the shaft?

4.45 A 5-hp electric motor delivers 3 hp and 2 hp, respectively, to accessory gears B and C. Determine **(a)** the maximum shear stress in the steel drive shaft and **(b)** the angle of twist between gears B and C. The angular speed of the shaft is $n = 1500$ rev/min (rpm).

Figure P4.45

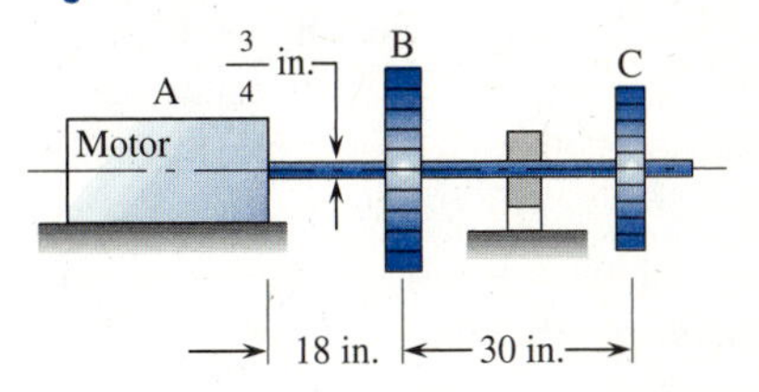

Section 4.7

4.40 A power takeoff shaft transmits 40 hp at 450 rpm. Determine the maximum shear stress in the 1-in. diameter solid shaft.

4.41 Determine the maximum shear stress in a steel shaft of 2-in. diameter that transmits 200 hp at 875 rpm.

4.42 A solid circular shaft is made of a material whose shear stress-strain diagram is shown. Determine the diameter of the shaft needed to transmit 100 hp at a speed of 1200 rpm if the maximum shear stress is not to exceed 10,000 psi and the angle of twist in a 100-in. length is not to exceed 1°.

Figure P4.42

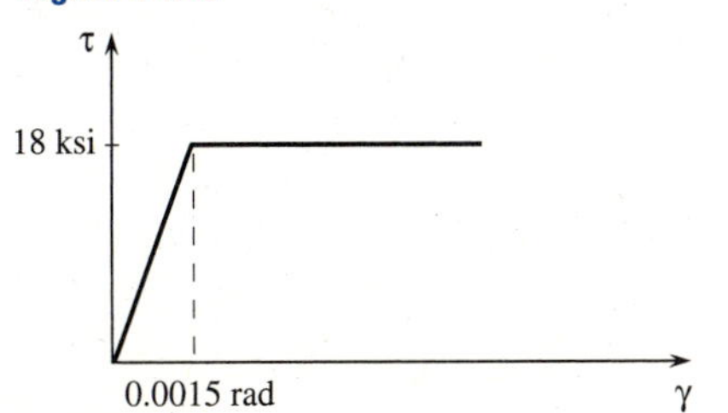

4.46 Determine the maximum shear stress in a 2.5-in. diameter solid power takeoff shaft that transmits 60 hp at 540 rpm.

4.47 A 4.5-kW electric motor drives a grinder through a 20-mm-diameter shaft turning at 1800 rpm. Determine the shear stress in the shaft.

4.48 Determine the diameter of a solid steel shaft to transmit 1 kW of power at a speed of 15 Hz, if the shear stress is not to exceed 30 MPa.

4.49 A 10-kW engine drives a solid shaft 1 m long by means of a V-belt and pulley at the midpoint of the shaft. At one end of the shaft 7 kW is removed and at the other end 3 kW is removed. The shaft is steel with a diameter of 15 mm and operates at 1000 rpm. Assuming that no power is lost in the V-belt drive, determine **(a)** the relative angle of twist between the two ends of the shaft and **(b)** the maximum shear stress in the shaft.

Section 4.8

4.50 A hollow shaft has an outer diameter of 3 in. and a wall thickness of 1⁄4 in. The material is mild steel with $G = 12 \times 10^6$ psi. Assuming that the material is elastic-plastic with a proportional limit of $\tau_p = 20$ ksi, determine the torque required to produce a fully plastic deformation.

4.51 A solid steel shaft of 2 in. diameter is made of a mild steel which is assumed to be elastic-plastic with $\tau_Y = 20$ ksi and $G = 12 \times 10^6$ psi. Determine the maximum shear stress due to an applied torque of **(a)** 10 kip · in. and **(b)** 50 kip · in.

4.52 A solid 3/4-in.-diameter shaft 15 in. long is twisted through an angle of 1°. The shear stress-strain diagram for the material is shown in Figure P4.52. Determine **(a)** the applied torque and **(b)** the maximum shear stress due to this torque.

Figure P4.52

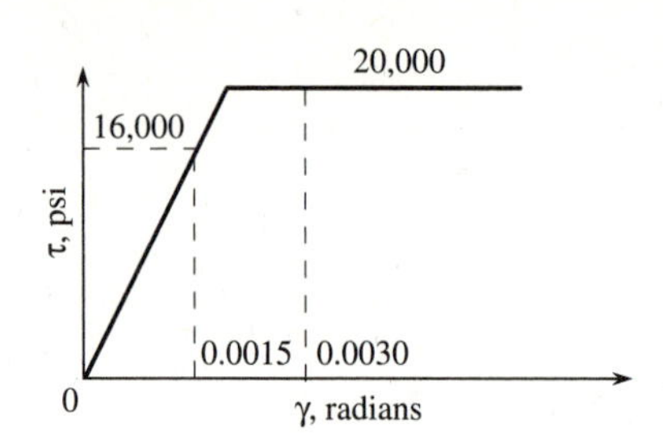

4.53 The same shaft as in Problem 4.52 is now twisted through an angle of 10°. Determine **(a)** the applied torque and **(b)** the maximum shear stress due to this torque.

4.54 A hollow circular shaft is made of a material whose shear stress-strain diagram is shown in Figure P4.54. The outer diameter is 3 in. and the inner diameter is 1.5 in. Let $\gamma_{max} = 0.004$ rad. Determine the applied torque.

Figure P4.54

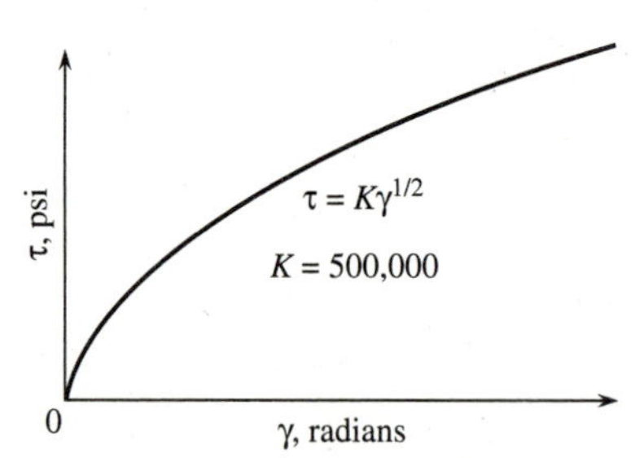

Section 4.9

4.55 Determine the maximum shear stress in an elliptical cross-section with a major diameter of 50 mm and a minor diameter of 25 mm. The shaft is subjected to a torque of 20 N · m.

4.56 Determine the maximum shear stress in a meter stick 6 mm × 24 mm in cross section subjected to a torque of 15 N · m.

4.57 A portion of an automobile suspension system is forged with an elliptical cross-section. If this member has a major diameter of 2 in. and a minor diameter of 1 in., determine the torque required to produce yielding. Assume a yield stress in shear $\tau_Y = 40$ ksi.

Figure P4.57

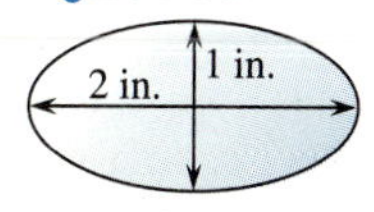

4.58 Determine the angle of twist per lb · ft of torque for the elliptical member of Problem 4.57 for a 6-in. length. Let $G = 11.5 \times 10^6$ psi.

4.59 An aluminum sheet having the dimensions of 1/4 in. thick × 8 in. long is formed into a 2 in. × 4 in. channel. Neglecting stress concentrations, determine **(a)** the maximum shear stress due to an applied torque of $T = 150$ lb · in. and **(b)** the angle of twist per unit length. Let $G = 3.0 \times 10^6$ psi.

Figure P4.59

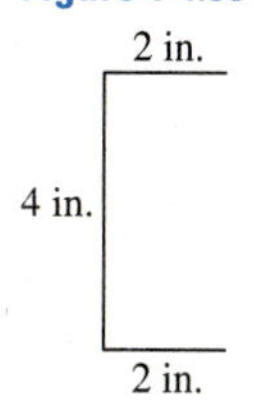

4.60 A power takeoff shaft transmits 10 kW at 2000 rpm. Compare the maximum shear stress of a 25-mm-diameter solid circular shaft with a 25-mm solid square shaft.

4.61 A power takeoff shaft transmits 40 hp at 540 rpm. Compare the maximum shear stress of a 1-in.-diameter solid circular shaft with a 1-in. solid square shaft.

Section 4.10

4.62 An extruded aluminum tube has the dimensions shown in Figure P4.62. Determine **(a)** the shear flow, **(b)** the maximum shear stress, and **(c)** the angle of twist per inch of length due to an applied torque of $T = 2000$ lb · in. Let $G = 3.0 \times 10^6$ psi.

Figure P4.62

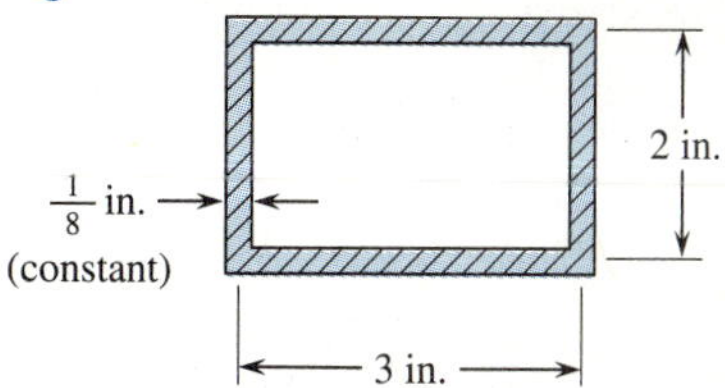

4.63 An extruded aluminum tube has the dimensions shown in Figure P4.63. Determine **(a)** the shear flow, **(b)** the maximum shear stress, and **(c)** the angle of twist per unit length due to an applied torque of $T = 250$ N · m. Let $G = 27$ GPa.

Figure P4.63

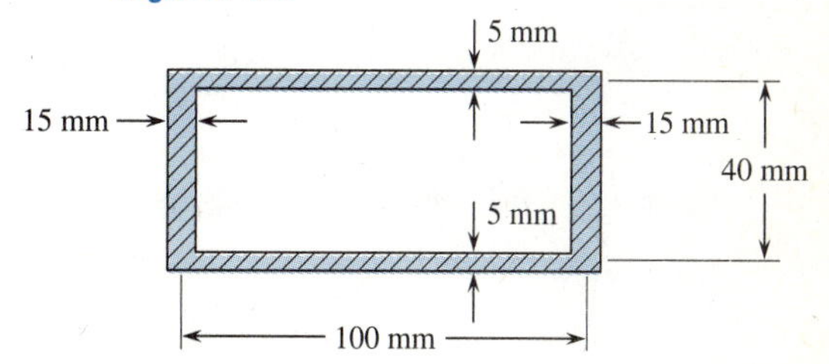

4.64 A thin-walled aluminum tube has an elliptical cross-section. Determine **(a)** the shear flow, **(b)** the maximum shear stress, and **(c)** the angle of twist per unit length due to an applied torque of $T = 5000$ lb · in. Let $G = 3.9 \times 10^6$ psi. Assume that the circumference of the ellipse equals 11 in.

Figure P4.64

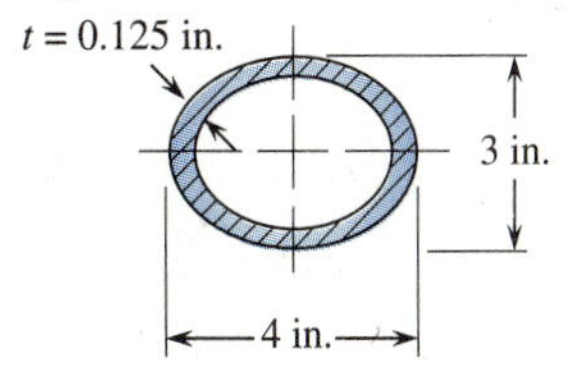

4.65 The two-cell box beam is subjected to a torque of $T = 6000$ lb · in. Determine **(a)** the shear flow in each web, **(b)** the shear stress in each web, and **(c)** the angle of twist per unit inch of length. Let $G = 12 \times 10^6$ psi.

Figure P4.65

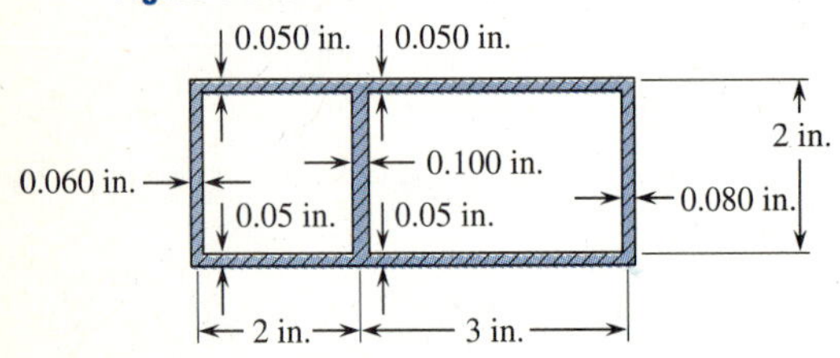

Section 4.11

For the problems in this section, consider stress concentrations.

4.66 The stepped shaft shown resists a torque $T = 500$ lb · in. Determine the maximum shear stress in the stepped shaft for a fillet radius r of **(a)** 1⁄16 in. and **(b)** 1⁄4 in.

Figure P4.66

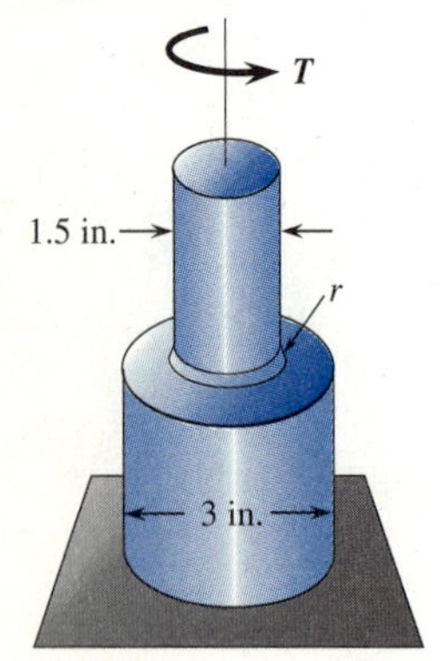

4.67 For the stepped shaft shown, the allowable shear stress is 10,000 psi. Determine the allowable magnitude of the torque T when the fillet radius r is **(a)** 1⁄16 in. and **(b)** 1⁄4 in.

Figure P4.67

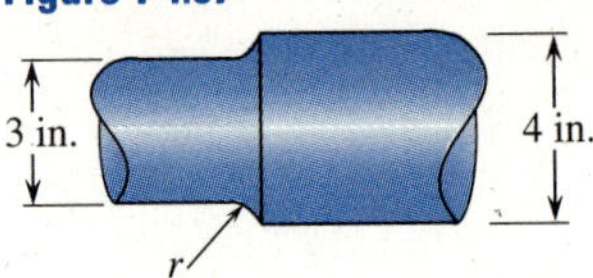

4.68 The stepped shaft shown rotates at 400 rpm. Determine the maximum horsepower that may be transmitted without exceeding an allowable shear stress of 50 MPa.

Figure P4.68

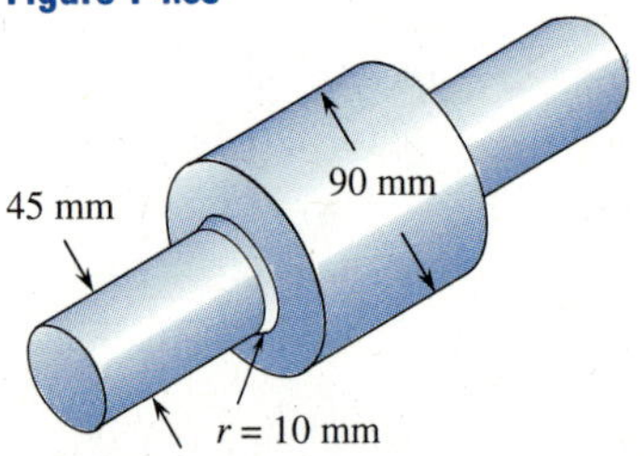

4.69 Determine the maximum shear stress in the stepped shaft shown. The shaft is rigidly connected to end A. The fillet radius is 0.5 in.

Figure P4.69

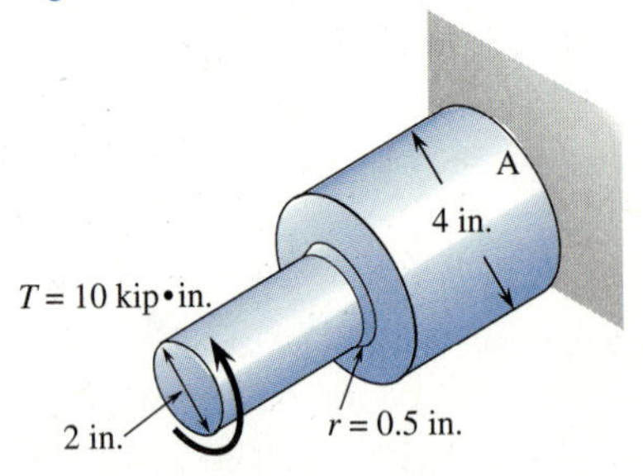

4.70 A fillet with radius r of 10 mm is used at the junction in a stepped shaft where the diameter is reduced from 125 mm to 100 mm. Determine the maximum shear stress in the fillet when the shaft is subjected to a torque of 10 kN · m.

Figure P4.70

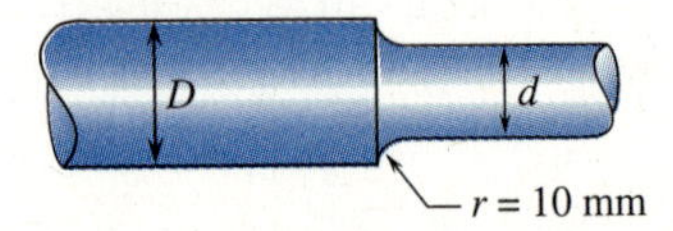

4.71 A transmission shaft has a semicircular groove with a 5-mm radius r machined into its 100 mm diameter D (d = 90 mm). If the maximum allowable shear stress in the shaft must not exceed 60 MPa, determine the maximum torque that the shaft can resist.

Figure P4.71

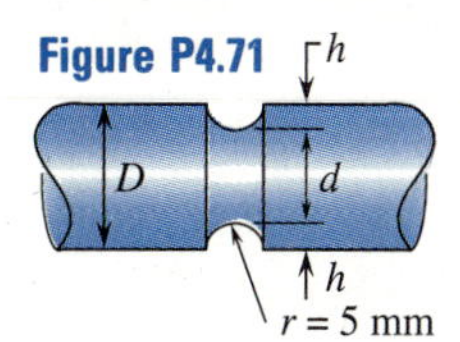

4.72 A shaft has a 100 mm diameter with a semicircular groove cut into it. If the maximum shear stress in the shaft must not exceed 60 MPa when transmitting a torque of 7000 N · m, determine the minimum radius needed for the groove.

4.73 The stepped shaft shown must transmit 40 kW of power at a speed of 600 rpm. Determine the minimum radius of the fillet if the allowable shear stress of 60 MPa is not to be exceeded.

Figure P4.73

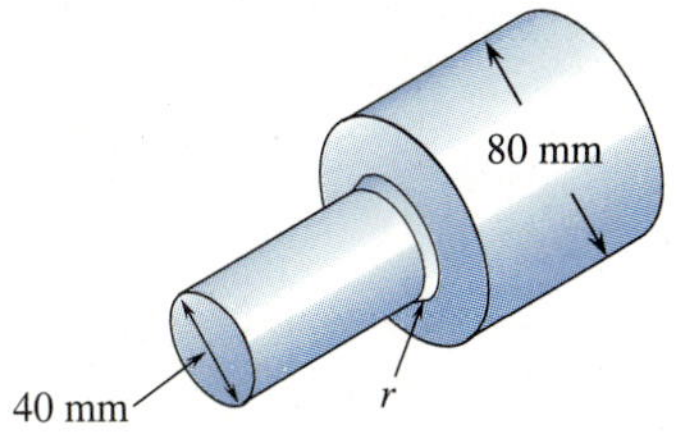

COMPUTER PROBLEMS

The following problems are suggested to be solved using a programmable calculator, microcomputer, or mainframe computer. It is suggested that you write a simple computer program in BASIC, FORTRAN, PASCAL, or some other language to solve these problems.

C4.1 Determine the maximum shear stresses and angles of twist of solid or hollow circular cylindrical shafts subjected to applied torque T. The input should include the torque T, the inner and outer diameters of the shaft, the length L of the shaft, and the shear modulus G of the shaft. Check your program using the following data: T = 50 kip · in.; diameter of solid shaft, d = 2 in.; length of shaft, L = 25 in.; and $G = 12 \times 10^3$ ksi.

Figure C4.1

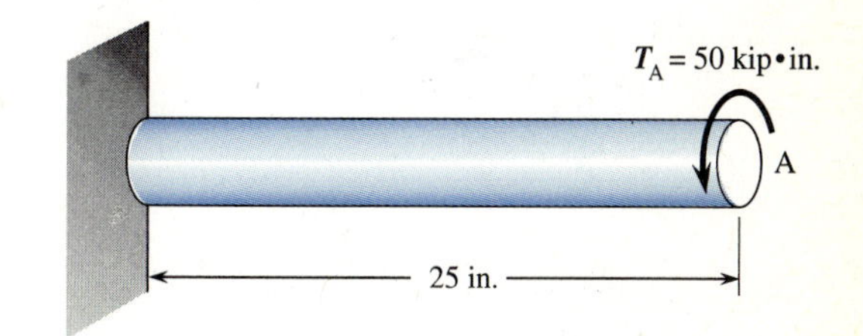

C4.2 Write a computer program to determine the maximum shear stress in each of the two materials, making up a composite circular shaft subjected to a known torque. Your program should include the radius of the inner core material, the inner and outer radii of the outer shell material, the shear moduli G_{core} and G_{shell}, and the resisting internal torque. Check your program using the following data: radius of core = 1 in., inner radius of shell = 1.25 in., outer radius of shell = 1.5 in., $G_{core} = 5.6 \times 10^6$ psi, $G_{shell} = 11.5 \times 10^6$ psi, and resisting torque = 20 kip · in.

Figure C4.2

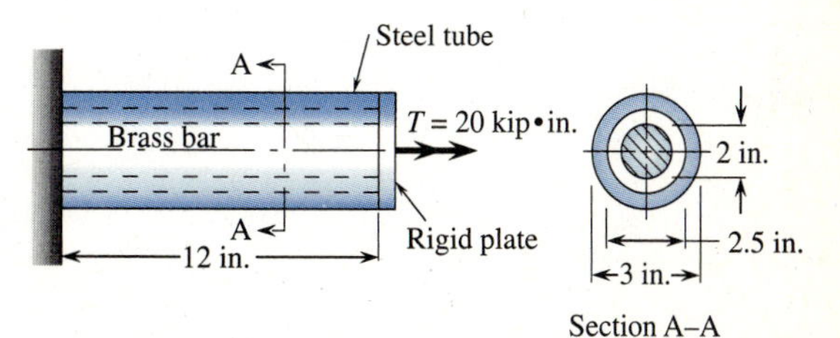

C4.3 Determine the maximum shear stress and angle of twist of a shaft of elliptical cross-section. The input should include the applied torque T, the major and minor axes, a and b, respectively, of the cross-section, the shear modulus G, and the length L. Check your program using the following data: $T = 20$ N · m, major axis $a = 50$ mm, minor axis $b = 25$ mm, and shear modulus $G = 27$ GPa.

Figure C4.3

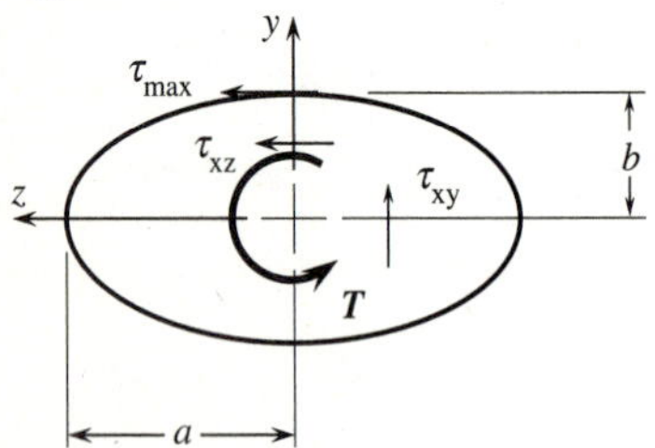

C4.4 Determine the maximum shear stress and angle of twist of a shaft of rectangular cross section. The input should include the applied torque T, the dimensions of the cross-section, the shear modulus G, the length L, and the constants k_1 and k_2. Check your program using the following data: $T = 15$ N · m, width $b = 24$ mm, and height $a = 6$ mm.

Figure C4.4

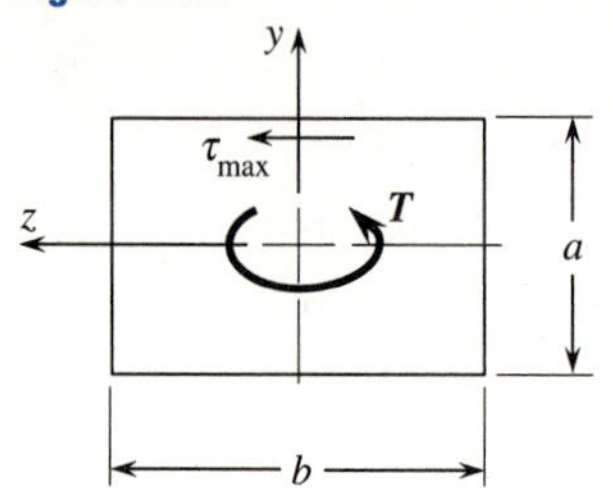

C4.5 Determine the shear stress and angle of twist of a shaft of closed, thin-walled, rectangular cross-section. The input should include the applied torque T, the thickness t of the wall, the dimensions of the rectangle, the shear modulus G, and the length of the shaft. Check your program using the following data: $T = 2000$ lb · in, average width of 3 in., average height of 2 in., thickness t of 1/8 in., and $G = 3.0 \times 10^6$ psi.

Figure C4.5

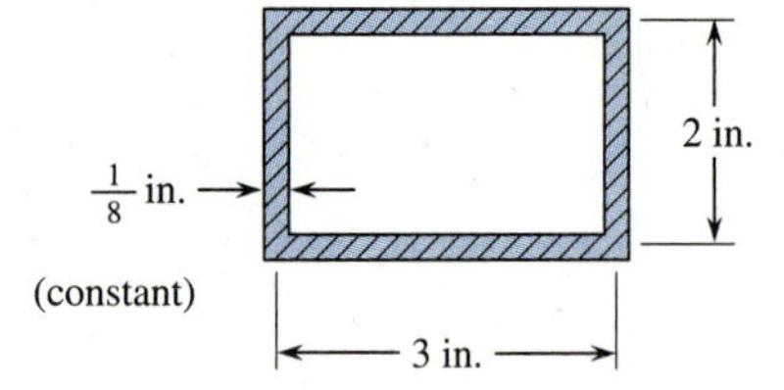

Beam Stresses

5.1 INTRODUCTION

In this chapter, we discuss the loading on beams and the subsequent normal and shear stress distributions that result. We develop the basic beam equations and describe how to draw shear force and bending moment diagrams. We develop the beam bending stress formula and beam shear stress formula. We then describe how to select or design prismatic beams. We describe how to analyze beams subjected to moving loads as when a vehicle moves across a bridge. Then we consider stress concentrations in bending members. Unsymmetric bending, along with principal moments of inertia, are considered. Finally, we consider shear flow and shear center in thin cross-sections, such as wide-flange and channel sections.

The beam is probably the most common type of structural element. A *beam* is generally defined as a member having cross-sectional dimensions significantly less than its length and loaded transversely as shown in Figures 5.1 and 5.2.

Although complicated combinations of loading and support conditions can exist, statically determinate beams are generally referred to as either cantilevered, simply supported, or overhanging. A *cantilevered beam* is fixed at one end and completely unsupported or free at the other end as shown in Figure 5.2(a). Recall from statics that the fixed end reactions may be forces in the x- and y-directions and a bending moment as shown in Figure 5.2(b). Simple beams are supported at each end by pins and/or rollers as shown in Figure 5.2(c). The reactions at a pin may be a force in the x-direction and a force in the y-direction, while the reaction at the roller support is a single force normal to the surface the roller acts on. Here this is a force in the y-direction. Figure 5.2(d) shows the reactions acting on a simple beam. Overhanging beams have portions of their length hanging over a support as shown in Figure 5.2(e). Figure 5.2(f) shows the reactions on an overhanging beam supported by a pin and a roller.

(a)

(b)

Figure 5.1 (a) Beam subjected to loading from a large air conditioning unit, and (b) steel beams being placed for interstate highway bridge.

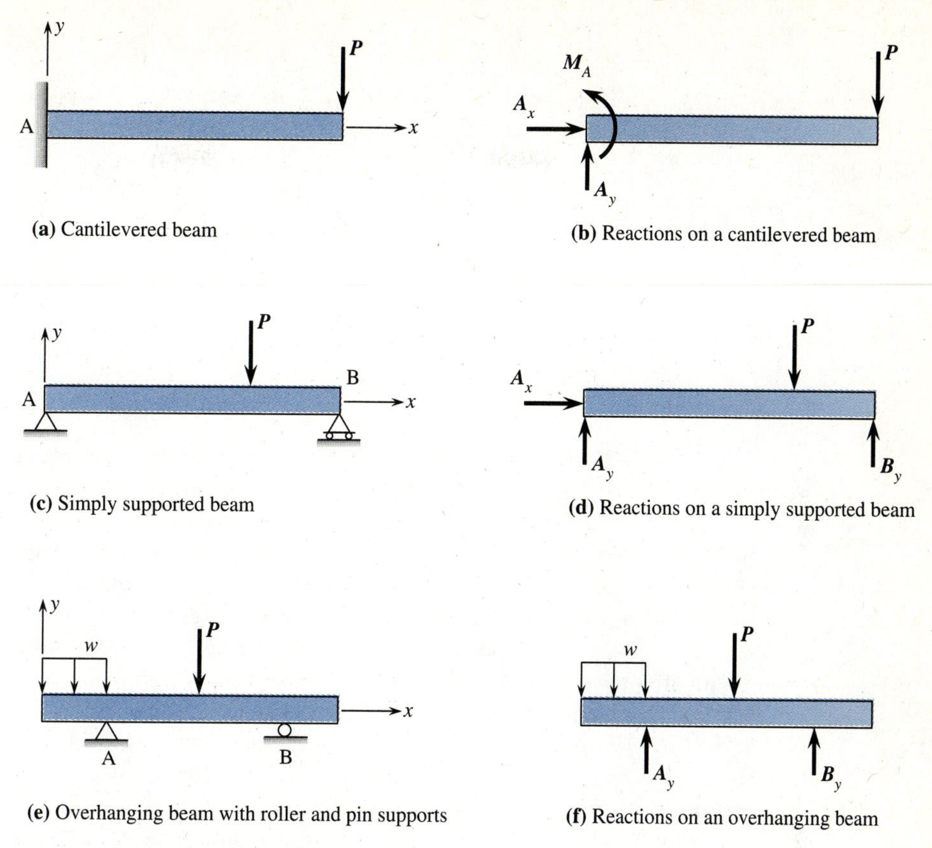

Figure 5.2 Beams with various types and arrangements of supports.

5.2 BASIC BEAM EQUATIONS

To determine the stresses in beams due to externally applied loads, we must first evaluate the shear force and bending moment at each cross section of the beam. These values of shear force and bending moment are then plotted over the span coordinate of the beam. The plots of shear force and bending moment as a function of the longitudinal coordinate x of the beam are called *shear force and bending moment diagrams* and are essential in the analysis of beams.

Consider the beam shown in Figure 5.3 loaded with distributed load $w(x)$ (expressed in units of force per unit length).

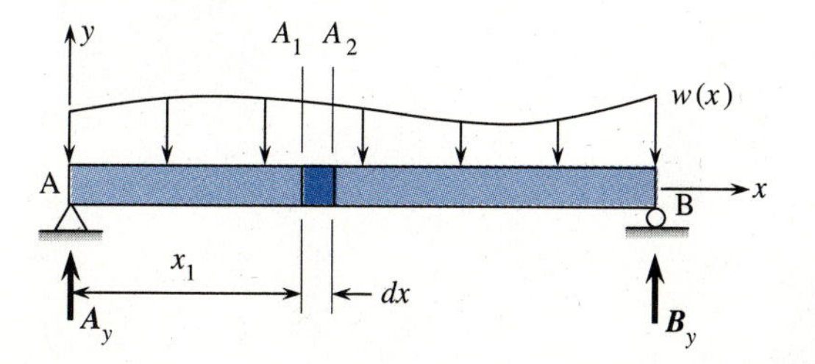

Figure 5.3 Simply supported beam with distributed load.

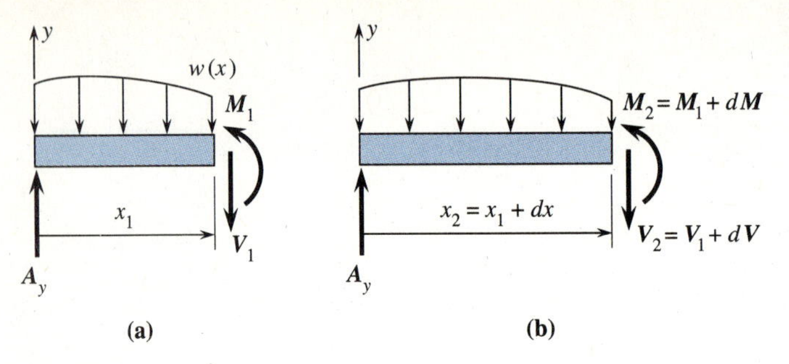

Figure 5.4 Cut sections of beam in equilibrium.

The vertical reactions of the supports, A_y and B_y, are obtained from the equations of static equilibrium. That is, $\Sigma F_y = 0$ and $\Sigma M = 0$.

We pass a plane through section A_1, located at distance x_1, and remove the right portion. The left portion is shown in Figure 5.4(a) with the shear force, V_1 and the bending moment, M_1, that are required to hold the cut section in equilibrium.

The unknown shear force, V_1, and bending moment, M_1, at this cut section located a distance x_1 from the left support are easily determined from the equations of static equilibrium after the distributed load function $w(x)$ is known.

Next, we cut the beam at section A_2 located a distance $x_2 = x_1 + dx$ from the left end (Figure 5.4(b)). The shear force and bending moment at section A_2 will be $V_2 = V_1 + dV$ and $M_2 = M_1 + dM$, which again can be determined from statics. Here dV and dM represent the changes in shear force and bending moment at distance dx from $x = x_1$.

If we now isolate the element dx and place it in equilibrium, we obtain the free-body diagram shown in Figure 5.5. Note that V_1 and M_1 are acting opposite to the directions that they were acting in Figure 5.4(a). This is because in Figure 5.5 they are acting on the left face instead of the right face. This is another illustration of Newton's third law of equal but opposite actions. Positive shear and positive moment are shown in Figure 5.5. That is, the shear is down on the right side (face) and up on the left side, and the moment is counterclockwise on the right face and clockwise on the left so as to produce compression in the top of the beam. These sign conventions will be used throughout our discussion on shear and moment diagrams in Section 5.3.

Since the element of Figure 5.5 must be in static equilibrium, we have

$$\Sigma F_y = 0: \quad V_1 - w(x)dx - (V_1 + dV) = 0 \tag{5.1}$$

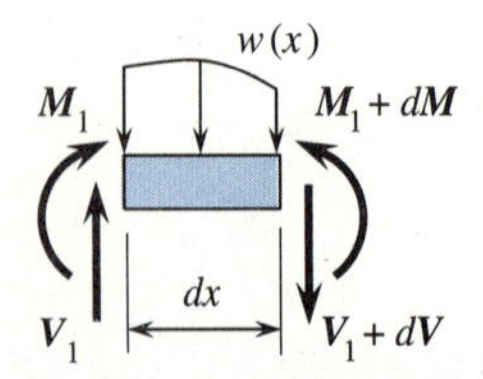

Figure 5.5 Free-body diagram of beam element *dx*.

where $w(x)$ can be considered constant over length dx as dx becomes small. Therefore, from Eq. (5.1), we obtain

$$\frac{dV}{dx} = -w(x) \tag{5.1a}$$

Equation (5.1a) indicates that for portions of the beam where the distributed load is a continuous function (no abrupt changes), the rate of change in the shear force is equal to the negative of the distributed load. The negative sign indicates that the shear force decreases with x when the distributed load acts downward. (As shown in Figure 5.3, $w(x)$ is considered to be positive when acting downward.)

The element of Figure 5.5 is also in moment equilibrium. Therefore, taking moments about any point (here point two) yields

$$\Sigma M_2 = 0: \quad -M_1 - V_1 dx + w(x)dx\frac{dx}{2} + (M_1 + dM) = 0 \tag{5.2}$$

Dividing Eq. (5.2) by dx and letting dx approach zero (then $w(x)dx$ approaches zero), we obtain

$$V_1 = \frac{dM}{dx} \tag{5.2a}$$

Since the location x_1 is arbitrary, the subscript can be dropped and we obtain the general expression

$$V = \frac{dM}{dx} \tag{5.2b}$$

Equation (5.2b) indicates that the shear force at any section of the beam is equal to the rate of change of the bending moment at that section with respect to the span coordinate, x. This concept is often useful when plotting the bending moment diagram as will be illustrated in Examples 5.1 through 5.6. The integral expression of Eq. (5.2b) also has a physical interpretation. If we integrate V between two points on the beam, say x_1 and x_2, we obtain

$$\int_{x_1}^{x_2} V dx = \int_{x_1}^{x_2} dM = M_2 - M_1 \tag{5.2c}$$

The physical interpretation of Eq. (5.2c) is that the total area under the shear force versus span diagram between two points is equal to the change in bending moment between those same two points.

The differential of Eq. (5.2b) is sometimes useful when plotting the shear force diagram. Differentiating Eq. (5.2b) with respect to x and using Eq. (5.1a), we obtain

$$\frac{dV}{dx} = \frac{d^2M}{dx^2} = -w(x) \tag{5.2d}$$

In all of the above definitions, it has been assumed that the shear force V, the bending moment M, and the distributed load, $w(x)$, have all been continuous functions of x. In the cases of discontinuous functions (resulting from abrupt changes in loading along the span) the beam can be treated by breaking it up into piecewise continuous portions or by the method of discontinuity functions. In Section 5.3, we illustrate the method of breaking up the beam into piecewise continuous portions to obtain the shear force and bending moment diagrams for beams subjected to abrupt changes in loading. The method of discontinuity functions is described in Section 6.6. In this section, we illustrate how to express loading functions, shear, bending moment, slope, and deflection in terms of discontinuity functions.

5.3 SHEAR FORCE AND BENDING MOMENT DIAGRAMS

Now that we have established some basic relationships between shear forces and bending moments on beams, we will utilize that information to learn how to construct basic shear force and bending moment diagrams. As we discussed in Section 5.2, the determination of the shear and moment diagrams is the first step in the analysis of a beam.

STEPS IN SOLUTION

The following steps are generally used to obtain the complete shear force and bending moment diagrams for beams subjected to general loading conditions.

1. Draw a free-body diagram of the beam and determine the reactions as needed from statics.
2. Place any original distributed loading on the beam and cut through a section and draw a free-body diagram of the section showing V and M on the cut. Remember if the cut is on the right side of the beam, show V down and M counterclockwise on the cut section in accordance with the shear and moment conventions introduced in Section 5.2. (If the cut section is on the left side of the beam, show V up and M clockwise on the cut.)
3. Determine V and M from statics, that is, from

$$\Sigma F_y = 0 \quad \text{and} \quad \Sigma M = 0$$

4. Repeat this process of cutting sections between discontinuous loads on the beam until the whole beam has been considered.
5. For cantilevered beams, it is normally best to start from the free end and proceed across the beam.

Examples 5.1 through 5.6 illustrate the steps used to obtain complete shear force and bending moment diagrams.

EXAMPLE 5.1

For the simply supported beam subjected to the concentrated force P, determine the shear force and bending moment diagrams (Figure 5.6).

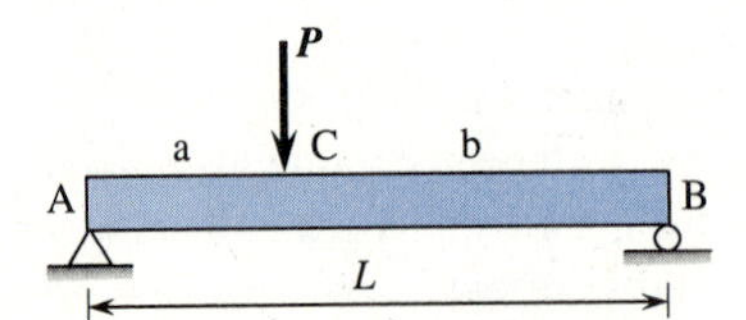

Figure 5.6 Simply supported beam subjected to concentrated load.

Solution

(a) Draw the free-body diagram of the beam.

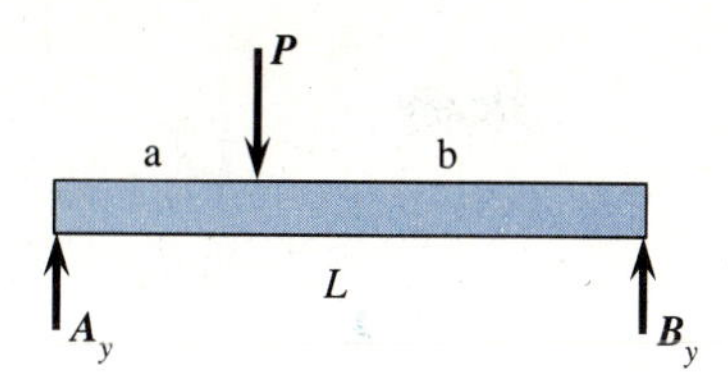

By statics, we have

$$\Sigma M_B = 0: \quad -A_y L + Pb = 0$$

$$A_y = \frac{Pb}{L}$$

$$\Sigma F_y = 0: \quad A_y - P + B_y = 0$$

$$B_y = P - A_y$$

$$= P - P\frac{b}{L}$$

$$= P\left(\frac{L - b}{L}\right)$$

$$= \frac{Pa}{L}$$

(b) Cut the beam between points A and C and draw a free-body diagram of the cut section showing V and M on the cut. Because the cut is on the right side of the section, by convention show V down and M counterclockwise.

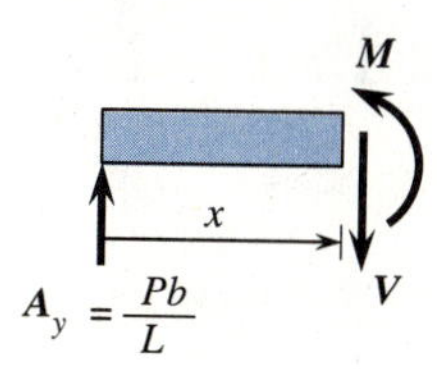

(c) By statics, we obtain

$$\Sigma F_y = 0: \quad \frac{Pb}{L} - V = 0$$

$$V = \frac{Pb}{L} \qquad \textbf{(a)}$$

$$\Sigma M = 0: \quad -\left(\frac{Pb}{L}\right)x + M = 0$$

$$M = \frac{Pbx}{L} \qquad \textbf{(b)}$$

Equations (a) and (b) hold for $0 \le x \le a$.

(d) Now cut the section between points C and B and draw the free-body diagram of the cut section.

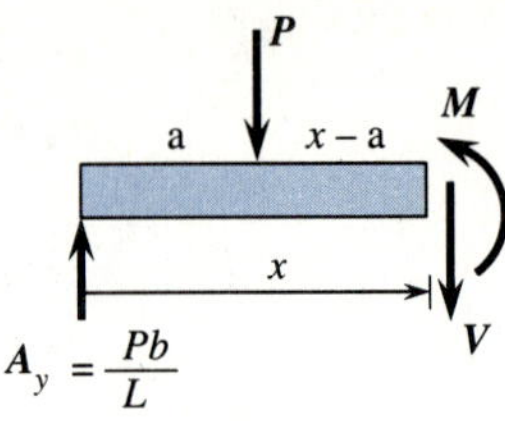

By statics, we have

$$\Sigma F_y = 0: \quad \frac{Pb}{L} - P - V = 0$$

$$V = \frac{Pb}{L} - P = P\left(\frac{b - L}{L}\right)$$

$$V = -\frac{Pa}{L} \tag{c}$$

$$\Sigma M = 0: \quad -\frac{Pb}{L}x + P(x - a) + M = 0$$

$$M = Px\left(\frac{b}{L} - 1\right) + Pa$$

$$= -\frac{Pax}{L} + Pa$$

$$M = Pa\left(\frac{L - x}{L}\right) \tag{d}$$

Equations (c) and (d) hold for $a \leq x \leq L$.

(e) We now plot Eqs. (a) and (c) to obtain the complete shear force diagram, and Eqs. (b) and (d) to obtain the complete bending moment diagram as shown in Figure 5.7.

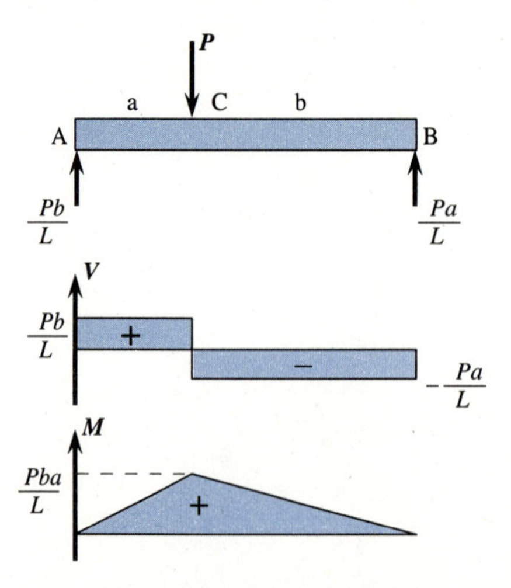

Figure 5.7 Shear force and bending moment diagrams for simple beam subjected to concentrated load.

EXAMPLE 5.2

For the simply supported beam subjected to the uniformly distributed load w, determine the shear force and bending moment diagrams (Figure 5.8).

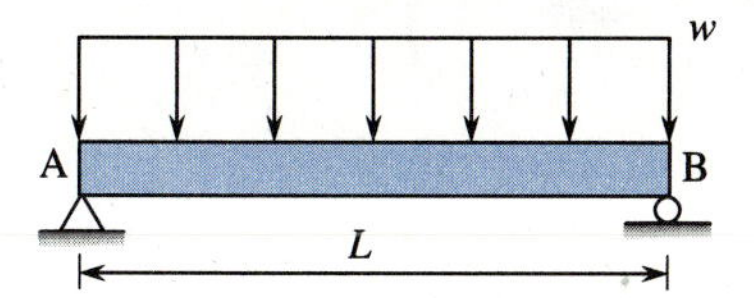

Figure 5.8 Simply supported beam subjected to uniformly distributed load.

Solution

(a) Draw the free-body diagram of the beam.

NOTE: We replace the uniformly distributed load by its statically equivalent force, F.

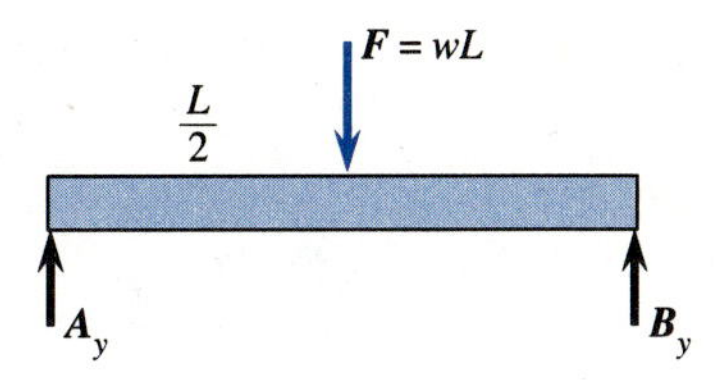

By statics, we have

$$\Sigma M_B = 0: \quad -A_y L + (wL)\frac{L}{2} = 0$$

$$A_y = \frac{wL}{2}$$

$$\Sigma F_y = 0: \quad A_y - wL + B_y = 0$$

$$B_y = \frac{wL}{2}$$

NOTE: Due to symmetry, we observe that one-half the total load is supported at each support as formally shown using the equations of statics above.

(b) Cut the beam to the right of point A and draw the free-body diagram of the section showing V and M on the cut. Remember to leave the original distributed load on the portion of the cut beam initially.

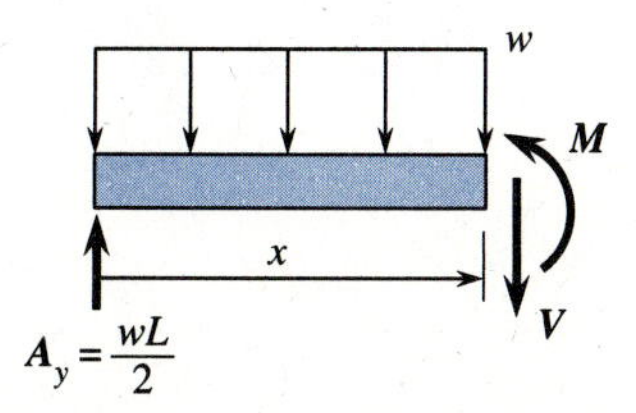

Replace the distributed load by its statically equivalent load.

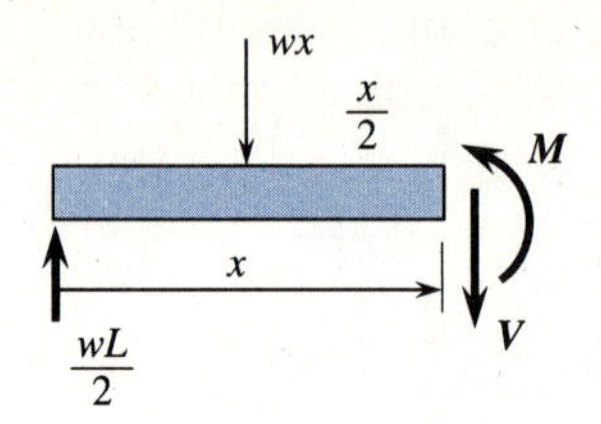

By statics, we obtain

$$\Sigma F_y = 0: \quad \left(\frac{wL}{2}\right) - wx - V = 0$$

$$V = \frac{wL}{2} - wx \tag{a}$$

$$\Sigma M = 0: \quad -\left(\frac{wL}{2}\right)x + wx\left(\frac{x}{2}\right) + M = 0$$

$$M = \left(\frac{wL}{2}\right)x - \frac{wx^2}{2} \tag{b}$$

Because the load is continuous across the whole beam, one cut is sufficient to obtain V and M expressions for the beam. We should note that differentiating Eq. (b) with respect to x yields Eq. (a) and therefore satisfies Eq. (5.2b), that is

$$V = \frac{dM}{dx}$$

(c) Plot Eqs. (a) and (b) to obtain the complete shear force and bending moment diagrams as shown in Figure 5.9.

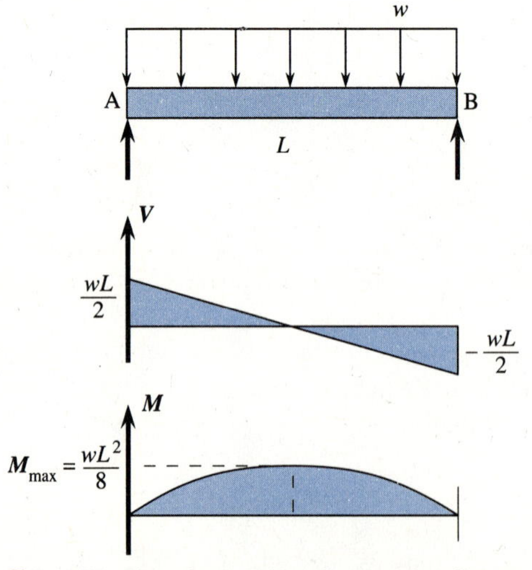

Figure 5.9 Shear force and bending moment diagrams for simple beam subjected to uniformly distributed load.

EXAMPLE 5.3

For the overhanging beam subjected to the linearly varying distributed load, determine the shear force and bending moment diagrams.

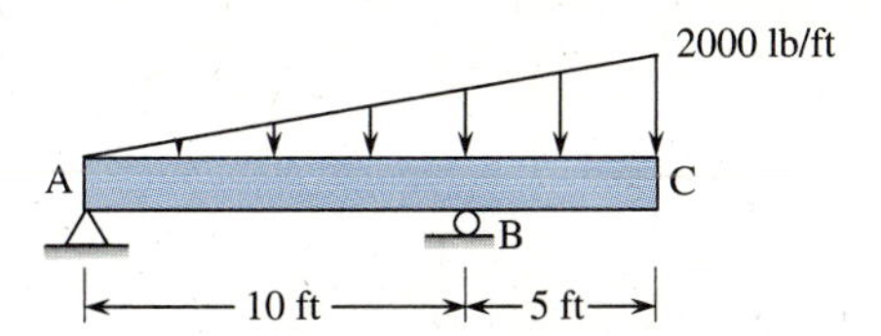

Figure 5.10 Overhanging beam subjected to linearly varying distributed load.

Solution

(a) Draw the free-body diagram of the beam. We replace the triangular load distribution with its statically equivalent force F, where

$$F = \frac{1}{2}(15 \text{ ft})\,(2000 \text{ lb/ft}) = 15{,}000 \text{ lb}$$

and F is located $\frac{2}{3}$ of the distance from A to C.

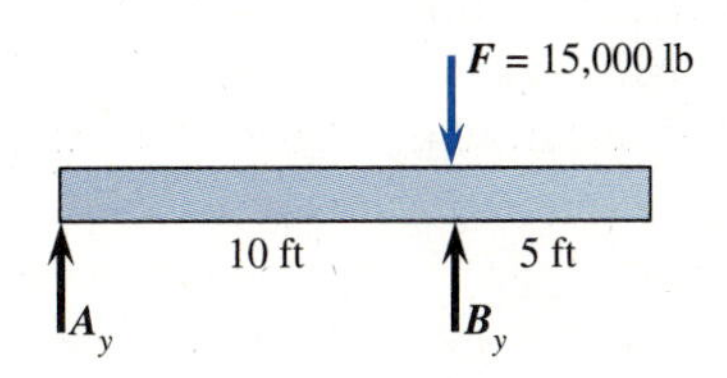

By statics, we obtain

$$\Sigma M_B = 0: \quad -A_y(10 \text{ ft}) = 0$$

$$A_y = 0$$

$$\Sigma F_y = 0: \quad B_y - F = 0$$

$$B_y = 15{,}000 \text{ lb}$$

(b) Cut the beam between points A and B and draw the free-body diagram of the section. Remember to place the original distributed load on the portion of the cut beam.

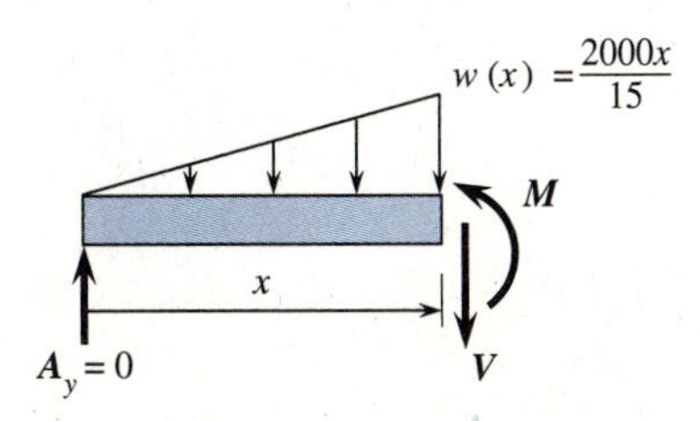

Replace the distributed load by its statically equivalent force F, where

$$F = \frac{1}{2}(x)\left(\frac{2000x}{15}\right) = \frac{1000x^2}{15} = 66.7\,x^2$$

located at 2/3 x.

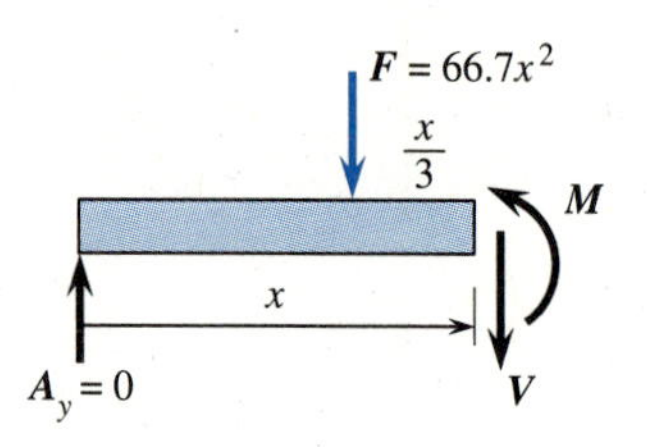

By statics, we have

$$\Sigma F_y = 0: \quad -F - V = 0$$

$$V = -66.7x^2 \qquad \textbf{(a)}$$

$$\Sigma M = 0: \quad (66.7x^2)\left(\frac{x}{3}\right) + M = 0$$

$$M = -22.2x^3 \qquad \textbf{(b)}$$

Equations (a) and (b) hold for $0 \leq x \leq 10$ ft.

Due to the reaction at B, another cut to the right of point B is needed to obtain the shear force and bending moment between points B and C.

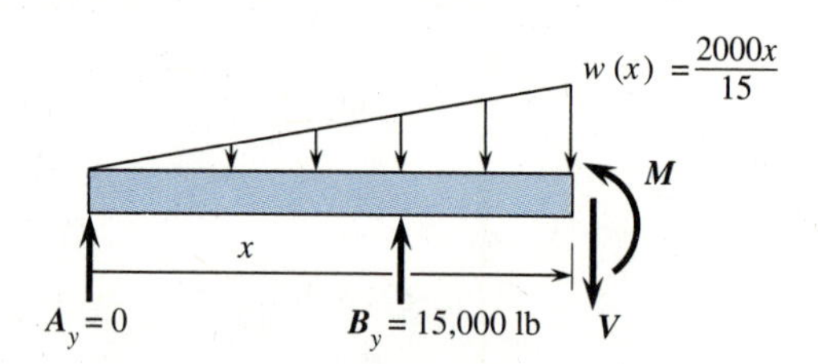

Replace the distributed load by its statically equivalent force F as shown below.

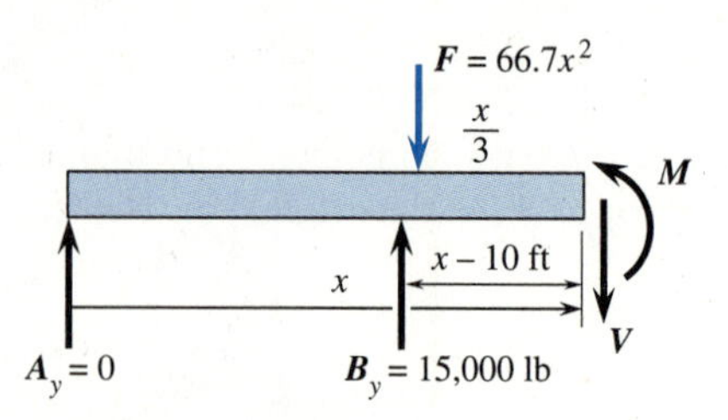

By statics, we have

$$\Sigma F_y = 0: \quad -66.7x^2 + 15{,}000 - V = 0$$

$$V = -66.7x^2 + 15{,}000 \qquad \textbf{(c)}$$

$$\Sigma M = 0: \quad (66.7x^2)\left(\frac{x}{3}\right) - 15{,}000(x - 10) + M = 0$$

$$M = -22.2x^3 + 15{,}000x - 150{,}000 \qquad \textbf{(d)}$$

Equations (c) and (d) hold for 10 ft $\leq x \leq$ 15 ft.

(c) Plot Eqs. (a) and (c) to obtain the complete shear force diagram, and Eqs. (b) and (d) to obtain the complete bending moment diagram as shown in Figure 5.11.

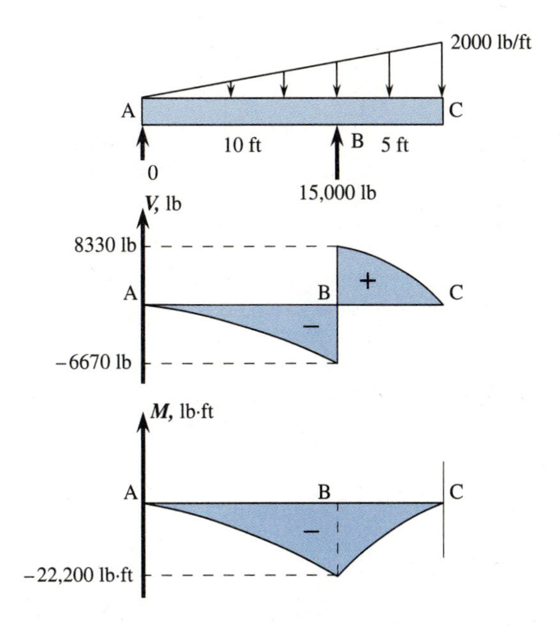

Figure 5.11 Complete shear force and bending moment diagrams for overhanging beam subjected to linearly varying load.

It is now of interest to describe how Eq. (5.1a) $\left(\frac{dV}{dx} = -w\right)$ can be used to aid in plotting the shear force diagram. In Figure 5.10, we observe that w is an increasing value as x increases from left to right (by convention positive w is taken as a downward value). Based on Eq. (5.1a), the shear diagram has a negatively increasing slope from A to B. The discontinuity in the shear diagram at B is due to the reaction force at B. The slope of the shear diagram has a negatively increasing value from B to C (based on Eq. (5.1a)). Also the shear force goes to zero at C as it should because there is not a concentrated load or reaction at C.

The moment diagram is then plotted with the aid of Eq. (5.2b) $\left(\frac{dM}{dx} = V\right)$. For instance, from A to B the shear is a negatively increasing value; therefore, the slope of the bending moment diagram must be a negatively increasing value (based on satisfying Eq. (5.2b)). From B to C the value of the shear force is a positive, but decreasing, value. Therefore, the slope of the bending moment diagram must be a positive but decreasing one as shown in Figure 5.11.

EXAMPLE 5.4

For the cantilevered beam shown in Figure 5.12 subjected to both a concentrated and a uniformly distributed load, determine the shear force and bending moment diagrams.

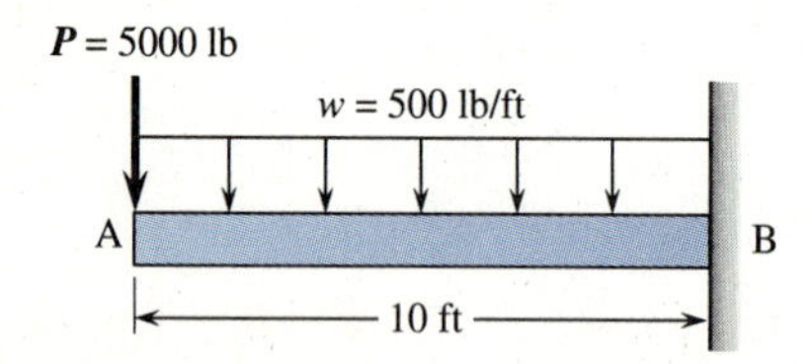

Figure 5.12 Cantilevered beam subjected to both concentrated and uniformly distributed loads.

Solution

(a) For the cantilevered beam, start from the free end and cut the section. The reactions at the fixed end B are then not needed in the analysis.

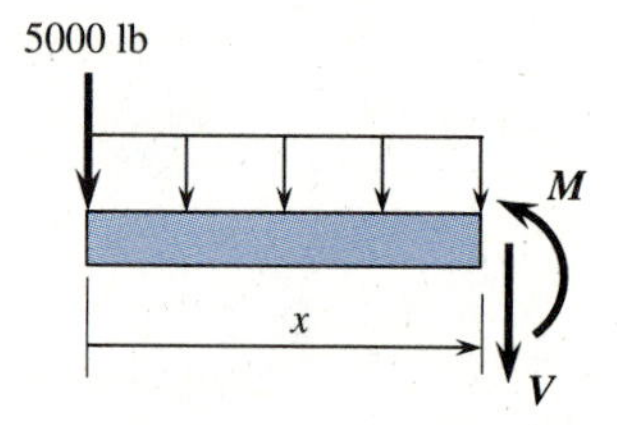

By statics, we have

$$\Sigma F_y = 0: \quad -5000 - 500x - V = 0$$

$$V = -5000 - 500x \qquad \textbf{(a)}$$

$$\Sigma M = 0: \quad 5000x + (500x)\left(\frac{x}{2}\right) + M = 0$$

$$M = -5000x - 250x^2 \qquad \textbf{(b)}$$

Equations (a) and (b) are sufficient to obtain V and M for the whole beam. This is because the loading is continuous over the whole beam.

(b) We now plot Eqs. (a) and (b) to obtain the complete V and M diagrams as shown in Figure 5.13.

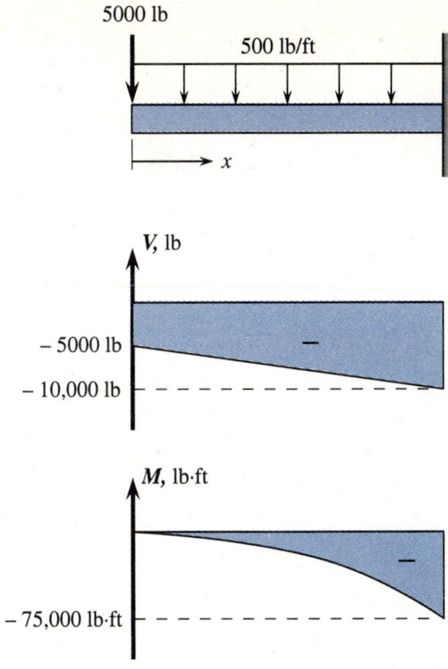

Figure 5.13 Complete shear and bending moment diagrams for cantilevered beam subjected to concentrated and uniformly distributed load.

EXAMPLE 5.5

For the overhanging beam shown in Figure 5.14 supporting the 100-kg mass, determine the shear force and bending moment diagrams.

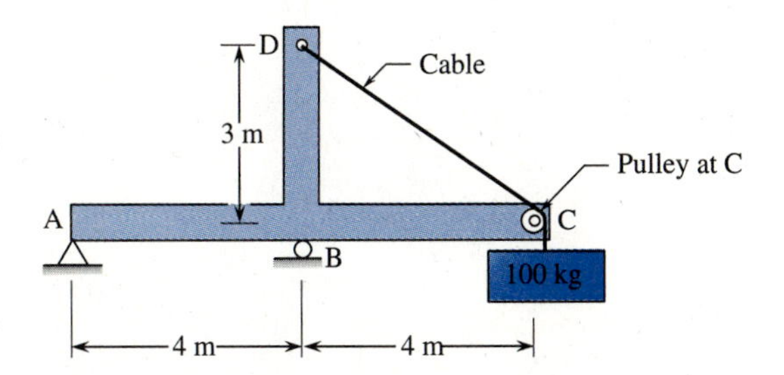

Figure 5.14 Overhanging beam supporting 100 kg mass.

Solution

(a) Determine the reactions at A and B and internal reactions from BD at B and from the cable at C.

First, draw the free-body diagram of pulley C.

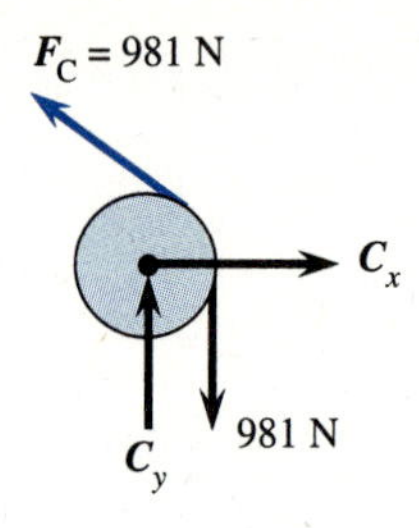

By statics, we obtain

$$\Sigma F_y = 0: \quad C_y + 981\left(\frac{3}{5}\right) - 981 = 0$$

$$C_y = 392 \text{ N}$$

$$\Sigma F_x = 0: \quad C_x - 981\left(\frac{4}{5}\right) = 0$$

$$C_x = 785 \text{ N}$$

Next, draw a free-body diagram of member BD.

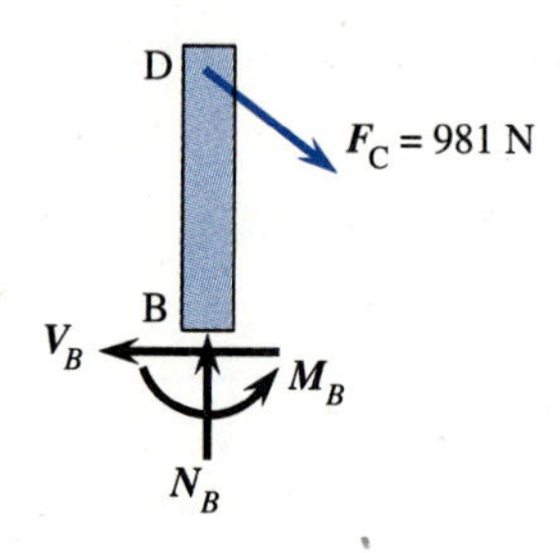

By statics, we obtain

$$\Sigma F_x = 0: \quad -V_B + 981\left(\frac{4}{5}\right) = 0$$

$$V_B = 785 \text{ N}$$

$$\Sigma F_y = 0: \quad N_B - 981\left(\frac{3}{5}\right) = 0$$

$$N_B = 589 \text{ N}$$

$$\Sigma M_B = 0: \quad M_B - 981\left(\frac{4}{5}\right)(3m) = 0$$

$$M_B = 2355 \text{ N} \cdot \text{m}$$

Now, by using Newton's third law, apply the opposite forces C_x and C_y at C, and N_B, V_B, and moment M_B at B to the beam ABC.

Finally, draw the free-body diagram of beam ABC.

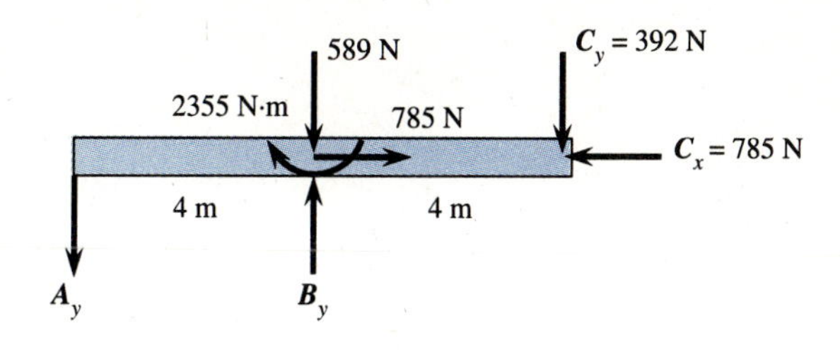

By statics, we have

$$\Sigma M_B = 0: \quad A_y(4m) - 2355 - 392(4m) = 0$$

$$A_y = 981 \text{ N}$$

$$\Sigma F_y = 0: \quad B_y - 981 - 589 - 392 = 0$$

$$B_y = 1962 \text{ N}$$

(b) Cut the beam between A and B and show V and M on the cut.

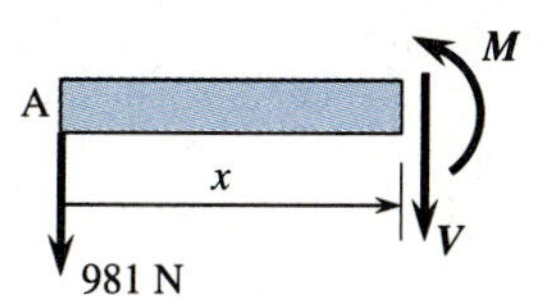

By statics, we obtain

$$\Sigma F_y = 0: \quad -981 - V = 0$$

$$V = -981 \text{ N} \qquad \textbf{(a)}$$

$$\Sigma M_{\text{cut}} = 0: \quad 981x + M = 0$$

$$M = -981x \qquad \textbf{(b)}$$

These equations (a) and (b) are valid for $0 \leq x \leq 4m$.

(c) Cut the beam between B and C and determine V and M again.

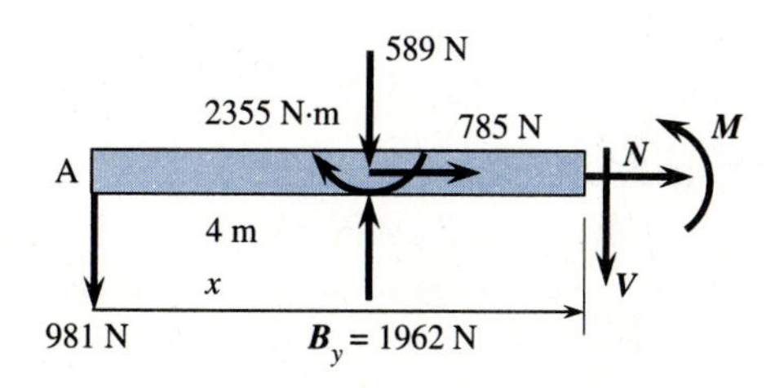

By statics, we write

$$\Sigma F_y = 0: \quad -981 + 1962 - 589 - V = 0$$

$$V = 392 \text{ N} \qquad \textbf{(c)}$$

$$\Sigma M_{cut} = 0: \quad 981x - 1962\,(x - 4) - 2355 + 589\,(x - 4) + M = 0$$

$$M = 981x - 7848 + 2355 - 589x + 2356$$

$$= 392x - 3137 \qquad \textbf{(d)}$$

These equations (c) and (d) are valid for $4m \leq x \leq 8m$.

Note that you could have obtained vertical reactions A_y and B_y by using the entire member ABCD as a free body. This serves as a check on these values as obtained above.

(d) Plot Eqs. (a) and (c) to obtain the complete shear force diagram, and Eqs. (b) and (d) to obtain the complete bending moment diagram as shown in Figure 5.15.

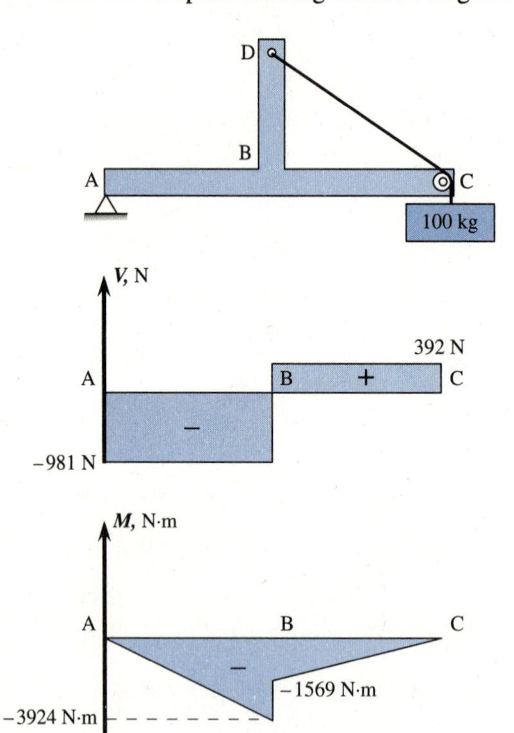

Figure 5.15 Shear force and bending moment diagrams for overhanging beam supporting 100 kg mass.

EXAMPLE 5.6

For the overhanging beam shown in Figure 5.16 subjected to the uniformly applied distributed load, determine the shear force and bending moment equations and draw the shear force and bending moment diagrams.

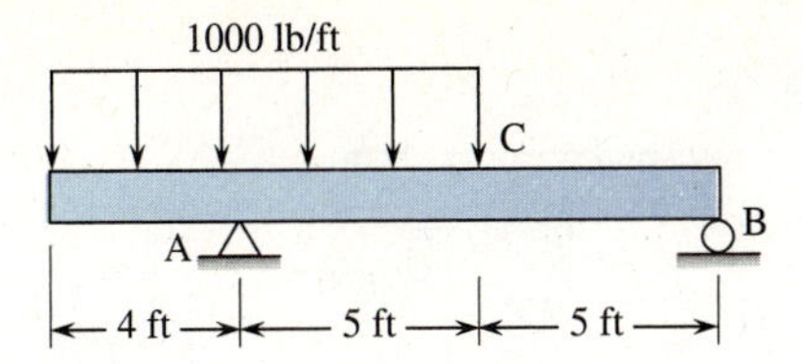

Figure 5.16 Overhanging beam subjected to distributed load.

Solution

(a) Draw a free-body diagram of the beam. Replace the uniformly distributed load by its statically equivalent force F as follows.

$$F = (1000 \text{ lb/ft})(9 \text{ ft}) = 9000 \text{ lb}$$

This force is located at 4.5 ft from the left end.

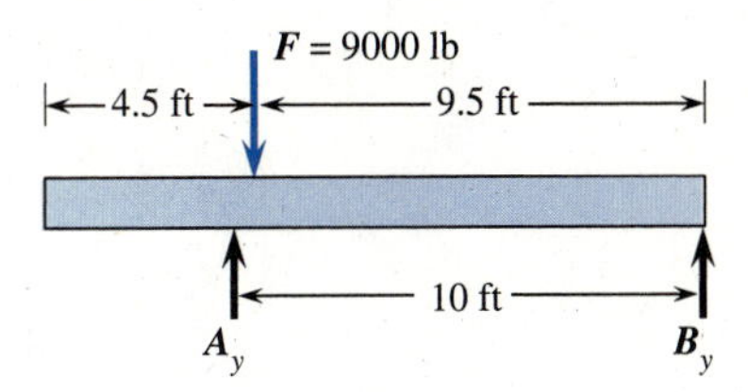

Determine the reactions at A and B by using statics as

$$\Sigma M_B = 0: \quad -A_y(10 \text{ ft}) + (9000 \text{ lb})(9.5 \text{ ft}) = 0$$

$$A_y = 8550 \text{ lb}$$

$$\Sigma F_y = 0: \quad A_y - 9000 + B_y = 0$$

$$B_y = 450 \text{ lb}$$

(b) Cut the beam at the sections needed to obtain the shear force and bending moment equations as follows:

For $0 \leq x \leq 4$ ft

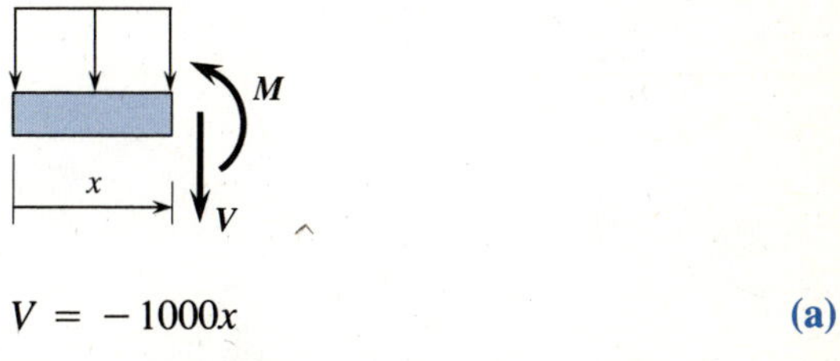

$$V = -1000x \qquad \textbf{(a)}$$

$$M = -1000\left(\frac{x^2}{2}\right) \qquad \textbf{(b)}$$

For 4 ft $\leq x \leq$ 9 ft

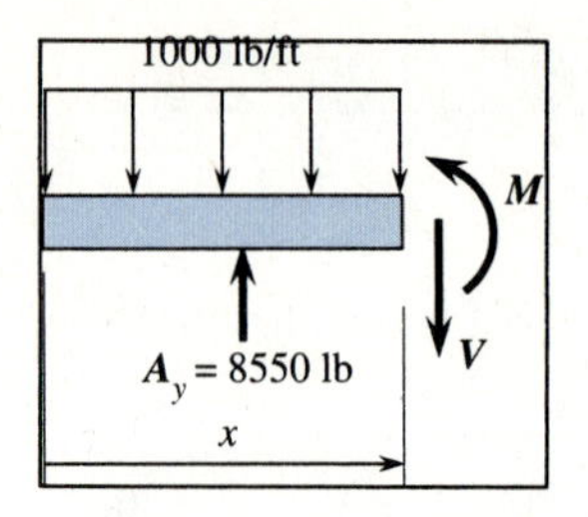

$$V = -1000x + 8550 \tag{c}$$

$$M = -1000\left(\frac{x^2}{2}\right) + 8550(x - 4) \tag{d}$$

Now start from the right end (indicated by using the y-coordinate).
For $0 \leq y \leq 5$ ft

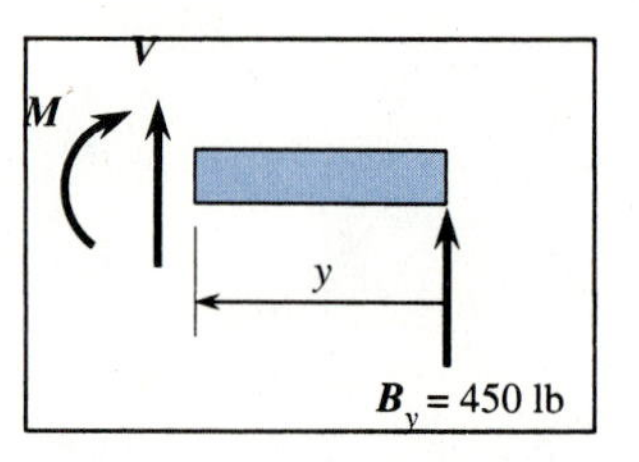

$$V = -450 \tag{e}$$

$$M = 450y \tag{f}$$

(c) Plot Eqs. (a) through (f) to obtain the complete shear force and bending moment diagrams as shown in Figure 5.17.

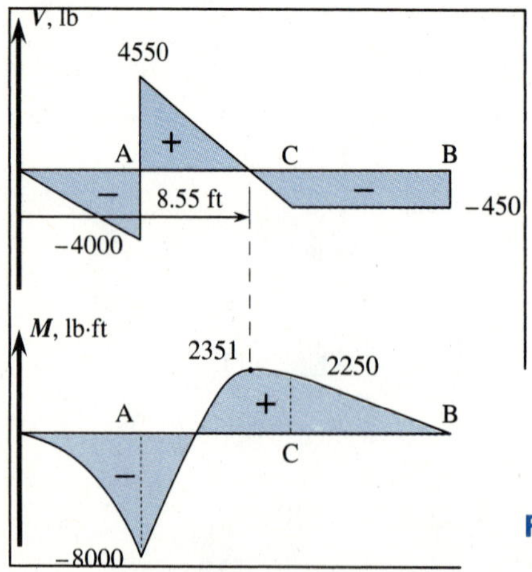

Figure 5.17 Complete shear force and bending moment diagrams for beam of Figure 5.16.

5.4 BEAM BENDING STRESS

In Section 5.3, we determined the bending moment for beams under various loading conditions. We will now determine the *bending* or *flexural stresses* on a cross-section of a beam. We will observe that this bending stress depends on the bending moment acting on the cross-section of the beam.

We make specific assumptions during the derivation of the bending stress equations. The following assumptions may seem to be somewhat restrictive; however, they are closely approximated in many practical engineering problems involving beam bending stress analysis. These assumptions are the following.

1. The loads are assumed to act on the beam so that no twisting of the beam occurs. This requires the loads to lie in a principal plane or along the principal axes of the cross-section as shown in Figure 5.18. This assumption is valid, for instance, whenever a cross-section has at least one axis or plane of symmetry and the loading acts through the centroid and in this plane of symmetry as shown in Figure 5.18.
2. The beam is initially straight before loading, that is, the beam is not curved.
3. Plane cross-sections perpendicular to the longitudinal centroidal axis of the beam before bending occurs remain plane and perpendicular to the longitudinal axis after bending occurs. This is illustrated in Figure 5.19 where a plane through vertical line ac is perpendicular to the longitudinal x-axis before bending and this same plane through a′c′ remains perpendicular to the bent x-axis after bending. This is also shown in Figures 5.20(a) and 5.20(b) for a beam with grid marks shown on it to illustrate better the concept that plane sections remain plane before and after being subjected to pure bending.

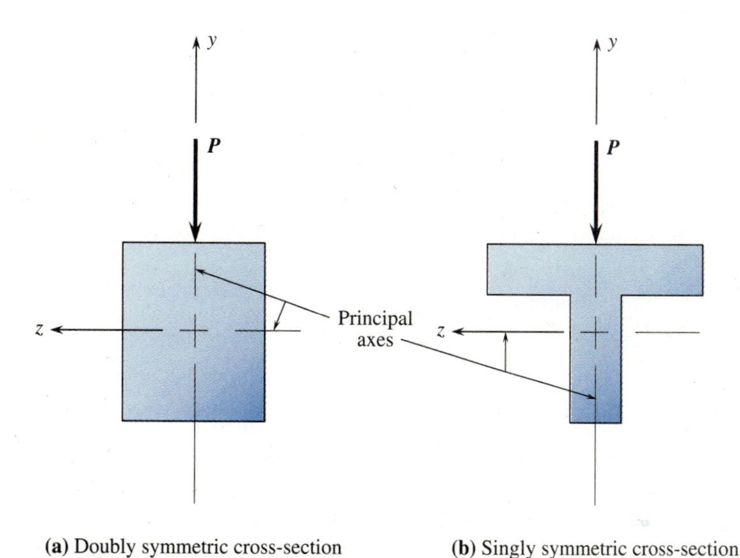

(a) Doubly symmetric cross-section **(b)** Singly symmetric cross-section

Figure 5.18 Illustration of assumption one showing location of transverse load on a beam cross section and showing loads acting along principal axes.

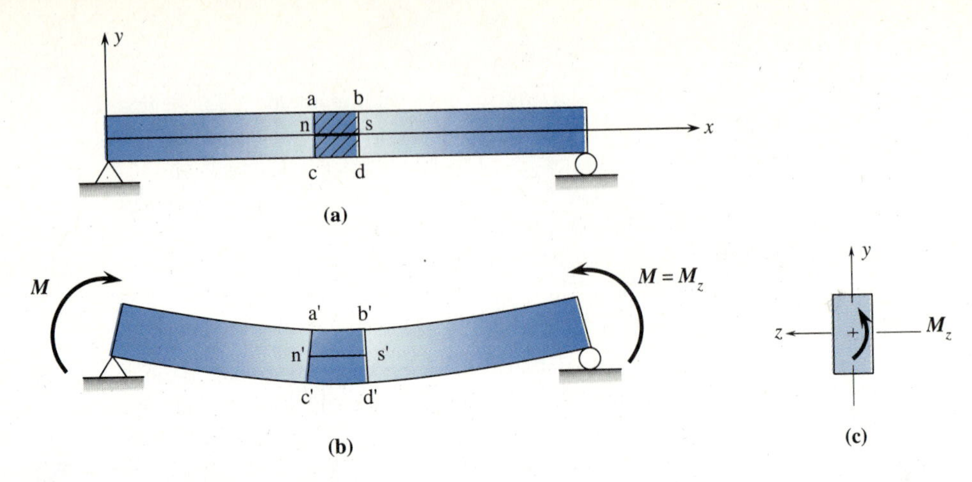

Figure 5.19 (a) Undeformed beam,
(b) deformed beam due to applied bending moments M, and
(c) cross section of the beam showing M applied through centroidal plane.

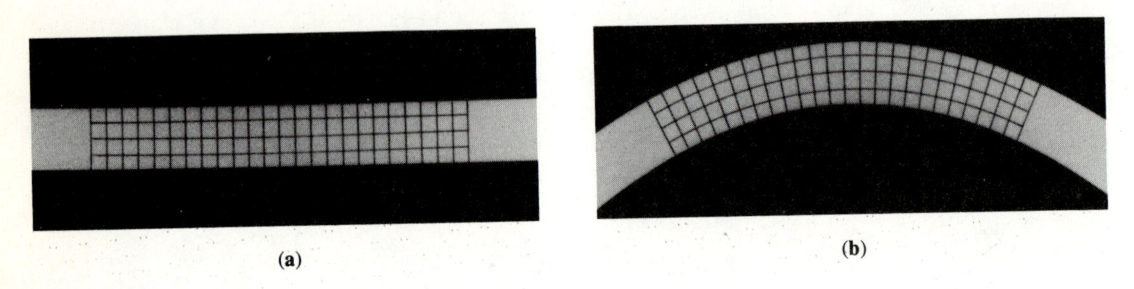

Figure 5.20 (a) Undeformed rubber beam and
(b) deformed rubber beam due to applied end moments.

4. The proportional limit stress of the material is not exceeded, that is, Hooke's law is obeyed, $\sigma = E\epsilon$.
5. The deformations and rotations are small.
6. The modulus of elasticity E of the material is the same in both tension and compression.
7. The material is homogeneous.

Kinematics of Bending

To derive the bending stress formula, we first consider the kinematics of bending or the geometry of deformation. We first consider the undeformed beam shown in Figure 5.19(a). Reference points a and b are on the top side of the beam, points n and s are at the neutral axis of the beam (also shown subsequently to be the centroidal z-axis of the beam), and points c and d are at the bottom side of the beam. Also remember that lines ac and bd are perpendicular to the longitudinal x-axis.

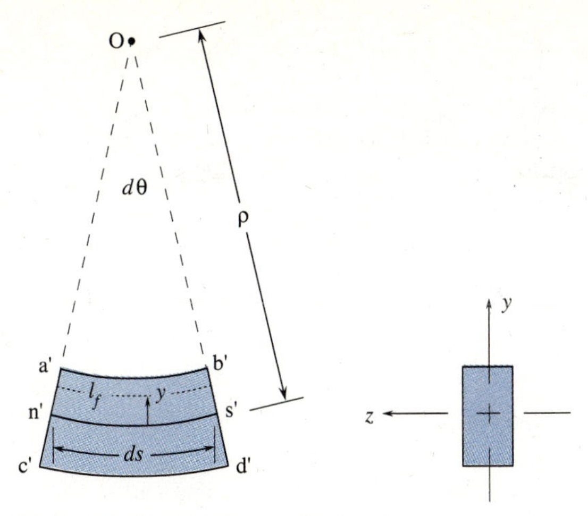

Figure 5.21 Portion of beam after bending.

It is then most convenient for the derivation to subject the beam to applied bending moments M. The deformed shape of the beam due to these moments is shown in Figure 5.19(b). These moments are applied in the principal vertical plane, as shown in Figure 5.19(c), so as to produce no twisting of the cross-section. A deformed section ds of this beam is shown in Figure 5.21 where the bending deformation has been magnified for clarity. We observe that according to assumption 3, the lines a′c′ and b′d′ are in planes passing perpendicular to the bent longitudinal axis and remain plane. These planes merely rotate such that the extension of these lines intersects at O, the origin or center of the radius of curvature measured to the arc n′s′. The angle between the intersecting lines is labeled $d\theta$. The y-axis is measured from the neutral axis or line n′s′.

As the beam is bent, the top fibers ab become compressed or shorten to a′b′, while the bottom fibers cd are tensioned or elongate to c′d′. The so-called neutral axis or surface, represented by ns, does not change length during or after the bending takes place, that is, length ns equals n′s′. In fact, all the material above n′s′ (in the positive y-direction or above the neutral axis) is now in compression, while all the material below the z-axis or in the negative y-direction is now in tension. You can verify this bending deformation by first drawing the vertical lines a-n-c and b-s-d perpendicular to the longitudinal axis of a soft rubber beam and measuring ab = cd, flexing the beam with both hands, and then measuring the new arc lengths a′b′, n′s′, and c′d′.

Now considering the geometry of the deformed portion *ds* in Figure 5.21, and using the definition of normal strain given in Chapter 2, we obtain the normal strain ϵ_x as follows:

$$\epsilon_x = \frac{\text{final length } (l_f) - \text{original length } (l_o)}{\text{original length } (l_o)} \tag{5.3}$$

$$= \frac{l_f - l_o}{l_o} \tag{a}$$

Considering a fiber at distance y above n′s′, we obtain the new or final length using the arc length formula as

$$l_f = (\rho - y)d\theta \tag{b}$$

The original length was

$$l_o = ds = n's' = \rho(d\theta) \tag{c}$$

Substituting Eqs. (b) and (c) into (a), yields

$$\epsilon_x = \frac{(\rho - y)d\theta - \rho d\theta}{\rho d\theta} = \frac{-y}{\rho} \tag{5.4}$$

By Eq. (5.4), the axial strain varies linearly from the neutral surface, n′s′. The negative sign arises because for positive y, the material is in compression. Equation (5.4) does not depend on any assumed stress distribution nor on any assumed material behavior. The material is not limited to elastic beam behavior by Eq. (5.4). Assumptions 1 through 3 are the only ones that apply to Eq. (5.4).

Moment-Bending Stress Relationship

We now develop the *moment-bending stress* relationship or the so-called *flexure formula* for a beam.

First, the relationship between normal stress and strain for a linear-elastic material subjected to a uniaxial loading also applies to the bending stress problem. This is also assumption 4. Therefore, using Eq. (5.4) in $\sigma = E\epsilon$, we obtain

$$\sigma_x = E\left(\frac{-y}{\rho}\right) \tag{5.4a}$$

Equation (5.4a) shows that the normal stress occurs in the x-direction and for linear-elastic materials varies linearly through the depth of the beam. Figure 5.22 shows the linear strain variation ϵ_x given by Eq. (5.4), and the stress variation given by Eq. (5.4a).

We now show that the neutral axis corresponds to the centroidal z-axis of the cross-section. This fact then will establish the origin of the y-axis and is needed to apply the bending stress formula, which will be subsequently developed.

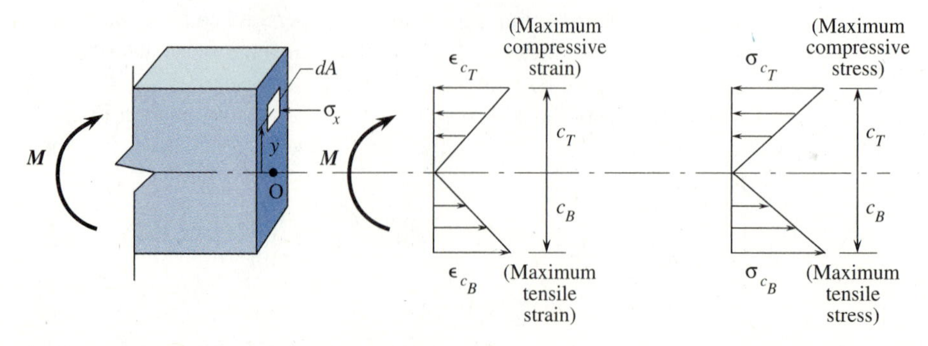

Figure 5.22 Strain and stress variation on section of beam.

To show that the neutral axis corresponds to the centroidal z-axis, we now consider an element of area dA in the cross-section at distance y from the neutral axis as shown in Figure 5.22. Because the stresses are normal to the cross-section, the force acting on this element is also normal to the cross-section and has magnitude σdA. The resultant normal force acting on the cross-section is zero because no axial forces are applied to the cross-section. Therefore, by summing forces in the x-direction on the cross-section, we obtain

$$\Sigma F_x = 0: \quad \int \sigma dA = 0 \tag{5.5}$$

Using Eq. (5.4a), in Eq. (5.5), we have

$$\int -\frac{E}{\rho} ydA = 0 \tag{5.5a}$$

Remember that both the modulus of elasticity E and the radius of curvature ρ are constants at the cross-section. Equation (5.5a) then becomes

$$\int ydA = 0 \tag{5.5b}$$

Equation (5.5b) also defines the location of the centroid if we recall the definition of the centroid $\bar{y}$ of an area as

$$\bar{y} = \int \frac{ydA}{A} \tag{5.5c}$$

Based on Eqs. (5.5b) and (5.5c), we conclude that

$$\bar{y} = 0$$

Hence, we have shown that the z-axis must pass through the centroid of the cross-section. The z-axis is also the neutral axis. Therefore, we conclude that the neutral axis passes through the centroid of the cross-section. This property is very important because it tells us the location of the neutral axis for a beam of any cross-sectional shape. That is, determine the centroid of the cross-section and this location corresponds to the z-neutral axis. Due to our assumption 1, we have symmetry of the cross-section about the y-axis and this axis is also a principal axis. For a beam satisfying Hooke's law (assumption 4) and subjected to pure bending, the y- and z-axes are both *principal centroidal axis*.

To develop the beam bending stress formula, we now express the differential moment about O due to the stress σ_x acting over differential area dA in Figure 5.22 as

$$dM = -y(\sigma_x dA) \tag{5.6}$$

On integrating Eq. (5.6), we obtain

$$M = -\int y\sigma_x dA \tag{5.6a}$$

where σ_x is the normal stress due to bending acting on area dA at a distance y from the neutral axis. The compressive stress distribution above the neutral axis (point O) together with the tensile stress distribution below the neutral axis are thus seen to produce a counterclockwise moment exactly equal and opposite to the applied moment M.

Substituting Eq. (5.4a) into Eq. (5.6a), yields

$$M = \int_A -\left(E\frac{-y}{\rho}\right) y \, dA \tag{5.6b}$$

Because E and ρ are not functions of the cross-sectional coordinates y and z, Eq. (5.6b) becomes

$$M = \frac{E}{\rho}\int_A y^2 dA \tag{5.6c}$$

The integral in Eq. (5.6c) is defined as the *area moment* of *inertia* I of the cross-section about the centroidal z-axis. Hence, I is

$$I = \int_A y^2 dA \tag{5.6d}$$

Therefore, using Eq. (5.6d) in Eq. (5.6c), we obtain

$$M = \frac{EI}{\rho} \tag{5.6e}$$

Equation (5.6e) indicates that the applied moment, M, is proportional to the bending stiffness defined by the product of $E \times I$, and inversely proportional to the radius of curvature, ρ.

Solving for E/ρ in Eq. (5.4a) and substituting into Eq. (5.6e), the radius of curvature and the modulus of elasticity are eliminated, and we obtain the beam bending stress formula or *flexure formula* for a beam as

$$\sigma_x = \frac{-My}{I} \tag{5.7}$$

Equation (5.7) is the highly useful equation for determining bending stress in a beam. This equation shows that the normal stress at a cross section due to bending (or bending stress), σ_x, depends on the applied moment, M, acting on that cross-section and the area moment of inertia, I, about the z-axis of the cross-section. The stress, σ_x, is also a linear function of the distance, y, measured from the centroidal z-axis or neutral axis. The negative sign indicates that for positive M and positive y, the stress σ_x will be compressive (negative). This is consistent with the previous convention of tension being positive and compression being negative. In Figure 5.22, for example, the maximum compressive stress would be obtained by substituting $y = c_T$ into Eq. (5.7), and the maximum tensile stress due to bending by substituting $y = -c_B$ into Eq. (5.7). That is,

$$\sigma_{\text{max comp}} = -\frac{Mc_T}{I} \tag{5.8a}$$

and

$$\sigma_{\text{max ten}} = -\frac{M(-c_B)}{I} = \frac{Mc_B}{I} \tag{5.8b}$$

By substituting $y = 0$ into Eq. (5.7), we verify that the bending stress equals zero at the neutral axis (where the normal strain is also zero).

Even though Eq. (5.7) was derived for a beam subjected to pure moments M (no shear forces V act on the cross-section), the flexure formula also applies to beams subjected to transverse loading. We merely determine the bending moment at the cross-section of interest, as described in Section 5.3, and substitute this value for M into Eq. (5.7).

However, we should understand that when shear forces are present they produce warping, or out-of-plane distortion, of the cross-section. Thus a section that was plane before bending is no longer plane after bending due to transverse loads resulting in shear forces. More detailed analysis has shown that the normal stresses calculated by the flexure formula are not significantly altered by the existence of the shear stresses and the associated warping that takes place. This analysis justifies the use of the flexure formula for determining normal stresses even when pure bending does not exist, such as when the more common transverse loading is applied to a beam.

Finally, another form of Eqs. (5.8a) and (5.8b) is often used, particularly when the section is symmetric about the principal z-axis. For these symmetric cross-sections ($c_T = c_B = c$), we define the *section modulus* S as

$$S = \frac{I}{c} \tag{5.8c}$$

where S has units of in.^{3} or m^3.

Using the definition of S in Eq. (5.8a), we have

$$\sigma = \frac{M}{S} \tag{5.8d}$$

For common hot-rolled cross-sections, such as wide flanges or channels, values of S are listed in Appendix C.

STEPS IN SOLUTION

The following steps are used to obtain the bending stress in a beam.

1. Draw the bending moment diagram for the beam. (Usually the maximum bending moment M_{max} is needed to obtain the largest tensile or compressive bending stress.) However, remember for cross-sections such as tees, that have only one axis of symmetry (the y-axis), the distances c_B and c_T to the bottom and top fibers from the neutral axis are not equal. Depending on the moment diagram (i.e., if there are both large positive and negative moments), the largest tensile and compressive stresses may be located at different positions along the beam based on the product of absolute maximum moment times c_B or c_T. Example 5.8 illustrates this concept.
2. Choose the maximum bending moment or another moment from the diagram where the stress is desired.
3. Calculate the moment of inertia, I.
4. Determine the maximum fiber distances, $y = -c_B$ and $y = c_T$. Remember the positive y direction is upward.
5. Determine the beam bending stress using Eq. (5.7) as

$$\sigma = -\frac{My}{I}$$

The largest stresses occur for the largest products of $|M|_{max} c_B$ and $|M|_{max} c_T$.

Examples 5.7 through 5.11 illustrate the steps used to obtain the beam bending stress.

EXAMPLE 5.7

Determine the maximum tensile and compressive bending stresses for the uniformly loaded simply supported wooden beam shown in Figure 5.23. The cross-section is a nominal 2 in. × 6 in.

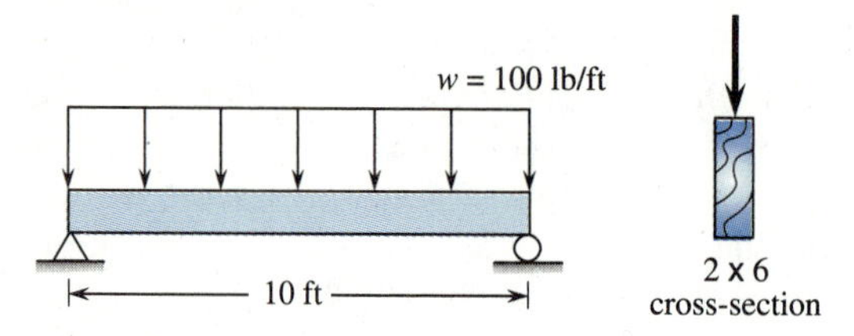

Figure 5.23 Uniformly loaded simply supported beam.

Solution

The actual dimensions of the 2 in. × 6 in. lumber are given in Appendix C as $1\frac{1}{2}$ in. by $5\frac{1}{2}$ in. The area moment of inertia is also listed in Appendix C as

$$I = 20.80 \text{ in.}^4$$

or verified for a solid rectangular-cross section as

$$I = \frac{bh^3}{12} = 20.80 \text{ in.}^4$$

The maximum moment occurs at the midspan ($x = 5$ ft) and was obtained in Example 5.2 as

$$M_{\max} = \frac{wL^2}{8} = \frac{(100 \text{ lb/ft})(10 \text{ ft})^2}{8} = 1250 \text{ lb} \cdot \text{ft}$$

Using Eq. (5.7), we obtain the bending stress in the beam. The maximum compressive stress occurs in the top of the beam cross-section. By substituting $y = c_T = 2.75$ in. into Eq. (5.7), we obtain

$$\sigma_{\max\text{ comp}} = \frac{-(1250 \text{ lb} \cdot \text{ft} \times 12 \text{ in./ft})(2.75 \text{ in.})}{20.80 \text{ in.}^4}$$

$$\sigma_{\max\text{ comp}} = -1983 \text{ psi (C)}$$

The negative sign indicates that the stress is compressive, as initially observed.

The maximum tensile stress occurs in the bottom of the beam by substituting $y = -c_B = -2.75$ in. into Eq. (5.7). Remember positive y is upward, hence the negative sign in c_B.

$$\sigma_{\max\text{ ten}} = -\frac{(1250 \text{ lb} \cdot \text{ft}) \times (12 \text{ in./ft})(-2.75 \text{ in.})}{20.80 \text{ in.}^4}$$

$$\sigma_{\max\text{ ten}} = 1983 \text{ psi (T)}$$

For cross-sections symmetric about the z-axis, as in this example, the maximum compressive and tensile stresses are the same in magnitude. For cross-sections that are not symmetric about the z-axis, such as tee sections, the maximum compressive and tensile stresses on a given cross-section will generally not be equal. Example 5.9 illustrates this concept.

EXAMPLE 5.8

Now turn the 2 × 6 of Example 5.7 such that the long dimension is horizontal as shown in Figure 5.24 and again determine the maximum compressive bending stress.

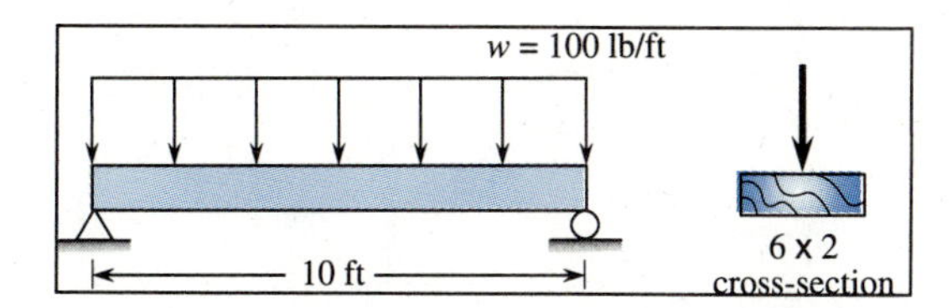

Figure 5.24 Uniformly loaded simply supported beam.

Solution

From Appendix C, we have

$$I = 1.547 \text{ in.}^4$$

$$M_{\max} = 1250 \text{ lb} \cdot \text{ft (from Example 5.7)}$$

$$y = c_T = 0.75 \text{ in.}$$

$$\sigma_{\text{max comp}} = -\frac{(1250 \times 12 \text{ in./ft})(0.75 \text{ in.})}{1.547 \text{ in.}^4}$$

$$\sigma_{\text{max comp}} = -7270 \text{ psi (compressive)}$$

The maximum tensile stress is then

$$\sigma_{\text{max ten}} = 7270 \text{ psi (tensile)}$$

The maximum bending stress is over three times that of Example 5.7. We conclude that by loading the beam on its narrow edge, as shown in Figure 5.23, rather than on its wider edge, as shown in Figure 5.24, the beam is more than three times as strong. This is the reason design engineers attempt to maximize the depth dimension in beam design.

EXAMPLE 5.9

For the tee cross-sectional beam loaded as shown in Figure 5.25, determine the largest tensile and compressive stresses.

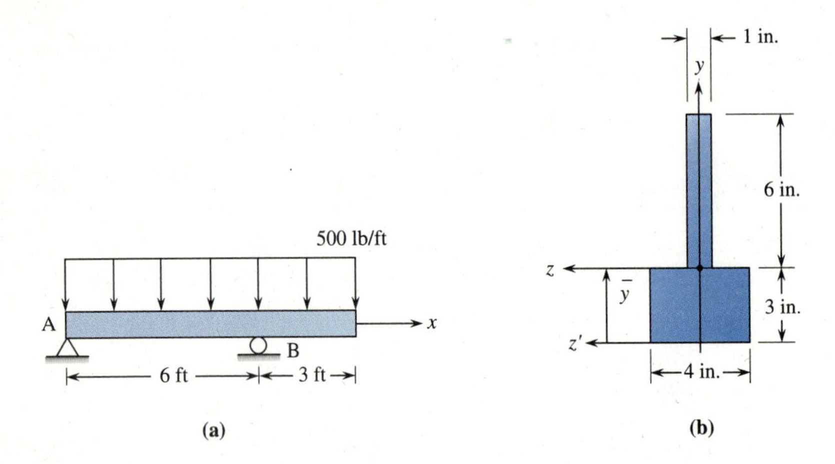

Figure 5.25 (a) Overhanging beam subjected to transverse load and (b) tee cross-section of the beam.

Solution

Although this cross-section is not symmetric about both centroidal axes, it is symmetric about the y-axis and the loading is in the centroidal x-y plane. Assumption 1 is then satisfied and we can use Eq. (5.7) to determine the bending stress.

We first determine the centroid of the cross-section. We arbitrarily, but for convenience, use the base axis z' as a reference. The centroidal distance $\bar{y}$ is determined using the concept for finding the centroid of a composite section. For the tee section, we use

$$\bar{y} = \frac{A_1\bar{y}_1 + A_2\bar{y}_2}{A_1 + A_2} \tag{a}$$

where the composite section is separated into two rectangular-shaped areas A_1 and A_2, (Figure 5.26), and $\bar{y}_1$ and $\bar{y}_2$ locate the centroids of these areas with respect to the base axis z'. Therefore, using Eq. (a), with $\bar{y}_1 = 3 \text{ in.}/2 = 1.5$ in. and $\bar{y}_2 = 3 \text{ in.} + 6 \text{ in.}/2 = 6$ in., we obtain

$$\bar{y} = \frac{(3 \text{ in.} \times 4 \text{ in.})(1.5 \text{ in.}) + (6 \text{ in.} \times 1 \text{ in.})(6 \text{ in.})}{(3 \text{ in.} \times 4 \text{ in.}) + (6 \text{ in.} \times 1 \text{ in.})} = 3 \text{ in.}$$

The centroid is then 3 in. up from the bottom and the origin of the y-z coordinates is located at that point.

The area moment of inertia I, about the z-axis is now calculated using the *parallel axis theorem* (sometimes called the *transfer axis theorem*). This theorem says that the moment of inertia I_z of an area with respect to the z-axis is equal to the moment of inertia $I_{\bar{z}}$ of the area with respect to the centroidal $\bar{z}$-axis parallel to the z-axis, plus the product

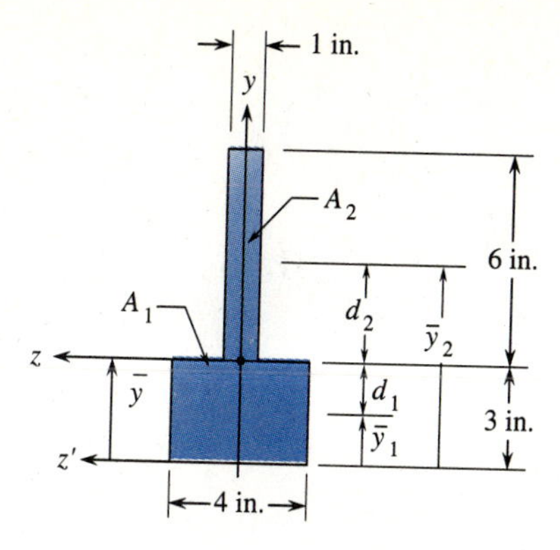

Figure 5.26 Cross-section of beam showing symbols for calculating geometric properties.

Ad^2 of the area A and of the perpendicular distance d between the two axes. The theorem is expressed by

$$I_z = I_{\bar{z}} + Ad^2$$

For composite areas, such as in Figure 5.26, we then apply this theorem to each regular area A_1 and A_2 and add the results. Having obtained the centroidal z-axis for the composite, the moment of inertia about the z-axis is then given by

$$I_z = (I_{\bar{z}})_1 + A_1 d_1^2 + (I_{\bar{z}})_2 + A_2 d_2^2$$

Using Appendix A for the rectangular cross-section, we have

$$I_z = \frac{1}{12} b_1 h_1^3 + A_1 d_1^2 + \frac{1}{12} b_2 h_2^3 + A_2 d_2^2$$

where b_1 and h_1 are the base and height of area A_1, and b_2 and h_2 are the base and height of area A_2. In Figure 5.26, we have $b_1 = 4$ in., $h_1 = 3$ in., $b_2 = 1$ in., $h_2 = 6$ in., $d_1 = 1.5$ in., and $d_2 = 3$ in. The moment of inertia is then

$$\begin{aligned} I_z &= \frac{1}{12}(4)(3)^3 + (4 \times 3)(1.5)^2 + \frac{1}{12}(1)(6)^3 + (1 \times 6)(3)^2 \\ &= 9 + 27 + 18 + 54 \\ &= 108 \text{ in.}^4 \end{aligned}$$

We then determine the beam reactions by drawing the free-body diagram shown in Figure 5.27(a) and writing the equations of equilibrium as follows

$$\Sigma M_B = 0: \quad -6A + 4500(1.5 \text{ ft}) = 0$$

$$A = 1125 \text{ lb}$$

$$\Sigma F_y = 0: \quad 1125 + B - 4500 = 0$$

$$B = 3375 \text{ lb}$$

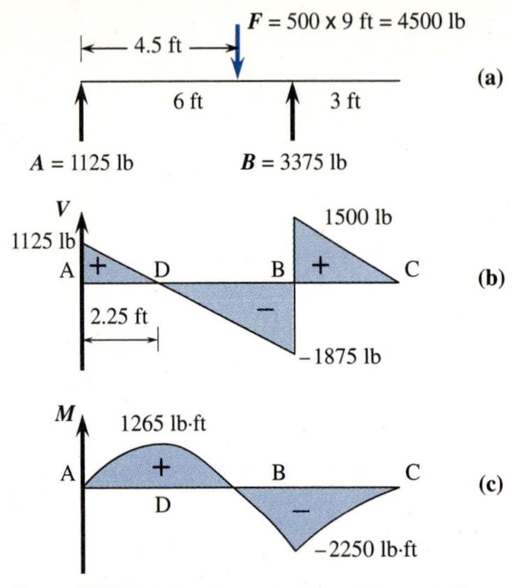

Figure 5.27 (a) Free-body diagram of beam in Figure 5.26,
(b) shear force diagram, and
(c) bending moment diagram.

Using the steps outlined in Section 5.3, we determine the shear force and bending moment diagrams shown in Figures 5.27(b) and 5.27(c). The verification of these diagrams is left to your discretion.

We now determine the largest tensile and compressive bending stresses in the beam. For beams that are not symmetric about the centroidal z-axis (such as the tee considered in this example), we must consider the product of $M \times y$ to obtain the maximum tensile and compressive stresses.

Based on the bending moment diagram in Figure 5.27(c), the largest moments are 1265 lb · ft at $x = 2.25$ ft and -2250 lb · ft at support B. Therefore, we will consider these two locations for the maximum bending stress.

First consider point B. The maximum tensile stress due to bending will occur in the top fibers of the beam at point B because the product of the negative moment times the distance to the top fibers is greatest there. Using Eq. (5.7), with $y = c_T = 6$ in., and $M = -2250$ lb · ft, we obtain

$$\sigma_{\text{max ten}} = -\frac{(-2250 \text{ lb} \cdot \text{ft})(12 \text{ in./ft})(6 \text{ in.})}{108 \text{ in.}^4}$$

$$= 1500 \text{ psi (tensile)} \qquad \textbf{(a)}$$

The maximum compressive stress due to bending at point B is obtained by using $y = -c_B = -3$ in. and $M = -2250$ lb · ft in Eq (5.7). This yields

$$\sigma_{\text{max comp}} = -\frac{(-2250 \times 12)(-3 \text{ in.})}{108 \text{ in.}^4}$$

$$= -750 \text{ psi (compressive)} \qquad \textbf{(b)}$$

Now consider point D at $x = 2.25$ ft. The maximum tensile bending stress at D occurs in the bottom of the beam, corresponding to $y = -3$ in. and $M = 1265$ lb · ft, and is given by

$$\sigma_{\text{max ten}} = -\frac{(1265 \times 12)(-3 \text{ in.})}{108 \text{ in.}^4} = 422 \text{ psi (tensile)} \quad \textbf{(c)}$$

The maximum compressive bending stress is in the top of the beam at D, corresponding to $y = 6$ in. and $M = 1265$ lb · ft, and is given by

$$\sigma_{\text{max comp}} = -\frac{(1265 \times 12)(6 \text{ in.})}{108 \text{ in.}^4} = -844 \text{ psi (compressive)} \quad \textbf{(d)}$$

These results indicate that the maximum tensile and compressive bending stresses may occur at different locations along the beam. This can occur whenever the beam cross-section is nonsymmetric about the centroidal z-axis, and hence, c_T is not equal to c_B.

We emphasize that in presenting values for bending stress, care must be taken in relying on the sign convention indicating whether a stress is tensile or compressive. It is strongly recommended that you use physical insight to reduce the chance of misinterpretation of the sign whenever possible and simply write after your answer either tension (T) or compression (C) as appropriate.

EXAMPLE 5.10

For a 9-m-long hollow circular cross-section flagpole fixed at the base and subjected to an end load of 2000 N (Figure 5.28), determine the maximum bending stress. The inner diameter is 125 mm and the outer diameter is 135 mm.

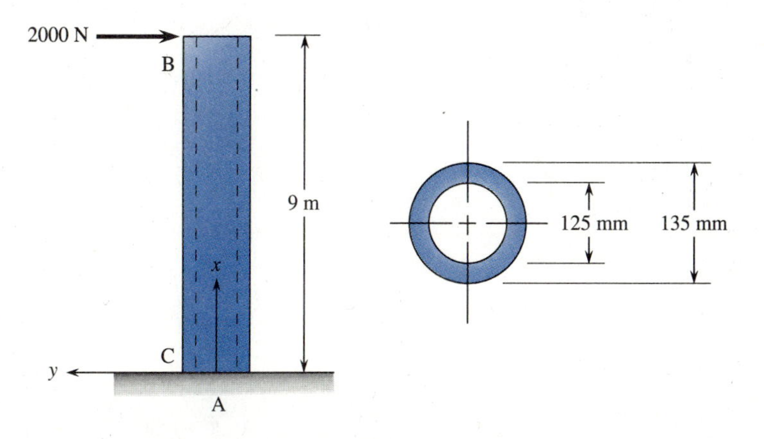

Figure 5.28 Flagpole subjected to end load.

Solution

Even though the beam does not have straight sides, we can use the flexure formula, Eq. (5.7), to determine the bending stress in the pipe.

First we determine the area moment of inertia of the hollow pipe. The area moment of inertia for a hollow circular cross-section is

$$I = \frac{\pi}{4}(r_o^4 - r_i^4) = \frac{\pi}{4}(67.5^4 - 62.5^4) = 4.32 \times 10^6 \text{ mm}^4 = 4.32 \times 10^{-6} \text{ m}^4$$

Next we determine the maximum bending moment. For the cantilevered beam subjected to end load, the maximum bending moment occurs at the fixed end and is given by

$$M_{max} = -(2000 \text{ N})(9 \text{ m}) = -18{,}000 \text{ N} \cdot \text{m}$$

Using Eq. (5.7), with $M_{max} = -18{,}000 \text{ N} \cdot \text{m}$ and $y_{max} = 67.5 \text{ mm} = 0.0675 \text{ m}$, we obtain the maximum bending stress as

$$\sigma_{max} = -\frac{(-18{,}000 \text{ N} \cdot \text{m})(0.0675 \text{ m})}{4.32 \times 10^{-6} \text{ m}^4}$$

$$= 281 \times 10^6 \text{ N/m}^2$$

$$= 281 \text{ MPa (T)}$$

This tensile stress occurs at the base of the pipe and on the left side of the outer surface of the pipe (labeled point C in Figure 5.28).

EXAMPLE 5.11

Determine the maximum tensile and compressive bending stresses for a beam resisting a maximum bending moment of 4000 lb · ft with the steel cross-sections shown in Figure 5.29. Based on bending strength, which cross-section would you recommend? Note that the cross-sectional areas are nearly equal for these four sections.

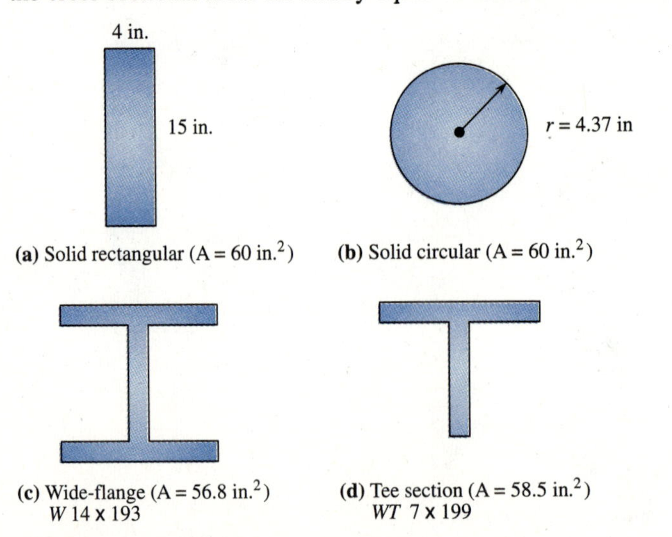

Figure 5.29 Cross-sections of beam for bending stress analysis.

Solution

We use Eqs. (5.8a) and (5.8b), to determine the maximum compressive and tensile stresses due to bending.

(a) First determine the moment of inertia for the solid rectangular cross-section. From Appendix A, we have

$$I = \frac{1}{12}bh^3 = \frac{1}{12}(4\text{ in.})(15\text{ in.})^3 = 1125\text{ in.}^4$$

The distances to the top and bottom from the neutral axis are

$$c_T = c_B = 7.5\text{ in.}$$

The maximum bending stresses are

$$\sigma_{\text{max comp}} = \frac{-(4000\text{ lb}\cdot\text{ft})(12\text{ in./ft})(7.5\text{ in.})}{1125\text{ in.}^4} = -320\text{ psi (C)}$$

$$\sigma_{\text{max ten}} = \frac{(4000\text{ lb}\cdot\text{ft})(12\text{ in./ft})(7.5\text{ in.})}{1125\text{ in.}^4} = 320\text{ psi (T)}$$

(b) From Appendix C, the moment of inertia for the circular cross-section is

$$I = \frac{\pi r^4}{4} = \frac{\pi(4.37)^4}{4} = 286.5\text{ in.}^4$$

The distances c_T and c_B are

$$c_T = c_B = r = 4.37\text{ in.}$$

The maximum bending stresses are

$$\sigma_{\text{max comp}} = \frac{-(4000\text{ lb}\cdot\text{ft})(12\text{ in./ft})(4.37\text{ in.})}{286.5\text{ in.}^4} = -732\text{ psi (C)}$$

$$\sigma_{\text{max ten}} = 732\text{ psi (T)}$$

(c) From Appendix C, the moment of inertia for the wide-flange is

$$I = 2400\text{ in.}^4$$

The distances c_T and c_B are

$$c_T = c_B = \frac{\text{depth}}{2} = \frac{15.48\text{ in.}}{2} = 7.74\text{ in.}$$

The maximum bending stresses are

$$\sigma_{\text{max comp}} = \frac{-(4000 \times 12)(7.74\text{ in.})}{2400\text{ in.}^4} = -155\text{ psi (C)}$$

$$\sigma_{\text{max ten}} = 155\text{ psi (T)}$$

(d) From Reference 3, the moment of inertia for the tee section is

$$I = 257\text{ in.}^4$$

The distances c_T and c_B are

$$c_T = 2.30 \text{ in.} \quad \text{and} \quad c_B = 9.145 \text{ in.} - 2.30 \text{ in.} = 6.845 \text{ in.}$$

The bending stresses are

$$\sigma_{\text{max comp}} = \frac{-(4000 \times 12)(2.30)}{257 \text{ in.}^4} = -428 \text{ psi (compressive)}$$

$$\sigma_{\text{max ten}} = \frac{(4000 \times 12)(6.845)}{257 \text{ in.}^4} = 1273 \text{ psi (tensile)}$$

The results indicate that the bending stress is smallest in the wide-flange section. It also has the smallest cross-sectional area and so the smallest weight. We conclude that the wide-flange is the most efficient section in bending because it has more of its area away from the neutral axis where it can resist bending.

5.5 BEAM SHEAR STRESS

In Section 5.4 we described how bending moments in beams produce bending stresses in the normal direction to the cross-section. These normal stresses were seen to vary from zero at the neutral axis to maximums at the extreme top and bottom fibers of the cross-section.

In this section, we will describe how the presence of a transverse shear force V on a beam causes shear stresses. We will develop a formula to obtain the shear stress at any point in a beam and will solve numerous example problems to illustrate the use of this formula.

When a beam is subjected to pure bending moments M, as shown in Figure 5.30, no shear forces V develop in the beam. (This is easily seen by summing forces in the y-direction.) That is, the beam of three smooth members resting one on top of the other deforms as three concentric arcs of circles.

Now consider the beam subjected to a transverse end load P as shown in Figure 5.31(a). Considering a section at distance x from the left end, the free-body diagram shows that a vertical shear force V develops on longitudinal planes.

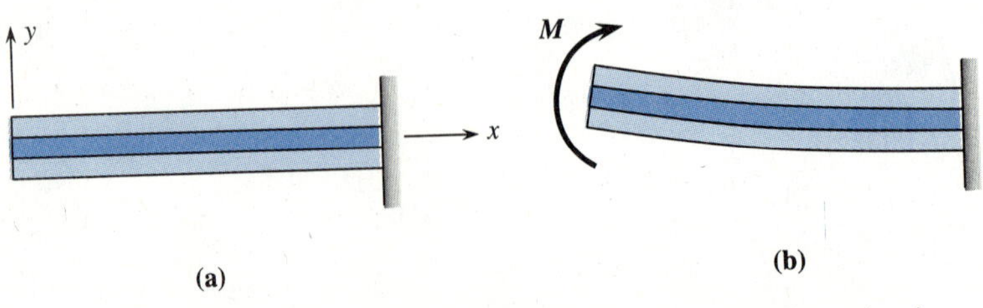

Figure 5.30 (a) Beam of three smooth members joined only at the right end and (b) deformed shape of the beam after subjected to pure bending moment M.

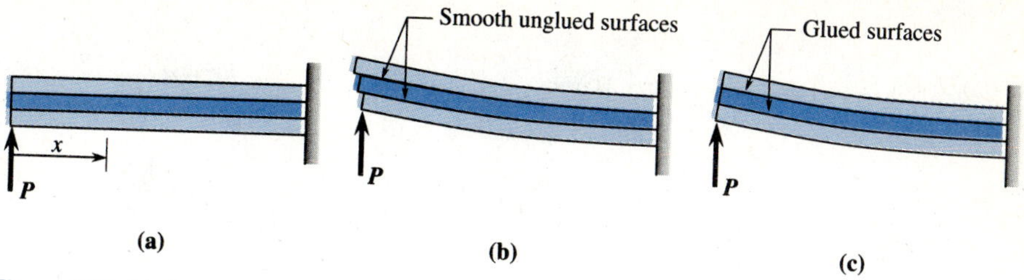

Figure 5.31 (a) Cantilevered beam of three smooth members subjected to end load, (b) deformed shape of beam due to load, and (c) deformed shape of beam for three members glued together.

By statics, we have

$$V = P$$

$$M = Px$$

The beam of three members resting on each other with smooth surfaces of contact deforms as shown in Figure 5.31(b). Each member has its top in compression and bottom in tension. They bend independent of each other and slide with respect to each other. Now when the members are glued together along their horizontal faces, sliding between the members is prevented due to shear stresses occurring on horizontal planes between the members. The three-member beam now behaves as a single beam with depth equal to the three members. The deformed shape is shown in Figure 5.31(c), where the upper fibers of the beam are in compression and the lower ones in tension.

To develop the shear stress formula for a beam, we now consider a single-member rectangular cross-section beam of width b subjected to end load P as shown in Figure 5.32(a). A section of length dx, shown in Figure 5.32(b), will have shear and bending moment at x (point L) equal to

$$V_{\mathrm{L}} = P$$

$$M_{\mathrm{L}} = Px$$

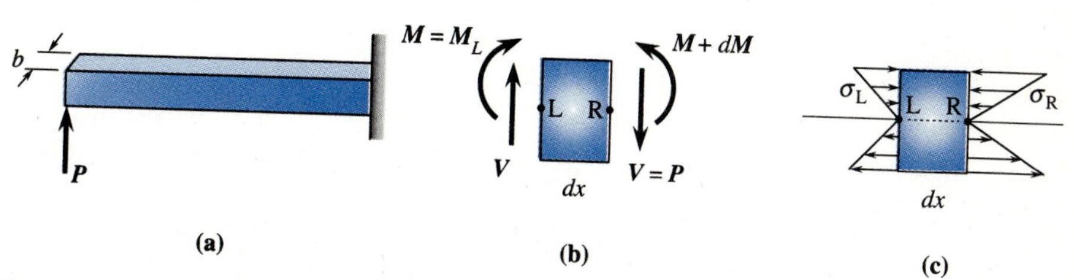

Figure 5.32 (a) Single-member cantilevered beam subjected to end load, (b) shear and moment on element of length dx, and (c) bending stress distributions on left and right sides.

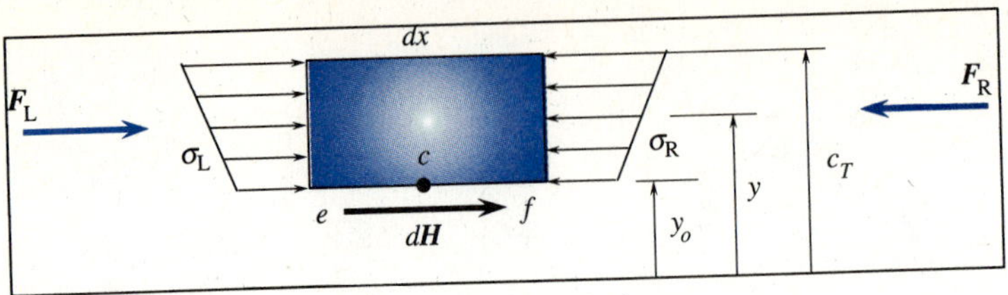

Figure 5.33 Free-body diagram of top section of a beam.

and at a small distance away at $x + dx$ (point R), we have

$$V_R = P$$

$$M_R = P(x + dx) = M_L + dM$$

From Section 5.4, we know that as a result of the applied moments, the bending stress distributions shown on the left and right sides of Figure 5.32(c) occur. By using Eq. (5.7), we find that these bending stress distributions are given by

$$\sigma_L = -\frac{My}{I} \tag{a}$$

and

$$\sigma_R = -\frac{(M + dM)y}{I} \tag{b}$$

We observe that σ_R will be numerically greater than σ_L since the moment is larger at point R than at L.

Because any portion of the beam must be in equilibrium, we now cut an element off the top section, as shown in Figure 5.33, to obtain the horizontal shear stress. Because the resultant force F_L on the left face is not equal to the force F_R on the right face, an additional force dH must show up on the horizontal face e-f for equilibrium of forces in the horizontal direction.

Summing forces in the horizontal direction yields

$$\Sigma F_x = 0: \quad F_L - F_R + dH = 0 \tag{c}$$

Now we have

$$F_L = \int_{A_u} \sigma_L \, dA \quad \text{and} \quad F_R = \int_{A_u} \sigma_R \, dA \tag{d}$$

where A_u is the cross sectional area above $y = y_0$ to $y = c_T$.

Using Eqs. (d) in (c), we obtain

$$\int_{A_u} \sigma_L dA - \int_{A_u} \sigma_R dA + dH = 0 \tag{5.9}$$

where dA is a differential area over which the varying bending stress acts. Substituting Eqs. (a) and (b) into Eq. (5.9) yields

$$\int_{A_u} \left(\frac{My}{I}\right) dA - \int_{A_u} \frac{(M + dM)y}{I} dA + dH = 0 \tag{5.9a}$$

The negative signs from Eqs. (a) and (b) were dropped since the stresses used in Eq. (5.9a) were compressive and the final correct signs on the forces are now accounted for. Since M and I are constant over the cross-section, Eq. (5.9a) becomes

$$dH = \frac{dM}{I} \int_{A_u} y dA \tag{5.10}$$

Now defining

$$Q = \int_{A_u} y dA \tag{5.11}$$

where Q is the first moment with respect to the neutral axis of that part of the cross-sectional area A_u that is located above the line $y = y_0$ in Figure 5.33.

Using Eq. (5.11) in Eq. (5.10), we have

$$dH = \frac{dM}{I} Q \tag{5.12}$$

Now dividing Eq. (5.12) by the length dx, we obtain

$$\frac{dH}{dx} = \frac{dM}{dx} \frac{Q}{I} \tag{5.13}$$

We define

$$q = \frac{dH}{dx}$$

as the horizontal shear force per unit length or *shear flow,* and also use Eq. (5.2b), to rewrite Eq. (5.13) as

$$q = \frac{VQ}{I} \quad \left(\frac{\text{force}}{\text{length}}\right) \tag{5.14}$$

Finally, dividing Eq. (5.14) by the width b of the cross-section where q acts, we obtain

$$\tau = \frac{VQ}{Ib} \tag{5.15}$$

Equation (5.15) is called the *beam shear stress formula.* In Eq. (5.15), τ is the horizontal shear stress on the beam at distance y_0 on the cross-section, V is the transverse shear force at the location along the beam axis where the shear stress is desired (V is normally determined from the shear force diagram), I is the area moment of inertia of the whole cross-section, and b is the width through the cross-section where τ acts.

Equation (5.15) is very important in the design of beams of materials weak in shear, such as timber beams. The equation indicates that shear stresses exist on a beam cross section whenever a transverse shear force is present. In Section 4.4 we indicated that shear stresses exist on mutually perpendicular faces to maintain moment equilibrium of an element. The beam shear stress τ at some point, such as c on the beam of Figure 5.33, is by convention, shown together with the normal bending stress σ at c on the stress element of Figure 5.34.

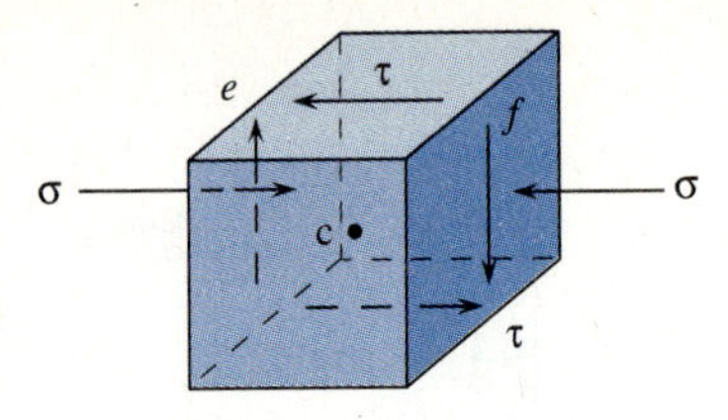

Figure 5.34 Element representing state of stress at point c in beam.

The distribution of the beam shear stress over the cross-section must be carefully understood. The shear stress at the top and bottom of the cross-section is zero because, based on Eq. (5.11), Q is zero at the top and bottom, while the shear stress will be a maximum at the neutral axis ($y = 0$) for uniform width beams. This distribution of shear stress is not generally linear. Bending shear stress distributions for some common-shaped cross-sections are shown in Figure 5.35. These distributions are based on evaluating Eq. (5.15) at various locations through the depth of the beam cross-section.

In the idealized flange-web of Figure 5.35(a), all of the effective area of the cross section that is resisting bending is considered to be concentrated at the flange areas, A_F, of the beam. The web, or narrow area separating the flanges, is assumed not effective in bending. Therefore, on neglecting the area of the web, by the parallel axis theorem, $I = 2A_F(h/2)^2$. These flanges are similarly assumed not effective in resisting a transverse shear force, and therefore the shear stress τ is simply the shear force V divided by the shear area $h \times t$. This type of construction is quite efficient from a weight standpoint and is commonly used in structural design where narrow-webbed sections, such as wide-flange cross-sections, as shown in Figure 5.35(e), are taken advantage of.

Except for the idealized flange-web, each of the other shapes in Figure 5.35 have shear stress distributions that are a maximum at the neutral axis and zero at the top and bottom. Using Eq. (5.15), we could show that for rectangular sections

$$\tau_{max} = \frac{3V}{2A} \tag{5.16}$$

and for circular sections

$$\tau_{max} = \frac{4V}{3A} \tag{5.17}$$

Equations (5.16) and (5.17) are easily verified by evaluating the shear stress at the neutral axis.

Also, it is common engineering practice to calculate the shear stress for narrow-webbed cross-sections, such as wide-flange and channel sections, as

$$\tau = \frac{V}{A_{web}} \tag{5.18}$$

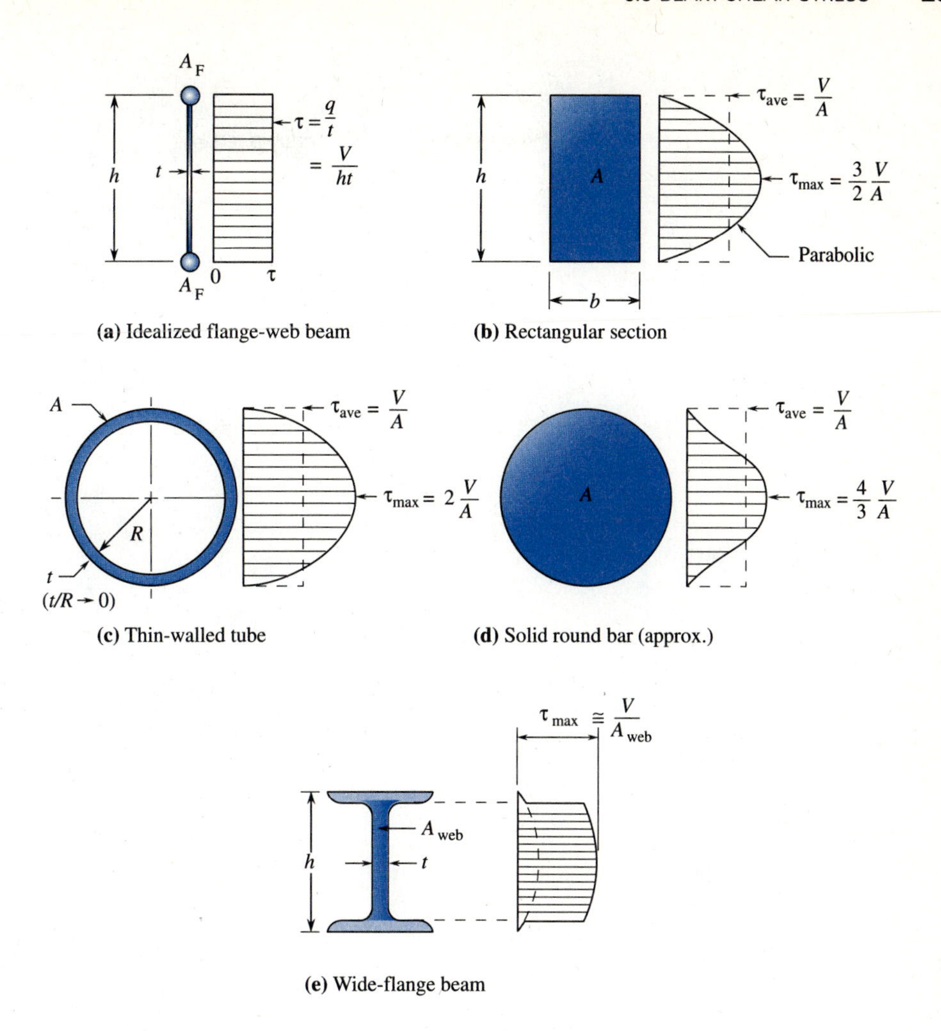

Figure 5.35 Shear stress distribution for different cross-sections.

From Figure 5.35, we observe that Eq. (5.18) will generally yield useful engineering results. (Also see Example 5.16.)

Equation (5.17) indicates that the shear stress for a solid circular cross-section is constant everywhere along the width at the neutral axis. The maximum values actually vary along the neutral axis from 1.38 V/A at the center of the circle to 1.23 V/A at the edge. This is shown using the theory of elasticity. (For instance, see Timoshenko and Goodier [1].) This variation is usually disregarded in most engineering analyses and will not be considered in our solutions.

The concept of beam shear stress is sometimes also referred to as *bending shear stress* and sometimes as *horizontal shear stress* (although it acts also on perpendicular vertical faces as shown in Figure 5.34).

STEPS IN SOLUTION

The following steps are used to obtain the shear stress in a beam due to a transverse shear force.

1. Draw the shear force diagram for the beam.
2. Choose the absolute maximum shear force V_{max}, or the shear force corresponding to the location along the beam where the shear stress is desired.
3. Calculate Q, the first moment of the area above the location on the beam cross-section where the shear stress τ is desired, evaluated about the neutral axis.
4. Calculate the moment of inertia I.
5. Determine the beam width b through the section where τ is being determined.
6. Determine the shear stress using Eq. (5.15) as

$$\tau = \frac{VQ}{Ib}$$

or use special equations, Eqs. (5.16) through (5.18), that apply to the rectangular, circular, and narrow webbed cross-sectional beams. That is,

$$\tau_{max} = \frac{3V}{2A} \quad \text{for a rectangular cross-section}$$

$$\tau_{max} = \frac{4V}{3A} \quad \text{for a solid circular cross-section}$$

$$\tau = \frac{V}{A_{web}} \quad \text{for a narrow webbed cross-section}$$

Examples 5.12 through 5.18 illustrate the steps used to obtain the beam shear stress.

EXAMPLE 5.12

At a section of a beam (Figure 5.36) the vertical shear force is $V = 16{,}000$ lb and the bending moment is $M = 568{,}000$ lb · in. Determine: **(a)** the maximum shear stress, **(b)** the shear stress in the web at the junction of the web and flange, **(c)** the variation in the shear stress through the cross-section, **(d)** the percentage of the shear force carried by the web alone, and **(e)** the percentage of the bending moment carried by the flanges alone.

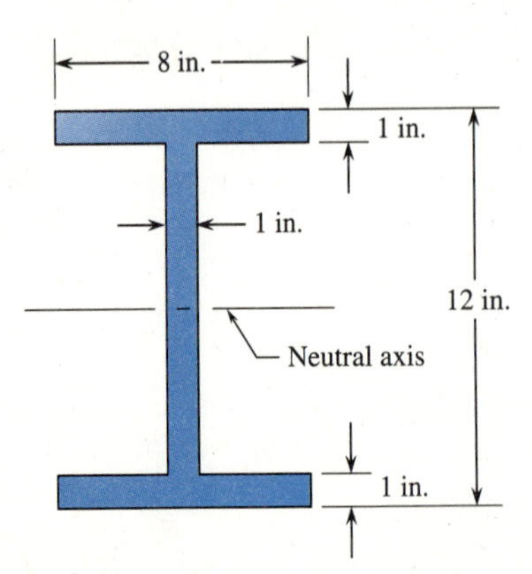

Figure 5.36 Cross-section of beam with $V = 16{,}000$ lb and $M = 568{,}000$ lb · in.

Solution

(a) First we determine the area moment of inertia I by evaluating I for a solid 8 in. × 12 in. rectangular area, and then subtracting the smaller rectangular areas (3.5 in. × 10 in.) that are not really part of the cross-section. Therefore, we obtain

$$I = \frac{1}{12}(8)(12)^3 - 2\left(\frac{1}{12}\right)(3.5)(10)^3 = 569 \text{ in.}^4$$

Next we determine Q. The maximum shear stress occurs at the neutral axis, therefore use the area above the neutral axis (Figure 5.37) to evaluate Q. By the definition of Q (Eq. (5.11)), we evaluate the moment of the area shown in Figure 5.37 about the neutral axis as follows:

$$Q = A_1 y_1 + A_2 y_2 = (8 \times 1)(5.5) + (1 \times 5)(2.5) = 56.5 \text{ in.}^3$$

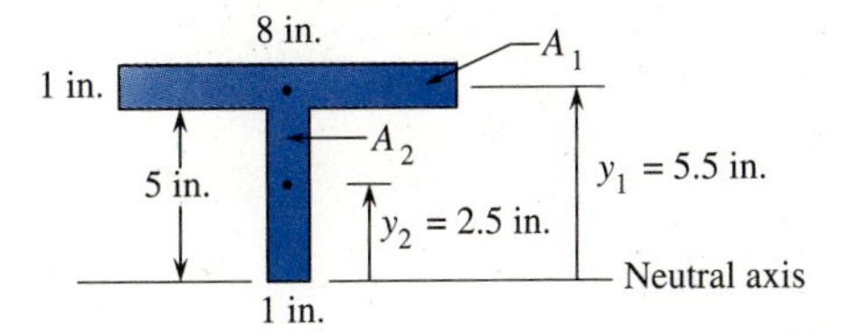

Figure 5.37 Area used to evaluate Q to obtain maximum shear stress.

The width b at the neutral axis is $b = 1$ in. The shear stress is then

$$\tau_{max} = \frac{VQ}{Ib} = \frac{(16{,}000)(56.5)}{(569)(1)} = 1590 \text{ psi}$$

(b) At the junction of the web and flange the area used to evaluate Q is the upper flange area A_1, as shown in Figure 5.37. Therefore

$$Q = (8 \times 1)(5.5) = 44 \text{ in.}^3$$

The other terms V, I, and b are as in part (a). Therefore, the shear stress is

$$\tau_{junc} = \frac{(16{,}000)(44)}{(569)(1)} = 1240 \text{ psi}$$

(c) To obtain the shear stress variation through the depth of the cross-section, we need to evaluate τ at the web-flange junction in the flange. The only change in the shear stress formula is that b is now $b = 8$ in., the width of the flange.

$$\tau_{flange} = \frac{(16{,}000)(44)}{(569)(8)} = 155 \text{ psi}$$

The shear stress variation through the beam depth is then shown in Figure 5.38.

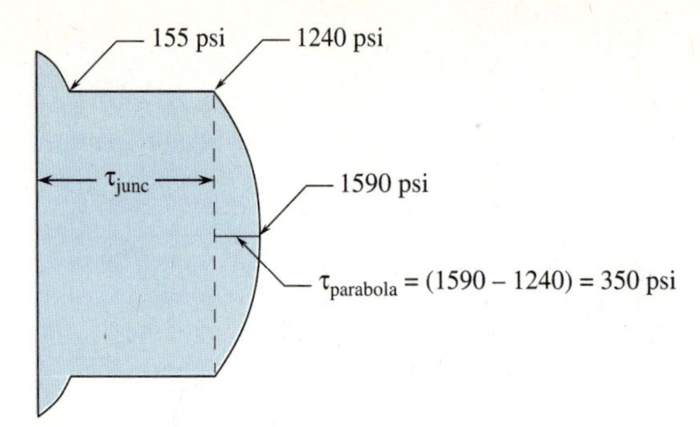

Figure 5.38 Shear stress distribution through beam depth.

(d) The total shear force carried by the web is obtained by multiplying the shear stress by the area of the web. That is,

$$V_{web} = \tau_{junc} A_{web} + \frac{2}{3} \tau_{parabola} A_{web}$$

$$V_{web} = (1240 \text{ psi})(1 \times 10) + \frac{2}{3}(350 \text{ psi})(1 \times 10) = 14{,}730 \text{ lb}$$

where the shear force V due to the portion of the shear stress in the web labeled $\tau_{parabola}$ in Figure 5.38 is found for the area of a parabola (see Appendix A where the area of a parabola is $A = [\frac{2}{3}]bh$).

The percentage of the shear force resisted by the web is then

$$\% \text{ Load} = \frac{14{,}730}{16{,}000} \times 100 = 92.1\%$$

(e) The total normal force resisted by each flange is obtained by multiplying the normal stress (Figure 5.39) by the area of the flange. That is,

$$\sigma_{max} = \frac{Mc_T}{I} = \frac{(568{,}000)(6)}{569} = 5990 \text{ psi}$$

and

$$\sigma_{\substack{\text{bottom of}\\ \text{flange}}} = \frac{(568{,}000)(5)}{569} = 4990 \text{ psi}$$

Then

$$\sigma_{ave} = \frac{5990 + 4990}{2} = 5490 \text{ psi}$$

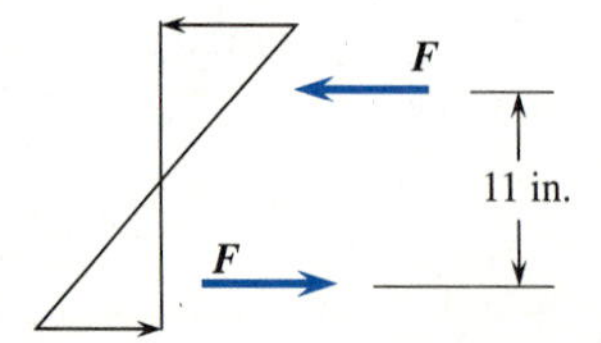

Figure 5.39 Normal bending stress distribution through beam depth.

The normal force resisted by a flange is then

$$F = (5490 \text{ psi})(8 \times 1) = 43{,}900 \text{ lb}$$

The moment resisted by the couple forces F and $-F$ is

$$M = d \times F = 11 \times 43{,}900 = 483{,}000 \text{ lb.} \cdot \text{in.}$$

where, for simplicity, d is taken as 11 in.

The percentage of the bending moment carried by the flanges is

$$\% \text{ moment} = \frac{483{,}000}{568{,}000} \times 100 = 85.0\%$$

EXAMPLE 5.13

A beam is constructed by gluing three 2 in. × 4 in. (nominal dimensions) boards together as shown in Figure 5.40. The beam must resist a transverse shear force of $V = 2000$ lb. Determine **(a)** the shear stress resisted by the glue and **(b)** the maximum shear stress on the cross-section.

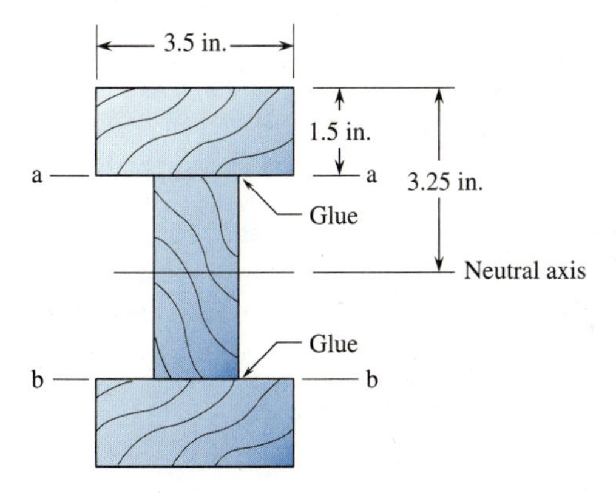

Figure 5.40 Beam constructed of 2 × 4 boards.

Solution

From Appendix C, the actual dimensions of a 2 × 4 are 1.5 in. by 3.5 in. The area moment of inertia is again easily calculated by evaluating I for a solid 3.5 in. × 6.5 in. rectangular area, and then subtracting the contributions of the smaller rectangular areas that are really not part of the cross-section as follows:

$$I = \frac{(3.5)(6.5)^3}{12} - 2\,\frac{(1.0)(3.5)^3}{12} = 73.0 \text{ in.}^4$$

(a) Since we are evaluating the shear stress at the glue line a-a, Q is the moment of the area above line a-a (see Figure 5.41) evaluated about the neutral axis of the whole cross-section. Therefore, we have

$$Q = (3.5 \text{ in.} \times 1.5 \text{ in.})(2.5 \text{ in.}) = 13.1 \text{ in.}^3$$

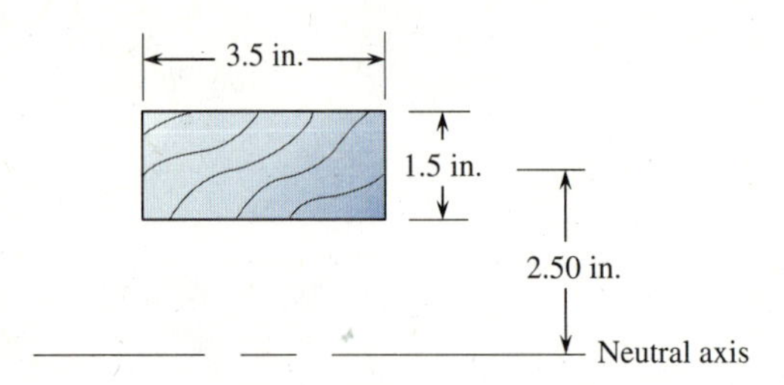

Figure 5.41 Area used to evaluate Q for shear stress at a-a.

Now using $V = 2000$ lb, $Q = 13.1$ in.^{3}, $I = 73.0$ in.^{4}, and $b = 1.5$ in. in Eq. (5.15), we obtain

$$\tau_{\text{a-a}} = \frac{(2000)(13.1)}{(73.0)(1.5)} = 239 \text{ psi}$$

This is the shear stress that must be resisted by the glue at section a-a and also section b-b. The determination of Q, the first moment of the area above a-a about the neutral axis, is the greatest source of error in calculating beam shear stress. We emphasize that Q is the area above the plane where the shear stress is desired but that the moment of this area is taken about the neutral axis of the beam. This is consistent with the definition of Q given by Eq. (5.11).

(b) The maximum beam shear stress will occur at the neutral axis where Q is largest and b is smallest. We now use V, I, and b as in part (a). We must evaluate Q for the area above the neutral axis as shown in Figure 5.42. Therefore, we obtain

$$\begin{aligned} Q &= (1.5 \text{ in.} \times 1.75 \text{ in.})\left(\frac{1.75}{2} \text{ in.}\right) + (3.5 \text{ in.} \times 1.5 \text{ in.})(2.5 \text{ in.}) \\ &= 15.4 \text{ in.}^3 \end{aligned}$$

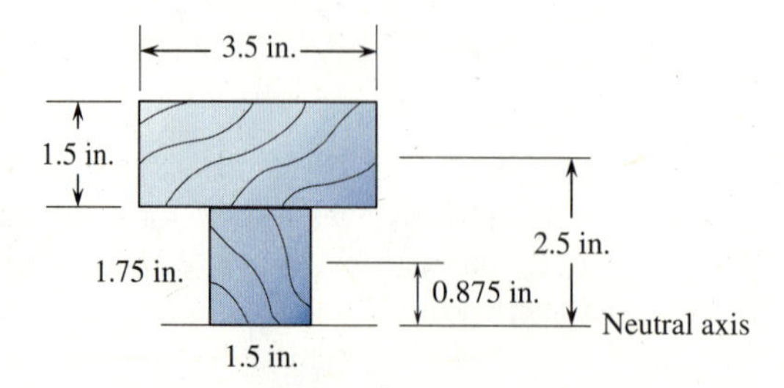

Figure 5.42 Area used to evaluate Q for maximum shear stress.

The shear stress is then

$$\tau_{max} = \frac{(2000)(15.4)}{(73.0)(1.5)} = 281 \text{ psi}$$

(For comparison, the allowable shear stress in a southern pine construction-grade lumber, as given by the *National Design Specification for Wood Construction* [2] is 105 psi. Therefore, for this species and grade of lumber the shear force V is unacceptably large. If the shear force cannot be reduced, a thicker board should be used for the web section.)

EXAMPLE 5.14

Determine the maximum beam shear stress for the beam and loading of Example 5.9 (See Figures 5.25, 5.26, and 5.27).

Solution

The maximum shear stress will occur at the neutral axis. Therefore, referring to Figure 5.26, we use the area above the neutral axis (Figure 5.43) to obtain Q as

$$Q = (6 \text{ in.} \times 1 \text{ in.})(3 \text{ in.}) = 18 \text{ in.}^3$$

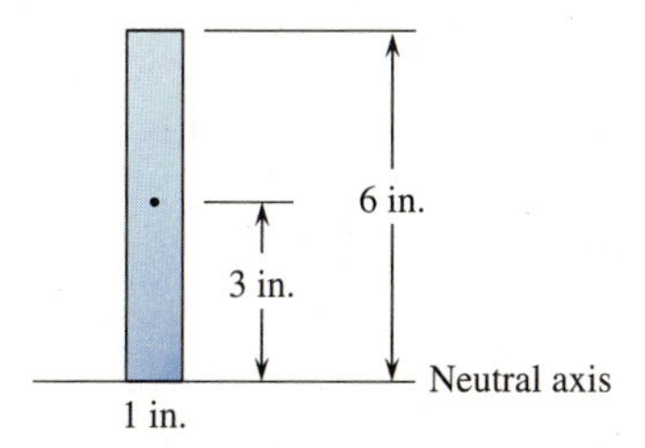

Figure 5.43 Area used to evaluate Q.

(We could have equivalently calculated Q by using the area below the neutral axis.) That is,

$$Q = (4 \text{ in.} \times 3 \text{ in.})(1.5 \text{ in.}) = 18 \text{ in.}^3$$

From the shear force diagram of Figure 5.27(b), the absolute maximum shear force is $V = 1875$ lb. Also from Example 5.9, the moment of inertia is $I = 108$ in.^{4} The width of the cross-section at the neutral axis is $b = 1$ in. Now using Eq. (5.15), we obtain the shear stress as

$$\tau_{max} = \frac{(1875)(18)}{(108)(1)} = 313 \text{ psi}$$

EXAMPLE 5.15

Determine the maximum beam shear stress for the beam and loading of Example 5.10 shown in Figure 5.28.

Solution

The maximum shear stress again occurs at the neutral axis. From Appendix A, the centroid of a semicircle is

$$\bar{y} = \frac{4r}{3\pi}$$

To obtain Q, we then subtract the moment of the semicircular hole of diameter 125 mm from the moment of the semicircular solid of diameter 135 mm, as shown in Figure 5.44.

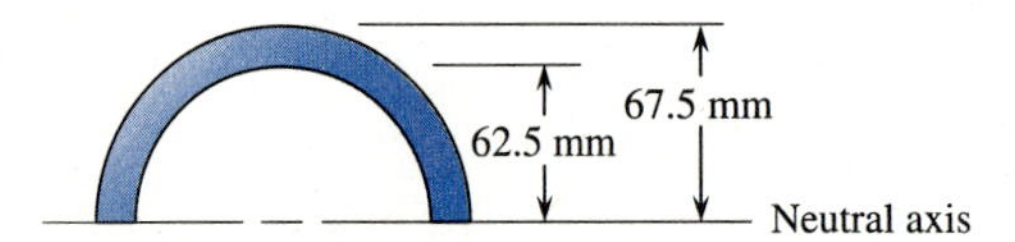

Figure 5.44 Cross-section used to evaluate Q.

We obtain for Q

$$Q = \left(\frac{\pi(67.5)^2}{2}\right)\left(\frac{4(67.5)}{3\pi}\right) - \left(\frac{\pi(62.5)^2}{2}\right)\left(\frac{4(62.5)}{3\pi}\right)$$

$$Q = 205{,}000 - 163{,}000 = 42{,}000 \text{ mm}^3$$

Using $b = 10$ mm, $I = 4.32 \times 10^6$ mm^4, and $V = 2000$ N in Eq. (5.15), we obtain

$$\tau_{max} = \frac{(2000)(42{,}000)}{(4.32 \times 10^6)(10)} = 1.96 \text{ N/mm}^2 = 1.96 \text{ MPa}$$

EXAMPLE 5.16

A W 8 × 24 wide-flange beam (see Appendix C for its properties) is reinforced by bolting $\frac{1}{2}$ in. × 8.0 in. plates to the flanges as shown in Figure 5.45. The bolts are spaced every 6 in. along the beam. The shear force is a constant $V = 20{,}000$ lb along the length of the beam. Determine: **(a)** the average shear force acting on each bolt and **(b)** the maximum shear stress in the beam.

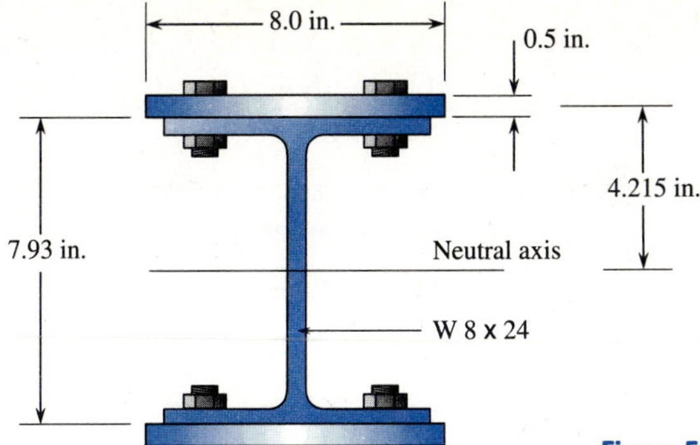

Figure 5.45 Reinforced wide-flange beam.

Solution

(a) From Appendix C, the area moment of inertia of the wide-flange is $I = 82.8$ in.^{4} and the web thickness at the neutral axis is $t_w = b = 0.245$ in. Including the reinforcing plates, the total moment of inertia using the parallel-axis theorem is

$$I = 82.8 + 2\left[\frac{(8.0)(0.5)^3}{12} + (8.0 \times 0.5)(4.215)^2\right]$$

$$I = 225 \text{ in.}^4$$

Because we want to determine the shear force in each bolt, we obtain Q as the moment of the area above the intersection of the upper plate and the top of the flange as

$$Q = (0.5 \text{ in.} \times 8 \text{ in.})(4.215 \text{ in.}) = 16.86 \text{ in.}^3$$

Therefore, the shear flow q along the bolted surfaces is obtained from Eq. (5.14) as

$$q = \frac{VQ}{I} = \frac{(20{,}000 \text{ lb})(16.86 \text{ in.}^3)}{225 \text{ in.}^4}$$

$$q = 1500 \text{ lb/in.}$$

This shear flow along each flange-plate intersection must be carried by two bolts every 6 in. (3 in. on each side of a bolt as shown in Figure 5.46).

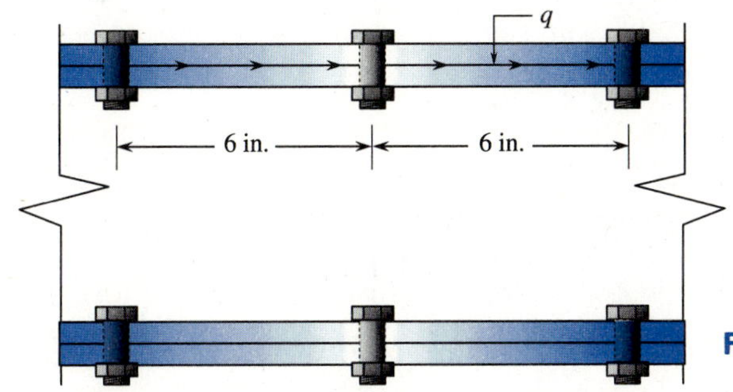

Figure 5.46 Side view showing shear flow q resisted by bolts.

The shear force per bolt is then

$$F_b = \frac{(1500\ \text{lb/in.})(6\ \text{in.})}{2\ \text{bolts}} = 4500\ \text{lb/bolt}$$

(b) The first moment of the area of the beam above the neutral axis consists of the contribution from the wide-flange beam plus the top plate. That is

$$Q = Q_w + Q_{plate}$$

Using the dimensions given in Appendix C for the W 8 × 24 cross-section, we have

$$Q_w = (0.40 \times 6.495)(3.765) + (0.245 \times 3.565)(1.7825) = 11.34\ \text{in.}^3$$

For the plate, we obtain

$$Q_{plate} = (0.5 \times 8.0)(4.215) = 16.86\ \text{in.}^3$$

Therefore, Q is

$$Q = 11.34 + 16.86 = 28.2\ \text{in.}^3$$

The maximum shear stress is obtained from Eq. (5.15) as

$$\tau_{max} = \frac{(20{,}000\ \text{lb})(28.2\ \text{in.}^3)}{(225\ \text{in.}^4)(0.245\ \text{in.})}$$

$$\tau_{max} = 10{,}230\ \text{psi}$$

For comparison, we now use the simple Eq. (5.18) to obtain the shear stress in the wide-flange beam as

$$\tau = \frac{V}{A_{web}} = \frac{20{,}000\ \text{lb}}{(0.245\ \text{in.})[7.93 - 2(0.4)]\ \text{in.}}$$

$$= 11{,}450\ \text{psi}$$

The approximate formula, Eq. (5.18), then gives a quick useful prediction of the shear stress for initial design purposes. However, remember Eq. (5.18) is only a good approximation for narrow-webbed cross-sections.

EXAMPLE 5.17

For the cantilevered beam with end load of 20 kips and cross-section shown in Figure 5.47, determine the maximum beam shear stress.

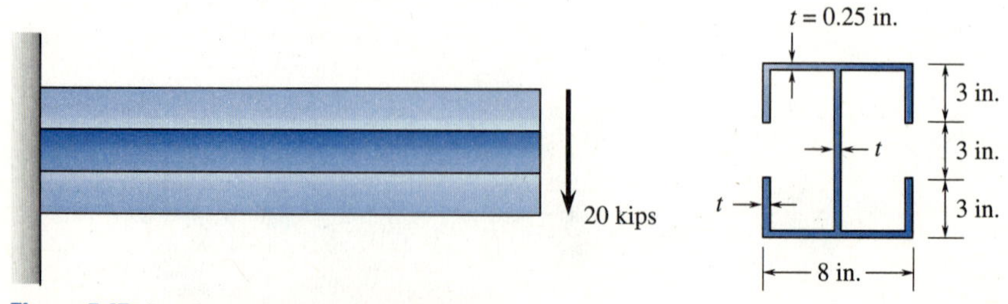

Figure 5.47 Cantilevered beam and cross-section.

Solution

First determine the area moment of inertia. This is easily done by using the areas shown in Figure 5.48. That is, evaluate I for the 8 in. × 9 in. rectangle and then subtract the contributions from rectangular areas labeled ② and ③ in Figure 5.48.

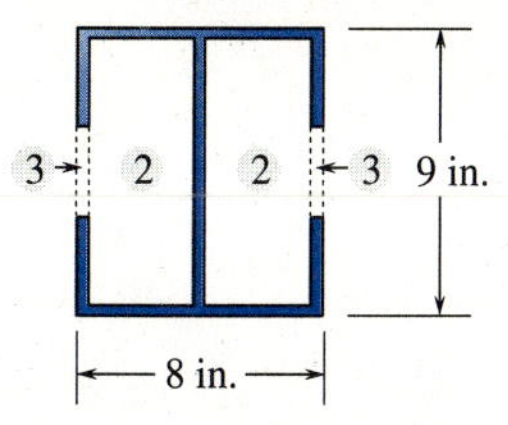

Figure 5.48 Areas used to determine I.

The moment of inertia is then

$$I = \frac{1}{12}(8)(9)^3 - \frac{1}{12}(7.25)(8.5)^3 - \frac{1}{12}(0.25)(3)^3 \times 2$$

$$I = 114 \text{ in.}^4$$

We want to determine the maximum beam shear stress, therefore, we now determine Q using the area above the neutral axis shown in Figure 5.49.

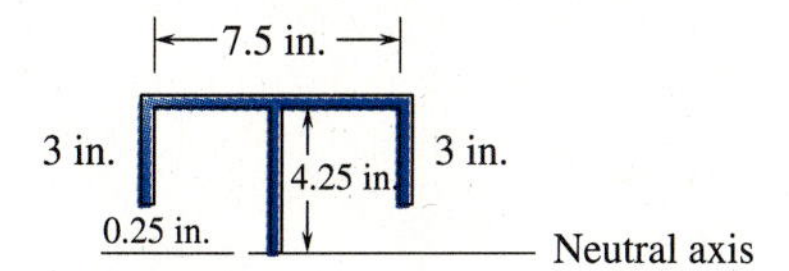

Figure 5.49 Area used to evaluate Q.

We then obtain Q as

$$Q = 2(3 \times 0.25)(3) + (7.5 \times 0.25)(4.375) + (4.25 \times 0.25)(2.125)$$

$$= 14.96 \text{ in.}^3$$

Using Eq. (5.15), we obtain the maximum beam shear stress as

$$\tau_{max} = \frac{(20 \text{ kips})(14.96 \text{ in.}^3)}{(114 \text{ in.}^4)(0.25 \text{ in.})} = 10.5 \text{ ksi}$$

EXAMPLE 5.18

A welded steel girder is fabricated of two 15 mm × 250 mm flange plates and a web plate 12 mm thick and 600 mm deep as shown in Figure 5.50. If the girder must resist a shear force of $V = 500$ kN, what force f per unit length of weld must be transmitted by each fillet weld?

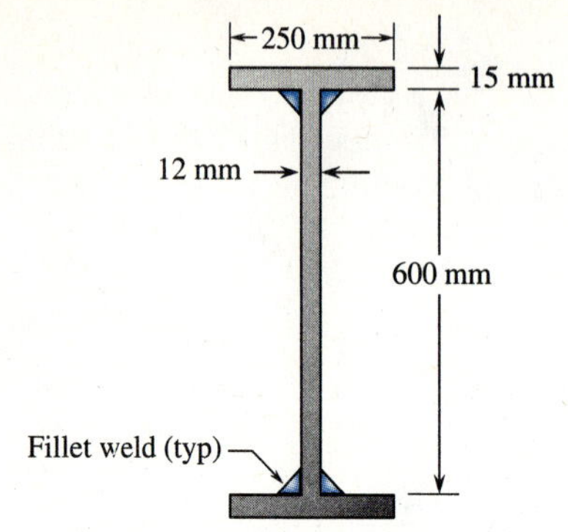

Figure 5.50 Welded steel girder.

Solution

We determine the moment of inertia using the solid rectangular area A_1 and subtracting the contribution from the smaller rectangular areas A_2 shown in Figure 5.51.

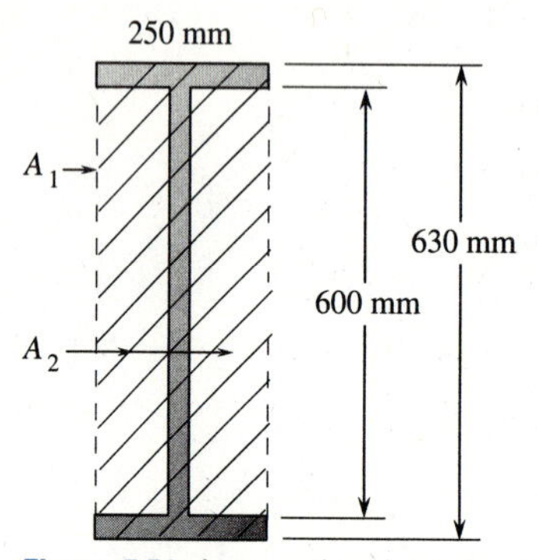

Figure 5.51 Area used to determine I.

The moment of inertia is

$$I = \frac{1}{12}(250)(630)^3 - \frac{1}{12}(238)(600)^3 = 930 \times 10^6 \text{ mm}^4$$

Next, we determine Q. We want to determine the shear resisted by the welds, therefore, we use the area above the weld line (the upper flange) shown in Figure 5.52.

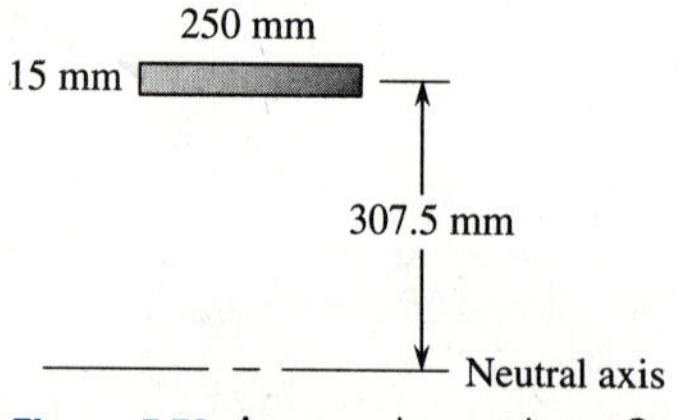

Figure 5.52 Area used to evaluate Q.

We obtain Q as

$$Q = \bar{y}A = (307.5)(15 \times 250) = 1150 \times 10^3 \text{ mm}^3$$

The shear flow at the weld line is

$$q = \frac{VQ}{I} = \frac{(500 \text{ kN})(1150 \times 10^3 \text{ mm}^3)}{930 \times 10^6 \text{ mm}^4}$$

$$q = 619 \text{ N/mm}$$

There is one weld each side of the web or two total welds resisting q, therefore the force per unit length resisted by one of the welds is

$$f = \frac{q}{2} = \frac{619 \text{ N/mm}}{2} = 309.5 \text{ kN/m}$$

5.6 DESIGN OF PRISMATIC BEAMS FOR STRENGTH

The general process of designing a prismatic (or constant cross-section) beam requires many of the same considerations described in Section 2.9 regarding selection of factor of safety. That is, we must consider the type of construction, the number and type of loads, the type of material to be used, and environmental effects.

However, we often reduce the problem to determining the shape and size of a beam such that the actual stresses do not exceed the allowable stresses. For practical purposes here, we will consider only the bending stress obtained from the bending stress equation, Eq. (5.7), and the shear stress obtained from the shear stress equation, Eq. (5.15). We will assume that local and lateral buckling are prevented. A complete design would require considerations of buckling, and deflection (see Chapter 6).

For various cross-sectional shapes of beams of the same material, we will consider the most economical design to be the beam with the smallest weight per unit length, since this beam should be the least expensive.

Standard beam sizes of steel, aluminum, and wood have been manufactured. The American Institute of Steel Construction (AISC)[3], Aluminum Association [4], and National Design Specification for Wood Construction [2] or American Institute of Timber Construction (AITC) [5] publish handbooks of properties of common shapes. This book provides abridged tables of structural steel, aluminum, and wood sections in Appendix C.

For structural steel, properties such as cross-sectional area, moment of inertia and section modulus for wide-flange beam (W), American Standard beam (S), channel (C), and angle (L) shapes are listed in Appendix C. For instance, a W 36 × 300 designation means the section is a W shape (called a wide-flange shape, although it really looks like an I shape) with a nominal or approximate depth of 36 in. and a weight of 300 lb/ft of length and a L $2 \times 2 \times \frac{1}{4}$ is an angle cross-section with legs of 2 in. length and thickness of $\frac{1}{4}$ in. These sections are manufactured by passing a hot billet of steel back and forth between rolls until it is formed into the desired shape. Figure 5.53 shows typical rolled steel sections.

Name of shape	Shape	Symbol	Example designation
Wide-flange beam		W	W36 x 300
American standard beam		S	S10 x 35
Channel		C	C15 x 40
Angle		L	L2 x 2 x $\frac{1}{4}$
Tee, cut from W shape		WT	WT18 x 150
Construction pipe			Standard 4 x 0.188

Figure 5.53 Typical rolled steel sections.

Aluminum structural sections are made by an extrusion process in which the hot billet is pushed (or extruded) through a shaped die. Common shapes are wide-flange, channel, and angle. For properties of common aluminum sections see Appendix C.

Wood beams are cut into rectangular cross-sections that are denoted by their rough-sawn cut size. If the rough surfaces are planed or surfaced to make them smooth (the usual case) then the actual dimensions are less than the rough ones. For instance, a 4 in. × 10 in. nominal size member is actually 3.5 in. × 9.25 in. in size if surfaced on all four sides. We use these actual dimensions in design work. For properties of common lumber sizes consult Appendix C.

STEPS IN SOLUTION

The following steps are normally used to design or select a beam.

1. Draw the shear force and bending moment diagrams corresponding to the specified loading conditions on the beam, and determine the maximum absolute values $|M|_{max}$ and $|V|_{max}$. These values normally control the beam size.
2. Assuming initially that the design is controlled by the value of the largest bending normal stress at $y = \pm c$ at the section where the maximum bending moment occurs, determine the minimum section modulus S (where $S = I/c$) as

$$S_{min} = \frac{|M|_{max}}{\sigma_{allow}} \tag{5.19}$$

where the allowable normal stress has been determined from a table of allowable stresses for the material (for instance, see AISC for steel, AITC, or National Design Specifications for Wood Construction for wood, or the Aluminum Association for aluminum). Or use Eq. (2.29) or (2.30), that is,

$$\sigma_{\text{allow}} = \frac{\sigma_{\text{u}}}{\text{FS}} \quad \text{or} \quad \sigma_{\text{allow}} = \frac{\sigma_{\text{Y}}}{\text{FS}}$$

3. Select a beam size (using Appendix C when appropriate) with an $S \geq S_{\text{min}}$, and with the smallest weight per unit length. This is considered to be the most economical section for which $\sigma \leq \sigma_{\text{allow}}$. The section with the smallest weight may not be the one with the smallest S, as you will notice by examining Appendix C.
4. Next, determine the shear stress in the section we have initially selected and compare this stress to the allowable one. In general, substitute $|V|_{\text{max}}$, Q, I, and b into Eq. (5.15) to obtain

$$\tau_{\text{max}} = \frac{|V|_{\text{max}} Q}{Ib} \tag{5.20}$$

For rectangular cross-sections, recall we can use Eq. (5.16) to obtain the maximum shear stress as

$$\tau_{\text{max}} = \frac{3|V|_{\text{max}}}{2A} \tag{5.21}$$

For W and S beams with long narrow webs, it is customary engineering practice to assume the shear stress to be uniformly distributed over the web (see Figure 5.35(e)), so we use Eq. (5.18) and obtain

$$\tau_{\text{max}} = \frac{|V|_{\text{max}}}{A_{\text{web}}} \tag{5.22}$$

For solid circular cross-sections, we use Eq. (5.17) to obtain

$$\tau_{\text{max}} = \frac{4|V|_{\text{max}}}{3A} \tag{5.23}$$

Then check if

$$\tau_{\text{max}} \leq \tau_{\text{allow}}$$

Examples 5.19 through 5.21 illustrate the steps used to obtain a beam size.

EXAMPLE 5.19

A timber beam of 4 in. nominal width is to be used as a floor joist on a span of 14 ft. The joist carries an effective uniformly distributed load of 250 lb/ft (Figure 5.54). What is the minimum nominal size timber required based on an allowable bending stress of 1200 psi and an allowable longitudinal shear stress of 70 psi?

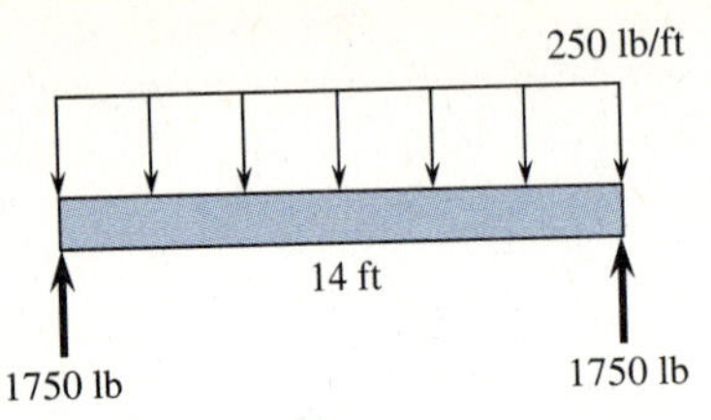

Figure 5.54 Floor joist carrying uniform load.

Solution

For a simply supported beam subjected to a uniform load, the maximum moment was determined in Example 5.2 (see Figure 5.9) as

$$M_{max} = \frac{wL^2}{8}$$

Therefore, we have

$$M_{max} = \frac{(250 \text{ lb/ft})(14 \text{ ft})^2}{8} = 6125 \text{ lb} \cdot \text{ft}$$

The section modulus for a rectangular cross-section is

$$S = \frac{bh^2}{6}$$

Using Eq. (5.19) allows us to determine the height of the joist based on the allowable bending stress as

$$\frac{bh^2}{6} = \frac{M_{max}}{\sigma_{allow}}$$

Solving for h, we obtain

$$h = \left[\frac{6(6125 \text{ lb} \cdot \text{ft})(12 \text{ in./ft})}{(3.5 \text{ in.})(1200 \text{ psi})}\right]^{1/2} = 10.24 \text{ in.}$$

Using Appendix C, the closest standard-sized lumber is a 4 × 12 nominal-sized timber.

We must then check the size based on the allowable shear stress. The maximum shear force was determined in Example 5.2 as

$$V_{max} = \frac{wL}{2} = \frac{(250 \text{ lb/ft})(14 \text{ ft})}{2} = 1750 \text{ lb}$$

For a solid rectangular cross-section, we have from Eq. (5.17)

$$\tau_{max} = \frac{3V}{2A} = \frac{3V}{2bh}$$

Solving for h, we obtain

$$h = \frac{3(1750 \text{ lb})}{2(3.5 \text{ in.})(70 \text{ psi})} = 10.71 \text{ in.}$$

To satisfy both the allowable bending stress and allowable shear stress criteria, we must use the 4 in. × 12 in. nominal-sized timber.

EXAMPLE 5.20

A floor of a warehouse is to be supported by wide-flange beams spaced 4 ft on centers. The beams span 20 ft as shown in Figure 5.55. The floor is a poured-in-place concrete slab, 4 in. thick. The design live load on the floor is 150 lb/ft². Determine a suitable wide-flange beam that will limit the bending stress to 22,000 psi.

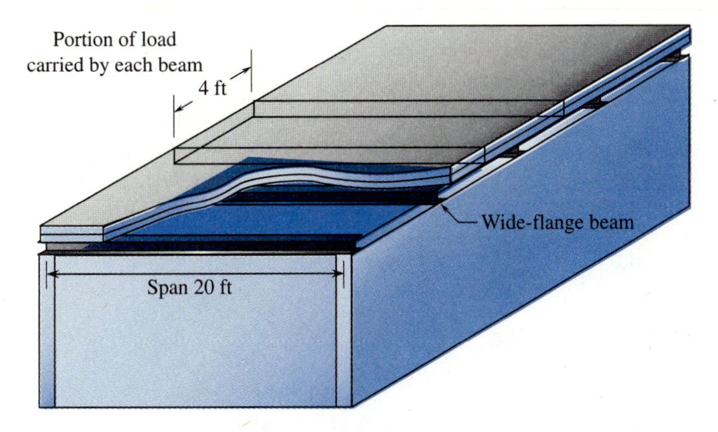

Figure 5.55 Warehouse floor supported by wide-flange beams.

Solution

First, determine the load on the typical beam. Dividing the load evenly among the adjacent beams results in each beam supporting a 4-ft-wide portion of the floor load (called the *tributary area*). The total floor load is the sum of the live load plus the dead load of the concrete. Assuming the concrete weighs 150 lb/ft³, then for a 4 in.-thick-slab every square foot weighs 150 lb/ft³ times (4 in./12 in./ft) = 50 lb/ft².

The total floor load on a beam is then

$$p = 150 + 50 = 200 \text{ lb/ft}^2$$

The distributed load w in lb/ft carried by each beam is then p times the spacing of the beams. Therefore

$$w = ps = (200 \text{ lb/ft}^2)(4 \text{ ft}) = 800 \text{ lb/ft}$$

Figure 5.56 shows the loaded beam, shear force, and bending moment diagrams.

Using Eq. (5.19), the required section modulus is

$$S_{\min} = \frac{(40{,}000 \text{ lb} \cdot \text{ft})(12 \text{ in./ft})}{22{,}000 \text{ psi}} = 21.8 \text{ in.}^3$$

From Appendix C, the following sections are admissible.

$$\text{W } 12 \times 22 \quad S = 25.4 \text{ in.}^3$$

$$\text{W } 10 \times 22 \quad S = 23.2 \text{ in.}^3$$

$$\text{W } 8 \times 28 \quad S = 24.3 \text{ in.}^3$$

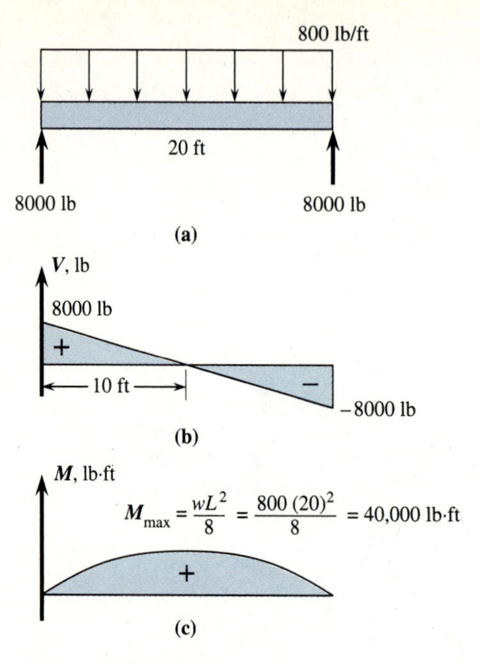

Figure 5.56 (a) Loaded beam,
(b) shear force diagram, and
(c) bending moment diagram.

The final selection may depend, for instance, on necessary clearances. The lightest section with minimum depth is the W 10 × 22. We choose it as the preferred section. (Recall that the number 22 refers to the weight per foot of beam length, i.e., 22 lb/ft.)

EXAMPLE 5.21

A solid circular shaft for a machine is subjected to the loads shown in Figure 5.57. The supports at each end are simple. The allowable bending stress in the shaft is limited to 90 MPa. Determine the diameter of shaft needed based on the allowable bending stress.

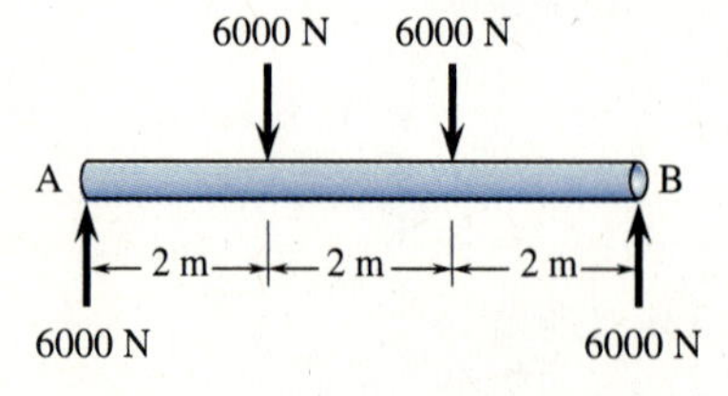

Figure 5.57 Solid shaft subjected to concentrated loads.

Solution

The shear force and bending moment diagrams are shown in Figure 5.58.

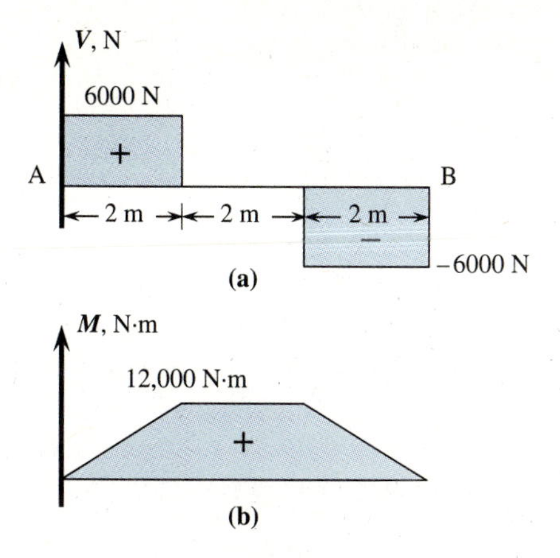

Figure 5.58 (a) Shear force diagram and (b) bending moment diagram of shaft.

The maximum bending moment is

$$M_{max} = (6000 \text{ N})(2 \text{ m}) = 12{,}000 \text{ N} \cdot \text{m} \tag{a}$$

We use Eq. (5.19) to size the shaft as follows:

$$S_{min} = \frac{M_{max}}{\sigma_{allow}} \tag{b}$$

For a solid circular shaft, S is

$$S = \frac{I}{c} = \frac{\frac{\pi}{4}r^4}{r} = \frac{\pi r^3}{4} = \frac{\pi d^3}{32} \tag{c}$$

Using Eq. (c) in Eq. (b), and solving for the diameter, we obtain

$$d = \left[\frac{32\, M_{max}}{\pi \sigma_{allow}}\right]^{1/3} = \left[\frac{32(12{,}000 \text{ N} \cdot \text{m})}{\pi(90 \times 10^6 \frac{\text{N}}{m^2})}\right]^{1/3} = 0.111 \text{ m} = 111 \text{ mm}$$

Use a standard 120-mm-diameter shaft.

5.7 MOVING LOADS ON BEAMS

In this section, we describe how to obtain the maximum shear force and maximum bending moment in beams subjected to moving loads. A truck or other vehicle moving across a beam constitutes a system of concentrated loads at fixed distances from one another. For example, a highway vehicle load designated by the American Association of State Highway and Transportation Officials (AASHTO) [6] as an HS 20-44 loading is

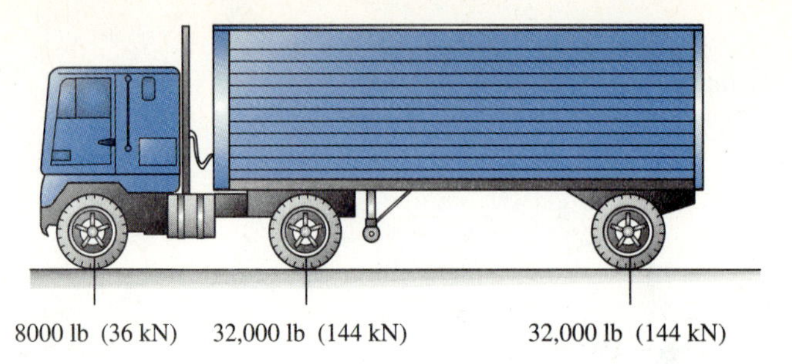

Figure 5.59 A standard vehicle load.

shown in Figure 5.59. The number 20 means there are 20 tons (40,000 lb or 180 kN) on the front two axles as shown. The number 44 is the year of adoption of the load. This loading is the standard truck load that has been adopted for interstate highway bridges.

Maximum Shear Force

For simply supported beams, the absolute maximum shear force will occur next to one of the reactions with the other loads on the beam as shown in Figure 5.60.

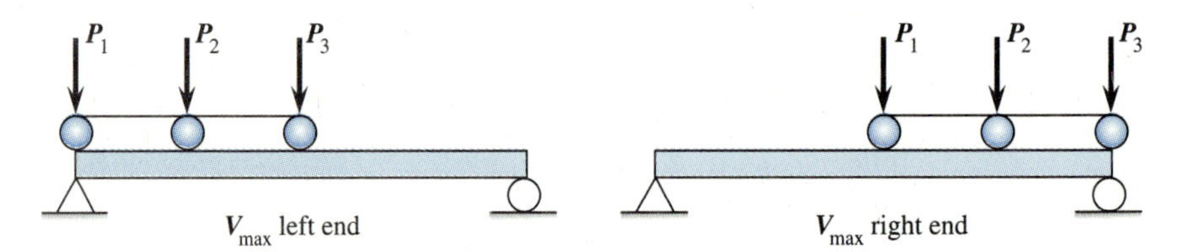

Figure 5.60 Position of loads for maximum shear force.

Maximum Bending Moment

For simply supported beams, the absolute maximum bending moment occurs under one of the concentrated forces such that this force is positioned on the beam so that it and the resultant force of the load system are equidistant from the beam's centerline.

Since we have a series of loads, we must apply this principle to each load in the series and compute the maximum moment for each case. However, the absolute maximum moment often occurs under the largest force lying nearest the resultant force of the system.

We now prove the statement regarding the location of the maximum moment. Consider the beam subjected to the moving loads shown in Figure 5.61. In Figure 5.61, P_1, P_2, P_3, and P_4 represent the system of loads at fixed distances a, b, and c from one another. These loads move as a single unit across the simply supported beam with span length L. We now locate the position of P_2 when the bending moment under this load is

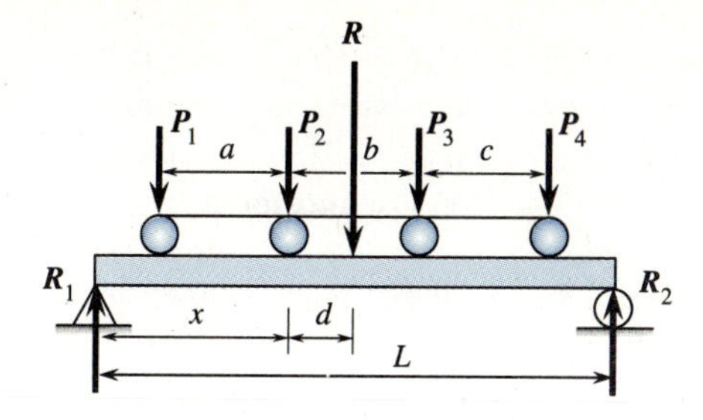

Figure 5.61 System of moving loads on a beam.

maximum. If we represent the resultant of the loads on the span by R and its position from P_2 by d, the value of the left reaction is

$$\Sigma M_2 = 0: \quad R_1 L = R(L - d - x)$$

or

$$R_1 = \frac{R}{L}(L - d - x)$$

From the free-body diagram of the beam up to load P_2, we observe that the bending moment under P_2 is then

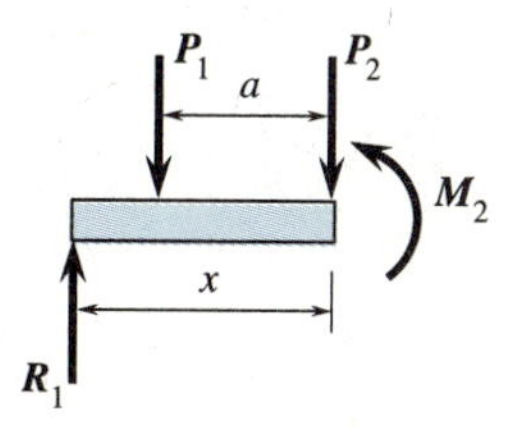

$$M_2 = R_1 x - P_1 a$$

or

$$M_2 = \frac{R}{L}(L - d - x)(x) - P_1 a$$

To obtain the value of x that will yield the maximum moment M_2, we set the derivative of M_2 with respect to x to zero and solve for x as

$$\frac{dM_2}{dx} = \frac{R}{L}(L - d - 2x) = 0$$

Therefore, x is

$$x = \frac{L}{2} - \frac{d}{2} \tag{5.24}$$

This value of x is independent of the number of loads to the left of P_2, since the derivative of all terms of the form $P_1 a$ with respect to x will be zero. We see that Eq. (5.24) locates the load P_2 and the resultant R equal distances from the center of the beam.

Equation (5.24) is explained by the following rule: *The bending moment under a particular load is a maximum when the center of the beam is midway between that load and the resultant of all loads presently on the beam span.* This rule allows us to locate the position of each load when the moment at that load is a maximum. We can then determine the moment using a section cut at the load.

The maximum shear force occurs at the maximum reaction and is equal to this maximum reaction. The maximum reaction for a group of moving loads on a span occurs either at the left reaction, when the leftmost load is over the reaction (here P_1) or at the right reaction when the rightmost load (here P_4) is over that reaction. That is, the maximum reaction occurs at the reaction that the resultant load is closest to.

EXAMPLE 5.22

A simply supported bridge beam shown carries one-half of an HS 20-44 semitrailer loading (Figure 5.62). Determine the maximum moment in the beam.

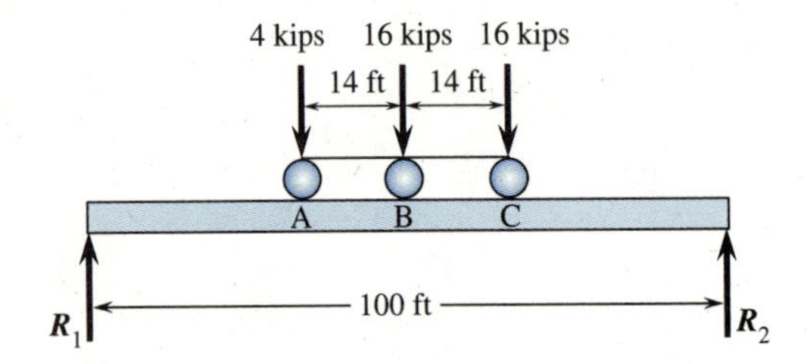

Figure 5.62 Simple beam supporting semitrailer load.

Solution

The maximum moment will occur under one of the largest loads. Try load C. First, determine the resultant force R.

$$R = 4 \text{ kips} + 16 \text{ kips} + 16 \text{ kips} = 36 \text{ kips}$$

Next determine the center x of the force distribution (location of the force resultant) as follows.

Summing moments of the force system at A yields

$$\Sigma M_A\text{:}\quad (16 \text{ kips})(14 \text{ ft}) + (16 \text{ kips})(28 \text{ ft}) = xR = x(36 \text{ kips})$$

$$x = \frac{(16 \text{ kips})(42 \text{ ft})}{36 \text{ kips}} = 18.67 \text{ ft from A}$$

The maximum moment occurs under C in the position shown in Figure 5.63(a), that is, when the load of 16 kips at C and the resultant R are equidistant from the center of the beam.

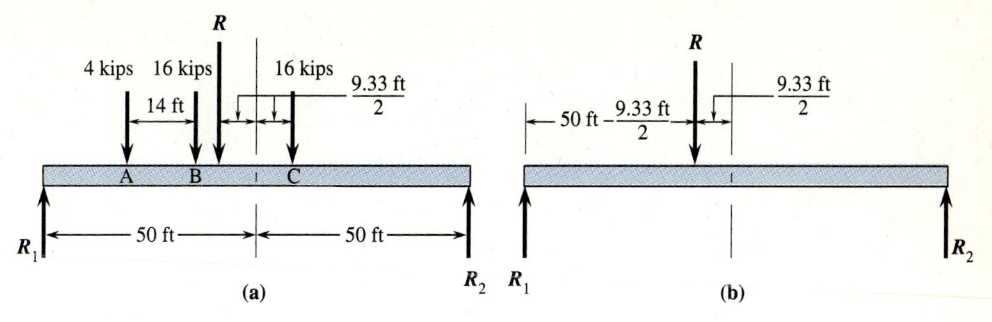

Figure 5.63 (a) Beam showing location of moving load for maximum moment under C and (b) beam with resultant for determining reaction at right end.

To obtain the moment under the load C, first determine the reaction at the right end by using Figure 5.63(b) as follows.

$$\Sigma M_{R_1} = 0: \quad -(36 \text{ kips})\left[50 \text{ ft} - \left(\frac{9.33 \text{ ft}}{2}\right)\right] + R_2(100 \text{ ft}) = 0$$

$$R_2 = 16.32 \text{ kips}$$

The moment at C is then determined by using a free-body diagram of the right section of the beam cut at C as shown below. The moment is obtained as

$$M_C = R_2\left[50 \text{ ft} - \frac{9.33}{2} \text{ ft}\right] = (16.32 \text{ kips})[45.33 \text{ ft}] = 740 \text{ kip} \cdot \text{ft}$$

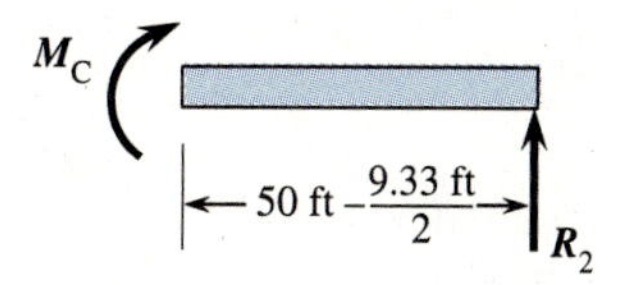

Next, we determine the maximum moment under point B. The maximum moment under B occurs when the resultant R and the 16-kip load at B are equidistant from the center of the beam as shown in Figure 5.64.

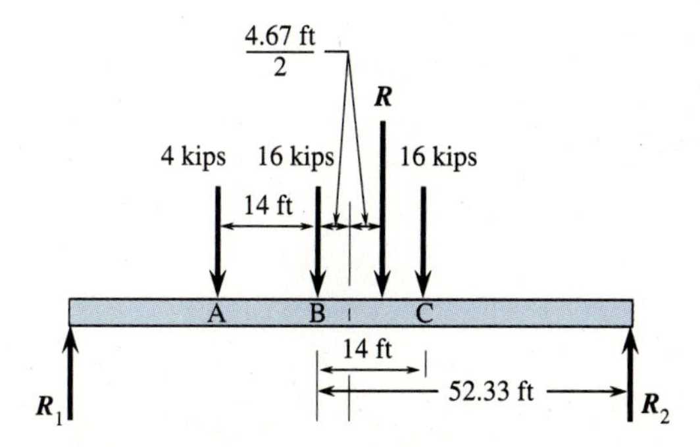

Figure 5.64 Location of loads for maximum moment at B.

We proceed in the same manner as used to obtain the maximum moment under C. First, using Figure 5.64 determine the reaction R_2 as

$$\Sigma M_{R_1} = 0: \quad R_2(100 \text{ ft}) - 36 \text{ kips}(52.33 \text{ ft}) = 0$$

$$R_2 = 18.84 \text{ kips}$$

Next, cut the right section of the beam up to B to obtain the moment at B as

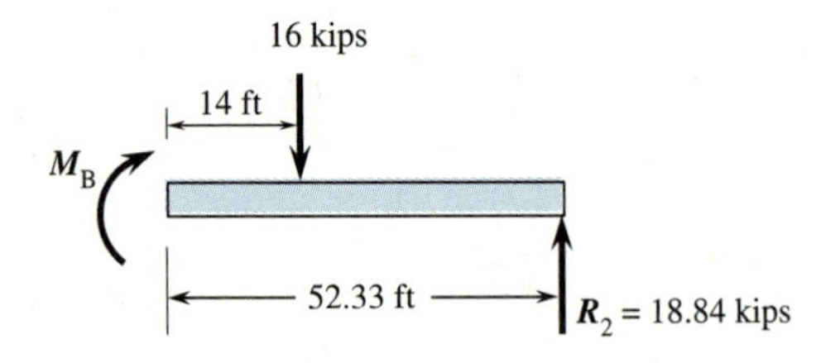

$$M_B = (18.84 \text{ kips})(52.33 \text{ ft}) - (16 \text{ kips})(14 \text{ ft}) = 762 \text{ kip} \cdot \text{ft}$$

Comparing the moments determined at C and B, the maximum one is

$$M_{max} = M_B = 762 \text{ kip} \cdot \text{ft}$$

This is the same moment you can find in Appendix A of the AASHTO specification for highway bridge design [6].

EXAMPLE 5.23

For the simple beam made of a W 30 × 99 wide flange, carrying a moving load from a grain truck, determine **(a)** the maximum moment, **(b)** the largest bending stress, and **(c)** the maximum shear force (Figure 5.65).

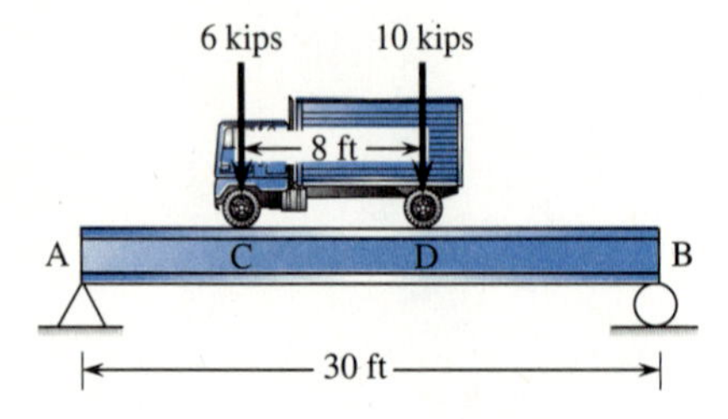

Figure 5.65 Simple beam subjected to moving load.

Solution

(a) Determine the resultant force R as

$$R = 6 \text{ kips} + 10 \text{ kips} = 16 \text{ kips}$$

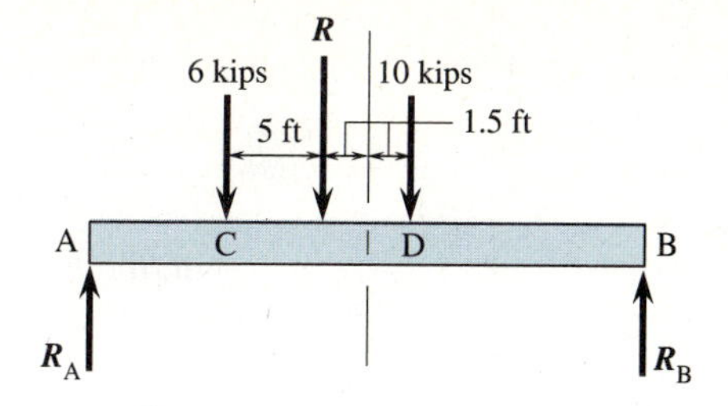

Figure 5.66 Position of loads for maximum moment under D.

Determine the location of the resultant R from the load at C as

$$Rx = (10 \text{ kips})(8 \text{ ft})$$

$$x = \frac{80}{16} = 5 \text{ ft}$$

The maximum moment will occur under the largest load at D when the resultant and the 10-kips load are equidistant from the center of the beam. The position of the loads for this case are shown in Figure 5.66. Next, we determine R_B by using the resultant load as follows.

$$\Sigma M_A = 0: \quad (-16 \text{ kips})(13.5 \text{ ft}) + R_B(30 \text{ ft}) = 0$$

$$R_B = 7.2 \text{ kips}$$

Now cutting the beam at D and using the right section, we have

$$M_D = (7.2 \text{ kips})(13.5 \text{ ft}) = 97.2 \text{ kip} \cdot \text{ft}$$

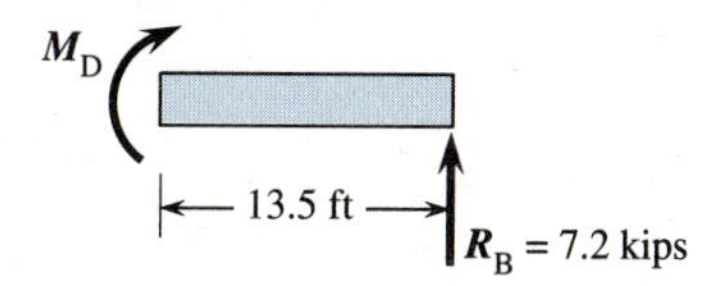

You can verify that this is the maximum moment in the beam by checking the moment under the 6-kip load at C when this load and the resultant are equidistant from the center of the beam span. This case is shown in Figure 5.67. You can show that the maximum moment for this case is $M_{max} = M_C = 83.3 \text{ kip} \cdot \text{ft}$.

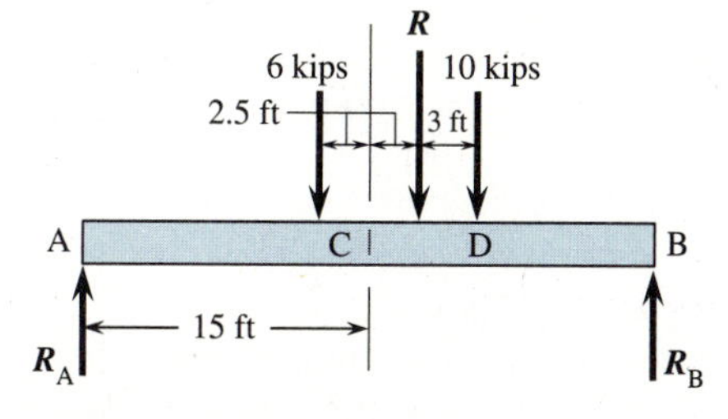

Figure 5.67 Position of loads for maximum moment under C.

(b) The bending stress in the wide-flange beam is obtained using the flexure formula as

$$\sigma = \frac{M}{S} = \frac{97.2 \text{ kip} \cdot \text{ft} \times 12 \text{ in./ft}}{269 \text{ in.}^3} = 4.34 \text{ ksi}$$

(c) The maximum shear force occurs at the left reaction with the leftmost load (6 kips) over the support and the other load (10 kips) on the beam (Figure 5.68(a)), or at the right reaction with the rightmost load (10 kips) on the support and the 6-kip load on the beam (Figure 5.68(b)).

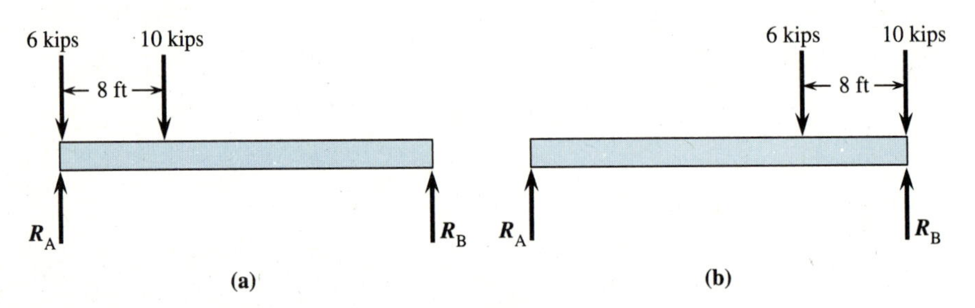

Figure 5.68 Locations of loads for maximum shear force.

Considering the load in the position shown in Figure 5.68(a), the reaction is determined as

$$\Sigma M_B = 0: \qquad -R_A(30 \text{ ft}) + (6 \text{ kips})(30 \text{ ft}) + (10 \text{ kips})(22 \text{ ft}) = 0$$

$$R_A = 6 + 10\left(\frac{22}{30}\right) = 13.33 \text{ kips}$$

Considering the load in the position shown in Figure 5.68(b), we have

$$\Sigma M_A = 0$$

$$R_B(30 \text{ ft}) - (6 \text{ kips})(22 \text{ ft}) - (10 \text{ kips})(30 \text{ ft}) = 0$$

$$R_B = 10 + 6\left(\frac{22}{30}\right) = 14.4 \text{ kips}$$

The largest shear force in the beam occurs at the right end with the truck loads in the position shown in Figure 5.68(b) and is equal to 14.4 kips.

5.8 COMPOSITE BEAMS (BEAMS OF MORE THAN ONE MATERIAL)

Numerous practical cases exist where beams are made of more than one material. For instance, it is common to reinforce a timber beam by adding steel plates to the bottom and top sides of the beam or plates to the edge sides of the beam (Figure 5.69(a)), or to reinforce concrete by adding reinforcing steel rods (Figure 5.69(b)). These beams consisting of more than one material are often called *composite beams*.

The advantage of composite beams is that we can use the lower modulus material in the lower stressed regions or in regions where that material can sufficiently resist the kind

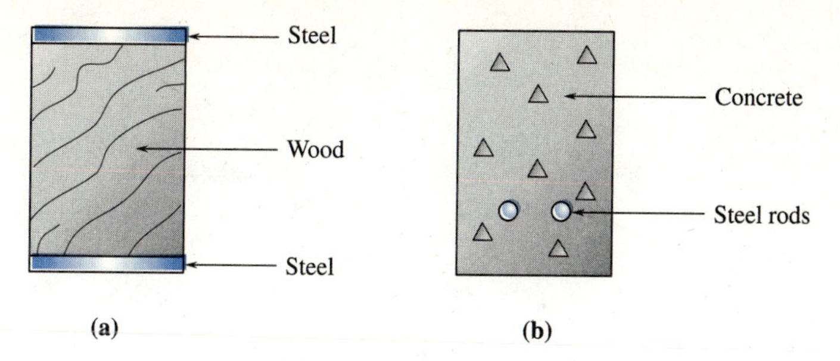

Figure 5.69 (a) Composite timber-steel beam and (b) reinforced concrete beam.

of stresses acting on it. As shown in Figure 5.69(a), the steel is used in the largest bending stress regions, as it is the higher-modulus and higher-strength material.

We can use the transformed or equivalent material approach to transform the beam to one of a single material so that the usual flexure formula, Eq. (5.7), can be used to obtain the stresses. For instance, Figure 5.70(a) shows a composite beam of two materials with moduli of elasticity, E_1 and E_2. We will explain how to transform the composite beam into one of a single material with only modulus of elasticity E_1 (Figure 5.70(b)) or E_2. This will require using an effective width b_e for the transformed material.

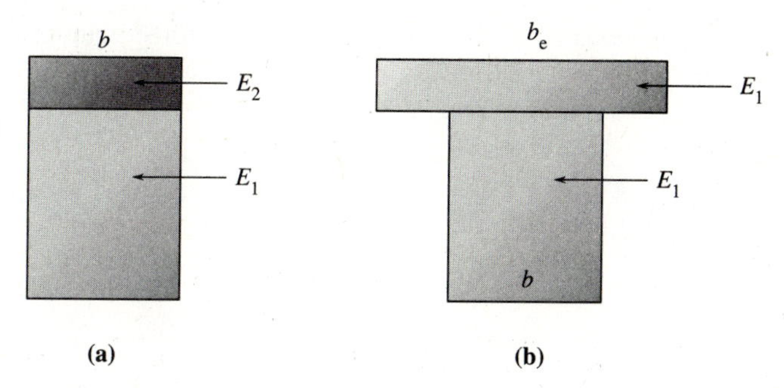

Figure 5.70 (a) Composite beam and (b) transformed beam of one material.

The new (transformed) beam must satisfy the same strain behavior as the actual beam, that is, we assume the strain to vary linearly over the depth y of the cross-section, and the transformed beam must resist the same forces as the actual beam.

Consider a beam of two different materials as shown in Figure 5.71(a). The strain distribution in the elastic beam varies linearly in the same manner as previously described in Section 5.4 for beams of homogeneous material (Figure 5.71(b)). However, since the moduli of elasticity E_1 and E_2 are different, the stresses in each material are now given by

$$\sigma_{x1} = E_1\epsilon_x = -E_1\frac{y}{\rho}$$

$$\sigma_{x2} = E_2\epsilon_x = -E_2\frac{y}{\rho}$$

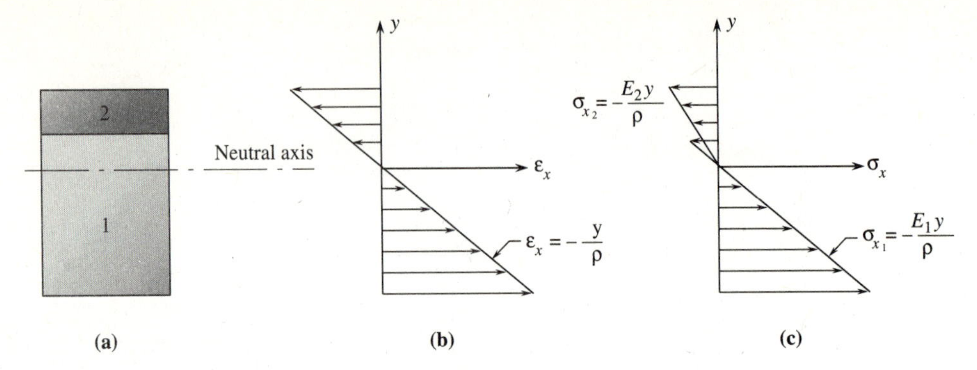

Figure 5.71 Strain and stress variations in two-material beam.

The stress distributions consist of two straight line segments shown in Figure 5.71(c). We would like to replace one of the materials with the other such that the same resistance to load occurs. When this is accomplished, we will have a homogeneous material and can use the usual procedure of beam analysis to obtain the bending stresses. That is, we will be able to find the neutral axis (corresponding to the centroid) and the bending stress using the flexure formula.

To maintain the same force resistance, consider replacing material of width b (Figure 5.72(a)) with a new (effective width) b_e (Figure 5.72(b)). This new dimension must be in a direction parallel to the neutral axis of the section, since we must maintain the same distance from the neutral axis as in the original beam. For force equilibrium, we must have the force F_e on the equivalent section (Figure 5.72(b)) equal to the force F_a on the actual section (Figure 5.72(a)). Therefore

$$F_a = F_e$$

or in terms of the normal stress acting on the area, we have

$$\sigma_x b \, \Delta y = \sigma_{xe} b_e \Delta y$$

Solving for the effective width, we obtain

$$b_e = \frac{\sigma_x b}{\sigma_{xe}} \tag{5.25}$$

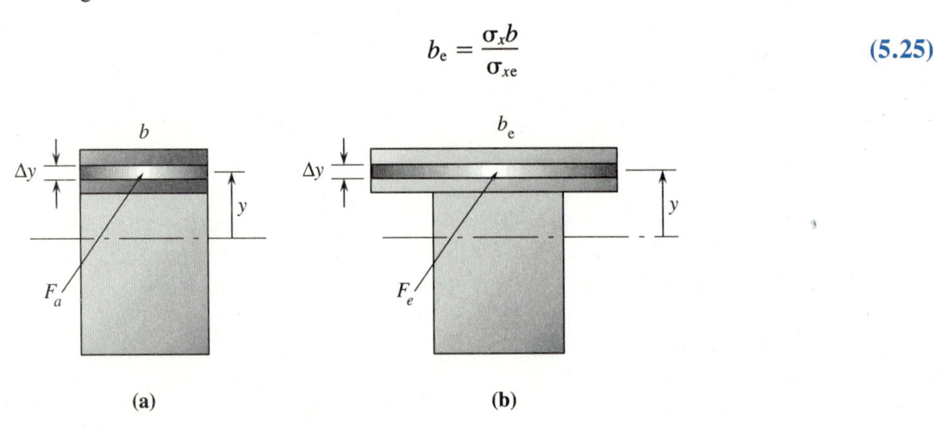

Figure 5.72 (a) Forces on actual beam and (b) forces on equivalent beam.

Since the strains at any distance y must be the same for the equivalent and the actual sections, we write

$$\epsilon_x = \epsilon_{xe}$$

or by Hooke's law, the original material with E_2 has strain, $\epsilon_x = \sigma_x/E_2$ and the equivalent section (now of material E_1) has strain, $\epsilon_{xe} = \sigma_{xe}/E_1$. Therefore

$$\epsilon_x = \frac{\sigma_x}{E_2} = \epsilon_{xe} = \frac{\sigma_{xe}}{E_1}$$

or

$$\frac{\sigma_x}{\sigma_{xe}} = \frac{E_2}{E_1} \tag{5.26}$$

Using Eq. (5.26) in Eq. (5.25), we obtain the effective width b_e as

$$b_e = \frac{E_2}{E_1}b \tag{5.27}$$

Therefore, the equivalent width of the transformed section (material two) is given by Eq. (5.27).

STEPS IN SOLUTION

The following steps are used in the transformed section method to determine the stresses in composite beams.

1. Transform the cross-section to one of a single material, say with modulus of elasticity E_1, by using Eq. (5.27) to obtain the effective width of the material 2 as

$$b_e = \frac{E_2}{E_1}b$$

2. Now the beam is of one material with E_1, so locate the neutral axis (equal to the centroid) in the usual manner for homogeneous materials.
3. Determine the moment of inertia about the neutral axis of the transformed cross-section.
4. Use the flexure formula to obtain the stresses. In the transformed section (now considered to be of only material one), we have

$$\sigma_e = \sigma_{\text{trans}} = \frac{-My}{I}$$

where y locates material in the transformed section.

5. The actual stress in the transformed material is then obtained by multiplying the transformed stress by E_2/E_1. That is,

$$\sigma_{\text{act}} = \frac{E_2}{E_1}\sigma_{\text{trans}} \tag{5.28}$$

EXAMPLE 5.24

A full-sized 2 in. × 4 in. timber beam is reinforced by bolting a $\frac{1}{4}$ in.-thick steel plate to the top and bottom of the beam as shown. The beam must resist a bending moment of 10,000 lb · in. Let E of timber be 1.5×10^6 psi and E of steel be 30×10^6 psi. Determine the largest bending stress in the steel and timber (Figure 5.73).

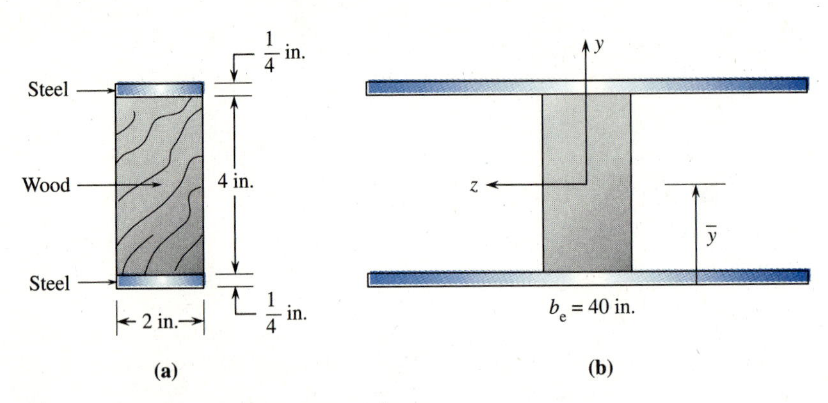

Figure 5.73 (a) Composite timber and steel beam and (b) transformed section.

Solution

Transform the cross-section to one material (here choose timber). Use Eq. (5.27) to obtain the equivalent width b_e of the steel plates (see Figure 5.73(b)) as

$$b_e = \left(\frac{E_2}{E_1}\right)b = \left(\frac{30}{1.5}\right)(2 \text{ in.}) = 40 \text{ in.}$$

By symmetry, we locate the centroid at one-half the depth, that is,

$$\bar{y} = 2.25 \text{ in.}$$

The moment of inertia about the neutral axis is

$$I_z = \frac{1}{12}(2 \text{ in.})(4 \text{ in.})^3 + 2[(40 \text{ in.} \times 0.25 \text{ in.})(2.125 \text{ in.})^2]$$

$$= 101 \text{ in.}^4$$

where the $(\frac{1}{12})bh^3$ terms for the plates are negligible. The largest stress in the transformed section (at the bottom of the beam) is

$$\sigma_{\text{trans}} = \frac{(10{,}000 \text{ lb} \cdot \text{in.})(2.25 \text{ in.})}{101 \text{ in.}^4} = 223 \text{ psi}$$

However, the steel was transformed, therefore the actual stress in the steel (at the bottom of the beam) is obtained using Eq. (5.28) as

$$\sigma_{\max} = \sigma_{\text{trans}} = \frac{E_2}{E_1}(223 \text{ psi}) = 20(223 \text{ psi}) = 4460 \text{ psi}$$

The maximum stress in the timber is at $y = 2$ in. and is

$$\sigma_{\text{timber}} = \frac{(10{,}000)(2)}{101} = 198 \text{ psi}$$

The stress in the steel at $y = 2$ in. is

$$\sigma_s = 20(198 \text{ psi}) = 3960 \text{ psi}$$

The bending stress distribution over the cross-section is shown below.

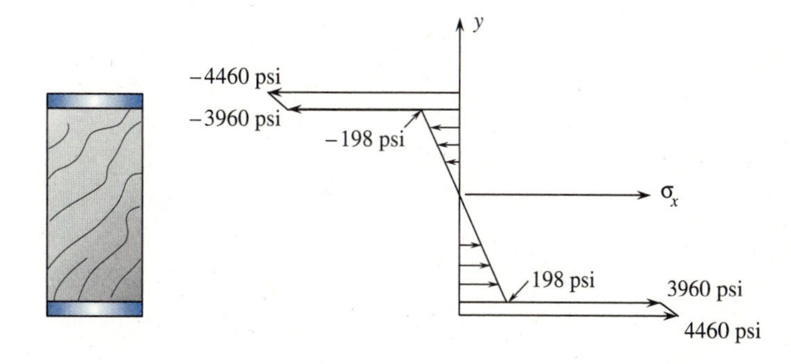

Note that without the steel reinforcement, the moment of inertia is $I = (1/12)(2 \text{ in.})(4 \text{ in.})^3 = 10.67 \text{ in.}^4$, and the largest bending stress in the timber is

$$\sigma_{\max} = \frac{(10{,}000)(2)}{10.67} = 1874 \text{ psi}$$

This stress is 9.5 times greater than with steel added and exceeds the allowable bending stress in most timber (see the National Design Specification for timber [2]).

EXAMPLE 5.25

For the full sized 2 in. × 4 in. timber beam reinforced with 1/4 in. thick steel side plates shown, determine the maximum bending stresses in both the timber and steel. The beam resists a bending moment of 10,000 lb · in. Let E of the timber be 1.5×10^6 psi and E of steel be 30×10^6 psi.

Solution

Transform the cross-section to one of all timber. Use Eq. (5.27) to obtain the equivalent width b_e of the steel plates (Figure 5.74(b)) as

$$b_e = \frac{E_2}{E_1}b = \frac{30}{1.5}\left(\frac{1}{4}\right) = 5 \text{ in.}$$

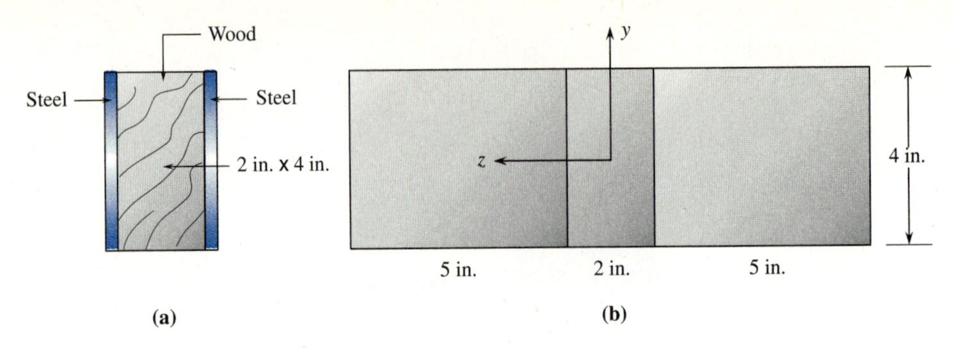

Figure 5.74 (a) Steel-reinforced timber beam and (b) transformed section.

By inspection, the centroid is at one-half the depth, that is,

$$\bar{y} = 2 \text{ in.}$$

The moment of inertia about the neutral axis is

$$I_z = \frac{1}{12}(12 \text{ in.})(4 \text{ in.})^3 = 64 \text{ in.}^4$$

The largest stress in the transformed steel section is

$$\sigma_{\text{trans}} = \frac{(10{,}000)(2)}{64} = 312.5 \text{ psi}$$

However, the steel was transformed. Therefore, the actual stress in the steel is given by using Eq. (5.28) as

$$\sigma_{\max} = \frac{30}{1.5}(312.5 \text{ psi}) = 6250 \text{ psi}$$

The maximum stress in the timber is

$$\sigma_{\text{timber}} = \frac{(10{,}000)(2)}{64} = 312.5 \text{ psi}$$

Note that in comparing results from the previous Example 5.24 and this one, it is more efficient to use steel on the top and bottom of the timber to reduce timber stresses. Less steel is used and a greater decrease in timber stress achieved by placing the steel plates on the top and bottom of the timber beam.

5.9 STRESS CONCENTRATIONS IN BENDING MEMBERS

In Section 2.10, we discussed stress concentrations in axially loaded members due to geometric changes such as holes, fillets, and notches in the members. Stress concentrations in torsionally loaded members were discussed in Section 4.11.

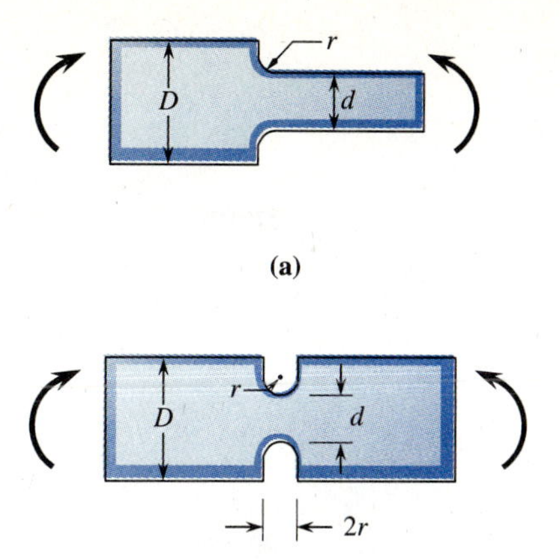

Figure 5.75 Flat bars with geometric discontinuities subjected to bending.

In this section, we discuss stress concentrations that result due to geometric changes, such as fillets or notches, in bending members as shown in Figure 5.75.

The bending stress in a uniform cross-section with a plane of symmetry can be determined using the flexure formula (see Section 5.4, Eq. (5.7)). To repeat, the flexure formula is

$$\sigma = -\frac{My}{I} \tag{5.29}$$

and

$$\sigma_{\text{max}} = \frac{Mc}{I} \tag{5.30}$$

Now if the bending member has a sudden change in geometry, stress concentrations occur near the abrupt change. This stress can be determined by including a stress concentration factor in Eq. (5.30) as

$$\sigma_{\text{max}} = K\frac{Mc}{I} \tag{5.31}$$

where again K is the stress concentration factor and I and c refer to the critical cross-section, that is, the section of smallest depth, d.

Plots of stress concentration factors for flat bars with fillets and notches (grooves) are shown in Figures 5.76(a) and (b) and for round bars with fillet and groove in Figures 5.76(c) and (d). It should be clear from these figures that the bigger the radius of the fillet or groove for a given d, the smaller the stress concentration factor.

More extensive tables of stress concentration factors are provided in Peterson's *Stress Concentration Factors* [7]. The values of these stress concentration factors K have been determined based on linear-elastic relations between the stresses and strains. Hence, in practice if the maximum stress obtained from Eq. (5.31) is greater than the yield stress of a ductile material, then the largest stress can be no greater than the yield stress. This topic was discussed in detail in Section 2.10.

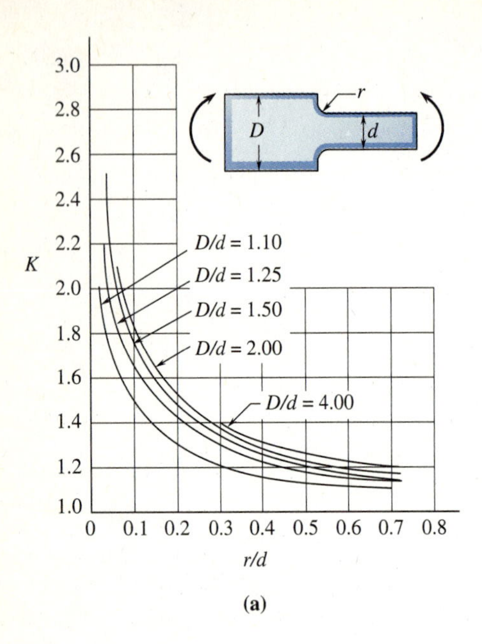

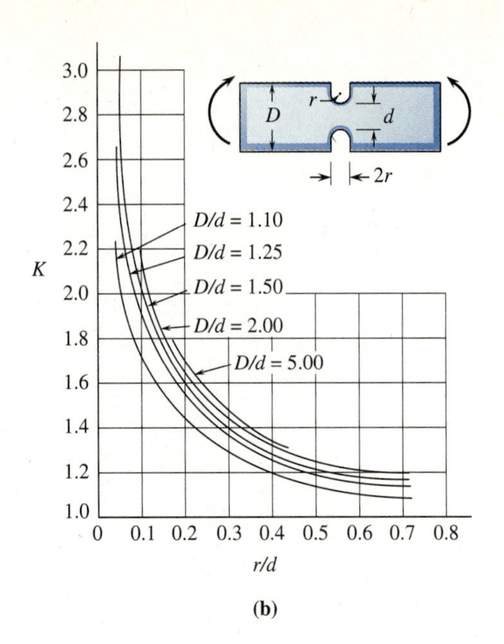

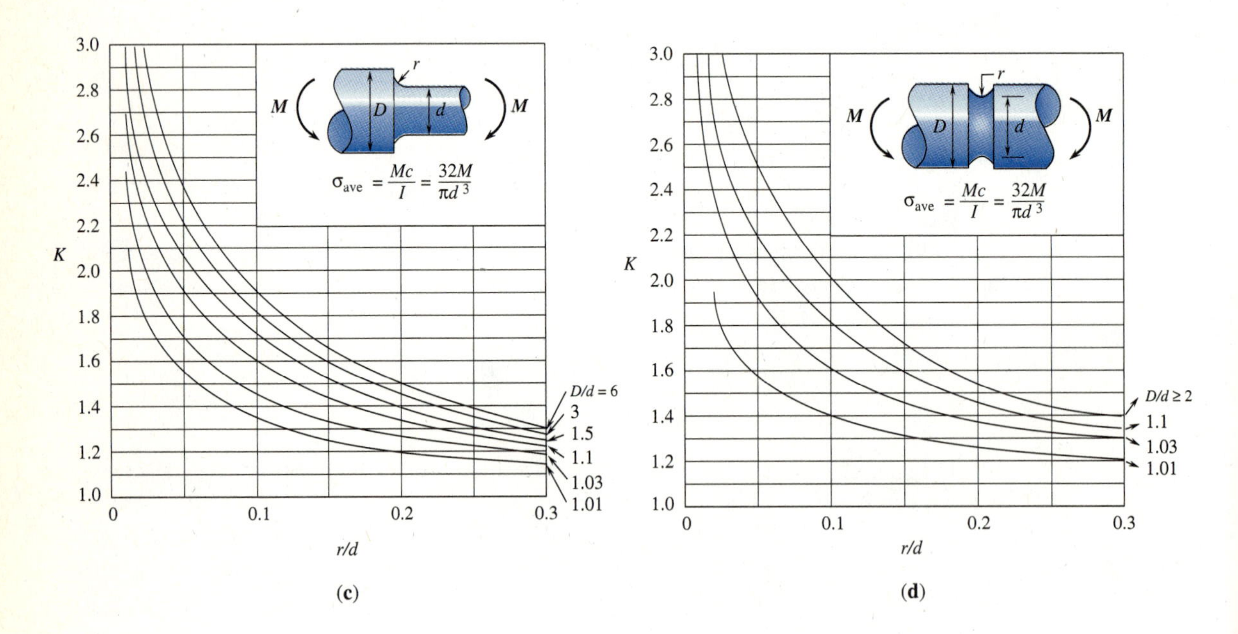

Figure 5.76 Stress concentration factors for a flat bar in bending
(a) with a fillet,
(b) with a groove and for a round bar in bending,
(c) with a fillet, and
(d) with a groove.

EXAMPLE 5.26

A machine part in bending has a moment of 25,000 lb · in. applied to it. The member is a flat bar of mild steel with yield stress of 36 ksi. The member has two different depths of 6 in. and 4 in. and has a thickness of 1 in. The radius of the fillet in the transition region is 1 in. Determine the maximum bending stress. Include stress concentration effects (Figure 5.77).

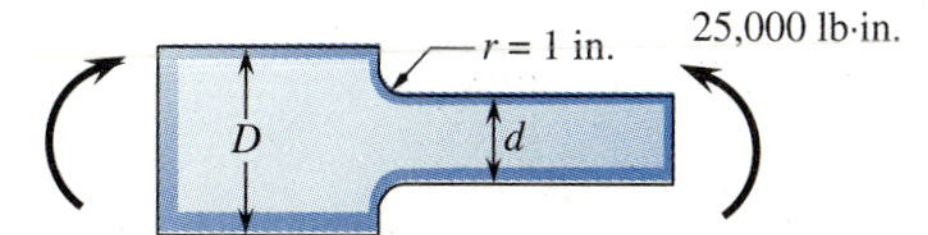

Figure 5.77 Machine part with fillets.

Solution

To obtain the stress concentration factor, we determine the ratios

$$\frac{r}{d} = \frac{1}{4} = 0.25 \quad \text{and} \quad \frac{D}{d} = \frac{6}{4} = 1.5$$

From Figure 5.76(a), with $r/d = 0.25$ and $D/d = 1.5$, we obtain

$$K = 1.4$$

The moment of inertia about the neutral axis of the smallest cross-section is

$$I = \frac{1}{12}(1 \text{ in.})(4 \text{ in.})^3 = 5.33 \text{ in.}^4$$

Using Eq. (5.31), the largest bending stress is

$$\sigma_{\max} = 1.4\frac{(25{,}000)(2)}{5.33}$$

$$= 13{,}130 \text{ psi}$$

This stress is less than the yield stress and so is a valid answer.

EXAMPLE 5.27

A flat steel bar is to have grooves cut into both edges. The whole depth of the bar is 100 mm and the thickness is 10 mm. The depth where the grooves are cut is required to be no less than 80 mm (Figure 5.78). The allowable bending stress is 150 MPa. The resisting moment is 1 kN · m. Determine the radius of the grooves.

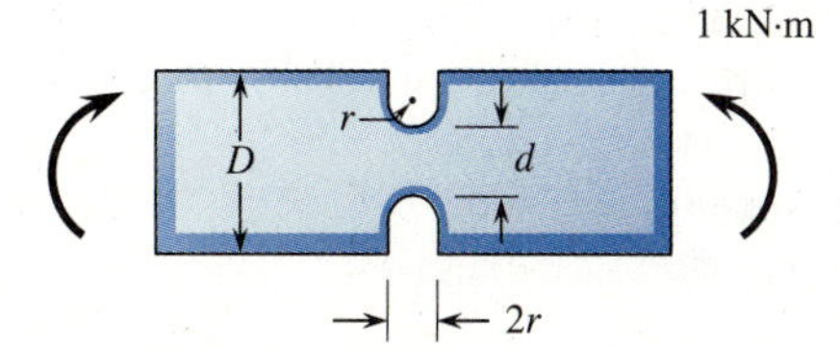

Figure 5.78 Bending member with grooves.

Solution

The moment of inertia about the neutral axis of the smallest cross-section is

$$I = \frac{1}{12}(10\text{ mm})(80\text{ mm})^3 = 426{,}700\text{ mm}^4 = 0.4267 \times 10^{-6}\text{ m}^4$$

The nominal bending stress without stress concentration is

$$\sigma = \frac{Mc}{I} = \frac{(1\text{ kN}\cdot\text{m})(0.040\text{ m})}{0.4267 \times 10^{-6}\text{ m}^4}$$

$$= 0.09374 \times 10^6\text{ kN/m}^2$$

$$= 93{,}740\text{ kN/m}^2 = 93.74\text{ MPa}$$

Using Eq. (5.31), we have

$$\sigma_{\max} = 150\text{ MPa} = K(93.74\text{ MPa})$$

or solving for K, we obtain

$$K = 1.6$$

We also have the ratio of D/d given by

$$\frac{D}{d} = \frac{100}{80} = 1.25$$

Using Figure 5.76(b), with the curve corresponding to $D/d = 1.25$ and $K = 1.60$, we need $r/d = 0.18$. Therefore, we can solve for r (the radius of the groove) as

$$r = 0.18d$$

or

$$r = 0.18d = 0.18(80\text{ mm}) = 14.4\text{ mm}$$

The width of the grooves must be

$$2r = 2(14.4\text{ mm}) = 28.8\text{ mm}$$

5.10 UNSYMMETRIC BENDING

So far we have analyzed beam bending for beams having a plane of symmetry acting through the longitudinal axis. The transverse loads acting on the beam were in the plane of symmetry. Figure 5.79 shows such a beam with a plane of symmetry (x-y) and transverse loading acting in this x-y plane such that the only deflection is in the y-direction. The y-axis is an axis of symmetry and hence a principal axis. The z-axis is the neutral axis perpendicular to the y-axis and is also a principal axis.

A very large number of beams in practice are under symmetric bending and the bending stress can be determined from the flexure formula given by Eq. (5.7) as $\sigma = -My/I$, where the bending takes place about the z-axis.

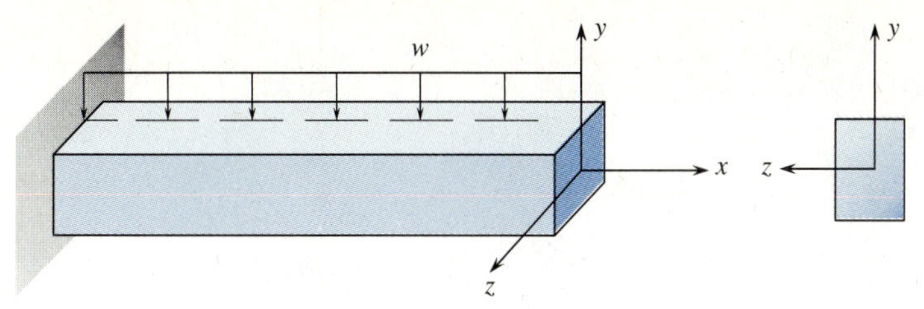

Figure 5.79 Beam under symmetric bending.

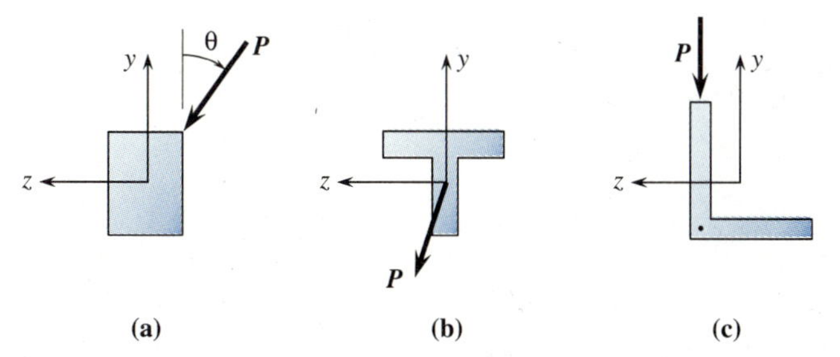

Figure 5.80 Cases of unsymmetric bending.

We will now consider unsymmetric bending of beams. Two common cases will be considered, one when the loads do not act in a plane of symmetry (Figure 5.80(a)) and the other when the cross-section is not symmetric (Figures 5.80(b) and (c)).

A beam, such as the rectangular cross-section of Figure 5.80(a), is doubly symmetric as it has two planes of symmetry. A beam, such as the tee beam (Figure 5.80(b)), is singly symmetric as it has one plane of symmetry. Finally, a beam is unsymmetric if it has no plane of symmetry as in the angle of Figure 5.80(c).

Doubly Symmetric Beams and Loads Not in Plane of Symmetry

First, consider a doubly symmetric beam with loads acting in directions that are not in either plane of symmetry as shown (Figure 5.81). Here the load must act through the centroid of the cross-section to avoid twisting about the x-axis.

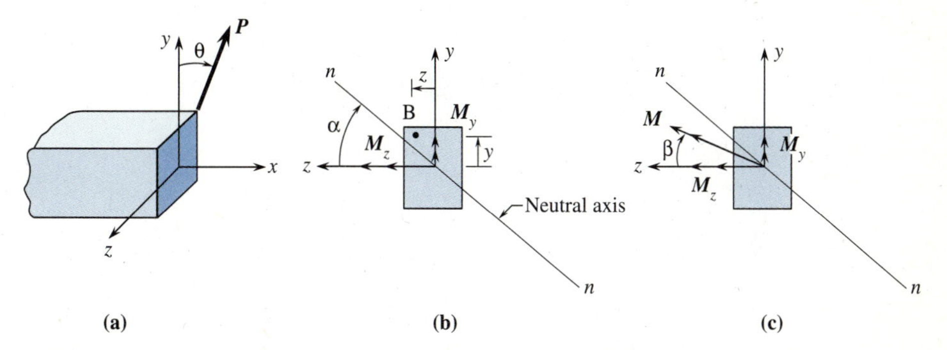

Figure 5.81 Doubly symmetric beam load not in either plane of symmetry.

STEPS IN SOLUTION

To determine the bending stress in an unsymmetrically loaded (skewed loaded) beam, the following steps can be used.

1. Resolve the load into two components, one in each plane of symmetry. That is,

$$P_y = P \cos \theta \quad \text{and} \quad P_z = P \sin \theta$$

2. Evaluate the bending moments about the z- and y-axis (see Figure 5.81(b)) as M_y and M_z. Here M_y and M_z are both positive for the direction of P shown.
3. Because these bending moments act in the planes of symmetry, apply the usual flexure formula for each moment to obtain the bending stress say at point B in the beam cross-section (see Figure 5.81(b)) as

$$\sigma_x = \frac{M_y z}{I_y} - \frac{M_z y}{I_z} \tag{5.32}$$

where the positive sign on the first term is due to the fact that the positive moment M_y produces a positive stress in the positive z-direction, while the positive moment M_z produces a negative stress in the positive y-direction. The moments of inertia I_y and I_z are those evaluated about the principal axes y and z, respectively.

To locate points of maximum stress, in general, we need to determine the location of the neutral plane or axis of the cross-section. The maximum perpendicular distances from this plane to points on the cross-section will be points of maximum tensile and compressive bending stresses.

The neutral axis is the location where the bending stress σ_x is zero. Therefore, using Eq. (5.32), we have

$$\frac{M_y z}{I_y} - \frac{M_z y}{I_z} = 0$$

or

$$\frac{y}{z} = \frac{M_y I_z}{M_z I_y} \tag{5.33}$$

We now define an angle α clockwise from the z-axis (see Figure 5.81(b)), such that

$$\tan \alpha = \frac{y}{z} = \frac{M_y I_z}{M_z I_y} \tag{5.34}$$

Depending on the directions and magnitudes of M_y and M_z, the angle may vary from $-90°$ to $+90°$. The ratio of the moments is

$$\frac{M_y}{M_z} = \tan \beta \tag{5.35}$$

Therefore, the resultant moment is at an angle β from the z-axis (see Figure 5.81(c)).

The moment vector M is perpendicular to the longitudinal plane containing the force P. In general, the resultant moment M is not in line with the neutral axis. That means the neutral axis is generally not perpendicular to the direction of the load P. As exceptions to this, we have three cases: (1) If $I_y = I_z$, the principal moments of inertia are equal and the plane of loading is a principal one and is perpendicular to the neutral axis as $\alpha = \beta$. (2) If $\alpha = 0$, the load is in the x-y plane and the z-axis becomes the neutral axis. The only bending moment is then M_z. (3) If $\alpha = 90°$, the load P is in the x-z plane and the y-axis is the neutral axis while the only bending moment is M_y.

Examples 5.28 and 5.29 illustrate how to determine the bending stresses in doubly symmetric beams with unsymmetric loading.

EXAMPLE 5.28

A simply supported beam with a rectangular cross-section is subjected to two unsymmetric loads of 800 lb (Figures 5.82(a) and (b)). Determine the largest tensile and compressive bending stresses.

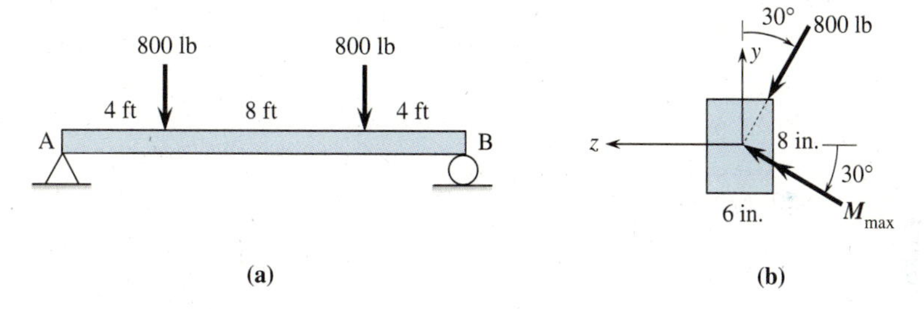

Figure 5.82 Simple beam with unsymmetric load.

Solution

First, a free-body diagram is drawn and then the bending moment diagram is obtained as shown.

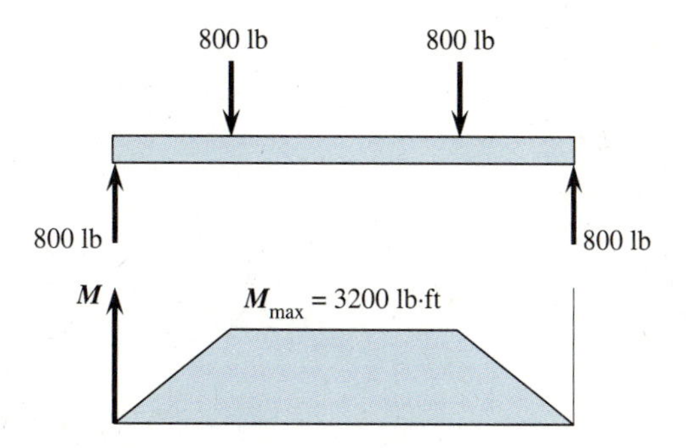

The maximum bending moment is 3200 lb · ft at an angle of 30° from the z-axis as shown in Figure 5.82(b). We can obtain the moments M_y and M_z as follows

$$M_y = (P \sin 30°)(4 \text{ ft})(12 \text{ in./ft}) = 19{,}200 \text{ lb} \cdot \text{in.}$$

and

$$M_z = (P \cos 30°)(4 \text{ ft})(12 \text{ in./ft}) = 33{,}300 \text{ lb} \cdot \text{in.}$$

The neutral axis is located by Eq. (5.34) as

$$\tan \alpha = \frac{M_y I_z}{M_z I_y}$$

where the principal moments of inertia are

$$I_y = \frac{1}{12}(8 \text{ in.})(6 \text{ in.})^3 = 144 \text{ in.}^4$$

and

$$I_z = \frac{1}{12}(6 \text{ in.})(8 \text{ in.})^3 = 256 \text{ in.}^4$$

Therefore

$$\tan \alpha = \frac{(19{,}200)(256)}{(33{,}300)(144)} = 1.025$$

and

$\alpha = 45.71°$ (the positive number means clockwise from the z-axis). The largest bending stresses occur at the farthest locations from the neutral axis. These are points C and D in the figure shown below.

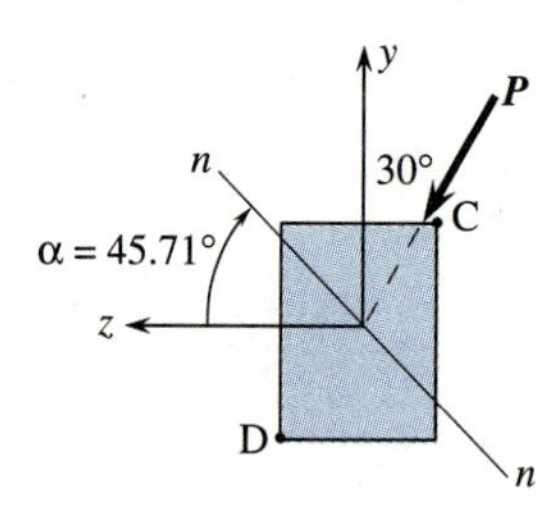

At C:

$$y = 4 \text{ in.}, \; z = -3 \text{ in.}$$

and the stress is

$$\sigma_x = \frac{M_y z}{I_y} - \frac{M_z y}{I_z}$$

$$= \frac{(19{,}200)(-3)}{144} - \frac{(33{,}300)(4)}{256}$$

$$= -400 \text{ psi} - 520 \text{ psi}$$

$$\sigma_x = -920 \text{ psi (C)}$$

At D:

$$y = -4 \text{ in.}, z = 3 \text{ in.}$$

and the stress is

$$\sigma_x = \frac{(19{,}200)(3)}{144} - \frac{(33{,}300)(-4)}{256}$$

$$= 920 \text{ psi (T)}$$

EXAMPLE 5.29

A wide-flange W 10 × 45 section is used as a cantilever beam 10 ft long in a mill building. It supports a combined inclined load of $P = 4$ kips at its free end. Determine the maximum tensile stress (Figure 5.83).

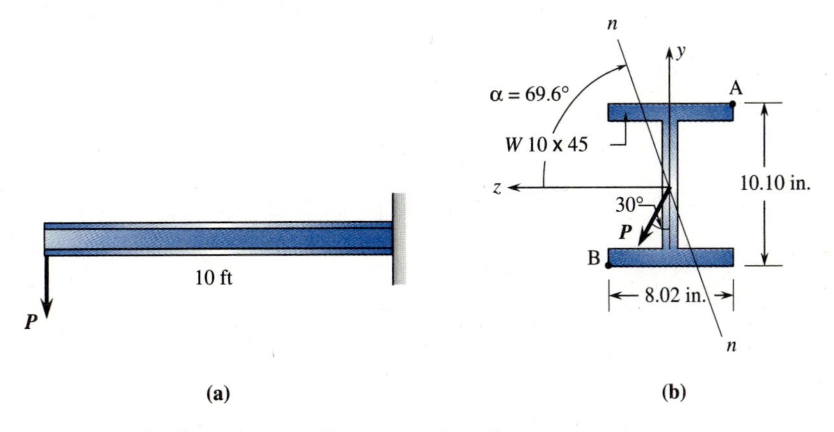

Figure 5.83 Cantilevered beam with unsymmetric load.

Solution

Determine the force components:

$$P_y = -P \cos\theta = -4(\cos 30°) = -3.46 \text{ kips}$$

$$P_z = P \sin\theta = 4(\sin 30°) = 2.00 \text{ kips}$$

The largest moments occur at the support and are given by

$$M_y = -P_z(10 \text{ ft}) = -(2.00 \text{ kips})(10 \text{ ft}) = -20.0 \text{ kip} \cdot \text{ft}$$

$$M_z = P_y(10 \text{ ft}) = -(3.46 \text{ kips})(10 \text{ ft}) = -34.6 \text{ kip} \cdot \text{ft}$$

By Appendix C, the depth d of the beam and the width b_f of the flange are

$$d = 10.10 \text{ in.} \quad \text{and} \quad b_f = 8.02 \text{ in.}$$

Also from Appendix C, the principal moments of inertia are

$$I_y = 53.4 \text{ in.}^4 \quad \text{and} \quad I_z = 248 \text{ in.}^4$$

The neutral axis is located using Eq. (5.34) as

$$\tan\alpha = \frac{M_y I_z}{M_z I_y} = \frac{(-20)(248)}{(-34.6)(53.4)}$$

$$\tan\alpha = 2.685$$

$$\alpha = 69.6° \text{ (as shown in Figure 5.83(b))}$$

The largest tensile stress occurs at point A with coordinates

$$y = \frac{10.10 \text{ in.}}{2} = 5.05 \text{ in.}$$

$$z = \frac{-8.02 \text{ in.}}{2} = -4.01 \text{ in.}$$

Using Eq. (5.32), this bending stress is

$$\sigma_x = \frac{(-20 \times 12 \text{ in./ft})(-4.01)}{53.4} - \frac{(-34.6 \times 12 \text{ in./ft})(5.05)}{248}$$

$$= 18.02 \text{ ksi} + 8.45 \text{ ksi}$$

$$= 26.47 \text{ ksi (T)}$$

The largest compressive stress is at B with coordinates

$$y = -5.05 \text{ in.} \quad \text{and} \quad z = 4.01 \text{ in.}$$

Therefore, the bending stress is

$$\sigma_x = -26.47 \text{ ksi (C)}$$

For comparison, if the force were all vertical, then the stress would be

$$\sigma_x = \frac{(4 \text{ kips} \times 10 \text{ ft})(12 \text{ in./ft})(5.05 \text{ in.})}{248 \text{ in.}^4}$$

$$= 9.77 \text{ ksi (tensile at top and compressive at bottom)}$$

Therefore, having a force out of line from the vertical yields a large increase in stress for a beam with much larger I_z than I_y. In practice, a cap channel is sometimes used to strengthen and stiffen the lateral direction (see Figure below). Also lateral support is often provided.

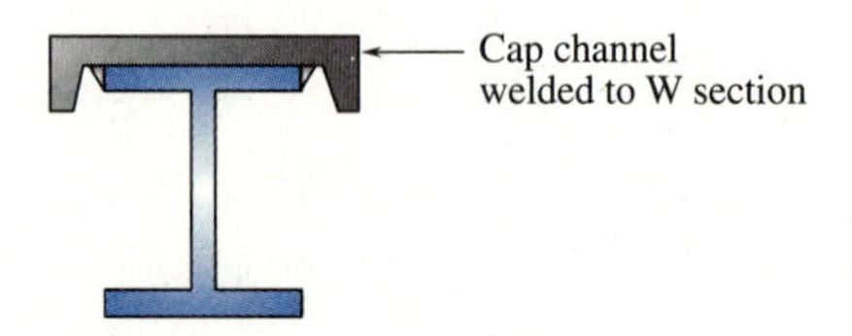

Pure Bending of Unsymmetric Beams

If a beam cross-section is unsymmetric, we can use the same formula, Eq. (5.32), to determine the resultant bending stress at any point A shown in Figure 5.84, provided that the beam is subjected to pure bending couples or that the load passes through the shear center so that no twisting of the cross-section occurs. (We discuss the shear center concept in Section 5.12.)

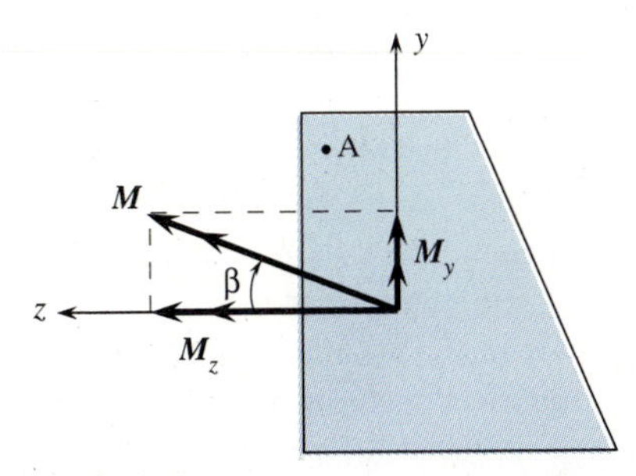

Figure 5.84 Unsymmetric beam under pure moment load.

STEPS IN SOLUTION

The following steps can be used to determine the bending stress in an unsymmetric beam loaded in pure bending or by a transverse load through the shear center.

1. First, we must locate the principal centroidal axes y and z of the cross-section and evaluate the principal moments of inertia. We will describe how to do this in Section 5.11.
2. Then resolve the couple M (Figure 5.84) into components

$$M_y = M \sin \beta \quad \text{and} \quad M_z = M \cos \beta$$

where β is the angle between M and the z-axis.

3. Then determine the stress at any point A as

$$\sigma_x = \frac{M_y z}{I_y} - \frac{M_z y}{I_z}$$

The equation to locate the neutral axis is again given by setting $\sigma_x = 0$ as follows:

$$\frac{M_y z}{I_y} - \frac{M_z y}{I_z} = 0$$

or

$$\frac{y}{z} = \frac{I_z M_y}{I_y M_z}$$

or by

$$\frac{y}{z} = \frac{I_z}{I_y} \tan \beta$$

or

$$\tan \alpha = \frac{I_z}{I_y}(\tan \beta)$$

Because α and β are generally not equal, the neutral axis is normally not perpendicular to the plane in which the applied moment M acts.

If transverse loading acts on unsymmetric beams, the possibility of twisting occurs unless the load acts through the shear center. The concept of shear center will be discussed in Section 5.12. In this section and Section 5.11, we assume that the load acts through the shear center so that the cross-section does not twist.

5.11 PRINCIPAL MOMENTS OF INERTIA

Perhaps the most difficult part of using Eq. (5.32) for unsymmetric sections is obtaining the principal moments of inertia. Since this concept is so important in analyzing bending of unsymmetric beams, we now describe how the principal moments of inertia can be determined. The method we describe is that based on transformation equations for moments of inertia. Similar transformation equations for plane stress are derived in Chapter 8. The method of Mohr's circle is a very straightforward approach for obtaining principal moments of inertia. However, discussion of the Mohr's circle concept will be reserved for Chapter 8 after we discuss transformation equations for stress. The Mohr's circle method for obtaining principal moments of inertia is then analogous to that used to obtain principal stresses.

We should understand that the moments of inertia of plane areas depend on the reference axes used. For instance, for a rectangular cross-section (Figure 5.85), we know that the moment of inertia about the centroidal axis $\bar{z}$ is

$$I_{\bar{z}} = \frac{1}{12}bh^3$$

while the moment of inertia about an axis through the base z is

$$I_z = \frac{1}{3}bh^3$$

These properties are also listed in Appendix A.

Also for a given origin of axes, the moments of inertia and the product of inertia vary as the axes are rotated about the origin. We now develop equations to give us these properties and the maximum and minimum (called principal) moments of inertia.

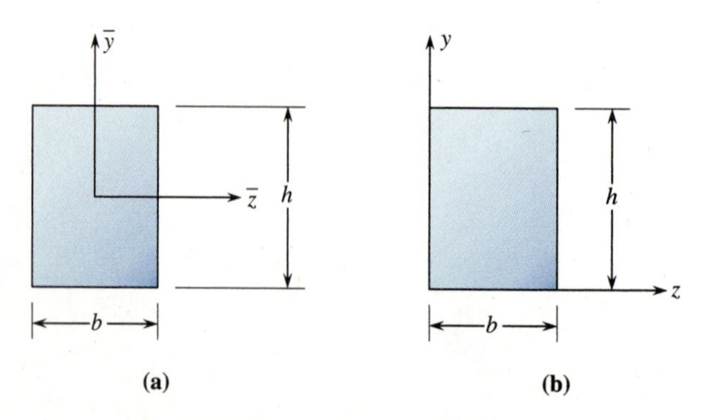

Figure 5.85 Rectangular cross-section with moments of inertia about different axes.

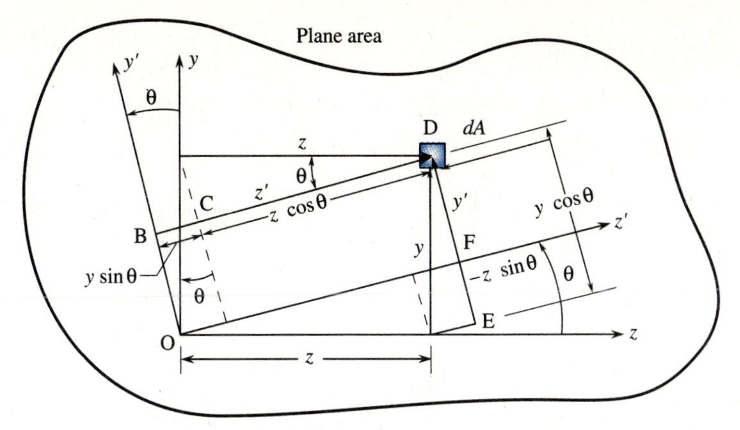

Figure 5.86 Plane area used to obtain expressions for moments and product of inertia.

To develop the general expressions for moments and product of inertia, we consider the plane area with a set of arbitrary coordinates y-z located at O (Figure 5.86). These axes are often chosen conveniently so that I_y, I_z, and I_{yz} can be easily determined. The moments of inertia about the y- and z-axes are

$$I_y = \int z^2 dA, \qquad I_z = \int y^2 dA, \qquad I_{yz} = \int yz dA \tag{a}$$

where y and z locate the coordinates of an element of area dA.

We now desire to determine the moments and product of inertia with respect to the y'-z' axes. These inertias are denoted by $I_{y'}$, $I_{z'}$, and $I_{y'z'}$. To obtain these inertias in terms of I_y, I_z, and I_{yz}, we express the coordinates y' and z' locating dA in terms of y and z and the angle θ. From Figure 5.86, we have

$$z' = CD + BC$$

or

$$z' = z \cos\theta + y \sin\theta \tag{b}$$

and

$$y' = DE - EF$$

or

$$y' = y \cos\theta - z \sin\theta \tag{c}$$

Using the usual definition for the moment of inertia, we obtain the moment of inertia with respect to the z' axis as

$$I_{z'} = \int (y')^2 dA \tag{d}$$

Substituting Eq. (c) into (d), we obtain

$$I_{z'} = \int (y \cos\theta - z \sin\theta)^2 dA \tag{e}$$

Simplifying Eq. (e) yields

$$I_{z'} = \cos^2\theta \int y^2 dA + \sin^2\theta \int z^2 dA - 2 \sin\theta \cos\theta \int zy dA \tag{f}$$

Using definitions, Eq. (a) in Eq. (f), we obtain

$$I_{z'} = I_z \cos^2 \theta + I_y \sin^2 \theta - 2I_{yz} \sin \theta \cos \theta \tag{g}$$

We now introduce the following trigonometric relations

$$\cos^2 \theta = \frac{1}{2}(1 + \cos 2\theta) \quad \sin^2 \theta = \frac{1}{2}(1 - \cos 2\theta)$$

$$2 \sin \theta \cos \theta = \sin 2\theta$$

Using these trigonometric relations in Eq. (g), we obtain

$$I_{z'} = \frac{I_z + I_y}{2} + \frac{I_z - I_y}{2} \cos 2\theta - I_{yz} \sin 2\theta \tag{5.36}$$

Similarly,

$$I_{y'} = \int (z')^2 dA \tag{h}$$

Using Eq. (b) in Eq. (h), we have

$$I_{y'} = \int (z \cos \theta + y \sin \theta)^2 dA$$

Using Eq. (a) in the above equation yields

$$I_{y'} = I_z \sin^2 \theta + I_y \cos^2 \theta + 2I_{yz} \sin \theta \cos \theta \tag{i}$$

Using the trigonometric relations in Eq. (i) yields

$$I_{y'} = \frac{I_z + I_y}{2} - \frac{I_z - I_y}{2} \cos 2\theta + I_{yz} \sin 2\theta \tag{5.37}$$

The product of inertia with respect to the z'-y' axes is

$$I_{z'y'} = \int z'y' dA \tag{j}$$

Using Eqs. (b) and (c) in Eq. (j) results in

$$I_{y'z'} = \int (z \cos \theta + y \sin \theta)(y \cos \theta - z \sin \theta) dA$$

Using Eq. (a) in the above equation yields

$$I_{y'z'} = (I_z - I_y) \sin \theta \cos \theta + I_{yz}(\cos^2 \theta - \sin^2 \theta) \tag{k}$$

Using the trigonometric identities in Eq. (k) yields

$$I_{y'z'} = \frac{I_z - I_y}{2} \sin 2\theta + I_{yz} \cos 2\theta \tag{5.38}$$

Equations (5.36), (5.37), and (5.38) express the moments and product of inertia with respect to rotated axes y'-z' in terms of the original y-z axes. These equations are called the *transformation equations for moments and product of inertia*. If we take the sum of $I_{z'}$ and $I_{y'}$, we obtain

$$I_{z'} + I_{y'} = I_z + I_y \tag{5.39}$$

Equation (5.39) indicates that the sum of moments of inertia with respect to a pair of orthogonal axes remains constant (invariant) as the axes are rotated about the origin. Also this sum is equal to J, the polar moment of inertia evaluated with respect to this origin. That is

$$J = I_z + I_y \tag{5.40}$$

To apply Eq. (5.32) to obtain the bending stresses for unsymmetric shapes, we need the principal moments of inertia. These are the maximum and minimum moments of inertia. To find these values, we must find the values of θ that maximize and minimize $I_{z'}$ and $I_{y'}$. Therefore, by the calculus, we take the derivative of Eq. (5.36) with respect to θ and set the result to zero to obtain

$$(I_z - I_y)\sin 2\theta + 2I_{yz}\cos 2\theta = 0 \tag{k}$$

Solving for θ in Eq. (k), we obtain

$$\tan 2\theta_P = -\frac{2I_{yz}}{I_z - I_y} \tag{5.41}$$

where θ_P denotes the angle defining a principal axis associated with a principal moment of inertia. Equation (5.41) gives two roots for $\tan 2\theta_P$ that are 180° apart. Therefore, θ_{P_1} and θ_{P_2} are 90° apart. One of these angles, when substituted into Eq. (5.36) for $I_{z'}$, yields the maximum moment of inertia, I_{max}, while the other angle yields the minimum moment of inertia, I_{min}.

By defining the sine and cosine of the double angle, we can obtain general expressions for I_{max} and I_{min} as follows. From Eq. (5.41), we develop a triangle, shown in Figure 5.87, to easily obtain the cosine and sine of the angle $2\theta_P$. The hypotenuse R of the triangle is

$$R = \sqrt{\left(\frac{I_z - I_y}{2}\right)^2 + I_{yz}^2} \tag{5.42}$$

Therefore

$$\cos 2\theta_P = \frac{I_z - I_y}{2R} \quad \text{and} \quad \sin 2\theta_P = -\frac{I_{yz}}{R} \tag{5.43}$$

When Eq. (5.43) is substituted into Eq. (5.36), we have

$$I_{max,min} = I_{1,2} = \frac{I_z + I_y}{2} \pm \sqrt{\left(\frac{I_z - I_y}{2}\right)^2 + I_{yz}^2} \tag{5.44}$$

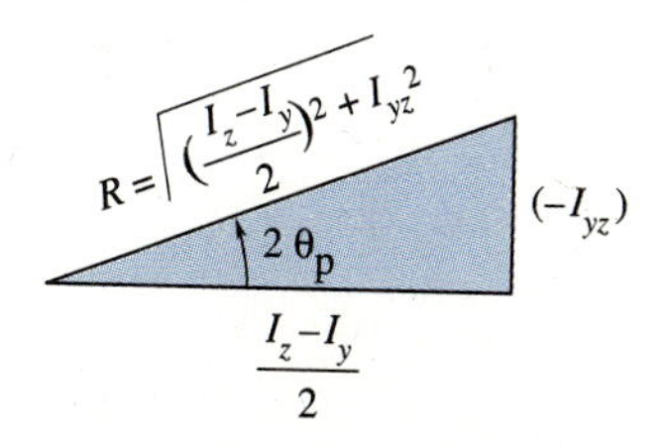

Figure 5.87 Triangle used to obtain cosine and sine of angle.

We assume that angle $2\theta_{p1}$ yields I_1 and $2\theta_{p2}$ yields I_2. Also the product of inertia is zero for the principal axes as can be seen by substituting Eqs. (5.43) into Eq. (5.38).

We now consider an example to illustrate how to obtain principal moments of inertia.

EXAMPLE 5.30

For the 8 in. × 8 in. × $\frac{3}{4}$ in. angle cross section shown, determine the principle axes and the principal moments of inertia (Figure 5.88).

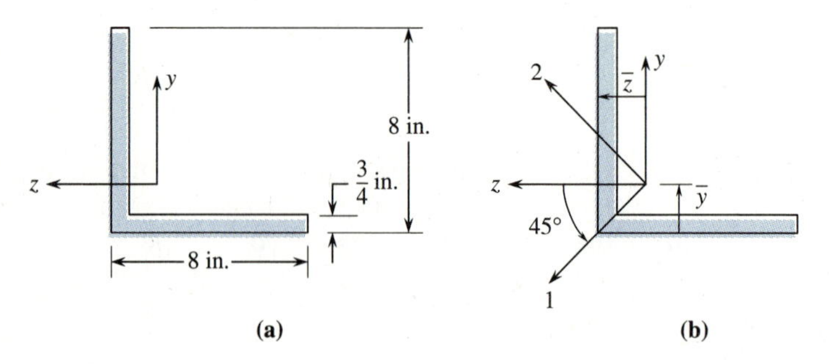

Figure 5.88 (a) Angle for principal moments of inertia, and (b) angle showing principal axes.

Solution

First, determine the centroid $\bar{y}$ of the cross section as

$$\bar{y} = \frac{\left(8 \times \frac{3}{4}\right)(4) + \left(7.25 \times \frac{3}{4}\right)\left(\frac{3}{8}\right)}{8 \times \frac{3}{4} + 7.25 \times \frac{3}{4}}$$

$$= 2.28 \text{ in.}$$

By symmetry

$$\bar{z} = 2.28 \text{ in.}$$

Determine moments of inertia I_z and I_y and product of inertia I_{yz} using the parallel-axis theorem as

$$I_z = \frac{\left(\frac{3}{4}\right)8^3}{12} + \left(8 \times \frac{3}{4}\right)(4 - 2.28)^2 + \frac{7.25\left(\frac{3}{4}\right)^3}{12} + \left(7.25 \times \frac{3}{4}\right)(2.28 - 0.375)^2$$

$$= 69.73 \text{ in.}^4$$

This result can also be found in Appendix C. By symmetry

$$I_y = 69.73 \text{ in.}^4$$

Finally

$$I_{yz} = (I_{yz})_1 + \bar{y}_1\bar{z}_1A_1 + (I_{yz})_2 + \bar{y}_2\bar{z}_2A_2$$

$$= 0 + (4 - 2.28)(2.28 - 0.375)\left(8 \times \frac{3}{4}\right) + 0 + (-1.905)(-2.095)\left(7.25 \times \frac{3}{4}\right)$$

$$I_{yz} = 41.63 \text{ in.}^4$$

Now use Eq. (5.44) to obtain the principal moments of inertia I_1 and I_2 as

$$I_{1,2} = \frac{I_z + I_y}{2} \pm \sqrt{\left(\frac{I_z - I_y}{2}\right)^2 + I_{yz}^2}$$

$$= \frac{69.73 + 69.73}{2} \pm \sqrt{0 + (41.63)^2}$$

$$I_{1,2} = 69.73 \text{ in.}^4 + 41.63 \text{ in.}^4$$

$$I_1 = 111.36 \text{ in.}^4, \qquad I_2 = 28.1 \text{ in.}^4$$

Using Eq. (5.41), the principal angle is

$$\tan 2\theta_p = -\frac{2I_{yz}}{I_z - I_y} = -\frac{2(41.63)}{69.73 - 69.73}$$

$$2\theta_p = 90°$$

$$\theta_p = 45° \text{ (counterclockwise from the z-axis)}$$

This angle locates one of the principal axes. By substituting the angle for θ_p into Eq. (5.36), the moment of inertia becomes $I_1 = 111.36$ in.^{4} Therefore, this angle is θ_{p1} and the axis it locates is the 1-axis. Figure 5.88(b) shows the principal axes, 1–2.

EXAMPLE 5.31

For the cantilevered beam made of the angle of Example 5.30, determine the largest tensile and compressive stresses. The cantilever is 50 in. long and subjected to a 6-kip vertical load as shown (Figure 5.89).

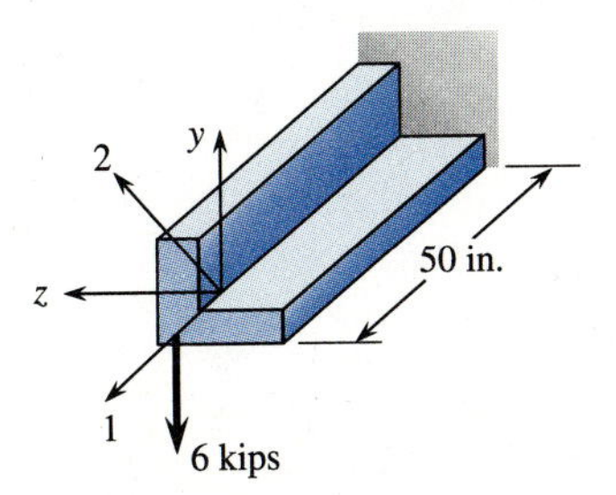

Figure 5.89 Angle for bending stress analysis.

Solution

The principal moments of inertia were obtained in Example 5.30 as

$$I_1 = 111.36 \text{ in.}^4 \quad \text{and} \quad I_2 = 28.10 \text{ in.}^4$$

The bending moment is

$$M_z = (-6 \text{ kips})(50 \text{ in.}) = -300 \text{ kip} \cdot \text{in.}$$

The moments about the principal axes are

$$M_1 = -300 \cos 45° = -212 \text{ kip} \cdot \text{in.}$$

$$M_2 = -300 \sin 45° = -212 \text{ kip} \cdot \text{in.}$$

The neutral axis, *n-n*, is

$$\tan \alpha = \frac{M_2 I_1}{M_1 I_2} = \frac{111.36}{28.1} = 3.96$$

or

$$\alpha = 75.8° \text{ (clockwise from the 1-axis as shown below)}$$

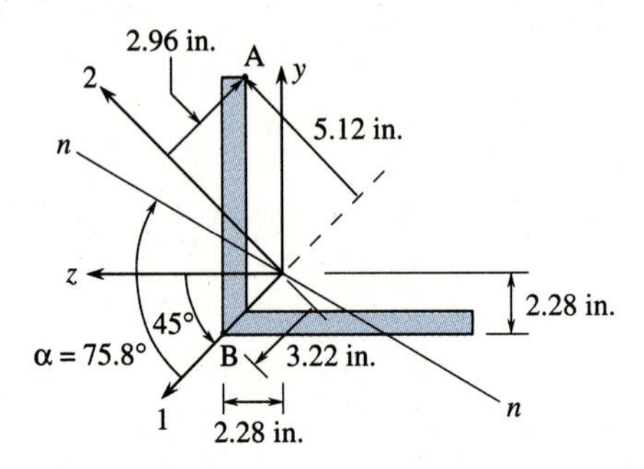

The largest stress occurs farthest from the neutral axis at points A and B on the cross-section. Using Eq. (5.32), with principal axes now 1 and 2, not z and y, we have

For point A

$$c_1 = -2.96 \text{ in.}, \; c_2 = 5.12 \text{ in.}$$

$$\sigma_A = \frac{M_2 c_1}{I_2} - \frac{M_1 c_2}{I_1} = \frac{(-212)(-2.96)}{28.10} - \frac{(-212)(5.12)}{111.36}$$

$$= 22.33 \text{ ksi} + 9.75 \text{ ksi}$$

$$= 32.1 \text{ ksi (T)}$$

For point B

$$c_1 = 3.22 \text{ in.}, \; c_2 = 0$$

$$\sigma_B = \frac{(-212)(3.22)}{28.10} - 0$$

$$= -24.3 \text{ ksi (C)}$$

Note that using the flexure formula incorrectly, as if we do not have unsymmetric bending yields

$$\sigma = -\frac{M_z y}{I_z} = -\frac{(-300)(8 - 2.28)}{69.73}$$

$$= 24.61 \text{ ksi (T)}$$

This tensile stress is much less than the actual stress of 32.1 ksi using the unsymmetric bending formula. Therefore, to obtain the correct stress in unsymmetric bending, we must use the proper equation.

5.12 SHEAR FLOW IN BEAMS WITH THIN SECTIONS AND SHEAR CENTER

In this section, we will describe how shear flows in a beam with a thin cross-section. The concept of shear flow will then be used to understand the concept of the shear center.

Shear Flow in Beam

Recall that in Section 5.5 we used the shear stress formula given by

$$\tau_{xy} = \frac{VQ}{Ib} \tag{5.15}$$

to obtain the transverse shear stress on a cross-section. This equation is based on assuming the shear stress to be uniform in the z-direction. For beams with thin walls, such as I beams, channels, angles, and boxes (Figure 5.90), we really do not have uniform τ_{xy} in the z-direction.

Equation (5.15) predicts that at any depth y (such as on the free edge of a flange, $y = \pm(\frac{h}{2} - t_f)$) τ_{xy} is uniform in the z-direction (Figure 5.91). Here t_f denotes the thickness of the flange. However, on the stress free surfaces τ_{xy} must be zero (Figure 5.91). Also Eq. (5.15) predicts larger shear stress τ_{xy} in flanges than really occurs, and so is not good for predicting τ_{xy} in flanges. However, τ_{xy} is resisted mostly in the web and Eq. (5.15) is accurate in the web except near the junction of the web and flange at $y = \pm(\frac{h}{2} - t_f)$.

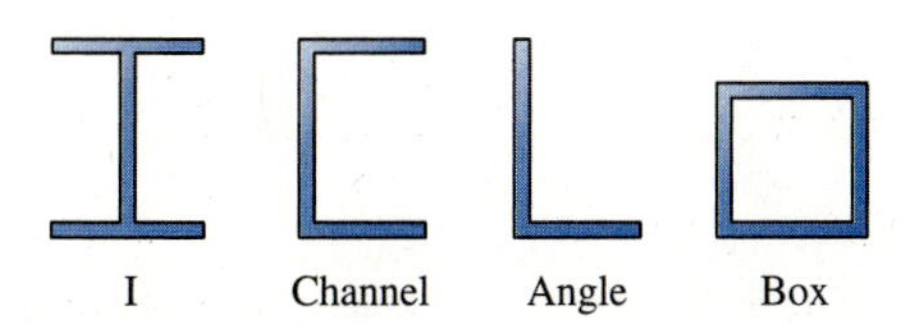

Figure 5.90 Thin-walled cross-sections.

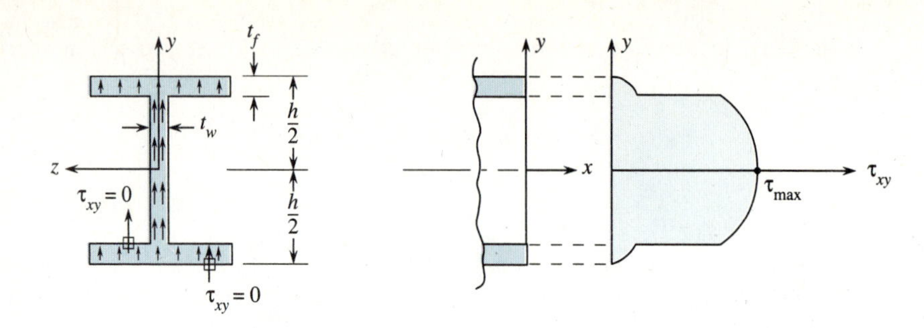

Figure 5.91 Shear stress distribution predicted by Eq. (5.15).

Now we determine τ_{zx} in the flange. This shear stress will occur on a vertical face of the flange. Consider the piece of material of length dx sectioned off with τ_{zx} acting on the vertical face as shown in Figure 5.92(b).

On summing forces in the x-direction, we have

$$\Sigma F_x = 0: \quad F_1 + F_H - F_2 = 0 \tag{a}$$

Expressing the forces in terms of stresses, we have

$$\int \sigma_{x_1} dA - \int \sigma_{x_2} dA + \tau_{zx}\, t_f\, dx = 0 \tag{b}$$

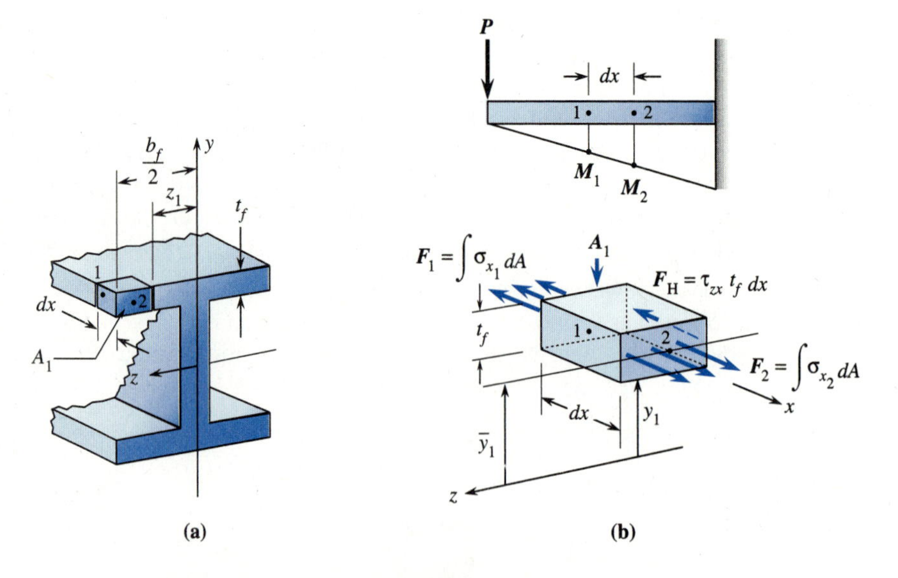

Figure 5.92 (a) Wide flange section subjected to vertical shear force and (b) material sectioned off to introduce the shear stress acting on a vertical plane.

Using the flexure formula Eq. (5.7) for bending stress σ_x and solving Eq. (b) for τ_{zx}, we obtain

$$\tau_{zx} t_f \, dx = \int \left(\frac{-M_1 y}{I} + \frac{M_2 y}{I} \right) dA \tag{c}$$

$$= \int \frac{(M_2 - M_1) y}{I} \, dA \tag{d}$$

Realizing that M_1 and M_2 are independent of the area dA, letting $dM = M_2 - M_1$, and dividing Eq. (d) by dx, we have

$$\tau_{zx} \, t_f = \frac{1}{I} \frac{dM}{dx} \int_{y_1}^{\frac{h}{2}} y dA \tag{e}$$

Dividing Eq. (e) by t_f and using Eq. (5.2b) $\left(V = \dfrac{dM}{dx} \right)$, we have

$$\tau_{zx} = \frac{VQ}{It_f} \tag{5.45}$$

where Q is the integral in Eq. (e).

Note that Q is determined for the area resisting the normal bending stress. In Figure 5.92(b), this area is A_1. Also, the shear flow on the vertical face is

$$q = \frac{VQ}{I} \tag{f}$$

For the area considered in Figure 5.92(a) and 5.92(b), Q is

$$Q = \int_{y_1}^{\frac{h}{2}} y dA = \bar{y}_1 A_1 \tag{g}$$

with the centroidal distance from the neutral axis to the centroid of area dA given by

$$\bar{y}_1 = \frac{h - t_f}{2} \tag{h}$$

and

$$A_1 = t_f \left(\frac{b_f}{2} - z_1 \right) \tag{i}$$

Therefore, using Eqs. (h) and (i) in (g), we have

$$Q = \left(\frac{h - t_f}{2} \right) \left[t_f \left(\frac{b_f}{2} - z_1 \right) \right] \tag{j}$$

Using Eq. (j) in Eq. (5.45), yields the shear stress on the vertical plane as

$$\tau_{zx} = \frac{V}{2I} (h - t_f) \left(\frac{b_f}{2} - z_1 \right) \tag{5.46}$$

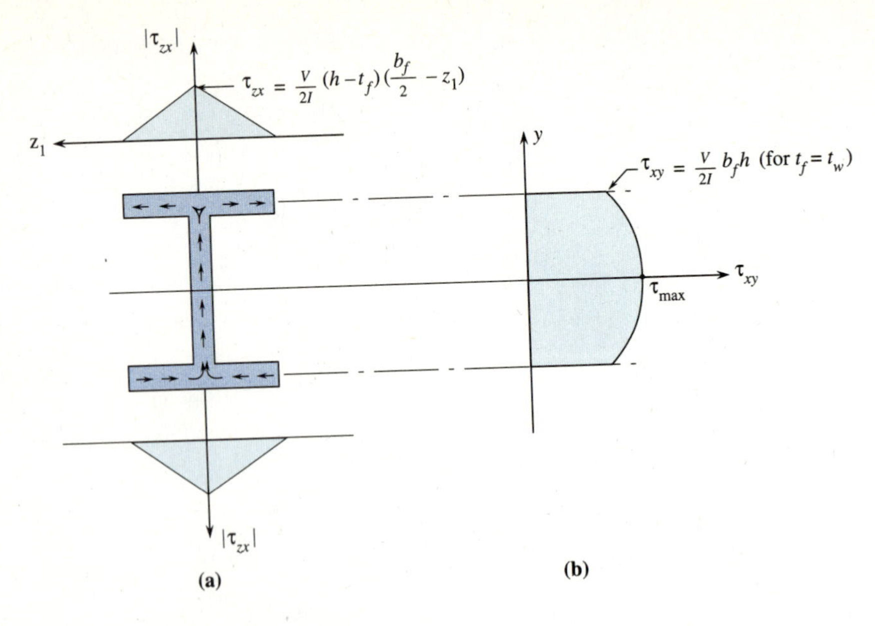

Figure 5.93 Shear stress distribution
(a) given by Eq. (5.46) and
(b) given by Eq. (5.15) for an I beam.

From Eq. (5.46) we observed that τ_{zx} is a linear function of coordinate z_1 shown in Figure 5.92(a), reaching a maximum when $z_1 = 0$. The shear stress distribution in the flanges given by Eq. (5.46) is shown in Figure 5.93(a). The shear stress distribution in the web given by Eq. (5.15) is shown in Figure 5.93(b). It is common practice to calculate the shear stress distributions in each segment of flange and web at the centerlines of the flanges and web so that the values in the shear stress are continuous. Example 5.32 illustrates this idea. Also from Figure 5.93(a), we see that the shear stress appears to flow through the section. This is why the term *shear flow* is commonly used to describe this behavior.

EXAMPLE 5.32

Determine the shear force per unit length in the welds for the two cases shown in Figure 5.94 when $V = 41.5$ kips.

Solution

The moment of inertia of both cross-sections is

$$I_z = \frac{1}{12}(6 \text{ in.})(8 \text{ in.})^3 - \frac{1}{12}(5 \text{ in.})(6 \text{ in.})^3 = 166 \text{ in.}^4$$

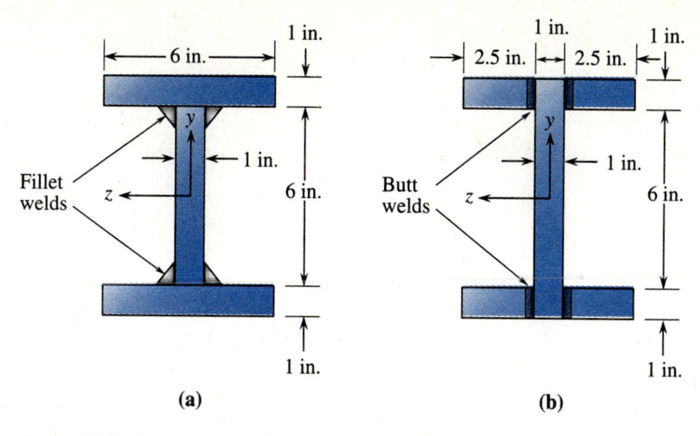

Figure 5.94 Cross-sections for determining weld shear.

(a) The area above the welds is used to obtain Q as shown below

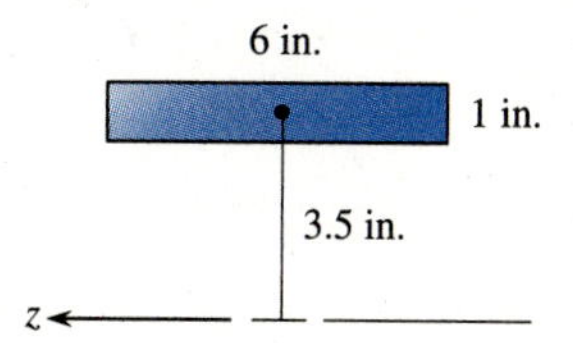

Then Q (the moment of the area about the neutral axis of the beam) is

$$Q = (6 \text{ in.} \times 1 \text{ in.})(3.5 \text{ in.}) = 21.0 \text{ in.}^3$$

The shear flow q is

$$q = \frac{VQ}{I} = \frac{(41.5 \text{ kips})(21.0 \text{ in.}^3)}{166 \text{ in.}^4} = 5.25 \text{ kips/in.}$$

There are two weld sides resisting this shear flow. Therefore, the shear per unit length in a weld is

$$f = \frac{q}{2} = 2.625 \text{ kips/in. per weld}$$

(b) The area to the left (or right) of the vertical weld is used to obtain Q as shown.

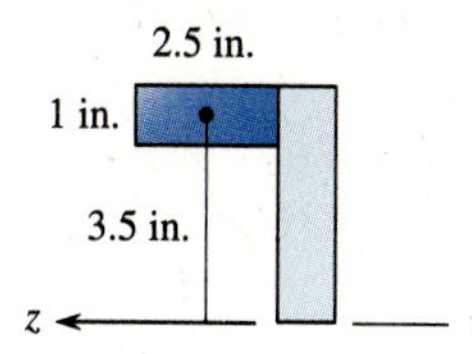

Then Q is determined as

$$Q = (2.5 \text{ in.} \times 1 \text{ in.})(3.5 \text{ in.}) = 8.75 \text{ in.}^3$$

The shear flow in the welds is

$$q = \frac{VQ}{I} = \frac{(41.5 \text{ kips})(8.75 \text{ in.}^3)}{166 \text{ in.}^4} = 2.188 \text{ kips/in. per weld}$$

Shear Center

The *shear center* is defined as the location where the transverse load should be placed to avoid twisting of the cross-section. Let us consider the channel shown in Figure 5.95, which will twist and bend under the load applied through the centroid C.

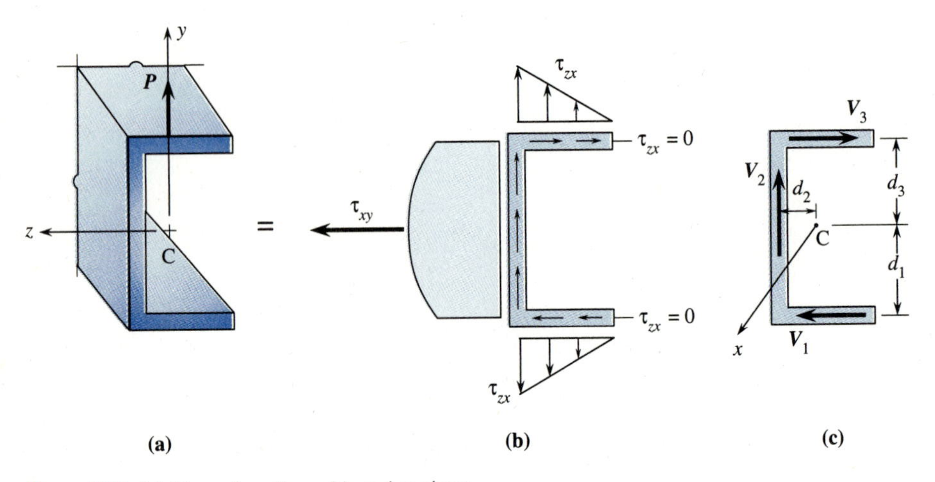

Figure 5.95 (a) Channel section subjected to shear, (b) shear stresses in the flanges and web, and (c) net shear forces on each surface.

To show that twisting must occur, we show the shear flow and net shear forces on each face in Figure 5.95(c). Then summing moments about the x-axis (Figure 5.95(c)), we have

$$\Sigma M_x = 0: \quad V_1 d_1 + V_2 d_2 + V_3 d_3 \neq 0$$

Therefore, a net torque occurs on the cross-section. There must be a torsional shear stress, as shown in Figure 5.96(a), to resist the net torque due to the shear flow predicted by Eqs. (5.15) and (5.45). However, if we place the force P, such that no net torque occurs in Figure 5.95(c), we will have found the location of the shear center (SC) (Figure 5.96(b)). We see from Figure 5.96(b) that the shear center generally does not coincide with the centroid of the cross-section.

The shear center is a particularly important concept when constructing with thin-walled open cross sections, as often used in the aerospace industry.

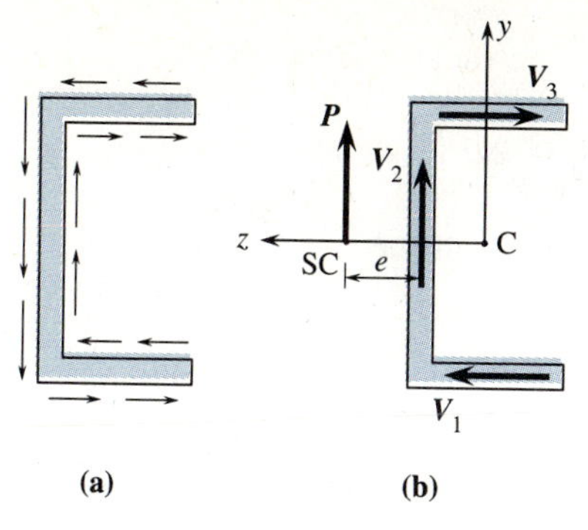

Figure 5.96 (a) Shear flows in section and
(b) net shear forces on section when P is located at shear center.

Examples 5.33 and 5.34 illustrate how to determine the shear center.

EXAMPLE 5.33

(a) Locate the shear center (SC) of the channel section shown in Figure 5.97(a). **(b)** Determine the distribution of the shear stress on the cross-section caused by a shear force of $V = 1000$ lb located at the shear center for $b = 4$ in., $a = 6$ in., and $t = \frac{1}{4}$ in. **(c)** Determine the maximum shear stress when the shear force is applied at the centroid C of the cross-section in Figure 5.97(c).

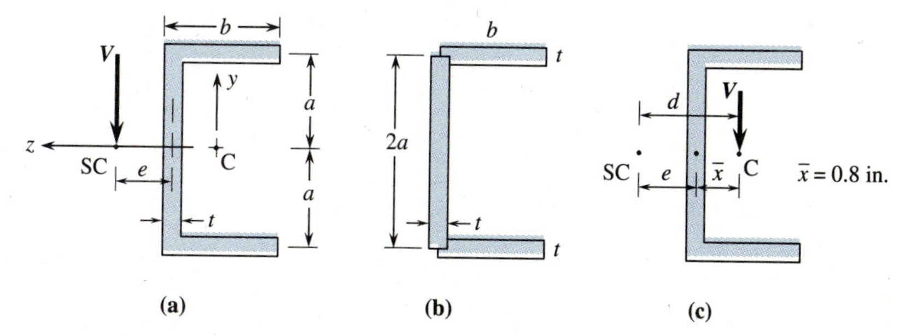

Figure 5.97 (a) Channel section for shear center (SC) calculation,
(b) channel split into sections for calculations, and
(c) channel with shear force through centroid C.

Solution

(a) Determine the moment of inertia of the cross-section as

$$I_z = \frac{1}{12}(t)(2a)^3 + 2\left(\frac{1}{12}bt^3 + (bt)a^2\right)$$

$$= \frac{2}{3}ta^3 + 2\left(\frac{1}{12}bt^3 + bta^2\right)$$

If t is small compared to a and b, then we approximate I_z by

$$I_z = 2ta^2\left(\frac{a}{3} + b\right)$$

Determine the shear flow in the upper flange as

$$q = \frac{VQ}{I_z} = \frac{V(st)a}{I_z} = \frac{Vsta}{2ta^2(\frac{a}{3} + b)} = \frac{Vs}{2a(\frac{a}{3} + b)}$$

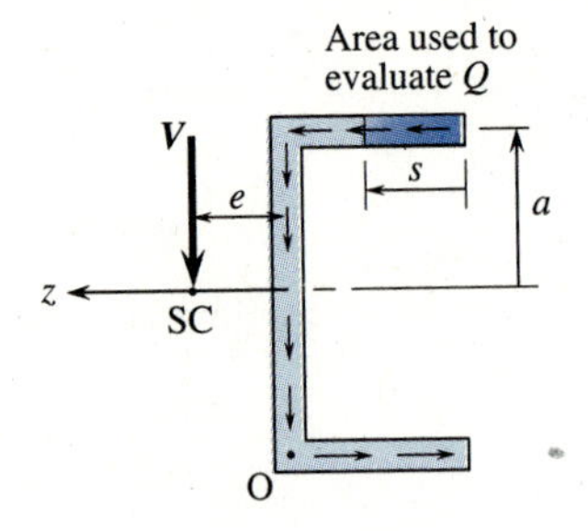

Determine the resultant shear force V_3 on the upper flange as

$$V_3 = \int_0^b q ds = \int_0^b \frac{Vs}{2a(\frac{a}{3} + b)} ds$$

$$= \frac{Vb^2}{2}\left(\frac{1}{2a(\frac{a}{3} + b)}\right).$$

Now sum moments at O as

$$\Sigma M_o = 0:$$

$$Ve = V_3(2a)$$

$$e = \frac{V_3}{V}(2a)$$

Substituting V_3 from above, we have

$$e = \frac{3b^2}{2(a + 3b)} \qquad \textbf{(a)}$$

Using the numeric values in Eq. (a), the shear center is located at

$$e = \frac{3(4 \text{ in.})^2}{2(6 \text{ in.} + 3(4 \text{ in.}))}$$

$$e = 1.33 \text{ in.}$$

(b) The distribution of shear stress is determined by using Eq. (5.15) for the web and Eq. (5.45) for the flange.

WEB SHEAR STRESS

By Eq. (5.15), the shear stress in the web is

$$\tau_{xy} = \frac{VQ}{It}$$

The area used to express Q is shown below as

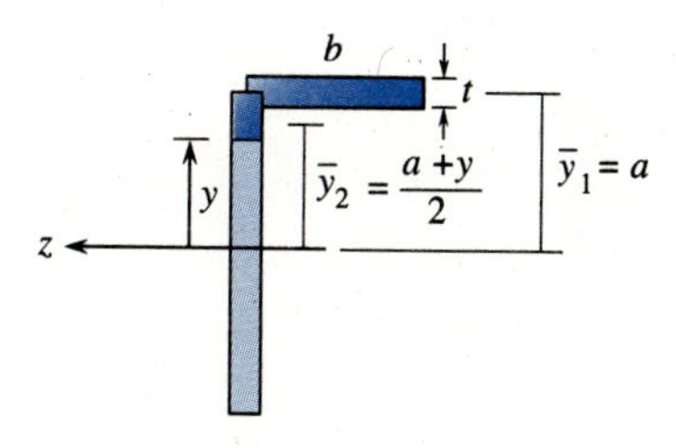

Then Q is determined as

$$Q = \Sigma \bar{y} A = a(bt) + \left(\frac{a+y}{2}\right)(a-y)(t)$$

$$= abt + \frac{t}{2}(a^2 - y^2)$$

Therefore, the shear stress in the web is

$$\tau_{xy} = \frac{V}{I}\left[ab + \frac{1}{2}(a^2 - y^2)\right] \quad \text{(b)}$$

which is a parabolic function as shown in Figure 5.98.

The moment of inertia of the whole cross-section is

$$I = 2\frac{1}{12}\left(4\tfrac{1}{8}\right)\left(\frac{1}{4}\right)^3 + \left(4\tfrac{1}{8}\right) \times \left(\frac{1}{4}\right)6^2 + \frac{1}{12}\left(\frac{1}{4}\right)\left(11\tfrac{3}{4}\right)^3$$

$$= 0.0107 + 37.125 + 33.80$$

$$= 70.9 \text{ in.}^4$$

Using numeric quantities in Eq. (b), we have

At $y = a = 6$ in.

$$\tau_{xy} = \frac{V}{I}(ab) = \frac{1000 \text{ lb}}{70.9 \text{ in}^4}[(6 \text{ in.})(4 \text{ in.})] = 338 \text{ psi} \quad \text{(c)}$$

At $y = 0$

$$\tau_{xy} = \frac{V}{I}\left(ab + \frac{a^2}{2}\right) = \frac{1000}{70.9}\left[(6)(4) + \frac{6^2}{2}\right] = 592 \text{ psi} \quad \text{(d)}$$

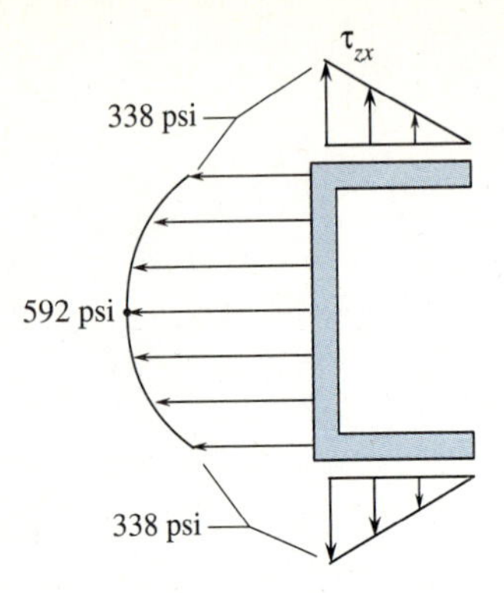

Figure 5.98 Shear stress distribution over cross-section.

FLANGE SHEAR STRESS

By Eq. (5.45), the shear stress in the vertical face of the top flange is

$$\tau_{zx} = \frac{VQ}{I_z t_f}$$

To obtain Q, the moment of the area at the cut shown below is used.

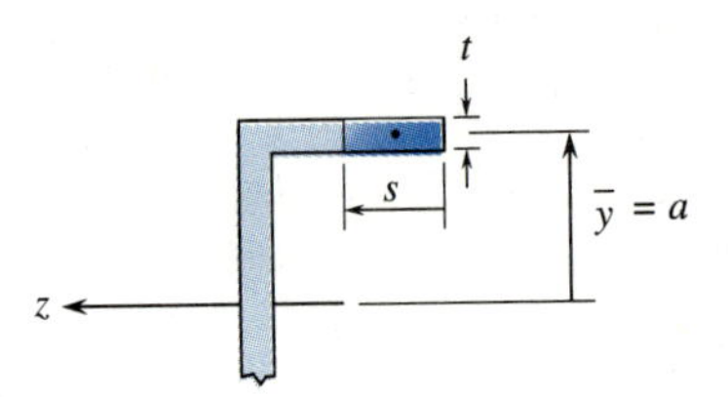

Therefore, Q is

$$Q = \bar{y}A = a(st)$$

Therefore, with $t = t_f$

$$\tau_{zx} = \frac{V}{I_z}(as) \quad \textbf{(e)}$$

The shear stress is a linear function in s. Using numeric values to evaluate Eq. (e), the largest value occurs at $s = b = 4$ in. Therefore

$$\tau_{zx} = \frac{(1000)(6)(4)}{70.9} = 338 \text{ psi} \quad \textbf{(f)}$$

This same stress distribution occurs in the lower flange. Figure 5.98 shows the shear stress distribution plot over the cross-section.

(c) If the shear force V is located at the centroid C, the cross-section will be subjected to twisting. We replace the shear V by an equivalent force-couple system at the shear center SC. This system consists of the shear force $V = 1000$ lb and a torque $T = V$

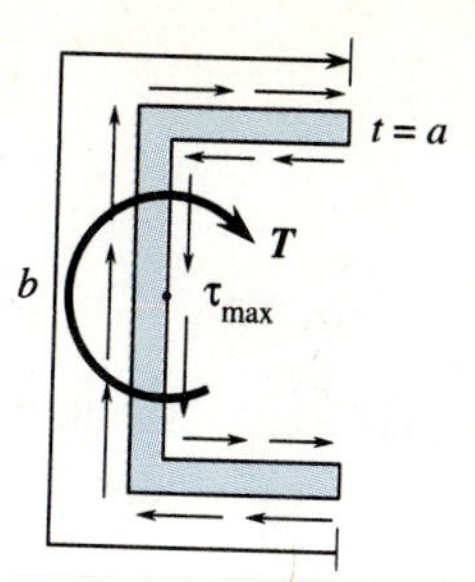

Figure 5.99 Shear stress distribution due to T.

times the distance from the centroid C to the shear center SC (Figure 5.97(c)) as

$$T = V(e + \bar{x}) = (1000 \text{ lb})(1.33 \text{ in.} + \bar{x}) = (1000)(1.33 + 0.8) = 2130 \text{ lb} \cdot \text{in.}$$

The shear force V causes the shear stress distribution shown in part (b) and Figure 5.98.

The torque T causes the member to twist. The shear stress distribution due to T is shown in Figure 5.99. We recall from Section 4.9, Eq. (4.33), that the shear stress due to twisting of a rectangular cross-section is

$$\tau_{max} = \frac{T}{k_2 a^2 b} \tag{4.33}$$

Using the numeric values in Eq. (4.33), we have

$$\tau_{max} = \frac{2130 \text{ lb} \cdot \text{in.}}{(0.333)(0.25)^2(20 \text{ in.})} = 5120 \text{ psi} \tag{g}$$

where $k_2 = 0.333$ (see Table 4.1) and a and b are as defined in Figure 4.27.

The maximum stress due to the combined actions of V and T occurs at the neutral surface, on the inside of the web, and is obtained by adding the results from Eqs. (d) and (g).

$$\tau_{max} = 592 \text{ psi} + 5120 \text{ psi}$$
$$= 5712 \text{ psi}$$

EXAMPLE 5.34

Find the shear center for the semicircular thin-walled cross-section shown in Figure 5.100.

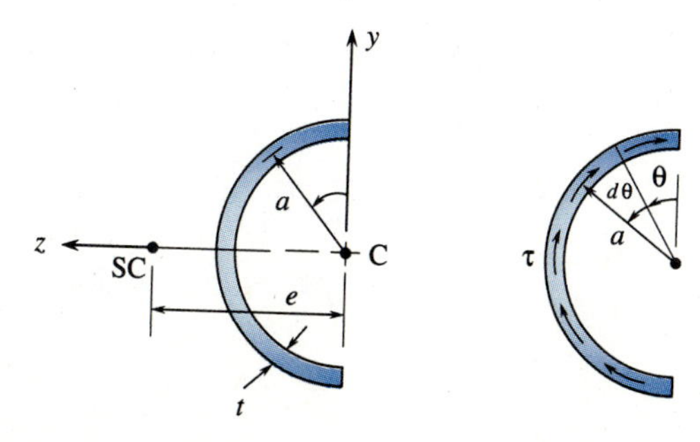

Figure 5.100 Semi-circular cross-section for shear center (SC) determination.

Solution

First determine the moment of inertia of the cross-section about the z-axis as

$$I = \frac{1}{2}\pi t a^3$$

Next summing moments of the shear force V and the internal resultant force dF acting on area dA, we have

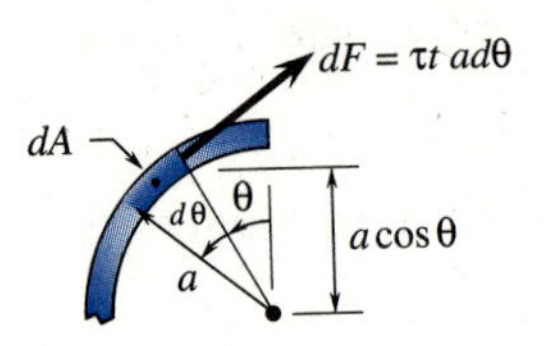

$$\Sigma M_c = 0: \qquad Ve = \int_0^{\pi} adF = \int_0^{\pi} a^2 \tau t d\theta$$

$$e = \frac{a^2 t}{V}\int_0^{\pi} \tau d\theta \qquad \textbf{(a)}$$

We must evaluate the integral in Eq. (a) as follows:

Expressing the shear stress as a function of angle θ, we have

$$dQ = (tad\theta)(a\cos\theta)$$

$$Q = \int_0^{\theta} ta^2 \cos\theta\, d\theta = ta^2 \sin\theta$$

$$\tau = \frac{VQ}{It} = \frac{Vta^2 \sin\theta}{(\frac{1}{2}\pi t a^3)t} = \frac{2V\sin\theta}{\pi a t} \qquad \textbf{(b)}$$

Substituting Eq. (b) into (a) yields

$$e = \frac{a^2 t}{V}\int_0^{\pi} \frac{2V\sin\theta}{\pi a t}\, d\theta$$

$$= \frac{2a}{\pi}[-\cos\theta]\Big|_0^{\pi}$$

$$e = \frac{4a}{\pi}$$

Finally, shear center locations of various thin-walled cross-sections are shown in Figure 5.101.

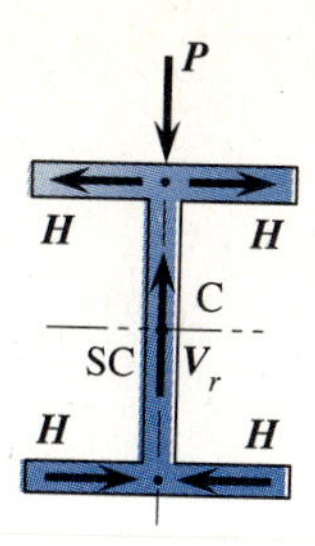

In sections having two axes of symmetry, the shear center SC coincides with the centroid C

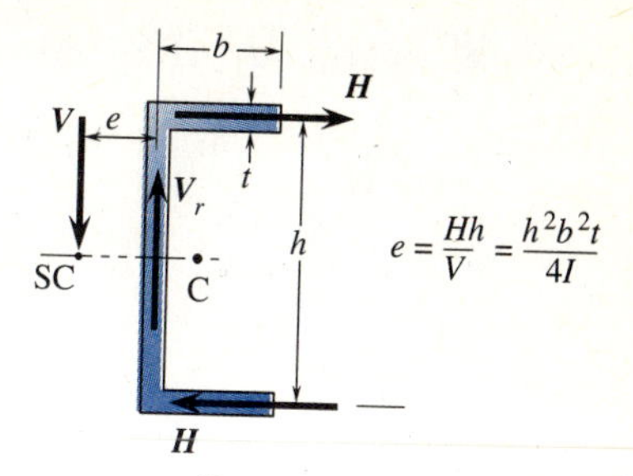

Channel section

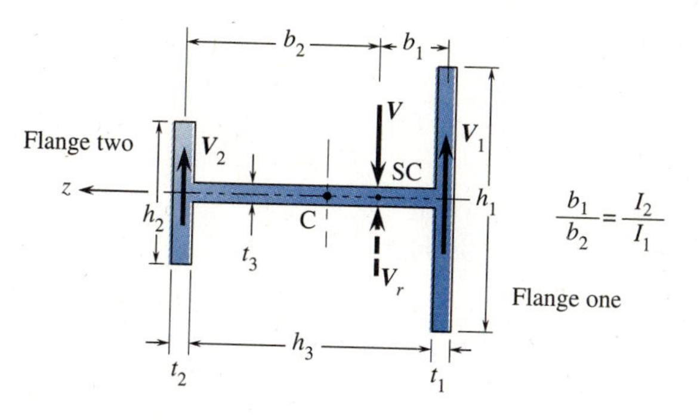

Wide flange turned sideways (two unequal flanges)
($I_1 = I$ of flange one about the z-axis, etc.)

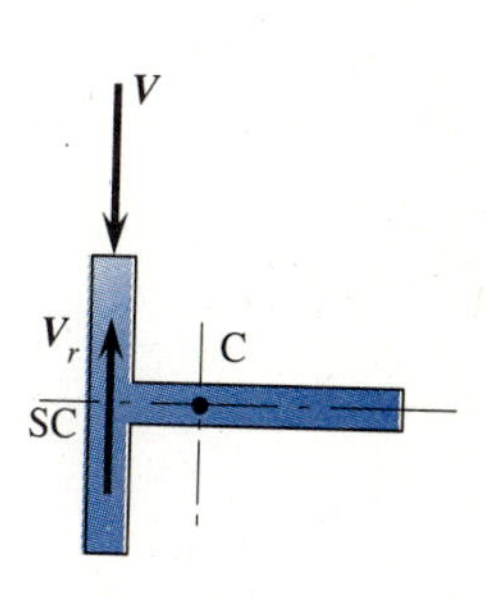

One-flange section
(web bending neglected)

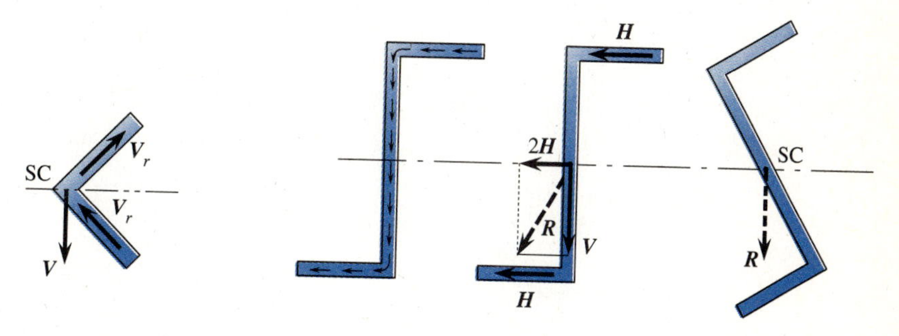

Two narrow rectangles
(shear center at intersection of rectangles)

Z-section
(shear center coincides with centroid)

Figure 5.101 Shear centers of various thin-walled cross sections.

5.13 SUMMARY OF IMPORTANT DEFINITIONS AND EQUATIONS

1. Relationship between shear and distributed load

$$\frac{dV}{dx} = -w(x) \qquad \textbf{(5.1a)}$$

2. Relationship between moment and shear

$$\frac{dM}{dx} = V \qquad \textbf{(5.2b)}$$

3. Relationship between bending moment and radius of curvature

$$M = \frac{EI}{\rho} \qquad \textbf{(5.6e)}$$

4. Maximum shear forces and bending moments for the most common beam supports and loadings

   a. Simply supported beam with concentrated load at center

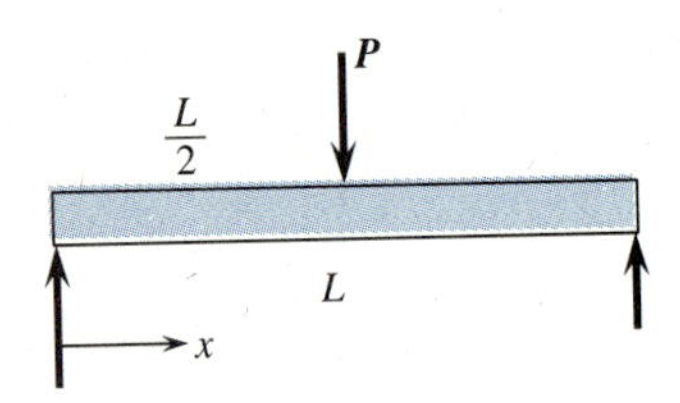

$V_{max} = \frac{P}{2}$ at 0 $\qquad M_{max} = \frac{PL}{4}$ at $\frac{L}{2}$

   b. Simply supported beam with uniformly distributed load

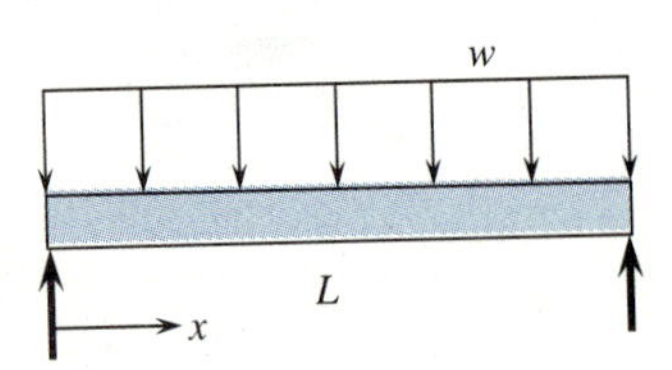

$V_{max} = \frac{wL}{2}$ at 0 $\qquad M_{max} = \frac{wL^2}{8}$ at $\frac{L}{2}$

   c. Cantilevered beam with free end load

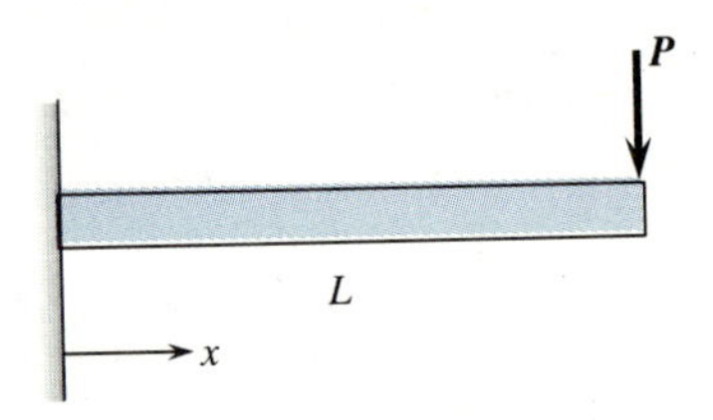

$V_{max} = P$ at 0 $\qquad M_{max} = PL$ at 0

   d. Cantilevered beam with uniformly distributed load

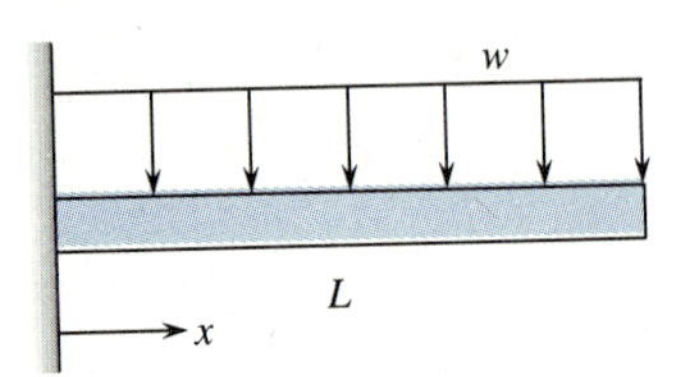

$V_{max} = wL$ at 0 $\qquad M_{max} = \frac{wL^2}{2}$ at 0

5. Beam bending stress (flexure formula)

$$\sigma_x = -\frac{My}{I} \qquad \textbf{(5.7)}$$

6. Beam shear flow

$$q = \frac{VQ}{I} \qquad \textbf{(5.14)}$$

7. Beam shear stress

$$\tau = \frac{VQ}{Ib} \qquad \textbf{(5.15)}$$

8. Maximum beam shear stress for solid rectangular cross-sections

$$\tau_{max} = \frac{3V}{2A} \qquad \textbf{(5.16)}$$

9. Maximum beam shear stress for solid circular cross-sections

$$\tau_{max} = \frac{4V}{3A} \tag{5.17}$$

10. Beam shear stress for narrow-webbed sections

$$\tau = \frac{V}{A_{web}} \tag{5.18}$$

11. Minimum section modulus based on bending stress

$$S_{min} = \frac{|M|_{max}}{\sigma_{allow}} \tag{5.19}$$

12. Moving loads on beams

The bending moment under a particular load is at a maximum when the center of the beam is midway between that load and the resultant of all loads presently on the beam span. This was shown by Eq. (5.24) as

$$x = \frac{L}{2} - \frac{d}{2} \tag{5.24}$$

13. Composite beams or beams of more than one material

The equivalent width of the transformed section (material two) is given by Eq. (5.27) as

$$b_e = \frac{E_2}{E_1} b \tag{5.27}$$

14. Stress concentrations in bending members

The bending stress is increased by the stress concentration factor K when abrupt changes in geometry occur. This stress is given by

$$\sigma_{max} = K\frac{Mc}{I} \tag{5.31}$$

15. Unsymmetric bending in beams

For beams subjected to unsymmetric bending, the bending stress is obtained by

$$\sigma_x = \frac{M_y z}{I_y} - \frac{M_z y}{I_z} \tag{5.32}$$

where I_y and I_z are principal moments of inertia.

The neutral axis is the location where the bending stress σ_x is zero. Therefore, using Eq. (5.32), we have

$$\frac{M_y z}{I_y} - \frac{M_z y}{I_z} = 0$$

or

$$\frac{y}{z} = \frac{M_y I_z}{M_z I_y} \tag{5.33}$$

The angle α between the z-axis and the neutral axis is defined by

$$\tan \alpha = \frac{y}{z} = \frac{M_y I_z}{M_z I_y} \tag{5.34}$$

16. Principal moments of inertia

The principal moments of inertia of a cross-section are given by

$$I_{max,min} = I_{1,2} = \frac{I_z + I_y}{2} \pm \sqrt{\left(\frac{I_z - I_y}{2}\right)^2 + I_{yz}^2} \tag{5.44}$$

where now I_z and I_y are moments of inertia about convenient perpendicular axes z and y.

17. Shear center

The *shear center* is defined as the location where the transverse load should be placed to avoid twisting of the cross-section. Figure 5.101 lists locations of shear centers for various cross-sections.

REFERENCES

1. Timoshenko SP., Goodier JN. *Theory of Elasticity,* 3rd ed. McGraw-Hill, New York, 1970.
2. National Forest Products Association. *National Design Specification for Wood Construction,* 1986 ed. 1619 Massachusetts Avenue, N.W., Washington, DC 20036, 1986.
3. American Institute of Steel Construction. *Manual of Steel Construction,* 9th ed., One East Wacker Dr., Chicago, IL 60601, 1989.
4. The Aluminum Association. *Specifications for Aluminum Structures,* 9th ed. 900 19th St., N.W., Washington, DC 20006, 1988.
5. American Institute of Timber Construction (333 West Hampden Avenue, Englewood, CO 80110). *Timber Construction Manual,* 3rd ed. New York, John Wiley & Sons, 1986.
6. American Association of State Highway and Transportation Officials. *Standard Specifications for Highway Bridges,* 14th ed. 440 North Capital St., N.W., Washington, DC, 20001, 1989.
7. Peterson RE. *Stress Concentration Factors.* New York, John Wiley & Sons, 1974.

PROBLEMS

Section 5.3

5.1–5.18 Draw shear and bending moment diagrams for the beam and loading shown. In each problem indicate clearly the location and the magnitude of the maximum shear and the maximum moment.

Figure P5.1

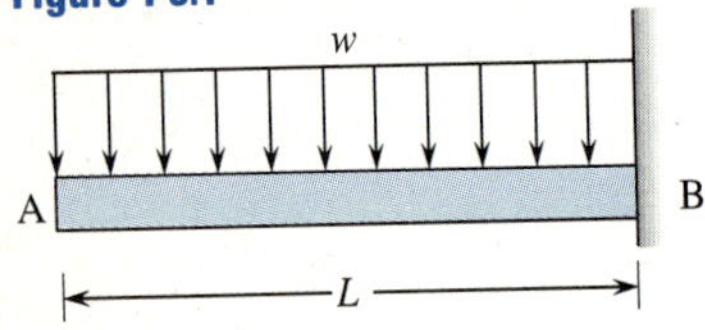

Figure P5.2

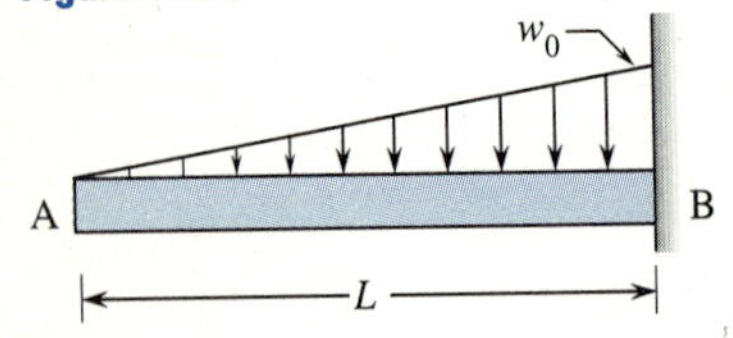

Figure P5.3

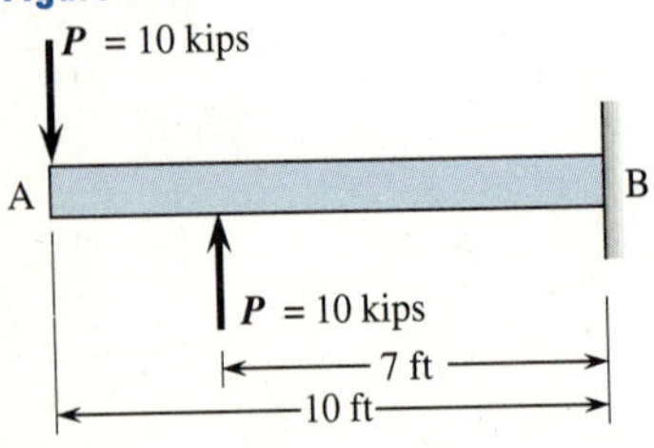

Figure P5.4

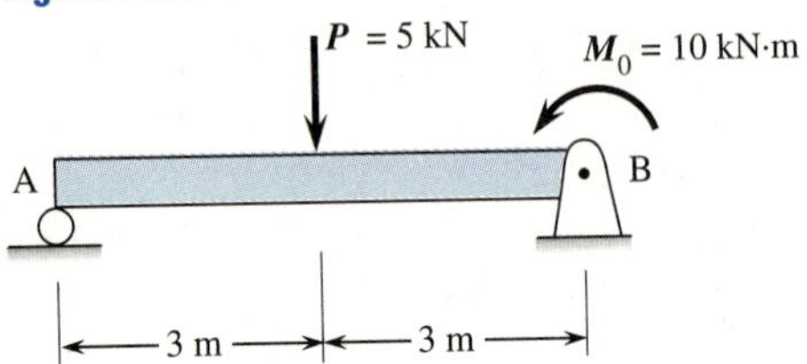

Figure P5.5

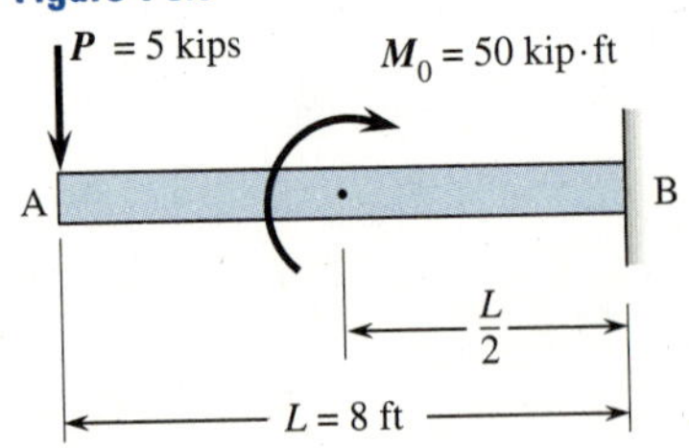

Figure P5.6

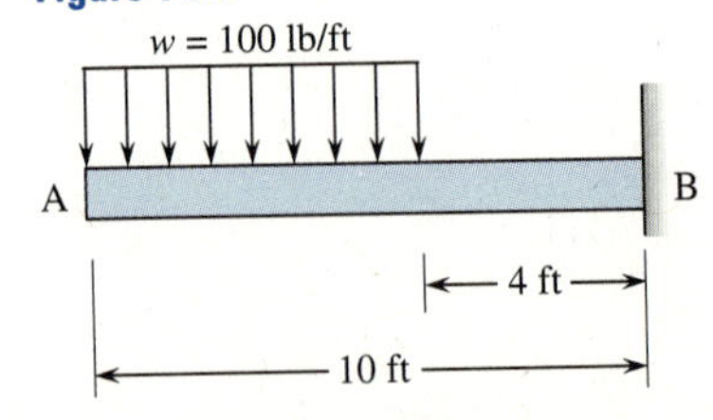

Figure P5.7

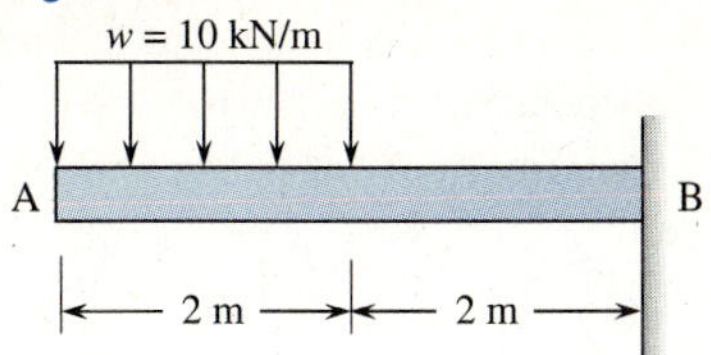

Figure P5.8

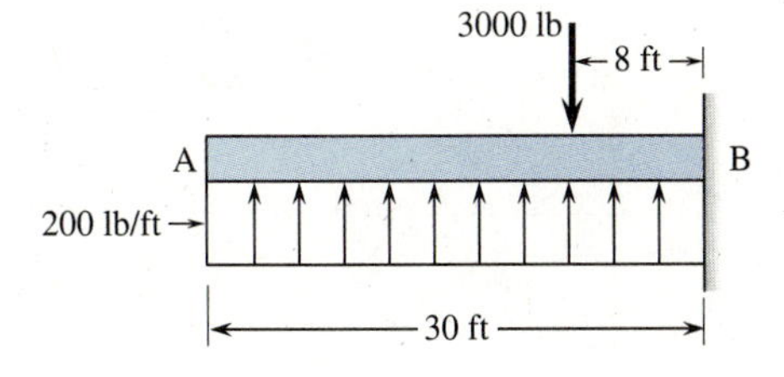

Figure P5.9

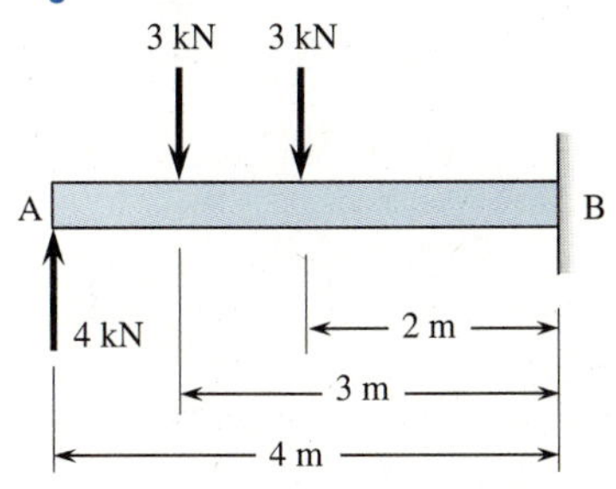

Figure P5.10

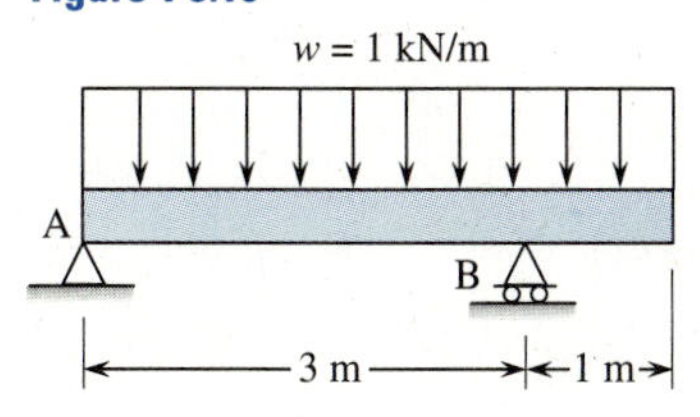

Figure P5.11

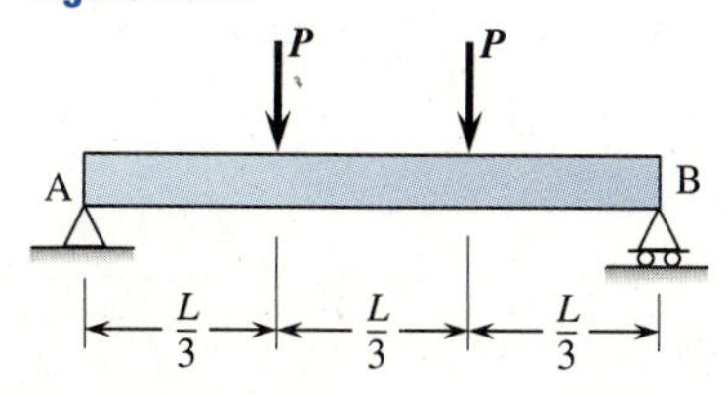

Figure P5.12

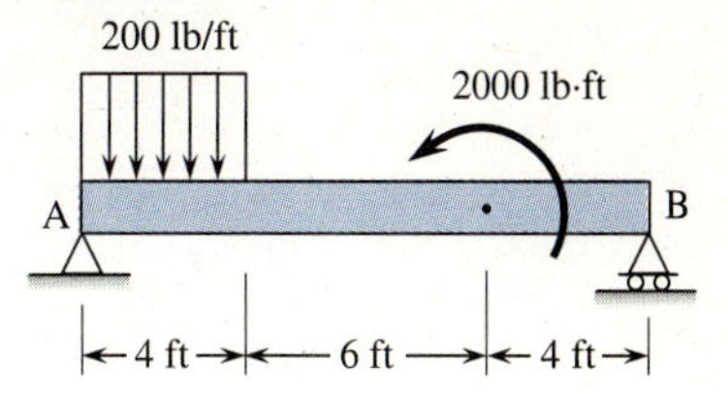

Figure P5.13

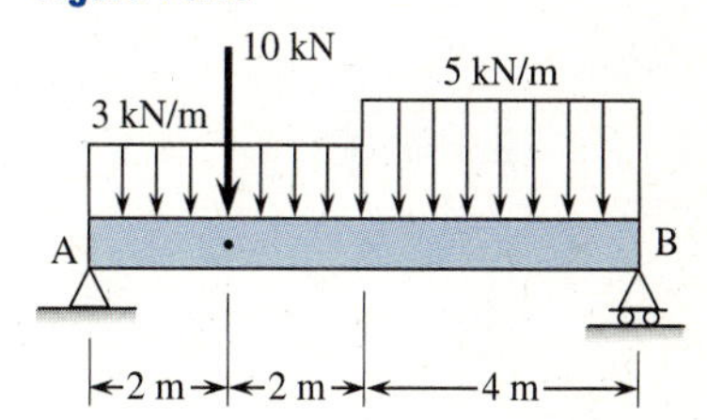

Figure P5.14

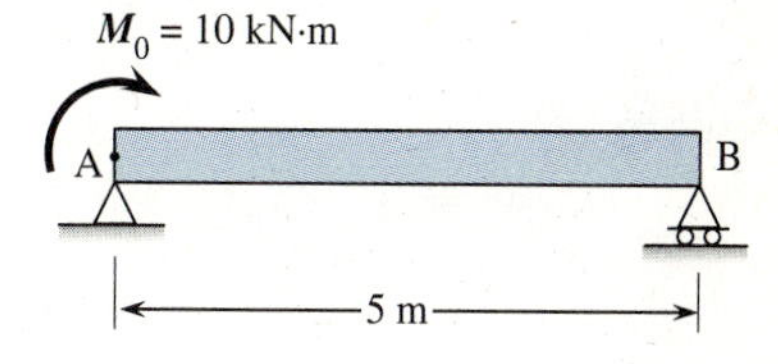

Figure P5.15

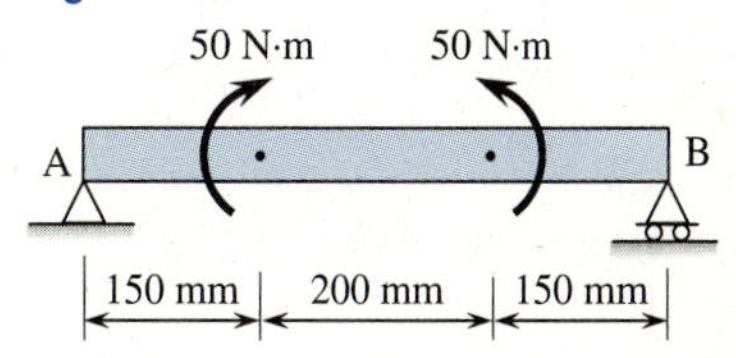

Figure P5.16

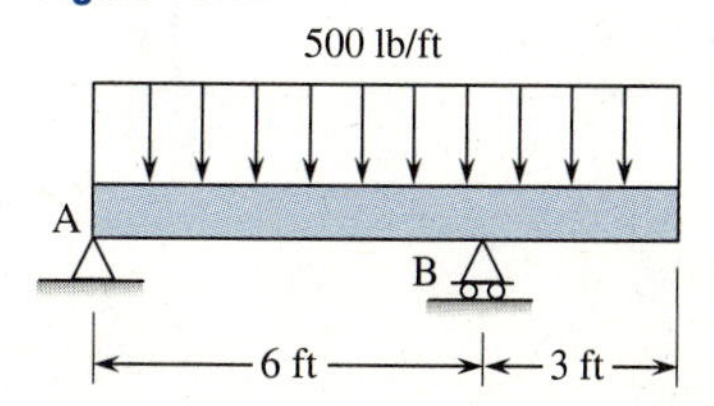

Figure P5.17

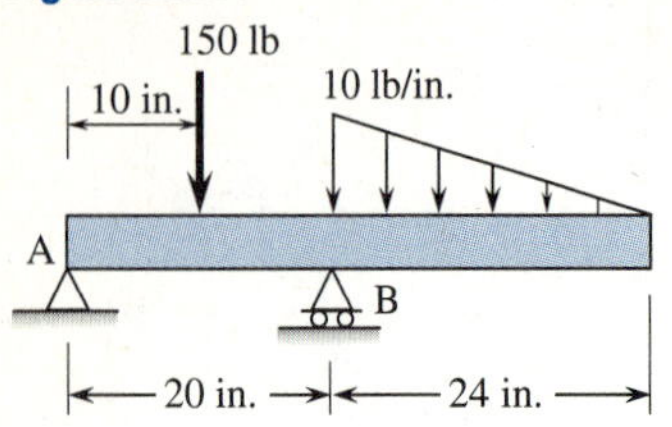

Figure P5.18

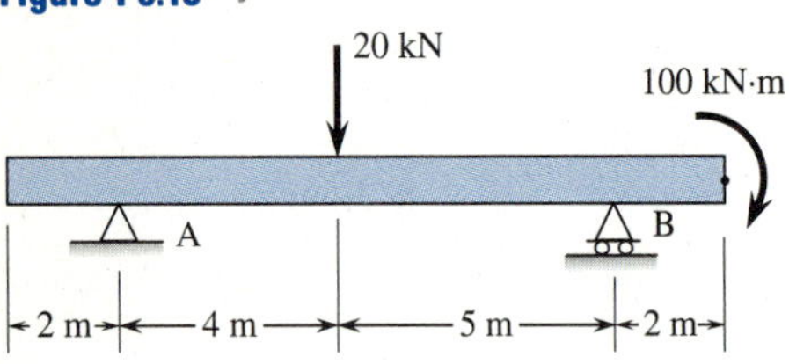

5.19 For the overhanging beam shown, draw **(a)** the shear force diagram and **(b)** the bending moment diagram.

Figure P5.19

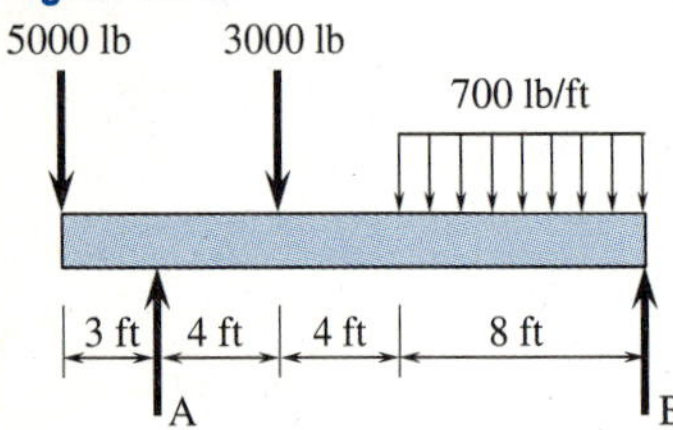

5.20 For the beam shown, draw the shear force and bending moment diagrams showing values at all significant points.

Figure P5.20

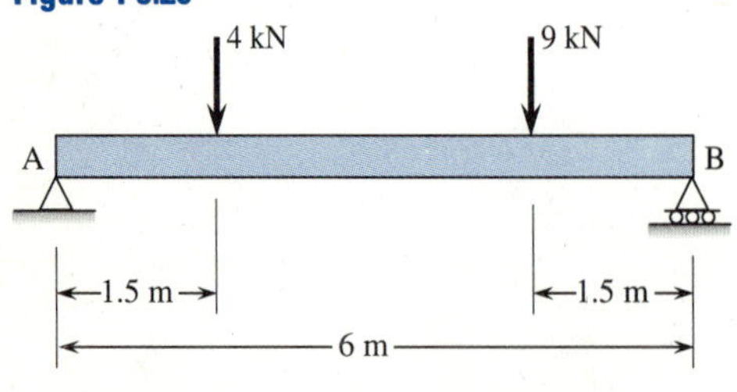

5.21 For the beam subjected to a concentrated load and couple shown, determine the shear force and bending moment diagrams. What are the maximum shear force and bending moments?

Figure P5.21

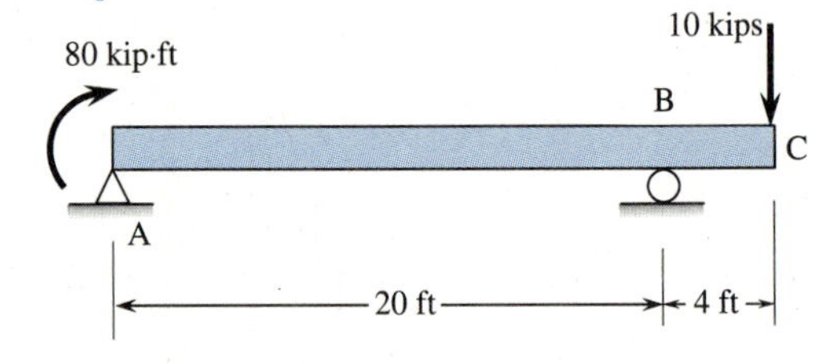

Section 5.4

5.22 A 12-ft header beam is made by nailing two 2 in. × 8 in. (actual dimensions) boards together. Determine the maximum bending stress.

Figure P5.22

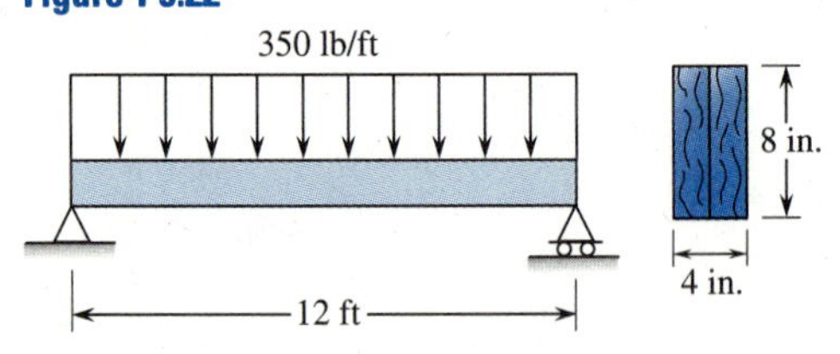

5.23 Determine the maximum compressive stress and the maximum tensile stress due to bending if $P = 6.0$ kN.

Figure P5.23

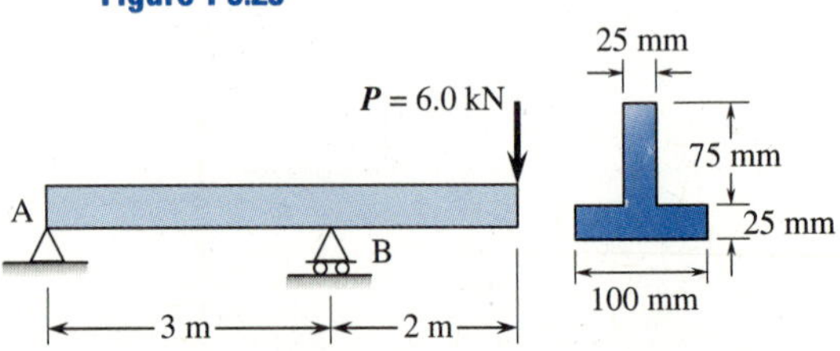

5.24 Two L5 × 5 × $\frac{1}{2}$ angle sections are bolted back-to-back (Figure P5.24) and used as the beam in Figure P5.22. Calculate the maximum tensile stress and the maximum compressive stress.

Figure P5.24

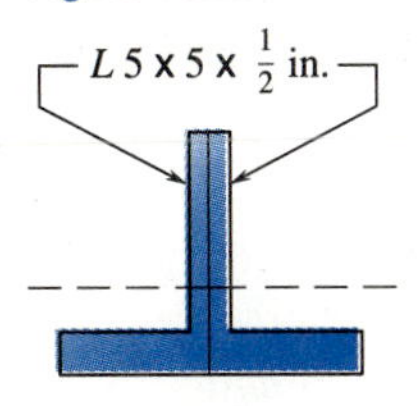

5.25 A 3-in. diameter hollow pipe with a 0.25-in. wall thickness is used as the beam in Figure P5.16. Determine the maximum bending stress.

5.26–5.32 The beams in Problems 5.3, 5.5, 5.6, 5.8, 5.12, 5.16, and 5.17 are 12 in. wide × 24 in. high. Determine the maximum tensile bending stress in each beam.

5.33–5.40 The beams in Problems 5.4, 5.7, 5.9, 5.10, 5.13 through 5.15, and 5.18 are 300 mm wide × 600 mm high. Determine the maximum tensile bending stress in each beam.

5.41 Determine the maximum bending stress for the beam of Problem 5.21. The beam cross-section is 6 in. wide × 8 in. deep.

5.42 A cantilevered beam is 4 m long and supports a uniform load of 1 kN/m over its entire length. The beam is 150 mm wide × 300 mm deep. Determine the maximum bending stress in the beam.

5.43 A 14-ft-long simple beam is uniformly loaded with 200 lb/ft over its entire length. If the beam is 3.5 in. wide by 7.5 in. deep, what is the maximum bending stress?

5.44 An axle is loaded in static bending with a moment of 2500 lb · in. The axle is steel with a diameter of 1 in. What is the maximum bending stress in the material?

5.45 For the beam with cross-section shown in Figure P5.45 (with I = 100 in.^{4}), the bending moment is M = 1000 lb · ft. Determine the largest positive bending stress.

Figure P5.45

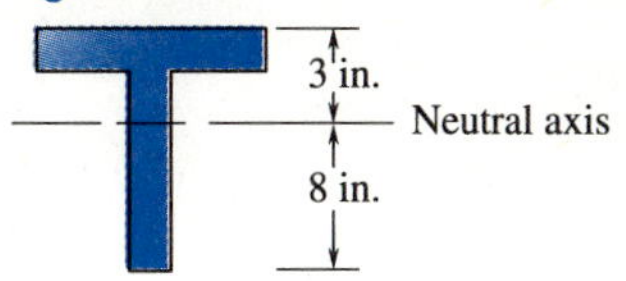

5.46 A square bar, 50 mm on a side, is used as a simply supported beam subjected to a bending moment of 400 N · m. Determine the maximum stress due to bending of the bar.

5.47 A bending moment of 6000 lb · in. is applied to a rectangular beam of dimensions 2 in. × 4 in. in cross-section. Determine the maximum bending stress in the beam **(a)** if the 2-in. side is placed vertical, and **(b)** if the 4-in. side is placed vertical.

5.48 A cylindrical tank is filled with water (the weight density of water is 62.4 lb/ft^3) and simply supported at each end. The tank has a 3-ft outer diameter and a wall thickness of t = 0.25 in. Calculate the maximum compressive stress in the tank wall. Neglect the weight of the tank but not the contents. For thin-walled vessels an approximate value of I about the neutral axis is given by $I = \pi r^3 t$ where r is the radius.

Figure P5.48

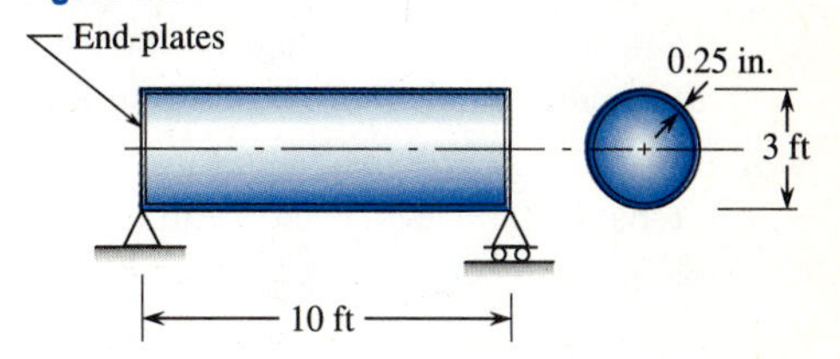

5.49 A sawmill is cutting rectangular wooden beams from circular logs. What should be the values of b and h to have the "strongest" beam (lowest bending stress for a given moment)?

Figure P5.49, P5.52

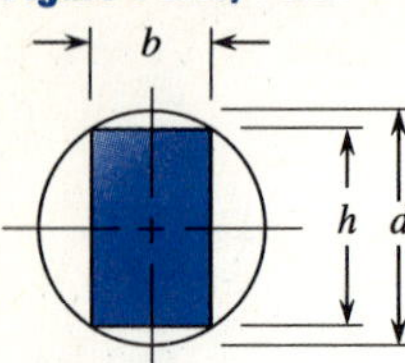

5.50 The beam of 20-ft span has the cross-section shown. The z-z centroidal axis of the cross-section is located as shown. The beam is supported and loaded as shown. The bending moment diagram is also shown. Determine the maximum tensile and compressive stress in the beam due to bending moment.

Figure P5.50

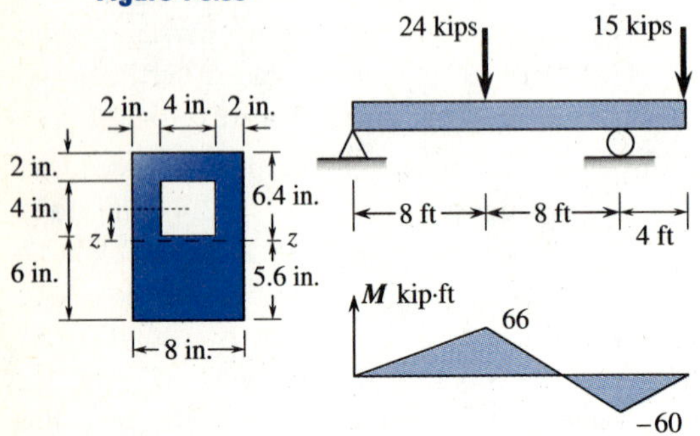

5.51 A simply supported wooden beam supports a uniform load of 100 lb/in. over its full length of 100 in. The beam cross-section is 5.50 in. wide and 11.25 in. high. Determine the maximum bending stress in the beam.

5.52 What should be the values of b and h in Problem 5.49 so that the beam would have minimum bending deflection? (HINT: I should be maximized.)

Sections 5.4 and 5.5

5.53 A box beam carries the loads shown. Determine **(a)** the maximum bending stress and **(b)** the shear stress at the neutral axis.

Figure P5.53

5.54 A structural steel tee, WT 12 × 51.5, is used as a simply supported beam of span 12 ft and subjected to a uniformly distributed load of 2000 lb/ft. Determine the maximum bending stress and maximum shear stress due to bending in the beam.

5.55 For the cantilevered beam loaded as shown, consider section n-n and determine **(a)** the maximum shear stress, **(b)** the shear stress at point a, and **(c)** the shear stress at point b in the web.

Figure P5.55

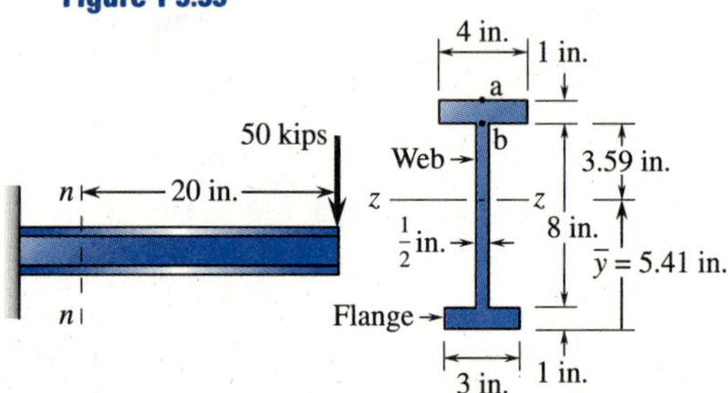

5.56–5.62 The beams in Problems 5.3, 5.5, 5.6, 5.8, 5.12, 5.16, and 5.17 are 12 in. wide × 24 in. high. Determine the largest transverse shear stress in each beam.

5.63–5.70 The beams in Problems 5.4, 5.7, 5.9, 5.10, 5.13 through 5.15, and 5.18 are 300 mm wide × 600 mm high. Determine the largest transverse shear stress in each beam.

5.71 Determine the maximum shear stress for the beam of Problem 5.21. The beam cross-section is 6 in. wide × 8 in. deep.

5.72 A 4 in. wide × 8 in. deep wood beam is simply supported and loaded as shown. What is the maximum shear stress in the beam?

Figure P5.40

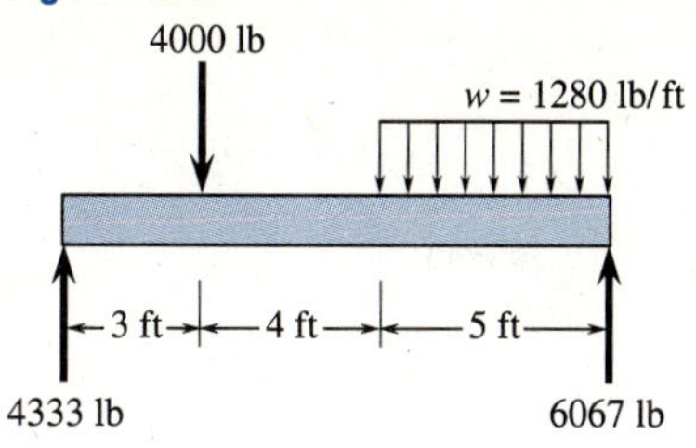

5.73 Calculate the maximum bending stress and the maximum shear stress at the neutral axis for the wide-flange beam shown in Figure P5.73.

Figure P5.73

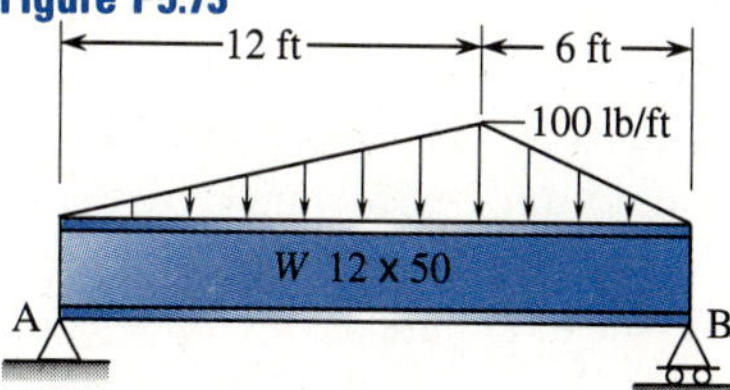

5.74 For the cantilevered beam shown determine **(a)** the maximum tension stress due to bending, **(b)** the maximum compression stress due to bending, **(c)** the shear stress at point C, and **(d)** the shear stress at the neutral axis. Let $I = 5.048$ in^4.

Figure P5.74

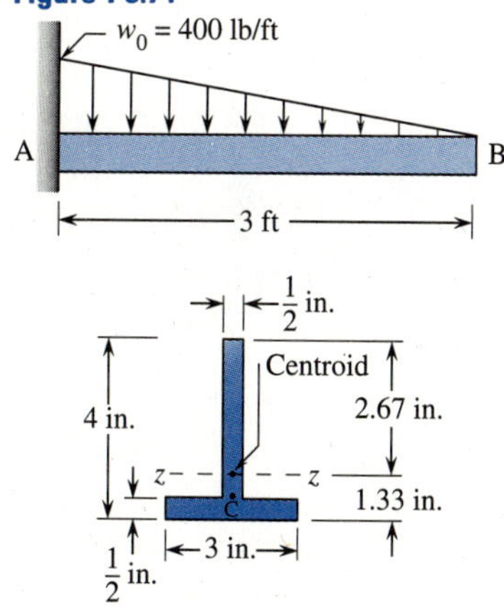

5.75 A W 10 × 12 wide flange is used for the beam shown. Determine the maximum bending stress and the maximum shear stress in the beam.

Figure P5.75

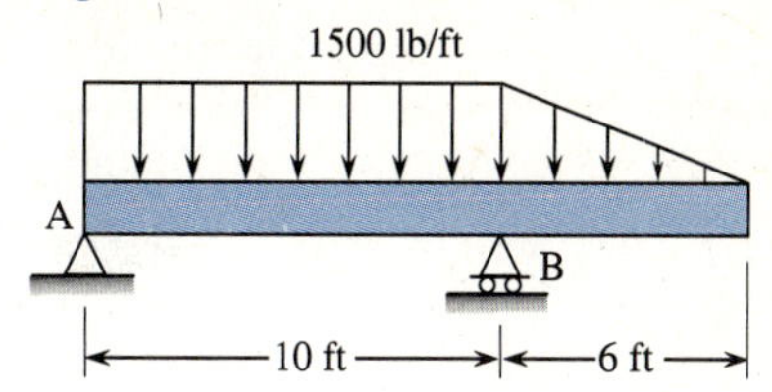

5.76 For the beam with overhang shown, determine **(a)** the maximum tensile bending stress. Indicate whether this stress occurs at the top or bottom of the cross-section. **(b)** Determine the maximum shear stress in the beam.

Figure P5.76

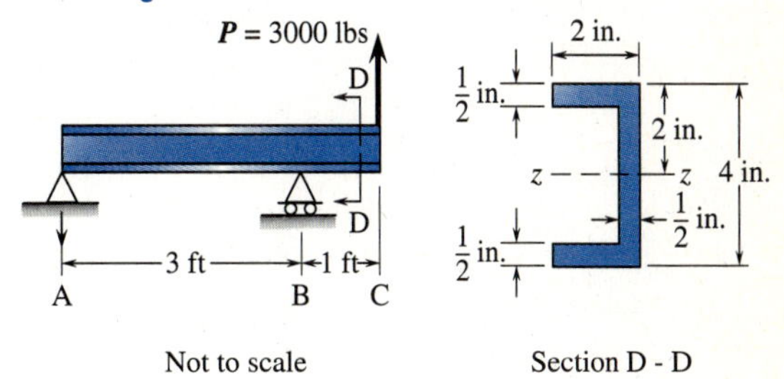

5.77 A steel girder shown in Figure P5.77 consists of two 1 in. × 16 in. flange plates welded to a $\frac{3}{8}$ in. × 60 in. web plate. Determine the allowable shear force V if each fillet weld has an allowable load in shear of $q = 2200$ lb per inch of weld.

Figure P5.77

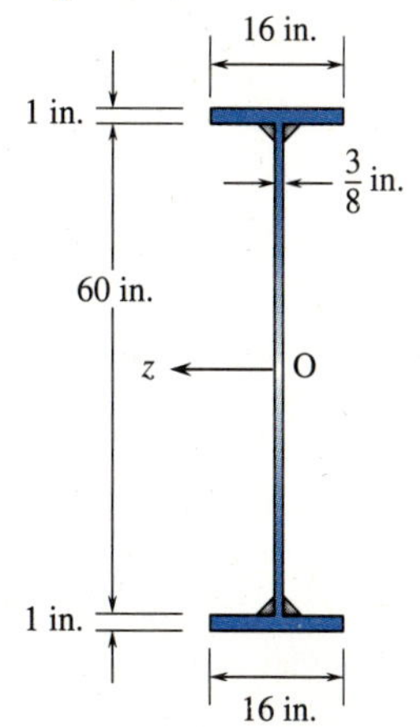

5.78 A beam is "built up" by gluing three 2 in. × 4 in. boards together (assume actual dimensions). Determine **(a)** the maximum bending stress, **(b)** the maximum shear stress, and **(c)** the shear stress on the glue joints.

Figure P5.78

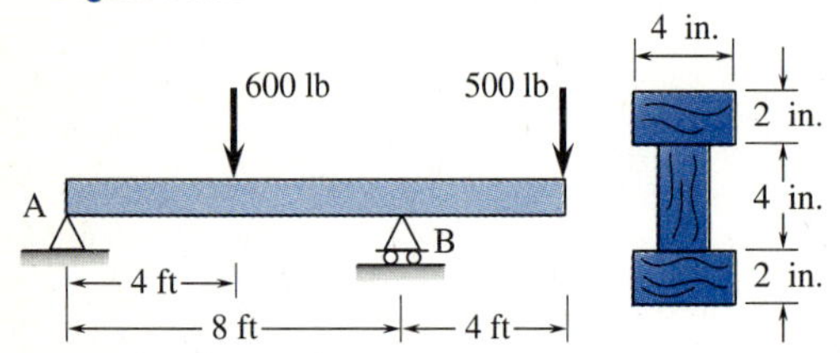

5.79 For the beam constructed of full-sized 2 in. × 4 in. boards nailed together with nail spacing of $s = 2.0$ in. and vertical shear force $V = 400$ lb, determine the shear force in a typical nail.

Figure P5.79

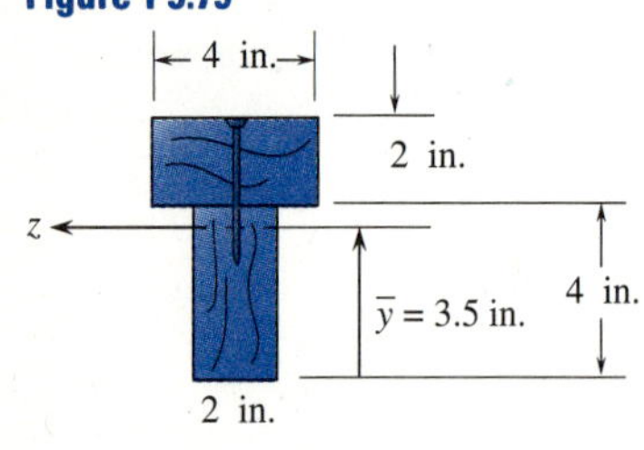

5.80 A box beam is constructed of wooden boards of size 1 in. × 8 in. (actual dimensions) as shown in Figure P5.80. The boards are joined together by screws that can resist an allowable load in shear of 300 lb per screw. Determine the maximum permissible longitudinal spacing s of the screws if the shear force V is 1000 lb.

Figure P5.80

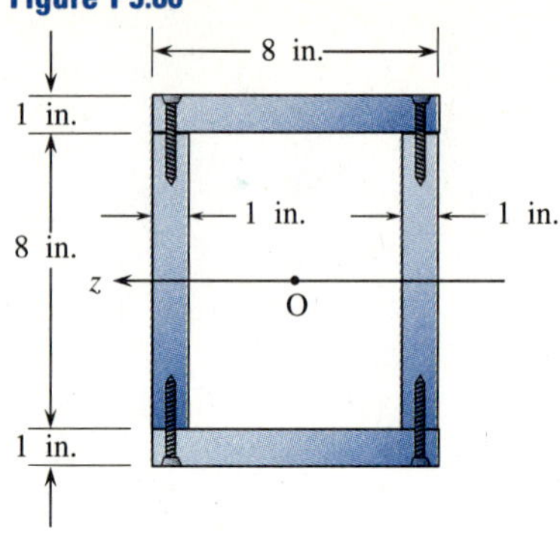

5.81 A wooden beam is constructed of two members of 25 mm × 200 mm cross-section that are joined by two 25 mm × 150 mm boards as shown in Figure P5.81. The boards are nailed to the beams at a longitudinal spacing $s = 75$ mm. If each nail has an allowable shear force of 1200 N, determine the maximum allowable shear force V.

Figure P5.81

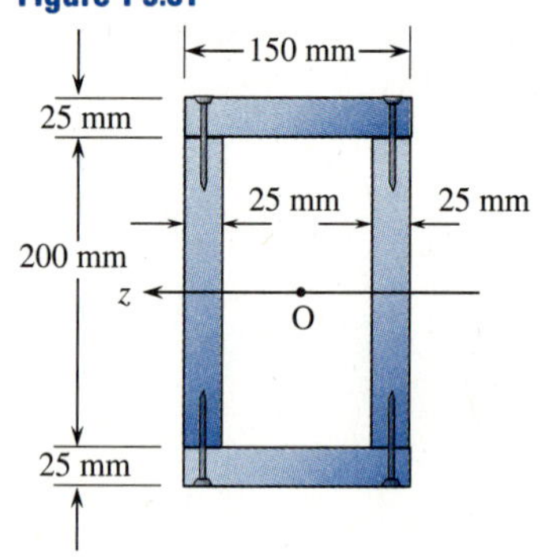

5.82 A beam of T cross-section is constructed by nailing together two boards having dimensions of 50 mm × 200 mm as shown in Figure P5.82. The shear force acting on the cross-section is $V = 2000$ N and each nail can carry 750 N in shear. Determine the maximum allowable nail spacing s.

Figure P5.82

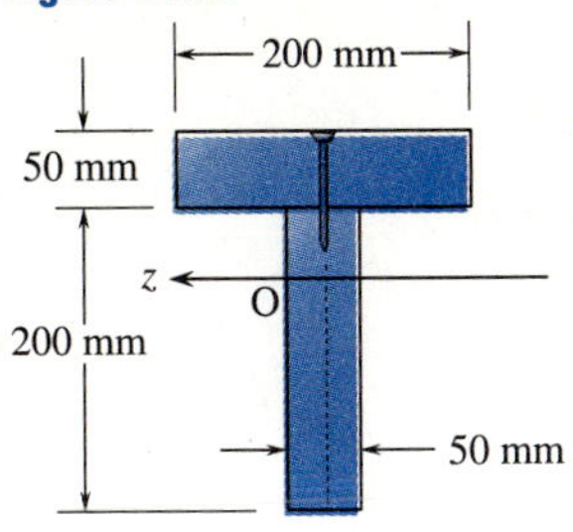

Section 5.6

5.83 A standard steel pipe is used to support a concentrated load of 300 lb at its middle. The pipe is 40 in. long and simply supported at its ends. Determine a suitable pipe size if the bending stress is limited to 12,000 psi. (See Appendix C.)

5.84 A wide-flange beam is used to support a uniformly distributed floor load of 750 lb/ft over a span of 20 ft. The beam is simply supported. Determine the lightest size of beam needed. Assume the beam is sufficiently supported against buckling. The allowable bending stress is 24 ksi and the allowable shear stress is 14.4 ksi.

5.85 A simple, solid axle for a trailer suspension is made as shown in Figure P5.85. A load of 4000 N is applied through each spring. Assume the supports at each wheel are simple supports. The allowable bending stress in the axle is limited to 100 MPa to account for dynamic and fatigue loading. Determine the size (diameter) of the shaft.

Figure P5.85

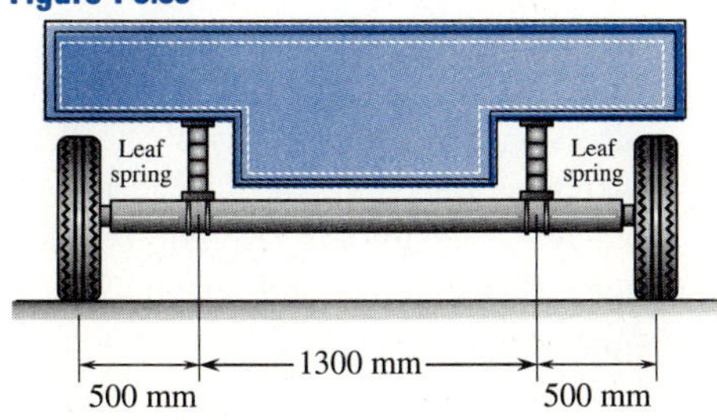

5.86 A cantilevered wooden roof system over a warehouse is 21 ft wide × 36 ft long, as shown in Figure P5.86. Four main beams each 21 ft long extend out from the building. Cross-beams connect the main beams and support 30 lb per square foot of uniformly distributed load. The crossbeams are spaced every 7 ft from the edge of the building. Assume the crossbeams to be simply supported at the main beams. The allowable bending stress in the wood beams is 1500 psi and the allowable shear stress is 80 psi. Determine the sizes of the main and cross-beams. (See Appendix C for standard sizes of lumber.)

Figure P5.86

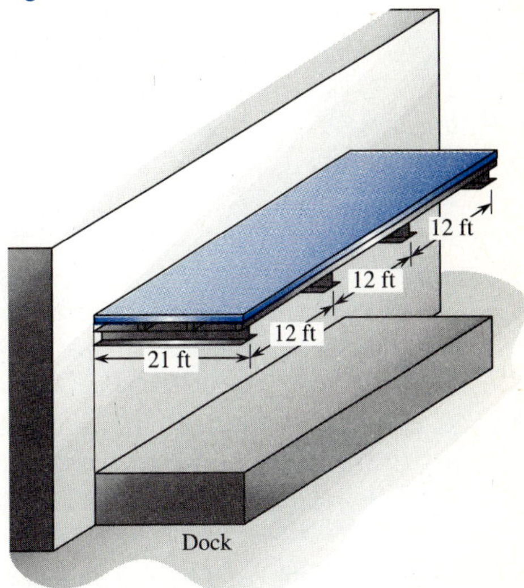

5.87 A frame for hoisting automobile engines consists of a beam supported by A-frames at the ends. The beam is 10 ft long and simply supported at its ends and must support a concentrated load of an engine weighing 600 lb. Determine the size of a structural tee section for the beam if the allowable bending stress is 20,000 psi.

5.88 A wide-flange beam is used as one of a series supporting the loads of a highway bridge. The loads on the beam represent those of a truck of standard design as shown in Figure P5.88. (Assume the position of the loads shown.) The beam is assumed to be simply supported. Determine the lightest W section for an allowable bending stress of 18,000 psi.

Figure P5.88

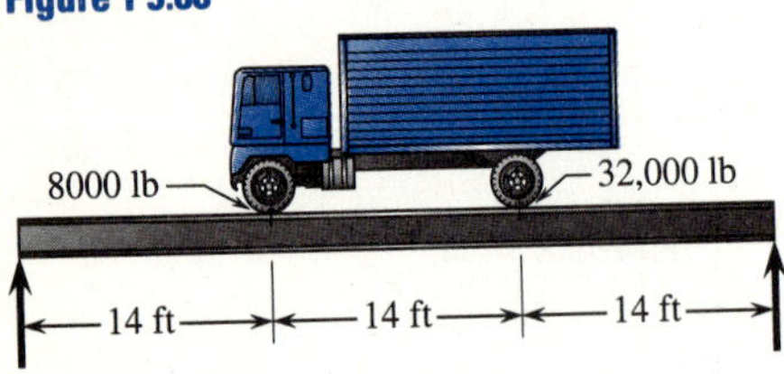

5.89 A wide-flange steel beam is needed to span 30 ft across a small stream as part of a new bridge. The largest load to be resisted has been simplified to 40 kips at the midspan. The allowable bending stress is 20 ksi and the allowable shear stress is 12 ksi. Determine the size of a wide-flange beam needed to support the load.

5.90 A simply supported, 10 ft long pipe must resist a concentrated load of 5 kips at its midspan. Determine the size of a standard construction pipe needed such that the bending stress in the pipe is less than 22 ksi.

5.91 A cantilevered beam 10 ft long is to be designed to support an end load of 10 kips. The allowable bending stress is 22 ksi. Determine the smallest **(a)** solid round bar, **(b)** solid square bar, **(c)** solid rectangular bar with height two times its width, and **(d)** lightest wide-flange beam. Then compare the magnitude of the cross-sectional areas. The smallest one is the lightest. Recommend the best section based on the lightest one.

5.92 A channel section is used as a stairway beam as shown. Two such channels spaced at 6 ft must support a uniform live load of 100 lb/ft^2. Assume the channels to be simply supported on their ends. Determine the minimum size of channel needed based on an allowable bending stress of 20 ksi.

Figure P5.92

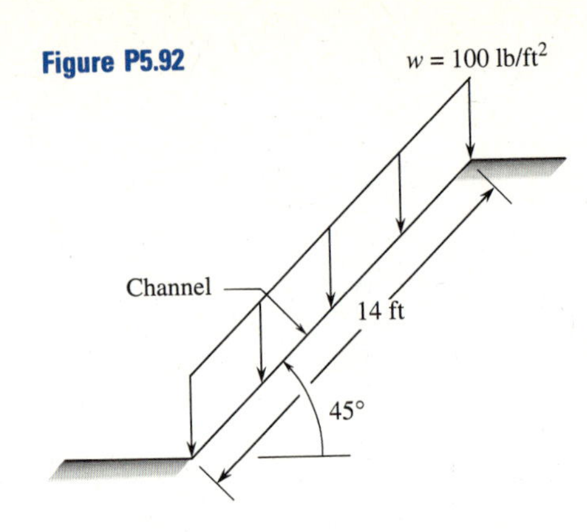

5.93 A simply supported southern pine wood beam of rectangular cross-section and span length of 10 ft is subjected to a uniform load of 1 kip/ft. Determine the size of the wood beam based on an allowable bending stress of 1200 psi and an allowable shear stress of 80 psi.

5.94 The same beam in Problem 5.93 is now subjected to a concentrated load at its center of 10 kips, instead of the uniform load. Determine the size of wood beam needed in this case. Use the allowable stresses from Problem 5.93.

5.95 A simply supported wood beam of span length 4 m is subjected to a concentrated load of 50 kN at its midspan. The allowable bending stress is 8 MPa and the allowable shear stress is 550 kPa. Determine the size of a rectangular cross-section with height two times the width to satisfy the allowable stresses.

Section 5.7

5.96 A truck weighing 8 kips exerts the wheel reactions shown on the deck of a girder bridge. Assume the truck can travel in either direction along the center of the deck, and therefore transfers half of its load to each of the two side girders. Determine the maximum shear force and maximum bending moment in the girder due to the moving load.

Figure P5.96

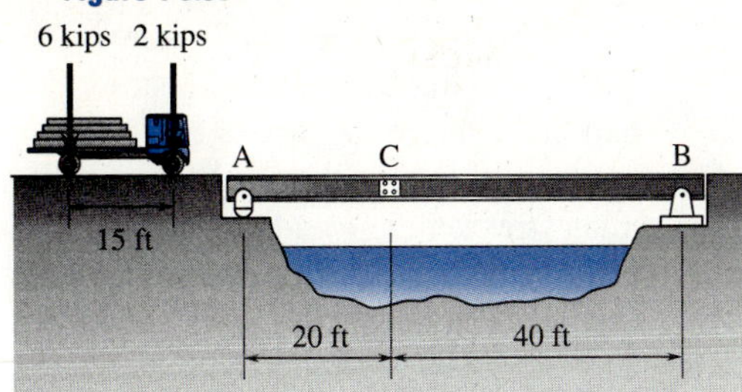

5.97 The semitruck shown transfers the loads shown to a girder of a bridge. Determine the largest bending moment due to the live load.

Figure P5.97

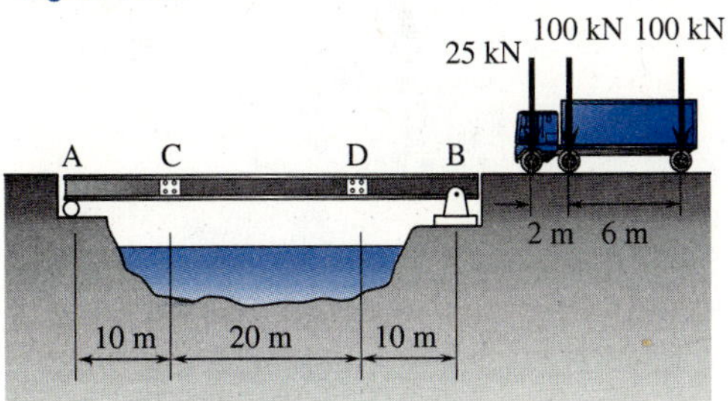

5.98 The semitruck shown transfers the loads shown to a girder of a bridge. Determine the maximum bending moment due to the live load.

Figure P5.98

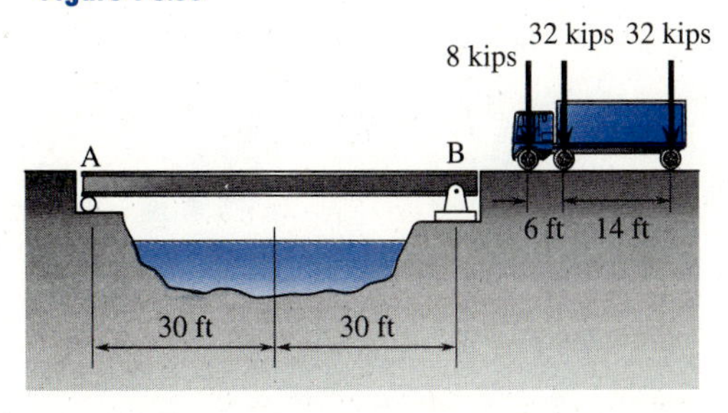

5.99 A car with a weight of 3000 lb and center of gravity at G moves across a girder bridge. Assume the car travels in either direction along the bridge deck, so that one-half of its load is transferred to each of two side girders. Determine the maximum shear force and maximum bending moment in the girder.

Figure P5.99

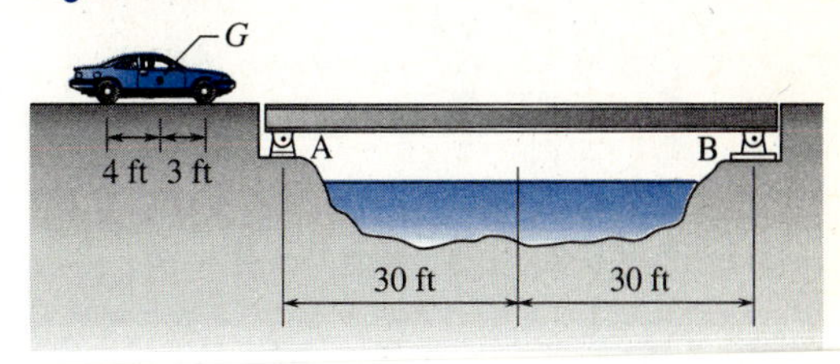

5.100 A car with a mass shown and center of mass at G moves across a bridge in either direction. Assume one-half of its mass is transferred to each of two girders. Determine the maximum shear force and maximum bending moment in the girder.

Figure P5.100

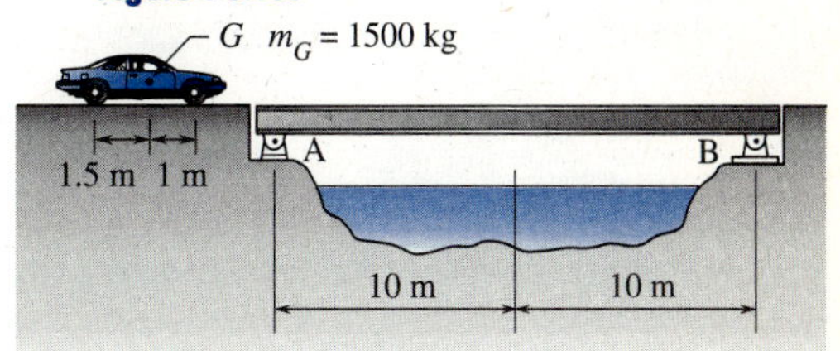

5.101 A four-wheel set shown moves slowly across a 40-ft-long simply supported beam. Determine the maximum bending moment and the position of the wheel set under this maximum moment.

Figure P5.101

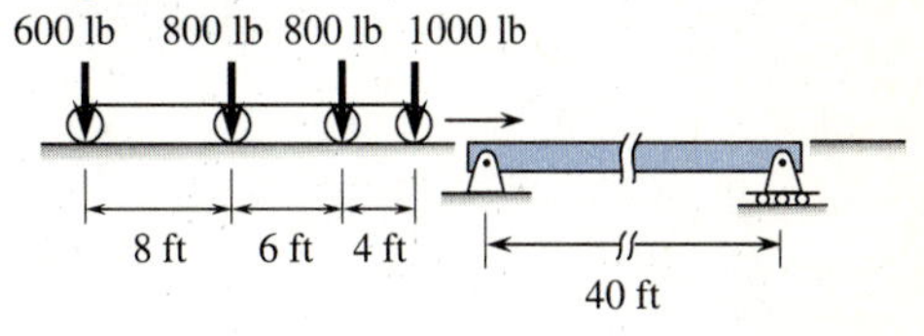

5.102 The four-wheel set moves across a 18-ft-long simply supported beam. Determine the maximum bending moment and the position of the wheel set under this maximum moment. Note that the maximum moment may not occur when all wheel loads are on the bridge.

Figure P5.102

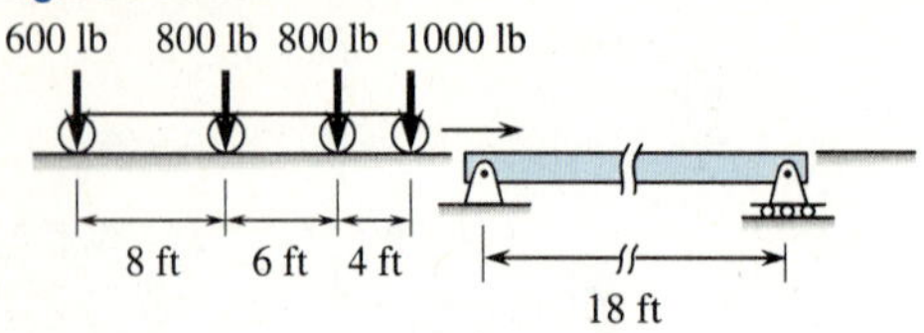

5.103 A grain truck with loads shown moves across a timber girder bridge. Assume half the load is transferred to each of two timber beams. Determine the largest bending moment due to this moving load.

Figure P5.103

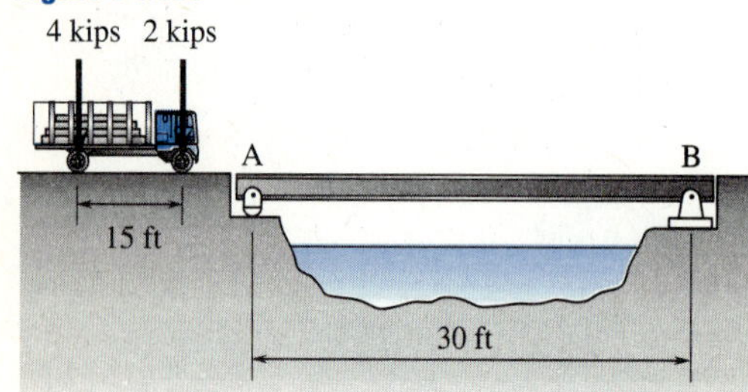

5.104 A three-wheel set moves across a 20-m-long simply supported beam. Determine the maximum bending moment due to this load.

Figure P5.104

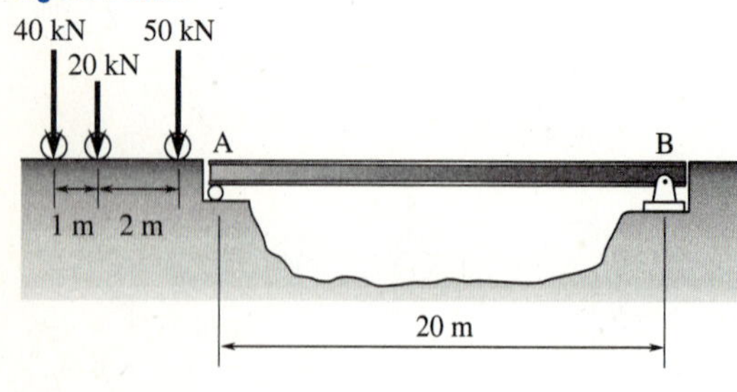

Section 5.8

5.105 A composite beam shown is made of nominal 2 in. × 6 in. southern pine with a modulus of elasticity of 1.2×10^6 psi and two $\frac{1}{4}$-in thick steel plates attached to the top and bottom of the wood section. The maximum bending moment about the z-axis is 10,000 lb · in. Determine the stress distribution in the wood and steel due to the moment. Compare the results for the maximum stress in the wood with the steel reinforcement to that of the wood without the steel plates.

Figure P5.105

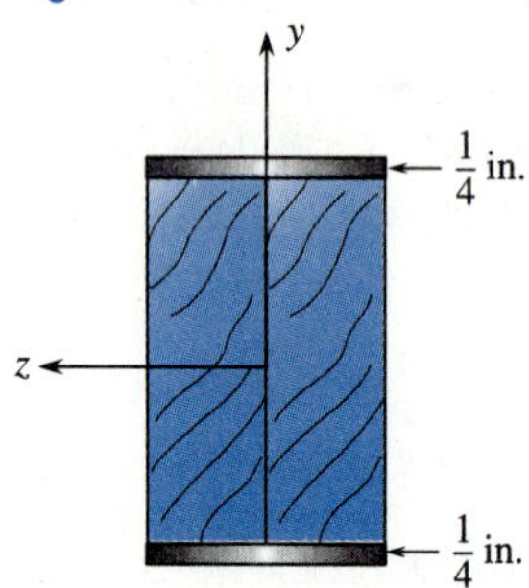

5.106 A timber beam 150 mm × 300 mm is reinforced on the bottom only with a steel plate 100 mm by 10 mm thick, as shown. Determine the maximum moment allowed if the stresses in the steel and wood are limited to 150 MPa and 10 MPa, respectively. Assume $E_s/E_w = 20$.

Figure P5.106

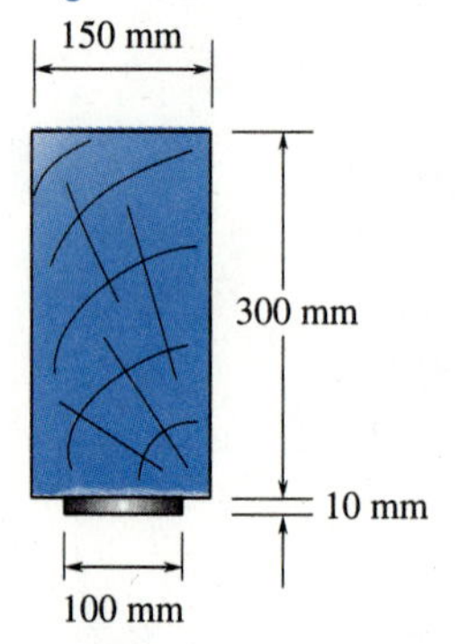

5.107 A timber beam 6 in. × 10 in. is reinforced with steel plates rigidly attached to its top and bottom as shown in Figure P5.107. Determine the largest concentrated vertical load that can be applied to the center of a simply supported beam with a span of 20 ft. The modulus of elasticity of the timber is 1.5×10^6 psi and that of steel is 29×10^6 psi. The allowable stresses in the timber and steel are 1200 psi and 20,000 psi, respectively.

Figure P5.107

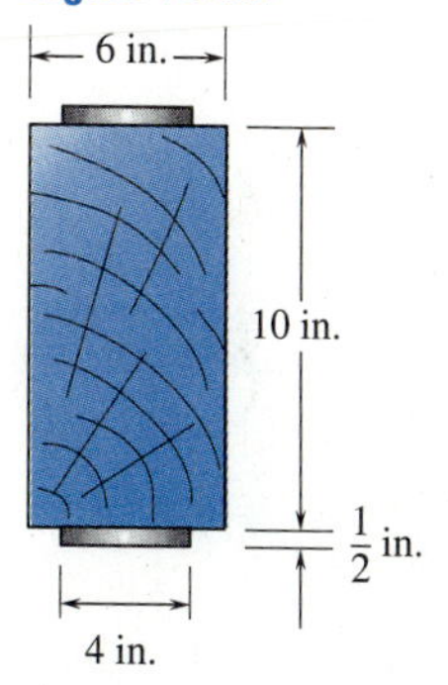

5.108 If the 20-ft-long beam of Problem 5.107 was cantilevered instead of simply supported, determine the allowable concentrated vertical free end load that could be applied.

5.109 A simply supported beam with a 30-ft span is subjected to a uniform load of 400 lb/ft. The beam is a composite timber and steel reinforced as shown in Figure P5.109. Determine the maximum stresses in the wood and steel. Assume $E_S/E_W = 20$.

Figure P5.109

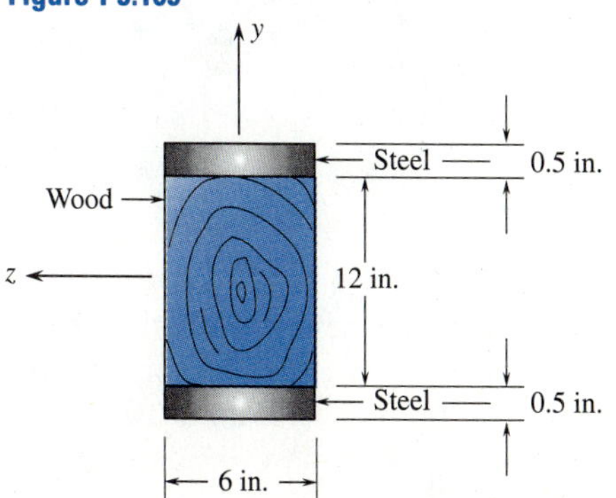

5.110 A timber beam is reinforced by two side vertical plates shown in Figure P5.110. The beam resists a maximum bending moment of 30 kN · m. Determine the stresses in the timber and steel. The steel/timber ratio of modulus of elasticity is 20.

Figure P5.110

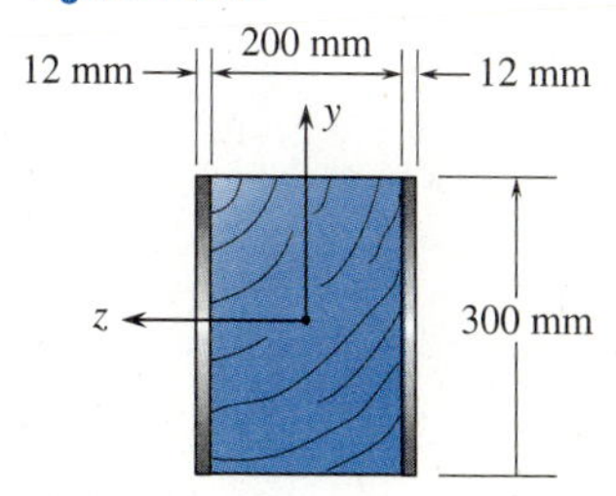

5.111 A pair of channel sections are securely bolted to a wood beam as shown in Figure P5.111. Determine the allowable bending moment if the allowable stresses are 20,000 psi in steel and 1200 psi in wood. Assume $E_s/E_w = 20$.

Figure P5.111

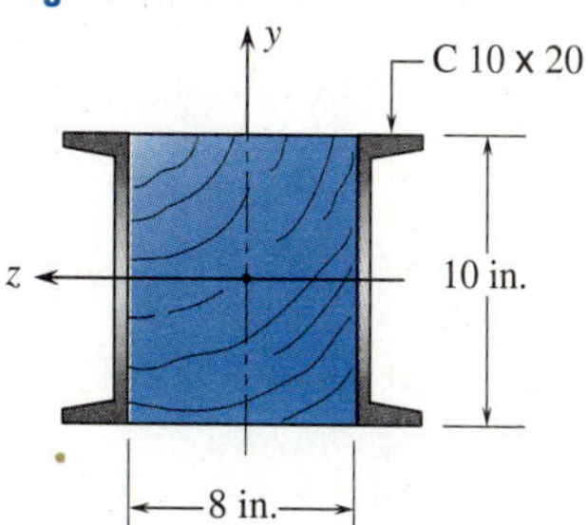

5.112 An aluminum beam is reinforced with a steel plate bonded to its bottom as shown in Figure P5.112. Determine the allowable bending moment if the allowable stresses in the aluminum and steel are 60 MPa and 120 MPa, respectively. Let E of aluminum be 70 GPa and E of steel be 200 GPa.

Figure P5.112

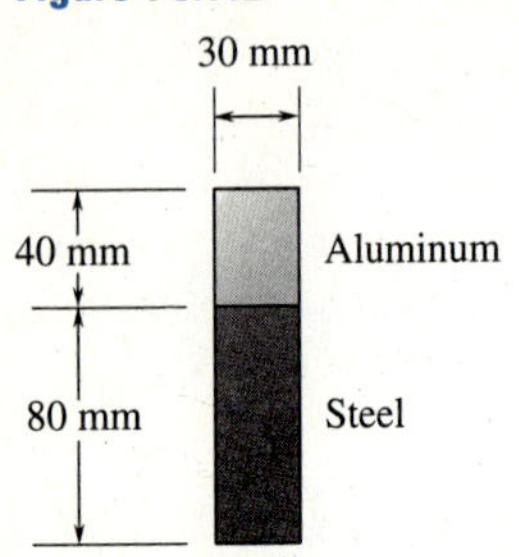

5.113 A composite steel and wood beam is subjected to a 15,000 lb · ft bending moment. The modulus of elasticity of steel is 30×10^6 psi and that of the wood is 1.5×10^6 psi. Determine the maximum bending stress in the steel and the wood.

Figure P5.113

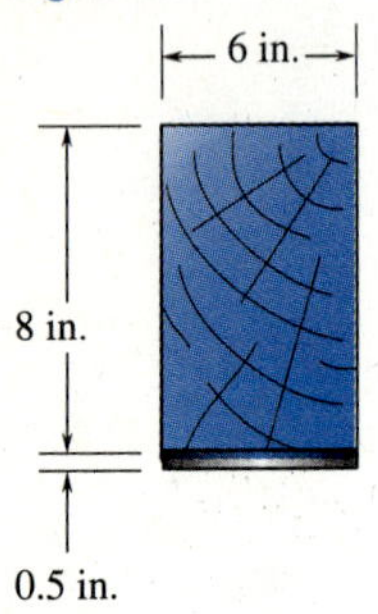

Section 5.9

In all problems in this section consider stress concentrations.

5.114 The cantilevered beam with fillets shown is subjected to a bending moment of $M = 10$ kip · in. Determine the maximum bending stress when the fillet radius is **(a)** 0.25 in. and **(b)** 0.50 in.

Figure P5.114

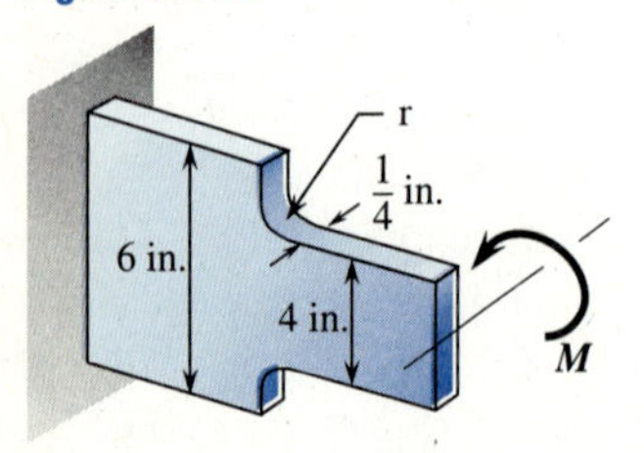

5.115 For the beam of Problem 5.114, if the allowable stress is 20,000 psi, determine the maximum moment that can be applied.

5.116 A cantilevered beam has grooves cut into each edge as shown. If the allowable bending stress is 15 ksi, determine the largest bending moment that can be applied.

Figure P5.116

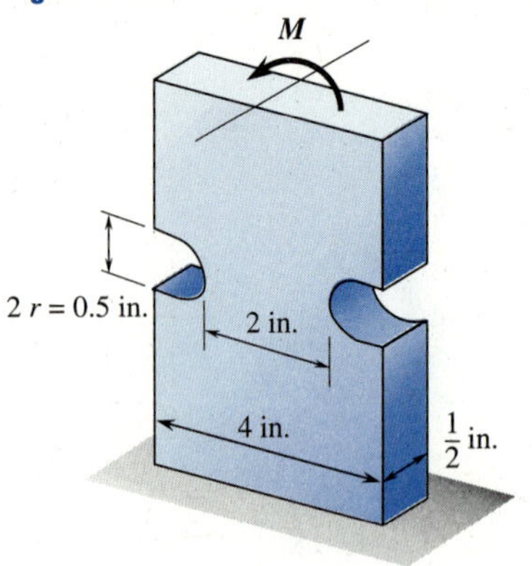

5.117 For the beam of Problem 5.116, if the bending moment is 5 kip · in, determine the bending stress at the grooves.

5.118 A part is machined with grooves shown. The bending moment resisted by the part is $M = 5$ kN · m. Determine the bending stress at the grooves.

Figure P5.118

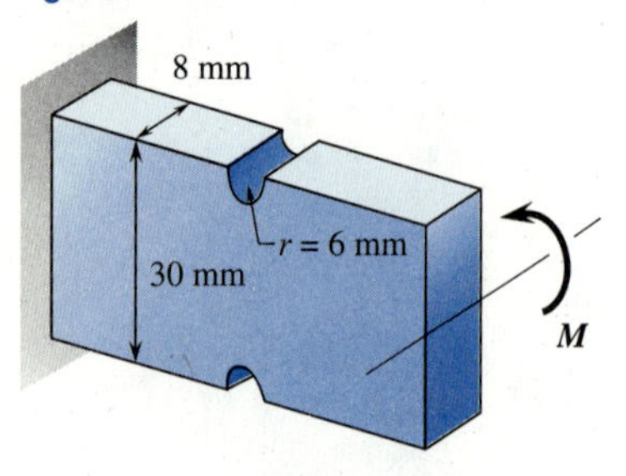

5.119 A simply supported beam with grooves at each edge is subjected to a uniform load of 1000 lb/ft. The beam is 10 ft long. Determine the largest bending stress at the grooves.

Figure P5.119

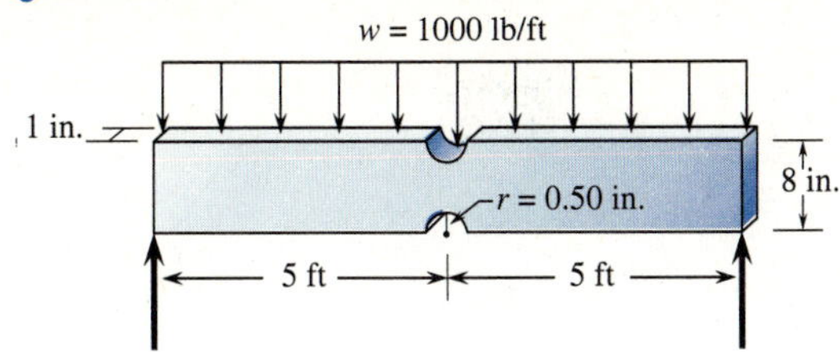

5.120 A round steel beam has 0.25-in. grooves machined into it as shown. The beam is built into a wall and is 5 ft long. Determine the largest end moment that can be applied to the beam if the allowable bending stress is 15 ksi.

Figure P5.120

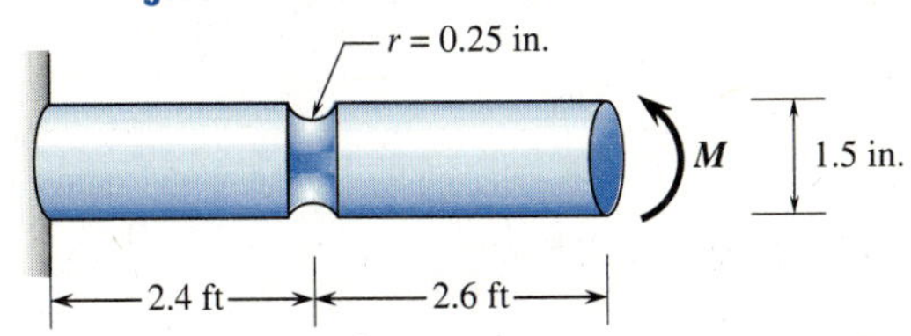

5.121 For the beam of Problem 5.120, determine the largest vertical end load only that can be applied to the beam.

5.122 A round steel beam has a fillet cut into it as shown. Determine the largest bending stress at the fillet.

Figure P5.122

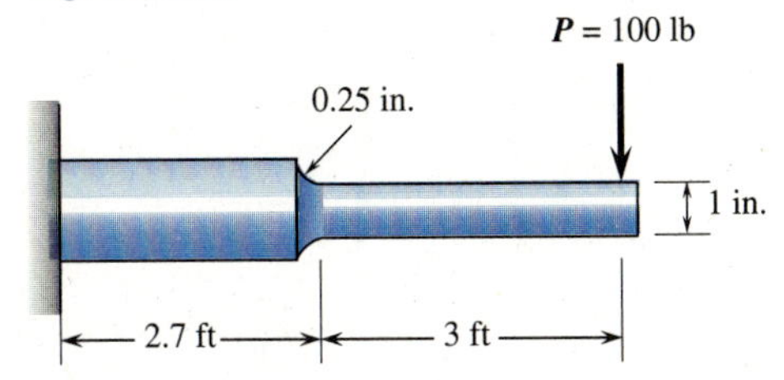

5.123 A round steel beam with a fillet machined into it as shown has an allowable bending stress of 160 MPa. Determine the largest bending moment that can be resisted by the beam.

Figure P5.123

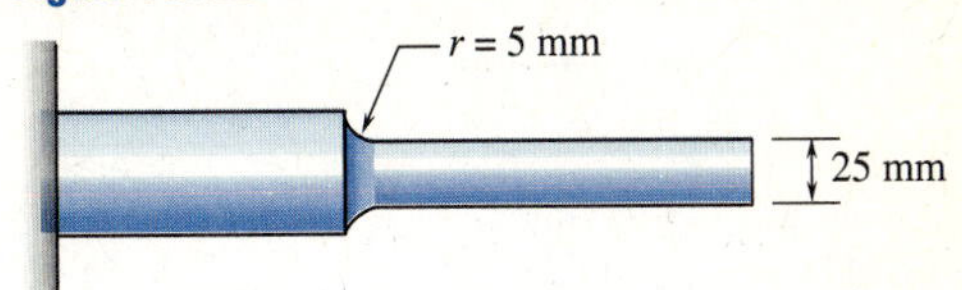

5.124 A lever used as a brake actuator for a piece of construction machinery has fillets cut into it as shown. Determine the bending stress at the fillet when a 20-lb force is applied at the top of the level. The cross-section of the smaller part of the lever is 0.25 in. × 1 in.

Figure P5.124

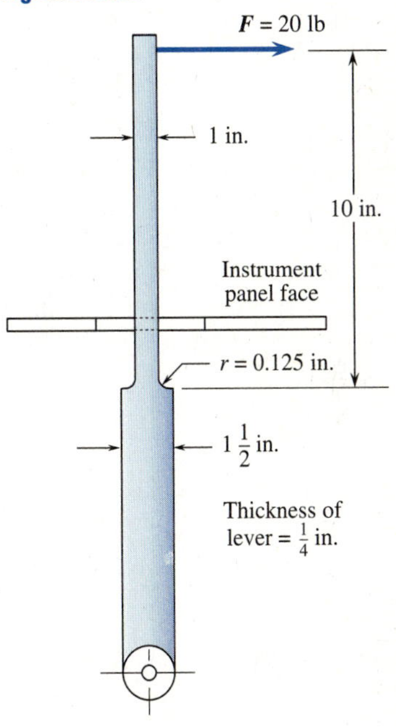

Section 5.10

5.125 A wood cantilevered beam with a rectangular cross-section supports an inclined load $P = 5$ kN at its free end. Determine the maximum tensile bending stress. Let $E = 12$ GPa, $b = 200$ mm, $h = 400$ mm, and $\theta = 30°$.

Figure P5.125

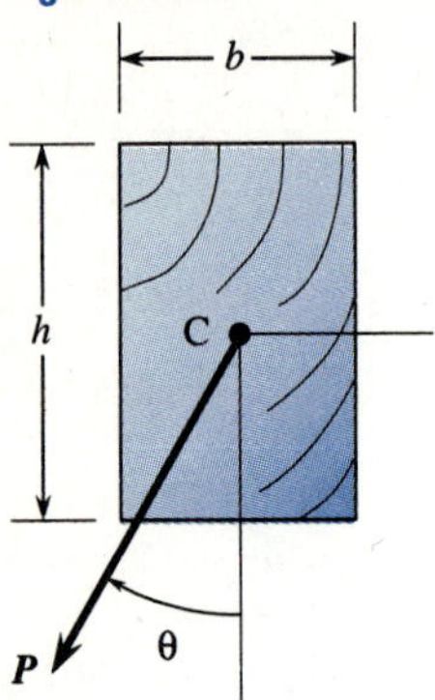

Figure P5.128

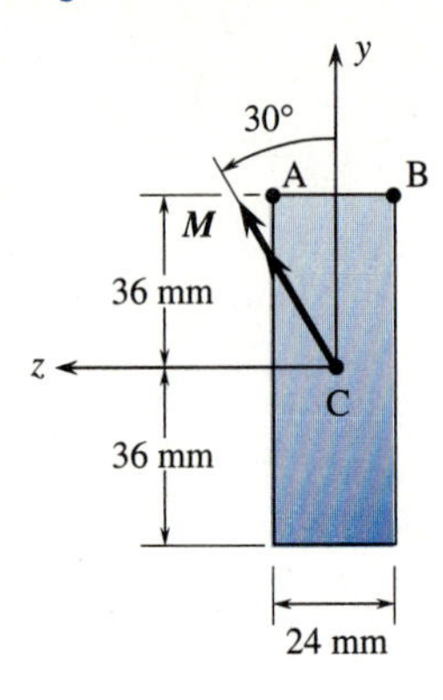

5.126 For the beam of Problem 5.125, if the allowable bending stress is 8 MPa, determine the allowable end load P.

5.127 A cantilevered beam 5 ft long resists a vertical load of $P = 200$ lb at the free end as shown. Determine the maximum bending stress at the corner A. The properties of the Z-section are $I_1 = 19.2$ in.^{4}, $I_2 = 9.0$ in.^{4}, $I_y = 2.95$ in.^{4}, and $I_z = 25.25$ in.^{4}

Figure P5.129

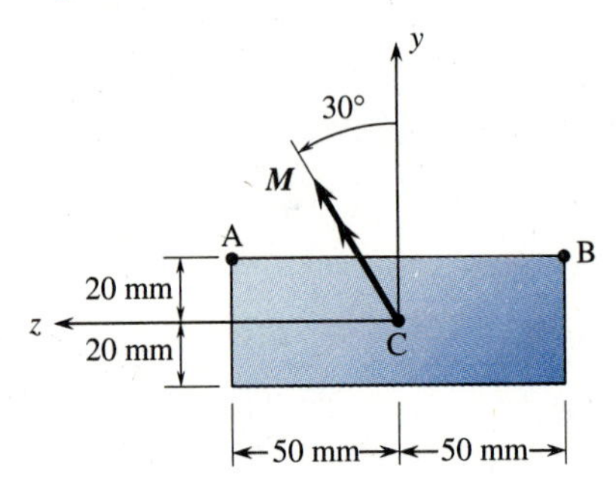

Figure P5.127

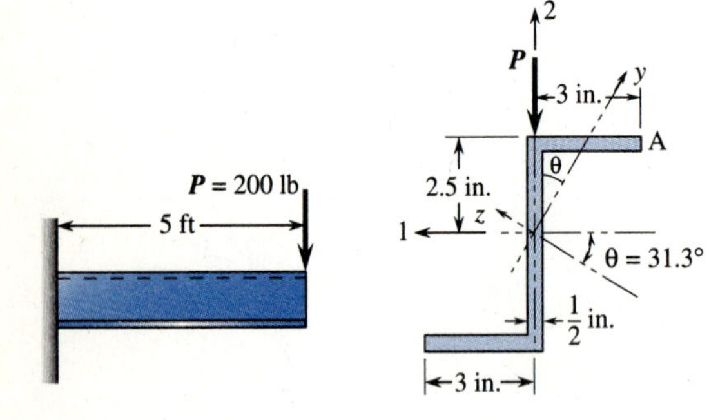

Figure P5.130

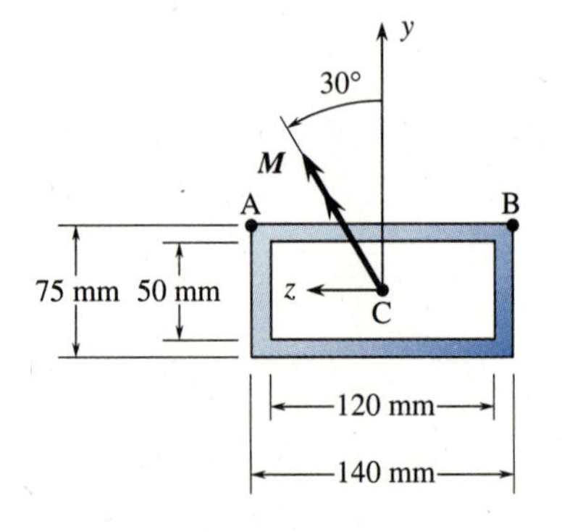

5.128–5.131 A beam of rectangular cross-section is subjected to a maximum bending moment of M = 5 kN · m making an angle of 30° with the vertical. Determine the stress at **(a)** point A and **(b)** point B.

Figure P5.131

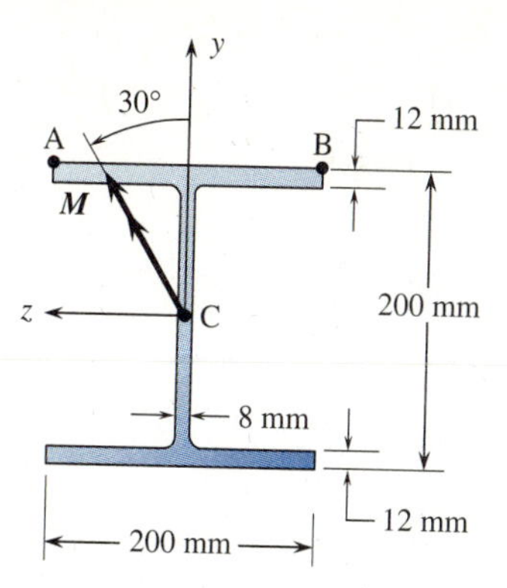

5.132 The couple moment M acts in a vertical plane about the z-axis on the angle cross-section shown. Determine the stress at point A.

Figure P5.132

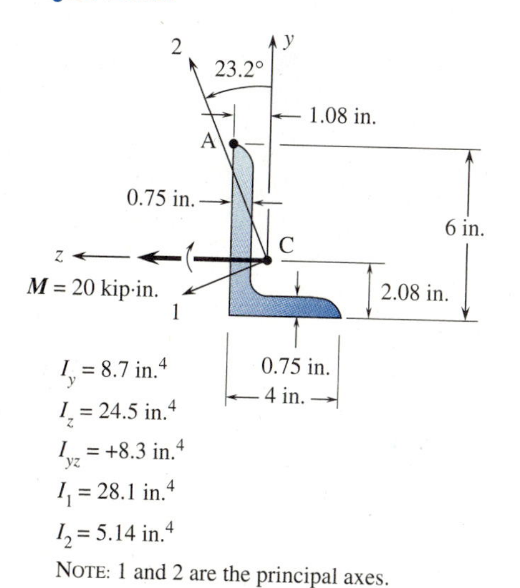

5.133 The cantilevered beam is a W 12 × 16 section with length of 10 ft. It supports a slightly inclined load of $P = 500$ lb at the free end. Determine the bending stress at point A for angle $\theta = 30°$.

Figure P5.133

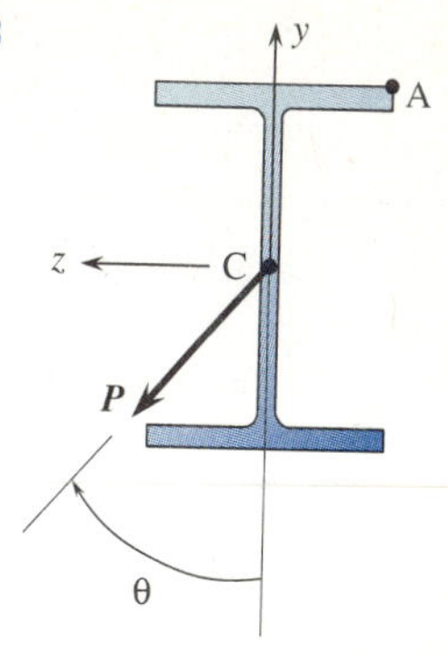

5.134 A simply supported beam has the cross-section and is loaded as shown. If the maximum bending stress is not to exceed 22 ksi, determine the allowable value of P.

Figure P5.134

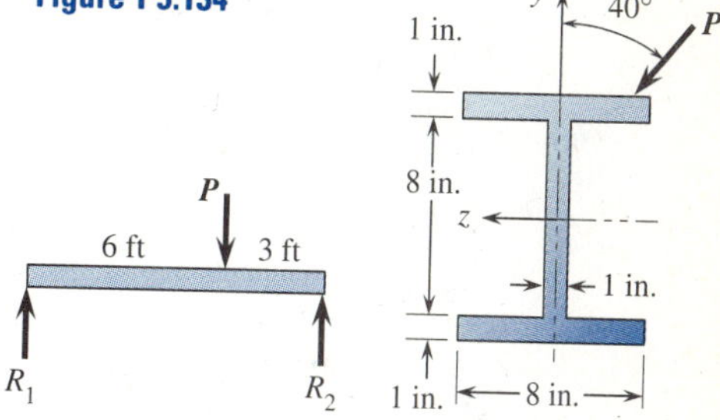

5.135 The T-beam shown is used as a simple beam 15 ft long. It supports a concentrated load inclined at 60° to the y-axis. The centroidal z-axis is 3.07 in. below the top of the section. Also $I_z = 112.6$ in.^{4} and $I_y = 18.7$ in.^{4} The allowable bending stress is 18 ksi. Determine the maximum load that can be supported.

Figure P5.135

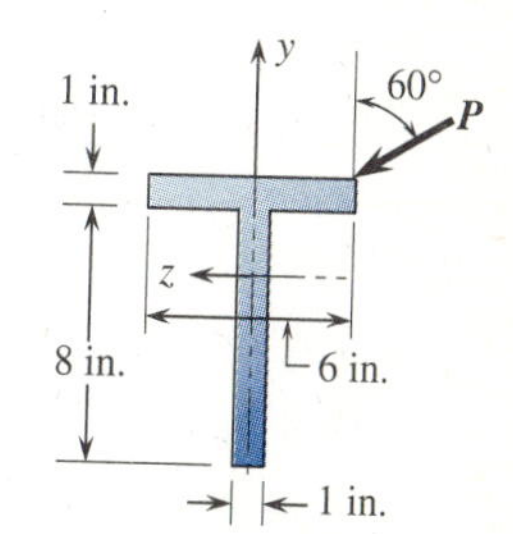

Section 5.11

5.136–5.141 Determine the principal moments of inertia for the cross-sections shown.

Figure P5.136

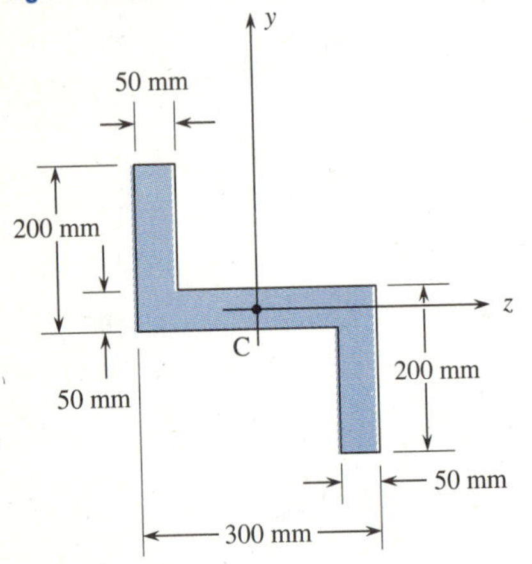

Figure P5.137

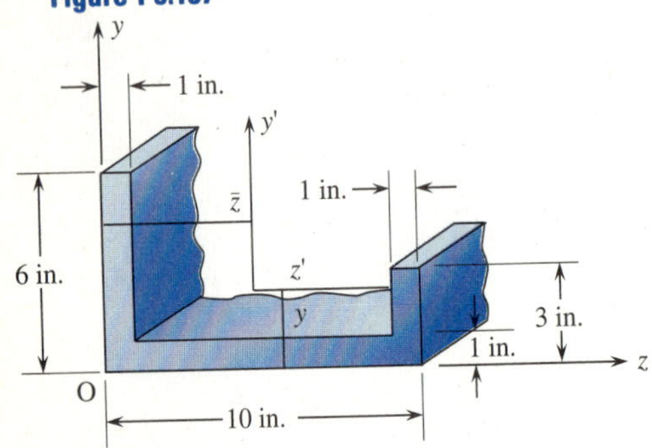

Figure P5.138

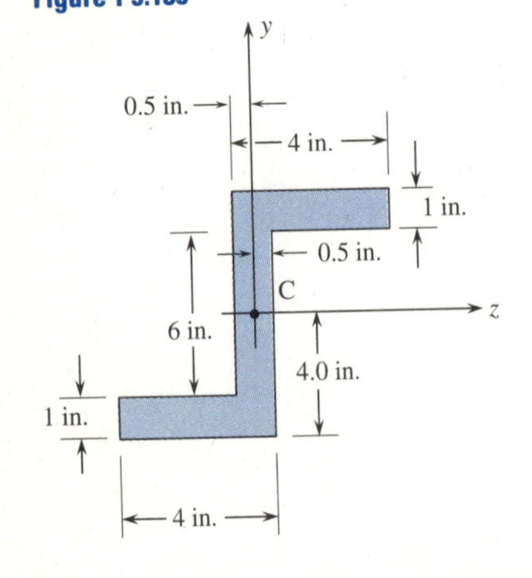

Figure P5.139

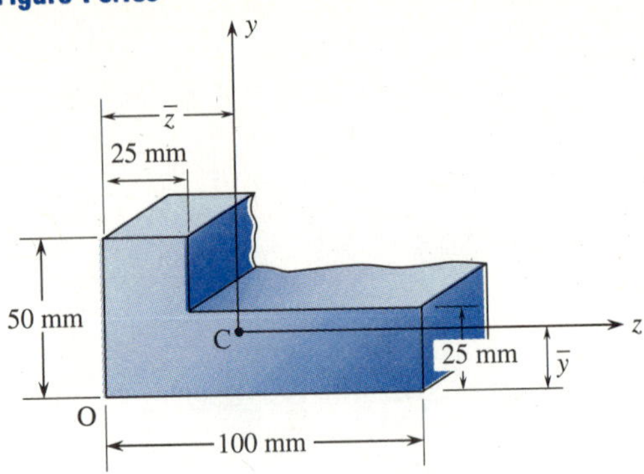

Figure P5.140

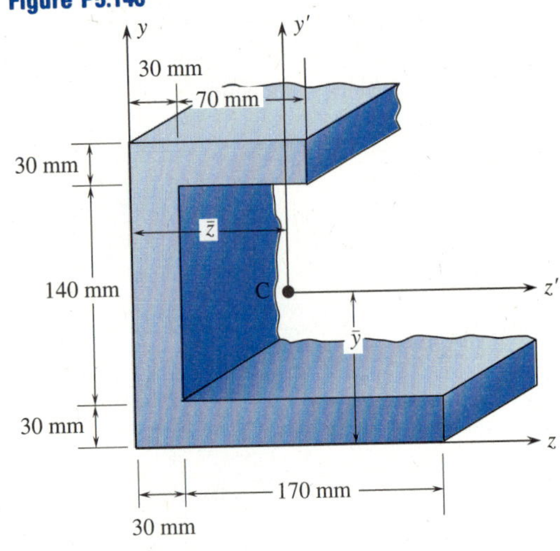

Figure P5.141

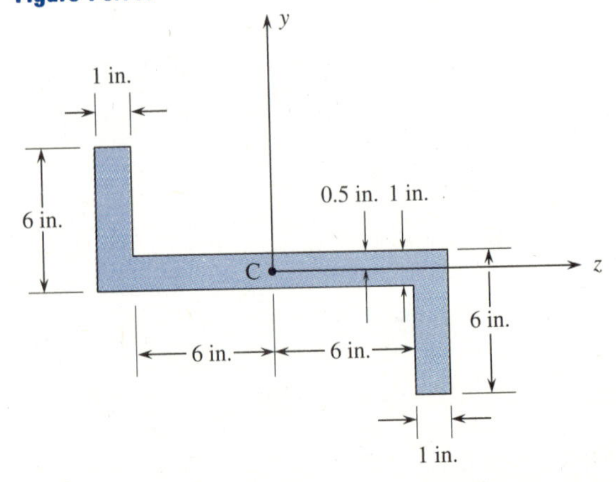

Section 5.12

5.142–5.147 Determine the location of the shear center SC of the thin-walled beams of uniform thickness having the cross-sections as shown.

Figure P5.142

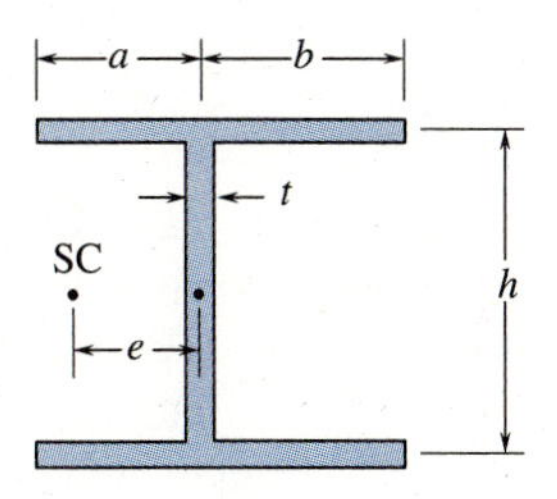

Figure P5.143

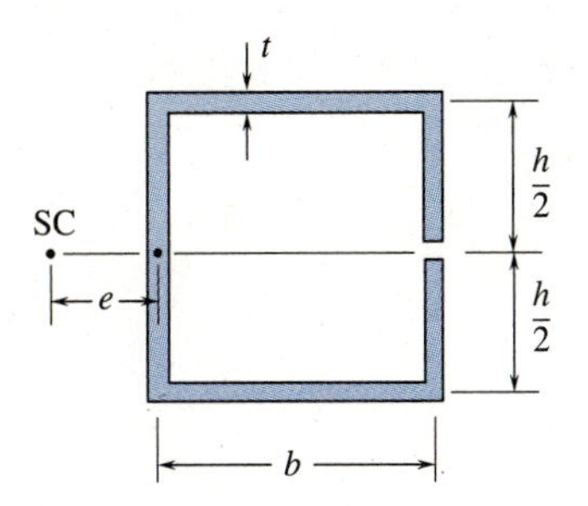

Figure P5.144

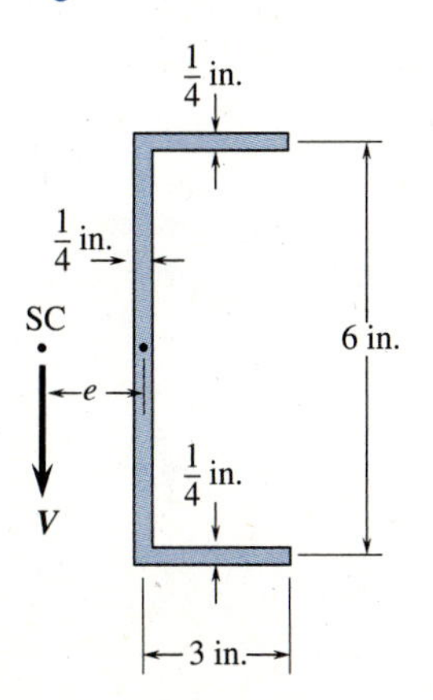

Figure P5.145

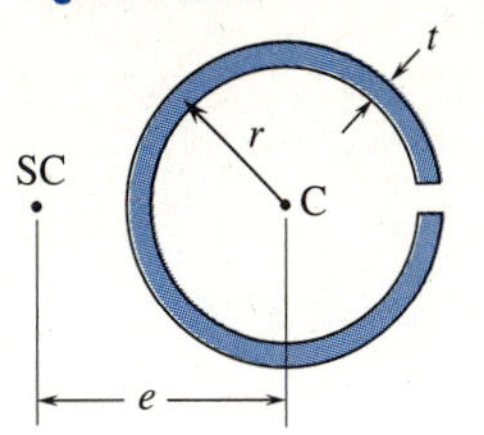

Figure P5.146

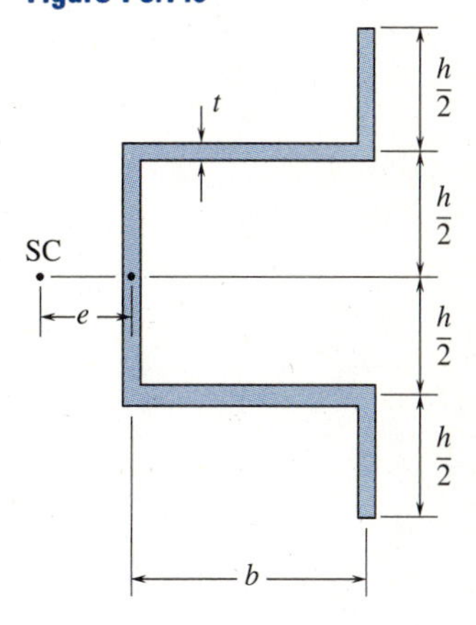

Figure P5.147

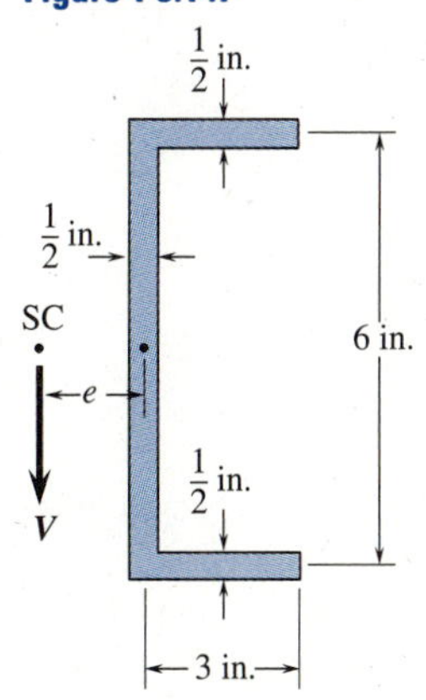

5.148 and 5.149 For the extruded beam having cross-sectional dimensions shown, determine **(a)** the location of the shear center SC and **(b)** the distribution of the shear stresses due to the 10 kN vertical shear force V applied at the shear center SC.

Figure P5.148

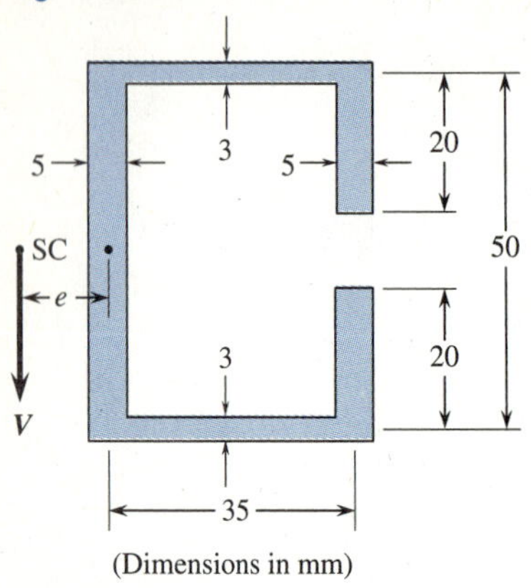

Figure P5.149

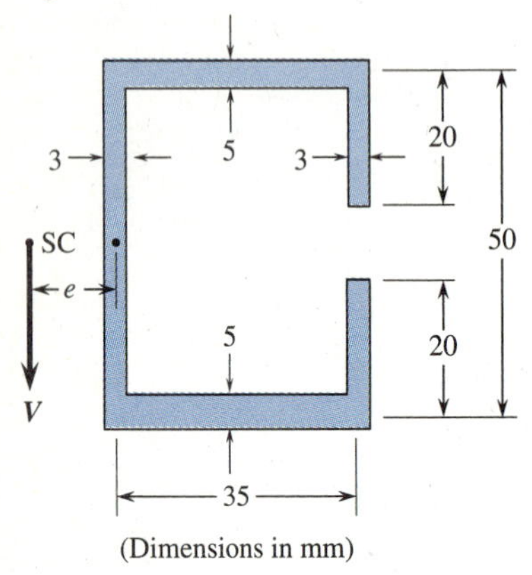

COMPUTER PROBLEMS

The following problems are suggested to be solved using a programmable calculator, microcomputer, or mainframe computer. It is suggested that you write a simple computer program in BASIC, FORTRAN, PASCAL, or some other language to solve these problems.

C5.1 A simply supported beam of length L is loaded by a downward uniform load of intensity w. Determine the shear force and bending moment in the beam. Start with $x = 0$ from the left end and use increments of, say, $L/10$. The program should input the span length L, the load intensity w (force/length units), and the increment of L for each solution. Check your program using the following data: $L = 20$ ft and $w = 1000$ lb/ft.

Figure C5.1

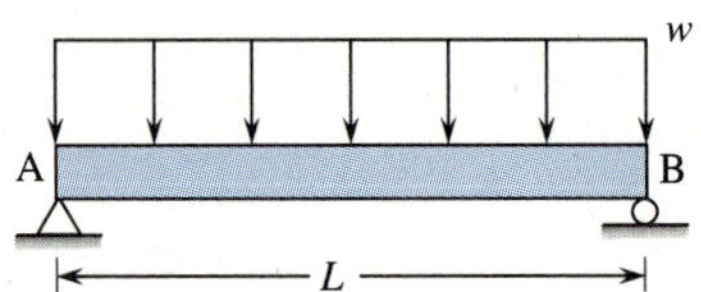

C5.2 A simply supported beam of length L is loaded by a concentrated downward force P at position a from the left end. Determine the shear force and bending moment in the beam. Start with $x = 0$ from the left end and use increments of, say, $L/10$. The program should input the span length L, the concentrated load P and its location, and the increment of L for each solution. Check your program using the following data: $L = 4$ m, $a = 2$ m and $P = 5$ kN.

Figure C5.2

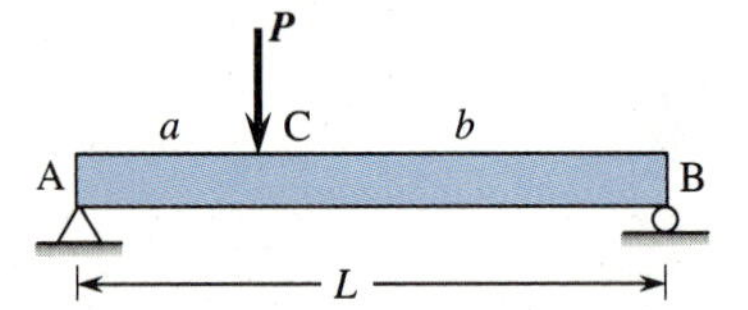

C5.3 A cantilevered beam of length L is subjected to a uniformly applied downward load of intensity w over the whole length. Determine the shear force and bending moment in the beam. Input the beam length L, the load intensity w, and the increment of L. Check your program using the following data: $L = 10$ ft and $w = 500$ lb/ft.

Figure C5.3

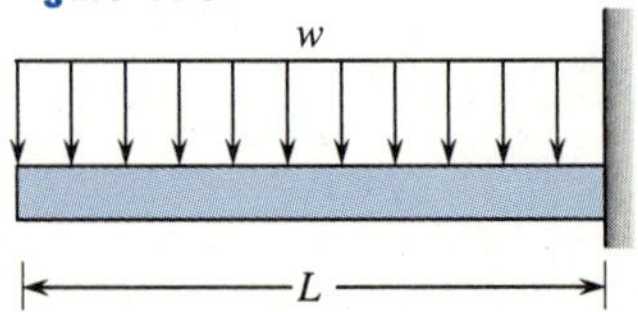

C5.4 Write a program to obtain the centroid of the T-beam cross-section shown and the principal centroidal moments of inertia. The program should be written to obtain the centroidal distance and the moments of inertia for any input values of b, h, t_f, and t_w. Check your program using the following data: $b = 200$ mm, $h = 200$ mm, $t_f = 50$ mm, and $t_w = 50$ mm.

Figure C5.4

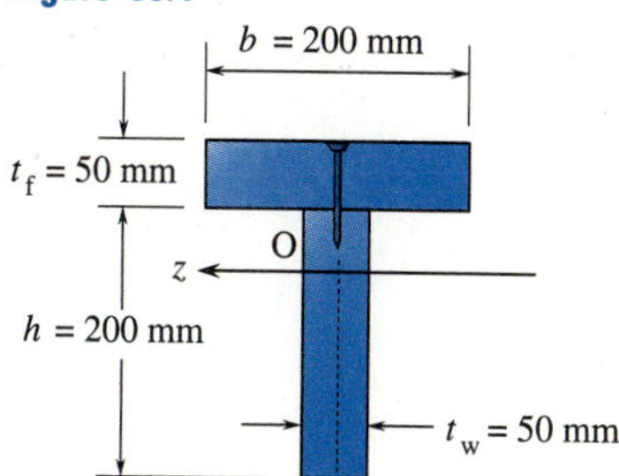

C5.5 A simply supported beam is subjected to a moving load through the carriage shown. Assume the load is equally distributed to each wheel of the carriage. The carriage rolls slowly along the beam. Determine the largest shear force and bending moment in the beam. The program should input the lifted load P, the beam span length L, and the distance d between the wheels of the carriage. Check your program using the following data: $P = 4000$ lb, $L = 20$ ft, and $d = 2$ ft.

Figure C5.5

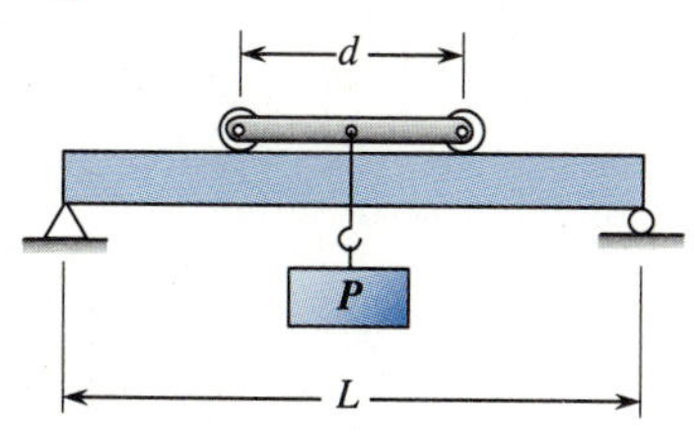

C5.6 Write a program to obtain the maximum bending moment in a simply supported beam subjected to a standard moving truck load shown. The program should input the wheel loads, the wheel load spacing, and the beam span length L. Check your program using the following data: Front wheel load of 8000 lb, rear wheel load of 32,000 lb, wheel spacing of 14 ft, and beam span of 42 ft.

Figure C5.6

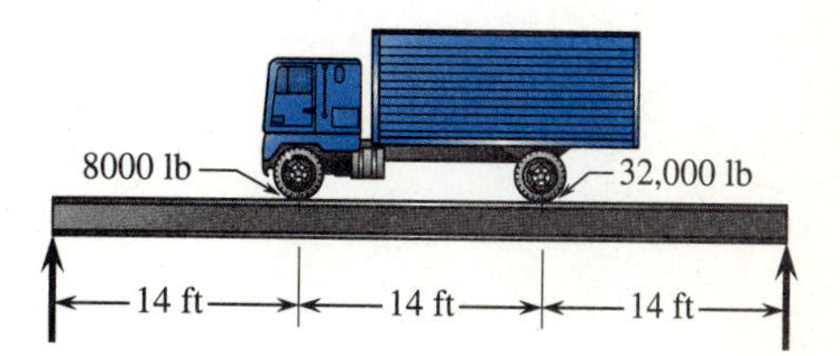

C5.7 Write a program to obtain the principal moments of inertia given the values of the moments of inertia I_y and I_z about the y- and z-axes, and the product of inertia I_{yz}. Also have the program locate the principal axes. Check your program using the following data: $I_z = I_y = 69.73$ in.^{4} and $I_{yz} = 41.63$ in.^{4}

Figure C5.7

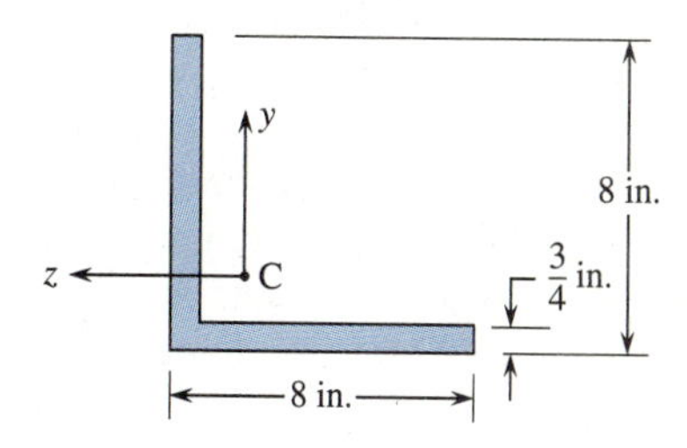

6 Beam Deflections

6.1 INTRODUCTION

In Chapter 5, we determined the bending and shear stresses in a beam. We then selected member sizes based on the allowable bending and shear stresses. The beam was considered safe when these allowable stresses were not exceeded.

In this chapter, we determine the bending deflection in a beam. It is possible that allowable deflections (not stresses) will be the basis for selecting beam sizes. This is often the case for long spanning beams. It is then possible for excessive bending deflections to cause a beam to be unsafe even when the beam stresses are less than the allowable ones. Our concern for deflections is addressed by design codes such as the American Institute for Steel Construction (AISC) code for steel building design [1], and the Crane Manufacturers Association of America (CMAA) code for overhead crane beam design [2]. For example, a simple beam in a building (Figure 6.1(a)) is normally limited to a deflection equal to the beam span length L divided by 360 ($L/360$) to reduce cracking in nonstructural components, such as plaster walls and ceilings. For proper functioning of the traveling overhead trolley, a crane beam in a mill building (Figure 6.1(b)) is often limited by code to a deflection of $L/600$. Also excessive bending deflections of shafts (Figure 6.1(c)) may result in the misalignment of the shaft in the bearings or of misalignment of gears, resulting in wear and possible malfunction.

In a beam the transverse deflections may be caused by both bending and shear. For long, slender beams (those with length much greater than width and depth) the deflection

Figure 6.1 Examples of beams where deflection may be of concern.
(a) beams in a steel-framed building, Indianapolis, IN.,
(b) overhead crane beam, and
(c) drive shaft in a bridge crane that drives the bridge.

(a)

(b)

(c)

due to bending is normally much greater than that due to shear. Since most beams are long and slender, in this chapter we consider only deflections due to bending to be significant.

We will first present the basic method of integration (sometimes referred to as the double-integration method or the method of successive integration) to determine beam deflections for beams subjected to various types of loadings and with different kinds of support conditions.

We then describe the method of superposition, which is a very practical one to obtain deflections in beams subjected to complex loadings that can be separated into simple loadings whose solutions for deflections have been previously tabulated.

Next, we introduce the moment-area method and develop the theorems used with this classic method based on the use of the differential relationship between moment, slope, and deflection. Then we consider the general method of finding deflections, using discontinuity functions. We learn to analyze statically indeterminate beams. Finally, we consider deflections in unsymmetric bending.

6.2 DERIVATION OF BEAM DEFLECTION EQUATION

In this section, we develop the basic beam deflection equation. To develop the deflection equation, we first introduce some basic concepts. These concepts are illustrated by considering the cantilevered beam subjected to the free end or tip load shown in Figure 6.2. The beam is shown in an exaggerated deflected state. The x-axis coincides with the neutral axis before deformation, while $y(x)$ represents the deflection of the neutral axis at any location x. We then define

$$y(x) = \text{deflection} \tag{6.1}$$

Remember that it is the deflection of the neutral axis that is considered in beam deflection analysis.

We now consider the detailed deflected curve (or elastic curve) at location x shown in Figure 6.3. The slope of the deflected curve at x is obtained by drawing a line tangent to the deflection at x.

In Figure 6.3, we observe that the arc length dL is given by the arc length formula as

$$dL = \rho d\theta \tag{a}$$

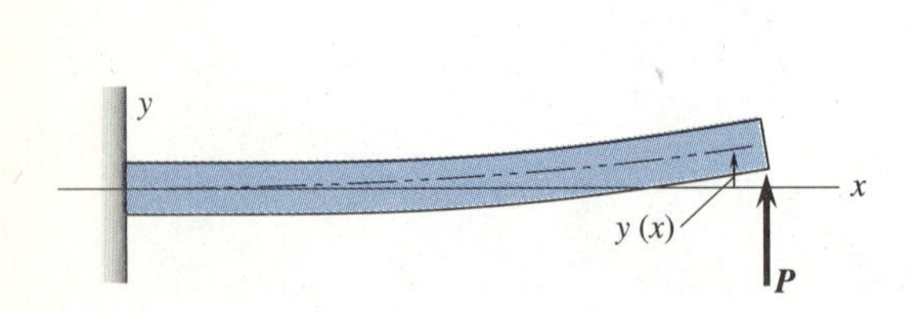

Figure 6.2 Cantilevered beam subjected to free end load.

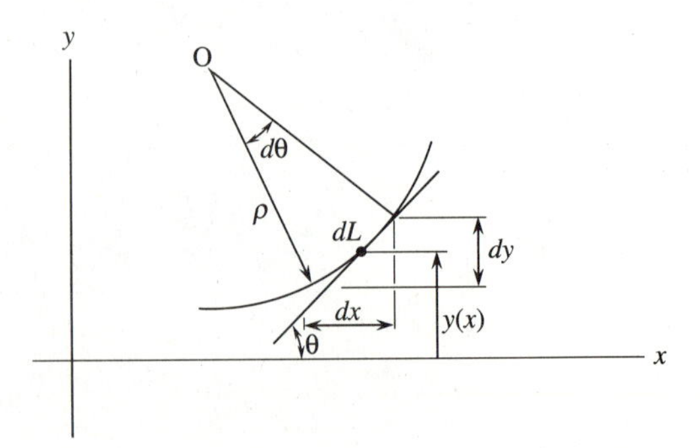

Figure 6.3 Detailed deflected curve.

where ρ is the radius of the curve representing the neutral axis and the arc length is obtained as

$$dL = \sqrt{(dx)^2 + (dy)^2} \tag{b}$$

$$= \sqrt{1 + (dy/dx)^2}\, dx \tag{c}$$

Also the slope θ is given by

$$\tan\theta = \frac{dy}{dx} \tag{d}$$

$$\theta(x) = \tan^{-1}\frac{dy}{dx} \tag{6.2}$$

The slopes in most beams in practical use are small, therefore, we can express Eq. (6.2) as

$$\theta(x) = \frac{dy}{dx} = \text{slope} \tag{6.3}$$

Differentiating Eq. (d) with respect to x, we obtain

$$\sec^2\theta\left(\frac{d\theta}{dx}\right) = \frac{d^2y}{dx^2} \tag{e}$$

Solving Eq. (e) for $d\theta$, we obtain

$$d\theta = \frac{d^2y}{dx^2}\frac{1}{\sec^2\theta}\, dx \tag{f}$$

Now recall the following trigonometric relationship

$$\sec^2\theta = 1 + \tan^2\theta \tag{g}$$

Substituting Eq. (d) into (g) and then into (f), we obtain

$$d\theta = \frac{d^2y/dx^2}{1 + (dy/dx)^2}\, dx \tag{h}$$

Next solving Eq. (a) for $d\theta$ and equating to Eq. (h), we have

$$d\theta = \frac{dL}{\rho} = \frac{d^2y/dx^2}{1 + (dy/dx)^2}\, dx \tag{i}$$

Solving Eq. (i) for $1/\rho$ and substituting Eq. (c) for dL into (i), yields

$$\frac{1}{\rho} = \frac{d\theta}{dL} = \frac{\dfrac{d^2y}{dx^2}}{\left[1 + \left(\dfrac{dy}{dx}\right)^2\right]^{3/2}} = \text{curvature} \tag{6.4}$$

Equation (6.4) relates the curvature of a line $\left(\text{defined by } \kappa = \dfrac{1}{\rho}\right)$, such as the neutral axis of a beam, to the deflection y.

For initially straight beams with small slopes, $\theta(= dy/dx)$ is small and hence $(dy/dx)^2$ is very small so that the denominator of Eq. (6.4) approaches unity. We then use the simpler approximate expression of Eq. (6.4) for the curvature given by

$$\kappa = \frac{1}{\rho} = \frac{d^2y}{dx^2} = \frac{d}{dx}(\theta) = \text{curvature} \tag{6.5}$$

Equation (6.5) indicates that the curvature κ is equal to the rate of change of the slope.

We now relate the bending moment to the deflection as follows. Recall from Eq. (5.6e) that the radius of curvature is equal to the bending stiffness, EI, divided by the moment (M) given by

$$\frac{1}{\rho} = \frac{M}{EI} \tag{6.6}$$

Equating Eqs. (6.5) and (6.6) yields the relationship between the bending moment and the deflection as

$$\kappa = \frac{d^2y}{dx^2} = \frac{M}{EI} \tag{6.7}$$

Equation (6.7) is the governing differential equation for the deflected or elastic curve of a beam and is often called the *moment-curvature* relationship.

The relationship between shear force and deflection is obtained by solving Eq. (6.7) for M and then differentiating with respect to x as follows.

$$\frac{dM}{dx} = \frac{d}{dx}\left[EI\frac{d^2y}{dx^2}\right] \tag{6.8}$$

Using Eq. (5.2b) $\left(V = \frac{dM}{dx}\right)$ in Eq. (6.8), we obtain

$$V = \frac{d}{dx}\left[EI\frac{d^2y}{dx^2}\right] \tag{6.9}$$

We now differentiate Eq. (6.9) and use Eq. (5.1a) $\left(w = -\frac{dV}{dx}\right)$ to obtain the relationship between distributed load and deflection as

$$w = -\frac{d^2}{dx^2}\left[EI\frac{d^2y}{dx^2}\right] \tag{6.10}$$

For constant *bending stiffness, EI,* which is often the case, Eqs. (6.9) and (6.10) become

$$V = EI\frac{d^3y}{dx^3} \tag{6.11}$$

and

$$w = -EI\frac{d^4y}{dx^4} \tag{6.12}$$

Equations (6.1), (6.3), (6.7), (6.11), and (6.12) yield a useful set of equations relating deflection, y, and derivatives of y to bending moment, shear force, and distributed load. In the derivation of these equations, we have assumed small slopes and small deflections.

TABLE 6.1 RELATIONSHIPS AMONG w, M, V, θ, AND y, AND THEIR SIGN CONVENTIONS

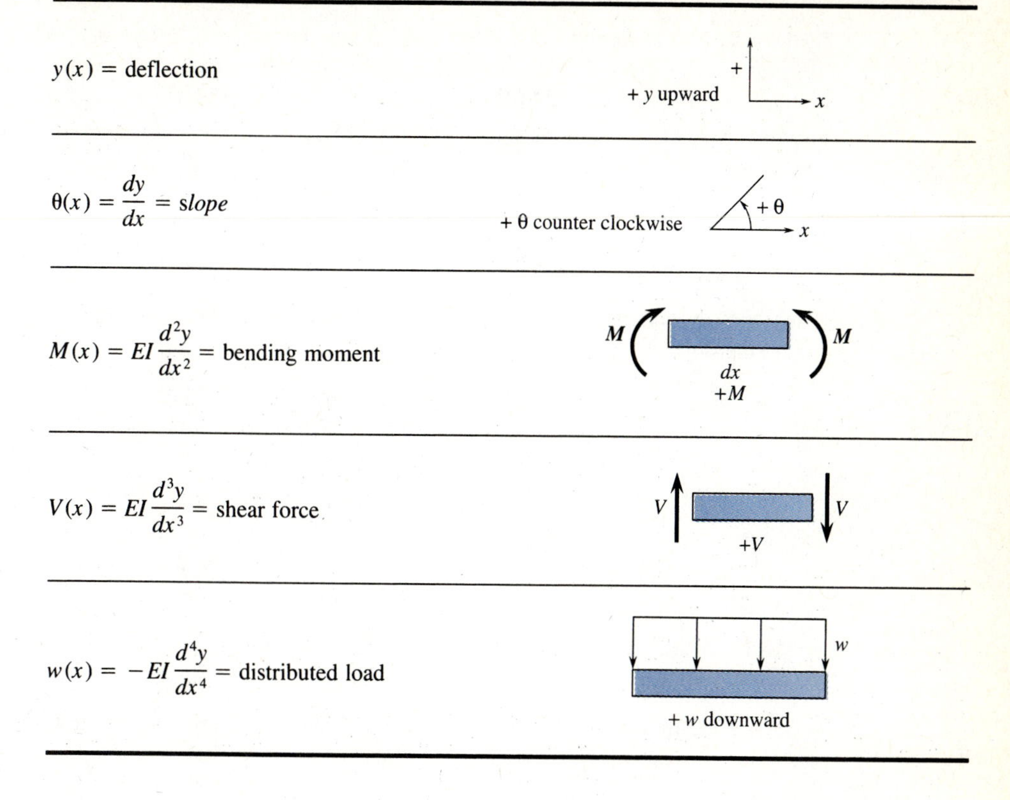

Relationship	Sign convention
$y(x) =$ deflection	$+y$ upward
$\theta(x) = \dfrac{dy}{dx} =$ *slope*	$+\theta$ counter clockwise
$M(x) = EI\dfrac{d^2y}{dx^2} =$ bending moment	$+M$
$V(x) = EI\dfrac{d^3y}{dx^3} =$ shear force	$+V$
$w(x) = -EI\dfrac{d^4y}{dx^4} =$ distributed load	$+w$ downward

These two assumptions are most common in the majority of practical load-bearing structures serving as beams. The relationships developed in this section are now summarized in Table 6.1.

From Table 6.1, we observe that if we know the deflection $y(x)$, for example, then successive differentiation would easily yield θ, M, V, and w. In practice, however, we often determine the moment function, as we did previously in Chapter 5 to aid in drawing the moment diagram. We then successively integrate to obtain the slope and deflection. The relationships from Table 6.1 are the basis for the methods of deflection analysis, called integration, described in Section 6.3, and moment-area, described in Section 6.5.

We should remember that the deflection equations developed in this section are valid for beams

1. with *small deflections* compared to the geometry of the beam (see Fig. 6.3 and derivation of Eq. (6.4))
2. with *small* slopes (Eq. (6.5))
3. that are *long* compared to the cross-sectional dimensions, that is, $L >> b$ or h (b and h representing base and height of a rectangular cross-section)
4. where the deflection due to bending moment M dominates (the deflection due to shear force V has not been considered). The influence of the shear force on the deflection is considered in Chapter 10, on energy methods.

6.3 INTEGRATION METHOD FOR DEFLECTION

In the *integration method* of deflection analysis (sometimes referred to as the *double-integration* method or the method of *successive integration*), we begin with a known moment function for a beam and then integrate once to obtain the slope and a second time to obtain the deflection. That is, we begin with Eq. (6.7) and integrate once to obtain the slope expression as

$$\frac{dy}{dx} = y'(x) = \theta(x) = \int_0^x \frac{M(x)}{EI}\,dx + C_1 \tag{6.13}$$

where the prime denotes differentiation with respect to x. We then integrate Eq. (6.13) to obtain the deflection expression as

$$y(x) = \int_0^x \left[\int_0^x \frac{M(x)}{EI}\,dx \right] dx + C_1 x + C_2 \tag{6.14}$$

In Eqs. (6.13) and (6.14), C_1 and C_2 are constants of integration, to be found by applying boundary conditions on the deflection and possibly slope of the beam based on support conditions, and possibly continuity conditions on deflection and slope within the beam.

Boundary Conditions

We now describe the boundary conditions for the common beam support conditions shown in Figure 6.4. For the clamped or fixed support, shown in Figure 6.4(a), both the deflection, y, and slope $y'(=\theta)$ are zero at the support. That is,

$$y = 0 \qquad y' = 0 \qquad \text{at fixed support.} \tag{6.15a}$$

For the roller (Figure 6.4(b)), and pin (or hinge) support (Figure 6.4(c)), the deflection is zero but the slope is generally not zero. That is,

$$y = 0 \qquad \text{at roller and pin support.} \tag{6.15b}$$

Continuity Conditions

Continuity conditions ensure that the beam remains together, that is, the beam does not separate or overlap at all positions along the beam. There are two common cases to be considered.

1. At a point on a *continuous* beam where a concentrated load exists, there is a deflection expression y_1 to the left of the load and another deflection expression y_2 to the right of the load. For the beam to have continuity, we must have both y_1 equal to y_2 and y_1' equal to y_2' at the load. This is shown in Figure 6.5(a). That is

$$\left.\begin{aligned} y_1(a) &= y_2(a) \\ y_1'(a) &= y_2'(a) \end{aligned}\right\} \quad \text{continuity conditions continuous beam} \tag{6.16a}$$

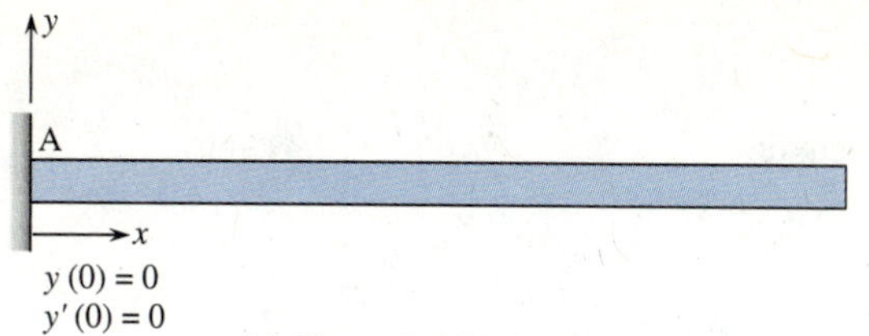

(**a**) Clamped or fixed support

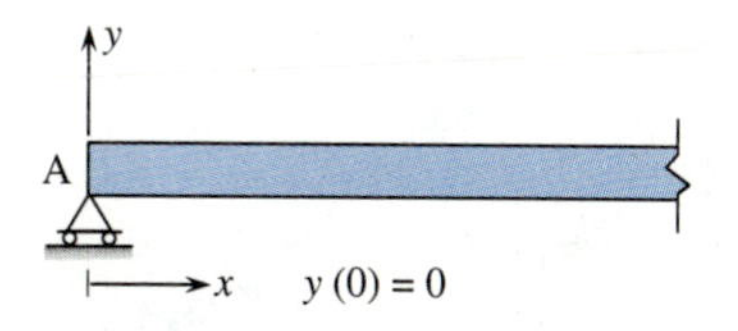

(**b**) Roller support

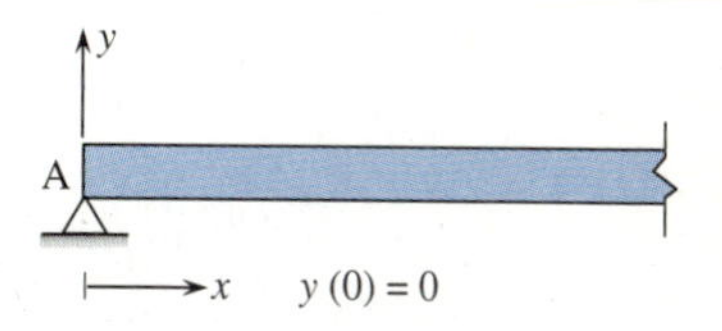

(**c**) Pin or hinge support

Figure 6.4 Common beam support boundary conditions.

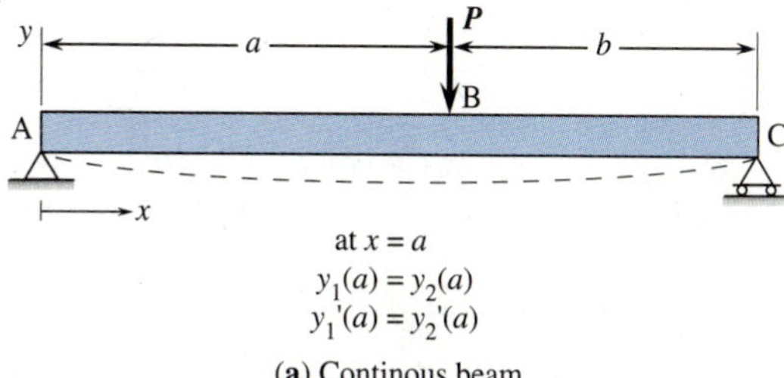

at $x = a$

$y_1(a) = y_2(a)$

$y_1'(a) = y_2'(a)$

(**a**) Continous beam

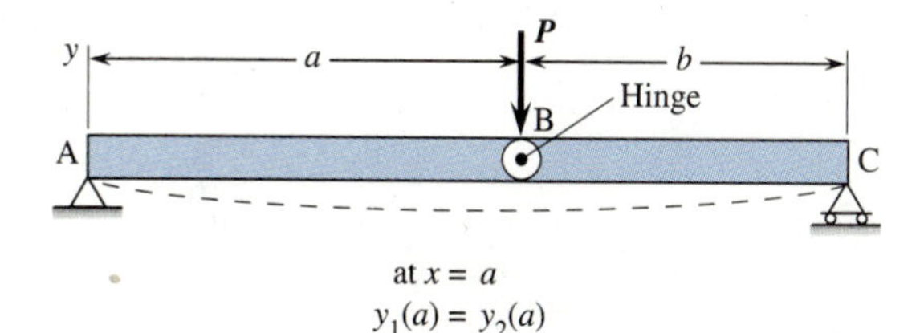

at $x = a$

$y_1(a) = y_2(a)$

(**b**) Internal hinge connection

(**c**) Actual pin connection in bridge beam

Figure 6.5 Common continuity conditions for beams.
(a) continuous beam,
(b) internal hinge or pin connection, and
(c) actual pin connection in bridge beam.

2. At a point on a beam with an internal hinge connection, the deflections y_1 and y_2 must be equal at the hinge. However, the slopes y_1' and y_2' evaluated at the hinge need not be equal for continuity of the beam. This is shown in Figures 6.5(b) and (c). That is

$$y_1(a) = y_2(a) \quad \text{Internal hinge} \tag{6.16b}$$

STEPS IN SOLUTION

The following steps are often used in solving for beam deflections using the integration method.

1. Evaluate the moment expression(s), $M(x)$, for the beam as you did in Chapter 5 to obtain the moment diagram. Remember there may be more than one moment expression.
2. Use the moment expression in Eqs. (6.13) and (6.14). Perform the explicit integrations to obtain expressions for slope and deflection. These expressions include constants of integration. You may instead choose to begin with the basic differential Equation (6.7) and substitute the moment expression into this equation.
3. Apply appropriate boundary and/or continuity conditions, given by Eqs. (6.15) and (6.16), to evaluate the constants of integration and hence obtain the explicit slope and deflection equations.

We now illustrate the integration method for some commonly encountered beam loadings and support conditions.

EXAMPLE 6.1

Determine the deflection expression for the cantilevered beam subjected to the vertical tip load P shown in Figure 6.6. The beam has a constant bending stiffness, EI. After obtaining the deflection expression in variable form, evaluate the numerically maximum downward deflection for a beam with $E = 30 \times 10^6$ psi, length $L = 10$ ft, rectangular cross-section of width $b = 4$ in. and height $h = 8$ in., and $P = 5000$ lb.

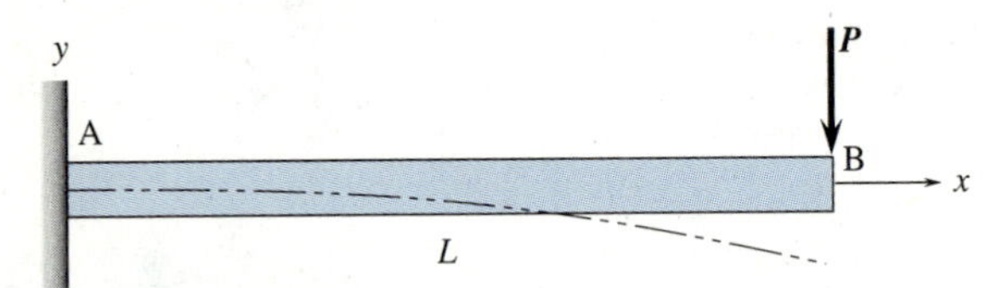

Figure 6.6 Cantilevered beam subjected to end load.

Solution

First draw the free-body diagram of the beam. Since we are starting from the fixed end, we must first determine the reactions at the cantilevered end as follows.

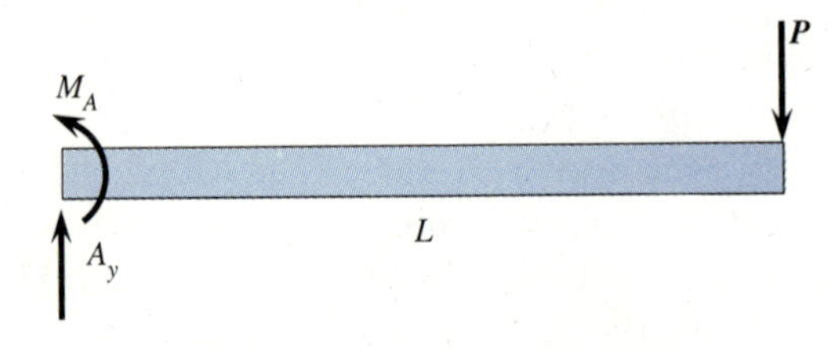

$$\Sigma F_y = 0: \qquad A_y = P$$

$$\Sigma M_A = 0: \qquad M_A = PL$$

Next we cut the beam at distance x from the left end and determine the bending moment at x.

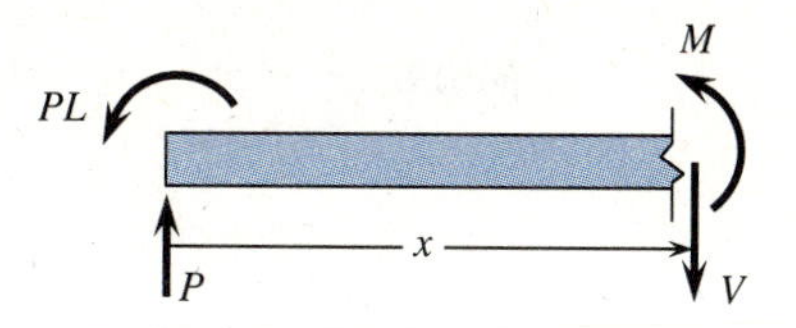

$$\Sigma M_{\text{cut}} = 0: \quad M + PL - Px = 0$$

$$M = -P(L - x) \tag{a}$$

Using Eq. (a) in Eq. (6.7), we have

$$\frac{d^2y}{dx^2} = \frac{-P(L - x)}{EI} \tag{6.17}$$

Integrating Eq. (6.17) with respect to x yields

$$\frac{dy}{dx} = \int_0^x \frac{-P(L - x)}{EI}\, dx + C_1 \tag{6.18}$$

where C_1 is a constant of integration. The specific integration of Eq. (6.18) yields the general slope equation as

$$y'(x) = \frac{-P}{EI}\left(Lx - \frac{1}{2}x^2\right) + C_1 \tag{6.19}$$

To obtain the general deflection equation, we now integrate Eq. (6.19) to yield

$$y(x) = \int_0^x \left[\frac{-P}{EI}\left(Lx - \frac{1}{2}x^2\right) + C_1\right] dx + C_2 \tag{6.20}$$

The specific integration of Eq. (6.20) yields

$$y(x) = \frac{-P}{EI}\left(\frac{Lx^2}{2} - \frac{x^3}{6}\right) + C_1x + C_2 \tag{6.21}$$

The constants of integration C_1 and C_2 resulting from the indefinite integrals are now evaluated on applying the boundary conditions for the fixed support given by Eq. (6.15). That is

$$y'(x = 0) = 0 \quad \text{and} \quad y(x = 0) = 0$$

Starting with $y'(x = 0) = 0$ in Eq. (6.19), we obtain

$$y'(0) = 0 = 0 + C_1$$

or

$$C_1 = 0 \tag{6.22}$$

Next, using $y(x = 0) = 0$ in Eq. (6.21), we obtain

$$y(0) = 0 = 0 + C_1(0) + C_2$$

or

$$C_2 = 0 \tag{6.23}$$

Substituting Eqs. (6.22) and (6.23) for C_1 and C_2 into Eqs. (6.19) and (6.20), we obtain the equations for the slope and deflection of the cantilevered beam as

$$y'(x) = \frac{-Px}{EI}\left(L - \frac{x}{2}\right) \tag{6.24}$$

and

$$y(x) = \frac{-Px^2}{2EI}\left(L - \frac{x}{3}\right) \tag{6.25}$$

The maximum slope and deflection both occur at the free end, $x = L$. Therefore, evaluating Eqs. (6.24) and (6.25) at $x = L$, we obtain

$$y'(L) = \frac{-PL^2}{2EI} \tag{6.26}$$

and

$$y(L) = \frac{-PL^3}{3EI} \tag{6.27}$$

The negative signs indicate clockwise rotation (or slope) and downward deflection. These signs are consistent with our previously stated sign conventions for slope and deflection. Equations (6.26) and (6.27) correspond to load case 12 in Appendix D.

Now substituting the numeric values into Eqs. (6.26) and (6.27), we obtain the maximum slope and deflection. First we evaluate I for the rectangular cross-section (see Appendix A) as

$$\begin{aligned} I &= \frac{1}{12} bh^3 \\ &= \frac{1}{12}(4)(8)^3 \\ &= 171 \text{ in.}^4 \end{aligned}$$

The maximum slope is given by

$$\begin{aligned} y'(L) &= \frac{-PL^2}{2EI} \\ &= \frac{-(5000 \text{ lb})(10 \text{ ft.} \times 12 \text{ in./ft})^2}{2(30 \times 10^6 \text{ psi})(171 \text{ in.}^4)} \\ &= -0.00702 \text{ rad} \\ &= -0.402^\circ \text{ (clockwise)} \end{aligned}$$

The maximum deflection is

$$\begin{aligned} y(L) &= \frac{-PL^3}{3EI} \\ &= \frac{-(5000 \text{ lb})(10 \text{ ft} \times 12 \text{ in./ft})^3}{3(30 \times 10^6 \text{ psi})(171 \text{ in.}^4)} \\ &= -0.561 \text{ in. (downward)} \end{aligned}$$

EXAMPLE 6.2

Determine the slope and deflection equations for the solid circular cross-section cantilevered beam subjected to the uniformly distributed load shown in Figure 6.7. The beam has a constant bending stiffness, EI. After obtaining the slope and deflection equations in variable form let $w = 2000$ N/m, $L = 5$ m, $r = 100$ mm, and determine the numeric slope and deflection at the free end.

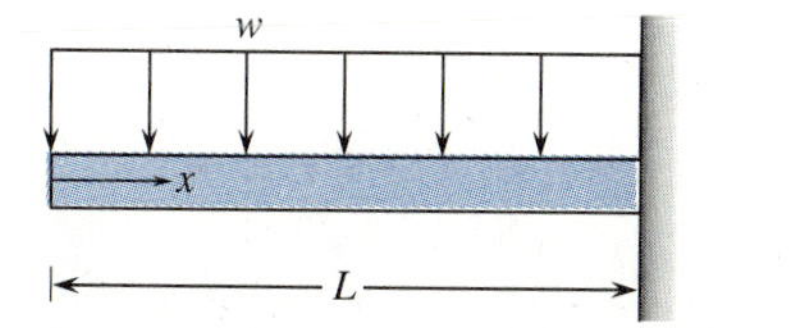

Figure 6.7 Cantilevered beam subjected to uniformly distributed loading and its circular cross-section.

Solution

We begin by considering a section of the beam and draw its free-body diagram as shown below.

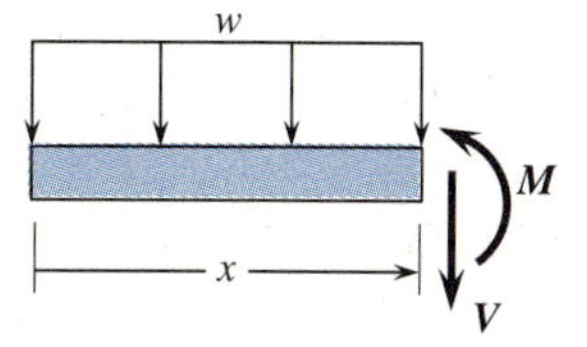

From the free-body diagram of the section, we obtain the bending moment M as

$$M = -(wx)\frac{x}{2} = -w\frac{x^2}{2} \tag{a}$$

Now use Eq. (a) for M in Eq. (6.13) to obtain the slope as

$$y' = \int_0^x -\frac{wx^2}{2EI}\,dx + C_1 \tag{6.28}$$

or on explicitly integrating Eq. (6.28), we obtain

$$y' = -\frac{wx^3}{6EI} + C_1 \tag{6.29}$$

Now integrating Eq. (6.29), we obtain the deflection as

$$y = \int_0^x \left(\frac{-wx^3}{6EI} + C_1\right) dx \tag{6.30}$$

Simplifying Eq. (6.30), we obtain

$$y = -\frac{wx^4}{24EI} + C_1x + C_2 \tag{6.31}$$

Next we impose the boundary condition Eq. (6.15a) for the fixed support. Starting with $y'(L) = 0$ in Eq. (6.29), we have

$$y'(L) = 0 = -\frac{wL^3}{6EI} + C_1$$

or

$$C_1 = \frac{wL^3}{6EI} \tag{6.32}$$

Now using $y(L) = 0$ in Eq. (6.31) along with C_1 from Eq. (6.32), we obtain

$$y(L) = 0 = -\frac{wL^4}{24EI} + \frac{wL^3}{6EI}L + C_2$$

or

$$C_2 = -\frac{wL^4}{8EI} \tag{6.33}$$

Using Eqs. (6.32) and (6.33) for C_1 and C_2 in Eqs. (6.29) and (6.31), the final expressions for the slope and deflection are

$$y' = -\frac{wx^3}{6EI} + \frac{wL^3}{6EI} = \frac{w}{6EI}(-x^3 + L^3) \tag{6.34}$$

and

$$y = -\frac{wx^4}{24EI} + \frac{wL^3}{6EI}x - \frac{wL^4}{8EI}$$

$$= \frac{w}{48EI}(-2x^4 + 8L^3x - 6L^4) \tag{6.35}$$

The maximum slope and deflection again occur at the free end $x = 0$ and are evaluated using $x = 0$ in Eqs. (6.34) and (6.35) as

$$y' = \frac{wL^3}{6EI} \tag{6.36}$$

$$y = -\frac{wL^4}{8EI} \tag{6.37}$$

Equations (6.36) and (6.37) correspond to Case 16 in Appendix D.

Substituting the numeric quantities into Eqs. (6.36) and (6.37), we obtain the maximum slope and deflection. We first calculate I for the solid circular cross-section (see Appendix A) as

$$I = \frac{\pi}{4}r^4 = \frac{\pi}{4}(0.1\text{ m})^4 = 0.0000785\text{ m}^4$$

The slope and deflection are then

$$y' = \frac{wL^3}{6EI} = \frac{(2000\text{ N/m})(5\text{ m})^3}{6(200 \times 10^9\text{ N/m}^2)(0.0000785\text{ m}^4)}$$

$$= 0.00265\text{ rad (counterclockwise)}$$

and

$$y = -\frac{wL^4}{8EI} = -\frac{(2000\ \text{N/m})(5\ \text{m})^4}{8(200 \times 10^9\ \text{N/m}^2)(0.0000785\ \text{m}^4)}$$

$$= -9.95 \times 10^{-3}\ \text{m}$$

$$= -9.95\ \text{mm (downward)}$$

EXAMPLE 6.3

Determine the maximum deflection in variable form for the simply supported beam subjected to the uniformly distributed load shown in Figure 6.8. Assume constant EI. Then let the beam be an American Society for Testing Materials (ASTM) A36 structural steel W 10 × 45 shape, with $w = 1000$ lb/ft and $L = 25$ ft, and determine the numerically maximum deflection. Is this deflection less than $L/360$, a criterion often used in steel building design for allowable deflection?

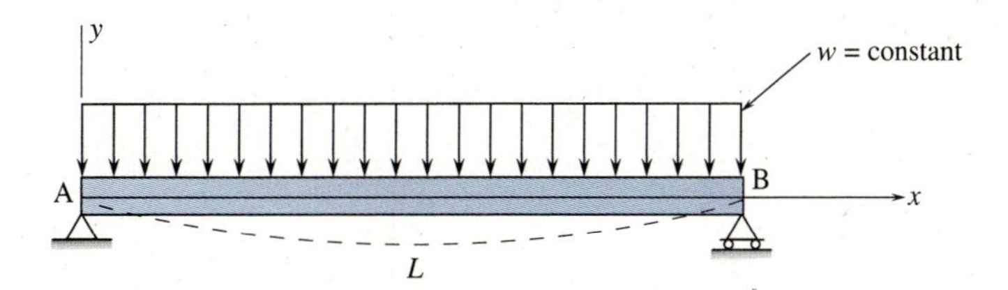

Figure 6.8 Simple beam subjected to uniformly distributed load.

Solution

By statics, we find the reaction at the left end as

$$A_y = \frac{wL}{2}$$

Then draw a free-body diagram of a section to the left of a cut at x to obtain the bending moment as

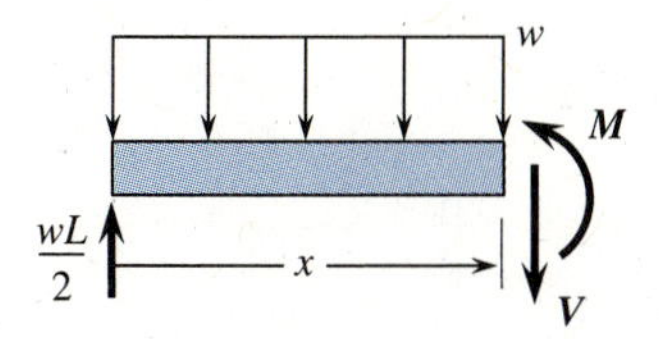

$$M = \left(\frac{wL}{2}\right)x - (wx)\left(\frac{x}{2}\right)$$

$$= \frac{w}{2}(Lx - x^2) \qquad \textbf{(a)}$$

Using Eq. (a) for M in Eq. (6.7), we obtain

$$\frac{d^2y}{dx^2} = \frac{w}{2EI}(Lx - x^2) \tag{6.38}$$

Explicitly integrating Eq. (6.38) once, we obtain the slope as

$$\frac{dy}{dx} = \frac{w}{2EI}\left(\frac{Lx^2}{2} - \frac{x^3}{3}\right) + C_1 \tag{6.39}$$

Integrating Eq. (6.39), we obtain the deflection as

$$y = \frac{w}{2EI}\left(\frac{Lx^3}{6} - \frac{x^4}{12}\right) + C_1x + C_2 \tag{6.40}$$

Now applying boundary condition Eq. (6.15b) for simple supports (the deflections are zero at both ends A and B), we have

$$y(0) = 0 \qquad y(L) = 0 \tag{6.41}$$

Using the first boundary condition of Eq. (6.41) in Eq. (6.40), we obtain

$$y(0) = 0 = C_2 \tag{6.42}$$

Using the second boundary condition of Eq. (6.41) in Eq. (6.40), we obtain

$$y(L) = 0 = \frac{w}{2EI}\left(\frac{L^4}{6} - \frac{L^4}{12}\right) + C_1L$$

or

$$C_1 = -\frac{wL^3}{24EI} \tag{6.43}$$

Substituting Eqs. (6.42) and (6.43) for C_1 and C_2 into Eq. (6.40), the deflection expression becomes

$$y(x) = \frac{w}{12EI}\left(Lx^3 - \frac{1}{2}x^4 - \frac{1}{2}L^3x\right) \tag{6.44}$$

The maximum deflection occurs at the midspan of the beam due to the uniform load over the whole span. In general, the maximum deflection of a simply supported beam will occur where the slope equals zero. Therefore, substituting Eq. (6.43) for C_1 into Eq. (6.39), setting the equation to zero, and solving for x, we obtain

$$0 = \frac{w}{2EI}\left(\frac{Lx^2}{2} - \frac{x^3}{3} - \frac{L^3}{12}\right)$$

or

$$x = \frac{L}{2}$$

Therefore, we conclude that the maximum deflection occurs at the midspan. The maximum deflection is obtained by evaluating Eq. (6.44) at $x = L/2$ as

$$y_{max} = y\left(\frac{L}{2}\right) = -\frac{5}{384}\frac{wL^4}{EI} \tag{6.45}$$

Equations (6.44) and (6.45) are found in Case 4 in Appendix D.

To determine the numerically maximum deflection, we substitute $w = 1000$ lb/ft, $L = 25$ ft, $E = 29 \times 10^6$ psi (see Appendix B for ASTM A36 steel), and $I = 248$ in.^{4} (see Appendix C for wide flange properties) into Eq. (6.45). This yields

$$y_{max} = -\frac{5\,(1000\text{ lb/ft})(1\text{ ft}/12\text{ in.})(25\text{ ft} \times 12\text{ in./ft})^4}{384\,(29 \times 10^6\text{ lb/in.}^2)(248\text{ in.}^4)}$$

$$= -1.22\text{ in. (downward)}$$

Now

$$\frac{L}{360} = \frac{(25\text{ ft} \times 12\text{ in./ft})}{360} = 0.833\text{ in.}$$

Based on this allowable deflection, the beam deflection is too large. We need to use a bigger section (one with a larger I) to satisfy this deflection criterion. (You might verify that a W 16 × 31 is satisfactory.)

EXAMPLE 6.4

Determine the maximum deflection and its location for the simply supported beam subjected to the concentrated load shown in Figure 6.9. Let EI be constant.

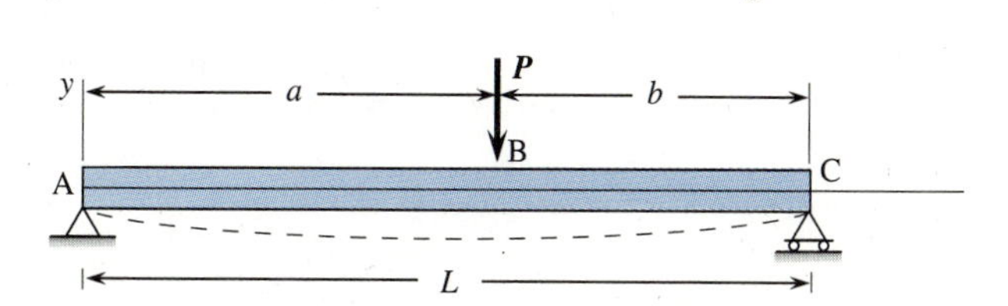

Figure 6.9 Simple beam subjected to concentrated load.

Solution

Using statics, we obtain the left and right support reactions as

$$A_y = P\frac{b}{L} \qquad C_y = P\frac{a}{L}$$

Due to the concentrated load P at B, the bending moment in section AB is different than the bending moment in section BC. To obtain M_{AB}, we cut a section between A and B. From the free-body diagram of section AB, we obtain

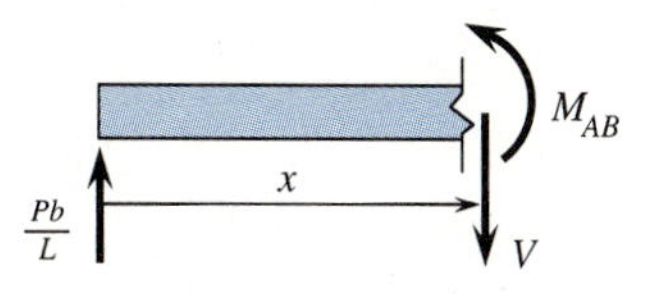

For $0 \leq x \leq a$

$$M_{AB} = \left(P\frac{b}{L}\right)x \qquad \textbf{(a)}$$

To obtain M_{BC}, we cut a section between B and C. From the free-body diagram, we obtain

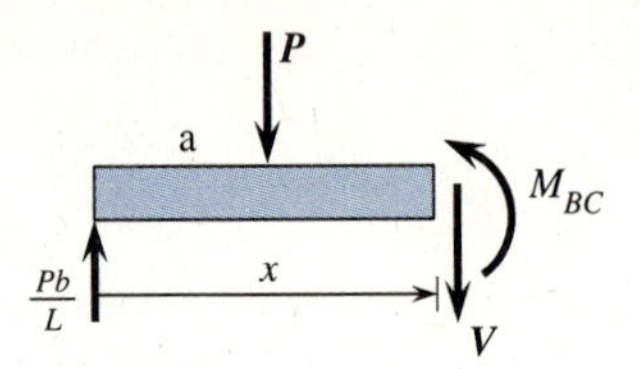

For $a \leq x \leq L$

$$M_{BC} = \left(P\frac{b}{L}\right)x - P(x - a) \tag{b}$$

Next using Eq. (6.7), we have

For section AB

$$\frac{d^2y_1}{dx^2} = \frac{bP}{LEI}x \tag{6.46a}$$

For section BC

$$\frac{d^2y_2}{dx^2} = \frac{P}{EI}\left(\frac{bx}{L} - x + a\right) \tag{6.46b}$$

where y_1 and y_2 are the deflections in sections AB and BC, respectively.

Integrating Eqs. (6.46a) and (6.46b), we obtain

For section AB

$$\frac{dy_1}{dx} = \frac{bP}{2LEI}x^2 + C_1 \tag{6.47a}$$

For section BC

$$\frac{dy_2}{dx} = \frac{P}{EI}\left(\frac{bx^2}{2L} - \frac{x^2}{2} + ax\right) + C_2 \tag{6.47b}$$

Integrating Eqs. (6.47a) and (6.47b), we have

For section AB

$$y_1 = \frac{bP}{6LEI}x^3 + C_1x + C_3 \tag{6.48a}$$

For section BC

$$y_2 = \frac{P}{EI}\left(\frac{bx^3}{6L} - \frac{x^3}{6} + \frac{ax^2}{2}\right) + C_2x + C_4 \tag{6.48b}$$

Now we must have four known conditions to obtain the four constants of integration C_1 through C_4. Because both supports are simple (pin and roller), we use Eq. (6.15b). That is,

$$y_1(0) = 0 \quad \text{and} \quad y_2(L) = 0 \tag{6.49}$$

Using Eq. (6.49), we obtain

$$y_1(0) = 0 = C_3$$

and

$$y_2(L) = 0 = \frac{P}{EI}\left(\frac{bL^2}{6} - \frac{L^3}{6} + \frac{aL^2}{2}\right) + C_2L + C_4$$

$$0 = \frac{P}{EI}\left(\frac{aL^2}{3}\right) + C_2L + C_4 \tag{6.50}$$

where $b = (L - a)$ has been used in the final form, Eq. (6.50).

Now from the continuity conditions, Eq. (6.16a) applies. That is, the slope at point B as given by y_1' must equal the slope at B given by y_2'. Therefore, we equate Eqs. (6.47a) and (6.47b) at the value $x = a$ (location B on the beam) to obtain

$$y_1'(a) = y_2'(a)$$

or

$$\frac{bP}{2LEI}(a^2) + C_1 = \frac{P}{EI}\left(\frac{ba^2}{2L} + \frac{a^2}{2}\right) + C_2 \tag{6.51}$$

Also the deflection at point B as given by both Eqs. (6.48a) or (6.48b) must give the same value. Therefore, equating Eqs. (6.48a) and (6.48b) at the value $x = a$ yields

$$y_1(a) = y_2(a)$$

or

$$\frac{bP}{6LEI}(a^3) + C_1a = \frac{P}{EI}\left(\frac{ba^3}{6L} + \frac{a^3}{3}\right) + C_2a + C_4 \tag{6.52}$$

Equations (6.50), (6.51), and (6.52) can be solved simultaneously for C_1, C_2, and C_4. The result is

$$C_1 = \frac{Pa}{EI}\left(\frac{a}{2} - \frac{L}{3} - \frac{a^2}{6L}\right)$$

$$C_2 = \frac{-Pa}{EI}\left(\frac{L}{3} + \frac{a^2}{6L}\right) \tag{6.53}$$

$$C_4 = \frac{Pa^3}{6EI}$$

Substituting the constants C_1, C_2, C_3, and C_4 into Eqs. (6.48a) and (6.48b), we obtain the deflection equations for both portions of the beam as

For section AB

$$y_1 = \frac{Pb}{6EIL}x^3 + \frac{Pa}{EI}\left(\frac{a}{2} - \frac{L}{3} - \frac{a^2}{6L}\right)x \tag{6.54}$$

or on simplifying Eq. (6.54), we have

$$y_1 = -\frac{Pbx}{6EIL}(L^2 - x^2 - b^2) \tag{6.55}$$

For section BC

$$y_2 = \frac{P}{EI}\left(\frac{bx^3}{6L} - \frac{x^3}{6} + \frac{ax^2}{2}\right) - \frac{Pa}{EI}\left(\frac{L}{3} + \frac{a^2}{6L}\right)x + \frac{Pa^3}{6EI} \tag{6.56}$$

or

$$y_2 = -\frac{Pb}{6EIL}\left[\frac{L}{b}(x-a)^3 + (L^2 - b^2)x - x^3\right] \tag{6.57}$$

Equations (6.55) and (6.57) correspond to Case 2 in Appendix D. For dimension a greater than b, the maximum deflection is located between A and B. This location is determined by setting Eq. (6.47a), for the slope y' (with the known expression for C_1), equal to zero. This yields

$$\frac{Pb}{2EIL}x^2 + \frac{Pa}{EI}\left(\frac{a}{2} - \frac{L}{3} - \frac{a^2}{6L}\right) = 0 \tag{6.58}$$

Solving Eq. (6.58) for x, we obtain

$$x = \left[\frac{2La}{b}\left(\frac{a^2}{6L} + \frac{L}{3} - \frac{a}{2}\right)\right]^{1/2} \tag{6.59}$$

Substituting Eq. (6.59) for x into Eq. (6.55) yields the maximum deflection as

$$y_{\text{max}} = \frac{P}{EI}\left[\frac{a^2}{6L} + \frac{L}{3} - \frac{a}{2}\right]^{3/2} \cdot \left[\frac{b}{6L}\left(\frac{2La}{b}\right)^{3/2} - \left(\frac{2La^3}{b}\right)^{1/2}\right] \tag{6.60}$$

For the special case of the load P located at midspan, $a = L/2$. Then Eq. (6.60) becomes

$$y_{\text{max}} = y\left(\frac{L}{2}\right) = -\frac{PL^3}{48EI} \tag{6.61}$$

Equation (6.61) corresponds to Case 1 in Appendix D.

In summary, the integration method for determining slopes and deflections in beams is a very general, straightforward method. The method provides us with the equations used to obtain the slope and the deflection over the entire beam. Maximum values of slope and deflection or values at any desired location can be evaluated by substituting the appropriate value of x (the axial coordinate) into the general equations.

The disadvantage of the integration method is that as the beam loading becomes more complicated, the evaluation of the integration constants becomes more involved. As illustrated in Example 6.4, when the moment functions have to be written in parts to account for sudden changes in the beam loading, the evaluation of integration constants becomes tedious. Much of this tedium is reduced using discontinuity functions as described in Section 6.6.

Fortunately most of the common loading situations occurring on beams have been analyzed previously and tabulated. Appendix D provides the slope and deflection formulas for many of the more common beam loadings and support conditions.

6.4 BEAM DEFLECTIONS BY SUPERPOSITION

In Section 6.3, we discussed the integration method for determining deflections and slopes in beams. Examples 6.1 through 6.4 illustrated the method. With this method we can determine the deflections for a variety of beam loadings and support conditions encountered in practice. For beams with numerous loadings, the integration method quickly becomes tedious.

Fortunately, most of the common beam loading cases have been previously analyzed and the expressions for deflections and slopes tabulated. Appendix D lists the deflections and slopes for the most common beam loadings and support conditions. Appendix D is most useful and can often considerably reduce our time to determine deflections and slopes for beams subjected to numerous loadings.

The purpose of this section is to illustrate how the superposition method, along with known solutions for deflections from Appendix D, is used to solve for deflections in more complicated beam problems.

The *principle of superposition* applied to deflections in beams is defined as follows.

The total deflection at a point in a beam subjected to several loads is equal to the sum of the individual deflections at that same point due to each load acting independently.

This principle applies whenever the deflections are linear functions of the applied loads. The linear assumption is satisfied whenever the material remains in the linear-elastic portion of the stress-strain curve due to the sum of all loads and for each individual load.

STEPS IN SOLUTION

The following steps are used in solving for beam deflections using the superposition method.

1. Separate the actual loading into individual loads that have known solutions such as those listed in Appendix D.
2. Determine the desired deflection or slope at the common location using each separate deflection or slope expression.
3. Use superposition to obtain the total deflection or slope. That is, add together the separate deflection or slope expressions to obtain the total deflection or slope. Examples 6.5 through 6.8 illustrate the steps in solution using superposition.

EXAMPLE 6.5

(a) Determine by superposition the maximum vertical deflection for the cantilevered beam shown in Figure 6.10 subjected to a uniformly distributed load w and a concentrated end load P. Let the bending stiffness EI be constant.

(b) Then let $w = 50$ lb/ft, $P = 200$ lb, $L = 6$ ft and calculate the numerically maximum downward deflection for a nominal 4 in. $\times$ 8 in. piece of southern pine (construction grade).

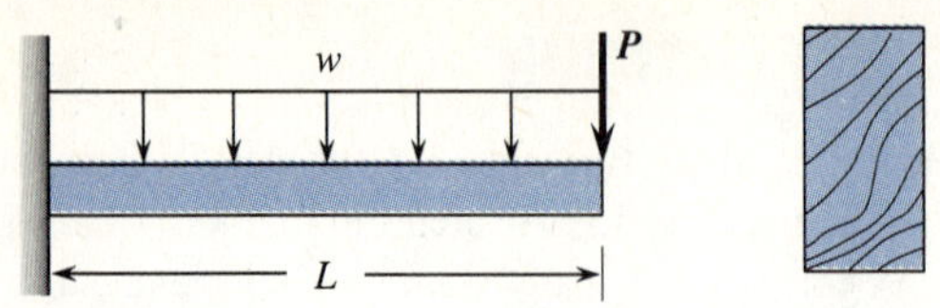

Figure 6.10 Cantilevered beam subjected to combined distributed and concentrated load.

Solution

(a) To use superposition, we consider the beam under separate loads that have known solutions. In this case, we use the cantilevered beam subjected to the separate loads of (1) a uniformly distributed load w (see Case 16 in Appendix D) and (2) a concentrated end load P (see Case 12 in Appendix D) (Figure 6.11).

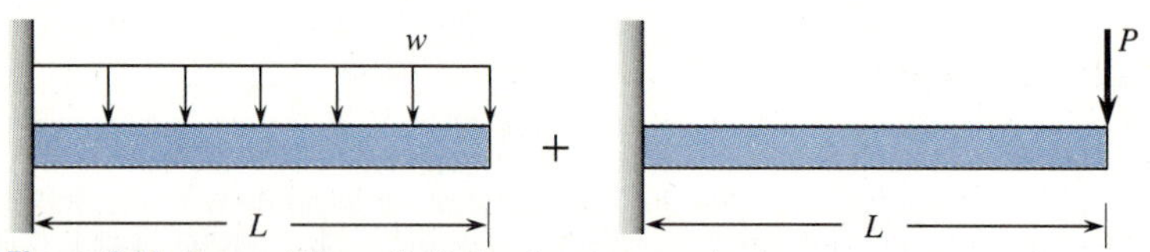

Figure 6.11 Beam of Figure 6.10 showing separate loads.

The maximum deflection is at the free end. Therefore, for the cantilevered beam subjected to the uniform load w, from Appendix D, Case 16, we have

$$y_1 = -\frac{wL^4}{8EI} \tag{a}$$

For the cantilevered beam subjected to the end load P, from Appendix D, Case 12, we have

$$y_2 = -\frac{PL^3}{3EI} \tag{b}$$

By superposition, the total deflection at the free end is

$$\begin{aligned} y &= y_1 + y_2 \\ &= -\frac{L^3}{EI}\left(\frac{wL}{8} + \frac{P}{3}\right) \end{aligned} \tag{c}$$

(b) From Appendix C, Properties of Structural Lumber, the moment of inertia of a nominal 4 in. $\times$ 8 in. piece of structural lumber is

$$I = 111.15 \text{ in.}^4$$

From Appendix B, the modulus of elasticity of the southern pine is $E = 1.2 \times 10^6$ psi. Using the numeric values in Eq. (c), the maximum deflection is

$$y = -\frac{(6 \text{ ft} \times 12 \text{ in./ft})^3}{(1.2 \times 10^6 \text{ psi})(111.15 \text{ in.}^4)}\left(\frac{50 \text{ lb/ft} \times 6 \text{ ft}}{8} + \frac{200 \text{ lb}}{3}\right)$$

$$y = -0.292 \text{ in. (downward)}$$

EXAMPLE 6.6

Determine the deflection at the midspan for the simply supported beam subjected to the trapezoidal load distribution shown in Figure 6.12. Let *EI* be constant.

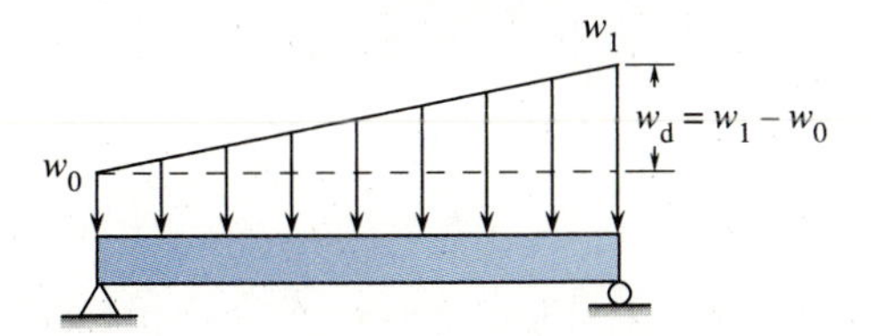

Figure 6.12 Simply supported beam subjected to trapezoidal load.

Solution

The trapezoidal loading can be separated into a uniform distribution w_0 plus a triangular distribution that varies from 0 to w_d. For the uniformly distributed load, we use Case 4 in Appendix D to obtain the midspan deflection as

$$y_1 = \frac{-5}{384}\frac{w_0 L^4}{EI}$$

For the triangular distribution, we use Case 7 in Appendix D to obtain the deflection equation as

$$y_2 = -\frac{w_d x}{360\,LEI}(7L^4 - 10L^2x^2 + 3x^4)$$

We now evaluate y_2 at $x = L/2$. Therefore, using superposition, the total midspan deflection is

$$\begin{aligned} y(x = L/2) &= y_1 + y_2 \\ &= -\left(\frac{5}{384}\frac{w_0 L^4}{EI} + \frac{5w_d L^4}{768EI}\right) = -\frac{L^4}{EI}\left(\frac{5w_0}{384} + \frac{5w_d}{768}\right) \end{aligned}$$

EXAMPLE 6.7

The overhanging beam shown in Figure 6.13 is subjected to a distributed load, *w*, over the middle section and a concentrated load, *P*, at *L*/2. Determine the vertical deflection expression at the center of the middle section ($x = L/2$). Let *EI* be constant.

Then using $w = 2000$ kN/m, $L = 5$ m, $a = 2$ m, $P = 5000$ kN, $E = 200$ GPa, and a circular cross-section with radius $r = 0.5$ m, determine the numeric deflection at $x = L/2 = 2.5$ m.

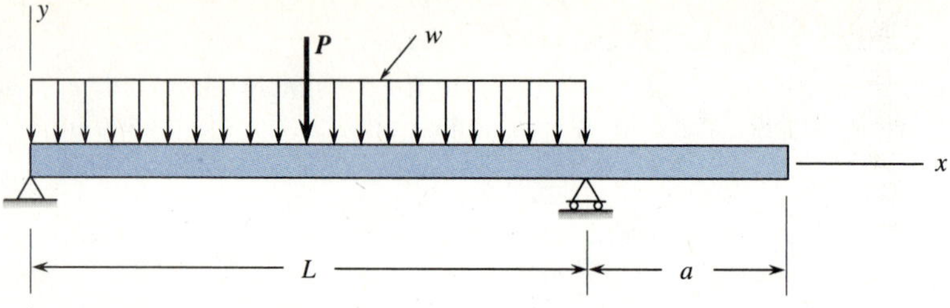

Figure 6.13 Beam with overhang.

Solution

To use superposition, we separate the actual load into load Cases 22 and 23 shown in Appendix D. For the uniformly distributed load, we use Case 22 to obtain a deflection at the midspan ($x = L/2$) of

$$y_1 = \frac{-5}{384}\frac{wL^4}{EI}$$

For the concentrated load at midspan, we use Case 23 and set $a = b = L/2$ in the expression for y to obtain

$$y_2 = \frac{-PL^3}{48EI}$$

The total deflection at $x = L/2$ is then

$$y = y_1 + y_2$$

$$y = \frac{-5wL^4}{384EI} - \frac{PL^3}{48EI}$$

$$y = \frac{-L^2}{384EI}(5wL^2 + 8PL)$$

Now using the numeric quantities (with $I = \frac{\pi}{4}r^4 = \frac{\pi}{4}(0.5 \text{ m})^4 = 0.0491 \text{ m}^4$ for the solid circular cross-section), we obtain the deflection at $x = L/2$ as

$$y = -\frac{(5 \text{ m})^2[5(2000 \text{ kN/m})(5 \text{ m})^2 + 8(5000 \text{ kN})(5 \text{ m})]}{384(200 \times 10^6 \text{ kN/m}^2)(0.0491 \text{ m}^4)}$$

$$= -2.98 \times 10^{-3}\text{m}$$

$$y = -2.98 \text{ mm (downward)}$$

EXAMPLE 6.8

It is common practice in mill building beams to stiffen a steel wide-flange section by welding a cap channel section to the top as shown in the cross-section of Figure 6.14. Here a W 27 × 94 has a C 12 × 20.7 welded to it. The beam is considered to be simply supported with a span length of 41 ft. Determine the maximum deflection for the beam subjected to the two equal concentrated forces symmetrically placed on the beam and due to the dead load from both the beam self weight and an attached rail.

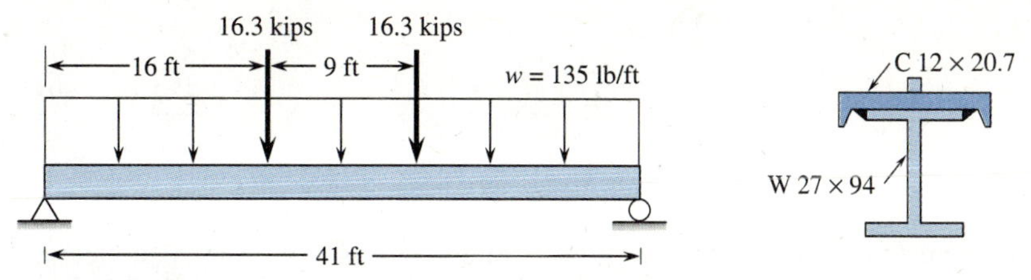

Figure 6.14 Beam subjected to two equal concentrated forces and uniform load.

Solution

We will use superposition to obtain the solution. The beam loading is then separated into two common loadings whose solutions are known as given in Appendix D. The separate load Cases 3 and 4 from Appendix D will be used.

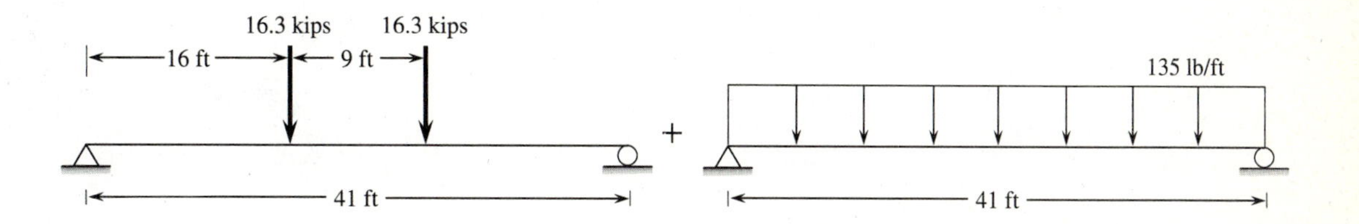

For load Case 3, we use $P = 16.3$ kips, $a = 16$ ft, $L = 41$ ft, and $E = 29 \times 10^3$ ksi (assume ASTM A36 steel and use Appendix B). We also must evaluate the moment of inertia I as follows.

From Appendix C, we obtain the properties needed for the W 27 × 94 and the C 12 × 20.7. For W 27 × 94

$$A = 27.7 \text{ in.}^2$$

$$I_x = 3270 \text{ in.}^4$$

$$d = 26.92 \text{ in.}$$

$$t_f = 0.745 \text{ in.}$$

For C 12 × 20.7

$$A = 6.09 \text{ in.}^2$$

$$I_x = 129 \text{ in.}^4$$

$$I_y = 3.88 \text{ in.}^4$$

$$t_w = 0.282 \text{ in.}$$

$$\bar{x} = 0.698 \text{ in.}$$

The centroid of the composite section is (noting that for the present application the x and y axes for the cap channel must be exchanged)

$$\bar{y} = \frac{A_1\bar{y}_1 + A_2\bar{y}_2}{A_1 + A_2}$$

$$= \frac{(27.7 \text{ in.}^2)\left(\frac{26.91 \text{ in.}}{2}\right) + (6.09 \text{ in.}^2)(26.92 + 0.282 - 0.698) \text{ in.}}{27.7 \text{ in.}^2 + 6.09 \text{ in.}^2}$$

$$\bar{y} = 15.8 \text{ in.}$$

The moment of inertia about the centroidal $\bar{x}$-$\bar{x}$ axis of the composite section is obtained by using the parallel axis theorem as

$$I_{\bar{x}} = I_{x_1} + A_1d_1^2 + I_{x_2} + A_2d_2^2$$

$$I_{\bar{x}} = 3270 \text{ in.}^4 + (27.7 \text{ in.}^2)\left(15.8 - \frac{26.92}{2}\right)^2 + 3.88 \text{ in.}^4 + (6.09 \text{ in.}^2)(26.49 - 15.8)^2$$

$$= 4122 \text{ in.}^4$$

The deflections at midspan are then obtained from Case 3 in Appendix D as

$$y_1 = -\frac{Pa}{24EI}(3L^2 - 4a^2)$$

$$= -\frac{(16.3 \text{ kips})(16 \text{ ft})[3(41 \text{ ft})^2 - 4(9 \text{ ft})^2]}{24(29 \times 10^3 \text{ ksi})(4122 \text{ in.}^4)} \times \frac{1728 \text{ in.}^3}{1 \text{ ft}^3}$$

$$= -0.741 \text{ in. (downward)}$$

From Case 4, Appendix D

$$y_2 = \frac{-5}{384}\frac{wL^4}{EI}$$

$$= -\frac{5(135 \text{ lb/ft})(41 \text{ ft})^4 (1728 \text{ in.}^3)}{384(29 \times 10^6 \text{ psi})(4122 \text{ in.}^4)(1 \text{ ft}^3)}$$

$$= -0.072 \text{ in. (downward)}$$

The total deflection at midspan is

$$y = y_1 + y_2 = -0.741 \text{ in.} - 0.072 \text{ in.}$$

$$= -0.813 \text{ in. (downward)}$$

In summary, the superposition method is a very straightforward method to determine beam deflections. However, we must remember that the system must be linear-elastic. That is, the load-deflection relationships must be linear for each load and for the total load acting on the beam. As long as this requirement is satisfied, we can superimpose the linear or angular deflections (as desired) at the same point due to several different loading conditions.

For commonly used loadings and support conditions, formulas, such as those provided in Appendix D can be used to obtain the solution. However, if formulas are not available, then one of the general methods of deflection analysis, such as the integration method or moment area (described in Section 6.5) must be used. Also for beams with varying *EI*, one of the general methods will need to be used to obtain solutions.

6.5 MOMENT-AREA METHOD FOR DEFLECTION

In this section we will develop and illustrate a classical method of beam deflection analysis called the *moment-area method* (sometimes called the area-moment method). This method has certain advantages over the integration method for specific kinds of problems as we will learn in this section. This method is sometimes used by structural analysts and machine designers to analyze beam deflections. This method is useful when the slope and deflection at only a few specific locations along the span of a beam are desired. When possible, these locations are chosen where the maximum slope and deflection will occur. Often times these locations can be determined by inspection based on the loading and support conditions. This method is easier to use than the integration method when the beam is subjected to several concentrated loads or abrupt changes in distributed loads (resulting in abrupt changes in the moment function), and when the beam has changes in its cross-sectional dimensions resulting in variable *EI*.

There are two theorems associated with the moment-area method. To develop these theorems, we consider the deflected shape of the simple beam shown in Figure 6.15. Reference locations are chosen at arbitrary points A and B.

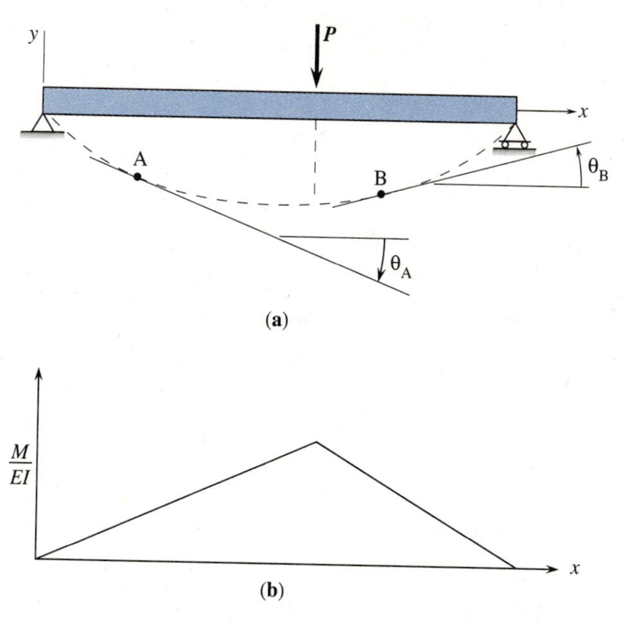

Figure 6.15 (a) Deflected curve used for derivation of moment-area theorems and (b) $\frac{M}{EI}$ diagram of the beam.

Derivation of First Moment-Area Theorem

To derive the first moment-area theorem, consider Figure 6.15. In Figure 6.15, θ_A and θ_B are the slopes of the deflection curve at points A and B. The slope θ_A is shown as a negative slope as the rotation of the tangent is clockwise. Equation (6.6) relates the beam radius of curvature, ρ, to the applied moment, M, and the beam stiffness, EI, by

$$\frac{1}{\rho} = \frac{M}{EI} \tag{6.62}$$

Equation (6.5) relates the radius of curvature to the rate of change of the slope as

$$\frac{1}{\rho} = \frac{d\theta}{dx} \tag{6.63}$$

Therefore, using Eqs. (6.62) and (6.63), we have

$$d\theta = \frac{M}{EI}\,dx \tag{6.64}$$

If we now integrate Eq. (6.64) between the two points A and B on the beam, we obtain

$$\int_A^B d\theta = \int_A^B \frac{M}{EI}\,dx$$

or

$$(\theta_B - \theta_A) = \int_A^B \frac{M}{EI}\,dx \tag{6.65}$$

Equation (6.65) is stated as the first moment-area theorem as follows.

First Moment-Area Theorem

The difference between the angles to the tangents to the deflected curve at A and B, denoted by θ_{BA}, is equal to the area of the moment divided by EI, (M/EI) diagram between A and B. This is called the *first moment-area theorem*.

From Eq. (6.65), we observe that the difference in angles corresponds to a positive M/EI diagram (see Figure 6.15(b)). Therefore, in moving from A to B along the deflected curve, the tangent to the curve rotates counterclockwise as shown in Figure 6.15(a).

Derivation of Second Moment-Area Theorem

To derive the second moment-area theorem, refer to Figure 6.16 which is an enlarged view of the deflection (elastic) curve of Figure 6.15(a) at point A.

In Figure 6.16, the tangents to the deflection curve are shown at points A and A′, where A′ is located distance, dx, from A. The difference in the slopes at A and A′ is the

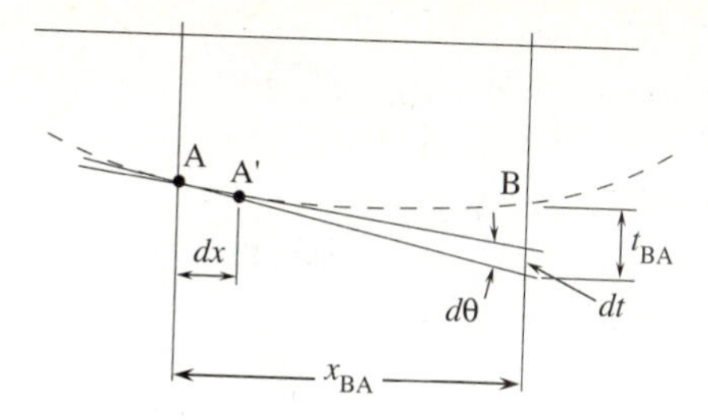

Figure 6.16 Enlarged view of deflection curve.

small angle $d\theta$. Using the arc length formula ($S = \theta r$) with $x = r$, since the slopes are small, we express the vertical segment dt by

$$dt = x_{BA} d\theta \tag{6.66}$$

where dt is actually the vertical deviation of the tangents between points A and A′, dx apart.

Using Eq. (6.64), Eq. (6.66) becomes

$$dt = x_{BA} \frac{M}{EI} dx \tag{6.67}$$

Integrating Eq. (6.67) between points A and B yields the vertical deviation of the tangent at B from that at A (denoted by t_{BA}) as

$$t_{BA} = \int_A^B x_{BA} \frac{M}{EI} dx \tag{6.68}$$

The right side of Eq. (6.68) is interpreted as follows. The term $\int_A^B \frac{M}{EI} dx$ represents the area of the M/EI diagram between A and B. Now recall from statics that $\bar{x} \int dA = \int x dA$ is used to determine the centroid $\bar{x}$ of an area. Therefore, the right side of Eq. (6.68) represents the moment of the M/EI diagram between points A and B computed about point B (as indicated by x_{BA}). In summary, Eq. (6.68) is stated as the *second moment-area* theorem.

Second Moment-Area Theorem

The vertical deviation from the deflection curve at B to the tangent to the deflection curve at A is equal to the first moment of the *M/EI* diagram between points A and B computed about point B, where the deviation is to be determined.

Using the sign conventions presented previously for positive moment diagrams, Eq. (6.68) indicates that for a positive M/EI area from A to B, the point B on the deflection curve is above the tangent to the deflected curve extended from A. Similarly, for a negative M/EI, the point B is below the tangent extended from A.

STEPS IN SOLUTION

The following steps can often be used to determine slopes and deflections in beams by the moment-area method.

1. Determine sufficient support reactions and draw the bending moment diagram and divide it by EI (M/EI diagram). When the beam is subjected to concentrated loads, the M/EI diagram will be a series of straight line segments, and the areas will be triangles and/or trapezoids. These areas and their moments required for the moment-area theorems are then relatively easy to determine. If the loading is a combination of concentrated and distributed loads, it is usually easier to use superposition, that is, to separate the loadings, and determine the M/EI diagram areas and moments for each separate load before applying the moment-area theorems. (See Example 6.13.) For beams subjected to distributed loads, the M/EI diagram will be a parabolic curve or possibly higher-order curve and the table on the inside back cover, or the more extensive table in Appendix A can be used to locate the area and centroid under each curve.
2. Sketch the approximate deflection curve. Remember that a point of zero slope always occurs at a fixed support, and zero deflection occurs at fixed, pinned, and roller supports. It may be helpful to use the moment diagram to aid in determining the curvature of the deflection curve. That is, a positive moment results in the beam bending concave downward so as to hold water (see Figure 6.15), whereas a negative moment bends the beam in a convex manner. Also an inflection point or change in curvature occurs where the moment in the beam is zero.
3. Apply the first moment-area theorem to determine the angle between any two tangents on the deflection curve. For cantilever beams and symmetrically loaded simple beams, the first moment-area theorem can be used to yield the slope at a desired location directly. (See Examples 6.9 through 6.11.)
4. Apply the second moment-area theorem to determine the tangential deviation. This theorem will directly yield the deflection for cantilever beams and will be used to obtain deflections for other beams. (See Examples 6.9 through 6.13.)

EXAMPLE 6.9

Determine the slope and deflection at the free end B of the cantilevered beam subjected to the concentrated load P located as shown in Figure 6.17. Let EI be constant.

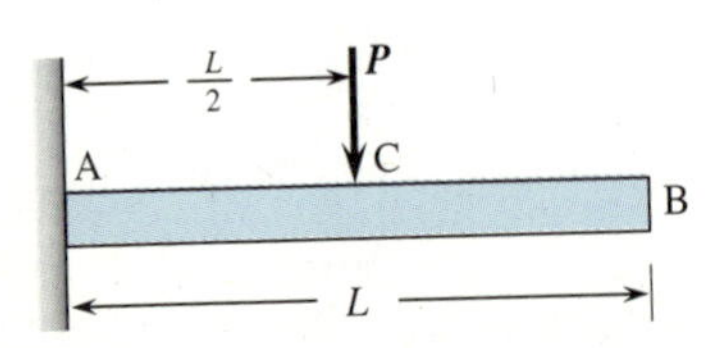

Figure 6.17 Cantilevered beam subjected to concentrated load.

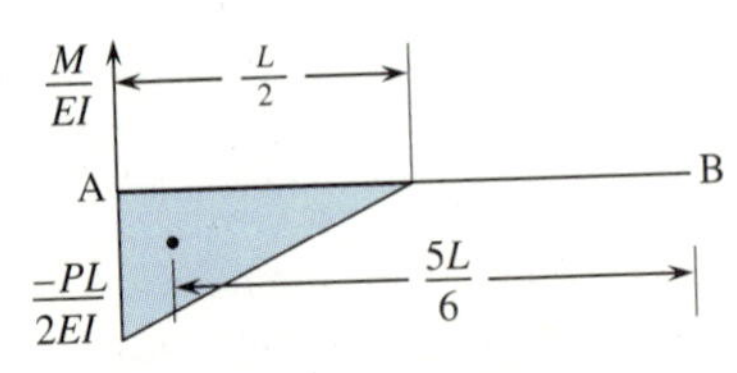

Figure 6.18 *M/EI* diagram of beam of Figure 6.17.

Solution

First we draw the M/EI diagram as shown in Figure 6.18. The centroid of the M/EI diagram is shown in Figure 6.18.

Now the first moment area theorem states that the difference in slopes of the deflection curve between points B and A is equal to the area of the M/EI diagram between A and B. From Figure 6.18, this area is determined and we have

$$\theta_B - \theta_A = \text{Area } M/EI\Big|_A^B$$

$$\theta_B - \theta_A = -\frac{1}{2}\left(\frac{PL}{2EI}\right)\left(\frac{L}{2}\right) \qquad \text{(a)}$$

Now because the beam is fixed at A, we have

$$\theta_A = 0 \qquad \text{(b)}$$

Therefore, using Eq. (b) in (a), we obtain the slope at B as

$$\theta_B = -\frac{PL^2}{8EI} \text{ rad} \quad \text{(clockwise)} \qquad \text{(6.69)}$$

The negative sign indicates that the slope at B is clockwise.

The second moment-area theorem states that the vertical deviation of the deflection at B to the tangent at A is equal to the moment of the M/EI diagram between A and B about B. From Figure 6.18, the moment of the area is determined and we have

$$t_{BA} = \text{Moment of } M/EI \text{ diagram between A and B about B}$$

$$t_{BA} = -\left(\frac{PL^2}{8EI}\right)\left(\frac{5L}{6}\right)$$

Since the tangent at A is zero, t_{BA} measures the absolute deflection of B. Therefore

$$t_{BA} = y_B = -\frac{5PL^3}{48EI} \qquad \text{(6.70)}$$

The negative sign indicates that the deflection at B is downward. Equations (6.69) and (6.70) are given as Case 13 in Appendix D (with $a = L/2$ in Case 13 formulas). The deflected curve and slope at B are shown in Figure 6.19.

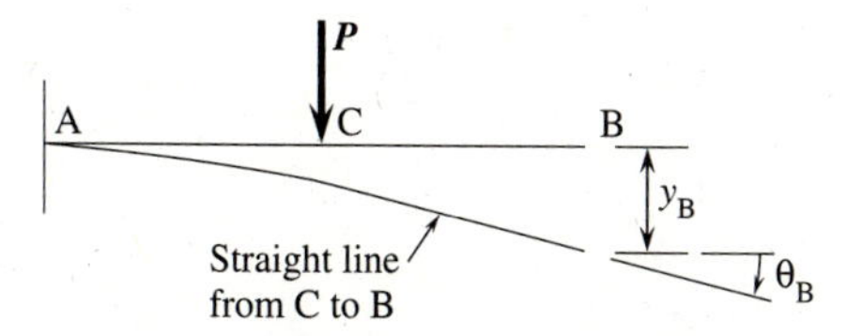

Figure 6.19 Deflected curve and slope at B.

EXAMPLE 6.10

Determine the slope and deflection at the free end B of the cantilevered beam subjected to a uniformly distributed load w acting over half the beam span as shown in Figure 6.20. Let EI be constant.

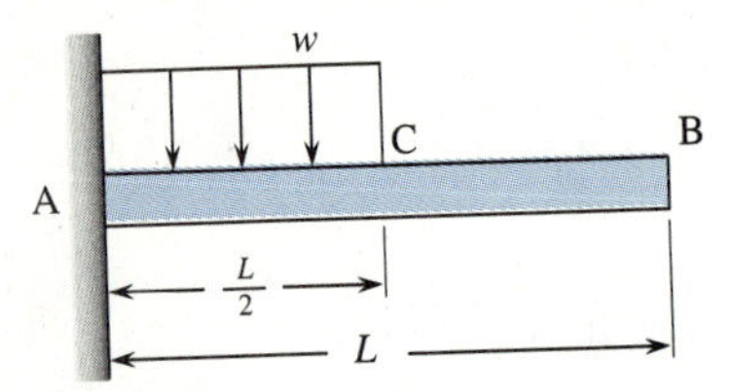

Figure 6.20 Cantilevered beam subjected to uniform load over portion of beam.

Solution

First draw the M/EI diagram (Figure 6.21).

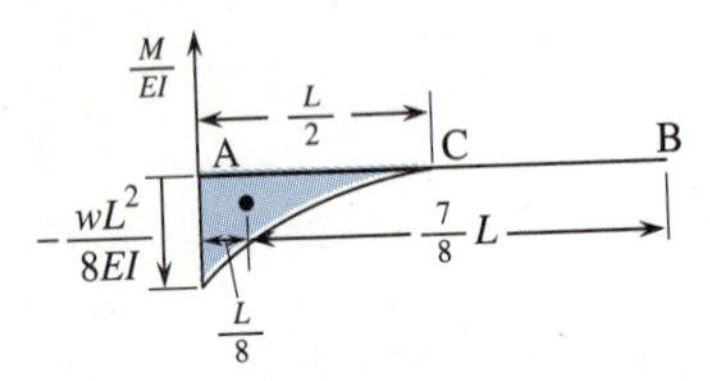

Figure 6.21 M/EI diagram of beam of Figure 6.20.

The M/EI diagram is a parabolic spandrel as shown in Figure 6.21. The centroid of a parabolic spandrel (see Appendix A) is at 1/4 of $L/2 = L/8$. The first moment-area theorem yields

$$\theta_B - \theta_A = \text{Area } M/EI \Big|_A^B$$

Now at fixed end A

$$\theta_A = 0$$

The area of the parabola is $\frac{1}{3}$ base times height (see Appendix A). This yields the area under the M/EI diagram as

$$\theta_B = -\frac{1}{3} \cdot \frac{wL^2}{8EI} \cdot \frac{L}{2}$$

$$\theta_B = -\frac{wL^3}{48EI} \quad \text{(clockwise)} \tag{6.71}$$

The second moment-area theorem yields

$$t_{BA} = \text{moment of } M/EI \Big|_A^B \text{ about B}$$

or

$$t_{BA} = \left(-\frac{wL^3}{48EI}\right)\left(\frac{7L}{8}\right)$$

As in Example 6.9, the tangent at A is zero, therefore, t_{BA} measures the absolute deflection of B.

$$t_{BA} = y_B = -\frac{7wL^4}{384EI} \tag{6.72}$$

Equations (6.71) and (6.72) are Case 17 in Appendix D if we let $a = L/2$ in Case 17 formulas.

EXAMPLE 6.11

Determine the slope at the left end and the maximum deflection of the simple beam subjected to the uniformly distributed load shown in Figure 6.22. Let EI be constant.

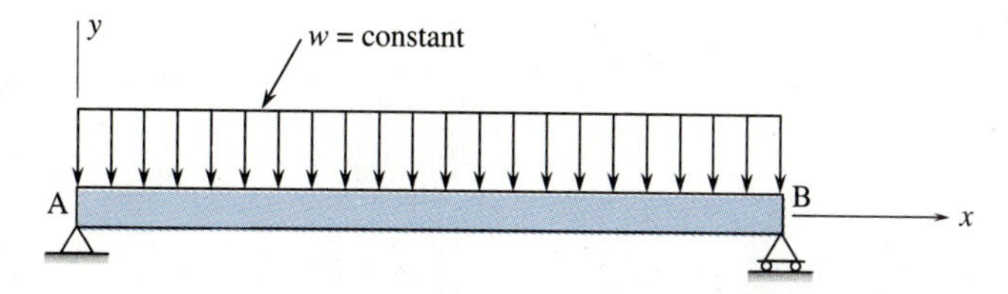

Figure 6.22 Simple beam with uniformly distributed load.

Solution

Due to the symmetry of the loading, the reactions will be equal. By statics they are

$$R_L = R_R = wL/2$$

The bending moment is determined using the free-body diagram of a section as follows.

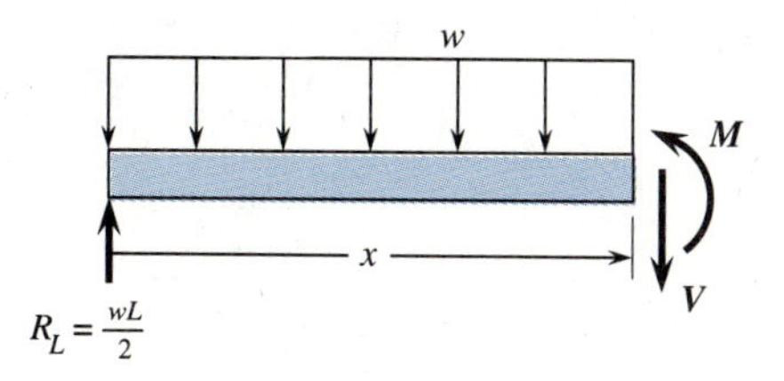

$$M = \left(\frac{wL}{2}\right)x - (wx)\left(\frac{x}{2}\right) \tag{a}$$

The deflection curve and the *M/EI* diagram are shown in Figure 6.23. Figure 6.23(b) shows the total *M/EI* diagram and Figure 6.23(c) shows the *M/EI* diagrams due to the left reaction (upper part) and the distributed load (lower part) separately, as based on Eq. (a). This separation will make the calculations easier and, along with superposition, will yield the solution.

Using the first moment-area theorem between points A and C, we have

$$\theta_C - \theta_A = \text{area of } M/EI\Big|_A^C \tag{b}$$

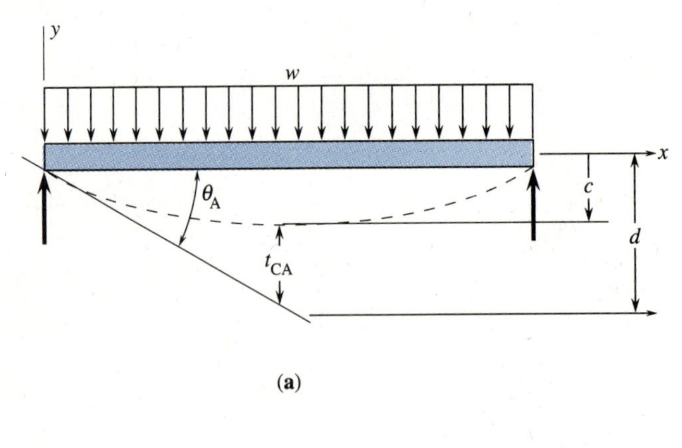

(a)

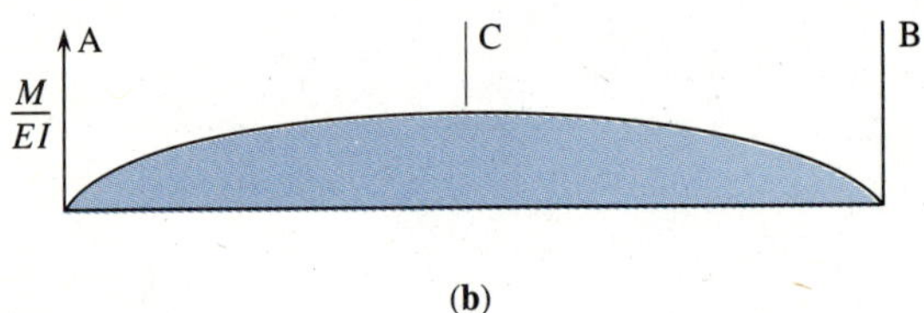

(b)

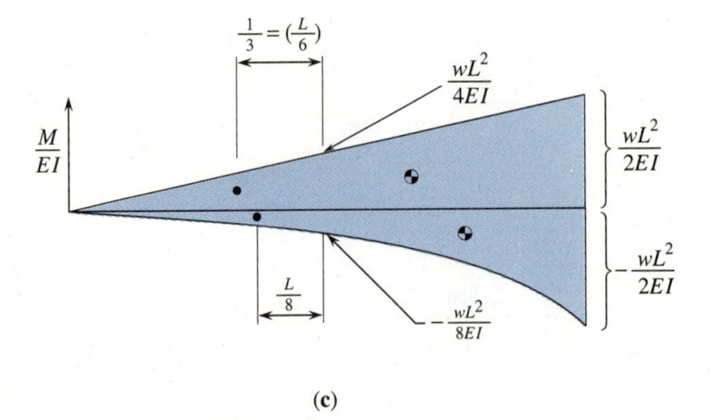

(c)

Figure 6.23 (a) Deflected curve,
(b) *M/EI* diagram for simple beam, and
(c) separated *M/EI* diagrams due to the left reaction and the uniform load.

The symmetric loading results in a maximum deflection at mid-span C and a corresponding zero slope. That is

$$\theta_C = 0$$

Therefore the slope at A will be equal to the area of the M/EI diagram between A and C. Using Figure 6.23(c), we evaluate the separate areas from A to C for the triangular and parabolic spandrel areas to obtain the slope at A as

$$-\theta_A = \frac{1}{2} \cdot \frac{wL^2}{4EI} \cdot \frac{L}{2} - \frac{1}{3} \cdot \frac{wL^2}{8EI} \cdot \frac{L}{2}$$

$$-\theta_A = \frac{wL^3}{16EI} - \frac{wL^3}{48EI}$$

$$\theta_A = -\frac{wL^3}{24EI} \text{ rad (clockwise)} \tag{6.73}$$

where again recall that the area of the parabolic spandrel is $\frac{1}{3}$ of the spandrel base times height, as given in Appendix A.

For small deflections, the vertical distance d in Figure 6.23(a) is given by

$$d = \theta_A\left(\frac{L}{2}\right) \tag{6.74}$$

Using Eq. (6.73) in (6.74), we have

$$d = \left(-\frac{wL^3}{24EI}\right)\left(\frac{L}{2}\right) = -\frac{wL^4}{48EI} \tag{6.75}$$

Using the second moment-area theorem, the tangential deviation, t_{CA}, is obtained as the first moment of the M/EI diagram between A and C evaluated at C. This yields (see Figure 6.23(c))

$$t_{CA} = \frac{wL^3}{16EI} \cdot \frac{L}{6} - \frac{wL^3}{48EI} \cdot \frac{L}{8} \tag{6.76}$$

where the moments due to the triangular and parabolic spandrel areas between A and C are shown as the first and second terms in Eq. (6.76). Simplifying, Eq. (6.76), we obtain

$$t_{CA} = \frac{3}{384} \cdot \frac{wL^4}{EI} \tag{6.77}$$

Therefore from Figure 6.23(a), the displacement, y_C, at the midspan will be

$$y_C = d + t_{CA} = \frac{wL^4}{EI}\left(\frac{-8}{384} + \frac{3}{384}\right)$$

$$y_C = -\frac{5}{384} \cdot \frac{wL^4}{EI} \tag{6.78}$$

Equations (6.73) and (6.78) are Case 4 in Appendix D.

EXAMPLE 6.12

Determine the slope and the deflection in variable form at B and at the free end C of the stepped cantilevered beam loaded as shown in Figure 6.24. Then let $E = 30 \times 10^6$ psi, $I = 200$ in.^{4}, $L = 10$ ft, a = b = 5 ft, $P_1 = 2000$ lb, and $P_2 = 2000$ lb and determine the numeric values for deflection at B and C.

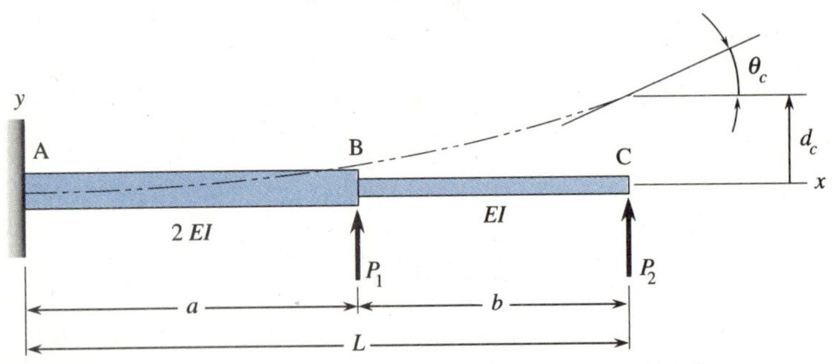

Figure 6.24 Stepped cantilevered beam subjected to concentrated loads.

Solution

This type of problem can be solved relatively easily using the moment-area method. Although the integration method could be used, it would have to be done in two parts to account for the two loads and the change in *EI*. Since we are only interested in points A and B, and then points A and C, the moment-area method will yield the solution faster than the integration method.

Figure 6.25 shows the *M* diagrams and the *M*/*EI* diagrams for the stepped beam. Figures 6.25(a) and 6.25(c) are due to P_1, and Figures 6.25(b) and (d) are due to P_2. (Remember we use 2EI in section AB).

Using the first moment-area theorem between A and B, the slope at B will be equal to the area of the *M*/*EI* diagram between points A and B, since the slope at A is zero. Therefore

$$\theta_B = \frac{1}{2}\left(\frac{P_1 a}{2EI}\right)\left(a\right) + \frac{1}{2}\left(\frac{P_2 a}{2EI}\right)\left(a\right) + \left(\frac{P_2 b}{2EI}\right)\left(a\right) \tag{a}$$

Simplifying Eq. (a), we have

$$\theta_B = \frac{1}{EI}\left(\frac{P_1 a^2}{4} + \frac{P_2 a^2}{4} + \frac{P_2 ab}{2}\right) \tag{6.79}$$

Similarly, the slope at C will be equal to the area of the *M*/*EI* diagram between points A and C, since the slope at A is zero. Therefore,

$$\theta_C = \theta_B + \frac{1}{2}\left(\frac{P_2 b}{EI}\right)\left(b\right) \tag{b}$$

Simplifying, Eq. (b), we have

$$\theta_C = \frac{1}{EI}\left[\frac{P_1 a^2}{4} + P_2\left(\frac{a^2}{4} + \frac{ab}{2} + \frac{b^2}{2}\right)\right] \tag{6.80}$$

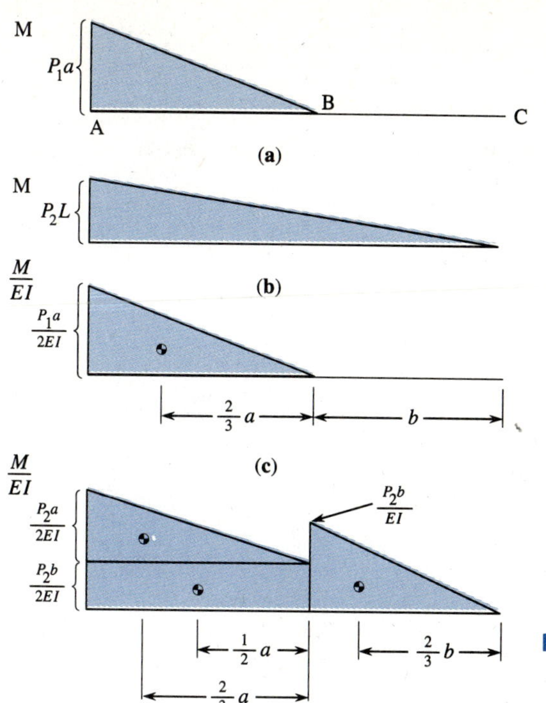

Figure 6.25 (a) M diagram for load P_1, (b) M diagram for load P_2, (c) M/EI diagram for load P_1, and (d) M/EI diagram for load P_2, all for stepped beam.

Now using the second moment-area theorem between A and B, we have t_{BA} = moment of $M/EI\Big|_A^B$ about B. Since the slope at A is zero, the deflection at B is equal to the deviation t_{BA}. Therefore

$$t_{BA} = y_B = \frac{P_1a^2}{4EI}\left(\frac{2}{3}a\right) + \frac{P_2a^2}{4EI}\left(\frac{2}{3}a\right) + \frac{P_2ba}{2EI}\left(\frac{1}{2}a\right)$$

$$y_B = \frac{1}{EI}\left(\frac{P_1a^3}{6} + \frac{P_2a^3}{6} + \frac{P_2ba^2}{4}\right) \tag{6.81}$$

Using the second moment-area theorem between A and C, the deflection at C is equal to the first moment of the M/EI diagram between A and C taken about C. Therefore using Figures 6.25(c) and (d), we obtain

$$y_C = \frac{P_1a^2}{4EI}\left(b + \frac{2}{3}a\right) + \frac{P_2a^2}{4EI}\left(b + \frac{2}{3}a\right) + \frac{P_2ab}{2EI}\left(b + \frac{1}{2}a\right) + \frac{P_2b^2}{2EI}\left(\frac{2}{3}b\right)$$

$$y_C = \frac{1}{EI}\left[P_1\left(\frac{a^2b}{4} + \frac{a^3}{6}\right) + P_2\left(\frac{a^2b}{2} + \frac{a^3}{6} + \frac{ab^2}{2} + \frac{b^3}{3}\right)\right] \tag{6.82}$$

Although these calculations may appear somewhat lengthy, the integration method would have been even more involved for a problem of this kind.

Substituting numeric values into Eqs. (6.81) and (6.82), we obtain

$$y_B = \frac{1}{(30 \times 10^6 \text{ psi})(200 \text{ in.}^4)}\left[\frac{2(2000 \text{ lb})(60 \text{ in.})^3}{6} + \frac{(2000 \text{ lb})(60 \text{ in.})(60 \text{ in.})^2}{4}\right]$$

$$y_B = 0.042 \text{ in. (upward)}$$

$$y_C = \frac{1}{(30 \times 10^6 \text{ psi})(200 \text{ in.}^4)}\left[(2000 \text{ lb})\left(\frac{(60 \text{ in.})^2(60 \text{ in.})}{4} + \frac{(60 \text{ in.})^3}{6}\right) + (2000 \text{ lb})\left((60 \text{ in.})^2(60 \text{ in.}) + \frac{(60 \text{ in.})^3}{2}\right)\right]$$

$$y_C = 0.138 \text{ in. (upward)}$$

EXAMPLE 6.13

For the overhanging beam shown in Figure 6.26, determine the vertical deflection at A and the slope at A. Use the moment-area method. Let $E = 30 \times 10^3$ ksi and $I = 288$ in.^{4}

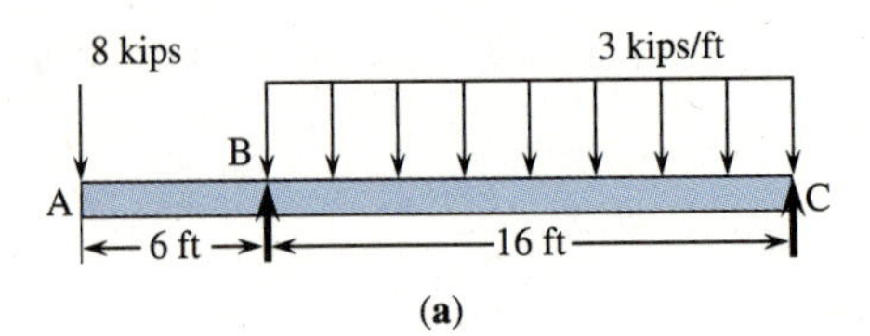

(a)

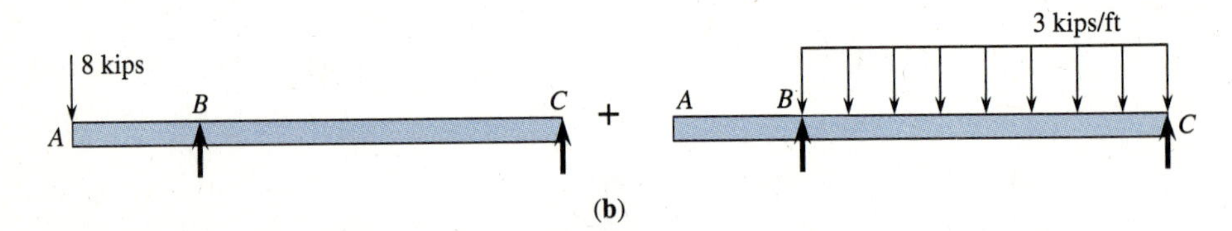

(b)

Figure 6.26 (a) Overhanging beam with two loadings and (b) beam separated into two for superposition.

Solution

This problem is best solved using superposition. We separate the problem into two parts. One beam with the 8-kip load and one beam with the 3 kips/ft uniform load as shown.

The moment diagrams for each loading can be found in the usual manner. The moment diagram from the separate loads is shown in Figure 6.27(a) and the superimposed or total moment diagram under the combined load is shown in Figure 6.27(b). We observe that the separate diagrams have regular shapes which make the calculations of these areas and their moments amenable for solution. We see that the superimposed diagram (Figure 6.27(b) is an irregular shape and hence difficult to use in calculations.

Figure 6.28 shows relevant reference slopes and tangential deviators of the deflected curve to be determined by the moment-area theorems and then used to obtain the slope and deflection at A.

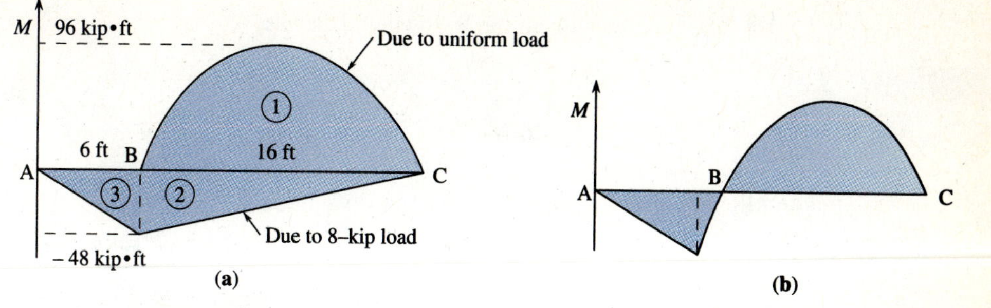

Figure 6.27 Moment diagrams
(a) separate loads and
(b) superimposed.

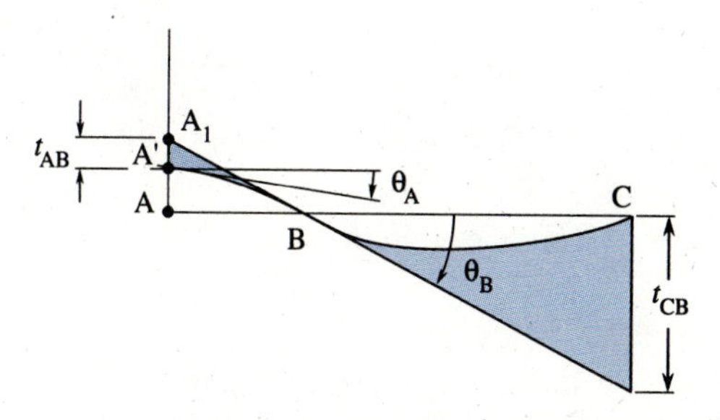

Figure 6.28 Slopes and tangential deviators on deflected curve of beam.

From the second moment-area theorem, we see that the deviation of point C relative to the tangent drawn at B is

$$t_{CB} = \text{moment of } M/EI \text{ area between B and C about C}$$

$$= \frac{1}{EI}(\text{moment of area 1 in Figure 6.27(a)} - \text{moment of area 2 in Figure 6.27(a)) about C})$$

$$= \frac{1}{EI}\left[\frac{2}{3}(96)(16)(8 \text{ ft}) - 48(16)\frac{1}{2}\left(\frac{2}{3}\,16\right)\right]$$

$$= \frac{1}{EI}(256 \times 32 - 256 \times 16)$$

$$t_{CB} = \frac{1}{EI}(256 \times 16) \tag{a}$$

Then, from Figure 6.28, we have

$$\theta_B = \frac{t_{CB}}{16} = \frac{1}{EI}\left(\frac{256 \times 16}{16}\right) = \frac{256}{EI} \text{ (clockwise)} \tag{b}$$

From Fig. 6.28, we obtain the following distances.

$$AA_1 = \theta_B \times 6 \text{ ft}$$

$$= \frac{256}{EI} \times 6 \qquad \text{(c)}$$

and by the second moment-area theorem, we obtain

$$A'A_1 = t_{AB} = \text{moment of } M/EI \text{ area between A and B about A}$$

$$= \frac{l}{EI}\left[\frac{(48)(6)}{2}\left(\frac{2}{3}(6)\right)\right] = \frac{96}{EI} \times 6 \qquad \text{(d)}$$

Remember $A'A_1$ is the deviation of A′ (not A), as we are considering the deflected curve, relative to the tangent at B. From Figure 6.28, we obtain the deflection at A as

$$y_A = AA' = AA_1 - A'A_1 \qquad \text{(e)}$$

Using (c) and (d) in (e), we have

$$= \frac{256 \times 6}{EI} - \frac{96 \times 6}{EI}$$

$$= \frac{160 \times 6}{EI} = \frac{960}{EI}$$

or

$$AA' = \frac{960 \times 1728 \text{ in.}^3/\text{ft}^3}{(30 \times 10^3)(288 \text{ in.}^4)}$$

$$AA' = y_A = 0.192 \text{ in. (upward)} \qquad \text{(f)}$$

The slope at A is

$$\theta_A = \theta_B - \theta_{BA} \qquad \text{(g)}$$

and by the first moment-area theorem

$$\theta_{BA} = \text{area of } M/EI \text{ diagram between A and B}$$

$$= \frac{48 \times 6}{2EI} = \frac{144}{EI} \qquad \text{(h)}$$

and using (b) and (h) in (g), we obtain

$$\theta_A = \frac{256 - 144}{EI} = \frac{112}{EI}$$

or

$$\theta_A = \frac{112 \times 144 \text{ in.}^2/\text{ft}^2}{30{,}000 \times 288}$$

$$= 1.867 \times 10^{-3} \text{ rad (clockwise)}$$

Recall that the integration method determined general expressions for the slope and deflection curves along an entire span of a beam. Specific values of x (locations along the span) could then be substituted into these general expressions wherever the slope or deflection was desired.

The moment-area method can be used to obtain general expressions for slope and deflection. However, this method is normally used to determine slopes and deflections at specific points and therefore is best utilized in problems where we already know the location of the maximum deflection. Although the moment-area method does require that we draw at least a portion of the M/EI diagram, it does have the advantage that changes in loads and bending stiffness, EI, along the span are easily included in the calculations as was illustrated in Example 6.12.

6.6 USE OF DISCONTINUITY FUNCTIONS

In Section 6.3, we introduced the integration method for obtaining deflections in beams. We observed that this method provides a very simple procedure for determining deflections and slopes at any point in a uniform cross-sectional beam, provided the bending moment can be expressed as a simple function in x. From Example 6.4, we observe that when the loading on a beam has discontinuities, such as due to concentrated point loads, more than one moment expression is needed. This results in a large number of constants of integration that must be solved simultaneously. The computation then becomes somewhat lengthy and tedious. These lengthy computations may be reduced using discontinuity functions.

Recall from Chapter 5 that the relations between loading $w(x)$, shear force V, and bending moment M are

$$\frac{dV}{dx} = -w(x) \tag{5.1a}$$

$$V = \frac{dM}{dx} \tag{5.2b}$$

$$\frac{dV}{dx} = \frac{d^2M}{dx^2} = -w(x) \tag{5.2d}$$

These expressions are valid as long as the loading is a continuous function of x.

Now we want to define some loading function that can physically represent concentrated point loads and concentrated moments and jump loading conditions while satisfying Eqs. (5.1a), (5.2b) and (5.2d). To do this, we introduce the *discontinuity functions* (sometimes called *singularity functions*). Some of these functions have special names, such as the Dirac delta function or unit impulse function, unit doublet function, and unit step or Heaveside function.

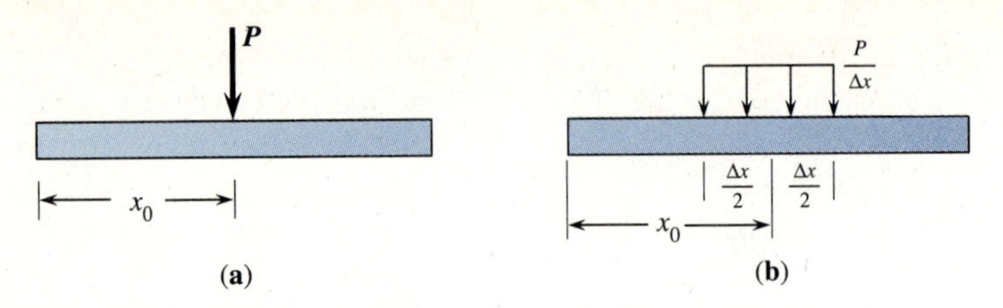

Figure 6.29 Representation of unit impulse function.

Unit Impulse Function

The *unit impulse function* is used to represent a unit concentrated force. To derive this function, consider Figure 6.29(a), where a concentrated force P is applied to a beam. We replace this force with a distributed one of intensity $P/\Delta x$ distributed over a very small length Δx so that as Δx approaches zero ($\Delta x \rightarrow 0$), the product of $P/\Delta x$ times Δx becomes P, as illustrated in Figure 6.29(b). For the unit load, we then define the unit impulse function or Dirac delta function (first developed by Paul Dirac in the early 1900s) as

$$\delta\langle x - x_0\rangle = \lim_{\Delta x \rightarrow 0} \begin{cases} 0 & \text{when } x < \left(x_0 - \dfrac{\Delta x}{2}\right) \\ \dfrac{1}{\Delta x} & \text{when } \left(x_0 - \dfrac{\Delta x}{2}\right) < x < \left(x_0 + \dfrac{\Delta x}{2}\right) \\ 0 & \text{when } x > \left(x_0 + \dfrac{\Delta x}{2}\right) \end{cases} \tag{6.83}$$

The distribution of load in Figure 6.29(b) can be represented by using the unit impulse function as

$$w(x) = P\delta\langle x - x_0\rangle \tag{6.84}$$

where the pointed brackets $\langle\ \rangle$ are used to denote this special kind of function. The unit impulse function is not a function in the usual mathematical definition as it is neither continuous nor differentiable at $x = x_0$. Nonetheless, Eq. (6.84) is a very convenient way to express a concentrated force as a load distribution function. Integration of Eq. (6.84) over the length of the beam yields the concentrated force P as

$$\int_0^L P\delta\langle x - x_0\rangle\, dx = \lim_{\Delta x \rightarrow 0} \int_{x_0 - \Delta x/2}^{x_0 + \Delta x/2} \frac{P}{\Delta x}\, dx = P \tag{6.85}$$

Unit Doublet Function

The *unit doublet function* is used to represent a concentrated moment or couple in terms of a load distribution function. To derive this function, consider Figure 6.30(b) where a pair of equal but opposite forces of magnitude $M_0/\Delta x$ are separated by a small distance Δx

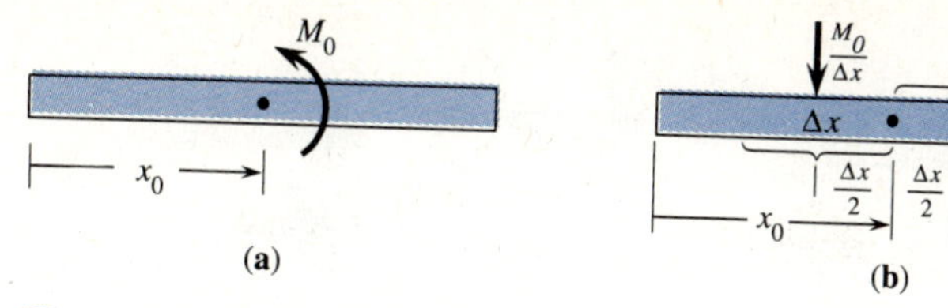

Figure 6.30 Unit doublet function.

apart so that the product of $M_0/\Delta x$ times Δx equals the moment M_0. For convenience, let $M_0 = 1$ and consider the forces to act over a small distance Δx as shown in Figure 6.30(b). The unit doublet function is defined by

$$D\langle x - x_0\rangle = \lim_{\Delta x \to 0} \begin{cases} 0 & \text{when } x < \left(x_0 - \frac{\Delta x}{2}\right) \\ \frac{1}{(\Delta x)^2} & \text{when } \left(x_0 - \frac{\Delta x}{2}\right) < x < x_0 \\ -\frac{1}{(\Delta x)^2} & \text{when } x_0 < x < \left(x_0 + \frac{\Delta x}{2}\right) \\ 0 & \text{when } x > \left(x_0 + \frac{\Delta x}{2}\right) \end{cases} \tag{6.86}$$

The load distribution corresponding to a couple M_0 is then given by

$$w(x) = M_0\, D\langle x - x_0\rangle \tag{6.87}$$

Integrating the unit doublet function in Eq. (6.86) yields

$$\int_0^x D\langle x - x_0\rangle dx = \lim_{\Delta x \to 0} \begin{cases} 0 & \text{when } x < \left(x_0 - \frac{\Delta x}{2}\right) \\ \frac{1}{\Delta x} & \text{when } x = x_0 \\ 0 & \text{when } x > \left(x_0 + \frac{\Delta x}{2}\right) \end{cases} \tag{6.88}$$

which is the unit impulse function.

Unit Step Function

The *unit step function* is used to represent a unit step or jump beginning at some point x_0. This function allows us to represent a uniformly distributed load. The function is represented by

$$u\langle x - x_0\rangle = \int_0^x \delta\langle x - x_0\rangle \mathrm{dx} \tag{6.89}$$

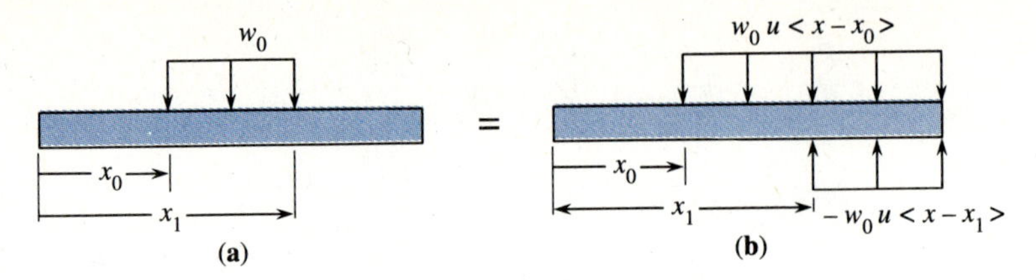

Figure 6.31 Uniform load represented by unit step function.

which from the definition of $\delta\langle x - x_0 \rangle$ given by Eq. (6.83), has the following property

$$u\langle x - x_0 \rangle = \begin{cases} 0 & \text{when } x < x_0 \\ 1 & \text{when } x \geq x_0 \end{cases} \tag{6.90}$$

and the integral of Eq. (6.90) becomes

$$\int_0^x u\langle x - x_0 \rangle dx = (x - x_0)u\langle x - x_0 \rangle \tag{6.91}$$

Using the unit step function, we represent the uniformly distributed load shown in Figure 6.31 as

$$w(x) = w_0 u\langle x - x_0 \rangle - w_0 u\langle x - x_1 \rangle = \begin{cases} 0 & \text{when } x < x_0 \\ w_0 & \text{when } x_0 \leq x \leq x_1 \\ 0 & \text{when } x > x_1 \end{cases} \tag{6.92}$$

Equation (6.92) is illustrated in Figure 6.31(b). Also the definite integral of the product of a continuous function $f(x)$ with the unit step function $u\langle x - x_0 \rangle$ is

$$\int_0^x f(x)u\langle x - x_0 \rangle dx = u\langle x - x_0 \rangle \int_{x_0}^x f(x)dx \tag{6.93}$$

and in particular, if $f(x) = (x - x_0)^n$

$$\begin{aligned} \int_0^x (x - x_0)^n u\langle x - x_0 \rangle dx &= u\langle x - x_0 \rangle \int_{x_0}^x (x - x_0)^n dx \\ &= u\langle x - x_0 \rangle \frac{(x - x_0)^{n+1}}{n + 1} \end{aligned} \tag{6.94}$$

for $n > -1$.

Unit Ramp Function

The *unit ramp function* is used to represent a linear function beginning at point x_0. This function allows us to represent a linear varying distributed load. The function is represented by

$$R\langle x - x_0 \rangle = \int_0^x (x - x_0)u\langle x - x_0 \rangle dx \tag{6.95}$$

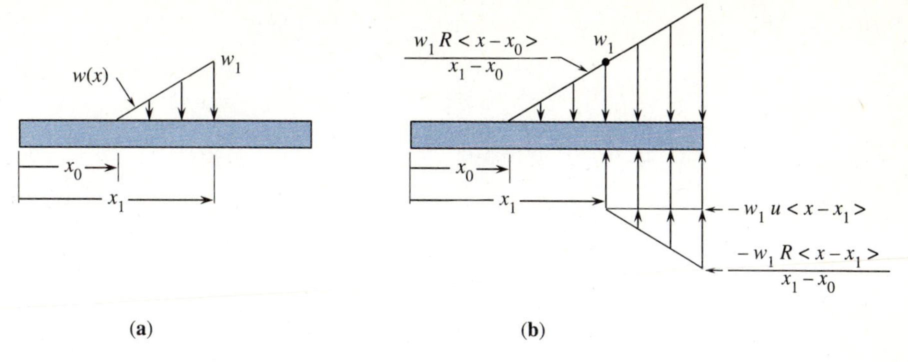

Figure 6.32 Linear varying load as a ramp function.

which from the definition of $u\langle x - x_0\rangle$, given by Eq. (6.90), has the following properties:

$$R\langle x - x_0\rangle = \begin{cases} 0 & \text{when } x < x_0 \\ x - x_0 & \text{when } x \geq x_0 \end{cases} \tag{6.96}$$

Using the unit ramp function, we represent the linear varying load shown in Figure 6.32(a) using the combination of loads shown in Figure 6.32(b) as

$$\begin{aligned} w(x) &= \frac{w_1}{x_1 - x_0} R\langle x - x_0\rangle - \frac{w_1}{x_1 - x_0} R\langle x - x_1\rangle - w_1 u\langle x - x_1\rangle \\ &= \begin{cases} 0 & \text{when } x < x_0 \\ \dfrac{w_1(x - x_0)}{x_1 - x_0} & \text{when } x_0 \leq x \leq x_1 \\ 0 & \text{when } x > x_1 \end{cases} \end{aligned} \tag{6.97}$$

When comparing the forms of the Eqs. (6.90) and (6.96), we can represent these functions in a more general form as

$$f_n(x) = \langle x - x_0\rangle^n = \begin{cases} 0 & \text{when } x < x_0 \\ (x - x_0)^n & \text{when } x \geq x_0 \end{cases} \tag{6.98}$$

for $n \geq 0$

For instance, the unit step function is represented when $n = 0$ in Eq. (6.98) as

$$f_0(x) = \langle x - x_0\rangle^0 = \begin{cases} 0 & \text{when } x < x_0 \\ 1 & \text{when } x \geq x_0 \end{cases} \tag{6.99}$$

or comparing Eq. (6.90) to (6.99)

$$u\langle x - x_0\rangle = \langle x - x_0\rangle^0 \tag{6.100}$$

The integral of Eq. (6.100) is

$$\int \langle x - x_0\rangle^0 dx = \langle x - x_0\rangle^1 \tag{6.101}$$

The unit ramp function is obtained when $n = 1$ as

$$f_1(x) = \langle x - x_0 \rangle^1 = \begin{cases} 0 & \text{when } x < x_0 \\ x - x_0 & \text{when } x \geq x_0 \end{cases} \tag{6.102}$$

Comparing Eq. (6.96) to (6.102)

$$R\langle x - x_0 \rangle = \langle x - x_0 \rangle^1 \tag{6.103}$$

The integral of Eq. (6.103) is

$$\int \langle x - x_0 \rangle^1 dx = \frac{\langle x - x_0 \rangle^2}{2} \tag{6.103a}$$

The unit second-degree function is obtained when $n = 2$ as

$$f_2(x) = \langle x - x_0 \rangle^2 = \begin{cases} 0 & \text{when } x < x_0 \\ (x - x_0)^2 & \text{when } x \geq x_0 \end{cases} \tag{6.104}$$

From the results of Eqs. (6.99), (6.102), and (6.104), we observe that any higher-order function can be expressed in terms of the unit step function as

$$f_n(x) = \langle x - x_0 \rangle^n = (x - x_0)^n \langle x - x_0 \rangle^0 \tag{6.105}$$

as

$$\langle x - x_0 \rangle^0 \begin{cases} = 0 & \text{when } x < x_0 \\ = 1 & \text{when } x \geq x_0 \end{cases}$$

and the integral of the function is

$$\int f_n(x)dx = \frac{f_{n+1}(x)}{n + 1} = \frac{\langle x - x_0 \rangle^{n+1}}{n + 1} \tag{6.106}$$

$$\text{for } n \geq 0$$

which, on using Eq. (6.105) in (6.106), is the same as Eq. (6.94).

British engineer W. Macauley in 1919 first suggested using these functions to solve beam problems, hence, these functions are sometimes called *Macauley functions*. The units of these discontinuity functions are the same as x^n, that is, f_0 is dimensionless, f_1 has units of x (in. or mm), f_2 has units of x^2 (in.^{2} or mm^2), and so forth. These functions are differentiable and integrable.

The other discontinuity functions (actually *singularity functions*), such as the unit impulse and doublet, can be represented as follows.

$$f_n(x) = \langle x - x_0 \rangle^n = \begin{cases} 0 & \text{when } x \neq x_0 \\ \pm\infty & \text{when } x = x_0 \end{cases} \tag{6.107}$$

$$\text{for } n = -1, -2, -3, \ldots$$

Note that singularity functions are represented for negative integer values, while the Macauley functions are defined for positive integers and for n equals zero.

For instance, using the notation of Eq. (6.107), the unit impulse is given by

$$w(x) = P\langle x - x_0 \rangle^{-1} \tag{6.108}$$

Note by comparing Eqs. (6.84) and (6.108) that

$$\delta\langle x - x_0\rangle = \langle x - x_0\rangle^{-1} \tag{6.109}$$

and the unit doublet is

$$w(x) = M_0\langle x - x_0\rangle^{-2} \tag{6.110}$$

Comparing Eqs. (6.87) and (6.110)

$$D\langle x - x_0\rangle = \langle x - x_0\rangle^{-2} \tag{6.111}$$

Even though the singularity functions are neither continuous nor differentiable, they can be integrated through the singularities. This integral formula is

$$\begin{aligned} \int f_n dx &= \int \langle x - x_0\rangle^n dx \\ &= \langle x - x_0\rangle^{n+1} = f_{n+1} \end{aligned} \tag{6.112}$$

for n negative, that is, $n = -1, -2, -3, \ldots$

Table 6.2 summarizes the discontinuity functions, along with derivatives and integrals of these. Table 6.3 summarizes the most common load intensities represented by the discontinuity functions.

Beam Shear Force And Bending Moment Equations

The following steps can be used to obtain the shear force and bending moment expressions using discontinuity functions.

1. Express the equivalent distributed load $w(x)$ in terms of discontinuity functions (including reactions in the total equivalent distributed load expression).
2. Using Eq. (5.1a), relate the shear to the distributed load by

$$V = -\int_0^x w(x)dx + C_1$$

Hence, integrate the load to obtain the shear. This integration produces one constant of integration C_1.

3. Use Eq. (5.2b) to relate the shear to the moment by

$$M = \int_0^x V dx + C_2$$

That is, integrate the V expression to obtain M. This integration produces a second constant of integration C_2. Each of these integration constants can be determined by applying known boundary conditions on V and M.

We observe that only a single expression for V and M results as we do not have to integrate a separate equation for each segment of the beam. The discontinuity functions allow us to integrate across discontinuities and singularities without introducing continuity conditions. Table 6.2 can be used to evaluate the integrals of the functions. Examples 6.14 through 6.16 illustrate the steps used to obtain shear force and bending moment equations using discontinuity functions.

TABLE 6.2 DISCONTINUITY FUNCTIONS USED FOR LOAD REPRESENTATIONS

Case	Name of Function	Function	Graph	Derivative and Integral
1	Unit doublet	$f_{-2} = \langle x - x_0\rangle^{-2} = \begin{cases} 0 & x \neq x_0 \\ \pm\infty & x = x_0 \end{cases}$ or $D\langle x - x_0\rangle$		$\int f_{-2}\,dx = f_{-1}$
2	Unit impulse	$f_{-1} = \langle x - x_0\rangle^{-1} = \begin{cases} 0 & x \neq x_0 \\ \pm\infty & x = x_0 \end{cases}$ or $\delta\langle x - x_0\rangle$		$\int f_{-1}\,dx = f_0$
3	General singularity	$f_n = \langle x - x_0\rangle^{n} = \begin{cases} 0 & x \neq x_0 \\ \pm\infty & x = x_0 \end{cases}$	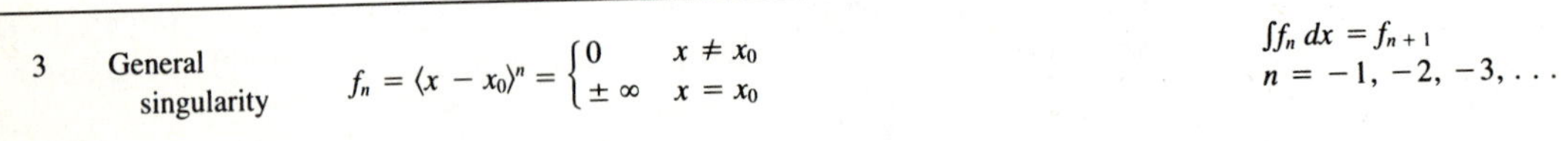	$\int f_n\,dx = f_{n+1}$ $n = -1, -2, -3, \ldots$
4	Unit step	$f_0 = \langle x - x_0\rangle^{0} = \begin{cases} 0 & x < x_0 \\ 1 & x \geq x_0 \end{cases}$ or $u\langle x - x_0\rangle$	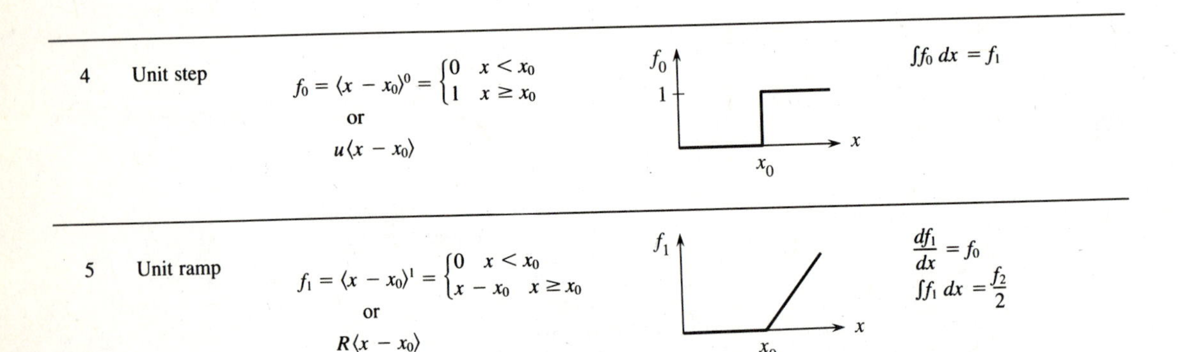	$\int f_0\,dx = f_1$
5	Unit ramp	$f_1 = \langle x - x_0\rangle^{1} = \begin{cases} 0 & x < x_0 \\ x - x_0 & x \geq x_0 \end{cases}$ or $R\langle x - x_0\rangle$		$\frac{df_1}{dx} = f_0$ $\int f_1\,dx = \frac{f_2}{2}$
6	Unit second degree	$f_2 = \langle x - x_0\rangle^{2} = \begin{cases} 0 & x < x_0 \\ (x - x_0)^2 & x \geq x_0 \end{cases}$	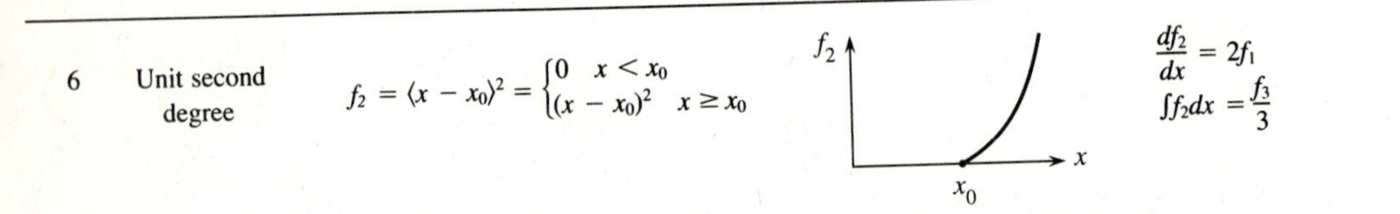	$\frac{df_2}{dx} = 2f_1$ $\int f_2 dx = \frac{f_3}{3}$
7	Unit n^{th} degree	$f_n = \langle x - x_0\rangle^{n} = \begin{cases} 0 & x < x_0 \\ (x - x_0)^n & x \geq x_0 \end{cases}$	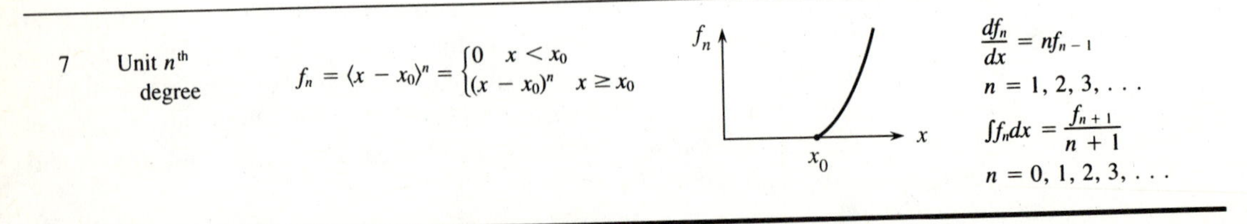	$\frac{df_n}{dx} = nf_{n-1}$ $n = 1, 2, 3, \ldots$ $\int f_n dx = \frac{f_{n+1}}{n+1}$ $n = 0, 1, 2, 3, \ldots$

TABLE 6.3 LOAD INTENSITIES EXPRESSED BY DISCONTINUITY FUNCTIONS

Case	Load on Beam	Load Intensity, $w(x)$ (Positive Downward)	Moment Function/Plot
1		$w(x) = M_0\langle x - x_0\rangle^{-2}$	$M(x) = -M_0\langle x - x_0\rangle^{0}$
2		$w(x) = P\langle x - x_0\rangle^{-1}$	$M(x) = -P\langle x - x_0\rangle^{1}$
3		$w(x) = w_0\langle x - x_0\rangle^{0}$	$M(x) = -w_0\dfrac{\langle x - x_0\rangle^2}{2}$
4		$w(x) = w_1\dfrac{\langle x - x_0\rangle^1}{x_1 - x_0}$	$M(x) = \dfrac{-w_1\langle x - x_0\rangle^3}{(x_1 - x_0)\cdot 2\cdot 3}$
5		$w(x) = w_0\langle x - x_0\rangle^{0}$ $-w_0\langle x - x_1\rangle^{0}$	$M(x) = -w_0\dfrac{\langle x - x_0\rangle^2}{2}$ $+ w_0\dfrac{\langle x - x_1\rangle^2}{2}$
6		$w(x) = w_1\dfrac{\langle x - x_0\rangle^1}{x_1 - x_0}$ $-w_1\dfrac{\langle x - x_1\rangle^1}{x_1 - x_0}$ $-w_1\langle x - x_1\rangle^0$	$M(x) = \dfrac{-w_1\langle x - x_0\rangle^3}{(x_1 - x_0)\cdot 2\cdot 3}$ $\dfrac{+w_1\langle x - x_1\rangle^3}{(x_1 - x_0)\cdot 2\cdot 3}$ $+w_1\dfrac{\langle x - x_1\rangle^2}{2}$
7	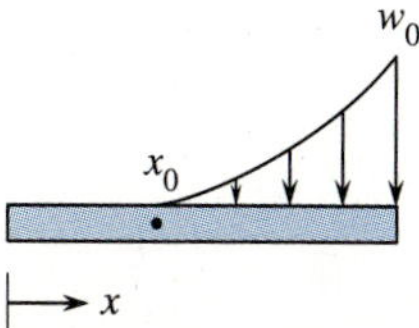	$w(x) = w_0\dfrac{\langle x - x_0\rangle^2}{(x_1 - x_0)^2}$	$M(x) = \dfrac{-w_0\langle x - x_0\rangle^4}{(x_1 - x_0)^2\cdot 3\cdot 4}$ 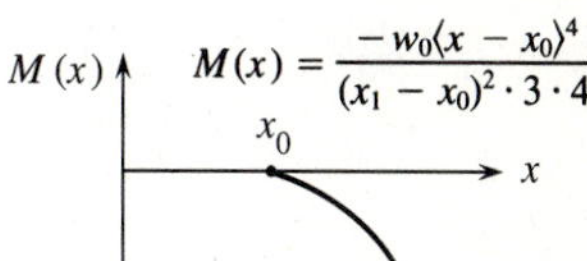

EXAMPLE 6.14

The beam shown is subjected to the concentrated force P and bending moment M_0. Express (a) the loading intensity for the beam including the reactions, (b) the shear force equation, and (c) the bending moment equation.

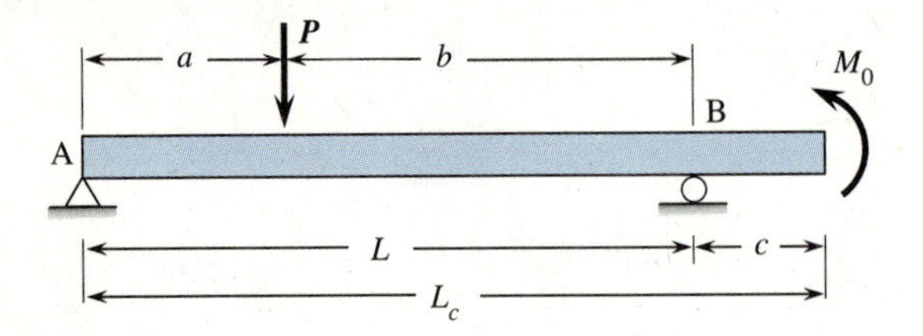

Figure 6.33 Beam for solution using discontinuity functions.

Solution

Find the reactions from the free-body diagram.

$$\Sigma M_B = 0: \quad -AL + Pb + M_0 = 0$$

$$A = \frac{Pb + M_0}{L}$$

$$\Sigma F_y = 0: \quad A - P + B = 0$$

$$B = P - \left(\frac{Pb + M_0}{L}\right)$$

$$= \frac{Pa - M_0}{L}$$

(a) Considering loadings directed downward to be positive and using the unit impulse and unit doublet functions, Eqs. (6.84) and (6.87), we have

$$w(x) = -\left(\frac{Pb + M_0}{L}\right)\delta\langle x\rangle + P\delta\langle x - a\rangle - \left(\frac{Pa - M_0}{L}\right)\delta\langle x - L\rangle + M_0 D\langle x - L_c\rangle \qquad \text{(a)}$$

or using the general notation of Eqs. (6.108) and (6.110), we have

$$w(x) = -\left(\frac{Pb + M_0}{L}\right)\langle x\rangle^{-1} + P\langle x - a\rangle^{-1} - \left(\frac{Pa - M_0}{L}\right)\langle x - L\rangle^{-1} + M_0\langle x - L_c\rangle^{-2} \qquad \text{(b)}$$

Equations (a) and (b) illustrate the equivalence of the two notations introduced for expressing discontinuity functions.

(b) Using Eq. (5.1a), the shear force expression is

$$V = -\int_0^x w(x)dx$$

Substituting Eq. (b) into the above integral and using Table 6.2 to aid in evaluating the integrals, we have

$$V = \left(\frac{Pb + M_0}{L}\right)\langle x\rangle^0 - P\langle x - a\rangle^0 + \left(\frac{Pa - M_0}{L}\right)\langle x - L\rangle^0 - M_0\,\langle x - L_c\rangle^{-1} + C_1 \quad \textbf{(c)}$$

The constant of integration can be evaluated from the known shear force at any point along the beam. For instance, when we use the shear force condition at the left end, the shear force equals the reaction, and we have

$$V(0^+) = A = \frac{Pb + M_0}{L} \quad \textbf{(d)}$$

Applying condition Eq. (d) to Eq. (c) at $x = 0$, we obtain

$$V(0^+) = \frac{Pb + M_0}{L} + C_1 \quad \textbf{(e)}$$

or

$$C_1 = 0 \quad \textbf{(f)}$$

where $V(0^+)$ means just to the right of zero at the reaction. To obtain the left side of Eq. (e), we used the following:

$$\langle x\rangle^0 = 1 \text{ for all } x$$

$$\langle x - a\rangle^0 = 0 \text{ for } x < a$$

$$\langle x - L\rangle^0 = 0 \text{ for } x < L$$

$$\langle x - L_c\rangle^{-1} = 0 \text{ for } x < L_c$$

Next, checking the shear at the $x = L_c$ value, we obtain

$$V(L_c) = \frac{Pb + M_0}{L} - P + \frac{Pa - M_0}{L}$$

$$= \frac{P(b + a)}{L} - P = 0$$

We observe that when the expression for the load, $w(x)$, includes all forces (both applied and reactive), the constant of integration, C_1, becomes zero.

(c) Next we integrate the shear force Eq. (c) to obtain the bending moment as

$$M = \int_0^x V dx$$

Using Table 6.2 to evaluate the integral, we obtain

$$M = \frac{Pb + M_0}{L}\langle x\rangle^1 - P\langle x - a\rangle^1 + \frac{Pa - M_0}{L}\langle x - L\rangle^1 - M_0\langle x - L_c\rangle^0 + C_2 \quad \textbf{(g)}$$

We now evaluate the constant C_2 from the known boundary condition for the bending moment at the left end of the beam. That is

$$M(0^-) = 0 = C_2$$

In evaluating $M(0^-)$, we interpret 0^- to mean just less than zero. Therefore,

$$\langle x \rangle^0 = 0 \quad \text{as} \quad x < 0$$

Again, the constant of integration, C_2, is zero because we have considered the load function to include all applied and reactive forces.

EXAMPLE 6.15

Use discontinuity functions to obtain the shear force and bending moment equations for the beam shown. This is the same beam solved in Example 5.6.

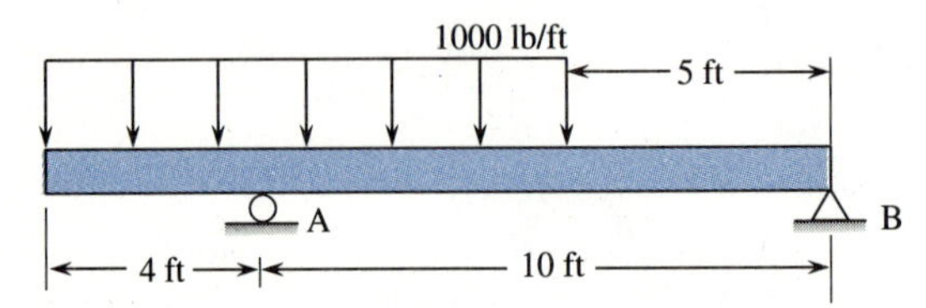

Figure 6.34 Beam analyzed using discontinuity functions.

Solution

First draw the free-body diagram and determine the reactions at A and B as

1000 × 9 = 9000 lb
4.5 ft
4 ft
10 ft
A_y
B_y

$$\Sigma M_B = 0: \quad -A_y\,(10\text{ ft}) + 9000\,(9.5\text{ ft}) = 0$$

$$A_y = 8550\text{ lb}$$

$$\Sigma F_y = 0: \quad 8550 - 9000 + B_y = 0$$

$$B_y = 450\text{ lb}$$

Next show the original beam with reactions.

We will use the general notation set forth in Eqs. (6.107) through (6.111) and summarized in Tables 6.2 and 6.3, to express the loadings. Considering loadings directed downward to be positive, we express each load (including reactions) in terms of discontinuity functions as follows.

1. For the uniformly distributed load of 1000 lb/ft which terminates at $x = 9$ ft, we use the unit step function (see Table 6.3, Case 3) to obtain

$$w(x) = 1000\langle x - 0\rangle^0 - 1000\langle x - 9\rangle^0 \quad \textbf{(a)}$$

2. For the left roller reaction located at $x = 4$ ft, we use the unit impulse function (see Table 6.3, Case 2) to obtain

$$w(x) = -8550\langle x - 4\rangle^{-1} \quad \textbf{(b)}$$

(The negative sign indicates that this load is upward.)

3. For the right pin reaction located at $x = 14$ ft, we use the unit impulse function (see Table 6.3, Case 2) to obtain

$$w(x) = -450\langle x - 14\rangle^{-1} \quad \textbf{(c)}$$

Therefore, for the whole beam we have

$$w(x) = 1000\langle x - 0\rangle^0 - 1000\langle x - 9\rangle^0 - 8550\langle x - 4\rangle^{-1} - 450\langle x - 14\rangle^{-1} \quad \textbf{(d)}$$

Now from Eq. (5.1a), we have

$$V = -\int_0^x w(x)dx$$

Substituting Eq. (d) into the integral, we obtain

$$\begin{aligned} V(x) = &-\int_0^x [1000\langle x - 0\rangle^0 - 1000\langle x - 9\rangle^0 - 8550\langle x - 4\rangle^{-1} \\ &-450\langle x - 14\rangle^{-1}]dx + C_1 \end{aligned} \quad \textbf{(e)}$$

Also by Eqs. (6.89) and (6.91) or (6.101) (or using Table 6.2, Cases 2 and 3 to evaluate the integral in Eq. (e) explicitly, we have

$$\begin{aligned} V(x) = &-1000\langle x - 0\rangle^1 + 1000\langle x - 9\rangle^1 + 8550\langle x - 4\rangle^0 \\ &+450\langle x - 14\rangle^0 + C_1 \end{aligned} \quad \textbf{(f)}$$

Now applying the boundary condition of $V(0) = 0$ to Eq. (f), we obtain

$$V(0) = 0 = C_1$$

as $\langle x - 9\rangle^1$ is zero for $x < 9$

$\langle x - 4\rangle^0$ is zero for $x < 4$

$\langle x - 14\rangle^0$ is zero for $x < 14$

Again, because we included the reactions in the load function, $C_1 = 0$. Next we evaluate V at $x = 14$ ft as follows.

$$V(14^+) = -1000(14) + 1000(14 - 9) + 8550 + 450 = 0$$

as $\langle x - 9\rangle^1 = 1$ for $x > 9$

$\langle x - 4\rangle^0 = 1$ for $x > 4$

$\langle x - 14\rangle^0 = 1$ for $x > 14$ $(x = 14^+)$

Here 14^+ means we consider the value just to the right of the pin reaction or just greater than 14 ft so that $\langle x - 14\rangle^0 = 1$ for $x = 14^+$. Therefore, Eq. (f) satisfies the shear force boundary conditions as required, that is,

$$V(0) = 0 \quad \text{at} \quad x = 0,\ x = 14^+$$

Now from Eq. (5.2b), we have

$$M(x) = \int_0^x V(x)dx \tag{g}$$

Using Eq. (f) in (g), the moment becomes

$$M(x) = \int_0^x [-1000\langle x - 0\rangle^1 + 1000\langle x - 9\rangle^1 + 8550\langle x - 4\rangle^0 + 450\langle x - 14\rangle^0]dx \tag{h}$$

Again, using Table 6.2, the integration in Eq. (h) becomes

$$M(x) = -1000\frac{\langle x - 0\rangle^2}{2} + 1000\frac{\langle x - 9\rangle^2}{2} + 8550\langle x - 4\rangle^1 + 450\langle x - 14\rangle^1 + C_2 \tag{i}$$

We see from Eq. (i) that only one equation is necessary to represent the moment in the beam, whereas in Example 5.6 three different equations were necessary to describe the bending moment over the whole beam.

Now applying the boundary condition that $M = 0$ at $x = 0$, we have

$$M(0) = 0 = C_2$$

We also evaluate M at $x = 14$ ft as follows.

$$M(14) = -1000\frac{14^2}{2} + 1000\frac{(14 - 9)^2}{2} + 8550(14 - 4)$$

$$= -98{,}000 + 12{,}500 + 85{,}500$$

$$M(14) = 0$$

This result verifies the boundary condition on the moment at $x = 14$ ft.

Next we evaluate M at $x = 8.55$ ft as follows.

$$M(8.55) = -1000\frac{8.55^2}{2} + 0 + 8550(8.55 - 4) + 0$$

$$= -36{,}551.25 + 38{,}902.5 = 2351.25 \text{ lb}\cdot\text{ft}$$

The moment is identical to that obtained in Example 5.6 using a series of cuts at appropriate sections of the beam. The extra significant figures were kept only to show the identity.

EXAMPLE 6.16

Determine the shear force and bending moment equations for the beam of Example 5.3.

Solution

The beam with the loading and reactions is first repeated as shown.

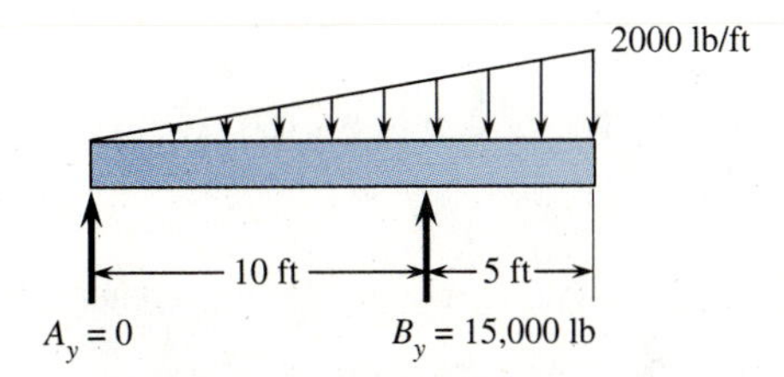

Figure 6.35 Beam analyzed using discontinuity functions.

Now using the general notation set forth by Eqs. (6.98) through (6.111), and summarized in Table 6.2 (with $x_0 = 0$), we express the loading as

$$\begin{aligned} w(x) &= \frac{2000}{15}\langle x - 0\rangle^1 - 15{,}000\langle x - 10\rangle^{-1} \\ &= \frac{2000x}{15} - 15{,}000\langle x - 10\rangle^{-1} \end{aligned} \tag{a}$$

The shear force is obtained by integrating Eq. (a) as

$$V = -\int_0^x \left[\frac{2000}{15}x - 15{,}000\langle x - 10\rangle^{-1}\right]dx \tag{b}$$

Explicit integration of Eq. (b), yields

$$V = -\frac{2000}{15}\frac{x^2}{2} + 15{,}000\langle x - 10\rangle^0 + C_1 \tag{c}$$

Next we evaluate the constant of integration C_1 and then verify that the boundary conditions on shear force are satisfied as follows.

$$V(0) = 0 = C_1$$

as $\langle x - 10\rangle^0 = 0$ for $x < 10$ and

$$V(15^+) = 15{,}000 - 15{,}000 = 0$$

as $\langle x - 10\rangle^0 = 1$ for $x > 10$

Next we evaluate the bending moment as

$$M = \int_0^x V(x)dx \tag{d}$$

Substituting Eq. (c) for V into (d), we have

$$M = \int_0^x \left[\frac{-1000}{15} x^2 + 15{,}000\langle x - 10 \rangle^0 \right] dx$$

Using Eq. (6.106) or Table 6.2 to perform the explicit integration, we obtain

$$\begin{aligned} M &= \frac{-1000}{15}\frac{x^3}{3} + 15{,}000\langle x - 10 \rangle^1 \\ &= -22.22x^3 + 15{,}000\langle x - 10 \rangle^1 + C_2 \end{aligned} \tag{e}$$

Applying the boundary condition of $M = 0$ at $x = 0$ to Eq. (e), we obtain

$$C_2 = 0$$

Again verifying that Eq. (e) satisfies the boundary conditions on moment, we have

$$\begin{aligned} M(15^+) &= -22.22(15^3) + 15{,}000(5) \\ &= 0 \end{aligned}$$

Also evaluating M at $x = 10$ ft, we have

$$M(10^-) = -22.22(10^3) = -22{,}220 \text{ lb} \cdot \text{ft}$$

This moment is the same as that found in Example 5.3 and, in part, verifies the correctness of our solution.

Beam Deflections

We now illustrate how to use the discontinuity functions to determine the slope and deflection in a beam.

Recall from Chapters 5 and 6 the following relations.

$$w(x) = -\frac{dV}{dx} \tag{5.1a}$$

$$V = \frac{dM}{dx} \tag{5.2b}$$

$$\frac{M}{EI} = \frac{d\theta}{dx} \tag{6.7}$$

$$\theta = \frac{dy}{dx} \tag{6.3}$$

From these relations we can first express the loading $w(x)$ in terms of the discontinuity functions. Then successively integrate each of these equations. Then apply appropriate boundary conditions on V, M, θ, and y to determine constants of integration. We finally obtain a single deflection equation for the entire beam. Using this method we will also have obtained the expressions for V and M.

Another variation of the method is to draw a free-body diagram and then express the moment equation for the entire beam in terms of discontinuity functions. Then integrate once to obtain the slope θ, and once again to obtain the deflection y.

However, the first method starting with the load function appears to be somewhat easier to use and so will be illustrated in Examples 6.17 and 6.18.

EXAMPLE 6.17

For the beam of Example 6.4 (repeated here including the reactions), express the slope and deflection equations using discontinuity functions. Also determine the deflection under the load P.

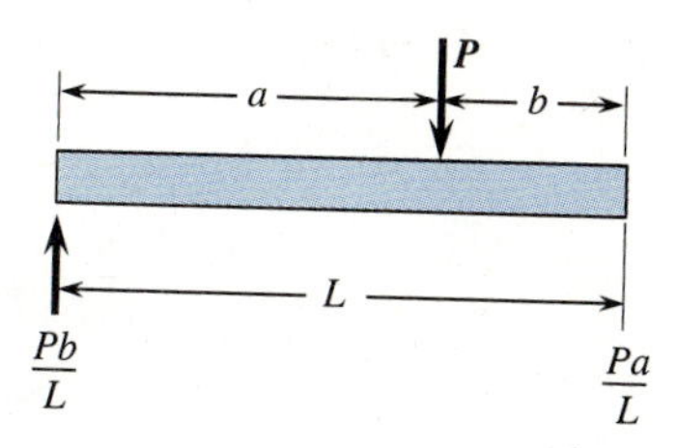

Figure 6.36 Beam for analysis using discontinuity functions.

Solution

First express the load in terms of the discontinuity functions as follows:

$$w(x) = -\frac{Pb}{L}\langle x - 0\rangle^{-1} + P\langle x - a\rangle^{-1} - \frac{Pa}{L}\langle x - L\rangle^{-1} \qquad \textbf{(a)}$$

Next express V as shown.

$$V(x) = -\int w(x)dx$$

Substituting $w(x)$ from Eq. (a) into the above and using Table 6.2 for the explicit integrations, we obtain

$$V = \frac{Pb}{L}\langle x - 0\rangle^{0} - P\langle x - a\rangle^{0} + \frac{Pa}{L}\langle x - L\rangle^{0} + C_1 \qquad \textbf{(b)}$$

Next express M as shown.

$$M = \int V dx$$

Substituting Eq. (b) into the above and using Table 6.2 for the evaluation of the integral, we obtain

$$M = \frac{Pb}{L}\langle x - 0\rangle^{1} - P\langle x - a\rangle^{1} + \frac{Pa}{L}\langle x - L\rangle^{1} + C_1x + C_2 \qquad \textbf{(c)}$$

On integrating the moment, we obtain the slope as

$$EI\theta = \int M dx$$

Integration of Eq. (c) results in

$$EI\theta = \frac{Pb}{L}\frac{\langle x - 0\rangle^2}{2} - P\frac{\langle x - a\rangle^2}{2} + \frac{Pa}{L}\frac{\langle x - L\rangle^2}{2} + C_1\frac{x^2}{2} + C_2x + C_3 \quad \text{(d)}$$

Finally, integrating the slope, we obtain the deflection as

$$EIy = \frac{Pb}{L}\frac{\langle x - 0\rangle^3}{6} - P\frac{\langle x - a\rangle^3}{6} + \frac{Pa}{L}\frac{\langle x - L\rangle^3}{6} + C_1\frac{x^3}{6} + C_2\frac{x^2}{2} + C_3x + C_4 \quad \text{(e)}$$

Applying the boundary conditions, we obtain the constants of integration as

$$V(0^+) = \frac{Pb}{L} = \frac{Pb}{L} + C_1, \quad C_1 = 0$$

$$M(0) = 0 = C_2$$

$$y(0) = 0 \quad \text{and} \quad y(L) = 0$$

$$EIy(0) = 0 = C_4$$

and

$$EIy(L) = 0 = \frac{PbL^3}{6L} - \frac{Pb^3}{6} + C_3L$$

or

$$C_3 = -\frac{Pb}{6L}(L^2 - b^2)$$

Therefore, the deflection is

$$y = \frac{P}{6EI}\left[\frac{b}{L}\langle x - 0\rangle^3 - \langle x - a\rangle^3 + \frac{a}{L}\langle x - L\rangle^3 - \frac{b}{L}(L^2 - b^2)x\right] \quad \text{(f)}$$

The third term has no effect on the results as it is zero everywhere along the axis of the beam up to the support. Also we can replace $\langle x - 0\rangle^3$ with x^3. Then the deflection becomes

$$y = \frac{P}{6EI}\left[\frac{b}{L}x^3 - \frac{bx}{L}(L^2 - b^2) - \langle x - a\rangle^3\right]$$

$$= \frac{Pb}{6EIL}\left[x^3 - x(L^2 - b^2) - \frac{L}{b}\langle x - a\rangle^3\right] \quad \text{(g)}$$

Equation (g) yields the deflection for the entire beam span. That is, this single equation represents the deflection for the entire beam span. Remember that for $0 < x < a$, the last term is zero. Equation (g) compares with Eq. (6.55) when $\langle x - a\rangle = 0$ (i.e., when $x < a$), and Eq. (6.57) when $x > a$.

EXAMPLE 6.18

For the beam of Example 6.15, determine (1) the equation of the deflection curve using discontinuity functions and (2) the deflection at points C and D. Let $E = 30 \times 10^6$ psi and $I = 200$ in.^{4}.

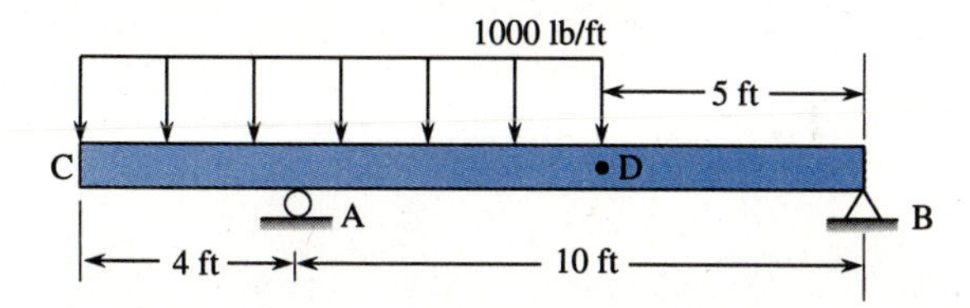

Figure 6.37 Beam for deflection analysis using discontinuity functions.

Solution

The bending moment expression was developed in Example 6.15, Eq. (i) as

$$M(x) = -1000\frac{\langle x\rangle^2}{2} + 1000\frac{\langle x-9\rangle^2}{2} + 8550\langle x-4\rangle^1 + 450\langle x-14\rangle^1 \quad \text{(a)}$$

Using Eq. (6.7), we integrate the moment Eq. (a) to obtain the slope as

$$EI\theta = \int_0^x \left[-1000\frac{\langle x\rangle^2}{2} + 1000\frac{\langle x-9\rangle^2}{2} + 8550\langle x-4\rangle^1 + 450\langle x-14\rangle^1\right]dx$$

Explicit integration yields

$$= -\frac{1000}{2\cdot 3}\langle x\rangle^3 + \frac{1000}{2\cdot 3}\langle x-9\rangle^3 + \frac{8550}{2}\langle x-4\rangle^2 + \frac{450}{2}\langle x-14\rangle^2 + C_1 \quad \text{(b)}$$

Integrating again, we have the deflection as

$$EIy = -\frac{1000}{2\cdot 3\cdot 4}\langle x\rangle^4 + \frac{1000}{2\cdot 3\cdot 4}\langle x-9\rangle^4 + \frac{8550}{2\cdot 3}\langle x-4\rangle^3 + \frac{450}{2\cdot 3}\langle x-14\rangle^3$$
$$+C_1x + C_2 \quad \text{(c)}$$

The boundary conditions are

$$y = 0 \text{ at } x = 4 \text{ ft} \quad \text{and} \quad y = 0 \text{ at } x = 14 \text{ ft}$$

Evaluating y at $x = 4$ ft, we have

$$EIy(4\text{ ft}) = 0 = -\frac{1000}{2\cdot 3\cdot 4}4^4 + C_1 4 + C_2$$

$$4C_1 + C_2 = 10{,}666.7 \quad \text{(d)}$$

Evaluating y at $x = 14$ ft, we have

$$EIy(14\text{ ft}) = 0 = -\frac{1000}{2\cdot 3\cdot 4}14^4 + \frac{1000}{2\cdot 3\cdot 4}(14-9)^4 + \frac{8550}{2\cdot 3}(14-4)^3 + C_1(14) + C_2$$

or

$$14C_1 + C_2 = 149{,}625 \tag{e}$$

Subtracting Eq. (d) from (e), we obtain C_1 as

$$10C_1 = 149{,}625 - 10{,}666.7$$

$$C_1 = 13{,}895.8$$

Then

$$C_2 = 10{,}666.7 - 4C_1 = 10{,}666.7 - 4(13{,}895.8)$$

$$= -44{,}916.6$$

Substituting C_1 and C_2 into Eq. (c) and simplifying, we obtain

$$y = \frac{1}{EI}[-41.67x^4 + 41.67\langle x - 9\rangle^4 + 1425\langle x - 4\rangle^3 + 75\langle x - 14\rangle^3 + 13{,}896x - 44{,}917] \tag{f}$$

The constant in Eq. (f) has units of lb · ft^3.
We now evaluate the deflection at $y = 0$ as

$$y_C = y(0) = \frac{1}{30 \times 10^6 \times 200}[-44{,}917 \text{ lb} \cdot \text{ft}^3](1728 \text{ in.}^3/\text{ft}^3)$$

$$= -0.0129 \text{ in. or downward}$$

Finally, evaluate the deflection at D as

$$y_D = y(9) = \frac{1}{30 \times 10^6 \times 200}[-41.67(9)^4 + 1425\langle 9 - 4\rangle^3 + 13{,}896(9) - 44{,}917](1728 \text{ in.}^3/\text{ft}^3)$$

$$= -0.00436 \text{ in. or downward}$$

6.7 STATICALLY INDETERMINATE BEAMS

In the previous sections of this chapter, we learned methods to obtain deflections in statically determinate beams. For instance, using the integration method we applied the equations of equilibrium to obtain the internal bending moment expression and then related this moment to the deflection using Eq. (6.7). On integrating twice and applying proper boundary conditions, the deflection was obtained. However many practical beams are supported by more than the minimum number of reactions required to maintain the beam in equilibrium as shown in Figure 6.38.

In this section we consider beams in which the equations of equilibrium alone do not provide enough equations to obtain the deflections in the beam. Hence, these beams are statically indeterminate and require the additional compatibility equations developed considering the geometry of the deformed structure as was previously done for the axially

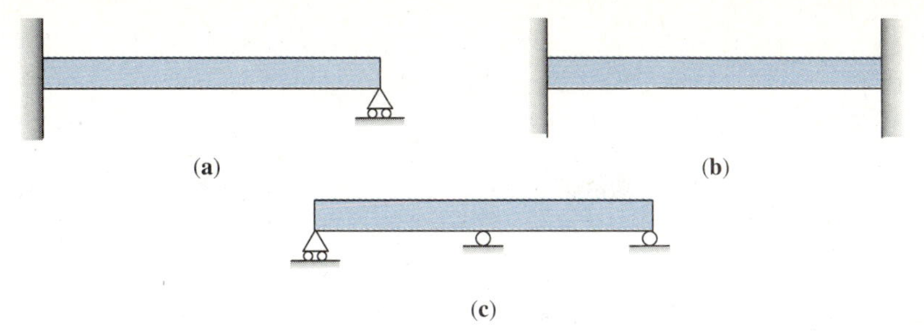

Figure 6.38 Examples of statically indeterminate beams
(a) propped cantilever,
(b) fixed-fixed,
(c) continuous beam over simple support.

loaded member in Section 2.7, and for the torsional member in Section 4.6. For beams, the compatibility equations usually result from the constraints that the supports place on the deflected shape of the beam.

STEPS IN SOLUTION

The following steps are used to analyze (determine reactions and then the deflection curve) statically indeterminate beams.

1. Draw the free-body diagram of the beam and write the equilibrium equations.
2. Apply a typical method for finding displacements, such as direct integration, to obtain the deflection equation.
3. Invoke the geometric boundary conditions based on the support conditions. Using these boundary conditions and the equilibrium equations, solve for the constants of integration and the reactions.
4. Substitute the constants of integration and the reactions back into the displacement expression to obtain the explicit displacement equation.

EXAMPLE 6.19

For the propped cantilevered beam shown in Figure 6.39, determine the reaction and the deflection expression using the integration method as described in Section 6.3. Let $E = 200$ GPa and $I = 248 \times 10^6$ mm^4.

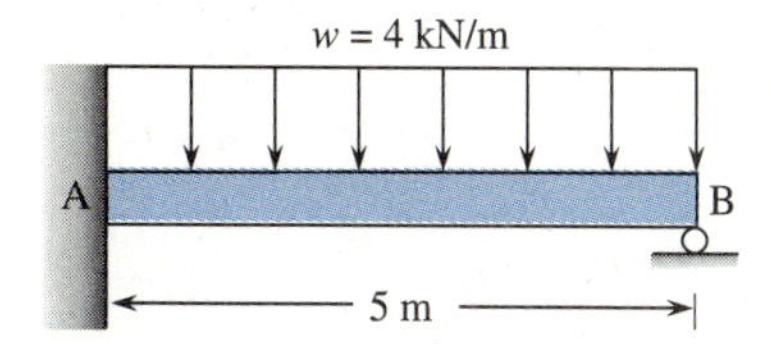

Figure 6.39 Propped cantilevered beam.

Solution

First, we draw the free-body diagram of the beam and then write the equilibrium equations as follows.

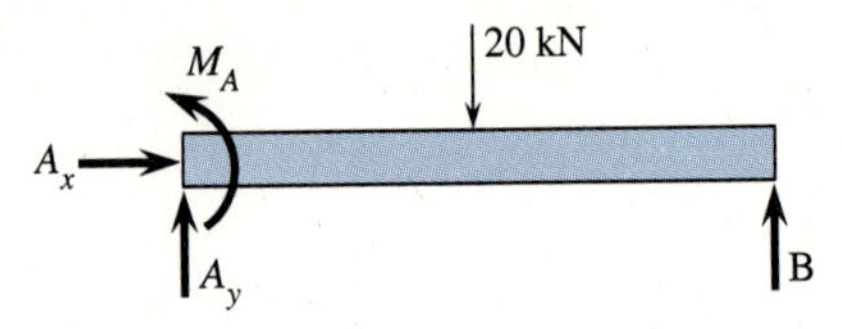

$$\Sigma F_x = 0: \quad A_x = 0$$

$$\Sigma F_y = 0: \quad A_y - 20 + \text{B} = 0 \tag{a}$$

$$\Sigma M_A = 0: \quad M_A - (20 \text{ kN})(2.5 \text{ m}) + \text{B}(5 \text{ m}) = 0$$

From the equilibrium equations (a), we have three unknowns A_y, M_A, and B, with only two equations. We need one more independent equation, which is obtained through considering the geometry of the deflected curve as follows.

From a free-body diagram of a portion of the original beam, we have

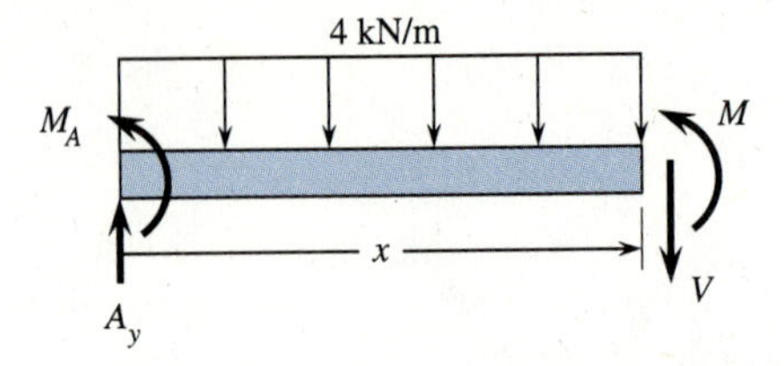

$$\Sigma M_{\text{cut}} = 0: \quad M + (4x)\left(\frac{x}{2}\right) - A_y x + M_A = 0 \tag{b}$$

or

$$M = -2x^2 + A_y x - M_A \tag{c}$$

Substituting Eq. (c) into the differential equation, Eq. (6.7), we obtain

$$EI\frac{d^2y}{dx^2} = -2x^2 + A_y x - M_A \tag{d}$$

Integrating Eq. (d) with respect to x yields an expression for the slope as

$$EI\theta = -\frac{2}{3}x^3 + A_y\frac{x^2}{2} - M_A x + C_1 \tag{e}$$

Integrating Eq. (e) with respect to x yields an expression for the deflection as

$$EIy = \frac{-x^4}{6} + A_y\frac{x^3}{6} - M_A\frac{x^2}{2} + C_1 x + C_2 \tag{f}$$

The geometric boundary conditions at the fixed support ($x = 0$) and the roller support ($x = L$) are

$$y(x = 0) = 0 \qquad \theta = y'(x = 0) = 0 \qquad y(x = L) = 0 \tag{g}$$

Using the first two boundary conditions of Eq. (g) in Eqs. (e) and (f), we have

$$EI\theta(0) = 0 = C_1 \tag{h}$$

and

$$EIy(0) = 0 = C_2 \tag{i}$$

Substituting (h) and (i) into (f), we have

$$EIy = -\frac{x^4}{6} + A_y\frac{x^3}{6} - M_A\frac{x^2}{2} \tag{j}$$

Now using the third boundary condition of Eq. (g) in (j), yields

$$EIy(x = 5m) = 0 = -\frac{5^4}{6} + A_y\frac{5^3}{6} - M_A\frac{5^2}{2}$$

Simplifying the above equation yields

$$0 = -104.16 + 20.83\, A_y - 12.5\, M_A \tag{k}$$

Equation (k) is the additional equation necessary to obtain the reactions. Solving Eqs. (a) and (k) simultaneously, we obtain the reactions at the supports as

$$A_x = 0,\ A_y = 12.5 \text{ kN},\ M_A = 12.5 \text{ kN}\cdot\text{m},\ B = 7.5 \text{ kN} \tag{1}$$

It is of interest to compare the reactions in Eq. (1) to those reactions at A for a cantilever without the roller support at B. For the cantilever, statics yields

$$\Sigma F_y = 0: \quad A_y = (4 \text{ kN/m})(5 \text{ m}) = 20 \text{ kN}$$

$$\Sigma M_A = 0: \quad M_A = (4 \text{ kN/m})(5 \text{ m})(2.5 \text{ m})$$

$$= 50 \text{ kN}\cdot\text{m}$$

Comparing these reactions with those of the propped cantilever, we see that the reactions at A are greatly reduced using the additional support at B.

Substituting Eq. (1) into (j), we obtain the deflection expression as

$$EIy = -\frac{x^4}{6} + \frac{12.5x^3}{6} - \frac{12.5x^2}{2} \tag{m}$$

or for $E = 200 \times 10^6$ kN/m^2 and $I = 248 \times 10^{-6}\, m^4$, we have

$$EI = 49{,}600 \text{ kN}/m^2$$

and the deflection expression, Eq. (m), becomes

$$y = \frac{1}{49{,}600}\left(-\frac{x^4}{6} + 2.08x^3 - 6.25x^2\right) \tag{n}$$

Differentiating Eq. (n) with respect to x, and setting the expression to zero, yields the value of x to obtain y_{max}. This value is $x = 2.91\ m$. Substituting $x = 2.91\ m$ into Eq. (n), yields

$$y_{max}(x = 2.91\ m) = \frac{1}{49{,}600}(-11.95 + 51.26 - 52.93)$$

$$= -0.275 \times 10^{-3}\ m$$

$$= -0.275 \text{ mm (downward)}$$

EXAMPLE 6.20

Determine the reactions, the deflection expression, and the maximum deflection of the fixed-fixed beam subjected to the uniform load shown in Figure 6.40. Use the integration method. Let *EI* be constant.

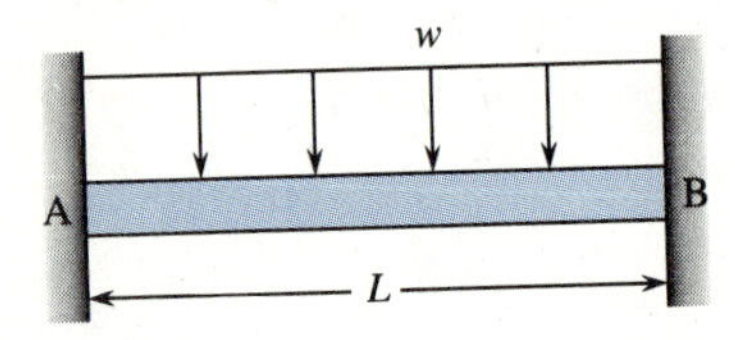

Figure 6.40 Fixed-fixed beam subjected to uniform load.

Solution

Draw the free-body diagram and write the equilibrium equations as

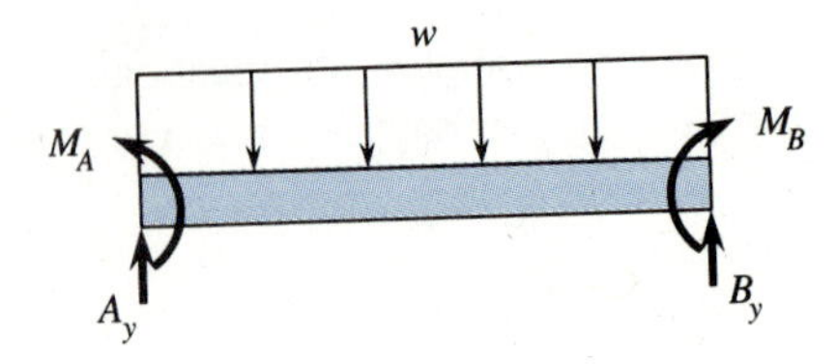

$$\Sigma F_y = 0: \quad A_y - wL + B_y = 0$$

$$\Sigma M_A = 0: \quad M_A - (wL)\left(\frac{L}{2}\right) + B_y L - M_B = 0 \tag{a}$$

There are four unknown reactions in Eqs. (a). We need two more independent equations which are obtained through considering the geometry of the deflected curve as follows.

Cut a portion of the beam to obtain the internal moment

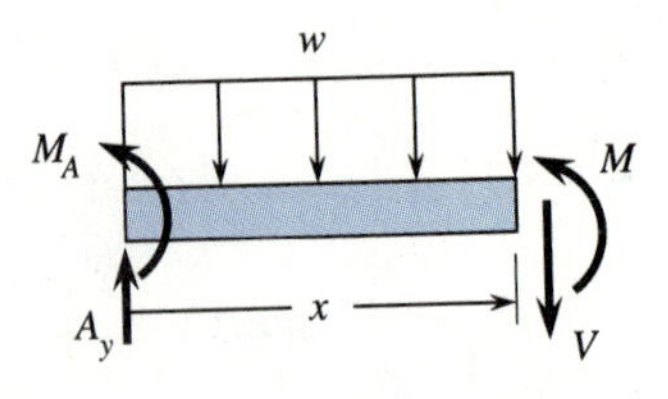

$$\Sigma M_{cut} = 0: \quad M + (wx)\left(\frac{x}{2}\right) - A_y x + M_A = 0$$

or

$$M = -\frac{wx^2}{2} + A_y x - M_A \tag{b}$$

Substituting Eq. (b) into the differential equation, Eq. (6.7), we obtain

$$EI\frac{d^2y}{dx^2} = -\frac{wx^2}{2} + A_y x - M_A \qquad \textbf{(c)}$$

Integrating Eq. (c) once, we obtain

$$EI\theta = -\frac{wx^3}{6} + A_y\frac{x^2}{2} - M_A x + C_1 \qquad \textbf{(d)}$$

Integrating Eq. (d), we obtain

$$EIy = -\frac{wx^4}{24} + A_y\frac{x^3}{6} - M_A\frac{x^2}{2} + C_1 x + C_2 \qquad \textbf{(e)}$$

The geometric boundary conditions for a fixed support are

$$\begin{aligned} y(x = 0) = 0 \qquad y(x = L) = 0 \\ \theta(x = 0) = 0 \qquad \theta(x = L) = 0 \end{aligned} \qquad \textbf{(f)}$$

Applying the boundary conditions (f) at $x = 0$ in (d) and (e) yields

$$C_1 = 0 \qquad C_2 = 0 \qquad \textbf{(g)}$$

Applying the boundary conditions (f) at $x = L$ in (d) and (e) yields

$$\begin{aligned} -\frac{wL^3}{6} + A_y\frac{L^2}{2} - M_A L = 0 \\ -\frac{wL^4}{24} + A_y\frac{L^3}{6} - M_A\frac{L^2}{2} = 0 \end{aligned} \qquad \textbf{(h)}$$

Solving Eqs. (h) simultaneously, we obtain

$$A_y = \frac{wL}{2} \qquad M_A = \frac{wL^2}{12} \qquad \textbf{(i)}$$

Then using the results from (i) in the equilibrium equations (a), we obtain

$$B_y = \frac{wL}{2} \qquad M_B = \frac{wL^2}{12} \qquad \textbf{(j)}$$

The results in (j) are expected due to the symmetry of the beam. Using Eqs. (g) and (i) in (e), the deflection equation becomes

$$EIy = -\frac{wx^4}{24} + \frac{wLx^3}{12} - \frac{wL^2x^2}{24} \qquad \textbf{(k)}$$

The maximum deflection occurs when $x = L/2$ is substituted into Eq. (k). This deflection becomes

$$y_{max} = \frac{1}{EI}\left[-\frac{wL^4}{24(16)} + \frac{wL^4}{12(8)} - \frac{wL^4}{24(4)}\right]$$

$$y_{max} = -\frac{wL^4}{384\,EI} \qquad \textbf{(l)}$$

Comparing this maximum deflection to that of the simply supported beam subjected to the same uniform load w, we observe, using Case 4 of Appendix D, that

$$y_{max} = -\frac{5wL^4}{384\,EI}$$

The maximum deflection of the simply supported beam is five times that of the fixed-fixed beam.

EXAMPLE 6.21

Use the method of superposition to determine the reactions and the deflection equation for the beam with a pin and two roller supports subjected to the uniform load w shown in Figure 6.41. Let EI be constant.

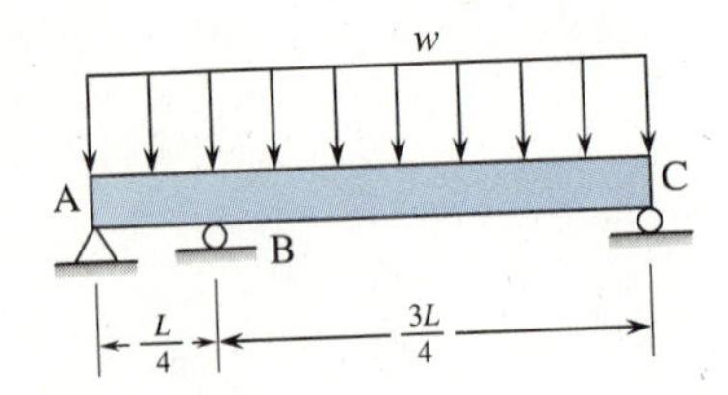

Figure 6.41 Beam with three supports.

Solution

Draw the free-body diagram and write the equilibrium equations shown below.

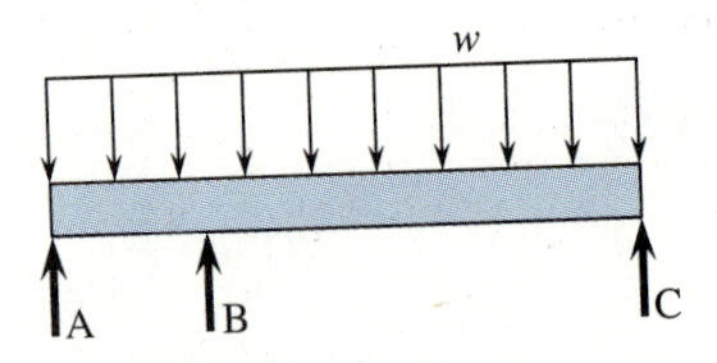

$$\Sigma M_A = 0: \quad B\left(\frac{L}{4}\right) + CL - (wL)\left(\frac{L}{2}\right) = 0$$

$$\Sigma F_y = 0: \quad A + B + C - wL = 0 \qquad \text{(a)}$$

Equations (a) are the only two independent equilibrium equations. However, there are three unknown reactions A, B, and C. The beam is then statically indeterminate to one degree.

Now use superposition and for convenience treat the reaction at B as the redundant force. That is, we separate the beam loading into two convenient loads for which deflection expressions are known from previous solutions such as tabulated in Appendix D.

These separate beams are shown below.

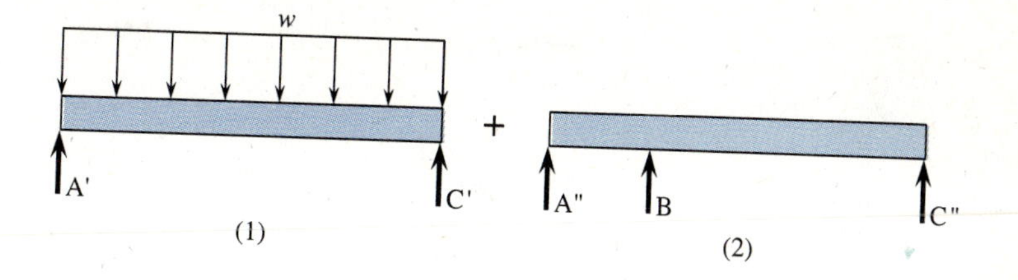

The deflection expression for the uniformly loaded simple beam is given in Appendix D, Case 4 as

$$EIy_1 = -\frac{wx^4}{24} + \frac{wLx^3}{12} - \frac{wL^3x}{24} \tag{b}$$

The deflection expression for the simply supported beam subjected to the upward load B at $x = L/4$ from the left support is known from Appendix D, load Case 2. For the portion of the beam from $0 \le x \le L/4$, the deflection expression is

$$EIy_2 = -\frac{B\left(\frac{3L}{4}\right)x}{6L}\left(-L^2\left(\frac{3L}{4}\right)^2 + x^2\right) \tag{c}$$

where $b = 3L/4$ has been used in the formula of Case 2. By superposition of Eqs. (b) and (c), we have the total deflection for $0 \le x \le L/4$ as

$$EIy = EIy_1 + EIy_2 \tag{d}$$

$$= -\frac{wx^4}{24} + \frac{wLx^3}{12} - \frac{wL^3x}{24} - \frac{Bx}{8}(x^2 - 0.4375L^2) \tag{e}$$

We now invoke the boundary condition that the deflection must be zero at the original support B. That is,

$$y = 0 \quad \text{at} \quad x = L/4 \tag{f}$$

Using Eq. (f) in (e), we solve for B as

$$0 = -\frac{w\left(\frac{L}{4}\right)^4}{24} + \frac{wL\left(\frac{L}{4}\right)^3}{12} - \frac{wL^3\left(\frac{L}{4}\right)}{24} - \frac{B}{8}\left(\frac{L}{4}\right)\left[\left(\frac{L}{4}\right)^2 - 0.4375L^2\right]$$

or

$$B = \frac{57}{72}wL \tag{g}$$

From the first of equilibrium Eqs. (a), we then have

$$C = \frac{87}{288}wL \tag{h}$$

From the second of Eqs. (a), we obtain

$$A = wL - B - C$$

$$= wL - \frac{57}{72} wL - \frac{87}{288} wL$$

or

$$A = -\frac{3}{32} wL$$

Substituting the value of B from Eq. (g) into Eq. (e), we obtain the final deflection expression for $0 \le x \le L/4$ as

$$EIy = -\frac{wx^4}{24} + \frac{wLx^3}{12} - \frac{wL^3x}{24} - \left(\frac{57}{72} wL\right) \frac{x}{8} \left(x^2 - \frac{7}{16} L^2\right)$$

A similar procedure can be used to obtain the final deflection expression for $L/4 \le x \le L$.

6.8 DEFLECTIONS IN UNSYMMETRIC BENDING MEMBERS

In the previous sections of this chapter, we assumed that the deflections were due to bending about only the principal z axis. However, if unsymmetric bending occurs, deflections must be calculated in each of the principal planes and these deflections added vectorially to obtain the total deflection. Figure 6.42 illustrates the vector addition of the deflections calculated first in the principal directions 1 and 2 (Δ_1 and Δ_2) for an angle cross-section and then added together vectorially to obtain the total deflection Δ.

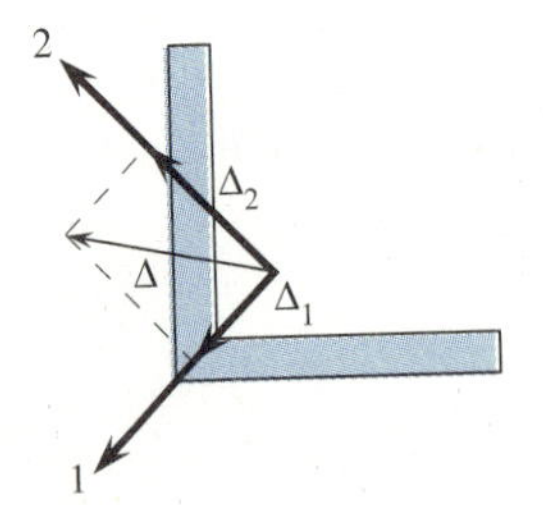

Figure 6.42 Deflection of a beam subjected to unsymmetric bending.

To prevent twisting of the cross section the applied force must act through the shear center of the cross section. The concept of shear center was discussed in Section 5.12. If the load is not through the shear center, then torsional stresses and deformations, treated in Chapter 4, must be considered.

Examples 6.22 and 6.23 illustrate how to determine deflections in beams under unsymmetric bending.

EXAMPLE 6.22

For the beam of Example 5.28, determine the maximum deflection. The beam is shown here as Figure 6.43. Assume the beam is made of steel with $E = 29 \times 10^6$ psi.

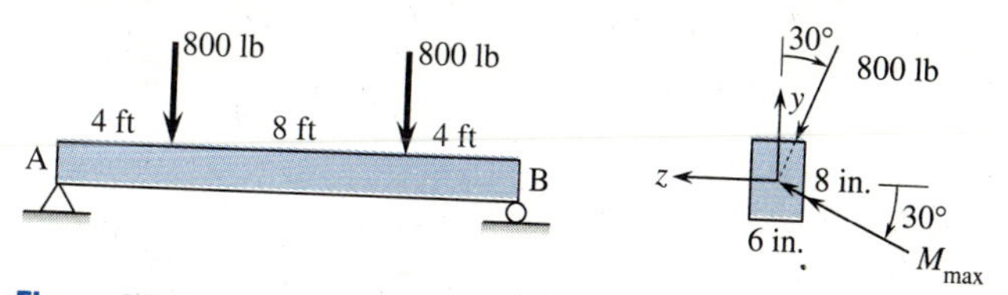

Figure 6.43 Beam subjected to unsymmetric bending.

Solution

From Appendix D, Case 3, the maximum deflection in the y-direction is given by

$$y_{max} = \frac{-Pa}{24\,EI}(3L^2 - 4a^2) \quad \text{(a)}$$

We use P_y for the maximum deflection in the y direction and P_z for the maximum deflection in the z direction. Here the y and z axes are the principal ones. Therefore

$$P_y = 800 \cos 30° = 693 \text{ lb (downward)}$$

$$P_z = 800 \sin 30° = 400 \text{ lb (to the left)}$$

The moments of inertia about the y and z axes are

$$I_z = \frac{1}{12}(6 \text{ in.})(8 \text{ in.})^3 = 256 \text{ in.}^4$$

$$I_y = \frac{1}{12}(8 \text{ in.})(6 \text{ in.})^3 = 144 \text{ in.}^4$$

Using Eq. (a) for both the y and z deflections, we have

$$y_{max} = -\frac{693}{24(256)E}[3(16 \times 12)^2 - 4(4 \times 12)^2]$$

$$= -\frac{11435}{E} \quad \text{(b)}$$

$$= -0.394 \times 10^{-3} \text{ in. or downward}$$

$$z_{max} = \frac{400}{24(144)E}[3(16 \times 12)^2 - 4(4 \times 12)^2]$$

$$= \frac{11733}{E} \quad \text{(c)}$$

$$= 0.405 \times 10^{-3} \text{ in. or to the left in positive } z \text{ direction)}$$

The final resultant deflection is obtained by adding vectorially the deflections from Eqs. (b) and (c) as shown in the figure below.

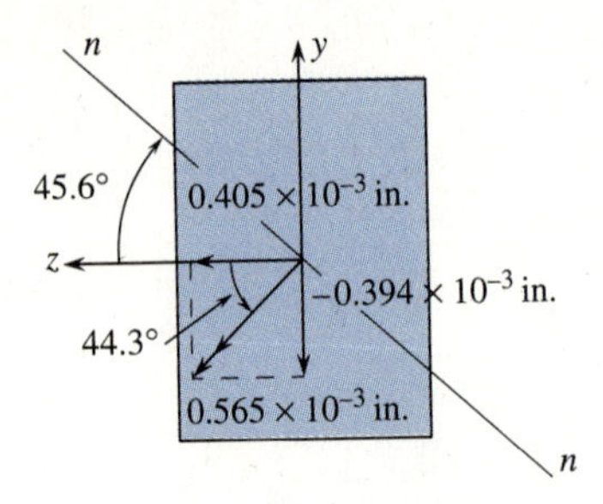

The resulting deflection (0.565×10^{-3} in.) is perpendicular to the neutral axis of the beam. The angle of 45.6° compares with the 45.7° angle found in Example 5.28, Section 5.10. The small difference is due to round-off error.

EXAMPLE 6.23

For the beam of Example 5.31, determine the resultant deflection due to the vertical load applied through the shear center of the angle cross-section. The angle is structural steel with $E = 29 \times 10^6$ psi. The beam is shown in Figure 6.44.

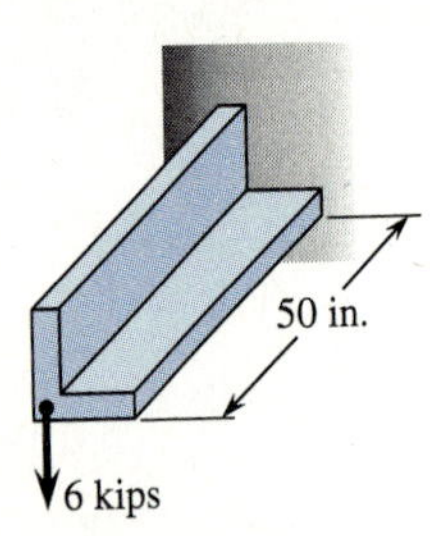

Figure 6.44 Angle beam for deflection analysis.

Solution

The beam is cantilevered and subjected to an end load. Therefore, the free-end deflection is given in Appendix D, Case 12 as

$$y_{\max} = \frac{-PL^3}{3EI} \tag{a}$$

The principal axes were determined in Example 5.30 of Section 5.11. The principal moments of inertia were obtained in Example 5.30 as

$$I_1 = 111.36 \text{ in.}^4 \quad \text{and} \quad I_2 = 28.10 \text{ in.}^4$$

The end loads in the 1 and 2 directions (see subsequent Figure) are

$$P_1 = 6 \cos 45° \text{ kips} = 4.24 \text{ kips (in the positive 1 direction)}$$

and

$$P_2 = 6 \sin 45° \text{ kips} = 4.24 \text{ kips (in the negative 2 direction)}$$

Using Eq. (a), the deflections in the one and two directions are then

$$\Delta_1 = \frac{4.24(50)^3}{3E(28.10)}$$

$$= 0.217 \text{ in. (in the positive 1 direction)} \quad \text{(b)}$$

$$\Delta_2 = \frac{-4.24(50)^3}{3E(111.36)}$$

$$= -0.0547 \text{ in. (in the negative 2 direction)} \quad \text{(c)}$$

The resultant deflection (0.224 in.) is obtained by vector addition of the deflections in the 1 and 2 directions as shown below.

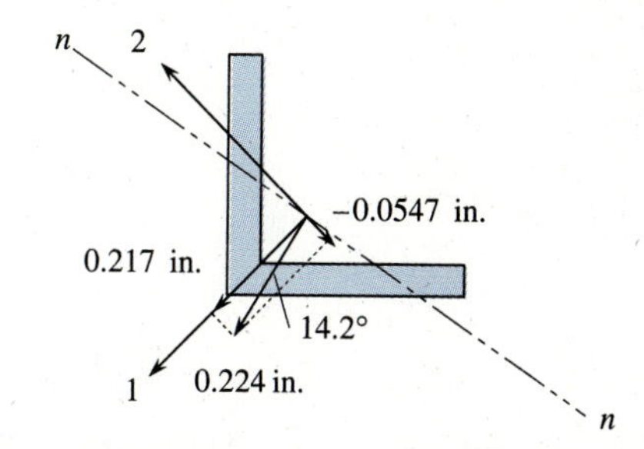

The direction of the resultant deflection is again perpendicular to the neutral axis as can be seen by comparing the neutral axis located in Example 5.31 for the angle.

6.9 SUMMARY OF IMPORTANT DEFINITIONS AND EQUATIONS

1. Curvature of a beam

$$\kappa = \frac{1}{\rho} = \frac{d^2y}{dx^2} = \frac{d}{dx}(\theta) \quad \text{(6.5)}$$

2. Basic second-order differential equation relating bending moment to deflection (also called the moment-curvature relation)

$$\kappa = \frac{d^2y}{dx^2} = \frac{M}{EI} \quad \text{(6.7)}$$

3. Integration method

For slope in beam

$$\frac{dy}{dx} = y'(x) = \theta(x) = \int_0^x \frac{M(x)}{EI}\,dx + C_1 \quad \text{(6.13)}$$

For deflection in beam

$$y(x) = \int_0^x \left[\int_0^x \frac{M(x)}{EI} \, dx \right] dx + C_1 x + C_2 \qquad \textbf{(6.14)}$$

4. Beam boundary conditions (to obtain constants of integration)

Clamped support

$$y = 0 \qquad y' = 0 \qquad \textbf{(6.15a)}$$

Roller or pin support

$$y = 0 \qquad \textbf{(6.15b)}$$

5. Beam continuity conditions (to obtain constants of integration)

Continuous beam

$$y_1(a) = y_2(a)$$
$$y_1'(a) = y_2'(a) \qquad \textbf{(6.16a)}$$

Internal hinge

$$y_1(a) = y_2(a) \qquad \textbf{(6.16b)}$$

6. Maximum deflection of some common beams

Cantilevered beam concentrated load at free end

$$y(L) = \frac{-PL^3}{3EI} \qquad \textbf{(6.27)}$$

Cantilevered beam uniform load

$$y(L) = -\frac{wL^4}{8EI} \qquad \textbf{(6.37)}$$

Simply supported beam uniform load

$$y_{max} = y\left(\frac{L}{2}\right) = -\frac{5}{384}\frac{wL^4}{EI} \qquad \textbf{(6.45)}$$

Simply supported beam concentrated load at midspan

$$y_{max} = y\left(\frac{L}{2}\right) = -\frac{PL^3}{48EI} \qquad \textbf{(6.61)}$$

7. Superposition principle for beam slopes and deflections

Used to obtain deflections in beams subjected to more than one load by separating the loads into common ones found in tables providing slope and deflection formulas such as in Appendix D.

8. First moment-area theorem

$$(\theta_B - \theta_A) = \int_A^B \frac{M}{EI}\,dx \tag{6.65}$$

9. Second moment-area theorem

$$t_{BA} = \int_A^B x_{BA}\frac{M}{EI}\,dx \tag{6.68}$$

10. Discontinuity functions used for load representations are given in Table 6.2.

11. Load intensities expressed by discontinuity functions are given in Table 6.3.

REFERENCES

1. American Institute of Steel Construction, Inc., Manual of Steel Construction, 9th Ed., One East Wacker Dr., Chicago, IL. 60601.
2. Crane Manufacturers Association of America, Inc., Specifications for Electric Overhead Traveling Cranes, CMAA Specification #70, Revised 1983.

PROBLEMS

Section 6.3

6.1 Use the integration method to determine the equations for the slope and deflection of a cantilevered beam loaded with an end moment M_o. Let EI be constant. Now let $L = 10$ ft, $E = 30 \times 10^6$ psi, $I = 200$ in.^{4}, and $M_o = 10$ kip·ft, and determine the numeric values for the slope and deflection at the free end.

Figure P6.1

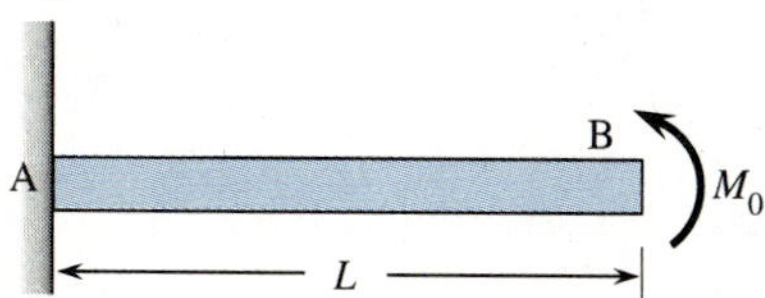

6.2 Use the integration method to determine the equations for the slope and deflection of a cantilevered beam loaded with a linearly varying load. Let EI be constant. Now let $L = 4$ m, $E = 205$ GPa, $I = 2 \times 10^{-6}$ m^4, and $w = 10$ kN/m and determine the numeric values of the maximum slope and deflection.

Figure P6.2

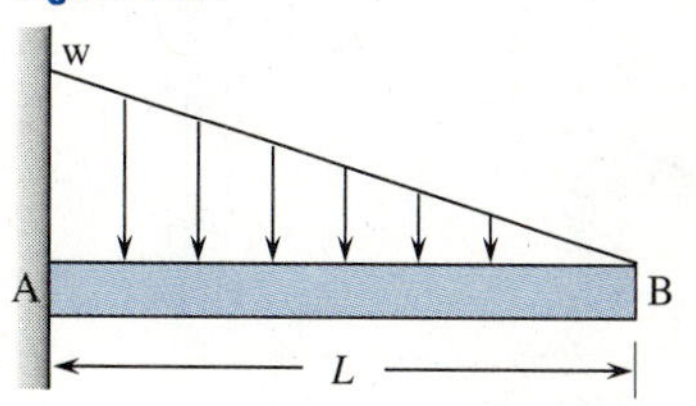

6.3 Use the integration method to determine the free end horizontal deflection of the vertical column subjected to the concentrated loads shown. Let $E = 30 \times 10^6$ psi and $I = 300$ in.^{4}

Figure P6.3

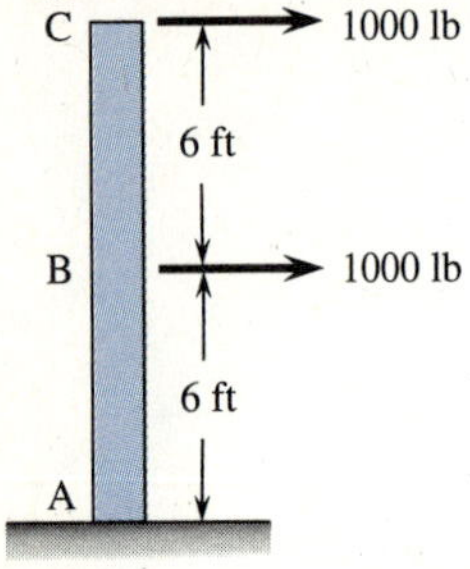

6.4 Determine the vertical deflection at the right end of the beam shown. Let EI be constant.

Figure P6.4

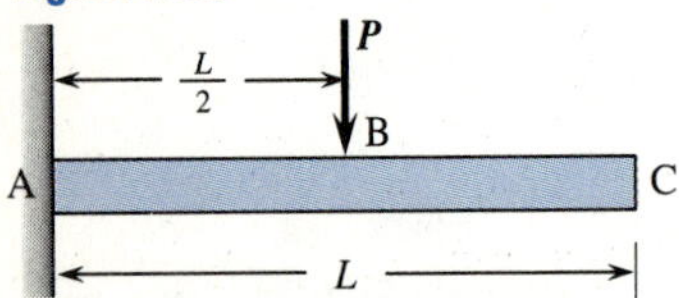

6.5 Use the integration method to determine the equations for the slope and deflection of a simple beam loaded with a linearly increasing distributed load. Let EI be constant. Determine the location and magnitude of the maximum deflection. Now let $L = 12$ ft, $E = 30 \times 10^6$ psi, $I = 200$ in.^{4}, and $w_0 = 1000$ lb/ft and determine the numeric value of the maximum deflection.

Figure P6.5

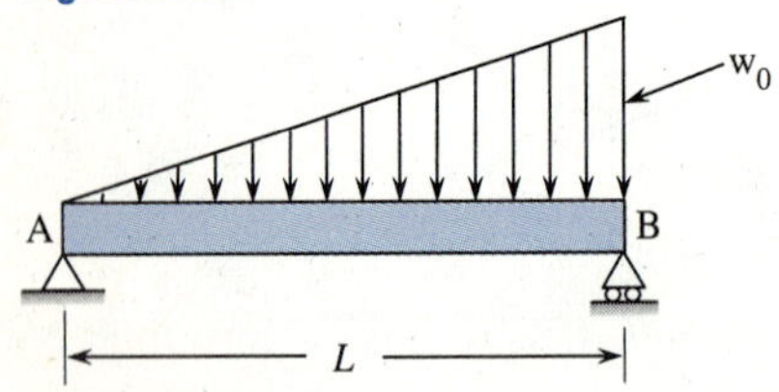

6.6 Use the integration method to determine the equations for the slope and deflection of a simple beam loaded with a moment on one end. Let EI be constant. Determine the location and the magnitude of the maximum deflection. Then let $M_L = 20$ kip · ft; $L = 10$ ft, $E = 30 \times 10^6$ psi, and $I = 100$ in.^{4} and determine the numeric value of the maximum deflection.

Figure P6.6

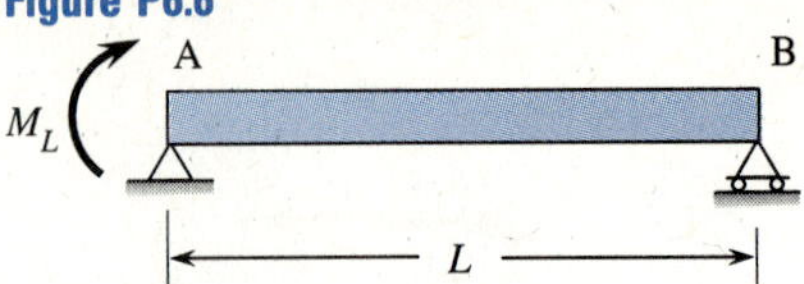

6.7 Use the integration method to determine the vertical deflection at B for the simply supported beam.

Figure P6.7

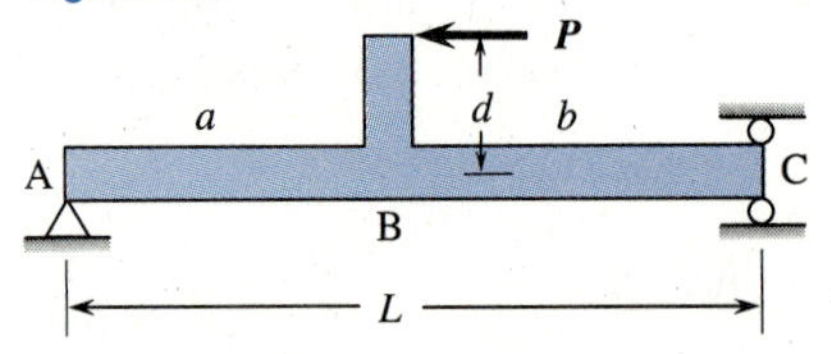

6.8 Use the integration method to determine the vertical deflection at midspan for the simple beam. Let EI be constant.

Figure P6.8

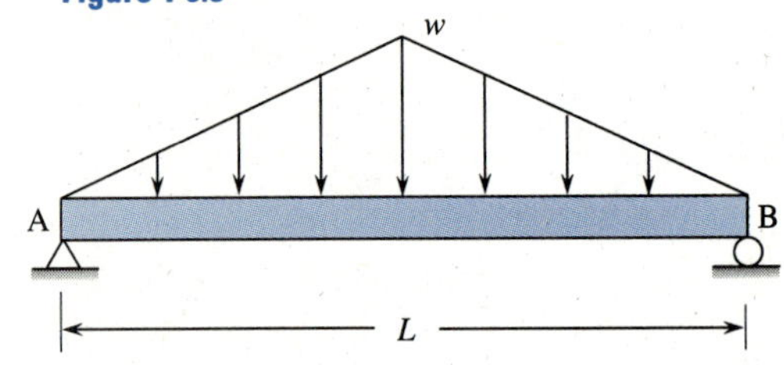

6.9 Use the integration method to obtain the equation of the elastic curve for the cantilevered beam AB carrying the uniform load of intensity w over one-half its length. Also determine the maximum deflection.

Figure P6.9

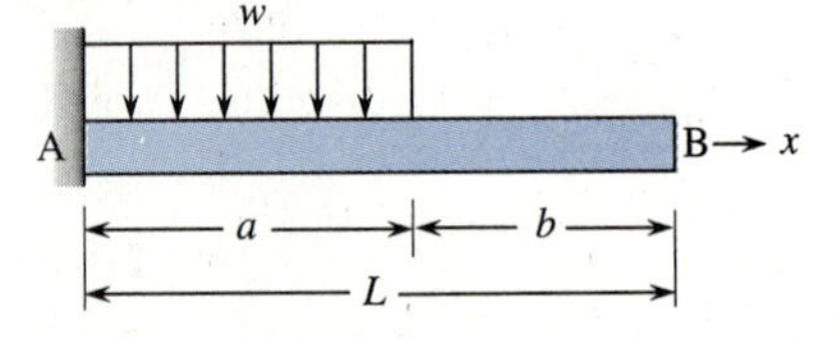

6.10 Use the integration method to obtain the elastic curve (deflected curve) for the cantilevered beam with uniform load acting over one-half the length as shown. Also determine the free end deflection and slope.

Figure P6.10

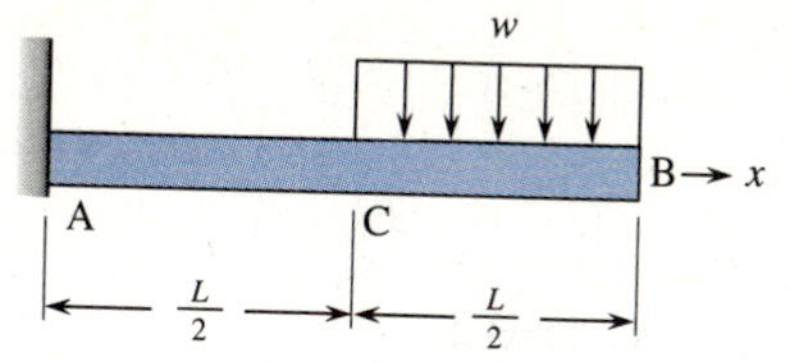

Figure P6.14

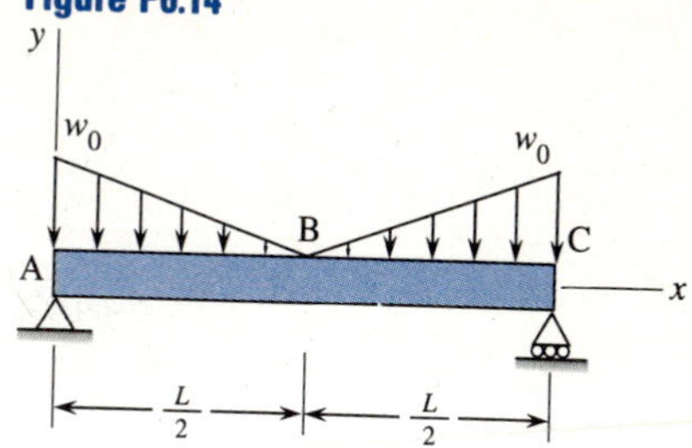

6.11–6.14 For the beams and loadings shown, determine **(a)** the equation(s) of the elastic curve (deflection equation) for section AB and **(b)** the deflection at the midpoint B.

Figure P6.11

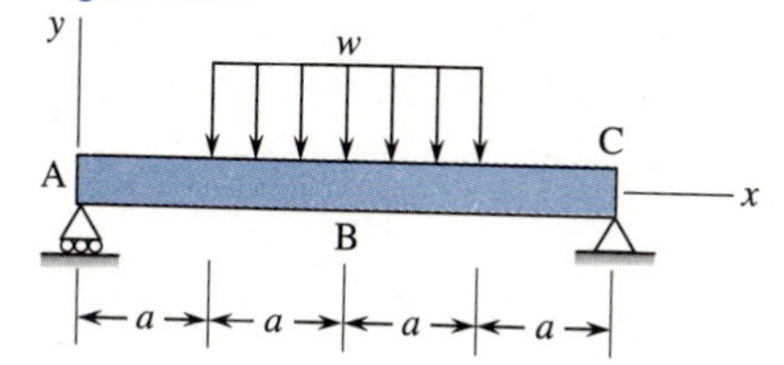

Figure P6.12

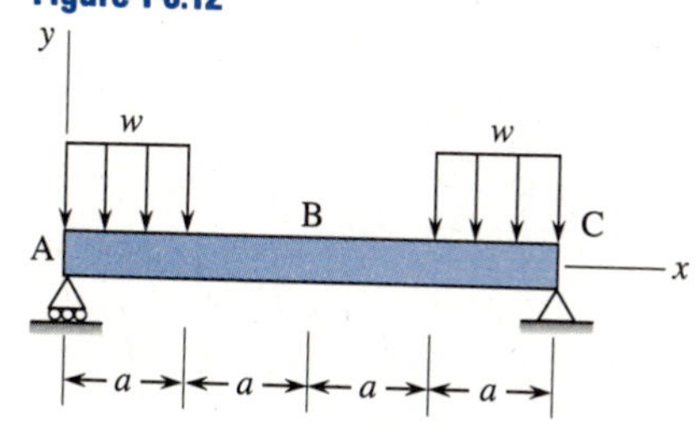

Figure P6.13

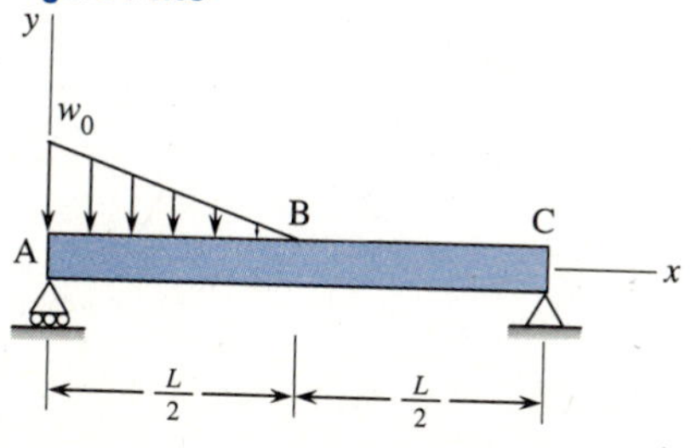

6.15 For the beam supported and loaded as shown, derive the equation of the elastic curve for section AB in terms of EI and x. The beam has constant EI. Use feet and kip units.

Figure P6.15

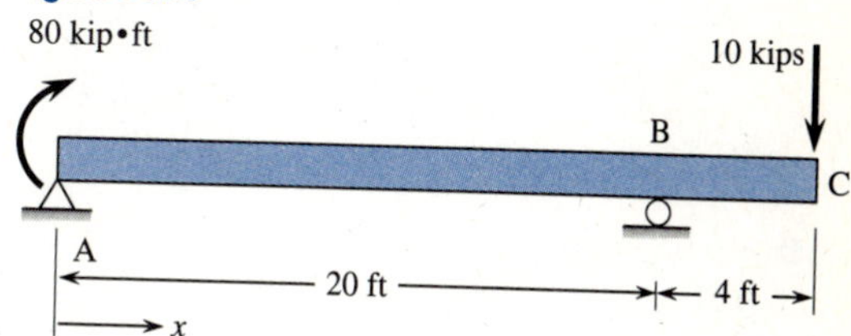

6.16 For the beam supported and loaded as shown, derive the equation of the elastic curve for section AB in terms of EI and x. The beam has constant EI. Use units of meters and kN.

Figure P6.16

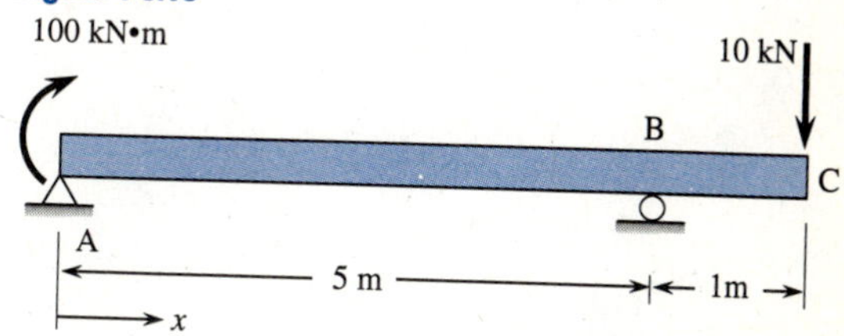

Section 6.4

6.17 Use the superposition method to determine the free end deflection of the cantilevered beam. Let EI be constant.

Figure P6.17

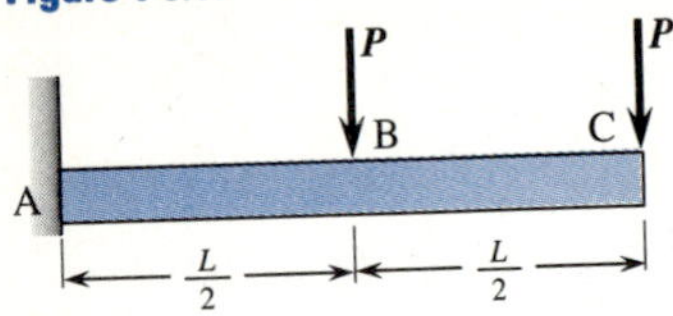

6.18 Use the superposition method to find the free end deflection and slope of the cantilevered beam. Let $E = 205$ GPa and $I = 186.1 \times 10^6$ mm^4.

Figure P6.18

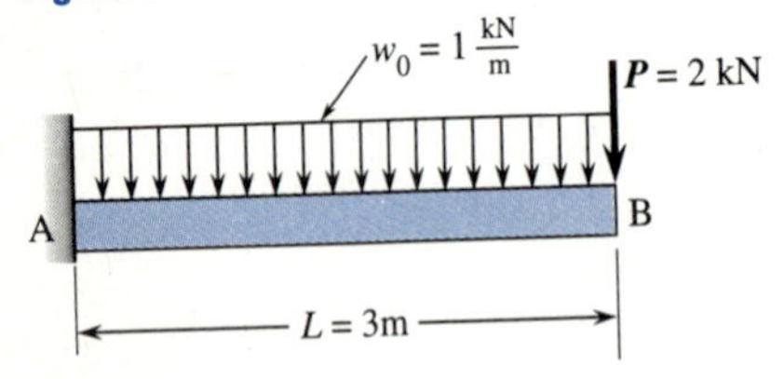

6.19 Use the superposition method to find the tip deflection for the cantilevered beam.

Figure P6.19

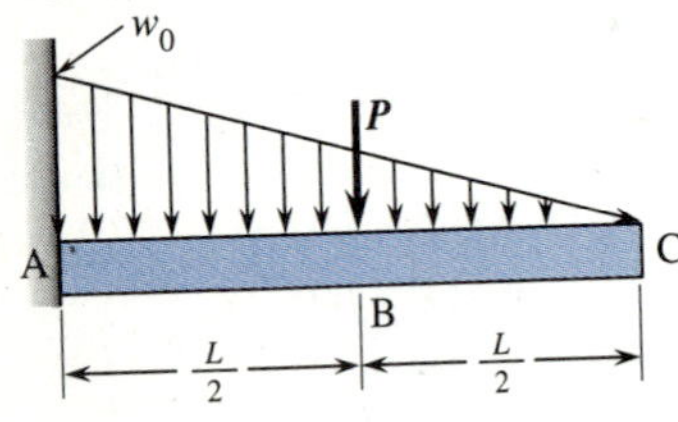

6.20 Use the superposition method to find the deflection at midspan of the simple beam. The beam is an S 12 × 35 of ASTM A36 steel. Let L = 10 ft.

Figure P6.20

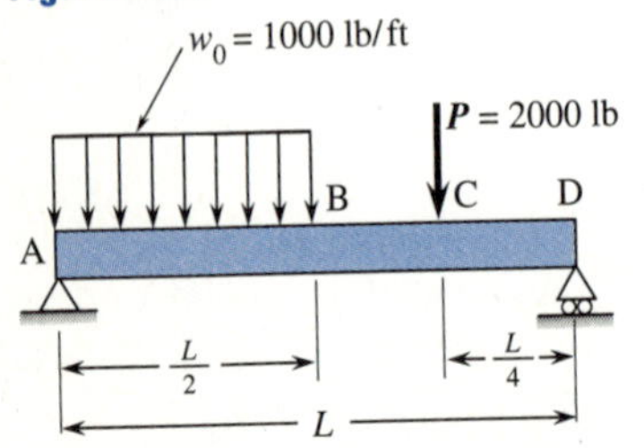

6.21 A steel wide flange (W 12 × 26) is used as a simple beam. The beam is subjected to a left end moment and uniform load. Use superposition to determine the midspan deflection.

Figure P6.21

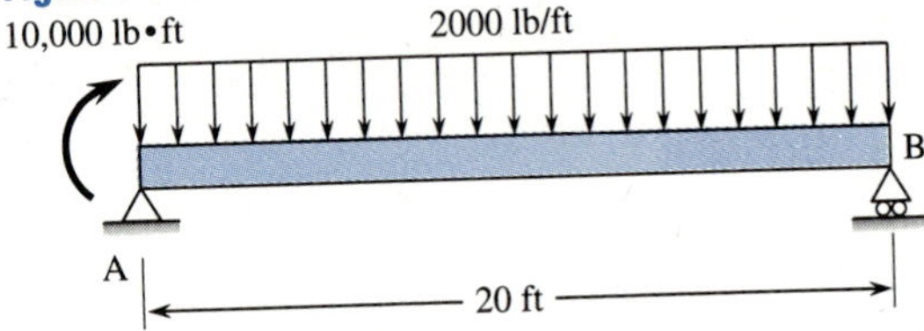

6.22–6.25 For the beams and loadings shown, use superposition to determine the deflection at the midspan. Use $E = 29 \times 10^6$ psi (200 GPa).

Figure P6.22

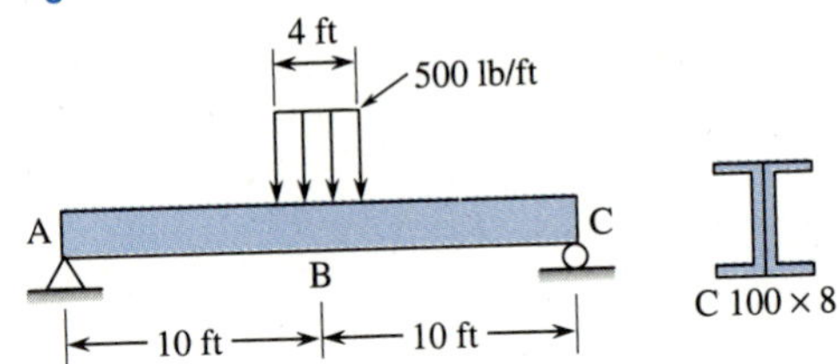

Figure P6.23

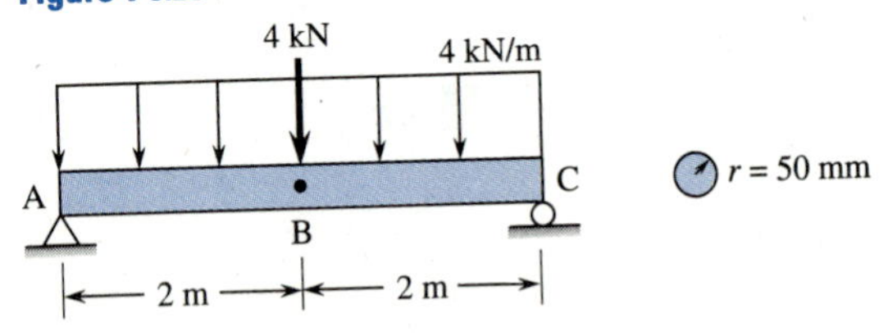

Figure P6.24

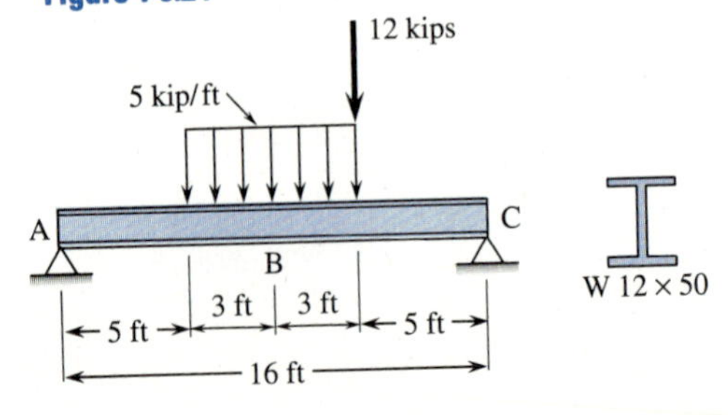

Figure P6.25

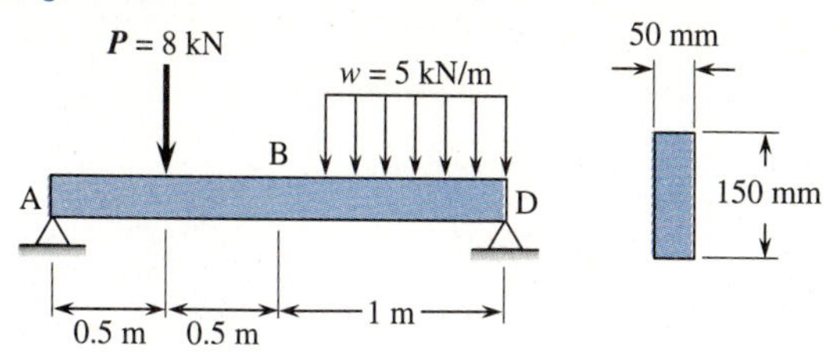

6.26 Use superposition to determine the vertical deflection at the free end of the cantilevered circular steel shaft. Let $E = 30 \times 10^6$ psi and the radius of the shaft be 1.0 in.

Figure P6.26

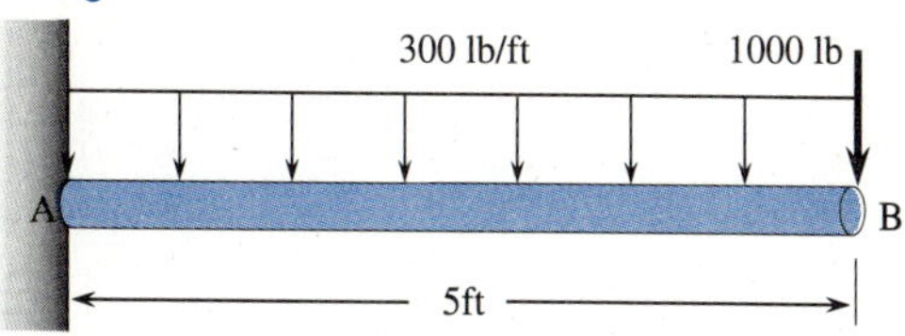

6.27 Use superposition to determine the deflection at point C on the steel cantilevered beam. Let $E = 200$ GPa and $I = 10^8$ mm^4.

Figure P6.27

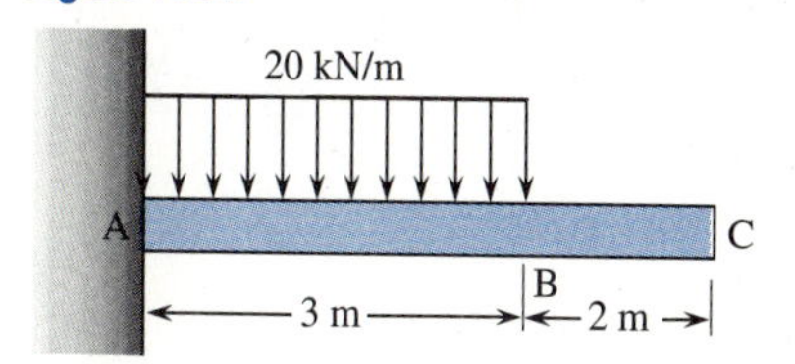

6.28 A nominal 2×8 Douglas fir joist (see Appendix C for actual dimensions) is used on a 14-ft span. Assuming that the joist is prevented from buckling laterally, determine the maximum deflection due to two 500-lb forces symmetrically placed on the beam 3 ft apart, and due to the uniform load of 60 lb/ft. Use superposition.

Figure P6.28

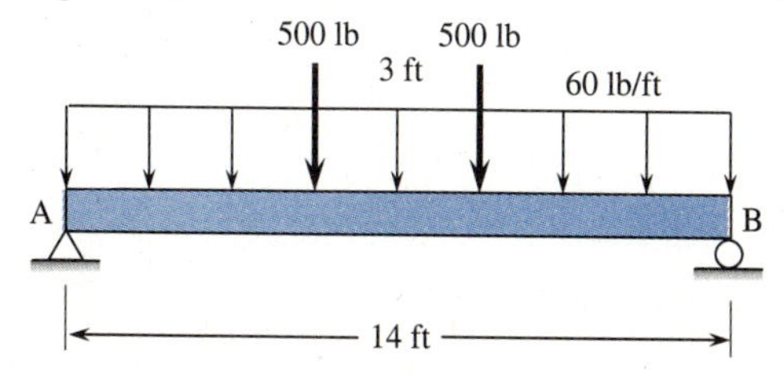

6.29 A simply supported crane runway box beam is subjected to the two loads shown as transfered from a trolley car lifting a roll of aluminum foil from below and from a uniform dead load of 100 lb/ft. Determine the maximum deflection. Is this deflection less than $L/600$ where L is the span length? Use superposition. Let $E = 29 \times 10^6$ and $I = 600$ in.^{4}

Figure P6.29

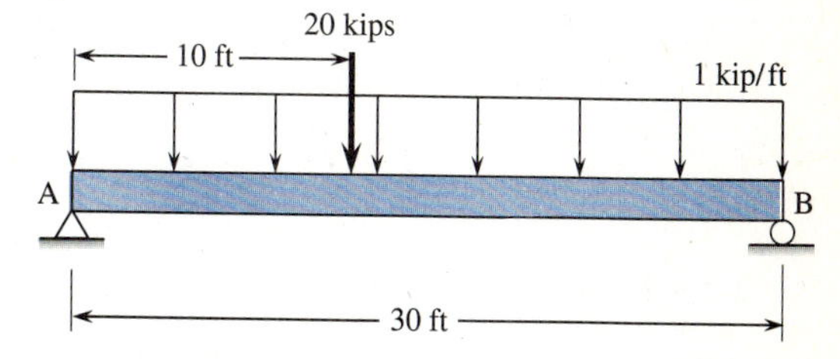

6.30 A simple beam (W14 × 68) in a steel building is subjected to a concentrated and a distributed load shown. Use superposition to determine the deflection at midspan. Let $E = 29 \times 10^6$ psi.

Figure P6.30

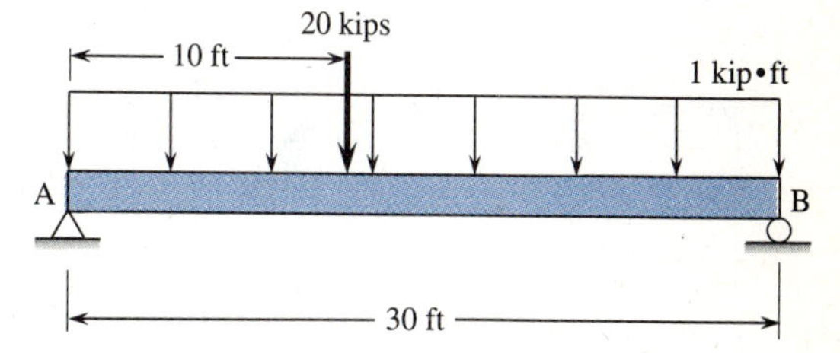

Section 6.5

6.31 Use the moment-area method to determine the deflection and the slope at the free end of the cantilever beam shown in Figure P6.1.

6.32 Use the moment-area method to determine the deflection and the slope at the free end of the beam in Figure P6.4.

6.33 Use the moment-area method to determine the deflection at $x = L/2$ on the cantilever in Figure P6.2.

6.34 Use the moment-area method to determine the deflection at the midspan of the simple beam in Figure P6.5.

6.35 Use the moment-area method to determine the midspan deflection of the simple beam in Figure P6.6.

6.36 A simple beam has an abrupt change in cross-section. Use the moment-area method to determine the deflection at $L/2$. Let $E = 200$ GPa, $I = 2 \times 10^{-4}$ m^4, P = 100 kN, $L = 6$ m, and $a = 2.5$ m.

Figure P6.36

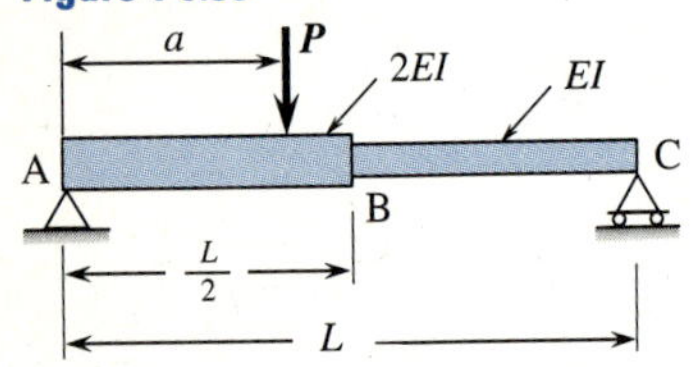

6.37 Use the moment-area method to solve for the deflection at point C on the overhanging beam. Let E = 30 × 10^6 psi, I = 300 in.^{4}, and L = 10 ft.

Figure P6.37

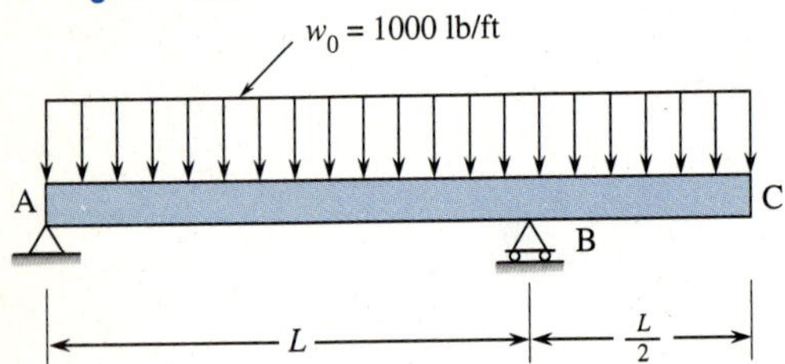

Section 6.6

6.38–6.45 For the beams shown in the figures, use discontinuity functions to express the intensity $w(x)$ of the equivalent distributed load acting on the beam, including all reactions. Assume $E = 29 \times 10^6$ psi (200 GPa).

Figure P6.38

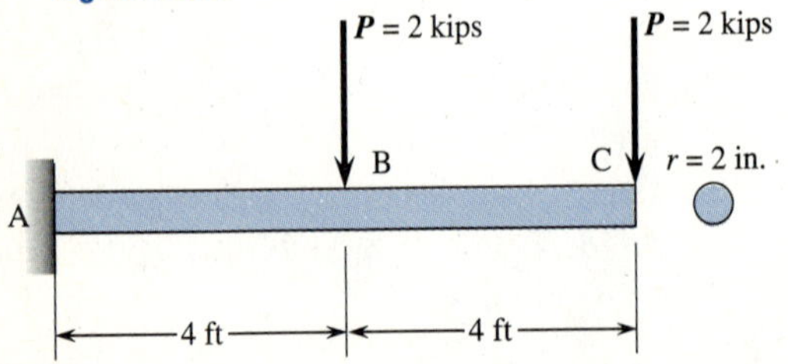

Figure P6.39

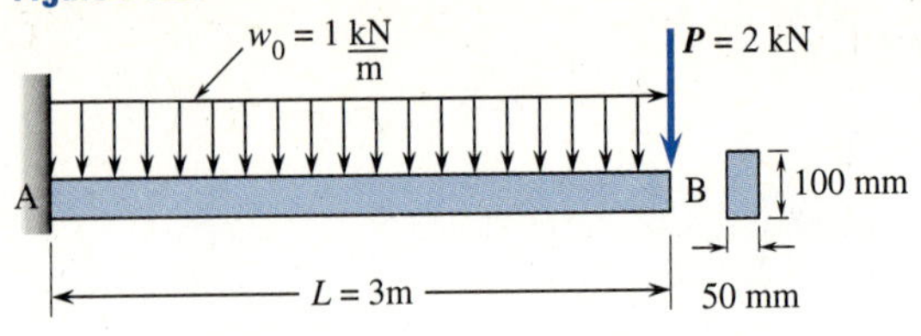

Figure P6.40

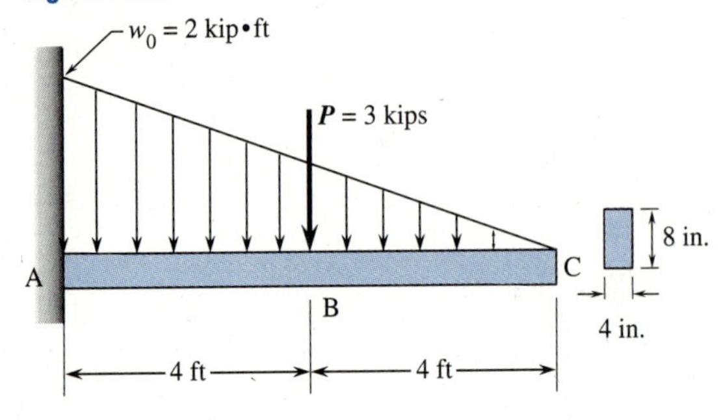

Figure P6.41

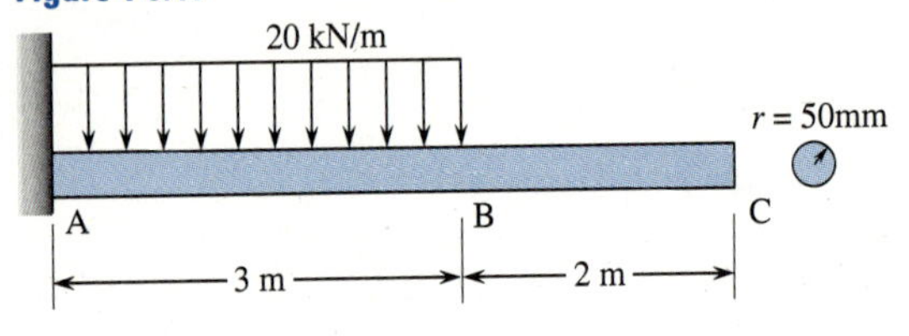

Figure P6.42

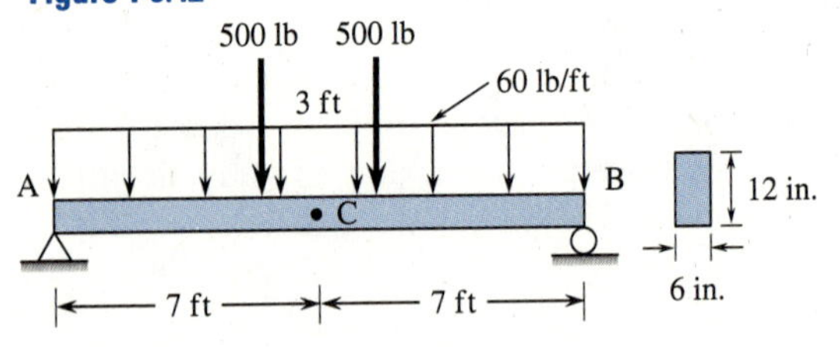

Figure P6.43

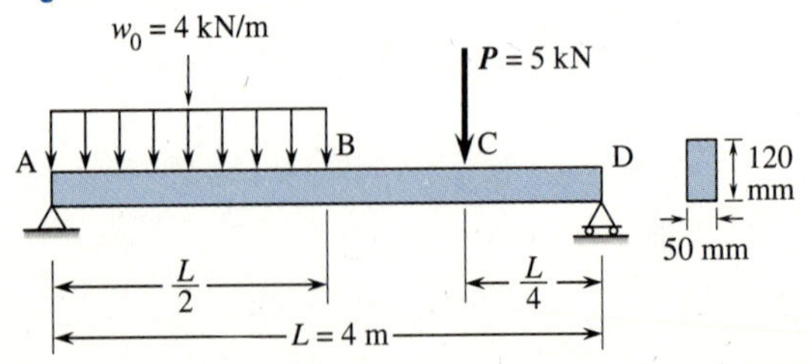

Figure P6.44

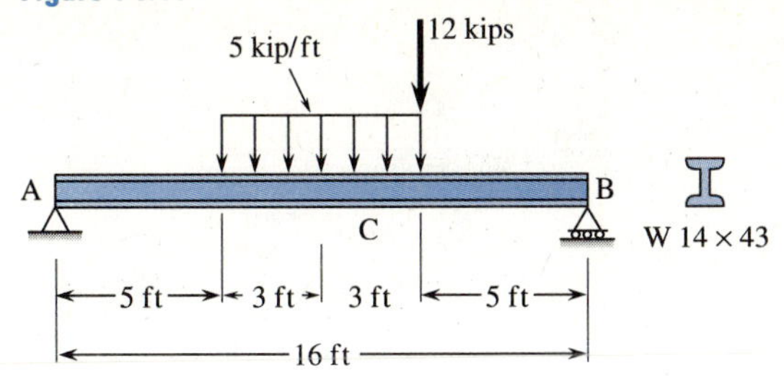

Figure P6.45

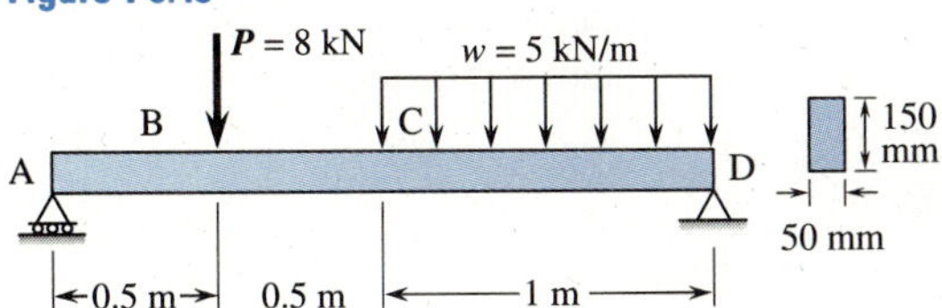

6.46–6.49 For the cantilever beams of problems 6.38 through 6.41, use discontinuity functions to determine the equation of the deflection curve. Also determine the deflection at the free end.

6.50–6.53 For the simply supported beams of problems 6.42 through 6.45, use discontinuity functions to determine the equation of the deflection curve. Also determine the deflection at point C.

6.54–6.56 For the beams with overhangs shown, use discontinuity functions to determine the equation of the deflection curve. Also determine the deflections at points C and D. Use $E = 29 \times 10^6$ psi (200 GPa) and $I = 12$ in.^{4} (5×10^{-6}m^4).

Figure P6.54

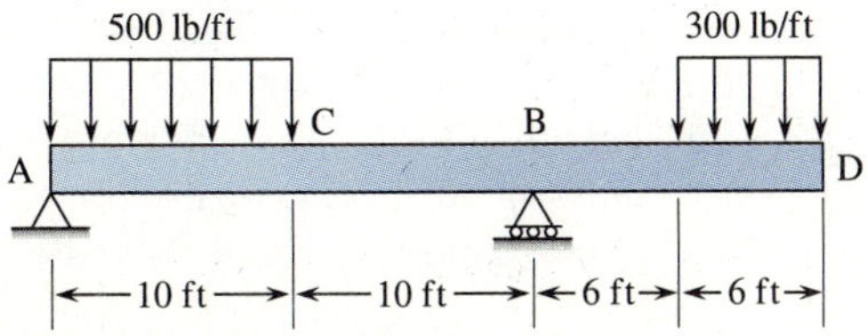

500 lb/ft 300 lb/ft A C B D 10 ft 10 ft 6 ft 6 ft

Figure P6.55

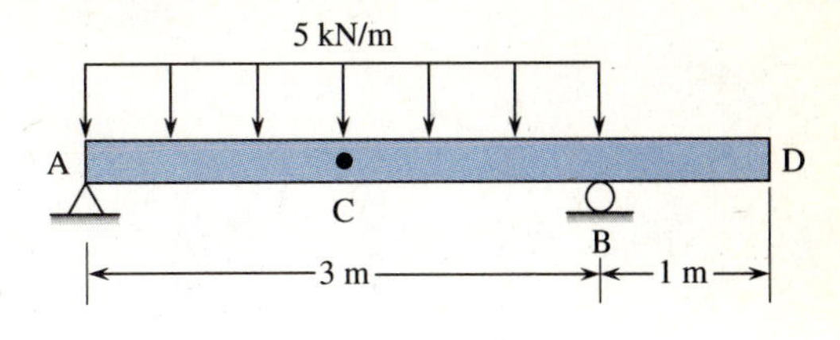

Figure P6.56

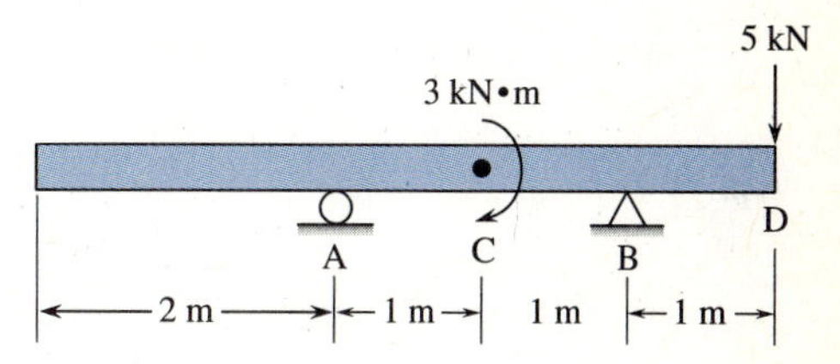

Section 6.7

6.57 For the statically indeterminate vertical beam shown, use the integration method to determine **(a)** reactions at B and **(b)** the midspan deflection. Let $E = 30 \times 10^6$ psi and $I = 200$ in.^{4}

Figure P6.57

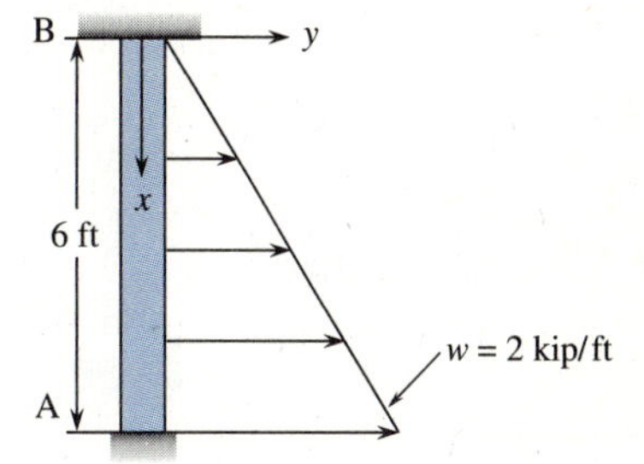

6.58 For the statically indeterminate beam shown, use the integration method to determine **(a)** the reactions, **(b)** the deflection curve, and **(c)** the deflection at B. Let $E = 30 \times 10^6$ psi, $I = 100$ in.^{4} and $L = 10$ ft.

Figure P6.58

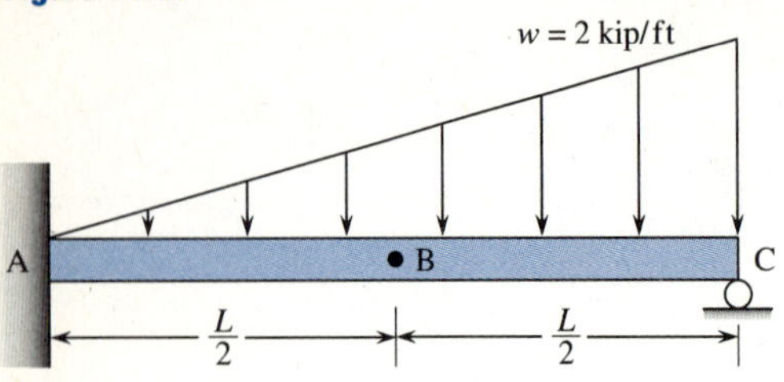

6.59–6.60 For the statically indeterminate beams shown, use the integration method to determine the deflection curve. Let EI be constant.

Figure P6.59

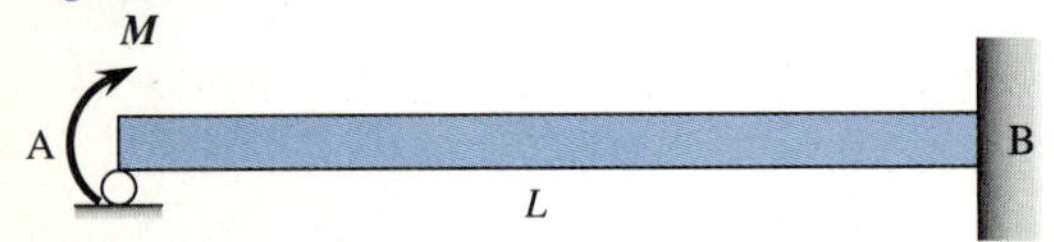

Figure P6.60

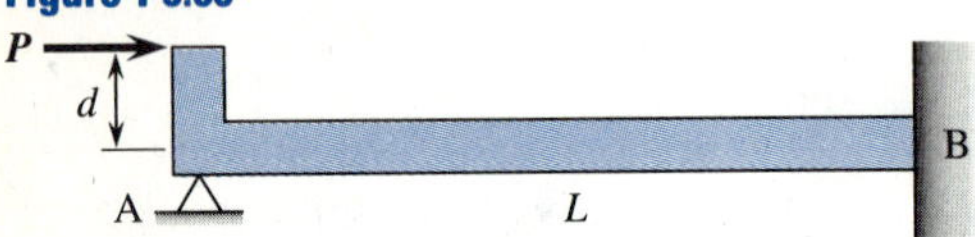

6.61 Solve Problem 6.57 for the reactions at A by the method of superposition.

6.62 Solve Problem 6.58 for the reaction at C by the method of superposition.

6.63 Solve Problem 6.60 for the reaction at A by the method of superposition.

6.64–6.65 For the statically indeterminate beams shown, use the method of superposition to obtain **(a)** the reactions at A and B, **(b)** the deflection curve, and **(c)** the deflection at C. Let $E = 200$ GPa and $I = 200 \times 10^{-6}$ m^4.

Figure P6.64

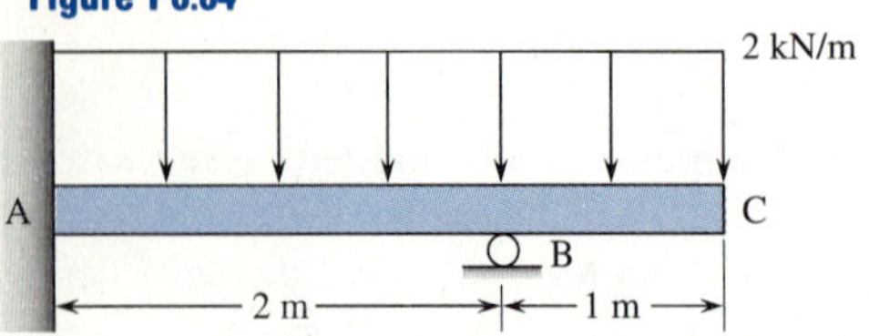

Figure P6.65

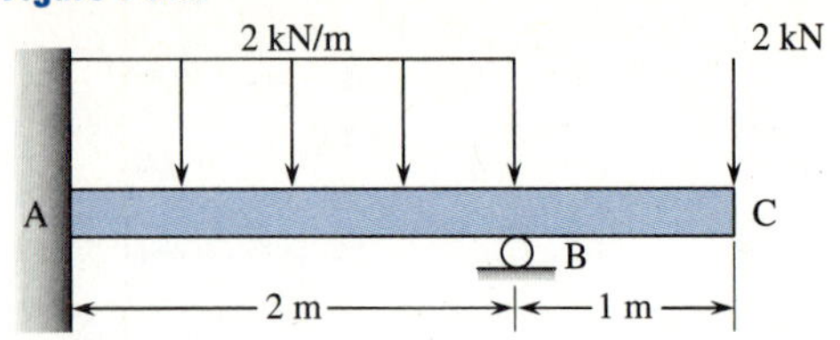

6.66 For the beam shown, there is a gap equal to d between the beam and the end support C before the load P is applied. After the load is applied, this gap is closed between the beam and support C so that part of the load is carried by support C. Determine the reactions at the three supports in terms of P, d, L, E, and I. Use the method of superposition.

Figure P6.66

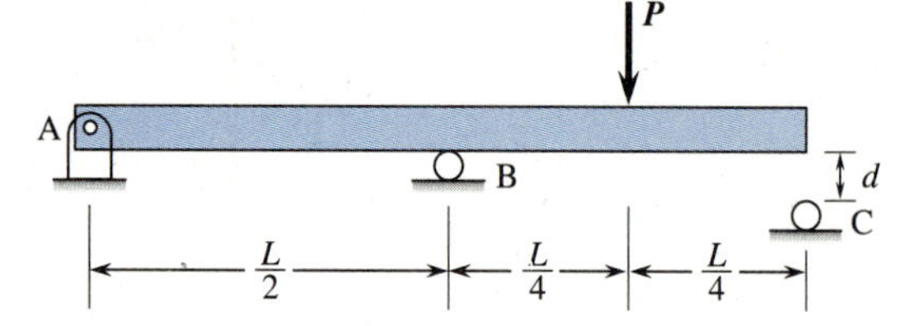

COMPUTER PROBLEMS

The following problems are suggested to be solved using a programmable calculator, microcomputer, or mainframe computer. It is suggested that you write a simple computer program in BASIC, FORTRAN, PASCAL, or some other language to solve these problems.

C6.1 A simply supported beam of length L is subjected to a concentrated downward load at distance a from the left end. Write a program to determine the deflection anywhere in the beam and the maximum deflection. The input should include the span length L, the load P, the principal moment of inertia I that bending is taking place about, the modulus of elasticity E, and the increment of L for each output. Check your program using the following data: $L = 4$ m, $a = 2$ m, $P = 50$ kN, $E = 200$ GPa, $I = 0.0491$ m^4.

Figure C6.1

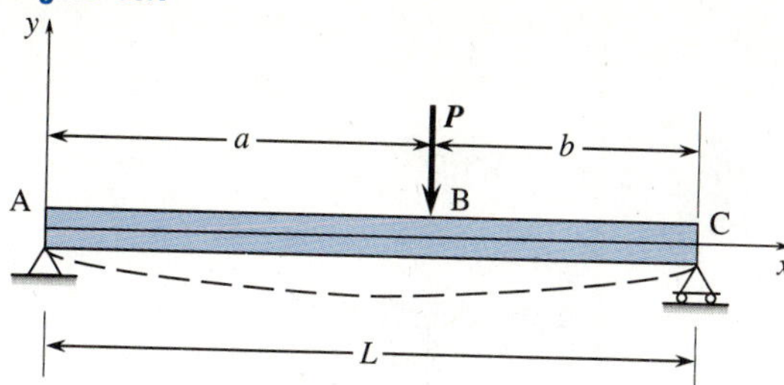

C6.2 Rework Problem C6.1 if the beam is now subjected to a uniformly applied load over the whole beam instead of the concentrated load P. Check your program using the following data: All data as in problem C6.1 except $w = 2000$ kN/m and $P = 0$.

Figure C6.2

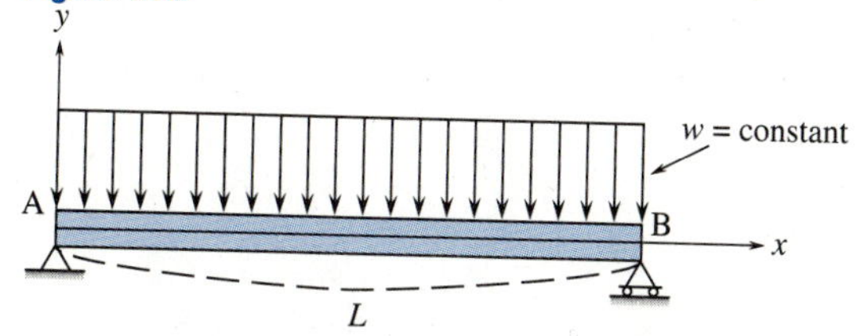

C6.3 A cantilevered beam of length L is fixed at its left end and free at the right end. It is subjected to a downward uniform load of intensity w over its entire length. Write a program to determine the deflection anywhere in the beam. The input should include the beam length L, the load intensity w, the principal moment of inertia that bending is taking place about, the modulus of elasticity, and the increment of L for each output. Check your program using the following data: $L = 6$ ft, $w = 50$ lb/ft, $E = 29 \times 10^6$ psi, and a rectangular cross-section with width 4 in. and depth 8 in.

Figure C6.3

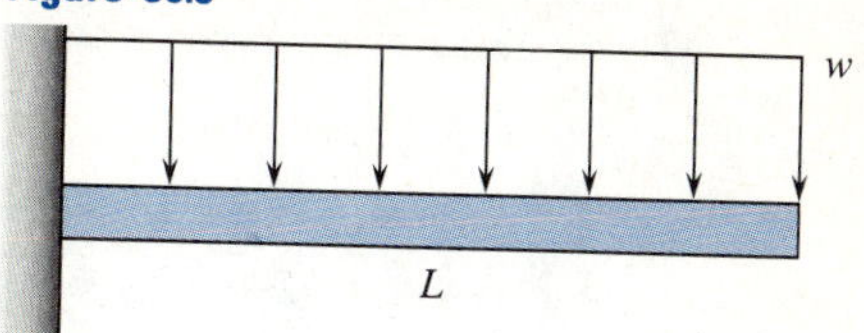

C6.4 Rework Problem C6.3 if the beam is now subjected to a concentrated load at its free end instead of the uniform load. Check your program using the following data: $L = 6$ ft, $P = 200$ lb, $E = 29 \times 10^6$ psi, and rectangular cross-section with width 4 in. and depth 8 in.

Figure C6.4

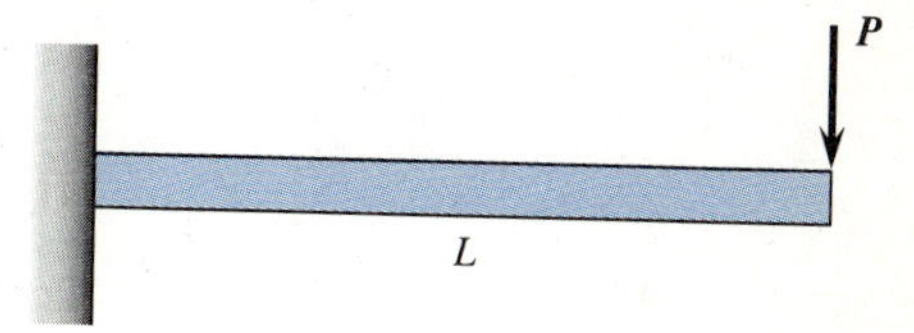

C6.5 Rework Problem C6.3 if the beam is now subjected to a linearly varying line load that increases from zero at the free end to a maximum value w at the fixed end. Check your program using the following data: $L = 6$ ft, $w = 500$ lb/ft, and rectangular cross-section of width 4 in. and depth 8 in.

Figure C6.5

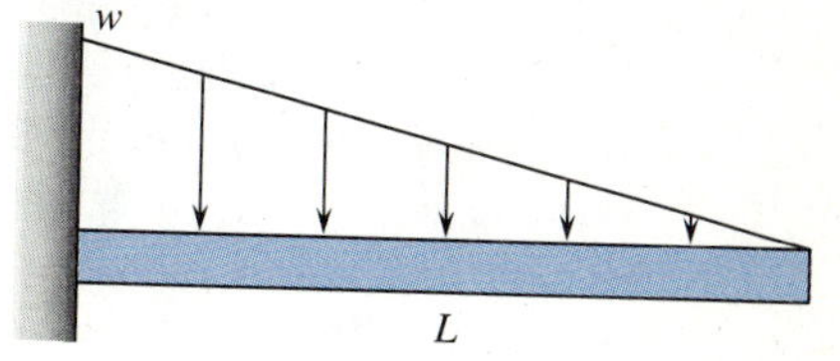

Stresses Under Combined Loadings, Pressure Vessels

7.1 INTRODUCTION

In the preceding chapters we developed the various formulas used to obtain the normal and shear stresses in bars, shafts, and beams. For instance, we used $\sigma = P/A$ to obtain the normal stress in a bar subjected to direct axial loading and $\tau = Tr/J$ to obtain the shear stress in a circular shaft subjected to a torsional moment. However, many times machine parts and structural members are simultaneously subjected to both normal and shear stresses on an element of the member. As an example, a shaft may be subjected to simultaneous axial loading and torque (Figure 7.1(a)) resulting in both normal and shear stresses acting on an element of the shaft. Or a shaft may be subjected to a transverse loading and a torque such that the transverse shear stress determined by $\tau = VQ/(Ib)$, and the torsional stress determined by $\tau = Tr/J$ are superimposed to obtain the largest shear stress at a point on the shaft (Figure 7.1(b)).

Similarly, a beam may be simultaneously subjected to transverse loading, resulting in a normal stress obtained from the flexure formula, $\sigma = Mc/I$, and a direct axial loading resulting in a normal stress $\sigma = P/A$, which are superimposed to obtain the largest normal stress at a point in the beam (Figure 7.2).

Therefore, our objectives in this chapter are to determine, through superposition, (1) the normal stresses that develop from the simultaneous action of axial force and bending moment and (2) the shear stresses due to simultaneous action of torque and direct shear force. Superposition of normal stresses or of shear stresses separately is allowed if

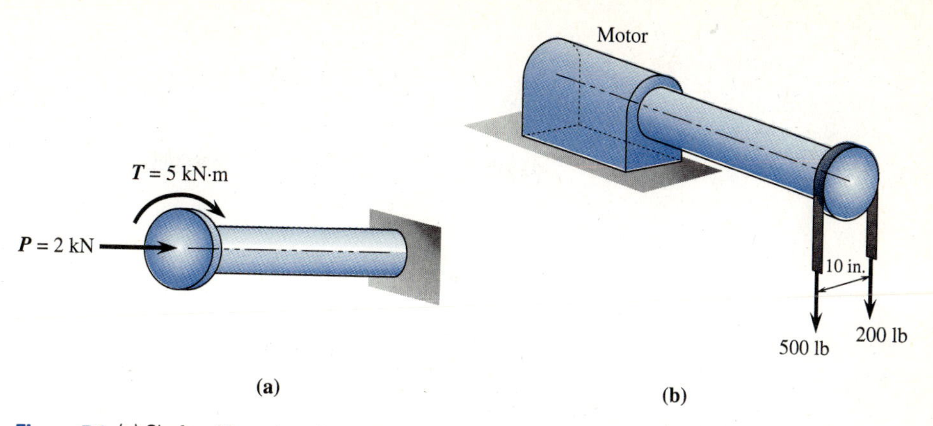

Figure 7.1 (a) Shaft subjected to simultaneous axial loading and torque and (b) shaft subjected to simultaneous transverse loading and torque.

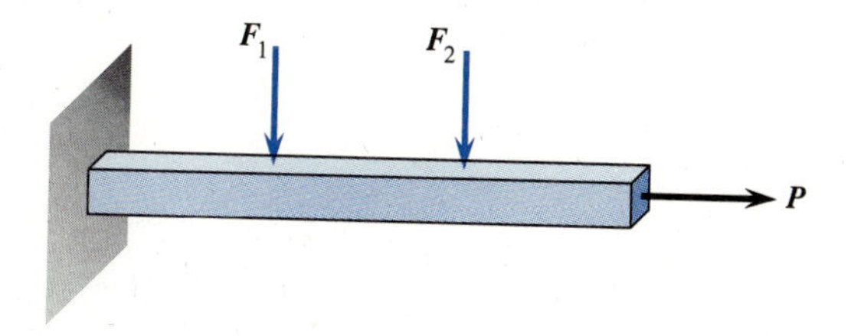

Figure 7.2 Beam simultaneously subjected to transverse and axial loading.

(1) the individual stresses are linear functions of the loads, that is, the stresses are below the elastic limit stress σ_e of the material, and the superimposed stresses are below σ_e of the material, and (2) there is no interaction effect among differently (independently) applied loads, that is, stresses due to one load are not affected by the presence of other loads. The material usually remains in the linear-elastic range if the deflections and rotations of the structure are small.

In the last two sections of this chapter we discuss stresses due to pressure inside thin-walled pressure vessels and solve problems of pressurized cylindrical and spherical tanks and combined loadings of tanks.

7.2 SUPERPOSITION OF NORMAL STRESSES

We illustrate the concept of superposition of normal stresses by first considering the simple beam subjected to simultaneous axial loading P and transverse loading w shown in Figure 7.3(a), with cross-section in Figure 7.3(b). The beam is shown subjected to the individual loadings P and w in Figures 7.3(c) and (d).

The normal stress due to P is given from Eq. (2.1) as

$$\sigma = \frac{P}{A} \tag{7.1}$$

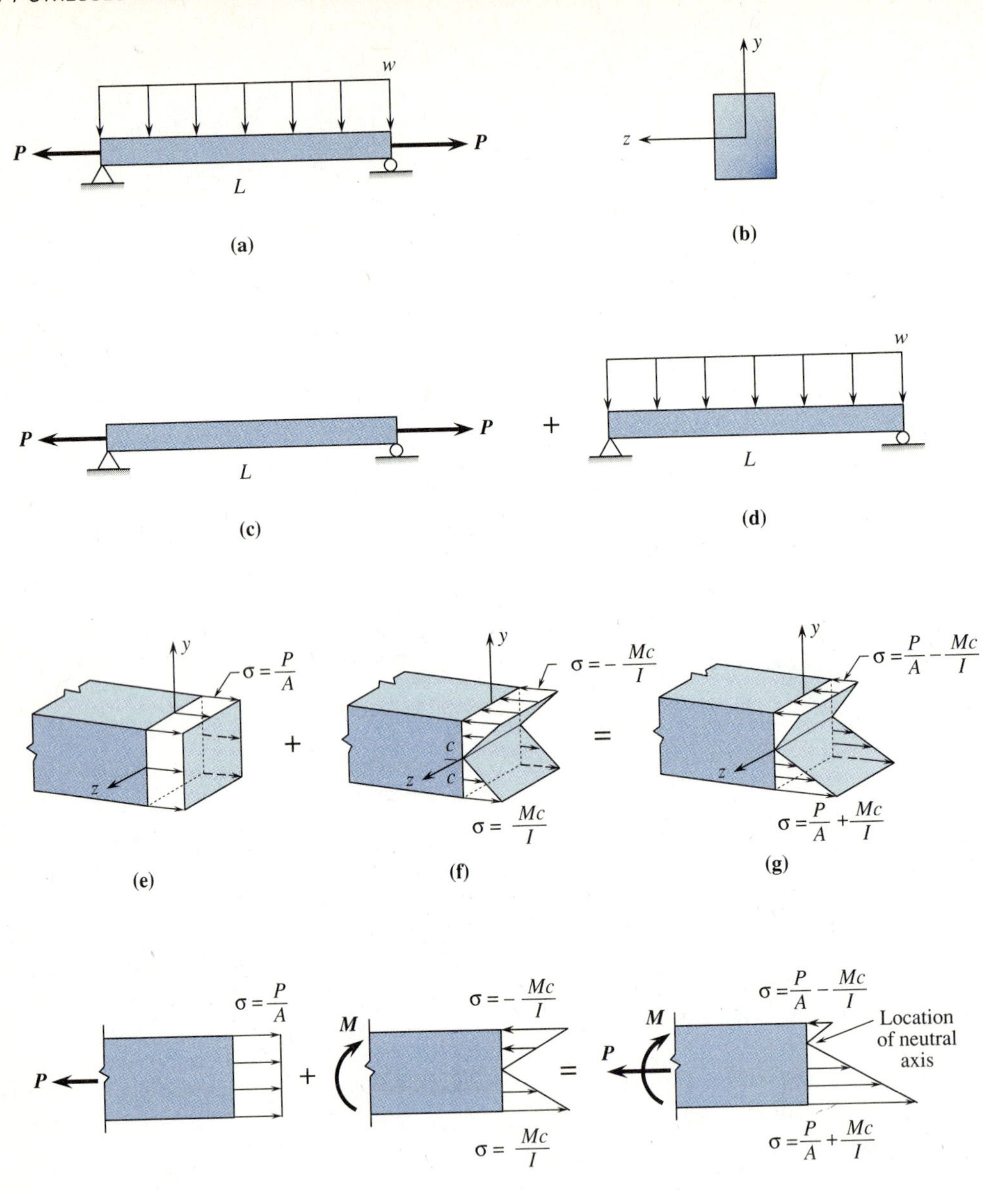

Figure 7.3 (a) Beam subjected to simultaneous transverse loading w and axial force P,
(b) cross-section of beam,
(c) beam subjected to axial load P, and
(d) beam subjected to transverse load w.
(e) Normal stress distribution due to direct axial force,
(f) normal stress distribution due to bending moment, and
(g) superposition of normal stresses due to direct axial force and bending moment.
(h) Side view representation of stress variations.

where A is the cross-sectional area of the beam. This normal stress acts uniformly over the cross-section as shown in Figure 7.3(e). The bending stress due to the transverse loading w is given from Eq. (5.7) as

$$\sigma = -\frac{My}{I} \tag{7.2}$$

where M is the bending moment due to w, y is the distance from the neutral axis, and I is the moment of inertia evaluated about the z-axis. The bending stress acts normal to the cross-section and varies linearly from the neutral axis as shown in Figure 7.3(f). To obtain the total normal stress at any point on the cross-section, we directly add the direct normal stress in Figure 7.3(e) to the bending normal stress in Figure 7.3(f). The result is shown in Figure 7.3(g). This total normal stress is given by

$$\sigma = \frac{P}{A} + \left(-\frac{My}{I}\right) \tag{7.3}$$

For simplicity, we often use side views to represent the stress variations through the depth (y-direction) of the beam for each load case as shown in Figure 7.3(h).

We see from Figure 7.3(h) that the line of zero stress, corresponding to the neutral axis of the beam, which is located at the centroid of the section for bending only, moves upward when the axial force also acts. In general, depending on the direction of M and P, the neutral axis may move up or down.

STEPS IN SOLUTION

The following steps are used to obtain the normal stress due to the simultaneous action of both direct axial force P and bending moment M.

1. Draw a free-body diagram of the desired section of the member showing the internal axial force P and bending moment M.
2. Write the equilibrium equations to obtain both P and M.
3. Use Eq. (7.3) to obtain the total normal stress due to both P and M. This requires
   (a) calculating the cross-sectional area A,
   (b) calculating the moment of inertia I, and
   (c) determining the appropriate coordinate value y through the depth of the member.

Examples 7.1 through 7.3 illustrate these steps to obtain the largest normal stress by superposition of normal stresses due to both normal force and bending moment.

EXAMPLE 7.1

A simply supported 10 ft long W 14 × 61 beam is loaded with a uniformly distributed load of 2 kips/ft and an axial tensile load of 120 kips (Figure 7.4). Determine the maximum normal stress. (The properties of the beam are found in Appendix C.)

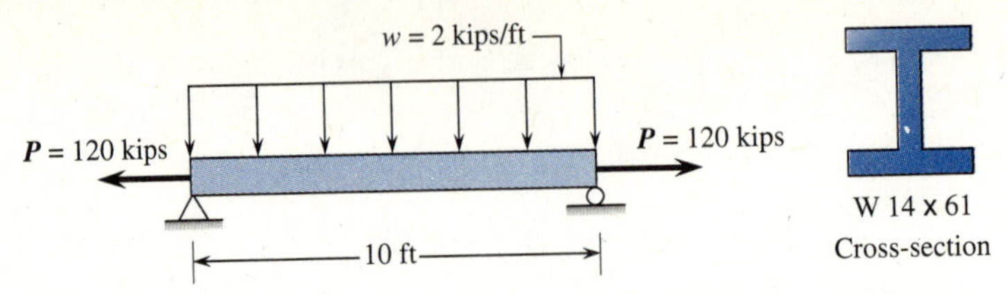

Figure 7.4 Simply supported beam subjected to a uniformly distributed load and axial load.

Solution

Using Appendix C, we obtain the cross-sectional area and moment of inertia of the W beam as

$$A = 17.9 \text{ in.}^2 \qquad I = 640 \text{ in.}^4$$

The largest stress occurs at the bottom of the beam where the direct and bending stress combine. In the flexure formula, we then use

$$y = -c = -\frac{h}{2} = -6.945 \text{ in.}$$

where $h = 13.89$ in. is the depth of the beam from Appendix C. Recall that y is negative in the downward direction. The maximum bending moment for simply supported beams subjected to a uniformly distributed load is

$$M_{\text{max}} = \frac{wL^2}{8}$$

at the midspan location (see Appendix D). Therefore

$$M_{\text{max}} = \frac{\left(2\frac{\text{kips}}{\text{ft}}\right)(10 \text{ ft})^2}{8} = 25 \text{ kip} \cdot \text{ft}$$

Using Eq. (7.3), the maximum normal stress is

$$\sigma_{\text{max}} = \frac{120 \text{ kips}}{17.9 \text{ in.}^2} + \frac{(25 \text{ kip} \cdot \text{ft})\left(12\frac{\text{in.}}{\text{ft}}\right)(6.945 \text{ in.})}{640 \text{ in.}^4}$$

$$= 6.70 \text{ ksi} + 3.26 \text{ ksi}$$

$$= 9.96 \text{ ksi} \quad \text{(tensile)}$$

EXAMPLE 7.2

A machine part for transmitting a pull of 20 kN is offset as shown in Figure 7.5. Determine the largest normal stress in the offset portion a-a of the member. The cross-section of the offset is triangular as shown.

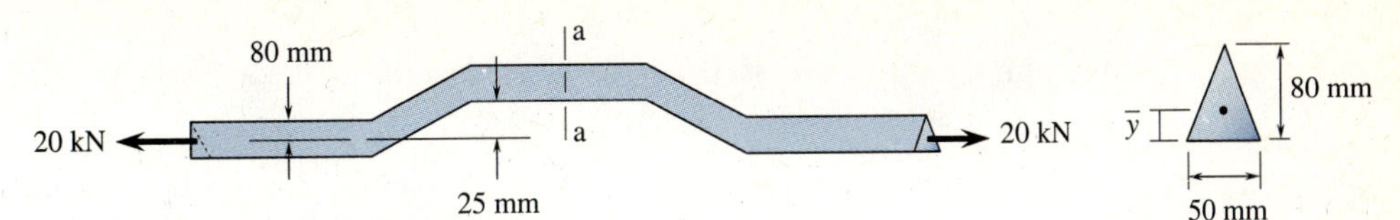

Figure 7.5 Machine part transmitting force.

Solution

From a free-body diagram of the left section to a-a, P and M are shown.

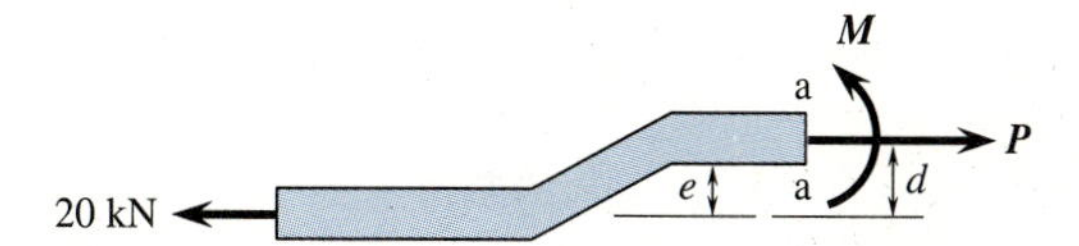

By statics

$$M = (20 \text{ kN})d = (20 \text{ kN})(0.0517 \text{ m}) = 1.034 \text{ kN} \cdot \text{m}$$

$$P = 20 \text{ kN}$$

where the distance d between the 20-kN forces is obtained as the eccentricity $e = 0.025$ m plus the distance $\bar{y}$ to the centroid of the triangular cross-section where the normal force $P = 20$ kN acts at a-a. That is

$$\bar{y} = \frac{h}{3} = \frac{0.080 \text{ m}}{3} = 0.0267 \text{ m}$$

The cross-sectional area and moment of inertia about the centroid z-axis for the triangular cross-section are

$$A = \frac{bh}{2} = \frac{(0.05 \text{ m})(0.08 \text{ m})}{2} = 0.002 \text{ m}^2$$

and

$$I = \frac{bh^3}{36} = \frac{(0.05 \text{ m})(0.08 \text{ m})^3}{36} = 0.711 \times 10^{-6} \text{ m}^4$$

Maximum Tensile Stress Using Eq. (7.3), we obtain the largest tensile normal stress at the bottom (negative y-direction), where the direct and bending stress add as

$$\sigma_{\text{bottom}} = \frac{P}{A} - \frac{M(-c_B)}{I}$$

$$= \frac{20 \text{ kN}}{0.002 \text{ m}^2} + \frac{(1.034 \text{ kN} \cdot \text{m})\left(\frac{0.08 \text{ m}}{3}\right)}{0.711 \times 10^{-6} \text{ m}^4}$$

$$= 10{,}000 \frac{\text{kN}}{\text{m}^2} + 38{,}800 \frac{\text{kN}}{\text{m}^2}$$

$$\sigma_{\text{bottom}} = 48{,}800 \frac{\text{kN}}{\text{m}^2} = 48.8 \frac{\text{MN}}{\text{m}^2} = 48.8 \text{ MPa} \quad \text{(tensile)}$$

where c_B is the distance from the neutral axis to the bottom of the cross-section.

Maximum Compressive Stress The largest compressive stress occurs at the top of the cross-section due to the bending stress. Again, using Eq. (7.3), we now have

$$\sigma_{\text{top}} = \frac{P}{A} - \frac{M(c_T)}{I}$$

where $y = c_T$ and $c_T = 0.0533\ m$ is the distance from the neutral axis to the top of the cross-section.

Therefore

$$\sigma_{\text{top}} = +10{,}000\,\frac{\text{kN}}{\text{m}^2} - \frac{(1.034\ \text{kN}\cdot\text{m})(0.0533\ \text{m})}{0.711 \times 10^{-6}\ \text{m}^4}$$

$$= 10{,}000\,\frac{\text{kN}}{\text{m}^2} - 77{,}500\,\frac{\text{kN}}{\text{m}^2}$$

$$\sigma_{\text{top}} = -67{,}500\,\frac{\text{kN}}{\text{m}^2} = 67.5\ \text{MPa} \quad \text{(compressive)}$$

The results show that the bending stress dominates with the large compressive stress attributed to the large distance c_T from the neutral axis to the uppermost fibers of the cross section.

EXAMPLE 7.3

A cast iron frame for a punch press has the dimensions shown in Figure 7.6. Determine the force P that may be applied to the frame controlled by the stresses in section a-a, if the allowable stresses are 4000 psi in tension and 12,000 psi in compression. The cross-section at a-a is also shown in Figure 7.6.

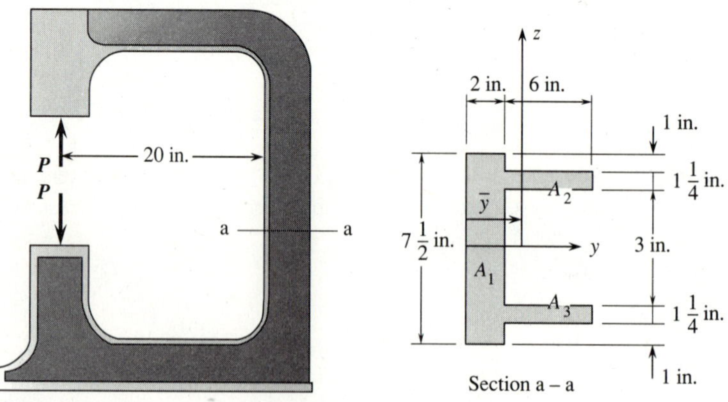

Figure 7.6 Cast iron frame for punch press.

Solution

First, we cut through a section at a-a and draw a free-body diagram of the upper portion of the frame with the internal normal force P and the bending moment M expressed in terms of P for equilibrium of the upper part.

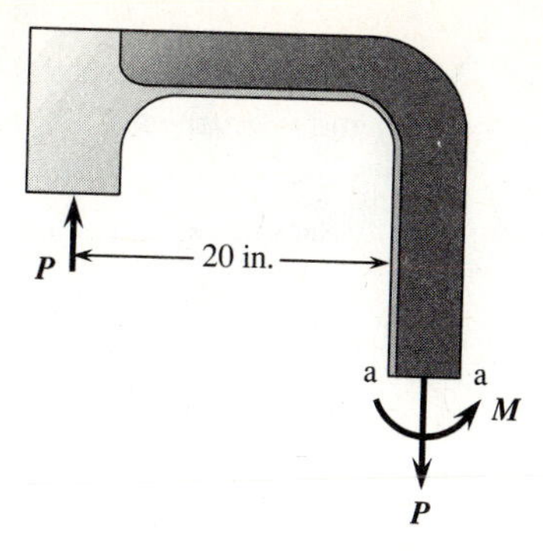

Figure 7.7 Upper portion of frame in equilibrium.

The centroid of the cross-section is obtained by the standard method using the three rectangular areas shown in Figure 7.6.

$$\bar{y} = \frac{\Sigma A_i \bar{y}_i}{\Sigma A_i} = \frac{A_1\bar{y}_1 + A_2\bar{y}_2 + A_3\bar{y}_3}{A_1 + A_2 + A_3} = \frac{A_1\bar{y}_1 + 2A_2\bar{y}_2}{A_1 + 2A_2}$$

since $A_2 = A_3$ and $\bar{y}_2 = \bar{y}_3$.

$$\bar{y} = \frac{(7.5 \text{ in.} \times 2 \text{ in.})(1 \text{ in.}) + 2(1.25 \text{ in.} \times 6 \text{ in.})(5 \text{ in.})}{(7.5 \text{ in.} \times 2 \text{ in.}) + 2(1.25 \text{ in.} \times 6 \text{ in.})}$$

$$\bar{y} = 3 \text{ in.} \quad \text{(from the left edge)}$$

The moment of inertia is evaluated about the centroidal z-axis since the bending moment is about the z-axis. Using $I_z = bh^3/12$ for a rectangular cross-section and the parallel axis theorem, we obtain

$$I_z = \frac{1}{12} b_1 h_1^3 + A_1 d_1^2 + 2\left(\frac{1}{12} b_2 h_2^3 + A_2 d_2^2\right)$$

$$= \frac{1}{12}(7.5 \text{ in.})(2 \text{ in.})^3 + (15 \text{ in.}^2)(2 \text{ in.})^2 + 2\left[\frac{1}{12}(1.25 \text{ in.})(6 \text{ in.})^3 + (7.5 \text{ in.}^2)(2 \text{ in.})^2\right]$$

$$I_z = 170 \text{ in.}^4$$

Maximum Tensile Stress Using Eq. (7.3), with $M = P(20 \text{ in.} + 3 \text{ in.})$, we observe from the direction of M that the maximum tensile stress occurs at the inside of section a-a and is given as 4000 psi. This allows us to calculate P, since we express the stress as

$$\sigma_{\text{inside}} = 4000 = \frac{P}{A} + \frac{Mc}{I_z}$$

$$= \frac{P}{30 \text{ in.}^2} + \frac{P(23 \text{ in.})(3 \text{ in.})}{170 \text{ in.}^4}$$

$$4000 = 0.439P$$

Solving for P, we have

$$P = 9110 \text{ lb}$$

Maximum Compressive Stress Using Eq. (7.3) again, we observe that the largest compressive stress occurs at the outer edge of section a-a and is given as 12,000 psi. Therefore

$$-12{,}000 = \frac{P}{A} - \frac{Mc}{I_z}$$

$$-12{,}000 = \frac{P}{30\text{ in.}^2} - \frac{P(23\text{ in.})(5\text{ in.})}{170\text{ in.}^4} = -0.643P$$

Solving for P, we obtain

$$P = 18{,}700\text{ lb}$$

We must use the smallest value of P to satisfy both the allowable tensile and compressive stresses. Therefore, the solution is

$$P = 9110\text{ lb}$$

7.3 SUPERPOSITION OF SHEAR STRESSES

We illustrate the superposition of shear stress due to simultaneous actions of torsion and transverse loading by considering the solid circular shaft of radius R subjected to torque T and transverse load P shown in Figure 7.8.

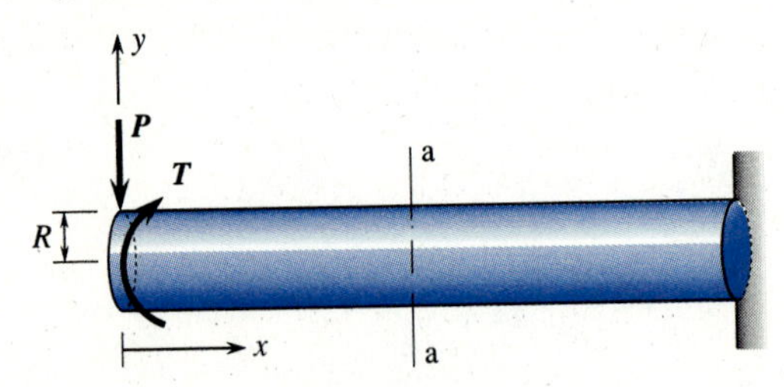

Figure 7.8 Solid circular shaft subjected to torque and transverse load.

We determine the shear stress due to the torque T and the direct shear force P by cutting through the shaft at a section a-a. The free-body diagram of the left portion of the shaft is shown in Figure 7.9.

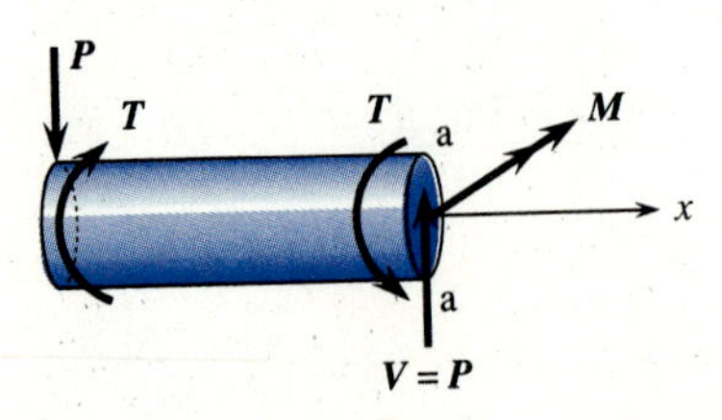

Figure 7.9 Free-body diagram of left portion of shaft.

Considering the circular cross-section at a-a, the shear stress due to the torque varies linearly from the center of the cross-section and reaches the maximum value as given by the torsion formula, Eq. (4.14), $\tau_{max} = TR/J$. These maximum shear stresses are shown at points 1, 2, 3, and 4 in Figure 7.10(a). The shear stress at 0 is zero.

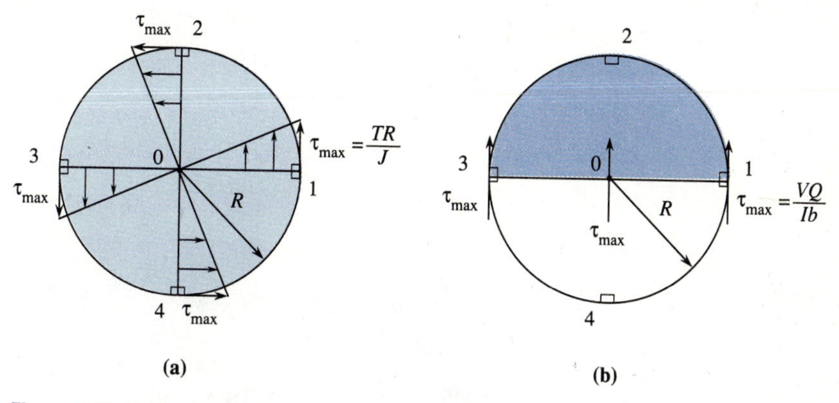

Figure 7.10 Maximum shear stress on shaft cross-section
(a) due to torque and
(b) due to shear force.

The beam shear stress due to the shear force V is obtained by Eq. (5.15), $\tau = VQ/(Ib)$. The maximum beam shear stress occurs along the line 3-0-1. For this location, Q becomes a maximum and we obtain Q from its definition as the moment of the upper semicircular area about the centroid. Here $\bar{y} = 4R/(3\pi)$ for a semicircle. Therefore

$$Q = \left(\frac{\pi R^2}{2}\right)\left(\frac{4R}{3\pi}\right) = \frac{2R^3}{3}$$

and using

$$b = 2R, \qquad I = \frac{J}{2} = \frac{\pi R^4}{4}$$

we obtain the maximum transverse shear stress as

$$\tau_{max} = \frac{VQ}{Ib} = \frac{V\left(\frac{2R^3}{3}\right)}{\left(\frac{\pi R^4}{4}\right)(2R)} = \frac{4V}{3\pi R^2} = \frac{4V}{3A}$$

This value of shear stress acts up at points 3, 0, and 1. This is consistent with the upward direction of the shear force V. The transverse shear stress is zero at points 2 and 4 because Q is zero for points 2 and 4.

Using superposition of the shear stresses, determined in Figures 7.10(a) and (b), we observe that the stresses add together at point 1 to yield

$$\tau_{max} = \frac{TR}{J} + \frac{4V}{3A} \tag{7.4}$$

Equation (7.4) holds for circular cross-sections only since $4V/(3A)$ is used. For other cross-sections we must use the appropriate transverse shear stress expression. For instance, as shown in Chapter 4 for rectangular cross-sections, we have

$$\tau_{max} = \frac{VQ}{Ib} = \frac{3V}{2A}$$

for the second term in Eq. (7.4).

There are no direct shear stresses at 2 and 4 while at 0 the torsional shear stress is zero. At point 3 the two shear stresses are of opposite sign. Therefore, the maximum combined shear stress is indeed at point 1. Stresses at other locations than points 0 through 4 cannot be treated by methods developed in this text. Finally, the normal bending stress at point 1 is zero because point 1 is located on the neutral axis.

STEPS IN SOLUTION

The following steps are used to obtain the shear stress due to the simultaneous actions of both a torsional moment T and a transverse shear force V.

1. Draw a free-body diagram of the desired section of the member showing the internal torque T and the transverse shear force V.
2. Write the equilibrium equations to obtain both T and V.
3. For circular cross-sections, directly use Eq. (7.4) to obtain the total shear stress due to both T and V. This requires calculating both the cross-sectional area A and torsional or polar moment of inertia J. For other cross-sectional shapes, replace the terms in Eq. (7.4) with the appropriate term for that cross section. For instance, use Eqs. (4.33) and (5.16) for a rectangular cross-section.

Examples 7.4 through 7.6 illustrate the steps above to obtain the total shear stress due to both torque and shear force.

EXAMPLE 7.4

A solid circular shaft of 2 in.-diameter is fixed to the wall at the right end and subjected to a downward transverse force $P = 100$ lb and a torque $T = 1000$ lb · in. as shown in Figure 7.11. Determine the combined shear stress at a-a, 50 in. from the free end due to P and T.

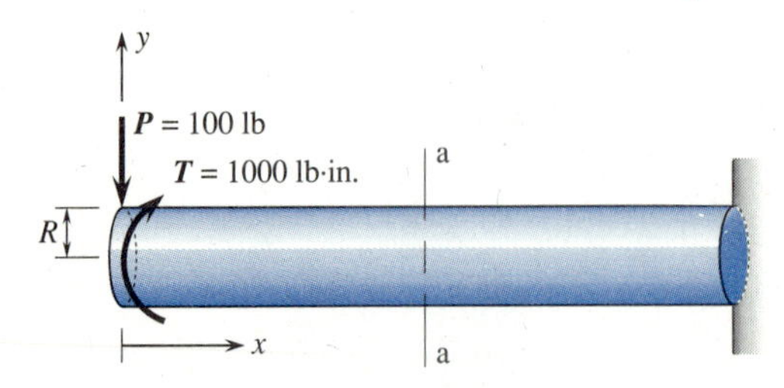

Figure 7.11 Solid circular shaft subjected to torque and transverse force.

Solution

First, we draw a free-body diagram of the section of the shaft to the left of a-a as shown in Figure 7.12.

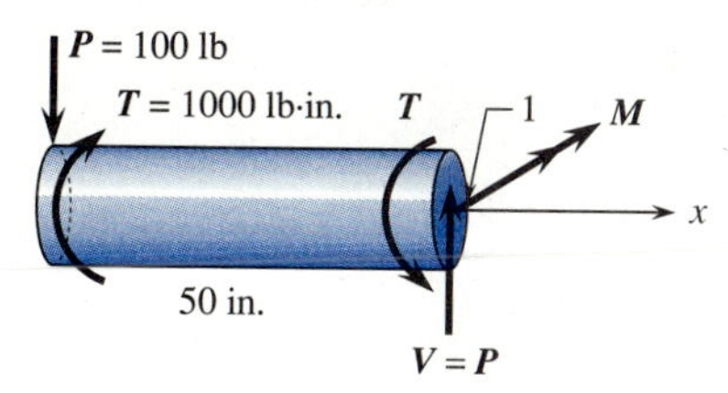

Figure 7.12 Free-body diagram of shaft.

From statics

$$T = 1000 \text{ lb} \cdot \text{in.} \qquad V = 100 \text{ lb} \qquad M = 5000 \text{ lb} \cdot \text{in.}$$

The largest combined shear stress occurs at point 1 on the cross-section where the torsional and transverse shear stresses are in the same direction and hence add together.

Using Eq. (7.4), which applies to solid circular cross-section, we have

$$\tau_{\max} = \frac{TR}{J} + \frac{4V}{3A}$$

with

$$R = 1 \text{ in.}$$

$$J = \frac{\pi R^4}{2} = \frac{\pi(1 \text{ in.})^4}{2} = \frac{\pi}{2} \text{ in.}^4$$

$$A = \pi R^2 = \pi(1 \text{ in.})^2 = \pi \text{ in.}^2$$

$$\tau_{\max} = \frac{(1000 \text{ lb} \cdot \text{in.})(1 \text{ in.})}{\frac{\pi}{2} \text{ in.}^4} + \frac{4(100 \text{ lb})}{3\pi \text{ in.}^2}$$

$$= 636 \text{ psi} + 43 \text{ psi}$$

$$\tau_{\max} = 679 \text{ psi}$$

EXAMPLE 7.5

A sign is supported by a pipe having outside diameter of 100 mm and inside diameter of 80 mm (Figure 7.13). The dimensions of the sign are 2 m × 0.75 m and its lower edge is 3 m above the support. The wind pressure against the sign is 1.5 kPa. Determine the combined shear stress due to the wind pressure on the sign at points A, B, and C located at the base of the pipe.

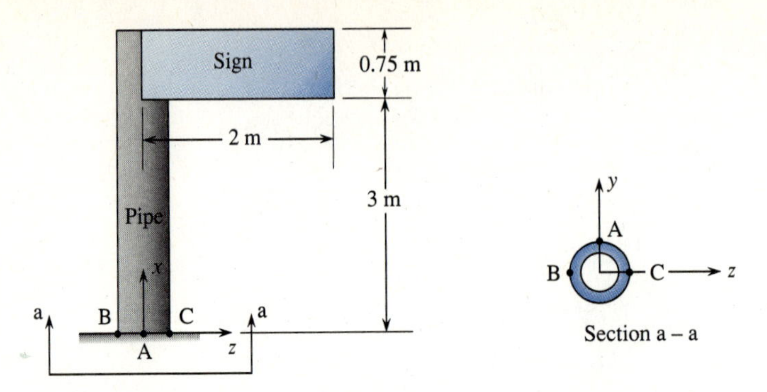

Figure 7.13 Sign supported by pipe and subjected to wind load.

Solution

First we draw a free-body diagram of the sign (Figure 7.14) showing reactions at the base and the total wind force F located at the centroid of the sign.

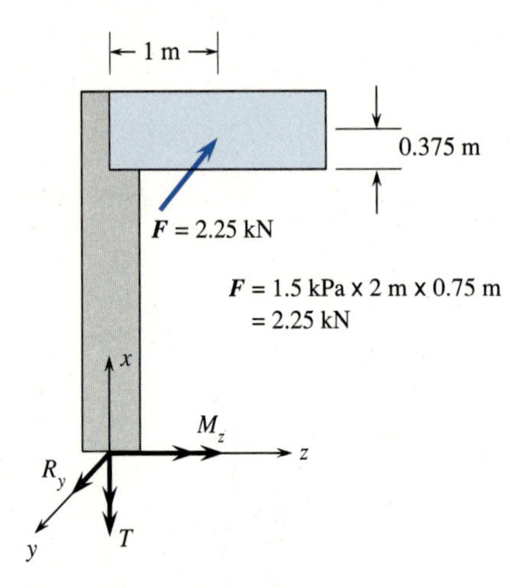

Figure 7.14 Free-body diagram of sign.

By statics, we obtain

$$\Sigma F_y = 0: \quad R_y - 2.25 \text{ kN} = 0 \qquad R_y = 2.25 \text{ kN}$$

$$\Sigma M_x = 0: \quad (2.25 \text{ kN})(1 \text{ m}) - T = 0 \qquad T = 2.25 \text{ kN} \cdot \text{m}$$

$$\Sigma M_z = 0: \quad -(2.25 \text{ kN})(3.375 \text{ m}) + M_z = 0 \qquad M_z = 7.59 \text{ kN} \cdot \text{m}$$

Next we calculate geometric properties as follows

$$J = \frac{\pi}{2}[(50 \text{ mm})^4 - (40 \text{ mm})^4]$$

$$J = 5.80 \times 10^6 \text{ mm}^4 = 5.80 \times 10^{-6} \text{ m}^4$$

$$I = \frac{J}{2} = 2.90 \times 10^{-6} \text{ m}^4$$

For a semicircular area $Q = A\bar{y}$ becomes

$$Q = \left(\frac{\pi R^2}{2}\right)\left(\frac{4R}{3\pi}\right) = \frac{2R^3}{3}$$

For a pipe

$$Q = \frac{2}{3}\left[R_o^3 - R_i^3\right] = \frac{2}{3}[(50 \text{ mm})^3 - (40 \text{ mm})^3]$$

$$Q = 40.67 \times 10^3 \text{ mm}^3 = 40.67 \times 10^{-6} \text{ m}^3$$

We then determine the shear stresses due to the torque T and shear force V shown in Figure 7.15(a).

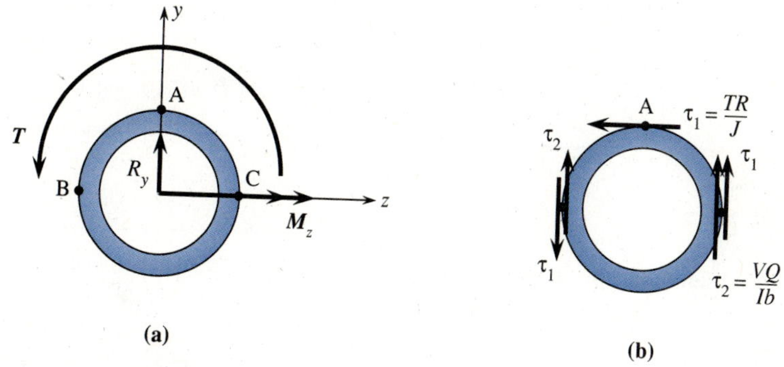

Figure 7.15 Section at base of pipe, (a) loads and (b) shear stresses.

At point A

Torsional stress

$$\tau_1 = \frac{TR}{J} = \frac{(2.25 \text{ kN} \cdot \text{m})(0.05 \text{ m})}{5.80 \times 10^{-6} \text{ m}^4} = 19.4 \text{ MPa}$$

Transverse shear stress

$$\tau_2 = \frac{VQ}{Ib} = 0$$

(since $Q = 0$ when evaluating direct shear stress at A)

Therefore

$$\tau_A = \tau_1 = 19.4 \text{ MPa} \quad \text{(to the left)}$$

At point B

Torsional stress

$$\tau_1 = 19.4 \text{ MPa} \quad \text{(down)}$$

Transverse shear stress

$$\tau_2 = \frac{VQ}{Ib} = \frac{(2.25 \text{ kN})(40.67 \times 10^{-6} \text{ m}^3)}{(2.90 \times 10^{-6} \text{ m}^4)(0.01 + 0.01) \text{ m}}$$

$$= 1.58 \text{ MPa} \quad \text{(up)}$$

From Figure 7.15(b) we see that τ_1 is directed downward while τ_2 is directed upward. Therefore

$$\tau_B = \tau_1 - \tau_2 = 19.4 - 1.58 = 17.8 \text{ MPa} \quad \text{(down)}$$

At point C

$$\tau_1 = 19.4 \text{ MPa} \quad \text{(up)}$$

$$\tau_2 = 1.58 \text{ MPa} \quad \text{(up)}$$

Therefore τ_1 and τ_2 add together to obtain

$$\tau_C = \tau_1 + \tau_2 = 19.4 + 1.58 = 21.0 \text{ MPa} \quad \text{(up)}$$

EXAMPLE 7.6

The axle of an automobile is subjected to the forces and torque shown in Figure 7.16. The diameter of the axle is 32 mm. Determine the shear stress at **(a)** point A on top of the axle, and **(b)** point B on the side of the axle.

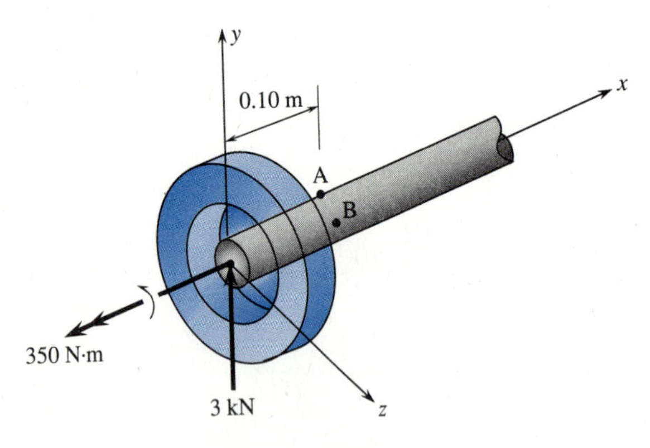

Figure 7.16 Axle subjected to forces and torque.

Solution

A free-body diagram of a section of the axle to point A and B is shown in Figure 7.17.

By statics

$$\Sigma F_y = 0: \quad 3 \text{ kN} - V = 0 \qquad V = 3 \text{ kN} = 3000 \text{ N}$$

$$\Sigma M_x = 0: \quad -350 \text{ N} \cdot \text{m} + T = 0 \qquad T = 350 \text{ N} \cdot \text{m}$$

$$\Sigma M_z = 0: \quad -3 \text{ kN}(0.10 \text{ m}) + M_z = 0 \qquad M_z = 0.30 \text{ kN} \cdot \text{m} = 300 \text{ N} \cdot \text{m}$$

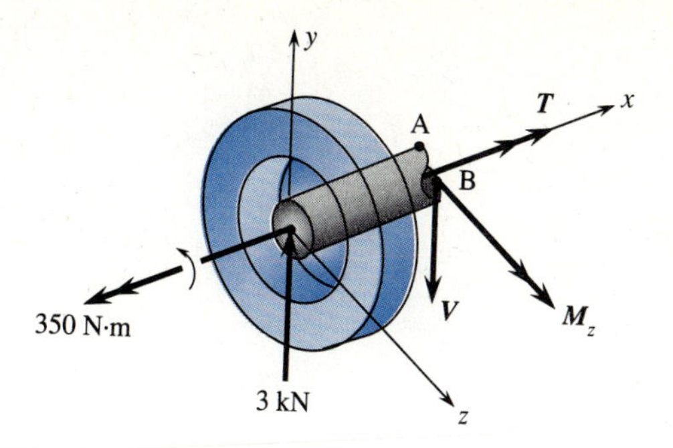

Figure 7.17 Section of axle in equilibrium.

Stresses

(a) At point A

Torsional shear stress

$$\tau_1 = \frac{TR}{J} = \frac{(350\ \text{N}\cdot\text{m})(0.016\ \text{m})}{\frac{\pi}{2}(0.016\ \text{m})^4}$$

$$= 54.4 \times 10^6 \frac{N}{\text{m}^2} = 54.4\ \text{MPa}$$

Transverse shear stress

$$\tau_2 = \frac{VQ}{Ib} = 0 \quad (\text{since } Q = 0 \text{ at A})$$

Therefore the combined shear stress at point A is

$$\tau_A = \tau_1 + \tau_2 = 54.4\ \text{MPa} + 0 = 54.4\ \text{MPa}$$

(b) At point B

Torsional shear stress

$$\tau_1 = 54.4\ \text{MPa} \quad (\text{same as for point A})$$

Transverse shear stress (for a solid circular shaft)

$$\tau_2 = \frac{4V}{3A} = \frac{4(3\ \text{kN})}{3\pi(0.016\ \text{m})^2} = 4974\frac{\text{kN}}{\text{m}^2} = 4.97\ \text{MPa}$$

NOTE: At point B the maximum direct shear stress occurs and is given by

$$\tau = \frac{4V}{3A}$$

for the solid circular shaft.

The combined shear stress at point B (Figure 7.18) is then the sum of torsional and transverse shear stresses given by

$$\tau_B = \tau_1 + \tau_2 = 54.4\ \text{MPa} + 4.97\ \text{MPa} = 59.37\ \text{MPa} \quad (\text{downward})$$

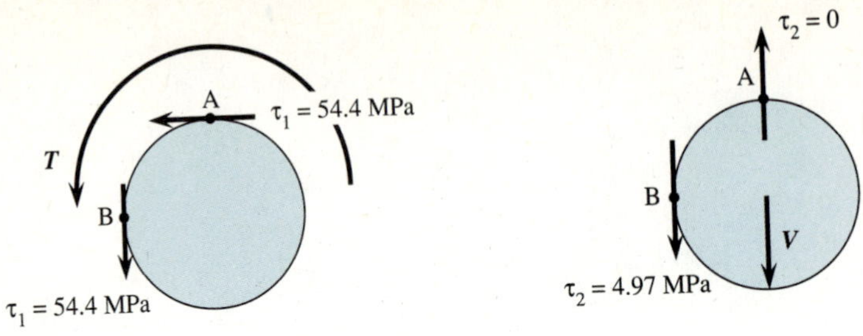

Figure 7.18 Shear stresses at section through A and B.

EXAMPLE 7.7

An L-shaped bracket ACD lying in a horizontal plane supports a load $P = 200$ lb (see Figure 7.19). The bracket is a structural rectangular cross-section tube with outside dimensions of 2 in. × 4 in. and a wall thickness of 0.125 in. Arm AC is 20 in. long while arm CD is 30 in. long. Determine the combined shear stress at (a) point A which is located on the top of the bracket at the support, and (b) point B which is located on the outer side of the bracket at the support.

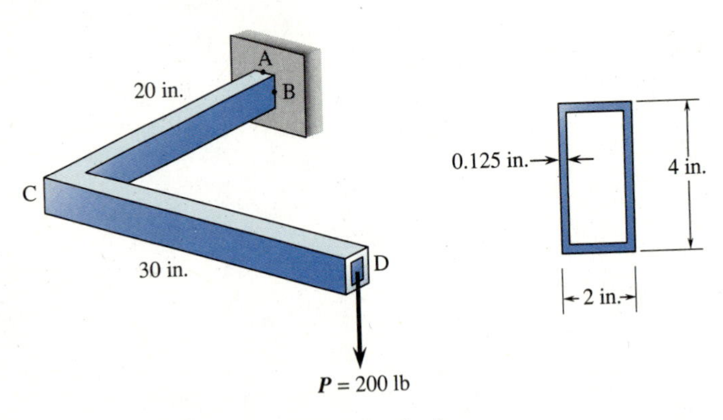

Figure 7.19 L-shaped bracket supporting load.

Solution

Figure 7.20 shows a free-body diagram of the bracket.

By statics, we obtain

$$\Sigma F_y = 0: \quad V - 200 \text{ lb} = 0 \qquad V = 200 \text{ lb}$$

$$\Sigma M_x = 0: \quad T - (200 \text{ lb})(30 \text{ in.}) = 0 \qquad T = 6000 \text{ lb} \cdot \text{in.}$$

$$\Sigma M_z = 0: \quad M - (200 \text{ lb})(20 \text{ in.}) = 0 \qquad M = 4000 \text{ lb} \cdot \text{in.}$$

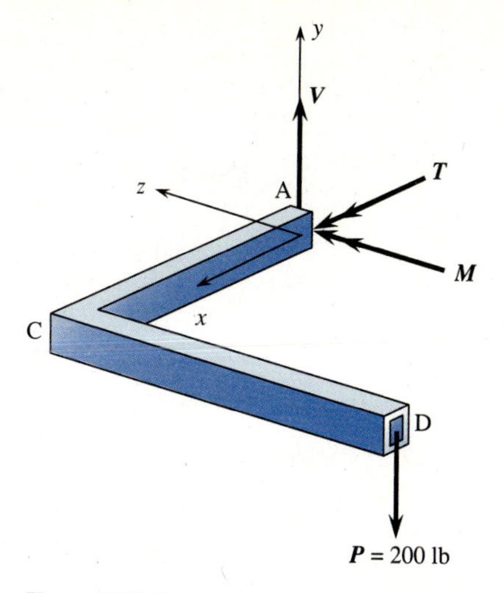

Figure 7.20 Free-body diagram of bracket in equilibrium.

Stresses

(a) At point A

Torsional shear stress (from Eq. (4.45)), for thin-walled cross-sections)

$$\tau_1 = \frac{T}{2tA_m} = \frac{6000 \text{ lb} \cdot \text{in.}}{2(0.125 \text{ in.})(1.875 \text{ in.} \times 3.875 \text{ in.})}$$

$$\tau_1 = 3300 \text{ psi}$$

where A_m is the area enclosed by the median line of the cross-section as shown in Figure 7.21.

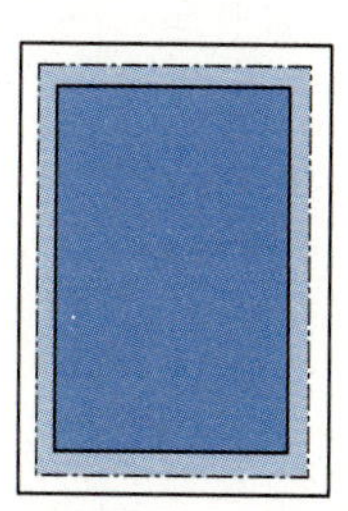

Figure 7.21 Enclosed area of thin-walled tube.

Transverse shear stress

$$\tau_2 = 0$$

(since point A is at the top of the bracket where $Q = 0$)

The combined shear stress at point A is

$$\tau_A = \tau_1 + \tau_2 = 3300 \text{ psi} + 0 = 3300 \text{ psi}$$

(b) At point B

Torsional shear stress (same as for point A)

$$\tau_1 = 3300 \text{ psi}$$

Transverse shear stress (Eq. 5.15)

$$\tau_2 = \frac{VQ}{Ib}$$

where

$$V = 200 \text{ lb}$$

$$b = 0.125 \text{ in.} + 0.125 \text{ in.} = 0.250 \text{ in.}$$

$$I = \frac{1}{12}(2 \text{ in.})(4 \text{ in.})^3 - \frac{1}{12}(1.75 \text{ in.})(3.75 \text{ in.})^3$$

$$I = 10.67 \text{ in.}^4 - 7.69 \text{ in.}^4 = 2.98 \text{ in.}^4$$

$$Q = (2 \text{ in.})(2 \text{ in.})(1 \text{ in.}) - (1.75 \text{ in.})(1.875 \text{ in.})\frac{(1.875 \text{ in.})}{2}$$

$$Q = 4 \text{ in.}^3 - 3.08 \text{ in.}^3 = 0.92 \text{ in.}^3$$

where Q is calculated from the cross-section shown in Figure 7.22.

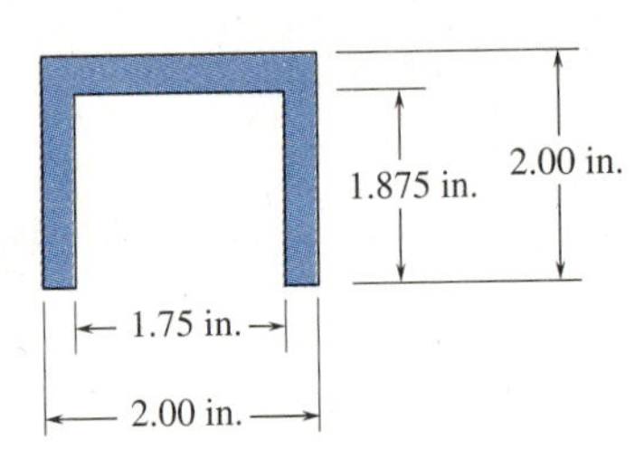

Figure 7.22 Cross-section to evaluate Q.

Therefore the transverse shear stress is

$$\tau_2 = \frac{(200 \text{ lb})(0.92 \text{ in.}^3)}{(2.98 \text{ in.}^4)(0.250 \text{ in.})} = 247 \text{ psi}$$

The combined shear stress at point B is

$$\tau_B = \tau_1 + \tau_2 = 3300 \text{ psi} + 247 \text{ psi} = 3550 \text{ psi}$$

7.4 STRESSES IN THIN-WALLED PRESSURE VESSELS

Thin-walled pressure vessels are closed containers that hold pressurized liquids or gases inside. Examples include cylindrical tanks containing compressed air, pressurized pipes, and water storage tanks (Figure 7.23). Thin-walled vessels have wall thicknesses that are small compared to the radius of the wall. The resulting stresses in the walls of the vessels

Figure 7.23 (a) Horizontal pressure tank for propane storage and (b) vertical tank for water storage.

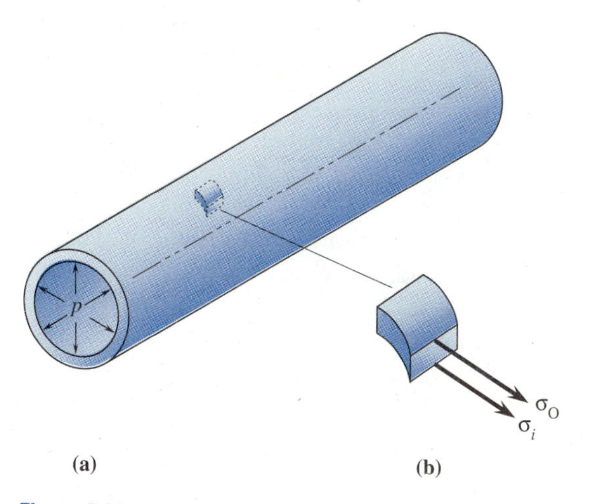

Figure 7.24 Stresses in wall of thin-walled vessel.

are nearly constant from the inner side to the outer side of the wall thickness as shown in Figure 7.24 for a cylindrical vessel under internal pressure p. That is, σ_i is nearly equal to σ_o in a thin-walled vessel.

However, in a thick-walled vessel, the difference between σ_i and σ_o can be quite large and of considerable importance in the design and analysis of the vessel. Hydraulic cylinders, deep submersibles, and arteries and veins are normally thick-walled vessels. A thin-walled vessel is generally defined as one with a ratio of radius r to wall thickness t greater than 10, that is,

$$\frac{r}{t} > 10 \tag{7.5}$$

Many practical vessels have r/t ratios of the order 50 to 200 and are classified as thin-walled.

We will limit our analysis of stresses in thin-walled pressure vessels to the most commonly encountered shapes, the cylindrical and spherical.

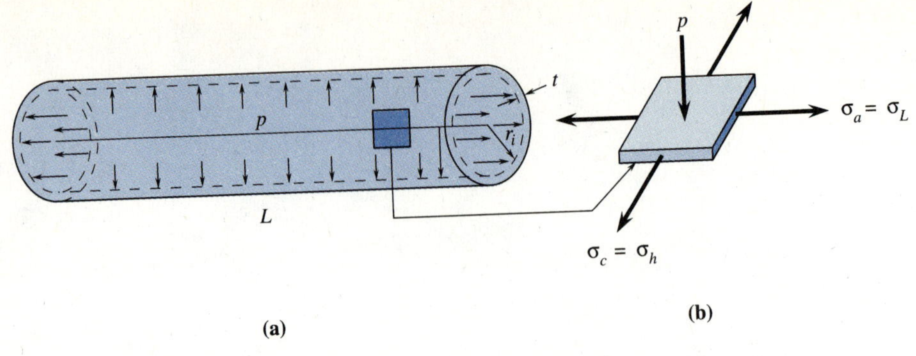

Figure 7.25 (a) Cylindrical tank showing pressure and geometry and (b) element of cylinder.

Cylindrical Pressure Vessels

We first consider the analysis of a thin-walled circular cylindrical tank with closed ends and subjected to internal pressure p (or gauge pressure, which is the pressure above atmospherical or above the external pressure). The tank has inner radius r_i, thickness t, and length L (Figure 7.25(a)).

Two types of stresses develop in the walls of the tank, a circumferential or hoop stress, σ_c or σ_h, and an axial or longitudinal stress, σ_a or σ_L. These stresses are shown on an element of the cylinder in Figure 7.25(b).

To determine the hoop stress σ_h, we consider one-half of the cylinder of length L obtained by passing a plane in the longitudinal direction (parallel to the xy-plane) through the whole cylinder and cut off the ends by passing planes perpendicular to the longitudinal axis at each end as shown in Figure 7.26(a).

The stress acting to pull on the cut wall is σ_h, while the pressure p acts on the curved surface. Summing forces in the z-direction requires us to multiply the stresses by the appropriate areas that they act over. For instance, letting F_h be the total hoop force acting on each cut face in the z-direction, we obtain

$$F_h = \sigma_h (Lt) \tag{7.6}$$

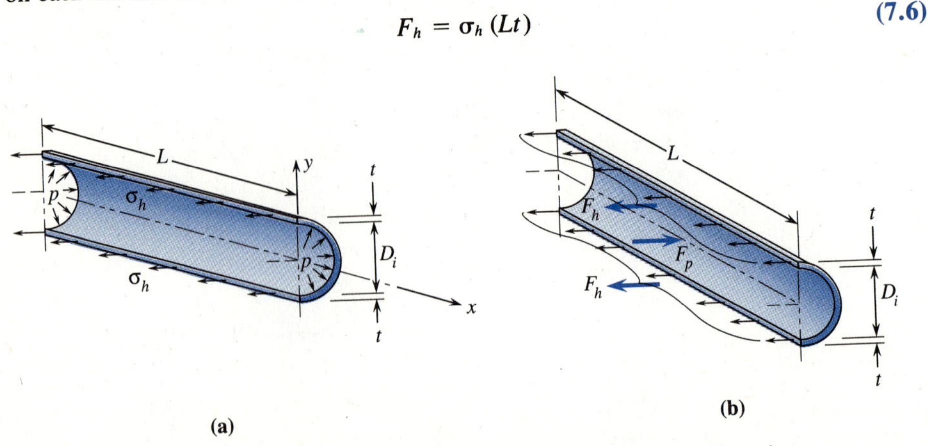

Figure 7.26 (a) One-half of cylinder showing hoop or circumferential stress in wall, and (b) equivalent forces from pressure and hoop stress.

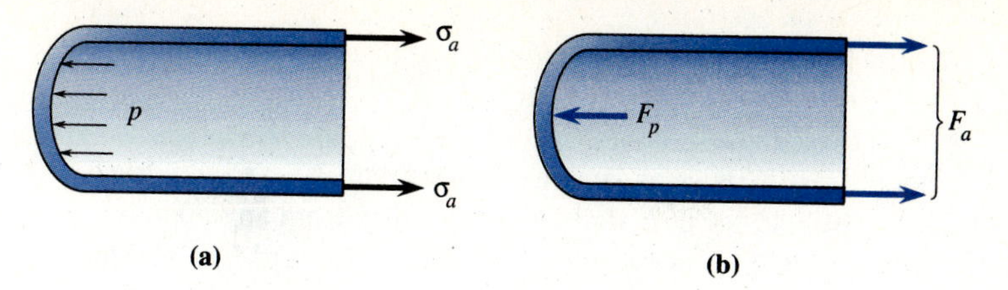

Figure 7.27 (a) One-half of cylinder showing axial stress in wall and (b) equivalent forces from pressure and axial stress.

and letting F_p be the total force on the curved surface in the z-direction, and recalling that this force is obtained by multiplying the pressure times the projected area of the curved surface (a vertical area equal to $2r_iL$) (Figure 7.26(b)), we obtain

$$F_p = p(2r_iL) \tag{7.7}$$

Now by statics

$$\Sigma F_z = 0: \quad 2F_h - F_p = 0 \tag{7.8}$$

and substituting Eqs. (7.6) and (7.7) into (7.8), we have

$$2\sigma_h Lt - p2r_iL = 0 \tag{7.9}$$

Solving Eq. (7.9) for the hoop stress yields

$$\sigma_h = \frac{pr_i}{t} \tag{7.10}$$

This stress is assumed to be uniform over the wall thickness provided the wall is very thin.

For a closed-ended vessel, an axial stress will also develop. The free-body diagram of one-half of the length of the tank, obtained by passing a plane through the tank perpendicular to the longitudinal x-axis, is shown in Figure 7.27(a). The stress pulling on the cut wall is σ_a while the relevant pressure p is that pushing against the end wall. The total force F_a in the x-direction due to the stress σ_a (Figure 7.27(b)) is

$$F_a = \sigma_a 2\pi r_i t \tag{7.11}$$

(Actually the average radius should be used in Eq. (7.11). However for thin-walled vessels r_i is satisfactory.) The total force F_p in the x-direction due to the pressure against the inside circular back wall area is

$$F_p = p\pi r_i^2 \tag{7.12}$$

Summing forces in the x-direction requires that

$$\Sigma F_x = 0: \quad F_a - F_p = 0 \tag{7.13}$$

Substituting Eqs. (7.11) and (7.12) into (7.13), yields

$$\sigma_a(2\pi r_i)t - p\pi r_i^2 = 0 \tag{7.14}$$

Solving Eq. (7.14) for the axial stress, we obtain

$$\sigma_a = \frac{pr_i}{2t} \tag{7.15}$$

Comparing Eqs. (7.10) and (7.15), we observe that

$$\sigma_h = 2\sigma_a \tag{7.16}$$

The hoop stress is twice as large as the axial stress for a cylindrical vessel. Equations (7.10) and (7.15) hold for thin-walled vessels away from any openings, attachments, and supports that normally exist. Example Problems 7.8 and 7.9 illustrate the solution of cylindrical pressure vessels.

EXAMPLE 7.8

A circular cylindrical pressure vessel is subjected to an internal pressure of $p = 1.8$ MPa (Figure 7.28). The vessel has inside radius $r_i = 0.75$ m. The working or allowable stress is 90 MPa. Determine the required wall thickness of the vessel.

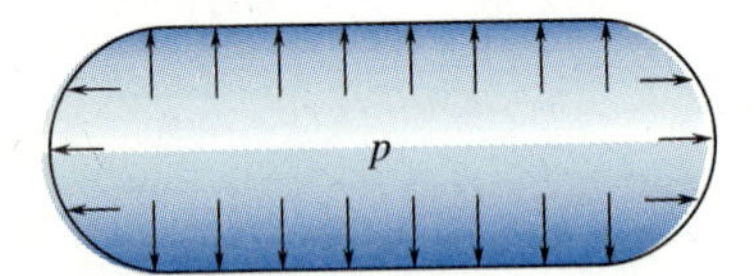

Figure 7.28 Circular cylindrical pressure vessel.

Solution

From Eq. (7.10), the hoop stress is given by

$$\sigma_h = \frac{pr_i}{t}$$

Substituting $\sigma_h = 90$ MPa (the allowable stress), $r_i = 0.75$ m, and $p = 1.8$ MPa into Eq. (7.10) and solving for t, we have

$$t = \frac{pr_i}{\sigma_h} = \frac{(1.8 \text{ MPa})(0.75 \text{ m})}{90 \text{ MPa}} = 0.015 \text{ m} = 15 \text{ mm}$$

Note that the allowable stress occurs in the circumferential direction as given by the hoop stress formula because the hoop stress is twice as large as the axial stress (see Eq. (7.16)).

EXAMPLE 7.9

A cylindrical water tank 30 ft high and 20 ft in diameter is made of 1/4-in. thick steel plate (Figure 7.29). Determine the maximum tensile stress in the tank when the tank is full.

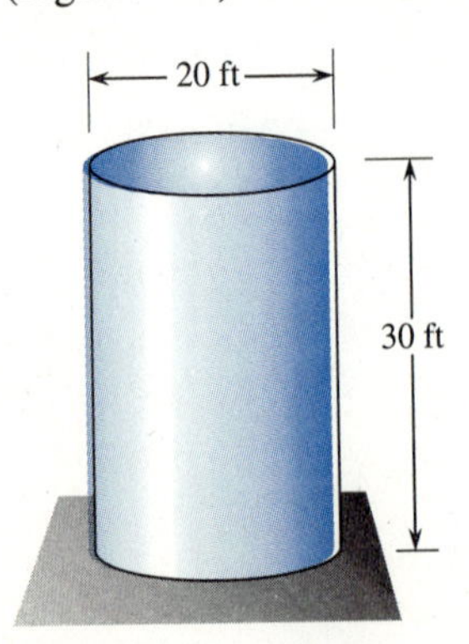

Figure 7.29 Cylindrical water tank.

Solution

Using Eq. (7.10), we can determine the largest tensile stress that is equal to the hoop stress. Therefore, we need to first determine the pressure. For a hydrostatic water pressure, the pressure varies linearly from zero at the top (neglecting atmospheric pressure) to a maximum at the bottom. The maximum pressure is given by multiplying the liquid weight density by the height of water as follows.

$$p = \gamma h = \left(62.4 \frac{\text{lb}}{\text{ft}^3}\right)(30 \text{ ft}) = 1872 \frac{\text{lb}}{\text{ft}^2}$$

Then using Eq. (7.10), the hoop stress is

$$\sigma_h = \frac{p r_i}{t} = \frac{\left(1872 \frac{\text{lb}}{\text{ft}^2}\right)(10 \text{ ft} \times 12 \text{ in./ft})}{(0.25 \text{ in.})(144 \text{ in.}^2/\text{ft}^2)} = 6240 \text{ psi}$$

Spherical Pressure Vessels

We consider the analysis of a thin-walled spherical tank subjected to internal pressure p. The tank has inner radius r_i and wall thickness t (Figure 7.30).

A single tensile stress develops in the walls of the tank. This is shown on an element of the sphere (see Figure 7.30) where the stresses on the four sides of the element are shown equal. These equal stresses are due to the symmetry of the geometry and the loading on the tank.

Figure 7.30 Spherical tank.

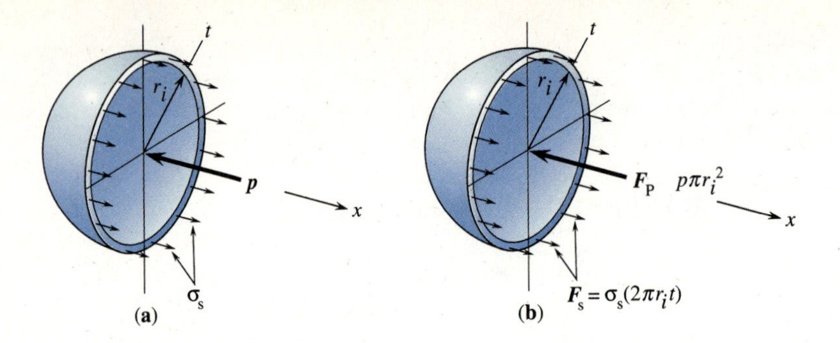

Figure 7.31 (a) One-half sphere and (b) equivalent forces from pressure and stress in wall.

To determine the tensile stress σ_S, we consider one-half of the sphere obtained by passing a vertical plane through half of the shell as shown in Figure 7.31(a). The stress acting to pull on the wall is σ_S, while the pressure p acts on the inner surface of the hemisphere. Summing forces in the x-direction requires us to multiply the stresses by the appropriate areas that they act over. Letting F_S (Figure 7.31(b)) be the total force acting on the cut edges, we obtain

$$F_S = \sigma_S(2\pi r_i t) \tag{7.17}$$

and letting F_p (Figure 7.31(b)) be the total force acting in the x-direction due to the pressure acting on the curved inner surface, we obtain

$$F_p = p(\pi r_i^2) \tag{7.18}$$

where the projected area πr_i^2 of the inner hemisphere has been used. Now by statics, we obtain

$$\Sigma F_x = 0: \qquad F_S - F_p = 0 \tag{7.19}$$

and substituting Eqs. (7.17) and (7.18) into (7.19), we have

$$\sigma_S(2\pi r_i t) - p(\pi r_i^2) = 0 \tag{7.20}$$

Solving Eq. (7.20) for σ_S yields

$$\sigma_S = \frac{p r_i}{2t} \tag{7.21}$$

This stress is assumed to be uniform over the wall thickness provided the wall is very thin. Comparing Eq. (7.21) with Eq. (7.15), we observe that the tensile stress due to internal pressure in a spherical vessel is one-half that of the hoop stress in a cylindrical vessel.

Example Problems 7.10 and 7.11 illustrate the analysis of spherical pressure vessels. Example 7.12 illustrates a cylindrical vessel with spherical end caps, and Example 7.13 is a combined loading problem involving a pressure load and an axial load.

EXAMPLE 7.10

A stainless steel spherical tank has an inside diameter of 500 mm and a wall thickness of 6.25 mm. It is used to contain a pressurized fuel. The allowable tensile stress is 240 MPa. Determine the maximum permissible pressure p inside the tank.

Solution

Using Eq. (7.21), and solving it for p, we have

$$p = \frac{\sigma_S 2t}{r_i} = \frac{(240\text{ MPa})(2)(6.25\text{ mm})}{250\text{ mm}} = 12\text{ MPa}$$

EXAMPLE 7.11

A spherical tank of inside diameter 96 in. and wall thickness of 1.25 in. contains compressed air at a pressure of 1000 psi (Figure 7.32). The tank is constructed of two hemispheres connected together by butt welding. Determine the tensile force f in pounds per inch of length resisted by the weld. Assume the tank is made of American Society for Testing Materials (ASTM)-A36 steel.

Figure 7.32 Spherical tank with compressed air.

Solution

The tensile force per unit length f is related to the tensile stress in the wall of the spherical tank by

$$f = \sigma_S t$$

where σ_S is given by Eq. (7.21). Therefore, we have

$$f = \left(\frac{pr_i}{2t}\right)t = \frac{pr_i}{2} = \frac{(1000 \text{ psi})(48 \text{ in.})}{2} = 24{,}000 \frac{\text{lb}}{\text{in.}}$$

A typical allowable value of f in a weld is obtained for, as an example, an E 70 electrode as follows.

An E 70 electrode has a tensile strength of 70 ksi (corresponding to the number to the right of the E). The allowable stress is then given in the American Welding Society code [1] as the lesser of 30% of the tensile strength or 60% of the yield strength of the base metal. Therefore

$$\sigma_{\text{allow}} = 0.3 \text{ (70 ksi)} = 21 \text{ ksi}$$

or

$$\sigma_{\text{allow}} = 0.6 \text{ (36 ksi)} = 21.6 \text{ ksi}$$

Then assuming a full-penetration weld, we have

$$f_{\text{allow}} = (\sigma_{\text{allow}})\, t = (21 \text{ ksi})\,(1.25 \text{ in.}) = 26.25 \frac{\text{kips}}{\text{in.}}$$

$$f_{\text{allow}} = 26{,}250 \frac{\text{lb}}{\text{in.}} > 24{,}000 \frac{\text{lb}}{\text{in.}}$$

Based on strength, the weld should be satisfactory.

EXAMPLE 7.12

A scuba diver's tank is 6 in. in diameter, 18 in. long, and has a wall thickness of 1/16 in. (Figure 7.33). The tank is a cylindrical pressure vessel with spherical end caps. Determine the maximum tensile stress in the tank wall for an internal gauge pressure of 75 psi.

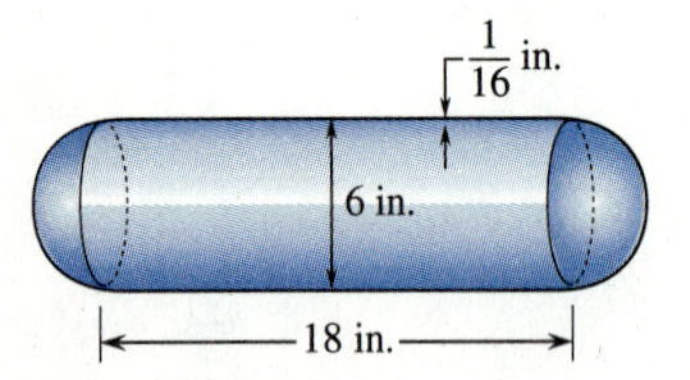

Figure 7.33 Scuba diver's tank.

Solution

For the cylindrical part:
Using Eq. (7.10), the hoop stress is

$$\sigma_h = \frac{pr_i}{t} = \frac{(75 \text{ psi})(3 \text{ in.})}{0.0625 \text{ in.}} = 3600 \text{ psi}$$

Using Eq. (7.15), the axial stress is

$$\sigma_a = \frac{pr_i}{2t} = \frac{\sigma_h}{2} = 1800 \text{ psi}$$

For the spherical end caps:
Using Eq. (7.21), the tensile stress is

$$\sigma_S = \frac{pr_i}{2t} = \frac{(75 \text{ psi})(3 \text{ in.})}{2(0.0625 \text{ in.})} = 1800 \text{ psi}$$

The largest tensile stress is then the hoop stress in the cylindrical portion of the tank, given by

$$\sigma_{max} = \sigma_h = 3600 \text{ psi}$$

EXAMPLE 7.13

A circular steel oil storage tank contains oil to its full height of 10 m. The lower portion of the tank has a wall thickness of 18 mm and a radius of 5 m. The mass density of the oil is $\rho_m = 950 \text{ kg/m}^3$. In addition, a uniform pressure of 2500 Pa is applied to the top of the tank. Determine **(a)** the axial stress and **(b)** the hoop stress in the tank.

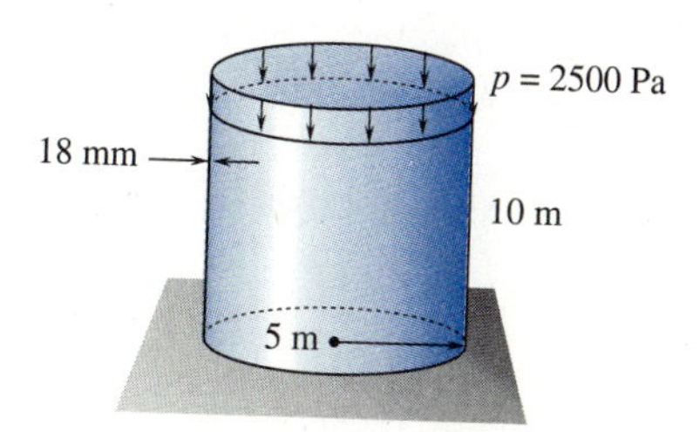

Figure 7.34 Oil storage tank with uniform load on top.

Solution

(a) The axial stress is due to the uniform live load P where

$$P = (2500 \text{ Pa})\pi(5 \text{ m})^2 = 196{,}000 \text{ N} = 196 \text{ kN}$$

Therefore, the axial stress is

$$\sigma_a = \frac{P}{A} = \frac{-196 \text{ kN}}{2\pi rt} = \frac{-196 \text{ kN}}{2\pi(5 \text{ m})(0.018 \text{ m})} = -347 \frac{\text{kN}}{\text{m}^2}$$

$$\sigma_a = -347 \text{ kPa} \quad \text{(compression)}$$

(b) The largest hoop stress occurs at the bottom of the tank because the pressure is greatest at the bottom. The pressure is

$$p = (\rho_m g)h$$

$$p = \left(950 \frac{\text{kg}}{\text{m}^3}\right)\left(9.81 \frac{\text{m}}{\text{s}^2}\right)(10 \text{ m})$$

$$p = 93{,}200 \frac{\text{kg} \cdot \text{m}}{\text{s}^2 \cdot \text{m}^2} = 93{,}200 \frac{\text{N}}{\text{m}^2} = 93.2 \frac{\text{kN}}{\text{m}^2}$$

Now using Eq. (7.10), we obtain the hoop stress as

$$\sigma_h = \frac{\left(93.2 \frac{\text{kN}}{\text{m}^2}\right)(5.0 \text{ m})}{0.018 \text{ m}} = 25{,}900 \frac{\text{kN}}{\text{m}^2} = 25.9 \text{ MPa}$$

7.5 CODE DESIGN OF PRESSURE VESSELS

The code most commonly used in the design of pressure vessels is the *ASME Boiler and Pressure Vessel Code* [2]. There are 11 sections of this voluminous code. The rules of Section VIII, Pressure Vessels apply to most vessels. Section VIII consists of two books, Division 1 and Division 2. We consider the rules of Division 1, which are based on fundamental mechanics of materials including equations developed in Section 7.4. Division 2 requires more advanced stress analysis methods beyond the scope of this text. These methods sometimes result in more economical designs.

Division 1 of the code specifies the minimum thickness t and maximum pressure p_{max} of shells subjected to only internal pressure loading as follows.

Cylindrical Shells

Circumferential stress longitudinal joints (see Figure 7.35).

$$t = \frac{pr_i}{(\sigma_{allow})j - 0.6p} \quad \text{or} \quad p_{max} = \frac{(\sigma_{allow})jt}{r_i + 0.6t} \tag{7.22}$$

Longitudinal stress circumferential joints (see Figure 7.35)

$$t = \frac{pr_i}{2(\sigma_{allow})j + 0.4p} \quad \text{or} \quad p_{max} = \frac{2\sigma_{allow}jt}{r_i - 0.4t} \tag{7.23}$$

Spherical Shells

$$t = \frac{pr_i}{2(\sigma_{allow})j - 0.2p} \quad \text{or} \quad p_{max} = \frac{2(\sigma_{allow})jt}{r_i + 0.2t} \tag{7.24}$$

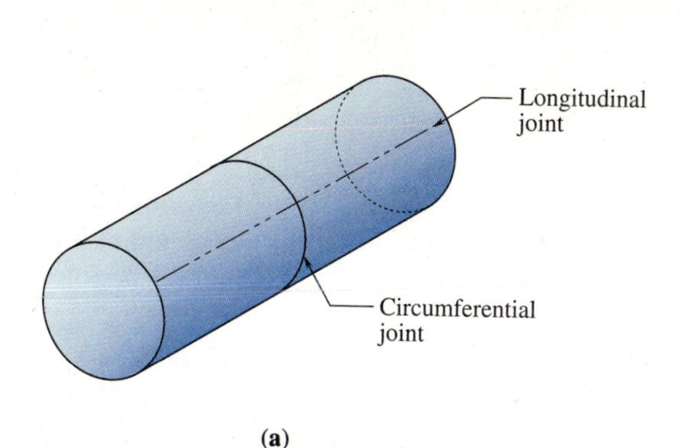

(a)

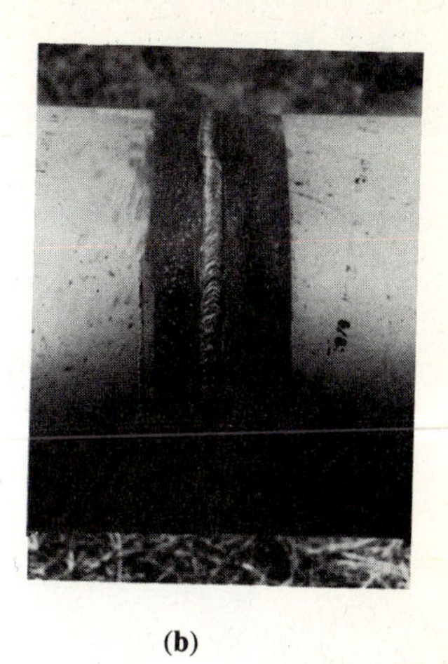

(b)

Figure 7.35 Cylindrical vessel showing
(a) longitudinal and circumferential joints and
(b) actual circumferential weld.

In Eqs. (7.22), (7.23) and (7.24), we have

t = required minimum thickness

p = design or working pressure

r_i = inner radius of shell

σ_{allow} = allowable stress at design temperature

j = joint efficiency factor

For the case of no joints, $j = 1$, and when the pressure is low with respect to the allowable stress, the second term in the denominators of Eqs. (7.22) through (7.24) is small. For this situation, Eqs. (7.22) through (7.24) closely approximate the membrane stress equations (7.10), (7.15), and (7.21), respectively.

We must remember that code Eqs. (7.22) through (7.24) are applicable for shells subjected to only internal pressure. External pressure causes compressive stresses in the vessel and shell buckling is then the likely mode of failure. In this case, another part of the code, not considered here, must be used.

The allowable stress used in Eqs. (7.21) through (7.24) is given by the lesser of

$$\sigma_{allow} = \frac{1}{4}\sigma_u \qquad \textbf{(7.25)}$$

or

$$\sigma_{allow} = \frac{2}{3}\sigma_Y \qquad \textbf{(7.26)}$$

In Eq. (7.25), the ultimate stress is used because sudden catastrophic bursting of the vessel is of greater concern than the slower general failure due to yielding of the material. We see that a factor of safety of 4 is used to account for variations in material properties, construction problems, uncertainty in loadings that might produce stress concentrations, and the serious nature of a sudden pressure vessel explosion.

Since these vessels may be used at elevated temperatures, values of allowable stress for the materials at different design temperatures are listed in the code. For instance, for SA285C boiler plate steel with a $\sigma_Y = 30$ ksi (207 MPa), the allowable stresses for different temperatures are

$$\sigma_{\text{allow}} = 13.8 \text{ ksi} \quad (-20°\text{F to } 650°\text{F})$$

$$\sigma_{\text{allow}} = 6.5 \text{ ksi} \quad (900°\text{F})$$

The joint efficiency factor j is used to account for uncertainties about the strength of welded joints used to join the plates making up the vessel. Most joints are butt welds. They may be multipass welds, machine deposited, using submerged arc or gas-shielded arc processes [1]. During this process cracks, voids, or inclusions of impurities may result in the weld and the defects must be removed and replaced if inspection dictates. Based on the thoroughness of the type of weld inspection used, different joint efficiency factors are used as follows:

$$j = 1 \quad \text{(if radiographic inspections used)}$$

$$j = 0.85 \quad \text{(if spot x-ray used)}$$

$$j = 0.70 \quad \text{(if not x-rayed)}$$

The higher value of j ($j = 1$) if radiographic inspection is used is due to this methods high reliability in discovering weld defects, which can then be eliminated.

Also, for carbon steel vessels, a corrosion allowance of 1/16 to 1/8 in. is usually added to the code minimum thickness. Finally, the wall around openings or nozzles in the vessel must be reinforced with extra thickness as described in the code.

EXAMPLE 7.14

A cylindrical vessel is to be designed to hold propane gas at a pressure of 100 psi. The material is ASTM A36 plate steel. There will be no x-ray inspection of the weld joints. The inner radius of the tank is 10 ft. Determine the thickness of the tank based on the ASME Pressure Vessel Code rules described in this section.

Solution

For ASTM A36 plate steel the ultimate and yield stresses are given in Appendix B as

$$\sigma_u = 58 \text{ ksi}$$

$$\sigma_Y = 36 \text{ ksi}$$

Using Eqs. (7.25) and (7.26), the allowable tensile stress is the lesser of

$$\sigma_{\text{allow}} = \frac{1}{4}(58) = 14.5 \text{ ksi}$$

or

$$\sigma_{\text{allow}} = \frac{2}{3}(36) = 24 \text{ ksi}$$

Use $\sigma_{\text{allow}} = 14.5$ ksi.

Now using Eq. (7.22) with $j = 0.7$, since there are no weld joint inspections, the shell thickness is

$$t = \frac{(100\text{ psi})(10\text{ ft} \times 12\text{ in./ft})}{(14{,}500\text{ psi})(0.7) - 0.6(100\text{ psi})}$$

$$= 1.19\text{ in.}$$

Use $t = 1.25$ in. thick plate. (This thickness also allows for 1/16 in. corrosion.)

7.6 SUMMARY OF IMPORTANT DEFINITIONS AND EQUATIONS

1. Combined normal stress
   Due to direct axial force and bending moment

$$\sigma = \frac{P}{A} - \frac{My}{I} \qquad \textbf{(7.3)}$$

2. Combined shear stress
   a. Due to torque in a solid circular shaft and transverse shear

$$\tau = \frac{Tr}{J} + \frac{VQ}{Ib} \qquad \left(\frac{VQ}{Ib}\right)_{\text{max}} = \frac{4V}{3A} \qquad \textbf{(7.4)}$$

   b. Due to torque in a closed thin-walled tube and transverse shear

$$\tau = \frac{T}{2A_{m}t} + \frac{VQ}{Ib} \qquad \textbf{(4.45),(5.15)}$$

   c. Due to torque in a solid rectangular shaft and transverse shear

$$\tau = \frac{T}{k_2a^2b} + \frac{VQ}{Ib} \qquad \left(\frac{VQ}{Ib}\right)_{\text{max}} = \frac{3V}{2A} \qquad \textbf{(4.33),(5.16)}$$

3. Stresses in thin-walled pressure vessels
   a. Thin-walled vessels

$$\frac{r}{t} > 10 \qquad \textbf{(7.5)}$$

   (1) Cylindrical vessels

$$\sigma_h = \frac{pr_i}{t} \qquad \textbf{(7.10)}$$

$$\sigma_a = \frac{pr_i}{2t} \qquad \textbf{(7.15)}$$

   b. Spherical vessels

$$\sigma_S = \frac{pr_i}{2t} \qquad \textbf{(7.21)}$$

REFERENCES

1. Blodgett OW. *Design of Welded Structures*. James F. Lincoln Arc Welding Foundation, 1966. Cleveland, Ohio
2. American Society of Mechanical Engineers. *ASME Boiler and Pressure Vessel Code*. 1989. 345 E. 47th St., New York, NY 10017

PROBLEMS

Section 7.2

7.1 A W 16 × 45 beam (see Appendix C) is loaded with a uniformly distributed load of 3 kips/ft, including its own weight, and an axial tensile force of 100 kips. Determine the maximum combined normal stress σ_x if the beam span is 10 ft.

Figure P7.1

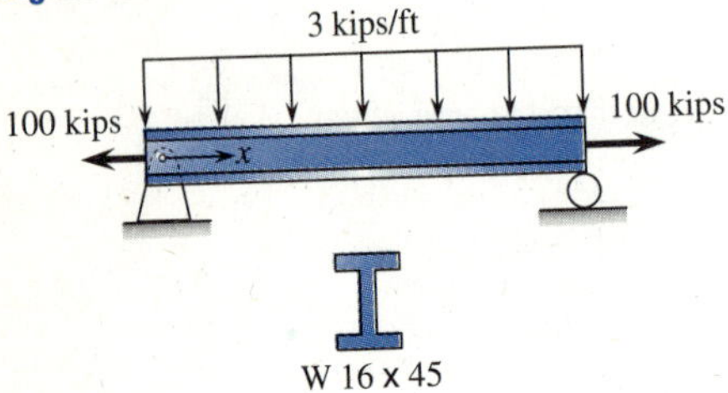

7.2 A W 12 × 45 beam 10 ft long is subjected to a pull of 90 kips as shown. At the ends where the pin connections are made the beam is reinforced with doubler plates welded to the beam as shown. Determine the maximum flange stress σ_x in the middle part of the beam span caused by the applied forces of 90 kips.

Figure P7.2

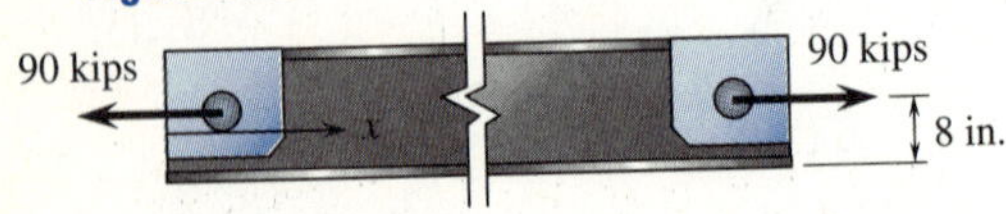

7.3 A $2\frac{1}{2}$ in. diameter (t = 0.203 in.) A36 steel construction pipe (see Appendix C) is used as a support for a basketball backboard as shown. It is securely fixed into the ground. Compute the combined normal stress σ_x that would be developed in the pipe if a 200-lb player hung on the base of the rim of the basket.

Figure P7.3

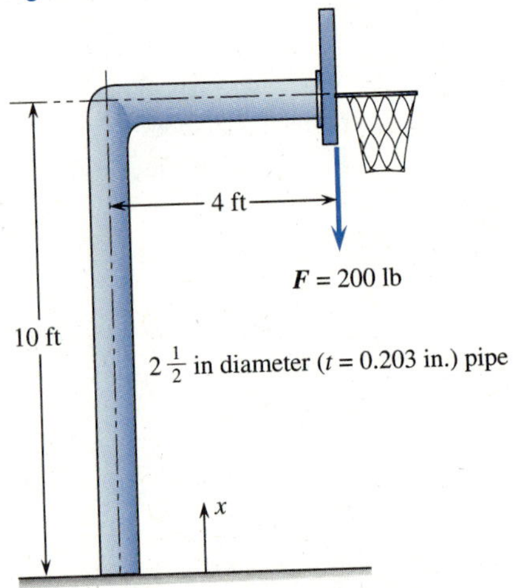

7.4 The bracket shown has a rectangular cross-section of 20 mm wide × 75 mm high. It is fixed to the wall at A. Determine the maximum normal stress σ_x in the bracket.

Figure P7.4

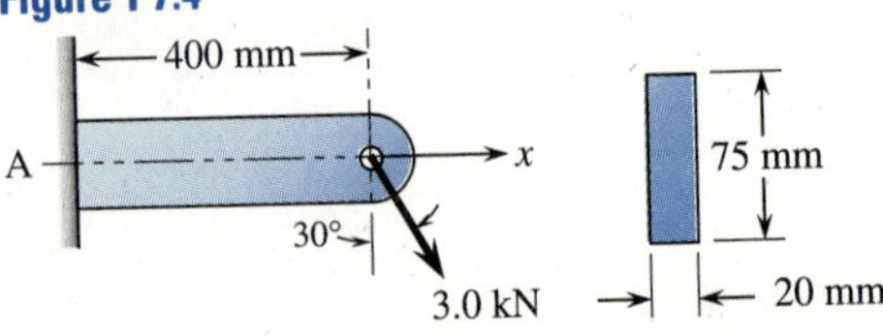

7.5 The beam shown supports a 9000-lb load attached to a bracket below the beam. Determine the normal stress σ_x at points A and B where it is attached to the column.

Figure P7.5

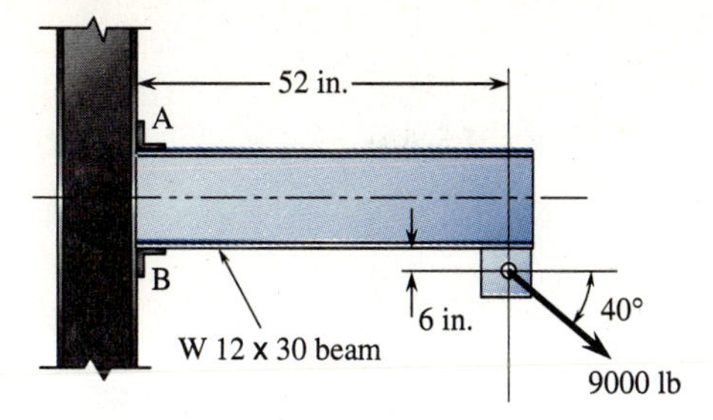

7.6 Determine the maximum normal stress σ_x in the crane beam shown when the load of 15 kN is applied at the middle of the beam.

Figure P7.6

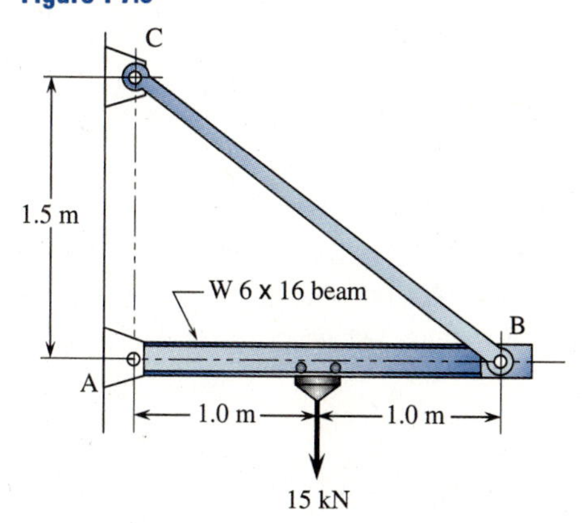

7.7 A machine part for transmitting a pull of 20 kN is offset as shown. Determine the largest normal stress σ_x in the offset portion of the member.

Figure P7.7

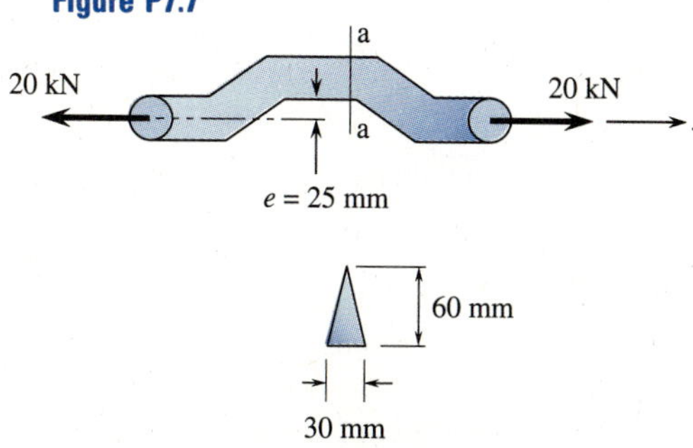

7.8 A large hook fabricated from a structural steel C 8 × 11.5 (see Appendix C) is loaded as shown. Determine the largest normal stress at the fixed end A.

Figure P7.8

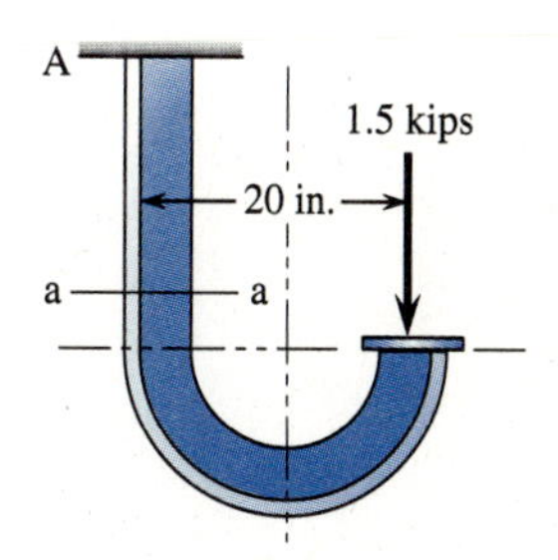

7.9 The frame of a metal cutting hacksaw is made of hollow tubing having an outside diameter of 12 mm and a wall thickness of 1.0 mm. The blade is pulled taut by the wing nut so that a tensile force of 180 N is applied to the blade. Determine the maximum normal stress in the top section of the tubular frame.

Figure P7.9

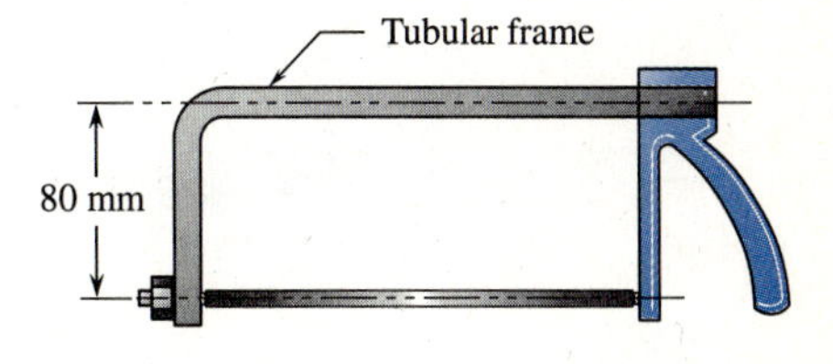

7.10 The C-clamp shown is made of cast malleable iron, ASTM A47 (see Appendix B). Determine the allowable clamping force that the clamp can exert if it is desired to have a factor of safety of 2 based on yield strength in either tension or compression.

Figure P7.10

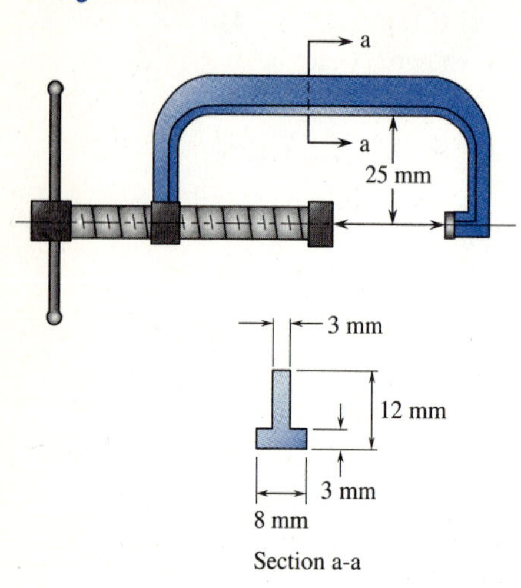

Figure P7.12

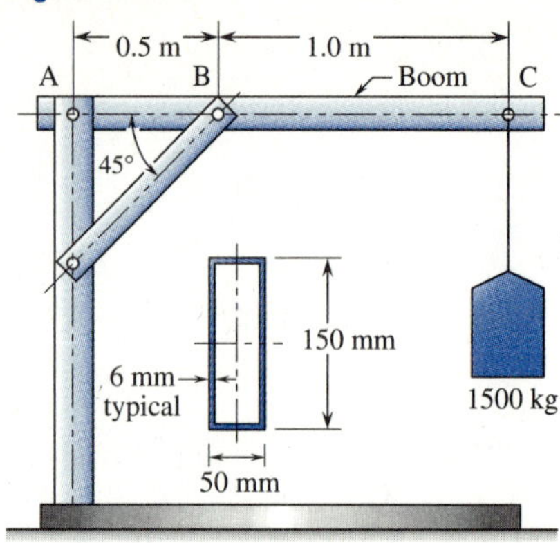

7.11 An S 6 × 12.5 American Standard I beam (see Appendix C) is subjected to the forces shown. The 5000-lb force acts directly in line with the axis of the beam. The 800-lb downward force at A produces the reactions shown at the supports B and C. Compute the maximum tensile and compressive stresses in the beam.

Figure P7.11

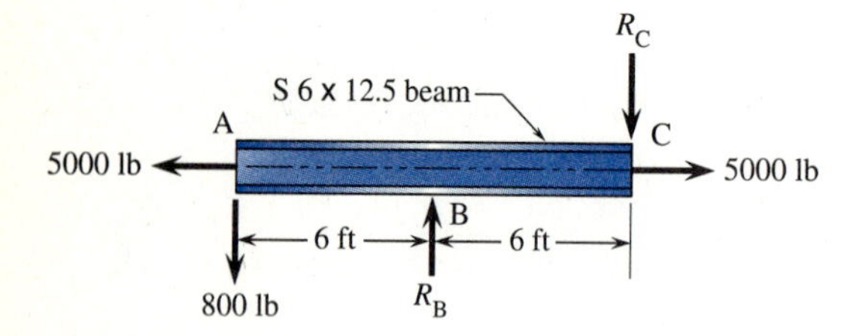

7.12 The horizontal crane boom shown is made of a hollow rectangular steel tube. Compute the stress in the boom just to the left of point B when a mass of 1500 kg is supported at the end.

7.13 For the crane boom shown in Figure P7.12, compute the load which could be supported if a factor of safety of 3 based on yield strength is desired. The boom is made of ASTM A36 structural steel. Analyze only sections where the full box section carries the load, assuming that sections at connections are adequately reinforced.

7.14 A malleable cast iron frame for a punch press has the proportions shown in the figure. What force P may be applied to this frame controlled by the stresses in the sections such as a-a, if the allowable stresses are 5000 psi tension and 18,000 psi in compression?

Figure P7.14

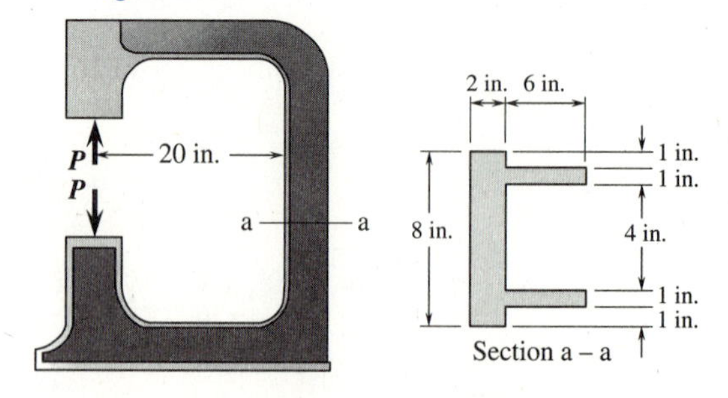

7.15 A short 100-mm square steel bar with a 50-mm-diameter axial hole is built in at the base and is loaded at the top as shown in the figure. Neglecting the weight of the bar, determine the value of the force P so that the maximum normal stress at the built-in end would not exceed 150 MPa.

Figure P7.15

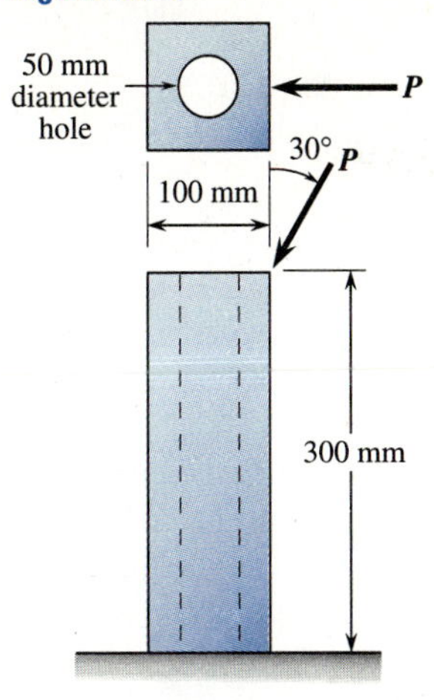

Figure P7.17

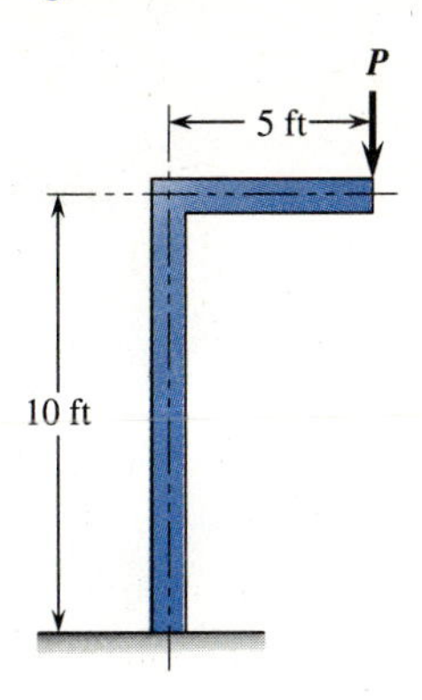

7.16 A jib crane is made from a W 8 × 21 section steel beam and a high-strength steel rod as shown in the figure. **(a)** Find the location of the movable load P that would cause the largest bending moment in the beam. Neglect the weight of the beam. **(b)** Using the load location found in **(a)**, how large may the load P be? Assume that the effect of shear in the beam is not significant, and let the allowable normal stress in the beam be 18,000 psi.

Figure P7.16

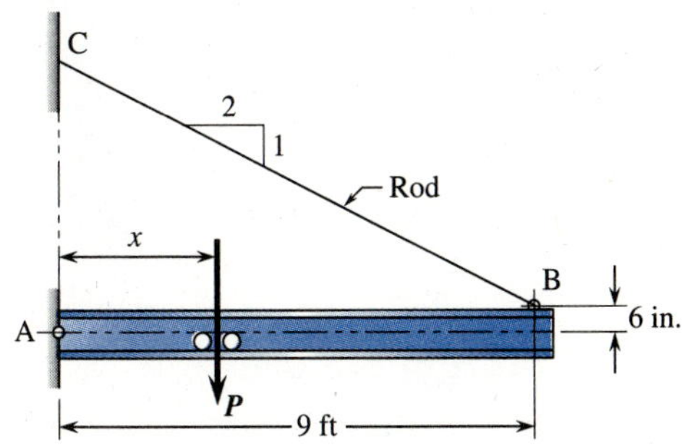

7.17 A steel frame fabricated from W 10 × 22 steel sections supports a load $P = 10$ kips at a distance of 5 ft from the center of the vertical column as shown in the figure. Determine the maximum normal stress in the vertical column.

7.18 A factory stairway having the centerline dimensions shown in the figure is made from two steel channels on edge separated by treads framing into them. The loading on each channel, including its own weight, is estimated to be 250 lb per foot of horizontal projection. Assuming that the lower end of the stairway is pinned and that the wall provides only horizontal support at the top, find the largest normal stress in the channels 5 ft above the floor level.

Figure P7.18

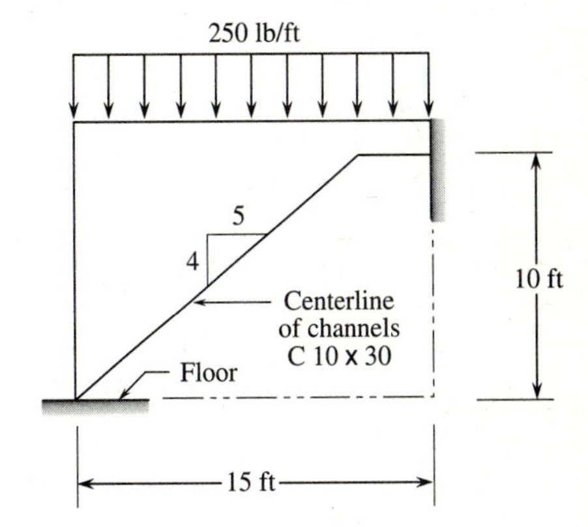

7.19 The post shown is a circular timber 0.30 m in diameter. What is the largest tensile stress acting normal to a section 2.5 m above the bottom of the post?

Figure P7.19

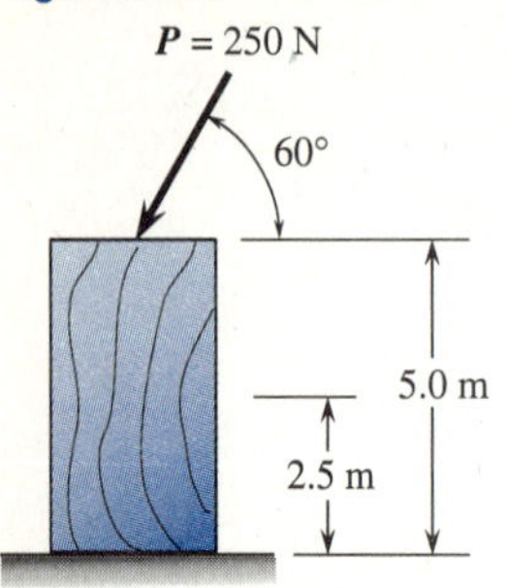

7.20 The beam shown supports a 40-kN load attached to a bracket below the beam. Determine the normal stress σ_x at points A and B where the beam is attached to the column.

Figure P7.20

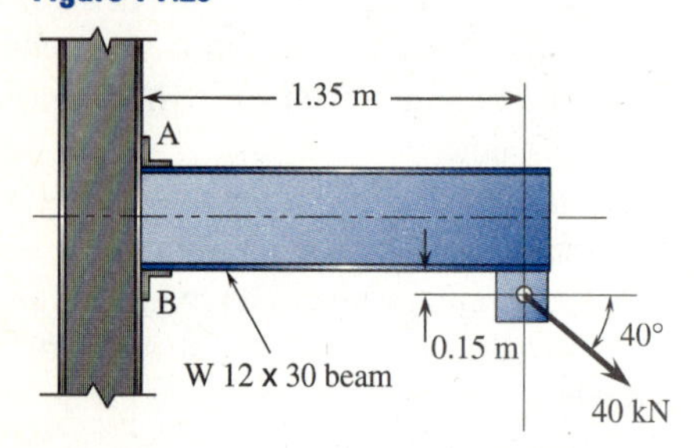

7.21 The crane beam shown is subjected to a lifted load of $P = 4000$ lb applied at the middle of the beam. Determine the maximum normal stress σ_x in the beam.

Figure P7.21

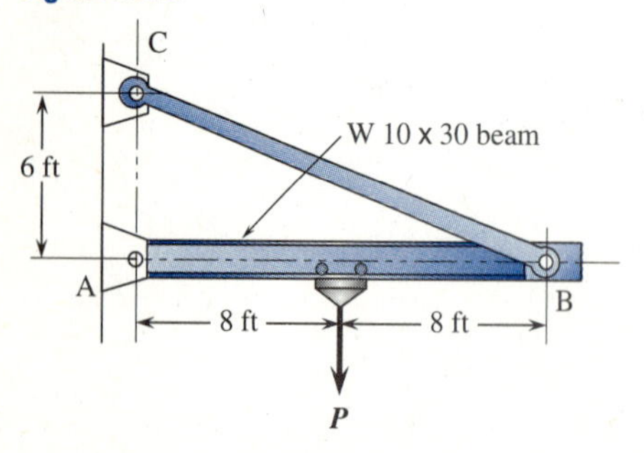

7.22 A stabilizing bar in a race car is bent as shown to provide clearance between it and the exhaust duct. The bar must resist loads at each end of 3000 lb as shown. Determine the largest normal stress in the bar at section a-a. The bar diameter is 3/4 in.

Figure P7.22

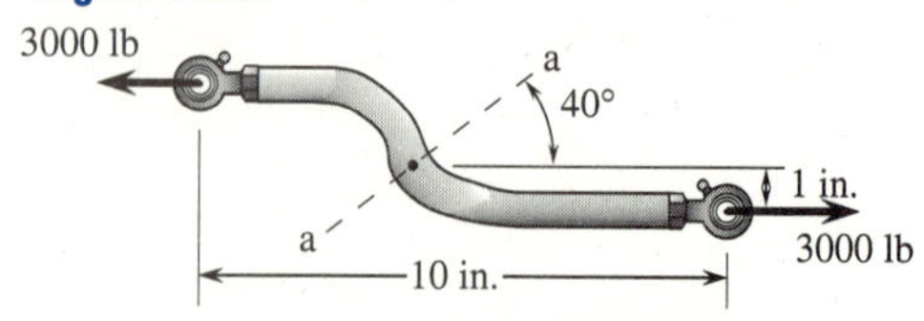

7.23 A machine part for transmitting a pull of 10,000 lb is offset as shown. Determine the largest normal stress σ_x in the offset portion a-a of the member. The cross section is a 2 in. × 2 in. square.

Figure P7.23

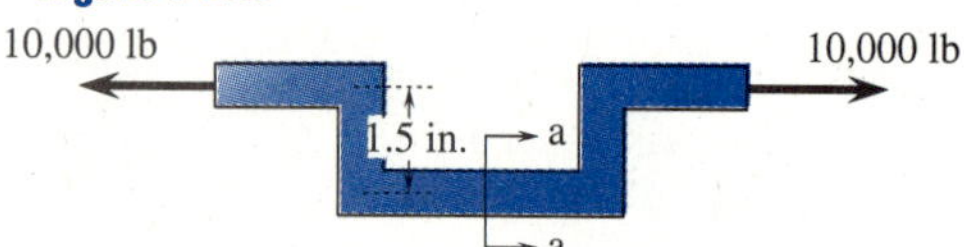

7.24 A lift ripper used to plow a tunnel for cable to be buried must be designed to resist a pulling force P of 40 kips and a lifting force of F = 10 kips. Determine the maximum normal stress at section a-a. The beam cross-section at a-a is 4 in. wide × 12 in. deep.

Figure P7.24

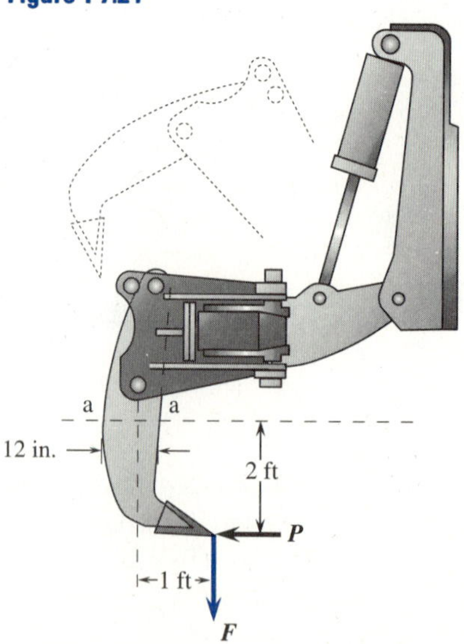

7.25 A hook for lifting bales of hay (assume a bale of hay weighs 25 lb) is shown. Determine the maximum normal stress σ_x in section a-a.

Figure P7.25

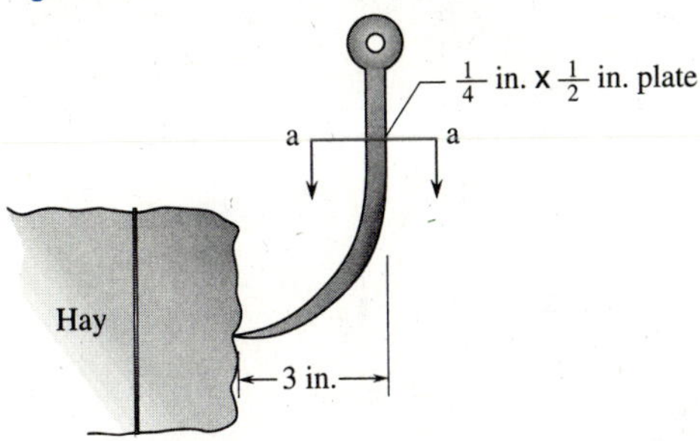

7.26 For the offset link shown, determine the maximum normal stress due to combined bending and tension at section a-a. The link resists a tensile load of $P = 16$ kN. The cross-section at a-a is 15 mm × 50 mm.

Figure P7.26

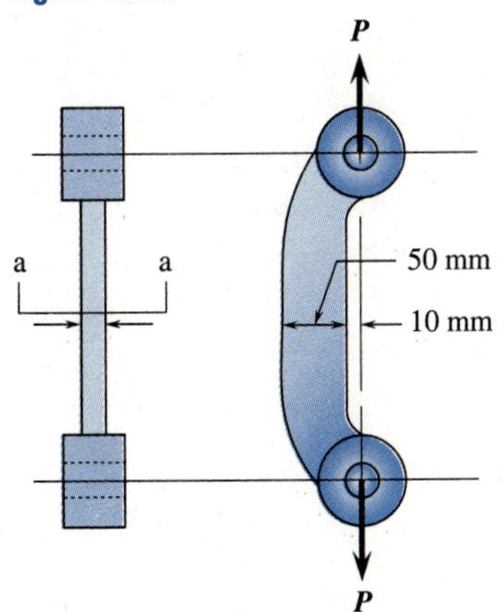

7.27 The crane beam shown is constructed of a wide flange with $A = 3060$ mm^2, $I = 13.36 \times 10^6$ mm^4, and depth $h = 160$ mm. The reactions are shown in the figure. Determine the maximum compressive normal stress in the beam.

Figure P7.27

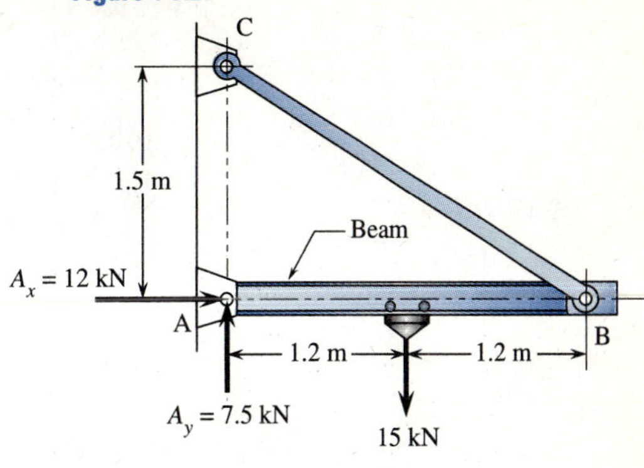

7.28 For the load frame shown, determine the normal stress at section a-a. The cross-sectional area at a-a is 75 mm wide × 50 mm thick.

Figure P7.28

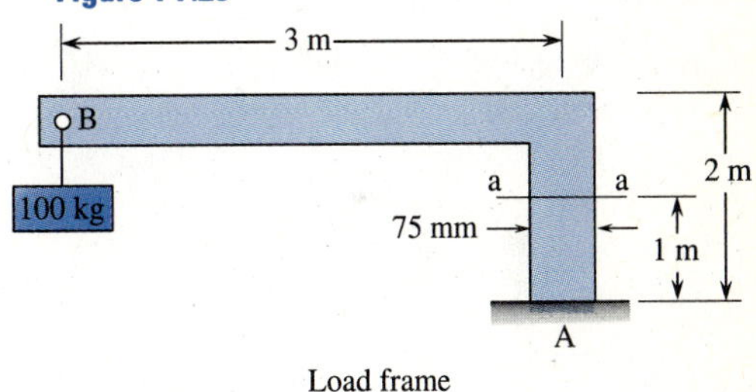

7.29 For the beam supported by the bar AB and pin C, determine the normal stress in section a-a. The beam is a rectangular cross-section with depth 6 in. and width 3 in.

Figure P7.29

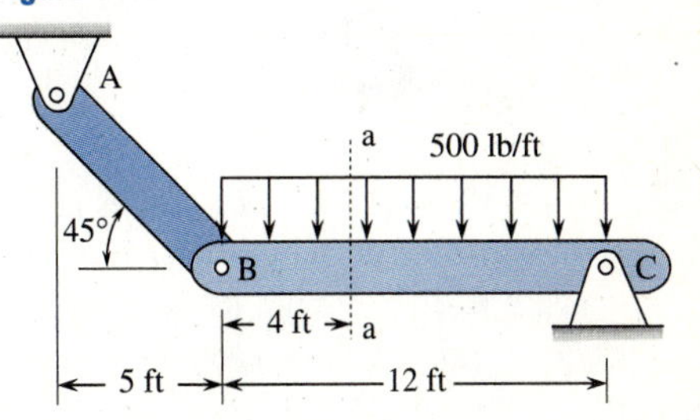

7.30 For the beam shown supported by the bar AB and pin at C, determine the normal stress at section a-a. The cross-section at a-a is rectangular with a depth of 200 mm and a width of 100 mm.

Figure P7.30

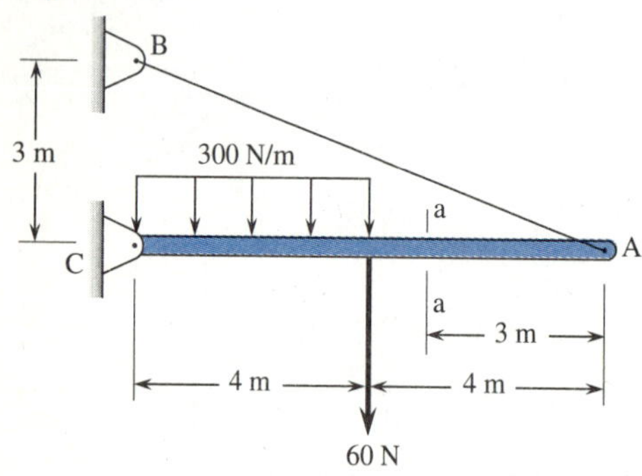

7.31 For the column subjected to the loads shown, determine the combined normal stress at the base point C.

Figure P7.31

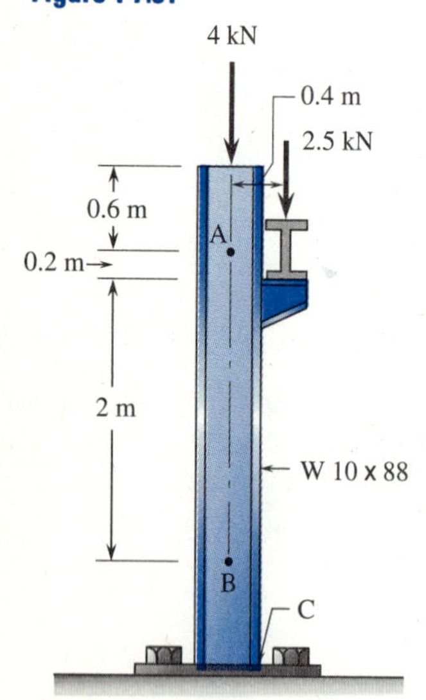

7.32 What must the thickness t of a rectangular concrete dam 1.5 m high be in order to retain a water level even with its crest, as shown in Figure P7.32, without causing tension on the foundation at the upstream face A? Assume the unit weight of concrete to be 25 kN/m^3 and that of water to be 10 kN/m^3.

Figure P7.32

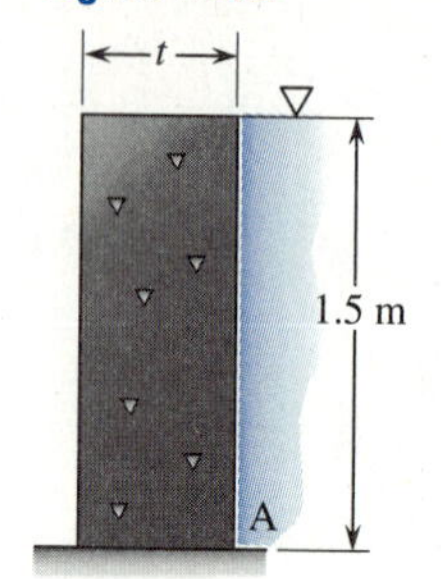

Section 7.3

7.33 A television antenna is mounted on a hollow aluminum tube as shown. A force of 30 lb is applied to the end of the antenna. Calculate the shear stress in the tube due to both the torque and transverse shear force. Assume the tube to be simply supported against bending at the clamps, but assume that rotation is not permitted at either clamp. If the tube is made of 6061-T6 extruded aluminum, would it be safe under this load? Use a factor of safety of 2.0 against yielding in shear. Assume the yield stress in shear is 27 ksi. The tube has an outside diameter of 1.5 in. and a wall thickness of 1/16 in.

Figure P7.33

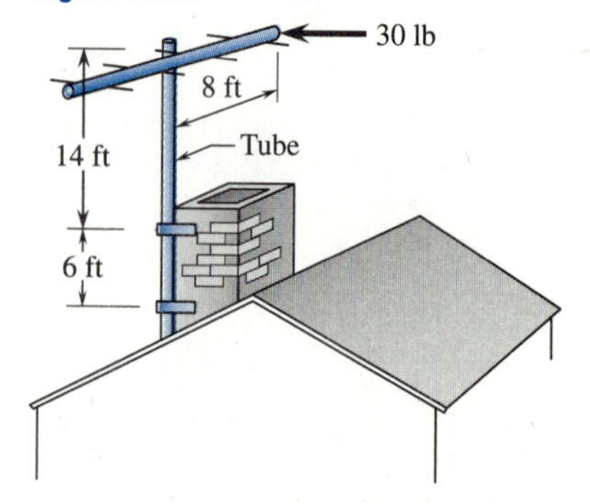

7.34 If the tube of Figure 7.33 was a 1 in. × 1 in. square tube with a 1/16 in. wall thickness, determine the shear stress due to both the torque and transverse shear force.

7.35 A crank has a force F of 1000 N applied to it. Determine the maximum shear stress in the circular portion of the crank due to torque and vertical shear.

Figure P7.35

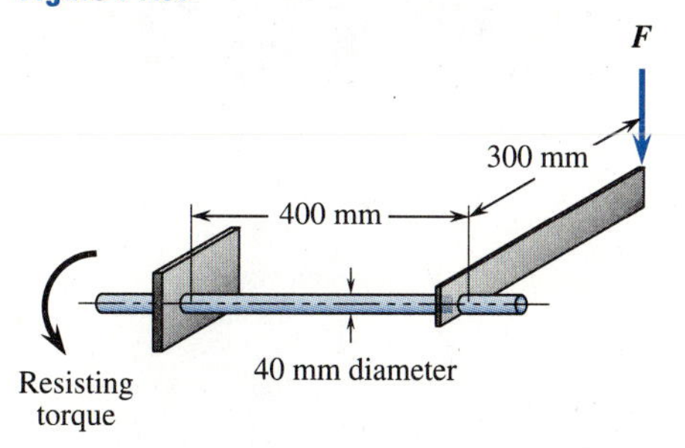

7.36 A standard steel construction pipe is used to support a bar carrying the load shown. Specify the size of pipe needed to keep the combined shear stress to 10,000 psi. (See Appendix C.)

Figure P7.36

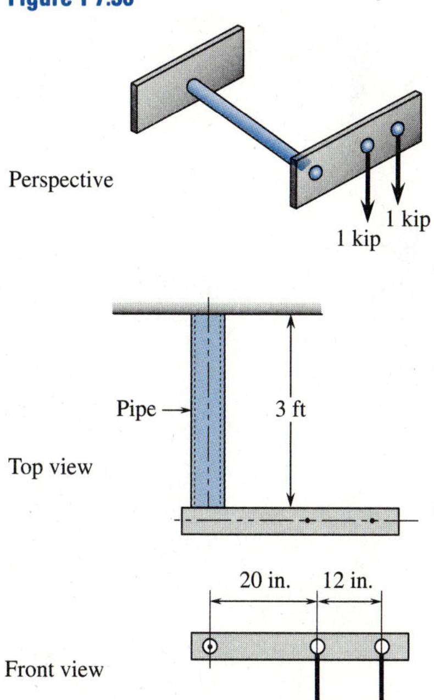

7.37 A simply supported hollow box beam 40 ft long is supporting a catwalk off one side as shown along its entire length. The dead load of the catwalk is 20 lb/ft² and the live load on the catwalk is 50 lb/ft². Determine the maximum shear stress in the beam due to the combined torque and vertical shear.

Figure P7.37

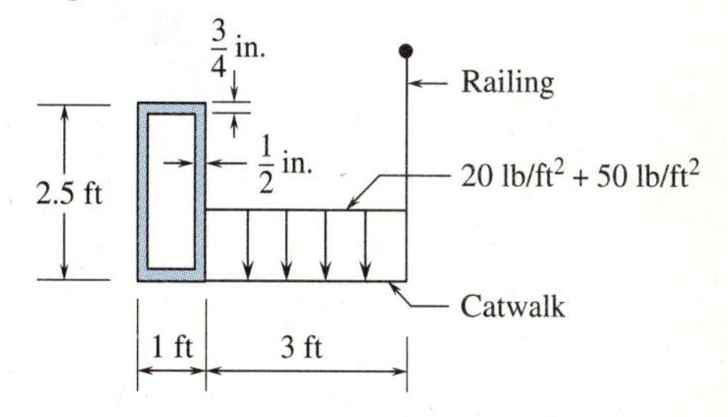

7.38 A crankshaft AB has an applied horizontal force acting on it of 700 lb and is held in equilibrium by the torque T and by reactions at A and B. The bearings at A and B are journal bearings and exert no couples on the shaft. Determine the shear stresses at points E, F, G, and H located at the top, bottom, and horizontal diameter of a vertical plane located 3.0 in. to the left of bearing B.

Figure P7.38

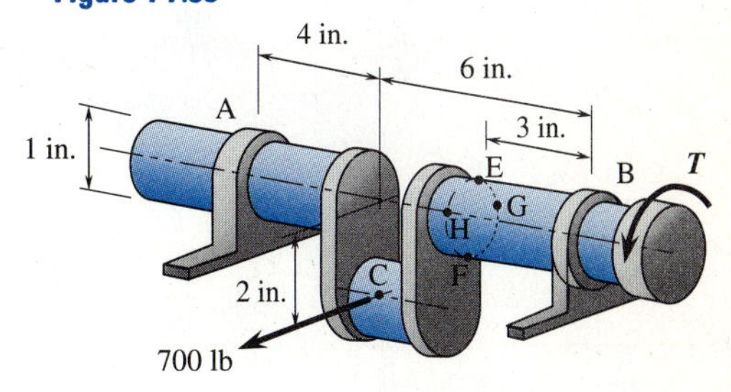

7.39 For the structural tube shown with uniform wall thickness of 0.25 in., determine the shear stresses at the three points A, B, and C indicated.

Figure P7.39

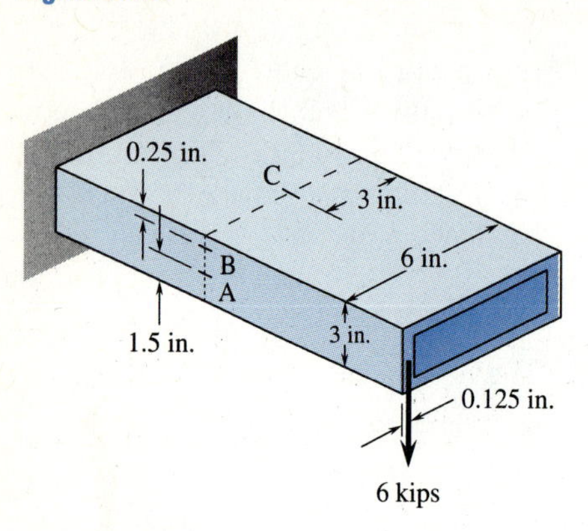

Figure P7.41

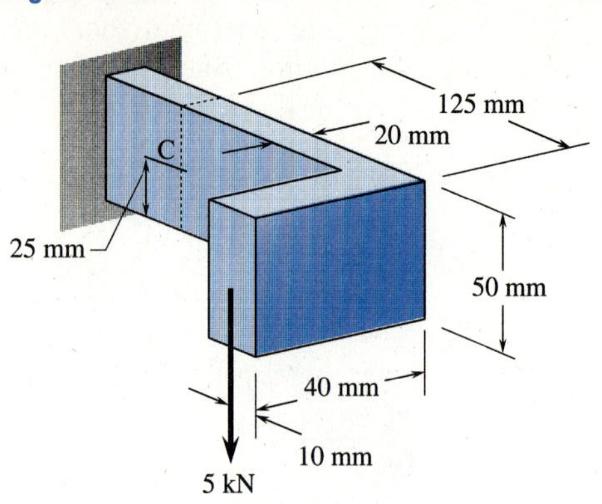

7.40 An extruded aluminum tube has a vertical force of 25 kN applied to it at the end A of bar AB, which is welded to the tube. The tube has a uniform wall thickness of 6 mm. Determine the shear stress at points C and D shown on the outer surface of the tube.

Figure P7.40

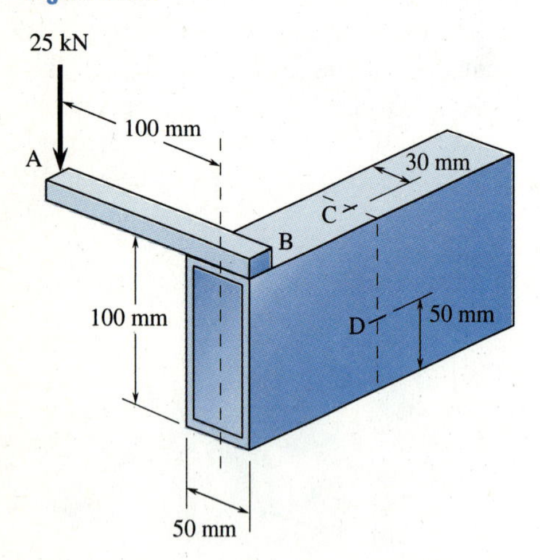

7.41 For the machine part made of a solid rectangular cross-section shown, determine the shear stress at point C.

7.42 For the torsion bar shown, determine the combined torsion stress at the fixed support C. The radius of the bar is 0.5 in. Let $L_1 = 20$ in. and $L_2 = 10$ in. Let $F = 100$ lb.

Figure P7.42

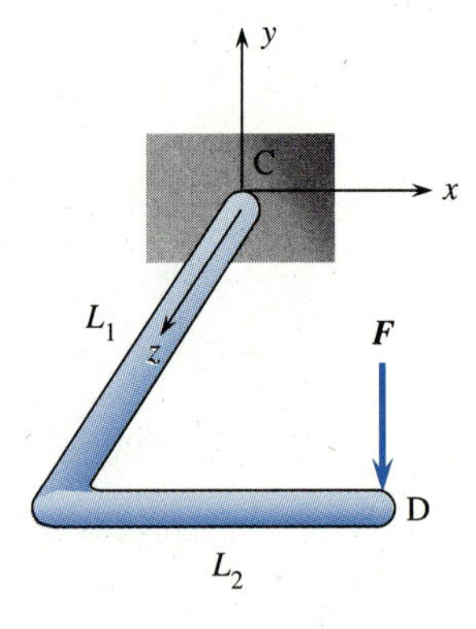

7.43 For the helical spring shown and section shown, determine the combined shear stress. Use the free-body diagram shown to determine the torque T and shear force V at the cut section of the spring. The radius r of the spring cross-section is 0.50 in. Let dimension $R = 2.5$ in., and force $F = 500$ lb.

Figure P7.43

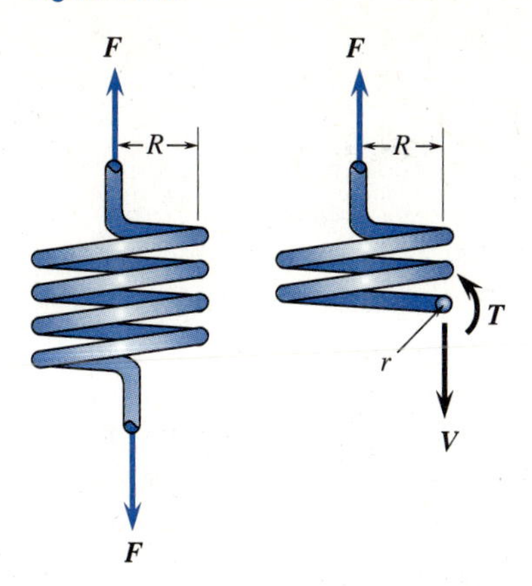

Figure P7.45

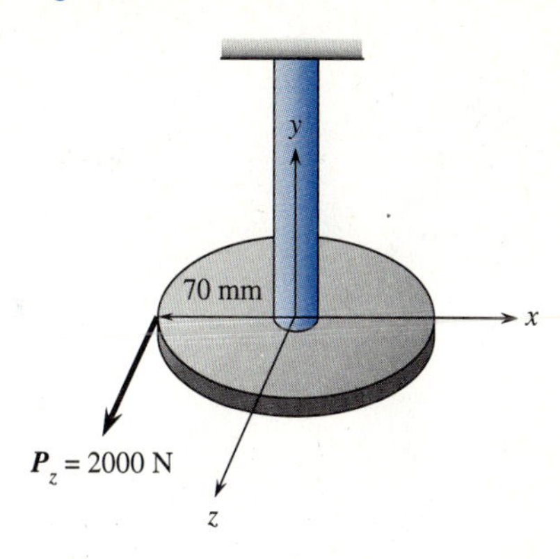

7.44 A vertical force is applied to the pipe wrench as shown. Determine the largest combined shear stress in the pipe at the wall B. The pipe has an outer diameter of 25 mm and an inner diameter of 22 mm.

Figure P7.44

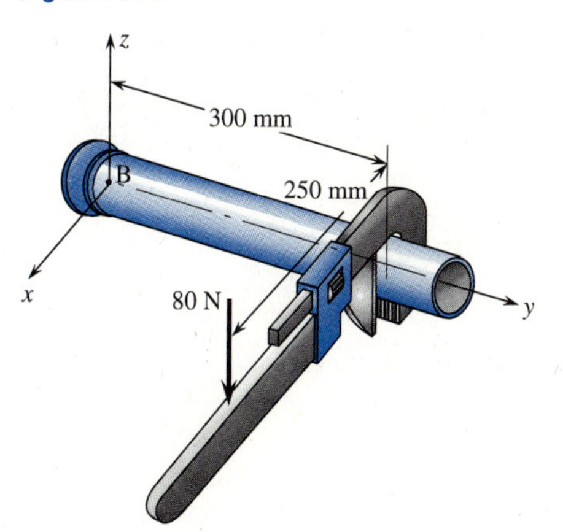

7.45 For the pendulum shown, determine the largest combined shear stress in the vertical shaft. The pendulum is subjected to a load of 2000 N in the z direction. The radius of the vertical member is 10 mm.

Section 7.4

7.46 A cylinder constructed of hot-rolled ASTM-A36 steel holds acetylene at an internal pressure of 1.8 MPa. The diameter of the tank is 350 mm. Using a factor of safety of 2.5 against yielding, determine the wall thickness.

7.47 A propane tank made of ASTM-A36 steel is used in a recreational vehicle. The tank diameter is 400 mm and its wall thickness is 3 mm. Determine the hoop and axial stress at the outer wall of the tank if the pressure inside is 800 kPa.

7.48 Determine the largest internal pressure that a cylindrical tank can hold if the ultimate stress in the steel is 125 ksi and a factor of safety of 5.0 is used. The tank has a wall thickness of 0.5 in. and a diameter of 60 in.

7.49 A water tank is filled to the top as shown. The tank has a wall thickness of 20 mm at the lower section and a diameter of 10 m. Determine the hoop stress at the base of the tank. (The mass density of water = 1000 kg/m^3.)

Figure P7.49

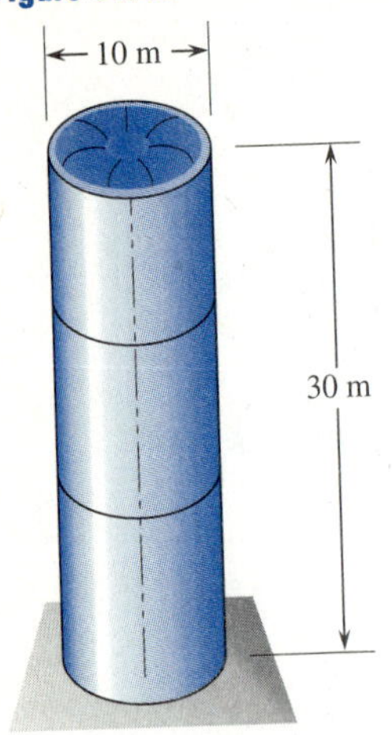

7.50 The cylindrical propane tank shown is 120 in. in diameter and is constructed of 3/4-in.-thick ASTM-A242 steel with all joints welded. The pressure inside the tank is 260 psi. Determine the largest normal stress in the wall of the tank and the factor of safety with respect to failure by fracture.

Figure P7.50

7.51 A hydraulic cylinder has a bore of 100 mm and a wall thickness of 2.5 mm. Determine the hoop stress in the cylinder wall if an internal pressure is 2.5 MPa.

7.52 Determine the normal stress in a basketball of 30 in. circumference and wall thickness of 0.1 in. that is inflated to a gauge pressure of 16 psi.

7.53 A large spherical storage tank used for a compressed gas in a chemical plant is 12 m in diameter and is made of ASTM-A242 hot-rolled steel plate 12 mm thick. Determine the internal pressure if a factor of safety of 2.5 is used based on the yield strength of the steel.

7.54 A spherical tank made of aluminum alloy 7075-T6 has a diameter of 50 in. and wall thickness of 1/4 in. For an internal pressure of 100 psi, determine the normal stress in the wall of the tank.

7.55 A scuba diver's tank is made of ASTM-A242 steel. The ends are hemispherical with wall thickness of 0.125 in. and diameter of 12 in. Based on a factor of safety against yielding of 4.0, determine the largest internal pressure that the tank can hold.

Figure P7.55

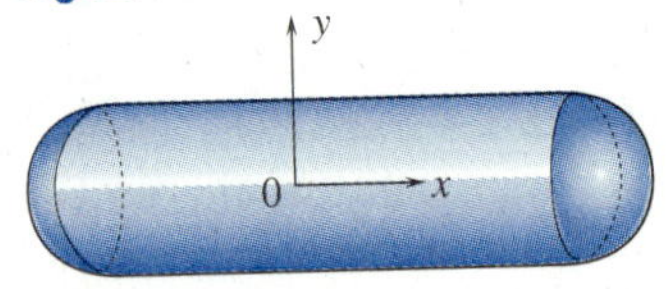

7.56 A spherical oxygen tank holds a pressure of 60 MPa. The tank has an outside diameter of 220 mm and a wall thickness of 15 mm. Determine the normal stress in the wall of the sphere.

Section 7.5

7.57 A vertical storage tank is to be constructed of ASTM A36 plate steel for an oil refinery. The tank is to be 70 ft high and have a diameter of 100 ft. The vessel is to have a spot x-ray on the vertical welds. The corrosion allowance is to be 1/8 in. Determine the thickness of the lowest course of plates for the tank. The tank will hold water.

7.58 A cylindrical tank of 15 in. diameter is to be designed to store hydrogen gas at a pressure of 200 psi. No corrosion allowance is necessary and no x-ray inspection of the welds will be made. Determine the required thickness of the tank. Assume A36 steel.

7.59 A cylindrical pipe ASTM A36 plate steel of 6 in. diameter must carry a pressure of 150 psi. The pipe should have a corrosion allowance of 1/8 in. There will not be any x-ray inspection of the welds. Determine the required thickness of the pipe.

7.60 A vertical cylindrical tank is to be designed to store oil. The tank will be 20 m high and have a diameter of 15 m. The tank is made of a steel with a yield stress of 250 MPa and an ultimate stress of 400 MPa. No corrosion allowance is necessary and no x-ray inspections will be made. Determine the thickness of the tank.

7.61 A spherical pressure vessel with a thickness of 0.225 in. and an outer diameter of 28.4 in. is known to be made of ASTM A36 steel plate. Determine the pressure rating of the tank in the service condition. Assume no x-ray inspections were performed on the welds of the tank during construction.

7.62 A scuba diver's tank is to be made of a high-strength, low-alloy ASTM-A514 steel. The tank is cylindrical with spherical ends. The cylindrical portion of the tank has a diameter of 2.5 m. The tank must hold a gas pressure at 6.0 MPa. The tank welds will be fully x-ray inspected. A corrosion allowance of 2 mm is used. Determine the required thickness of the tank.

COMPUTER PROBLEMS

The following problems are suggested to be solved using a programmable calculator, microcomputer, or mainframe computer. It is suggested that you write a simple computer program in BASIC, FORTRAN, PASCAL, or some other language to solve these problems.

C7.1 Write a program to determine the thickness of a rectangular concrete dam in order to retain a water level even with its crest (top), as shown in the figure, without causing tension in the foundation at the upstream side. Your input should include the height of the dam, the unit weight of concrete, and the unit weight of water or other liquid. Check your program using the following data: height of dam = 1.5 m, unit weight of concrete = 25 kN/m^2, and unit weight of water = 10 kN/m^2.

Figure C7.1

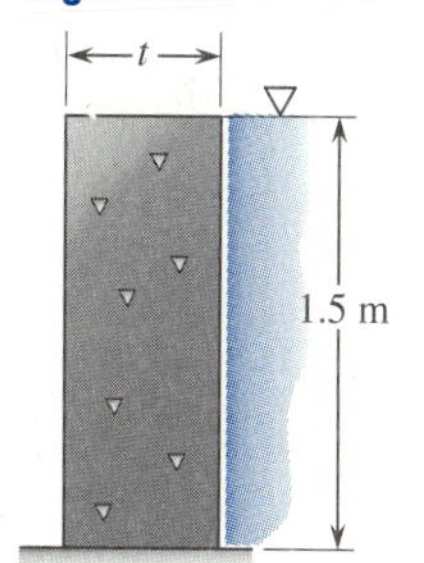

C7.2 Write a program to determine the maximum combined shear stress in a box beam due to the combined torque and vertical shear from a supporting catwalk with its live load. The input should include the catwalk dead load and live load in force/unit length units, the width of the catwalk, and the size of the box beam (including length, width, height, and thickness of the box beam cross-section). Check your program with the following data: catwalk dead load = 20 lb/ft^2, catwalk live load = 50 lb/ft^2, catwalk width 3 ft, dimensions of the box beam (length = 40 ft) and cross-section (height = 2.5 ft, width = 1 ft, and thickness of wall = 0.5 in.). Let the beam length = 40 ft.

Figure C7.2

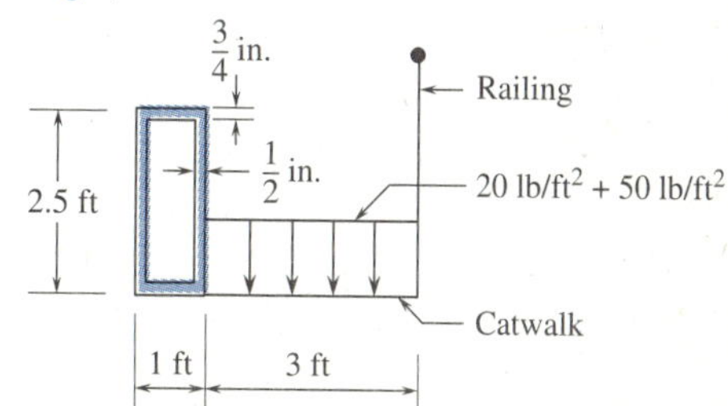

C7.3 Write a program to determine the thickness of a cylindrical pressure vessel based on the ASME Pressure Vessel Code (see Eqs. (7.22) and (7.23)). Your input should include the yield strength and the ultimate strength of the material, the internal pressure, the inner radius of the vessel, and the joint efficiency j. Check your program using the following data: yield stress of 36 ksi, ultimate stress of 58 ksi, internal pressure of 260 psi, inner radius of 60 in., and joint efficiency $j = 0.70$.

C7.4 Write a program to determine the thickness of a spherical pressure vessel based on the ASME Pressure Vessel Code (see Eq. (7.24)). The input should be the same as that of Problem C7.3. Use the data of problem C7.3 to check your program.

Analysis of Plane Stress and Strain

8.1 INTRODUCTION

In this chapter we develop the stress transformation equations to describe the state of stress at a point in a body. We also use the conventional stress formulas developed in Chapters 2 through 5 to obtain stresses on known reference planes (as also illustrated for combined loading in Chapter 7). These stresses are then used in the stress transformation equations to obtain the stresses acting on any plane through a given point in a body. We then obtain equations for the maximum normal and shear stresses and the planes that they act on because these stresses are important design considerations.

We then introduce the concept of Mohr's circle and show it to be an invaluable tool for rapidly determining stresses on arbitrarily oriented planes and for determining maximum normal and shear stresses and their associated planes.

Numerous problems involving combined loads, resulting in, for instance, simultaneous axial, torsional, and bending stresses (Figures 8.1 and 8.2), illustrate the transformation equations and the role of the Mohr's circle concept in analyzing combined stress problems.

Next we consider the concept of absolute maximum shear stress. Then we describe common failure theories used to predict actual failure for brittle or ductile materials.

We then analyze strain using the transformation equations for strain and the Mohr's circle concept. Next, we introduce strain rosettes. These rosettes are shown to be a practical method for obtaining the strains used in the stress-strain equations that follow. The plane stress-strain equations are then developed and used to determine stresses in a number of engineering problems.

Finally, we consider the three-dimensional stress-strain equations.

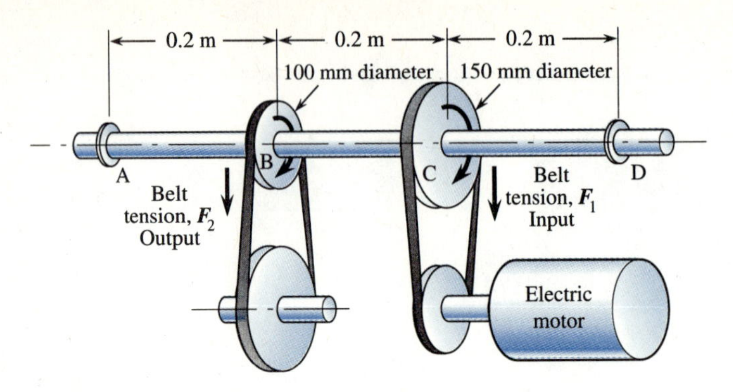

Figure 8.1 Machine shaft subjected to combined bending and torsion.

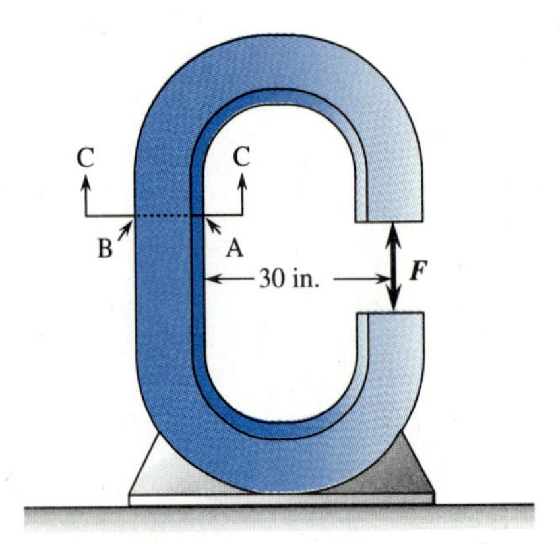

Figure 8.2 Press frame subjected to axial force and bending.

8.2 PLANE STRESS AND EQUATIONS FOR THE TRANSFORMATION OF PLANE STRESS

We now consider the common state of stress called plane stress. The axially loaded bar, the torsionally loaded shaft, the beam, thin-walled pressure vessels, and members subjected to combinations of loads, such as the bar with axial and torsional loads acting together, can often be treated as plane stress problems.

To describe plane stress more completely, we begin by examining the general three-dimensional state of stress at a given point Q under general loading conditions (Figure 8.3). In Figure 8.3, six components of stress are possible. Three of these components, σ_x, σ_y, and σ_z, define the normal stresses acting perpendicular to the faces of a small cubic element centered at Q. Here we show the normal stresses acting on the positive faces x, y, and z. These normal stresses, as shown, are tensile stresses because they pull away on the faces they act on. The other three components, τ_{xy}, τ_{xz}, and τ_{yz}, represent the shear stresses on the same element where, for instance, positive τ_{xy} represents the shear stress acting on

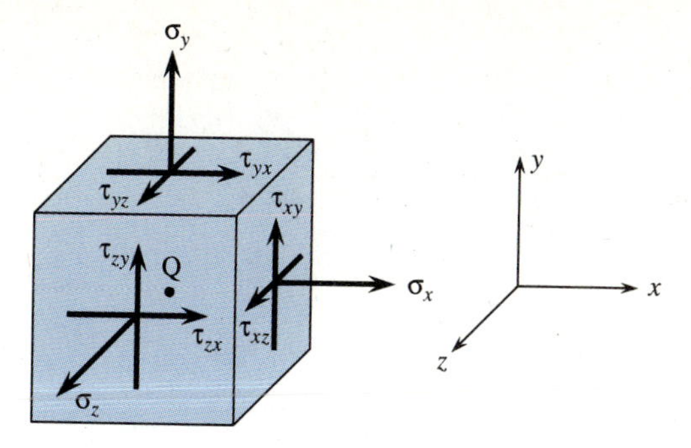

Figure 8.3 General three dimensional state of stress.

the positive x-face and in the positive y-direction. (Recall that there are three independent shear stresses since we have previously shown that $\tau_{xy} = \tau_{yx}$, $\tau_{xz} = \tau_{zx}$, and $\tau_{yz} = \tau_{zy}$.) This state of stress is called *triaxial*.

In plane stress, two faces of the cubic element are free of stress. It is conventional to assume the stress-free faces to be the front and back faces of the cubic element. Therefore, we have $\sigma_z = \tau_{zx} = \tau_{zy} = 0$. The stresses σ_x, σ_y, and τ_{xy} remain. These stresses act on the x- and y-faces and, for convenience, are commonly represented by the plane as shown in Figure 8.4, even though the element is really a solid one with a constant thickness perpendicular to the plane of the figure. Thin plates subjected to forces acting in the midplane of the plate and the free surfaces of structural and machine components (Figure 8.5) are typical examples of states of plane stress.

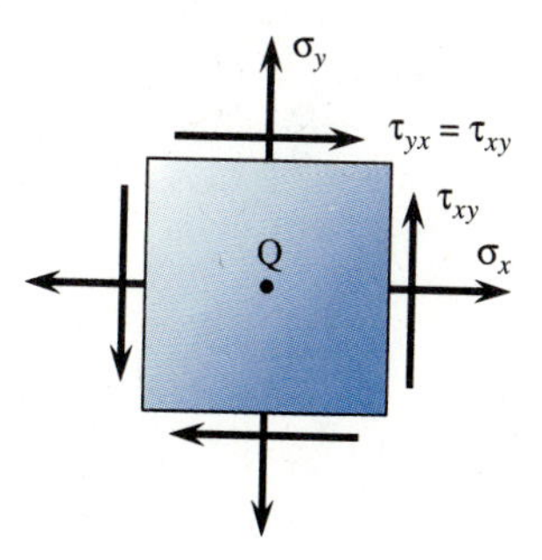

Figure 8.4 Plane stress element.

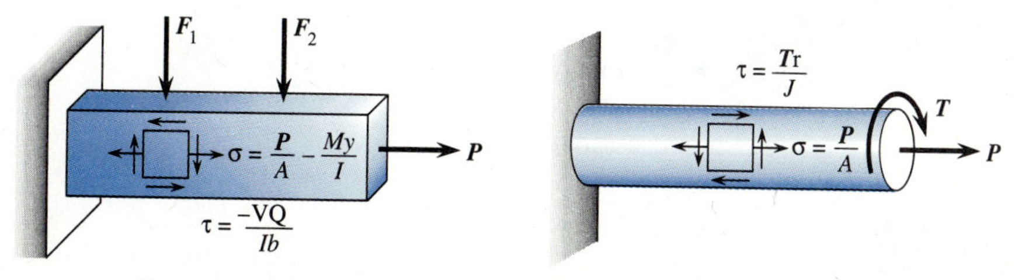

Figure 8.5 Examples of plane stress.

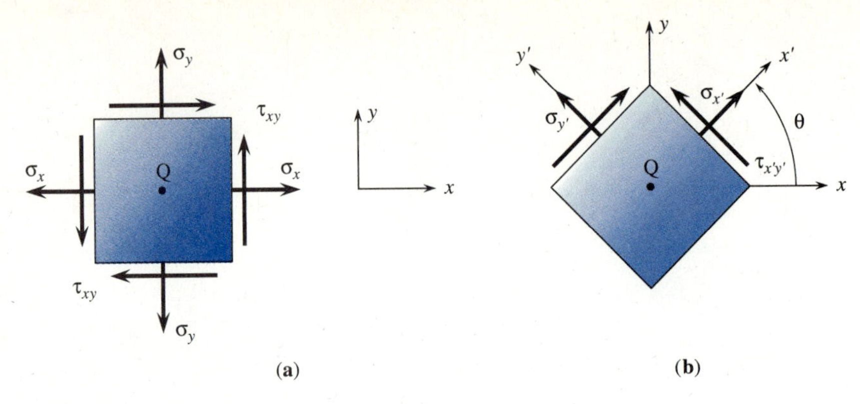

Figure 8.6 (a) State of plane stress at point Q and (b) element rotated about the z axis.

To develop the stress transformation equations, we assume a state of plane stress exists at a point Q in a body and that stresses σ_x, σ_y, and τ_{xy} exist on planes associated with an element representing the state of stress at Q (Figure 8.6(a)). We desire to obtain the stress components $\sigma_{x'}$, $\sigma_{y'}$, and $\tau_{x'y'}$ acting on an element that has been rotated through an angle θ about the z axis (Figure 8.6(b)). These stresses can be expressed in terms of σ_x, σ_y, τ_{xy}, and θ.

To obtain the stresses $\sigma_{x'}$ and $\tau_{x'y'}$ acting on the face perpendicular to the x'-axis, we consider a wedge-shaped element consisting of faces perpendicular to the x-, y-, and x'-axes as shown in Figure 8.7(a), where the thickness of the element is included, or equivalently, by the plane element (Figure 8.7(b)) with uniform thickness implied. We arbitrarily assign the inclined x'-face to have area dA. Then using trigonometry, the vertical face has area $dA(\cos\theta)$ and the horizontal face has area $dA(\sin\theta)$. Multiply these areas by the respective stresses acting on each area, we obtain the forces acting on each of the three faces as shown in Figure 8.8(a).

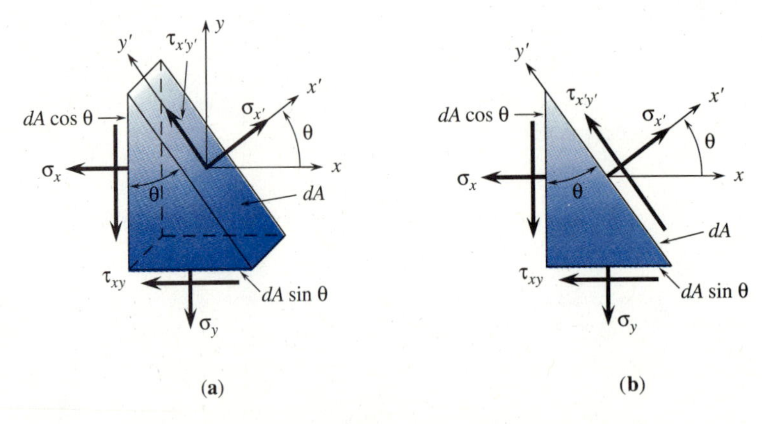

Figure 8.7 (a) Wedge shaped stress element with thickness shown and (b) two dimensional representation of wedge.

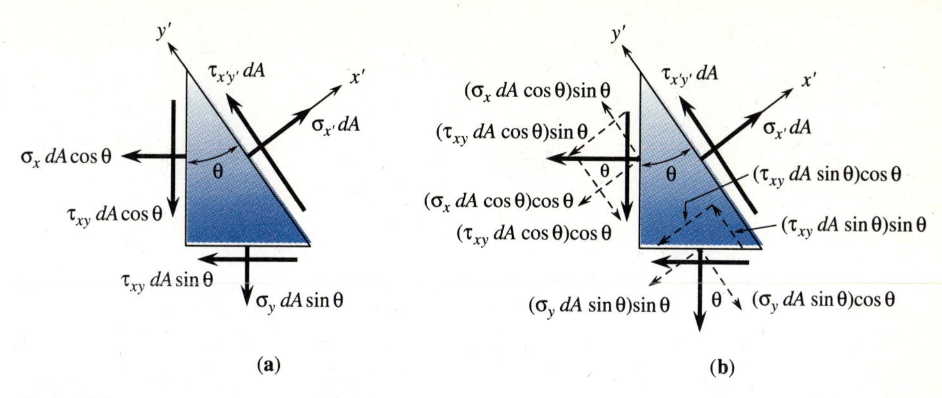

Figure 8.8 (a) Forces acting on wedge and
(b) components of forces along x' and y' axes.

Summing force components along the x'- and y'-axes (Figure 8.8(b)), we obtain the following equilibrium equations:

$$\Sigma F_{x'} = 0: \quad \sigma_{x'}dA - \sigma_x(dA\cos\theta)\cos\theta - \tau_{xy}(dA\cos\theta)\sin\theta$$
$$- \sigma_y(dA\sin\theta)\sin\theta - \tau_{xy}(dA\sin\theta)\cos\theta = 0$$

$$\Sigma F_{y'} = 0: \quad \tau_{x'y'}\,dA + \sigma_x(dA\cos\theta)\sin\theta - \tau_{xy}(dA\cos\theta)\cos\theta$$
$$- \sigma_y(dA\sin\theta)\cos\theta + \tau_{xy}(dA\sin\theta)\sin\theta = 0$$

Solving the first equation for $\sigma_{x'}$ and the second for $\tau_{x'y'}$, we obtain

$$\sigma_{x'} = \sigma_x\cos^2\theta + \sigma_y\sin^2\theta + 2\tau_{xy}\sin\theta\cos\theta \quad \textbf{(8.1)}$$

$$\tau_{x'y'} = -(\sigma_x - \sigma_y)(\sin\theta\cos\theta) + \tau_{xy}(\cos^2\theta - \sin^2\theta) \quad \textbf{(8.2)}$$

Using the trigonometric relations

$$\cos^2\theta = \frac{1+\cos 2\theta}{2} \qquad \sin^2\theta = \frac{1-\cos 2\theta}{2} \quad \textbf{(8.3)}$$

and

$$\sin 2\theta = 2\sin\theta\cos\theta \qquad \cos 2\theta = \cos^2\theta - \sin^2\theta \quad \textbf{(8.4)}$$

in Eqs. (8.1) and (8.2) and rearranging terms, we have

$$\sigma_{x'} = \frac{\sigma_x + \sigma_y}{2} + \frac{\sigma_x - \sigma_y}{2}\cos 2\theta + \tau_{xy}\sin 2\theta \quad \textbf{(8.5)}$$

$$\tau_{x'y'} = -\frac{\sigma_x - \sigma_y}{2}\sin 2\theta + \tau_{xy}\cos 2\theta \quad \textbf{(8.6)}$$

Equations (8.5) and (8.6) are the general *stress transformation equations* used to obtain the normal and shear stress, respectively, on any plane located by the angle θ between the x and x′ axes, and caused by the initially known stresses σ_x, σ_y, and τ_{xy} that are often obtained using the conventional stress formulas developed in Chapters 2 through 5.

Remember in using Eqs. (8.5) and (8.6) that positive σ_x and σ_y are tensile or pulling away on the faces of the element, τ_{xy} is positive if it acts in the positive y-direction on the positive x-face (to produce a counterclockwise rotation of the element) and θ is positive when moving counterclockwise from x to x'. The expression for $\sigma_{y'}$ can be obtained by replacing θ in Eq. (8.5) by the angle $\theta + 90°$ formed by the y'-axis with the x-axis and by making use of the trigonometric relations

$$\cos(2\theta + 180°) = -\cos 2\theta \qquad \sin(2\theta + 180°) = -\sin 2\theta \tag{8.7}$$

to obtain

$$\sigma_{y'} = \frac{\sigma_x + \sigma_y}{2} - \frac{\sigma_x - \sigma_y}{2}\cos 2\theta - \tau_{xy}\sin 2\theta \tag{8.8}$$

Example Problems 8.1 and 8.2 illustrate the application of the stress transformation equations. In these problems it is assumed that the stresses on the x- and y-faces have been determined using the conventional stress formulas introduced in Chapters 2 through 5. Later we will illustrate their use in obtaining σ_x, σ_y, and τ_{xy}. However, at this stage our intent is to illustrate the direct use of Eqs. (8.5) and (8.6).

EXAMPLE 8.1

For the state of plane stress shown in Figure 8.9, determine the normal and shear stresses acting on a plane parallel to the line a-a indicated.

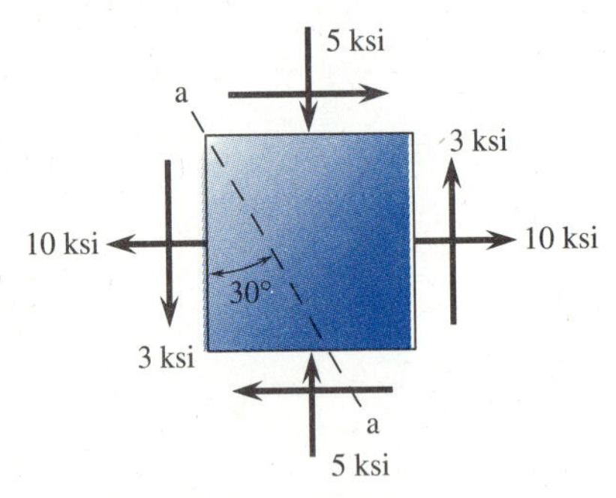

Figure 8.9 Element in plane stress.

Solution

We begin by constructing the wedge of material showing the known stresses on the vertical and horizontal faces and the stresses to be obtained on the face parallel to the a-a line.

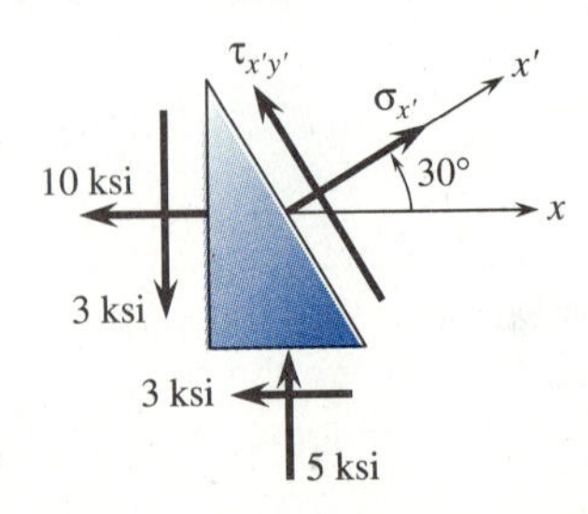

Using Eqs. (8.5) and (8.6), with $\sigma_x = 10$ ksi, $\sigma_y = -5$ ksi, $\tau_{xy} = 3$ ksi, and $\theta = 30°$, we obtain

$$\sigma_{x'} = \frac{10 + (-5)}{2} + \frac{10 - (-5)}{2} \cos 60° + 3 \sin 60° = 8.85 \text{ ksi}$$

$$\tau_{x'y'} = -\frac{10 - (-5)}{2} \sin 60° + 3 \cos 60° = -4.00 \text{ ksi}$$

The positive number indicates that $\sigma_{x'}$ is acting away as a tensile stress on the face a-a, while the negative number indicates that $\tau_{x'y'}$ tends to create a clockwise rotation of the element as it acts downward and to the right opposite to that initially shown on the wedge.

EXAMPLE 8.2

The grain of a wooden block forms an angle of 15° with the vertical (Figure 8.10(a)). For the state of stress shown, determine the normal stress perpendicular to the grain and the shear stress in the plane of the grain.

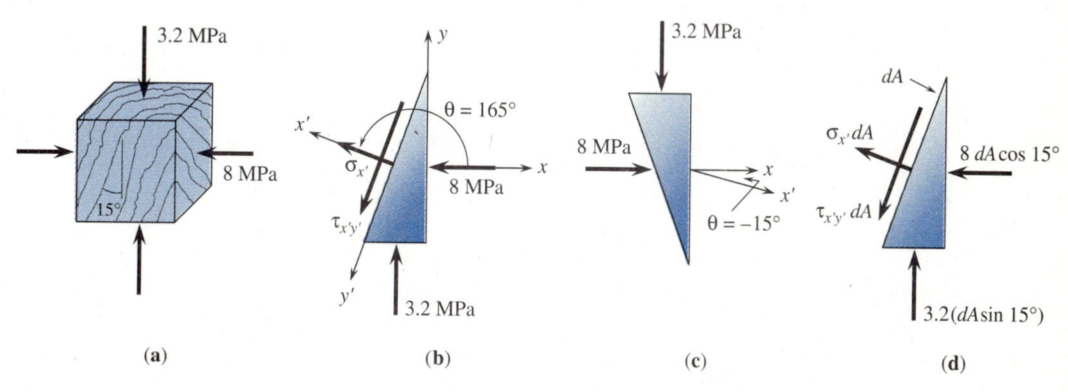

Figure 8.10 (a) Wooden block in plane stress, (b) wedge of material showing stresses, (c) alternative wedge and (d) forces acting on wedge taken as in (b).

Solution

We begin by constructing a wedge of material showing the known stresses on the vertical and horizontal faces, and the stresses to be obtained on the face parallel to the grain (Figure 8.10(b)). Note that positive $\sigma_{x'}$ is still directed away from the face parallel to the grain, and $\tau_{x'y'}$ still acts on the x'-face in the positive y'-direction (or produces a counterclockwise rotation of the wedge). Alternatively, we could choose a wedge shown in Figure 8.10(c) with $\theta = -15°$ since the angle from x to x' is in the clockwise direction. Using Eqs. (8.5) and (8.6), with $\sigma_x = -8$ MPa, $\sigma_y = -3.2$ MPa, $\tau_{xy} = 0$, and

$\theta = 165° = (90° + 75°)$, we obtain

$$\sigma_{x'} = \frac{-8 + (-3.2)}{2} + \frac{-8 - (-3.2)}{2} \cos 330° + 0 = -7.68 \text{ MPa}$$

$$\tau_{x'y'} = \frac{-8 - (-3.2)}{2} \sin 330° + 0 = -1.20 \text{ MPa}$$

The negative signs mean that $\sigma_{x'}$ is acting into or pushing on the plane parallel to the grain, while $\tau_{x'y'}$ is tending to rotate the wedge clockwise. These stresses are then in the opposite directions from those shown on the wedge. We readily observe that $\sigma_{x'}$ must be pushing into the plane parallel to the grain (in the negative x' direction) to obtain force equilibrium since the components of the forces due to the 8 MPa and 3.2 MPa stresses are both directed in the positive x'-direction.

An alternative method to determine $\sigma_{x'}$ and $\tau_{x'y'}$ is to sum forces directly in the x'- and y'-directions for the wedge. This is done by first converting the stresses to forces by multiplying the stresses by the areas they act over, and then using trigonometry to obtain the components of each force in the x'- and y'-directions. Figure 8.10(d) shows the forces acting on the wedge. Here the area that $\sigma_{x'}$ and $\tau_{x'y'}$ act on is arbitrarily defined as dA. Then, based on trigonometry, the horizontal and vertical faces have areas $dA \cos 15°$ and $dA \sin 15°$. On summing forces in the x'- and y'-directions, we obtain

$$\Sigma F_{x'} = 0, \quad \sigma_{x'} dA + 8(dA \cos 15°)\cos 15° + 3.2(dA \sin 15°)\sin 15° = 0$$

$$\sigma_{x'} = -8 \cos^2 15° - 3.2 \sin^2 15°$$

$$\sigma_{x'} = -7.68 \text{ MPa}$$

$$\Sigma F_{y'} = 0, \quad \tau_{x'y'} dA + 8(dA \cos 15°)\sin 15° - 3.2(dA \sin 15°)\cos 15° = 0$$

$$\tau_{x'y'} = -8\cos 15° \sin 15° + 3.2\sin 15° \cos 15°$$

$$\tau_{x'y'} = -1.20 \text{ MPa}$$

These stresses are identical to those obtained by direct use of Eqs. (8.5) and (8.6). We are then reminded that Eqs. (8.5) and (8.6) are really obtained by writing equilibrium equations for a wedge of material.

In summary, this method, often called the *wedge method,* because we are directly summing forces acting on a wedge, is somewhat easier to apply than the direct application of the stress transformation equations. Using the stress transformation equations requires remembering that θ is the angle between x and x' and is positive if we move from x to x' through angle θ in the counterclockwise direction and is negative if we move from x to x' in the clockwise direction. We must also be careful to identify θ properly if the element is not oriented exactly as shown in Figure 8.7.

8.3 PRINCIPAL STRESSES

The transformation equations, Eqs. (8.5) and (8.6), indicate that normal stress $\sigma_{x'}$ and shear stress $\tau_{x'y'}$ vary with angle θ. That is, as we investigate the state of stress at point Q in a body, as the element representing that state of stress is rotated to different orientations different values of normal and shear stress appear on the planes of the element. For design purposes, we normally seek the largest positive and negative normal stresses and largest shear stress. To determine the maximum and minimum normal stresses, called the *principal stresses,* we differentiate Eq. (8.5) with respect to θ and set the expression to zero, and solve for the values of θ. These values of θ make $\sigma_{x'}$ a maximum or a minimum. This procedure yields

$$\frac{d\sigma_{x'}}{d\theta} = \frac{d}{d\theta}\left(\frac{\sigma_x + \sigma_y}{2} + \frac{\sigma_x - \sigma_y}{2}\cos 2\theta + \tau_{xy}\sin 2\theta\right) = 0$$

or

$$-(\sigma_x - \sigma_y)\sin 2\theta + 2\tau_{xy}\cos 2\theta = 0$$

which can be solved for θ as

$$\tan 2\theta_p = \frac{2\tau_{xy}}{\sigma_x - \sigma_y} \tag{8.9}$$

The subscript p is used to indicate that the angle θ_p defines the *principal planes*. Two angles θ_{p1} and θ_{p2} are obtained from Eq. (8.9). These values differ by 90°. One angle is between 0° and 90°, while the other is between 90° and 180°. The stresses, evaluated on back-substituting θ_{p1} and θ_{p2} into Eq. (8.5), are called the *principal stresses*. One of the angles θ_{p1} or θ_{p2} yields the largest normal stress in the x-y plane, while the other angle yields the smallest normal stress. Because these angles differ by 90°, the principal stresses are 90° apart and act on mutually perpendicular planes. The principal angles then give the directions of the normals to the principal planes.

Example Problems 8.3 and 8.4 now illustrate how to use Eq. (8.9) to determine principal angles and how to back-substitute these resulting angles from Eq. (8.9) into Eq. (8.5) to determine principal stresses and, hence, how to establish which principal angle corresponds to which principal stress.

EXAMPLE 8.3

A circular shaft is subjected to a combined tension and torsion load producing the state of stress shown in Figure 8.11. Determine the principal stresses and their corresponding principal planes.

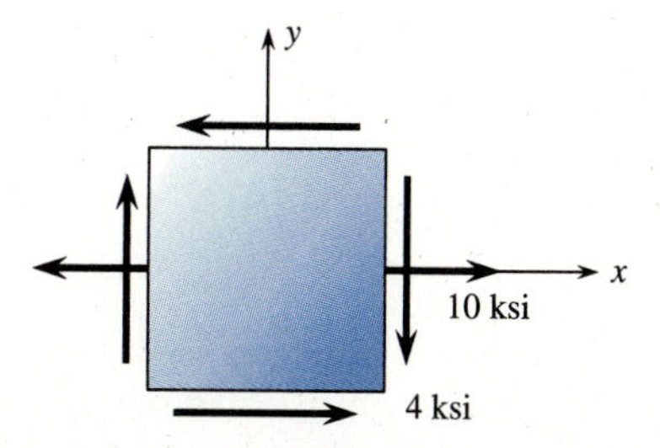

Figure 8.11 Element from circular shaft subjected to tension and torsion.

Using Eq. (8.9), with $\sigma_x = 10$ ksi, $\sigma_y = 0$, and $\tau_{xy} = -4$ ksi, ($\tau_{xy} = -4$ ksi, since it acts on the x-face in the negative y-direction), we obtain

$$\tan 2\theta_p = \frac{2(-4)}{10 - 0} = -\frac{8}{10}$$

$$2\theta_p = 321.3^\circ \quad \text{and} \quad \theta_p = 160.6^\circ$$

or

$$2\theta_p = 141.3^\circ \quad \text{and} \quad \theta_p = 70.6^\circ$$

The principal stresses are obtained by substituting the two values of $2\theta_p$ into Eq. (8.5) for $\sigma_{x'}$. Substituting $2\theta_p = 321.3^\circ$ into Eq. (8.5), along with σ_x, σ_y, and τ_{xy} above, we obtain

$$\begin{aligned}\sigma_{x'} &= \frac{10 + 0}{2} + \frac{10 - 0}{2}\cos 321.3^\circ + (-4)\sin 321.3^\circ \\ &= 5 + 5(0.781) - 4(-0.625) \\ &= 11.4 \text{ ksi}\end{aligned}$$

Similarly, substituting $2\theta_p = 141.3^\circ$ into Eq. (8.5), we obtain

$$\begin{aligned}\sigma_{x'} &= \frac{10 + 0}{2} + \frac{10 - 0}{2}\cos 141.3^\circ + (-4)\sin 141.3^\circ \\ &= 5 + 5(-0.780) - 4(0.625) \\ &= -1.40 \text{ ksi}\end{aligned}$$

Therefore, the principal stresses and their corresponding angles are

$$\sigma_1 = 11.4 \text{ ksi} \quad \text{and} \quad \theta_{p1} = 160.6^\circ$$

$$\sigma_2 = -1.4 \text{ ksi} \quad \text{and} \quad \theta_{p2} = 70.6^\circ$$

Note that θ_{p1} and θ_{p2} differ by 90°. The principal stresses and their directions are shown on the element in Figure 8.12.

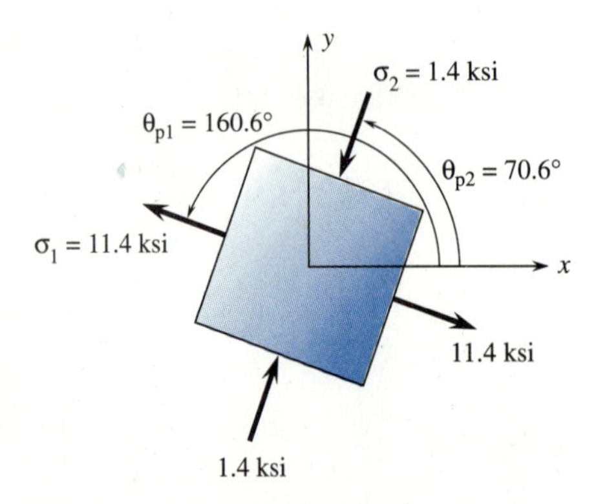

Figure 8.12 Principal stresses and their directions.

No shear stresses act on the element with principal stresses acting on it. We show this by substituting the principal angles (one at a time) into Eq. (8.6) as follows.

$$\tau_{x'y'} = -\frac{10-0}{2}\sin 321.3^\circ + (-4)\cos 321.3^\circ$$

$$= -5(-0.625) - 4(0.781)$$

$$= 0$$

Similarly, when $2\theta_{p2} = 141.3^\circ$ is substituted into Eq. (8.6), we obtain

$$\tau_{x'y'} = 0$$

We have now shown that for the element oriented with the principal stresses acting on it no shear stresses act on these principal planes.

We can obtain general formulas for the principal stresses by substituting Eq. (8.9) into Eq. (8.5) as follows.

$$\sigma_{x'} = \frac{\sigma_x + \sigma_y}{2} + \frac{\sigma_x - \sigma_y}{2}\cos 2\theta_p + \tau_{xy}\left(\frac{2\tau_{xy}}{\sigma_x - \sigma_y}\right)\cos 2\theta_p$$

Simplifying, we have

$$\sigma_{x'} = \frac{\sigma_x + \sigma_y}{2} + \left[\frac{\sigma_x - \sigma_y}{2} + \frac{(\tau_{xy})^2}{(\sigma_x - \sigma_y)/2}\right]\cos 2\theta_p$$

Factoring out $1/((\sigma_x - \sigma_y)/2)$, we obtain

$$\sigma_{x'} = \frac{\sigma_x + \sigma_y}{2} + \left[\left(\frac{\sigma_x - \sigma_y}{2}\right)^2 + (\tau_{xy})^2\right]\frac{\cos 2\theta_p}{(\sigma_x - \sigma_y)/2} \qquad \textbf{(8.10)}$$

Now using Fig. 8.13, which represents Eq. (8.9) for $2\theta_p$, we express $\cos 2\theta_p$ as

$$\cos 2\theta_p = \frac{\left(\frac{\sigma_x - \sigma_y}{2}\right)}{R} \qquad \textbf{(8.11)}$$

where R is the length of the hypotenuse of the triangle in Figure 8.13 given by

$$R = \sqrt{\left(\frac{\sigma_x - \sigma_y}{2}\right)^2 + (\tau_{xy})^2} \qquad \textbf{(8.12)}$$

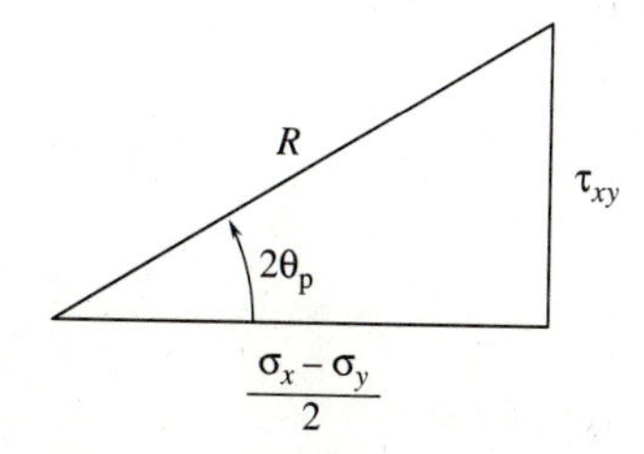

Figure 8.13 Relationship between θ_p and in-plane stresses.

Substituting Eqs. (8.11) and (8.12) into (8.10), yields

$$\sigma_{x'} = \sigma_1 = \frac{\sigma_x + \sigma_y}{2} + \sqrt{\left(\frac{\sigma_x - \sigma_y}{2}\right)^2 + (\tau_{xy})^2} \quad \textbf{(8.13)}$$

which is the principal stress σ_1, where $2\theta_p$ was assumed to be $2\theta_{p1}$ to yield σ_1. Substituting $2\theta_{p2} = 2\theta_{p1} + 180°$, which is the negative of Eq. (8.11), and $\sin 2\theta_p$, obtained from Figure 8.13, into Eq. (8.5) and evaluating as above, we obtain

$$\sigma_2 = \frac{\sigma_x + \sigma_y}{2} - \sqrt{\left(\frac{\sigma_x - \sigma_y}{2}\right)^2 + (\tau_{xy})^2} \quad \textbf{(8.14)}$$

Due to the similarities in Eqs. (8.13) and (8.14), we combine them into one formula for the principal stresses as

$$\sigma_{1,2} = \frac{\sigma_x + \sigma_y}{2} \pm \sqrt{\left(\frac{\sigma_x - \sigma_y}{2}\right)^2 + (\tau_{xy})^2} \quad \textbf{(8.15)}$$

The positive sign on the radical term gives the algebraically larger principal stress σ_1 and the negative sign gives the smaller principal stress σ_2. Making use of Eq. (8.9) for $2\theta_p$ and Eq. (8.6), we obtain the shear stress acting on the principal planes as follows. First, we divide Eq. (8.6) by $\cos 2\theta_p$ to yield

$$\frac{\tau_{x'y'}}{\cos 2\theta_p} = -\frac{\sigma_x - \sigma_y}{2} \tan 2\theta_p + \tau_{xy}$$

Now using Eq. (8.9) for $\tan 2\theta_p$, we get

$$\frac{\tau_{x'y'}}{\cos 2\theta_p} = -\left(\frac{\sigma_x - \sigma_y}{2}\right)\left(\frac{2\tau_{xy}}{\sigma_x - \sigma_y}\right) + \tau_{xy}$$

On simplifying, we have

$$\tau_{x'y'} = 0$$

which says that the shear stresses are zero on the principal planes. (We recall that this was also illustrated by Example 8.3.) The orientation of an element with the principal stresses acting on it is shown in Figure 8.14.

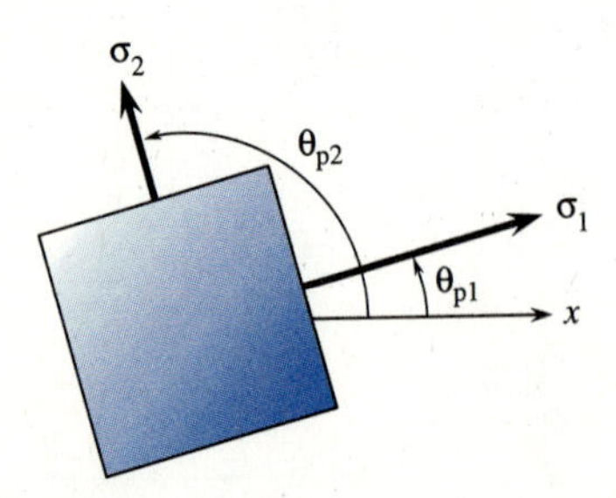

Figure 8.14 Orientation of element with principal stresses acting on it.

We also see that by adding together Eqs. (8.13) and (8.14), we obtain

$$\sigma_x + \sigma_y = \sigma_1 + \sigma_2$$

That is, the sum of the normal stresses acting on any two orthogonal planes passing through a point in a plane body is a constant or invariant.

EXAMPLE 8.4

The axle of an automobile is subjected to the normal forces and a torque producing the state of stress shown in Figure 8.15. Determine the principal stresses and principal angles. Show a sketch of the element oriented with the principal stresses acting on it.

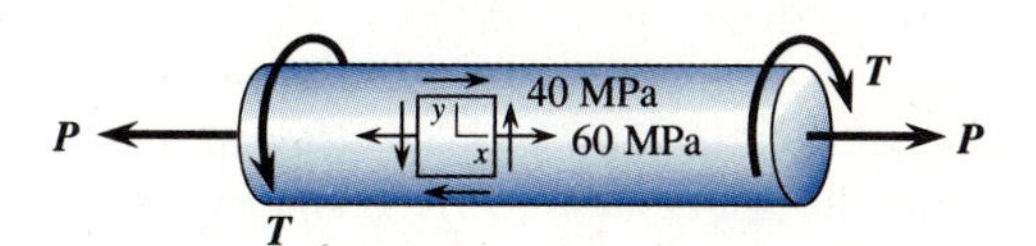

Figure 8.15 State of stress in element of an axle.

With $\sigma_x = 60$ MPa, $\sigma_y = 0$, and $\tau_{xy} = 40$ MPa, we use Eq. (8.15) to obtain the principal stresses directly as

$$\sigma_{1,2} = \frac{60 + 0}{2} \pm \sqrt{\left(\frac{60 - 0}{2}\right)^2 + 40^2}$$

$$= 30 \pm 50$$

$$\sigma_1 = 80 \text{ MPa} \qquad \sigma_2 = -20 \text{ MPa}$$

We use Eq. (8.9) to obtain θ_p as

$$\tan 2\theta_p = \frac{2(40)}{60 - 0} = 1.33$$

$$2\theta_p = 53.1^\circ \quad \text{and} \quad 2\theta_p = 53.1^\circ + 180^\circ = 233.1^\circ$$

Substituting $2\theta_p = 53.1^\circ$ into Eq. (8.5) for $\sigma_{x'}$, we obtain

$$\sigma_{x'} = \frac{60 + 0}{2} + \frac{60 - 0}{2} \cos 53.1^\circ + 40 \sin 53.1^\circ$$

$$\sigma_{x'} = 80 \text{ MPa} = \sigma_1$$

Therefore, $2\theta_{p1} = 53.1^\circ$ and $2\theta_{p2} = 233.1^\circ$, because when substituting 53.1° into Eq. (8.5), we obtained σ_1.

The properly oriented element associated with the principal stresses is shown in Figure 8.16.

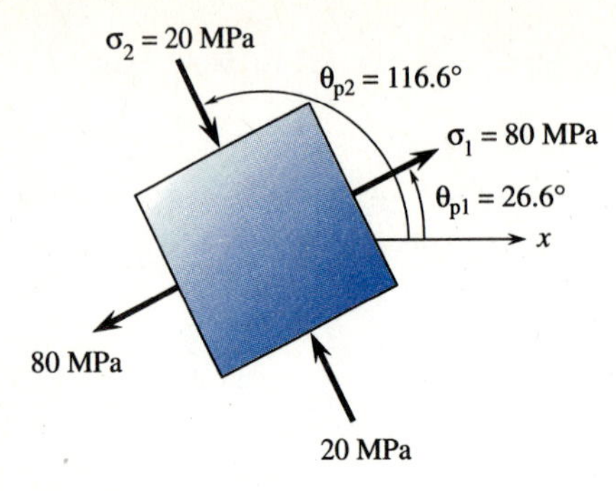

Figure 8.16 Orientation of element with principal stresses acting on it.

8.4 MAXIMUM SHEAR STRESS

The maximum shear stress on an element in plane stress is determined by differentiating Eq. (8.6) with respect to θ, setting the expression to zero, and solving for the angle θ. These values of θ make $\tau_{x'y'}$ a maximum or a minimum. This procedure yields

$$\frac{d\tau_{x'y'}}{d\theta} = \frac{d}{d\theta}\left(-\frac{\sigma_x - \sigma_y}{2}\sin 2\theta + \tau_{xy}\cos 2\theta\right) = 0$$

or

$$-(\sigma_x - \sigma_y)\cos 2\theta - 2\tau_{xy}\sin 2\theta = 0$$

Solving for θ, we obtain

$$\tan 2\theta_s = -\frac{\sigma_x - \sigma_y}{2\tau_{xy}} \tag{8.16}$$

The subscript s is used to indicate that the angle θ_s defines the orientation of the planes that the maximum shear stresses act on.

Equation (8.16) yields two values of θ_s. These two values, θ_{s1} and θ_{s2}, differ by 90° with one value between 0° and 90° and the other between 90° and 180°. The angles θ_{s1} and θ_{s2} then yield the maximum and minimum values of $\tau_{x'y'}$, upon substitution of these angles into Eq. (8.6). Also $\tau_{max} = -\tau_{min}$ since these shear stresses act on perpendicular planes (i.e., recall $\tau_{xy} = \tau_{yx}$). A formula for the maximum shear stress is obtained by substituting Eq. (8.16) into Eq. (8.6) as follows.

$$\tau_{max} = -\frac{\sigma_x - \sigma_y}{2}\sin 2\theta_{s1} + \tau_{xy}\cos 2\theta_{s1} \tag{8.17}$$

Now using Figure 8.17, which represents Eq. (8.16) for $2\theta_{s1}$, we express $\cos 2\theta_{s1}$ and $\sin 2\theta_{s1}$ as

$$\cos 2\theta_{s1} = \frac{\tau_{xy}}{R} \qquad \sin 2\theta_{s1} = \frac{-1/2(\sigma_x - \sigma_y)}{R} \tag{8.18}$$

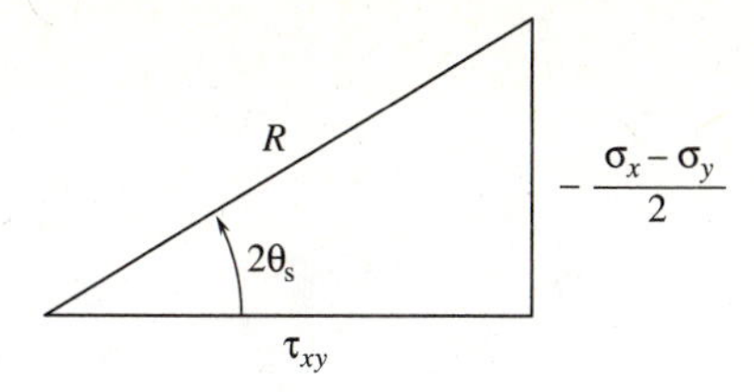

Figure 8.17 Graphical representation of Eq. (8.16).

where R is then

$$R = \sqrt{\left(-\frac{\sigma_x - \sigma_y}{2}\right)^2 + (\tau_{xy})^2} \tag{8.19}$$

Note that Eq. (8.19) is the same as Eq. (8.12). Substituting Eqs. (8.18) and (8.19) into (8.17), we obtain

$$\tau_{max} = \frac{\left(\frac{\sigma_x - \sigma_y}{2}\right)^2 + (\tau_{xy})^2}{R}$$

$$\tau_{max} = \sqrt{\left(\frac{\sigma_x - \sigma_y}{2}\right)^2 + (\tau_{xy})^2} = R \tag{8.20}$$

Similarly, substitution of $2\theta_{s2}$ into Eq. (8.6) yields

$$\tau_{min} = -\sqrt{\left(\frac{\sigma_x - \sigma_y}{2}\right)^2 + (\tau_{xy})^2} \tag{8.21}$$

Equations (8.20) and (8.21) yield what are called the *maximum* and *minimum in-plane shear stresses* because these stresses are associated with the in-plane (x-y plane) stresses.

In developing Eqs. (8.20) and (8.21), we have assumed that $2\theta_{s1}$ produces τ_{max} and $2\theta_{s2}$ produces τ_{min}. In a numeric example we simply substitute, one at a time, the two angles θ_{s1} and θ_{s2} into Eq. (8.6) and a positive shear stress τ_{max} and a negative shear stress τ_{min} result corresponding to either θ_{s1} or θ_{s2}. Of course, only one angle θ_s needs to be substituted into Eq. (8.6) since $\tau_{max} = -\tau_{min}$. We also observe on comparing Eqs. (8.9) and (8.16) that

$$\tan 2\theta_s = -\frac{1}{\tan 2\theta_p} = -\cot 2\theta_p \tag{8.22}$$

and by trigonometry

$$\tan (\theta \pm 90°) = -\cot \theta \tag{8.23}$$

Therefore, we must have

$$2\theta_s = 2\theta_p \pm 90° \tag{8.24}$$

or

$$\theta_s = \theta_p \pm 45° \tag{8.25}$$

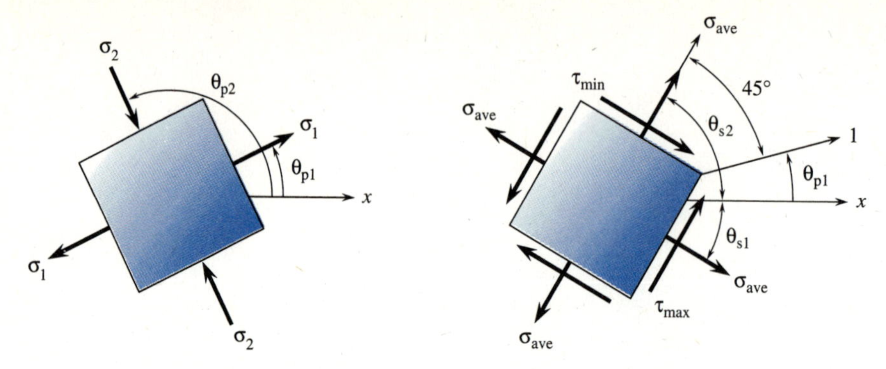

Figure 8.18 Elements oriented with principal stresses and maximum shear stress.

to satisfy Eq. (8.22). Thus we see that the planes of maximum and minimum shear stress occur at 45° orientations from the planes that the principal stresses act on. In fact, from Eqs. (8.18) and (8.11) with $\theta_p = \theta_{p1}$, we observe that

$$\theta_{s1} = \theta_{p1} - 45°$$

Substituting Eqs. (8.18) into Eq. (8.5) results in

$$\sigma_{x'} = \frac{\sigma_x + \sigma_y}{2} = \sigma_{\text{ave}} \tag{8.26}$$

which is the normal stress acting on the planes that also have the maximum and minimum shear stress acting on them. As seen by Eq. (8.26), this stress is the average of the normal stresses on the x- and y-planes. Figure 8.18 shows a general representation of elements oriented with maximum principal stresses, and with maximum and minimum shear stresses, along with the normal stress σ_{ave} given by Eq. (8.26).

Finally, a wedge can be used to indicate the principal stresses and the maximum in-plane shear stress in one sketch, as shown in Figure 8.19 for a general state of plane stress. Two sides AB and BC of the wedge show the principal stresses and the principal planes. Since $\theta_{s1} = \theta_{p1} - 45°$, the diagonal plane AC represents one of the planes of maximum shear stress. Plane AC then has τ_{max} and σ_{ave} acting on it.

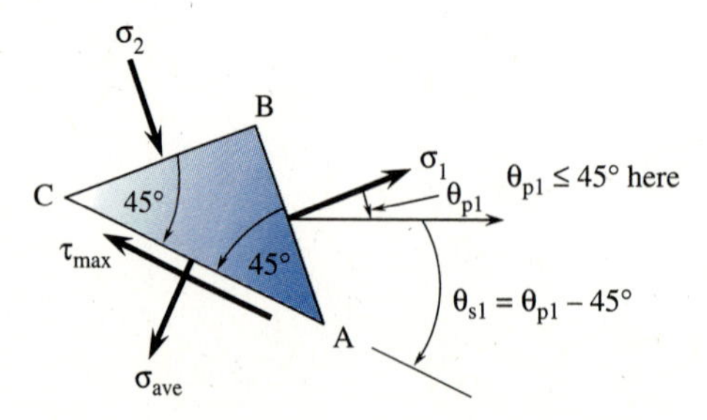

Figure 8.19 Wedge indicating principal stresses and maximum shear stress.

From Figure 8.19, some general conclusions regarding the directions of the principal stresses and maximum shear stress follow.

1. The numerically greater principal stress will act on the plane that makes an angle of 45° or less with the plane of the numerically larger of the two given normal stresses σ_x and σ_y. In Figure 8.19, we have shown σ_1 such that σ_x must be numerically larger than σ_y.

2. The direction of the maximum shear stress in the plane must tend to oppose the larger of the two principal stresses. In Figure 8.19, σ_1 is primarily directed to the right, therefore τ_{max} is directed primarily to the left as shown on plane AC.

The maximum shear stress in the x-y plane can be related to the principal stresses σ_1 and σ_2 by using Eq. (8.15). Subtracting σ_2 from σ_1 as given by Eq. (8.15), and then comparing the result to Eq. (8.20), we have

$$\tau_{max} = \frac{\sigma_1 - \sigma_2}{2} \tag{8.27}$$

Therefore, we see that the maximum shear stress is equal to one-half the difference of the principal stresses. Example Problems 8.5 and 8.6 illustrate how to determine principal stresses and maximum shear stress with its corresponding normal stress.

EXAMPLE 8.5

For the state of plane stress shown in Figure 8.20, determine **(a)** the principal planes, **(b)** the principal stresses, **(c)** the planes the maximum and minimum shear stresses act on, and **(d)** the maximum shear stress and its corresponding normal stress.

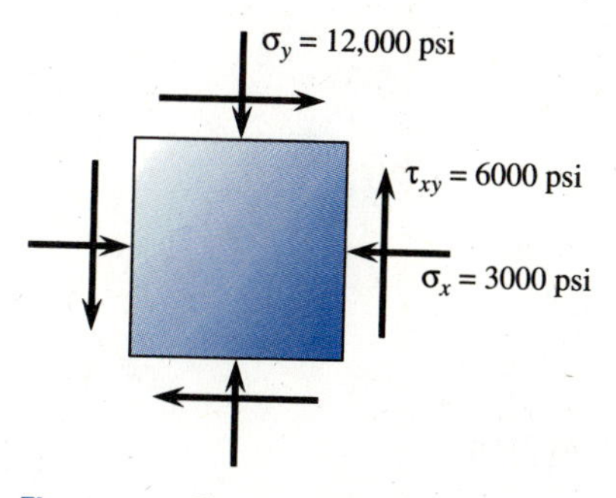

Figure 8.20 Element in state of plane stress.

Solution

(a) Principal planes

Substituting into Eq. (8.9), we obtain θ_p as

$$\tan 2\theta_p = \frac{2\tau_{xy}}{\sigma_x - \sigma_y} = \frac{2(6000)}{-3000 - (-12{,}000)} = \frac{12}{9} = \frac{4}{3}$$

$$2\theta_p = 53.1^\circ \qquad \theta = 26.6^\circ$$

and

$$2\theta_p = 53.1^\circ + 180^\circ = 233.1^\circ \qquad \theta_p = 116.6^\circ$$

(b) Principal stresses

Using Eq. (8.15), we obtain the principal stresses as

$$\sigma_{1,2} = \frac{\sigma_x + \sigma_y}{2} \pm \sqrt{\left(\frac{\sigma_x - \sigma_y}{2}\right)^2 + (\tau_{xy})^2}$$

$$\sigma_{1,2} = \frac{-3000 - 12{,}000}{2} \pm \sqrt{\left(\frac{-3000 + 12{,}000}{2}\right)^2 + (6000)^2}$$

$$\sigma_{1,2} = -7500 \pm 7500$$

$$\sigma_1 = 0 \qquad \sigma_2 = -15{,}000 \text{ psi}$$

Substituting $2\theta_p = 53.1°$ into Eq. (8.5), we obtain

$$\sigma_{x'} = \sigma_1 = 0$$

Therefore

$$2\theta_{p1} = 53.1° \qquad \theta_{p1} = 26.6°$$

and

$$2\theta_{p2} = 233.1° \qquad \theta_{p2} = 116.6°$$

The principal planes and principal stresses are sketched in Figure 8.21. Remember the shear stress is zero on the principal planes.

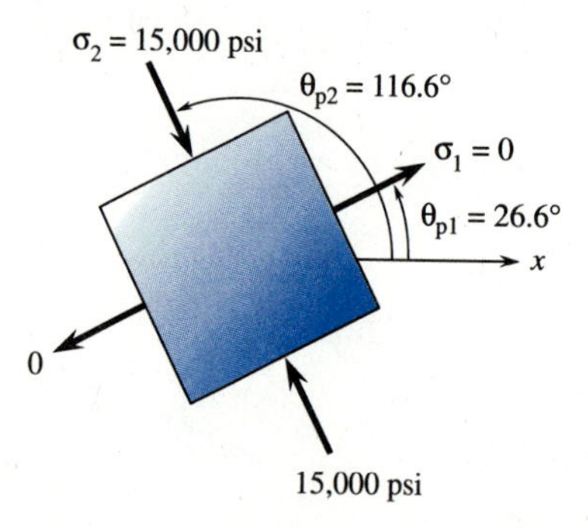

Figure 8.21 Element with principal stresses acting on it.

(c) Planes of maximum shear stress

Using Eq. (8.16), we obtain $2\theta_s$ as

$$\tan 2\theta_s = -\frac{\sigma_x - \sigma_y}{2\tau_{xy}} = -\frac{-3000 + 12{,}000}{2(6000)} = -\frac{3}{4}$$

$$2\theta_s = -36.9° \; (= 2\theta_p - 90°) \qquad \theta_s = -18.4°$$

and

$$2\theta_s = 2\theta_p + 90° = 53.1° + 90° = 143.1°$$

$$\theta_s = 71.6°$$

(d) Maximum shear stress and corresponding normal stress
Using Eqs. (8.20) and (8.21), we obtain

$$\tau_{max} = \sqrt{\left(\frac{\sigma_x - \sigma_y}{2}\right)^2 + (\tau_{xy})^2}$$

$$= \sqrt{\left(\frac{-3000 + 12,000}{2}\right)^2 + (6000)^2}$$

$$\tau_{max} = 7500 \text{ psi and } \tau_{min} = -7500 \text{ psi}$$

We verify that $\theta_{s1} = -18.4°$ by substituting $\theta_s = -18.4°$ into Eq. (8.6) to obtain

$$\tau_{x'y'} = 7500 \text{ psi} = \tau_{max}$$

Using Eq. (8.26), we obtain the normal stress acting on the planes of maximum and minimum shear stress as

$$\sigma_{ave} = \frac{\sigma_x + \sigma_y}{2} = \frac{-3000 - 12,000}{2} = -7500 \text{ psi}$$

Figure 8.22 is the sketch of the element with τ_{max}, τ_{min}, and σ_{ave} acting on it.

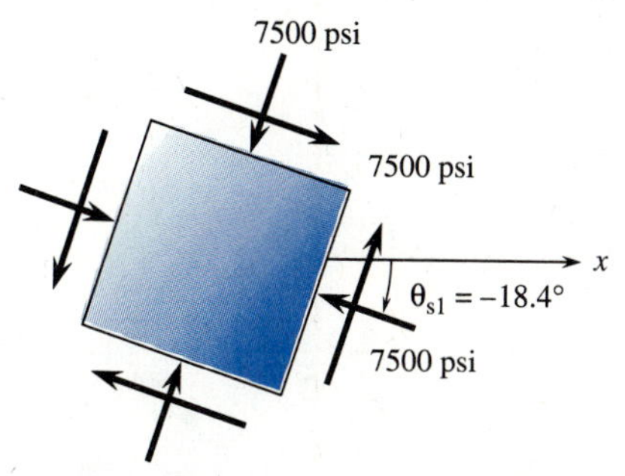

Figure 8.22 Element with maximum shear stress acting on it.

Finally, we use the wedge shown in Figure 8.23 to represent both the principal stresses and the orientation of their planes from Figure 8.21, and the maximum shear stress and its corresponding normal stress σ_{ave} from Figure 8.22.

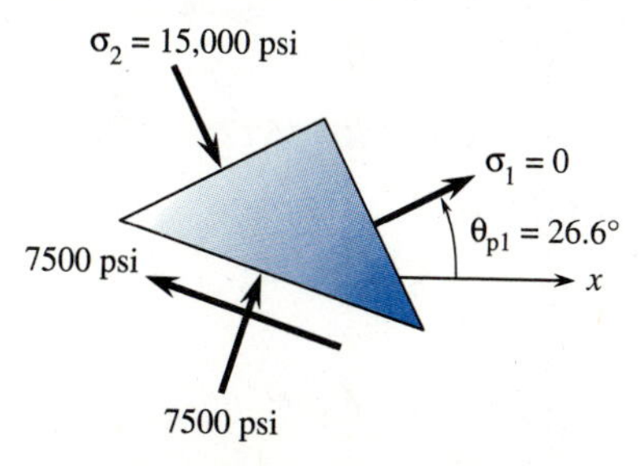

Figure 8.23 Wedge with both principal stresses and maximum shear stresses shown.

In Figure 8.23, we observe that (1) σ_1 makes an angle less than or equal to 45° with the numerically larger normal stress $\sigma_x = -3000$ psi, that is, $\theta_{p1} = 26.6°$, and (2) the maximum shear stress tends to oppose the numerically larger principal stress $\sigma_1 = 0$ (remember, numerically $\sigma_2 = -15{,}000$ psi).

EXAMPLE 8.6

At a point in a steel beam, the stresses shown in Figure 8.24 occur. Determine **(a)** the principal stresses and show them on a properly oriented element, and **(b)** the maximum shear stresses and show them on a sketch of a properly oriented element.

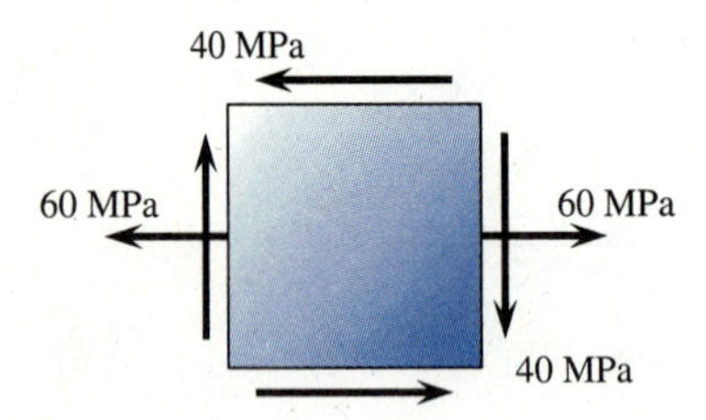

Figure 8.24 Stresses acting at a point in a steel beam.

Solution

(a) Principal stresses

Using Eq. (8.15), we obtain the principal stresses

$$\sigma_{1,2} = \frac{60+0}{2} \pm \sqrt{\left(\frac{60-0}{2}\right)^2 + (-40)^2}$$

$$= 30 \pm 50$$

$$\sigma_1 = 80 \text{ MPa} \qquad \sigma_2 = -20 \text{ MPa}$$

Using Eq. (8.9), we obtain the principal angle as

$$\tan 2\theta_p = \frac{2(-40)}{60-0} = -\frac{4}{3}$$

$$2\theta_p = -53.1° \qquad \theta_p = -26.6°$$

and

$$2\theta_p = -53.1 + 180° = 126.9°$$

Substituting $2\theta_p = -53.1°$ into Eq. (8.5), we obtain

$$\sigma_{x'} = \sigma_1 = 80 \text{ MPa}$$

Therefore

$$2\theta_{p1} = -53.1° \qquad \theta_{p1} = -26.6°$$

The principal planes, along with the principal stresses, are sketched in Figure 8.25.

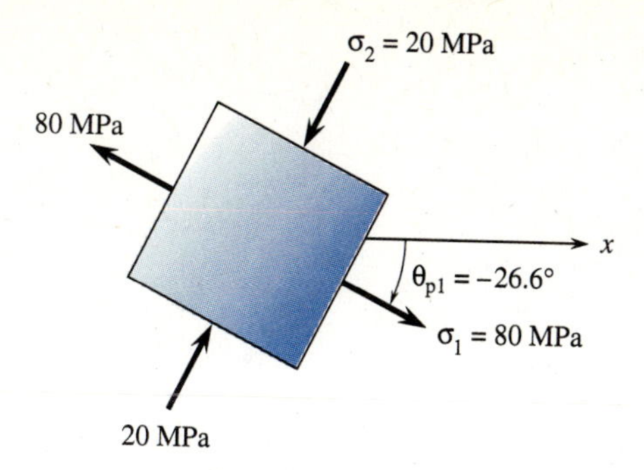

Figure 8.25 Element with principal stresses acting on it.

(b) Maximum shear stress
Using Eqs. (8.20) and (8.21), we obtain

$$\tau_{max} = \sqrt{\left(\frac{60-0}{2}\right)^2 + (-40)^2}$$

$$\tau_{max} = 50 \text{ MPa} \quad \text{and} \quad \tau_{min} = -50 \text{ MPa}$$

From Eq. (8.16), we obtain

$$\tan 2\theta_s = -\frac{60-0}{2(-40)} = \frac{3}{4}$$

$$2\theta_s = 36.9° \qquad \theta_s = 18.4°$$

Using Eq. (8.26), we obtain the normal stress acting on the plane of maximum and minimum shear stress as

$$\sigma_{ave} = \frac{60+0}{2} = 30 \text{ MPa}$$

Figure 8.26 shows the orientation of the element with τ_{max}, τ_{min}, and σ_{ave} acting on it.

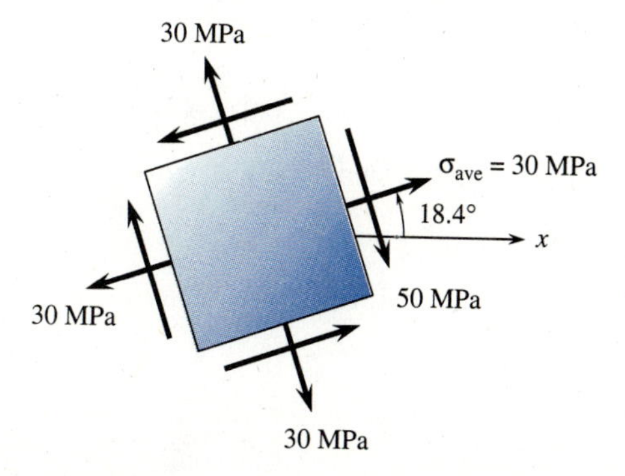

Figure 8.26 Element with maximum shear stress acting on it.

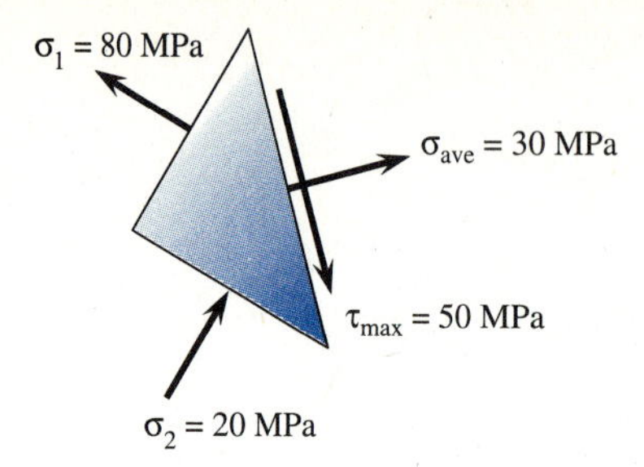

Figure 8.27 Wedge with both principal stresses and maximum shear stresses shown.

We now use the wedge shown in Figure 8.27 to represent both the principal stresses and their orientations and the maximum shear stress and its corresponding normal stress σ_{ave}. That is, we combine Figures 8.25 and 8.26 into a single figure to represent both principal stresses and maximum shear stress, as shown in Figure 8.27.

Conclusions similar to those given for Figure 8.23 also apply to Figure 8.27.

8.5 MOHR'S CIRCLE FOR PLANE STRESS ANALYSIS

We will show that the transformation Eqs. (8.5) and (8.6) used to obtain the normal and shear stress components on an arbitrary plane through a point can be represented as points on a circle. This circle representation was developed by O. Mohr and is called Mohr's circle. To develop this circle representation, we transpose the first term on the right side of Eq. (8.5) to the left side, and along with Eq. (8.6), we have

$$\begin{aligned} \sigma_{x'} - \frac{\sigma_x + \sigma_y}{2} &= \frac{\sigma_x - \sigma_y}{2}\cos 2\theta + \tau_{xy}\sin 2\theta \\ \tau_{x'y'} &= -\frac{\sigma_x + \sigma_y}{2}\sin 2\theta + \tau_{xy}\cos 2\theta \end{aligned} \quad \textbf{(8.28)}$$

We then square Eqs. (8.28) and add them together to obtain

$$\left(\sigma_{x'} - \frac{\sigma_x + \sigma_y}{2}\right)^2 + (\tau_{x'y'} - 0)^2 = \left(\frac{\sigma_x - \sigma_y}{2}\right)^2 + (\tau_{xy})^2 \quad \textbf{(8.29)}$$

Recalling Eqs. (8.12) and (8.26) for R and σ_{ave}, we rewrite Eq. (8.29) as

$$(\sigma_{x'} - \sigma_{ave})^2 + (\tau_{x'y'} - 0)^2 = R^2 \quad \textbf{(8.30)}$$

which is the equation of a circle. To see this, recall the equation of a circle in rectangular coordinates is given in general form by

$$(x - a)^2 + (y - b)^2 = r^2$$

where the center of the circle is located at (a, b) with a radius of r. Equation (8.30) then represents the equation of a circle for stresses in which the horizontal x-axis is normal

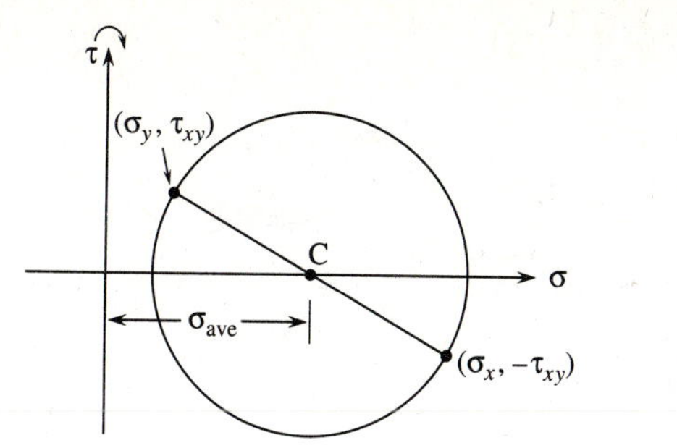

Figure 8.28 Mohr's circle for state of plane stress.

stress $\sigma_{x'}$, the vertical y-axis is shear stress $\tau_{x'y'}$, the center of the circle is at

$$a = \frac{\sigma_x + \sigma_y}{2} = \sigma_{\text{ave}}, \qquad b = 0 \tag{8.31}$$

with radius of

$$r = R = \sqrt{\left(\frac{\sigma_x - \sigma_y}{2}\right)^2 + (\tau_{xy})^2} \tag{8.32}$$

Since $b = 0$, the center of the circle always lies on the horizontal axis $x = \sigma_{x'}$. Figure 8.28 shows the Mohr's circle for a state of plane stress given at a point by σ_x, σ_y, and τ_{xy}. To construct the Mohr's circle, we adopt the following sign conventions:

1. Normal stresses σ are plotted horizontally with tensile stresses plotted positively (to the right of the origin) and compressive stresses plotted negatively (to the left of the origin).

2. Shear stresses, τ, are plotted vertically with shear stresses that apply a clockwise couple (moment) to the stressed element plotted positively (above the origin), and shear stresses that apply a counterclockwise couple to the element plotted negatively (below the origin). In Figure 8.29a, the shear stress τ_{xy} acting on the x-face would be plotted as a positive value since it applies a clockwise moment to the stressed element. In Figure 8.29b, τ_{xy} would be plotted as a negative value since it applies a counterclockwise moment to the stressed element. The coordinates of each point on the circle correspond to the stresses acting on some plane passing through a stressed point in a body. You should notice that the positive sign convention for τ_{xy} for Mohr's circle is opposite that used for the basic stress element of Figures 8.3 and 8.4.

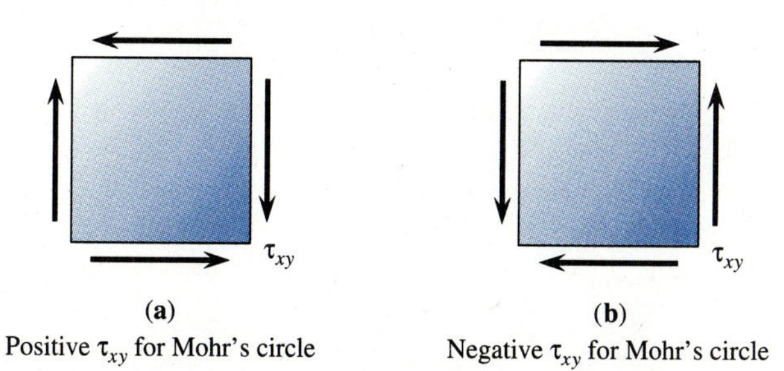

Figure 8.29 Element showing sign conventions for shear stresses on Mohr's circle.

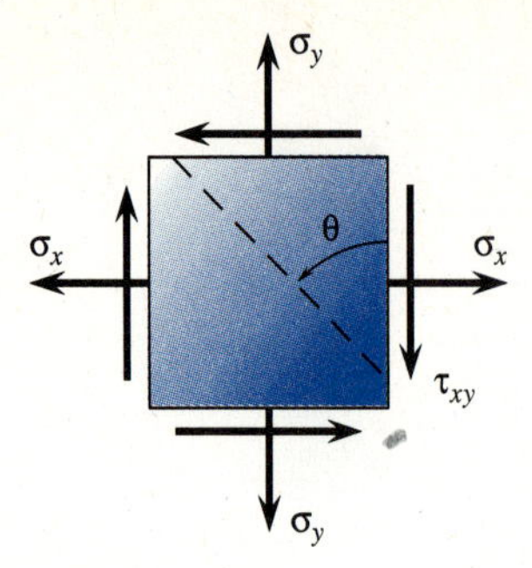

Figure 8.30 Element in plane stress.

Given the state of plane stress shown in Figure 8.30 the steps involved in constructing Mohr's circle are (σ_x, σ_y, τ_{xy}, all assumed positive as shown in Figure 8.30.)

1. Write down the two stress coordinates X, Y to be plotted. For instance, remembering that τ_{xy} on the x-face will be plotted as a positive value, since it causes a clockwise moment on the stressed element, and tensile normal stresses are plotted positively, we have

$$X = (\sigma_x, \tau_{xy}) \qquad Y = (\sigma_y, -\tau_{xy})$$

where X and Y represent the stresses on the x- and y-faces, respectively.

2. Determine the center C of the circle from

$$C = \sigma_{\text{ave}} = \frac{\sigma_x + \sigma_y}{2}$$

where C is always on the horizontal σ axis.

3. Sketch the circle using the points X, Y, and C, as shown in Figure 8.31.
4. Determine the radius R of the circle by observing from the circle that

$$R = \sqrt{(CD)^2 + (XD)^2}$$

where distance

$$CD = OD - OC = \sigma_x - \left(\frac{\sigma_x + \sigma_y}{2}\right) = \frac{\sigma_x - \sigma_y}{2}$$

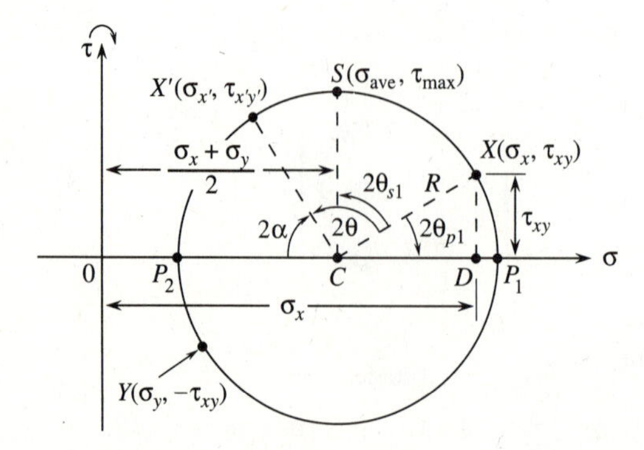

Figure 8.31 Mohr's circle for plane stress.

and distance

$$XD = \tau_{xy}$$

Therefore

$$R = \sqrt{\left(\frac{\sigma_x - \sigma_y}{2}\right)^2 + (\tau_{xy})^2}$$

which is equivalent to Eq. (8.32). Note that

$$\tau_{\max} = R$$

since the highest vertical point S on the circle has a vertical coordinate equal to R. The stresses at point S are then

$$S = (\sigma_{\text{ave}}, \tau_{\max})$$

Remember that $\tau_{\max}$ from the circle applies a clockwise moment to an element which is rotated through an angle $2\theta_{s1}$ counterclockwise from the x-face as seen from the circle in Figure 8.31. This angle is given by

$$\tan 2\theta_{s1} = \frac{CD}{XD}$$

or from the circle it is a counterclockwise angle given by

$$2\theta_{s1} = 90° - 2\theta_{p1}$$

where $2\theta_{p1}$ is determined in step 5.

5. Determine the principal stresses, the largest and smallest normal stresses on the circle, points P_1 and P_2 in Figure 8.31. From the circle the stresses corresponding to P_1 and P_2 are given by

$$\sigma_1 = C + R \qquad \sigma_2 = C - R \qquad (C = OC \text{ (distance from } O \text{ to } C))$$

6. Determine the principal planes from the circle, where any angle θ associated with the element is 2θ on the circle. This is seen by remembering that the stresses on the x- and y-faces of an element are 90° apart, while these stresses are 180° apart on the circle. To determine $2\theta_{p1}$ from the circle, we have

$$\tan 2\theta_{p1} = \frac{XD}{CD} = \frac{\tau_{xy}}{\left(\frac{\sigma_x - \sigma_y}{2}\right)}$$

and $2\theta_{p1}$ is seen to be clockwise from the x-face to the principal plane P_1. Figure 8.32 shows the principal plane P_1. Principal plane P_2 is then 180° counterclockwise from P_1 on the circle or 90° counterclockwise from P_1 on the stress element.

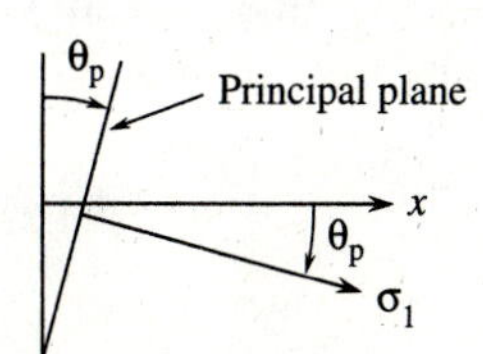

Figure 8.32 Principal plane

7. If desired, determine the normal and shear stress on any other plane at an angle θ from the vertical plane (x-face) or from the horizontal plane (y-face) using the circle to obtain the appropriate trigonometric relations and distances needed. For instance, assume the normal and shear stress are desired on a weld seam making an angle θ between the x-face and the seam as shown in Figure 8.30. We then use the circle starting from point X, or equivalently use the CX radius in Figure 8.31, and rotate counterclockwise 2θ on the circle about point C. This is point X' on the circle. The normal and shear stress at X' are then obtained from the circle by

$$\sigma_{x'} = C - R\cos 2\alpha$$

$$\tau_{x'y'} = R\sin 2\alpha$$

where

$$2\alpha = 180° - 2\theta - 2\theta_{pl}$$

is easily obtained on the circle. We see that the Mohr's circle is a semianalytical approach in which the circle merely guides us in determining proper formulas (distances) and angles and their directions to obtain, for instance, principal stresses, principal planes, maximum shear stress, or stresses on any arbitrary plane.

Example Problems 8.7 through 8.11 illustrate the use of Mohr's circle to solve problems in plane stress.

EXAMPLE 8.7

A shaft is subjected to simultaneous torsion and tension loads, which produce the state of stress shown in Figure 8.33. Construct Mohr's circle for this state of stress and determine (1) the maximum shear stress in the plane and (2) the principal stresses and principal planes.

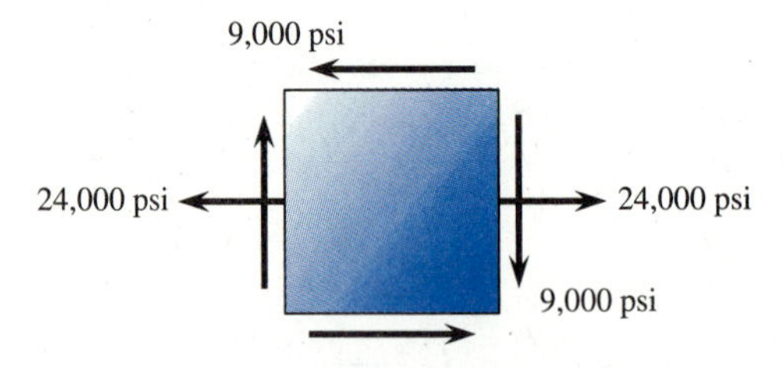

Figure 8.33 State of stress on a shaft.

Solution

We follow the steps previously outlined for construction and use of Mohr's circle to obtain the maximum shear stress and principal stresses.

1. The two coordinates to be plotted are the stresses on the x- and y-faces. Using $\sigma_x = 24{,}000$ psi, $\sigma_y = 0$, and $\tau_{xy} = 9000$ psi, we have

$$X = (24{,}000,\ 9000) \qquad Y = (0,\ -9000)$$

2. The center of the circle is

$$C = \sigma_{ave} = \frac{24{,}000 + 0}{2} = 12{,}000$$

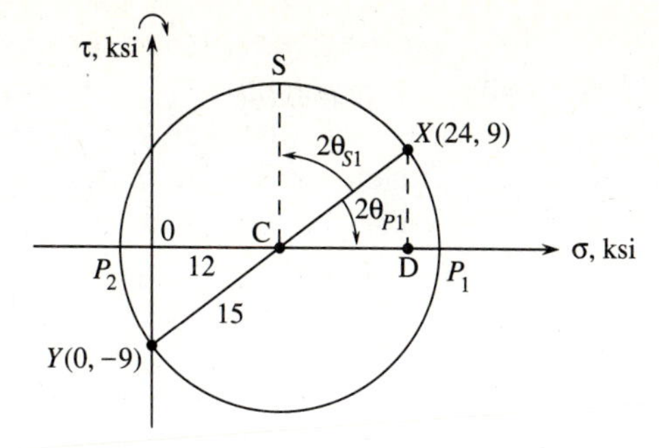

Figure 8.34 Sketch of Mohr's circle.

3. A sketch of the circle is now possible using points X, Y, and C and is shown in Figure 8.34.
4. The radius of the circle is determined from the distances CD and XD obtained from the circle as

$$R = \sqrt{(CD)^2 + (XD)^2} = \sqrt{(12)^2 + (9)^2} = 15$$

Hence, the maximum shear stress is

$$\tau_{\max} = R = 15 \text{ ksi} = 15{,}000 \text{ psi}$$

$$\tan 2\theta_{s1} = \frac{CD}{XD} = \frac{12}{9}$$

$$2\theta_{s1} = 53.1°$$

(counterclockwise from point X to S on the circle)

$$\theta_{s1} = 26.6°$$

(counterclockwise from the x-axis on the element)

5. The principal stresses are points P_1 and P_2 on the circle. These points are given by

$$\sigma_1 = C + R = 12 + 15 = 27 \text{ ksi} = 27{,}000 \text{ psi (tensile)}$$

$$\sigma_2 = C - R = 12 - 15 = -3 \text{ ksi} = -3000 \text{ psi (compressive)}$$

6. The orientation of the principal planes is obtained by determining $2\theta_{p1}$ on Figure 8.34 as

$$\tan 2\theta_{p1} = \frac{XD}{CD} = \frac{9}{12}$$

$$2\theta_{p1} = \tan^{-1}\frac{9}{12} = 36.8°$$

(clockwise from point X to P_1 on the circle)

$$\theta_{p1} = 18.4°$$

(clockwise from the x-axis) or

$$2\theta_{p1} = 360° - 36.8° = 323.2°$$

(counterclockwise from the x-axis)

Hence, the principal plane, having the principal stress σ_1 on it, is oriented at an angle of $\theta_{p1} = 18.4°$ clockwise from the x-face moving from point X to P_1 on the circle, while the principal plane having the principal stress σ_2 acting on it is oriented 18.4° clockwise from the y-face (moving from point Y to P_2 on the circle). Figure 8.35 shows the orientation of the element with the principal stresses acting on it.

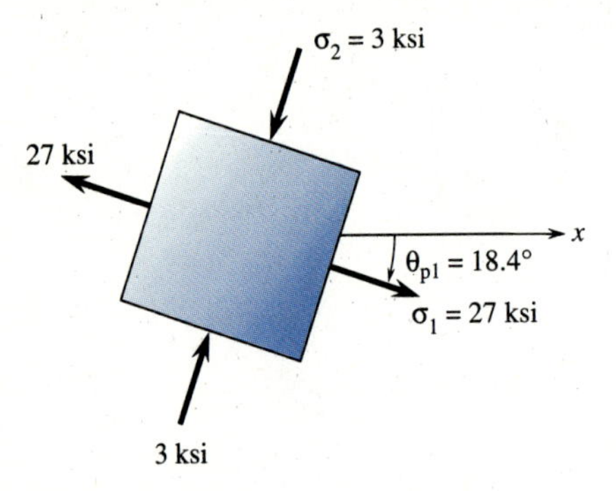

Figure 8.35 Orientation of element with principal stresses.

EXAMPLE 8.8

A crane hook shown in Figure 8.36(a) is loaded such that at point A the state of stress shown in Figure 8.36(b) results. Determine (1) the principal stresses and principal angle θ_{p1}, and show the principal stresses on a sketch of a properly oriented element, and (2) the maximum shear stresses in the plane. Solve using Mohr's circle method.

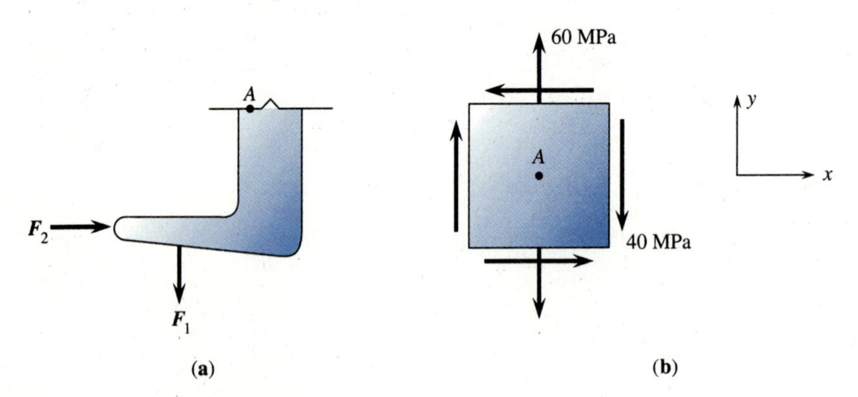

Figure 8.36 (a) Crane hook with loading and (b) stresses acting at point A.

Solution

1. $X = (0, 40) \qquad Y = (60, -40)$

2. $C = \dfrac{0 + 60}{2} = 30 \text{ MPa}$

3. Sketch the circle using points X, Y, and C.

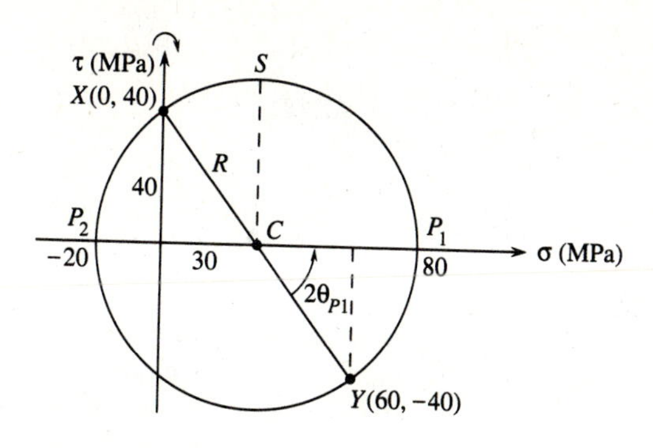

4. $R = \sqrt{30^2 + 40^2} = 50 \text{ MPa} = \tau_{max}$

5. $\sigma_1 = C + R = 30 + 50 = 80 \text{ MPa}$

   $\sigma_2 = C - R = 30 - 50 = -20 \text{ MPa}$

6. $\tan 2\theta_{p1} = \dfrac{4}{3} \qquad 2\theta_{p1} = 53.13° \qquad \theta_{p1} = 26.6°$

(counterclockwise from the y-plane to the σ_1-plane as shown on the element below)

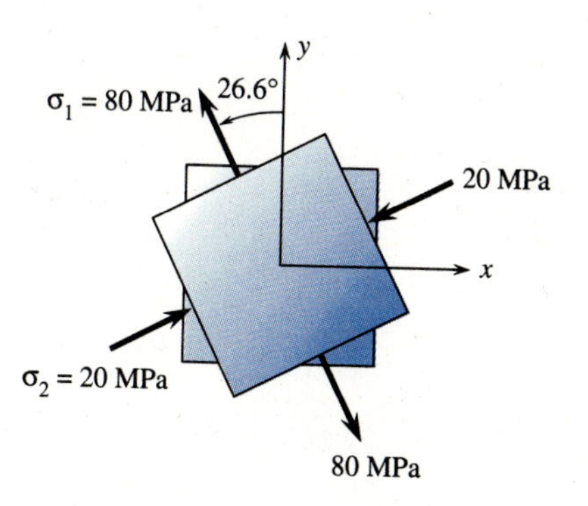

EXAMPLE 8.9

The propeller shaft of a ship is a hollow circular tube with outside diameter of 18 in. and inside diameter of 10 in. The shaft is subjected to simultaneous torque $T = 2000$ kip-in. and axial compression load $P = 100$ kips. Determine **(a)** the maximum tensile stress, **(b)** maximum compressive stress, and **(c)** maximum shear stress in the shaft. Use Mohr's circle method (Figure 8.37).

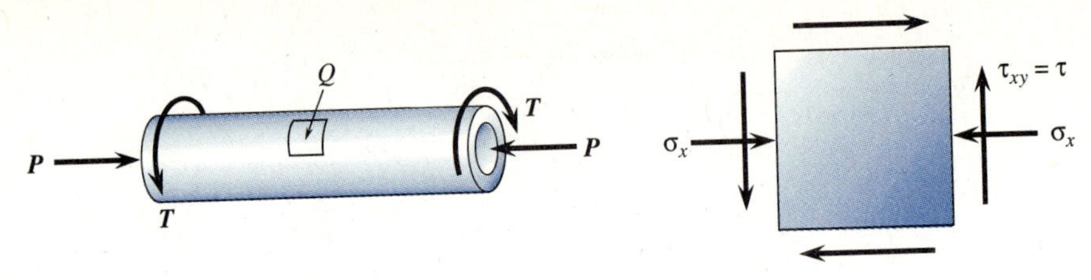

Figure 8.37 Propeller shaft subjected to torque and axial load.

Solution

The critical stresses occur at the outer surface Q of the shaft as the torsional shear stress is greatest there. The stresses on an element of material at point Q on the surface of the shaft are now obtained. First, we calculate the cross-sectional area A and polar moment of inertia J as

$$A = \frac{\pi}{4}(D_0^2 - D_i^2) = \frac{\pi}{4}[(18 \text{ in.})^2 - (10 \text{ in.})^2] = 176 \text{ in.}^2$$

$$J = \frac{\pi}{32}[D_0^4 - D_i^4] = \frac{\pi}{32}[(18 \text{ in.})^4 - (10 \text{ in.})^4] = 9324 \text{ in.}^4$$

By Eq. (2.1)

$$\sigma_x = -\frac{P}{A} = \frac{-100 \text{ kips}}{176 \text{ in.}^2} = -0.568 \text{ ksi}$$

By Eq. (4.14)

$$\tau = \frac{TR_0}{J} = \frac{(2000 \text{ kip-in.})(9 \text{ in.})}{9324 \text{ in.}^4} = 1.93 \text{ ksi}$$

Using the steps in the Mohr's circle method, we have

1. $X = (-0.568, -1.93), \qquad Y = (0, 1.93)$
2. $C = \dfrac{-0.568 + 0}{2} = -0.284$
3. Sketch the circle.

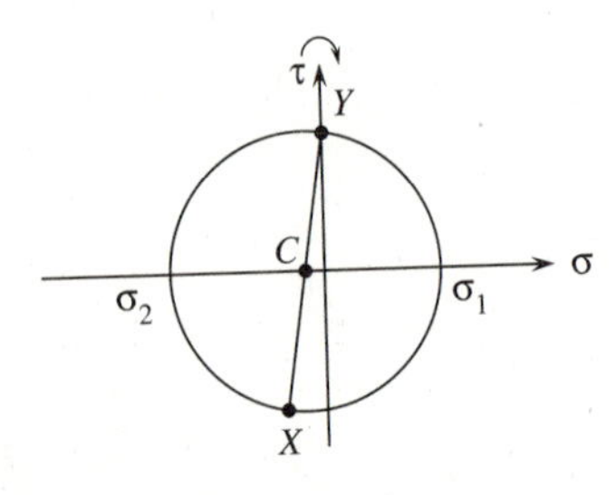

4. $R = \sqrt{(-0.284)^2 + (1.93)^2} = 1.95 \text{ ksi} = \tau_{max}$
5. $\sigma_1 = C + R = -0.284 + 1.95 = 1.67$ ksi (tensile)

   $\sigma_2 = C - R = -0.284 - 1.95 = -2.23$ ksi (compressive)

(a) The maximum tensile stress is

$$\sigma_1 = 1.67 \text{ ksi}$$

(b) The maximum compressive stress is

$$\sigma_2 = -2.23 \text{ ksi}$$

(c) The maximum shear stress is

$$\tau_{max} = 1.95 \text{ ksi}$$

EXAMPLE 8.10

The compressed-air tank AB has an inside diameter of 18 in. and a uniform wall thickness of 0.25 in. The gauge pressure inside the tank is 150 psi. Determine **(a)** the maximum normal stress and **(b)** the maximum in-plane shear stress at point D on top of the tank (Figure 8.38).

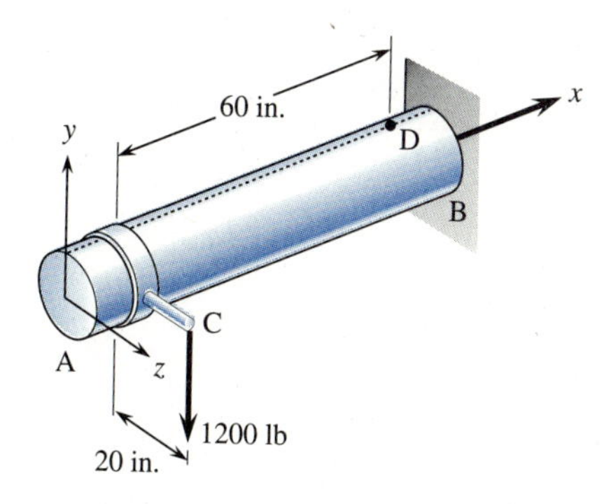

Figure 8.38 Compressed-air tank.

Solution

Draw a free-body diagram of the tank section AD.

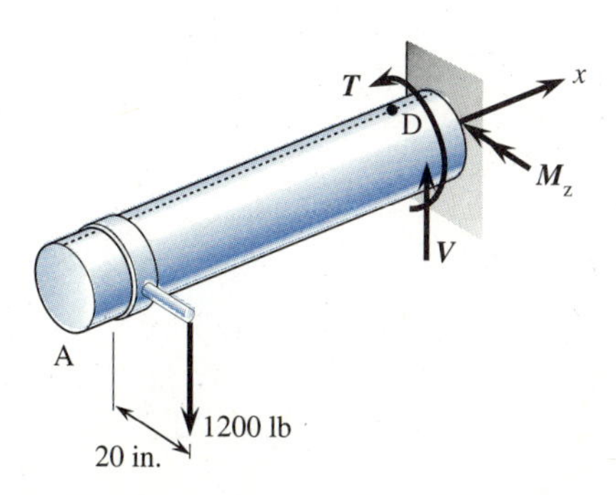

By statics

$$\Sigma F_y = 0: \quad -1200 + V = 0 \qquad V = 1200 \text{ lb}$$

$$\Sigma M_x = 0: \quad (20 \text{ in.})(1200 \text{ lb}) - T = 0 \qquad T = 24{,}000 \text{ lb-in.}$$

$$\Sigma M_z = 0: \quad (60 \text{ in.})(1200 \text{ lb}) - M_z = 0 \qquad M_z = 72{,}000 \text{ lb-in.}$$

Geometric properties

$$I_z = \frac{\pi}{4}(r_0^4 - r_i^4) = \frac{\pi}{4}(9.25 \text{ in.}^4 - 9 \text{ in.}^4) = 597 \text{ in.}^4$$

or for thin-walled tanks

$$I_z = \pi r_{\text{ave}}^3 t = \pi(9.125 \text{ in.})^3(0.25 \text{ in.}) = 597 \text{ in.}^4$$

$$J = 2I_z = 2(597 \text{ in.}^4) = 1194 \text{ in.}^4$$

Basic normal stresses

Due to bending, by Eq. (5.7)

$$\sigma_b = \frac{M_z c}{I_z} = \frac{(72{,}000 \text{ lb-in.})(9.25 \text{ in.})}{597 \text{ in.}^4} = 1120 \text{ psi}$$

Due to pressure, by Eqs. (7.15) and (7.10), the axial and hoop stresses are

$$\sigma_a = \frac{pr_i}{2t} = \frac{(150 \text{ psi})(9 \text{ in.})}{2(0.25 \text{ in.})} = 2700 \text{ psi}$$

$$\sigma_h = \frac{pr_i}{t} = 5400 \text{ psi}$$

NOTE: σ_b and σ_a are in the x-direction. σ_h is in the z-direction.

Basic shear stresses

Due to torsion, by Eq. (4.14)

$$\tau = \frac{Tr_0}{J} = \frac{(24{,}000 \text{ lb-in.})(9.25 \text{ in.})}{1194 \text{ in.}^4} = 186 \text{ psi}$$

Due to vertical shear, by Eq. (5.15)

$$\tau = \frac{VQ}{Ib} = 0 \quad (\text{since } Q = 0 \text{ at top surface})$$

The basic stress element at point D is then

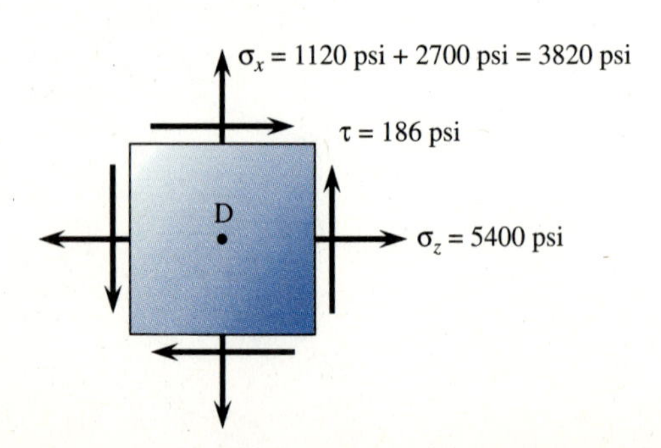

where superposition of stresses, as previously discussed in Chapter 7, has been used to obtain σ_x.

Mohr's circle method

1. $X(3820, 186) \qquad Z(5400, -186)$

2. $C = \dfrac{3820 + 5400}{2} = 4610$ psi

3. Sketch of circle

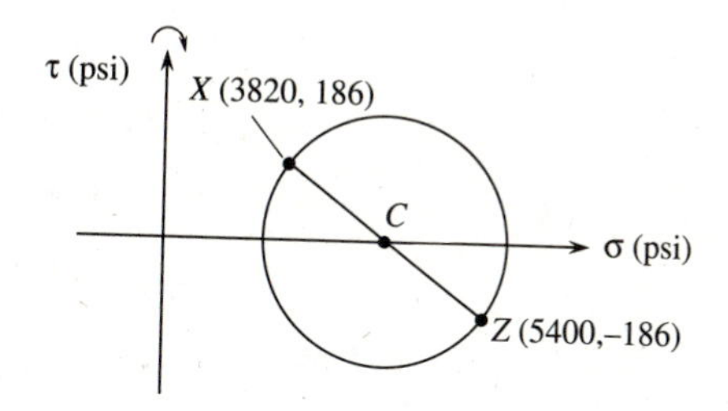

4. $R = \sqrt{(5400 - 4610)^2 + 186^2} = 812$ psi

5. $\sigma_1 = C + R = 4610 + 812 = 5420$ psi

   (a) The maximum normal stress at point D is

$$\sigma_1 = 5420 \text{ psi}$$

   (b) The maximum in-plane shear stress at point D is

$$\tau_{max} = 812 \text{ psi}$$

EXAMPLE 8.11

A pressurized cylindrical steel tank has inner radius 600 mm and thickness 16 mm. The tank is subjected to a gauge pressure $p = 1750$ kPa and an axial force $F = 125$ kN. The butt weld seams form an angle of 54° with the longitudinal axis of the tank. Determine (1) the normal stress perpendicular to the weld σ_w and (2) the in-plane shear stress parallel to the weld τ_w.

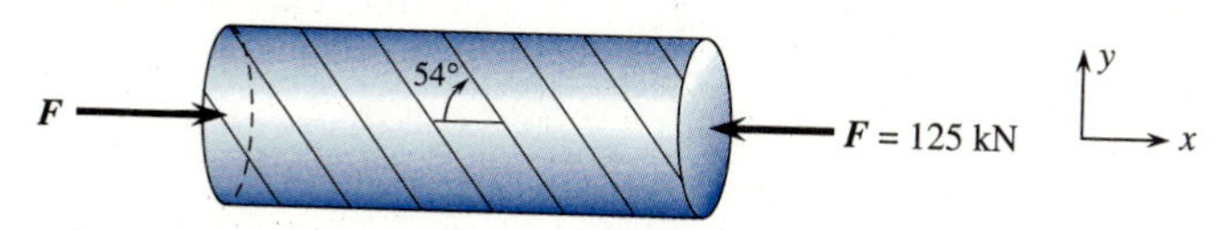

Figure 8.39 Pressurized cylindrical tank.

Solution

Basic normal stresses

By Eq. (2.1)

$$\sigma_F = \frac{F}{A} = \frac{F}{2\pi rt} = \frac{-125 \text{ kN}}{2\pi(0.6 \text{ m})(0.016 \text{ m})} = -2070 \text{ kPa} = -2.07 \text{ MPa}$$

By Eq. (7.15)

$$\sigma_a = \frac{pr_i}{2t} = \frac{(1750 \text{ kPa})(0.6 \text{ m})}{2(0.016 \text{ m})} = 32{,}800 \text{ kPa} = 32.8 \text{ MPa}$$

By Eq. (7.10)

$$\sigma_h = 2\sigma_a = 2(32{,}800 \text{ kPa}) = 65{,}600 \text{ kPa} = 65.6 \text{ MPa}$$

Using superposition, the total axial stress is

$$\sigma_x = \sigma_F + \sigma_a = -2.07 \text{ MPa} + 32.8 \text{ MPa} = 30.7 \text{ MPa}$$

A stress element on the surface is given by

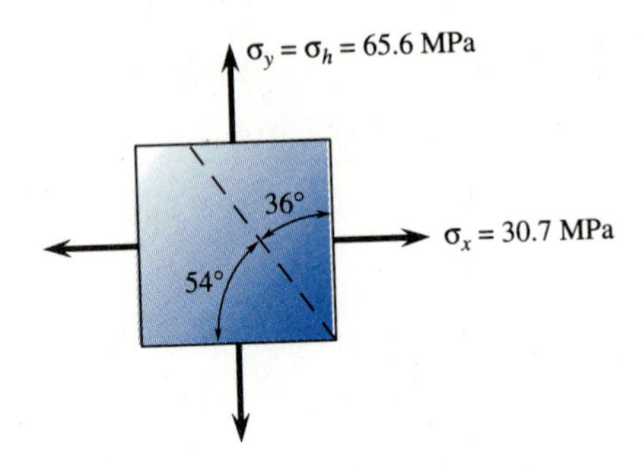

Mohr's circle method

1. $X(30.7, 0)$ $\quad Y(65.6, 0)$

2. $C = \dfrac{30.7 + 65.6}{2} = 48.2 \text{ MPa}$

3. Sketch of circle

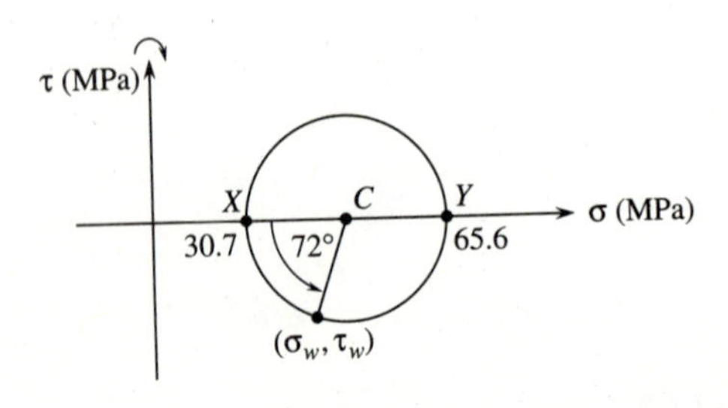

4. $R = \sqrt{\left(\frac{65.6 - 30.7}{2}\right)^2} = 17.5 \text{ MPa}$

5. The normal and shear stresses acting on the weld are obtained using trigonometry as aided by the circle above.

$$\sigma_w = C - R\cos 72^\circ = 48.2 - 17.5(\cos 72^\circ) = 42.8 \text{ MPa}$$

$$\tau_w = -R\sin 72^\circ = -17.5(0.951) = -16.6 \text{ MPa}$$

The negative sign means τ_w causes a counterclockwise moment on the wedge shown below.

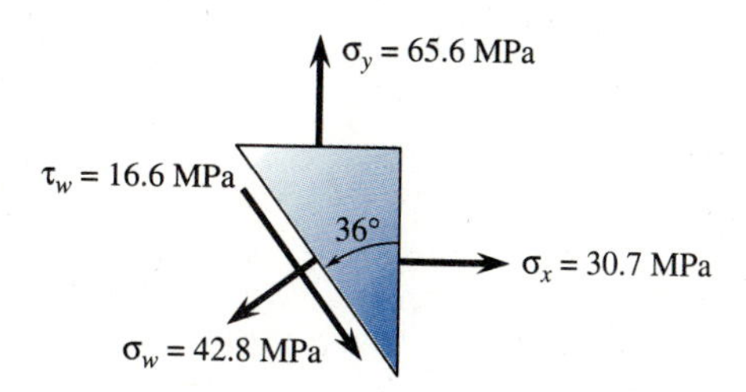

8.6 ABSOLUTE MAXIMUM SHEAR STRESS

In the preceding sections we considered only the shear stress associated with the in-plane (x-y plane) stresses. This is why we have called this shear stress the maximum in-plane shear stress. The ***absolute maximum shear stress*** may occur out of this x-y plane. To obtain the absolute maximum shear stress we must look at a three-dimensional analysis of our basic stress element. This element really is a three-dimensional element but for plane stress it was convenient to represent the element by a plane element.

Recall that for plane stress the maximum in-plane shear stress is given by Eq. (8.27) (repeated here for convenience as Eq. (8.33)) as

$$\tau_{\max} = \frac{\sigma_1 - \sigma_2}{2} \tag{8.33}$$

and acts in a plane perpendicular to the x-y plane and acts at 45° from the principal planes, as shown in Figure 8.40(a), or equivalently in the two-dimensional representation Figure 8.40(b). Here, σ_1 and σ_2 are the principal stresses determined in a plane stress problem.

When the principal stresses from a plane stress problem both have the same sign, that is, $\sigma_1 > \sigma_2 > 0$ or $\sigma_1 < \sigma_2 < 0$, then the absolute maximum shear stress is not given by Eq. (8.33), as we will now show.

Remember for plane stress $\sigma_z = 0$, $\tau_{zx} = \tau_{zy} = 0$, and on principal planes the shear stresses are zero. Since $\tau_{zx} = \tau_{zy} = 0$, then the z-axis must be a principal axis.

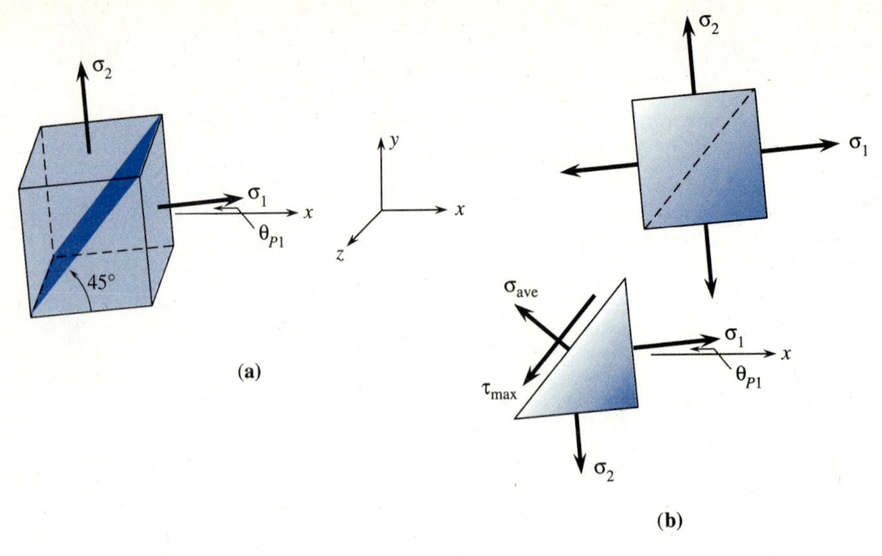

Figure 8.40 (a) Maximum in-plane shear stress and (b) its two-dimensional representation.

The element in Figure 8.40 can be viewed from the three different principal stress directions (1) looking in from the z-axis, (2) looking in from the axis corresponding to the σ_1 direction (called the 1-axis), and (3) looking in from the σ_2 direction (called the 2-axis).

These three cases of two-dimensional stress are represented in Figure 8.41. For instance, Figure 8.41(a) represents an element lying in the 1-2 plane.

The Mohr's circles for each case are shown in Figure 8.42. First assuming $\sigma_1 > \sigma_2 > 0$ (Figure 8.41(a)), we have the Mohr's circle shown in Figure 8.42 and labeled the 1-2 plane. The circles for cases (b) and (c) of Figure 8.41 are similarly drawn.

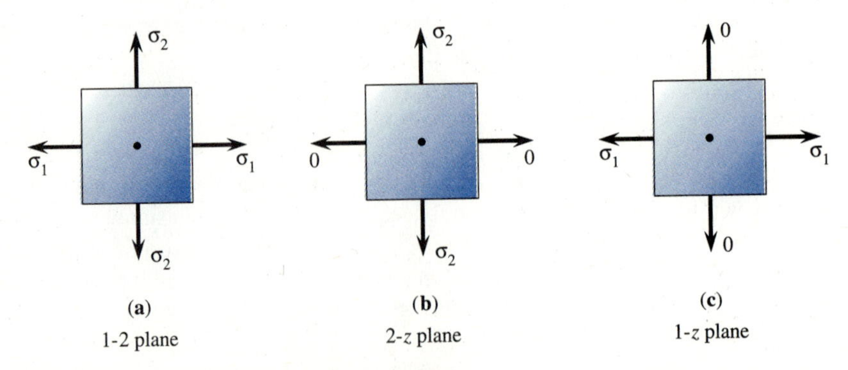

Figure 8.41 (a) Stresses on element in 1-2 plane, (b) stresses on element in 2-z plane and (c) stresses on element in 1-z plane.

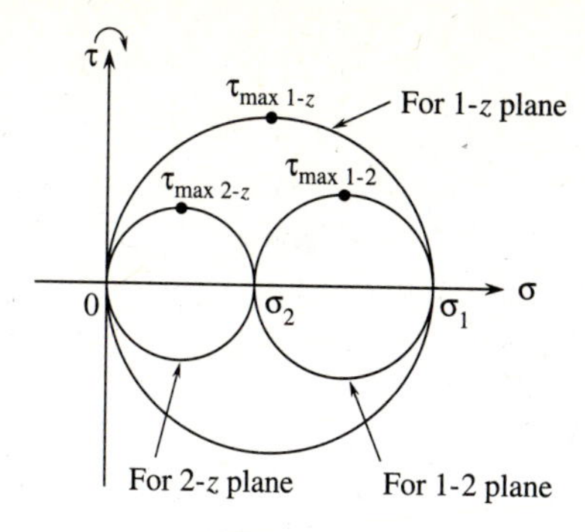

Figure 8.42 Mohr's circles for three stress states shown in Figure 8.41.

The maximum shear stress in each plane is given by the radius of the respective circle. For the 1-2 plane

$$R = \tau_{\max\ 1\text{-}2} = \frac{\sigma_1 - \sigma_2}{2} \tag{8.34}$$

for the 2-z plane

$$R = \tau_{\max\ 2\text{-}z} = \frac{\sigma_2 - 0}{2} = \frac{\sigma_2}{2} \tag{8.35}$$

and for the 1-z plane

$$R = \tau_{\max\ 1\text{-}z} = \frac{\sigma_1 - 0}{2} = \frac{\sigma_1}{2} \tag{8.36}$$

Comparing Eqs. (8.34), (8.35), and (8.36), for the case $\sigma_1 > \sigma_2 > 0$, the largest shear stress is given by Eq. (8.36), (and not by Eq. (8.33)) which corresponds to the circle of largest radius. This shear stress is called the *absolute maximum shear stress* and occurs on a plane rotated 45° from the principal planes 1 and z, and perpendicular to the 1-z plane as shown in Figure 8.43(a). The maximum shear stress $\tau_{\max\ 2\text{-}z}$ is shown in Figure 8.43(b).

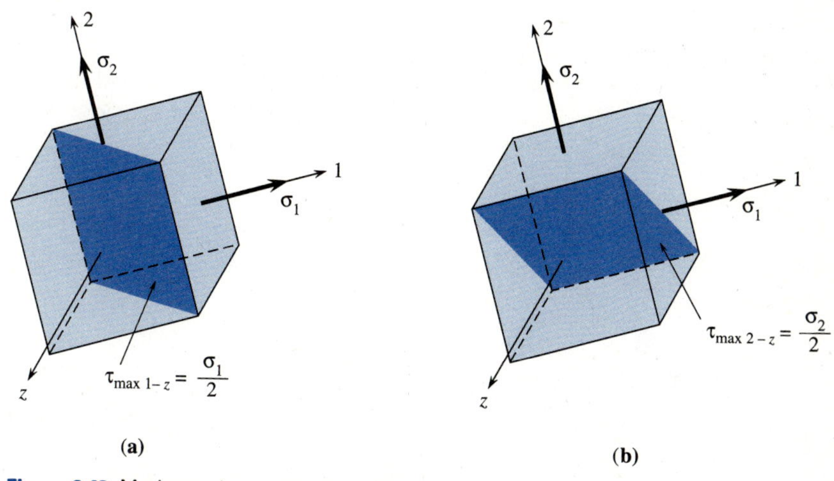

Figure 8.43 Maximum shear stresses on different planes.

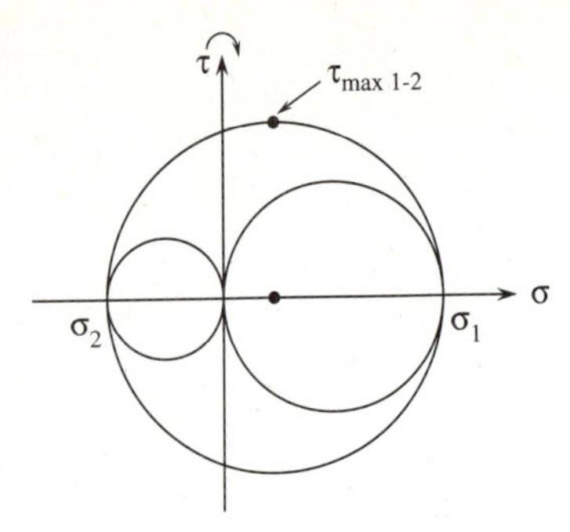

Figure 8.44 Mohr's circles for stress state $\sigma_1 > 0$ and $\sigma_2 < 0$.

For the case when $\sigma_1 > 0$ and $\sigma_2 < 0$, the absolute maximum shear stress is then given by Eqs. (8.33) or (8.34), as the largest radius of a Mohr's circle corresponds to this stress state, as shown in Figure 8.44. In conclusion, when σ_1 and σ_2 have different algebraic signs, the absolute maximum shear stress is the in-plane shear stress given by Eq. (8.33), but when σ_1 and σ_2 have the same algebraic sign, the absolute maximum shear stress is given by Eq. (8.36). In the general three-dimensional stress case σ_z does not equal zero and a third principal stress σ_3 occurs. If we let the algebraically largest principal stress be σ_{max} and the algebraically smallest stress be σ_{min}, the absolute maximum shear stress is always given by

$$\tau_{max} = \frac{\sigma_{max} - \sigma_{min}}{2} \tag{8.37}$$

Example Problems 8.12 and 8.13 illustrate cases where the in-plane maximum shear stress may not be the absolute maximum.

EXAMPLE 8.12

For the plane stress element shown determine the absolute maximum shear stress when **(a)** $\sigma_x = 14$ ksi, $\sigma_y = 2$ ksi, $\tau_{xy} = 8$ ksi, and **(b)** $\sigma_x = 14$ ksi, $\sigma_y = 14$ ksi, $\tau_{xy} = 8$ ksi.

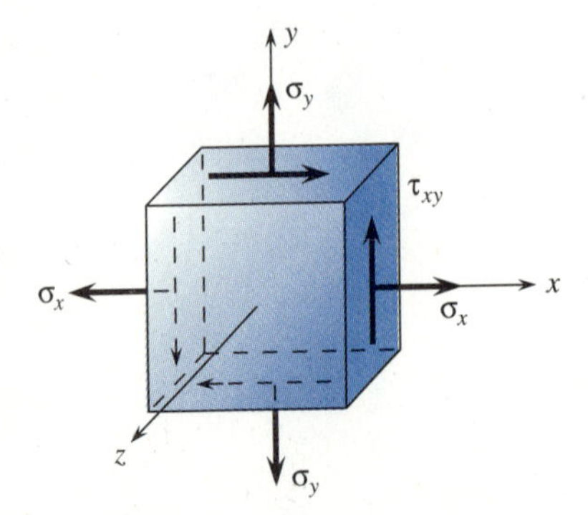

Figure 8.45 Plane stress element.

Solution

(a) For the *x*-*y* plane, we have

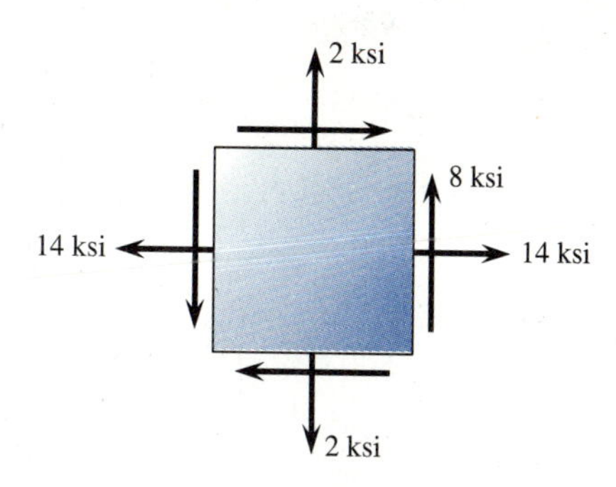

Following the steps outlined for use of the Mohr's circle method, we have

1. $X = (14, -8) \qquad Y = (2, 8)$

2. $C = \sigma_{ave} = \dfrac{14 + 2}{2} = 8$

3. Sketch of circle

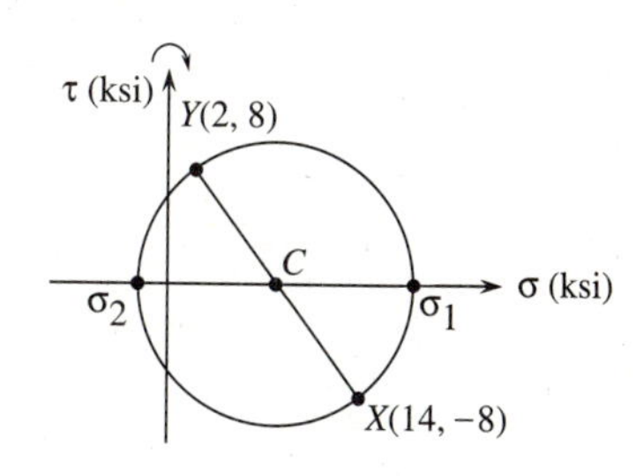

4. $R = \tau_{max\ 1\text{-}2} = \sqrt{(14 - 8)^2 + (8)^2} = 10 \text{ ksi}$

5. $\sigma_1 = C + R = 8 + 10 = 18 \text{ ksi} = \sigma_{max}$

$\sigma_2 = C - R = 8 - 10 = -2 \text{ ksi} = \sigma_{min}$

$\sigma_3 = 0 = \sigma_z$

Using Eq. (8.33), since σ_1 and σ_2 have opposite algebraic signs, or Eq. **(8.37)**, we obtain the absolute maximum shear stress as

$$\tau_{max} = \frac{\sigma_1 - \sigma_2}{2} = \frac{\sigma_{max} - \sigma_{min}}{2} = \frac{18 - (-2)}{2} = 10 \text{ ksi}$$

Hence, the maximum in-plane shear stress corresponds to the absolute maximum shear stress.

(b) For the *x-y* plane, we have

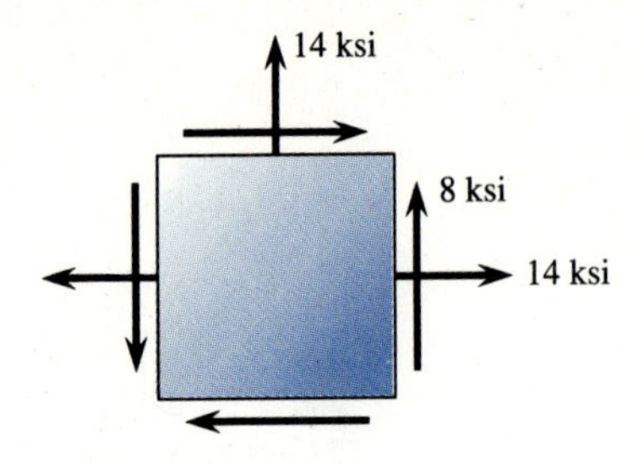

1. $X = (14, -8) \qquad Y = (14, 8)$

2. $C = \sigma_{ave} = \dfrac{14 + 14}{2} = 14$

3. Sketch of circles

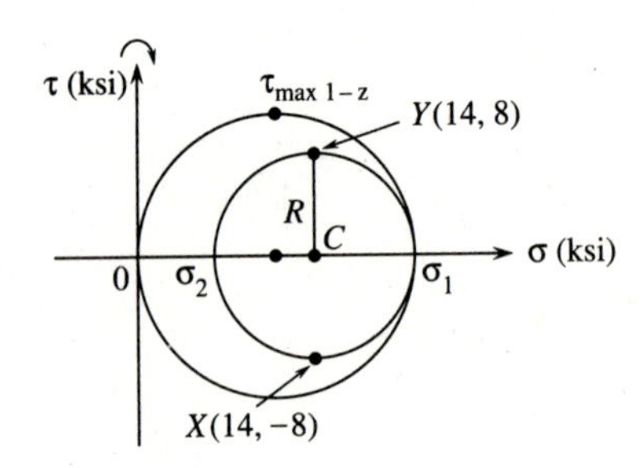

4. $R = \tau_{max\ 1\text{-}2} = 8$ ksi

5. $\sigma_1 = C + R = 14 + 8 = 22$ ksi

$\sigma_2 = C - R = 14 - 8 = 6$ ksi

Now since σ_1 and σ_2 have the same algebraic sign, absolute τ_{max} is given by Eq. (8.36) as

$$\tau_{max\ 1\text{-}z} = \frac{\sigma_1}{2} = \frac{22}{2} = 11 \text{ ksi}$$

or letting $\sigma_1 = \sigma_{max} = 22$ ksi and $\sigma_{min} = 0$, we use Eq. (8.37) to obtain the same value (11 ksi) for the absolute maximun shear stress, which is shown in the above Mohr's circle.

EXAMPLE 8.13

A cylindrical tank with closed ends contains a gas pressure $p = 2000$ kPa. The inside diameter of the tank is 1000 mm and the wall thickness is 6 mm. Determine **(a)** the principal stresses in the outer wall of the cylinder, **(b)** the maximum in-plane shear stress, and

(c) the absolute maximum shear stress in the cylinder. Show sketches of the stresses in parts **(a)**, **(b)**, and **(c)** on properly oriented elements.

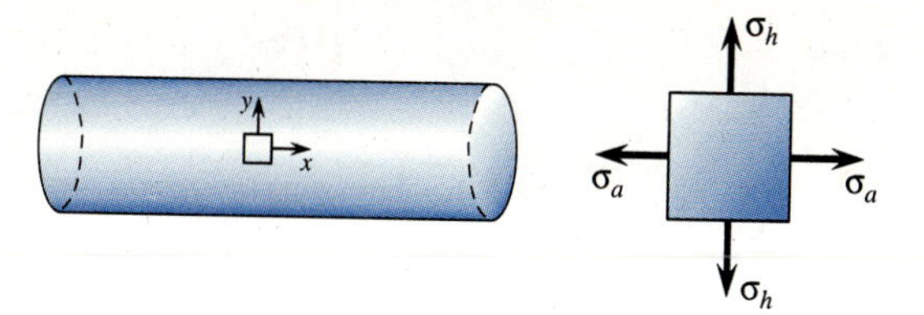

Figure 8.46 Cylindrical tank containing gas pressure.

Solution

(a) The hoop stress (in the y-direction) and axial stress (in the x-direction) are determined using Eqs. (7.10) and (7.15) as

$$\sigma_y = \sigma_h = \frac{pr_i}{t} = (2000 \text{ kPa})\left(\frac{500 \text{ mm}}{6 \text{ mm}}\right) = 167{,}000 \text{ kPa}$$

$$= 167 \text{ MPa}$$

$$\sigma_x = \sigma_a = \frac{pr_i}{2t} = \frac{\sigma_h}{2} = \frac{167 \text{ MPa}}{2} = 83.5 \text{ MPa}$$

For a vessel subjected to pressure loading, the shear stress $\tau_{xy} = 0$, therefore the hoop stress and axial stress are the principal stresses given by

$$\sigma_1 = \sigma_h = 167 \text{ MPa} \quad \sigma_2 = \sigma_a = 83.5 \text{ MPa}$$

A sketch of an element with the principal stresses acting on it is shown as

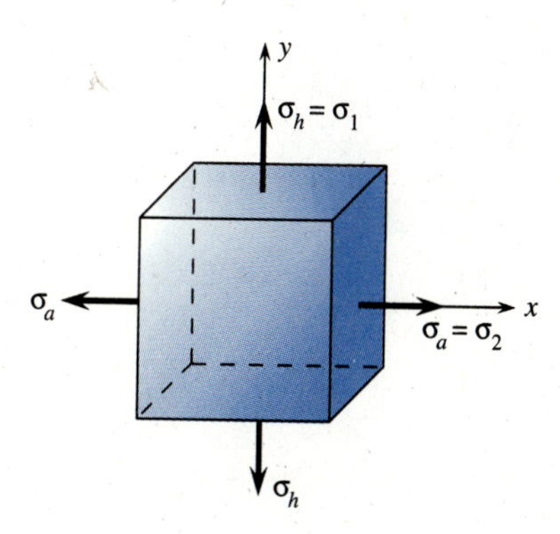

(b) The maximum in-plane shear stress is given by Eq. (8.27) as

$$\tau_{\text{max}} = \frac{\sigma_1 - \sigma_2}{2} = \frac{167 - 83.5}{2} = 41.8 \text{ MPa}$$

A sketch of an element showing the maximum in-plane shear stress is shown on the following page.

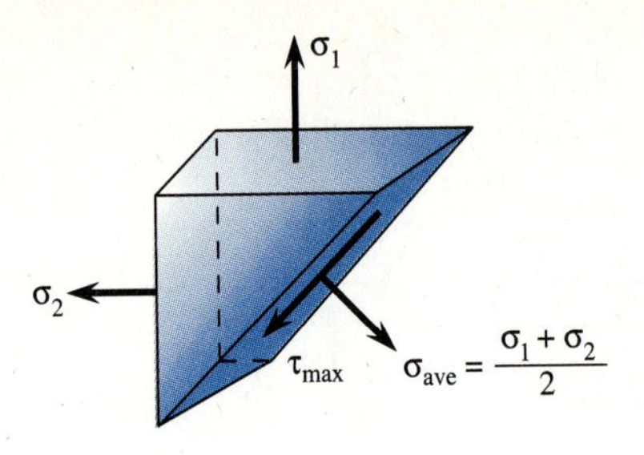

By Eq. (8.25), with $\theta_{p1} = 90°$ and $\theta_{p2} = 0°$, we obtain

$$\theta_s = \theta_p \pm 45°$$

$$\theta_{s1} = 45° \quad \text{and} \quad \theta_{s2} = -45°$$

(c) Since $\sigma_1 > \sigma_2 > 0$, the absolute maximum shear stress is given by Eq. (8.36) as

$$\tau_{max\ 1\text{-}z} = \frac{\sigma_1}{2} = \frac{167 \text{ MPa}}{2} = 83.5 \text{ MPa}$$

A sketch of an element with the absolute maximum shear stress is shown below.

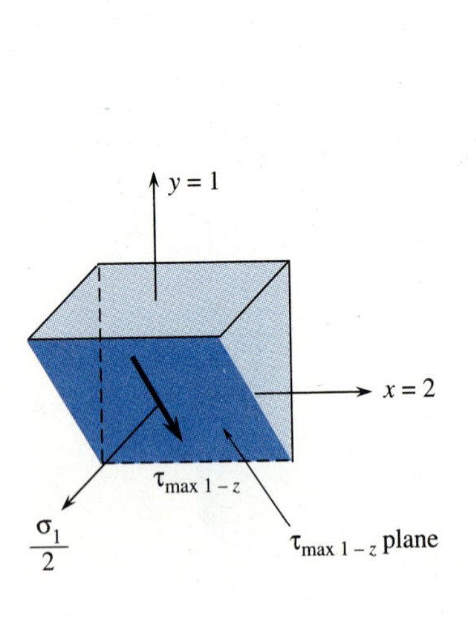

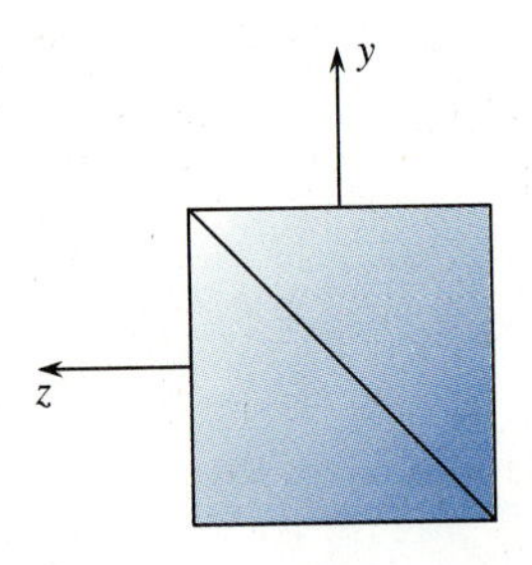

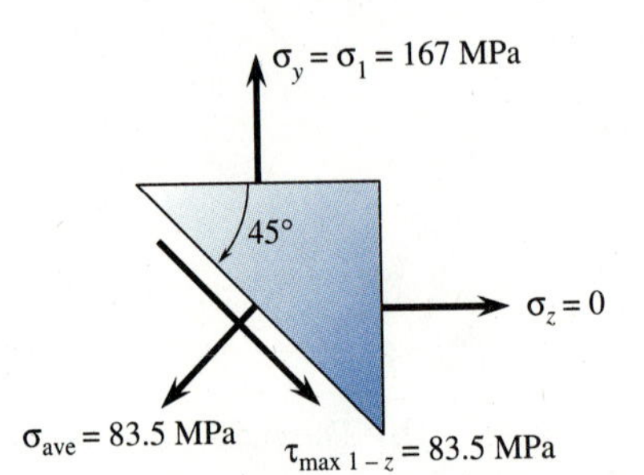

8.7 FAILURE THEORIES

In this section, we present the most common failure theories used to predict yielding in ductile materials and to predict fracture in brittle materials. These theories only apply when static loads exist. Other methods must be used when repeated or fatigue loads exist (see Chapter 11).

Failure theories attempt to relate a complex stress state, such as a biaxial or plane one (Figure 8.47; also see Figure 8.4), studied in this chapter to a single strength, normally the yield strength σ_Y or the ultimate strength σ_u based on a simple tension test of the same material.

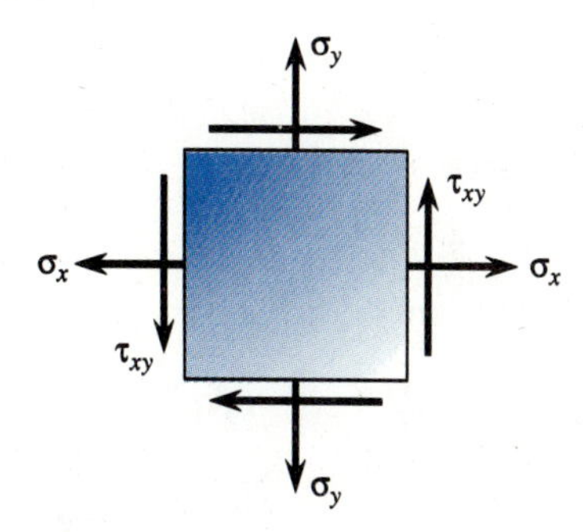

Figure 8.47 Biaxial stress state.

Several theories have been proposed, some dating back to the 1770s, for predicting failure of different kinds of materials subjected to numerous combinations of loads. Unfortunately, no one best theory has evolved that agrees with test data for all types of materials and load combinations. Those theories most often used in practice will be presented in this section. We divide these into theories primarily used for ductile materials and those used for brittle materials.

Ductile Material Failure

We will consider only the two most useful strength theories for ductile materials.

Maximum Shear Stress Theory

This theory, also called the *Tresca yield theory,* is a conservative theory in that it is on the safe side of test results. It is also simple to apply and so is often used in design to predict failure of ductile materials.

The maximum shear stress theory says yielding failure begins when the maximum shear stress in a material under any general loading conditions equals the maximum shear stress in a tension test specimen of the same material when that specimen begins to yield, that is, when $\sigma_1 = \sigma_Y$ in the tension test.

From the tension test (Figure 8.48(a)), when we reach the yield stress σ_Y, the Mohr's circle for the tension test (when $\sigma_1 = \sigma_Y$) is shown in Figure 8.48(b).

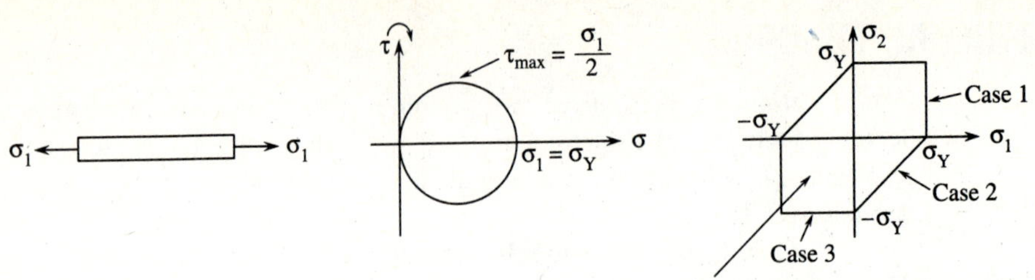

Figure 8.48 (a) Tension test to yielding,
(b) resulting Mohr's circle and
(c) failure envelope based on maximum shear stress theory.

The maximum shear stress in the tension test when $\sigma_1 = \sigma_Y$ is

$$\tau_{max} = \frac{\sigma_1}{2} = \frac{\sigma_Y}{2} \tag{8.38}$$

This is considered to be the largest shear stress allowed in the material subjected to any load combination.

For an actual biaxial stress state, as described numerous times in this chapter (see, for instance, the machine shaft of Figure 8.1 or the press frame of Figure 8.2, and Examples 8.1 through 8.10), the absolute maximum shear stress is given by Eq. (8.37) as

$$\tau_{max} = \frac{\sigma_{max} - \sigma_{min}}{2} \tag{8.39}$$

The maximum shear stress theory then predicts that failure by yielding along slip planes will occur when the maximum shear stress from Eq. (8.39) reaches the maximum shear stress given by Eq. (8.38). That is

$$\tau_{max} = \frac{\sigma_{max} - \sigma_{min}}{2} = \frac{\sigma_Y}{2} \tag{8.40}$$

or simplifying Eq. (8.40), we have

$$\sigma_{max} - \sigma_{min} = \sigma_Y \tag{8.41}$$

Specifically looking at the possible combinations of principal stresses, we have three separate cases to consider.

Case 1

If $\sigma_1 > \sigma_2 > 0$, then $\sigma_3 = 0$ is σ_{min} and failure occurs when $\sigma_1 - 0 = \sigma_Y$.

Case 2

If $\sigma_1 > 0 > \sigma_2$, then failure occurs when $\sigma_1 - \sigma_2 = \sigma_Y$.

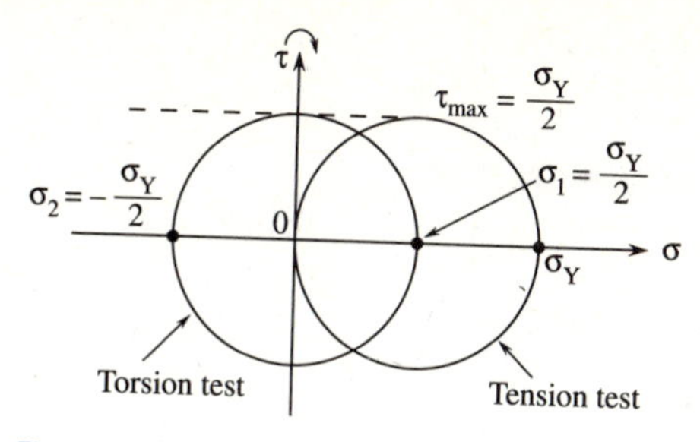

Figure 8.49 Mohr's circle for pure shear.

Case 3

If $\sigma_2 < \sigma_1 < 0$, then failure occurs when $\sigma_2 = \sigma_Y$.

The failure envelope, based on the maximum shear stress theory is shown in Figure 8.48(c). The other lines in the envelope occur if we switch σ_1 and σ_2 in Cases 1, 2, and 3.

In pure shear, such as the case of a shaft subjected to only torque, from the Mohr's circle in pure shear (Figure 8.49), the largest shear stress allowed according to the maximum shear stress theory is

$$\tau_{max} = \frac{\sigma_1 - \sigma_2}{2} = \frac{\sigma_Y/2}{2} - \frac{(-\sigma_Y/2)}{2} = \frac{\sigma_Y}{2} \qquad \textbf{(8.42)}$$

as $\sigma_1 = \sigma_Y/2$ and $\sigma_2 = -\sigma_Y/2$ in pure torsion and the largest shear stress can be no greater than that produced during the tension test that yields $\sigma_1 = \sigma_Y$.

Maximum Distortion Energy Theory

This theory is also known as *Von Mises's theory* or *Von Mises's-Hencky's theory* or the *maximum octahedral shear stress theory*. The theory was developed based on the observations that ductile materials stressed hydrostatically (with uniform pressure acting all around the material) yielded at stresses much greater than the values given by the simple tension test. It was then thought that yielding was not a simple tension or compression phenomenon, but probably was related to the angular distortion and hence, distortional energy of the stressed material element.

For ductile materials, this theory more closely matches the experimental data on yielding failure than the maximum shear stress theory. It is often employed in design and has been programmed, for convenient use, into computer codes dealing with stress analysis, such as ANSYS [1].

The theory says that yielding failure occurs when the distortional energy u_d in a material under general loading is equal to the distortional energy from a simple tension test of the same material when $\sigma_1 = \sigma_Y$ in the tension test.

We will show in Section 10.8, Eq. (10.59), that the distortional energy in biaxial stress is expressed in terms of the principal stresses by

$$u_d = \frac{1+\nu}{3E}\left[\sigma_1^2 - \sigma_1\sigma_2 + \sigma_2^2\right] \qquad \textbf{(8.43)}$$

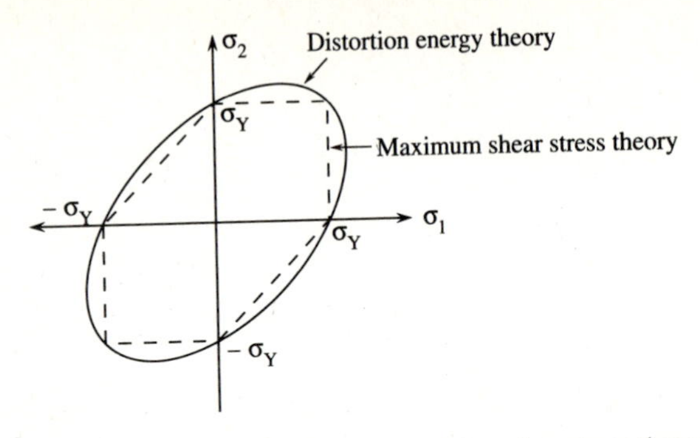

Figure 8.50 Failure envelope based on distortion energy theory and maximum shear stress theory shown for comparison.

Under tension test load only (with $\sigma_1 = \sigma_Y$ and $\sigma_2 = \sigma_3 = 0$), the distortional energy, Eq. (8.43), becomes

$$u_d = \frac{1+\nu}{3E}\,\sigma_Y^2 \tag{8.44}$$

The maximum distortion energy theory then predicts that yielding failure will occur when u_d from Eq. (8.43) equals u_d from Eq. (8.44). That is

$$\sigma_1^2 - \sigma_1\sigma_2 + \sigma_2^2 = \sigma_Y^2 \tag{8.45}$$

or defining the effective stress or Von Mises stress as

$$\sigma_e = (\sigma_1^2 - \sigma_1\sigma_2 + \sigma_2^2)^{1/2} \tag{8.46}$$

yielding occurs when

$$\sigma_e \geq \sigma_Y \tag{8.47}$$

Equation (8.45) is an equation of an ellipse that intersects the σ_1 axis at $\pm\sigma_Y$ and the σ_2 axis at $\pm\sigma_Y$ as shown in Figure 8.50. Figure 8.50 is the distortion energy envelope.

The name *octahedral shear stress theory* is a consequence of the following. The failure is assumed to occur when the octahedral shear stress τ_{oct} for any stress state equals the octahedral shear stress for the simple tension test specimen at failure (when $\sigma_1 = \sigma_Y$). The octahedral shear stress in biaxial stress is given by

$$\tau_{oct} = \frac{1}{3}[2\sigma_1^2 - 2\sigma_1\sigma_2 + 2\sigma_2^2]^{1/2} \tag{8.48}$$

In the simple tension test, the octahedral shear stress is then

$$\tau_{oct} = \frac{\sigma_Y}{3}(2)^{1/2} \tag{8.49}$$

Setting Eq. (8.48) equal to Eq. (8.49), yielding is assumed to occur when

$$(2\sigma_1^2 - 2\sigma_1\sigma_2 + 2\sigma_2^2)^{1/2} = \sigma_Y(2)^{1/2} \tag{8.50}$$

or simplifying and squaring both sides of Eq. (8.50), we obtain

$$\sigma_1^2 - \sigma_1\sigma_2 + \sigma_2^2 = \sigma_Y^2 \qquad (8.51)$$

Equation (8.51) is identical to Eq. (8.45).

In pure shear, when $\tau = \tau_{max}$ from the Mohr's circle, then $\sigma_1 = \tau_{max}$ and $\sigma_2 = -\tau_{max}$. Therefore, from Eq. (8.46), the effective stress is

$$\sigma_e = (3\tau_{max}^2)^{1/2} = \tau_{max}\sqrt{3}$$

and failure occurs according to Eq. (8.47) when

$$\sigma_e = \sigma_Y$$

or

$$\tau_{max}\sqrt{3} = \sigma_Y$$

or failure occurs when

$$\tau_{max} = \frac{\sigma_Y}{\sqrt{3}} = 0.577\sigma_Y \qquad (8.52)$$

EXAMPLE 8.14

The ice auger shown in Figure 8.51 is made of a steel with a yield strength of 276 MPa. Assume that the ice auger can resist a biaxial stress state of $\sigma_x = -105$ MPa and $\tau_{xy} = 105$ MPa. Based on **(a)** the maximum shear stress theory and **(b)** the maximum distortion energy theory, will the auger fail?

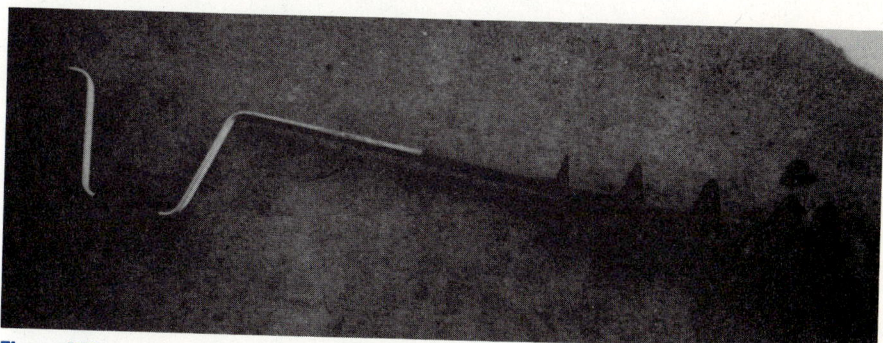

Figure 8.51 Ice auger under biaxial stress state.

Solution

(a) By the maximum shear stress theory, failure occurs when τ_{max} in the auger reaches one-half the yield stress in the material as given by Eq. (8.40). The principal stresses are obtained using Eq. (8.15) or from the Mohr's circle as

$$\sigma_{1,2} = \frac{-105 + 0}{2} \pm \sqrt{\left(\frac{-105 - 0}{2}\right)^2 + 105^2} = -52.5 \pm 117.4$$

$$\sigma_1 = 64.9 \text{ MPa}, \ \sigma_2 = -170 \text{ MPa}$$

The maximum shear stress is then

$$\tau_{max} = \frac{\sigma_1 - \sigma_2}{2} = \frac{64.9 - (-170)}{2} = 117 \text{ MPa}$$

and

$$\frac{\sigma_Y}{2} = \frac{276}{2} \text{ MPa} = 138 \text{ MPa}$$

Since $\tau_{max} = 117$ MPa is less than one-half the yield stress of 138 MPa, failure does not occur.

(b) From Eq. (8.46), the equivalent stress is

$$\sigma_e = (\sigma_1^2 - \sigma_1\sigma_2 + \sigma_2^2)^{1/2}$$

$$\sigma_e = \sqrt{64.9^2 - (64.9)(-170) + (-170)^2}$$

$$\sigma_e = 210 \text{ MPa}$$

Based on Eq. (8.47), since $\sigma_e = 210$ MPa is less than $\sigma_Y = 276$ MPa, failure does not occur.

Brittle Material Failure

Brittle materials have much larger compressive strengths than tensile strengths. For instance, see Section 2.5 and Figure 8.52.

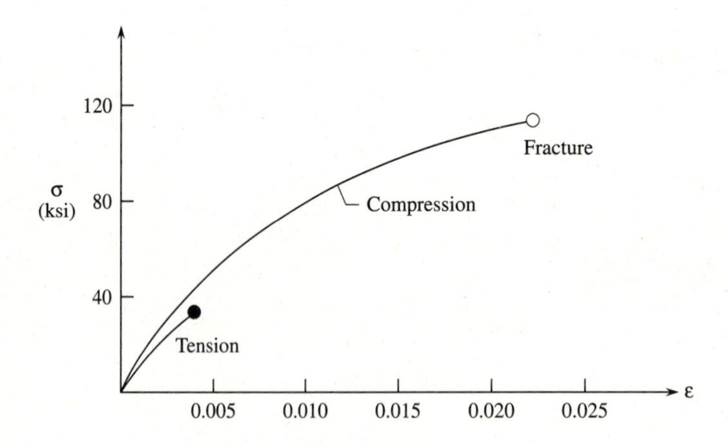

Figure 8.52 Stress-strain diagram for brittle material cast iron.

We present two theories for predicting brittle fracture under static loading.

Maximum Normal Stress Theory

The *maximum normal stress theory* is probably the simplest of all failure theories, although generally not one of the better ones. It states that brittle failure by rupture occurs

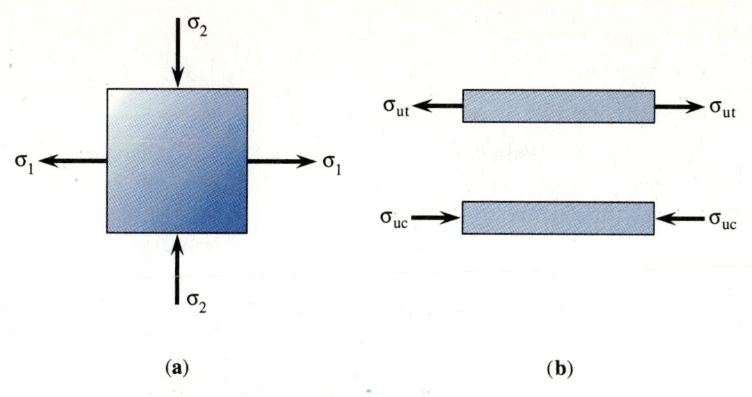

Figure 8.53 (a) Member in biaxial stress state and (b) uniaxial stress state.

whenever the greatest tensile stress in the actual structural or machine part exceeds the uniaxial tensile strength or whenever the largest compressive strength exceeds the uniaxial compressive strength. Figure 8.53(a) shows a member in a biaxial stress state with principal stresses assumed as shown and Figure 8.53(b) shows the uniaxial tensile or compressive specimen at ultimate strength in tension σ_{ut} and compression σ_{uc}. For σ_1 positive and σ_2 negative, failure then occurs when

$$\sigma_1 = \sigma_{ut} \quad \text{or} \quad \sigma_2 = \sigma_{uc} \tag{8.53}$$

Mohr's circle illustrates this concept in Figure 8.54(a) where the uniaxial tensile and compressive circles are drawn. The Mohr's circle for the actual stressed part must then be bound within these vertical lines of σ_{uc} and σ_{ut} to be safe. On the plot of σ_1 versus σ_2 (shown by the failure envelope, Figure 8.54(b), for biaxial stress, a material is safe as long as the (σ_1, σ_2) point is inside the shaded region. This theory correlates reasonably well with test data for brittle material fracture when the material is subjected to tensile loads. This is why we show σ_{uc} and σ_{ut}. Also for brittle materials, recall that σ_{uc} is

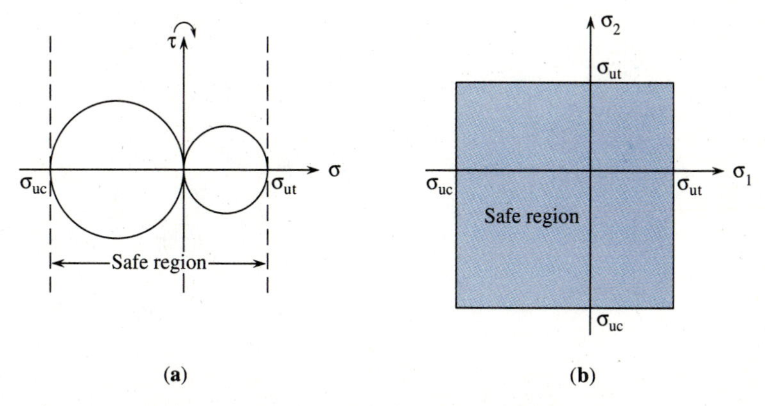

Figure 8.54 (a) Mohr's circles for maximum normal stress theory and (b) failure envelope.

generally much larger than σ_{ut}. For instance, from Appendix B and Figure 8.52, a typical gray cast iron has a $\sigma_{uc} = 95$ ksi (650 MPa) and a $\sigma_{ut} = 25$ ksi (170 MPa).

Coulomb-Mohr Theory

Brittle materials fail in tension under tension loads but fail in shear at higher compressive loads than tensile loads. A theory was developed by Mohr to account for this behavior of brittle materials. It is based on a modification of the maximum shear stress theory that was applied to ductile materials and retains part of the maximum normal stress theory for failure. The theory is called the *Coulomb-Mohr* or internal friction theory.

Consider a biaxial state of stress with principal stresses σ_1 and σ_2. The theory is presented for the three possible cases.

Case 1: $\sigma_1 > \sigma_2 > 0$

Then failure occurs when

$$\sigma_1 = \sigma_{ut} \tag{8.54}$$

Case 2: $\sigma_1 > 0 > \sigma_2$

Then failure occurs when

$$\frac{\sigma_1}{\sigma_{ut}} + \frac{\sigma_2}{\sigma_{uc}} = 1 \tag{8.55}$$

where σ_{uc} is used as a negative number in Eq. (8.55).

Case 3: $0 > \sigma_1 > \sigma_2$

Failure occurs when

$$\sigma_2 = \sigma_{uc} \tag{8.56}$$

It is Case 2, Eq. (8.55), that makes the Coulomb-Mohr theory different than the maximum normal stress theory. The theory is based on test results of a material, both the tension and compression test are used. The Mohr's circles of the tensile test and the compression test at failure are plotted as shown in Figure 8.55(a). The Coulomb-Mohr theory says that fracture occurs for any stress state which yields a Mohr's circle with points outside of the tangents to the two test circles shown in Figure 8.55(a). The failure envelope is shown in Figure 8.55(b).

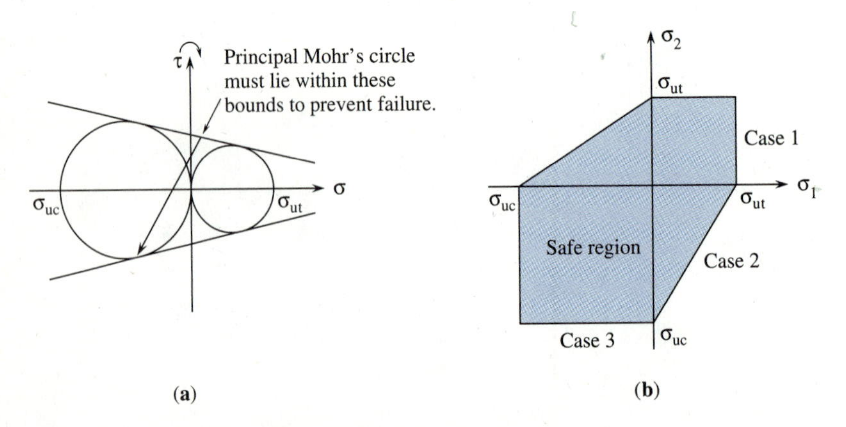

Figure 8.55 (a) Mohr's circle bounds for Coulomb-Mohr theory and (b) failure envelope.

EXAMPLE 8.15

A brittle cast iron, with $\sigma_{uc} = 95$ ksi and $\sigma_{ut} = 25$ ksi, is subjected to the biaxial stress state producing $\sigma_x = 10$ ksi, $\sigma_y = -4$ ksi, and $\tau_{xy} = 0$. Use the Coulomb-Mohr theory for brittle materials to determine whether the cast iron is safe.

Solution

Because $\tau_{xy} = 0$

$$\sigma_1 = \sigma_x = 10 \text{ ksi} \quad \text{and} \quad \sigma_2 = \sigma_y = -4 \text{ ksi}$$

Now since $\sigma_1 > 0$ and $\sigma_2 < 0$, use Case 2. Therefore, using Eq. (8.55), we have

$$\frac{\sigma_1}{\sigma_{ut}} + \frac{\sigma_2}{\sigma_{uc}} < 1$$

to be safe. Or on evaluation

$$\frac{10}{25} + \frac{(-4)}{(-95)} = 0.40 + 0.042 = 0.442 < 1$$

Based on the Coulomb-Mohr theory, the cast iron is safe.

8.8 PLANE STRAIN, THE EQUATIONS FOR THE TRANSFORMATION OF PLANE STRAIN, AND MOHR'S CIRCLE FOR PLANE STRAIN

Plane Strain

Recall that normal strain ϵ was defined in Chapter 2 for axially loaded members and shear strain γ was defined in Chapters 3 and 4 for members subjected to direct shear and torque, respectively. In this section, we consider strains in the x-y plane. Specifically, we treat the case of plane strain in which at most the three strains ϵ_x, ϵ_y, and γ_{xy} exist, while $\epsilon_z = \gamma_{zx} = \gamma_{zy} = 0$. These three strains are shown separately in Figure 8.56.

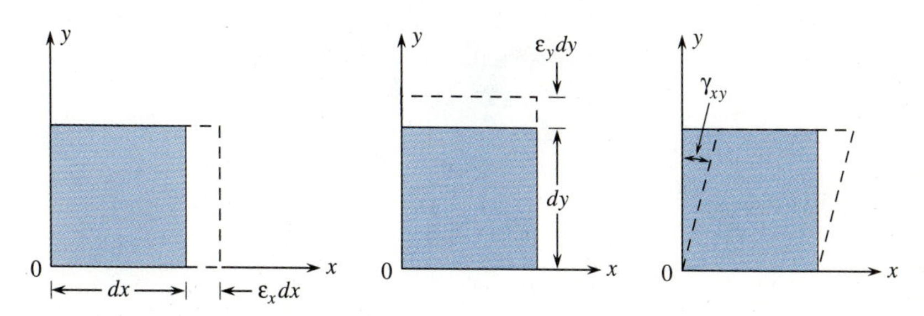

Figure 8.56 Three separate strains existing in plane strain.

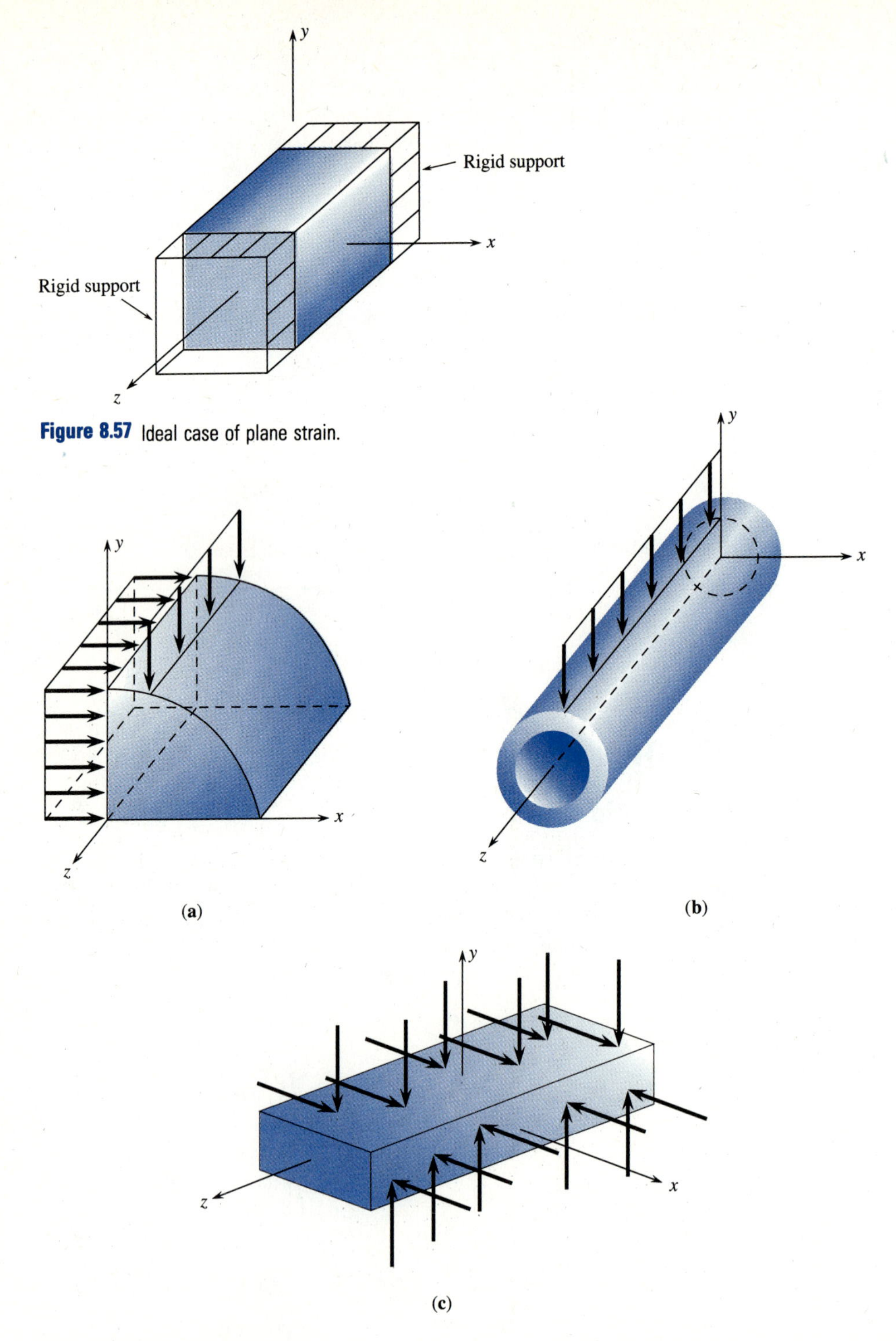

Figure 8.57 Ideal case of plane strain.

Figure 8.58 Examples of plane strain.

The case of plane strain exists when the deformations of the member occur within parallel planes and are the same in each parallel plane. This ideal situation is shown (Figure 8.57) for the block prevented from expanding or contracting by the rigid support plates.

Examples of plane strain occur in long structures subjected to distributed loads acting only in the x- and/or y-directions in which each unit slice of material in the z-direction has the same loading. The dam subjected to hydrostatic and vertical loads (Figure 8.58(a)), the pipe subjected to the line loading (Figure 8.58(b)), and the footing (Figure 8.58(c)) are specific examples of plane strain, in which the loads do not vary as we move from one position to another along the z-direction and we stay away from the ends of each structure.

Equations for Transformation of Plane Strain

We now develop the strain transformation equations in a manner similar to that used to derive the plane stress transformation equations in Section 8.2. We assume that a state of plane strain exists at some point in a body, in which ϵ_x, ϵ_y, and γ_{xy} associated with the x-y axes are known, and want to first determine the normal strain $\epsilon_{x'}$ along some new x′-axis, as shown in Figure 8.59. We will follow this derivation with that for $\gamma_{x'y'}$. (The strain $\epsilon_{y'}$ is obtained by substituting $\theta + 90°$ for θ in the equation for $\epsilon_{x'}$, and hence, need not be formally derived.)

Consider the plane strain, undeformed, rectangular-shaped element with dimensions dx and dy shown in Figure 8.60, in which the diagonal length dl coincides with the x′-axis. The strains ϵ_x, ϵ_y, and γ_{xy} produce an elongation of $\epsilon_x dx$ in the x-direction (Figure 8.60(a)), an elongation of $\epsilon_y dy$ in the y-direction (Figure 8.60(b)), and a decrease in the 90° angle between the x- and y-directions equal to γ_{xy} (Figure 8.60(c)). The three deformations in Figure 8.60 contribute to the change in the diagonal length dl given by Δd. For instance, the deformation $\epsilon_x dx$ causes the diagonal dl to increase in length by $\epsilon_x dx \cos\theta$ (Figure 8.60(a)). Similarly, $\epsilon_y dy$ causes an increase in dl of $\epsilon_y dy \sin\theta$ (Figure 8.60(b)), and $\gamma_{xy} dy$ causes an increase in dl of $\gamma_{xy} dy \cos\theta$ (Figure 8.60(c)). The total change in length Δd is then given by

$$\Delta d = \epsilon_x dx \cos\theta + \epsilon_y dy \sin\theta + \gamma_{xy} dy \cos\theta \tag{8.57}$$

The normal strain $\epsilon_{x'}$ is then given by

$$\epsilon_{x'} = \frac{\Delta d}{dl} = \epsilon_x \frac{dx}{dl} \cos\theta + \epsilon_y \frac{dy}{dl} \sin\theta + \gamma_{xy} \frac{dy}{dl} \cos\theta \tag{8.58}$$

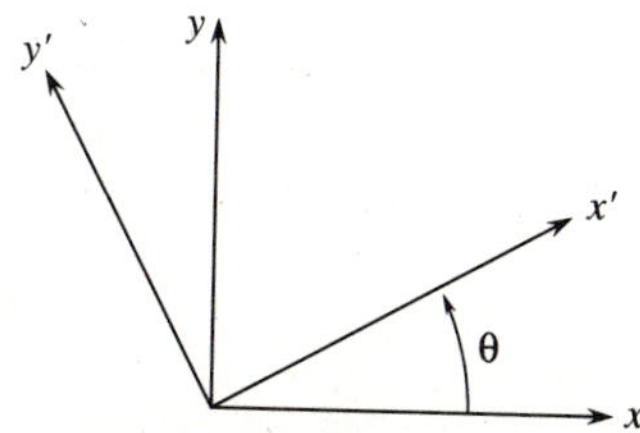

Figure 8.59 Direction x′ in which normal strain is to be determined.

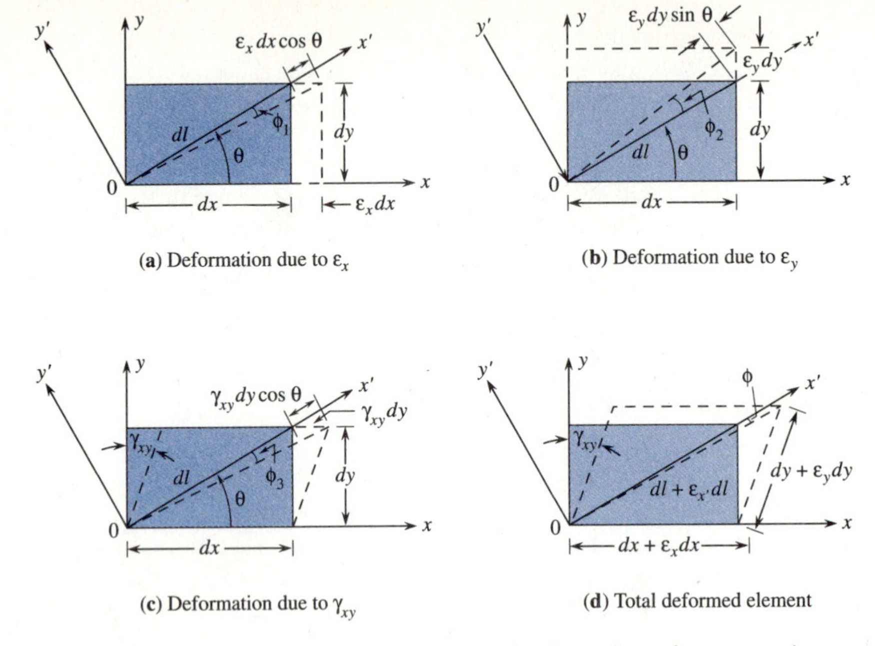

Figure 8.60 Elements showing the deformations occurring due to the strains ϵ_x, ϵ_y, and γ_{xy}.

From Figure 8.60, we also observe that

$$\frac{dx}{dl} = \cos\theta \qquad \frac{dy}{dl} = \sin\theta \tag{8.59}$$

Using Eq. (8.59) in Eq. (8.58), we obtain the transformation equation for the normal strain as

$$\epsilon_{x'} = \epsilon_x \cos^2\theta + \epsilon_y \sin^2\theta + \gamma_{xy} \sin\theta \cos\theta \tag{8.60}$$

Using trigonometric relations, Eqs. (8.3 and 8.4), we write Eq. (8.60) as

$$\epsilon_{x'} = \frac{\epsilon_x + \epsilon_y}{2} + \frac{\epsilon_x - \epsilon_y}{2} \cos 2\theta + \frac{\gamma_{xy}}{2} \sin 2\theta \tag{8.61}$$

Replacing θ by $\theta + 90°$, we obtain the normal strain along the y'-axis.

$$\epsilon_{y'} = \frac{\epsilon_x + \epsilon_y}{2} + \frac{\epsilon_x - \epsilon_y}{2} \cos(2\theta + 180°) + \frac{\gamma_{xy}}{2} \sin(2\theta + 180°) \tag{8.62}$$

From the trigonometric relations, we have

$$\cos(2\theta + 180°) = -\cos 2\theta \qquad \sin(2\theta + 180°) = -\sin 2\theta$$

Using these relations in Eq. (8.62), we obtain

$$\epsilon_{y'} = \frac{\epsilon_x + \epsilon_y}{2} - \frac{\epsilon_x - \epsilon_y}{2} \cos 2\theta - \frac{\gamma_{xy}}{2} \sin 2\theta \tag{8.63}$$

Adding Eqs. (8.61) and (8.63), we obtain

$$\epsilon_{x'} + \epsilon_{y'} = \epsilon_x + \epsilon_y \qquad \textbf{(8.64)}$$

Therefore, as in the case of plane stress, we now see from Eq. (8.64) that for plane strain, the sum of normal strains is independent of the orientation of the element.

We now derive the transformation equation for the shear strain $\gamma_{x'y'}$. Since the shear strain is the change in right angle, to obtain $\gamma_{x'y'}$, we need to determine the change in right angle between the x'- and y'-axes.

From Figure 8.60(d), we see that the line dl in the x$'$-direction underwent a change in relative orientation (or a rotation) ϕ, as well as a change in length $\epsilon_x \cdot dl$. As in determining $\epsilon_{x'}$, there are three contributions to the angle ϕ produced by ϵ_x, ϵ_y, and γ_{xy}. These angles are shown in Figures 8.60(a) through (c) as ϕ_1, ϕ_2, and ϕ_3. Since ϕ_1, ϕ_2, and ϕ_3 are small (small rotations assumed, i.e., $\tan \phi \approx \phi$), we have

$$\phi_1 = \frac{\epsilon_x dx \sin \theta}{dl} \qquad \phi_2 = \frac{\epsilon_y dy \cos \theta}{dl} \qquad \phi_3 = \frac{\gamma_{xy} dy \sin \theta}{dl} \qquad \textbf{(8.65)}$$

Using Eqs. (8.59) in (8.65), we obtain

$$\phi_1 = \epsilon_x \cos \theta \sin \theta \qquad \phi_2 = \epsilon_y \sin \theta \cos \theta \qquad \phi_3 = \gamma_{xy} \sin^2\theta \qquad \textbf{(8.66)}$$

Since the decrease in right angle is a positive $\gamma_{x'y'}$, we see that ϕ is a contribution to $\gamma_{x'y'}$ and that rotations of dl counterclockwise are then positive. Therefore

$$\phi = -\phi_1 + \phi_2 - \phi_3 \qquad \textbf{(8.67)}$$

Using Eqs. (8.66) in (8.67), we have

$$\phi = -(\epsilon_x - \epsilon_y)\sin \theta \cos \theta - \gamma_{xy} \sin^2\theta \qquad \textbf{(8.68)}$$

The other contribution to $\gamma_{x'y'}$ comes from the rotation of a line element dt, originally along the y'-axis and perpendicular to dl, as shown in Figure 8.61. The rotation of dt is given by Eq. (8.68), with θ replaced by $\theta + 90°$, as dt is 90° from dl. Therefore

$$\begin{aligned}\phi_{\theta + 90°} = &-(\epsilon_x - \epsilon_y)\sin(\theta + 90°)\cos(\theta + 90°) \\ &- \gamma_{xy} \sin^2(\theta + 90°)\end{aligned} \qquad \textbf{(8.69)}$$

Using the trigonometric identities

$$\sin(\theta + 90°) = \cos \theta \qquad \cos(\theta + 90°) = -\sin \theta$$

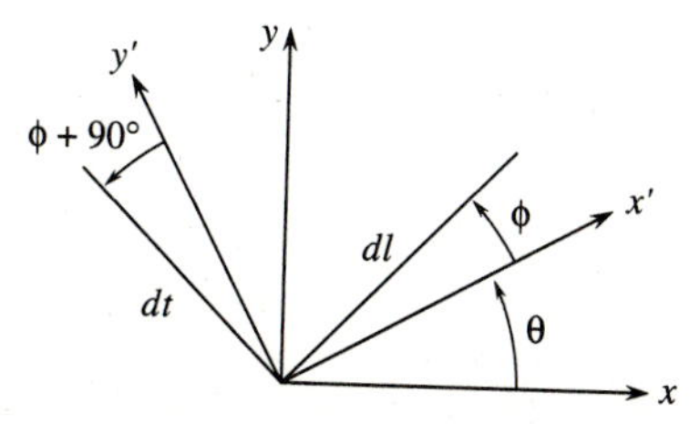

Figure 8.61 Rotation of line element dt.

in Eq. (8.69), we have

$$\phi_{\theta+90^\circ} = (\epsilon_x - \epsilon_y)\cos\theta\sin\theta - \gamma_{xy}\cos^2\theta \qquad \textbf{(8.70)}$$

Thus the total change in right angle between x' and y', which is $\gamma_{x'y'}$, is given by subtracting Eq. (8.70) from Eq. (8.68) as

$$\gamma_{x'y'} = \phi - \phi_{\theta+90^\circ} = -2(\epsilon_x - \epsilon_y)\sin\theta\cos\theta + \gamma_{xy}(\cos^2\theta - \sin^2\theta) \qquad \textbf{(8.71)}$$

Using the trigonometric identities

$$\cos^2\theta = \frac{(1+\cos 2\theta)}{2} \qquad \sin^2\theta = \frac{(1-\cos 2\theta)}{2} \qquad \sin\theta\cos\theta = \frac{(\sin 2\theta)}{2}$$

in Eq. (8.71), we express the shear strain as

$$\gamma_{x'y'} = -(\epsilon_x - \epsilon_y)\sin 2\theta + \gamma_{xy}\cos 2\theta \qquad \textbf{(8.72)}$$

In summary, given ϵ_x, ϵ_y, γ_{xy}, and θ, Eq. (8.72) yields the change in right angle between the x'- and y'-axes, where x' is at angle θ from the x-axis. A positive θ is counterclockwise between x and x' and a positive value for $\gamma_{x'y'}$ means the original 90° angle between x' and y' has decreased. Equations (8.61) and (8.72) for the transformation of plane strain are analogous to Eqs. (8.5) and (8.6) for plane stress. That is, comparing Eqs. (8.5) and (8.6) with (8.61) and (8.72), we see that

$$\begin{aligned} &\sigma_x \text{ corresponds to } \epsilon_x \\ &\sigma_y \text{ corresponds to } \epsilon_y \\ &\sigma_{x'} \text{ corresponds to } \epsilon_{x'} \\ &\tau_{xy} \text{ corresponds to } \frac{\gamma_{xy}}{2} \\ &\tau_{x'y'} \text{ corresponds to } \frac{\gamma_{x'y'}}{2} \end{aligned} \qquad \textbf{(8.73)}$$

Using these correspondences we can readily obtain expressions for principal strains and maximum shear strains as follows.

Using Eqs. (8.9) and (8.73), the principal angle θ_p is now

$$\tan 2\theta_p = \frac{\gamma_{xy}}{\epsilon_x - \epsilon_y} \qquad \textbf{(8.74)}$$

The principal strains, corresponding to Eq. (8.15), are calculated from

$$\epsilon_{1,2} = \frac{\epsilon_x + \epsilon_y}{2} \pm \sqrt{\left(\frac{\epsilon_x - \epsilon_y}{2}\right)^2 + \left(\frac{\gamma_{xy}}{2}\right)^2} \qquad \textbf{(8.75)}$$

The shear strain is zero in the directions of the principal strains.

The maximum shear strain in the x-y plane, corresponding to Eq. (8.20), is given by

$$\frac{\gamma_{max}}{2} = \sqrt{\left(\frac{\epsilon_x - \epsilon_y}{2}\right)^2 + \left(\frac{\gamma_{xy}}{2}\right)^2} \qquad \textbf{(8.76)}$$

The minimum shear strain is the negative of γ_{max}. In the directions of γ_{max} and γ_{min}, the normal strain, corresponding to Eq. (8.26), is

$$\epsilon_{ave} = \frac{\epsilon_x + \epsilon_y}{2} \tag{8.77}$$

The angle associated with the maximum shear strain is

$$\tan 2\theta_s = -\frac{\epsilon_x - \epsilon_y}{\gamma_{xy}} \tag{8.78}$$

and θ_s is associated with axes at 45° to the directions of the principal strains.

The plane strain transformation equations (Eqs. (8.61) and (8.72)) are actually valid even if strain ϵ_z exists since ϵ_z does not influence the in-plane strain transformations. Similarly, the transformation equations derived for plane stress can be used even if σ_z is not zero because σ_z does not enter the in-plane equilibrium equations, which were used to derive the transformation equations.

Furthermore, plane stress and plane strain do not occur simultaneously except when an element in plane stress is subjected to equal but opposite normal stresses (i.e., $\sigma_x = -\sigma_y$). In this case, the normal strain $\epsilon_z = 0$, as can be shown easily by considering the plane stress-strain equations given in Section 8.10. Therefore, the element is also in plane strain in addition to plane stress. Figure 8.62 summarizes the comparison of plane stress and plane strain.

Plane stress element	Plane strain element
σ_x, σ_y, τ_{xy} nonzero, in general, $\sigma_z = 0$, $\tau_{xz} = 0$, $\tau_{yz} = 0$ ϵ_x, ϵ_y, ϵ_z, γ_{xy} nonzero, in general, $\gamma_{xz} = 0$, $\gamma_{yz} = 0$	σ_x, σ_y, σ_z, τ_{xy} nonzero, in general, $\tau_{xz} = 0$, $\tau_{yz} = 0$ ϵ_x, ϵ_y, γ_{xy} nonzero, in general, $\epsilon_z = 0$, $\gamma_{xz} = 0$, $\gamma_{yz} = 0$

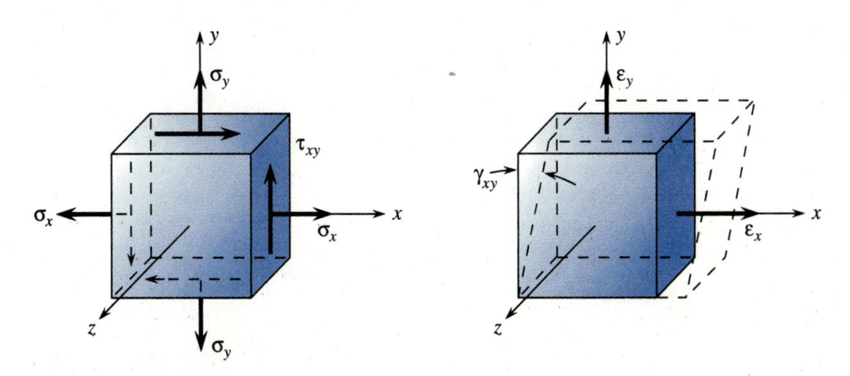

Figure 8.62 Summarization of comparison of plane stress and plane strain.

Example Problems 8.14 and 8.15 illustrate the application of the equations developed in this section.

EXAMPLE 8.14

At a point on the surface of a steel member under plane strain the following strains are known: $\epsilon_x = 0.001150$ in./in., $\epsilon_y = -0.000250$ in./in., and $\gamma_{xy} = 0.000900$ rad. These strains are shown highly exaggerated on a unit element in Figure 8.63. Determine **(a)** the strains for an element rotated 30° counterclockwise, **(b)** the principal strains, and **(c)** and the maximum in-plane shear strain.

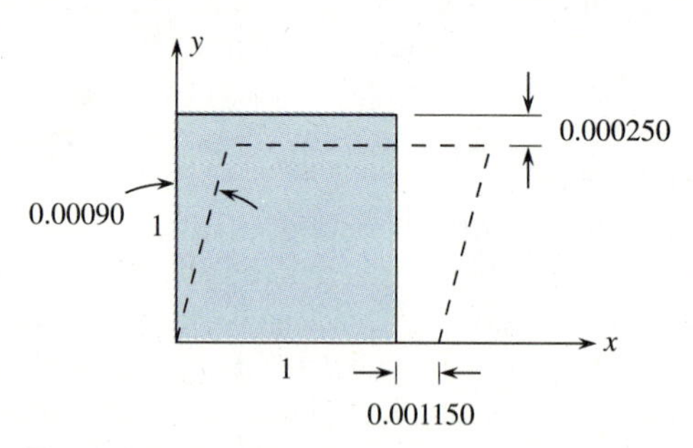

Figure 8.63 Unit element showing strains.

Solution

(a) Using the transformation Eqs. (8.61) and (8.72), we obtain

$$\epsilon_{x'} = \frac{\epsilon_x + \epsilon_y}{2} + \frac{\epsilon_x - \epsilon_y}{2}\cos 2\theta + \frac{\gamma_{xy}}{2}\sin 2\theta$$

$$= \left(\frac{1150 - 250}{2}\right) \times 10^{-6} + \left(\frac{1150 + 250}{2}\right) \times 10^{-6}(\cos 60°)$$

$$+ \left(\frac{900}{2}\right) \sin 60° \times 10^{-6}$$

$$= (450 + 350 + 390) \times 10^{-6} = 1190 \times 10^{-6} \text{ in./in.}$$

and

$$\gamma_{x'y'} = -(\epsilon_x - \epsilon_y)\sin 2\theta + \gamma_{xy}\cos 2\theta$$

$$= -(1150 + 250) \times 10^{-6}(\sin 60°) + \frac{900}{2} \times 10^{-6}(\cos 60°)$$

$$= (-1212 + 225) \times 10^{-6}$$

$$\gamma_{x'y'} = -987 \times 10^{-6} \text{ rad}$$

Using Eq. (8.64) for $\epsilon_{y'}$

$$\epsilon_{y'} = \epsilon_x + \epsilon_y - \epsilon_{x'}$$

$$= (1150 - 250 - 1190) \times 10^{-6}$$

$$= -290 \times 10^{-6} \text{ in./in.}$$

A sketch of the unit element at 30° and the strains associated with x' and y' is shown below.

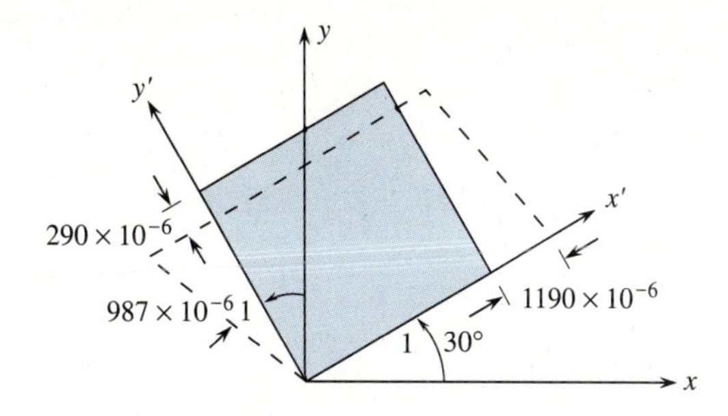

(b) Using Eq. (8.75), the principal strains are

$$\epsilon_{1,2} = \frac{\epsilon_x + \epsilon_y}{2} \pm \sqrt{\left(\frac{\epsilon_x - \epsilon_y}{2}\right)^2 + \left(\frac{\gamma_{xy}}{2}\right)^2}$$

$$= 450 \times 10^{-6} \pm \sqrt{(700 \times 10^{-6})^2 + (450 \times 10^{-6})^2}$$

$$= (450 \pm 832) \times 10^{-6}$$

$$\epsilon_1 = 1282 \times 10^{-6} \text{ in./in.} \qquad \epsilon_2 = -382 \times 10^{-6} \text{ in./in.}$$

By Eq. (8.74), the principal angle is

$$\tan 2\theta_p = \frac{\gamma_{xy}}{\epsilon_x - \epsilon_y} = \frac{900 \times 10^{-6}}{(1150 + 250) \times 10^{-6}} = 0.643$$

$$2\theta_p = 32.7° \qquad \theta_p = 16.4°$$

Substituting $2\theta_p = 32.7°$ into Eq. (8.61), we obtain

$$\epsilon_{x'} = 450 \times 10^{-6} + (700 \times 10^{-6})\cos 32.7°$$

$$+ (450 \times 10^{-6})\sin 32.7°$$

$$= (450 + 589 + 243) \times 10^{-6}$$

$$\epsilon_{x'} = 1282 \times 10^{-6} \text{ in./in.}$$

This value corresponds to $\epsilon_1 = 1282 \times 10^{-6}$ in./in. Therefore

$$\theta_{p1} = 16.4°$$

The element oriented with principal strains is shown below.

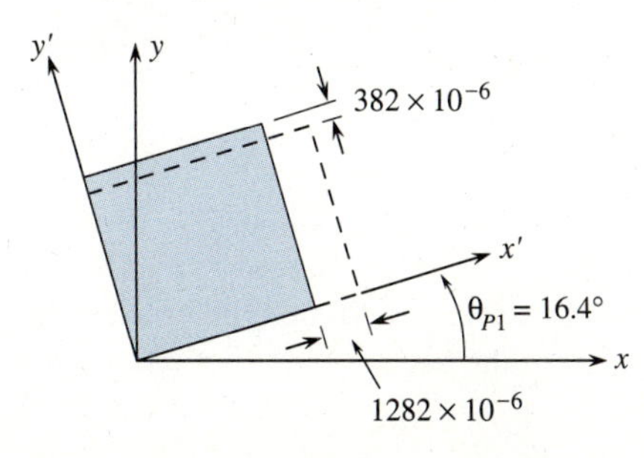

(c) Using Eq. (8.76)

$$\frac{\gamma_{max}}{2} = \sqrt{\left(\frac{\epsilon_x - \epsilon_y}{2}\right)^2 + \left(\frac{\gamma_{xy}}{2}\right)^2}$$

$$= \sqrt{(700 \times 10^{-6})^2 + (450 \times 10^{-6})^2}$$

$$\frac{\gamma_{max}}{2} = 832 \times 10^{-6}$$

$$\gamma_{max} = 1664 \times 10^{-6} \text{ rad}$$

From Eq. (8.77)

$$\epsilon_{ave} = \frac{\epsilon_x + \epsilon_y}{2} = 450 \times 10^{-6} \text{ in./in.}$$

and by Eq. (8.78)

$$\tan 2\theta_s = -\frac{\epsilon_x - \epsilon_y}{\gamma_{xy}} = -\frac{(1150 - (-250)) \times 10^{-6}}{900 \times 10^{-6}} = -\frac{14}{9}$$

$$2\theta_s = -57.3° \qquad \theta_s = -28.6°$$

The angle $\theta_{s1} = \theta_{p1} - 45° = 16.4° - 45° = -28.6°$ and $\theta_{s2} = \theta_{p1} + 45° = 61.4°$. The angle θ_s found above is indeed θ_{s1}. We could substitute $2\theta_s = -57.3°$ into Eq. (8.72), this yields $\gamma_{max} = 1664 \times 10^{-6}$ rad and confirms that $\theta_{s1} = -28.6°$. An element rotated with the maximum shear strains on it is shown below.

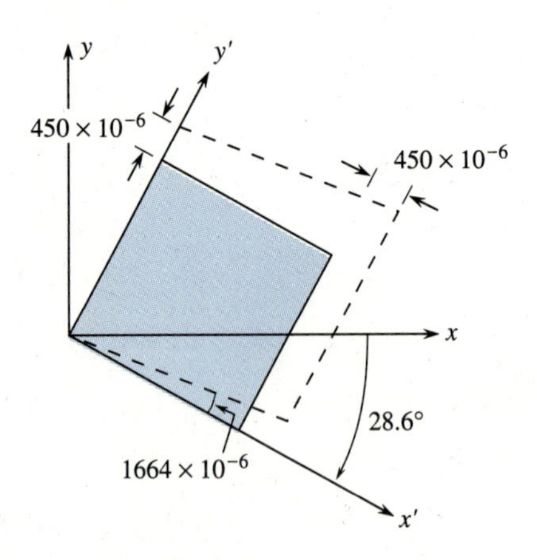

EXAMPLE 8.15

If $\epsilon_x = -800 \times 10^{-6}$ in./in., $\epsilon_y = -200 \times 10^{-6}$ in./in., and $\gamma_{xy} = 800 \times 10^{-6}$ rad, what value of strain would an axial strain gauge indicate that was oriented 45° counterclockwise from the x-direction? (See Figure 8.64.)

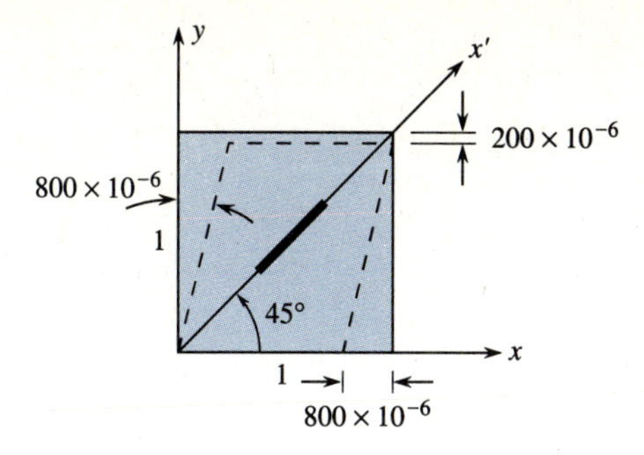

Figure 8.64 Element with in-plane strain.

Solution

Using Eq. (8.61), we determine $\epsilon_{x'}$ at 45°

$$\begin{aligned}\epsilon_{x'} &= \frac{\epsilon_x + \epsilon_y}{2} + \frac{\epsilon_x - \epsilon_y}{2}\cos 2\theta + \frac{\gamma_{xy}}{2}\sin 2\theta \\ &= \left[\frac{-800 - 200}{2} + \frac{-800-(-200)}{2}\cos 90° + \frac{800}{2}\sin 90°\right] \times 10^{-6} \\ &= (-500 + 0 + 400) \times 10^{-6} \\ &= -100 \times 10^{-6} \text{ in./in. (compression)}\end{aligned}$$

Mohr's Circle For Plane Strain

We observe that the transformation Equations (8.61), (8.63) and (8.72) are analogous to those for plane stress. Also the correspondence between the stresses and strains was shown in Eq. (8.73). Therefore we can extend the concept of Mohr's circle to the analysis of plane strain.

Given the plane strain components ϵ_x, ϵ_y, and γ_{xy} at a point in a body the following steps are used in constructing Mohr's circle for strain.

1. Write down the two strain coordinates X, Y to be plotted. Positive normal strain is associated with an extension of a line element while negative normal strain is associated with a shortening of a line element. Remember, we plot $\gamma_{xy}/2$ (not γ_{xy}) and a positive γ_{xy} is associated with a shear deformation causing a given side of an element to rotate clockwise and is plotted above the horizontal axis (analogous to positive shear stress causing a clockwise rotation of the element). If the deformation causes the side to rotate counterclockwise, the corresponding point is plotted below the horizontal axis. Figure 8.65 shows a deformed element with positive γ_{xy} associated with ϵ_y as the vertical side has rotated clockwise while negative γ_{xy} is associated with ϵ_x as the horizontal side has rotated counterclockwise. In other words, a positive γ_{xy} means the original 90° angle between the x and y axes has decreased to $90° - \gamma_{xy}$. For the element of Figure 8.65, we then plot

$$X = \left(\epsilon_x, -\frac{\gamma_{xy}}{2}\right) \qquad Y = \left(\epsilon_y, \frac{\gamma_{xy}}{2}\right)$$

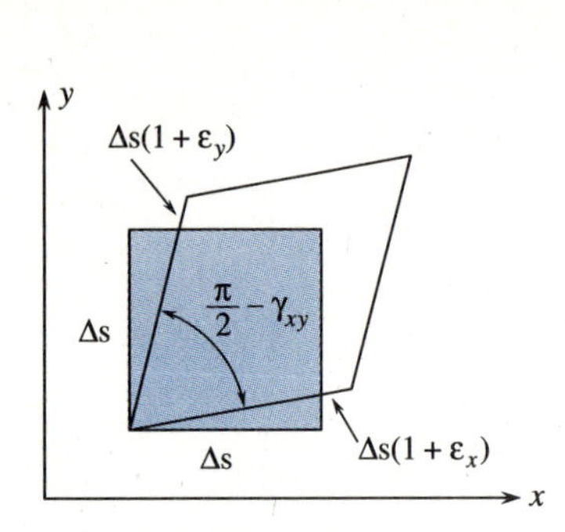

Figure 8.65 Deformed element.

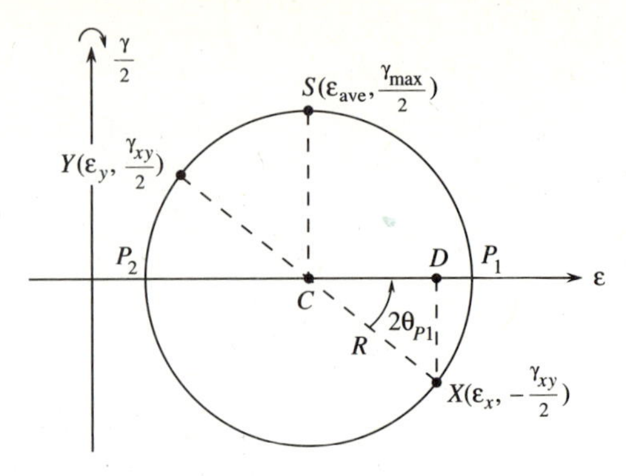

Figure 8.66 Mohr's circle for plane strain.

2. Determine the center C of the circle from

$$C = \epsilon_{ave} = \frac{\epsilon_x + \epsilon_y}{2}$$

where C is always on the horizontal ϵ axis.

3. Sketch the circle using the points X, Y and C as shown in Figure 8.66.
4. Determine the radius R of the circle by observing from the circle

$$R = \sqrt{(CD)^2 + (XD)^2}$$

$$= \sqrt{\left(\frac{\epsilon_x - \epsilon_y}{2}\right)^2 + \left(-\frac{\gamma_{xy}}{2}\right)^2}$$

Note that the maximum in-plane shear strain is

$$\frac{\gamma_{max}}{2} = R$$

as we have plotted $\gamma/2$ on the vertical axis. The strains at point S are then

$$S = (\epsilon_{ave}, \gamma_{max}/2)$$

5. Determine the principal strains, the largest and smallest normal strains on the circle, points P_1 and P_2 in Figure 8.66. These strains are then

$$\epsilon_1 = C + R \qquad \epsilon_2 = C - R$$

6. Determine the principal axes from the circle, where again angle 2θ on the circle is associated with angle θ on the element. From Figure 8.66 the principal angle is

$$\tan 2\theta_{p1} = \frac{XD}{CD} = \frac{\gamma_{xy}/2}{\frac{\epsilon_x - \epsilon_y}{2}} = \frac{\gamma_{xy}}{\epsilon_x - \epsilon_y}$$

and is seen to be counterclockwise from the x axis to the principal axis associated with ϵ_1. The principal axis associated with ϵ_2 is then 180° from that associated with ϵ_1 on the circle or the principal axes are 90° apart on the element.

7. Finally, if desired, the normal and shear strain in any other direction at angle θ from the x direction can be obtained by rotating the radius CX 2θ degrees about C in the same sense on the Mohr's circle and using appropriate trigonometric relations and distances needed.

Example Problem 8.18 illustrates how to obtain principal strains using Mohr's circle.

EXAMPLE 8.18

For the deformed element shown in Figure 8.67 a state of plane strain with $\epsilon_x = 400\mu$, $\epsilon_y = 300\mu$, and $\gamma_{xy} = 600\mu$ exists. Determine (1) the principal strains and (2) the maximum in-plane shear strain. Use Mohr's circle method.

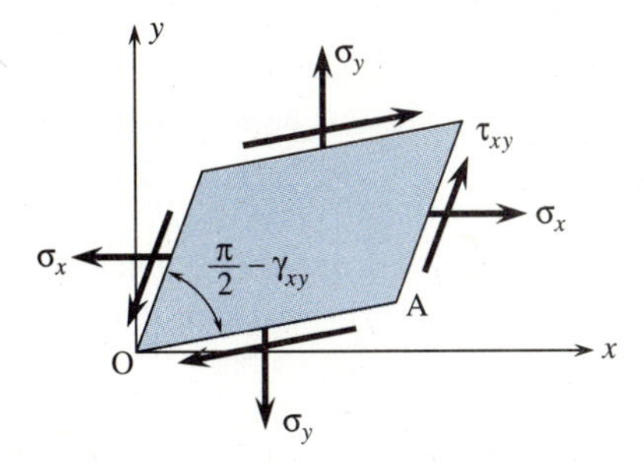

Figure 8.67 Deformed element and typical stresses that would cause this deformation.

Solution

Applying Mohr's circle for strain, we have

1. $X = \left(400, -\frac{600}{2}\right) \qquad Y = \left(300, \frac{600}{2}\right)$

where since the line OA associated with ϵ_x rotates counterclockwise the X point is plotted below the horizontal axis. Also all units are micro-inch per inch (μ).

2. $C = \epsilon_{\text{ave}} = \frac{\epsilon_x + \epsilon_y}{2} = \frac{400 + 300}{2} = 350$

3. Plot X, Y, and C points and sketch the circle.

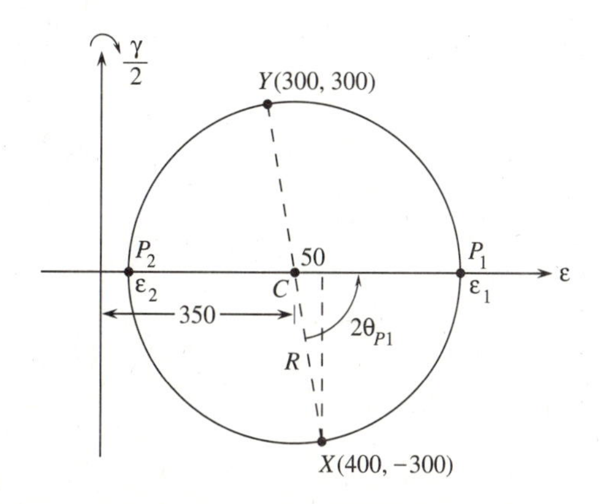

4. The maximum in-plane shear strain is

$$\frac{\gamma_{max}}{2} = R = \sqrt{50^2 + 300^2} = 304$$

$$\gamma_{max} = 608$$

5. The principal strains and principal angle are

$$\epsilon_1 = C + R = 350 + 304 = 654$$

$$\epsilon_2 = C - R = 350 - 304 = 46$$

$$2\theta_{p1} = \tan^{-1}\frac{300}{50} = 80.5°$$

$\theta_{p1} = 40.25°$ (counterclockwise from x-axis as shown in the sketch below)

$$\theta_{s1} = 45° - \theta_{p1} = 45° - (-40.25°) = 85.25°$$

where θ_{p1} is actually considered negative as the angle is counterclockwise from X to P_1 on the circle. The element below showing maximum in-plane shear strain also has average normal strains $\epsilon_{ave} = 350\mu$ and so has enlarged.

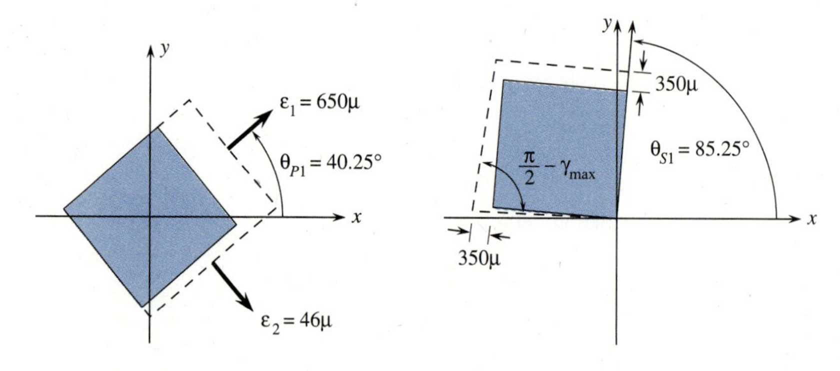

Principal strain element

Element with maximum in–plane shear strain

We should note that it is probably easier to visualize the proper deformed shape of the element associated with the maximum in-plane shear strain by considering the shear stresses associated with these strains. We should recall that from homogeneous, isotropic materials by Hooke's law $\tau_{max} = G\gamma_{max}$. Therefore the angle $\theta_{s1} = 85.25°$ also corresponds to the normal plane that has τ_{max} acting on it and by Mohr's circle convention this positive τ_{max} is the one that causes clockwise rotation of the element. Therefore we show the element with the maximum in-plane shear stresses below and this helps us to visualize the proper shape of the element associated with the maximum shear strain.

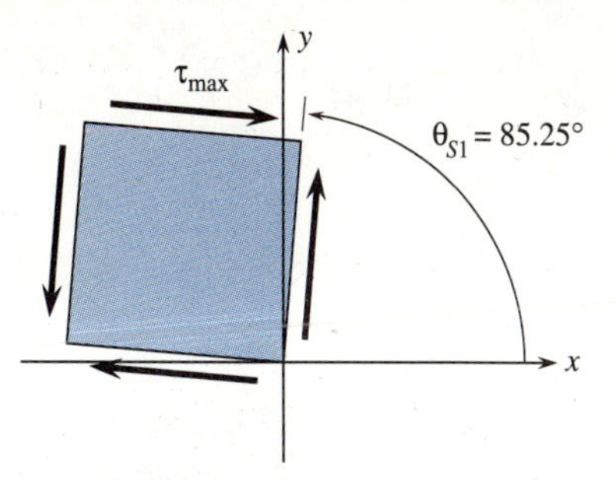

8.9 STRAIN MEASUREMENTS USING ROSETTES

Up to this point we have assumed that the normal or axial strains ϵ_x and ϵ_y and the shear strain γ_{xy} were given for any two-dimensional analysis. In theory, the normal strain may be determined in any given direction on the surface of a stressed structural member or machine part by marking a line segment, say AB, in a desired direction before loading (Figure 8.68). After loading, we then measure the line segment AB again. The normal strain is then

$$\epsilon_{AB} = \frac{\delta}{L}$$

where δ is the change in length of line segment AB whose original length was L. Similarly, by marking two lines perpendicular to each other, say in the x- and y-directions, as shown in Figure 8.69 on the surface of a member before loading and then measuring the change in the right angle after loading, the shear strain is determined as this change in right angle. This method of marking lines on the member to measure normal and shear strain is not very practical or very accurate.

A more accurate method for measuring normal strains is to use electrical resistance strain gauges. These gauges are generally made of a length of thin wire bent, as shown in

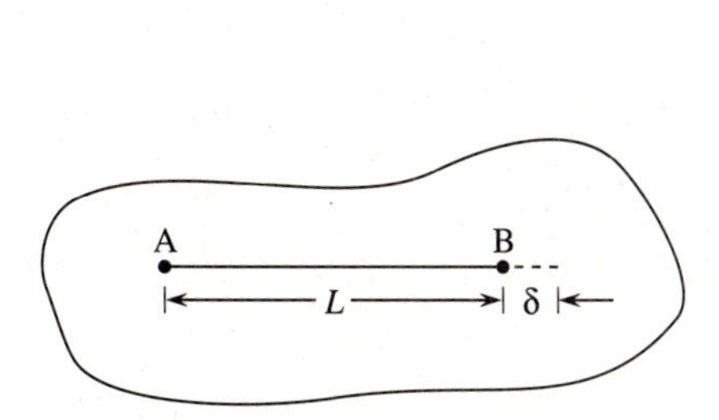

Figure 8.68 Line segment for normal strain measurement.

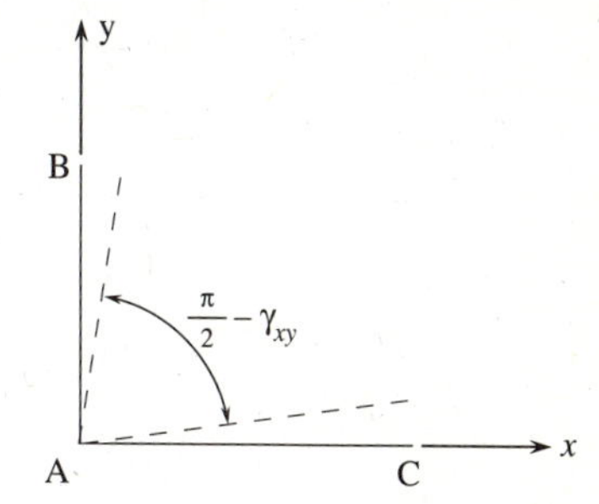

Figure 8.69 Line segments for measuring shear strain.

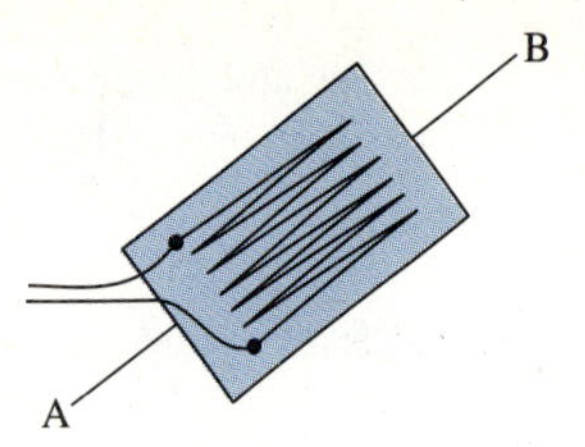

Figure 8.70 Electrical resistance strain gauge.

Figure 8.70, and sandwiched between two pieces of paper. The gauge is glued to the surface of the member in a direction that the normal strain is desired. As the member is loaded, the wire elongates or shortens causing the electrical resistance of the gauge to change. A current is passed through the gauge and the change in resistance is measured and converted into a strain measurement. These gauges are very sensitive and can record strains as small as 1×10^{-6}.

To obtain strain components ϵ_x and ϵ_y, we merely glue a gauge on the member in the x- and y-directions. Measuring the normal strain in a third different direction is sufficient to obtain γ_{xy}. The arrangement of the gauges to form a cluster to measure three normal strains is known as a *strain rosette*. For instance, assume we have three gauges in a cluster mounted on the surface of a member that make angles θ_a, θ_b, and θ_c with the x-axis as shown in Figure 8.71. Because the rosette is mounted to the free surface of the member, a state of plane stress exists. The same transformations for plane strain exist as for plane stress. Therefore, using the strain transformation Eq. (8.60), we write three equations as

$$\begin{aligned}
\epsilon_a &= \epsilon_x \cos^2\theta_a + \epsilon_y \sin^2\theta_a + \gamma_{xy} \sin\theta_a \cos\theta_a \\
\epsilon_b &= \epsilon_x \cos^2\theta_b + \epsilon_y \sin^2\theta_b + \gamma_{xy} \sin\theta_b \cos\theta_b \\
\epsilon_c &= \epsilon_x \cos^2\theta_c + \epsilon_y \sin^2\theta_c + \gamma_{xy} \sin\theta_c \cos\theta_c
\end{aligned} \tag{8.79}$$

The gauges record ϵ_a, ϵ_b, and ϵ_c. Therefore, Eqs. (8.79) may be solved simultaneously for ϵ_x, ϵ_y, and γ_{xy} and these strains can then be used, for instance, to construct the Mohr's circle of strain. The most common and practical arrangement of the three strain gauges is (1) the so-called *rectangular,* or 45° strain rosette, in which the gauges are arranged 45°

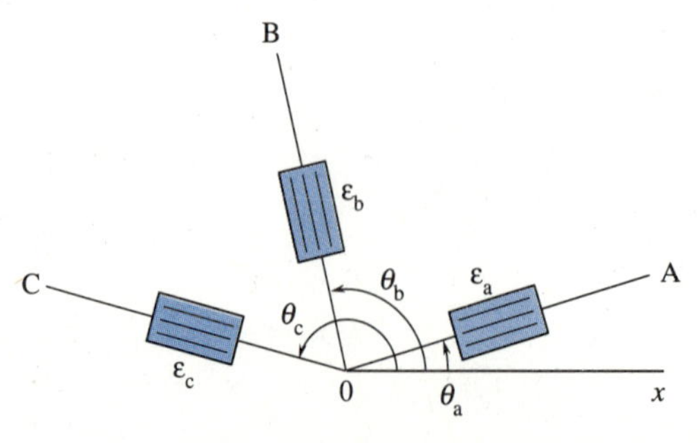

Figure 8.71 Strain rosette mounted on surface of a member.

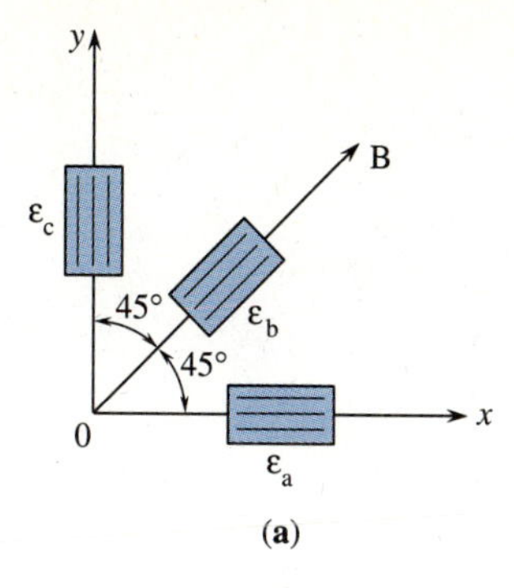

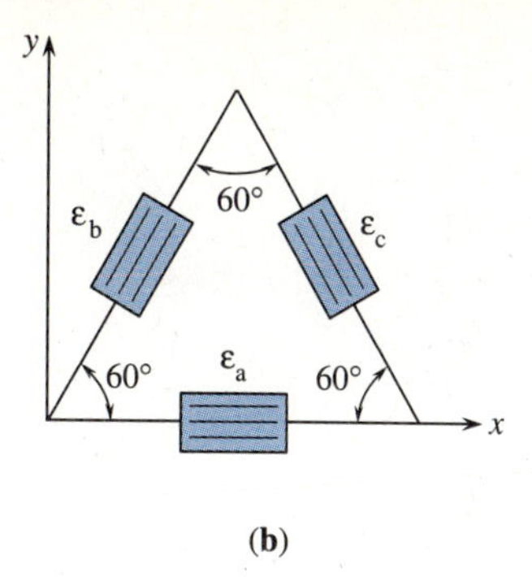

Figure 8.72 (a) Rectangular and (b) equiangular strain rosettes.

apart normally starting at 0° with the x-axis, as shown in Figure 8.72(a); or (2) the *equiangular*, or the *delta*, or the 60°, strain rosette of Figure 8.72(b).

For the rectangular rosette, the gauges measure normal strains $\epsilon_a = \epsilon_{0°}$, $\epsilon_b = \epsilon_{45°}$, and $\epsilon_c = \epsilon_{90°}$, where $\theta_a = 0°$, $\theta_b = 45°$, and $\theta_c = 90°$. Substituting these angles into Eqs. (8.79), we obtain ϵ_x, ϵ_y, and γ_{xy} as

$$\begin{aligned} \epsilon_x &= \epsilon_{0°} \\ \epsilon_y &= \epsilon_{90°} \\ \epsilon_{45°} &= \frac{\epsilon_x}{2} + \frac{\epsilon_y}{2} + \frac{\gamma_{xy}}{2} \end{aligned} \tag{8.80}$$

The third of Eqs. (8.80) is then solved for γ_{xy} as

$$\gamma_{xy} = 2\epsilon_{45°} - \epsilon_{0°} - \epsilon_{90°} \tag{8.81}$$

For the equiangular rosette, the measured normal strains are now $\epsilon_a = \epsilon_{0°}$, $\epsilon_b = \epsilon_{60°}$, and $\epsilon_c = \epsilon_{120°}$ with $\theta_a = 0°$, $\theta_b = 60°$, and $\theta_c = 120°$. We can now use Eqs. (8.79) and solve simultaneously for ϵ_x, ϵ_y, and γ_{xy} as

$$\begin{aligned} \epsilon_x &= \epsilon_{0°} \\ \epsilon_y &= \frac{1}{3}(2\epsilon_{60°} + 2\epsilon_{120°} - \epsilon_{0°}) \\ \gamma_{xy} &= \frac{2}{\sqrt{3}}(\epsilon_{60°} - \epsilon_{120°}) \end{aligned} \tag{8.82}$$

Example Problems 8.19 and 8.20 illustrate the use of strain rosette data, along with Mohr's circle, to obtain principal strains.

EXAMPLE 8.19

A rectangular or 45° strain rosette is mounted on an airplane wing as shown in Figure 8.73. Under the loadings applied to the wing, the following strains are recorded from the gauges: $\epsilon_a = 0.0005$, $\epsilon_b = 0.0002$, and $\epsilon_c = 0.0003$. Determine the principal strains and maximum in-plane shear strain.

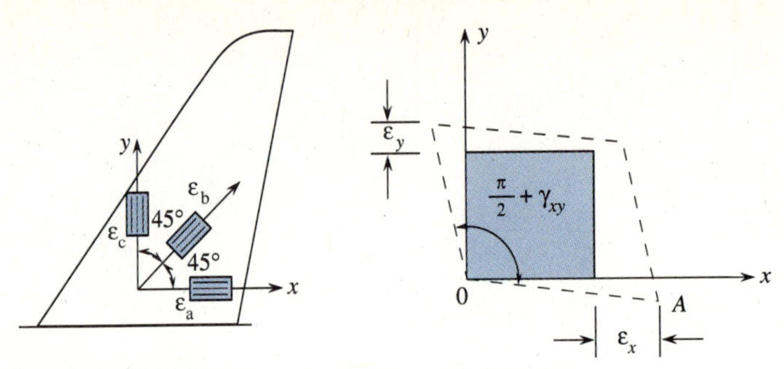

Figure 8.73 Rectangular strain rosette mounted to wing.

Solution

Using Eqs. (8.80) and (8.81), we obtain

$$\epsilon_x = \epsilon_a = \epsilon_{0°} = 0.0005$$

$$\epsilon_y = \epsilon_c = \epsilon_{90°} = 0.0003$$

$$\begin{aligned}\gamma_{xy} &= 2\epsilon_b - \epsilon_a - \epsilon_c = 2\epsilon_{45°} - \epsilon_{0°} - \epsilon_{90°}\\ &= 2(0.0002) - 0.0005 - 0.0003\\ &= -0.0004\end{aligned}$$

The negative sign for γ_{xy} means the original 90° angle between x and y axes has increased. Applying Mohr's circle, we have

1. $X = \left(0.0005, \dfrac{0.0004}{2}\right) \qquad Y = \left(0.0003, -\dfrac{0.0004}{2}\right)$

   where since the line OA associated with ϵ_x rotates clockwise the X point is plotted above the horizontal axis. Also remember we plot $\gamma_{xy}/2$ as $\gamma_{xy}/2$ is analogous to τ_{xy} in Eq. (8.73).

2. $C = \epsilon_{\text{ave}} = \dfrac{0.0005 + 0.0003}{2} = 0.0004$

3. Plot X, Y, and C points and sketch circle.

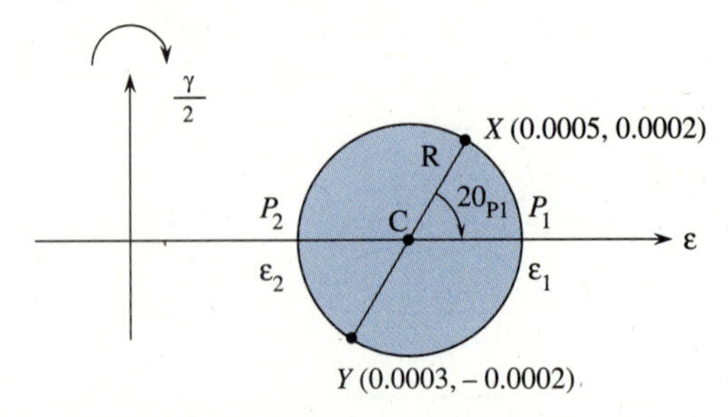

4. The maximum in-plane shear strain is

$$\frac{\gamma_{\max}}{2} = R = \sqrt{(0.0001)^2 + (0.0002)^2} = 0.000224$$

$$\gamma_{\max} = 0.000448 \text{ rad}$$

A properly oriented element showing the largest in-plane shear strain is shown below, where θ_s from step 5 has been used.

5. The principal strains are

$$\epsilon_1 = C + R = 0.0004 + 0.000224 = 0.000624$$

$$\epsilon_2 = C - R = 0.0004 - 0.000224 = 0.000176$$

$$2\theta_{p1} = \tan^{-1}\frac{0.0002}{0.0001} = 63.4°$$

$$\theta_{p1} = 31.7° \text{ (clockwise from the } x\text{-axis as shown below)}$$

$$\theta_s = 45° - \theta_{p1} = 45° - 31.7° = 13.3° \text{ counterclockwise}$$

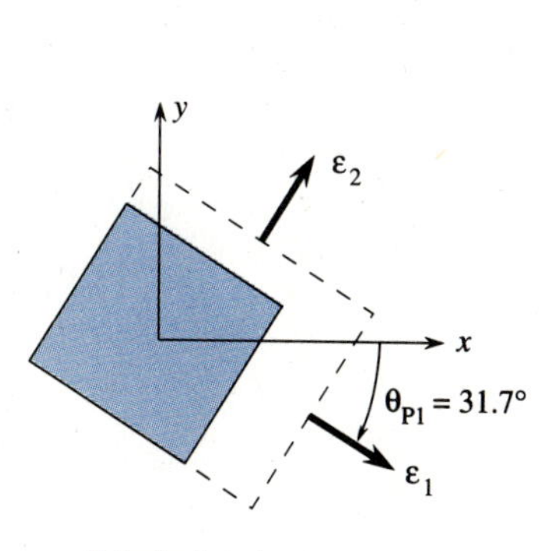

Principal strain element

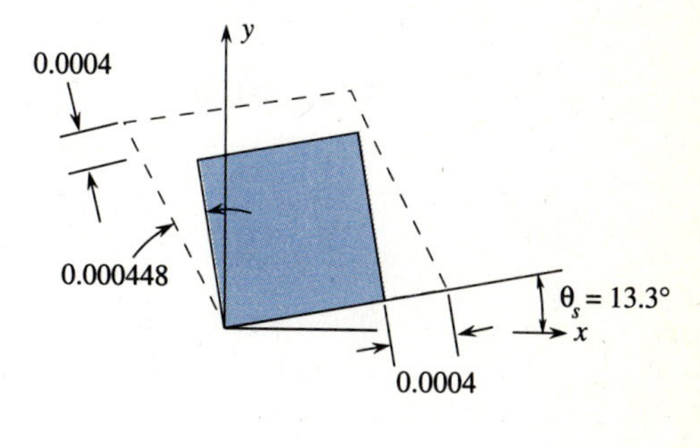

Element with maximum in–plane shear strain

EXAMPLE 8.20

An equiangular strain rosette is attached to the surface of a crane hook as shown in Figure 8.74. Under the maximum allowable lifted load of 10,000 lb, the gauges record the following strains $\epsilon_{0°} = \epsilon_a = 600 \times 10^{-6}$, $\epsilon_{60°} = \epsilon_b = 500 \times 10^{-6}$, $\epsilon_{120°} = \epsilon_c = -200 \times 10^{-6}$. Determine the principal strains and maximum in-plane shear strain (Figure 8.74).

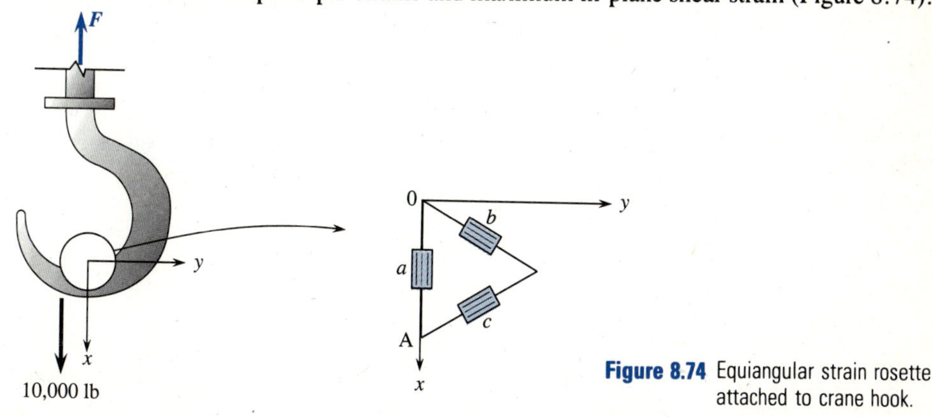

Figure 8.74 Equiangular strain rosette attached to crane hook.

Solution

Using Eqs. (8.82) for a 60° equiangular rosette, we obtain

$$\epsilon_x = \epsilon_a = 600 \times 10^{-6} \text{ in./in.}$$

$$\epsilon_y = \left(\frac{1}{3}\right)(2\epsilon_{60°} + 2\epsilon_{120°} - \epsilon_{0°})$$

$$= \left(\frac{1}{3}\right)[2(500) + 2(-200) - 600] \times 10^{-6} = 0$$

$$\gamma_{xy} = \left(\frac{2}{\sqrt{3}}\right)(\epsilon_{60°} - \epsilon_{120°})$$

$$= \left(\frac{2}{\sqrt{3}}\right)[500 - (-200)] \times 10^{-6} = 800 \times 10^{-6} \text{ rad}$$

Using Mohr's circle method (where for convenience all numbers are assumed to be in microinches per inch), we have

1. $X = \left(600, -\frac{808}{2}\right) \qquad Y = \left(0, \frac{808}{2}\right)$

   the X point is plotted below the horizontal axis, where $\gamma_{xy}/2$ is plotted as a negative value because the line OA associated with the x-direction has rotated counterclockwise. (The original 90° angle between x and y axes has decreased.)

2. $C = \dfrac{600 + 0}{2} = 300$

3. Plot the X, Y, and C points and sketch circle.

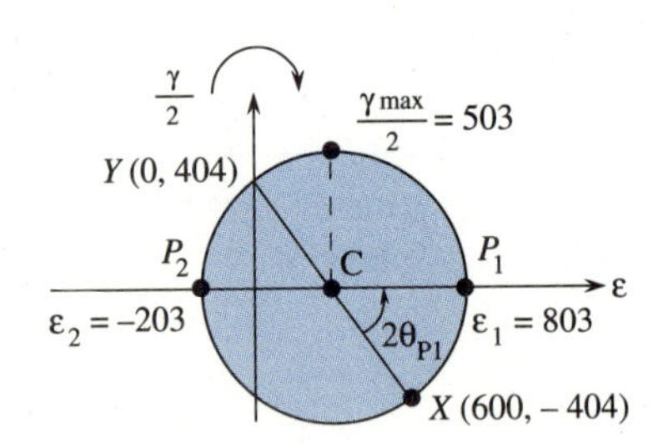

4. $\dfrac{\gamma_{max}}{2} = R = \sqrt{(300)^2 + (404)^2} = 503$

   $\gamma_{max} = 1006$

5. $\epsilon_1 = C + R = 300 + 503 = 803$

   $\epsilon_2 = C - R = 300 - 503 = -203$

$$\tan 2\theta_{p1} = \frac{404}{300} \qquad 2\theta_{p1} = 53.4°$$

$$\theta_{p1} = 26.7° \text{ counterclockwise from } x\text{-direction}$$

8.10 PLANE STRESS-STRAIN EQUATIONS

Recall from Chapter 2 that for the one-dimensional case, the stress and strain were related through Hooke's law by $\sigma = E\epsilon$, and in Chapter 4, for the case of pure shear, the shear stress was related to the shear strain by $\tau = G\gamma$. We now derive the equations that relate the stresses to the strains for the case of plane stress and for a material that is homogeneous, isotropic (the material has the same mechanical response regardless of orientation; see Chapter 2) and obeys Hooke's law. Recall that a material that obeys Hooke's law behaves in a linear-elastic manner. These stress-strain relationships are often called *constitutive equations*.

We begin by considering the element of material subjected to plane stresses σ_x, σ_y, and τ_{xy} shown in Figure 8.75.

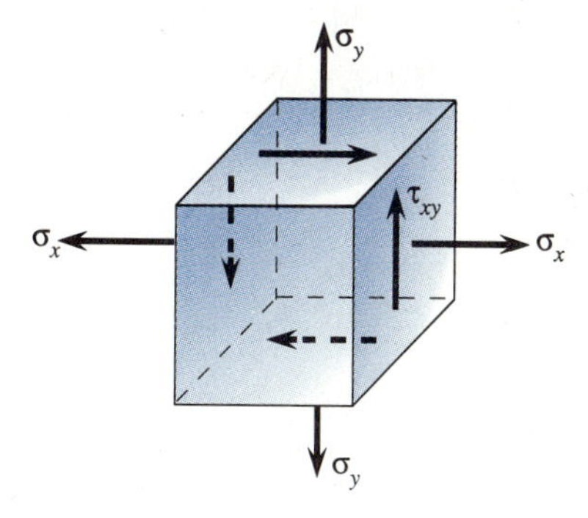

Figure 8.75 Element subjected to plane stress.

The separate effects of the positive normal stresses causing deformations is shown in Figure 8.76 where, for convenience, a unit cube is used.

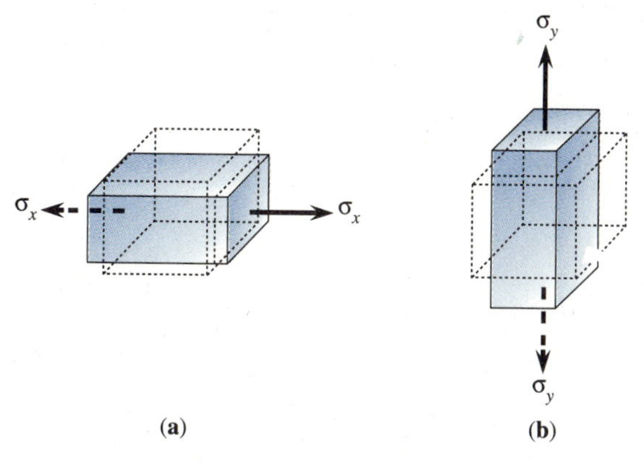

Figure 8.76 (a) Unit cube subjected to stress σ_x and (b) unit cube subjected to stress σ_y.

In Figure 8.76(a), the positive stress σ_x produces a strain ϵ_x equal to σ_x/E because the loading is uniaxial, and a strain ϵ_y equal to $-\nu\epsilon_x$ due to the Poisson effect. That is, due to σ_x, we have

$$\epsilon_x = \frac{\sigma_x}{E} \qquad \epsilon_y = -\nu\epsilon_x = -\nu\frac{\sigma_x}{E} \tag{8.83}$$

Similarly, the stress σ_y (Figure 8.76(b)) produces

$$\epsilon_y = \frac{\sigma_y}{E} \qquad \epsilon_x = -\nu\epsilon_y = -\nu\frac{\sigma_y}{E} \tag{8.84}$$

In Eqs. (8.83) and (8.84), the same E and ν are used because the material is assumed to be isotropic. Also the axial stresses do not distort the cube, that is, the original right angles between the sides before stress remain right angles after the cube is stressed. Since we are considering a linear-elastic material behavior, even under the superimposed stress states, we can add the axial strains in the x-direction from Eqs. (8.83) and (8.84) to obtain

$$\epsilon_x = \frac{\sigma_x}{E} - \nu\frac{\sigma_y}{E} \tag{8.85}$$

Similarly, in the y-direction, we obtain

$$\epsilon_y = \frac{\sigma_y}{E} - \nu\frac{\sigma_x}{E} \tag{8.86}$$

The shear stress τ_{xy} does not cause any axial strains ϵ_x or ϵ_y.

Considering now the effect of the shear stress τ_{xy}, we observe from the unit cube in Figure 8.77 that τ_{xy} distorts the element such that each z-face becomes a rhombus. The x- and y-faces remain square as no other shear stresses act on those sides.

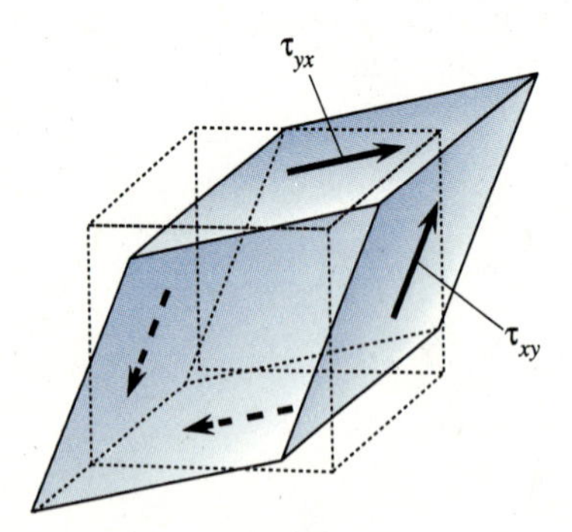

Figure 8.77 Unit cube subjected to shear stress τ_{xy}.

By Hooke's law for shear, we obtain the shear strain in terms of the shear stress as

$$\gamma_{xy} = \frac{\tau_{xy}}{G} \tag{8.87}$$

The normal stresses σ_x and σ_y do not effect the shear strain γ_{xy}. Equations (8.85), (8.86), and (8.87) are the final strains due to all plane stresses (σ_x, σ_y, and τ_{xy}) acting simultaneously.

We can now solve Eqs. (8.85) and (8.86) simultaneously for the normal stresses in terms of the strains, and rewrite Eq. (8.87) in terms of the shear stress to obtain

$$\sigma_x = \frac{E}{1-\nu^2}(\epsilon_x + \nu\epsilon_y)$$

$$\sigma_y = \frac{E}{1-\nu^2}(\epsilon_y + \nu\epsilon_x) \qquad (8.88)$$

$$\tau_{xy} = G\gamma_{xy}$$

All other stresses (σ_z, τ_{xz}, and τ_{yz}) are zero. Equations (8.88) are the final stress-strain or Hooke's law equations for an isotropic material under plane stress. Equations (8.88), written for the x- and y-directed stresses, also apply for principal stresses. That is, replacing say subscript x by 1 and subscript y by 2, the principal stresses and strains are related by

$$\sigma_1 = \frac{E}{1-\nu^2}(\epsilon_1 + \nu\epsilon_2)$$

$$\sigma_2 = \frac{E}{1-\nu^2}(\epsilon_2 + \nu\epsilon_1) \qquad (8.89)$$

Example Problems 8.21 through 8.23 illustrate use of Hooke's law and Mohr's circle to solve problems in stress analysis.

EXAMPLE 8.21

A torsion bar with a diameter of 50 mm used in an automobile suspension system has a strain gauge attached to it at a 60° angle from its axis as shown (Figure 8.78). Determine the torque T transmitted by the shaft if the gauge reads 250 μ (μ = 10^{-6}). The bar is steel with G = 75 GPa.

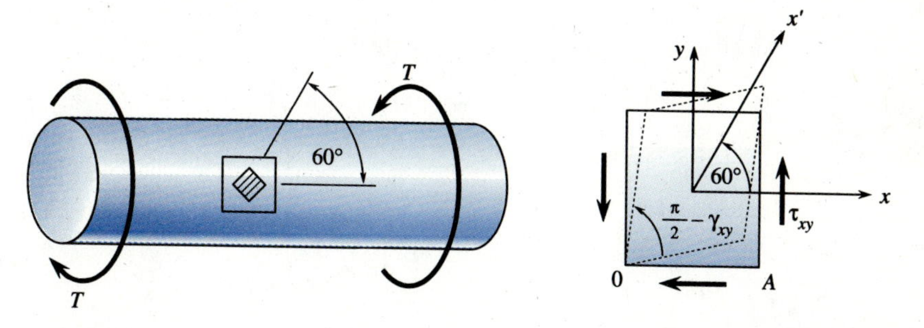

Figure 8.78 Torsion bar with strain gauge attached to it.

Solution

Following the steps for constructing Mohr's circle for strain, we have

1. $X = \left(0, -\frac{\gamma_{xy}}{2}\right) \qquad Y = \left(0, \frac{\gamma_{xy}}{2}\right)$

   where X is plotted below the horizontal axis because the side OA associated with ϵ_x rotates counterclockwise. Also using $\tau_{xy} = G\gamma_{xy}$, we observe from the stressed element that τ_{xy} on the x-face causes a counterclockwise moment and would be plotted as a negative value on the Mohr's circle for stress.

2. $C = \epsilon_{ave} = 0$

3. Plot the X, Y, and C points and sketch Mohr's circle.

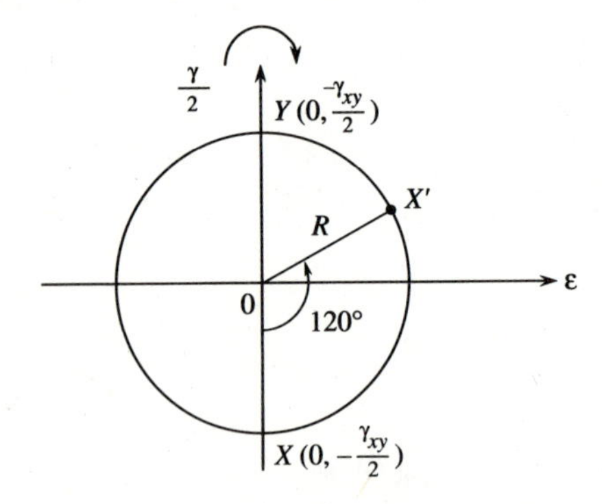

We must rotate 120° (60° × 2) on the circle to go from $0X$ (ϵ_x direction) to $0X'$ ($\epsilon_{x'}$ direction). From the given $\epsilon_{x'}$ and use of the circle, we have

$$\epsilon_{x'} = 250\ \mu = R\cos 30°$$

or

$$R = 289\ \mu$$

Therefore, since

$$\frac{\gamma_{xy}}{2} = R = 289\ \mu$$

$$\gamma_{xy} = 578\ \mu$$

By Hooke's law, Eq. (8.88), we obtain

$$\tau_{xy} = G\gamma_{xy} = (75\ \text{GPa})(578\ \mu) = 43.3\ \text{MPa}$$
$$= 43.3 \times 10^6\ \text{N/m}^2$$

Using the torsion formula, Eq. (4.14)

$$\tau_{xy} = \frac{Tr}{J}$$

$$T = \frac{\tau_{xy}J}{r}$$

and

$$J = (\pi/2)r^4 = (\pi/2)(0.025)^4 = 6.14 \times 10^{-7}\ \text{m}^4$$

$$T = (43.3 \times 10^{+6})(6.14 \times 10^{-7})/(0.025)$$

$$T = 1063\ \text{N} \cdot \text{m}$$

$$T = 1.063\ \text{kN} \cdot \text{m}$$

EXAMPLE 8.22

A single gauge forming an angle of 15° with the horizontal plane is used to determine the gauge pressure in a cylindrical tank shown (Figure 8.79). The tank is 6 mm thick and 500 mm in diameter. It is made of steel with an $E = 200$ GPa and $\nu = 0.29$. The strain gauge reading is 350 μ. Determine the pressure inside the tank.

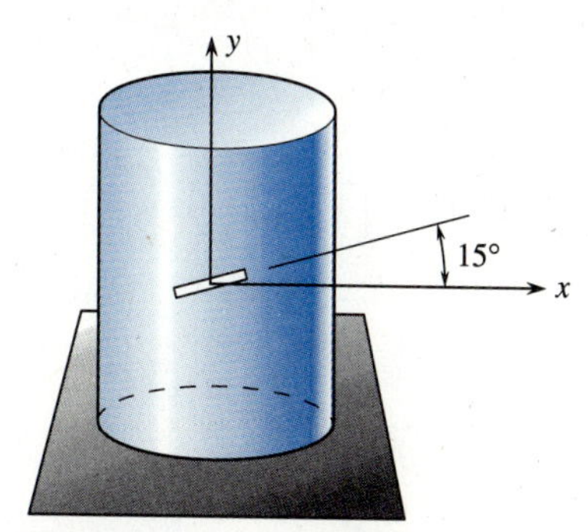

Figure 8.79 Strain gauge attached to pressurized tank.

Solution

For a cylindrical pressure vessel, the principal stresses are related by

$$\sigma_2 = \frac{1}{2}\sigma_1$$

From Hooke's law, Eq. (8.89), relating principal stresses to strains, we have

$$\epsilon_x = \epsilon_1 = \frac{\sigma_1 - \nu\sigma_2}{E} = \frac{\sigma_1 - \nu\left(\frac{\sigma_1}{2}\right)}{E} = \frac{\sigma_1}{E}\left(1 - \frac{\nu}{2}\right)$$

$$= \frac{\sigma_1}{E}\left(1 - \frac{0.29}{2}\right) = 0.855\frac{\sigma_1}{E}$$

$$\epsilon_y = \epsilon_2 = \frac{\sigma_2 - \nu\sigma_1}{E} = \frac{\frac{\sigma_1}{2} - \nu\sigma_1}{E} = \frac{\sigma_1}{E}\left(\frac{1}{2} - \nu\right)$$

$$= \frac{\sigma_1}{E}\left(\frac{1}{2} - 0.29\right) = 0.210\frac{\sigma_1}{E}$$

$\gamma_{xy} = 0$ because x and y are principal directions. Using Mohr's circle for strain, we have

1. $X = (\epsilon_x, 0) = \left(0.855\dfrac{\sigma_1}{E}, 0\right)$

   $Y = (\epsilon_y, 0) = \left(0.210\dfrac{\sigma_1}{E}, 0\right)$

2. $C = \epsilon_{\text{ave}} = \dfrac{1}{2}(\epsilon_x + \epsilon_y) = \dfrac{1}{2}(0.855 + 0.210)\dfrac{\sigma_1}{E} = 0.533\dfrac{\sigma_1}{E}$

3. Plot X, Y, and C and sketch the circle.

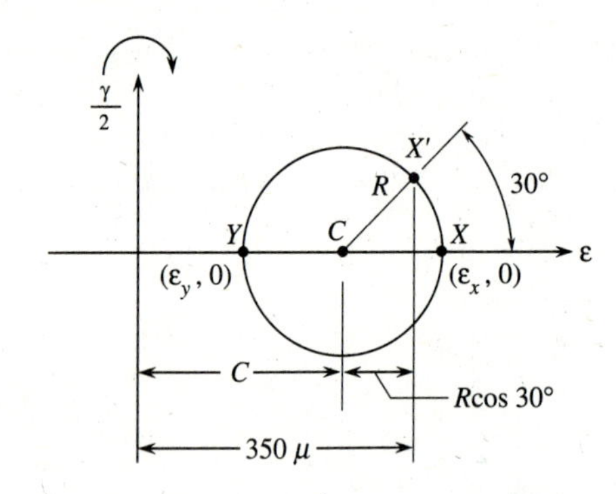

From the circle, we have

4. $R = \dfrac{1}{2}(\epsilon_x - \epsilon_y) = \dfrac{1}{2}(0.855 - 0.210)\dfrac{\sigma_1}{E} = 0.323\dfrac{\sigma_1}{E}$

5. $\epsilon_{x'} = 350\ \mu = C + R\cos 30°$

   or

$$350\ \mu = 0.533\frac{\sigma_1}{E} + 0.323\frac{\sigma_1}{E}\cos 30°$$

$$0.812\sigma_1 = (350 \times 10^{-6})E$$

$$\sigma_1 = (431 \times 10^{-6})(200 \times 10^9)$$

$$\sigma_1 = 86.23 \text{ MPa}$$

By Eq. (7.10) we have

$$\sigma_1 = \frac{pr_i}{t} \qquad p = \frac{\sigma_1 t}{r_i} = (86.23 \text{ MPa})\frac{6 \text{ mm}}{250 \text{ mm}} = 2.07 \text{ MPa}$$

EXAMPLE 8.23

The safety of a bridge across a river is investigated by inspectors who determine strains $\epsilon_x = 1200\ \mu$, $\epsilon_y = -200\ \mu$, and $\gamma_{xy} = 1000\ \mu$. The safety code indicates that the maximum allowable stresses are $\sigma = 20{,}000$ psi and $\tau = 15{,}000$ psi. Based on this strength criteria, is the bridge safe? Use $E = 29 \times 10^6$ psi, $G = 12 \times 10^6$ psi, and $\nu = 0.25$.

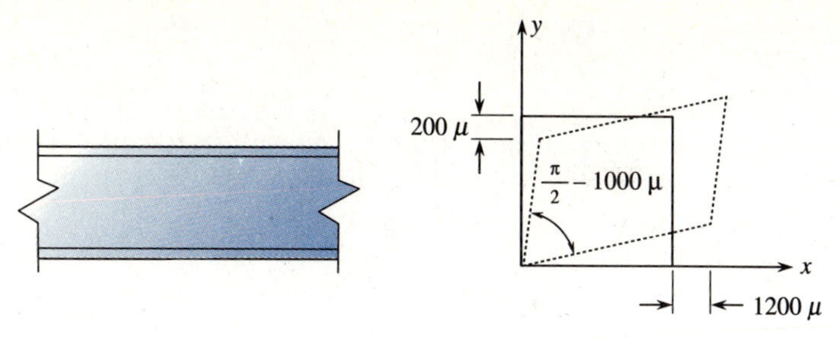

Figure 8.80 Bridge beam with in-plane strains shown.

Solution

Using Mohr's circle for strain, we have (all numbers are times 10^{-6})

1. $X = (1200, -500) \qquad Y = (-200, 500)$

2. $C = \epsilon_{ave} = \dfrac{(1200 - 200)}{2} = 500$

3. Plot X, Y, and C and sketch the circle.

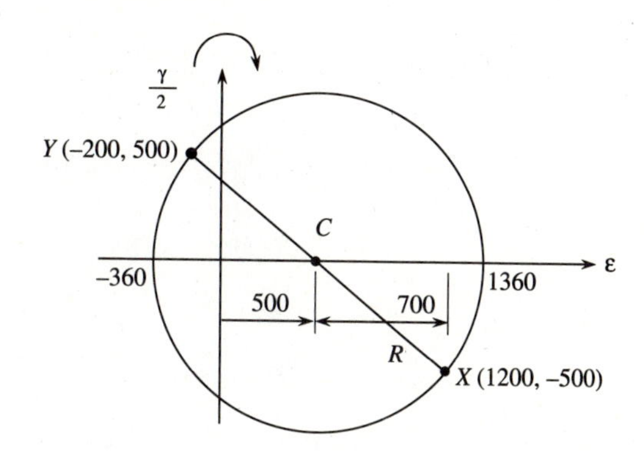

From the circle, we have

4. $R = \sqrt{700^2 + 500^2} = 860$

5. $\epsilon_1 = \epsilon_{max} = C + R = 500 + 860 = 1360$

   $\epsilon_2 = \epsilon_{min} = C - R = 500 - 860 = -360$

Now using Eq. (8.89), we obtain the maximum principal stress as

$$\sigma_{max} = \frac{E}{1 - \nu^2}(\epsilon_1 + \nu\epsilon_2)$$

$$= \frac{29 \times 10^6}{1-(0.25)^2}[1360 - 0.25(360)] \times 10^{-6}$$

$$\sigma_{max} = 39{,}285 \text{ psi} > 20{,}000 \text{ psi}$$

Therefore, the bridge is not safe.

8.11 THREE-DIMENSIONAL STRESS-STRAIN EQUATIONS AND DILATATION

For the general three-dimensional state of stress at a point, represented by Figure 8.81, we can use the same procedure as in Section 8.10 to obtain the stress-strain relations or constitutive equations for an isotropic, elastic material.

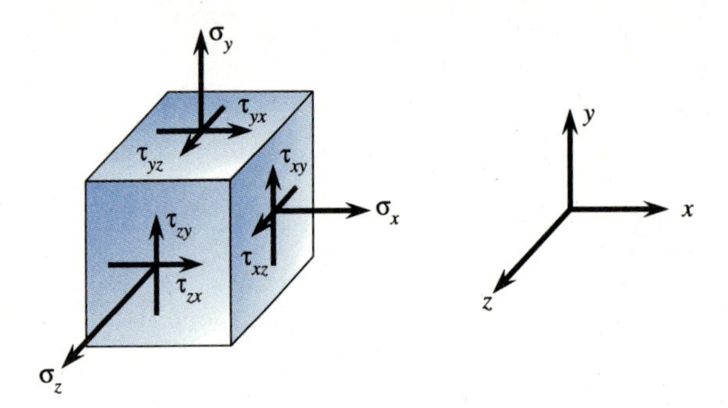

Figure 8.81 Body subjected to general state of stress.

For instance, by merely adding in the σ_z contribution to Eq. (8.88), we obtain

$$\begin{aligned}\sigma_x &= \frac{E}{1-\nu^2}(\epsilon_x + \nu\epsilon_y + \nu\epsilon_z)\\ \sigma_y &= \frac{E}{1-\nu^2}(\nu\epsilon_x + \epsilon_y + \nu\epsilon_z)\\ \sigma_z &= \frac{E}{1-\nu^2}(\nu\epsilon_x + \nu\epsilon_y + \epsilon_z)\end{aligned} \tag{8.90}$$

The additional shear stress-strain equations result in two additional equations to the one in Eq. (8.88), such that

$$\begin{aligned}\tau_{xy} &= G\gamma_{xy}\\ \tau_{xz} &= G\gamma_{xz}\\ \tau_{yz} &= G\gamma_{yz}\end{aligned} \tag{8.91}$$

Equations (8.90) and (8.91) represent the general three-dimensional stress-strain equations for isotropic materials behaving elastically. These equations are also called the *generalized Hooke's law equations*.

We will now determine an expression for *dilatation*, which is defined to be the change in volume per unit volume. This concept is particularly important in helping to understand the limits on Poisson's ratio and in determining the distortion energy (see Section 10.8 for the formal development of the distortion energy) which is used in the development of the maximum distortion energy failure theory in Section 8.7.

Consider, for convenience, a unit cube of material (Figure 8.82).

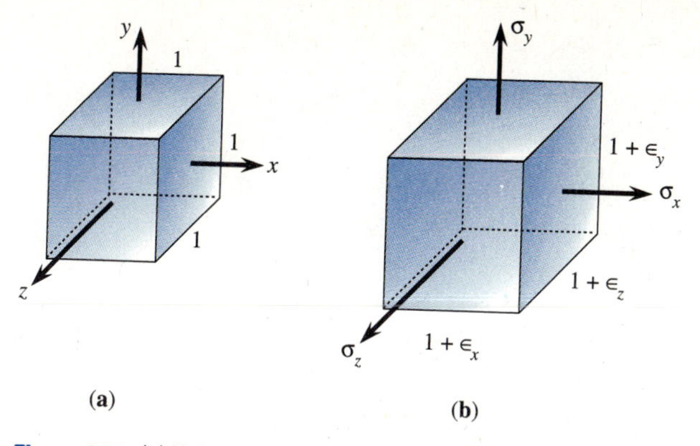

Figure 8.82 (a) Unit cube and
(b) deformed unit cube after subjected to normal stresses.

Now subject the cube to only normal stresses σ_x, σ_y, and σ_z. The sides of the cube now become $1 + \epsilon_x$, $1 + \epsilon_y$, and $1 + \epsilon_z$, where ϵ_x, ϵ_y, and ϵ_z are the normal strains, respectively, in the x-, y-, and z-directions (see Figure 8.82). The cube now deforms into a parallelepiped of volume v given by

$$v = (1 + \epsilon_x)(1 + \epsilon_y)(1 + \epsilon_z) \tag{8.92}$$

Multiplying the three terms together and considering the strains to be small quantities (much smaller than one), we can neglect products of the strains such as ϵ_x, ϵ_y, etc., to obtain

$$v = 1 + \epsilon_x + \epsilon_y + \epsilon_z \tag{8.93}$$

Now defining the dilatation e as the change in volume per unit volume, we have

$$e = \frac{v - 1}{1} = \epsilon_x + \epsilon_y + \epsilon_z \tag{8.94}$$

where the original volume was that of unity, as we started with a cube of unit sides.

Next, we use Eq. (8.90) and express the normal strains in terms of the normal stresses as

$$\begin{aligned}
\epsilon_x &= \frac{1}{E}(\sigma_x - \nu\sigma_y - \nu\sigma_z) \\
\epsilon_y &= \frac{1}{E}(-\nu\sigma_x + \sigma_y - \nu\sigma_z) \\
\epsilon_z &= \frac{1}{E}(-\nu\sigma_x - \nu\sigma_y + \sigma_z)
\end{aligned} \tag{8.95}$$

On substituting these normal strains into Eq. (8.94), the dilatation is expressed in terms of the normal stresses as

$$e = \frac{1-2\nu}{E}(\sigma_x + \sigma_y + \sigma_z) \tag{8.96}$$

In the special case of $\sigma_x = \sigma_y = \sigma_z = -p$, we have

$$e = -\frac{3(1-2\nu)}{E}p \tag{8.97}$$

This case corresponds to a hydrostatic pressure all around the body, such as when a body is immersed in fluid. We define another constant K_b as

$$K_b = \frac{E}{3(1-2\nu)} \tag{8.98}$$

Using Eq. (8.98) in Eq. (8.97), the dilatation becomes

$$e = -\frac{p}{K_b} \tag{8.99}$$

The constant K_b is called the *bulk modulus* or modulus of compression of the material of the body. Since e has no units, K_b has the same units as p, that is, psi or pascals.

The concept of bulk modulus can be used to determine the upper limit of Poisson's ratio. A normal material subjected to hydrostatic pressure can only decrease in volume. Therefore, the dilatation e in Eq. (8.99) must be negative and the bulk modulus K_b must then be positive since p is considered to be a positive number. Then from Eq. (8.98), since K_b is positive, the value of Poisson's ratio must be less than 1/2, that is, $1 - 2\nu$ must be greater than zero. Also by the definition of Poisson's ratio (see Eq. (2.12)), we know it is a positive quantity. Therefore, the limits on Poisson's ratio are

$$0 < \nu < 1/2 \tag{8.100}$$

An ideally incompressible material would have no volume change ($e = 0$). Therefore, from Eq. (8.99), $K_b = \infty$ and from Eq. (8.98), $\nu = 1/2$. Finally, from Eq. (8.96), if we stretch a material with $\sigma_x > 0$ and $\sigma_y = \sigma_z = 0$, the dilatation will be positive and the volume will increase (see Eq. (8.94)).

8.12 SUMMARY OF IMPORTANT DEFINITIONS AND EQUATIONS

1. Plane stress transformation equations

$$\sigma_{x'} = \frac{\sigma_x + \sigma_y}{2} + \frac{\sigma_x - \sigma_y}{2}\cos 2\theta + \tau_{xy}\sin 2\theta \tag{8.5}$$

$$\tau_{x'y'} = -\frac{\sigma_x - \sigma_y}{2}\sin 2\theta + \tau_{xy}\cos 2\theta \tag{8.6}$$

2. Principal stresses

$$\sigma_{1,2} = \frac{\sigma_x + \sigma_y}{2} \pm \sqrt{\left(\frac{\sigma_x - \sigma_y}{2}\right)^2 + (\tau_{xy})^2} \tag{8.15}$$

3. Principal angle

$$\tan 2\theta_p = \frac{2\tau_{xy}}{\sigma_x - \sigma_y} \quad \textbf{(8.9)}$$

4. Maximum shear stress

$$\tau_{max} = \frac{\left(\frac{\sigma_x - \sigma_y}{2}\right)^2 + (\tau_{xy})^2}{R}$$

$$\tau_{max} = \sqrt{\left(\frac{\sigma_x - \sigma_y}{2}\right)^2 + (\tau_{xy})^2} = R \quad \textbf{(8.20)}$$

5. Angle for maximum shear stress

$$\tan 2\theta_s = -\frac{\sigma_x - \sigma_y}{2\tau_{xy}} \quad \textbf{(8.16)}$$

6. Absolute maximum shear stress

$$\tau_{max} = \frac{\sigma_{max} - \sigma_{min}}{2} \quad \textbf{(8.37)}$$

7. Maximum shear stress failure theory

Yielding occurs when

$$\sigma_{max} - \sigma_{min} = \sigma_Y \quad \textbf{(8.41)}$$

Maximum distortion energy theory

Yielding occurs when

$$\sigma_e = (\sigma_1^2 - \sigma_1\sigma_2 + \sigma_2^2)^{1/2} \geq \sigma_Y \quad \textbf{(8.46)–(8.47)}$$

Maximum normal stress theory

Brittle failure occurs when

$$\sigma_1 = \sigma_{ut} \quad \text{or} \quad \sigma_2 = \sigma_{uc} \quad \textbf{(8.53)}$$

Coulomb-Mohr theory

Brittle failure occurs when

For $\sigma_1 > \sigma_2 > 0$ $\quad \sigma_1 = \sigma_{ut}$ **(8.54)**

For $\sigma_1 > 0 > \sigma_2$ $\quad \dfrac{\sigma_1}{\sigma_{ut}} + \dfrac{\sigma_2}{\sigma_{uc}} = 1$ **(8.55)**

For $0 > \sigma_1 > \sigma_2$ $\quad \sigma_2 = \sigma_{uc}$ **(8.56)**

8. Plane strain transformation equations

$$\epsilon_{x'} = \frac{\epsilon_x + \epsilon_y}{2} + \frac{\epsilon_x - \epsilon_y}{2}\cos 2\theta + \frac{\gamma_{xy}}{2}\sin 2\theta \quad \textbf{(8.61)}$$

$$\gamma_{x'y'} = -(\epsilon_x - \epsilon_y)\sin 2\theta + \gamma_{xy}\cos 2\theta \quad \textbf{(8.72)}$$

9. Plane stress-strain equations

$$\sigma_x = \frac{E}{1 - \nu^2}(\epsilon_x + \nu\epsilon_y)$$

$$\sigma_y = \frac{E}{1 - \nu^2}(\epsilon_y + \nu\epsilon_x) \quad \textbf{(8.88)}$$

$$\tau_{xy} = G\gamma_{xy}$$

10. Three-dimensional stress-strain equations

$$\sigma_x = \frac{E}{1 - \nu^2}(\epsilon_x + \nu\epsilon_y + \nu\epsilon_z)$$

$$\sigma_y = \frac{E}{1 - \nu^2}(\nu\epsilon_x + \epsilon_y + \nu\epsilon_z) \quad \textbf{(8.90)}$$

$$\sigma_z = \frac{E}{1 - \nu^2}(\nu\epsilon_x + \nu\epsilon_y + \epsilon_z)$$

$$\tau_{xy} = G\gamma_{xy}$$

$$\tau_{xz} = G\gamma_{xz} \quad \textbf{(8.91)}$$

$$\tau_{yz} = G\gamma_{yz}$$

11. Dilatation

$$e = \frac{1 - 2\nu}{E}(\sigma_x + \sigma_y + \sigma_z) \quad \textbf{(8.96)}$$

12. Bulk modulus

$$K_b = \frac{E}{3(1 - 2\nu)} \quad \textbf{(8.98)}$$

REFERENCE

1. Swanson, J. A., *ANSYS—Engineering Analysis System User's Manual,* Swanson Analysis Systems, Inc., Elizabeth, Pennsylvania.

PROBLEMS

Section 8.2

8.1–8.6 For the states of plane stress shown in Figures P8.1–P8.6, use the stress transformation equations to determine the normal and shear stress acting on plane *a-a*.

Figure P8.1

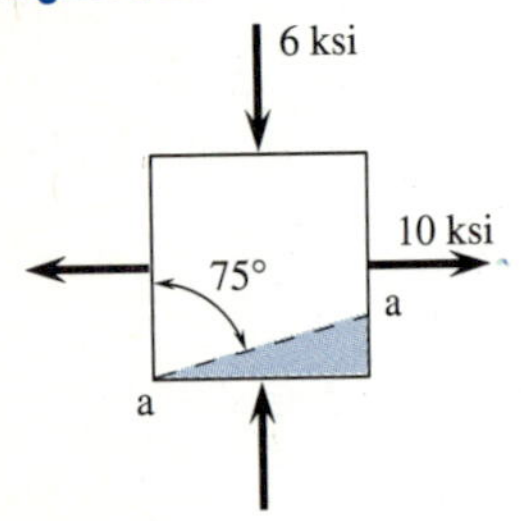

Figure P8.2

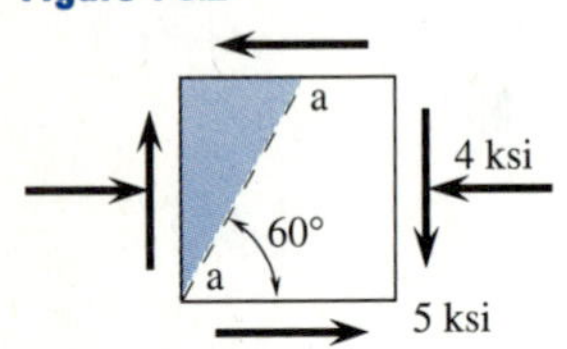

Figure P8.3

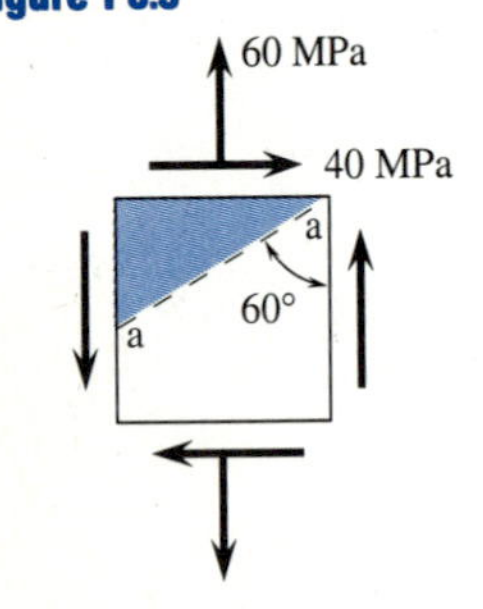

Figure P8.4

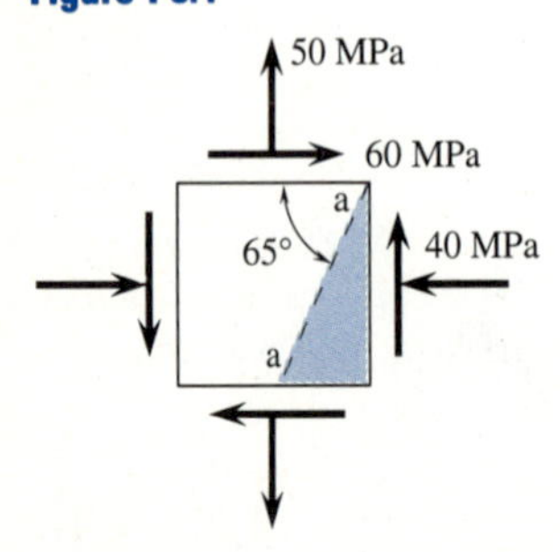

Figure P8.5

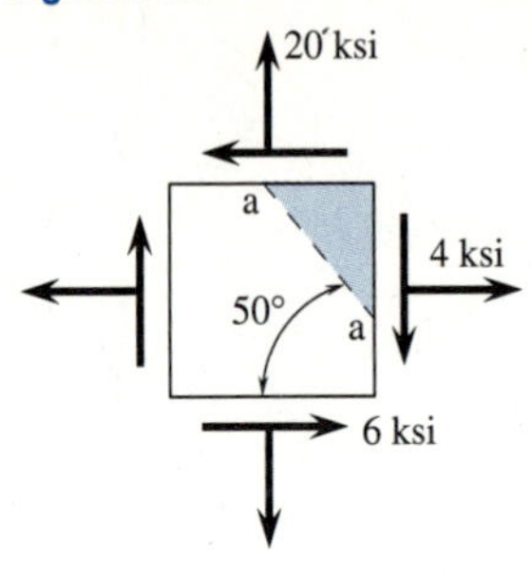

Figure P8.6

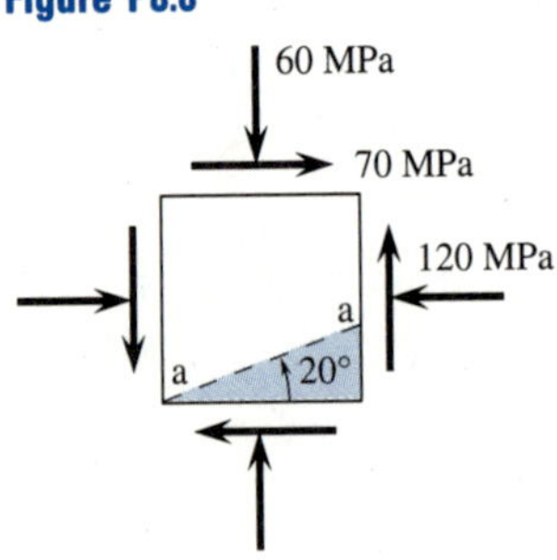

8.7–8.12 For the states of plane stress shown in Figures P8.1–P8.6, use the wedge method to determine the normal and shear stresses acting on plane *a-a*.

8.13–8.20 The elements in plane stress shown in Figures P8.13–P8.20 are subjected to stresses σ_x, σ_y, and τ_{xy}. Determine **(a)** the principal stresses and show them on a properly oriented element, and **(b)** the maximum in-plane shear stress and the corresponding normal stress and show them on a properly oriented element.

Figure P8.13 **Figure P8.14**

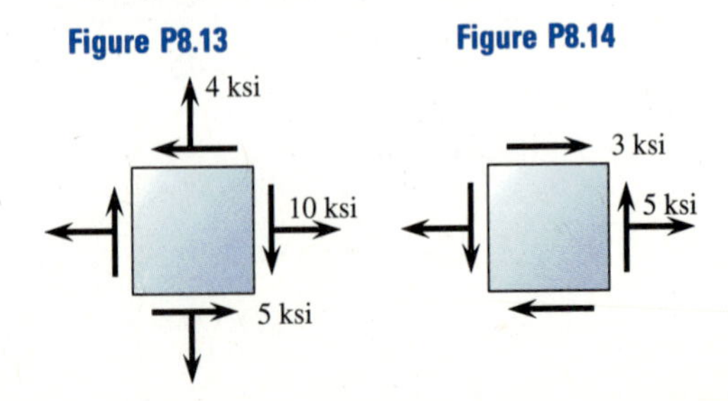

Figure P8.15 **Figure P8.16**

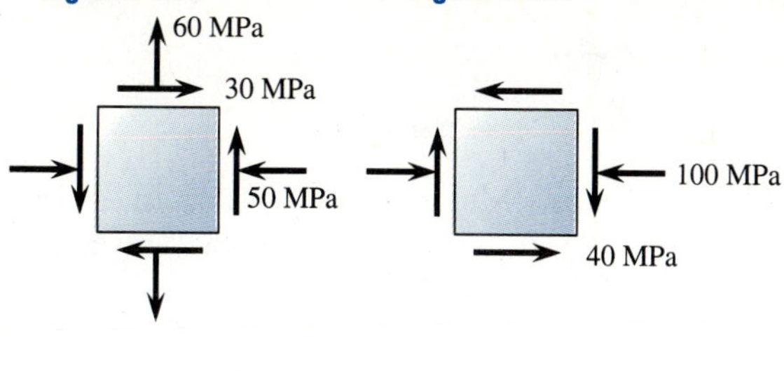

Figure P8.17 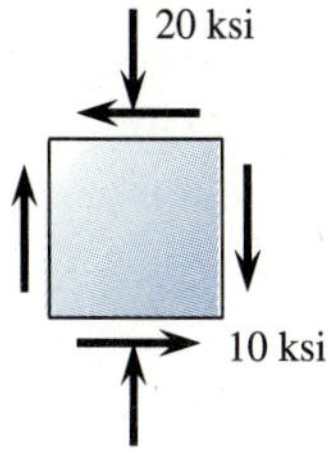 **Figure P8.18**

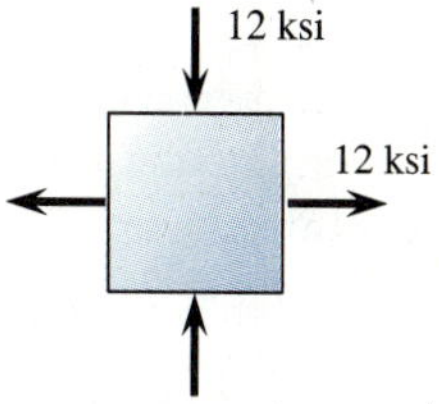

20 ksi
10 ksi
12 ksi
12 ksi

Figure P8.19 **Figure P8.20**

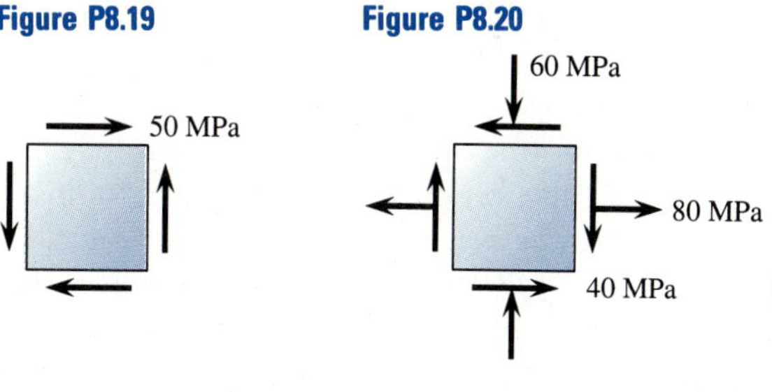

8.21–8.22 The grain of a wooden member forms an angle of 20° with the vertical. For the state of plane stress shown, determine **(a)** the normal stress perpendicular to the grain, and **(b)** the shear stress parallel to the grain.

Figure P8.21 **Figure P8.22**

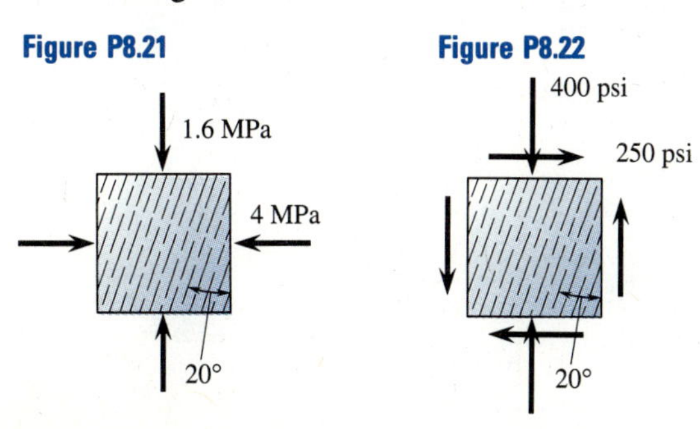

8.23–8.24 For the states of plane stress shown, determine the normal and in-plane shear stresses acting on a plane rotated 30° counterclockwise from the vertical.

Figure P8.23

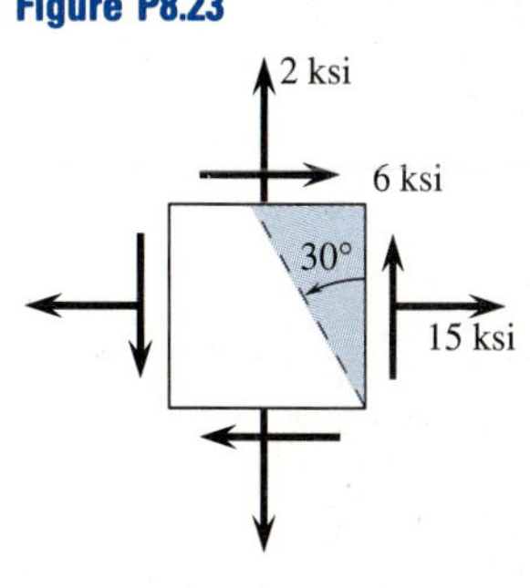

Figure P8.24

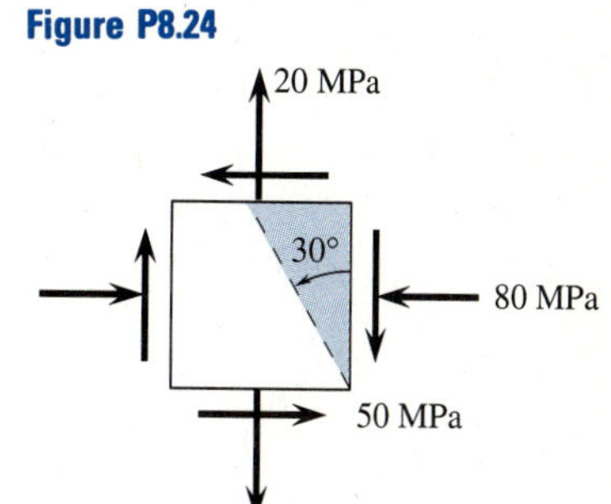

8.25 Two members of uniform cross-section 3 in. × 3 in. are glued together along a plane *a-a*, which forms an angle of 30° with the horizontal. The allowable normal stress in the glued joint is 110 psi and the allowable shear stress is 75 psi. Determine the largest axial force *P* that may be applied.

Figure P8.25

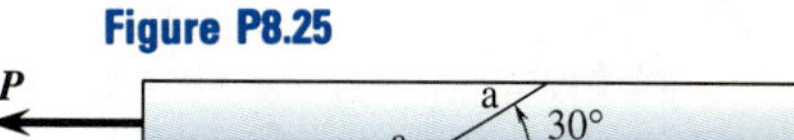

8.26 A steel pipe of 250 mm outside diameter is fabricated of 6-mm-thick plate by welding along a helix that forms an angle of 25° with a horizontal plane. An axial force $P = 300$ kN and a torque $T = 15$ kN · m are applied to the pipe. Determine the normal and shear stresses in the weld.

Figure P8.26

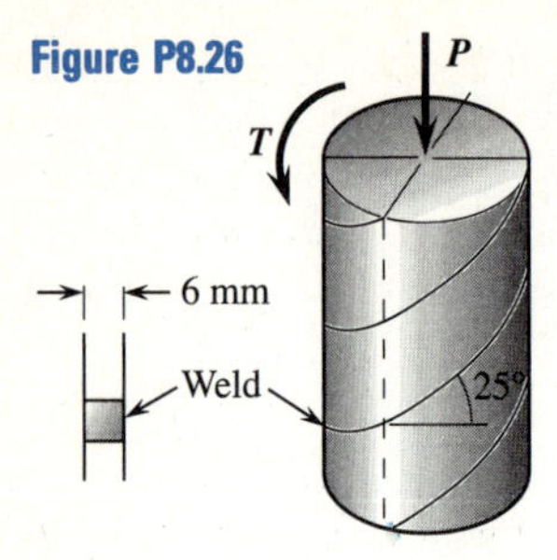

8.27 Solve Problem P8.26 with the direction of the torque reversed.

Section 8.5

8.28 A drive shaft of an automobile is subjected to the vertical forces and torque shown. The diameter of the solid shaft is 25 mm, determine the principal stresses at point B located on the top of the shaft and at point C located on the side of the shaft.

Figure P8.28

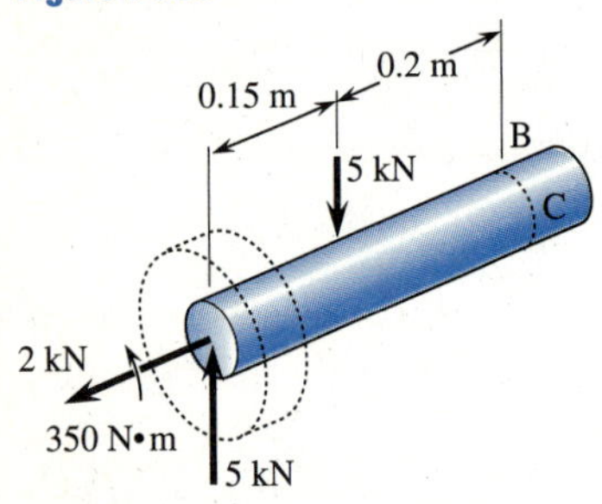

8.29–8.34 Solve Problems P8.1–P8.6 using Mohr's circle method.

8.35–8.42 Solve Problems P8.13–P8.20 using Mohr's circle.

8.43 Construct Mohr's circle for an element in uniaxial stress **(a)** from the circle, derive the following stress transformation equations for uniaxial loading.

$$\sigma_{x'} = \left(\frac{\sigma_x}{2}\right)(1 + \cos 2\theta) \quad \tau_{x'y'} = \left(\frac{\sigma_x}{2}\right)\sin 2\theta$$

Then use the circle to determine **(b)** the principal stresses and **(c)** the maximum shear stress and show them on a properly oriented element.

Figure P8.43

σ_x σ_x

8.44 Construct Mohr's circle for an element in pure shear. **(a)** From the circle derive the following stress transformation equations for pure shear.

$$\sigma_{x'} = \tau_{xy} \sin 2\theta \quad \text{and} \quad \tau_{x'y'} = \tau_{xy} \cos 2\theta$$

(b) Use the circle to obtain the principal stresses and show them on a properly oriented element. **(c)** From the circle, determine the maximum and minimum shear stresses.

Figure P8.44

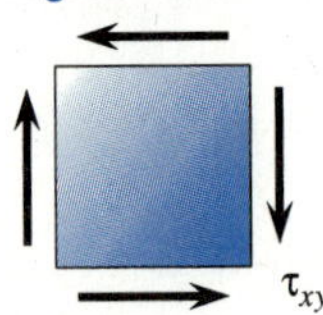

8.45–8.50 An element in plane stress is subjected to stresses σ_x, σ_y, and τ_{xy} as shown in the figures below. Use Mohr's circle method to determine the stresses acting on an element rotated through an angle $\theta = 30°$ clockwise.

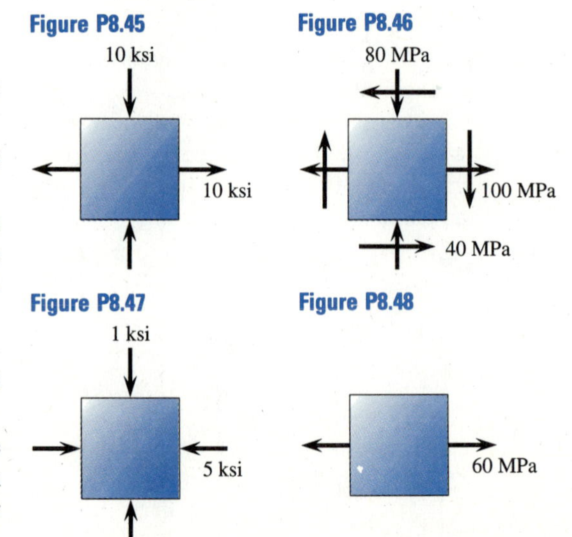

Figure P8.49

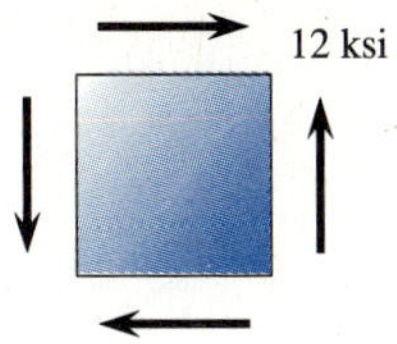

12 ksi

Figure P8.50

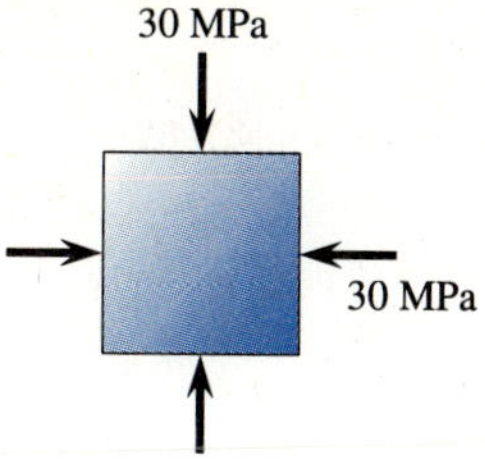

30 MPa

30 MPa

8.51 A solid circular bar is subjected simultaneously to an axial tensile force $P = 30$ kips and a torque $T = 40$ kip · in. as shown. Calculate the maximum tensile, maximum compressive, and maximum in-plane shear stress in the bar. The bar has a diameter of 3 in.

Figure P8.51

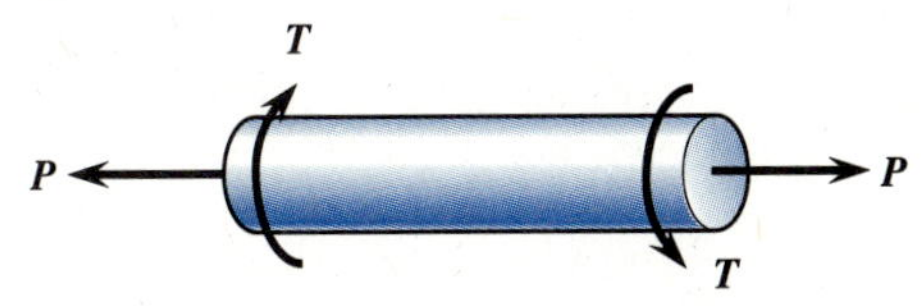

8.52 A socket wrench with an extension shaft attached to it is subjected to a force $F = 50$ lb perpendicular to the socket wrench and a vertical force $P = 30$ lb as shown. Determine the maximum tensile, compressive, and in-plane shear stresses in the extension. The extension is a hollow shaft with outer diameter of 1/2 in. and inner diameter of 3/8 in.

Figure P8.52

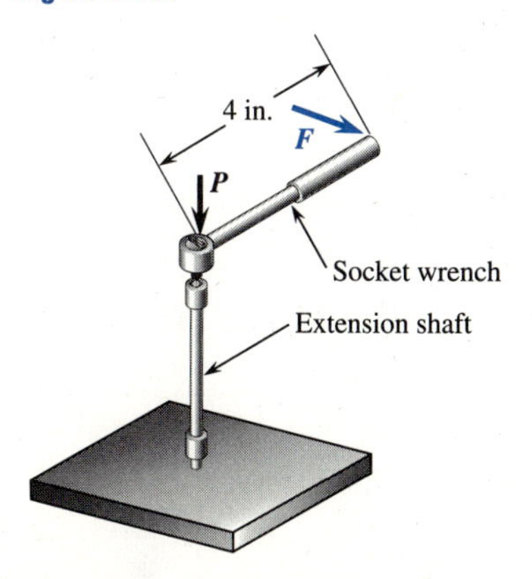

8.53 A closed-ended cylinder of inner radius 8 ft and thickness of 0.25 in. stores natural gas at an internal pressure of 100 psi. Determine **(a)** the hoop stress, **(b)** the longitudinal stress, and **(c)** the maximum in-plane shear stress.

8.54 The cylindrical pressure tank is made of 8-mm steel plate. The pressure inside is 1.2 MN/m^2 and the axial compressive force $P = 1$ MN. The diameter of the tank is 1 m. Determine the maximum tensile and compressive stresses in the tank wall.

Figure P8.54

8.55 A cylindrical tank is fabricated by butt welding 3/4 in. plate with a spiral seam, as shown. The pressure inside the tank is 300 psi, and an axial force of 30,000 lb is applied to the end of the tank through a rigid bearing plate. Determine the maximum normal and shear stresses on the plane of the weld.

Figure P8.55

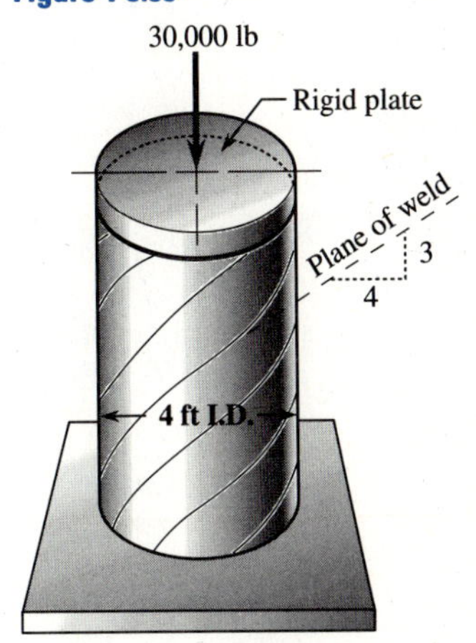

8.56 A cylindrical pressure vessel is subjected to an internal pressure of 50 psi and also an external torque about its longitudinal axis of 10,000 lb · ft. Determine the maximum tensile stress at a point on the outer wall in the cylindrical portion of the tank. The tank outer diameter is 40 in. and tank wall thickness is 0.125 in. Use $E = 30 \times 10^6$ psi. Also use the thin-walled approximation equation for J ($J = 2\pi r^3 t$).

Section 8.6

8.57 For the state of plane stress shown, determine the absolute maximum shear stress when **(a)** $\sigma_x = 10$ ksi, $\sigma_y = 3$ ksi, and $\tau_{xy} = 6$ ksi and **(b)** $\sigma_y = 0$ with the other stresses remaining the same. Consider both in-plane and out of plane shear.

Figure P8.57

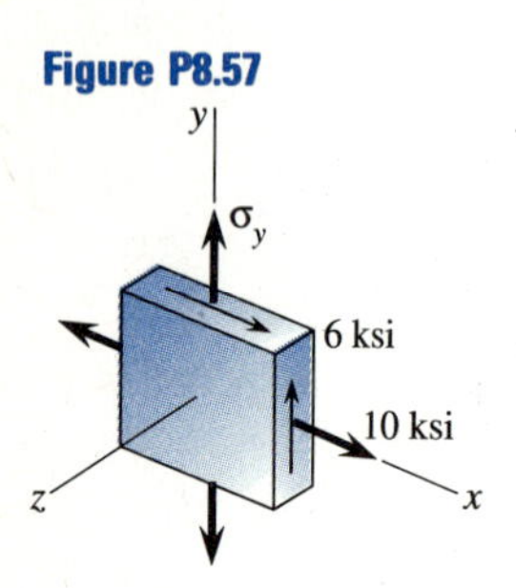

8.58 Solve Problem 8.57 when **(a)** $\sigma_y = 9$ ksi and **(b)** $\sigma_y = -4$ ksi.

8.59–8.64 For the states of plane stress shown, determine the absolute maximum shear stress.

Figure P8.59 **Figure P8.60**

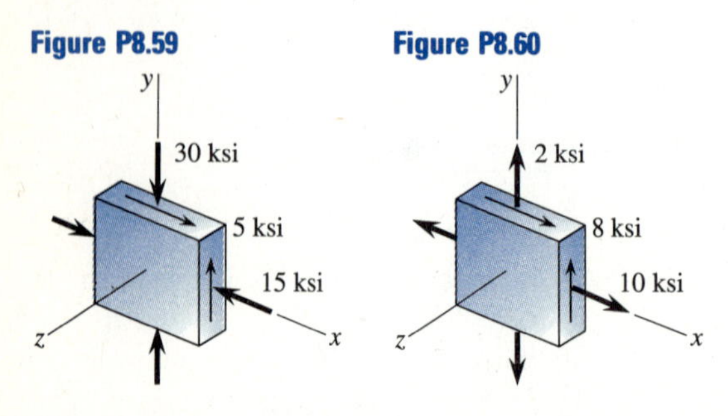

Figure P8.61 **Figure P8.62**

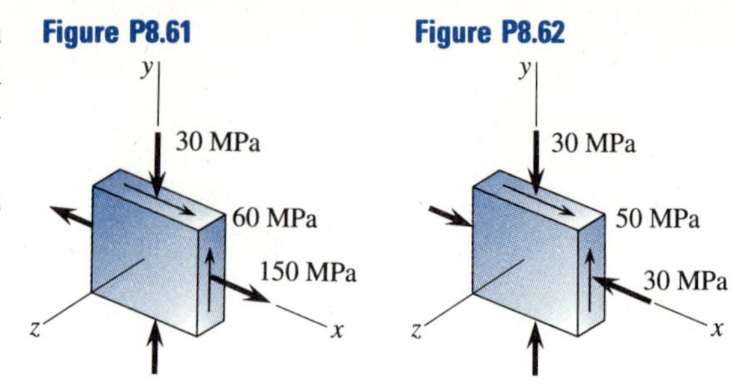

Figure P8.63 **Figure P8.64**

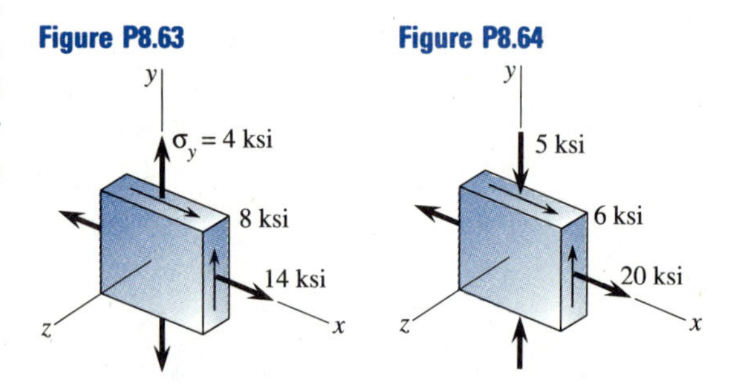

8.65 A standpipe is a thin-walled cylindrical tank holding water. The standpipe is 15 m high and 4 m in inside diameter and has a wall thickness of 20 mm. The density of water is 1000 kg/m^3. Determine the largest normal stress and absolute maximum shear stress at the bottom of the tank when it is full of water.

Figure P8.65

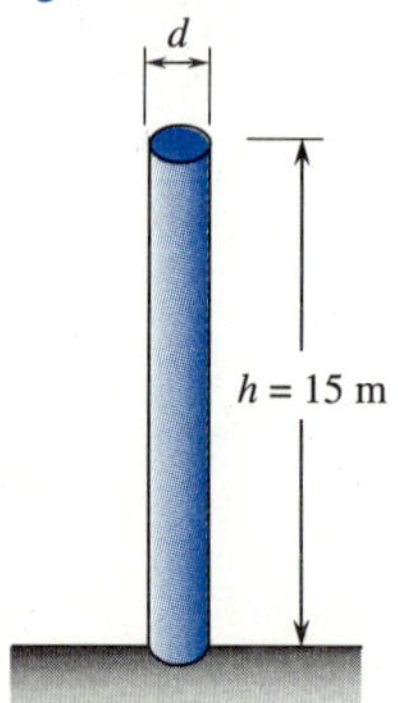

8.66 A thin-walled cylindrical container is subjected to an internal pressure of 300 psi and to an axial tensile force of $P = 75$ kips applied through rigid end plates. The inside diameter is 30 in. and the wall thickness is 1/2 in. Determine the principal stresses and the absolute maximum shear stress on the outer surface of the tank.

Figure P8.66

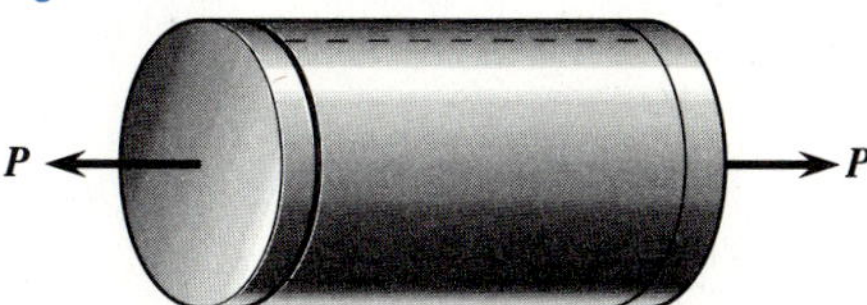

Section 8.7

8.67 A ductile steel has a yield strength of 36 ksi. Determine if it will fail based on (1) the maximum shear stress theory and (2) the distortion-energy theory, for each of the following stress states:

(a) $\sigma_x = 10$ ksi, $\sigma_y = -3$ ksi, $\tau_{xy} = 0$
(b) $\sigma_x = 10$ ksi, $\sigma_y = 3$ ksi, $\tau_{xy} = 0$
(c) $\sigma_x = 0$, $\sigma_y = 0$, $\tau_{xy} = 8$ ksi
(d) $\sigma_x = 15$ ksi, $\sigma_y = 0$, $\tau_{xy} = 5$ ksi

8.68 A ductile machine part has a yield strength of 350 MPa. It is loaded so that the following stresses below result. Determine if the machine part will fail based on (1) the maximum shear stress theory and (2) based on the distortion energy theory.

(a) $\sigma_x = 70$ MPa, $\sigma_y = -70$ MPa, $\tau_{xy} = 0$
(b) $\sigma_x = 0$, $\sigma_y = 0$, $\tau_{xy} = 100$ MPa
(c) $\sigma_x = 70$ MPa, $\sigma_y = 0$, $\tau_{xy} = 100$ MPa
(d) $\sigma_x = -100$ MPa, $\sigma_y = -100$ MPa, $\tau_{xy} = 0$

8.69 A thin-walled cylindrical pressure vessel of American Society for Testing Metals (ASTM) A36 steel has an outer diameter of 100 mm and a wall thickness of 5 mm. Determine the internal pressure that would fail the material by yielding based on the distortion energy theory.

8.70 A 1.5-in.-diameter shaft is made of steel with a yield strength of 36 ksi. Using the maximum shear stress theory, determine the magnitude of the torque T at which yielding first occurs.

8.71 Solve Problem 8.70 using the distortion energy theory.

8.72 A lever is subjected to a downward force of 500 lb. It is keyed to a round bar of 3/4 in. diameter as shown. The round bar is made of 36 ksi yield strength material. Based on the maximum shear stress theory will the bar fail?

Figure P8.72

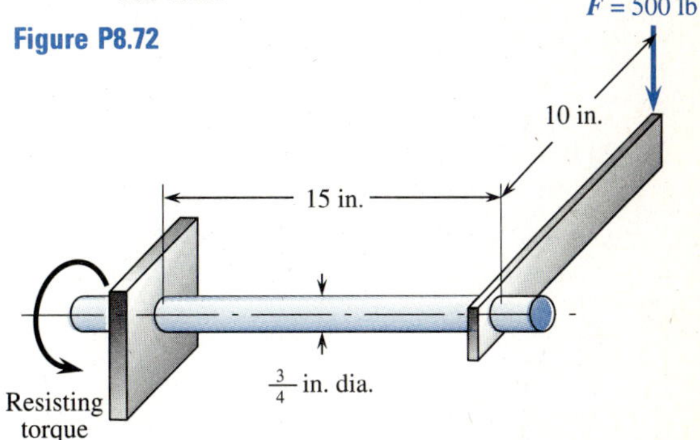

8.73 A solid aluminum bar of 10 mm diameter is built into a wall at one end and subjected to a torsional moment $T = 20$ N·m and an axial load $P = 5$ kN at the free end. Based on the maximum shear stress theory will the bar yield?

Figure P8.73

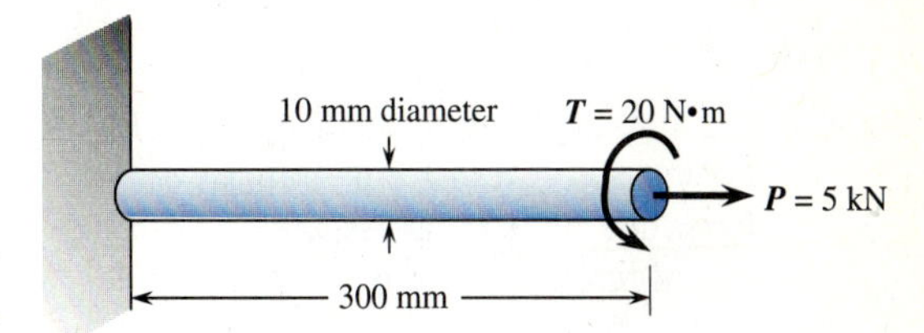

8.74 A thin-walled pressure vessel is made of an aluminum alloy tubing with a yield strength of 138 MPa. The vessel has an outer diameter of 60 mm and a wall thickness of 1.5 mm. What internal pressure would cause the material to yield based on the distortion energy theory?

8.75 A cast iron press, with ultimate strength in tension of 60 MPa and ultimate strength in compression of 170 MPa, is subjected to loadings that result in the following states of plane stress. Based on the Coulomb-Mohr theory, will the press rupture?

(a) $\sigma_x = 10$ MPa, $\sigma_y = 0$, $\tau_{xy} = 10$ MPa
(b) $\sigma_x = 10$ MPa, $\sigma_y = -10$ MPa, $\tau_{xy} = 0$
(c) $\sigma_x = 40$ MPa, $\sigma_y = 40$ MPa, $\tau_{xy} = 0$
(d) $\sigma_x = 0$, $\sigma_y = 0$, $\tau_{xy} = 50$ MPa

8.76 A cast aluminum machine part, with ultimate tensile strength of 10 ksi and ultimate compressive strength of 20 ksi, is subjected to a state of plane stress shown. Determine the normal stress σ_x at which failure may be expected based on the Coulomb-Mohr theory.

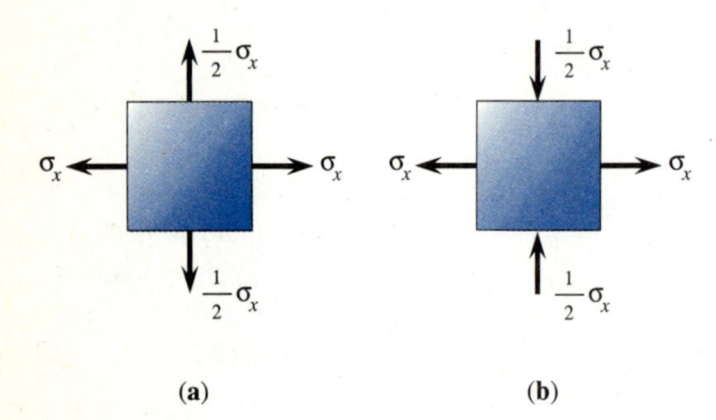

Section 8.8

8.77–8.82 For the states of plane strain listed, use the strain transformation equations to determine the strains associated with axes rotated through angle θ from the x-axis.

	ϵ_x	ϵ_y	γ_{xy}	θ
8.77	500 μ	500 μ	0 μ	30°
8.78	−300 μ	200 μ	180 μ	20°
8.79	1000 μ	500 μ	100 μ	20°
8.80	300 μ	−300 μ	−200 μ	30°
8.81	0 μ	100 μ	−400 μ	15°
8.82	800 μ	500 μ	0 μ	45°

8.83–8.88 Use Mohr's circle method to solve Problems 8.77–8.82.

8.89–8.94 Use the Mohr's circle method to determine the principal strains for Problems 8.77–8.82.

Section 8.9

8.95 During the testing of a beam, the following strains were recorded from a 45° strain rosette.

$\epsilon_a = 700\ \mu \quad \epsilon_b = 50\ \mu \quad \epsilon_c = -50\ \mu$

Determine (1) the principal strains and (2) the in-plane maximum shear strain.

8.96 During the pressure testing of a tank, the following strains were recorded from a rectangular strain rosette.

$\epsilon_a = 400\ \mu \quad \epsilon_b = 600\ \mu \quad \epsilon_c = 900\ \mu$

Determine (1) the principal strains, and (2) the maximum in-plane shear strain.

8.97 A machine part has a delta strain rosette mounted to it before loading. After the maximum load was applied the strains were recorded as

$\epsilon_a = 500\ \mu \quad \epsilon_b = 700\ \mu \quad \epsilon_c = 300\ \mu$

Determine the principal strains and maximum in-plane shear strain.

Section 8.10

8.98 A thin rectangular plate is subjected to normal stresses as shown in Figure P8.98. The strain gauges record $\epsilon_x = 1000\ \mu$ and $\epsilon_y = 400\ \mu$. Assuming that $E = 200$ GPa and $\nu = 0.30$, determine the stresses σ_x and σ_y.

Figure P8.98

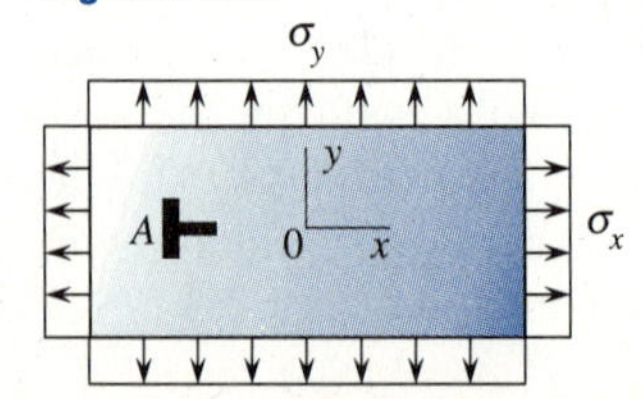

8.99 The strains measured on the outside surface of the cylindrical pressure vessel of Figure P8.99 are:

$$\epsilon_a = 500\ \mu \text{ and } \epsilon_b = 240\ \mu.$$

The angle θ is not known. The diameter of the vessel is 24 in. and the thickness is 1/8 in. The material is ASTM-A242 high-strength steel. Determine **(a)** the stresses σ_a and σ_b in the shell, **(b)** the principal stresses at the outer wall surface, and **(c)** the pressure in the vessel. Notice that when two orthogonal strains are measured the angle θ is not needed to determine the normal stresses.

Figure P8.99

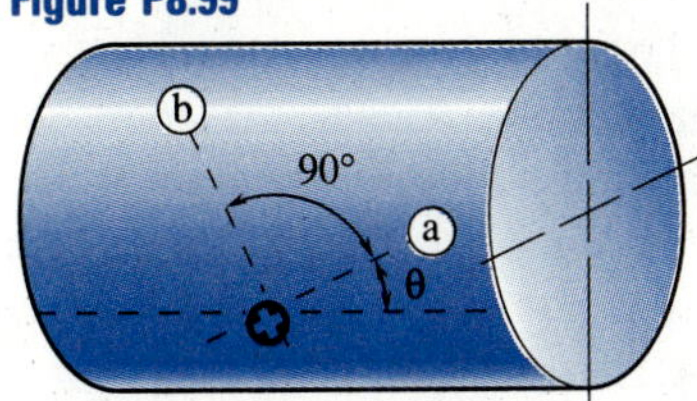

8.100 A light pressure cylinder is made of an aluminum alloy. The cylinder has a 3.5-in. outer diameter, a wall thickness of 0.065 in., a modulus of elasticity of 10.3×10^6 psi, and a Poisson's ratio of 0.334. If the internal pressure is 1860 psi, what is the principal normal strain on an element on the circumference?

8.101 An axle from a large earth moving tractor has a delta strain rosette mounted to it before loading. After loading the strains were recorded as:

$$\epsilon_a = 1000\ \mu \qquad \epsilon_b = 500\ \mu \qquad \epsilon_c = 200\ \mu$$

Determine the principal stresses in the shaft. Let $E = 30 \times 10^6$ psi and $\nu = 0.30$.

8.102 The safety of a bridge across a river is investigated by inspectors who determine strains $\epsilon_x = 1000\ \mu$, $\epsilon_y = -300\ \mu$, and $\gamma_{xy} = 800\ \mu$. The safety code indicates that the maximum allowable stresses are $\sigma = 20{,}000$ psi and $\tau = 15{,}000$ psi. Based on these allowable stresses, is the bridge safe? Use $E = 29 \times 10^6$ psi, $G = 12 \times 10^6$ psi, and $\nu = 0.30$.

Section 8.11

8.103 The following normal strains have been determined on a body:

$$\epsilon_x = 500\ \mu \qquad \epsilon_y = 400\ \mu \qquad \epsilon_z = -200\ \mu$$

Determine the dilatation of the body.

8.104 A submerged tank is subjected to a hydrostatic pressure of $p = 80$ psi. The material of the tank is steel with $E = 29 \times 10^6$ psi and $\nu = 0.30$. Determine the bulk modulus and dilatation of the material.

8.105 A solid steel sphere is subjected to a hyrdostatic pressure p that reduces the volume by 0.03%. (1) Determine the pressure p and (2) determine the bulk modulus K_b for the steel. The steel has $E = 29 \times 10^6$ psi and $\nu = 0.30$.

COMPUTER PROBLEMS

The following problems are suggested to be solved using a programmable calculator, microcomputer, or mainframe computer. It is suggested that you write a simple computer program in BASIC, FORTRAN, PASCAL, or some other language to solve these problems.

C8.1 Write a program to obtain (1) the normal and shear stresses on an arbitrary plane, (2) the principal stresses and the principal angles, and (3) the maximum in-plane shear stress and the angle associated with this shear stress for a plane stress problem. The program should be written to evaluate the normal and shear stress for small increments of θ. The input should include σ_x, σ_y, τ_{xy}, angle θ, and the increment in θ. Check your program using the following data: $\sigma_x = -3000$ psi, $\sigma_y = -12{,}000$ psi, and $\tau_{xy} = 6000$ psi.

Figure C8.1

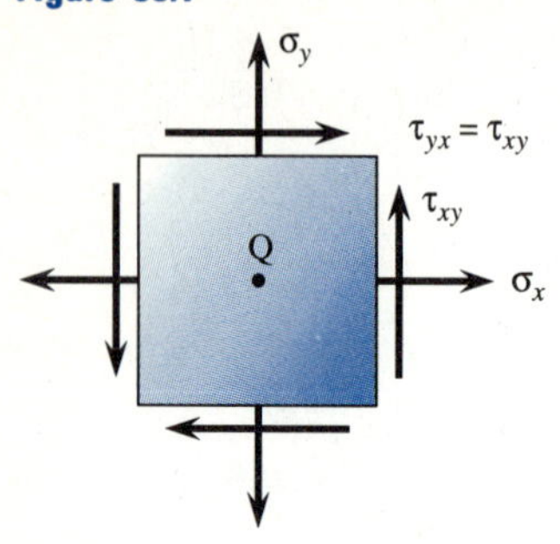

C8.2 Write a program to obtain (1) the normal and shear strain on an arbitrary plane, (2) the principal strains and principal angles, and (3) the maximum in-plane shear strain and the angle associated with this shear strain for a two dimensional stress state. The program should be written to evaluate the normal and shear strain for small increments of θ. The input should include ϵ_x, ϵ_y, γ_{xy}, angle θ, and increment in θ. Check your program using the following data $\epsilon_x = 0.00115$ in./in., $\epsilon_y = -0.00025$ in./in., and $\gamma_{xy} = 0.0009$ rad.

C8.3 Write a program that uses the strain rosette measurements to obtain the normal strains in the x- and y-directions and the shear strain γ_{xy}. Then calculate the stresses σ_x, σ_y, and τ_{xy}. Check your program using the following data: From an equiangular rosette $\epsilon_{0°} = \epsilon_a = 600\ \mu$, $\epsilon_{60°} = \epsilon_b = 500\ \mu$, and $\epsilon_{120°} = \epsilon_c = -200\ \mu$.

C8.4 Write a program to predict the failure of a ductile material. The program should be based on both the maximum shear stress theory and the distortion energy theory criteria for failure. The input should include the in-plane stresses, σ_x, σ_y, and τ_{xy}, and the yield strength of the material. Check your program using the following data: $\sigma_x = -105$ MPa, $\sigma_y = 0$, $\tau_{xy} =$ 105 MPa, and yield strength of 276 MPa.

C8.5 Write a program to predict the failure of a brittle material. The program should be based on both the maximum normal stress theory and the Coulomb-Mohr theory criteria for failure. The input should include the in-plane stresses σ_x, σ_y, and τ_{xy}, and the ultimate strengths in tension and compression. Check your program using the following data: $\sigma_x = 10$ ksi, $\sigma_y = -4$ ksi, and $\tau_{xy} = 0$, and ultimate tensile and compressive strengths of 25 ksi and 95 ksi, respectively.

Column Buckling

9.1 INTRODUCTION

In the previous chapters, we have considered both stress and deflection analysis of structural members. In this chapter, we will introduce the concept of column instability or buckling as a possible mode of failure. In particular, we will discuss a very important type of structural element called a *column*. A column is generally considered to be a long slender member that is loaded in axial compression. We will see that the failure of a column is usually influenced more by its geometrical configuration, the modulus of elasticity E, and boundary support conditions than by its allowable compressive stress.

We first develop Euler's elastic column formula for predicting buckling of pin-ended columns. Then we consider Euler's formula for other end conditions. We treat inelastic column behavior using the tangent modulus and then using the empirical Johnson's formula. Finally, we consider some code formulas used to design columns.

Columns or compression members are frequently found in load-resisting structures. Examples are water tower legs, truss members, bridge piers, building columns, steering linkages, push rods, and frame members in hoisting rigs (Figure 9.1).

The concept of stability is illustrated by the following simple example used in Ref. [1]. Consider a ball supported in succession by a concave surface, a flat horizontal surface, and a convex surface as shown in Figure 9.2.

The concave surface of Figure 9.2(a) represents stable equilibrium. If the ball is perturbed a small amount by some disturbing force from its equilibrium position (the lowest point in the trough), it will return to that equilibrium position on removal of the disturbing force. The horizontal surface of Figure 9.2(b) represents neutral equilibrium since the ball will remain in any displaced position. The convex surface of Figure 9.2(c), however, represents a situation of unstable equilibrium. As long as the ball is exactly balanced at the top it will remain there. However, the slightest perturbation from this precarious position will result in instability failure or a sudden movement of the ball.

Figure 9.1 (a) Water tower legs in compression,
(b) crane truss lower chord members and some diagonals in compression,
(c) bridge piers in bridge under construction,
(d) compression members in small crane hoist, and
(e) buckled column in load frame.

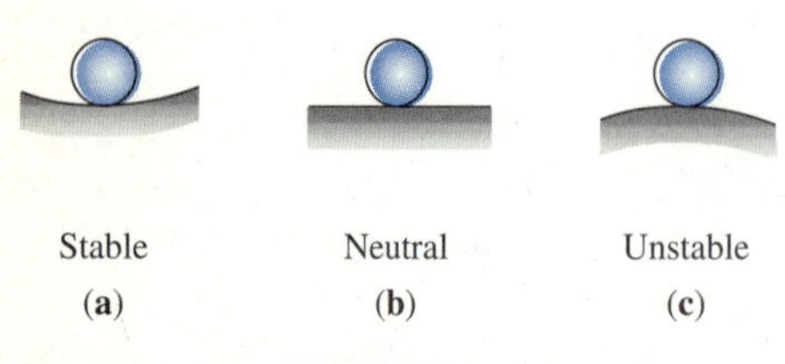

Figure 9.2 Conditions of stability
(a) stable,
(b) neutral and,
(c) unstable.

The buckling failure of a column is analogous to the unstable equilibrium case. That is, when the column compressive load reaches a certain value, called the *critical buckling load,* the lateral displacement increases suddenly (buckling occurs) until permanent deformation (see Figure 9.1(e)) or possible fracture occurs. Therefore, we will observe that the analysis of columns essentially becomes that of determining the applied loads resulting in the so-called neutral equilibrium position just before buckling occurs.

9.2 EULER'S COLUMN FORMULA FOR PINNED ENDS

In this section we will develop a well-known formula for predicting the buckling load of a column. Consider a slender column of uniform cross-section with pinned ends loaded with an axial compressive load P which acts through the centroid of the cross-section (Figure 9.3(a)). The column is assumed to be initially straight and to behave in a linear-elastic manner.

When the load P is small, the column will remain straight. Any sideways or lateral perturbation of the column will result in a return to the initially straight configuration. This situation is analogous to the stable case of Figure 9.2(a).

As the load P is gradually increased, a critical value of P will be reached where the column will be stable in any position, analogous to Figure 9.2(b). Notice that the load remains vertical at all times.

If the load is further increased even slightly beyond this critical value, the column will suddenly deflect catastrophically or snap through (called buckle) when subjected to even the slightest perturbation, analogous to Figure 9.2(c).

You can demonstrate very nicely for yourself this phenomenon by standing a yardstick vertically on a smooth surface and applying an increasing downward compressive force with your fingers at the top of the stick.

In the analysis of columns, we desire to predict the critical axial load, P_{Cr}. To determine P_{Cr}, we must find the load under which the column can be in equilibrium both in the

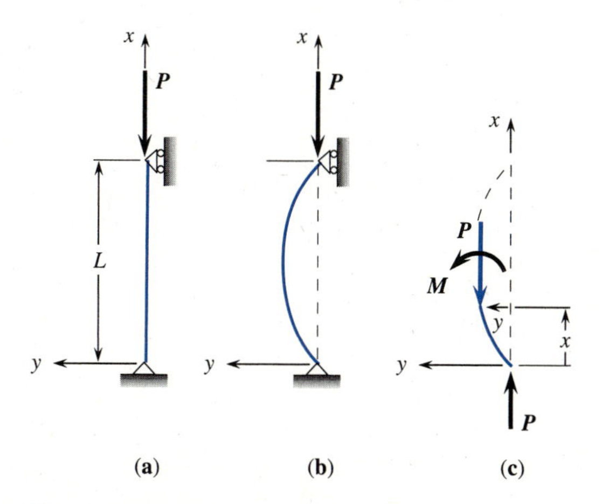

Figure 9.3 Pinned column in equilibrium.

straight and slightly deformed positions. This is the load associated with the neutral stability equilibrium and is called the method of neutral equilibrium. We begin by considering a free-body diagram of an arbitrary section of the laterally deflected pinned column in Figure 9.3(c). The column at some location x has a deflection y as shown. The direction of the deflection is arbitrary. For simplicity, we have assumed the deflection to be positive. For moment equilibrium of the column, the moment at this section will be

$$M = -Py \tag{9.1}$$

The negative sign is consistent with conventions described in Section 5.2. That is, a positive deflection y will produce a negative moment according to the convention of Section 5.2.

From Eq. (6.7), we recall the basic relationship between bending moment and deflection curve given by

$$\frac{M}{EI} = \frac{d^2y}{dx^2} \tag{9.2}$$

Combining Eqs. (9.1) and (9.2) yields

$$\frac{d^2y}{dx^2} + \frac{P}{EI}y = 0 \tag{9.3}$$

Equation (9.3) is the basic second-order, linear, homogeneous differential equation for column buckling. We will now solve this equation for the value of the axial load P which corresponds to the case of neutral stability or impending buckling.

To simplify the solution, we define

$$\beta^2 = \frac{P}{EI} \tag{9.4}$$

Equation (9.3) then becomes

$$\frac{d^2y}{dx^2} + \beta^2 y = 0 \tag{9.5}$$

The solution to Eq. (9.5) is shown from differential equations to be

$$y = C_1 \sin \beta x + C_2 \cos \beta x \tag{9.6}$$

where C_1 and C_2 are constants which are determined from the column end or boundary conditions. For a column with pinned ends, the transverse deflection, y, will be zero at $x = 0$ and at $x = L$. The boundary conditions for both ends pinned are then

$$y(0) = 0 \qquad y(L) = 0 \tag{9.7}$$

Substituting these boundary conditions into Eq. (9.6) yields

$$y(0) = 0 = C_2 \tag{9.8}$$

and

$$y(L) = 0 = C_1 \sin \beta L \tag{9.9}$$

In Eq. (9.9), if C_1 is zero, we will have a trivial solution, $y = 0$, and β, and therefore by Eq. (9.4), P can have any value. This result confirms that a column is in equilibrium under any axial load P as long as the column remains perfectly straight. Therefore, C_1, in general, will not be zero and we have

$$\sin \beta L = 0 \quad \text{or} \quad \beta L = n\pi \tag{9.10}$$

where n is an integer. Therefore, from Eqs. (9.4) and (9.10), we have

$$\beta^2 = \frac{P}{EI} = \frac{n^2\pi^2}{L^2} \tag{9.11}$$

Solving Eq. (9.11) for P gives us the critical value of the applied load as

$$P_{Cr} = n^2 \frac{\pi^2 EI}{L^2} \quad n = 1,2,3, \ldots \tag{9.12}$$

The value of $n = 0$ is disregarded since from Eq. (9.11) this would imply $\beta = P = 0$, a trivial case. At the loads predicted by Eq. (9.12), the column can be in equilibrium in the slightly deformed position. The shape of this deformation is given by

$$y = C_1 \sin \beta x \tag{9.13}$$

(This shape is also called the buckling mode shape.) However, the amplitude cannot be determined since C_1 can have any value when $\sin \beta L = 0$.

Equation (9.13) implies that there are an infinite number of solutions for P_{Cr} corresponding to $n = 1,2,3$, and so on. Although the solutions corresponding to n greater than one do have physical significance, it is the smallest value of P_{Cr} corresponding to $n = 1$ that we are interested in. This equation is

$$P_{Cr} = \frac{\pi^2 EI}{L^2} \tag{9.14}$$

Equation (9.14) is referred to as *Euler's formula* or the *Euler load* and is named after the famous Swiss mathematician Leonhard Euler (1707–1783) and is one of the more significant formulas in structural mechanics. This equation predicts the value of the axial load P at which a state of neutral equilibrium is possible. It is then the load for which the column is no longer in stable equilibrium. It is often called the *buckling load*.

Since columns are loaded axially, the critical stress corresponding to the critical load will be obtained by divided the critical load by the cross-sectional area A of the column. This yields

$$\sigma_{Cr} = \frac{P_{Cr}}{A} = \frac{\pi^2 EI}{L^2 A} \tag{9.15}$$

We can express Eq. (9.15) in an alternate form often used in practice. To do so we introduce the concept of the radius of gyration r of a cross-section given by

$$r = \sqrt{\frac{I}{A}} \tag{9.16}$$

where I is the minimum moment of inertia and A is the cross-sectional area of the column.

On substituting Eq. (9.16) into Eq. (9.15), we obtain

$$\sigma_{Cr} = \frac{\pi^2 E}{\left(\frac{L}{r}\right)^2} \tag{9.17}$$

The nondimensional term (L/r) is called the *slenderness ratio* and is an important characteristic of columns. A typical qualitative plot of Eq. (9.17) is shown in Figure 9.4. The critical column stress is seen to be inversely proportional to the square of the slenderness ratio. Euler's formula predicts that *the critical buckling stress approaches infinity as the slenderness ratio approaches zero*. This last condition of course is impossible and indicates that Euler's formula is not accurate for smaller slenderness ratios. These shorter column conditions will be considered in Section 9.5.

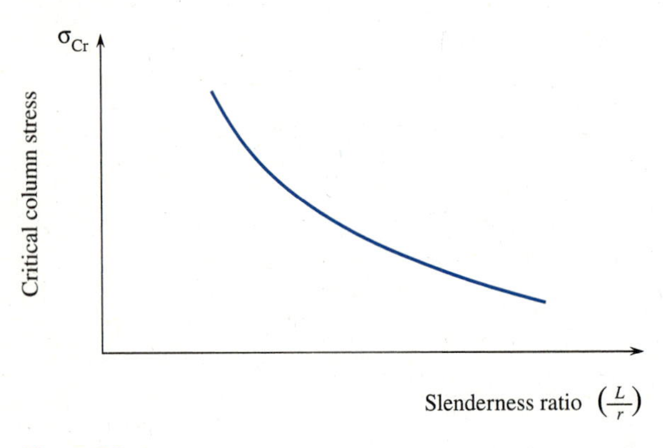

Figure 9.4 Euler's critical column stress versus slenderness ratio.

EXAMPLE 9.1

Determine the critical column load, P_{Cr}, and the critical column stress, σ_{Cr}, for an acrylic plastic meter stick, assuming that its ends are pinned. Assume that the cross-section of the stick is 8 mm × 24 mm and that $E = 3.1$ GPa. Use Euler's formula.

Solution

Using the cross-sectional properties of the stick, we obtain the cross-sectional area as

$$A = 8 \text{ mm} \times 24 \text{ mm} = 192 \text{ mm}^2$$

We next calculate the minimum value of I, since the column will bend about its weakest axis.

$$I = \frac{bh^3}{12} = \frac{(24 \text{ mm})\,(8 \text{ mm})^3}{12} = 1024 \text{ mm}^4$$

From Eq. (9.16), we calculate the radius of gyration as

$$r = \sqrt{\frac{I}{A}} = \sqrt{\frac{1024 \text{ mm}^4}{192 \text{ mm}^2}} = 2.31 \text{ mm}$$

The slenderness ratio is therefore

$$\left(\frac{L}{r}\right) = \frac{1\ \text{m}}{0.00231\ \text{m}} = 433$$

Using Eqs. (9.14) and (9.17) which apply for simply supported columns, the critical column load and the critical column stress are therefore

$$P_{\text{Cr}} = \frac{\pi^2(3.1\ \text{GPa})(1024\ \text{mm}^4) \times 10^{-12}\ \text{m}^4/\text{mm}^4}{(1\ \text{m})^2} = 31.3\ \text{N}$$

$$\sigma_{\text{Cr}} = \frac{\pi^2(3.1\ \text{GPa})}{(433)^2} = 163\ \text{kPa}$$

Additional Buckling Loads and Modes

Equation (9.12) indicates that multiple solutions exist for the critical buckling load corresponding to $n = 1,2,3, \ldots$. We saw that $n = 1$ corresponded to the lowest critical buckling load, P_{Cr}, and this is the value that we are normally the most interested in from a practical consideration.

The buckled mode shape corresponding to the critical buckling mode can be obtained by substituting $\beta L = n\pi$ from Eq. (9.10) into the general solution for y given by Eq. (9.13). When we use $n = 1$, we obtain the familiar sine wave first buckling mode shown in Figure 9.5(a) corresponding to the buckling load of Eq. (9.14).

Now if we let $n = 2$, the second buckling mode will look like Figure 9.5(b) and the second critical buckling load from Eq. (9.12) will be

$$P_{\text{Cr}_2} = 4\frac{\pi^2 EI}{L^2} \tag{9.18}$$

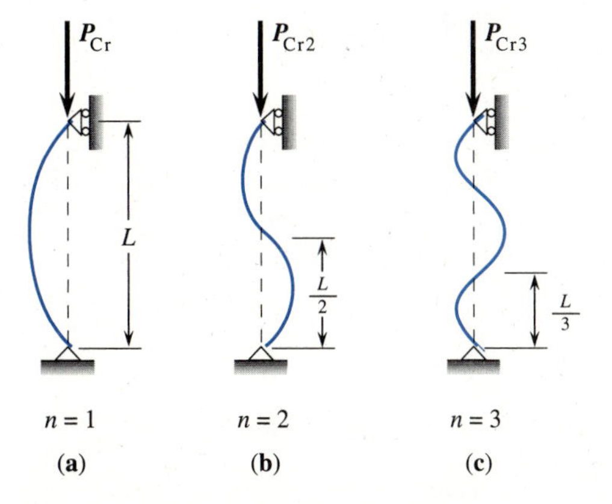

Figure 9.5 Euler buckling curves for higher modes.

If we let $n = 3$, the third mode will look like Figure 9.5(c) and the third buckling load will be

$$P_{Cr_3} = 9\frac{\pi^2 EI}{L^2} \tag{9.19}$$

Higher modes for $n = 4,5, \ldots$ are similarly obtained.

Since P_{Cr} corresponding to $n = 1$ is the lowest value, it is the critical load of the most practical use. Although the higher modes are theoretically possible to obtain, they are impossible to obtain in practice unless intermediate supports are used to prevent sideways motion at the points where the curves of Figure 9.5 have $y = 0$.

For instance, the possibility of increasing the critical load by a factor of 4 (Eq. (9.18)) is an important design consideration. Design engineers often provide lateral bracing to columns to force the deflection into the curve of Figure 9.5(b). This can be done by providing support at the midpoint $L/2$ or node ($y = 0$ location) of the column. The stiffness of the support necessary at the midpoint to produce this second deflection mode is surprisingly small (see reference [1]). You can demonstrate the difference between the critical column buckling loads corresponding to $n = 1$ and $n = 2$ by lightly supporting a yardstick (column) at its midpoint with your thumb and forefinger while applying a downward load at the top with your other hand.

9.3 EULER'S COLUMN FORMULA FOR OTHER END CONDITIONS

The Euler column formulas for the critical buckling load, Eq. (9.12), and the critical column stress, Eq. (9.17), were derived for the case of pinned ends (see Figure 9.3). Because numerous other types of column end conditions occur in practice, we will now develop a procedure for predicting the minimum loads (and stresses) that will cause buckling in those columns.

We use the formula for a column with pinned ends as the basis for developing other column formulas. The smallest critical buckling load for the column with pinned ends was given by Eq. (9.14) as

$$P_{Cr} = \frac{\pi^2 EI}{L^2} \tag{9.14}$$

In this section we will introduce the concept of effective length, L_e, for columns with different end conditions and substitute these values of L_e for L in Eq. (9.12). The *effective length*, L_e, is defined as the length of a column with pinned ends that would have the same critical load P_{Cr} as the actual column under consideration with its support conditions.

We now evaluate the effective length, L_e, for some common end conditions used in practice.

Column with One End Fixed and One End Free

Figure 9.6(a) shows a column that is fixed at the bottom and loaded axially at the unsupported top. The top is free to deflect laterally as well as vertically, similar to a flagpole.

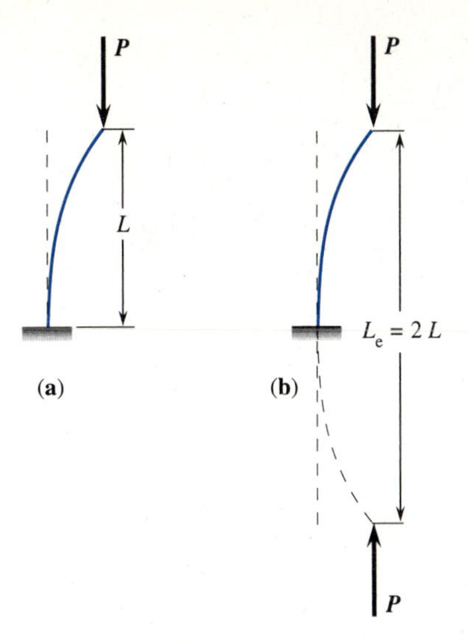

Figure 9.6 (a) Fixed-free column and (b) first buckled mode shape.

The buckled shape (or mode shape) corresponding to the lowest critical load, P_{Cr}, for this column is shown in Figure 9.6(a). Now notice that this mode is exactly the same as the top half of Figure 9.5(a), which shows the first buckled mode of the column with pinned ends. Figure 9.6(b) indicates that for a column with one end fixed and one end free, we will have an effective length of

$$L_e = 2L \tag{9.20}$$

and the lowest critical load will be obtained by substituting L_e into the more general expression for Euler's column formula

$$P_{Cr} = \frac{\pi^2 EI}{L_e^2} \tag{9.21}$$

That is, substituting Eq. (9.20) into Eq. (9.21), the critical load for a fixed-free column is

$$P_{Cr} = \frac{\pi^2 EI}{4L^2} \tag{9.22}$$

Column with Both Ends Fixed

Figure 9.7(a) shows a column that is fixed at both the top and bottom. The ends may move in the axial direction but cannot deflect laterally and cannot rotate. Figure 9.7(b) shows a free-body diagram of the column and the buckled shape corresponding to the lowest critical load, P_{Cr}. The direction of the deflection is, of course, arbitrary.

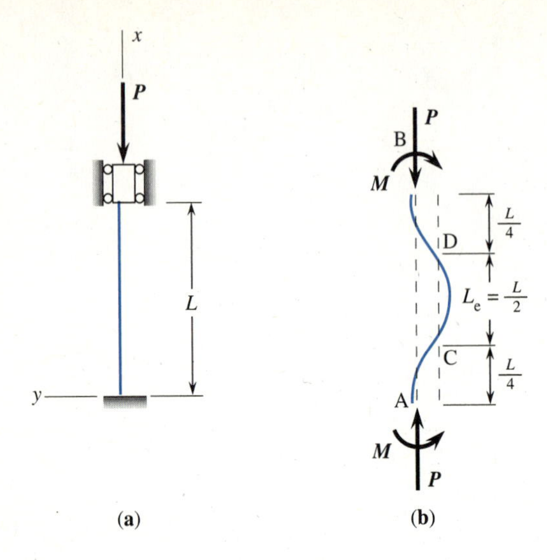

Figure 9.7 (a) Column with both ends fixed and (b) first buckling mode.

Due to the symmetry of this loading, points C and D are inflection points. That is, the moments are zero at C and D. Therefore, the section of the column C-D can be considered to be pinned at C and D and behave as a column with pinned ends having a length of $L/2$. The equivalent length of the column with fixed ends will then be

$$L_e = \frac{L}{2} \tag{9.23}$$

By substituting Eq. (9.23) into Eq. (9.21), the lowest critical column load is obtained as

$$P_{Cr} = \frac{4\pi^2 EI}{L^2} \tag{9.24}$$

Column with One End Fixed and One End Pinned

Figure 9.8(a) shows a column that is fixed at the bottom and pinned at the top. The top is restrained from lateral movement but can deflect in the axial direction and can also rotate. Figure 9.8(b) shows a free-body diagram of the column and the buckled mode shape corresponding to the lowest critical load P_{Cr}.

Figure 9.8(c) shows an arbitrary section of the displaced column in equilibrium. For equilibrium of moments about this section, we find that the moment, $M(x)$, necessary to hold the column in equilibrium at any location, x, must be

$$M(x) = M_A + V_x - Py \tag{9.25}$$

where $M(x)$ is shown positive for positive y consistent with the convention of Section 5.2.

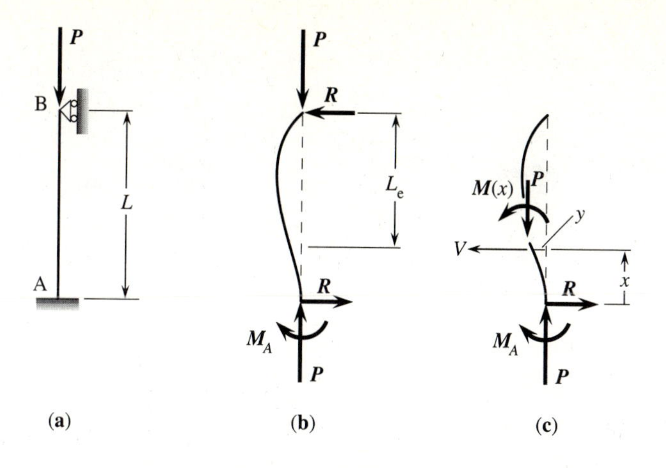

Figure 9.8 (a) Column with one end fixed and one end pinned, (b) first buckled mode shape, and (c) displaced column in equilibrium.

Now from Figure 9.8(b), we see that for equilibrium the end-moment, M_A, must be

$$M_A = -RL = -VL \tag{9.26}$$

Therefore, substituting Eq. (9.26) for M_A into Eq. (9.25), we have

$$M(x) = -VL + Vx - Py \tag{9.27}$$

Recall from Eq. (6.7), the basic relationship between bending moment and deflection is

$$\frac{M}{EI} = \frac{d^2y}{dx^2} \tag{6.7}$$

Using Eq. (6.7) in Eq. (9.27) and simplifying, we obtain

$$\frac{d^2y}{dx^2} + \frac{P}{EI}y = \frac{V}{EI}(x - L) \tag{9.28}$$

Using Eq. (9.4), Eq. (9.28) becomes

$$\frac{d^2y}{dx^2} + \beta^2 y = \frac{V}{EI}(x - L) \tag{9.29}$$

Equation (9.29) is similar to Eq. (9.5) except that the differential equation is not homogeneous because the right-hand side is not zero. Recall from differential equations we then have a homogeneous solution and a particular solution to Eq. (9.29).

The general solution to Eq. (9.29) will be

$$y = C_1\sin\beta x + C_2\cos\beta x + C_3(x - L) \tag{9.30}$$

where the particular solution of Eq. (9.29) is of the form

$$y_p = C_3(x - L)$$

The boundary conditions are that y and y' are zero at the fixed end ($x = 0$) and y is zero at the pinned end ($x = L$). These are written as

$$y(0) = 0 \qquad y(L) = 0 \qquad y'(0) = 0 \tag{9.31}$$

The constants C_1, C_2, and C_3 are obtained by applying boundary conditions. The coefficient C_3 is obtained by substituting y_P into Eq. (9.29). The result is

$$C_3 = \frac{V}{P}$$

Using C_3 in Eq. (9.30), we obtain

$$y = C_1 \sin \beta x + C_2 \cos \beta x + \frac{V}{P}(x - L) \tag{9.32}$$

where the definition of β from Eq. (9.4) has been used.

Now using the boundary conditions of Eq. (9.31) in Eq. (9.32), we obtain

$$0 = C_2 - \frac{VL}{P}$$

$$0 = C_1 \sin \beta L + C_2 \cos \beta L \tag{9.33}$$

$$0 = C_1 \beta + \frac{V}{P}$$

Solving Eqs. (9.33) simultaneously yields the transcendental equation (an equation having an infinite number of solutions) as

$$\tan \beta L = \beta L \tag{9.34}$$

The values of β that satisfy Eq. (9.34) will give us the values of the critical buckling load corresponding to each of the possible buckling modes. Since we are interested in the smallest buckling load, the smallest nonzero solution is obtained by trial and error (or by plotting $\tan \beta L$ and βL and looking for intersecting points) as

$$\beta L = 4.4934 \tag{9.35}$$

Using Eq. (9.35) in Eq. (9.4), we obtain the lowest buckling load for this column as

$$P_{Cr} = (4.4934)^2 \frac{EI}{L^2} \tag{9.36}$$

Equating Eq. (9.36) with the general expression for Euler's formula, Eq. (9.21), and solving for the equivalent column length, we obtain

$$L_e = 0.6992\, L \approx 0.7\, L \tag{9.37}$$

for the fixed-pinned column. Using Eq. (9.37) in Eq. (9.21), we obtain

$$P_{Cr} = 2.041\pi^2(EI/L^2) \tag{9.38}$$

Summary of Euler's Columns

Observing Eqs. (9.14), (9.22), (9.24), and (9.38) for the buckling formulas of the four column end conditions, we see they are all of the form

$$P_{Cr} = \frac{C\pi^2 EI}{L^2} \quad \textbf{(9.39)}$$

where $C = 1$ for pinned-pinned columns, $C = 1/4$ for fixed-free columns, $C = 4$ for fixed-fixed columns, and $C = 2.041$ for fixed-pinned columns.

Letting $L_e = L/\sqrt{C}$ in Eq. (9.39), we have Eq. (9.21) again as

$$P_{Cr} = \frac{\pi^2 EI}{L_e^2} \quad \textbf{(9.40)}$$

In Equation (9.40), L_e is the equivalent column length which depends on the end conditions of the column being analyzed. Values of L_e and C for the four common types of end-supports for columns are listed in Figure 9.9.

Often it is more convenient to solve for the critical buckling stress rather than the load. In this case we simply use L_e in Eq. (9.17) and obtain

$$\sigma_{Cr} = \frac{\pi^2 E}{\left(\frac{L_e}{r}\right)^2} \quad \textbf{(9.41)}$$

where again r is the radius of gyration as defined in Eq. (9.16) and L_e is obtained from Figure 9.9.

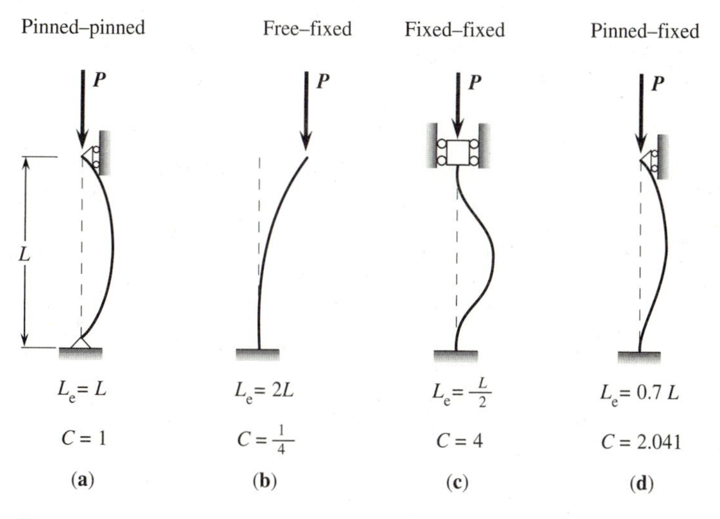

Figure 9.9 Equivalent column lengths and coefficients C for different end conditions.

EXAMPLE 9.2

Determine the critical column load for the plastic meter stick of Example 9.1 assuming that it is fixed at both ends.

Solution

From Figure 9.9, the effective column length, L_e, for a fixed-fixed column is $L_e = L/2$. Therefore, from Eq. (9.40), we obtain

$$P_{Cr} = \frac{\pi^2 EI}{\left(\frac{L}{2}\right)^2}$$

$$P_{Cr} = \frac{\pi^2\,(3.1\text{ GPa})(1024\text{ mm}^4) \times 10^{-12}\text{ m}^4/\text{mm}^4}{\left(\frac{1}{2}\text{ m}\right)^2} = 125.2\text{ N}$$

This load is four times that for the column pinned at both ends.

EXAMPLE 9.3

A W 8 × 15 wide-flange shape made of American Society for Testing Materials (ASTM) A36 structural steel is 20 ft long and is to be used as a vertical column with the bottom end fixed and the top pinned. Use Euler's formula to obtain the critical buckling stress and load.

Solution

From Appendix B, we find that for ASTM A36 structural steel, $E = 29 \times 10^6$ psi. From Appendix C, we find for the wide flange the radius of gyration, r, is given for two different axes. Since bending will naturally take place about the axis having the minimum moment of inertia, we use the minimum value of r. For the W 8 × 15, we find $r_y = r = 0.876$ in. (Buckling will take place about the y-axis.)

Referring to Figure 9.9, a column with one end fixed and one end free has an effective length of $L_e = 0.7L$. Therefore

$$L_e = (0.7)(20\text{ ft} \times 12\text{ in./ft}) = 168\text{ in.}$$

The effective slenderness ratio is

$$\left(\frac{L_e}{r_y}\right) = \frac{168}{0.876} = 192$$

The critical buckling stress is obtained from Eq. (9.41) as

$$\sigma_{Cr} = \frac{\pi^2 E}{\left(\frac{L_e}{r}\right)^2}$$

$$\sigma_{Cr} = \frac{\pi^2(29 \times 10^6\text{ psi})}{(192)^2} = 7764\text{ psi}$$

The critical column load is obtained by using $A = 4.44$ in.^{2} in Eq. (9.15) as

$$P_{Cr} = \sigma_{Cr} A$$

$$P_{Cr} = (7764 \text{ psi})(4.44 \text{ in.}^2) = 34{,}470 \text{ lb}$$

9.4 INELASTIC COLUMN BEHAVIOR/TANGENT MODULUS

Euler's column buckling formulas described in the previous sections have assumed that the modulus of elasticity, E, is constant. For examples, see Eqs. (9.40) and (9.41).

This implies that the average stress on the column remains below the proportional limit, or in the straight-line portion of the stress-strain curve.

It is possible that the stress in a loaded column may exceed the proportional limit before buckling actually occurs. In this case, the value of the modulus, E, will be reduced and Euler's formula will no longer be accurate. However, a modification to Euler's formula which corrects for the reduced modulus was suggested by the German engineer F. Engesser in 1889 [2]. Engesser suggested that E in Euler's formula be replaced by the tangent modulus, E_t, defined as the local slope of the stress-strain curve or

$$E_t = \frac{d\sigma}{d\epsilon} \tag{9.42}$$

For any given value of the stress, σ, the tangent modulus is then determined as shown in Figure 9.10. Euler's formula, using E_t instead of E, becomes

$$\sigma_{Cr} = \frac{\pi^2 E_t}{\left(\dfrac{L_e}{r}\right)^2} \tag{9.43}$$

This important modification to Euler's formula is called the *tangent-modulus formula* or the *Euler-Engesser formula* and provides a simple method by which the elastic column curves can be modified to account for inelastic effects.

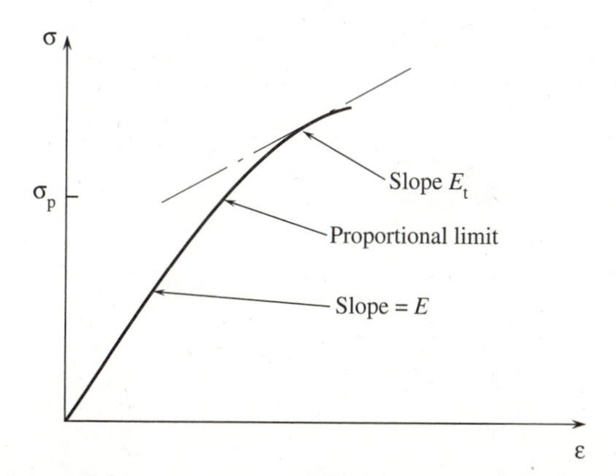

Figure 9.10 Column stress-strain diagram showing tangent modulus.

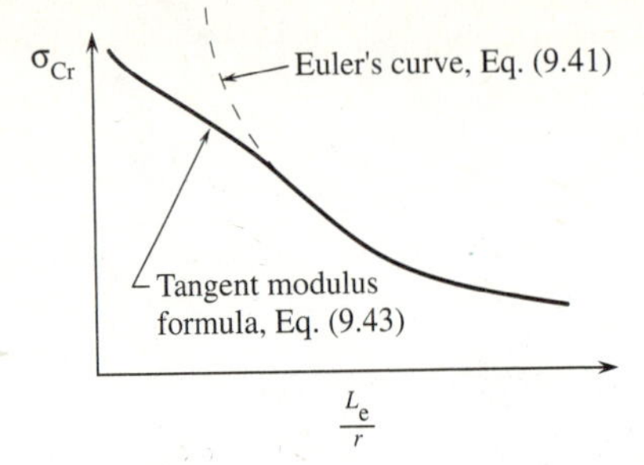

Figure 9.11 Allowable column stress versus critical slenderness ratio.

The procedure for using Eq. (9.43) to predict column buckling assumes that a compression stress-strain diagram is available. After selecting a value of (L_e/r), a value E_t is assumed and then σ_{Cr} is solved from Eq. (9.43). The value of E_t corresponding to this σ_{Cr} is then determined from the stress-strain diagram. If this value E_t does not correspond to the assumed value, then the process is repeated.

An alternative to this iterative procedure is to solve for the allowable or critical slenderness ratio from Eq. (9.43). That is,

$$\left(\frac{L_e}{r}\right)_{Cr} = \pi\sqrt{\frac{E_t}{\sigma}} \tag{9.44}$$

This equation gives us the effective nondimensional column length (or slenderness ratio) at which a stress, σ, would become critical. In other words, if we have the stress-strain curve for the material under consideration, an allowable column curve similar to Figure 9.11 could be plotted from Eq. (9.43).

EXAMPLE 9.4

A 2-ft-long aluminum tube with a 1 in. outside diameter and a wall thickness of 0.120 in. is fixed at both ends and loaded as a column. A compressive stress-strain curve and a plot of the tangent modulus, E_t, for this aluminum are shown in Figure 9.12.

Determine the critical column load, P_{Cr}, and the critical column stress, σ_{Cr}.

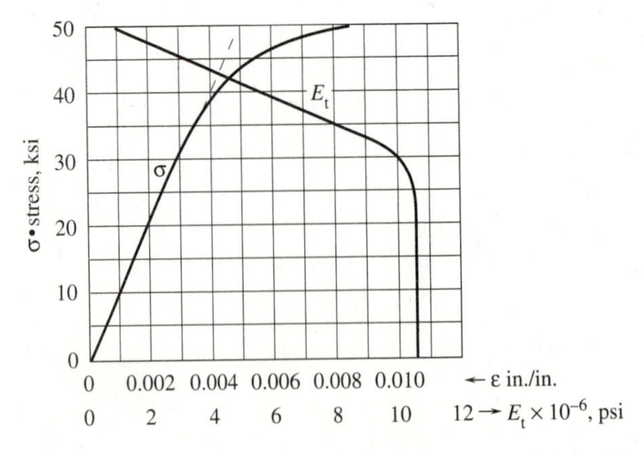

Figure 9.12 Stress-strain and tangent-modulus curves for aluminum.

Solution

The iteration procedure will be illustrated with this example. The area of the tube cross-section is

$$A = \pi(0.50^2 - 0.38^2) = 0.332 \text{ in.}^2$$

The moment of inertia is

$$I = \frac{\pi}{4}(0.50^4 - 0.38^4) = 0.0327 \text{ in.}^4$$

Therefore, the radius of gyration is

$$r = \sqrt{\frac{I}{A}} = 0.314 \text{ in.}$$

Since this column is fixed at both ends, the effective column length, L_e, is seen from Figure 9.9 to be $L_e = L/2$. The effective slenderness ratio is

$$\left(\frac{L_e}{r}\right) = \frac{\frac{24}{2} \text{ in.}}{0.314 \text{ in.}} = 38.2$$

Now we assume initially that the modulus is $E_t = 10.5 \times 10^6$ psi (see Figure 9.12). Therefore, Eq. (9.43) yields

$$\sigma_{Cr} = \frac{\pi^2 E_t}{\left(\frac{L_e}{r}\right)^2} = \frac{\pi^2 (10.5 \times 10^6 \text{ psi})}{(38.2)^2} = 71{,}000 \text{ psi}$$

Referring to Figure 9.12, we observe that a stress this high would be impossible for the assumed E_t. Therefore, we assume a lower value, say $E_t = 8 \times 10^6$ psi. The value of σ_{Cr} corresponding to this is

$$\sigma_{Cr} = \frac{\pi^2 (8 \times 10^6 \text{ psi})}{(38.2)^2} = 54{,}100 \text{ psi}$$

Again referring to Figure 9.12, we see that this σ_{Cr} is still too high. Assuming $E_t = 6 \times 10^6$ psi gives

$$\sigma_{Cr} = \frac{\pi^2 (6 \times 10^6 \text{ psi})}{(38.2)^2} = 40{,}600 \text{ psi}$$

This σ_{Cr} is still slightly high (yielding an E_t of about 4.5×10^6 psi from Figure 9.12). Now assume $E_t = 5.7 \times 10^6$ psi. This gives

$$\sigma_{Cr} = \frac{\pi^2 (5.7 \times 10^6 \text{ psi})}{(38.2)^2} = 38{,}500 \text{ psi}$$

Figure 9.12 shows that this σ_{Cr} value corresponds closely with $E_t = 5.7 \times 10^6$ psi. Therefore, we can assume that this will be the critical buckling stress for this tube with the material given in Figure 9.12.

The critical column load can be obtained from

$$P_{Cr} = \sigma_{Cr}A$$

$$P_{Cr} = (38{,}500 \text{ psi})(0.322 \text{ in.}^2) = 12{,}400 \text{ lb}$$

As previously explained, the solution to this problem could also have been obtained by plotting Eq. (9.43), using the data in Figure 9.12, and then finding the value of σ_{Cr} corresponding to the given value of the effective slenderness ratio (L_e/r). This figure would be similar to Figure 9.11. Although the work involved would have been more, the plot obtained could then have been used for other values of (L_e/r) as well.

The tangent modulus or Engesser method as discussed here is a useful method to predict column buckling in the inelastic stress range. It extends the range of validity of Euler's formula by simply altering the value of E to correspond to the stress level.

The tangent-modulus theory does not account for the fact that during bending one side of the cross-section is subjected to a decreasing stress (the side that is bowing out) which would tend to increase the modulus E_t. This inconsistency led to suggested additional modifications by Engesser [3], Bleich [4], and von Kármán [5], and was finally resolved by Shanley [6]. Although the Shanley theory has been verified by test results, the tangent-modulus theory is recommended in this text for inelastic behavior for stresses above the yield point since it predicts buckling loads only slightly lower than the Shanley theory and is easier to use.

Although the method is not exact, it has been found to compare closely to experimental test results and can be used for preliminary analysis. The major disadvantage to the method is that it requires a compressive stress-strain curve for the material being analyzed. This disadvantage is overcome with the method to be discussed in the Section 9.5.

9.5 INELASTIC BEHAVIOR/JOHNSON'S FORMULA

In the preceding sections on column buckling, we described Euler's formula for predicting the critical buckling load when the compressive stress in the material is less than the proportional limit. That is, for columns with large slenderness ratios (L_e/r), Euler's formula applies. Such columns are usually referred to as *long columns*.

In the Section 9.4 we learned that by using the tangent modulus, E_t, instead of Young's modulus, E, in Euler's formula, the buckling load or stress could be reasonably approximated for smaller slenderness ratios where the material began to behave inelastically. The disadvantage of this tangent modulus approach is that the column material stress-strain curve must be available.

In this section, we will describe an empirical method for predicting column buckling in the inelastic or short column range that is easier to use than the tangent modulus method and which has been shown from experimental tests to give good results. Figure 9.13 shows the critical column stress according to Euler's formula. As the column slenderness ratio, (L_e/r), becomes small, Euler's formula predicts that the critical column stress, σ_{Cr}, increases without bounds. We know, of course, that this is impossible. If we use yielding

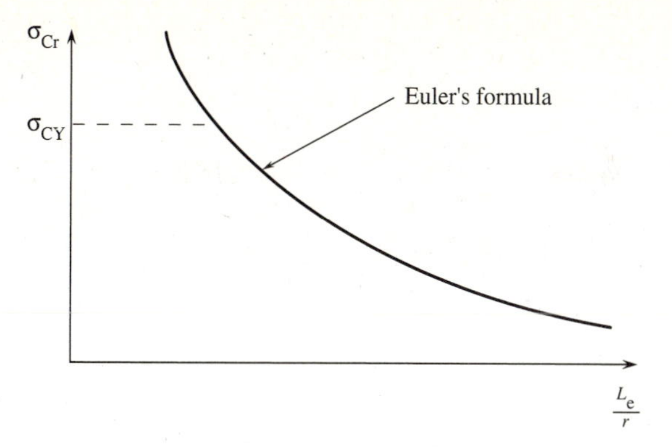

Figure 9.13 Limitations on Euler's formula.

as a practical upper stress limit, then the maximum permissible stress attainable would be that corresponding to the compressive yield stress, σ_{CY}, of the material. This is sometimes referred to as *block compression allowable,* corresponding to a very short block of material where compressive stability is not of concern.

This compressive yield stress, σ_{CY}, establishes an upper limit to σ_{Cr}, no matter how small the value of the slenderness ratio, (L_e/r), will be. This limit is shown by the horizontal dotted line in Figure 9.13. A plot of actual column buckling test data would follow closely Euler's curve for long columns and then would deviate from the Euler curve in the intermediate range and eventually intersect the ordinate at σ_{CY}. It will be the purpose of this discussion to establish an empirical curve agreeing with test data that plot smoothly from Euler's curve to the ordinate at σ_{CY}.

Many different empirical formulas have been suggested to represent this intermediate to short column range, but the parabolic formula due to J. B. Johnson is widely used in the machine, automotive, aerospace, and structural-steel construction fields and is the approach that will be used here. We will refer to this as the *Johnson formula* for column stress. This formula is given by

$$\sigma_{Cr} = \sigma_{CY} - \left(\frac{\sigma_{CY}}{2\pi}\right)^2 \frac{1}{E}\left(\frac{L_e}{r}\right)^2 \tag{9.45}$$

where σ_{CY} is the material compressive yield stress and (L_e/r) is the slenderness ratio as before.

A plot of Eq. (9.45) in Figure 9.14 shows the smooth transition from Euler's curve to σ_{CY} at the ordinate.

Figure 9.14 is a complete representation of the critical column buckling stress for values of the slenderness ratios from zero to approaching infinity. For long columns (large values of the slenderness ratio), the theoretical Euler's curve will give accurate predictions of σ_{Cr}. For intermediate to short columns, the curve due to the empirical Johnson's formula will predict σ_{Cr} with reasonable accuracy. These two curves become tangent to each

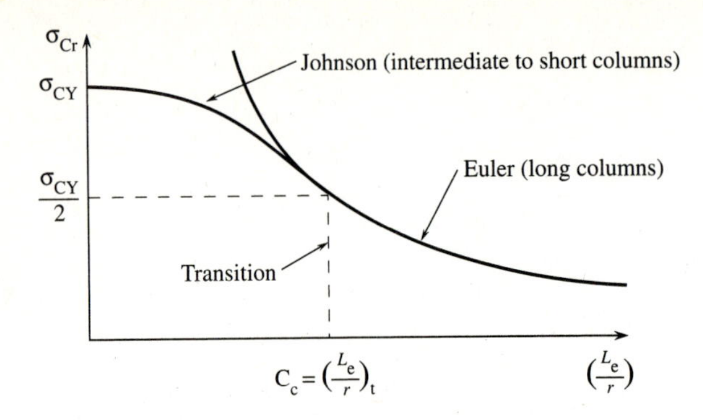

Figure 9.14 Johnson's column formula plotted with Euler's formula.

other at some value of the slenderness ratio that we will call the *transition slenderness ratio,* $C_c = (L_e/r)_t$. At this value of slenderness ratio, the value of σ_{Cr} from both Euler's and Johnson's formulas must be the same.

One of the most widely used versions of Johnson's formula requires that its plot become tangent to Euler's curve (at the transition slenderness ratio) at a critical stress of

$$\sigma_{Cr} = \frac{\sigma_{CY}}{2} \tag{9.46}$$

as shown in Figure 9.14.

Therefore, substituting Eq. (9.46) into Eq. (9.45), we can solve for the transition effective slenderness ratio as

$$C_c = \left(\frac{L_e}{r}\right)_t = \sqrt{\frac{2\pi^2 E}{\sigma_{CY}}} \tag{9.47}$$

The procedure for column analysis that is recommended in this chapter is to first calculate the transition slenderness ratio from Eq. (9.47). The actual slenderness ratio (L_e/r) for the column being analyzed is then calculated, being careful to use appropriate L_e from Figure 9.9. If (L_e/r) is greater than $(L_e/r)_t$, then we have a long column and Euler's formula, Eq. (9.41), should be used to calculate σ_{Cr}. That is,

$$\sigma_{Cr} = \frac{\pi^2 E}{\left(\frac{L_e}{r}\right)^2} \tag{9.41}$$

If the actual (L_e/r) is less than $(L_e/r)_t$, then we have an intermediate or short column and Johnson's formula, Eq. (9.45), should be used to calculate σ_{Cr}. That is,

$$\sigma_{Cr} = \sigma_{CY} - \left(\frac{\sigma_{CY}}{2\pi}\right)^2 \frac{1}{E}\left(\frac{L_e}{r}\right)^2 \tag{9.45}$$

STEPS IN SOLUTION

The steps to be followed in analyzing a column for the critical buckling load or critical buckling stress are as follows.

1. For the column being analyzed, calculate the cross-sectional area, A.
2. Calculate the minimum moment of inertia of the cross-section, I.
3. Calculate the minimum radius of gyration, r, as

$$r = \sqrt{\frac{I}{A}} \tag{9.16}$$

4. Establish the end-fixity conditions for the column being analyzed. Then from Figure 9.9 and the actual length of the column, determine the effective length, L_e.
5. Calculate the effective slenderness ratio

$$\left(\frac{L_e}{r}\right)$$

6. Determine the compressive yield strength, σ_{CY}, and the modulus of elasticity, E, for the column material. Appendix B will often be helpful in obtaining σ_{CY} and E.
7. Calculate the transition slenderness ratio from Eq. (9.47).

$$C_c = \left(\frac{L_e}{r}\right)_t = \sqrt{\frac{2\pi^2 E}{\sigma_{CY}}} \tag{9.47}$$

8. If $(L_e/r) > (L_e/r)_t$, use Euler's formula to find the critical column buckling stress

$$\sigma_{Cr} = \frac{\pi^2 E}{\left(\frac{L_e}{r}\right)^2} \tag{9.41}$$

9. If $(L_e/r) < (L_e/r)_t$, then use the Johnson formula to find the critical column buckling stress

$$\sigma_{Cr} = \sigma_{CY} - \left(\frac{\sigma_{CY}}{2\pi}\right)^2 \frac{1}{E}\left(\frac{L_e}{r}\right)^2 \tag{9.45}$$

10. The critical column buckling load is obtained from

$$P_{Cr} = \sigma_{Cr} A \tag{9.15}$$

EXAMPLE 9.5

Determine the critical buckling load of a standard nominal 2 in. × 4 in. by 8 ft long wooden stud of southern pine. The actual cross-sectional dimensions are 1.5 in. × 3.5 in. The compressive yield stress is 5 ksi and $E = 1.5 \times 10^6$ psi. Assume that the ends are pinned.

Solution

The area of the cross-section is

$$A = 1.5 \text{ in.} \times 3.5 \text{ in.} = 5.25 \text{ in.}^2$$

The minimum moment of inertia is

$$I = \frac{(3.5 \text{ in.})(1.5 \text{ in.})^3}{12} = 0.984 \text{ in.}^4$$

The radius of gyration is

$$r = \sqrt{\frac{I}{A}} = \sqrt{\frac{0.984 \text{ in.}^4}{5.25 \text{ in.}^2}} = 0.433 \text{ in.}$$

From Figure 9.9, the effective length, L_e, is the same as the length, L, for pinned ends. Therefore

$$L_e = (8 \text{ ft.} \times 12 \text{ in./ft.}) = 96 \text{ in.}$$

The effective slenderness ratio is

$$\left(\frac{L_e}{r}\right) = \left(\frac{96 \text{ in.}}{0.433 \text{ in.}}\right) = 222$$

The transition effective slenderness ratio is found from Eq. (9.47) as

$$\left(\frac{L_e}{r}\right)_t = \sqrt{\frac{2\pi^2 E}{\sigma_{CY}}}$$

$$\left(\frac{L_e}{r}\right)_t = \sqrt{\frac{2\pi^2(1.5 \times 10^6 \text{ psi})}{5000 \text{ psi}}} = 77$$

Since the wooden column has a slenderness ratio greater than the transition value, this indicates that we have a long column and should use Euler's formula, Eq. (9.41), to solve for the critical buckling stress. Therefore the critical stress is

$$\sigma_{Cr} = \frac{\pi^2 E}{\left(\frac{L_e}{r}\right)^2}$$

$$\sigma_{Cr} = \frac{\pi^2(1.5 \times 10^6 \text{ psi})}{(222)^2} = 300 \text{ psi}$$

The critical buckling load (from Eq. (9.15)) is equal to

$$P_{Cr} = \sigma_{Cr}A = (300 \text{ psi})(5.25 \text{ in.}^2) = 1575 \text{ lb}$$

EXAMPLE 9.6

Determine the critical buckling load for a W 8 × 15 structural steel column 5 ft long. The compressive yield stress for the steel is 36 ksi and $E = 29 \times 10^6$ psi. Assume pinned ends.

Solution

From Appendix C, the area of the W cross-section is seen to be 4.44 in.^{2} and minimum radius of gyration is given directly as $r = 0.876$ in. From Figure 9.9, we find that $L_e = L$ for pinned ends and therefore

$$L_e = (5 \text{ ft} \times 12 \text{ in./ft}) = 60 \text{ in.} \tag{a}$$

The actual column slenderness ratio is therefore

$$\left(\frac{L_e}{r}\right) = \left(\frac{60 \text{ in.}}{0.876 \text{ in.}}\right) = 68.5 \tag{b}$$

From Eq. (9.47), the transition slenderness ratio is

$$C_c = \left(\frac{L_e}{r}\right)_t = \sqrt{\frac{2\pi^2\, 29 \times 10^6}{36{,}000}} = 126 \tag{c}$$

Since the actual slenderness ratio, Eq. (b), is less than the transition value, Eq. (c), this is a short column and we need to use the Johnson formula.

Therefore, from Eq. (9.45), the critical column stress is

$$\sigma_{Cr} = 36 \text{ ksi} - \left(\frac{36 \text{ ksi}}{2\pi}\right)^2\left(\frac{1}{29 \times 10^3 \text{ ksi}}\right)(68.5)^2$$

$$\sigma_{Cr} = 30.7 \text{ ksi} \tag{d}$$

From Eq. (9.15), the critical column load is

$$P_{Cr} = (30.7 \text{ ksi})(4.44 \text{ in.}^2)$$

$$P_{Cr} = 136.3 \text{ kips} \tag{e}$$

EXAMPLE 9.7

Determine the critical buckling load of the W section in Example 9.6 if the column was built in (fixed) at both ends.

Solution

First, the equivalent length from Figure 9.9 is now

$$L_e = \frac{L}{2} = \frac{5 \text{ ft}}{2} = 2.5 \text{ ft}$$

$$= 30 \text{ in.}$$

The effective slenderness ratio is now

$$\left(\frac{L_e}{r}\right) = \left(\frac{30 \text{ in.}}{0.876 \text{ in.}}\right) = 34.2$$

Since this is less than the transition slenderness ratio of

$$\left(\frac{L_e}{r}\right)_t = 126$$

this is also a short column and we should use Johnson's formula. Therefore, from Eq. (9.45), we have

$$\sigma_{Cr} = 36 \text{ ksi} - \left(\frac{36 \text{ ksi}}{2\pi}\right)^2 \left(\frac{1}{29 \times 10^3 \text{ ksi}}\right)(34.2)^2$$

$$= 36 - 1.32$$

$$\sigma_{Cr} = 34.68 \text{ ksi}$$

The critical buckling load is now

$$P_{Cr} = (34.68 \text{ ksi})(4.44 \text{ in.}^2)$$

$$= 154 \text{ kips}$$

9.6 CODE DESIGN FORMULAS FOR COLUMNS UNDER CENTRIC LOAD

In the preceding sections, we described how to determine the critical (buckling) load of an elastic column using Euler's formula. We extended the procedure to inelastic columns by introducing the tangent modulus formula. Then we introduced the empirical formula, called the Johnson formula, used for intermediate length to short columns. Using a reasonable factor of safety of around 2 to 3, these formulas could be used to design acceptable columns.

However, as we have observed, a single formula is normally not adequate for all values of effective slenderness ratio, L_e/r, nor for the various materials used in column construction. Therefore, different formulas have been developed based in part on experimental data of these materials.

We now consider specific code formulas for the design of columns under centric loading for the three common materials, structural steel, aluminum, and timber. We emphasize that the formulas introduced here are intended solely to introduce you to a few of the design formulas used in practice. For a more complete description of, and limitations associated with, the numerous formulas available consult the specifications cited in the following discussions on steel, aluminum, and timber columns.

Structural Steel

The design of steel building columns under centric load is governed by the specifications of the American Institute of Steel Construction (AISC) [7]. There are two basic formulas used depending on the L_e/r of the column. The formula for long columns is based on Euler's formula, while that for intermediate to short columns is based on Johnson's formula. The design formulas used are based on the transition slenderness ratio $(L_e/r)_t$.

For

$$\left(\frac{L_e}{r}\right) < \left(\frac{L_e}{r}\right)_t = C_c$$

the allowable compressive stress is given by

$$\sigma_{allow} = \frac{\sigma_Y}{FS}\left[1 - \frac{(L_e/r)^2}{2C_c^2}\right] \tag{9.48}$$

where the transition slenderness ratio is

$$C_c = \left(\frac{L_e}{r}\right)_t = \sqrt{\frac{2\pi^2 E}{\sigma_{CY}}} \tag{9.49}$$

and the factor of safety (FS) is

$$FS = \frac{5}{3} + \frac{3(L_e/r)}{8C_c} - \frac{(L_e/r)^3}{8C_c^3} \tag{9.50}$$

For $L_e/r > C_c$ but $L_e/r < 200$

$$\sigma_{allow} = \frac{12\pi^2 E}{23\left(\frac{L_e}{r}\right)^2} \tag{9.51}$$

where the FS is

$$FS = \frac{23}{12} = 1.92$$

For structural steel with $E = 29 \times 10^3$ ksi, Eq. (9.51) becomes

$$\sigma_{allow} = \frac{149 \times 10^3}{(L_e/r)^2} \text{ ksi} \tag{9.52}$$

NOTE: that according to the AISC, L_e/r must be less than 200 to be acceptable. For σ_Y = 36-ksi yield strength steel and $E = 29 \times 10^3$ ksi, Eq. (9.49) yields $C_c = 126.1$ and therefore for columns with $L_e/r > 126.1$, the column is considered to be long.

For various grades (yield strengths) of steel, the AISC *Manual of Steel Construction* [7] has tabulated values of allowable stress for values of L_e/r from 1 to 200. Figure 9.15 shows the variation of σ_{allow} for various L_e/r for 36-ksi and 50-ksi yield strength steels.

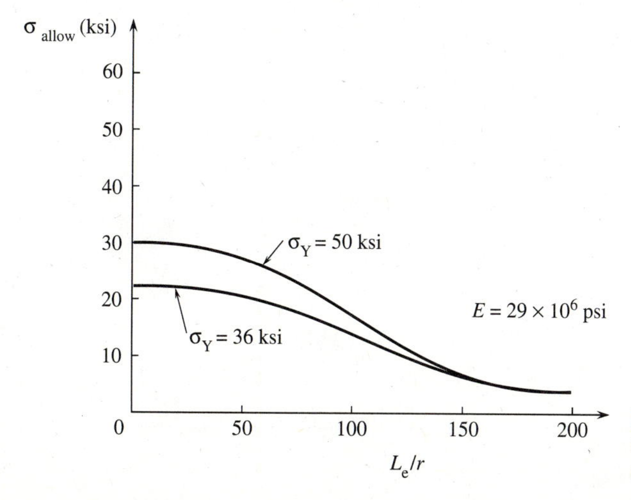

Figure 9.15 Variation of allowable compressive stress with various slenderness ratios for different steels.

Design formulas similar to Eqs. (9.48) and (9.51) have been also adopted for steel highway bridges by the American Association of State Highway and Transportation Officials (AASHTO) [8].

EXAMPLE 9.8

For the water tank shown in Figure 9.1(a), determine the allowable buckling stress and load. Assume the legs are made of A36 steel pipe. The pipes are 12 in. outer diameter and 0.5 in. thick, and are located at 50-ft intervals (between bracing).

Solution

Calculate the cross-sectional area A and moment of inertia I as

$$A = \pi(6 \text{ in.}^2 - 5.5 \text{ in.}^2) = 18.06 \text{ in.}^2$$

$$I = \frac{\pi}{4}(6^4 - 5.5^4)$$

$$= (1296 - 915.06)$$

$$= 299 \text{ in.}^4$$

Calculate the radius of gyration as

$$r = \sqrt{\frac{299}{18.06}} = 4.07 \text{ in.}$$

Assume a pinned-ended column and determine the slenderness ratio as

$$L_e/r = (50 \text{ ft} \times 12 \text{ in./ft})/4.07$$

$$= 147$$

Determine the transition slenderness ratio as

$$C_c = \sqrt{\frac{2\pi^2 E}{\sigma_Y}} = \sqrt{\frac{2\pi^2 \times 29 \times 10^3}{36}} = 126$$

Since $L_e/r > C_c$, we have an Euler column. Therefore, use Eq. (9.51) to obtain the allowable stress as

$$\sigma_{\text{allow}} = \frac{\pi^2 \times 29 \times 10^3}{(147)^2(1.92)} \text{ ksi}$$

$$\sigma_{\text{allow}} = 6.90 \text{ ksi}$$

The allowable buckling load is then

$$P_{\text{allow}} = 6.90 \times 18.06$$

$$= 125 \text{ kips}$$

Aluminum

The design of structural and machine components using aluminum alloys is governed by the specification of The Aluminum Association [9].

For each of the numerous alloys, three different formulas for the allowable stress in columns under centric loading are provided. The three formulas based on limits of the slenderness ratio are of the following general form.

For short columns

$$\sigma_{allow} = \frac{\sigma_Y}{FS} \tag{9.53}$$

(yield stress with a factor of safety)

For intermediate columns

$$\sigma_{allow} = \frac{C_1 - C_2(L/r)}{FS} \tag{9.54}$$

(a straight line with C_1 and C_2 constants)

For long columns

$$\sigma_{allow} = \frac{\pi^2 E}{FS(L/r)^2} \tag{9.55}$$

(Euler's formula with a factor of safety)

Formulas used for two commonly used alloys are listed as follows:

1. Alloy 6061-T6

   Alloy 6061-T6 is used in building construction and for machine parts in the form of sheets, plates, extrusion, structural shapes, rods, and tubes. It has good corrosion resistance, formability, and weldability.

   Short columns: $L/r \leq 9.5$

$$\sigma_{allow} = 19 \text{ ksi } (131 \text{ MPa}) \tag{9.56}$$

   Intermediate columns: $9.5 < L/r < 66$

$$\sigma_{allow} = (20.2 - 0.126\, L/r) \text{ ksi}$$

$$\sigma_{allow} = (139 - 0.869\, L/r) \text{ MPa} \tag{9.57}$$

   Long columns: $L/r \geq 66$

$$\sigma_{allow} = \frac{51{,}000}{(L/r)^2} \text{ ksi}$$

$$= \frac{351 \times 10^3}{(L/r)^2} \text{ MPa} \tag{9.58}$$

2. Alloy 2014-T6 (Alclad). Alloy 2014-T6 is used for aircraft skins and structures. It is heat-treatable with high mechanical properties.

   Short columns: $L/r \leq 12$

$$\sigma_{allow} = 28 \text{ ksi } (193 \text{ MPa}) \tag{9.59}$$

Intermediate columns: $12 < L/r < 55$

$$\sigma_{\text{allow}} = (30.7 - 0.23\, L/r)\ \text{ksi}$$
$$= (212 - 1.585\, L/r)\ \text{MPa} \tag{9.60}$$

Long columns: $L/r \geq 55$

$$\sigma_{\text{allow}} = \frac{54{,}000}{(L/r)^2}\ \text{ksi} = \frac{372 \times 10^3}{(L/r)^2}\ \text{MPa} \tag{9.61}$$

For aluminum alloys, the length L is defined as the distance between points of lateral support (hence $L_e = L$) except for a cantilevered column where two times the real length of the column is used ($L_e = 2L$).

EXAMPLE 9.9

A solid circular bar of aluminum alloy 2014-T6 is used as a strut in an airplane wing. Its length is 2 m and it must carry a compressive force of 10 kN. Assume the bar to be pinned-ended. Determine the diameter of the bar.

Solution

Assume a long column $L/r \geq 55$, then by Eq. (9.61), the allowable stress is

$$\sigma_{\text{allow}} = \frac{372 \times 10^3}{(L/r)^2}\ \text{MPa}$$

The radius of gyration is

$$r = \sqrt{\frac{I}{A}} = \sqrt{\frac{\frac{\pi}{64}d^4}{\frac{\pi}{4}d^2}} = \frac{d}{4}$$

Setting the allowable buckling stress (based on Eq. (9.61)) equal to the applied direct compressive stress, we have

$$\frac{372 \times 10^3}{[2/(d/4)]^2} = \frac{0.010}{\frac{\pi}{4}d^2}\ \text{MN}$$

Solving for d, we obtain

$$\frac{\pi d^2}{4} \times \frac{372 \times 10^3 (d/4)^2}{(4)} = 0.01$$

$$d^4 = 2.19 \times 10^{-6}$$

$$d = 0.0385\ \text{m}$$

$$d = 38.5\ \text{mm}$$

Now verify the assumption that $L/r \geq 55$.

The radius of gyration is

$$r = \frac{0.0385}{4}\text{ m} = 0.00962\text{ m}$$

Therefore, the slenderness ratio is $\frac{L}{r} = \frac{2\text{ m}}{0.00962\text{ m}} = 208 > 55$ and the column is long as assumed.

Timber

For the design of timber columns, the *National Design Specification for Wood Construction (NDS),* Section 3.7.3, [10], provides design formulas based on three categories of columns as follows.
(All formulas are for square or rectangular solid columns.)

1. Short columns: $L_e/d \leq 11$

$$\sigma_{\text{allow}} = \sigma_c \tag{9.62}$$

where σ_c is the design value in compression parallel to the grain of the wood and d is the least dimension of the cross-section.

2. Intermediate columns: $11 < L_e/d < K$

$$\sigma_{\text{allow}} = \sigma_c\left[1 - \frac{1}{3}\left(\frac{L_e/d}{K}\right)^4\right] \tag{9.63}$$

where

$$K = 0.671\sqrt{\frac{E}{\sigma_c}} \tag{9.64}$$

3. Long columns: $L_e/d \geq K$ but $L_e/d < 50$

$$\sigma_{\text{allow}} = \frac{\pi^2 E}{2.727(L_e/r)^2} = \frac{0.3E}{(L_e/d)^2} \tag{9.65}$$

NOTE: L_e/d must always be less than 50 to prevent having a column that is too slender.

EXAMPLE 9.10

A square timber column 8 ft long made of southern pine must be designed to support a floor load of 7500 lb. Determine the smallest cross-sectional dimension d of the column. The allowable compressive stress parallel to the grain is 1400 psi.

Solution

Assume that the column is long. Then we can use Eq. (9.65) for the allowable stress given by

$$\sigma_{\text{allow}} = \frac{0.3E}{(L_e/d)^2}$$

From Appendix B, the modulus of elasticity is $E = 1.6 \times 10^6$ psi. Assume pinned ends, then the effective length is $L_e = L = 8$ ft.

The direct compressive stress is

$$\sigma = \frac{P}{A} = \frac{P}{d^2}$$

Setting the direct stress equal to the allowable stress, we have

$$\frac{P}{d^2} = \frac{0.3E}{(L/d)^2}$$

Solving for d, we obtain

$$d = \sqrt[4]{\frac{PL^2}{0.3E}}$$

Substituting the numeric quantities, we have

$$d = \sqrt[4]{\frac{(7500 \text{ lb})(96 \text{ in.})^2}{0.3 \times 1.6 \times 10^6 \text{ psi}}} = 3.46 \text{ in.}$$

A nominal 4 × 4 has actual dimensions of 3.50 in. × 3.50 in. and is satisfactory. Now verify the assumption of a long column. The slenderness ratio is

$$\frac{L}{d} = \frac{96 \text{ in.}}{3.5 \text{ in.}} = 27.4 < 50$$

Also, from Eq. (9.64), the transition slenderness ratio is

$$K = 0.671\sqrt{\frac{E}{\sigma_c}}$$

$$= 0.671\sqrt{\frac{1.6 \times 10^6}{1400}} = 22.7$$

Since the slenderness ratio is greater than K and less than 50, the column is long as assumed.

Finally, check that the compressive stress is less than the allowable parallel to the grain.

$$\sigma = \frac{P}{A} = \frac{7500 \text{ lb}}{3.5 \times 3.5} = 612 \text{ psi} < 1400 \text{ psi}$$

The column is satisfactory.

In summary, the column design formulas for steel, aluminum, and timber provided in this section are intended to make you aware of just a few of the many formulas used in designing columns under centric loading (for columns under eccentric loading see, for instance, the AISC [7]). These formulas make up only a part of the complete design process for columns. There are numerous code limitations that have not been presented here. Hence all design formulas taken from codes and specifications must be applied by knowledgeable people trained in their use.

9.7 SUMMARY OF IMPORTANT DEFINITIONS AND EQUATIONS

1. Euler's formula for critical buckling load of column

$$P_{Cr} = \frac{\pi^2 EI}{L_e^2} \qquad \textbf{(9.21, 9.40)}$$

2. Effective column length

L_e values from Figure 9.9

3. Critical column stress

$$\sigma_{Cr} = \frac{P_{Cr}}{A} \qquad \textbf{(9.15)}$$

4. Radius of gyration

$$r = \sqrt{\frac{I}{A}} \qquad \textbf{(9.16)}$$

5. Euler's formula for critical buckling stress

$$\sigma_{Cr} = \frac{\pi^2 E}{\left(\frac{L_e}{r}\right)^2} \qquad \textbf{(9.41)}$$

6. Effective slenderness ratio

$$\left(\frac{L_e}{r}\right)$$

7. Tangent-modulus formula for critical buckling stress

$$\sigma_{Cr} = \frac{\pi^2 E_t}{\left(\frac{L_e}{r}\right)^2} \qquad \textbf{(9.43)}$$

8. Tangent-modulus critical slenderness ratio

$$\left(\frac{L_e}{r}\right)_{Cr} = \pi\sqrt{\frac{E_t}{\sigma}} \qquad \textbf{(9.44)}$$

9. Johnson's formula for critical buckling stress

$$\sigma_{Cr} = \sigma_{CY} - \left(\frac{\sigma_{cY}}{2\pi}\right)^2 \frac{1}{E}\left(\frac{L_e}{r}\right)^2 \qquad \textbf{(9.45)}$$

10. Transition effective slenderness ratio

$$C_c = \left(\frac{L_e}{r}\right)_t = \sqrt{\frac{2\pi^2 E}{\sigma_{CY}}} \qquad \textbf{(9.47)}$$

11. Design formulas for allowable stresses in columns are provided in codes. These columns must be designed using codes developed for the particular material such as steel, aluminum alloys, or timber. Section 9.6 provides some formulas used to design columns made of steel, aluminum alloys, and timber.

REFERENCES

1. Timoshenko SP, Gere JM. *Theory of Elastic Stability,* McGraw-Hill, New York, 1961.
2. Engesser F. Ueber die Knickfestigkeit gerader Stabe. *Zeitschrift fur Architektur und Ingenieurwesen,* 35, 1889.
3. Engesser F. Knickfragen. *Schweitzerische Bauzeitung,* 26, 1895.
4. Bleich F. *Buckling Strength of Metal Structures.* McGraw-Hill, New York, 1952.
5. von Kármán T, Untersuchungen uber knickfestigkeit. In *Mitteilungen uber Forschungsarbeiten aug dem Gebiete des Ingenieurwesens,* No. 81. Berlin, 1910.
6. Shanley FR. Inelastic column theory. *Journal of the Aeronautical Sciences* 14(5), 1947.
7. American Institute of Steel Construction, Inc., *Manual of Steel Construction,* 9th ed., 1 E. Wacker Dr., Chicago, IL 60601.
8. American Association of State Highway and Transportation Officials. *Standard Specifications for Highway Bridges,* 14th ed. Washington, DC, 20001, 1989.
9. The Aluminum Association. *Specifications for Aluminum Structures,* latest ed., 900 Nineteenth St., NW Washington, DC, 20006.
10. National Forest Products Association. *National Design Specification for Wood Construction,* Washington, DC, 1986.

PROBLEMS

Section 9.2 and 9.3

Problems 9.1 through 9.23 can be assumed to be Euler or long columns.

9.1 Determine the buckling load for a reinforced wooden column constructed by nailing two studs back-to-back such that a 3.0 in. × 3.5 in. cross-section results. Assume the studs are southern pine with $E = 1.5 \times 10^6$ psi. Compare your answer with that of Example 9.5.

9.2 Two wooden meter sticks each 8 mm × 24 mm in cross-section are glued back-to-back. Determine the critical column buckling stress and critical load when both ends of the column are pinned. Assume the wood has an $E = 11.0$ GPa.

9.3 Repeat Problem 9.2 if the column now has one end pinned and one end fixed.

9.4 A structural steel wide flange (W 6 × 25) 18 ft long is used as a column. Assume both ends are pinned. Determine the critical column buckling load.

9.5 A 160-lb acrobat wants to use a 30-ft-long hollow aluminum pole to perform an aerial handstand. The pole has a 4 in. outer diameter and a 0.1-in.-thick wall. Assume the pole to be fixed at the bottom and free at the top. Will the pole be safe (based on buckling) if a factor of safety of 3 is required against buckling? (That is, the pole should support 480 lb.) Use $I = \pi r^3 t$ (thin walled approximation for the pole). Let $E = 10.5 \times 10^6$ psi.

9.6 Using a factor of safety of 3 against buckling, determine the load capacity P of the jib crane shown based on buckling of the W 6 × 20 wide flange steel beam AB. Neglect the weight of the beam. Assume member AB is pinned at both ends.

Figure P9.6

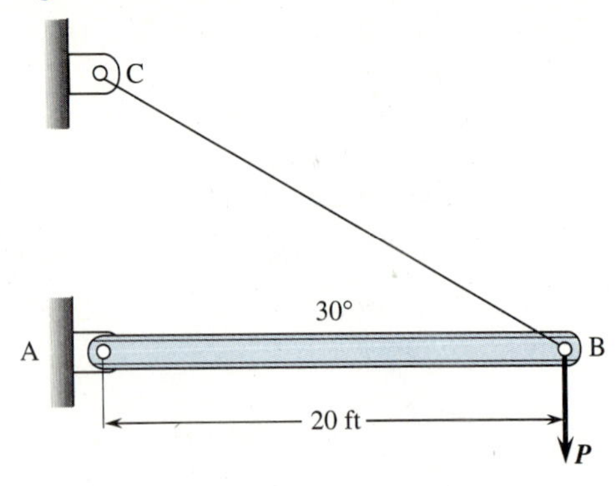

9.7 A 12-ft-long structural steel wide flange W 6 × 25 is used as a spreader bar AB in the sling arrangement shown. Based on buckling of the bar and a factor of safety of 3, determine the load capacity W.

Figure P9.7

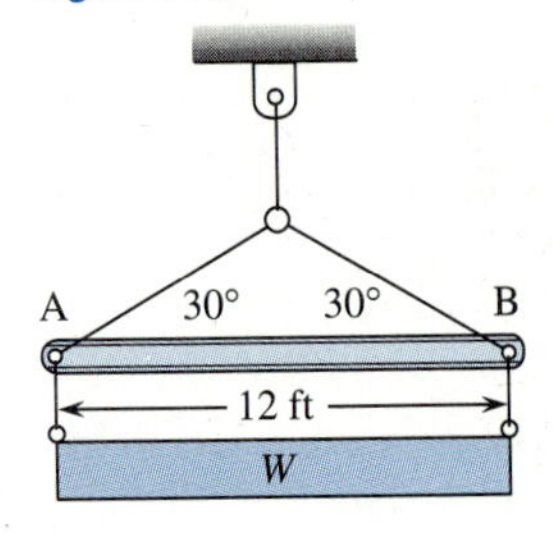

9.8 Repeat Problem 9.7 except use a nominal sized 4 × 4 (actual dimensions 3.5 in. × 3.5 in.) southern pine wooden member instead of the wide flange for member AB. Let $E = 1.2 \times 10^6$ psi for the southern pine.

9.9 A 3/8-in.-diameter wooden dowel 36 in. long has its bottom 4 in. clamped in a vise and an empty coffee can is attached to the top. If the dowel is initially perfectly vertical, how many pounds of sand can be poured into the coffee can before the dowel fails by buckling? Let $\sigma_Y = 8000$ psi and $E = 1.7 \times 10^6$ psi for the dowel.

9.10 A column with fixed ends is made from 6061-T6 aluminum. The column has a 40-mm-square cross-section. Determine the critical buckling load if the column is 4 m long.

9.11 Determine the maximum load F that can be applied to the truss shown in Figure P9.11 without causing buckling in member CD. Each of the members is made from A36, 3-in.-diameter, 0.30-in.-thick construction pipe. All joints are pinned.

Figure P9.11

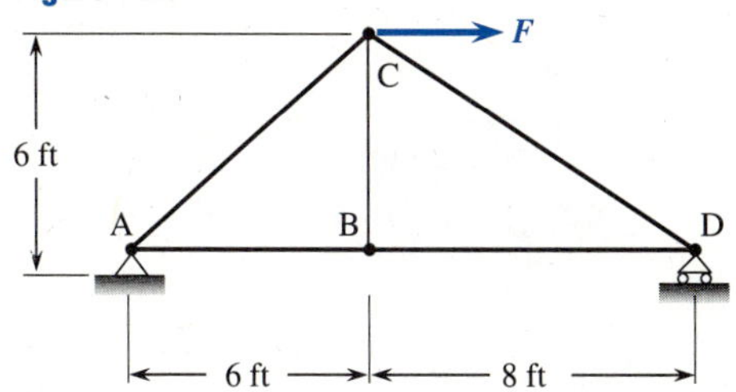

9.12 The solid square and solid circular cross-sections shown are used for a column 100 in. long. Determine the slenderness ratio of each if dimension d = 2 in.

Figure P9.12

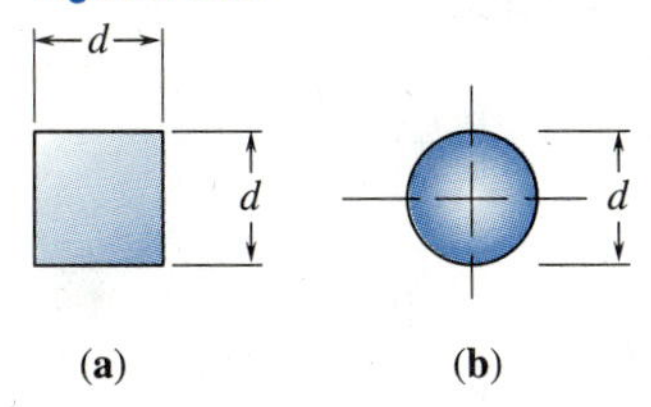

9.13 Calculate the critical buckling stress and the critical buckling load for both of the columns in Problem 9.12 if the ends are pinned and the material is aluminum. Which of these two is the more "efficient" column? HINT: Let efficiency in this case be defined as critical buckling load divided by total weight.

9.14 Determine the load P that will cause the column AB (Figure P9.14) to buckle. Assume the top B is also pinned. Let $E = 29 \times 10^6$ psi for the column.

Figure P9.14

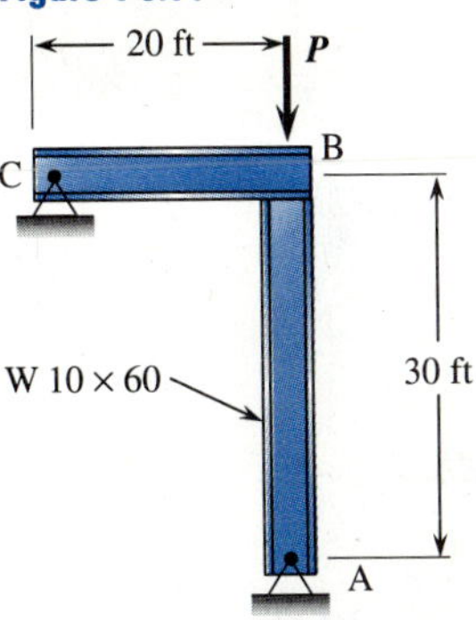

9.15 A column is made by nailing nominal 2 in. × 6 in. boards together as shown. The column length is 8 ft. Use $E = 1.5 \times 10^6$ psi. Determine the critical column load. Assume pinned ends.

Figure P9.15

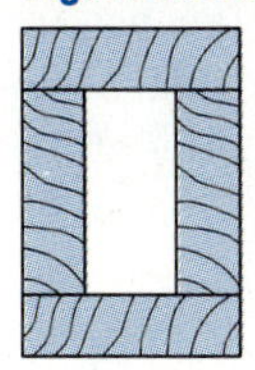

9.16 A column is made of a solid round bar of steel with a 30 mm diameter. The column is 2 m long. Use E = 200 GPa. Determine the critical column load. Assume pinned ends.

9.17 The column of Problem 9.16 is now hollowed out so that the outer diameter is 30 mm and the inner diameter is 15 mm. Determine the critical column load. Assume pinned ends. Compare the answer with that of Problem 9.16. Which column is more efficient based on the load per weight?

9.18 A hollow square column is made by welding 100 mm long by 5 mm thick plates together as shown. Determine the critical column load if the column is 3 m long and pinned at its ends. Use $E = 200$ GPa.

Figure P9.18

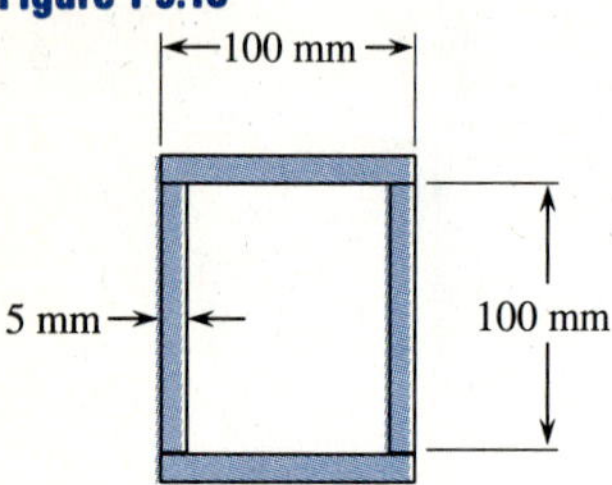

9.19 A vehicle jack is to be designed to support a maximum weight W of 5000 lb. The opposite-handed threads on the two ends of the screw AC are cut to allow the link angle θ to vary from 15° to 75°. The links are hot-rolled steel bars with minimum yield strength of 50 ksi. Each of the four links is to consist of two bars, one on each side of the screw. The bars are 12 in. long and have a width of 1 in. Assume the bars are pin-connected. Determine the thickness of the bars based on a factor of safety of 3.

Figure P9.19

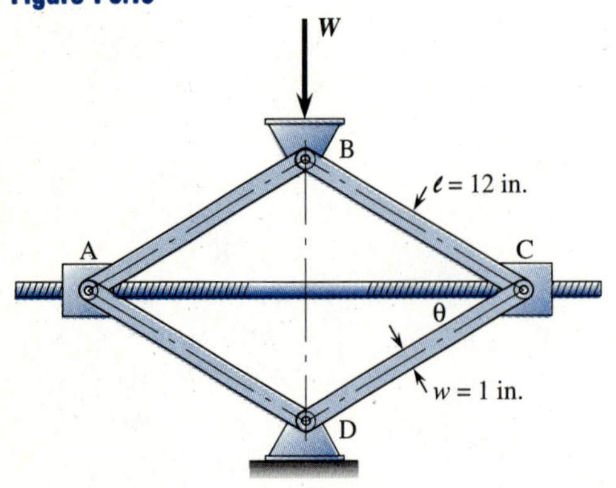

9.20 A hand-operated toggle press is subjected to a load $F = 50$ lb. Determine the size of a square cross-section member CD to prevent buckling, if the yield strength of the member CD is 36 ksi and a factor of safety of 2 against buckling is used. Let $\theta = 30°$.

Figure P9.20

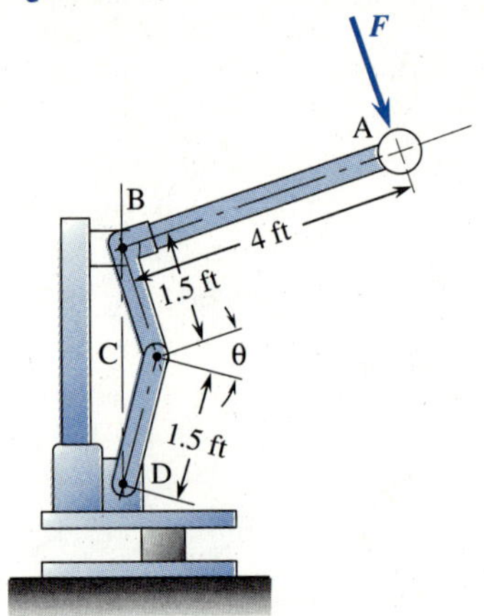

9.21 Compare the column buckling loads for **(a)** the I shape, **(b)** the box shape, and **(c)** the hollow circular shape shown. All have nearly the same cross-sectional area. Assume the columns are pinned at each end. Recommend the most efficient one based only on material savings. Let $b = 6$ in. and $t = 0.5$ in.

Figure P9.21

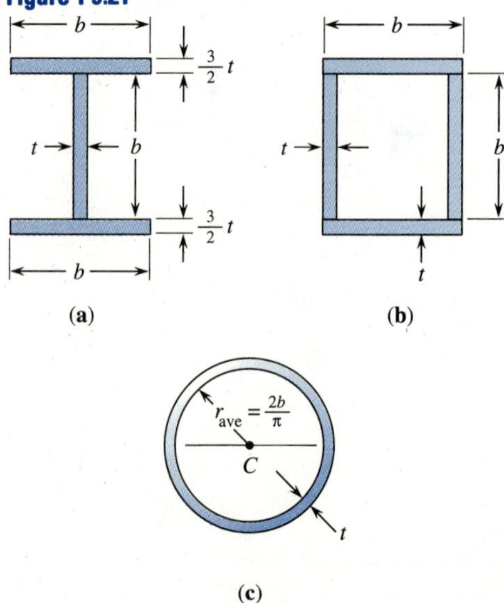

9.22 A steel column is used as an automobile lift as shown. The column is a hollow round pipe (a 4-in.-diameter pipe with wall thickness of 0.337 in.). The maximum vehicle weight to be supported is 6000 lb. Using a factor of safety of 4, determine if the pipe is safe. Here G denotes the center of gravity of the vehicle. Assume the pipe is fixed at the base and free at the top. Let $E = 29 \times 10^6$ psi.

Figure P9.22

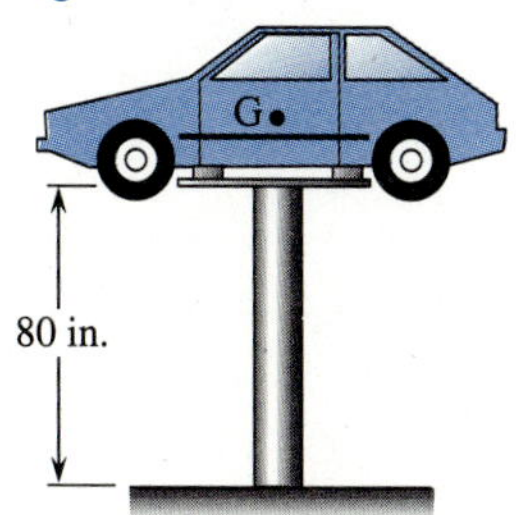

9.23 A vertical steel bar is built in at both ends while unstressed at 70°F. The bar is 20 ft long with a square cross section 2 in. × 2 in. Determine the temperature at which the bar will buckle. Let $E = 29 \times 10^6$ psi and the coefficient of thermal expansion be 6.5×10^{-6}/°F.

Section 9.4

9.24 Figure P9.24 shows an approximate stress-strain diagram for a particular steel. Use Eq. (9.43) to plot the critical stress vs. the effective slenderness ratio and then determine the critical buckling stress for a 2-in. diameter solid rod 30 in. long made from this material. Assume ends are pinned.

Figure P9.24

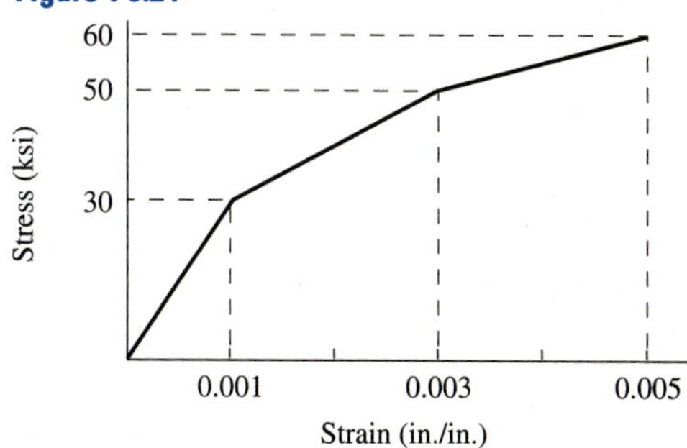

9.25 Use the critical stress versus slenderness ratio plot constructed for Problem 9.24 to determine the critical buckling stress of a 8-ft-long column made from the same material. The column has one end fixed and one end free and has the cross-section of a wide-flange W6 × 25 beam (see Appendix C).

9.26 Using the tabulated stainless steel stress-strain data shown, construct a column design curve, that is, a curve of tangent modulus stress versus slenderness ratio.

σ (ksi)	ϵ (in./in.)
10.0	0.00004
20.0	0.00008
25.0	0.00010
27.5	0.000115
30.0	0.000145
32.5	0.00018
35.0	0.00025
37.5	0.00075

9.27 The stress-strain diagram for an aluminum alloy is shown below. Plot the column buckling stress versus slenderness ratio curve.

Figure P9.27

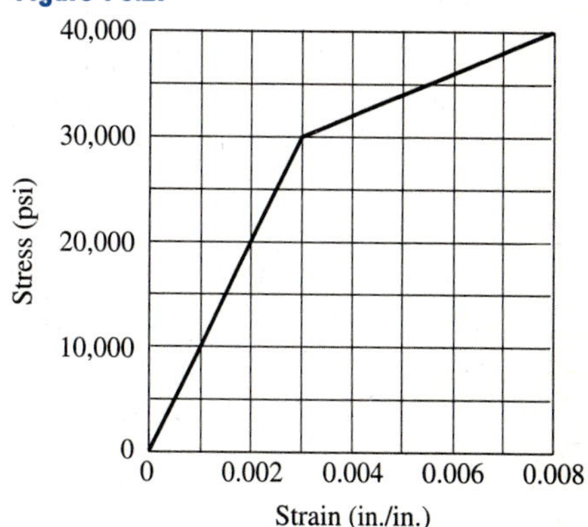

Sections 9.2, 9.3, and 9.5

Problems 9.28 through 9.42 may be Euler or Johnson columns. Use the appropriate formulas for each problem.

9.28 Determine the critical buckling load for a simply supported reinforced column 5 ft long made by bolting two 4 × 4 × 0.5 in. steel angles together back-to-back (see Appendix C). Use $\sigma_{cY} = 36$ ksi and $E = 29 \times 10^6$ psi.

9.29 A column with both ends rounded (pinned) is made from structural steel ($\sigma_{CY} = 250$ MPa, $E = 200$ GPa) with a 10 mm × 25 mm cross-section. Calculate the buckling load for column lengths of **(a)** 100 mm, **(b)** 300 mm, and **(c)** 500 mm.

9.30 A rectangular bar 0.5 × 1.5 in. is loaded as a column with one end fixed and one end pinned. Calculate the buckling stress and buckling load for lengths of **(a)** 20 in. and **(b)** 40 in. Use $\sigma_Y = 36$ ksi and $E = 29 \times 10^6$ psi.

9.31 Calculate the actual length of a structural steel post corresponding to the transition slenderness ratio. The post has a 4-in. outside diameter (OD) and a 0.5-in. thick wall. Use $\sigma_Y = 36$ ksi and $E = 29 \times 10^6$ psi. Assume both ends pinned.

9.32 A 600 mm length is cut out of the meter stick of Example 9.1 and loaded axially with both ends fixed. Calculate the critical column buckling load. $\sigma_{CY} = 14$ MPa.

9.33 Determine the magnitude of the force F that will cause buckling in either of the members of the pinned frame shown. Both of the aluminum members are 50 mm × 50 mm in cross-section. Use $E = 70$ GPa and $\sigma_{cY} = 255$ MPa.

Figure P9.33

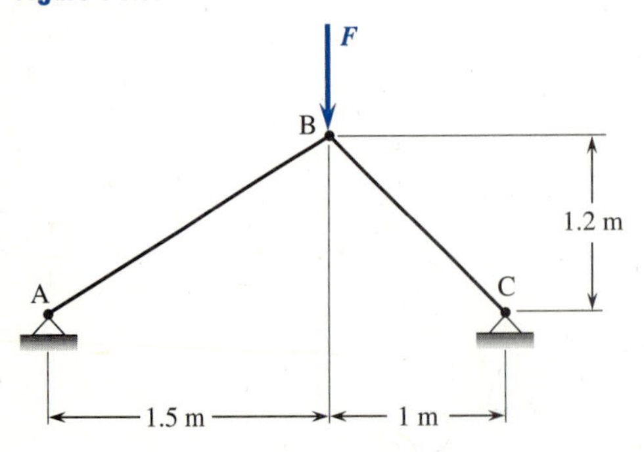

9.34 A home-made hoist is made from two 12-ft-long pine 4 × 4 inch posts (actual dimensions 3½ × 3½ in.) as shown. Determine the load F that will cause the posts to buckle. Let $\sigma_{CY} = 2000$ psi and $E = 1.7 \times 10^6$ psi. Assume ends pinned.

Figure P9.34

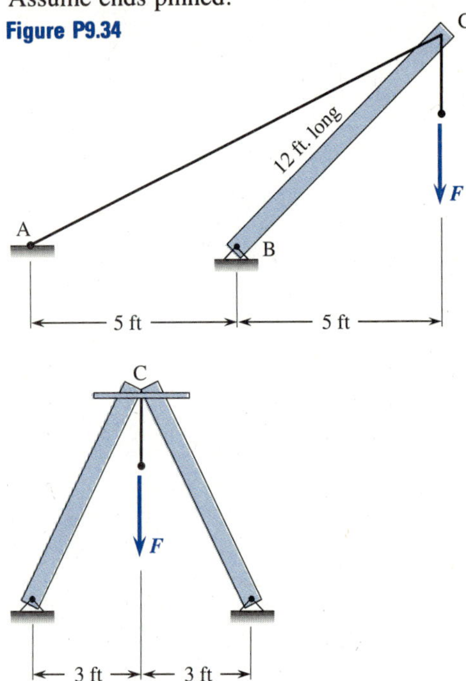

9.35 Each of the members of the truss shown is made from A36 steel construction pipe (2-in.-diameter × 0.154 in. wall). Determine the maximum load F possible without any of the compression members buckling as a column. All joints are pinned. Let $\sigma_Y = 36$ ksi and $E = 29 \times 10^6$ psi.

Figure P9.35

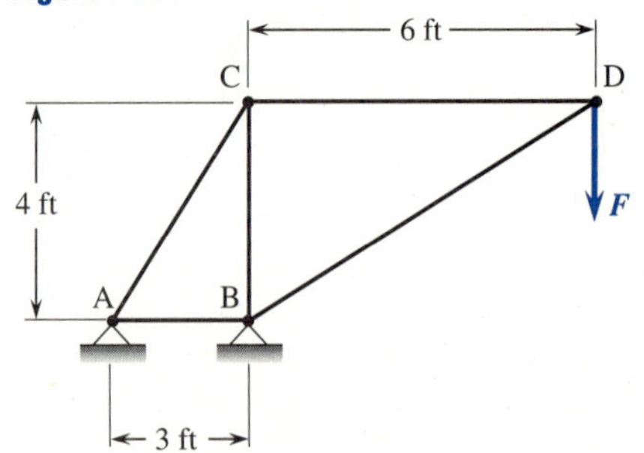

9.36 A 50-ft column is made by welding together two steel S 20 × 86 I beams as shown. Assuming pinned ends, calculate the Euler buckling load.

Figure P9.36

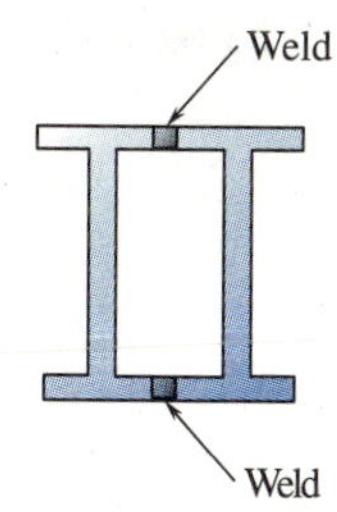

9.37 If the column of Problem 9.36 is only 10 ft long, determine the buckling load.

9.38 If the column of Problem 9.36 is only 5 ft long, determine the buckling load.

9.39 Determine the critical buckling load for a W 8 × 40 A36 structural steel column pinned at each end. The column length is 6 ft.

9.40 Determine the critical buckling load for a W 8 × 40 A36 structural steel column fixed at each end. The column length is 6 ft. Compare this answer to that of Problem 9.39.

9.41 For the W 8 × 40 column of Problem 9.39, determine the buckling load if the length is now 20 ft.

9.42 For the W 8 × 40 column of Problem 9.39, determine the buckling load if the length is now 30 ft.

Section 9.6

For Problems 9.43 through 9.61, assume that the axial loads are centric and applied at the ends of the column. The columns are supported only at the ends and they may buckle in any direction. Use the appropriate design formula in Section 9.6 to solve these problems.

9.43 A pin-ended column 20 ft long of ASTM A36 steel must carry a centric load of 100 kips. Select the lightest wide-flange shape to resist the load. Let $E = 29 \times 10^6$ psi.

9.44 A pin-ended column 12 ft long of ASTM A36 steel must carry a centric load of 100 kips. Select the lightest wide-flange shape to resist the load. Let $E = 29 \times 10^6$ psi.

9.45 A pin-ended column 4 m long resists a centric load of 500 kN. Determine the size of a solid square steel column to resist the load. Let $E = 200$ GPa and $\sigma_Y = 250$ MPa.

9.46 A 4-in.-diameter construction pipe with wall thickness of 0.237 in. is used as a pin-ended column to resist an axial load of 30 kips. Determine the longest permissible length L of the column. Let $E = 29 \times 10^6$ psi and $\sigma_Y = 36$ ksi.

9.47 A solid circular steel rod with pinned ends and length 3 m must support an axial load of 400 kN. Assume $E = 200$ GPa and $\sigma_Y = 250$ MPa. Determine the minimum permissible diameter d of the column.

9.48 For the bar BC of A36 steel supporting the beam shown, determine the size of a square bar to prevent buckling.

Figure P9.48

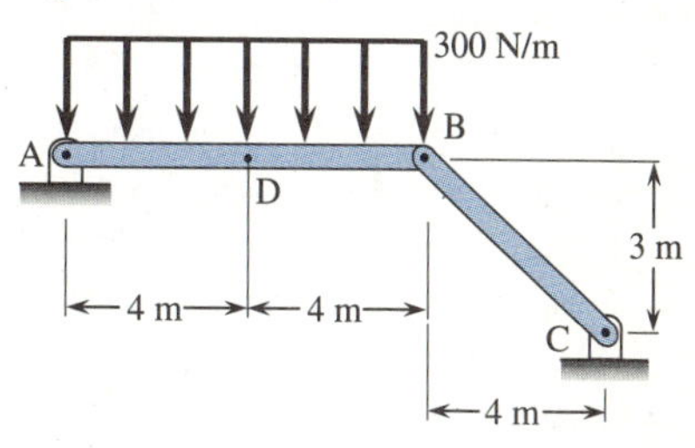

9.49 For the load frame shown, determine the size of a round bar of A36 steel to prevent buckling of member CF.

Figure P9.49

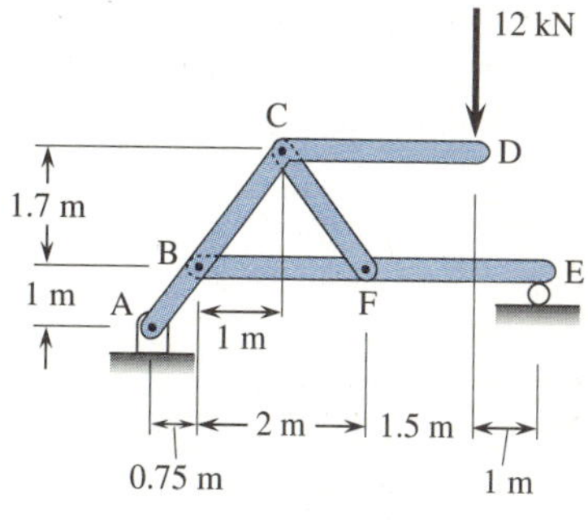

Load frame

9.50 For the vise loaded as shown, determine the thickness of bar AB needed to prevent buckling. Bar AB of A36 steel is a rectangular cross-section with a width of 1 in.

Figure P9.50

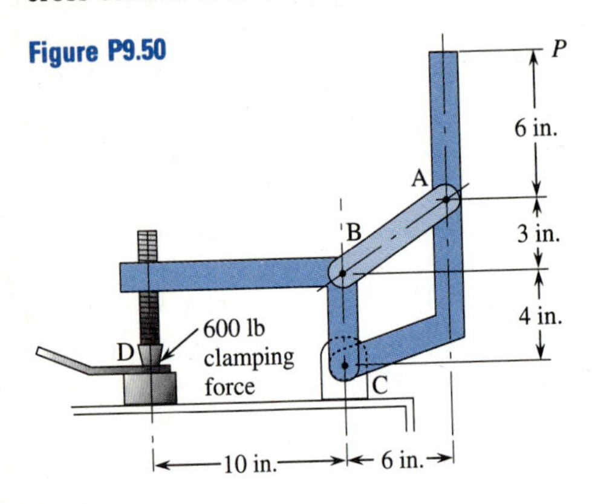

9.51 A 2014-T6 aluminum alloy is used as a pin-ended column 60 in. long. The column is a circular cross-section and supports a centric load of 10 kips. Determine the minimum permissible diameter of the column.

9.52 A 6061-T6 aluminum alloy is used as a pin-ended column of length 2 m. The column is a hollow square cross-section 60 mm wide with a thickness of 7.5 mm. Determine the maximum allowable centric load.

9.53 A square rod of 2014-T6 aluminum alloy 1 m long must resist an axial load of 150 kN. The column is fixed at one end and free at the other. Determine the size of the column.

9.54 An aluminum alloy (6061-T6) tube with a 3-in. outside diameter is used to resist an axial compressive force of 20 kips. The tube is 50 in. long with pinned ends. The stock of tubes available have wall thicknesses in increments of 1/16 in. Determine the lightest tube to resist the load.

9.55 A square timber column with pinned ends must support an axial load of 30 kips. Its length is 8 ft. Assume $E = 1.6 \times 10^6$ psi and the allowable stress parallel to the grain is 1000 psi. Determine the size of the column.

9.56 A square timber column with pinned ends and length 4 m must resist an axial compressive load of 100 kN. Assume $E = 12$ GPa and the allowable stress parallel to the grain is 9 MPa. Determine the size of the column in millimeters.

9.57 A square timber column with pinned ends and length 24 ft must resist an axial compressive load of 90 kips. Assume $E = 1.6 \times 10^6$ psi and allowable compressive stress parallel to the grain of 1200 psi. Determine the size of the column.

9.58 A square timber column of Douglas fir with $E = 1.5 \times 10^6$ psi and allowable compressive stress parallel to the grain of 1400 psi must resist a load of 35 kips. The column length is 12 ft. Determine the size of a square column to resist the load.

9.59 A barn is to be restored as a restaurant. In its basement are southern pine timbers 11.5×11.5 in. and 8 ft long. They must support axial loads of 10 kips. Are they safe? (The allowable compressive stress parallel to the grain is 1200 psi.)

9.60 Three wooden planks, each 5 mm $\times$ 10 mm cross-section are securely nailed together to form a column of dimensions 10×15 mm. The length of the column is 3 m. The wood is southern pine with $E = 12$ GPa and allowable stress in compression parallel to the grain of 10 MPa. Determine the maximum allowable centric load for the column.

COMPUTER PROBLEMS

The following problems are suggested to be solved using a programmable calculator, microcomputer, or mainframe computer. It is suggested that you write a simple computer program in BASIC, FORTRAN, PASCAL, or some other language to solve these problems.

C9.1 Write a program to determine the critical buckling load and stress in a column based on either Euler's or Johnson's formula. The program should check which formula is appropriate. The input should include the cross-sectional area A and moment of inertia I of the column, the type of fixities (support conditions), the actual length L of the column, the modulus of elasticity E, and the compressive yield strength σ_{CY} of the material of the column. Check your program using the following data: $A = 192$ mm^2, $I = 1024$ mm^4, $L = 1$ m, $E = 3.1$ GPa, and $\sigma_{CY} = 8$ ksi. Assume the column to be pinned at each end.

C9.2 Write a program to analyze a steel column based on the AISC Code formulas of Section 9.6. The input should include the yield strength of the steel, the modulus of elasticity of the steel, the cross-sectional area A and moment of inertia I of the column, the length of the column, and the end-fixity conditions. Check your program using the following data: $A = 18.06$ in.^{2}, $I = 299$ in.^{4}, $L = 30$ ft, $E = 29 \times 10^6$ psi, $\sigma_Y = 36$ ksi, and pinned ends.

C9.3 Write a program to analyze an aluminum alloy column based on the formulas in Section 9.6. The input should include the choice of the two aluminum alloys included in Section 9.6, the area A and moment of inertia I of the column, the length L of the column, and the allowable stresses for the different alloys. Check your program using the following data: $A = 1735$ mm^2, $I = 240 \times 10^3$ mm^4, aluminum alloy 2014-T6, and $L = 3$ m.

C9.4 Write a program to analyze a timber column based on the formulas in Section 9.6. The input should include the allowable compressive stress parallel to the grain σ_c, the modulus of elasticity E, the length of the column, the end fixities, and the dimensions of the column. Check your program using the following data: $\sigma_c = 1400$ psi, $E = 1.6 \times 10^6$ psi, $L = 8$ ft, and 3.5 in. $\times$ 3.5 in. column cross-section dimensions. Assume pinned ends.

10 Energy Methods

10.1 INTRODUCTION

In the preceding chapters, we used the equilibrium equations and relations between forces (and moments) and deformation due to different kinds of loading conditions to solve basic problems in mechanics of materials. These solutions usually involved finding stresses, strains, and deformations in a load-bearing structure or machine. Specifically, to determine deformations in beams, we used the geometric methods of double integration and moment area presented in Chapter 6. These methods have the advantage of yielding analytical expressions for deflection and slope throughout the beam. However, as we will see in this chapter, energy methods provide an attractive alternative to obtain deformations in structures. When we desire the deflection or slope at a specific point in a structure, such as a truss, beam, or frame, energy methods provide a faster way to obtain these deformations.

In this chapter, we then introduce energy methods to obtain solutions to problems in mechanics of materials. These energy methods are based on the principle of conservation of energy. To apply energy methods for the solution of problems, we must understand the concept of external work and internal work or strain energy. We introduce these concepts first. We will observe that on application of the principle of conservation of energy and equating external work to strain energy (also called the work-energy principle), we can obtain deformations in structures such as trusses, frames, and beams.

The direct solution of problems by the work-energy method is only feasible for cases where a single load is applied to the structure. Therefore, we introduce generalized methods of virtual work (or dummy unit load) and Castigliano's theorem to solve for deflections and slopes for trusses, beams, and frames subjected to a number of loadings. We then illustrate the use of Castigliano's theorem to solve statically indeterminate problems.

Finally, using energy methods, we can analyze members subjected to not only slowly applied (static) loads, but also to impact or energy loads. Hence, we consider the special topic of structures subjected to impact loads.

10.2 EXTERNAL WORK AND STRAIN ENERGY

To understand the energy methods, we must first introduce the concepts of external work and strain energy for various kinds of loadings and members.

External Work Due to Force

The external work due to a force F is most easily understood if we consider the bar BC of uniform cross-sectional area A attached to a rigid support at B and subjected to a slowly applied increasing load F at C as shown in Figure 10.1. On plotting load F versus deformation x, the bar will have a certain load-deformation diagram shown in Figure 10.2.

When the force moves through a differential displacement dx in the same direction as the force, the differential work done by the force is

$$dW = Fdx \tag{10.1}$$

and is equal to the area width dx and height F under the load-deformation diagram shown in Figure 10.3.

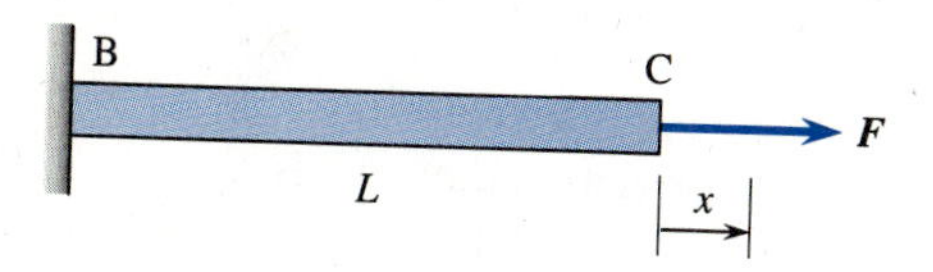

Figure 10.1 Bar subjected to axial force.

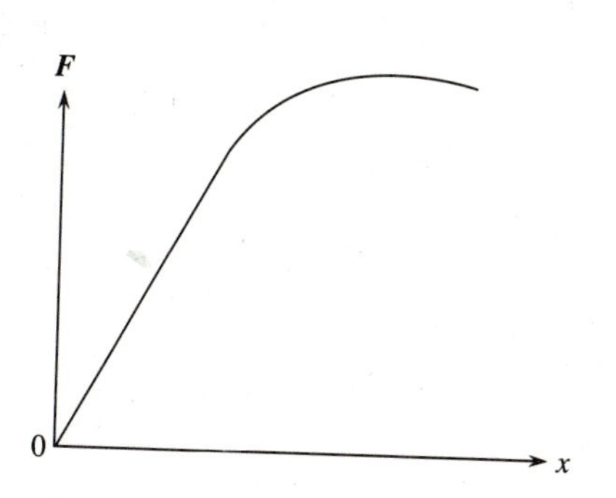

Figure 10.2 Load-deformation diagram of the bar.

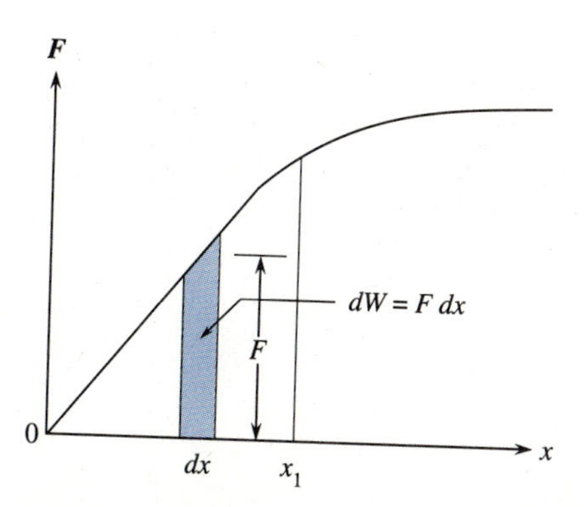

Figure 10.3 Load-deformation diagram.

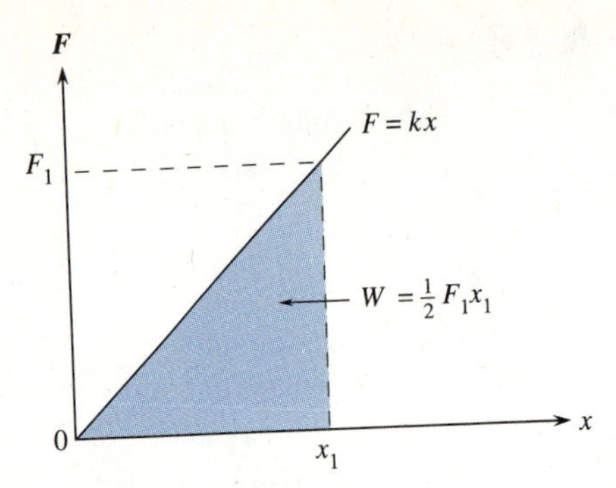

Figure 10.4 Load-deformation diagram for bar with linear-elastic behavior.

The total work W done by the force as the bar deforms an amount x_1, is then

$$W = \int_0^{x_1} F dx \tag{10.2}$$

and is equal to the area under the load-deformation diagram between $x = 0$ and $x = x_1$.

Consider now the effect of the axial force applied to the end of a linear and elastic deformable bar. In this case, the load-deformation diagram is a straight line shown by Figure 10.4. The load, being directly proportional to the deformation, allows us to express a load-deformation equation as

$$F = kx \tag{10.3}$$

where k is the proportionality constant. Substituting Eq. (10.3) for F into Eq. (10.2), the work becomes

$$W = \int_0^{x_1} kx\, dx = \frac{1}{2}kx_1^2 \tag{10.4}$$

or using Eq. (10.3) in Eq. (10.4), we have

$$W = \frac{1}{2}F_1x_1 \tag{10.5}$$

where F_1 is the value of the force at the deformation x_1. The units of work are then ft · lb or in. · lb in the U.S. Customary System and N · m or joules (J) (1 J = 1 N · m) in the SI system of units. We also note from Eq. (10.5) that as the force is gradually applied to the bar, its magnitude increases from zero to some value F_1 and the work is equal to the average force magnitude ($F_1/2$) times the total deformation x_1.

Now assume that a force F_1 is already acting on a bar and another force F' is slowly applied to the bar so that an additional deformation x' occurs. The work done by the force F_1 already acting on the bar (not the work due to F') due to the additional deformation x' is

$$W' = F_1x' \tag{10.6}$$

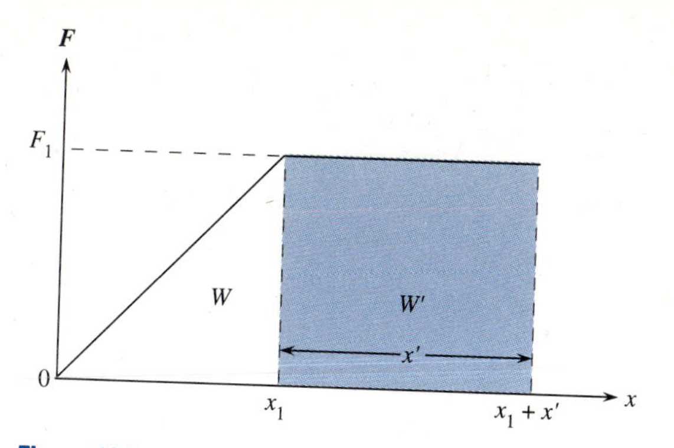

Figure 10.5 Load-deformation diagram showing additional work due to a force already on a bar when additional deformation occurs.

This is shown in Figure 10.5, where the shaded rectangular area represents this additional work. In this case, the force F_1 being already on the bar when the additional deformation due to another force F' was applied, results in additional work given simply by the force magnitude times the deformation due to F'.

External Work Due To a Moment

The external work due to a moment is defined by the product of the magnitude of the moment M and the angle through which the moment rotates. To understand this concept, consider Figure 10.6 where a beam is pinned at A and a force applied at B. The length AB rotates to AB′ through angle $d\theta$.

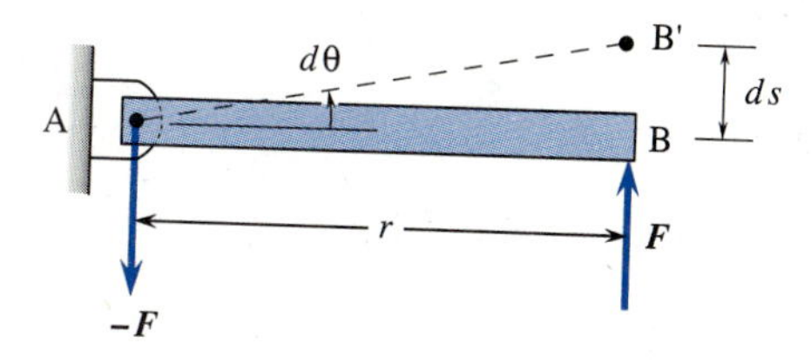

Figure 10.6 Body subjected to a couple moment.

The work of F moving through displacement ds is

$$dW = Fds$$

and $ds = rd\theta$. The moment is

$$M = Fr$$

Therefore, the work of M moving through angle $d\theta$ is

$$dW = Md\theta$$

The total work of the moment moving through total angle θ is

$$W = \int_0^{\theta} M d\theta$$

If the moment is applied gradually, we have

$$W = \frac{1}{2} M\theta \tag{10.7}$$

If the moment is already applied to the structure and other loadings additionally distort the structure so that M moves through a rotation θ', the work is

$$W' = M\theta' \tag{10.8}$$

Strain Energy and Strain Energy Density

Consider again the bar of Figure 10.1 subjected to the slowly applied loading F. The force F results in an axial stress in the bar and produces strains in the bar. These strains increase the energy level of the bar. This energy of the strain due to stress is called *strain energy* and is denoted by U. This strain energy is then equal to the external work done by the force F.

We can obtain an expression for the strain energy in terms of the axial normal stress and the normal strain as follows. Since the work given by Eq. (10.2) is equal to the strain energy, we have

$$W = U = \int_0^{x_1} F dx \tag{10.9}$$

We divide the force by the cross-sectional area A and the deformation by the length L of the bar in Eq. (10.9) to obtain

$$\frac{U}{AL} = \int_0^{x_1} \frac{F}{A} \frac{dx}{L} \tag{10.10}$$

Now recall that normal stress is $\sigma_x = F/A$ and differential normal strain is $d\epsilon_x = dx/L$, we write Eq. (10.10) as

$$\frac{U}{AL} = \int_0^{\epsilon_1} \sigma_x d\epsilon_x \tag{10.11}$$

where ϵ_1 is the value of the strain corresponding to x_1. Noting that $AL = V$, the volume of the bar, we express Eq. (10.11) as

$$\frac{U}{V} = \int_0^{\epsilon_1} \sigma_x d\epsilon_x \tag{10.12}$$

We now define the strain energy per unit volume, U/V, as the *strain energy density* u. Equation (10.12) then becomes

$$u = \int_0^{\epsilon_1} \sigma_x d\epsilon_x \tag{10.13}$$

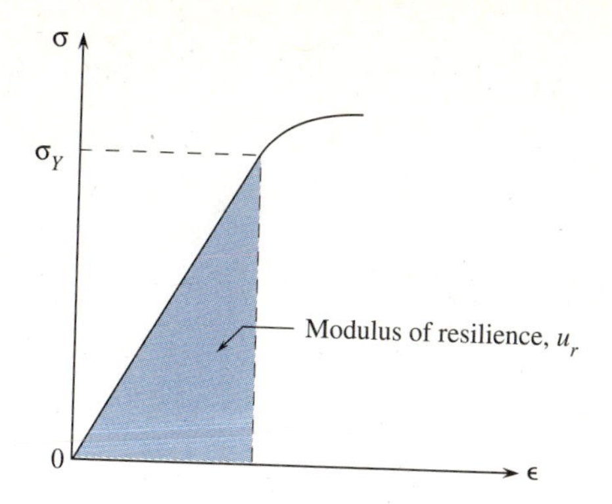

Figure 10.7 Modulus of resilience.

For linear-elastic material, the stress is proportional to the strain and we recall

$$\sigma_x = \epsilon_x E \tag{10.14}$$

Substituting Eq. (10.14) into Eq. (10.13), we obtain

$$u = \int_0^{\epsilon_1} E\epsilon_x d\epsilon_x = \frac{E\epsilon_1^2}{2} \tag{10.15}$$

Using Eq. (10.14), we also have $\sigma_1 = E\epsilon_1$, and using this relation in Eq. (10.15), Eq. (10.15) becomes

$$u = \frac{\sigma_1^2}{2E} \tag{10.16}$$

The strain energy density when the stress is equal to the proportional limit is called the *modulus of resilience, u_r* or *MR* (as described in Section 2.5). In practice, the proportional limit is normally difficult to obtain and the yield strength is then used in Eq. (10.16) for σ_1. The modulus of resilience is then

$$u_r = \frac{\sigma_Y^2}{2E} \tag{10.17}$$

As shown in Figure 10.7, u_r is the area under the stress-strain diagram up to the yield stress σ_Y. This is often taken as the area under the straight-line portion of the diagram. The modulus of resilience measures the capacity of a material to absorb energy in its linear-elastic range. Based on Eq. (10.17), materials with higher yield stress and lower modulus of elasticity can absorb greater elastic energy.

A related term used to describe the ability to absorb energy is *toughness*. Toughness is the ability to absorb energy up to fracture. Toughness is measured by the *modulus of toughness u_t* or *MT* (as described in Section 2.5), which is the area under the entire stress-strain diagram.

In general we can express the strain energy density for a differential element of material as

$$u = \frac{dU}{dV} \tag{10.18}$$

and the total strain energy by integrating Eq. (10.18) as

$$U = \int u dV \tag{10.19}$$

For a linear-elastic material, with $\sigma_x = E\epsilon_x$, we then have

$$u = \frac{\sigma_x^2}{2E} \tag{10.20}$$

by replacing σ_1 with σ_x in Eq. (10.16).

Substituting Eq. (10.20) into Eq. (10.19), a useful form of the elastic strain energy is

$$U = \int \frac{\sigma_x^2}{2E} dV \tag{10.21}$$

EXAMPLE 10.1

Determine the modulus of resilience for **(a)** American Society for Testing Materials (ASTM) A36 structural steel and **(b)** aluminum alloy 6061-T6.

Solution

Using the properties of the materials table in Appendix B, we obtain

(a) For the structural steel

$$E = 29 \times 10^6 \text{ psi} \quad \text{and} \quad \sigma_Y = 36{,}000 \text{ psi}$$

Using Eq. (10.17), the modulus of resilience is

$$u_r = \frac{\sigma_Y^2}{2E} = \frac{(36{,}000)^2}{2 \times 29 \times 10^6} = 22.34 \text{ in.} \cdot \text{lb}$$

(b) For the aluminum alloy

$$E = 10 \times 10^6 \text{ psi} \quad \text{and} \quad \sigma_Y = 37{,}000 \text{ psi}$$

The modulus of resilience is

$$u_r = \frac{\sigma_Y^2}{2E} = \frac{(37{,}000)^2}{2 \times 10 \times 10^6} = 68.45 \text{ in.} \cdot \text{lb}$$

The modulus of resilience is about three times as great for the aluminum alloy as for the structural steel.

10.3 ELASTIC STRAIN ENERGY DUE TO AXIAL FORCE

Consider a uniform cross sectional area bar subjected to a centric axial force F as shown in Figure 10.8. This problem was treated in Chapter 2. For this case, the normal stress σ_x was assumed to be uniformly distributed over the cross-sectional area.

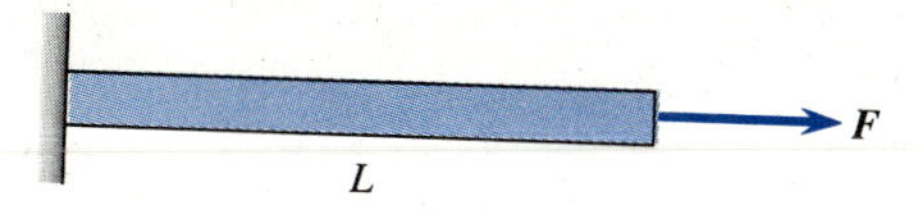

Figure 10.8 Uniform bar subjected to centric axial force.

In this case, the internal force is $P = F$ acting over the whole length L of the bar and $\sigma_x = P/A$. Therefore, Eq. (10.21) becomes

$$U = \int \frac{P^2}{2EA^2} dV \quad \textbf{(10.22)}$$

Recognizing that $dV = Adx$ and integrating Eq. (10.22), we have

$$U = \frac{P^2L}{2AE} \quad \textbf{(10.23)}$$

EXAMPLE 10.2

Determine the elastic strain energy **(a)** in the uniform cross-sectional area bar and **(b)** the stepped bar shown in Figure 10.9 in terms of the centric axial force P, cross-sectional area A, length L, and modulus of elasticity E.

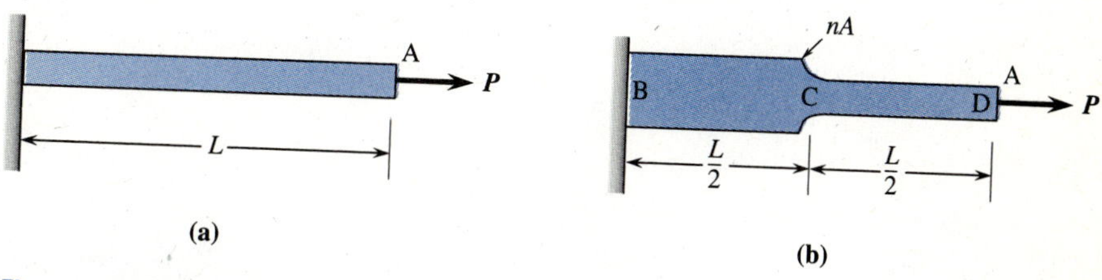

Figure 10.9 (a) Uniform bar and (b) stepped bar.

Solution

(a) Using Eq. (10.23), the strain energy in the uniform bar is

$$U_1 = \frac{P^2L}{2AE} \quad \textbf{(a)}$$

(b) Using Eq. (10.23) for the two portions of the stepped bar, the strain energy becomes

$$U_n = \frac{P^2(L/2)}{2AE} + \frac{P^2(L/2)}{2(nA)E}$$

$$= \frac{P^2L}{4AE}\left(1 + \frac{1}{n}\right)$$

or

$$U_n = \frac{P^2L}{2AE}\left(\frac{1+n}{2n}\right) \qquad \textbf{(b)}$$

We note for $n = 1$ (the uniform bar case), we obtain Eq. (a) again and for $n > 1$, $U_n < U_1$. For instance, letting $n = 2$, we have

$$U_n = \frac{P^2L}{2AE}\left(\frac{3}{4}\right)$$

Since the maximum stress occurs in the smaller section CD of the stepped bar and is equal to $\sigma_{max} = P/A$, we conclude that for a given allowable stress, increasing the diameter of portion BC of the bar results in a decrease of the total energy-absorbing capacity of the bar. For bars subjected to impact or energy loads, this result is very important. We conclude that changes in cross-sectional area should be avoided whenever possible if a member is going to be subjected to impact or energy loads.

10.4 ELASTIC STRAIN ENERGY DUE TO SHEAR STRESS

The strain energy density for an element of material subjected to pure shear stresses in the x-y plane (Figure 10.10) is obtained in a manner analogous to that of Eq. (10.13) for axial loading. The strain energy density is

$$u = \int_0^{\gamma_{xy}} \tau_{xy}\, d\gamma_{xy} \qquad \textbf{(10.24)}$$

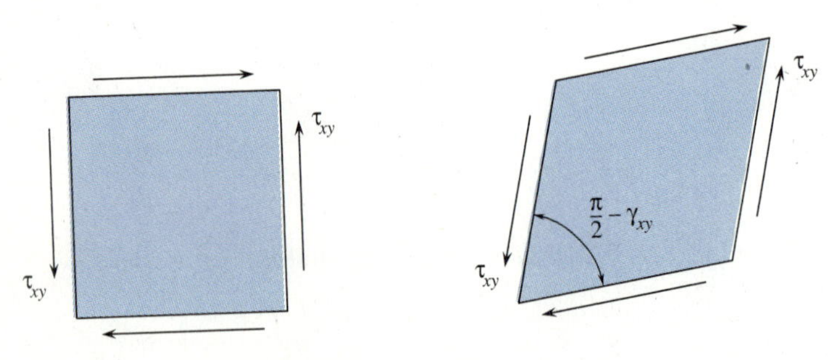

Figure 10.10 Element subjected to pure shear.

When the shear stress is within the linear-elastic range of the material, we have

$$\tau_{xy} = G\gamma_{xy} \tag{10.25}$$

where G is the modulus of rigidity of the material. Substituting Eq. (10.25) for τ_{xy} into Eq. (10.24), and integrating, the strain energy density becomes

$$u = \frac{1}{2}G\gamma_{xy}^2 \tag{10.26}$$

Using Eq. (10.25) in Eq. (10.26), we write

$$u = \frac{\tau_{xy}^2}{2G} \tag{10.27}$$

Equation (10.27) is analogous to Eq. (10.20) for the axially loaded bar.

The strain energy for an element subjected to pure shear stress is then obtained using Eq. (10.19) as

$$U = \int u dV = \int \frac{\tau_{xy}^2}{2G} dV \tag{10.28}$$

Strain Energy in Direct Shear

For the special case of a member subjected to direct uniform shear force V acting over area A (Figure 10.11), we have

$$\tau_{xy} = \frac{V}{A} \tag{10.29}$$

Substituting Eq. (10.29) into Eq. (10.28), the strain energy becomes

$$U = \int \frac{V^2}{2A^2G} A dx \tag{10.30}$$

Integrating and simplifying Eq. (10.30), we obtain

$$U = \frac{V^2L}{2AG} \tag{10.31}$$

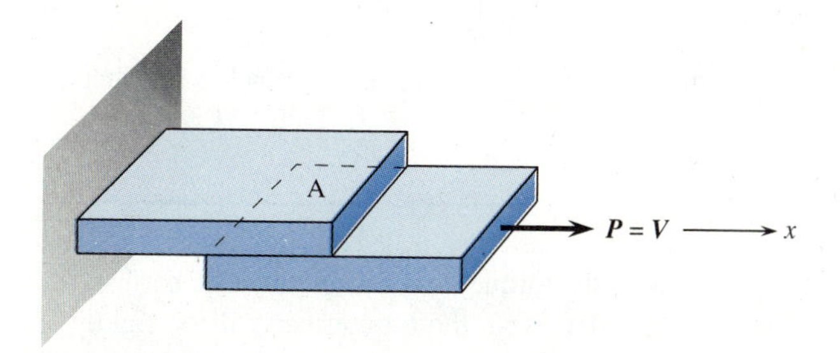

Figure 10.11 Member subjected to direct shear.

10.5 ELASTIC STRAIN ENERGY DUE TO TORSION

Consider the circular shaft BC of length L subjected to a twisting moment T at end C and fixed at end B (Figure 10.12). Recall that the shear stress in any cross-section is given by

$$\tau_{xy} = \frac{Tr}{J} \tag{10.32}$$

where J is the polar moment of inertia of the cross section and T is the internal torque in the section.

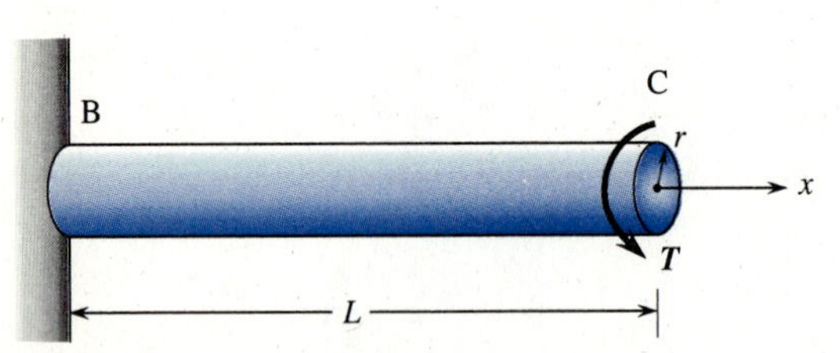

Figure 10.12 Circular shaft subjected to torque.

Substituting for τ_{xy} from Eq. (10.32) into Eq. (10.28), the strain energy becomes

$$U = \int_0^L \frac{T^2 r^2}{2GJ^2} dV \tag{10.33}$$

Now $dV = dAdx$, as τ_{xy} now varies with a differential area dA of the cross-sectional area. Also noting that T, G, and J are at most functions only of x, we have for U

$$U = \int_0^L \frac{T^2}{2GJ^2} (\int r^2 dA) dx \tag{10.34}$$

Recalling that $\int r^2 dA = J$, Eq. (10.34) becomes

$$U = \int_0^L \frac{T^2}{2GJ} dx \tag{10.35}$$

For constant cross-sectional area, constant G, and constant internal torque over length L, Eq. (10.35) becomes

$$U = \frac{T^2 L}{2GJ} \tag{10.36}$$

Equation (10.36) is analogous to Eq. (10.23) for axially loaded members.

EXAMPLE 10.3

For the torsion bar in the automotive suspension system shown in Figure 10.13, determine the elastic strain energy U due to the torque $T = 10$ kN · m. The bar has length $L = 2$ m, radius of 10 mm, and $G = 79$ GPa. Note the torsion bar will be rigidly attached to the frame and as the lower control arm moves up and down, it twists the torsion bar which then resists the twisting motion and returns the control arm to its normal position.

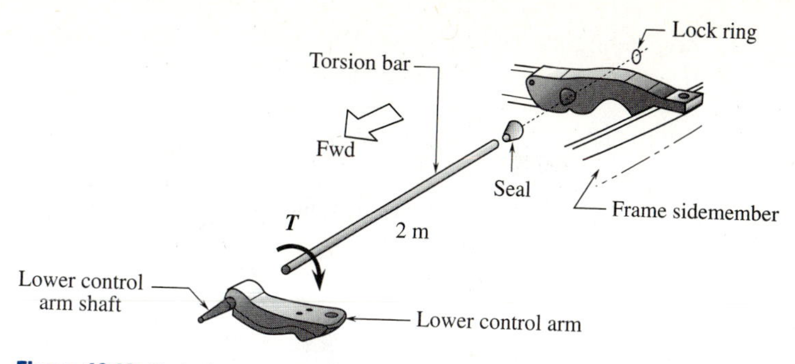

Figure 10.13 Typical torsion bar suspension.

Solution

The elastic strain energy will be obtained using Eq. (10.36) as

$$U = \frac{T^2L}{2GJ}$$

First evaluating J, we have

$$J = \frac{\pi}{2}r^4 = \frac{\pi}{2}(0.01 \text{ m})^4$$

$$= 1.571 \times 10^{-8} \text{ m}^4$$

Using the numeric quantities in Eq. (10.36), we obtain

$$U = \frac{(10 \text{ kN} \cdot \text{m})^2(2 \text{ m})}{2(79 \times 10^6 \text{ kN/m}^2)(1.571 \times 10^{-8}\text{m}^4)}$$

$$U = 80.6 \text{ kN} \cdot \text{m}$$

10.6 ELASTIC STRAIN ENERGY DUE TO BENDING

The strain energy due to bending moment in a member such as a beam or frame member can be obtained in a manner similar to that used to obtain the elastic strain energy for a uniaxially loaded member. Consider a beam subjected to a transverse loading (Figure 10.14).

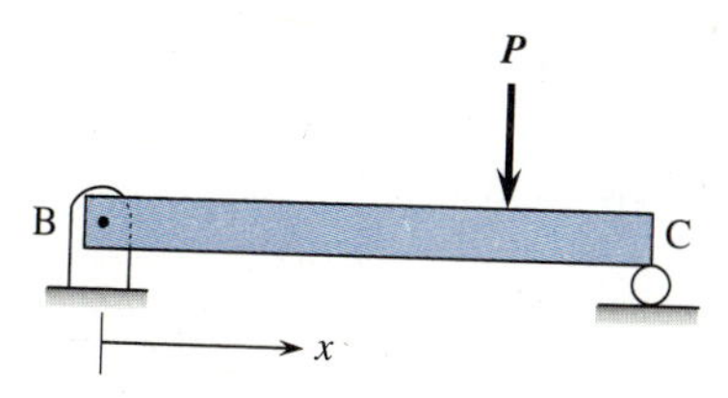

Figure 10.14 Beam subjected to transverse loading.

We assume the bending moment is M at any distance x from the left end. The normal stress due to bending moment M was derived in Chapter 5 as

$$\sigma_x = -\frac{My}{I} \tag{10.37}$$

Substituting Eq. (10.37) into Eq. (10.21), we obtain the strain energy as

$$U = \int \frac{\sigma_x^2}{2E} dV = \int \frac{M^2 y^2}{2EI^2} dV \tag{10.38}$$

We express the differential volume as $dV = dAdx$. Since M, E, and I are functions only of x, we rewrite Eq. (10.38) as

$$U = \int_0^L \frac{M^2}{2EI^2} (\int y^2 dA) dx \tag{10.39}$$

Recalling that the moment of inertia I is

$$I = \int y^2 dA \tag{10.40}$$

we write Eq. (10.39) as

$$U = \int_0^L \frac{M^2}{2EI} dx \tag{10.41}$$

EXAMPLE 10.4

For the simply supported rectangular beam shown in Figure 10.15 subjected to a concentrated force P at midspan, determine the elastic strain energy due to bending.

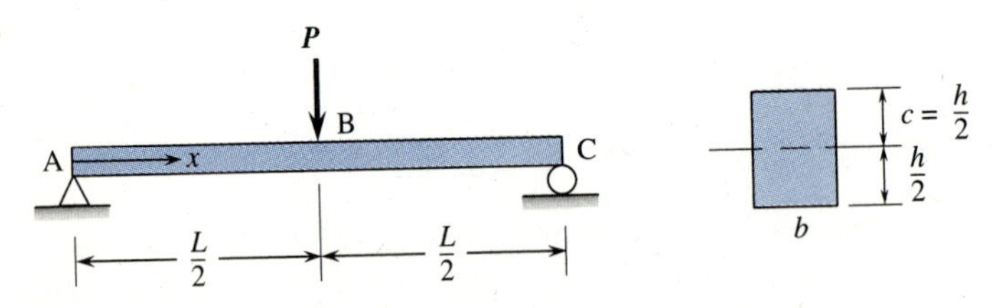

Figure 10.15 Beam subjected to midspan force.

Solution

Due to symmetry, the strain energy in section AB is equal to that in section BC. Therefore, we consider only one-half of the beam.

First, determine the bending moment in section AB.

For $0 \leq x \leq \frac{L}{2}$

$$M = \frac{Px}{2}$$

The strain energy is then obtained using Eq. (10.41) as

$$U = \int_0^L \frac{M^2}{2EI}dx = 2\int_0^{L/2} \frac{\left(\frac{Px}{2}\right)^2}{2EI}\,dx$$

$$= \frac{P^2x^3}{12EI}\bigg|_0^{L/2}$$

$$U = \frac{P^2L^3}{96EI} \tag{10.42}$$

10.7 ELASTIC STRAIN ENERGY DUE TO TRANSVERSE SHEAR STRESS IN BEAM

The strain energy due to transverse shear stress in a beam is, in general, given by Eq. (10.28). To obtain more useful expressions for the strain energy in shear due to transverse shear force V, we can express the strain energy in shear as

$$U = \int_0^L \frac{fV^2}{2GA}\,dx \tag{10.43}$$

where f is a *form factor* that depends on the shape of the cross-section of the beam. We observe that Eq. (10.43) is an extension of Eq. (10.30). For long beams with length/depth ratios of $L/d > 10$, the strain energy due to transverse shear stress (or shear force V) can be less than 1% of that due to the strain energy due to bending moment. The exception to this would be for a material with a very low shear modulus G compared to modulus of elasticity E.

For V constant, Eq. (10.43) becomes

$$U = \frac{fV^2L}{2GA} \tag{10.44}$$

The form factor for the following cross-sectional shapes are

solid circular, $f = 10/9$

thin tube, $f = 2$

rectangular, $f = 6/5$

cross sections with thin-walled webs, such as I-sections or box sections, $f = A/A_{web}$

where A is the total cross-sectional area of the section and A_{web} is the area of the web. Recall for these narrow-webbed cross-sections, the shear stress is nearly uniformly distributed over the height of the web and is approximately equal to the shear force divided by the area of the web (see Section 5.5, Example 5.12).

Example 10.5 illustrates how the form factor is obtained for a rectangular cross-section beam.

EXAMPLE 10.5

Determine **(a)** the form factor for a rectangular cross-section beam, **(b)** the strain energy due to shear force for the simply supported rectangular beam subjected to the midspan force P as given in Example 10.4, and **(c)** compare the amount of strain energy due to bending and due to shear.

Solution

The shear force diagram is shown below as

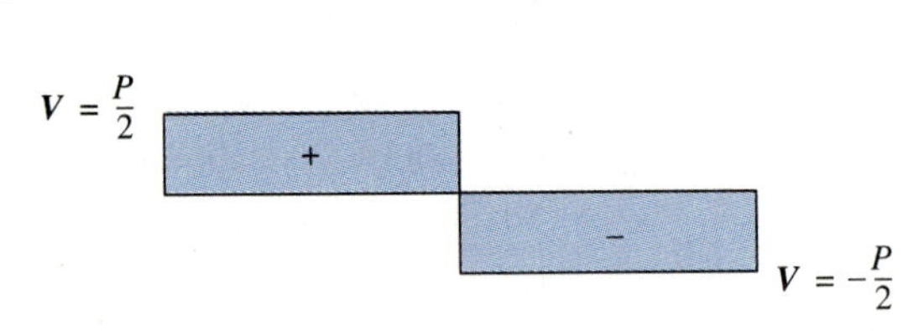

Therefore, the shear force is

$$V = \frac{P}{2} \tag{a}$$

The shear stress is given by

$$\tau_{xy} = \frac{3V}{2A}\left(1 - \frac{y^2}{c^2}\right) \tag{b}$$

Substituting Eq. (b), along with $V^2 = (P/2)^2$, into Eq. (10.28), we obtain the strain energy as

$$U_s = 2\left\{\frac{1}{2G}\int_{-h/2}^{c=h/2}\left[\frac{3(P/2)}{2A}\left(1 - \frac{y^2}{c^2}\right)\right]^2 dy \int_0^{L/2} b dx\right\}$$

$$= \frac{9P^2}{16Gb^2h^2}\int_{-h/2}^{h/2}\left(1 - \frac{2y^2}{c^2} + \frac{y^4}{c^4}\right) dybL$$

$$U_s = \frac{3P^2L}{20GA} \tag{10.45}$$

(a) If we set Eq. (10.45) equal to Eq. (10.44) and solve for the form factor, we obtain

$$\frac{f(P/2)^2L}{2GA} = \frac{3P^2L}{20GA}$$

$$f = \frac{6}{5}$$

(b) For the rectangular cross-section, $A = 12I/(h^2)$ is substituted into Eq. (10.45) to yield the strain energy in shear as

$$U_s = \frac{3P^2L(h^2)}{20G(12I)}$$

or

$$U_s = \frac{P^2Lh^2(6/5)}{96GI} \tag{c}$$

(c) Using Eq. (10.42) (obtained in Example 10.4), we write the strain energy in shear in terms of that in bending as

$$U_s = \frac{P^2L^3}{96EI}\left(\frac{6Eh^2}{5GL^2}\right)$$

or

$$U_s = U_b\left(\frac{6Eh^2}{5GL^2}\right) \tag{d}$$

Now letting $G = 0.4E$, a typical relationship for structural steel, Eq. (d) becomes

$$U_s = U_b\left(\frac{3h^2}{L^2}\right) \tag{e}$$

Long beams are considered to have a depth/length ratio of $h/L \leq 1/10$. Using this ratio in Eq. (e), we obtain the strain energy in shear as

$$U_s = 0.03U_b$$

We conclude that for long beams, the strain energy in shear is negligible compared to the strain energy in bending. This is why deflection due to shear is neglected for long beams and only that due to bending is important.

10.8 STRAIN ENERGY FOR A GENERAL STATE OF STRESS

In Sections 10.3 through 10.7, we determined the strain energy in a body due to separate kinds of loadings, such as axial, direct shear, torsion, bending, and transverse shear. We now want to express the strain energy for a body subjected to a general three-dimensional state of stress as shown in Figure 10.16.

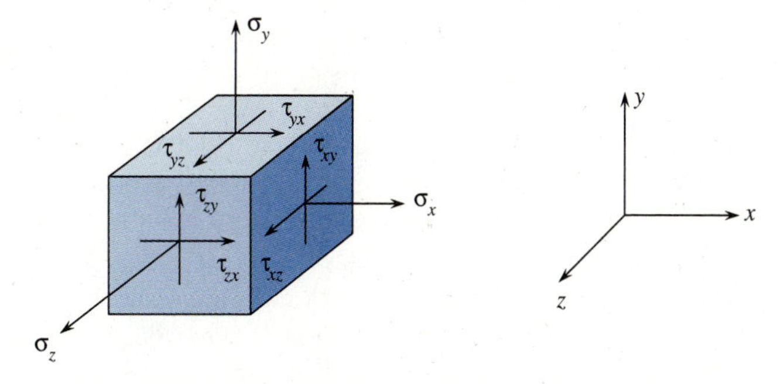

Figure 10.16 Body subjected to general state of stress.

The strain energy density for an elastic deformation of an isotropic body under a general stress state is obtained by adding together terms like Eq. (10.20) due to σ_x and Eq. (10.27) due to τ_{xy}, and including similar terms due to σ_y, σ_z, τ_{xz}, and τ_{yz}. The resulting strain energy expression is given by

$$u = \frac{1}{2}(\sigma_x\epsilon_x + \sigma_y\epsilon_y + \sigma_z\epsilon_z + \tau_{xy}\gamma_{xy} + \tau_{yz}\gamma_{yz} + \tau_{xz}\gamma_{xz}) \tag{10.46}$$

Using the three-dimensional stress-strain equations in Section 8.11 (Eqs. (8.90) and (8.91)), Eq. (10.46) becomes

$$u = \frac{1}{2E}[\sigma_x^2 + \sigma_y^2 + \sigma_z^2 - 2\nu(\sigma_x\sigma_y + \sigma_y\sigma_z + \sigma_x\sigma_z)] + \frac{1}{2G}(\tau_{xy}^2 + \tau_{yz}^2 + \tau_{xz}^2) \quad \textbf{(10.47)}$$

If we use the principal axes 1, 2, and 3, instead of x, y, and z for the stresses, then Eq. (10.47) becomes

$$u = \frac{1}{2E}[\sigma_1{}^2 + \sigma_2{}^2 + \sigma_3{}^2 - 2\nu(\sigma_1\sigma_2 + \sigma_2\sigma_3 + \sigma_1\sigma_3)] \quad \textbf{(10.48)}$$

where σ_1, σ_2, and σ_3 are the principal stresses at the given point.

The maximum distortion energy theory, used to predict failure by yielding of ductile materials, is based on the distortion energy of the body. This *distortion energy* is the amount of energy needed to distort or change the shape of the body. It is then important that we separate the total strain energy density into two parts, a part associated with a change in volume of the body, called u_v, and a part associated with the distortion of the body, called u_d.

The total strain energy at a point in a body is then

$$u = u_v + u_d \quad \textbf{(10.49)}$$

It is easiest to obtain the expression for u_v in terms of the principal stresses and then subtract this quantity from u, given by Eq. (10.48). First, we introduce the average stress $\overline{\sigma}$ of the principal stresses at the point in the body as

$$\overline{\sigma} = \frac{\sigma_1 + \sigma_2 + \sigma_3}{3} \quad \textbf{(10.50)}$$

Then we define the principal stresses to be made up of this average stress and an additional stress as

$$\begin{aligned} \sigma_1 &= \overline{\sigma} + \sigma_1{}' \\ \sigma_2 &= \overline{\sigma} + \sigma_2{}' \\ \sigma_3 &= \overline{\sigma} + \sigma_3{}' \end{aligned} \quad \textbf{(10.51)}$$

Based on Eq. (10.51), the actual stress state can be thought of as a superposition of the two stress states shown in Figure 10.17. The state of stress given by Figure 10.17(b) is often called a *hydrostatic stress state* in that all faces of the element are subjected to the same stress $\overline{\sigma}$. As these stresses are the same on each face, this state of stress changes the volume of the element but not its shape. For instance, if $\overline{\sigma}$ is tensile, the volume increases and if $\overline{\sigma}$ is compressive the volume decreases but no change in shape will occur. We might recall that in Section 8.12, the unit volume change or dilatation e was given by

$$e = \frac{1 - 2\nu}{E}(\sigma_x + \sigma_y + \sigma_z) \quad \textbf{(10.52)}$$

Now comparing Eqs. (10.50) and the sum of the three equations, Eq. (10.51), we see that

$$\sigma_1{}' + \sigma_2{}' + \sigma_3{}' = 0 \quad \textbf{(10.53)}$$

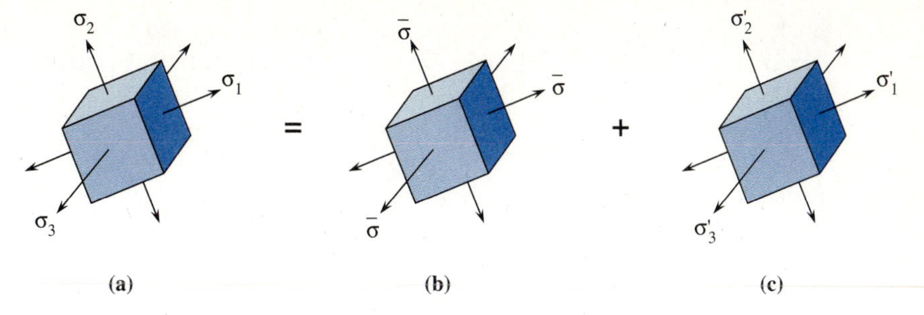

Figure 10.17 (a) Principal stress state in a body, (b) hydrostatic stress state resulting in u_v, and (c) additional stress state resulting in u_d.

Since one or more of these stresses in Eq. (10.53) must then be negative, the stress state in Figure 10.17(c) tends to change the shape of the body. However, since we could replace the sum $\sigma_x + \sigma_y + \sigma_z$ with $\sigma_1' + \sigma_2' + \sigma_3'$ in Eq. (10.52), as the sum of the normal stresses on three mutually perpendicular planes is invariant, we observe that this stress state does not change the volume of the body as e would now be zero. It then follows that the volume change is only due to the stress state in Figure 10.17(b) and therefore u_v is associated with Figure 10.17(b) and u_d with Figure 10.17(c). Therefore, substituting the stress state Figure 10.17(b) into Eq. (10.48), we have

$$u_v = \frac{1}{2E}[3\bar{\sigma}^2 - 2\nu(3\bar{\sigma}^2)]$$

$$u_v = \frac{3\bar{\sigma}^2(1 - 2\nu)}{2E} \tag{10.54}$$

Substituting Eq. (10.50) for $\bar{\sigma}$ in Eq. (10.54), u_v becomes

$$u_v = \frac{1 - 2\nu}{6E}(\sigma_1 + \sigma_2 + \sigma_3)^2 \tag{10.55}$$

Now that we have expressions for u, Eq. (10.48) and u_v, Eq. (10.55), we can obtain the strain energy due to the distortion of the element. Using Eq. (10.49), we solve for u_d and then substitute Eqs. (10.48) and (10.55) for u and u_v as

$$u_d = u - u_v = \frac{1}{6E}[3(\sigma_1^2 + \sigma_2^2 + \sigma_3^2) - 6\nu(\sigma_1\sigma_2 + \sigma_2\sigma_3 + \sigma_1\sigma_3) - (1 - 2\nu)(\sigma_1 + \sigma_2 + \sigma_3)^2] \tag{10.56}$$

On expanding the squared term three and rearranging terms for simplification purposes, we obtain

$$u_d = \frac{1 + \nu}{6E}[(\sigma_1^2 - 2\sigma_1\sigma_2 + \sigma_2^2) + (\sigma_2^2 - 2\sigma_2\sigma_3 + \sigma_3^2) + (\sigma_3^2 - 2\sigma_1\sigma_3 + \sigma_1^2)] \tag{10.57}$$

Each of these terms in the parentheses in Eq. (10.57) is a perfect square. Therefore, we can simplify Eq. (10.57) to obtain the distortion energy per unit volume as

$$u_d = \frac{1+\nu}{6E}[(\sigma_1 - \sigma_2)^2 + (\sigma_2 - \sigma_3)^2 + (\sigma_3 - \sigma_1)^2] \tag{10.58}$$

For the case of biaxial or plane stress, assuming that the three-axis is the perpendicular to the plane of stress, we have $\sigma_3 = 0$ and reduce Eq. (10.58) to

$$u_d = \frac{1+\nu}{3E}[\sigma_1^2 - \sigma_1\sigma_2 + \sigma_2^2] \tag{10.59}$$

We have used Eq. (10.59) in Section 8.7 as a basis for the prediction of failure for ductile materials. This failure theory was presented as the maximum distortion energy theory. See, for instance, Eqs. (8.43) and (8.45).

10.9 PRINCIPLE OF WORK-ENERGY

In Section 10.2, we formulated the work expressions due to an applied force and an applied moment. In Section 10.3 through 10.7, we formulated strain energy expressions for uniaxial load, torsional moment, bending moment, and transverse shear force. We now apply the *principle of work-energy* or *conservation of energy* to obtain the displacement at a point in a member or a structure. This principle states that the external work W done by a single applied force or moment is equal to the total strain energy in the member or structure. This principle is only applicable if the member is subjected to a single concentrated force or moment. Examples now illustrate the principle.

EXAMPLE 10.6

Using the work-energy principle, determine the axial displacement of the free end of the elastic bar of constant cross sectional area A, constant modulus of elasticity E, and of length L subjected to a gradually applied axial force F at the free end (Figure 10.18).

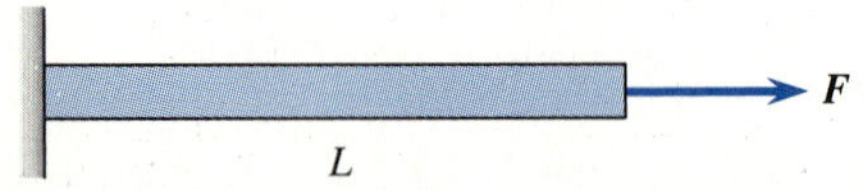

Figure 10.18 Uniform bar subjected to axial force.

Solution

The force is gradually applied to the bar. Therefore, the external work is given by Eq. (10.5) as

$$W = \frac{1}{2}F\Delta \tag{a}$$

where Δ is the displacement of the bar. The strain energy for an elastic bar is given by Eq. (10.23) as

$$U = \frac{F^2L}{2AE} \tag{b}$$

By the work-energy principle, $W = U$, we obtain

$$\frac{1}{2}F\Delta = \frac{F^2L}{2AE}$$

$$\Delta = \frac{FL}{AE} \qquad \text{(c)}$$

which is the same deflection given by Eq. (2.21).

EXAMPLE 10.7

For the truss shown in Figure 10.19, determine the vertical deflection at point C. Use the work-energy principle. Let $E = 29 \times 10^6$ psi and $A = 5$ in.^{2} for each member.

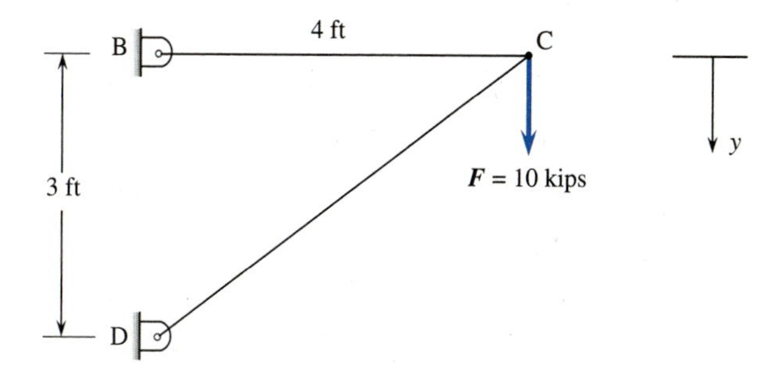

Figure 10.19 Truss subjected to vertical force.

Solution

From statics, we obtain the internal force in each bar as

$$F_{CB} = \frac{4}{3}F \quad \text{and} \quad F_{CD} = -\frac{5}{3}F$$

The work of the vertical force is

$$W = \frac{1}{2}Fy_c$$

The strain energy in the truss is the sum of the strain energy in each bar. This is given by

$$U = \frac{F_{CB}^2L_{CB}}{2AE} + \frac{F_{CD}^2L_{CD}}{2AE}$$

$$= \frac{\left(\frac{4}{3}F\right)^2 L_{CB}}{2AE} + \frac{\left(-\frac{5}{3}F\right)^2 L_{CD}}{2AE}$$

$$U = \frac{1}{2AE}\left[\frac{16}{9}F^2(4\text{ ft}) + \frac{25}{9}F^2(5\text{ ft})\right]$$

Setting $W = U$, we obtain

$$\frac{1}{2}Fy_c = \frac{1}{2AE}\left[\frac{16}{9}F^2(4) + \frac{25}{9}F^2(5)\right]$$

$$y_c = \frac{21F}{AE}$$

$$= \frac{21(10 \text{ kips-ft}) \times 12 \text{ in./ft}}{(5 \text{ in.}^2)(29{,}000 \text{ ksi})}$$

$$y_c = 0.0174 \text{ in.} \quad \text{(down)}$$

EXAMPLE 10.8

For the simply supported beam of Example 10.4, determine the deflection under the concentrated force P located at midspan (Figure 10.20). Consider only deflection due to bending. Use the work-energy principle.

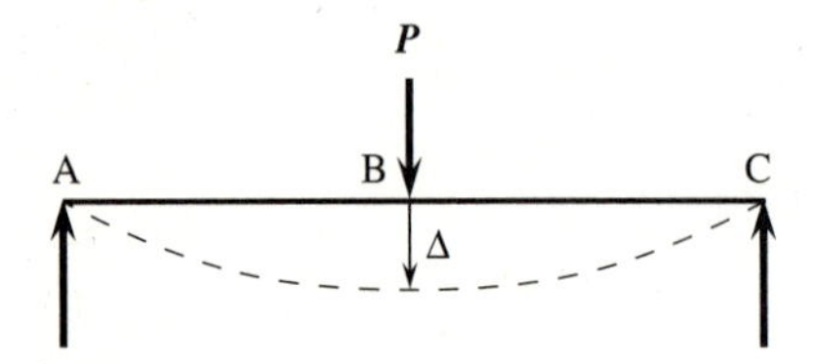

Figure 10.20 Simple beam under concentrated force.

Solution

The force is gradually applied to the beam. Therefore, the work is given by

$$W = \frac{1}{2}P\Delta \qquad \text{(a)}$$

From Example 10.4, Eq. (10.42), we have the strain energy given by

$$U = \frac{P^2L^3}{96EI} \qquad \text{(b)}$$

Setting the work equal to the strain energy, we obtain

$$W = U$$

$$\frac{1}{2}P\Delta = \frac{P^2L^3}{96EI} \qquad (10.60)$$

$$\Delta = \frac{PL^3}{48EI}$$

It appears that the work-energy principle is useful due to its direct manner in which a displacement is obtained. However, application of this method is limited to only a few special problems. We note that only one force (or moment) may be applied to the member or structure. If more than one force is applied, there would be an unknown displacement associated with each force. However, it will only be possible to obtain one work-energy equation for the structure. This method then fails when more than one force is applied to the structure. To determine the deflection or slope at an arbitrary point of a structure subjected simultaneously to several concentrated forces, distributed loads, or moments, we will subsequently present two commonly used methods, the method of virtual work (Section 10.10) and Castigliano's theorem (Section 10.13).

10.10 VIRTUAL WORK METHOD FOR DEFLECTIONS

The application of the work-energy method, that is, the equating of the external work to the strain energy, has the disadvantage that normally only the deflection caused by a single force can be obtained. The *method of virtual work* (sometimes referred to as the *dummy unit load method* or *method of virtual forces*) provides a general procedure to determine deflections and slopes (rotations) at a point in a structure, such as a truss, a beam, or a frame, subjected to any number of loadings.

To develop the virtual work method in a general way, we consider a body or structure of arbitrary shape (later this body will represent a specific truss, beam, or frame) shown in Figure 10.21(a).

Let us assume that we want to determine the deflection Δ of a point A along line AB on the body caused by a number of real or actual forces P_1, P_2, and P_3 shown in Figure 10.21(b). We assume for the present that the supports can resist the forces and maintain equilibrium of the body. The method of virtual work to find Δ is easily employed by using the following orderly steps:

1. Place a virtual force (here we conveniently use a virtual unit force) on the body at point A in the same direction AB that Δ is to be found (as shown in Figure 10.21(a)). The term virtual force is used to indicate the force is an imaginary one and does not actually exist as part of the real forces. This unit force, however, causes internal virtual forces throughout the body. A typical internal virtual force u acting on a representative element of the body is shown in Figure 10.21(a).

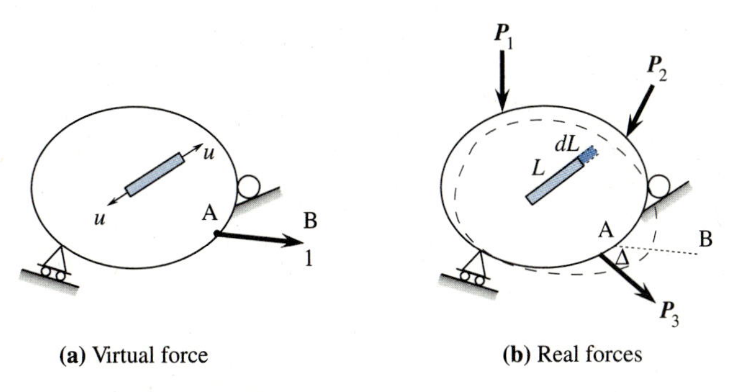

(a) Virtual force **(b)** Real forces

Figure 10.21 (a) Arbitrary body with virtual force and (b) with real forces.

2. Next place the real forces, P_1, P_2, and P_3 on the body (Figure 10.21(b)). These forces cause the point A to deform an amount Δ along line AB, while the representative element now deforms an amount dL. While these deformations occur, the external virtual unit force already on the body before P_1, P_2, and P_3 are applied, moves through the displacement Δ and the internal virtual force u acting on the element before P_1, P_2, and P_3 are applied, moves through the displacement dL. These forces, moving through displacements, result in work.

3. The external virtual unit force moving through displacement Δ performs external virtual work given by 1 times Δ on the body as in Eq. (10.6). Similarly, the internal virtual force u moving through displacement dL performs internal virtual work given by $u \times dL$ on the element. Since the external virtual work is equal to the internal work done on all elements making up the body, we express the virtual work equation as

Virtual forces

$$1 \cdot \Delta = \Sigma u \cdot dL \tag{10.61}$$

Real deformations

where

1 = the external virtual unit force acting in the direction of Δ

u = the internal virtual force acting on the element in the direction of dL.

Δ = the external deflection caused by the real forces

dL = the internal deflection of an element caused by the real forces

The summation sign in Eq. (10.61), indicates that all internal virtual work in the whole body must be included. Equation (10.61) then yields the deflection along the line of the virtual unit force. A positive result indicates that the deflection is in the same direction as the unit force.

In expressing Eq. (10.61), remember that the full values of the virtual forces (unit force and internal force u) were already on the body when the real deformations were imposed. Therefore, no one-half appears in any terms of the equation.

In a similar manner, we can determine the rotation or slope at a point in a body by applying a virtual unit moment or couple instead of a unit force at the point where the rotation is desired (Figure 10.22.)

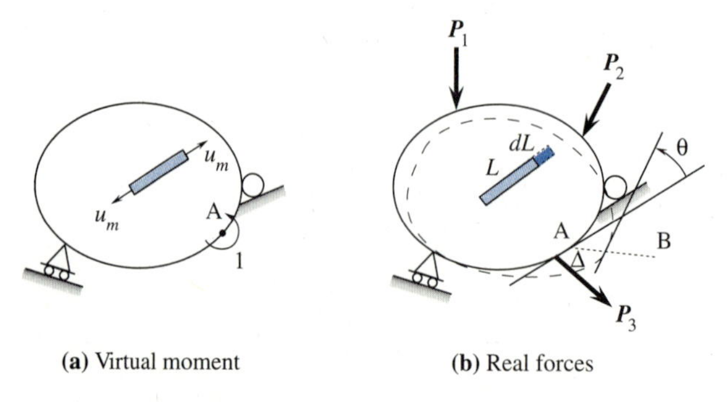

Figure 10.22 (a) Arbitrary body with virtual moment and (b) with real forces.

This unit moment would result in a virtual unit force u_m in an element. Again, assuming that the application of the real forces would deform the element an amount dL, the rotation is obtained from the virtual work equation of the form

$$\underbrace{1}_{\text{Virtual unit moment}} \cdot \underbrace{\theta}_{\text{Real slope}} = \Sigma \underbrace{u_m}_{\text{Virtual forces}} \cdot \underbrace{dL}_{\text{Real deformation}} \tag{10.62}$$

where

1 = the external virtual unit moment acting in the θ direction

u_m = the internal virtual force acting on the element in the direction of dL

θ = the external slope caused by the real forces

dL = the internal deflection of an element caused by the real forces

10.11 METHOD OF VIRTUAL WORK FOR TRUSSES

Deflection Due to External Forces

We can determine the deflection of a truss joint due to an applied load by applying Eq. (10.61). To apply Eq. (10.61), each member of the truss represents an element L from Figure 10.21. The deflection dL in Eq. (10.61) is now obtained for each truss element from Eq. (2.21) as $dL = PL/(AE)$. The virtual work equation for a truss is then given by

$$1 \cdot \Delta = \sum_{i=1}^{n} \frac{u_i P_i L_i}{A_i E_i} \tag{10.63}$$

where

1 = the external virtual unit force acting on the truss joint in the direction of Δ that you want to find

Δ = the external joint deflection due to the real forces acting on the truss

u = the internal virtual force in a truss member due to the external virtual unit force

P = the internal force in a truss member due to the real forces

L = the length of a truss member

A = the cross sectional area of a truss member

E = the modulus of elasticity of a truss member

The summation in Eq. (10.63) indicates that all members must be considered. In Eq. (10.63), the external virtual unit force causes an internal virtual force u in each member. The real forces cause the joint where the deflection is desired to deflect amount Δ along the same direction as the unit force, while each member deflects an amount of $PL/(AE)$. The internal virtual forces u in each member then move through the deflection $PL/(AE)$ of each member. Again, the external virtual work, $1 \times \Delta$, equals the internal virtual work or strain energy stored in all the truss members, $\Sigma uPL/(AE)$.

Deflection Due to Temperature Change

We can determine the deflection in a truss when one or more members of the truss is subjected to a uniform temperature change. When a member is subjected to temperature change, the change in length of a member is

$$dL = \alpha(\Delta T)L \tag{10.64}$$

where α is the coefficient of thermal expansion and ΔT the temperature change of a member (an increase in temperature is assumed positive). Using Eq. (10.64) in Eq. (10.61), we determine the deflection of a selected truss joint due to this temperature change as

$$1 \cdot \Delta = \Sigma[u_i \alpha_i (\Delta T_i) L_i] \tag{10.65}$$

where

1 = the external virtual unit force acting on the truss joint in the direction where Δ is desired

Δ = the external joint deflection caused by the temperature change

u = the internal virtual force in a truss member due to the external virtual unit force

α = the coefficient of thermal expansion of the member

ΔT = the change in temperature of the member

L = the length of a member

STEPS IN SOLUTION

The following steps are used to determine the deflection of any joint on a truss using the virtual work method.

1. Use the method of joints and/or method of sections to determine the force P in each member due to only the real forces acting on the truss. Assume tensile forces positive and compressive forces negative.
2. Apply a unit force on the truss at the joint where the deflection is desired. This unit force must be placed on the joint in the direction along the line of action that the deflection is desired. Remove all real loads so that the unit force is the only one acting on the truss. Then use the method of joints and/or method of sections to determine the internal force u in each truss member. Again, assume tensile forces positive and compressive forces negative.
3. Apply the virtual work equation, Eq. (10.63) for deflection due to external forces, or Eq. (10.65) for deflection due to a temperature change in one or more members, to determine the desired deflection Δ. Remember to keep the algebraic signs for each P and u force from steps 1 and 2 when substituting into the equation. If the resultant sum, $\Sigma PLu/(AE)$, is positive, the final deflection Δ is in the same direction as the unit force. If the resultant is negative, the final deflection is opposite that of the unit force.

We can use both Eqs. (10.63) and (10.65) together to obtain the deflection in a truss joint due to simultaneous external forces and temperature change of one or more members.

Examples 10.9 and 10.10 illustrate the method of virtual work applied to determine deflections in trusses.

EXAMPLE 10.9

Determine the horizontal deflection of point B of the truss shown in Figure 10.23. The cross-sectional areas of the members are indicated in Figure 10.23 and $E = 30{,}000$ ksi. Use the method of virtual work.

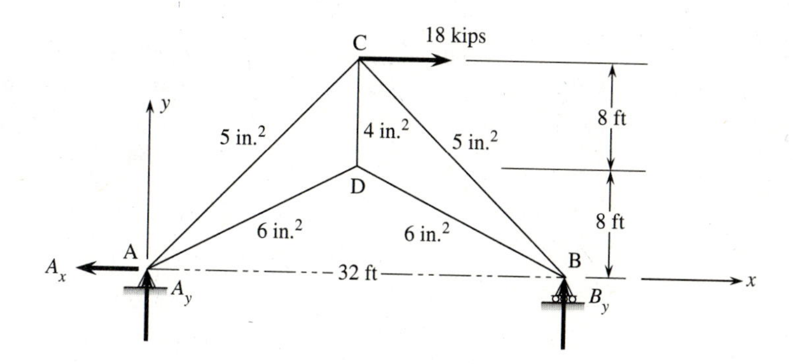

Figure 10.23 Truss subjected to horizontal force.

Solution

First, the reactions are found by using equilibrium equations for the whole truss as follows:

$$\Sigma M_A = 0: \quad B_y(32 \text{ ft}) - (18 \text{ kips})(16 \text{ ft}) = 0$$

$$B_y = 9 \text{ kips} \uparrow$$

$$\Sigma F_y = 0: \quad A_y + B_y = 0$$

$$A_y = -9 \text{ kips} \downarrow$$

$$\Sigma F_x = 0: \quad -A_x + 18 \text{ kips} = 0$$

$$A_x = 18 \text{ kips} \leftarrow$$

Next, the axial force in each member is obtained using the standard method of sections and/or joints. These forces are those caused only by the real force of 18 kips. For instance, using the method of joints, we draw a free-body diagram of joint B.

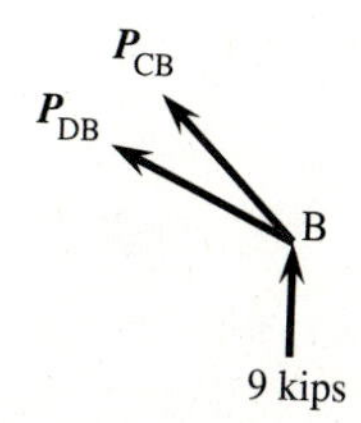

Next, using force equilibrium equations, we have

$$\Sigma F_x = 0: \quad -P_{CB}\frac{1}{\sqrt{2}} - P_{DB}\frac{2}{\sqrt{5}} = 0 \tag{a}$$

$$\Sigma F_y = 0: \quad P_{CB}\frac{1}{\sqrt{2}} + P_{DB}\frac{1}{\sqrt{5}} + 9 = 0 \tag{b}$$

Solving Eqs. (a) and (b) simultaneously

$$P_{CB} = -25.5 \text{ kips}$$

$$P_{DB} = 20.0 \text{ kips}$$

Other forces can be obtained in a similar manner. A summary of all the forces is shown on Figure 10.24, where positive numbers indicate tensile forces and negative numbers compressive forces.

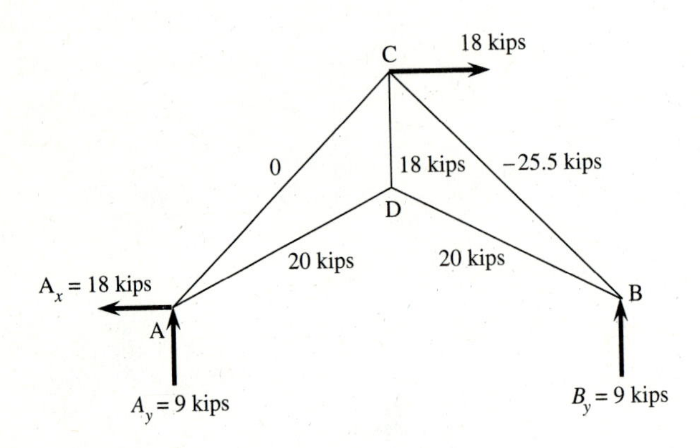

Figure 10.24 Internal force in each truss member.

Next, the real force is removed and only a horizontal unit force is applied at joint B. The force in each member due to the unit force is obtained using the method of joints and/or sections, whichever is convenient. These forces are shown in Figure 10.25.

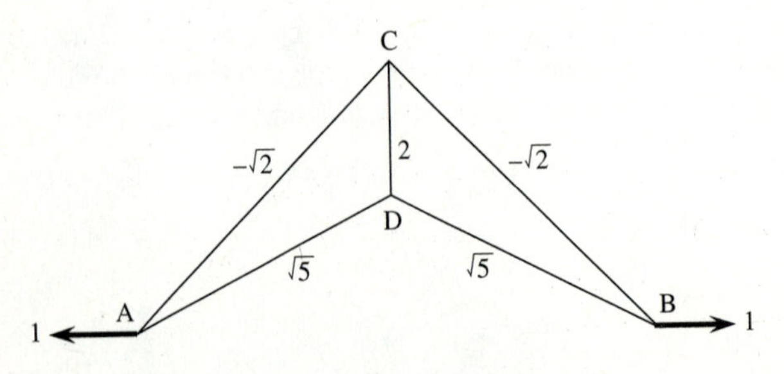

Figure 10.25 Forces in truss members due to unit load.

Finally, we arrange the data in tabular form to facilitate using the virtual work equation, Eq. (10.63).

Member	P (kips)	L (ft)	A (in.^{2})	u	PLu/A
AC	0	22.6	5	$-\sqrt{2}$	0
CB	−25.5	22.6	5	$-\sqrt{2}$	163.0
AD	20.0	17.9	6	$\sqrt{5}$	133.4
DB	20.0	17.9	6	$\sqrt{5}$	133.4
CD	18.0	8.0	4	2	72.0
					Σ 501.8

Therefore, the horizontal deflection at B is

$$1 \cdot \Delta_B = \Sigma\frac{PLu}{AE} = \frac{501.8 \text{ kips} \cdot \text{ft/in.}^2 \times 12 \text{ in./ft}}{30{,}000 \text{ ksi}}$$

$$\Delta_B = 0.201 \text{ in.}$$

The positive sign indicates the deflection is in the same direction as the unit force applied at B, that is, to the right.

EXAMPLE 10.10

For the truss shown previously in Figure 10.23, remove the 18-kip force and subject member BC to a temperature rise of 100°F. Determine the horizontal deflection of joint B due to this rise in temperature. Use the method of virtual work. Let $\alpha = 6.5 \times 10^{-6}$/°F (Figure 10.26).

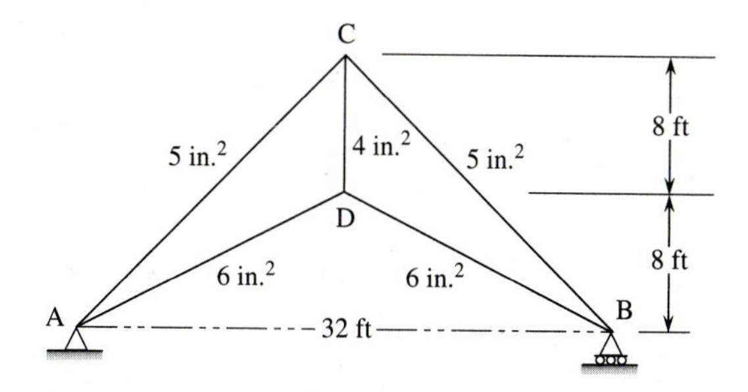

Figure 10.26 Truss member subjected to temperature rise.

Solution

The real forces in the members are zero as no external forces are applied to the truss. To obtain the horizontal deflection at B, we apply a unit horizontal force in the horizontal direction at B (as was done in Example 10.9). The internal forces u due to the unit horizontal force at B are shown in Figure 10.25 (see Example 10.9).

Finally, we apply the method of virtual work, Eq. (10.65), to obtain the deflection as:

$$1 \cdot \Delta = \Sigma u\alpha(\Delta T)L = u_{BC}\alpha(\Delta T)L_{BC}$$

$$= -\sqrt{2}(6.5 \times 10^{-6}/°\text{F})(100°\text{F})(22.63 \text{ ft})$$

$$= -0.0208 \text{ ft} = -0.249 \text{ in.}$$

The negative sign in the answer indicates that the deflection is opposite the direction of the unit load or to the left. Also member BC only contributes to the deflection as only BC is subjected to temperature change.

10.12 METHOD OF VIRTUAL WORK FOR BEAMS AND FRAMES

The general equations developed in Section 10.10 to obtain deflections and slopes can be extended to problems involving bending of beams and frames. To develop the virtual work equation for a beam or a frame, consider the beam in Figure 10.27(a). Assume we want to determine the deflection Δ at a point B on the beam. We determine the deflection Δ at B as follows.

1. Apply a unit vertical load at B (Figure 10.27(b)) and determine the internal virtual moment m (Figure 10.27(d)) at location x along the beam using the method of sections.

2. Apply the real load (here w). This load causes the point B to deform an amount Δ (Figure 10.27(a)), while a differential element dx of the beam rotates by an amount (Figure 10.27(c)) given by

$$d\theta = (M/EI)dx \tag{10.66}$$

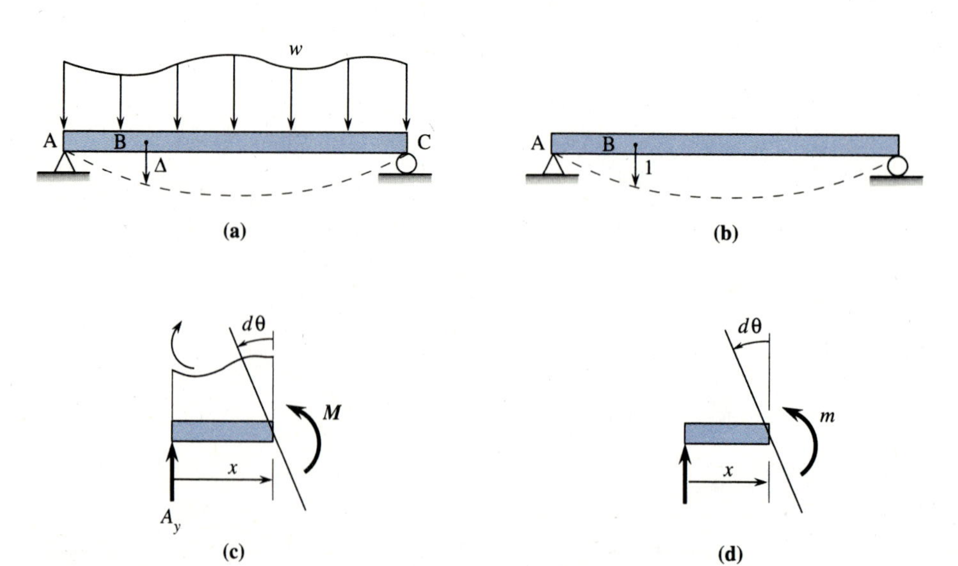

Figure 10.27 Beam for deflection analysis.

(Recall from Chapter 5 that $y'' = \theta' = M/EI$). While these deformations occur, the external virtual unit force, already acting on the beam before w was applied, moves through displacement Δ and the internal virtual moment m acting on the element before w was applied, moves through the rotation $d\theta$ resulting in virtual work.

3. The external virtual unit force moving through displacement Δ performs external virtual work of $1 \times \Delta$ on the beam. Similarly, the internal moment m moving through $d\theta$ performs strain energy of amount $md\theta = m(M/EI)dx$, by Eq. (10.66). Since the external work is equal to the strain energy done on all elements making up the beam, we have

$$1 \cdot \Delta = \int_0^L \frac{mM}{EI} dx \tag{10.67}$$

where

1 = the external virtual force acting on the beam or frame

Δ = the external deflection at the point caused by the real load acting on the beam or frame

m = the internal virtual moment in the beam or frame (expressed as a function of x) due to the virtual unit force

M = the internal moment in the beam or frame (expressed as a function of x) due to the real forces

E = the modulus of elasticity of the material of the beam or frame

I = the moment of inertia of the cross-sectional area determined about the neutral axis

Similarly, the rotation or slope at a point on the beams centerline can be determined by applying a virtual unit moment at the point. We would then determine the internal virtual moment m in the beam due to the unit moment. The resulting equation developed similar to that of Eq. (10.67) is

$$1 \cdot \theta = \int_0^L \frac{mM}{EI} dx \tag{10.68}$$

If concentrated forces or moments or discontinuous distributed loads act on the beam, a single integration over the beam length will not be possible. The usual procedure employed to obtain moment expressions in Chapter 5 must be used here. That is, various moment expressions in each section of the beam that has continuity in loading must be developed. For example, in Figure 10.28 moments M_1, M_2, and M_3 must be expressed as functions of x. We must also use the same origin or same regions for evaluating the real moment M and the virtual moment m. For example, for the beam in Figure 10.28, to determine the deflection at point C on the beam, we use coordinate x_1 to determine the internal virtual work in region AB, x_2 in region BC, x_3 in region CD, and x_4 in region DE (or possibly start x_4 from E back to the left within DE if it appears that easier expressions develop for M and m).

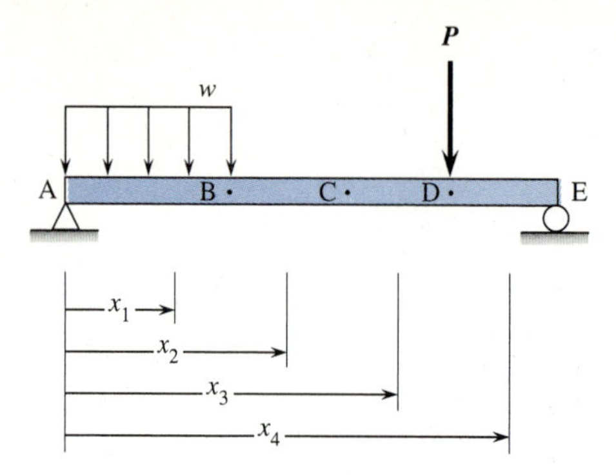

Figure 10.28 Beam for deflection analysis.

STEPS IN SOLUTION

The following steps are used to determine deflections in beams and frames using the virtual work method.

1. Use the method of sections to determine the internal moment M due to real loads acting on the beam or frame.
2. Apply a unit force at the point on the beam or frame where deflection is desired. This unit force must be placed on the beam or frame in the direction along the line of action that the deflection is desired. Remove all real loads so that the unit force is the only one acting on the beam or frame. Then use the method of sections to determine the internal moment m throughout the beam or frame. Use the same sign conventions to determine M in step 1 and m in step 2.
3. Apply the virtual work equation, Eq. (10.67), to determine the desired deflection.

If the slope is desired at a point on a beam or frame, apply a unit moment in step 2 instead of a unit force and then proceed as outlined in the steps above. Examples 10.11, 10.12, and 10.13 now illustrate the method of virtual work applied to beams and frames.

EXAMPLE 10.11

For the simply supported beam shown in Figure 10.29, determine the vertical deflection at point C. Let $E = 30{,}000$ ksi and $I = 500$ in.^{4}

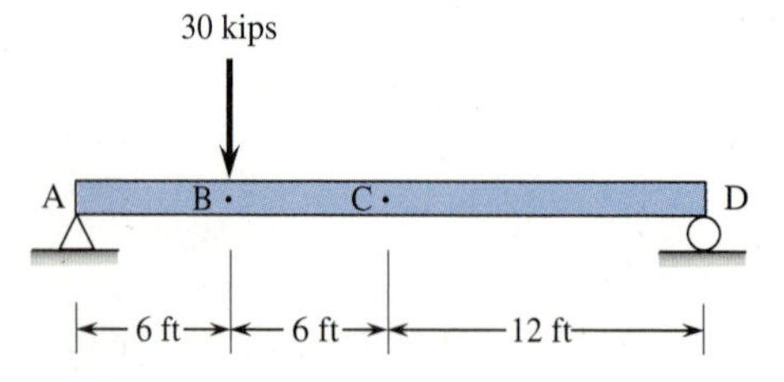

Figure 10.29 Beam subjected to concentrated force.

Solution

Draw the free-body diagram of the beam and determine the reactions as shown below.

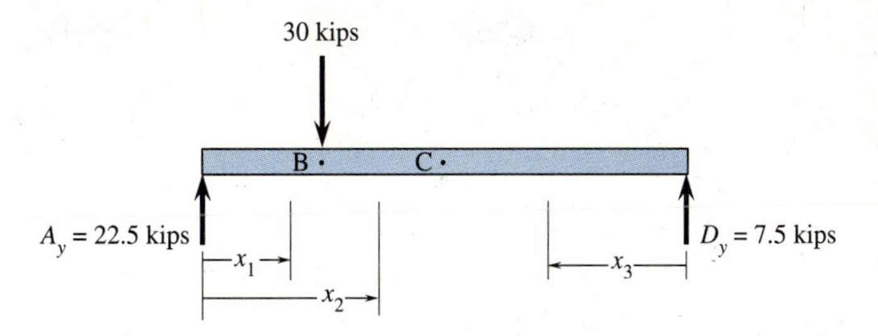

Determine the bending moment M due to the real load throughout the beam in the usual manner as done in drawing bending moment diagrams in Chapter 5. For instance, for $0 \leq x_1 \leq 6$ ft.

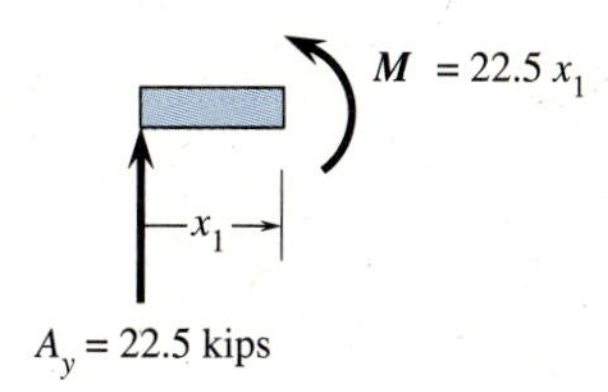

The table below summarizes these results. Note, that since the deflection at C is desired, we needed a portion BC. That is, we needed coordinates that cover all portions where no discontinuities in either real or virtual unit load occur.

Determine the bending moment m due to the vertical unit load.

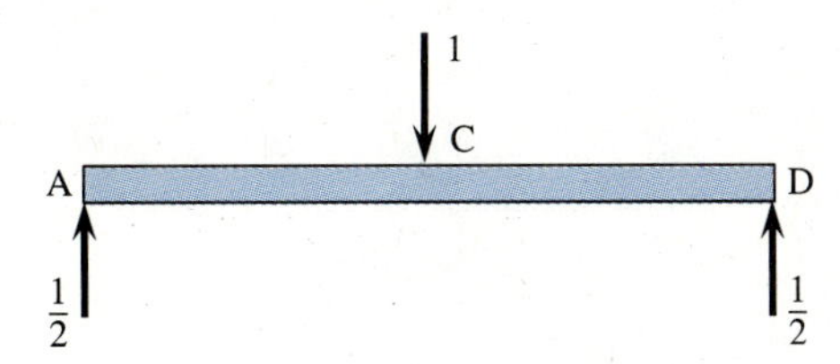

These moments are summarized in the following table.

Portion	Origin	Limits	M	m
AB	A	0–6 ft	$22.5x_1$	$0.5x_1$
BC	A	6–12 ft	$22.5x_2 - 30(x_2 - 6)$	$0.5x_2$
DC	D	0–12 ft	$7.5x_3$	$0.5x_3$

Using the virtual work equation, we obtain deflection at c as

$$1 \cdot \Delta_c = \int_0^L \frac{Mm}{EI}\,dx$$

$$= \int_0^{6\text{ ft}} \frac{(22.5x_1)}{EI}\left(\frac{x_1}{2}\right)dx_1 + \int_{6\text{ ft}}^{12\text{ ft}} \frac{[22.5x_2 - 30(x_2 - 6)]}{EI}\left(\frac{x_2}{2}\right)dx_2$$

$$+ \int_{0\text{ ft}}^{12\text{ ft}} \frac{7.5x_3}{EI}\left(\frac{x_3}{2}\right)dx_3$$

Upon integrating and simplifying, we obtain

$$\Delta_c = \frac{5940\text{ kips}\cdot\text{ft}^3 \times 1728\text{ in.}^3/\text{ft}^3}{(30{,}000\text{ ksi})(500\text{ in.}^4)}$$

$$\Delta_c = 0.684\text{ in.} \quad \text{(down)}$$

Remember, a positive answer indicates a deflection in the same direction as the virtual unit load.

EXAMPLE 10.12

For the stepped shaft with two times larger moment of inertia in the middle portion than the outer portions, determine the rotation at end A (Figure 10.30).

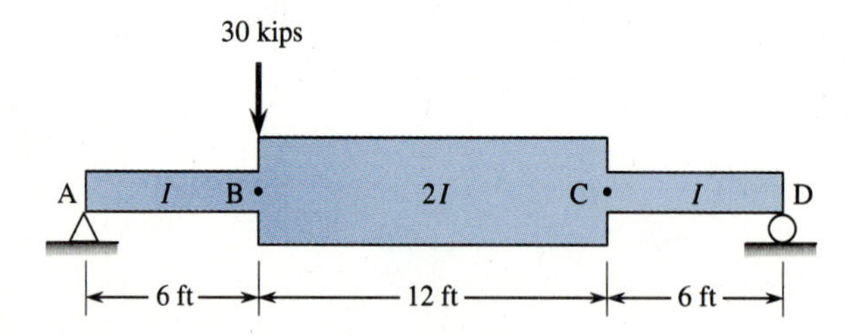

Figure 10.30 Stepped shaft subjected to concentrated load.

Solution

Determine the moment M in the beam due to the real load of 30 kips. Here three portions AB, BC, and CD are needed. The table below lists the moments in each portion.

To obtain the rotation at A, a virtual unit moment must be applied at A. Then determine the virtual moments m in the beam due to a unit moment applied at A. The table shows the resulting internal moments m.

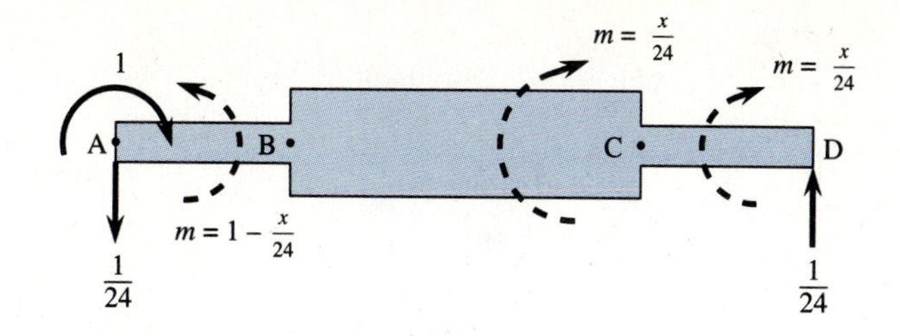

Portion	Origin	Limits	*M*	*m*	*I*
AB	A	0–6 ft	22.5x	1-x/24	I
CB	D	6–18 ft	7.5x	x/24	2I
DC	D	0–6 ft	7.5x	x/24	I

Using the virtual work equation, we obtain the rotation at A as

$$1 \cdot \theta_A = \int \frac{Mm}{EI} dx$$

$$= \int_0^6 \frac{22.5x}{EI}\left(1 - \frac{x}{24}\right) dx + \int_6^{18} \frac{7.5x}{2EI}\left(\frac{x}{24}\right) dx + \int_0^6 \frac{7.5x}{EI}\left(\frac{x}{24}\right) dx$$

On integrating and simplifying, we obtain

$$\theta_A = 6.046 \times 10^{-3} \text{ rad}$$

The positive result indicates that the rotation is in the same direction as the unit moment, that is, clockwise.

EXAMPLE 10.13

For the frame shown (Figure 10.31), determine the vertical deflection at free end C. The frame is subjected to a moment of 5 kN · m at C. Let $E = 200$ GPa and $I = 2.5 \times 10^6$ mm^4.

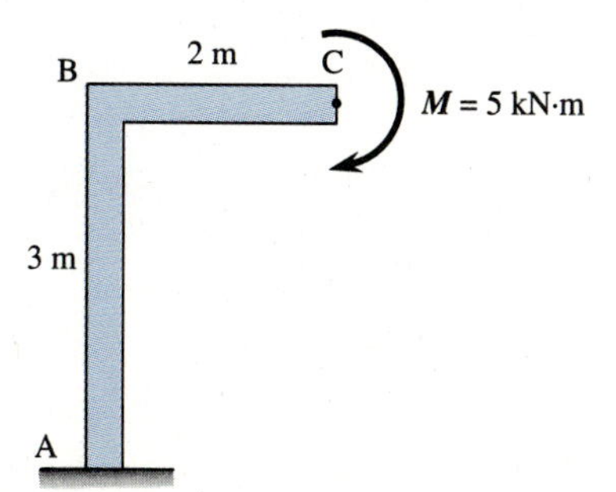

Figure 10.31 Frame subjected to end moment.

Solution

Determine the moment M throughout the frame.

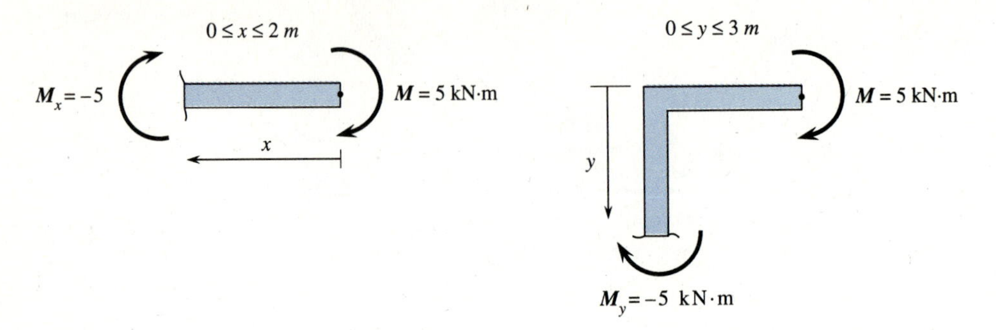

Apply a unit vertical force at C and determine the moment m throughout the frame.

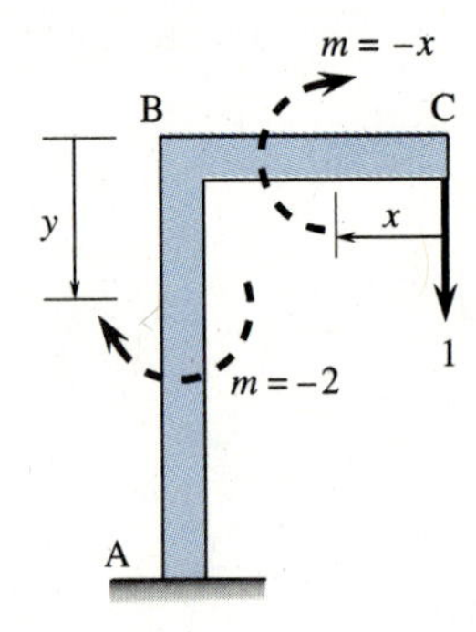

The following table summarizes the results.

Portion	Origin	Limits	M (kN · m)	m
CB	C	0–2 m	−5	$-x$
BA	B	0–3 m	−5	−2

Apply the virtual work equation to obtain the deflection as follows:

$$\Delta_c = \int \frac{Mm}{EI} dx$$

$$= \int_0^2 \frac{(-5)(-x)}{EI} dx + \int_0^3 \frac{(-5)(-2)}{EI} dy$$

$$= \frac{1}{EI}[10 + 30] = \frac{40 \text{ kN} \cdot \text{m}^3}{200 \times 10^6 \dfrac{\text{kN}}{\text{m}^2} \times 2.5 \times 10^{-6} \text{ m}^4}$$

$$= 0.08 \text{ m} = 80 \text{ mm} \quad \text{(down)}$$

The positive result indicates that the deflection is in the same direction as the unit force, that is, down.

10.13 CASTIGLIANO'S THEOREM

Castigliano's theorem is a method to determine the deflection or slope at a point in an elastic body, such as a truss, beam, or frame. The theorem described in this section is actually Castigliano's second theorem as developed by Alberto Castigliano in 1879 and is based on the strain energy in a body.

To derive the theorem, we consider for convenience, the beam of Figure 10.32 subjected to loads P_1 and P_2. The strain energy of the beam is equal to the work of the two forces P_1 and P_2 as they are gradually applied to the beam. To obtain the work due to these forces, we must express the deflections x_1 and x_2 under each force in terms of the forces P_1 and P_2.

To begin, apply only P_1 to the beam (Figure 10.33). The deflections under force P_1 (at point B_1) and under the point where P_2 was applied (at point B_2) are given by

$$x_{11} = f_{11}P_1 \quad \text{and} \quad x_{21} = f_{21}P_1 \tag{10.69}$$

where f_{11} and f_{21} are constants called *influence coefficients*. These constants give the influence of, or represent, the deflections of points B_1 and B_2, respectively, when a unit force is applied at B_1. These coefficients are functions of properties of the beam. For our discussion, these specific functions are not necessary. Equations (10.69) indicate that the deflections are proportional to the force P_1. This is consistent with the elastic behavior of the beam.

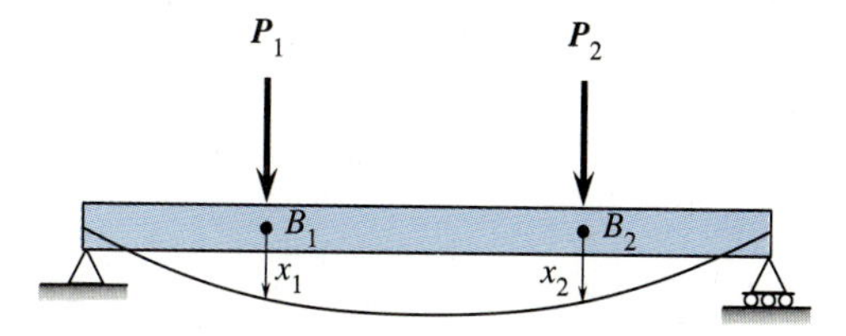

Figure 10.32 Beam subjected to two forces.

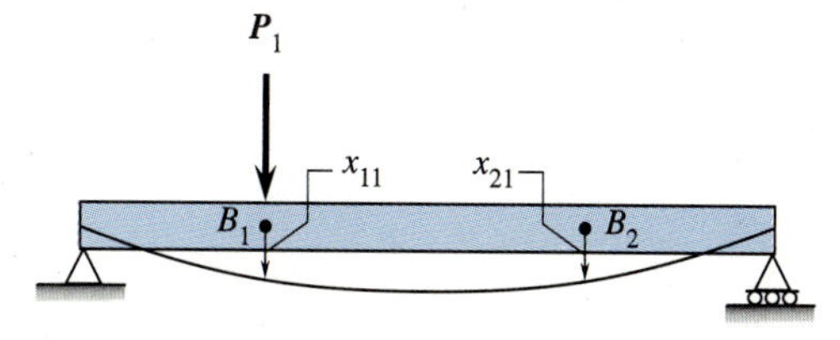

Figure 10.33 Beam with only P_1 acting.

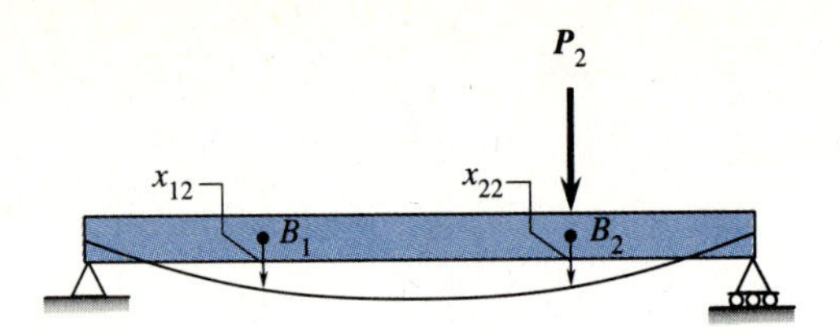

Figure 10.34 Beam with only P_2 acting.

Next, we remove force P_1 and apply force P_2 to the beam (Figure 10.34). The deflections under points B_1 and B_2 are now

$$x_{12} = f_{12}P_2 \quad \text{and} \quad x_{22} = f_{22}P_2 \tag{10.70}$$

where f_{12} and f_{22} are the influence coefficients representing the deflections of points B_1 and B_2, respectively, when a unit force is applied at B_2.

The total deflections at B_1 and B_2 are then obtained by superimposing the influences of each deflection from Eqs. (10.69) and (10.70) as

$$x_1 = x_{11} + x_{12} = f_{11}P_1 + f_{12}P_2 \tag{10.71}$$

$$x_2 = x_{21} + x_{22} = f_{21}P_1 + f_{22}P_2 \tag{10.72}$$

Now we want to determine the work done by P_1 and P_2, and thus the strain energy in the beam due to these forces. It is convenient to first apply force P_1 at B_1 (Figure 10.35(a)). The work of this force is

$$W_{11} = \frac{1}{2}P_1x_{11} \tag{10.73}$$

Using Eq. (10.69) in Eq. (10.73), we obtain

$$W_{11} = \frac{1}{2}P_1(f_{11}P_1) = \frac{1}{2}f_{11}P_1^2 \tag{10.74}$$

Remember that P_2 does not do any work as it is not acting on the beam.

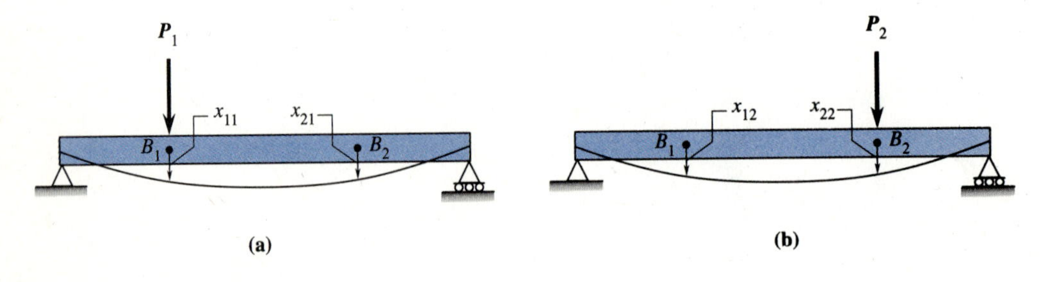

Figure 10.35 Beam with
(a) P_1 applied first and
(b) P_2 applied second.

Now we apply P_2 at B_2 (Figure 10.35(b)) while leaving P_1 on the beam. The work of P_2 is

$$W_{22} = \frac{1}{2}P_2x_{22} \tag{10.75}$$

Using Eq. (10.70) in Eq. (10.75), we have

$$W_{22} = \frac{1}{2}f_{22}P_2^2 \tag{10.76}$$

Remember that P_1 is acting on the beam while P_2 is slowly applied. Therefore, P_1 moves through displacement x_{12}, and does additional work given by

$$W_{12} = P_1x_{12} \tag{10.77}$$

Using Eq. (10.70), we rewrite Eq. (10.77) as

$$W_{12} = f_{12}P_1P_2 \tag{10.78}$$

The total work is then obtained by adding the expressions given by Eqs. (10.74), (10.76), and (10.78) as

$$W = U = \frac{1}{2}(f_{11}P_1^2 + 2f_{12}P_1P_2 + f_{22}P_2^2) \tag{10.79}$$

Now if we reverse the order that the forces are applied (Figure 10.36), we can obtain the total work done in a manner similar to that used to obtain Eq. (10.79). The result is

$$W = U = \frac{1}{2}(f_{22}P_2^2 + 2f_{21}P_2P_1 + f_{11}P_1^2) \tag{10.80}$$

Since the total work is the same whether we first apply P_1 and then P_2, or whether we first apply P_2 and then P_1, we can equate Eqs. (10.79) and (10.80) to obtain

$$f_{12} = f_{21} \tag{10.81}$$

Equation (10.81) indicates that the deflection at B_1 due to a unit force applied at B_2 (f_{12}) is equal to the deflection at B_2 due to a unit force applied at B_1 (f_{21}). This property is known as *Maxwell's law of reciprocity* (named after James C. Maxwell who derived it in 1864).

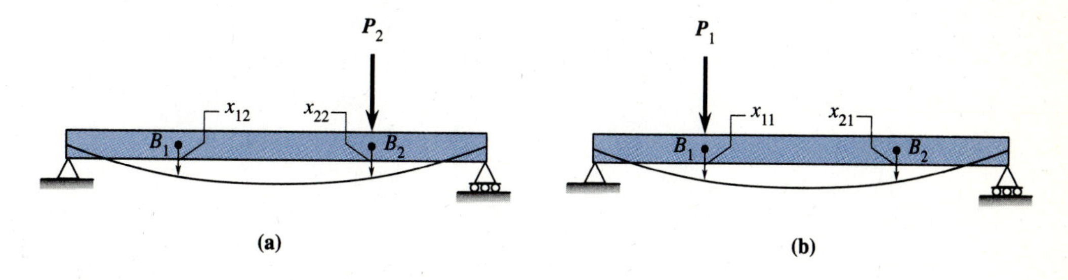

Figure 10.36 Reversing order that forces P_1 and P_2 are applied.

To derive Castigliano's theorem, we differentiate Eq. (10.79) with respect to P_1 and make use of Eq. (10.71) to obtain

$$\frac{\partial U}{\partial P_1} = f_{11}P_1 + f_{12}P_2 = x_1 \tag{10.82}$$

Now differentiating Eq. (10.79) with respect to P_2, and making use of Eq. (10.72), we obtain

$$\frac{\partial U}{\partial P_2} = f_{12}P_1 + f_{22}P_2 = x_2 \tag{10.83}$$

We can conclude from Eqs. (10.82) and (10.83), that for an elastic structure subjected to any number of forces, $P_1, P_2, \ldots, P_n$, the deflection x_i at the point of the applied force P_i, measured along the line of action of P_i, can be expressed as the partial derivative of the strain energy of the structure with respect to the force P_i. We express the general form of Castigliano's theorem as

$$x_i = \frac{\partial U}{\partial P_i} \tag{10.84}$$

Similarly, we can determine the slope of a beam or frame at the point of application of a moment M_i by using Castigliano's theorem in the following form:

$$\theta_i = \frac{\partial U}{\partial M_i} \tag{10.85}$$

STEPS IN SOLUTION (TRUSSES)

The following steps may be used to determine the displacement at a point in a truss using Castigliano's theorem.

1. If a concentrated force exists in the direction of, and at the joint where the deflection is desired, then use it in the analysis. (For convenience, call it F if it is a numeric value so that you can more easily keep track of it when you apply Castigliano's theorem.) If a force does not exist at a point where the deflection is desired, then apply an imaginary or fictitious force F_i at the joint and in the direction that the deflection is desired. After formulating the needed expressions used in Castigliano's theorem, set the imaginary force F_i to zero.
2. In analyzing a truss, determine the internal force P in each member in terms of the real and the imaginary force (if it was necessary as explained in step 1).
3. Determine the partial derivative expressions

$$\frac{\partial P}{\partial F} \tag{10.86}$$

for each member of the truss. Replace real force F with imaginary force F_i if an imaginary force was used in step 1.

4. Apply Castigliano's theorem to obtain the deflection at i in the direction of F or F_i as

$$\Delta_i = \frac{\partial U}{\partial F} = \frac{\partial}{\partial F}\left[\sum_{i=1}^{n}\left(\frac{P_n^2 L_n}{2A_n E_n}\right)\right] = \sum_{i=1}^{n}\frac{P_n(\partial P_n/\partial F)L_n}{A_n E_n} \tag{10.87}$$

where n = the total number of members. Again, replace F with F_i in Eq. (10.87) if an imaginary force was used.

The theorem is best applied if we perform the differentiation as indicated in Eq. (10.86) before the summation in Eq. (10.87). Example 10.14 illustrates Castigliano's theorem to determine the deflection at a joint in a truss.

EXAMPLE 10.14

For the truss shown (Figure 10.37), determine **(a)** the horizontal and **(b)** the vertical deflection of joint C using Castigliano's theorem. Let $E = 29 \times 10^3$ ksi and $A = 5$ in.^{2} for each member.

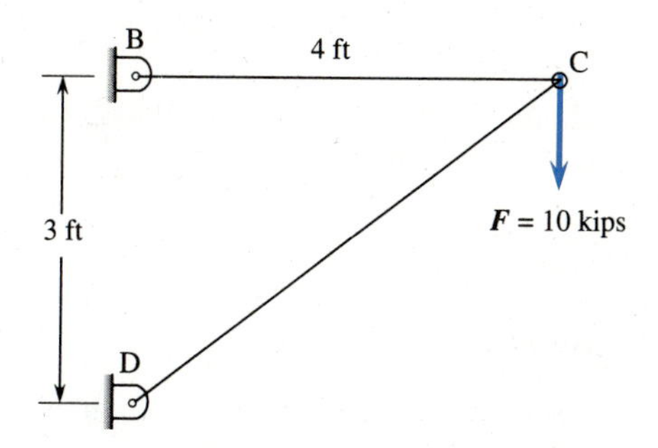

Figure 10.37 Truss analyzed for deflection.

Solution

(a) To obtain the vertical deflection at C, use the existing vertical force F in Castigliano's theorem.

Next, determine the internal force P in each member in terms of F. In this example, use a free-body diagram of joint C to obtain the internal forces as shown below.

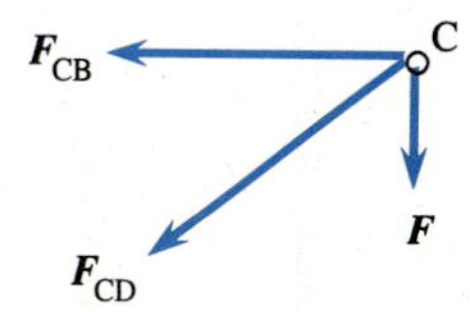

$$\Sigma F_y = 0: \quad -\frac{3}{5}F_{CD} - F = 0$$

$$F_{CD} = -\frac{5}{3}F$$

$$\Sigma F_x = 0: \quad -F_{CB} - \frac{4}{5}F_{CD} = 0$$

$$F_{CB} = -\frac{4}{5}\left(-\frac{5}{3}\right)F$$

$$F_{CB} = \frac{4}{3}F$$

Next, determine the partial derivatives of each internal force with respect to F as follows.

$$\frac{\partial F_{CB}}{\partial F} = \frac{4}{3} \qquad \frac{\partial F_{CD}}{\partial F} = -\frac{5}{3}$$

Finally, use Castigliano's theorem Eq. (10.87) to obtain the vertical deflection of point C as

$$y_c = \frac{(4/3)F(4/3)(4 \text{ ft})}{AE} + \frac{(-5/3)F(-5/3)(5 \text{ ft})}{AE}$$

$$= \frac{21F}{AE} \text{ ft}$$

Converting feet to inches and substituting in $F = 10$ kips, $E = 29 \times 10^3$ ksi, and $A = 5$ in.^{2}, we obtain the numeric value for the vertical deflection at C as

$$y_c = \frac{21(10 \text{ kips})(12 \text{ in./ft})(1 \text{ ft})}{(5 \text{ in.}^2)(29{,}000 \text{ ksi})}$$

$$y_c = 0.0174 \text{ in.}$$

The positive sign indicates that the deflection is in the same direction as F, that is, downward. This solution is identical to that obtained by the work-energy method in Example 10.7.

(b) To obtain the horizontal deflection at C, we must apply an imaginary force F_i in the horizontal direction at C because a real force in the horizontal direction does not exist at C.

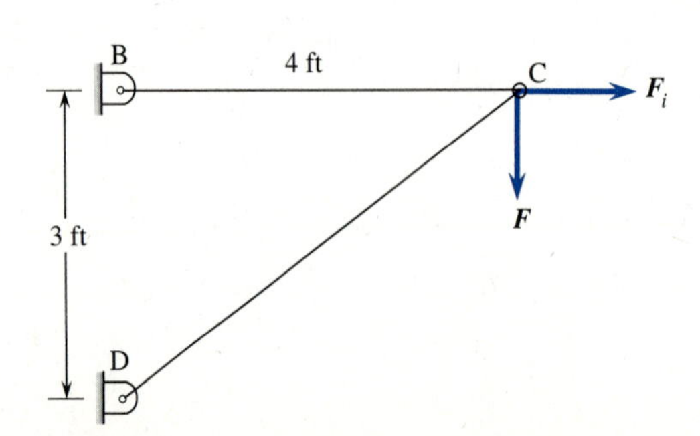

The internal forces P in each member are now determined in terms of F_i and F.

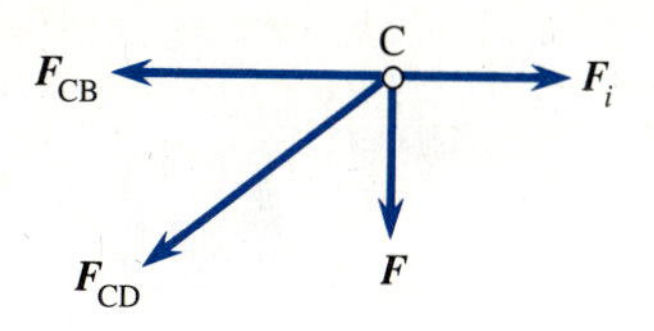

$$\Sigma F_y = 0: \quad -\frac{3}{5}F_{CD} - F = 0$$

$$F_{CD} = -\frac{5}{3}F$$

$$\Sigma F_x = 0: \quad -F_{CB} - \frac{4}{5}F_{CD} + F_i = 0$$

$$F_{CB} = -\frac{4}{5}\left(-\frac{5}{3}F\right) + F_i$$

$$F_{CB} = \frac{4}{3}F + F_i$$

Next, obtain the partial derivatives of each internal force with respect to F_i as

$$\frac{\partial F_{CB}}{\partial F_i} = 1, \quad \frac{\partial F_{CD}}{\partial F_i} = 0$$

Finally, use Castigliano's theorem Eq. (10.87), to obtain the horizontal deflection at C as

$$x_c = \frac{[(4/3)F + F_i](1)(4\text{ ft})}{AE} + 0$$

Now set $F_i = 0$.

$$x_c = \frac{(4/3)F(4\text{ ft})}{AE}$$

$$= \frac{(4/3)(10\text{ kips})(4\text{ ft})(12\text{ in./ft})}{(5\text{ in.}^2)(29{,}000\text{ ksi})}$$

$$x_c = 0.00441\text{ in.}$$

Beams and Frames

For a beam or a frame, Castigliano's theorem is applied after we recall the strain energy due to bending Eq. (10.41) is

$$U = \int_0^L \frac{M^2}{2EI}\,dx$$

To obtain the deflection at a point i, we use the force P at point i or introduce an imaginary force P_i at i (if a real force is not located at i). Castigliano's theorem then yields the deflection as

$$\Delta_i = \frac{\partial U}{\partial P_i} = \int_0^L \frac{M}{EI}\frac{\partial M}{\partial P}\,dx \tag{10.88}$$

where M is now the internal moment expressed as a function of P or P_i.

If we want the rotation at a point in a beam, then a real moment or an imaginary moment must be applied at the desired point. Castigliano's theorem then becomes

$$\theta_i = \frac{\partial U}{\partial M_i} = \int_0^L \frac{M}{EI}\frac{\partial M}{\partial M_i}\,dx \tag{10.89}$$

STEPS IN SOLUTION

The following steps may be used to obtain the deflection at a point in a beam or a frame.

1. If a concentrated force exists in the direction of and at the point where the deflection is desired, then use it in the analysis. For convenience, call it P even though it may have a numeric value. If a force does not exist at a joint where the deflection is desired, then apply an imaginary or fictitious force P_i at the point and in the direction that the deflection is desired. After formulating the needed expressions used in Castigliano's theorem, then set the imaginary force P_i to zero.
2. Determine the bending moment throughout the beam or frame as a function of the real forces, and possibly the imaginary forces if needed, as explained in step 1.
3. Determine the partial derivative of the moment with respect to the real force or imaginary one as appropriate; that is

$$\frac{\partial M}{\partial P} \quad \text{or} \quad \frac{\partial M}{\partial P_i}$$

4. Apply Castigliano's theorem to obtain the deflection at i in the direction of P or P_i as

$$\Delta_i = \frac{\partial U}{\partial P} = \int_0^L \frac{M}{EI}\frac{\partial M}{\partial P}\,dx$$

To obtain the rotation or slope at a point i, we simply apply a moment M_i at i, instead of a force P_i. The rotation is then given by

$$\theta_i = \frac{\partial U}{\partial M_i} = \int_0^L \frac{M}{EI}\frac{\partial M}{\partial M_i}\,dx$$

Examples 10.15 through 10.19 illustrate how to find deflections and slopes in beams and frames.

EXAMPLE 10.15

For the simply supported beam subjected to the concentrated force P at midspan (Figure 10.38), determine the deflection at midspan B. Let EI be constant. Use Castigliano's theorem.

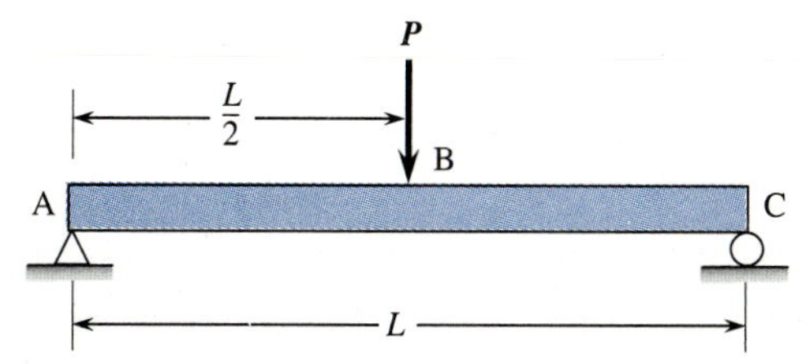

Figure 10.38 Beam subjected to midspan force.

Solution

To determine the deflection at B, we can use the real force P located at B. Determine the moment throughout the beam.

For $0 \leq x \leq \frac{L}{2}$

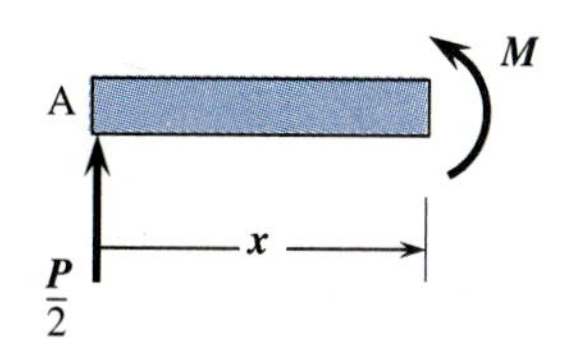

$$M = \frac{P}{2}x \tag{a}$$

For $\frac{L}{2} \leq x \leq L$

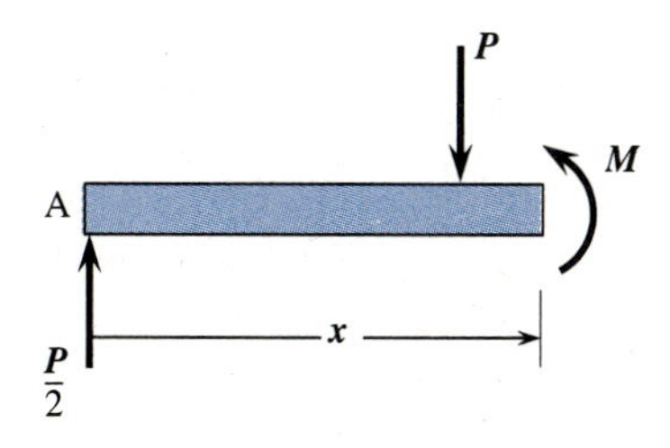

$$M = \frac{P}{2}x - P\left(x - \frac{L}{2}\right) = \frac{P(L - x)}{2} \tag{b}$$

Next, determine the derivatives of M with respect to P as follows.

For $0 \leq x \leq \frac{L}{2}$, we differentiate Eq. (a) to obtain

$$\frac{\partial M}{\partial P} = \frac{x}{2} \tag{c}$$

For $\frac{L}{2} \leq x \leq L$, we differentiate Eq. (b) to obtain

$$\frac{\partial M}{\partial P} = \frac{L - x}{2} \tag{d}$$

Finally, using Castigliano's theorem Eq. (10.88), we obtain

$$\Delta_B = \frac{1}{EI}\left\{\int_0^{L/2}\left(\frac{P}{2}x\right)\left(\frac{x}{2}\right)dx + \int_{L/2}^{L}\frac{P(L-x)}{2}\frac{(L-x)}{2}dx\right\} = \frac{PL^3}{4EI}\left(\frac{1}{24} + \frac{1}{24}\right)$$

$$\Delta_B = \frac{PL^3}{48EI} \quad \text{(down)}$$

We should notice in this example that due to symmetry, we could have used one-half of the beam and multiplied the result by 2. This would have greatly simplified the integration process.

EXAMPLE 10.16

For the cantilevered beam subjected to a uniformly distributed load w acting over one-half its length, determine **(a)** the vertical deflection at point A, and **(b)** the slope at point A using Castigliano's theorem (Figure 10.39).

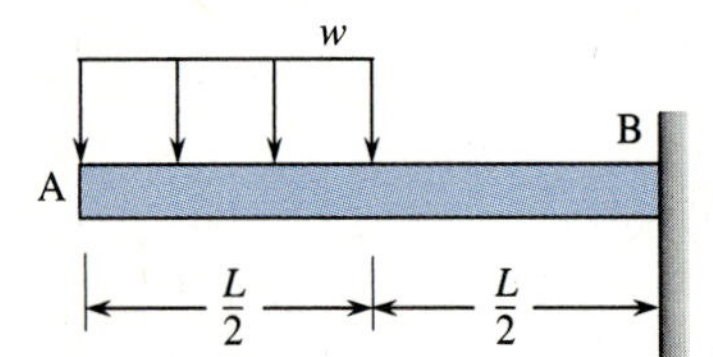

Figure 10.39 Cantilevered beam for deflection analysis.

Solution

(a) To determine the deflection at A, we must apply a fictitious or imaginary concentrated force P_i at A because a real concentrated force does not exist at A.

First, determine the moment M throughout the beam with the imaginary force P_i applied at A.

For $0 \leq x \leq \frac{L}{2}$

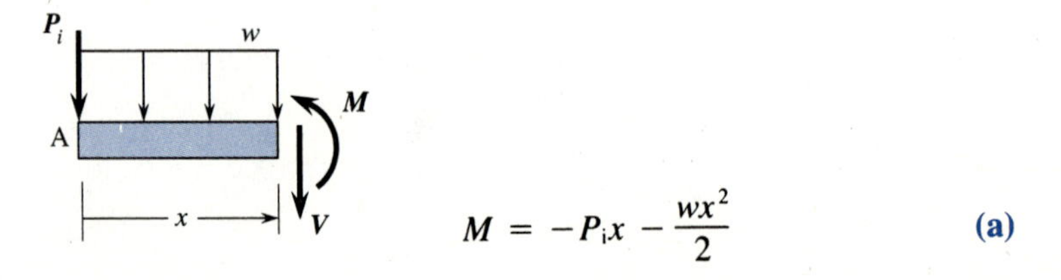

$$M = -P_i x - \frac{wx^2}{2} \tag{a}$$

For $\frac{L}{2} \leq x \leq L$

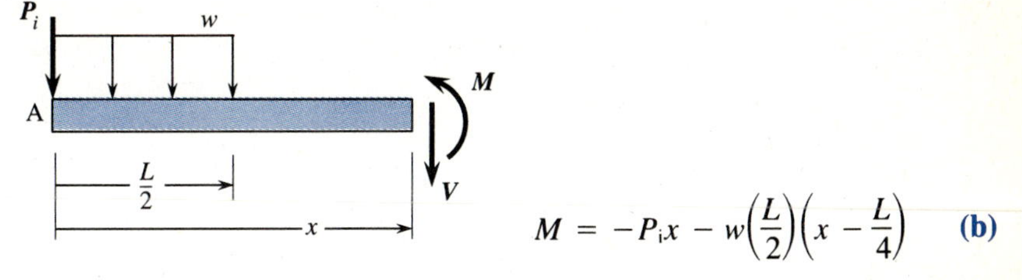

$$M = -P_i x - w\left(\frac{L}{2}\right)\left(x - \frac{L}{4}\right) \quad \textbf{(b)}$$

Next, determine the derivatives of M with respect to P_i as follows.

For $0 \leq x \leq \frac{L}{2}$, we differentiate Eq. (a) with respect to P_i to obtain

$$\frac{\partial M}{\partial P_i} = -x \quad \textbf{(c)}$$

For $\frac{L}{2} \leq x \leq L$, we differentiate Eq. (b) with respect to P_i to obtain

$$\frac{\partial M}{\partial P_i} = -x \quad \textbf{(d)}$$

Finally, use Castigliano's theorem, Eq. (10.88), to obtain the deflection as

$$\Delta_A = \frac{\partial U}{\partial P_i} = \int_0^L \frac{M}{EI}\frac{\partial M}{\partial P_i}\,dx$$

$$= \frac{1}{EI}\left\{\int_0^{L/2}\left(-P_i x - \frac{wx^2}{2}\right)(-x)dx + \int_{L/2}^{L}\left[-P_i x - \frac{wL}{2}\left(x - \frac{L}{4}\right)\right](-x)dx \quad \textbf{(e)}\right.$$

Now, for convenience, set $P_i = 0$ before integrating. On integrating Eq. (e), we obtain

$$\Delta_A = \frac{1}{EI}\left\{\frac{wx^4}{8}\bigg|_0^{L/2} + \left(\frac{wL}{2}\frac{x^3}{3} - \frac{wL^2}{8}\frac{x^2}{2}\right)\bigg|_{L/2}^{L}\right\}$$

$$\Delta_A = \frac{41}{384}\frac{wL^4}{EI} \quad \text{(down)}$$

The positive result indicates that the deflection is in the same direction as the imaginary force P_i, which was applied in the downward direction.

(b) To obtain the slope at point A, we apply an imaginary moment M_i at A. Then determine the moment throughout the beam with the imaginary moment applied at A.

For $0 \leq x \leq \frac{L}{2}$

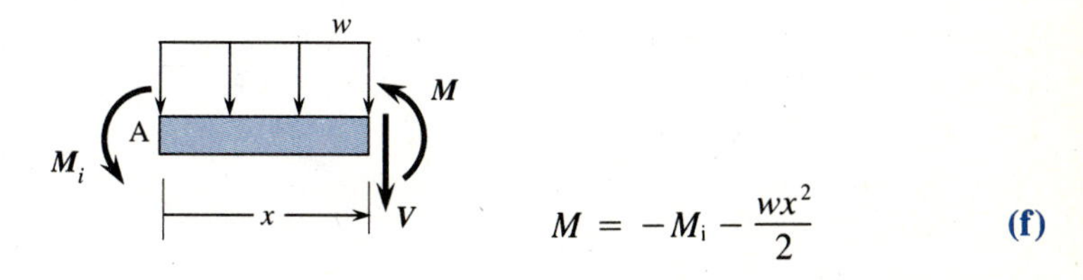

$$M = -M_i - \frac{wx^2}{2} \quad \textbf{(f)}$$

For $\frac{L}{2} \le x \le L$

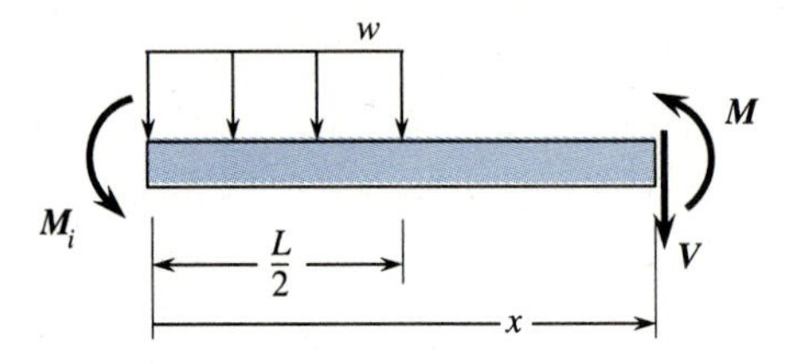

$$M = -M_i - \frac{wL}{2}\left(x - \frac{L}{4}\right) \tag{g}$$

Next, determine the derivative of M with respect to M_i as follows.

For $0 \le x \le \frac{L}{2}$, we differentiate Eq. (f) with respect to M_i to obtain

$$\frac{\partial M}{\partial M_i} = -1 \tag{h}$$

For $\frac{L}{2} \le x \le L$, we differentiate Eq. (g) with respect to M_i to obtain

$$\frac{\partial M}{\partial M_i} = -1 \tag{i}$$

Applying Castigliano's theorem, Eq. (10.89), we obtain the slope at A as

$$\begin{aligned}\theta_A &= \frac{\partial U}{\partial M_i} = \int_0^L \frac{M}{EI}\left(\frac{\partial M}{\partial M_i}\right)dx \\ &= \frac{1}{EI}\left\{\int_0^{L/2}\left(-M_i - \frac{wx^2}{2}\right)(-1)dx + \int_{L/2}^{L}\left[-M_i - \frac{wL}{2}\left(x - \frac{L}{4}\right)\right](-1)dx\right\}\end{aligned}$$

Setting M_i to zero and then integrating, we obtain

$$\begin{aligned}\theta_A &= \frac{1}{EI}\left\{\frac{wx^3}{6}\bigg|_0^{L/2} + \left(\frac{wL}{2}\frac{x^2}{2} - \frac{wL^2}{8}x\right)\bigg|_{L/2}^{L}\right\} \\ &= \frac{7}{48}\frac{wL^3}{EI} \quad \text{(counterclockwise)}\end{aligned}$$

The positive sign indicates that the rotation is in the same direction as the imaginary moment, that is, the tangent to the curve at A rotates counterclockwise with respect to the horizontal as shown.

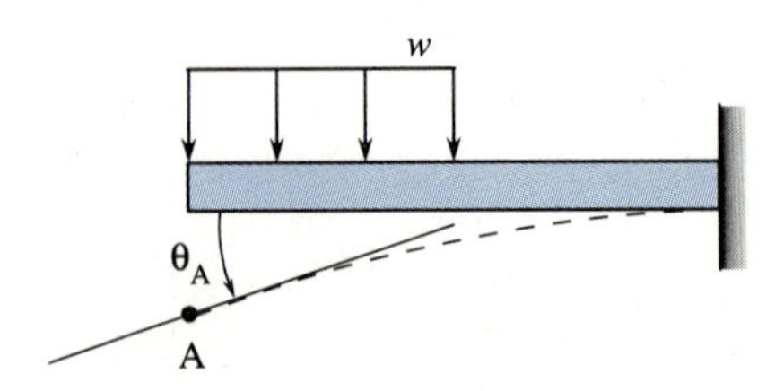

EXAMPLE 10.17

The semicircular curved frame with radius of $r = 120$ in. is subjected to the force P at its apex B. Determine the vertical deflection at B (Figure 10.40). The cross-section is a wide-flange with $I = 214$ in.^{4}, total cross-sectional area $A = 8.23$ in.^{2}, and web area $A_w = 2.7$ in.^{2}. Consider the strain energy due to bending, normal force, and shear force. Compare deflection contributions from the three parts. Use Castigliano's theorem.

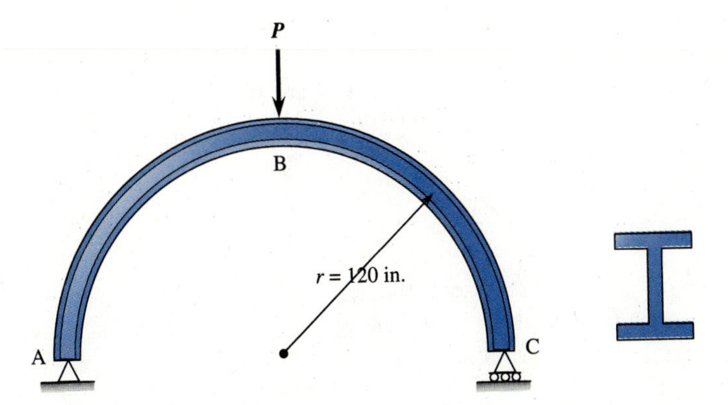

Figure 10.40 Curved frame for deflection analysis.

Solution

We want to determine the deflection under the real force P, so we use P in Castigliano's theorem. First, determine the internal bending moment, normal force, and shear force in the frame. From the free-body diagram of a portion of the frame, we have

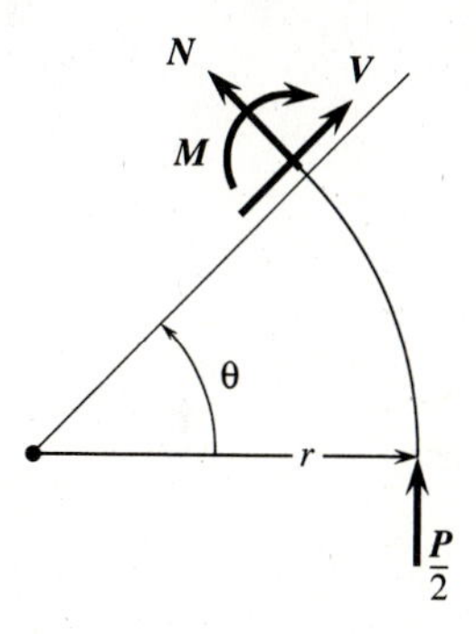

For $0 \leq \theta \leq 90°$

$$\Sigma F_n = 0: \quad N = -\frac{P}{2}\cos\theta \tag{a}$$

$$\Sigma F_v = 0: \quad V = -\frac{P}{2}\sin\theta \tag{b}$$

$$\Sigma M = 0: \quad M = \frac{P}{2}(r - r\cos\theta) \tag{c}$$

Due to symmetry, we formulate the solution for only one-half the frame and multiply this formulation by 2.

Next, take the derivatives of Eqs. (a), (b), and (c) with respect to P, to obtain

$$\frac{\partial N}{\partial P} = -\frac{\cos\theta}{2} \tag{d}$$

$$\frac{\partial V}{\partial P} = -\frac{\sin\theta}{2} \tag{e}$$

$$\frac{\partial M}{\partial P} = \frac{r}{2}(1 - \cos\theta) \tag{f}$$

Using Castigliano's theorem, we obtain the vertical deflection at B as

$$\Delta_B = 2\left[\int_A^B \frac{M}{EI}\frac{\partial M}{\partial P}\,ds + \int_A^B \frac{N}{AE}\frac{\partial N}{\partial P}\,ds + \int_A^B \frac{V}{A_w G}\frac{\partial V}{\partial P}\,ds\right] \tag{g}$$

In Eq. (g), the first integral represents the deflection due to bending moment, the second integral is that due to normal force (assuming N is constant over the whole area A), and the third integral is that due to shear force. In the third integral, we use a simplified approximation based on assuming that the shear force is resisted uniformly only by the area of the web of the cross section. Recall from Chapter 5, Example 5.12, that for narrow-webbed cross-sections, the shear force is primarily resisted by the web and the shear stress is nearly uniform over the web area. Therefore, the third integral in Eq. (g) is a very acceptable approximation.

Substituting Eqs. (a) through (f) into (g), we have

$$\Delta_B = 2\left[\int_0^{\pi/2} \frac{Pr(1-\cos\theta)}{2EI}\frac{r(1-\cos\theta)}{2}\,r\,d\theta \right.$$

$$\left. + \int_0^{\pi/2} \frac{P\cos\theta}{2AE}\left(\frac{\cos\theta}{2}\right) r\,d\theta + \int_0^{\pi/2} \frac{P\sin\theta}{2A_w G}\left(\frac{\sin\theta}{2}\right) r\,d\theta\right] \tag{h}$$

where $ds = r\,d\theta$ is the arc length.

Using calculus formulas for the integrals, we recall

$$\int \cos^2\theta\,d\theta = \frac{1}{2}\sin\theta\cos\theta + \frac{\theta}{2}$$

$$\int \sin^2\theta\,d\theta = -\frac{1}{2}\sin\theta\cos\theta + \frac{\theta}{2} \tag{i}$$

$$\int \cos\theta\,d\theta = \sin\theta$$

$$\int \sin\theta\,d\theta = -\cos\theta$$

Using the formulas, Eq. (i), in Eq. (h), we obtain

$$\Delta_B = \frac{0.178Pr^3}{EI} + \frac{0.393Pr}{AE} + \frac{0.393Pr}{A_w G} \tag{j}$$

Now letting $r = 120$ in., $I = 214$ in.^{4}, $A = 8.23$ in.^{2}, $A_w = 2.7$ in.^{2}, and assuming structural steel with $G = 0.4E$, we obtain

$$\Delta_B = \frac{1437P}{E} + \frac{6P}{E} + \frac{44P}{E} = \frac{1487P}{E} \tag{k}$$

The result shows that for long members (here large radius r) the major contribution to the deflection is from the bending moment (strain energy due to bending dominates). The deflection due to normal force and shear force is negligible. That is, in this example

$$\frac{\Delta_M}{\Delta_{\text{total}}} \times 100 = \frac{1437}{1487} \times 100 = 96.6\%$$

or 96.6% of the deflection is due to bending.

EXAMPLE 10.18

The frame shown (Figure 10.41) is supported by a fixed end at G and is free at A. Use Castigliano's theorem to determine the horizontal deflection at A. Let $E = 30 \times 10^3$ ksi and $I = 500$ in.^{4}.

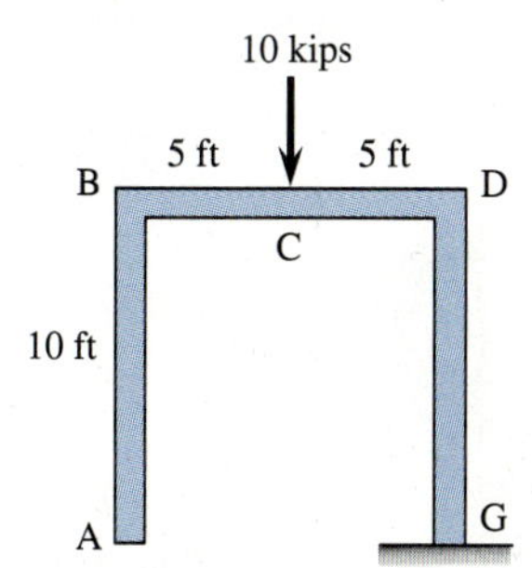

Figure 10.41 Rigid frame for deflection analysis.

Solution

First, apply an imaginary force P_i at A in the horizontal direction as shown.

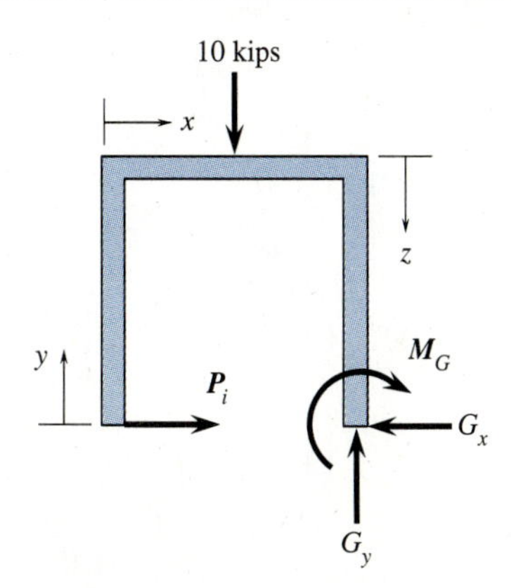

Determine the bending moment throughout the frame and then the partial derivative of M with respect to P_i. This is best accomplished by using the coordinates x, y and z shown above for each member.

For $0 \leq y \leq 10$ ft

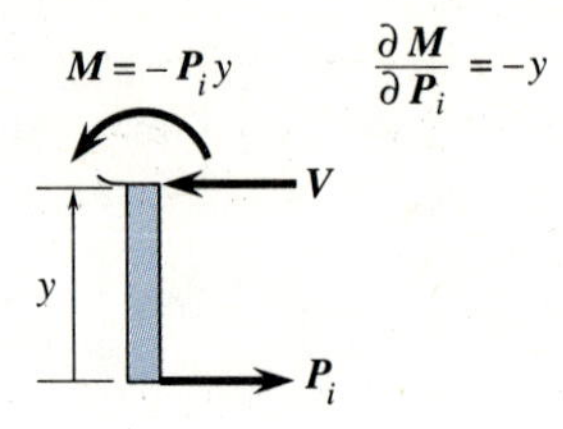

For $0 \leq x \leq 5$ ft

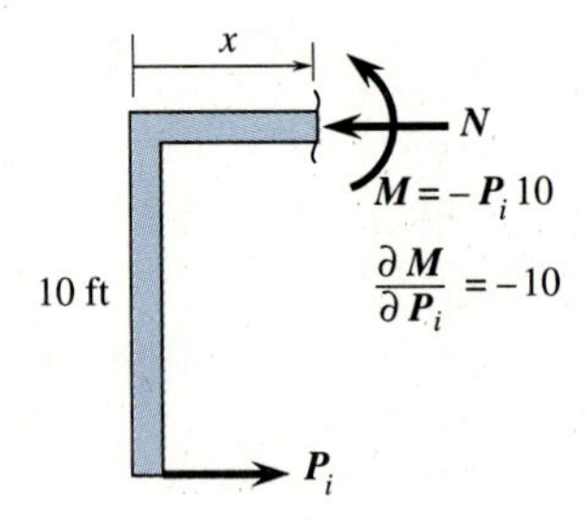

For $5 \leq x \leq 10$ ft

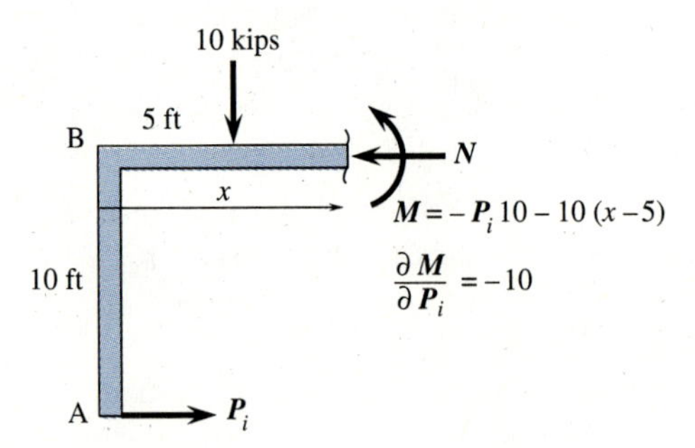

For $0 \leq z \leq 10$ ft

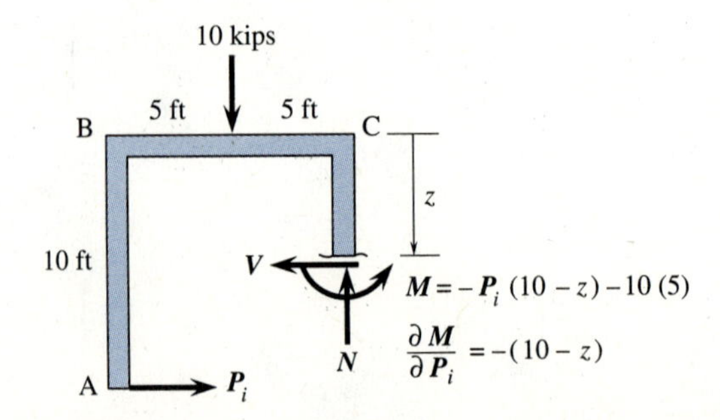

The following table summarizes the results needed to apply Castigliano's theorem.

Member	Origin	Limits	M	$\partial M/\partial P_i$
AB	A	0–10 ft	$-P_i y$	$-y$
BC	B	0–5 ft	$-P_i 10$	-10
CD	B	5–10 ft	$-P_i 10 - 10(x - 5)$	-10
DG	D	0–10 ft	$-P_i(10 - z) - 50$	$z - 10$

Using Castigliano's theorem, we obtain the horizontal deflection at A as

$$\Delta_H = \frac{1}{EI}\left\{\int_0^{10}(-P_i y)(-y)\,dy + \int_0^{5}(-P_i 10)(-10)\,dx + \int_5^{10}\left[-P_i 10 - 10(x - 5)\right](-10)dx + \int_0^{10}\left[-P_i(10 - z) - 50\right](z - 10)dz\right\} \quad \textbf{(a)}$$

Now setting $P_i = 0$ in Eq. (a) before integrating, we obtain

$$\Delta_H = \frac{1}{EI}\left[\left(\frac{100x^2}{2} - 500\,x\right)\bigg|_5^{10} + 500z\bigg|_0^{10} - 50\frac{z^2}{2}\bigg|_0^{10}\right]$$

$$= \frac{1}{EI}\left[50(100 - 25) - 500(10 - 5) + 500(10) - 50\left(\frac{100}{2}\right)\right]$$

$$= \frac{3750 \text{ kip-ft}^3 \times 1728 \text{ in.}^3/\text{ft}^3}{(30 \times 10^3 \text{ ksi})(500 \text{ in.}^4)}$$

$$\Delta_H = 0.43 \text{ in. (to the right)}$$

The positive result indicates that the horizontal deflection is in the same direction as P_i, that is, to the right.

EXAMPLE 10.19

For the torsion bar spring BC shown subjected to a vertical force F, determine **(a)** the vertical deflection at D and **(b)** the angle of twist at C. Use Castigliano's theorem. Let the bar properties be E, G, I, and J. Neglect any influence from the shear force.

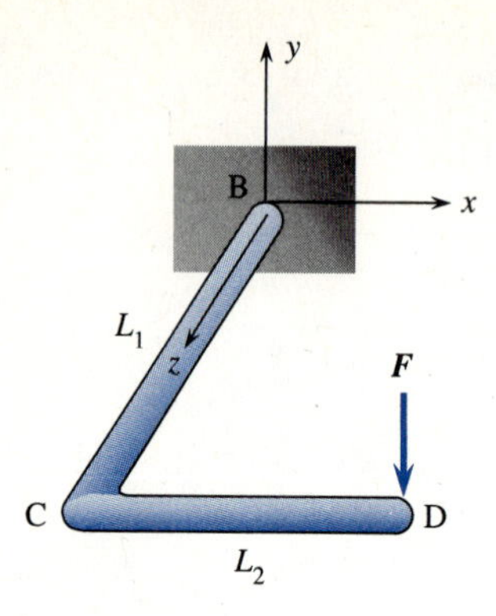

Figure 10.42 Torsion bar subjected to vertical load.

Solution

(a) First, determine the bending and twisting moments in sections BC and CD. The following free-body diagrams show the moments.

For section BC:

Section BC has a bending moment M_{BCx} and a torque T_{BC} acting on it.

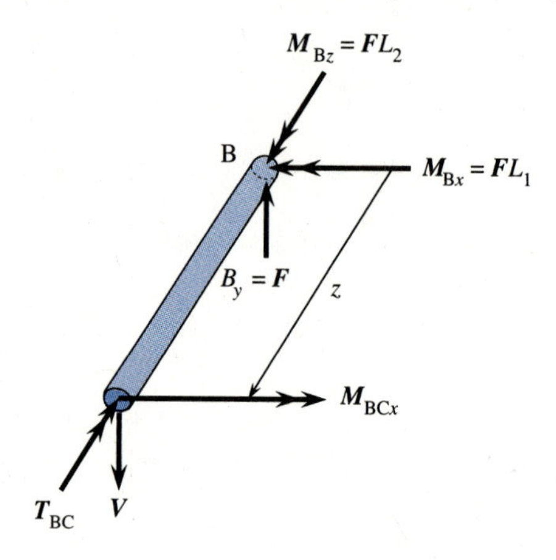

$$M_{BCx} = FL_1 - Fz$$
$$T_{BC} = FL_2 \qquad \textbf{(a)}$$

For section CD:

Section CD has only a bending moment acting on it

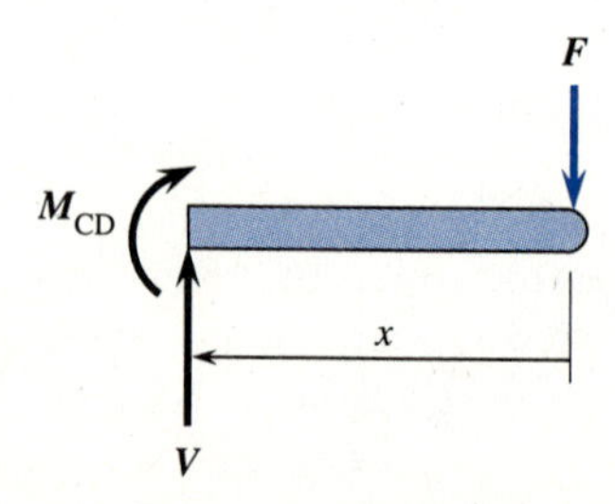

$$M_{CD} = -Fx \tag{b}$$

The partial derivatives of the moment expressions, Eqs. (a) and (b), with respect to F are:

$$\frac{\partial M_{BCx}}{\partial F} = L_1 - z \qquad \frac{\partial T_{BC}}{\partial F} = L_2 \qquad \frac{\partial M_{CD}}{\partial F} = -x \tag{c}$$

The strain energy due to both bending, Eq. (10.41), and twisting, Eq. (10.35), moments is considered. This strain energy is given by

$$U = U_{BC} + U_{CD}$$

$$U = \int_0^{L_1} \frac{M_{BCx}{}^2 dz}{2EI_{BC}} + \int_0^{L_1} \frac{T_{BC}{}^2}{2GJ_{BC}} dz + \int_0^{L_2} \frac{M_{CD}^2}{2EI_{CD}} dx \tag{d}$$

Using Castigliano's theorem, the vertical deflection at D is

$$\Delta_D = \frac{\partial U}{\partial F} = \int_0^{L_1} \frac{M_{BCx}}{EI_{BC}} \left(\frac{\partial M_{BCx}}{\partial F}\right) dz + \int_0^{L_1} \frac{T_{BC}}{GJ_{BC}} \left(\frac{\partial T_{BC}}{\partial F}\right) dz + \int_0^{L_2} \frac{M_{CD}}{EI_{CD}} \left(\frac{\partial M_{CD}}{\partial F}\right) dx \tag{e}$$

Substituting Eqs. (a), (b), and (c) into (e), we have

$$\Delta_D = \int_0^{L_1} \frac{(FL_1 - Fz)(L_1 - z)}{EI} dz + \int_0^{L_1} \frac{(FL_2)L_2}{GJ} dz + \int_0^{L_2} \frac{Fx(x)}{EI} dx \tag{f}$$

Performing the integration in Eq. (f) and evaluating, the deflection is

$$\Delta_D = \frac{F}{EI}\left(L_1^3 - L_1^3 + \frac{L_1^3}{3}\right) + \frac{FL_2^2 L_1}{GJ} + \frac{FL_2^3}{3EI}$$

or

$$\Delta_D = \frac{FL_1^3}{3EI} + \frac{FL_2^2 L_1}{GJ} + \frac{FL_2^3}{3EI}$$

(b) To obtain the angle of twist of CB, place an imaginary torque T_i acting at C.

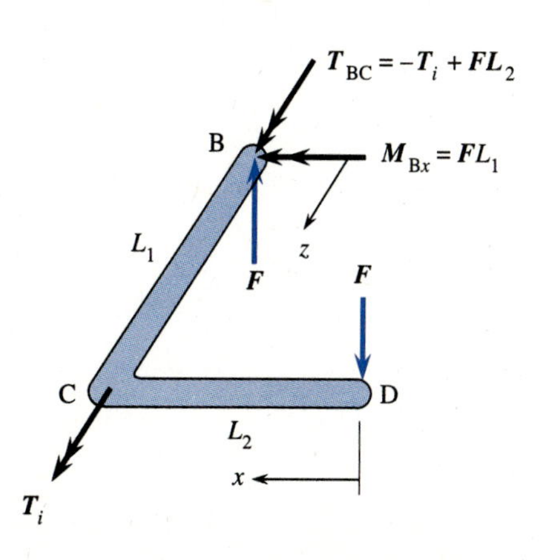

The internal torque and bending moment in the bar are

$$T_{BC} = -T_i + FL_2$$

$$T_{CD} = 0$$

$$M_{BCx} = FL_1 - Fz$$

$$M_{CD} = -Fx$$

The strain energy in the bar is now

$$U = U_{BC} + U_{CD}$$

$$= \frac{T_{BC}{}^2 L_1}{2GJ_{BC}} + \int_0^{L_2} \frac{M_{CD}{}^2}{2EI_{CD}}\,dx + \int_0^{L_1} \frac{M_{BCx}{}^2}{2EI_{BC}}\,dz$$

Taking the derivative of the strain energy with respect to T_i and setting $T_i = 0$, we obtain the angle of twist as

$$\frac{\partial U}{\partial T_i} = \theta_c = \frac{T_{BC}L_1}{GJ}\frac{\partial T_{BC}}{\partial T_i} + 0 + 0$$

$$= \frac{(-T_i + FL_2)L_1(-1)}{GJ}$$

$$= -\frac{FL_2 L_1}{GJ}$$

10.14 ANALYSIS OF STATICALLY INDETERMINATE STRUCTURES USING CASTIGLIANO'S THEOREM

We can use Castigliano's theorem to obtain the reactions at the supports of a statically indeterminate elastic structure. For example, consider the propped cantilever beam subjected to a uniformly distributed load of w shown in Figure 10.43. The beam has four reaction components, while three equilibrium equations exist. The beam is then statically indeterminate to the first degree because it has one more reaction component than equations of equilibrium.

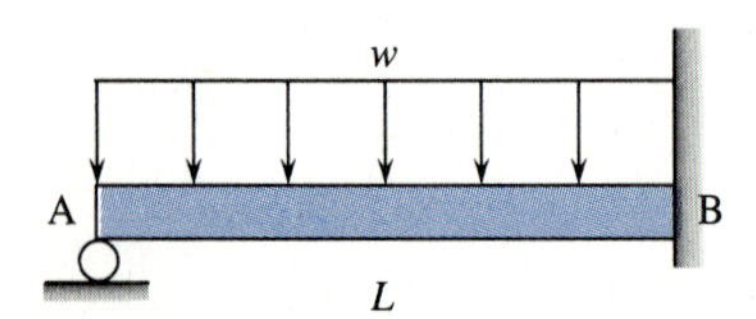

Figure 10.43 Propped cantilever beam.

STEPS IN SOLUTION

The following steps are used to obtain the reactions at the supports of a statically indeterminate structure by Castigliano's theorem.

1. Begin by drawing a free-body diagram of the beam and designate one of the reaction components as redundant. In Figure 10.44, we designate reaction R_A as the redundant. This reaction is then considered to be an unknown load, along with the other known load w.

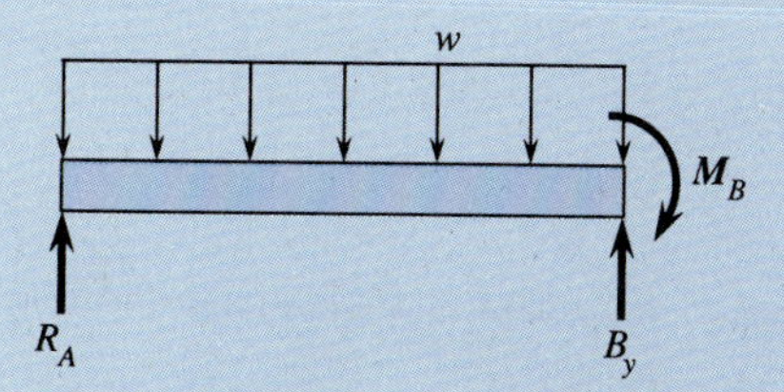

Figure 10.44 Propped cantilever with redundant

2. Then determine the internal reactions, such as the bending moment, due to the redundant and the known real load. The strain energy in the structure is then expressed in terms of the internal reactions, which includes the redundant.
3. Next, take the partial derivative of U with respect to the redundant. Remember, that by Castigliano's theorem this derivative represents the deflection (or slope if a moment was chosen as the redundant) at the support. However, at an unyielding (rigid) support, such as the roller at A, the deflection y_A is zero. We then set this derivative to zero as shown by Eq. (10.90) as

$$\frac{\partial U}{\partial R_A} = y_A = 0 \tag{10.90}$$

4. Finally, solve Eq. (10.90) for the redundant reaction. The remaining reaction components may be obtained from the equations of static equilibrium.

For a structure with two degrees of static indeterminacy, introduce two redundants. The internal reactions, such as the bending moment, are obtained due to both the two redundants and the real load. The strain energy in the structure is then a function of both redundants. Then take the partial derivative of U separately with respect to each redundant. Two equations result that can be solved simultaneously for the two redundants. In general, introduce as many redundants as degrees of static indeterminacy exist (say n), express the strain energy in terms of these redundants, take the partial derivative of U with respect to each redundant, and solve the n equations simultaneously.

Examples 10.20 through 10.23 illustrate how to analyze statically indeterminate structures by Castigliano's theorem.

EXAMPLE 10.20

For the propped cantilever beam subjected to the uniform load w (Figure 10.45), determine the reactions. Let EI be constant.

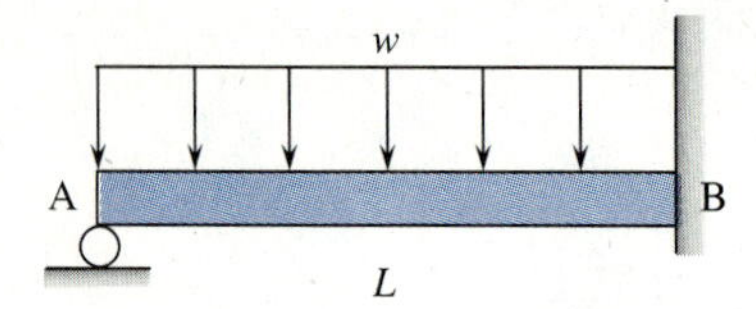

Figure 10.45 Propped cantilever for analysis.

Solution

First draw the free-body diagram of the beam and designate one reaction component as redundant. Here designate R_A as redundant.

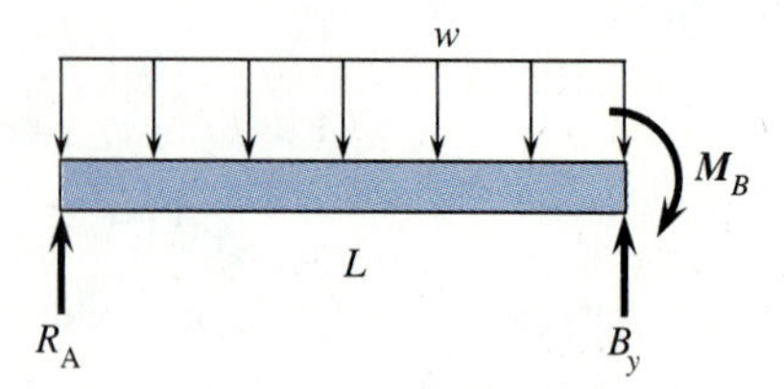

Using the free-body diagram, determine the bending moment M in the beam due to the redundant and the real load.

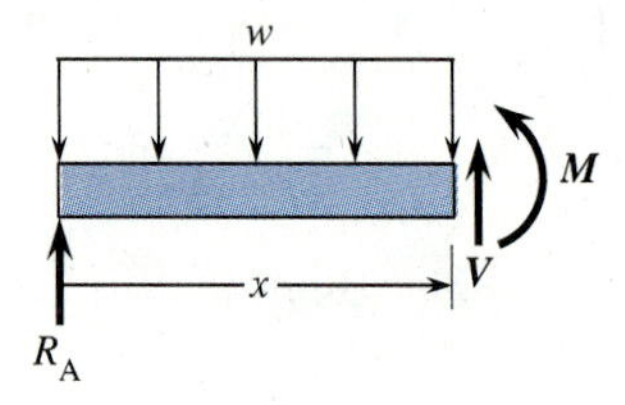

The bending moment is

$$M = R_A x - \frac{wx^2}{2} \tag{a}$$

The derivative of M with respect to R_A is

$$\frac{\partial M}{\partial R_A} = x \tag{b}$$

Substituting Eqs. (a) and (b) into Eq. (10.90), where the strain energy only in bending is considered, we have

$$y_A = 0 = \int_0^L \frac{M}{EI}\left(\frac{\partial M}{\partial R_A}\right)dx = \int_0^L \frac{\left(R_A x - \frac{wx^2}{2}\right)}{EI} x\,dx \tag{c}$$

Integrating Eq. (c) and evaluating, we obtain

$$0 = \left[R_A \frac{x^3}{3}\bigg|_0^L - \frac{wx^4}{8}\bigg|_0^L \right] \frac{1}{EI} \quad \textbf{(d)}$$

Solving Eq. (d) for the redundant R_A, we obtain

$$R_A = \frac{3wL}{8}$$

Using equations of equilibrium, the reaction force and moment at B are

$$R_B = \frac{5wL}{8} \quad \text{and} \quad M_B = \frac{wL^2}{8}$$

EXAMPLE 10.21

A cantilevered beam is supported by the spring at end B. The spring constant is k. Let EI be constant in the beam. Determine the reaction at B. (See Figure 10.46.)

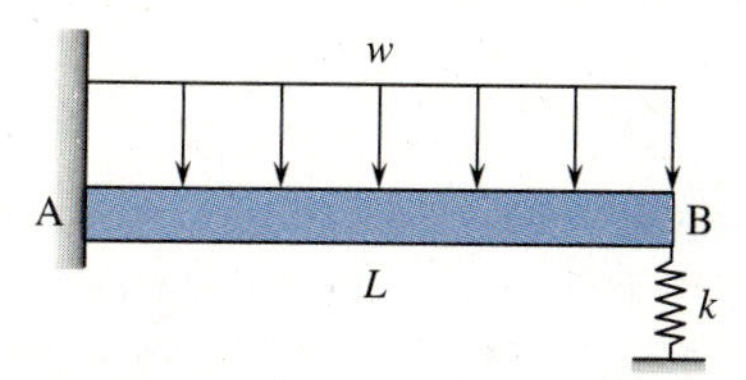

Figure 10.46 Cantilever beam with spring support.

Solution

Draw the free-body diagram of the beam and designate the spring support reaction as the redundant.

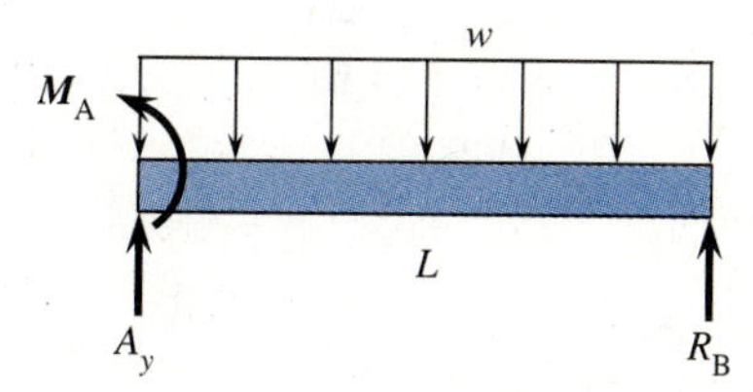

Determine the bending moment in terms of the redundant and the real load.

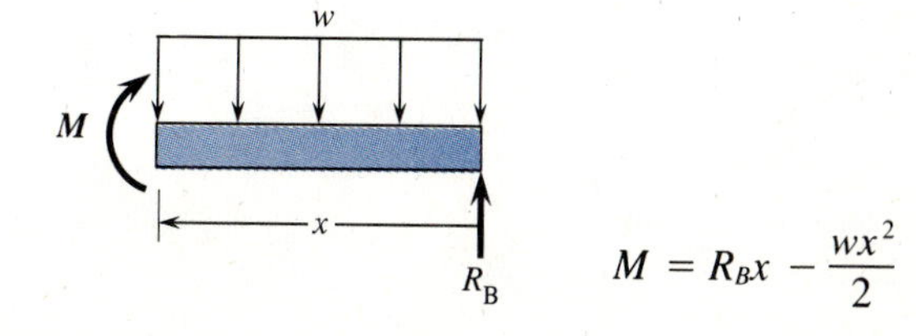

$$M = R_B x - \frac{wx^2}{2}$$

Take the partial derivative of M with respect to the redundant as follows:

$$\frac{\partial M}{\partial R_B} = x$$

Use Castigliano's theorem to express the deflection at B in terms of the partial derivative of the strain energy with respect to R_B.

$$\delta_B = \frac{\partial U}{\partial R_B} = \frac{1}{EI}\int_0^L \left(R_B x - \frac{wx^2}{2}\right) x\, dx$$

$$= \frac{1}{EI}\left(R_B \frac{L^3}{3} - \frac{wL^4}{8}\right)$$

Set this result equal to the deflection at B expressed in terms of R_B and k, where for a linear spring $F = k\delta$. Therefore

$$\frac{1}{EI}\left(R_B \frac{L^3}{3} - \frac{wL^4}{8}\right) = -\frac{R_B}{k}$$

Solve for the redundant as

$$R_B = \frac{wL/8}{\frac{1}{3} + \frac{EI}{L^3 k}} = \frac{kwL/8}{\frac{k}{3} + \frac{EI}{L^3}}$$

The result indicates that if the spring stiffness approaches infinity, the support becomes rigid and the reaction at B becomes that obtained in Example 10.20. That is, $R_B = 3wL/8$. If the spring stiffness approaches zero, then $R_B = 0$, as expected.

EXAMPLE 10.22

The beam with a pin and two roller supports is subjected to the uniform load w shown (Figure 10.47). Determine the reactions. Let EI be constant. (This is the same problem as Example 6.21 solved using superposition.)

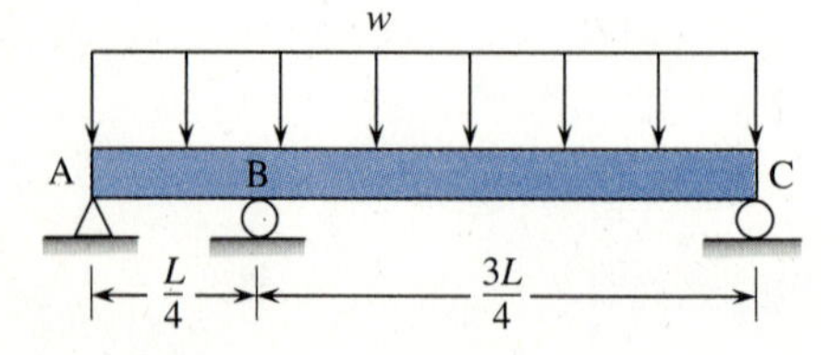

Figure 10.47 Beam with three supports.

Solution

First draw a free-body diagram of the beam and designate one reaction as the redundant. Here designate R_B as the redundant.

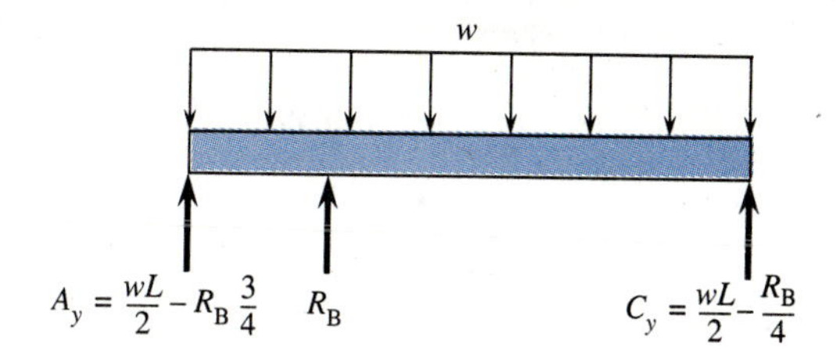

Using the free-body diagrams shown, determine the bending moment M in the beam due to the redundant and real load. For convenience, choose two different coordinate systems, x from the left and y from the right, to determine the bending moment in the beam. Also express the reactions at A and C in terms of the redundant R_B and the real load w. The bending moment expressions are as follows.

For $0 \leq x \leq \frac{L}{4}$

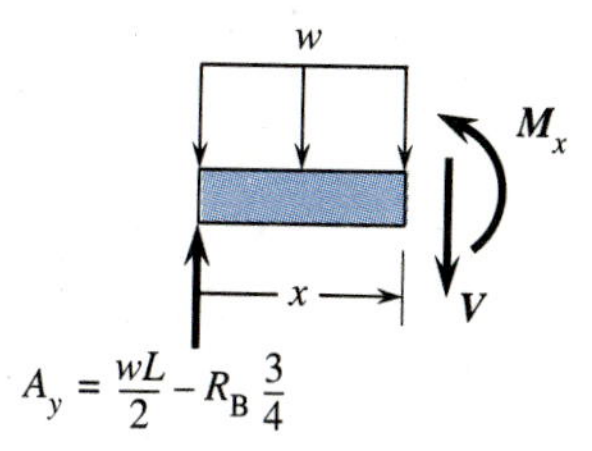

$$M_x = \left(\frac{wL}{2} - \frac{3}{4}R_B\right)x - \frac{wx^2}{2} \tag{a}$$

For $0 \leq y \leq \frac{3L}{4}$

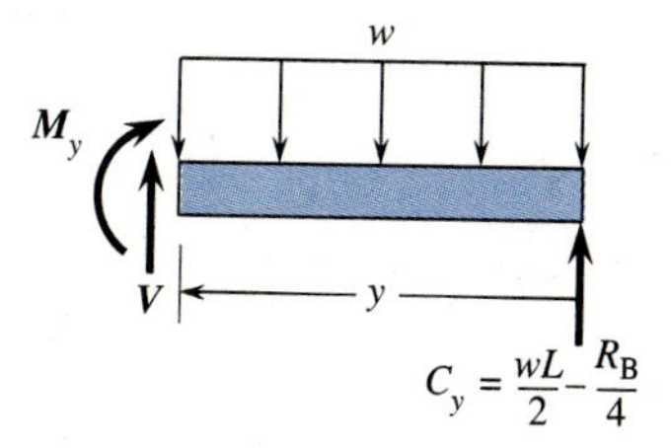

$$M_y = \left(\frac{wL}{2} - \frac{R_B}{4}\right)y - \frac{wy^2}{2} \tag{b}$$

The derivatives of M_x and M_y with respect to R_B are

$$\frac{\partial M_x}{\partial R_B} = -\frac{3x}{4} \qquad \frac{\partial M_y}{\partial R_B} = -\frac{y}{4} \tag{c}$$

Substituting Eqs. (a), (b), and (c) into Eq. (10.90), we obtain

$$y_B = 0 = \frac{\partial U}{\partial R_B} = \frac{1}{EI}\int_0^{L/4}\left(\frac{wLx}{2} - \frac{3}{4}R_B x - \frac{wx^2}{2}\right)\left(-\frac{3x}{4}\right)dx$$
$$+ \frac{1}{EI}\int_0^{3L/4}\left(\frac{wLy}{2} - \frac{R_B y}{4} - \frac{wy^2}{2}\right)\left(-\frac{y}{4}\right)dy \qquad \textbf{(d)}$$

Performing the integration in Eq. (d), we obtain

$$0 = \frac{wL}{2}\left(-\frac{3}{4}\right)\frac{x^3}{3}\bigg|_0^{3L/4} + \frac{9}{16}R_B\frac{x^3}{3}\bigg|_0^{L/4} + \frac{3w}{8}\frac{x^4}{4}\bigg|_0^{L/4}$$
$$+ \frac{wL}{2}\left(-\frac{1}{4}\right)\frac{y^3}{3}\bigg|_0^{3L/4} + \frac{1}{16}R_B\frac{y^3}{3}\bigg|_0^{3L/4} + \frac{w}{8}\frac{y^4}{4}\bigg|_0^{3L/4} \qquad \textbf{(e)}$$

Evaluating each expression in Eq. (e) and solving for R_B, we obtain

$$R_B = \frac{57}{72}wL$$

This is the same result as obtained in Example 6.21.

The reactions at A and C are determined using equilibrium equations as

$$A = -\frac{3}{32}wL$$

and

$$C = \frac{87}{288}wL$$

EXAMPLE 10.23

A crate of weight $W = 500$ lb is suspended by the three bars shown. Determine the force in each bar. Use Castigliano's theorem. Assume all bars have the same cross-sectional area $A = 2$ in.^{2} and modulus of elasticity $E = 30 \times 10^6$ psi.

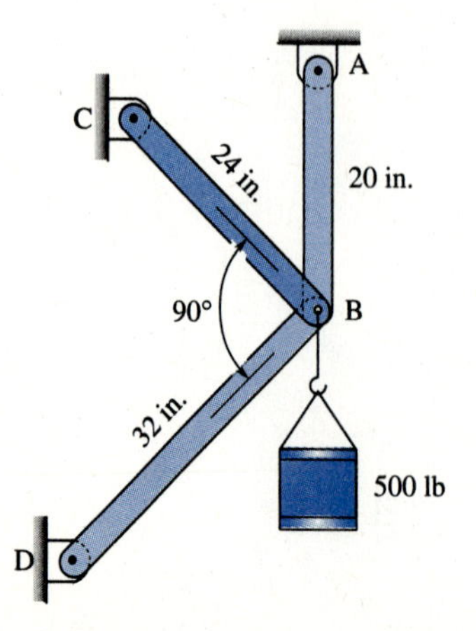

Figure 10.48 Crate supported by three bars.

Solution

From a free-body diagram of the crate, the problem is indeterminate to the first degree.

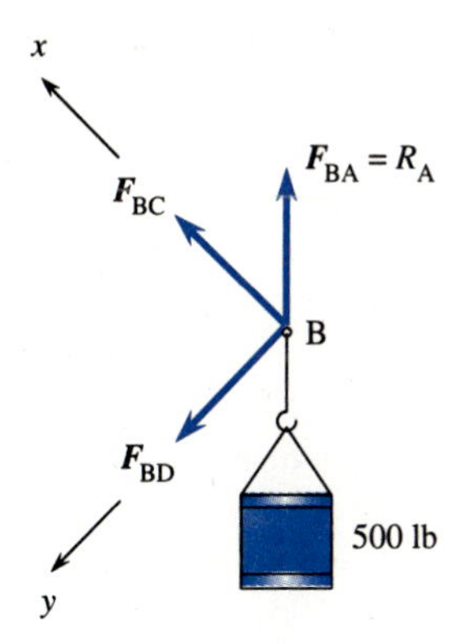

Consider the reaction at A as the redundant. The reaction at A is then considered as an unknown force. Then by using statics, express the forces in the other bars in terms of the redundant R_A and the weight of the crate as follows:

$$\Sigma F_x = 0: \quad F_{BC} = 0.6W - 0.6R_A$$
$$\Sigma F_y = 0: \quad F_{BD} = 0.8R_A - 0.8W \tag{a}$$

Next take the derivative of each bar force with respect to the redundant as follows:

$$\frac{\partial F_{BC}}{\partial R_A} = -0.6 \qquad \frac{\partial F_{BD}}{\partial R_A} = 0.8 \qquad \frac{\partial F_{BA}}{\partial R_A} = 1 \tag{b}$$

Recalling Eq. (10.87) for truss deflections by Castigliano's theorem, we obtain

$$\delta_A = 0 = \frac{F_{BC}L_{BC}}{AE}\frac{\partial F_{BC}}{\partial R_A} + \frac{F_{BD}L_{BD}}{AE}\frac{\partial F_{BD}}{\partial R_A} + \frac{F_{BA}L_{BA}}{AE}\frac{\partial F_{BA}}{\partial R_A} \tag{c}$$

Substituting Eqs. (a) and (b) into (c), we obtain

$$\delta_A = 0 = \frac{1}{AE}[(0.6W - 0.6R_A)(24)(-0.6) + (0.8R_A - 0.8W)(32)(0.8)$$
$$+ R_A(20)(1)]$$
$$0 = -8.64W + 8.64R_A + 20.48R_A - 20.48W + 20R_A \tag{d}$$

Solving Eq. (d) for R_A, we have

$$49.12R_A = 29.12W$$

or

$$R_A = 0.593W = 0.593(500 \text{ lb})$$
$$= 296 \text{ lb} \quad \text{(tensile)} \tag{e}$$

Substituting R_A from Eq. (e) into Eqs. (a), we have

$$\begin{aligned} F_{BC} &= 0.6(500 \text{ lb}) - 0.6(296 \text{ lb}) \\ &= 122 \text{ lb} \quad \text{(tensile)} \\ F_{BD} &= 0.8(296 \text{ lb}) - 0.8(500 \text{ lb}) \\ &= -163 \text{ lb} \quad \text{(compressive)} \end{aligned}$$

10.15 IMPACT PROBLEMS USING ENERGY METHODS

A number of problems involving *impact forces* (suddenly applied forces that act for a short duration of time) can be solved using energy methods. For example, the collision of an automobile with a guard rail or the collision of a pile driver with the pile being driven into the ground, or the dropping of a weight onto a floor.

When the velocity of impact of the load is low, (much lower than the velocity of propagating strain waves), we can use the same relations between stress and strain and load and deformation as developed for static problems, along with the conservation of energy principle to solve impact problems. Under impact loading, if we have elastic action, the loaded member will vibrate, until equilibrium is established. This is shown in Figure 10.49 for a beam subjected to a weight dropped onto it. We might recall from our dynamics class that we can treat a dynamic force using Newton's second law relating force to the product of mass times acceleration of the mass center or using the impulse-momentum law, where the force is the time rate of change of the momentum, or in terms of the change in kinetic energy of the body by using the work-energy principle.

We now illustrate the work-energy method to determine the maximum stress occurring in a structure subjected to impact loading. In using this approach, we make certain simplifying assumptions as follows:

1. First, we assume that all of the kinetic energy, $KE = 1/2\ mv_i^2$, of the striking mass is transferred entirely to the structure as strain energy within the deformable body. Low impact velocities allow the structure to behave in the same manner as it would under a static loading. This allows us to use the same equations relating stress to strain and load to deflection as already developed for static loading.

For all the kinetic energy to be transferred to the structure as strain energy, we have

$$U = \frac{1}{2} mv_i^2 \tag{10.91}$$

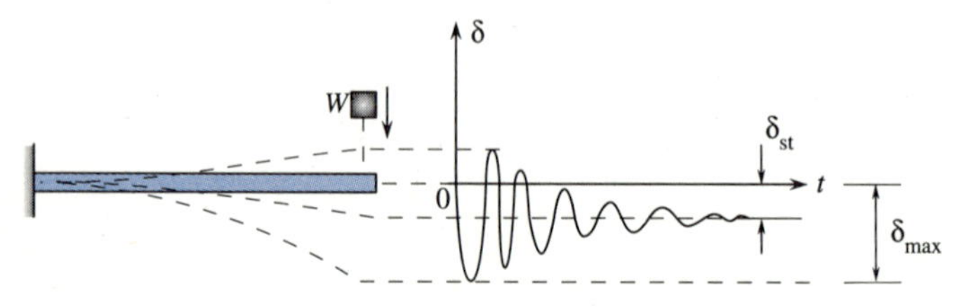

Figure 10.49 Motion of beam subjected to impact loading.

This means the striking mass should not bounce off the structure so as to retain some of its energy.

2. Therefore, we assume that the striking mass has a much larger inertia than the structure.

3. Also, we assume that no energy is lost in the form of heat, sound, or permanent deformation of the striking mass.

In reality, some or all of these assumptions are violated. However, these assumptions will yield a conservative design of the structure and will serve as a basis for preliminary design.

Elastic Rod

We first consider an elastic bar subjected to an impact loading (Figure 10.50). Since we assume elastic behavior of the material even under impact loading, we express the strain energy for axial stress (See Eq. (10.21)) as

$$U = \int \frac{\sigma_x^2}{2E} dV \tag{10.92}$$

For a uniform bar, Eq. (10.92) becomes

$$U = \frac{\sigma_x^2 V}{2E} \tag{10.93}$$

Solving Eq. (10.93) for σ_x, we have

$$\sigma_x = \sqrt{\frac{2EU}{V}} \tag{10.94}$$

Using Eq. (10.91) for U in Eq. (10.94), we express the stress in terms of the mass of the impacting body and the velocity at impact as

$$\sigma_x = \sqrt{\frac{mv_i^2 E}{V}} \tag{10.95}$$

where $V = AL$ is the volume of the bar.

We conclude that for a uniform rod, the stress can be reduced by choosing a material with a low modulus of elasticity E (generally softer materials such as rubber) or a large volume V.

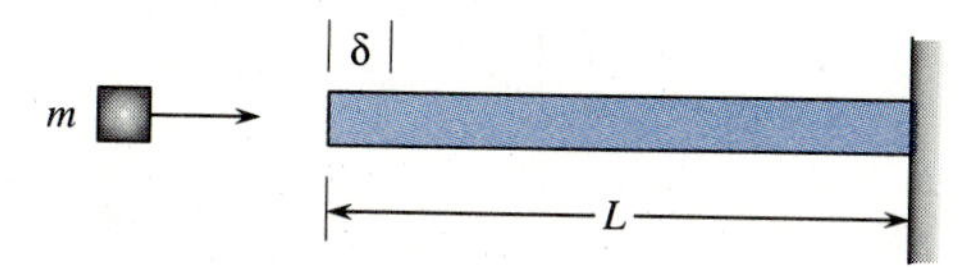

Figure 10.50 Bar subjected to impact load.

Also U can be expressed in terms of the deflection δ of the bar by substituting $\sigma_x = E\epsilon_x$ and $\epsilon_x = \dfrac{\delta}{L}$ into Eq. (10.93). The result is

$$U = \frac{EA\delta^2}{2L} \tag{10.96}$$

We set U equal to the kinetic energy to obtain

$$\frac{mv_i^2}{2} = \frac{EA\delta^2}{2L}$$

and solving for the deflection, we have

$$\delta = \sqrt{\frac{mv_i^2 L}{EA}} \tag{10.97}$$

Equation (10.97) gives the maximum deflection of the end of the bar. Remember that Eqs. (10.95) and (10.97) are based on ideal conditions that the stress is assumed uniform throughout the bar. In practice this often is not the case as stress concentrations and nonuniform loading on the impacting surface can occur. Also remember the mass of the impacted bar or the inertia from it has been neglected in this analysis. This inertia results in the bar having a greater local deflection and stress than when neglecting inertia of the bar. The influence of the bar inertia is considered in more advanced books [1].

As another example, based on our assumptions, consider the elastic spring subjected to an impact load from a falling weight (Figure 10.51). The spring has an elastic spring constant of k. Let δ be the maximum deflection due to impact. From conservation of energy, we have the potential energy of the falling mass turned into kinetic energy and this energy is transferred to the spring.

Therefore

$$W(h + \delta) = \frac{1}{2}\delta F_e \tag{10.98}$$

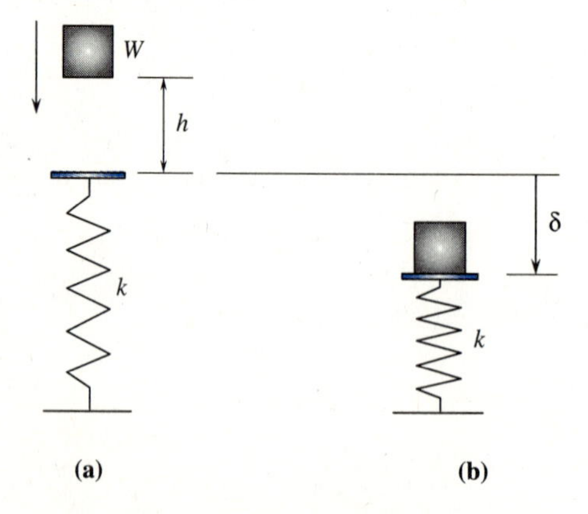

Figure 10.51 (a) Undeformed spring and (b) spring deformed due to impact load.

But $F_e = k\delta$ is the elastic spring force. Therefore, substituting for F_e into Eq. (10.98), we have

$$Wh + W\delta = \frac{1}{2}k\delta^2$$

or

$$\frac{1}{2}k\delta^2 - W\delta - Wh = 0 \tag{10.99}$$

The solution to the quadratic Eq. (10.99) is

$$\delta = \frac{W \pm \sqrt{W^2 - 4\left(\frac{1}{2}k\right)(-Wh)}}{k} = \frac{W \pm \sqrt{W^2 + 2kWh}}{k} \tag{10.100}$$

Now the static deflection, δ_{st}, from the weight slowly lowered onto the spring is

$$\delta_{st} = \frac{W}{k} \tag{10.101}$$

Therefore, we rewrite Eq. (10.100) as

$$\delta = \delta_{st} \pm \sqrt{\left(\frac{W}{k}\right)^2 + 2\frac{W}{k}h} \tag{10.102}$$

or as

$$\delta = \delta_{st} \pm \sqrt{\delta_{st}^2 + 2h\delta_{st}} \tag{10.103}$$

or finally factoring out the δ_{st} term in Eq. (10.103), we obtain

$$\delta = \delta_{st} \pm \delta_{st}\sqrt{1 + \frac{2h}{\delta_{st}}} = \delta_{st}\left(1 + \sqrt{1 + \frac{2h}{\delta_{st}}}\right) \tag{10.104}$$

The plus sign must be used for the downward deflection. Since $F_e = k\delta$, the spring force is given by

$$F_e = W + W\sqrt{1 + \frac{2h}{\delta_{st}}} = W\left(1 + \sqrt{1 + \frac{2h}{\delta_{st}}}\right) \tag{10.105}$$

The term in the parentheses in Eqs. (10.104) and (10.105) is called the *impact factor*. This is the factor that increases the force and deflection in the spring due to the dynamically applied weight W over that which would occur if the weight was slowly applied to the spring. For δ_{st} small compared to the height h of the drop, Eq. (10.103) can be approximated by

$$\delta = \delta_{st} + \sqrt{2h\delta_{st}} \cong \sqrt{2h\delta_{st}} \tag{10.106}$$

We can express the deflection δ and force F_e in terms of the velocity at impact v_i instead of the height h of the fall of weight W. The kinetic energy of the falling weight at impact is equal to the change in potential energy of the weight at impact. Therefore

$$\frac{1}{2}mv_i^2 = Wh$$

and $W = mg$. Solving for h, we obtain

$$h = \frac{v_i^2}{2g} \tag{10.107}$$

where g is the acceleration of gravity. Now substituting Eq. (10.107) for h into Eq. (10.104), we have

$$\delta = \delta_{st}\left(1 + \sqrt{1 + \frac{v_i^2}{g\delta_{st}}}\right) \tag{10.108}$$

and

$$F_e = W\left(1 + \sqrt{1 + \frac{v_i^2}{g\delta_{st}}}\right) \tag{10.109}$$

Note that if $h = 0$ and we suddenly release the weight from the top of the uncompressed spring, from Eqs. (10.108) and (10.109), we get

$$\delta = 2\delta_{st} \tag{10.110}$$

and

$$F_e = 2W \tag{10.111}$$

The impact factor is then 2. The factor of two has sometimes been used as a "rule of thumb" when using a static analysis to analyze a problem with impact loading. We will observe, by examples, that this is not recommended and can yield large errors in results.

EXAMPLE 10.24

Compare **(a)** the elastic energy absorbing capacities, **(b)** the deflections, and **(c)** the axial stresses in two different bars when a sliding mass of weight $W = 50$ lb drops from the top of the bar and hits the bottom end flange. One bar has a uniform cross-section with diameter $d = 1$ in. and length $L = 30$ in. The other bar has upper-half diameter of 2 in. and lower-half diameter of 1 in. with same total length $L = 30$ in. Assume both bars of steel with yield stress of 100 ksi and $E = 29 \times 10^3$ ksi (Figure 10.52).

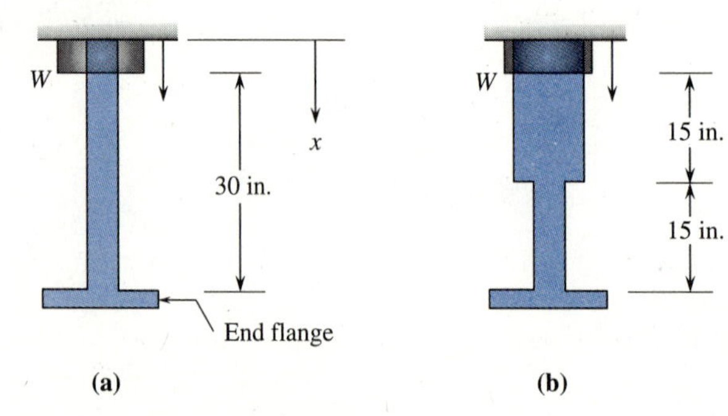

Figure 10.52 (a) Uniform bar and (b) stepped bar.

Solution

(a) Based on Eq. (10.93), the elastic energy absorbing capacities of the bars can be determined as follows.

For the uniform bar:
Assuming maximum elastic energy reached when $\sigma_x = \sigma_Y$

$$U = \frac{\sigma_Y^2 V}{2E} = \frac{(100 \text{ ksi})^2\,(\pi/4)(1 \text{ in.})^2(30 \text{ in.})}{2 \times 29 \times 10^3 \text{ ksi}}$$

$$U = 4.062 \text{ in.} \cdot \text{kip}$$

For the stepped bar:
The lower half reaches σ_Y under uniform load while the upper half will only reach $\sigma_Y/4$ as its area is four times the lower half. Therefore

$$U = \frac{(\sigma_Y/4)^2(\pi/4)(2d)^2(L/2)}{2E} + \frac{(\sigma_Y)^2(\pi/4)d^2(L/2)}{2E}$$

$$= \frac{(100/4)^2(\pi/4)(2)^2(15) + 100^2(\pi/4)(1^2)(15)}{2 \times 29 \times 10^3}$$

$$= 0.5078 + 2.031$$

$$U = 2.539 \text{ in.} \cdot \text{kip}$$

The energy capacity of the stepped bar is only $2.539/4.062 = 5/8$ that of the uniform bar. Also the stepped bar has 2.5 times the volume and weight of the uniform bar. Therefore, the energy capacity per pound is four times $(2.5/(5/8))$ as great with the uniform bar. In addition, the stress concentration in the middle of the stepped bar would reduce its energy absorbing capacity even more and would tend to enhance brittle fracture. Therefore, uniform cross-sections should be used whenever possible for energy absorbing or dynamic loadings.

For static loads, however, remember that the maximum stress, and not the energy absorbing capability, is most important in design.

(b) Equation (10.104) applies for a bar subjected to impact load.

For the uniform bar:
We first calculate the static deflection as

$$\delta_{st} = \frac{WL}{AE} = \frac{(50 \text{ lb})(30 \text{ in.})}{(\pi/4)(1 \text{ in.})^2(29 \times 10^6 \text{ psi})}$$

$$= 0.0000660 \text{ in.}$$

This deflection is substituted into Eq. (10.104) to obtain the deflection due to the dynamic loading as

$$\delta = \delta_{st}\left(1 + \sqrt{1 + \frac{2h}{\delta_{st}}}\right)$$

$$\delta = 0.000066 \text{ in.}\left(1 + \sqrt{1 + \frac{2\,(30 \text{ in.})}{0.000066 \text{ in.}}}\right)$$

$$= 0.000066(1 + 953) = 0.0630 \text{ in.}$$

Note, we could have used the approximate Eq. (10.106) (since h is much larger than δ_{st}) to obtain

$$\delta = \sqrt{2(30 \text{ in.})(0.000066 \text{ in.})} \tag{a}$$
$$\delta = 0.0630 \text{ in.}$$

From the calculations above, the impact factor is 954.

For the stepped bar:
To use equations developed in this section for the stepped bar, we must obtain an equivalent stiffness for the bar as follows. Consider the bar to have a static load P acting on the free end. The resulting elongation is

$$\delta = \frac{P(L/2)}{AE} + \frac{P(L/2)}{4AE} = \frac{5PL}{8AE} \tag{b}$$

where $A = \pi d^2/4$ is the cross-sectional area of the lower half of the bar. From Eq. (b), we obtain the equivalent stiffness k as

$$k = \frac{P}{\delta} = \frac{8AE}{5L} = \frac{8(\pi/4)(1)^2(29 \times 10^3)}{5(30)} \tag{c}$$

or

$$k = 1.215 \times 10^3 \text{ kips/in.} \tag{d}$$

The static deflection is then

$$\delta_{st} = \frac{W}{k} = \frac{50 \text{ lb}}{1.215 \times 10^6 \text{ lb/in.}}$$
$$= 0.0000412 \text{ in.}$$

Using Eq. (10.104), the deflection due to dynamic loading is

$$\delta = 0.0000412\left(1 + \sqrt{1 + \frac{2(30)}{0.0000412}}\right) = 0.000042(1208) \tag{e}$$
$$\delta = 0.0498 \text{ in.}$$

The deflection is smaller than for the uniform bar, but the impact factor is 1208 compared to 954 for the uniform bar.

(c) The maximum stresses are

For the uniform bar:

$$\sigma = E\frac{\delta}{L} = (29 \times 10^3 \text{ ksi})\left(\frac{0.0630 \text{ in.}}{30 \text{ in.}}\right) \tag{f}$$
$$= 60.9 \text{ ksi}$$

For the stepped bar:
First find the static load that yields the deflection $\delta = 0.0498$ in. as follows:

$$P = k\delta$$

Using k from Eq. (d) and δ from Eq. (e), we have

$$P = (1.215 \times 10^3 \text{ kips/in.})(0.0498 \text{ in.})$$

$$P = 60.5 \text{ kips}$$

The normal stress is largest in the smaller cross-section and is then

$$\sigma = \frac{60.5 \text{ kips}}{(\pi/4)(1 \text{ in.})^2} = 77.0 \text{ ksi} \qquad \text{(g)}$$

Comparing Eqs. (f) and (g), we observe that the normal stress is larger in the stepped bar than the uniform bar with smaller cross section.

Torsion Bar

A torsion bar subjected to sudden torsional impact T can be analyzed in a manner analogous to that used for the axially loaded bar. Using Eq. (10.36) for the uniform torsion bar, the strain energy is

$$U = \frac{T^2L}{2GJ} \qquad (10.112)$$

Also the torsional stiffness is

$$K = \frac{T}{\phi} = \frac{GJ}{L} \qquad (10.113)$$

Using Eq. (10.113) in (10.112), we have

$$U = \frac{T^2}{2K} \qquad (10.114)$$

and

$$T = \sqrt{2UK} \qquad (10.115)$$

Also recall from Eqs. (4.14) and (4.19), the torsional shear stress and angle of twist of the bar are

$$\tau = \frac{Tr}{J} \qquad (4.14)$$

and

$$\phi = \frac{TL}{GJ} \qquad (4.19)$$

We now write ϕ in terms of U as

$$\phi = \sqrt{\frac{2U}{K}} \qquad (10.116)$$

For a circular disk with mass moment of inertia $I_m = \frac{1}{2} mr^2$ and rotating with an angular velocity of ω, the kinetic energy of the disk is

$$KE = \frac{1}{2} I_m \omega^2 \tag{10.117}$$

If this kinetic energy is imparted to a torsion bar, then setting the strain energy from Eq. (10.116) equal to the kinetic energy (Eq. (10.117)), we have

$$\phi = \sqrt{\frac{I_m \omega^2}{K}} \tag{10.118}$$

EXAMPLE 10.25

A solid circular steel shaft of 10 mm diameter in a grinding machine is subjected to a suddenly applied torque that stops the shaft instantly (Figure 10.53). The shaft pulley (with diameter = 100 mm and thickness = 10 mm) was turning at 2000 rpm when it stopped suddenly. Let G = 80 GPa and mass density of the pulley ρ = 2000 kg/m^3. Determine the torsional shear stress in the shaft.

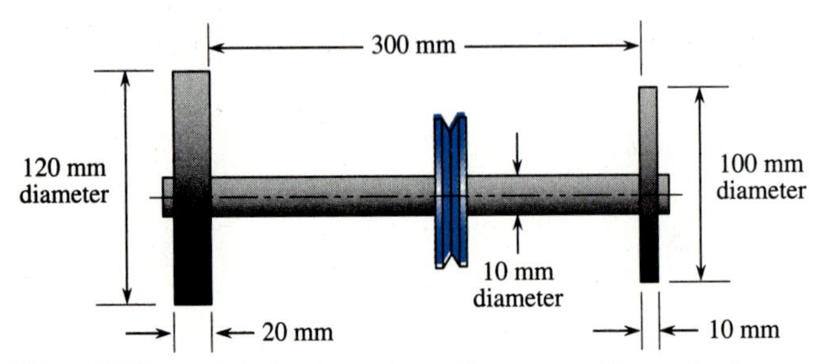

Figure 10.53 Steel shaft subjected to sudden torsional impact loading.

Solution

From Eq. (10.113), the torsional stiffness of the bar is

$$K = \frac{GJ}{L} = \frac{(80 \times 10^9 \text{ N/m}^2)(\pi/2)(0.005 \text{ m})^4}{0.3 \text{ m}}$$

$$= 262 \text{ N} \cdot \text{m}$$

The mass of the disk is

$$m = \pi r^2 t \rho = \pi(0.05 \text{ m})^2(0.01 \text{ m})(2000 \text{ kg/m}^3)$$

$$= 0.157 \text{ kg}$$

The mass moment of inertia of the disk is

$$I_m = \frac{1}{2} mr^2 = \frac{1}{2}(0.157 \text{ kg})(0.05 \text{ m})^2$$

$$= 0.0001963 \text{ kg} \cdot \text{m}^2$$

The kinetic energy of the pulley is

$$KE = \frac{1}{2} I_m \omega^2$$

$$= \frac{1}{2}(0.0001963 \text{ kg} \cdot \text{m}^2)\left[(2000 \text{ rpm})\left(2\pi\frac{\text{rad}}{\text{rev}}\right)\left(\frac{1 \text{ min.}}{60 \text{ sec}}\right)\right]^2$$

$$= 4.29 \text{ N} \cdot \text{m}$$

The kinetic energy is imparted to the bar as strain energy U. Therefore, using Eq. (10.116), we obtain the angle of twist of the bar as

$$\phi = \sqrt{\frac{2(4.29 \text{ N} \cdot \text{m})}{262 \text{ N} \cdot \text{m}}} = 0.181 \text{ rad}$$

and solving Eq. (10.113) for T, we have

$$T = \frac{\phi GJ}{L} = (0.181)(262 \text{ N} \cdot \text{m}) = 47.4 \text{ N} \cdot \text{m}$$

The maximum shear stress in the shaft is then

$$\tau = \frac{Tr}{J} = \frac{(47.4 \text{ N} \cdot \text{m})(0.005 \text{ m})}{9.82 \times 10^{-10} \text{ m}^4}$$

$$= 241 \text{ MPa}$$

Remember, this solution is based on equations developed for stresses in the linear-elastic range. Therefore, the yield stress in shear for the shaft must be greater than 241 MPa.

Beam Deflection and Stress Due to Impact

We use the same work-energy approach as used for the analysis of the axially loaded bar and torsion bar due to impact loads and torques, respectively. That is, we will equate the work done by the falling mass with the strain energy in the beam. In doing this, we make the following assumptions:

1. The falling weight sticks to the beam (no rebound) and moves with the beam.
2. The beam behaves linear-elastic.
3. No energy is lost during the impact, such as due to heat and noise.
4. The deflected shape is assumed the same under dynamic load as under static load.
5. The mass of the beam is small compared to the mass of the falling weight.

To illustrate the work-energy approach based on the above assumptions, consider the simple beam that is subjected to an impact force at the middle of the span due to a falling weight W (Figure 10.54). All the work done by the weight moving through distance $h + \delta$ is equated to the elastic strain energy of the beam as follows.

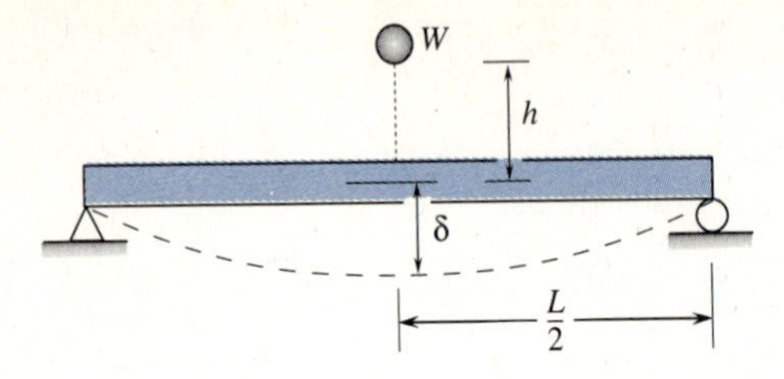

Figure 10.54 Deflection of simple beam due to impact from falling weight.

The work done by the falling weight is

$$W_e = W(h + \delta) \tag{10.119}$$

The work of the force P corresponding to when the displacement δ is a maximum is

$$W_p = \frac{P\delta}{2} \tag{10.120}$$

We then relate P to δ through the equation for beam deflection (see Chapter 6 or Case 1, Appendix D) as

$$\delta = \frac{PL^3}{48EI} \tag{10.121}$$

or

$$P = \frac{48EI\delta}{L^3} \tag{10.122}$$

Now the work of P moving through δ is equal to the strain energy of the beam. Therefore,

$$W_p = \frac{P\delta}{2} = \frac{48EI\delta^2}{2L^3} \tag{10.123}$$

Equating work done by the falling weight W_e to W_p, we have

$$W(h + \delta) = \frac{24EI\delta^2}{L^3} \tag{10.124}$$

The quadratic equation can be solved for δ as

$$\delta = \frac{WL^3}{48EI} + \left[\left(\frac{WL^3}{48EI}\right)^2 + 2h\frac{WL^3}{48EI}\right]^{1/2} \tag{10.125}$$

The static deflection is denoted by

$$\delta_{st} = \frac{WL^3}{48EI} \tag{10.126}$$

Using Eq. (10.126), we write Eq. (10.125) as

$$\delta = \delta_{st} \pm \sqrt{(\delta_{st}^2 + 2h\delta_{st})} \tag{10.127}$$

or

$$\delta = \delta_{st} \pm \delta_{st}\sqrt{1 + \frac{2h}{\delta_{st}}} \tag{10.128}$$

which is identical in form to Eq. (10.104) used to analyze a bar under impact force.

The deflection δ given by Eq. (10.128) is normally an upper limit because we assumed that there were no losses of energy, such as due to heat during the impact. The falling weight was assumed to stick to the beam (and not rebound from the beam) and not deform. Also the mass (or inertia) of the beam would tend to reduce the deflection. Again for h large, we can express Eq. (10.128) as

$$\delta = \sqrt{2h\,\delta_{st}} \tag{10.129}$$

The maximum bending stress occurs at the middle of the simple beam where the bending moment is maximum. The maximum bending moment is

$$M_{max} = \frac{PL}{4} = \left(\frac{48EI\delta}{L^3}\right)\left(\frac{L}{4}\right) \tag{10.130}$$

Therefore, the maximum bending stress from the flexure formula is

$$\sigma_{max} = \frac{M_{max}c}{I} \tag{10.131}$$

The strain energy during deformation is equal to the kinetic energy at impact. Therefore, we can express the force in terms of the velocity of impact as follows:

$$U = \frac{1}{2}mv_i^2 = \frac{P\delta}{2} \tag{10.132}$$

For the simply supported beam, δ given by Eq. (10.121) is substituted into Eq. (10.132) to obtain

$$U = \frac{1}{2}mv_i^2 = \frac{1}{2}\frac{P^2L^3}{48EI} \tag{10.133}$$

Solving Eq. (10.133) for P, we have

$$P = \sqrt{\frac{96UEI}{L^3}} = \sqrt{\frac{48\,mv_i^2EI}{L^3}} \tag{10.134}$$

Using Eqs. (10.130), and (10.134) for P, we obtain the maximum bending stress as

$$\sigma_{max} = \frac{PL}{4}\left(\frac{c}{I}\right) = \sqrt{\frac{3\,mv_i^2\,Ec^2}{LI}} \tag{10.135}$$

EXAMPLE 10.26

A simply supported ASTM-A36 steel beam is struck at its midspan by a falling weight of 50 lb (Figure 10.55). The weight drops through 3 in. onto the beam. The beam is 2 in. wide × 4 in. deep and is 100 in. long. Determine **(a)** the maximum deflection and **(b)** the maximum bending stress.

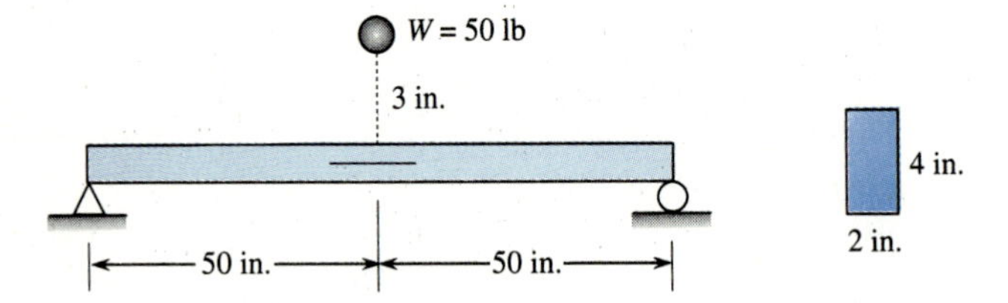

Figure 10.55 Beam subjected to impact force.

Solution

(a) Using Eq. (10.126), the static deflection is

$$\delta_{st} = \frac{WL^3}{48EI}$$

$$= \frac{(50 \text{ lb})(100 \text{ in.})^3}{48 \times 30 \times 10^6 \text{ psi} \times 10.67 \text{ in.}^4}$$

$$= 0.00325 \text{ in.}$$

Since $h = 3$ in. is much greater than 0.00325 in., we use the simple Eq. (10.129) to obtain the deflection as

$$\delta = \sqrt{2h\delta_{st}}$$

$$\delta = \sqrt{2(3)(0.00325)} = 0.140 \text{ in.}$$

(b) By Eq. (10.130), the maximum bending moment is

$$M_{max} = \frac{48EI\delta}{L^3}\left(\frac{L}{4}\right) = \frac{48 \times 30 \times 10^6 \times 10.67 \times 0.140 \times 100}{4(100)^3}$$

$$M_{max} = 53{,}780 \text{ lb} \cdot \text{in.}$$

Using Eq. (10.131), the maximum bending stress is

$$\sigma_{max} = \frac{(53{,}780 \text{ lb} \cdot \text{in.})(2 \text{ in.})}{10.67 \text{ in.}^4}$$

$$\sigma_{max} = 10{,}080 \text{ psi} < 36{,}000 \text{ psi}$$

Therefore, the beam is elastic and the solution is valid. For comparison, if the weight of 50 lb was slowly applied to the beam, we have

$$M_{max} = \frac{PL}{4} = \frac{(50)(100)}{4} = 1250 \text{ lb} \cdot \text{in.}$$

and the maximum static stress is

$$\sigma_{max} = \frac{(1250 \text{ lb} \cdot \text{in.})(2 \text{ in.})}{10.67 \text{ in.}^4} = 234 \text{ psi}$$

which is much smaller than the stress due to impact.

In summary, we notice from the results of Example 10.24 that a stepped rod cannot absorb as much energy as a uniform rod of the same volume and same material. That is, the lowest stresses will be in the uniform rod. Similarly, a beam is not as efficient at resisting impact loads as is the uniform rod. This is because the stresses vary linearly through the depth of the beam and along the length of the beam. The torsion member is also not as efficient as the uniform rod as the shear stresses vary linearly through the radial direction.

In general, the most efficient way to withstand impact loading is to try to achieve the following:

1. A large volume (see Eq. (10.93))
2. A uniform cross-section with stresses distributed the same throughout the member
3. A low modulus of elasticity and a high yield stress (see Eq. (10.93))

10.16 SUMMARY OF IMPORTANT DEFINITIONS AND EQUATIONS

1. Work of a slowly applied increasing force

$$W = \frac{1}{2}Fx \qquad \textbf{(10.5)}$$

2. Work of a slowly applied increasing moment

$$W = \frac{1}{2}M\theta \qquad \textbf{(10.7)}$$

3. Strain energy density for uniaxial stress state

$$u = \frac{\sigma_1^2}{2E} \qquad \textbf{(10.16)}$$

4. Modulus of resilience (Area under the linear portion of the stress-strain diagram)

$$u_r = \frac{\sigma_Y^2}{2E} \qquad \textbf{(10.17)}$$

5. Modulus of toughness (Area under the whole stress-strain diagram)

6. Elastic strain energy in a bar due to axial force

$$U = \frac{P^2L}{2AE} \qquad \textbf{(10.23)}$$

7. Elastic strain energy due to direct shear force

$$U = \frac{V^2L}{2AG} \qquad \textbf{(10.31)}$$

8. Elastic strain energy due to torsion in a circular bar

$$U = \frac{T^2L}{2GJ} \qquad \textbf{(10.36)}$$

9. Elastic strain energy due to bending

$$U = \int_0^L \frac{M^2}{2EI}dx \qquad \textbf{(10.41)}$$

10. Elastic strain energy due to transverse shear force in beam

$$U = \frac{fV^2L}{2GA} \quad (10.44)$$

11. Volumetric strain energy density

$$u_v = \frac{1 - 2\nu}{6E}(\sigma_1 + \sigma_2 + \sigma_3)^2 \quad (10.55)$$

12. Distortion strain energy density

$$u_d = \frac{1 + \nu}{6E}[(\sigma_1^2 - 2\sigma_1\sigma_2 + \sigma_2^2) + (\sigma_2^2 - 2\sigma_2\sigma_3 + \sigma_3^2) + (\sigma_3^2 - 2\sigma_1\sigma_3 + \sigma_1^2)] \quad (10.57)$$

13. Principle of work-energy (Conservation of energy). The external work W done by a single applied force or moment is equal to the total strain energy U in the member or structure.

14. Method of virtual work (dummy unit load method)

$$1 \cdot \Delta = \Sigma u \cdot dL \quad (10.61)$$

Virtual forces (1, u); Real deformations (Δ, dL)

$$1 \cdot \theta = \Sigma u_m \cdot dL \quad (10.62)$$

Virtual unit moment (1); Virtual forces (u_m); Real deformation (dL); Real slope (θ)

15. Method of virtual work

Truss deflections due to external forces

$$1 \cdot \Delta = \sum_{i=1}^{n} \frac{u_i P_i L_i}{A_i E_i} \quad (10.63)$$

Truss deflections due to temperature change

$$1 \cdot \Delta = \Sigma[u_i \alpha_i (\Delta T_i) L_i] \quad (10.65)$$

Beam and frame deflections

$$1 \cdot \Delta = \int_0^L \frac{mM}{EI} dx \quad (10.67)$$

Beam and frame slopes

$$1 \cdot \theta = \int_0^L \frac{mM}{EI} dx \quad (10.68)$$

16. Castigliano's theorem

$$x_i = \frac{\partial U}{\partial P_i} \quad (10.84)$$

$$\theta_i = \frac{\partial U}{\partial M_i} \quad (10.85)$$

Truss deflections

$$\Delta_i = \frac{\partial U}{\partial F} = \frac{\partial}{\partial F}\left[\sum_{i=1}^{n}\left(\frac{P_n^2 L_n}{2A_n E_n}\right)\right] = \sum_{i=1}^{n} \frac{P_n(\partial P_n / \partial F)L_n}{A_n E_n} \quad (10.87)$$

Beam and frame deflections

$$\Delta_i = \frac{\partial U}{\partial P_i} = \int_0^L \frac{M}{EI}\frac{\partial M}{\partial P} dx \quad (10.88)$$

Beam and frame slopes

$$\theta_i = \frac{\partial U}{\partial M_i} = \int_0^L \frac{M}{EI}\frac{\partial M}{\partial M_i} dx \quad (10.89)$$

Statically indeterminate structures to obtain redundant R_A set

$$\frac{\partial U}{\partial R_A} = y_A = 0 \quad (10.90)$$

and solve for R_A.

17. Impact and energy load problems

Stress in a bar due to horizontal impacting mass

$$\sigma_x = \sqrt{\frac{2EU}{V}} \quad (10.94)$$

$$\sigma_x = \sqrt{\frac{mv_i^2E}{V}} \quad (10.95)$$

Deflection in a bar due to horizontal impacting mass

$$\delta = \sqrt{\frac{mv_i^2L}{EA}} \quad (10.97)$$

Deflection in a spring due to a falling mass

$$\delta = \delta_{st} \pm \delta_{st}\sqrt{1 + \frac{2h}{\delta_{st}}}$$

$$= \delta_{st}\left(1 + \sqrt{1 + \frac{2h}{\delta_{st}}}\right) \quad (10.104)$$

Force in a spring due to a falling mass

$$F_e = W + W\sqrt{1 + \frac{2h}{\delta_{st}}}$$

$$= W\left(1 + \sqrt{1 + \frac{2h}{\delta_{st}}}\right) \quad (10.105)$$

Impact factor

$$1 + \sqrt{1 + \frac{2h}{\delta_{st}}}$$

Angle of twist in a torsion member due to energy load

$$\phi = \sqrt{\frac{I_m\omega^2}{K}} \quad (10.118)$$

Maximum bending stress in simple beam due to falling mass

$$\sigma_{max} = \frac{PL}{4}\left(\frac{c}{I}\right) = \sqrt{\frac{3mv_i^2Ec^2}{LI}} \quad (10.135)$$

REFERENCE

1. Paz, M., *Structural Dynamics Theory and Computation,* Van Nostrand Reinhold, New York, 1985

PROBLEMS

Section 10.3

10.1 A bar is made of a steel that has a yield strength of 36 ksi. The modulus of elasticity is $E = 29 \times 10^6$ psi. Determine the maximum strain energy in the bar before permanent deformation would occur.

Figure P10.1

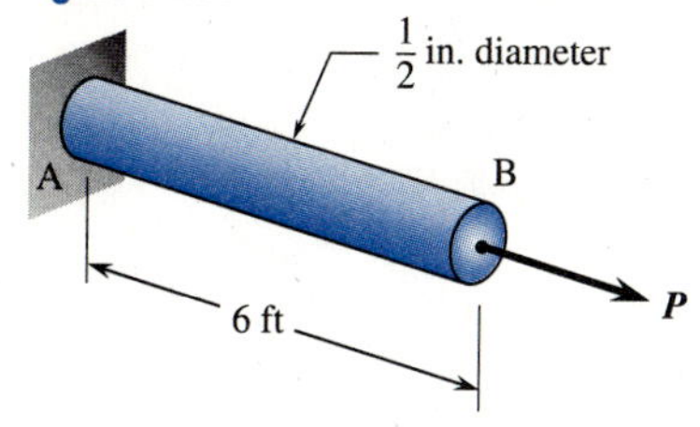

10.2 A bar is made of a steel that has a yield strength of 250 MPa. The modulus of elasticity is $E = 200$ GPa. Determine the maximum elastic strain energy in the bar.

Figure P10.2

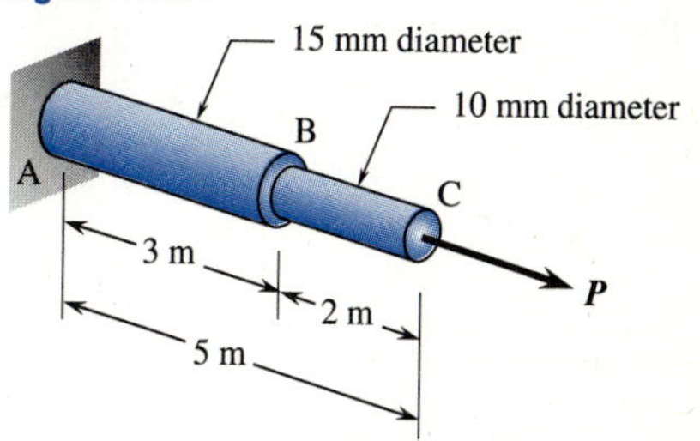

10.3 An aluminum bar with a yield strength of 225 MPa and $E = 75$ GPa is used for the stepped bar shown. Determine the maximum strain energy in the bar before permanent deformation occurs.

Figure P10.3

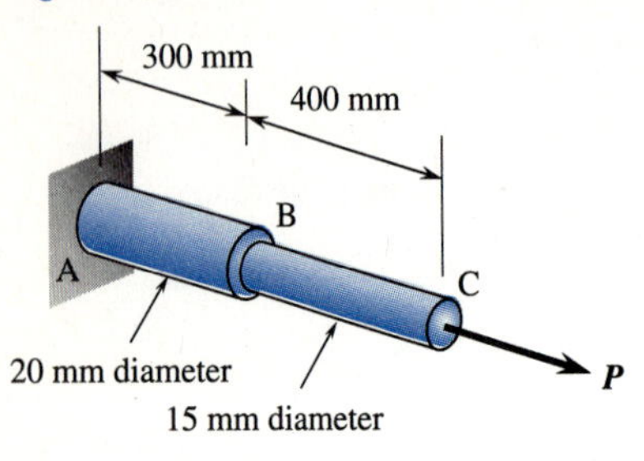

Figure P10.6

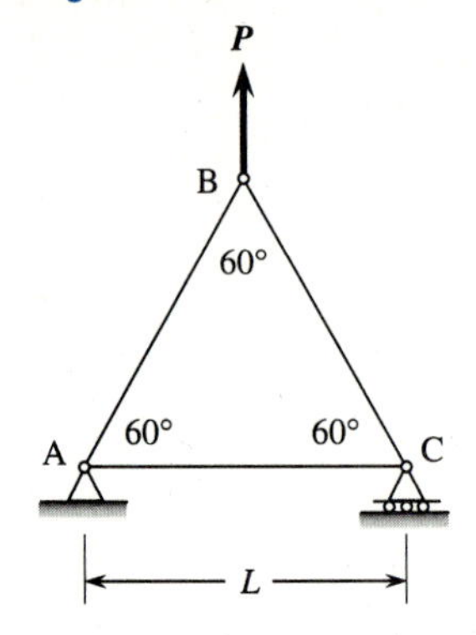

10.4 A uniform bar must absorb a strain energy of 20 J when the axial load P is applied. The yield strength of the bar is 250 MPa and $E = 200$ GPa. Determine the length of the bar just before yielding would occur.

Figure P10.4

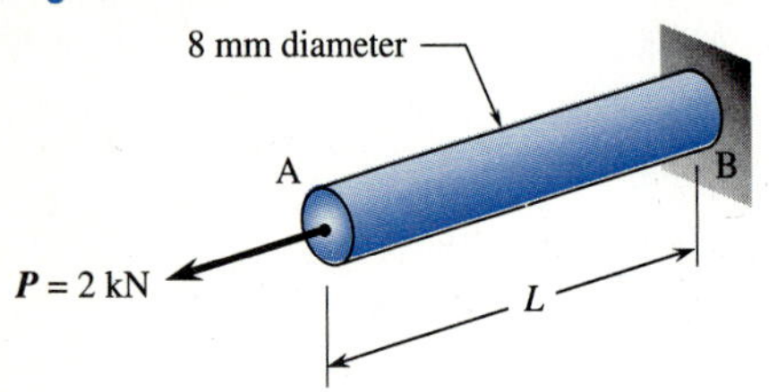

Figure P10.7

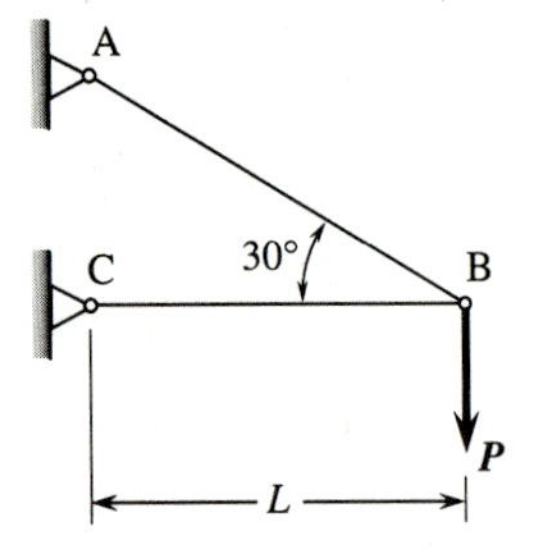

Figure P10.8

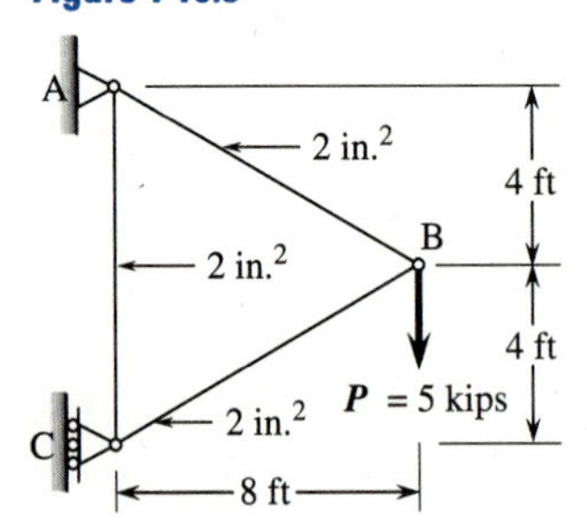

10.5–10.9 Determine the strain energy in the trusses shown when subjected to the load P. All members have a uniform cross-section A and modulus of elasticity E.

Figure P10.5

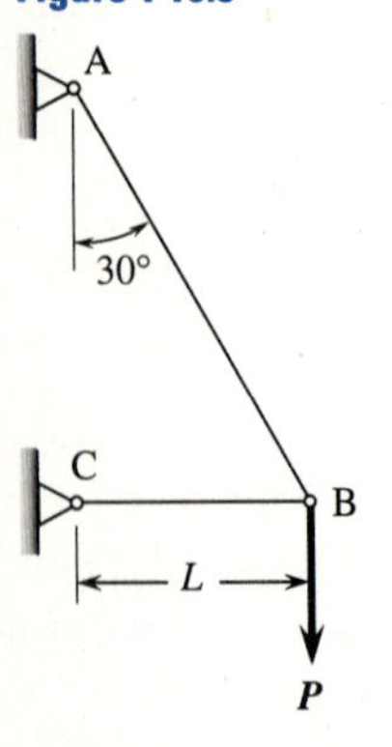

Figure P10.9

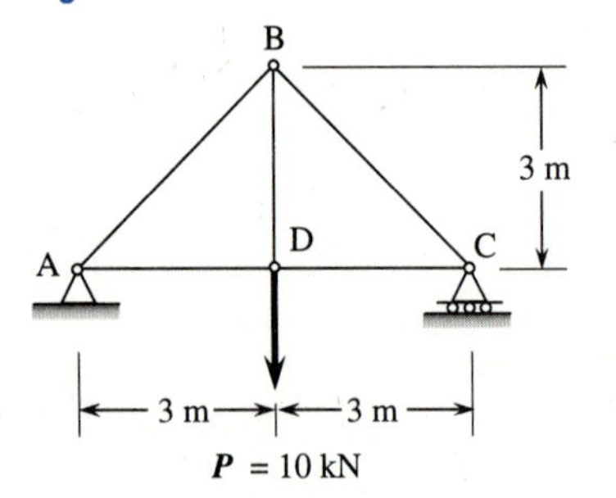

Section 10.5

10.10 A uniform torsion bar is subjected to an end torque T as shown. The bar has a diameter of 1 in. Determine the strain energy in the bar for a maximum shear stress of 12 ksi. Let $G = 12 \times 10^6$ psi.

Figure P10.10

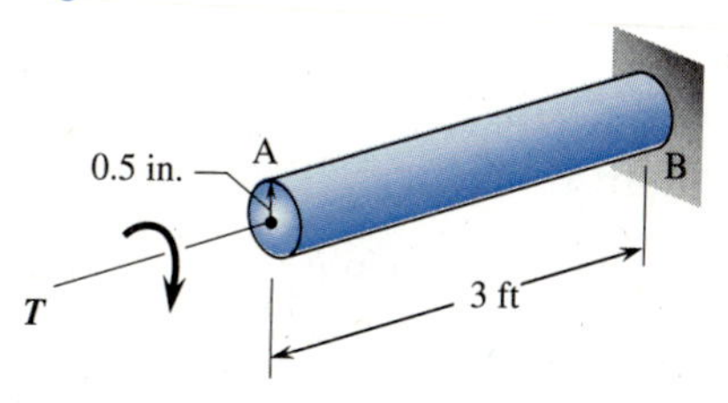

10.11 The stepped shaft shown is subjected to the end torque $T = 5.0$ kip · in. Determine the strain energy in the shaft. Let $G = 12 \times 10^6$ psi.

Figure P10.11

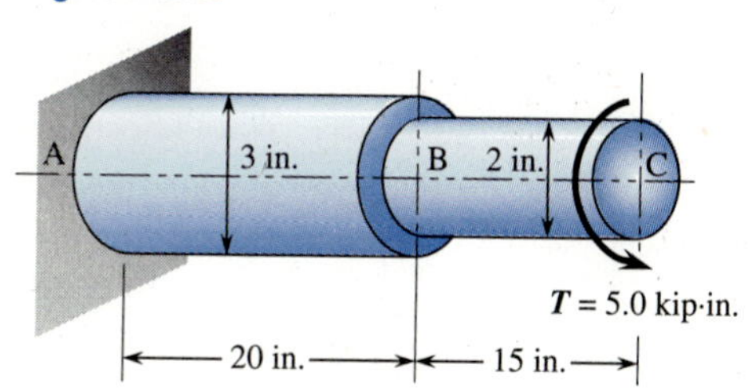

10.12 A hollow shaft is subjected to the torque shown. Determine the strain energy in the shaft for a maximum shear stress of 15 ksi. Let $G = 6 \times 10^6$ psi.

Figure P10.12

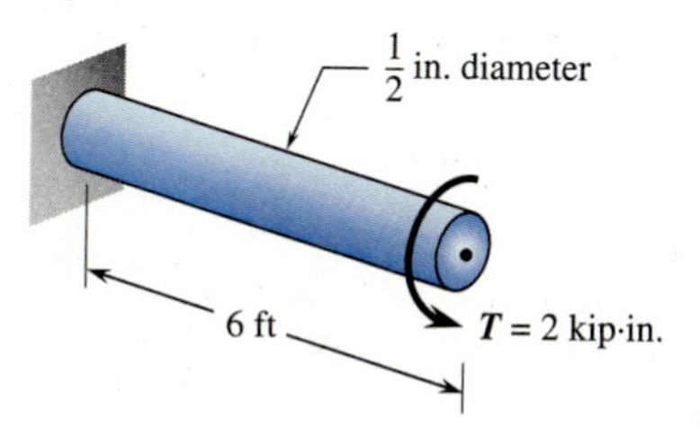

10.13 The pendulum shown is subjected to a force P resulting in a torque in the vertical shaft of the pendulum. The maximum shear stress in the shaft cannot exceed 90 MPa. Determine the maximum strain energy and the force P. Let $G = 79$ GPa.

Figure P10.13

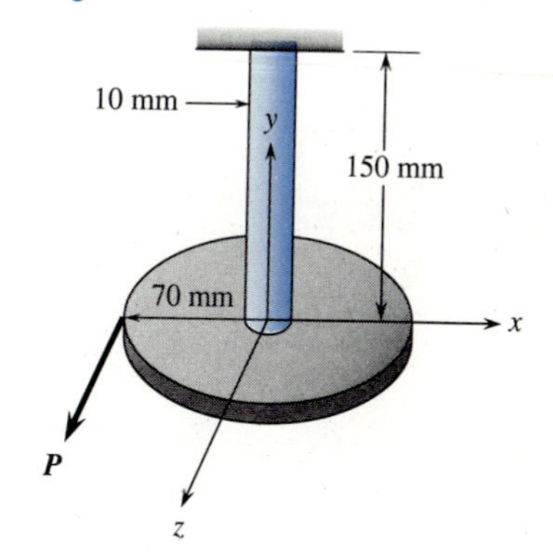

Section 10.6

10.14–10.21 Determine the strain energy due to bending stress only in the uniform cross-section beam shown. Let $E = 29 \times 10^6$ psi (200 GPa).

Figures P10.14, 22

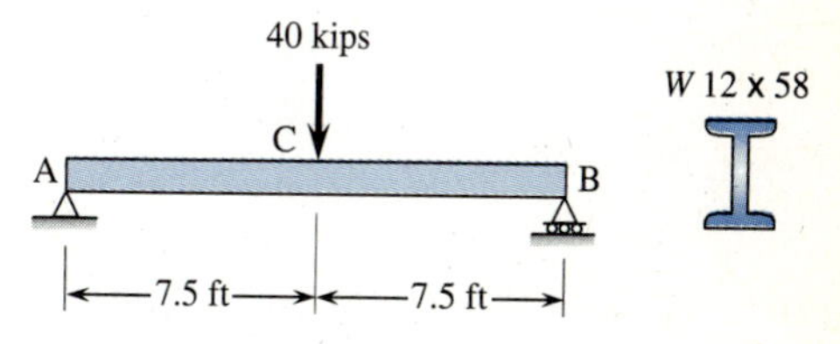

Figures P10.15, 23

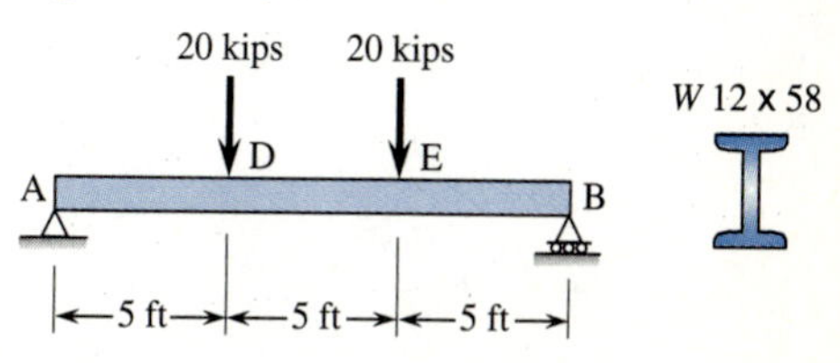

Figures P10.16, 24

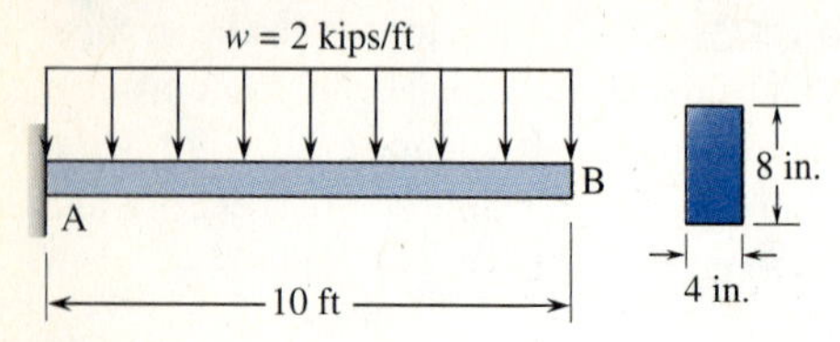

Figures P10.17, 25

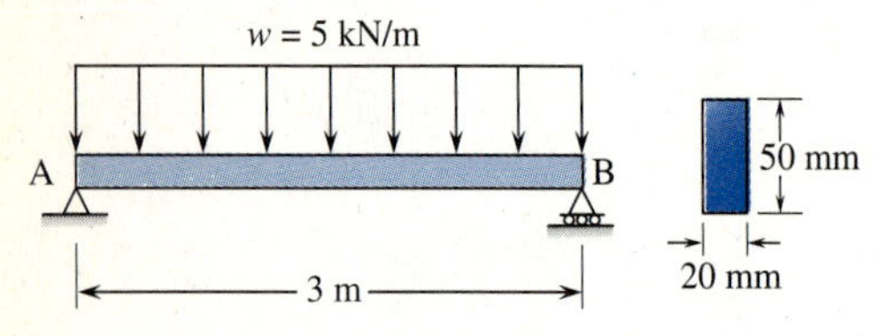

Figures P10.18, 26

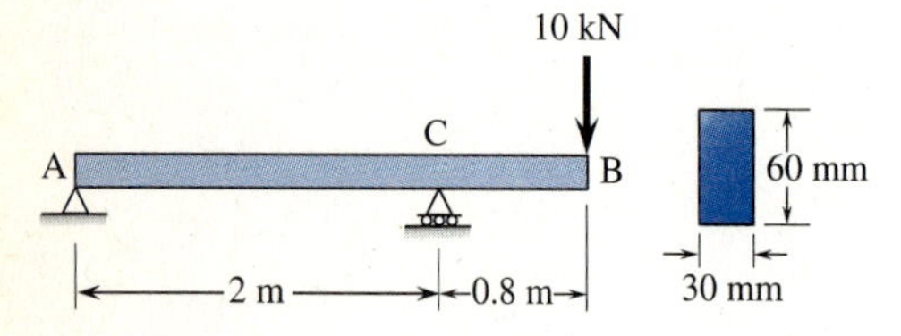

Figures P10.19, 27

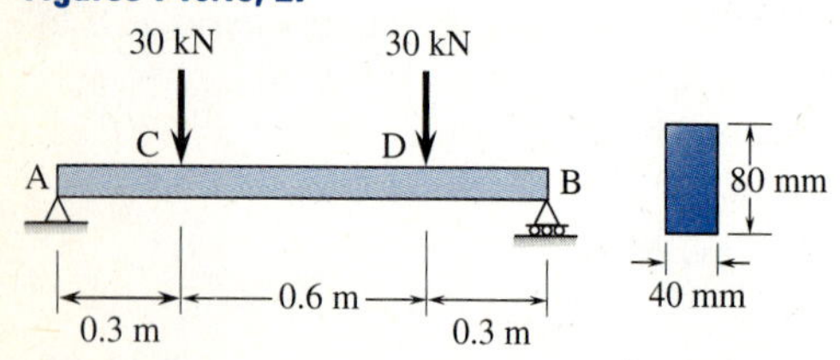

Figures P10.20, 28

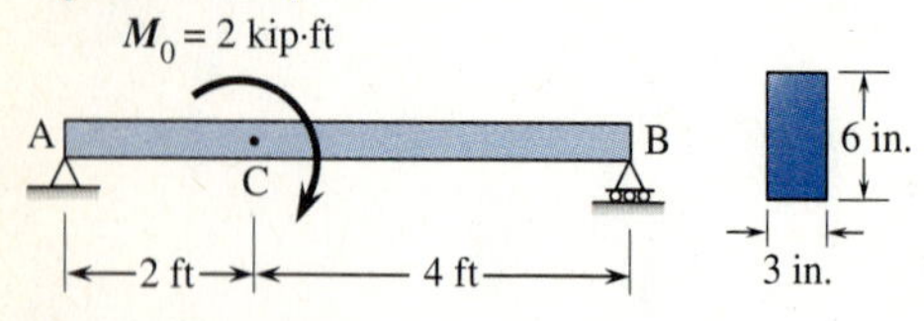

Figures P10.21, 29

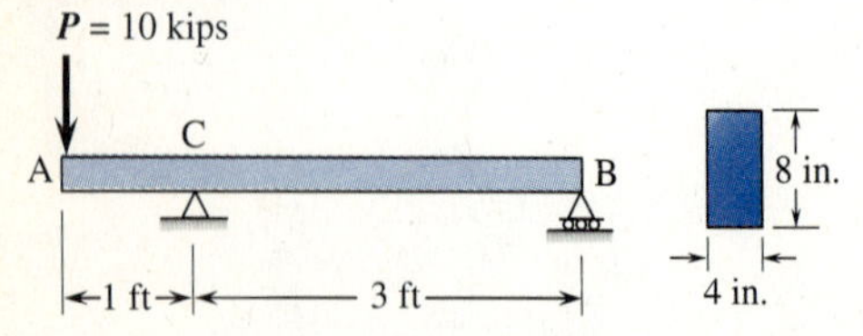

Section 10.7

10.22–10.29 Determine the strain energy due to transverse shear stress in the uniform cross-section beams of Problems 10.14 through 10.21. Compare the strain energy due to bending stress with that due to transverse shear stress. Let $G = 11.5 \times 10^6$ psi (80 GPa).

Section 10.9

10.30–10.36 Using the principle of work-energy, determine the vertical deflection at point B for the trusses shown. The cross-sectional areas of the bars are indicated on the figures. Let $E = 29 \times 10^6$ psi (200 GPa).

Figure P10.30

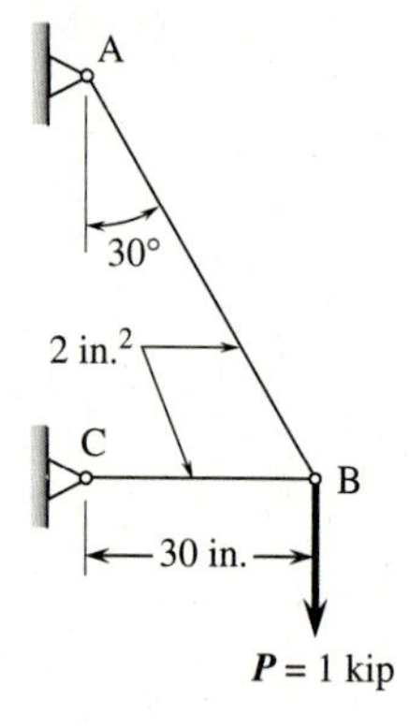

Figure P10.31

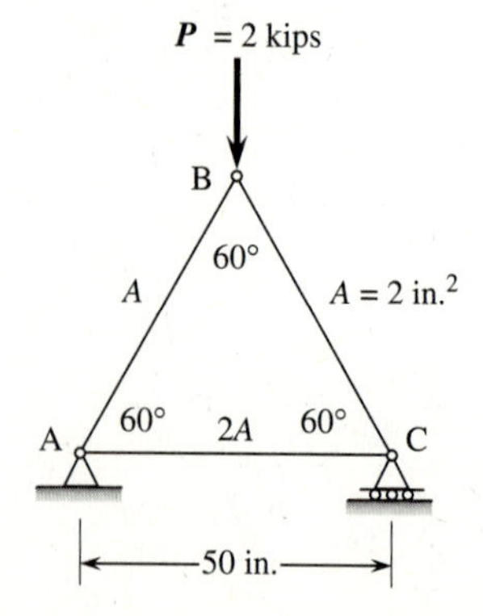

Figure P10.32

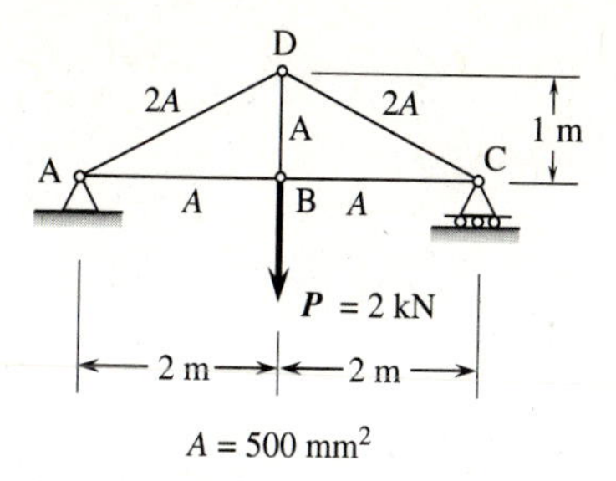

Figure P10.33

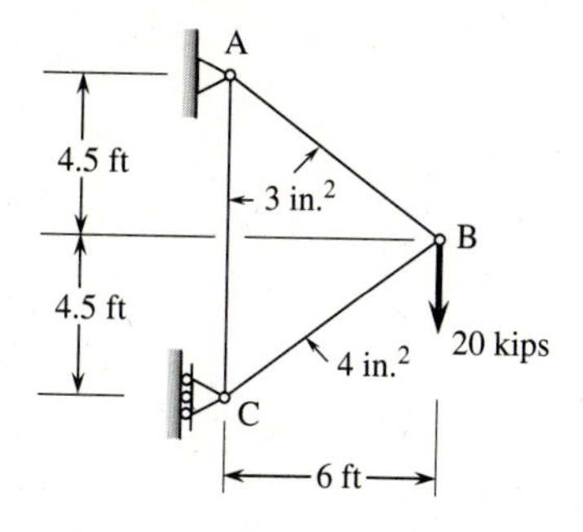

Figure P10.34

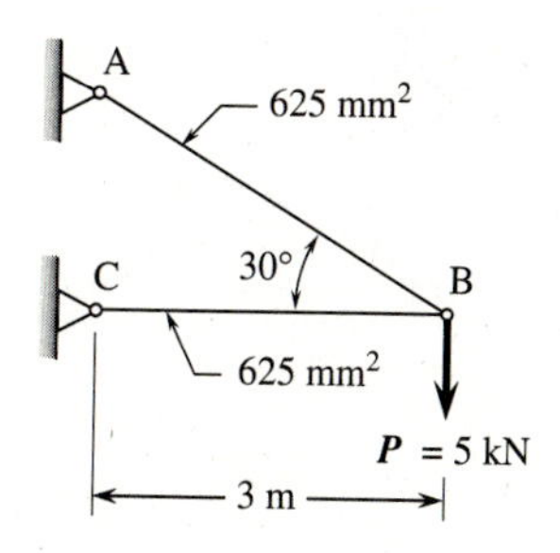

Figure P10.35

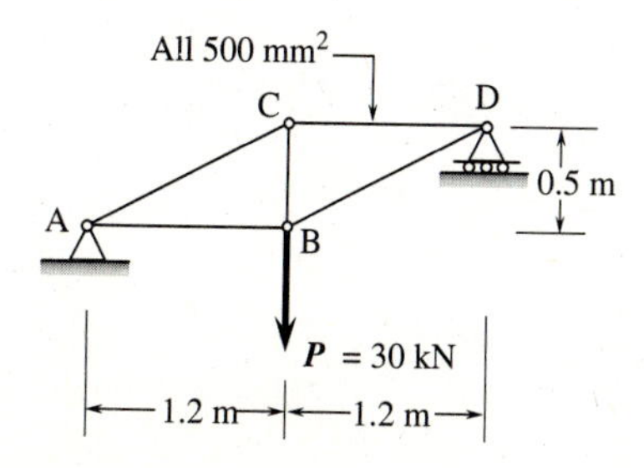

Figure P10.36

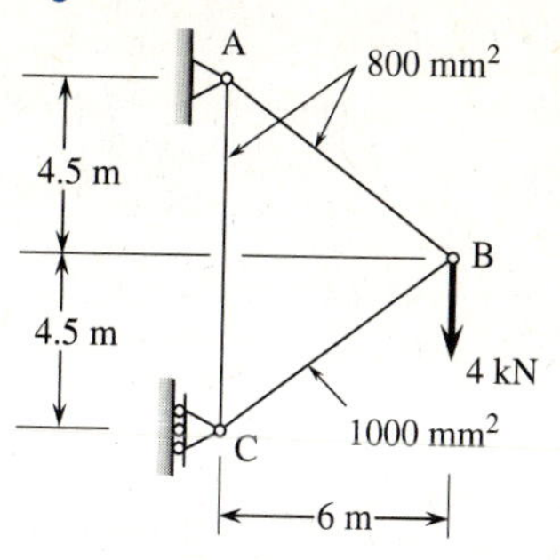

10.37–10.40 Using the principle of work-energy, determine the deflection under the load at point B for the beams shown. Let EI be constant.

Figure P10.37

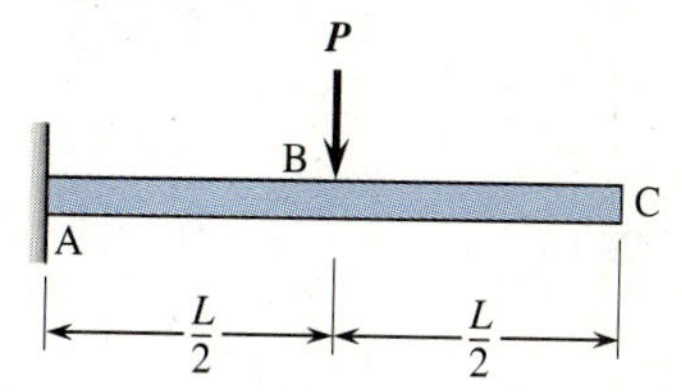

Figure P10.38

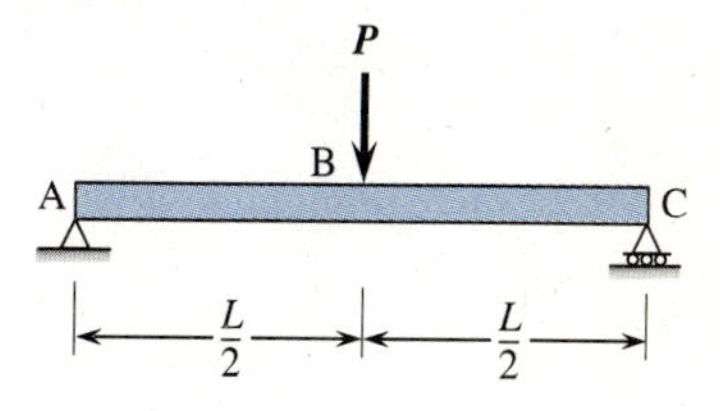

Figure P10.39

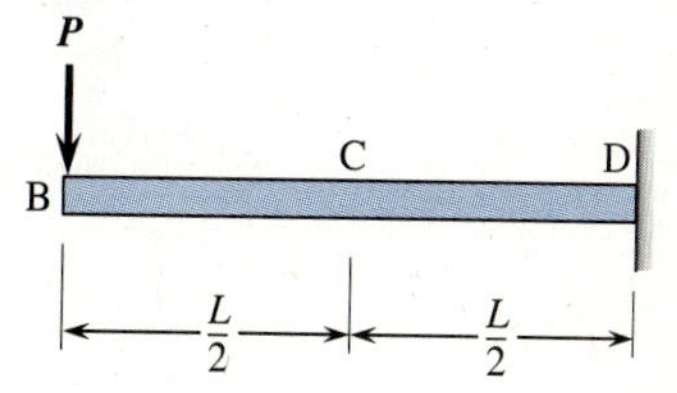

Figure P10.40

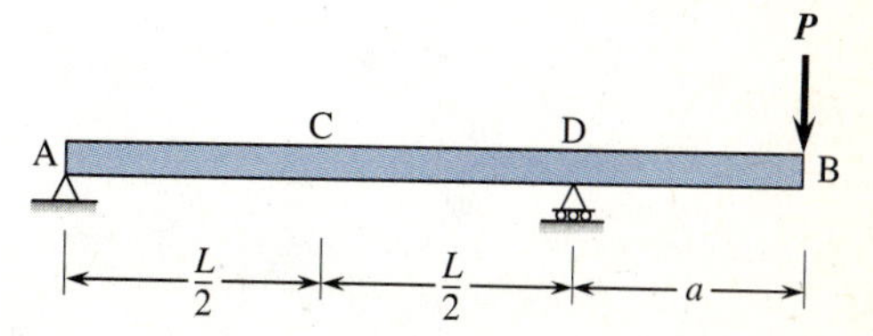

10.41–10.43 Using the principle of work-energy, determine the slope at the moment point C for the beams shown. Let EI be constant.

Figure P10.41

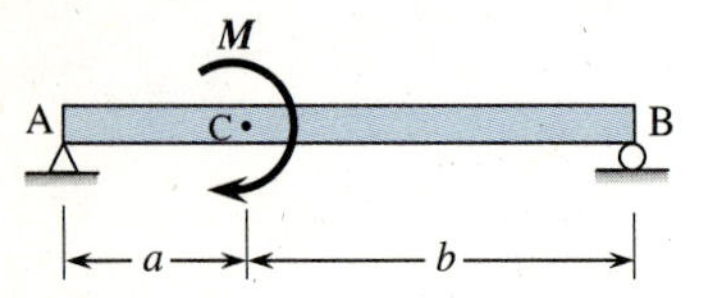

Figure P10.42

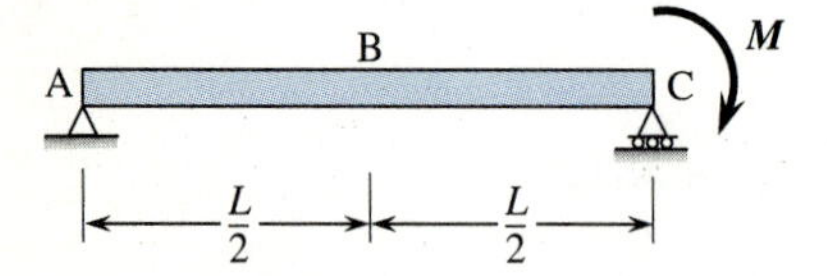

Figure P10.43

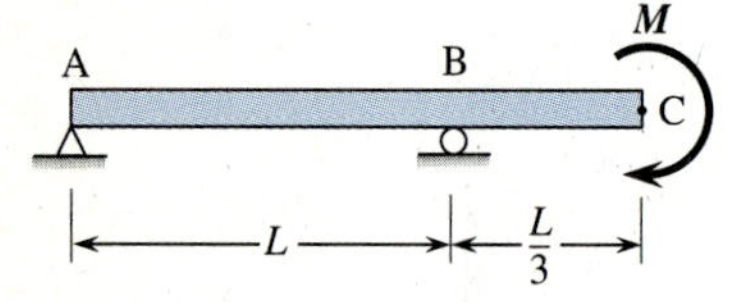

Section 10.11

10.44–10.52 Using the method of virtual work (dummy unit load method), determine the vertical deflection at point C in the trusses shown. Let EA be constant.

Figures P10.44, 53

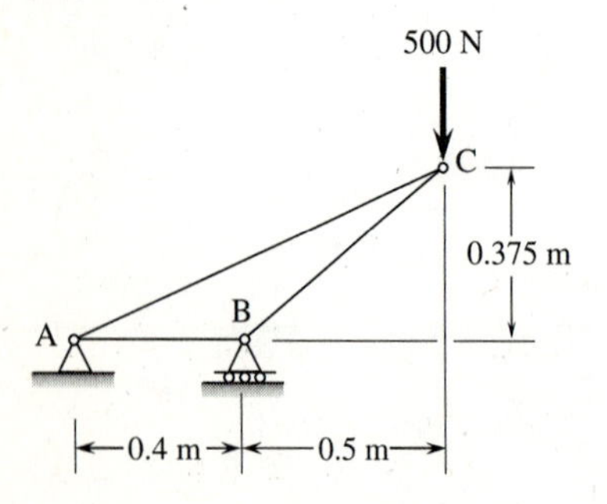

Figures P10.45, 54, 78

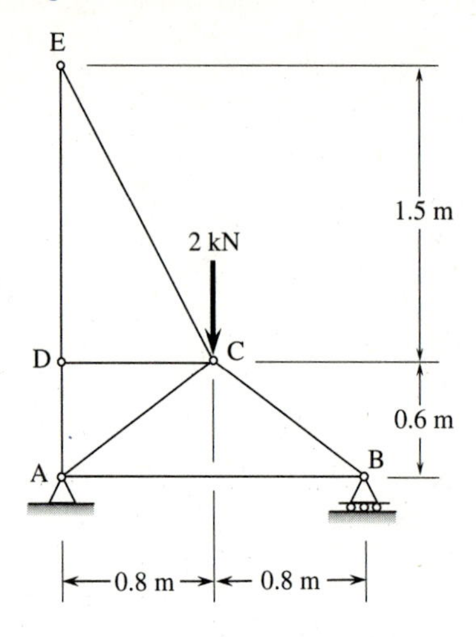

Figures P10.46, 55, 79

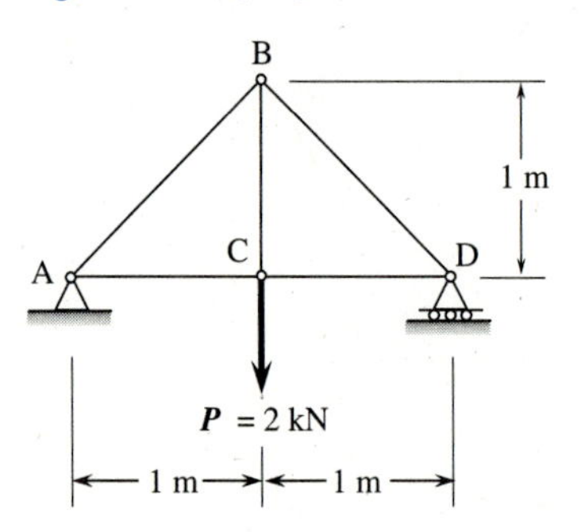

Figures P10.47, 56, 59, 80

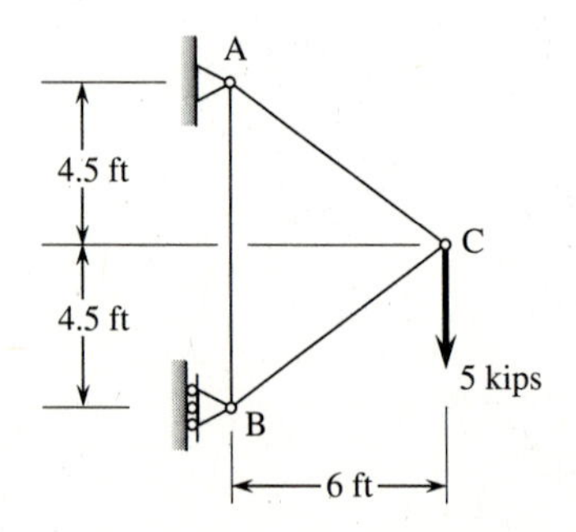

Figures P10.48, 57, 60, 81

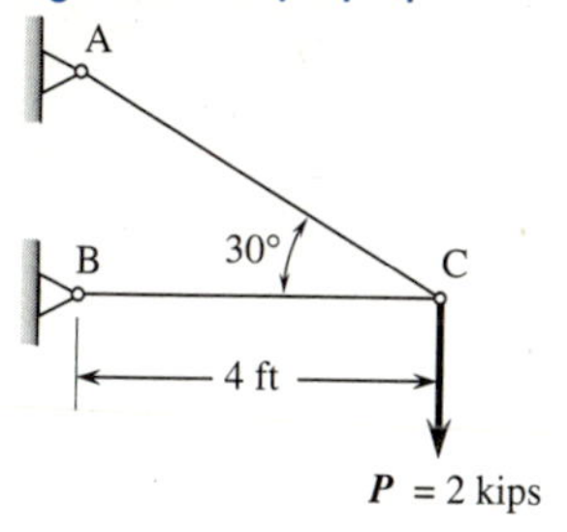

Figures P10.49, 58, 61, 82

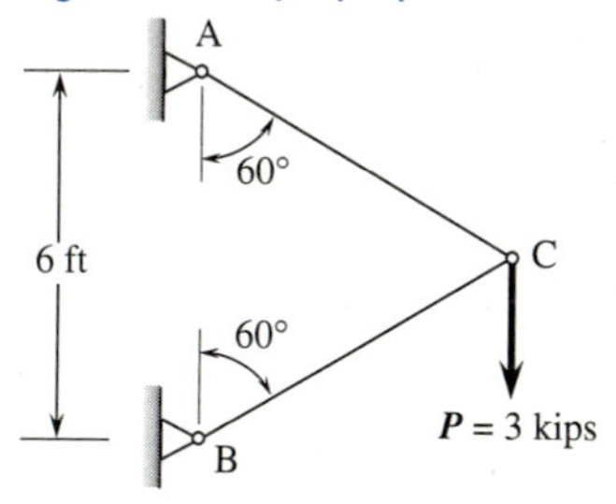

Figures P10.50, 83

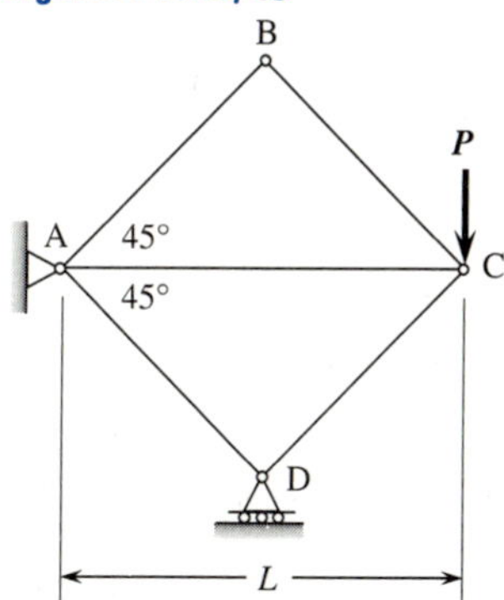

Figures P10.51, 84

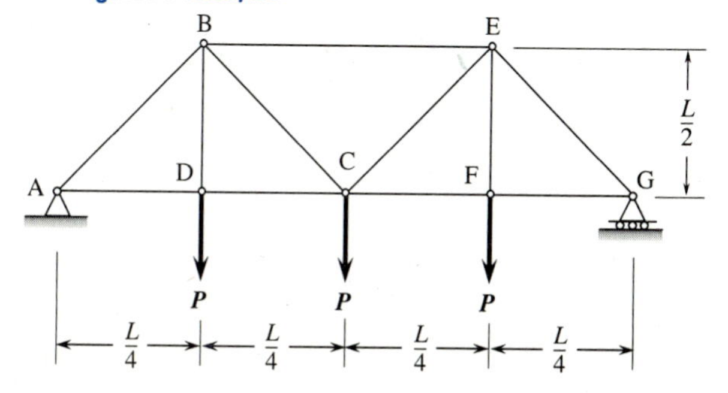

Figures P10.52, 85

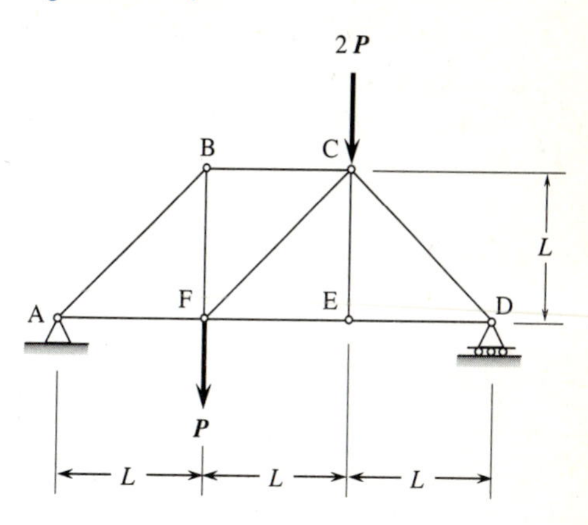

10.53–10.55 Using the method of virtual work (dummy unit load method), determine the vertical deflection at point C in the trusses of Problems 10.44 through 10.46 when the load is removed and only the member BC is subjected to a 30°C temperature decrease. Let $\alpha = 11.7 \times 10^{-6}$/°C.

10.56–10.58 Using the method of virtual work (dummy unit load), determine the vertical deflection at point C in the trusses of Problems 10.47 through 10.49 when the load is removed and only member BC is subjected to a uniform temperature increase of 100°F. Let $\alpha = 6.5 \times 10^{-6}$/°F.

10.59–10.61 Using the method of virtual work (dummy unit load method), determine the horizontal deflection at point C in the trusses of Problems 10.47 through 10.49 when the load is removed and only member BC is subjected to a uniform temperature increase of 100°F. Let $\alpha = 6.5 \times 10^{-6}$/°F.

Section 10.12

10.62–10.71 Using the method of virtual work (dummy unit load), determine the vertical deflection at point C for the beams with the loads shown. Let EI be constant.

Figures P10.62, 86

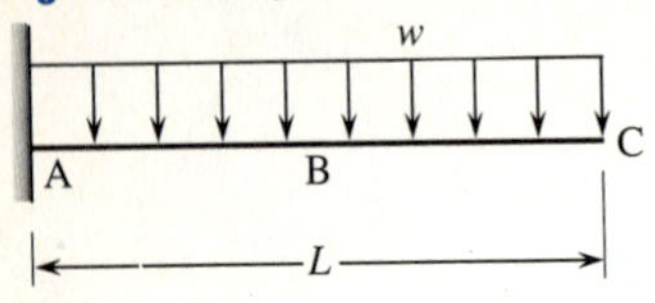

Figures P10.63, 87

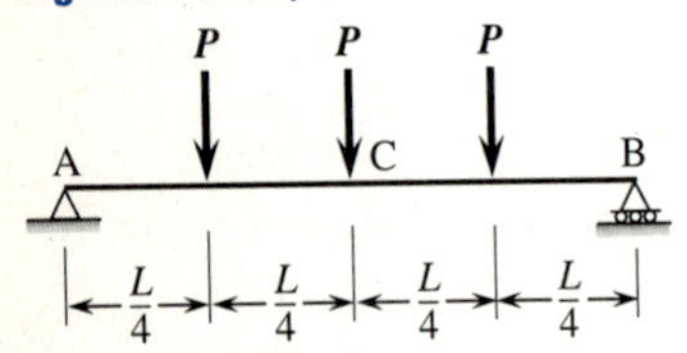

Figures P10.64, 88

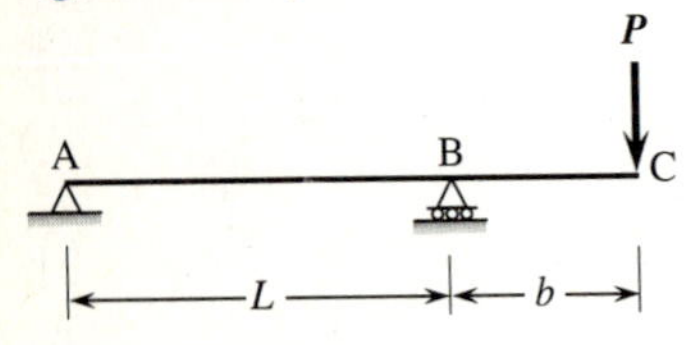

Figures P10.65, 89

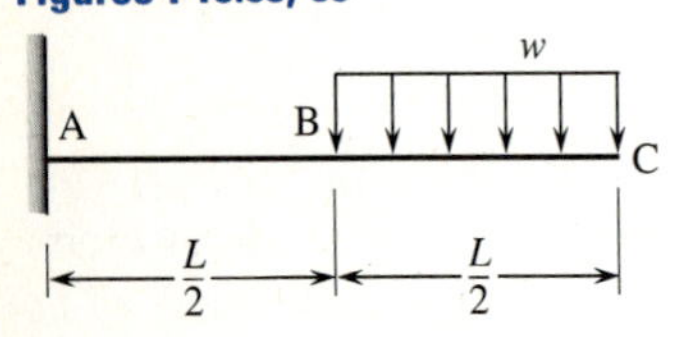

Figures P10.66, 90

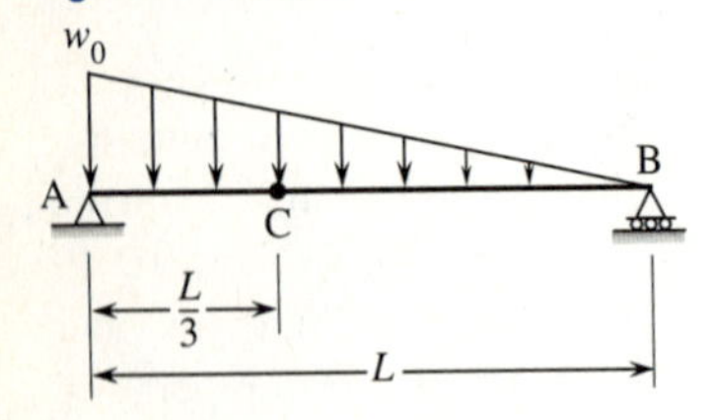

Figures P10.67, 91

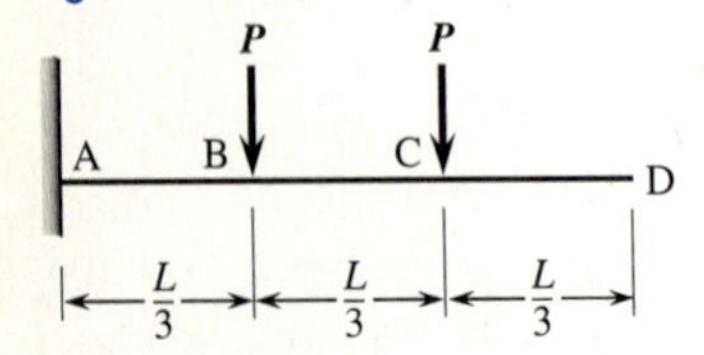

Figures P10.68, 92

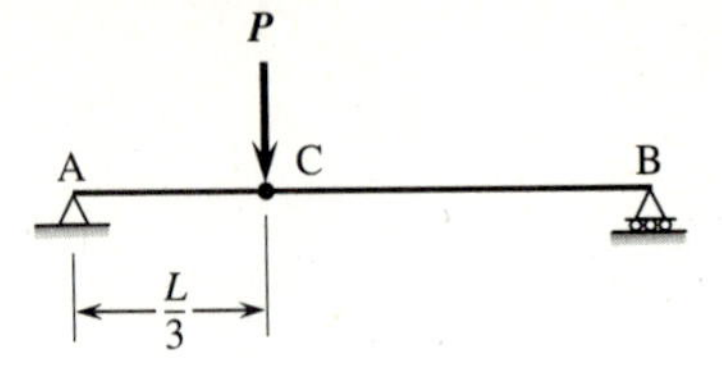

Figures P10.69, 93

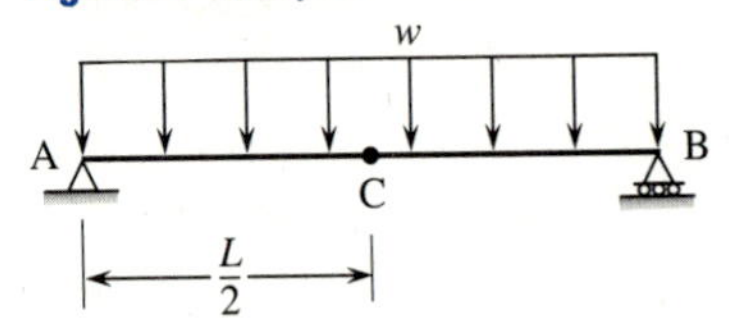

Figures P10.70, 94

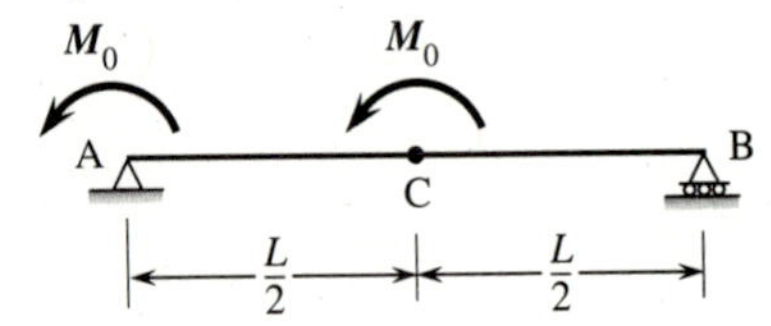

Figures P10.71, 95

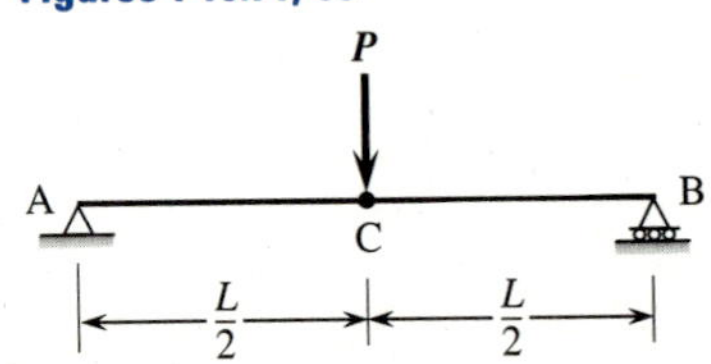

10.72–10.77 Using the method of virtual work (dummy unit load method), determine the vertical deflection at point C for the rigid frames shown.

Figures P10.72, 96

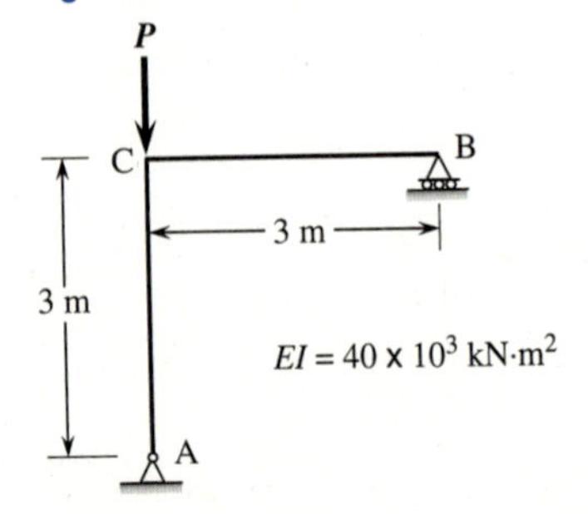

Figures P10.73, 97

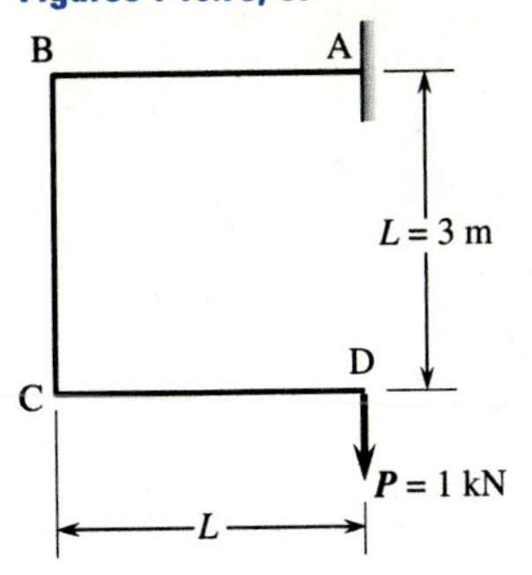

Figures P10.74, 98

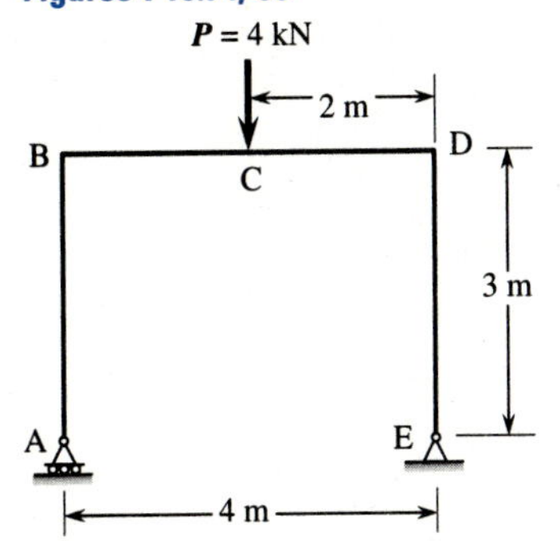

Figures P10.75, 99

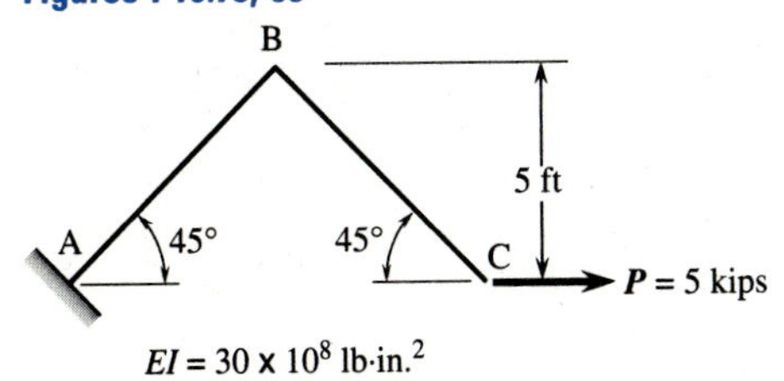

Figures P10.76, 100

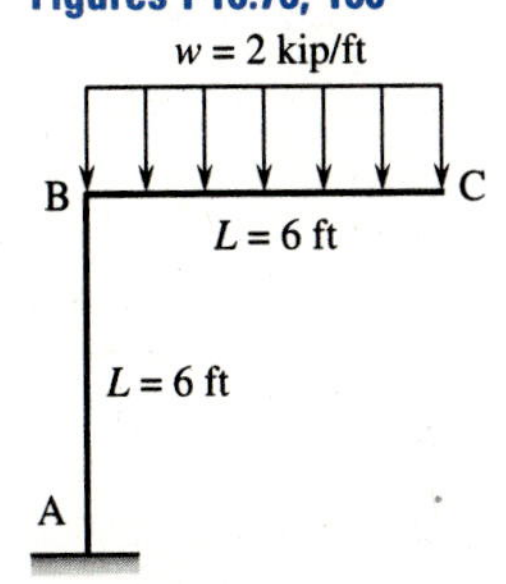

Figures P10.77, 101

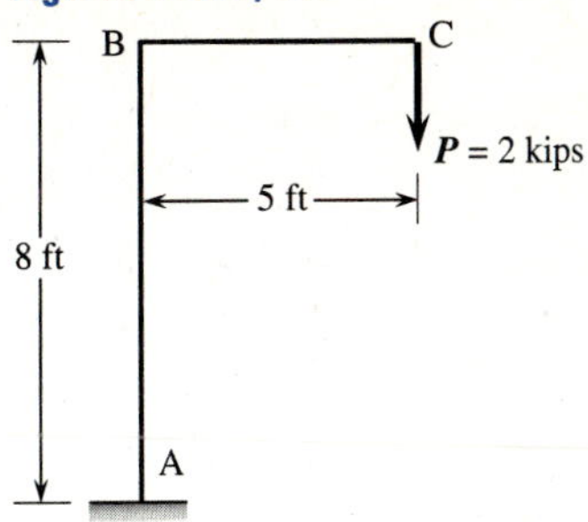

Section 10.13

Use Castigliano's theorem to solve the problems of this section.

10.78–10.85 Solve the truss problems of Section 10.11 (Problems 10.45 through 10.52).

10.86–10.95 Solve the beam problems of Section 10.12 (Problems 10.62 through 10.71).

10.96–10.101 Solve the frame problems of Section 10.12 (Problems 10.72 through 10.77).

10.102–10.104 For the rigid frames shown, determine **(a)** the slope and **(b)** the vertical deflections at C. Let *EI* be constant.

Figure P10.102

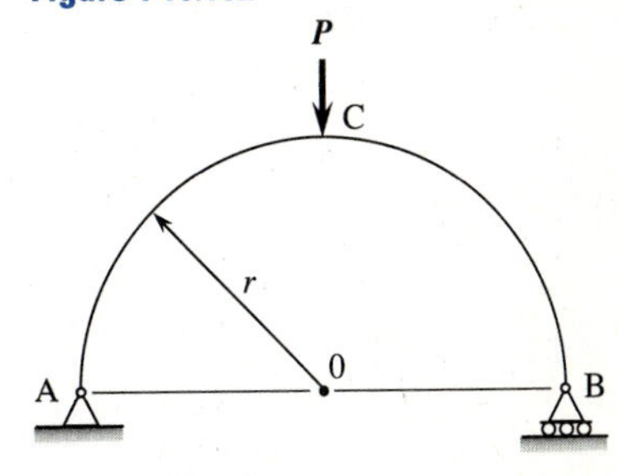

Figure P10.103

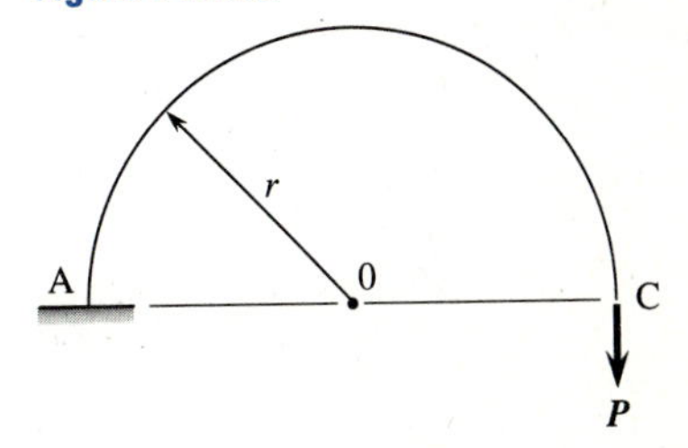

Figure P10.104

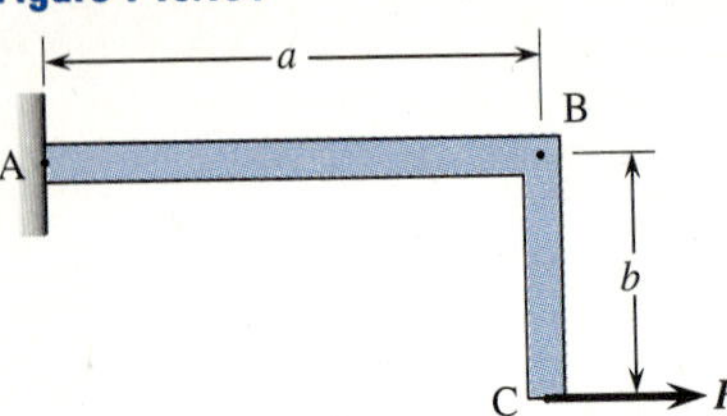

10.105–10.108 For the torsion members shown, determine the angle of twist at C. Let $G = 12 \times 10^6$ psi.

Figure P10.105

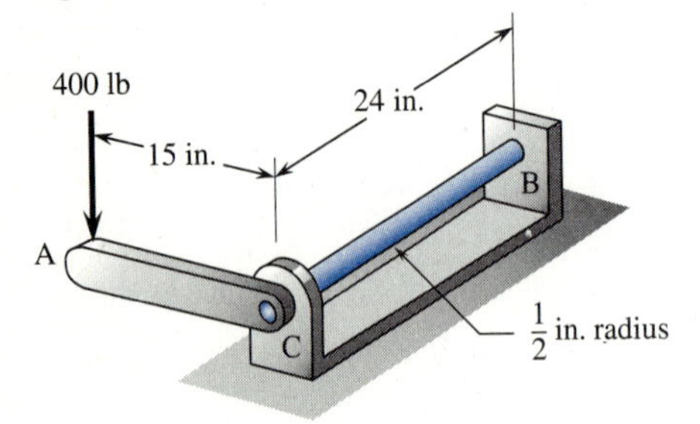

Figure P10.106

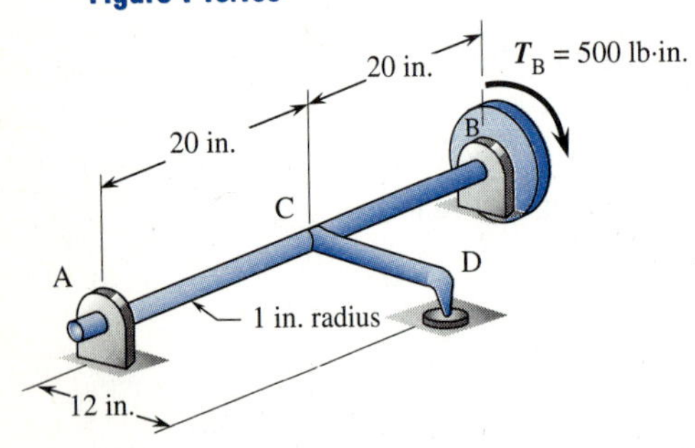

Figure P10.107

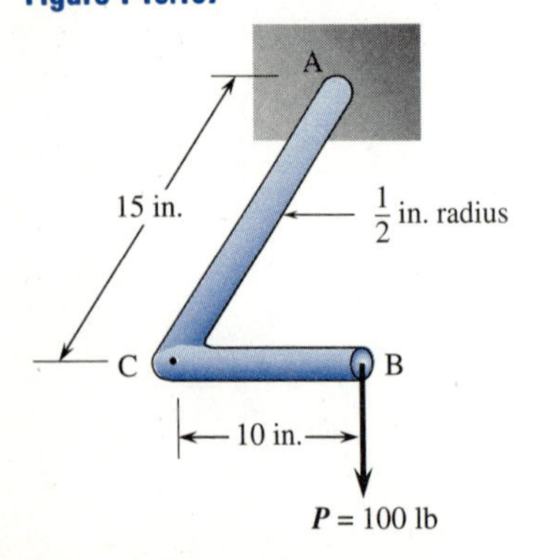

Figure P10.108

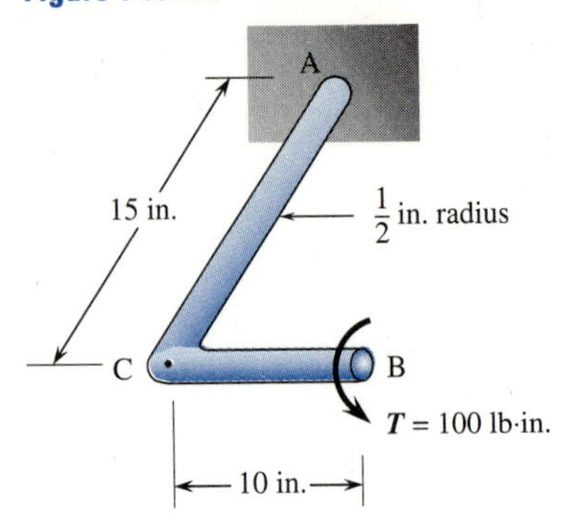

Section 10.14

Use Castigliano's theorem to solve the statically indeterminate problems of this section.

10.109–10.113 For the statically indeterminate beams shown, determine the reactions at each support. Let *EI* be constant.

Figure P10.109

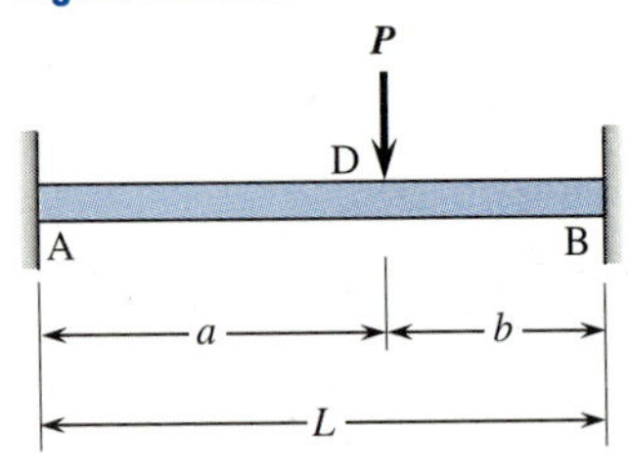

Figure P10.110

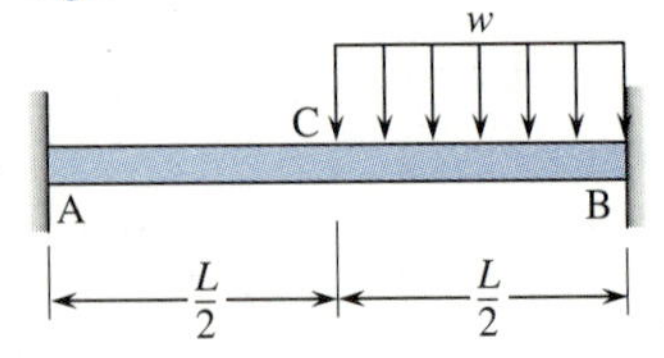

Figure P10.111

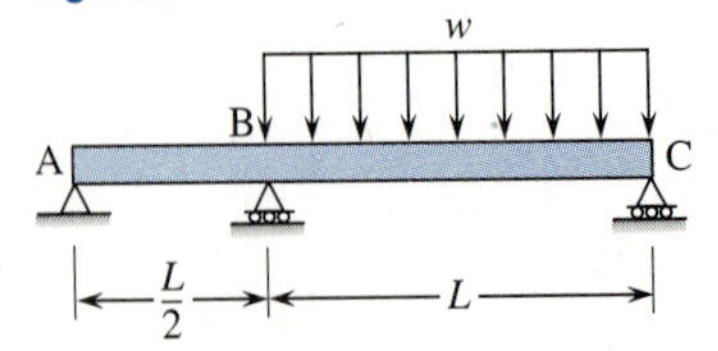

Figure P10.112

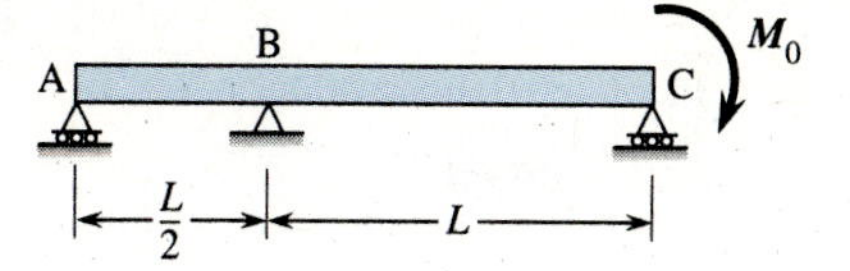

Figure P10.113

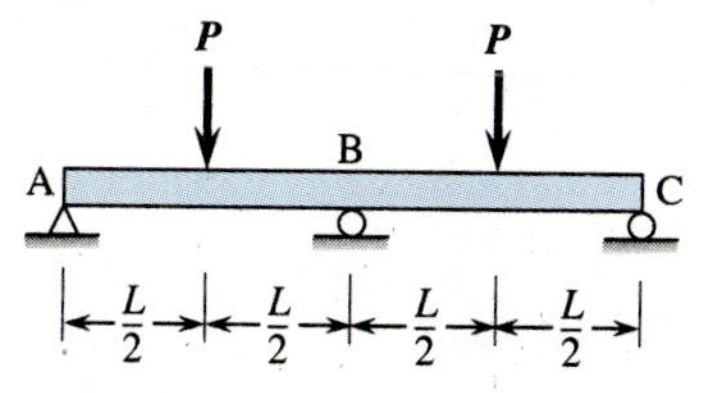

10.114–10.116 For the statically indeterminate trusses shown, determine the forces in each member. Let $EA = 30 \times 10^6$ lb.

Figure P10.114

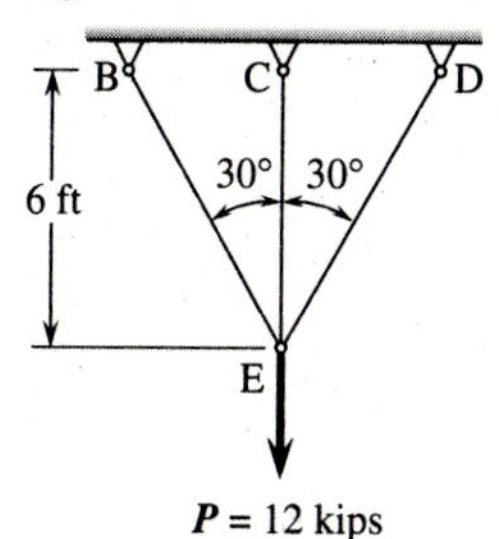

Figure P10.115

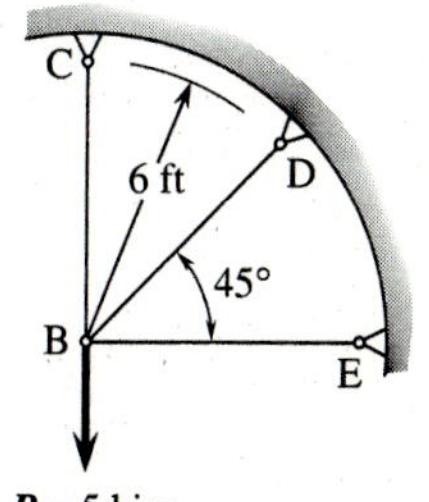

Figure P10.116

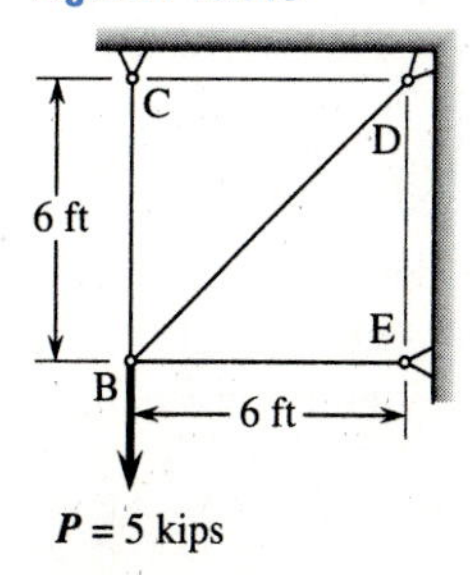

Section 10.15

Neglect energy losses during impact in all problems in this section.

10.117 A bumper stop for a trolley car moving across a crane beam runway is constructed of a spring with stiffness $k = 2000$ lb/in. If the car weighs 5000 lb and is traveling at 5 miles per hour when it strikes the spring, determine the maximum deflection of the spring and the maximum force in the spring.

Figure P10.117

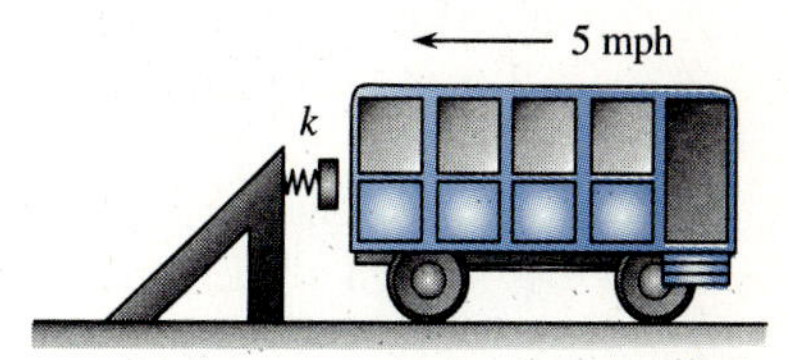

10.118 A bumper stop for a boat has a spring stiffness of 3000 lb/in. If a small boat weighing 1000 lb strikes the spring at 10 miles per hour, determine the maximum deflection in the spring and the maximum force in the spring.

10.119 A bumper stop for a coal car has a spring stiffness of 200 N/mm. If the car with a mass of 400 kg is traveling at 3 m/s, determine the maximum deflection and maximum force in the spring.

Figure P10.119

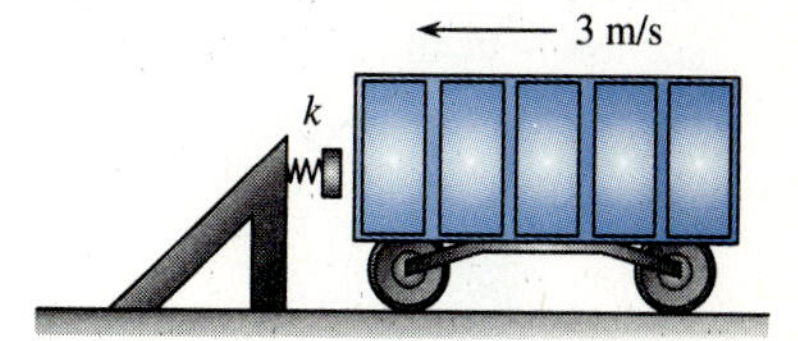

10.120 A mass of 20 kg is dropped from a height of 50 mm onto a spring that has a spring constant of 15 kN/m. Determine **(a)** the maximum deflection of the spring and **(b)** the static load that would produce the same deflection.

Figure P10.120

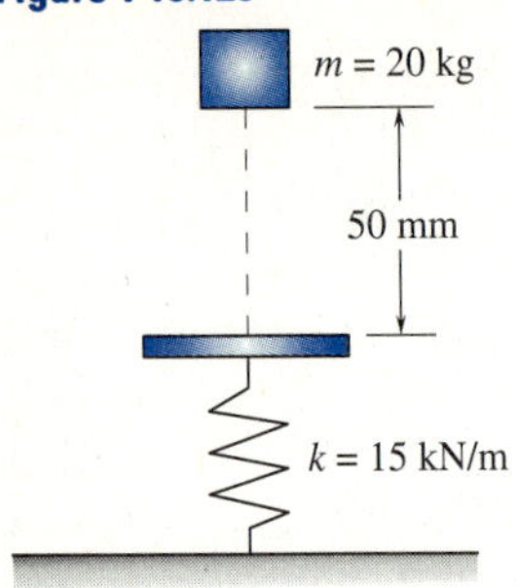

10.121 A collar of weight W is dropped from a height of 10 in. onto the end of a steel bar shown. If the maximum stress in the bar is not to exceed 20 ksi, determine the maximum allowable weight of the collar. The cross-section of the bar is 2 in.^{2}.

Figure P10.121

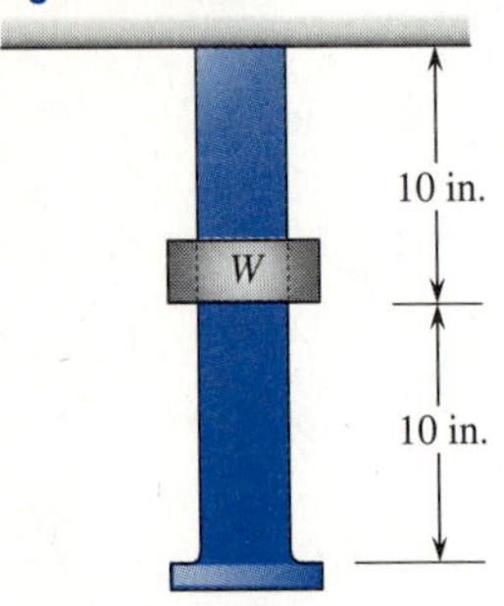

10.122 Solve Problem 10.121 if the bar is now stepped with a 4 in.^{2} cross-section over the upper half of the bar and a 2 in.^{2} cross-section over the lower half of the bar.

Figure P10.122

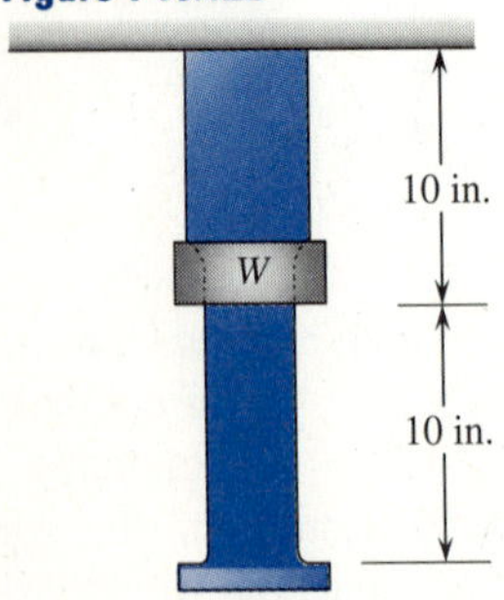

10.123 A 25-kg collar is released from a height h from rest and is stopped by a plate attached to the vertical bar at B. The modulus of elasticity of the bar is 200 GPa. Determine the height h for which the maximum stress in the bar is 240 MPa.

Figure P10.123

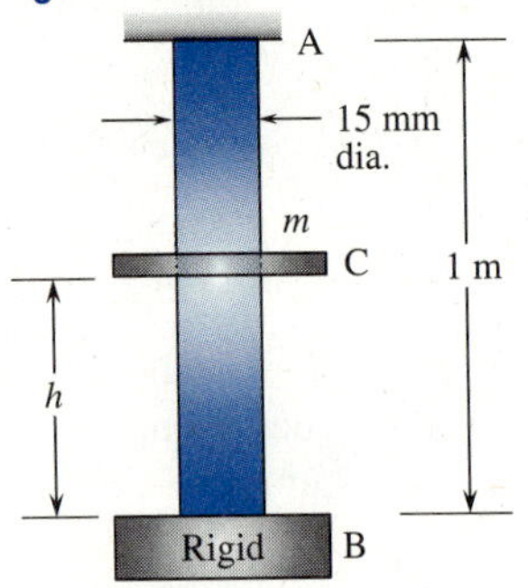

10.124 The uniform bar AB is made of steel which has a yield strength of 100 ksi and modulus of elasticity of 30×10^6 psi. Collar C moves along the bar with a constant speed of $v_i = 10$ ft/s as it strikes the plate attached to the bar at B. Using a factor of safety of 5, determine the largest allowable weight of the collar if the rod is not to be permanently deformed.

Figure P10.124

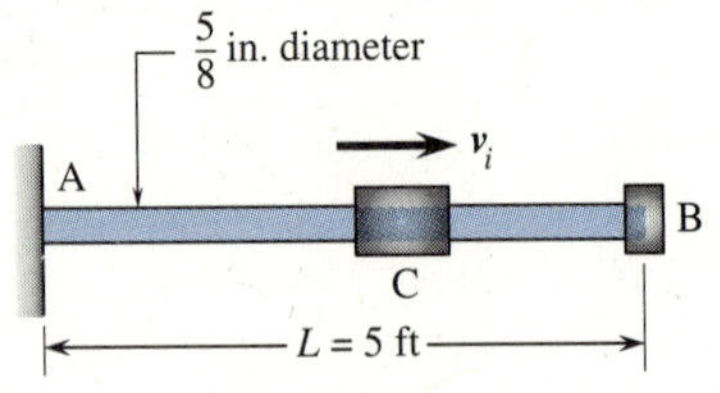

10.125 Solve Problem 10.124 assuming the bar length is increased to 10 ft.

10.126 The uniform bar AB is made of aluminum with a yield strength of 255 MPa and modulus of elasticity of 69 GPa. Collar C moves along the bar with constant speed

$v_i = 3$ m/s as it strikes the plate attached to the bar at B. Using a factor of safety of 5, determine the largest allowable weight of the collar if the rod is not to be permanently deformed.

Figure P10.126

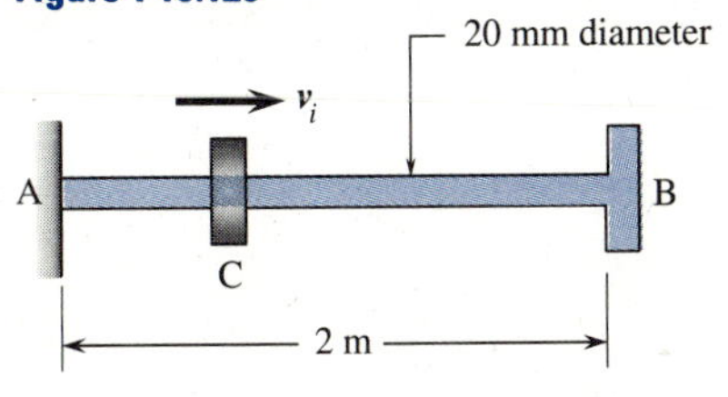

10.127 Solve Problem 10.126 assuming the bar length is increased to 4 m.

10.128 A 10-kg weight is dropped from a height $h = 300$ mm onto the midspan of a simply supported beam AB. The beam is made of steel with $E = 200$ GPa. Determine **(a)** the maximum deflection of point C and **(b)** the maximum stress in the beam. Let $L = 2$ m.

Figure P10.128

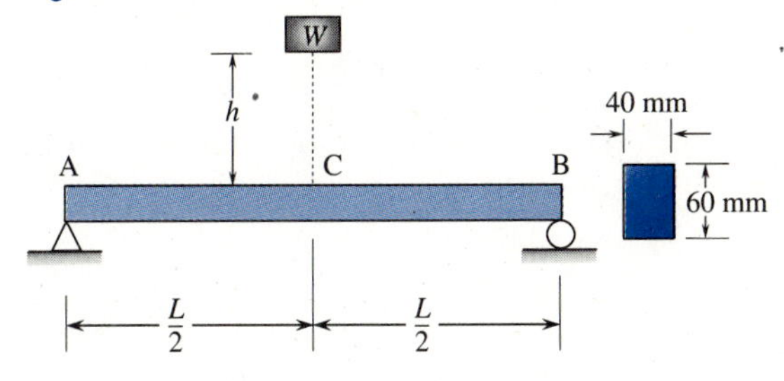

10.129 A steel beam used as a guard rail is struck at its midpoint C by a 50 lb block moving horizontally with a speed of 5 ft/s. The modulus of elasticity of the steel is 30×10^6 psi. Determine **(a)** the equivalent static load to cause the same deflection as the impact, **(b)** the maximum deflection of the midpoint C, and **(c)** the maximum stress in the beam.

Figure P10.129

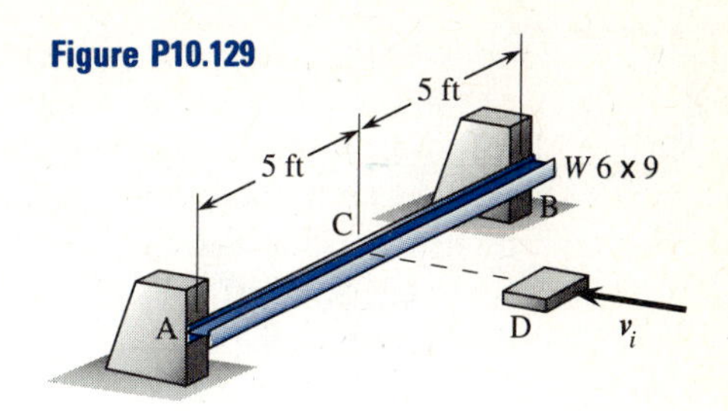

10.130 A 5-kg block is dropped from a height h onto an aluminum beam AB. The modulus of elasticity of the beam is 70 GPa and the allowable stress is 50 MPa. Determine the maximum height of drop of the block.

Figure P10.130

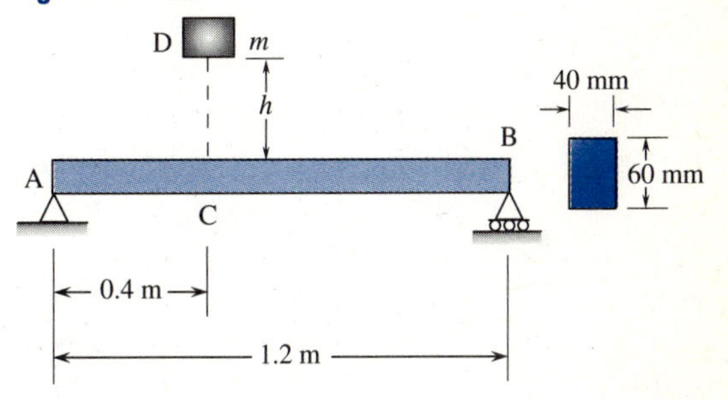

10.131 A steel post used as a bumper is a pipe of 100 mm outer diameter and wall thickness of 10 mm. A 5-kg mass C moving horizontally at 2 m/s strikes the pipe squarely at the top B. Use $E = 200$ GPa. Determine **(a)** the maximum deflection of the free end B, **(b)** the maximum bending moment in the pipe, and **(c)** the maximum stress in the pipe.

Figure P10.131

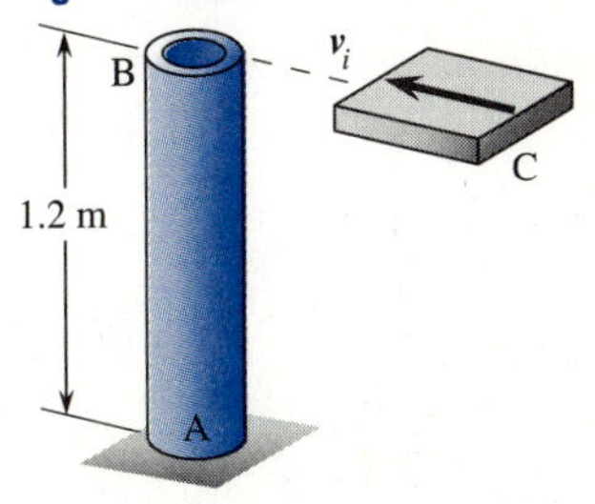

10.132 A 3-lb block is dropped from a height of 2 in. onto the free end of a cantilevered beam which is a 4 in. × 4 in. cross-section. Use $E = 30 \times 10^6$ psi, and determine **(a)** the maximum deflection of the free end A, **(b)** the maximum bending moment in the beam, and **(c)** the maximum stress in the beam.

Figure P10.132

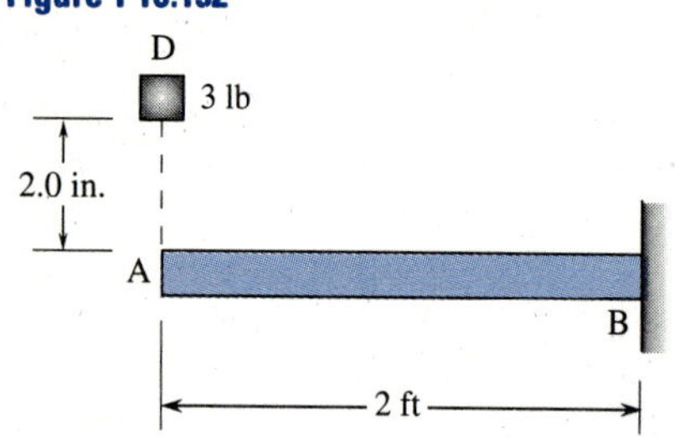

10.133 A diver weighing 150 lb jumps from a height of 2 in. onto the end of a diving board. Determine **(a)** the maximum end deflection at A and **(b)** the maximum bending stress in the diving board. The board is a rectangular cross-section with width $c = 12$ in. and height $d = 2$ in. and $E = 2.0 \times 10^6$ psi. Compare the impact values for maximum deflection and bending stress with the corresponding static values when the diver's weight is applied to the end of the board.

Figure P10.133

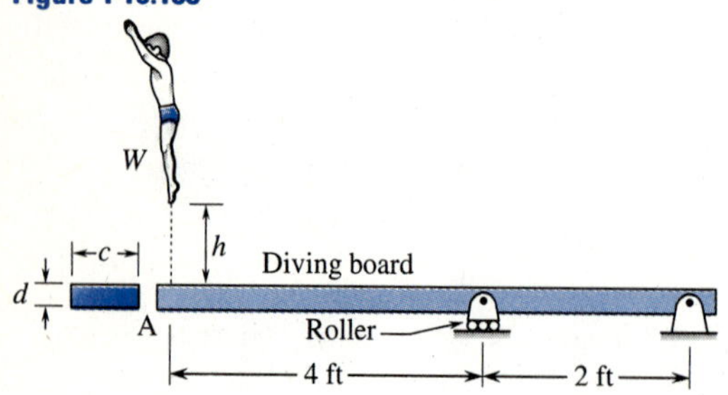

10.134 A diver weighing 670 N jumps from a height of 50 mm onto the end of a diving board. Determine **(a)** the maximum deflection at A and **(b)** the maximum bending stress in the board. The board is a rectangular cross-section with width $c = 250$ mm and height $d = 50$ mm and $E = 13$ GPa. Compare the impact results for maximum deflection and stress with the corresponding static values due to the diver's weight at end A.

Figure P10.134

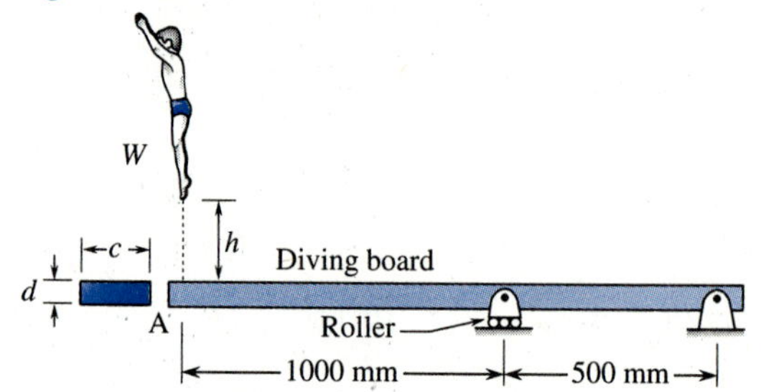

10.135 A solid circular shaft with a flywheel attached to one end rotates at 10 rad/s. The shaft is suddenly stopped at the opposite end. Assume all the kinetic energy of the flywheel is transformed into elastic strain energy in the shaft. The flywheel is idealized as a disk of radius 0.2 m and weight of 400 N. The shaft has a length of 1.0 m, a diameter of 50 mm, and a shear modulus $G = 79$ GPa. Determine **(a)** the maximum torsional shear stress and **(b)** the maximum angle of twist of the shaft.

Figure P10.135

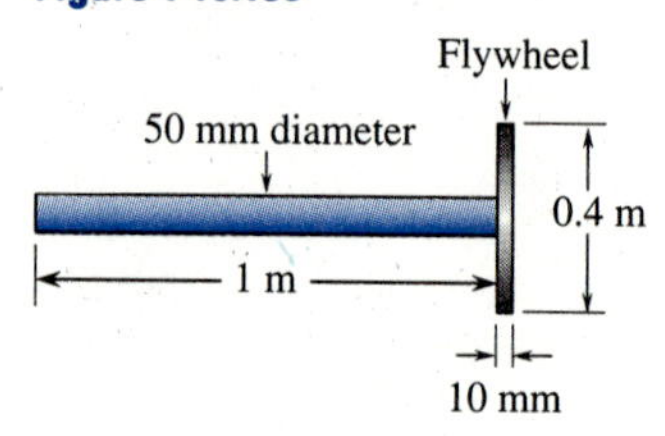

10.136 A solid circular shaft with a pulley attached to one end rotates at 10 rad/s. The shaft is suddenly stopped at the opposite end and the kinetic energy of the pulley is transformed into elastic strain energy in the shaft. The pulley is idealized as a thin disk of radius 10 in. and weight 50 lb. The length of the shaft is 75 in., the diameter is 2.5 in., and $G = 12.0 \times 10^6$ psi. Determine **(a)** the maximum torsional shear stress and **(b)** the maximum angle of twist of the shaft.

Figure P10.136

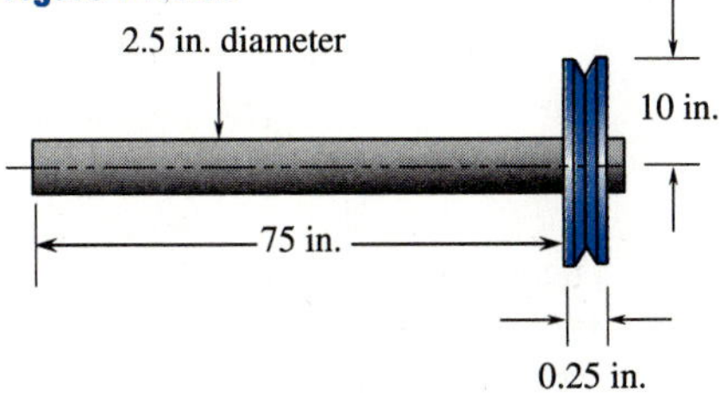

10.137 A solid steel shaft with 5 in. diameter and 12 ft length has an allowable shear stress of 10 ksi. The shear modulus is $G = 12 \times 10^6$ psi. Determine **(a)** the torsional energy U that the shaft can resist and **(b)** the angle of twist due to this energy load.

Figure P10.137

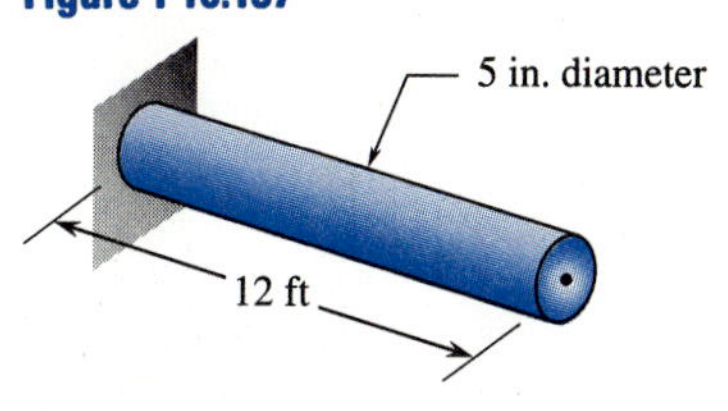

10.138 A torsional energy load of 30 N · m is applied to a steel shaft of diameter 75 mm and 4 m length. Use $G = 79$ GPa for the shaft. Determine **(a)** the maximum shear stress due to the energy load and **(b)** the angle of twist of the shaft.

Figure P10.138

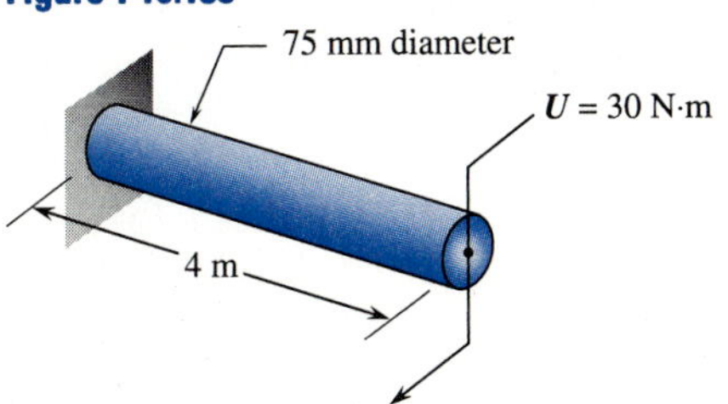

10.139 Solve Problem 10.138 if a stepped solid shaft with 75 mm diameter over half the length and 50 mm diameter over the other half-length replaces the solid shaft of uniform diameter of 75 mm.

COMPUTER PROBLEMS

The following problems are suggested to be solved using a programmable calculator, microcomputer, or mainframe computer. It is suggested that you write a simple computer program in BASIC, FORTRAN, PASCAL, or some other language to solve these problems.

C10.1 Write a program to determine the maximum deflection of a spring and the maximum force in the spring when a trolley car moving across a runway beam impacts the spring. Your input should include the weight of the car, the velocity of the car at impact, and the spring stiffness, k. Check your program using the following data: car weight = 5000 lb, car velocity = 5 miles/h, and $k = 2000$ lb/in.

Figure C10.1

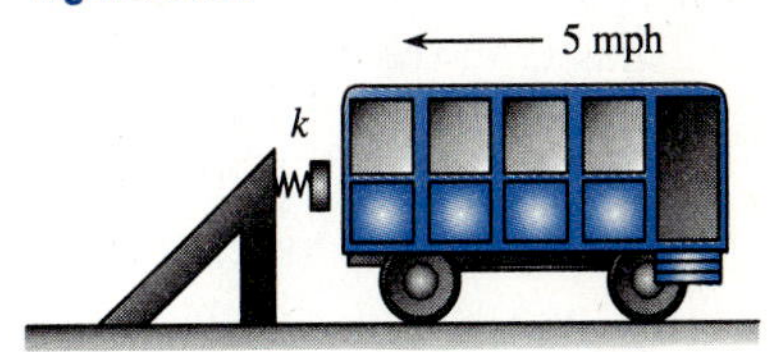

C10.2 Write a program to determine the maximum deflection and maximum axial stress in a vertical bar subjected to an impact from a collar dropped from a given height onto the lower end of the bar. Your input should include the weight of the collar, the height of drop of the collar, the cross-sectional area A of the bar, the modulus of elasticity E of the bar, and the yield strength of the bar. The program should check if the yield strength of the bar material has been exceeded. Check your program using the following data: collar weight = 50 lb, height of collar drop = 30 in., diameter of bar = 1 in., $E = 29 \times 10^3$ ksi, and $\sigma_Y = 100$ ksi.

Figure C10.2

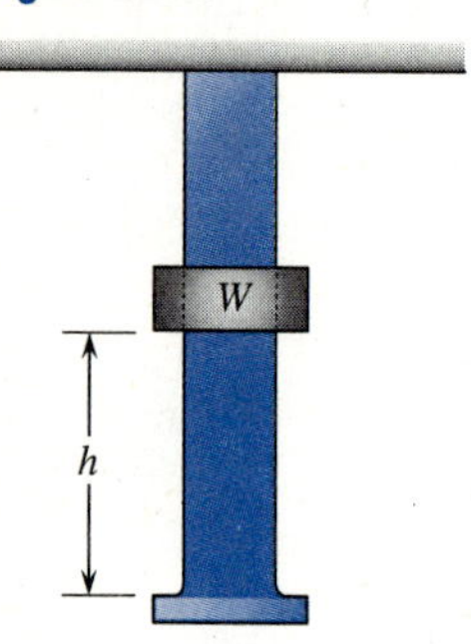

C10.3 Write a program to determine the shear stress in a circular shaft with an attached pulley that was turning at a constant angular velocity when suddenly stopped. The input should include the diameter of the shaft, the speed of the shaft pulley, the diameter and thickness of the pulley, the mass density of the pulley, and the shear modulus G of the shaft. Check your program using the following data: Shaft diameter of 10 mm, shaft pulley speed of 2000 rpm, pulley diameter of 100 mm, and pulley thickness of 10 mm, mass density of pulley of 2000 kg/m^3, and shear modulus of the shaft equal to 80 GPa.

Figure C10.3

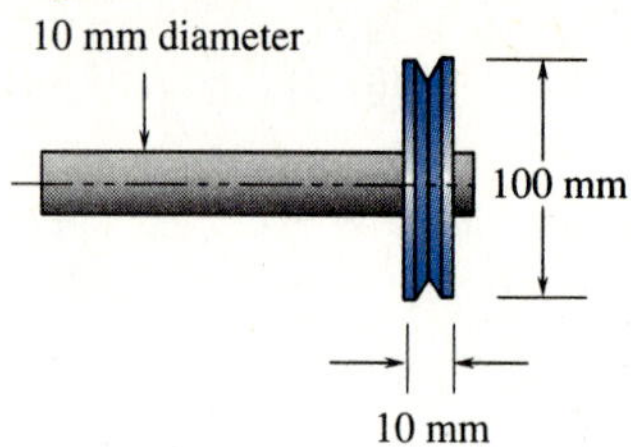

C10.4 Write a program to determine the maximum deflection and maximum bending stress in a simply supported beam subjected to an impact load at its midspan. The input should include the weight of the impacting object, the height of drop of the weight, the cross-sectional dimensions of the beam, the length of the beam, and the modulus of elasticity of the beam. Check your program using the following data: falling weight of 50 lb, height of weight drop of 3 in., beam cross-section of 2 in. wide × 4 in. deep and beam span of 100 in., and modulus of elasticity of 30×10^6 psi.

Figure C10.4

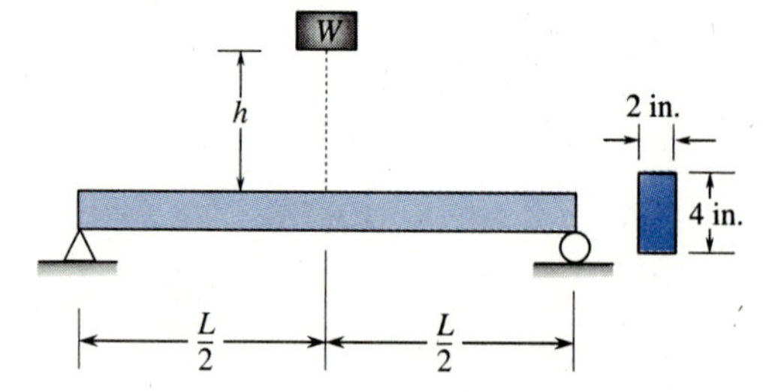

11 Fatigue

11.1 INTRODUCTION

In previous chapters we considered members subjected to only static loads or to a single suddenly applied dynamic load. That is, the loads were considered to be applied only occasionally to the member.

In this chapter we consider members subjected to repeated or fatigue loads. We consider only some of the basic concepts of fatigue and refer to advanced texts ([1–3]) for more details of fatigue design. An example of fatigue loading is a rotating shaft in a machine subjected to bending loads. The shaft will be subjected to tension and compression for each rotation of the shaft. If the shaft rotates at 1000 rpm, then the shaft will be subjected to tension and compression 1000 times per minute.

Fatigue failure, then, is the failure of a part due to the repeated or cyclic loading and unloading or fluctuating reversal in loading after a large number of cycles. This kind of failure is also called *progressive fracture*. The fatigue strength of a member is generally much less than either the yield or ultimate strength of the member.

Fatigue failure begins with a small crack, usually at a point of discontinuity in the material, such as at a change in cross-section, at a keyway in a shaft, at a hole in the member, or at any flaw, internal or external. The crack progresses and eventually stresses become large over a small area until, on loading with another load cycle, a sudden, or instantaneous, brittle type of failure occurs. The surface of a failed section shows the progressive crack with the beach mark (wavylike) appearance that one would observe along a beach at a lake. The final failure, being brittlelike, gives no warning. It is sudden (like a brittle material failure under static load). The fracture surface usually shows a smoother slightly wavier region (beach mark–like region) where the crack was progressing under successive load applications (often called the *fatigue zone*) and a region of rough surface where this final fracture occurred under the last cycle of load (often called the *instantaneous zone*). However, the stress at failure is usually below the yield strength of the

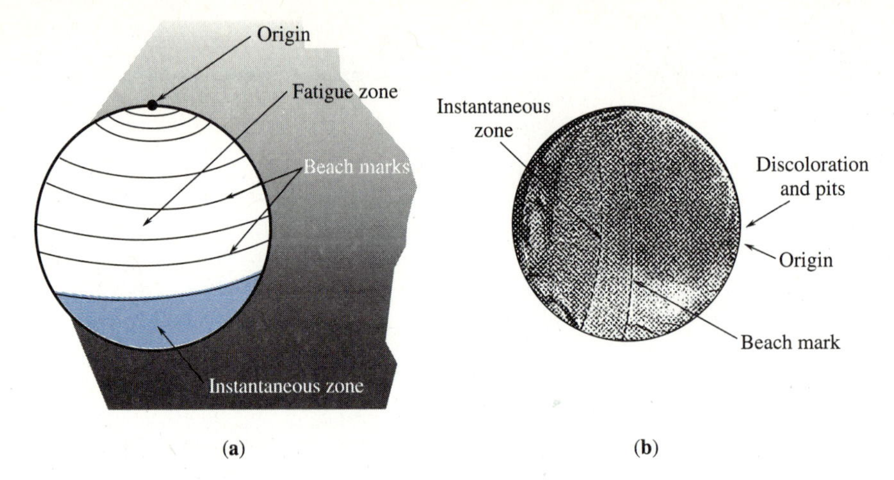

Figure 11.1 (a) Diagram of fatigue fracture surface showing origin of cracks, fatigue zone, and instantaneous zone and
(b) Cross-section of failed bolt in engine connecting rod due to fatigue loading.

material. Figure 11.1(a) shows a diagram of a fatigue fracture surface with its typical features—origin of crack, fatigue zone showing progressive cracks or beach marks, and instantaneous fracture zone. This cross-section is typical of the kind of failure in one-way bending fatigue. Figure 11.1(b) shows the cross-section of a bolt in an engine connecting rod that failed due to fatigue loading.

11.2 THE S-N DIAGRAM

The *S-N diagram* is used in fatigue strength analysis. The S-N diagram is a plot showing fatigue strength (S) versus number of cycles to a failure (N). The fatigue strength is plotted on a log scale on the vertical axis and the number of cycles to failure is plotted on a log scale on the horizontal axis. (Plotting on log paper clearly shows the bend in the curve.) A typical S-N diagram for a steel is shown in Figure 11.2.

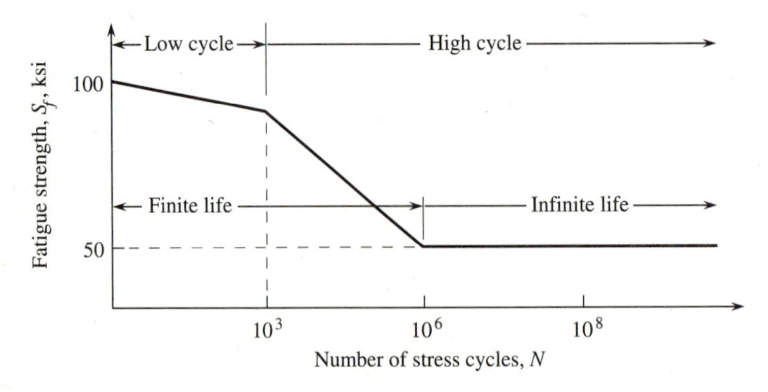

Figure 11.2 Typical S-N diagram for a steel.

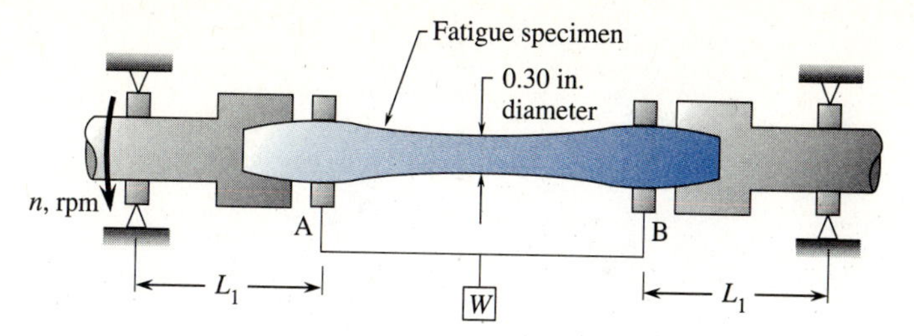

Figure 11.3 Rotating beam fatigue test.

A common test using the high-speed rotating beam machine was developed by R. R. Moore. This rotating beam test is shown in Figure 11.3. Here a specimen with 0.3 in. diameter at the center span is subjected to a load W, which is applied over two points A and B, such as to result in a constant moment in region AB. The specimen is then rotated with a specific constant angular speed such that the specimen is subjected to fatigue or cyclic loading. Since the load W is stationary and the shaft is rotating, at any point on the surface of the specimen the stress undergoes a variation from a positive normal stress to a negative normal stress ($\sigma = \pm\sigma_{max}$) for each cycle of load. A typical cycle of load is shown in Figure 11.4. The specimen is rotated until the specimen fractures. The number of cycles N to fracture is recorded. The value of σ_{max} corresponding to fracture for a specific number of cycles to failure is normally called the fatigue strength S_f. The test is usually stopped at 10^6 or 10^7 cycles if a fracture does not occur. A number of tests are performed and the results plotted on log-log paper. The test data are normally somewhat scattered and either statistical analysis or minimum base curves are used. For most steels the S-N diagram has a sharp knee in the diagram, which occurs at approximately 10^6 cycles of loading. Beyond the knee, failure will not occur no matter how many cycles are used. The strength at this point is called the *endurance limit* S_e'. Therefore, if a member is subjected to a bending stress below the endurance limit stress, the member will not fail no matter how many cycles of loading occur. Most steels have an endurance limit, while many nonferrous metals (aluminum, etc.) do not have a well-defined endurance limit. In Figure 11.2, we also observe a region of low cycle fatigue when N is less than or equal to 10^3 number of cycles and a region of high cycle fatigue for N greater than 10^3 cycles.

For most steels the relationship between log S_f and log N is linear from 10^3 to about 10^6 cycles, whereas from about 10^6 cycles and greater, S_f remains constant at a value S_e'. Usually under 10^3 cycles, the loading is considered to be static and at 10^3 cycles the failure occurs at about 0.8–0.9 times the ultimate strength of the material.

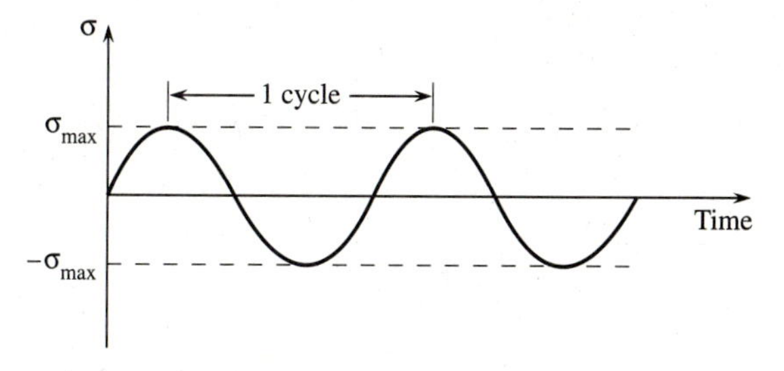

Figure 11.4 Typical cycle of loading.

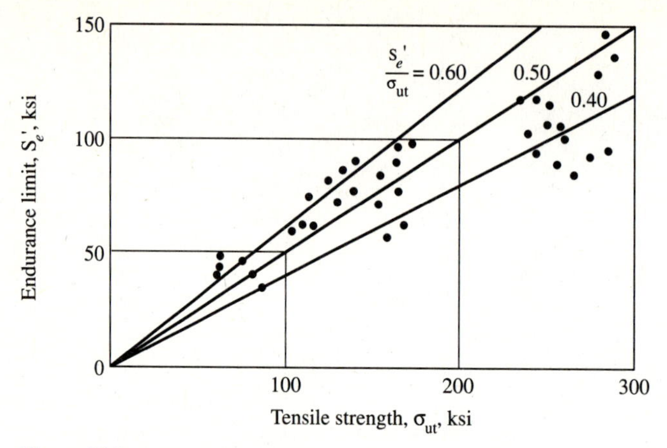

Figure 11.5 Typical graph of endurance limits versus tensile strengths from typical test results for a large number of steels. Ratios of $S_e'/\sigma_{ut} = 0.60$, 0.50, and 0.40 are shown by the solid lines.

For a large number of steels, Figure 11.5 shows a graph of endurance limits versus tensile strengths. These results have been used to arrive at the endurance limits given by Eq. (11.1). Similar tests on other materials have been used to obtain endurance limits for other materials given by Eqs. (11.2) through (11.6).

The generally accepted endurance limits based on experimental data for rotating beam specimens of various materials are listed below as follows:

For steel:

To estimate the endurance limit from the rotating beam test for rotating beam specimens, we use

$$
\begin{aligned}
S_e' &= 0.5\sigma_{ut} \quad \text{For steels with } \sigma_{ut} \leq 200 \text{ ksi (1400 MPa)} \\
S_e' &= 100 \text{ ksi (700 MPa)} \quad \text{For } \sigma_{ut} > 200 \text{ ksi (1400 MPa)}
\end{aligned}
\tag{11.1}
$$

For cast iron and cast steels:

$$
\begin{aligned}
S_e' &= 0.45\sigma_{ut} \quad \text{For cast irons with } \sigma_{ut} \leq 88 \text{ ksi (607 MPa)} \\
S_e' &= 40 \text{ ksi (280 MPa)} \quad \text{For } \sigma_{ut} > 88 \text{ ksi (607 MPa)}
\end{aligned}
\tag{11.2}
$$

For steels we do not want to stress beyond these endurance limits if a life cycle of 10^6 or more cycles is required.

For nonferrous alloys, even though a true endurance limit does not exist, we use the following:

For aluminum:

$$
\begin{aligned}
S_e' &= 0.30\sigma_{ut} \quad \text{(cast aluminum)} \\
S_e' &= 0.40\sigma_{ut} \quad \text{(wrought aluminum)}
\end{aligned}
\tag{11.3}
$$

(if a life of 5×10^8 cycles or more is desired)

For magnesium alloys:

$$S_e' = 0.35\sigma_{ut} \tag{11.4}$$

(if a life of 10^8 cycles or more is desired)

For copper alloys:

$$S_e' = 0.25\text{–}0.5\sigma_{ut} \tag{11.5}$$

(if a life of 10^8 cycles or more is desired)

For titanium and its alloys:

$$S_e' = 0.45\text{–}0.65\sigma_{ut} \tag{11.6}$$

(For titanium alloys a true endurance limit in the 10^6–10^7 cycle range exists.)

11.3 HIGH CYCLE FATIGUE UNDER COMPLETELY REVERSED LOADING

In this section we consider the common case of high cycle fatigue. *High cycle* fatigue is defined as that fatigue that exists when the number of cycles of loading is greater than 10^3 cycles.

Based on experimental results (from rotating beam specimens) for steels (Figure 11.6), we have the following equation to express the fatigue strength versus cycles to failure for 10^3–10^6 cycles of loading:

$$\log S_f' = b \log N + c \tag{11.7}$$

Now this line intersects the $N = 10^6$ cycles at S_e' and the 10^3 cycles at $0.8\sigma_{ut}$. This yields

$$b = -\frac{1}{3}\log\left(\frac{0.8\sigma_{ut}}{S_e'}\right) \tag{11.8}$$

and

$$c = \log\left[\frac{(0.8\sigma_{ut})^2}{S_e'}\right] \tag{11.9}$$

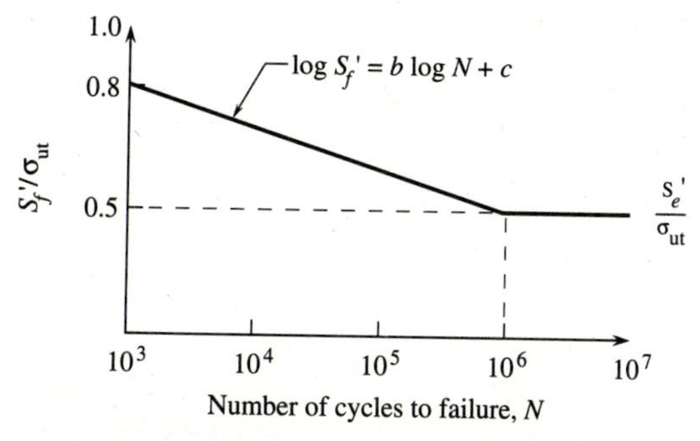

Figure 11.6 Typical average S-N diagram for different steels.

Also by Eq. (11.7), we have

$$S_f' = 10^c N^b \quad \text{for} \quad 10^3 \leq N \leq 10^6 \tag{11.10}$$

and

$$N = 10^{-c/b} S_f'^{1/b} \quad \text{for} \quad 10^3 \leq N \leq 10^6 \tag{11.11}$$

Remember in Eqs. (11.7) through (11.11) that S_f' and S_e' are the fatigue strength and endurance limit, respectively, for a rotating beam specimen of the material. The actual machine part or structural member will have a different endurance limit, which will be affected by such factors as its surface finish, its size, the type of load, the temperature, and stress concentrations. These considerations will be described in Section 11.4.

EXAMPLE 11.1

A steel part with an endurance limit of $S_e' = 48$ ksi and ultimate strength of 100 ksi is subjected to a standard rotating beam test (see Figure 11.3) using a weight $W = 186$ lb. The distance from the support to the load at A is $L_1 = 2.0$ in. Determine the number of cycles to failure.

Solution

The maximum stress in the rotating beam specimen is

$$\sigma_{max} = \frac{Mc}{I} = \frac{32\,M}{\pi d^3} = \frac{32\,WL_1/2}{\pi(0.30 \text{ in.})^3}$$

$$= 188.6\,WL_1 = 188.6(186 \text{ lb})(2.0 \text{ in.})$$

$$= 70.2 \text{ ksi}$$

This stress is then the fatigue strength of the part. Therefore, $S_f' = 70.2$ ksi.

Assuming high cycle fatigue ($N > 10^3$ cycles), we use Eq. (11.11) to obtain N, with b obtained using Eq. (11.8) and c obtained using Eq. (11.9) as follows:

$$b = -\frac{1}{3}\log\frac{0.8\sigma_{ut}}{S_e'} = -\frac{1}{3}\log\left(\frac{0.8(100)}{48}\right)$$

$$= -0.07395$$

$$c = \log\frac{(0.8\sigma_{ut})^2}{S_e'} = \log\frac{[0.8(100)]^2}{48}$$

$$= 2.125$$

Therefore the number of cycles to failure is

$$N = (10^{-2.125/(-0.07395)})(70.2^{1/(-0.07395)})$$

$$= 5857 \text{ cycles}$$

11.4 FACTORS AFFECTING ENDURANCE LIMIT OF MEMBERS SUBJECTED TO COMPLETELY REVERSED LOADING

From the rotating beam test, the endurance limit S_e' of a specimen of material is obtained. A real machine part has factors that reduce the endurance limit S_e' given by

$$S_e = k_{su} k_{si} k_l k_t S_e' \tag{11.12}$$

where

S_e = the endurance limit of the part

S_e' = the endurance limit of the rotating beam specimen of the same material as the part

k_{su} = the surface factor

k_{si} = the size factor

k_l = the load factor

k_t = the temperature factor

All the k's in Eq. (11.12) are less than or equal to 1. We apply Eq. (11.12) for high cycle fatigue. This means that for ferrous metals, with a known endurance limit, Eq. (11.12) is valid. For those materials with no well-defined endurance limit, such as aluminum alloys, we apply these factors to the 10^8 or 5×10^8 cycle strength as relevant.

We now consider each of these factors separately.

1. *Surface finish,* k_{su} In general smoother surface finishes of members result in members more resistant to fatigue loading. For a smoothly polished surface, $k_{su} = 1$. For members that are ground, machined or cold-drawn, hot-rolled, or forged, surface finish factor values are given in Figure 11.7. The mechanical surface treatment called *shot*

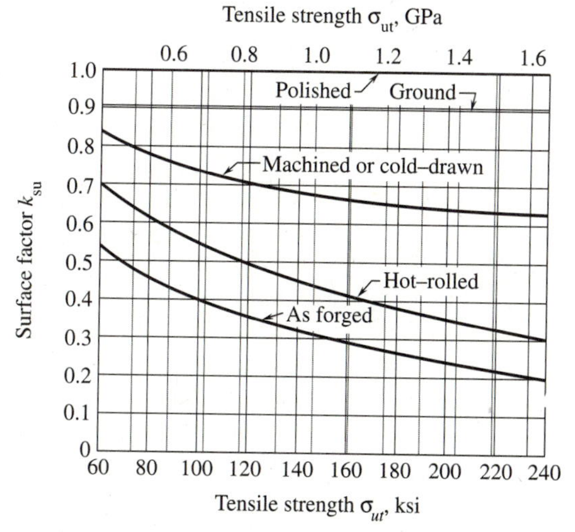

Figure 11.7 Surface effect influence on surface factor from [4]. Printed with permission of McGraw-Hill from Shigley, J.E. and Mitchell, L.D. *Mechanical Engineering Design,* 1983.

peening is often used to improve the surface finish. Springs, gears, shafts, and connecting rods are typical machine parts that are often shot peened. Shot peening involves bombarding the surface with high-velocity iron or steel shot discharged from a rotating wheel or pneumatic nozzle. This hammering effect tends to reduce the thickness and therefore increase the area of the exposed surface layer. This outer layer is then placed in residual compression. This residual compression works to the advantage of the material as the material is subjected to cycles of fatigue loading. If specific data are not available on shot peened parts, it is usually considered appropriate to use a surface finish factor of 1 when a part has been shot peened.

2. *Size effect,* k_{si} In general, there is a greater likelihood that members with larger cross-sections will contain flaws. From experimental testing of round bars in reversed bending and torsion, Eq. (11.13) has been developed [4] to obtain the size factor as

$$k_{si} = \begin{cases} 0.869d^{-0.097} & 0.3 \text{ in.} \leq d \leq 10 \text{ in.} \\ 1 & d \leq 0.3 \text{ in.} \quad \text{or} \quad d \leq 8 \text{ mm} \\ 1.189d^{-0.097} & 8 \text{ mm} < d \leq 250 \text{ mm} \end{cases} \tag{11.13}$$

For square bars, for simplicity we use the effective d (the d obtained for a circular bar of the same cross-sectional area although other more accurate methods are available).

For axially loaded bars, we use

$$k_{si} = 0.71 \text{ (if the mechanical properties are well known from actual tests)}$$

or (11.14)

$$k_{si} = 0.60 \text{ (if no test results are available)}$$

3. *Load factor,* k_l The load factor is given for the different kinds of loads by the following equation:

$$\begin{aligned} k_l &= 0.9 && \text{axial load, for } \sigma_{ut} \leq 220 \text{ ksi (1520 MPa)} \\ k_l &= 1 && \text{axial load, for } \sigma_{ut} > 220 \text{ ksi (1520 MPa)} \\ &= 1 && \text{bending} \\ &= 0.58 && \text{torsion} \end{aligned} \tag{11.15}$$

The endurance limit in reversed torsion is about 58% of the endurance limit in reversed bending. This result has also been predicted by the distortion energy theory (see Chapter 8, Eq. (8.52)). Therefore, the load factor of 0.58 is used for reversed torsion in Eq. (11.15).

4. *Temperature factor,* k_t The temperature affects the endurance limit of a member. For low temperatures, brittle fracture is a possibility and should be considered before fatigue. For steels with temperatures up to around 450°C (840°F) a factor of $k_t = 1$ is often used. For higher temperature, the factor may approach 0.5. For more details on studies regarding the temperature factor consult references [1–3]. In our work, we will assume the members to be at room temperature and use $k_t = 1$.

EXAMPLE 11.2

A 1 in. diameter bar of steel with an ultimate tensile strength of 100 ksi has a machined finish and will be used in reversed bending. Determine the endurance limit S_e of the mechanical part made of this steel bar.

Solution

Using Eq. (11.12), we obtain the endurance limit as

$$S_e = k_{su}k_{si}k_l k_t S_e'$$

We begin by using Figure 11.7 to obtain the surface finish factor. For a steel with tensile strength of 100 ksi and machined finish, Figure 11.7 yields

$$k_{su} = 0.73$$

Next we obtain the size factor from Eq. (11.13) as

$$k_{si} = 0.869d^{-0.097} = 0.869(1\text{ in.})^{-0.097} = 0.869$$

The load factor is given by Eq. (11.15). The member is subjected to reversed bending, therefore the load factor is

$$k_l = 1.0$$

The temperature is assumed to be room temperature, therefore $k_t = 1$. The rotating beam endurance limit is obtained from Eq. (11.1) as

$$S_e' = 0.5\sigma_{ut} = 0.5(100\text{ ksi}) = 50\text{ ksi}$$

Using Eq. (11.12), the endurance limit is

$$S_e = (0.73)(0.869)(1.0)(1.0)(50\text{ ksi})$$
$$= 31.7\text{ ksi}$$

EXAMPLE 11.3

A flat bar of wrought aluminum alloy is used in reversed axial loading of 500 lb as shown (Figure 11.8). The bar has a machined finish. The bar width is 1 in. and its thickness is 3/8 in. The ultimate tensile strength is 60 ksi. Determine the endurance limit of the bar. Compare the endurance limit to the axial stress in the bar.

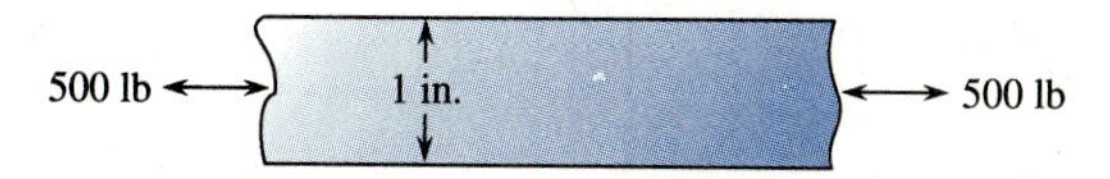

Figure 11.8 Bar subjected to reversed axial loading.

Solution

Using Eq. (11.12), we obtain the endurance limit of the bar as

$$S_e = k_{su} k_{si} k_l k_t S_e'$$

Using Eq. (11.3), the endurance limit of the rotating beam specimen of wrought aluminum is

$$S_e' = 0.4(60 \text{ ksi}) = 24 \text{ ksi}$$

The surface finish factors in Figure 11.7 are for steel. However, since we do not have any better information available for aluminum, we will use Figure 11.7 and apply it to the aluminum (realizing that we really need fatigue tests involving the material and surface being used to establish the surface factor). The surface finish factor from Figure 11.7 is then

$$k_{su} = 0.83$$

From Eq. (11.15), the load factor is

$$k_l = 0.9$$

All other factors are 1. The endurance limit is then

$$S_e = 0.83(0.9)(24 \text{ ksi})$$

$$= 17.9 \text{ ksi}$$

The axial stress in the bar is

$$\sigma = \frac{500 \text{ lb}}{(3/8 \text{ in.})(1 \text{ in.})} = 1333 \text{ psi} = 1.33 \text{ ksi} < 17.9 \text{ ksi}$$

EXAMPLE 11.4

A uniform shaft with diameter of 7/8 in. is made of a heat treated steel with an ultimate tensile strength of 190 ksi. The shaft surface is ground. The shaft is subjected to the completely reversed bending loads shown (Figure 11.9). Determine the endurance limit and compare it to the bending stress in the shaft.

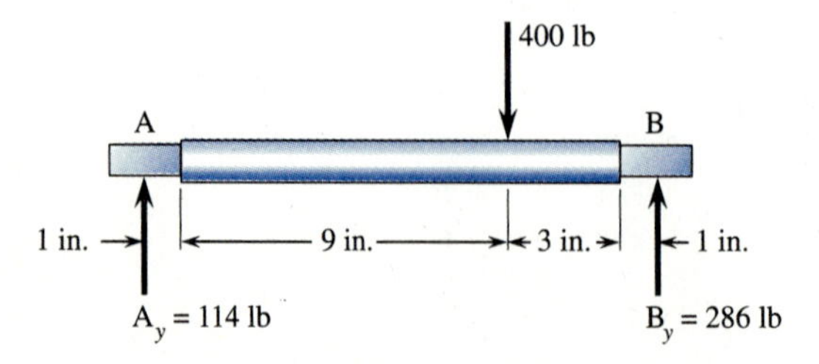

Figure 11.9 Shaft subjected to completely reversed loading.

Solution

By Eq. (11.1), the endurance limit of the rotating beam of the same material is

$$S_e' = 0.5(190 \text{ ksi})$$
$$= 95 \text{ ksi}$$

For a ground surface, using Figure 11.7, the surface factor is

$$k_{su} = 0.89$$

For a diameter of 7/8 in., the size factor is obtained from Eq. (11.13) as

$$k_{si} = 0.869(0.875)^{-0.097}$$
$$= 0.88$$

The load factor and temperature factor are equal to 1. Using Eq. (11.12), the endurance limit of the shaft is

$$S_e = (0.89)(0.88)(95 \text{ ksi})$$
$$= 74.4 \text{ ksi}$$

The bending stress is

$$\sigma = \frac{\text{M}}{\text{S}} = \frac{(114 \text{ lb} \cdot 10 \text{ in.})}{\dfrac{\pi d^3}{32}} = \frac{1140}{\dfrac{\pi(0.875)^3}{32}} = 17{,}330 \text{ psi}$$

11.5 INFLUENCE OF STRESS CONCENTRATION ON FATIGUE UNDER COMPLETELY REVERSED FATIGUE LOADING

We considered stress concentration effects under static tensile loads in Chapter 2, static torsional loads in Chapter 4, and static bending loads in Chapter 5. We now consider the influence of stress concentrations on the strength of a member subjected to completely reversed fatigue loading.

Under fatigue, loading tests have shown that some materials are not very sensitive to the presence of notches or discontinuities. Therefore, the full value of the stress concentration factors K described in previous chapters need not be used. Figures 11.10(a) and (b) show an unnotched and a notched specimen of the same material. The specimen dimensions are the same at the section where the notch occurs and in the unnotched specimen. Any load then yields the same nominal stress in either case. Figure 11.10(c) shows the two S-N curves for the specimens. The ratio of the unnotched to notched endurance limit is defined as the fatigue stress concentration factor, K_f. That is

$$K_f = \frac{\text{Unnotched endurance limit}}{\text{Notched endurance limit}} = \frac{S_{eu}'}{S_{en}'} \qquad \textbf{(11.16)}$$

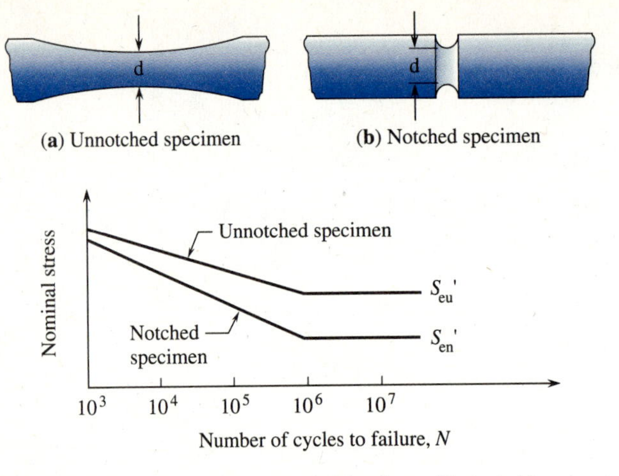

Figure 11.10 Fatigue stress concentration factor illustrated by resulting tests of notched and unnotched members.

Tests have shown that K_f is often less than K for a material. This is likely due to internal irregularities in the structure of a material. These irregularities, if severe enough, cause local points of high stress. The addition of a geometric stress concentration, such as from a groove or hole machined into a member, may then have little influence on the endurance limit of the unnotched or notched member. For instance, ordinary gray cast iron has numerous internal stress raisers due to the graphite flakes in the matrix. Hence, the addition of a geometric stress raiser has a small effect on the endurance limit.

The notch sensitivity is included in the definition of K, as

$$K_f = 1 + (K - 1)q \tag{11.17}$$

where q is the notch sensitivity factor. Values of q range from between zero (making $K_f = 1$ and the material has no sensitivity to notches) and unity (making $K_f = K$ and the material has full sensitivity to notches).

For steels and aluminum alloys subjected to reversed bending or axial loads, Figure 11.11 shows how notch sensitivity q varies with notch radius and hardness and strength of the steel. For larger notch radii, and for higher strength materials, the sensitivity q approaches 1. Therefore, for these situations, K_f approaches K. For cast iron, notch sensitivity is low. Based on experimental data, it is recommended that $q = 0.20$ be used for all grades of cast iron.

After obtaining K_f, it can be used as a stress concentration factor or as a strength reducing factor. Experts differ on this point. For simplicity, we will consider K_f as a factor that reduces the strength of the member. This means that the modification to the fatigue strength due to stress concentration is now defined by

$$k_{sc} = \frac{1}{K_f} \tag{11.18}$$

We then use the factor k_{sc} to further reduce the endurance limit of a member and multiply k_{sc} by the endurance limit S_e obtained from Eq. (11.12).

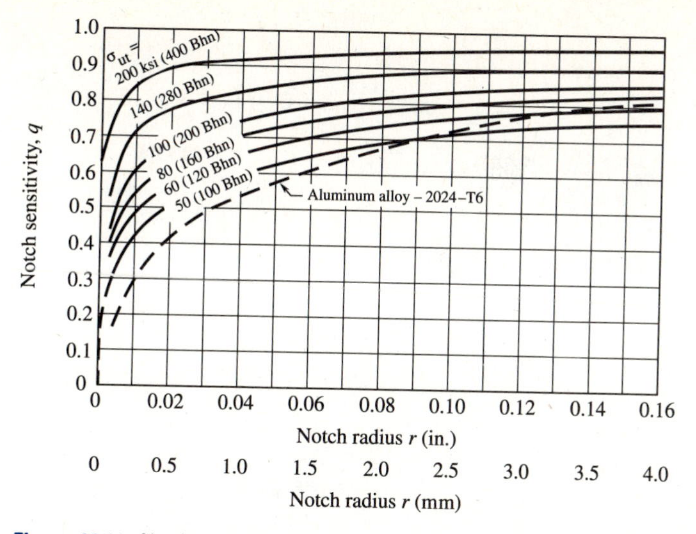

Figure 11.11 Notch sensitivity curves for steels and wrought aluminum alloys subjected to reversed bending or axial loads (r is the radius at the point of likely fatigue crack origin). (For $r > 0.16$ in. extrapolate.) From [5] with permission from John Wiley and Sons, Inc.

EXAMPLE 11.5

For Example 11.4, assume the shaft is now stepped with a fillet radius of 1/16 in. between the 7/8 in. diameter shaft and the larger shaft of 1-1/8 in. diameter (Figure 11.12). Now determine the endurance limit of the shaft.

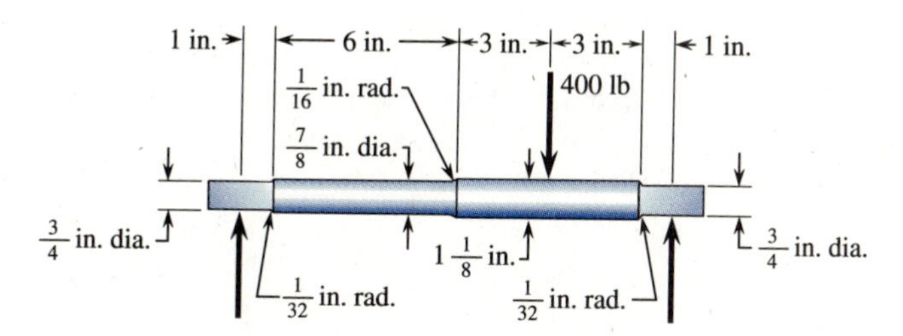

Figure 11.12 Stepped shaft to illustrate stress concentration in fatigue.

Solution

Stress reduction effects from surface finish and size were considered in Example 11.4. They are

$$k_{su} = 0.89 \quad \text{and} \quad k_{si} = 0.88$$

We now determine the geometric stress concentration factor. Using

$$\frac{D}{d} = \frac{1.125}{0.875} = 1.28$$

and

$$\frac{r}{d} = \frac{1/16}{0.875} = 0.0714$$

we obtain from Figure 5.76(c) of Chapter 5

$$K = 1.8$$

For a notch radius $r = 0.0625$ in. and $\sigma_{ut} = 190$ ksi, from Figure 11.11, we obtain $q = 0.90$.

From Eq. (11.17), we obtain

$$K_f = 1 + 0.90(1.8 - 1) = 1.72$$

and from Eq. (11.18), we have

$$k_{sc} = \frac{1}{1.72} = 0.581$$

The endurance limit is then

$$S_e = (0.89)(0.88)(0.581)(95) = 43.2 \text{ ksi}$$

11.6 FATIGUE STRENGTH UNDER FLUCTUATING STRESSES

Often machine and structural parts are subjected to stresses that are a combination of static and fluctuating stress which is not a complete reversal. To design for fatigue due to fluctuating load or stress, we will use the *modified Goodman diagram* as it is the most widely accepted one.

A typical fluctuating stress is shown in Figure 11.13. The stress may or may not reverse. In Figure 11.13 the stress is shown not reversing.

Figure 11.13 shows the maximum stress σ_{max}, the minimum stress σ_{min}, the stress range σ_r, the stress amplitude σ_a, and the steady or static or mean stress σ_m. The mean stress and stress amplitude are defined by

$$\sigma_m = \frac{\sigma_{max} + \sigma_{min}}{2} \tag{11.19}$$

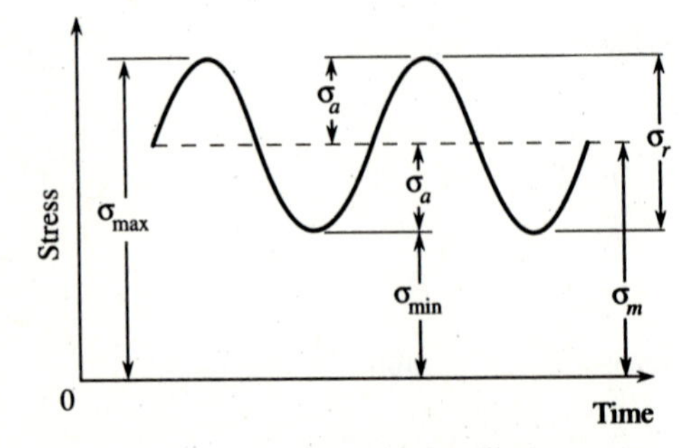

Figure 11.13 Sinusoidal fluctuating stress variation.

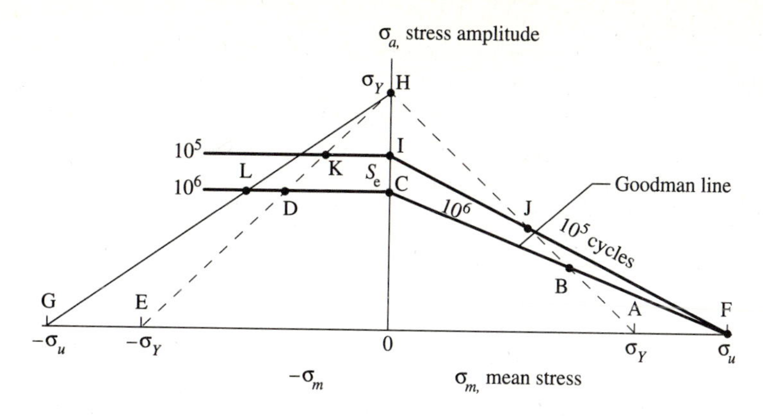

Figure 11.14 Plot of mean stress versus stress amplitude.

and

$$\sigma_a = \frac{\sigma_{max} - \sigma_{min}}{2} \tag{11.20}$$

To construct the Goodman diagram, we plot the mean stress versus stress amplitude as shown in Figure 11.14. Two lines are shown plotted on the diagram on the positive side of the mean stress axis. A line from the yield stress σ_Y on the mean stress axis (point A) to σ_Y on the stress amplitude axis (point H). The other line is from the ultimate stress σ_u on the mean stress axis (point F) to the endurance limit S_e of the part on the stress amplitude axis (point C). On the negative or compression side of the mean stress line, a line is drawn from the yield stress in compression of the material (point E) up to the yield stress plotted on the vertical stress amplitude axis. Here we will assume that the yield stress in compression is equal to that in tension. If we want to prevent failure by yielding and want a life of at least 10^6 cycles, we must have a stress state inside of these lines or inside of the area labeled ABCDEA. If no yielding is required but less than 10^6 cycles of life are required (such as 10^5 cycles), then we can also be inside the area above ABCDEA, that is, area DCBJIKD is considered safe. If 10^6 cycles are required, but yielding is acceptable, area ABFA may be included in the safe region and area GEDLG to the left may be included, as well as the area ABCDEA. The area above the segment HBF corresponds to yielding on the first load application and fatigue fracture before 10^6 cycles of loading.

In summary, most designers use the curve ABCDE in Figure 11.14 in design work when a part is subjected to fluctuating stresses. The modified Goodman line CBF can be put in equation form as

$$\frac{\sigma_m}{\sigma_{ut}} + \frac{\sigma_a}{S_e} = 1 \tag{11.21}$$

Example 11.6 illustrates how to design a part for fluctuating loads based on the modified Goodman diagram.

EXAMPLE 11.6

A rectangular link with a transverse hole is loaded by forces F acting at each end of the link shown. The forces F vary so as to produce axial tension ranging from 70 kN to 30 kN. The link is made of a steel with a yield stress of 920 MPa and an ultimate stress of 1010 MPa. The steel has been heat-treated and all surfaces are finished by grinding. Using a factor of safety of 1.5, determine the thickness t of the link.

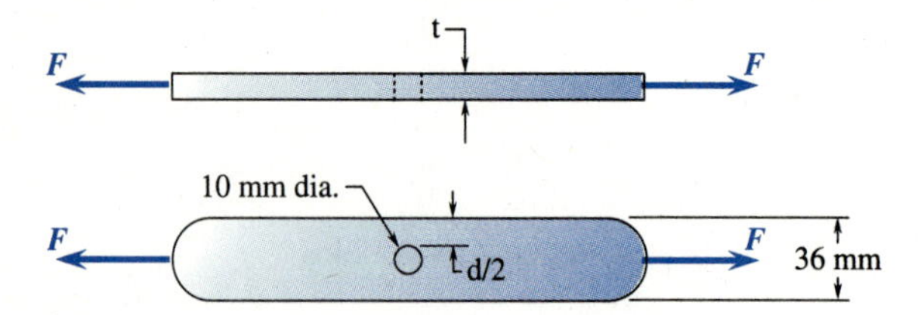

Figure 11.15 Link subjected to fluctuating axial load.

Solution

First, we determine the endurance limit of the link as follows: The surface is ground, therefore using Figure 11.7, the surface factor is

$$k_{su} = 0.89$$

The load is axial and assuming no test results available for the part, therefore from Eq. (11.14), the size factor is

$$k_{si} = 0.60$$

The load factor is obtained from Eq. (11.15) as

$$k_l = 0.90$$

The stress concentration reduction factor is obtained as follows

$$\frac{r}{d} = \frac{5 \text{ mm}}{26 \text{ mm}} = 0.192$$

Therefore, from Figure 2.58(d) of Chapter 2, we have

$$K = 2.3$$

For a notch radius of 5 mm and $\sigma_{ut} = 146$ ksi (1010 MPa), we obtain the notch sensitivity q as (see Figure 11.11)

$$q = 0.92$$

Using Eq. (11.17), we obtain K_f as

$$K_f = 1 + q(K - 1) = 1 + 0.92(2.3 - 1) = 2.20$$

Using Eq. (11.18), we obtain the stress concentration reduction factor as

$$k_{sc} = \frac{1}{2.20} = 0.455$$

The endurance limit of the link is then obtained using Eq. (11.12) as

$$S_e' = 0.5(1010 \text{ MPa}) = 505 \text{ MPa}$$

and

$$S_e = (0.89)(0.60)(0.90)(0.455)(505 \text{ MPa}) = 110 \text{ MPa}$$

The force amplitude is

$$F_a = \frac{70 - 30}{2} = 20 \text{ kN}$$

The mean force is

$$F_m = \frac{70 + 30}{2} = 50 \text{ kN}$$

The stress amplitude is then

$$\sigma_a = \frac{20 \text{ kN}}{(0.036 - 0.01)t} = \frac{769}{t} \text{ MPa}$$

The mean stress is

$$\sigma_m = \frac{50 \text{ kN}}{(0.036 - 0.01)t} = \frac{1923}{t} \text{ MPa}$$

We now construct the modified Goodman diagram as shown below. The ratio of mean stress to stress amplitude is

$$\frac{\sigma_m}{\sigma_a} = \frac{1923}{769} = 2.5$$

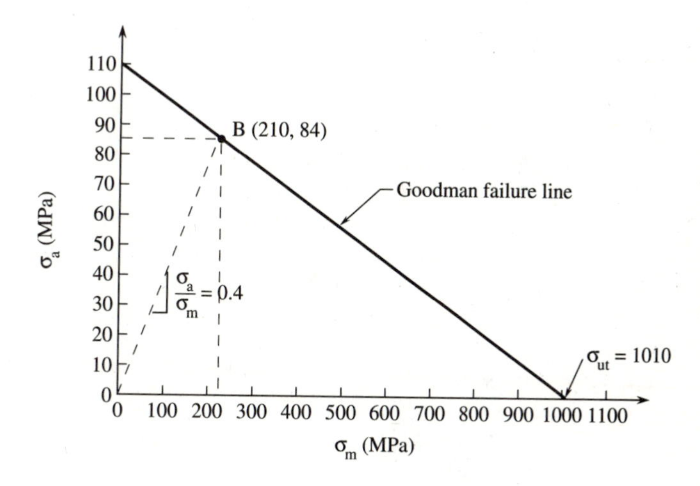

This results in an allowable mean stress of $S_m = 210$ MPa and allowable stress amplitude of $S_a = 84$ MPa (see point B on the adjacent diagram).

Using a factor of safety of 1.5, the mean stress is expressed in terms of the thickness as

$$\sigma_m = \frac{S_m}{n}$$

or

$$\frac{1923}{t} = \frac{210}{1.5}$$

Solving for t, we obtain

$$t = 13.7 \text{ mm}$$

The same result occurs if we use the allowable stress amplitude instead of the mean stress.

For a comparison, we determine the thickness based on the assumption of a static load only. This static load is $F = 70$ kN. The allowable static stress is given by

$$\sigma_{allow} = \frac{\sigma_Y}{1.5} = \frac{920 \times 10^3}{1.5} = \frac{70 \text{ kN}}{(0.026)t}$$

or solving for t, we obtain

$$t = 4.39 \text{ mm (much less than 13.7 mm obtained using the fatigue load assumption and the Goodman diagram)}$$

11.7 SUMMARY OF IMPORTANT DEFINITIONS AND EQUATIONS

1. *Fatigue failure:* the failure of a part due to repeated or cyclic loading and unloading or fluctuating reversal in loading after a large number of cycles of load

2. *S-N diagram:* a plot of fatigue strength versus number of cycles to failure

3. *Endurance limit:* the stress level below which a part will not fail no matter how many cycles of loading are used

4. Suggested endurance limits for steels, cast irons, aluminums, magnesium, copper, and titanium alloys are given by Eqs. (11.1)–(11.6)

5. Fatigue strength for high-cycle fatigue is given by

$$\log S_f' = b \log N + c \qquad \textbf{(11.7)}$$

where

$$b = -\frac{1}{3}\log\left(\frac{0.8\sigma_{ut}}{S_e'}\right) \qquad \textbf{(11.8)}$$

and

$$c = \log\left[\frac{(0.8\sigma_{ut})^2}{S_e'}\right] \qquad \textbf{(11.9)}$$

The number of cycles to failure for $10^3 \leq N \leq 10^6$ cycles is

$$N = 10^{-c/b} S_f'^{\,1/b} \qquad \textbf{(11.11)}$$

6. Endurance limit of actual machine parts is

$$S_e = k_{su} k_{si} k_1 k_t S_e' \qquad \textbf{(11.12)}$$

with factors affecting endurance limit of actual part given by

(a) Surface finish factor (see Figure 11.7)

(b) Size effect [see Eqs. (11.13) and (11.14)]

(c) Load factor [see Eq. (11.15)]

7. Influence of stress concentration factor

$$K_f = 1 + (K - 1)q \qquad \textbf{(11.17)}$$

and

$$k_{sc} = \frac{1}{K_f} \qquad \textbf{(11.18)}$$

8. Fatigue strength under fluctuating stresses [see modified Goodman diagram (Figure 11.14)]

REFERENCES

1. Fuchs HO, Stephens RI. *Metal Fatigue in Engineering*. John Wiley & Sons, New York, 1980.
2. Graham JA (ed). *Fatigue Design Handbook*. Society of Automotive Engineers, New York, 1968.
3. Sines G, Waisman JL (eds). *Metal Fatigue*. McGraw-Hill, New York, 1959.
4. Shigley JE, Mitchell LD, *Mechanical Engineering Design*, 4th Ed., McGraw-Hill, New York, 1983.
5. Juvinall RC, *Fundamentals of Machine Component Design*, John Wiley and Sons, New York, 1983.

PROBLEMS

Sections 11.2 and 11.3

11.1 A rotating beam specimen with a diameter of 0.3 in. is made of a steel with an ultimate strength of 120 ksi using a weight of 186 lb. The distance from the support to the load is $L_1 = 2.0$ in. Determine the number of cycles to failure.

11.2 Estimate the fatigue strength of a rotating beam specimen made of a steel with an ultimate strength of 60 ksi and having a life of 13,000 cycles of stress reversal.

11.3 For the steel specimen in Problem 11.2, determine the life of the specimen corresponding to a stress amplitude of 40 ksi.

11.4 Estimate the fatigue strength of a rotating beam specimen made of a steel with an ultimate strength of 1500 MPa and having a life of 10,000 cycles of reverse bending stress.

11.5 A uniform steel drive shaft is subjected to an alternating bending stress of 55 ksi with a mean stress of zero. Assuming that the ultimate tensile strength of the material is 80 ksi, determine the number of rotations of the shaft before failure.

Section 11.4

11.6 Determine the endurance limit of a 1-in.-diameter steel rod with a machined finish. The rod has an ultimate strength of 100 ksi. The rod has no points of stress concentration.

11.7 Two connecting rods each having a diameter of 1 in. are made as forgings. They are made of two different steels; one is an expensive high-strength material (with $\sigma_u = 250$ ksi) heat-treated and tempered to 600°F. The other is a less expensive lower-strength steel (with $\sigma_u = 110$ ksi) heat-treated and tempered in the same manner as the high-strength one. Determine both endurance limits. Which one would you recommend in fatigue loading? Why?

11.8 A machined steel drive shaft is subjected to an alternating bending stress of 50 ksi with a mean stress of zero. Assume that the ultimate tensile strength of the material is 80 ksi. Determine the endurance limit of the shaft.

Figure P11.8

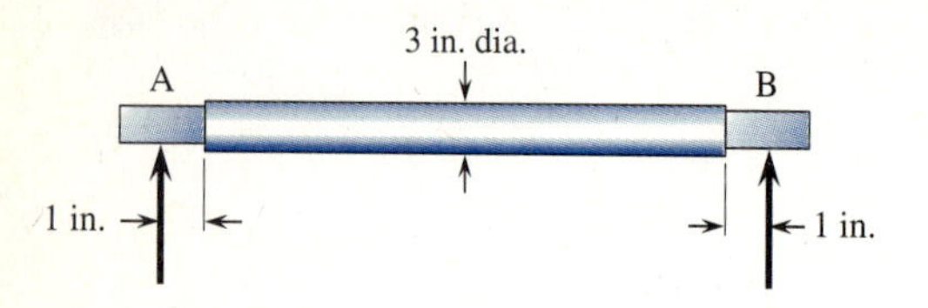

11.9 The bar shown is subjected to completely reversed tensile loading. The bar is made of a steel with a tensile strength of 80 ksi. Determine the endurance limit of the bar.

Figure P11.9

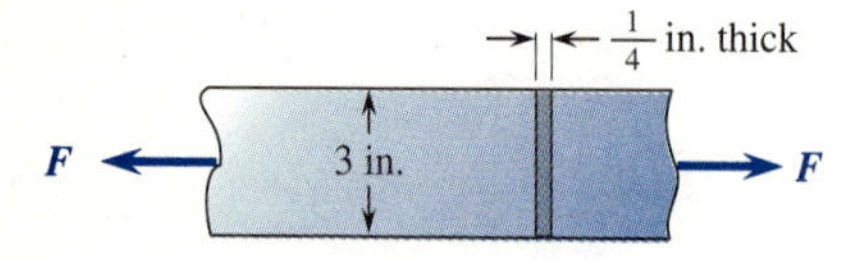

11.10 A shaft is subjected to a completely reversed bending stress from the 500-lb load shown. The shaft is made of a heat-treated steel that is shot peened and has a Brinell hardness of 400. Determine the endurance limit of the shaft.

Figure P11.10

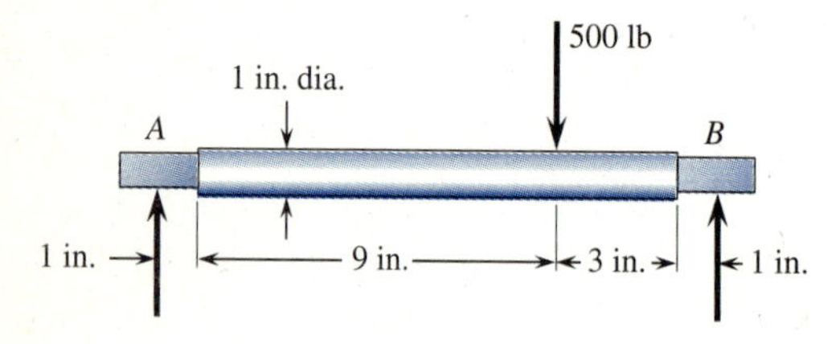

Section 11.5

11.11 The stepped shaft shown has a fillet radius of 3 mm between the 20-mm-diameter shaft and the 25-mm-diameter shaft. The shaft is made of a heat-treated steel with an ultimate strength of 825 MPa. The surface of the shaft is ground. Determine the endurance limit of the shaft.

Figure P11.11

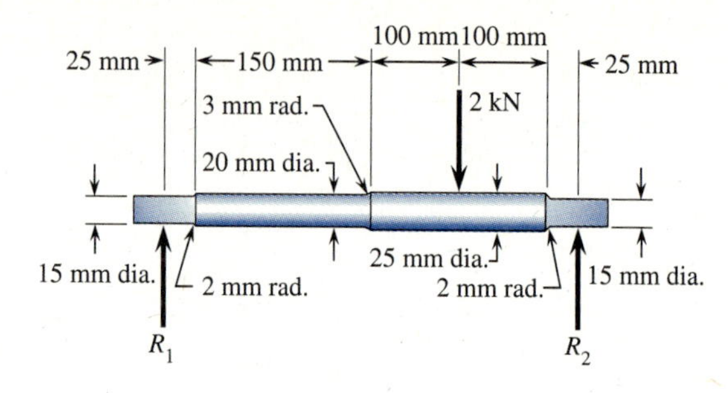

11.12 A flat bar shown with fillet and hole is machined from a steel with a tensile strength of 825 MPa. The axial load is completely reversed. Determine the endurance limit of the shaft.

Figure P11.12

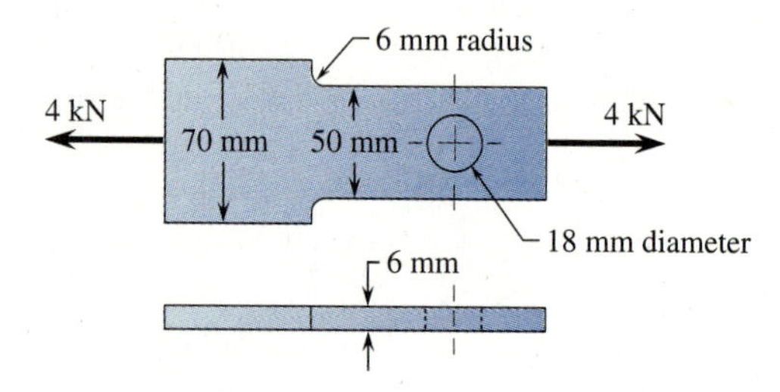

11.13 A shaft of diameter $D = 1.75$ in. has grooves machined into it to hold snap-ring retainers (one on each side of a pulley (not shown) attached to the shaft). The radius of the groove is $r = 0.25$ in. and the depth of the groove is $h = 0.375$ in. The shaft is made of a steel with an ultimate strength of 60 ksi. The shaft will be shot peened. Determine the endurance limit of the shaft. The size factor is to be determined based on the gross diameter D. If the shaft must support a transverse force causing a maximum bending moment of 2100 lb · in., is the shaft safe?

Figure P11.13

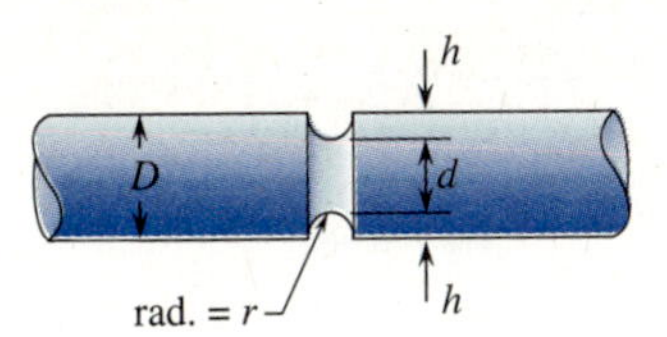

11.14 A stepped shaft is subjected to the transverse load shown. The shaft is made of a steel with an ultimate strength of 1000 MPa. The shaft is machined. The fillet radius is to be 1/10 times the diameter of the shaft. Determine the diameter of the shaft based on a factor of safety of 2.0. All dimensions are in millimeters.

Figure P11.14

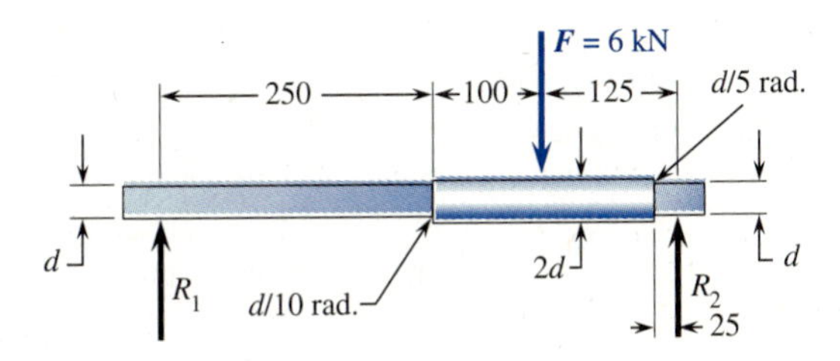

11.15 A bar shown is made of wrought aluminum alloy with a fatigue strength of $S_e' = 18$ ksi at 5×10^8 cycles of reversed axial loading. Determine the factor of safety against fatigue failure if the load is completely reversed. Assume the surface finish factor to be 1.

Figure P11.15

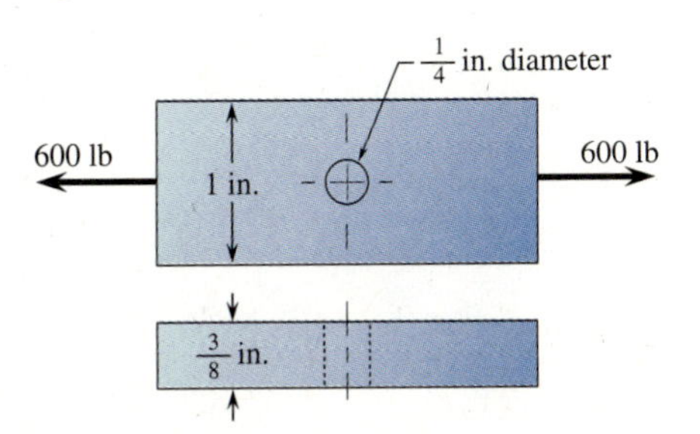

11.16 A bar shown is made of wrought aluminum alloy with a tensile strength of 255 MPa for 5×10^6 cycles of reversed axial loading. Determine the factor of safety against fatigue failure if the load is completely reversed. Assume the surface finish factor to be 1.

Figure P11.16

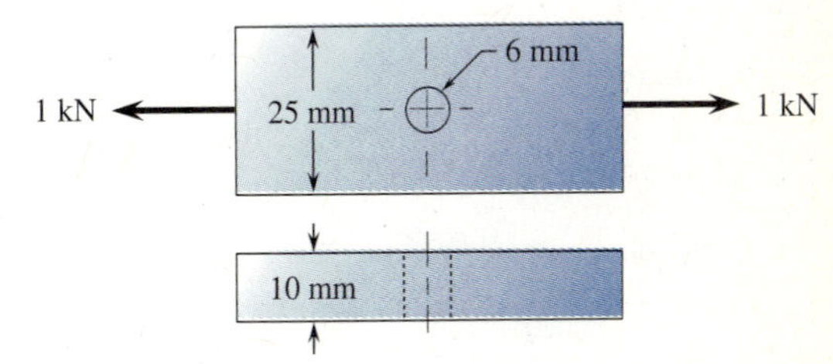

Section 11.6

11.17 A machine part is made of a steel with ultimate strength of 100 ksi, yield strength of 70 ksi, and endurance limit of 30 ksi. Determine the factor of safety (allowable maximum mean stress/actual mean stress) for the following stress state: a bending stress alternating between 6 ksi and 15 ksi.

11.18 Rework Problem 11.17 for a bending stress alternating between 0 and 25 ksi.

11.19 Rework Problem 11.17 for a bending stress alternating between −10 ksi and 10 ksi.

11.20 A machine part is made of a steel with an ultimate strength of 600 MPa, a yield strength of 470 MPa, and an endurance limit of 200 MPa. Determine the factor of safety (allowable maximum mean stress/actual mean stress) for the following stress state: an axial stress alternating between 75 MPa and 10 MPa.

11.21 Rework Problem 11.20 for an alternating axial stress between 100 MPa and 0 MPa.

11.22 Rework Problem 11.20 for an alternating axial stress between 75 MPa and −75 MPa.

11.23 A connecting rod has stress concentrations at two locations shown. The forces F fluctuate between a tension of 16 kips and a compression of 4 kips. The material is cold drawn steel with an ultimate strength of 60 ksi. Assume some uncertainty in the mechanical properties. Determine if the part is safe against fatigue failure based on an infinite life.

Figure P11.23

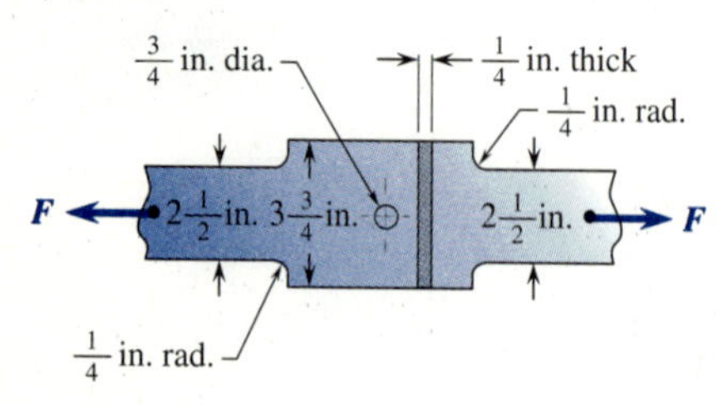

11.24 A cantilevered bar of circular cross section with diameter of 0.5 in. is used as a spring and is subjected to a fluctuating end force of maximum and minimum values given by F_{max} = 40 lb and F_{min} = 20 lb. The surface finish is hot rolled. Neglect stress concentrations. Determine if the spring is safe based on infinite life. Use ASTM-A514 steel.

Figure P11.24

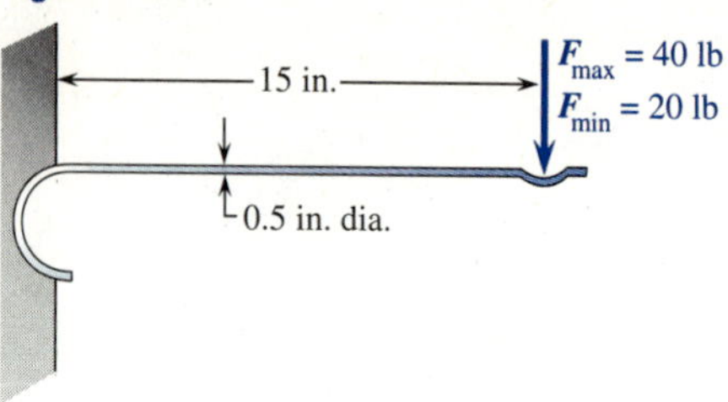

COMPUTER PROBLEM

The following problem is suggested to be solved using a programmable calculator, microcomputer, or mainframe computer. It is suggested that you write a simple computer program in BASIC, FORTRAN, PASCAL, or some other language to solve this problem.

C11.1 Write a program to solve Eqs. (11.7)–(11.11), that is, to determine the number of cycles to failure, or the fatigue strength of a rotating beam part. The input should include the ultimate strength of the material part, the fatigue strength of the rotating beam part of the same material as the actual part, and an option to include the desired number of cycles to failure, or the fatigue failure strength of the rotating beam part. Check your program using the following data: S_e' = 48 ksi, σ_{ut} = 100 ksi, and S_f' = 70.2 ksi.

Properties of Plane Areas

Shape	Area, A	Area Moment of Inertia for Centroidal Axes ($J_{\bar{z}} = I_{\bar{x}} + I_{\bar{y}}$)
Rectangle	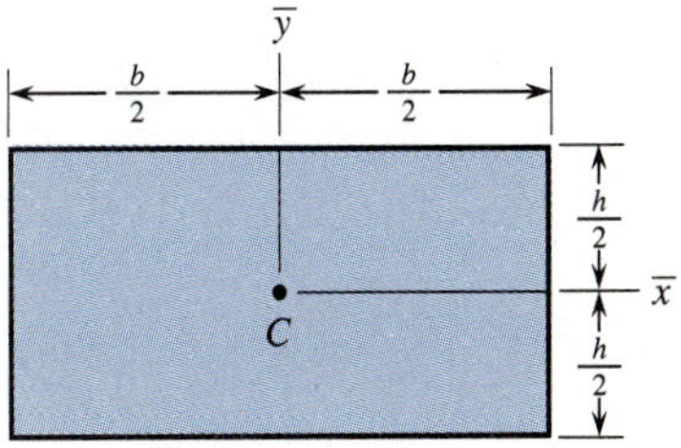 $A = bh$	$I_{\bar{x}} = \frac{1}{12}bh^3$ $I_{\bar{y}} = \frac{1}{12}b^3h$
Right triangle	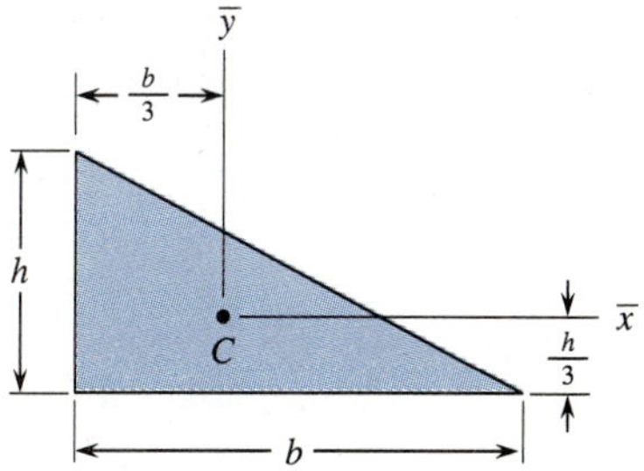 $A = \frac{1}{2}bh$	$I_{\bar{x}} = \frac{1}{36}bh^3$ $I_{\bar{y}} = \frac{1}{36}b^3h$ $I_{\bar{x}\bar{y}} = -\frac{1}{72}b^2h^2$
Scalene triangle	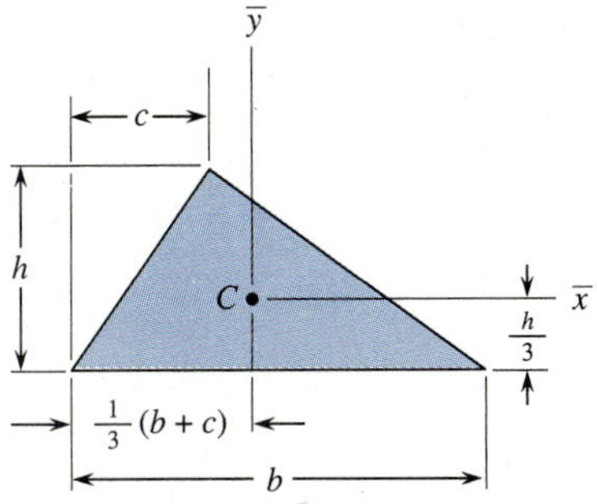 $A = \frac{1}{2}bh$	$I_{\bar{x}} = \frac{1}{36}bh^3$ $I_{\bar{y}} = \frac{1}{36}bh(b^2 + c^2 - bc)$ $I_{\bar{x}\bar{y}} = \frac{1}{72}bh^2(2c - b)$

Shape		Area, A	Area Moment of Inertia for Centroidal Axes ($J_{\bar{z}} = I_{\bar{x}} + I_{\bar{y}}$)
Circle	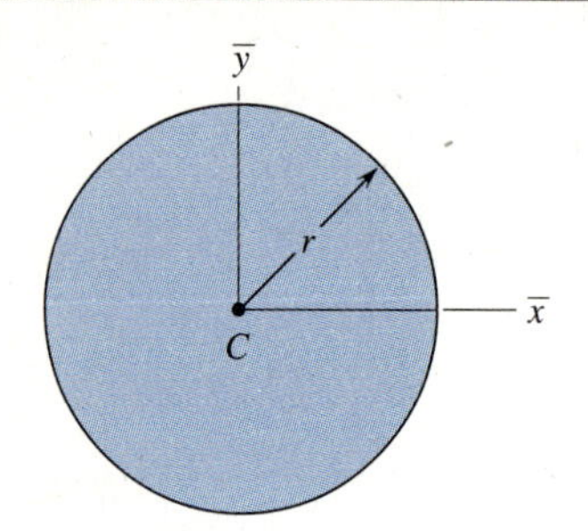	$A = \pi r^2$	$I_{\bar{x}} = I_{\bar{y}} = \frac{1}{4}\pi r^4$ $J_z = J_G = \frac{1}{2}\pi r^4$
Quarter circle	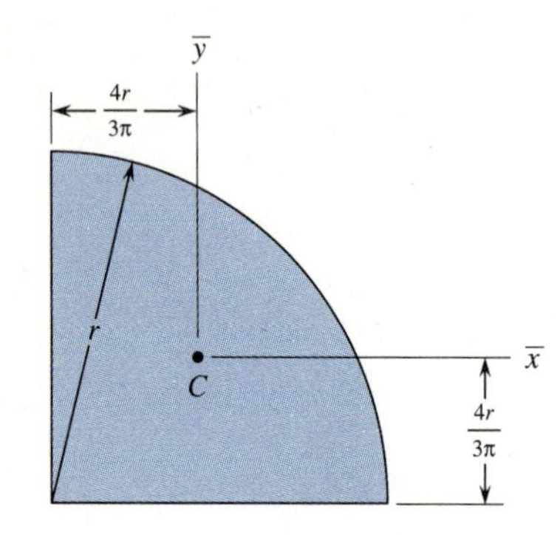	$A = \frac{1}{4}\pi r^2$	$I_{\bar{x}} = I_{\bar{y}} = \left(\frac{9\pi^2 - 64}{144\pi}\right)r^4$ $I_{\bar{x}\bar{y}} = \left(\frac{9\pi - 32}{72\pi}\right)r^4$
Circular sector	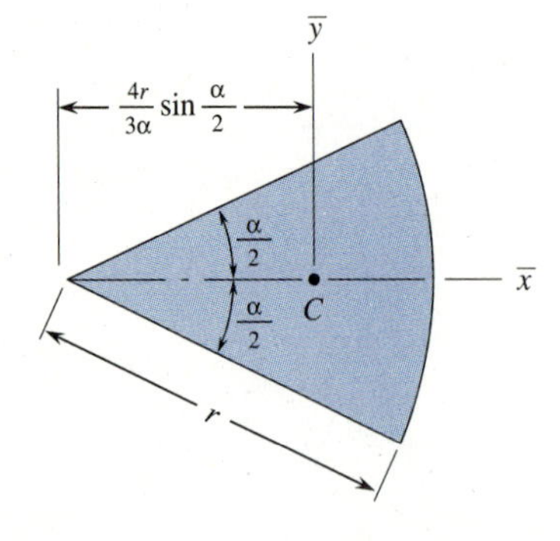	$A = \frac{1}{2}\alpha r^2$	$I_{\bar{x}} = \frac{1}{8}(\alpha - \sin\alpha)r^4$ $I_{\bar{y}} = \left[\frac{\alpha + \sin\alpha}{8} - \frac{4}{9\alpha}(1 - \cos\alpha)\right]r^4$

Shape		Area, A	Area Moment of Inertia for Centroidal Axes $(J_{\bar{z}} = I_{\bar{x}} + I_{\bar{y}})$
Circular arc	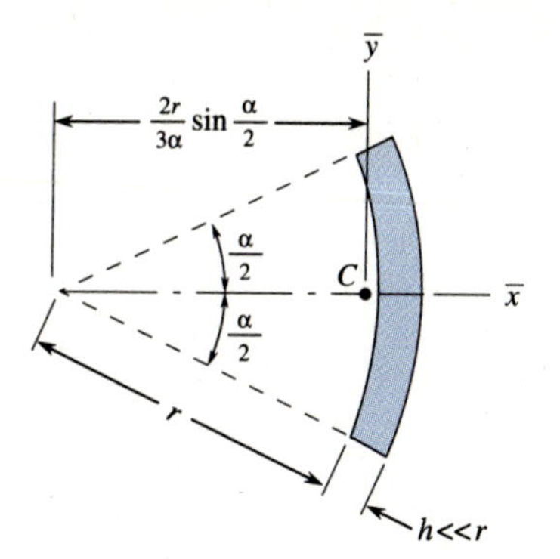	$A = \alpha rh$	$I_{\bar{x}} = \frac{1}{2}(\alpha - \sin\alpha)r^3h$ $I_{\bar{y}} = \left[\frac{\alpha + \sin\alpha}{2} - \frac{2}{\alpha}(1 - \cos\alpha)\right]r^3h$
Ellipse	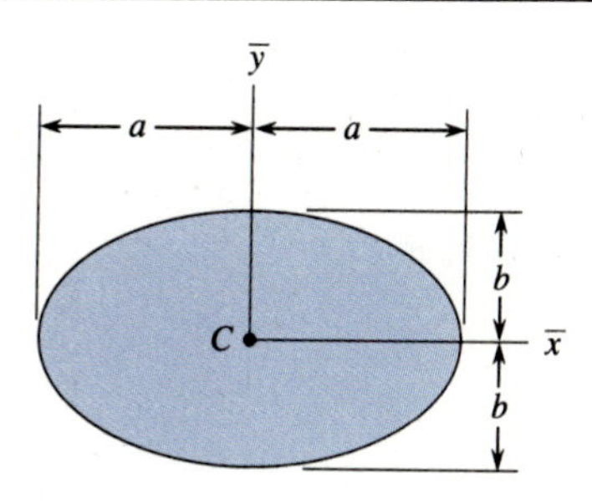	$A = \pi ab$	$I_{\bar{x}} = \frac{\pi}{4}ab^3$ $I_{\bar{y}} = \frac{\pi}{4}a^3b$
Quarter-ellipse	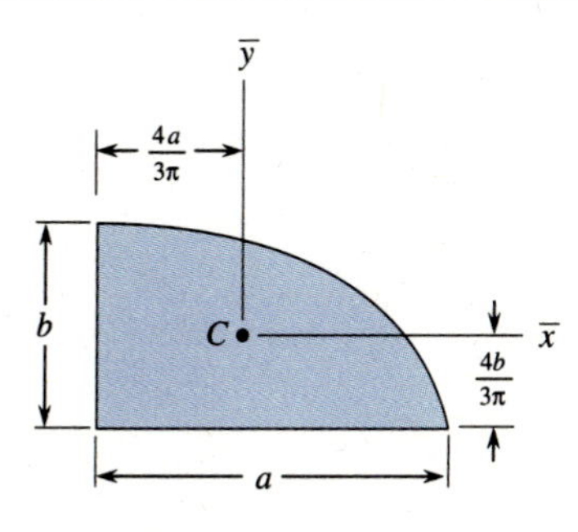	$A = \frac{1}{4}\pi ab$	$I_{\bar{x}} = \left(\frac{9\pi^2 - 64}{144\pi}\right)ab^3$ $I_{\bar{y}} = \left(\frac{9\pi^2 - 64}{144\pi}\right)a^3b$ $I_{\bar{x}\bar{y}} = \left(\frac{9\pi - 32}{72\pi}\right)a^2b^2$

Shape	Area, A	Area Moment of Inertia for Centroidal Axes ($J_{\bar{z}} = I_{\bar{x}} + I_{\bar{y}}$)
Parabolic section 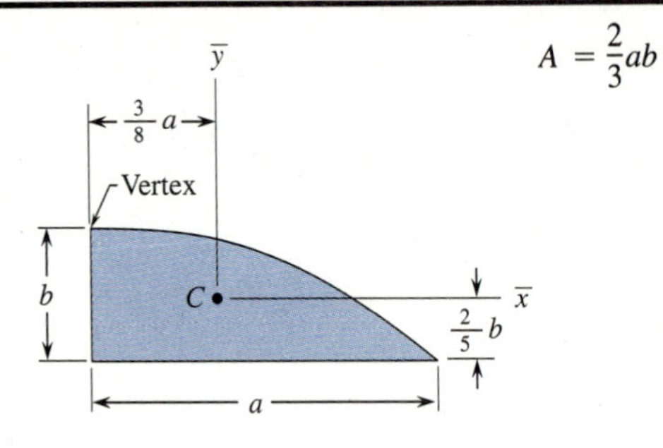	$A = \frac{2}{3}ab$	$I_{\bar{x}} = \frac{8}{175}ab^3$ $I_{\bar{y}} = \frac{19}{480}a^3b$ $I_{\bar{x}\bar{y}} = -\frac{1}{60}a^2b^2$
Parabolic spandrel 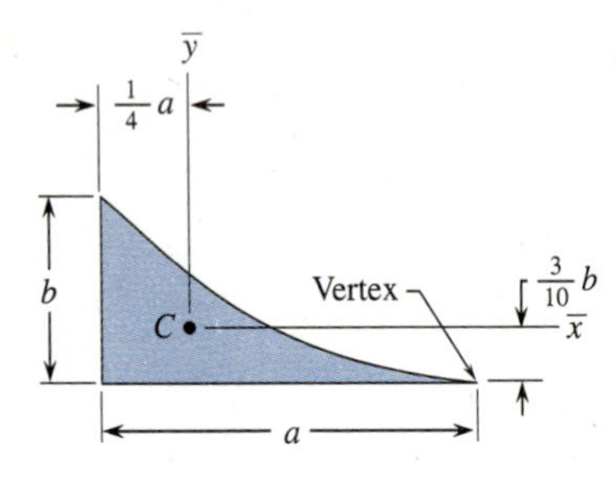	$A = \frac{1}{3}ab$	$I_{\bar{x}} = \frac{19}{1050}ab^3$ $I_{\bar{y}} = \frac{1}{80}a^3b$ $I_{\bar{x}\bar{y}} = -\frac{1}{120}a^2b^2$
Thin ring 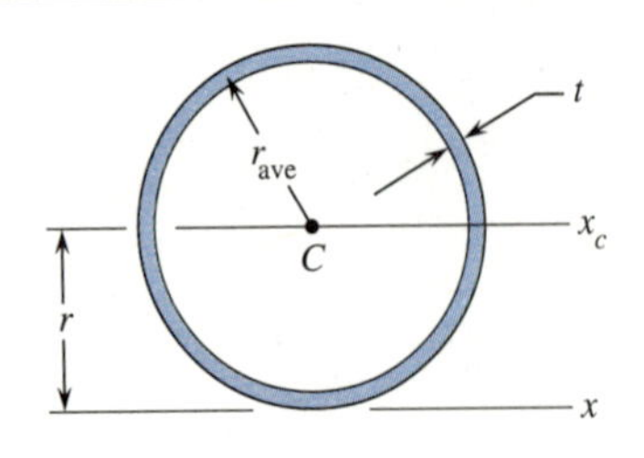	($r \gtrsim 10t$)	$A = 2\pi r_{ave} t$ $J_c = 2\pi r^3_{ave} t$ $I_{x_c} = \pi r^3_{ave} t$

Typical Properties of Selected Engineering Materials

Material	Ultimate Strength σ_u		0.2% Yield Strength σ_Y		Modulus of Elasticity E		Shear Modulus G	Coefficient of Thermal Expansion, α		Density, ρ	
	ksi	MPa	ksi	MPa	(10^6 psi	GPa)	(10^6 psi)	$10^{-6}/°F$	$10^{-6}/°C$	lb/in.^{3}	kg/m^3
Aluminum											
Alloy 1100-H14 (99% Al)	14	110(T)	14	95	10.1	70	3.7	13.1	23.6	0.098	2710
Alloy 2024-T3 (sheet and plate)	70	480(T)	50	340	10.6	73	4.0	12.6	22.7	0.100	2763
Alloy 6061-T6 (extruded)	42	260(T)	37	255	10.0	69	3.7	13.1	23.6	0.098	2710
Alloy 7075-T6 (sheet and plate)	80	550(T)	70	480	10.4	72	3.9	12.9	23.2	0.101	2795
Yellow brass (65% Cu, 35% Zn)											
Cold-rolled	78	540(T)	63	435	15	105	5.6	11.3	20.0	0.306	8470
Annealed	48	330(T)	15	105	15	105	5.6	11.3	20.0	0.306	8470
Phosphor bronze											
Cold-rolled (510)	81	560(T)	75	520	15.9	110	5.9	9.9	17.8	0.320	8860
Spring-tempered (524)	122	840(T)	—	—	16	110	5.9	10.2	18.4	0.317	8780
Cast iron											
Gray, 4.5%C, ASTM A-48	25	170(T)	—	—	10	70	4.1	6.7	12.1	0.260	7200
	95	650(C)									
Malleable, ASTM A-47	50	340(T)	33	230	24	165	9.3	6.7	12.1	0.264	7300
	90	620(C)	—	—							

The values given in this table are average mechanical properties. Further verification may be necessary for final design or analysis. For ductile materials, the compressive strength is normally assumed to equal the tensile strength. *Abbreviations:* C, compressive strength; T, tensile strength. For an explanation of the numbers associated with the aluminums, cast irons, and steels, see Appendix E.

Material	Ultimate Strength σ_u		0.2% Yield Strength σ_Y		Modulus of Elasticity E		Shear Modulus G	Coefficient of Thermal Expansion, α		Density, ρ	
	ksi	MPa	ksi	MPa	(10^6 psi	GPa)	(10^6 psi)	10^{-6}/°F	10^{-6}/°C	lb/in.^{3}	kg/m^3
Copper and its alloys											
CDA 145 copper, hard	48	331(T)	44	303	16	110	16	9.9	17.8	0.323	8940
CDA 172 beryllium copper, hard	175	1210(T)	240	965	19	131	19	9.4	17.0	0.298	8250
CDA 220 bronze, hard	61	421(T)	54	372	17	117	17	10.2	18.4	0.318	8800
CDA 260 brass, hard	76	524(T)	63	434	16	110	16	11.1	20.0	0.308	8530
Magnesium alloy (8.5% Al)	55	380(T)	40	275	4.5	45	2.4	14.5	26.0	0.065	1800
Monel alloy 400 (Ni-Cu)											
Cold-worked	98	675(T)	85	580	26	180	—	7.7	13.9	0.319	8830
Annealed	80	550(T)	32	220	26	180	—	7.7	13.9	0.319	8830
Steel											
Structural (ASTM-A36)	58	400(T)	36	250	29	200	11.5	6.5	11.7	0.284	7860
High-strength low-alloy ASTM-A242	70	480(T)	50	345	29	200	11.5	6.5	11.7	0.284	7860
Quenched and tempered alloy ASTM-A514	120	825(T)	100	690	29	200	11.5	6.5	11.7	0.284	7860
Stainless, (302)											
Cold-rolled	125	860(T)	75	520	28	190	10.6	9.6	17.3	0.286	7920
Annealed	90	620(T)	40	275	28	190	10.6	9.6	17.3	0.286	7920
Titanium alloy (6% Al, 4% V)	130	900(T)	120	825	16.5	114	6.2	5.3	9.5	0.161	4460
Concrete											
Medium strength	4.0	28(C)	—	—	3.5	25	—	5.5	10.0	0.084	2320
High strength	6.0	40(C)	—	—	4.5	30	—	5.5	10.0	0.084	2320
Granite	35	240(C)	—	—	10	69	—	4.0	7.0	0.100	2770
Glass, 98% silica	7	50(C)	—	—	10	69	—	44.0	80.0	0.079	2190
Melamine	6	41(T)	—	—	2.0	13.4	—	17.0	30.0	0.042	1162
Nylon, molded	8	55(T)	—	—	0.3	2	—	45.0	81.0	0.040	1100

The values given in this table are average mechanical properties. Further verification may be necessary for final design or analysis. For ductile materials, the compressive strength is normally assumed to equal the tensile strength. *Abbreviations:* C, compressive strength; T, tensile strength. For an explanation of the numbers associated with the aluminums, cast irons, and steels, see Appendix E.

Material	Ultimate Strength σ_u ksi	Ultimate Strength σ_u MPa	0.2% Yield Strength σ_Y ksi	0.2% Yield Strength σ_Y MPa	Modulus of Elasticity E (10^6 psi	GPa)	Shear Modulus G (10^6 psi)	Coefficient of Thermal Expansion, α $10^{-6}/°F$	Coefficient of Thermal Expansion, α $10^{-6}/°C$	Density, ρ lb/in.^{3}	Density, ρ kg/m^3
Polystyrene	7	48(T)	—	—	0.45	3	—	40.0	72.0	0.038	1050
Rubbers											
Natural	2	14(T)	—	—	—	—	—	90.0	162.0	0.033	910
Neoprene	3.5	24(T)	—	—	—	—	—			0.045	1250
Timber, air dry, parallel to grain											
Douglas fir, construction grade	7.2	50(C)	—	—	1.5	10.5	—	varies	varies	0.019	525
Eastern spruce	5.4	37(C)	—	—	1.3	9	—	1.7–	3–	0.016	440
Southern pine, construction grade	7.3	50(C)	—	—	1.2	8.3	—	3.0	5.4	0.022	610

The values given in this table are average mechanical properties. Further verification may be necessary for final design or analysis. For ductile materials, the compressive strength is normally assumed to equal the tensile strength. *Abbreviations:* C, compressive strength; T, tensile strength. For an explanation of the numbers associated with the aluminums, cast irons, and steels, see Appendix E.

Properties of Structural Steel Shapes, Aluminum, and Timber

WIDE FLANGE SHAPES (W SHAPES)*

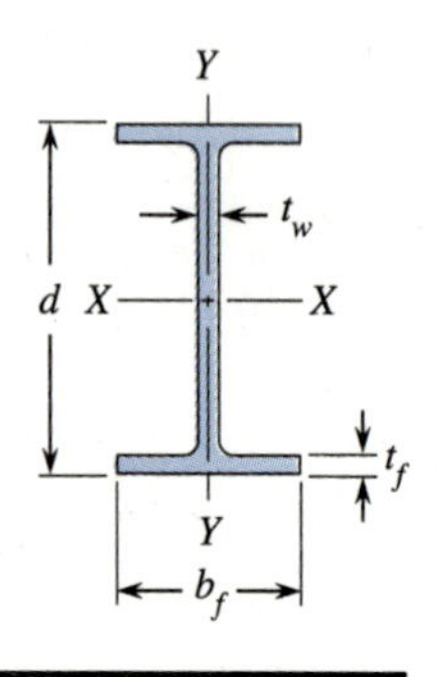

THEORETICAL DIMENSIONS AND PROPERTIES FOR DESIGNING

Section Number	Weight per Foot (lb)	Area of Section A (in.²)	Depth of Section d (in.)	Flange: Width b_f (in.)	Flange: Thickness t_f (in.)	Web Thickness t_w (in.)	Axis X-X: I_x (in.⁴)	Axis X-X: S_x (in.³)	Axis X-X: r_x (in.)	Axis Y-Y: I_y (in.⁴)	Axis Y-Y: S_y (in.³)	Axis Y-Y: r_y (in.)
W36 ×	300	88.3	36.74	16.655	1.680	0.945	20,300	1,110	15.2	1,300	156	3.83
	280	82.4	36.52	16.595	1.570	0.885	18,900	1,030	15.1	1,200	144	3.81
	260	76.5	36.26	16.550	1.440	0.840	17,300	953	15.0	1,090	132	3.78
	245	72.1	36.08	16.510	1.350	0.800	16,100	895	15.0	1,010	123	3.75
	230	67.6	35.90	16.470	1.260	0.760	15,000	837	14.9	940	114	3.73
W36 ×	210	61.8	36.69	12.180	1.360	0.830	13,200	719	14.6	411	67.5	2.58
	194	57.0	36.49	12.115	1.260	0.765	12,100	664	14.6	375	61.9	2.56
	182	53.6	36.33	12.075	1.180	0.725	11,300	623	14.5	347	57.6	2.55
	170	50.0	36.17	12.030	1.100	0.680	10,500	580	14.5	320	53.2	2.53
	160	47.0	36.01	12.000	1.020	0.650	9,750	542	14.4	295	49.1	2.50
	150	44.2	35.85	11.975	0.940	0.625	9,040	504	14.3	270	45.1	2.47
	135	39.7	35.55	11.950	0.790	0.600	7,800	439	14.0	225	37.7	2.38
W33 ×	241	70.9	34.18	15.860	1.400	0.830	14,200	829	14.1	932	118	3.63
	221	65.0	33.93	15.805	1.275	0.775	12,800	757	14.1	840	106	3.59
	201	59.1	33.68	15.745	1.150	0.715	11,500	684	14.0	749	95.2	3.56

*A W section is designated by the letter W followed by the nominal depth in inches and the weight in pounds per foot.

WIDE FLANGE SHAPES (W SHAPES): THEORETICAL DIMENSIONS AND PROPERTIES FOR DESIGNING

				Flange			Axis X-X			Axis Y-Y		
Section Number	Weight per Foot (lb)	Area of Section A (in.^{2})	Depth of Section d (in.)	Width b_f (in.)	Thickness t_f (in.)	Web Thickness t_w (in.)	I_x (in.^{4})	S_x (in.^{3})	r_x (in.)	I_y (in.^{4})	S_y (in.^{3})	r_y (in.)
W33 ×	152	44.7	33.49	11.565	1.055	0.635	8,160	487	13.5	273	47.2	2.47
	141	41.6	33.30	11.535	0.960	0.605	7,450	448	13.4	246	42.7	2.43
	130	38.3	33.09	11.510	0.855	0.580	6,710	406	13.2	218	37.9	2.39
	118	34.7	32.86	11.480	0.740	0.550	5,900	359	13.0	187	32.6	2.32
W30 ×	211	62.0	30.94	15.105	1.315	0.775	10,300	663	12.9	757	100	3.49
	191	56.1	30.68	15.040	1.185	0.710	9,170	598	12.8	673	89.5	3.46
	173	50.8	30.44	14.985	1.065	0.655	8,200	539	12.7	598	79.8	3.43
W30 ×	132	38.9	30.31	10.545	1.000	0.615	5,770	380	12.2	196	37.2	2.25
	124	36.5	30.17	10.515	0.930	0.585	5,360	355	12.1	181	34.4	2.23
	116	34.2	30.01	10.495	0.850	0.565	4,930	329	12.0	164	31.3	2.19
	108	31.7	29.83	10.475	0.760	0.545	4,470	299	11.9	146	27.9	2.15
	99	29.1	29.65	10.450	0.670	0.520	3,990	269	11.7	128	24.5	2.10
W27 ×	178	52.3	27.81	14.085	1.190	0.725	6,990	502	11.6	555	78.8	3.26
	161	47.4	27.59	14.020	1.080	0.660	6,280	455	11.5	497	70.9	3.24
	146	42.9	27.38	13.965	0.975	0.605	5,630	411	11.4	443	63.5	3.21
W27 ×	114	33.5	27.29	10.070	0.930	0.570	4,090	299	11.0	159	31.5	2.18
	102	30.0	27.09	10.015	0.830	0.515	3,620	267	11.0	139	27.8	2.15
	94	27.7	26.92	9.990	0.745	0.490	3,270	243	10.9	124	24.8	2.12
	84	24.8	26.71	9.960	0.640	0.460	2,850	213	10.7	106	21.2	2.07
W24 ×	162	47.7	25.00	12.955	1.220	0.705	5,170	414	10.4	443	68.4	3.05
	146	43.0	24.74	12.900	1.090	0.650	4,580	371	10.3	391	60.5	3.01
	131	38.5	24.48	12.855	0.960	0.605	4,020	329	10.2	340	53.0	2.97
	117	34.4	24.26	12.800	0.850	0.550	3,540	291	10.1	297	46.5	2.94
	104	30.6	24.06	12.750	0.750	0.500	3,100	258	10.1	259	40.7	2.91
W24 ×	94	27.7	24.31	9.065	0.875	0.515	2,700	222	9.87	109	24.0	1.98
	84	24.7	24.10	9.020	0.770	0.470	2,370	196	9.79	94.4	20.9	1.95
	76	22.4	23.92	8.990	0.680	0.440	2,100	176	9.69	82.5	18.4	1.92
	68	20.1	23.73	8.965	0.585	0.415	1,830	154	9.55	70.4	15.7	1.87
W24 ×	62	18.2	23.74	7.040	0.590	0.430	1,550	131	9.23	34.5	9.80	1.38
	55	16.2	23.57	7.005	0.505	0.395	1,350	114	9.11	29.1	8.30	1.34
W21 ×	147	43.2	22.06	12.510	1.150	0.720	3,630	329	9.17	376	60.1	2.95
	132	38.8	21.83	12.440	1.035	0.650	3,220	295	9.12	333	53.5	2.93
	122	35.9	21.68	12.390	0.960	0.600	2,960	273	9.09	305	49.2	2.92
	111	32.7	21.51	12.340	0.875	0.550	2,670	249	9.05	274	44.5	2.90
	101	29.8	21.36	12.290	0.800	0.500	2,420	227	9.02	248	40.3	2.89

WIDE FLANGE SHAPES (W SHAPES): THEORETICAL DIMENSIONS AND PROPERTIES FOR DESIGNING

Section Number	Weight per Foot (lb)	Area of Section A (in.^{2})	Depth of Section d (in.)	Flange		Web Thickness t_w (in.)	Axis X-X			Axis Y-Y		
				Width b_f (in.)	Thickness t_f (in.)		I_x (in.^{4})	S_x (in.^{3})	r_x (in.)	I_y (in.^{4})	S_y (in.^{3})	r_y (in.)
W21 ×	93	27.3	21.62	8.420	0.930	0.580	2,070	192	8.70	92.9	22.1	1.84
	83	24.3	21.43	8.355	0.835	0.515	1,830	171	8.67	81.4	19.5	1.83
	73	21.5	21.24	8.295	0.740	0.455	1,600	151	8.64	70.6	17.0	1.81
	68	20.0	21.13	8.270	0.685	0.430	1,480	140	8.60	64.7	15.7	1.80
	62	18.3	20.99	8.240	0.615	0.400	1,330	127	8.54	57.5	13.9	1.77
W21 ×	57	16.7	21.06	6.555	0.650	0.405	1,170	111	8.36	30.6	9.35	1.35
	50	14.7	20.83	6.530	0.535	0.380	984	94.5	8.18	24.9	7.64	1.30
	44	13.0	20.66	6.500	0.450	0.350	843	81.6	8.06	20.7	6.36	1.26
W18 ×	119	35.1	18.97	11.265	1.060	0.655	2,190	231	7.90	253	44.9	2.69
	106	31.1	18.73	11.200	0.940	0.590	1,910	204	7.84	220	39.4	2.66
	97	28.5	18.59	11.145	0.870	0.535	1,750	188	7.82	201	36.1	2.65
	86	25.3	18.39	11.090	0.770	0.480	1,530	166	7.77	175	31.6	2.63
	76	22.3	18.21	11.035	0.680	0.425	1,330	146	7.73	152	27.6	2.61
W18 ×	71	20.8	18.47	7.635	0.810	0.495	1,170	127	7.50	60.3	15.8	1.70
	65	19.1	18.35	7.590	0.750	0.450	1,070	117	7.49	54.8	14.4	1.69
	60	17.6	18.24	7.555	0.695	0.415	984	108	7.47	50.1	13.3	1.69
	55	16.2	18.11	7.530	0.630	0.390	890	98.3	7.41	44.9	11.9	1.67
	50	14.7	17.99	7.495	0.570	0.355	800	88.9	7.38	40.1	10.7	1.65
W18 ×	46	13.5	18.06	6.060	0.605	0.360	712	78.8	7.25	22.5	7.43	1.29
	40	11.8	17.90	6.015	0.525	0.315	612	68.4	7.21	19.1	6.35	1.27
	35	10.3	17.70	6.000	0.425	0.300	510	57.6	7.04	15.3	5.12	1.22
W16 ×	100	29.4	16.97	10.425	0.985	0.585	1,490	175	7.10	186	35.7	2.52
	89	26.2	16.75	10.365	0.875	0.525	1,300	155	7.05	163	31.4	2.49
	77	22.6	16.52	10.295	0.760	0.455	1,110	134	7.00	138	26.9	2.47
	67	19.7	16.33	10.235	0.665	0.395	954	117	6.96	119	23.2	2.46
W16 ×	57	16.8	16.43	7.120	0.715	0.430	758	92.2	6.72	43.1	12.1	1.60
	50	14.7	16.26	7.070	0.630	0.380	659	81.0	6.68	37.2	10.5	1.59
	45	13.3	16.13	7.035	0.565	0.345	586	72.7	6.65	32.8	9.34	1.57
	40	11.8	16.01	6.995	0.505	0.305	518	64.7	6.63	28.9	8.25	1.57
	36	10.6	15.86	6.985	0.430	0.295	448	56.5	6.51	24.5	7.00	1.52
W16 ×	31	9.12	15.88	5.525	0.440	0.275	375	47.2	6.41	12.4	4.49	1.17
	26	7.68	15.69	5.500	0.345	0.250	301	38.4	6.26	9.59	3.49	1.12
W14 ×	730	215	22.42	17.890	4.910	3.070	14,300	1,280	8.17	4,720	527	4.69
	665	196	21.64	17.650	4.520	2.830	12,400	1,150	7.98	4,170	472	4.62
	605	178	20.92	17.415	4.160	2.595	10,800	1,040	7.80	3,680	423	4.55
	550	162	20.24	17.200	3.820	2.380	9,430	931	7.63	3,250	378	4.49
	500	147	19.60	17.010	3.500	2.190	8,210	838	7.48	2,880	339	4.43
	455	134	19.02	16.835	3.210	2.015	7,190	756	7.33	2,560	304	4.38

WIDE FLANGE SHAPES (W SHAPES): THEORETICAL DIMENSIONS AND PROPERTIES FOR DESIGNING

Section Number	Weight per Foot (lb)	Area of Section A (in.^{2})	Depth of Section d (in.)	Flange		Web Thickness t_w (in.)	Axis X-X			Axis Y-Y		
				Width b_f (in.)	Thickness t_f (in.)		I_x (in.^{4})	S_x (in.^{3})	r_x (in.)	I_y (in.^{4})	S_y (in.^{3})	r_y (in.)
W14 ×	426	125	18.67	16.695	3.035	1.875	6,600	707	7.26	2,360	283	4.34
	398	117	18.29	16.590	2.845	1.770	6,000	656	7.16	2,170	262	4.31
	370	109	17.92	16.475	2.660	1.655	5,440	607	7.07	1,990	241	4.27
	342	101	17.54	16.360	2.470	1.540	4,900	559	6.98	1,810	221	4.24
	311	91.4	17.12	16.230	2.260	1.410	4,330	506	6.88	1,610	199	4.20
	283	83.3	16.74	16.110	2.070	1.290	3,840	459	6.79	1,440	179	4.17
	257	75.6	16.38	15.995	1.890	1.175	3,400	415	6.71	1,290	161	4.13
	233	68.5	16.04	15.890	1.720	1.070	3,010	375	6.63	1,150	145	4.10
	211	62.0	15.72	15.800	1.560	0.980	2,660	338	6.55	1,030	130	4.07
	193	56.8	15.48	15.710	1.440	0.890	2,400	310	6.50	931	119	4.05
	176	51.8	15.22	15.650	1.310	0.830	2,140	281	6.43	838	107	4.02
	159	46.7	14.98	15.565	1.190	0.745	1,900	254	6.38	748	96.2	4.00
	145	42.7	14.78	15.500	1.090	0.680	1,710	232	6.33	677	87.3	3.98
W14 ×	132	38.8	14.66	14.725	1.030	0.645	1,530	209	6.28	548	74.5	3.76
	120	35.3	14.48	14.670	0.940	0.590	1,380	190	6.24	495	67.5	3.74
	109	32.0	14.32	14.605	0.860	0.525	1,240	173	6.22	447	61.2	3.73
	99	29.1	14.16	14.565	0.780	0.485	1,110	157	6.17	402	55.2	3.71
	90	26.5	14.02	14.520	0.710	0.440	999	143	6.14	362	49.9	3.70
W14 ×	82	24.1	14.31	10.130	0.855	0.510	882	123	6.05	148	29.3	2.48
	74	21.8	14.17	10.070	0.785	0.450	796	112	6.04	134	26.6	2.48
	68	20.0	14.04	10.035	0.720	0.415	723	103	6.01	121	24.2	2.46
	61	17.9	13.89	9.995	0.645	0.375	640	92.2	5.98	107	21.5	2.45
W14 ×	53	15.6	13.92	8.060	0.660	0.370	541	77.8	5.89	57.7	14.3	1.92
	48	14.1	13.79	8.030	0.595	0.340	485	70.3	5.85	51.4	12.8	1.91
	43	12.6	13.66	7.995	0.530	0.305	428	62.7	5.82	45.2	11.3	1.89
W14 ×	38	11.2	14.10	6.770	0.515	0.310	385	54.6	5.88	26.7	7.88	1.55
	34	10.0	13.98	6.745	0.455	0.285	340	48.6	5.83	23.3	6.91	1.53
	30	8.85	13.84	6.730	0.385	0.270	291	42.0	5.73	19.6	5.82	1.49
W14 ×	26	7.69	13.91	5.025	0.420	0.255	245	35.3	5.65	8.91	3.54	1.08
	22	6.49	13.74	5.000	0.335	0.230	199	29.0	5.54	7.00	2.80	1.04
W12 ×	190	55.8	14.38	12.670	1.735	1.060	1,890	263	5.82	589	93.0	3.25
	170	50.0	14.03	12.570	1.560	0.960	1,650	235	5.74	517	82.3	3.22
	152	44.7	13.71	12.480	1.400	0.870	1,430	209	5.66	454	72.8	3.19
	136	39.9	13.41	12.400	1.250	0.790	1,240	186	5.58	398	64.2	3.16
	120	35.3	13.12	12.320	1.105	0.710	1,070	163	5.51	345	56.0	3.13
	106	31.2	12.89	12.220	0.990	0.610	933	145	5.47	301	49.3	3.11
	96	28.2	12.71	12.160	0.900	0.550	833	131	5.44	270	44.4	3.09
	87	25.6	12.53	12.125	0.810	0.515	740	118	5.38	241	39.7	3.07
	79	23.2	12.38	12.080	0.735	0.470	662	107	5.34	216	35.8	3.05
	72	21.1	12.25	12.040	0.670	0.430	597	97.4	5.31	195	32.4	3.04
	65	19.1	12.12	12.000	0.605	0.390	533	87.9	5.28	174	29.1	3.02

WIDE FLANGE SHAPES (W SHAPES): THEORETICAL DIMENSIONS AND PROPERTIES FOR DESIGNING

Section Number	Weight per Foot (lb)	Area of Section A (in.^{2})	Depth of Section d (in.)	Flange: Width b_f (in.)	Flange: Thickness t_f (in.)	Web Thickness t_w (in.)	Axis X-X: I_x (in.^{4})	Axis X-X: S_x (in.^{3})	Axis X-X: r_x (in.)	Axis Y-Y: I_y (in.^{4})	Axis Y-Y: S_y (in.^{3})	Axis Y-Y: r_y (in.)
W12 ×	58	17.0	12.19	10.010	0.640	0.360	475	78.0	5.28	107	21.4	2.51
	53	15.6	12.06	9.995	0.575	0.345	425	70.6	5.23	95.8	19.2	2.48
W12 ×	50	14.7	12.19	8.080	0.640	0.370	394	64.7	5.18	56.3	13.9	1.96
	45	13.2	12.06	8.045	0.575	0.335	350	58.1	5.15	50.0	12.4	1.94
	40	11.8	11.94	8.005	0.515	0.295	310	51.9	5.13	44.1	11.0	1.93
W12 ×	35	10.3	12.50	6.560	0.520	0.300	285	45.6	5.25	24.5	7.47	1.54
	30	8.79	12.34	6.520	0.440	0.260	238	38.6	5.21	20.3	6.24	1.52
	26	7.65	12.22	6.490	0.380	0.230	204	33.4	5.17	17.3	5.34	1.51
W12 ×	22	6.48	12.31	4.030	0.425	0.260	156	25.4	4.91	4.66	2.31	0.848
	19	5.57	12.16	4.005	0.350	0.235	130	21.3	4.82	3.76	1.88	0.822
	16	4.71	11.99	3.990	0.265	0.220	103	17.1	4.67	2.82	1.41	0.773
	14	4.16	11.91	3.970	0.225	0.200	88.6	14.9	4.62	2.36	1.19	0.753
W10 ×	112	32.9	11.36	10.415	1.250	0.755	716	126	4.66	236	45.3	2.68
	100	29.4	11.10	10.340	1.120	0.680	623	112	4.60	207	40.0	2.65
	88	25.9	10.84	10.265	0.990	0.605	534	98.5	4.54	179	34.8	2.63
	77	22.6	10.60	10.190	0.870	0.530	455	85.9	4.49	154	30.1	2.60
	68	20.0	10.40	10.130	0.770	0.470	394	75.7	4.44	134	26.4	2.59
	60	17.6	10.22	10.080	0.680	0.420	341	66.7	4.39	116	23.0	2.57
	54	15.8	10.09	10.030	0.615	0.370	303	60.0	4.37	103	20.6	2.56
	49	14.4	9.98	10.000	0.560	0.340	272	54.6	4.35	93.4	18.7	2.54
W10 ×	45	13.3	10.10	8.020	0.620	0.350	248	49.1	4.33	53.4	13.3	2.01
	39	11.5	9.92	7.985	0.530	0.315	209	42.1	4.27	45.0	11.3	1.98
	33	9.71	9.73	7.960	0.435	0.290	170	35.0	4.19	36.6	9.20	1.94
W10 ×	30	8.84	10.47	5.810	0.510	0.300	170	32.4	4.38	16.7	5.75	1.37
	26	7.61	10.33	5.770	0.440	0.260	144	27.9	4.35	14.1	4.89	1.36
	22	6.49	10.17	5.750	0.360	0.240	118	23.2	4.27	11.4	3.97	1.33
W10 ×	19	5.62	10.24	4.020	0.395	0.250	96.3	18.8	4.14	4.29	2.14	0.874
	17	4.99	10.11	4.010	0.330	0.240	81.9	16.2	4.05	3.56	1.78	0.845
	15	4.41	9.99	4.000	0.270	0.230	68.9	13.8	3.95	2.89	1.45	0.810
	12	3.54	9.87	3.960	0.210	0.190	53.8	10.9	3.90	2.18	1.10	0.785
W8 ×	67	19.7	9.00	8.280	0.935	0.570	272	60.4	3.72	88.6	21.4	2.12
	58	17.1	8.75	8.220	0.810	0.510	228	52.0	3.65	75.1	18.3	2.10
	48	14.1	8.50	8.110	0.685	0.400	184	43.3	3.61	60.9	15.0	2.08
	40	11.7	8.25	8.070	0.560	0.360	146	35.5	3.53	49.1	12.2	2.04
	35	10.3	8.12	8.020	0.495	0.310	127	31.2	3.51	42.6	10.6	2.03
	31	9.13	8.00	7.995	0.435	0.285	110	27.5	3.47	37.1	9.27	2.02

WIDE FLANGE SHAPES (W SHAPES): THEORETICAL DIMENSIONS AND PROPERTIES FOR DESIGNING

Section Number	Weight per Foot (lb)	Area of Section A (in.^{2})	Depth of Section d (in.)	Flange: Width b_f (in.)	Flange: Thickness t_f (in.)	Web Thickness t_w (in.)	Axis X-X: I_x (in.^{4})	Axis X-X: S_x (in.^{3})	Axis X-X: r_x (in.)	Axis Y-Y: I_y (in.^{4})	Axis Y-Y: S_y (in.^{3})	Axis Y-Y: r_y (in.)
W8 ×	28	8.25	8.06	6.535	0.465	0.285	98.0	24.3	3.45	21.7	6.63	1.62
	24	7.08	7.93	6.495	0.400	0.245	82.8	20.9	3.42	18.3	5.63	1.61
W8 ×	21	6.16	8.28	5.270	0.400	0.250	75.3	18.2	3.49	9.77	3.71	1.26
	18	5.26	8.14	5.250	0.330	0.230	61.9	15.2	3.43	7.97	3.04	1.23
W8 ×	15	4.44	8.11	4.015	0.315	0.245	48.0	11.8	3.29	3.41	1.70	0.876
	13	3.84	7.99	4.000	0.255	0.230	39.6	9.91	3.21	2.73	1.37	0.843
	10	2.96	7.89	3.940	0.205	0.170	30.8	7.81	3.22	2.09	1.06	0.841
W6 ×	25	7.34	6.38	6.080	0.455	0.320	53.4	16.7	2.70	17.1	5.61	1.52
	20	5.87	6.20	6.020	0.365	0.260	41.4	13.4	2.66	13.3	4.41	1.50
	15	4.43	5.99	5.990	0.260	0.230	29.1	9.72	2.56	9.32	3.11	1.45
W6 ×	16	4.74	6.28	4.030	0.405	0.260	32.1	10.2	2.60	4.43	2.20	0.967
	12	3.55	6.03	4.000	0.280	0.230	22.1	7.31	2.49	2.99	1.50	0.918
	9	2.68	5.90	3.940	0.215	0.170	16.4	5.56	2.47	2.20	1.11	0.905
W5 ×	19	5.54	5.15	5.030	0.430	0.270	26.2	10.2	2.17	9.13	3.63	1.28
	16	4.68	5.01	5.000	0.360	0.240	21.3	8.51	2.13	7.51	3.00	1.27
W4 ×	13	3.83	4.16	4.060	0.345	0.280	11.3	5.46	1.72	3.86	1.90	1.00

MISCELLANEOUS SHAPE (M SHAPES)

THEORETICAL DIMENSIONS AND PROPERTIES FOR DESIGNING

Section Number	Weight per Foot (lb)	Area of Section A (in.^{2})	Depth of Section d (in.)	Flange Width b_f (in.)	Flange Thickness t_f (in.)	Web Thickness t_w (in.)	Axis X-X I_x (in.^{4})	Axis X-X S_x (in.^{3})	Axis X-X r_x (in.)	Axis Y-Y I_y (in.^{4})	Axis Y-Y S_y (in.^{3})	Axis Y-Y r_y (in.)
M14 ×	18	5.10	14.00	4.000	0.270	0.215	148	21.1	5.38	2.64	1.32	0.719
M12 ×	11.8	3.47	12.00	3.065	0.225	0.177	71.9	12.0	4.55	0.980	0.639	0.532
M12 ×	10.8	3.18	11.97	3.065	0.210	0.160	65.0	10.9	4.55	0.905	0.591	0.537
M12 ×	10	2.94	11.97	3.250	0.180	0.149	61.6	10.3	4.57	0.994	0.612	0.576
M10 ×	9	2.65	10.00	2.690	0.206	0.157	38.8	7.76	3.83	0.609	0.453	0.480
M10 ×	8	2.35	9.95	2.690	0.182	0.141	34.5	6.94	3.82	0.537	0.399	0.427
M10 ×	7.5	2.21	9.99	2.690	0.173	0.130	32.8	6.57	3.85	0.498	0.370	0.474
M 8 ×	6.5	1.92	8.00	2.281	0.189	0.135	18.5	4.62	3.10	0.343	0.301	0.423
M 6 ×	4.4	1.29	6.00	1.844	0.171	0.114	7.20	2.40	2.36	0.165	0.179	0.358
M 5 ×	18.9	5.55	5.00	5.003	0.416	0.316	24.1	9.63	2.08	7.86	3.14	1.19

AMERICAN STANDARD SHAPES (S SHAPES)

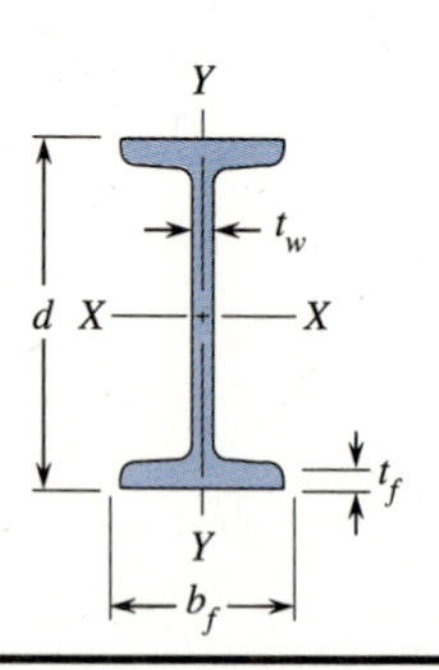

THEORETICAL DIMENSIONS AND PROPERTIES FOR DESIGNING

Section Number	Weight per Foot (lb)	Area of Section A (in.^{2})	Depth of Section d (in.)	Flange Width b_f (in.)	Flange Average Thickness t_f (in.)	Web Thickness t_w (in.)	Axis X-X I_x (in.^{4})	Axis X-X S_x (in.^{3})	Axis X-X r_x (in.)	Axis Y-Y I_y (in.^{4})	Axis Y-Y S_y (in.^{3})	Axis Y-Y r_y (in.)
S24 ×	121.0	35.6	24.50	8.050	1.090	0.800	3,160	258	9.43	83.3	20.7	1.53
	106.0	31.2	24.50	7.870	1.090	0.620	2,940	240	9.71	77.1	19.6	1.57

AMERICAN STANDARD SHAPES (S SHAPES):
THEORETICAL DIMENSIONS AND PROPERTIES FOR DESIGNING

Section Number	Weight per Foot (lb)	Area of Section A (in.^{2})	Depth of Section d (in.)	Flange: Width b_f (in.)	Flange: Average Thickness t_f (in.)	Web Thickness t_w (in.)	Axis X-X: I_x (in.^{4})	Axis X-X: S_x (in.^{3})	Axis X-X: r_x (in.)	Axis Y-Y: I_y (in.^{4})	Axis Y-Y: S_y (in.^{3})	Axis Y-Y: r_y (in.)
S24 ×	100.0	29.3	24.00	7.245	0.870	0.745	2,390	199	9.02	47.7	13.2	1.27
	90.0	26.5	24.00	7.125	0.870	0.625	2,250	187	9.21	44.9	12.6	1.30
	80.0	23.5	24.00	7.000	0.870	0.500	2,100	175	9.47	42.2	12.1	1.34
S20 ×	96.0	28.2	20.30	7.200	0.920	0.800	1,670	165	7.71	50.2	13.9	1.33
	86.0	25.3	20.30	7.060	0.920	0.660	1,580	155	7.89	46.8	13.3	1.36
S20 ×	75.0	22.0	20.00	6.385	0.795	0.635	1,280	128	7.62	29.8	9.32	1.16
	66.0	19.4	20.00	6.255	0.795	0.505	1,190	119	7.83	27.7	8.85	1.19
S18 ×	70.0	20.6	18.00	6.251	0.691	0.711	926	103	6.71	24.1	7.72	1.08
	54.7	16.1	18.00	6.001	0.691	0.461	804	89.4	7.07	20.8	6.94	1.14
S15 ×	50.0	14.7	15.00	5.640	0.622	0.550	486	64.8	5.75	15.7	5.57	1.03
	42.9	12.6	15.00	5.501	0.622	0.411	447	59.6	5.95	14.4	5.23	1.07
S12 ×	50.0	14.7	12.00	5.477	0.659	0.687	305	50.8	4.55	15.7	5.74	1.03
	40.8	12.0	12.00	5.252	0.659	0.462	272	45.4	4.77	13.6	5.16	1.06
S12 ×	35.0	10.3	12.00	5.078	0.544	0.428	229	38.2	4.72	9.87	3.89	0.980
	31.8	9.35	12.00	5.000	0.544	0.350	218	36.4	4.83	9.36	3.74	1.00
S10 ×	35.0	10.3	10.00	4.944	0.491	0.594	147	29.4	3.78	8.36	3.38	0.901
	25.4	7.46	10.00	4.661	0.491	0.311	124	24.7	4.07	6.79	2.91	0.954
S 8 ×	23.0	6.77	8.00	4.171	0.425	0.441	64.9	16.2	3.10	4.31	2.07	0.798
	18.4	5.41	8.00	4.001	0.425	0.271	57.6	14.4	3.26	3.73	1.86	0.831
S 7 ×	20.0	5.88	7.00	3.860	0.392	0.450	42.4	12.1	2.69	3.17	1.64	0.734
	15.3	4.50	7.00	3.662	0.392	0.252	36.7	10.5	2.86	2.64	1.44	0.766
S 6 ×	17.25	5.07	6.00	3.565	0.359	0.465	26.3	8.77	2.28	2.31	1.30	0.675
	12.5	3.67	6.00	3.332	0.359	0.232	22.1	7.37	2.45	1.82	1.09	0.705
S 5 ×	14.75	4.34	5.00	3.284	0.326	0.494	15.2	6.09	1.87	1.67	1.01	0.620
×	10	2.94	5.00	3.004	0.326	0.214	12.3	4.92	2.05	1.22	0.809	0.643
S 4 ×	9.5	2.79	4.00	2.796	0.293	0.326	6.79	3.39	1.56	0.903	0.646	0.569
×	7.7	2.26	4.00	2.663	0.293	0.193	6.08	3.04	1.64	0.764	0.574	0.581
S 3 ×	7.5	2.21	3.00	2.509	0.260	0.349	2.93	1.95	1.15	0.586	0.468	0.516
×	5.7	1.67	3.00	2.330	0.260	0.170	2.52	1.68	1.23	0.455	0.390	0.522

AMERICAN STANDARD CHANNELS

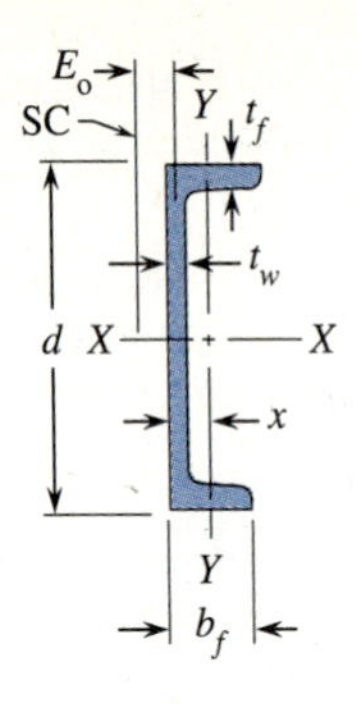

THEORETICAL DIMENSIONS AND PROPERTIES FOR DESIGNING

Section Number	Weight per Foot (lb)	Area of Section A (in.^{2})	Depth of Section d (in.)	Flange Width b_f (in.)	Flange Average Thickness t_f (in.)	Web Thickness t_w (in.)	Axis X-X I_x (in.^{4})	Axis X-X S_x (in.^{3})	Axis X-X r_x (in.)	Axis Y-Y I_y (in.^{4})	Axis Y-Y S_y (in.^{3})	Axis Y-Y r_y (in.)	Axis Y-Y x (in.)	Shear Center Location E_o (in.)
C15 ×	50.0	14.7	15.00	3.716	0.650	0.716	404	53.8	5.24	11.0	3.78	0.867	0.799	0.941
	40.0	11.8	15.00	3.520	0.650	0.520	349	46.5	5.44	9.23	3.36	0.886	0.778	1.03
	33.9	9.96	15.00	3.400	0.650	0.400	315	42.0	5.62	8.13	3.11	0.904	0.787	1.10
C12 ×	30.0	8.82	12.00	3.170	0.501	0.510	162	27.0	4.29	5.14	2.06	0.763	0.674	0.873
	25.0	7.35	12.00	3.047	0.501	0.387	144	24.1	4.43	4.47	1.88	0.780	0.674	0.940
	20.7	6.09	12.00	2.942	0.501	0.282	129	21.5	4.61	3.88	1.73	0.799	0.698	1.01
C10 ×	30.0	8.82	10.00	3.033	0.436	0.673	103	20.7	3.42	3.94	1.65	0.669	0.649	0.705
	25.0	7.35	10.00	2.886	0.436	0.526	91.2	18.2	3.52	3.36	1.48	0.676	0.617	0.757
	20.0	5.88	10.00	2.739	0.436	0.379	78.9	15.8	3.66	2.81	1.32	0.691	0.606	0.826
	15.3	4.49	10.00	2.600	0.436	0.240	67.4	13.5	3.87	2.28	1.16	0.713	0.634	0.916
C9 ×	20.0	5.88	9.00	2.648	0.413	0.448	60.9	13.5	3.22	2.42	1.17	0.642	0.583	0.739
	15.0	4.41	9.00	2.485	0.413	0.285	51.0	11.3	3.40	1.93	1.01	0.661	0.586	0.824
	13.4	3.94	9.00	2.433	0.413	0.233	47.9	10.6	3.48	1.76	0.962	0.668	0.601	0.859
C8 ×	18.75	5.51	8.00	2.527	0.390	0.487	44.0	11.0	2.82	1.98	1.01	0.599	0.565	0.674
	13.75	4.04	8.00	2.343	0.390	0.303	36.1	9.03	2.99	1.53	0.853	0.615	0.553	0.756
	11.5	3.38	8.00	2.260	0.390	0.220	32.6	8.14	3.11	1.32	0.781	0.625	0.571	0.807
C7 ×	14.75	4.33	7.00	2.299	0.366	0.419	27.2	7.78	2.51	1.38	0.779	0.564	0.532	0.651
	12.25	3.60	7.00	2.194	0.366	0.314	24.2	6.93	2.60	1.17	0.702	0.571	0.525	0.695
	9.8	2.87	7.00	2.090	0.366	0.210	21.3	6.08	2.72	0.968	0.625	0.581	0.541	0.752
C6 ×	13	3.83	6.00	2.157	0.343	0.437	17.4	5.80	2.13	1.05	0.642	0.525	0.514	0.599
	10.5	3.09	6.00	2.034	0.343	0.314	15.2	5.06	2.22	0.866	0.564	0.529	0.499	0.643
	8.2	2.40	6.00	1.920	0.343	0.200	13.1	4.38	2.34	0.693	0.492	0.537	0.511	0.699
C5 ×	9	2.64	5.00	1.885	0.320	0.325	8.90	3.56	1.83	0.632	0.450	0.489	0.478	0.590
	6.7	1.97	5.00	1.750	0.320	0.190	7.49	3.00	1.95	0.479	0.378	0.493	0.484	0.647

AMERICAN STANDARD CHANNELS: THEORETICAL DIMENSIONS AND PROPERTIES FOR DESIGNING

Section Number	Weight per Foot (lb)	Area of Section A (in.^{2})	Depth of Section d (in.)	Flange Width b_f (in.)	Flange Average Thickness t_f (in.)	Web Thickness t_w (in.)	Axis X-X I_x (in.^{4})	Axis X-X S_x (in.^{3})	Axis X-X r_x (in.)	Axis Y-Y I_y (in.^{4})	Axis Y-Y S_y (in.^{3})	Axis Y-Y r_y (in.)	Axis Y-Y x (in.)	Shear Center Location E_o (in.)
C4 ×	7.25	2.13	4.00	1.721	0.296	0.321	4.59	2.29	1.47	0.433	0.343	0.450	0.459	0.547
	5.4	1.59	4.00	1.584	0.296	0.184	3.85	1.93	1.56	0.319	0.283	0.449	0.457	0.594
C3 ×	6	1.76	3.00	1.596	0.273	0.356	2.07	1.38	1.08	0.305	0.268	0.416	0.455	0.500
	5	1.47	3.00	1.498	0.273	0.258	1.85	1.24	1.12	0.247	0.233	0.410	0.438	0.521
	4.1	1.21	3.00	1.410	0.273	0.170	1.66	1.10	1.17	0.197	0.202	0.404	0.436	0.546

MISCELLANEOUS CHANNELS

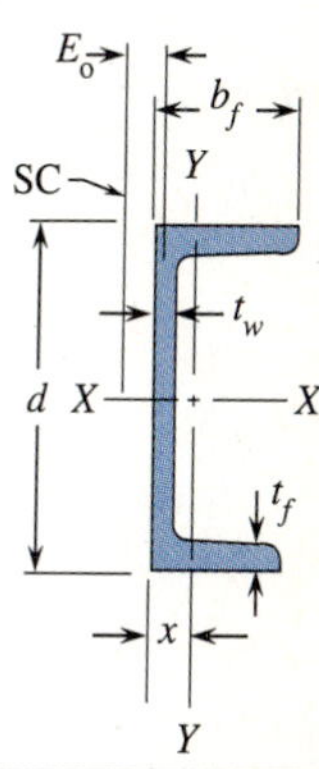

THEORETICAL DIMENSIONS AND PROPERTIES FOR DESIGNING

Section Number	Weight per Foot (lb)	Area of Section A (in.^{2})	Depth of Section d (in.)	Flange Width b_f (in.)	Flange Average Thickness t_f (in.)	Web Thickness t_w (in.)	Axis X-X I_x (in.^{4})	Axis X-X S_x (in.^{3})	Axis X-X r_x (in.)	Axis Y-Y I_y (in.^{4})	Axis Y-Y S_y (in.^{3})	Axis Y-Y r_y (in.)	Axis Y-Y x (in.)	Shear Center Location E_o (in.)
MC18 ×	58.0	17.1	18.00	4.200	0.625	0.700	676	75.1	6.29	17.8	5.32	1.02	0.862	1.04
	51.9	15.3	18.00	4.100	0.625	0.600	627	69.7	6.41	16.4	5.07	1.04	0.858	1.10
	45.8	13.5	18.00	4.000	0.625	0.500	578	64.3	6.56	15.1	4.82	1.06	0.866	1.16
	42.7	12.6	18.00	3.950	0.625	0.450	554	61.6	6.64	14.4	4.69	1.07	0.877	1.19
MC13 ×	50.0	14.7	13.00	4.412	0.610	0.787	314	48.4	4.62	16.5	4.79	1.06	0.974	1.21
	40.0	11.8	13.00	4.185	0.610	0.560	273	42.0	4.82	13.7	4.26	1.08	0.964	1.31
	35.0	10.3	13.00	4.072	0.610	0.447	252	38.8	4.95	12.3	3.99	1.10	0.980	1.38
	31.8	9.35	13.00	4.000	0.610	0.375	239	36.8	5.06	11.4	3.81	1.11	1.00	1.43

MISCELLANEOUS CHANNELS: THEORETICAL DIMENSIONS AND PROPERTIES FOR DESIGNING

				Flange			Axis X-X			Axis Y-Y				
Section Number	Weight per Foot (lb)	Area of Section A (in.^{2})	Depth of Section d (in.)	Width b_f (in.)	Average Thickness t_f (in.)	Web Thickness t_w (in.)	I_x (in.^{4})	S_x (in.^{3})	r_x (in.)	I_y (in.^{4})	S_y (in.^{3})	r_y (in.)	x (in.)	Shear Center Location E_0 (in.)
MC12 ×	50.0	14.7	12.00	4.135	0.700	0.835	269	44.9	4.28	17.4	5.65	1.09	1.05	1.16
	45.0	13.2	12.00	4.012	0.700	0.712	252	42.0	4.36	15.8	5.33	1.09	1.04	1.20
	40.0	11.8	12.00	3.890	0.700	0.590	234	39.0	4.46	14.3	5.00	1.10	1.04	1.25
	35.0	10.3	12.00	3.767	0.700	0.467	216	36.1	4.59	12.7	4.67	1.11	1.05	1.30
	31.0	9.12	12.00	3.670	0.700	0.370	203	33.8	4.71	11.3	4.39	1.12	1.08	1.36
MC10 ×	41.1	12.1	10.00	4.321	0.575	0.796	158	31.5	3.61	15.8	4.88	1.14	1.09	1.26
	33.6	9.87	10.00	4.100	0.575	0.575	139	27.8	3.75	13.2	4.38	1.16	1.08	1.35
	28.5	8.37	10.00	3.950	0.575	0.425	127	25.3	3.89	11.4	4.02	1.17	1.12	1.43
MC10 ×	25.0	7.35	10.00	3.405	0.575	0.380	110	22.0	3.87	7.35	3.00	1.00	0.953	1.22
	22.0	6.45	10.00	3.315	0.575	0.290	103	20.5	3.99	6.50	2.80	1.00	0.990	1.27
MC9 ×	25.4	7.47	9.00	3.500	0.550	0.450	88.0	19.6	3.43	7.65	3.02	1.01	0.970	1.21
	23.9	7.02	9.00	3.450	0.550	0.400	85.0	18.9	3.48	7.22	2.93	1.01	0.981	1.24
MC8 ×	22.8	6.70	8.00	3.502	0.525	0.427	63.8	16.0	3.09	7.07	2.84	1.03	1.01	1.26
	21.4	6.28	8.00	3.450	0.525	0.375	61.6	15.4	3.13	6.64	2.74	1.03	1.02	1.28
MC8 ×	20.0	5.88	8.00	3.025	0.500	0.400	54.5	13.6	3.05	4.47	2.05	0.872	0.840	1.04
	18.7	5.50	8.00	2.978	0.500	0.353	52.5	13.1	3.09	4.20	1.97	0.874	0.849	1.07
MC7 ×	22.7	6.67	7.00	3.603	0.500	0.503	47.5	13.6	2.67	7.29	2.85	1.05	1.04	1.26
	19.1	5.61	7.00	3.452	0.500	0.352	43.2	12.3	2.77	6.11	2.57	1.04	1.08	1.33
MC6 ×	18.0	5.29	6.00	3.504	0.475	0.379	29.7	9.91	2.37	5.93	2.48	1.06	1.12	1.36
MC6 ×	15.3	4.50	6.00	3.500	0.385	0.340	25.4	8.47	2.38	4.97	2.03	1.05	1.05	1.33
MC6 ×	16.3	4.79	6.00	3.000	0.475	0.375	26.0	8.68	2.33	3.82	1.84	0.892	0.927	1.12
	15.1	4.44	6.00	2.941	0.475	0.316	25.0	8.32	2.37	3.51	1.75	0.889	0.940	1.14
MC6 ×	12.0	3.53	6.00	2.497	0.375	0.310	18.7	6.24	2.30	1.87	1.04	0.728	0.704	0.880

STRUCTURAL TEES CUT FROM S SHAPES

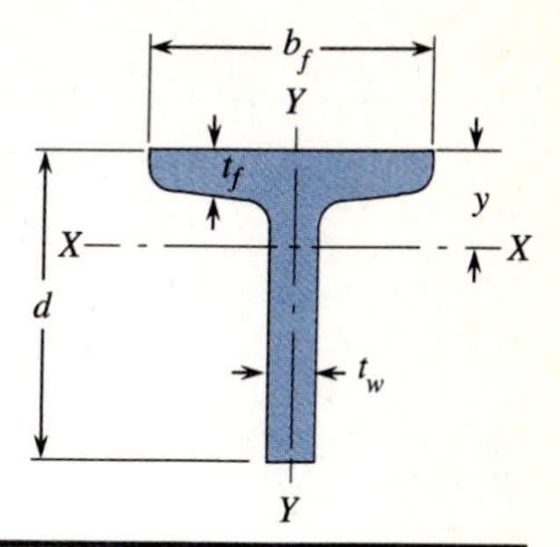

THEORETICAL DIMENSIONS AND PROPERTIES FOR DESIGNING

Section Number	Weight per Foot	Area of Section A	Depth of Tee d	Flange		Stem Thickness t_w	Axis X-X				Axis Y-Y		
				Width b_f	Average Thickness t_f		I_x	S_x	r_x	y	I_y	S_y	r_y
	(lb)	(in.^{2})	(in.)	(in.)	(in.)	(in.)	(in.^{4})	(in.^{3})	(in.)	(in.)	(in.^{4})	(in.^{3})	(in.)
ST12 ×	60.5	17.8	12.250	8.050	1.090	0.800	259	30.1	3.82	3.63	41.7	10.4	1.53
	53	15.6	12.250	7.870	1.090	0.620	216	24.1	3.72	3.28	38.5	9.80	1.57
ST12 ×	50	14.7	12.000	7.245	0.870	0.745	215	26.3	3.83	3.84	23.8	6.58	1.27
	45	13.2	12.000	7.125	0.870	0.625	190	22.6	3.79	3.60	22.5	6.31	1.30
	40	11.7	12.000	7.000	0.870	0.500	162	18.7	3.72	3.29	21.1	6.04	1.34
ST10 ×	48	14.1	10.150	7.200	0.920	0.800	143	20.3	3.18	3.13	25.1	6.97	1.33
	43	12.7	10.150	7.060	0.920	0.660	125	17.2	3.14	2.91	23.4	6.63	1.36
ST10 ×	37.5	11.0	10.000	6.385	0.795	0.635	109	15.8	3.15	3.07	14.9	4.66	1.16
	33	9.70	10.000	6.255	0.795	0.505	93.1	12.9	3.10	2.81	13.8	4.43	1.19
ST9 ×	35	10.3	9.000	6.251	0.691	0.711	84.7	14.0	2.87	2.94	12.1	3.86	1.08
	27.35	8.04	9.000	6.001	0.691	0.461	62.4	9.61	2.79	2.50	10.4	3.47	1.14
ST7.5 ×	25	7.35	7.500	5.640	0.622	0.550	40.6	7.73	2.35	2.25	7.85	2.78	1.03
	21.45	6.31	7.500	5.501	0.622	0.411	33.0	6.00	2.29	2.01	7.19	2.61	1.07
ST6 ×	25	7.35	6.000	5.477	0.659	0.687	25.2	6.05	1.85	1.84	7.85	2.87	1.03
	20.4	6.00	6.000	5.252	0.659	0.462	18.9	4.28	1.78	1.58	6.78	2.58	1.06
ST6 ×	17.5	5.14	6.000	5.078	0.544	0.428	17.2	3.95	1.83	1.65	4.93	1.94	0.980
	15.9	4.68	6.000	5.000	0.544	0.350	14.9	3.31	1.78	1.51	4.68	1.87	1.00
ST5 ×	17.5	5.15	5.000	4.944	0.491	0.594	12.5	3.63	1.56	1.56	4.18	1.69	0.901
	12.7	3.73	5.000	4.661	0.491	0.311	7.83	2.06	1.45	1.20	3.39	1.46	0.954
ST4 ×	11.5	3.38	4.000	4.171	0.425	0.441	5.03	1.77	1.22	1.15	2.15	1.03	0.798
	9.2	2.70	4.000	4.001	0.425	0.271	3.51	1.15	1.14	0.941	1.86	0.932	0.831
ST3.5 ×	7.65	2.25	3.500	3.662	0.392	0.252	2.19	0.816	0.987	0.817	1.32	0.720	0.766
ST3 ×	8.625	2.53	3.000	3.565	0.359	0.465	2.13	1.02	0.917	0.914	1.15	0.648	0.675
	6.25	1.83	3.000	3.332	0.359	0.232	1.27	0.552	0.833	0.691	0.911	0.547	0.705

STRUCTURAL TEES CUT FROM S SHAPES: THEORETICAL DIMENSIONS AND PROPERTIES FOR DESIGNING

Section Number	Weight per Foot (lb)	Area of Section A (in.^{2})	Depth of Tee d (in.)	Flange Width b_f (in.)	Flange Average Thickness t_f (in.)	Stem Thickness t_w (in.)	Axis X-X I_x (in.^{4})	Axis X-X S_x (in.^{3})	Axis X-X r_x (in.)	Axis X-X y (in.)	Axis Y-Y I_y (in.^{4})	Axis Y-Y S_y (in.^{3})	Axis Y-Y r_y (in.)
ST2.5 ×	7.375	2.17	2.500	3.284	0.326	0.494	1.27	0.740	0.764	0.789	0.833	0.507	0.620
	5	1.47	2.500	3.004	0.326	0.214	0.681	0.353	0.681	0.569	0.608	0.405	0.643
ST2 ×	4.75	1.40	2.000	2.796	0.293	0.326	0.470	0.325	0.580	0.553	0.451	0.323	0.569
	3.85	1.13	2.000	2.663	0.293	0.193	0.316	0.203	0.528	0.448	0.382	0.287	0.581
ST1.5 ×	3.75	1.10	1.500	2.509	0.260	0.349	0.204	0.191	0.430	0.432	0.293	0.234	0.516
	2.85	0.835	1.500	2.330	0.260	0.170	0.118	0.101	0.376	0.329	0.227	0.195	0.522

STRUCTURAL TEES CUT FROM W SHAPES

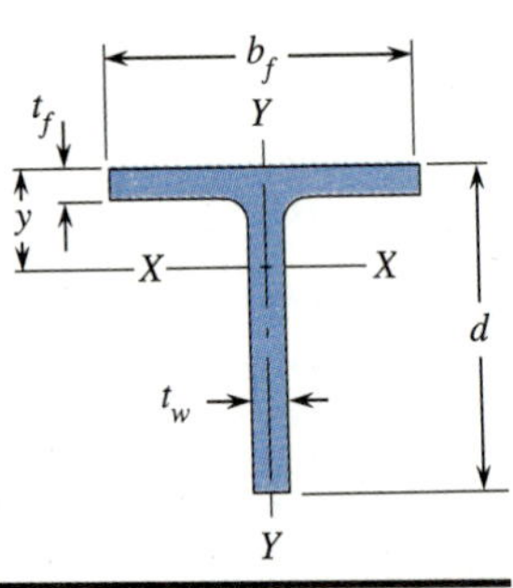

DIMENSIONS AND PROPERTIES FOR DESIGNING

Section Number	Weight Per Foot	Area of Section A (in.^{2})	Depth of Tee d (in.)	Flange Width b_f (in.)	Flange Thickness t_f (in.)	Stem Thickness t_w (in.)	Axis X-X I_x (in.^{4})	Axis X-X S_x (in.^{3})	Axis X-X r_x (in.)	Axis X-X $\bar{y}$ (in.)	Axis Y-Y I_y (in.^{4})	Axis Y-Y S_y (in.^{3})	Axis Y-Y r_y (in.)
WT18	× 150	44.1	18.36	16.655	1.680	0.945	1220	86.0	5.27	4.13	648	77.8	3.83
	× 140	41.2	18.25	16.595	1.570	0.885	1130	80.0	5.25	4.06	599	72.2	3.81
	× 130	38.2	18.12	16.551	1.440	0.841	1060	75.1	5.26	4.05	545	65.9	3.77
	× 122.5	36.1	18.03	16.512	1.350	0.802	995	71.1	5.25	4.03	507	61.4	3.75
	× 115	33.8	17.94	16.471	1.260	0.761	933	67.0	5.25	4.00	470	57.1	3.73
WT18	× 97	28.6	18.24	12.117	1.260	0.770	905	67.4	5.63	4.81	188	31.0	2.56
	× 91	26.8	18.16	12.072	1.180	0.725	845	63.1	5.61	4.77	174	28.8	2.55
	× 85	25.0	18.08	12.027	1.100	0.680	786	58.8	5.60	4.73	160	26.6	2.53
	× 80	23.6	18.00	12.000	1.020	0.653	742	56.0	5.61	4.75	147	24.6	2.50
	× 75	22.1	17.92	11.972	0.940	0.625	698	53.1	5.62	4.78	135	22.5	2.47
	× 67.5	19.9	17.78	11.945	0.794	0.598	636	49.5	5.65	4.94	113	18.9	2.39

STRUCTURAL TEES CUT FROM W SHAPES: DIMENSIONS AND PROPERTIES FOR DESIGNING

Section Number	Weight Per Foot	Area of Section A (in.²)	Depth of Tee d (in.)	Flange Width b_f (in.)	Flange Thickness t_f (in.)	Stem Thickness t_w (in.)	Axis X-X I_x (in.⁴)	Axis X-X S_x (in.³)	Axis X-X r_x (in.)	Axis X-X $\bar{y}$ (in.)	Axis Y-Y I_y (in.⁴)	Axis Y-Y S_y (in.³)	Axis Y-Y r_y (in.)
WT16.5	× 177	52.1	17.77	16.100	2.090	1.160	1320	96.8	5.03	4.16	729	90.6	3.74
	× 159	46.7	17.58	15.985	1.890	1.040	1160	85.8	4.99	4.02	645	80.7	3.71
	× 145.5	42.8	17.42	15.905	1.730	0.960	1050	78.3	4.97	3.94	581	73.1	3.69
	× 131.5	38.7	17.26	15.805	1.570	0.870	943	70.2	4.94	3.84	517	65.5	3.66
	× 120.5	35.4	17.09	15.860	1.400	0.830	871	65.8	4.96	3.85	466	58.8	3.63
	× 110.5	32.5	16.96	15.805	1.275	0.775	799	60.8	4.96	3.81	420	53.2	3.59
	× 100.5	29.5	16.84	15.745	1.150	0.715	725	55.5	4.95	3.78	375	47.6	3.56
WT16.5	× 84.5	24.8	16.91	11.500	1.220	0.670	649	51.1	5.12	4.21	155	27.0	2.50
	× 76	22.4	16.74	11.565	1.055	0.635	592	47.4	5.14	4.26	136	23.6	2.47
	× 70.5	20.8	16.65	11.535	0.960	0.605	552	44.7	5.15	4.29	123	21.3	2.43
	× 65	19.2	16.54	11.510	0.855	0.580	513	42.1	5.18	4.36	109	18.9	2.39
	× 59	17.3	16.43	11.480	0.740	0.550	469	39.2	5.20	4.47	93.6	16.3	2.32
WT15	× 117.5	34.5	15.65	15.055	1.500	0.830	674	55.1	4.42	3.42	427	56.8	3.52
	× 105.5	31.0	15.47	15.105	1.315	0.775	610	50.5	4.43	3.40	378	50.1	3.49
	× 95.5	28.1	15.34	15.040	1.185	0.710	549	45.7	4.42	3.35	336	44.7	3.46
	× 86.5	25.4	15.22	14.985	1.065	0.655	497	41.7	4.42	3.31	299	39.9	3.43
WT15	× 74	21.7	15.33	10.480	1.180	0.650	466	40.6	4.63	3.84	113	21.7	2.28
	× 66	19.4	15.15	10.545	1.000	0.615	421	37.4	4.66	3.90	98.0	18.6	2.25
	× 62	18.2	15.08	10.515	0.930	0.585	396	35.3	4.66	3.90	90.4	17.2	2.23
	× 58	17.1	15.00	10.495	0.850	0.565	373	33.7	4.67	3.94	82.1	15.7	2.19
	× 54	15.9	14.91	10.475	0.760	0.545	349	32.0	4.69	4.01	73.0	13.9	2.15
	× 49.5	14.5	14.82	10.450	0.670	0.520	322	30.0	4.71	4.09	63.9	12.2	2.10
WT13.5	× 108.5	31.9	14.215	14.115	1.500	0.830	502	45.2	3.97	3.11	352	49.9	3.32
	× 97	28.5	14.055	14.035	1.340	0.750	444	40.3	3.95	3.03	309	44.1	3.29
	× 89	26.1	13.905	14.085	1.190	0.725	414	38.2	3.98	3.05	278	39.4	3.26
	× 80.5	23.7	13.795	14.020	1.080	0.660	372	34.4	3.96	2.99	248	35.4	3.24
	× 73	21.5	13.690	13.965	0.975	0.605	336	31.2	3.95	2.95	222	31.7	3.21
WT13.5	× 64.5	18.9	13.815	10.010	1.100	0.610	323	31.0	4.13	3.39	92.2	18.4	2.21
	× 57	16.8	13.645	10.070	0.930	0.570	289	28.3	4.15	3.42	79.4	15.8	2.18
	× 51	15.0	13.545	10.015	0.830	0.515	258	25.3	4.14	3.37	69.6	13.9	2.15
	× 47	13.8	13.460	9.990	0.745	0.490	239	23.8	4.16	3.41	62.0	12.4	2.12
	× 42	12.4	13.355	9.960	0.640	0.460	216	21.9	4.18	3.48	52.8	10.6	2.07
WT12	× 88	25.8	12.625	12.890	1.340	0.750	319	32.2	3.51	2.74	240	37.2	3.04
	× 81	23.9	12.500	12.955	1.220	0.705	293	29.9	3.50	2.70	221	34.2	3.05
	× 73	21.5	12.370	12.900	1.090	0.650	264	27.2	3.50	2.66	195	30.3	3.01
	× 65.5	19.3	12.240	12.855	0.960	0.605	238	24.8	3.52	2.65	170	26.5	2.97
	× 58.5	17.2	12.130	12.800	0.850	0.550	212	22.3	3.51	2.62	149	23.2	2.94
	× 52	15.3	12.030	12.750	0.750	0.500	189	20.0	3.51	2.59	130	20.3	2.91

STRUCTURAL TEES CUT FROM W SHAPES: DIMENSIONS AND PROPERTIES FOR DESIGNING

Section Number		Weight Per Foot	Area of Section A (in.²)	Depth of Tee d (in.)	Flange Width b_f (in.)	Flange Thickness t_f (in.)	Stem Thickness t_w (in.)	Axis X-X I_x (in.⁴)	Axis X-X S_x (in.³)	Axis X-X r_x (in.)	Axis X-X $\bar{y}$ (in.)	Axis Y-Y I_y (in.⁴)	Axis Y-Y S_y (in.³)	Axis Y-Y r_y (in.)
WT12	×	51.5	15.1	12.26	9.000	0.980	0.550	204	22.0	3.67	3.01	59.7	13.3	1.99
	×	47	13.8	12.155	9.065	0.875	0.515	186	20.3	3.67	2.99	54.5	12.0	1.98
	×	42	12.4	12.050	9.020	0.770	0.470	166	18.3	3.67	2.97	47.2	10.5	1.95
	×	38	11.2	11.960	8.990	0.680	0.440	151	16.9	3.68	3.00	41.3	9.18	1.92
	×	34	10.0	11.865	8.965	0.585	0.415	137	15.6	3.70	3.06	35.2	7.85	1.87
WT12	×	31	9.11	11.870	7.040	0.590	0.430	131	15.6	3.79	3.46	17.2	4.90	1.38
	×	27.5	8.10	11.785	7.005	0.505	0.395	117	14.1	3.80	3.50	14.5	4.15	1.34
WT10.5	×	83	24.4	11.240	12.420	1.360	0.750	226	25.5	3.04	2.39	217	35.0	2.98
	×	73.5	21.6	11.030	12.510	1.150	0.720	204	23.7	3.08	2.39	188	30.0	2.95
	×	66	19.4	10.915	12.440	1.035	0.650	181	21.1	3.06	2.33	166	26.7	2.93
	×	61	17.9	10.840	12.390	0.960	0.600	166	19.3	3.04	2.28	152	24.6	2.92
	×	55.5	16.3	10.755	12.340	0.875	0.550	150	17.5	3.03	2.23	137	22.2	2.90
	×	50.5	14.9	10.680	12.290	0.800	0.500	135	15.8	3.01	2.18	124	20.2	2.89
WT10.5	×	46.5	13.7	10.810	8.420	0.930	0.580	144	17.9	3.25	2.74	46.4	11.0	1.84
	×	41.5	12.2	10.715	8.355	0.835	0.515	127	15.7	3.22	2.66	40.7	9.75	1.83
	×	36.5	10.7	10.620	8.295	0.740	0.455	110	13.8	3.21	2.60	35.3	8.51	1.81
	×	34	10.0	10.565	8.270	0.685	0.430	103	12.9	3.20	2.59	32.4	7.83	1.80
	×	31	9.1	10.495	8.240	0.615	0.400	93.8	11.9	3.21	2.58	28.7	6.97	1.77
WT10.5	×	28.5	8.37	10.530	6.555	0.650	0.405	90.4	11.8	3.29	2.85	15.3	4.67	1.35
	×	25	7.36	10.415	6.530	0.535	0.380	80.3	10.7	3.30	2.93	12.5	3.82	1.30
	×	22	6.49	10.330	6.500	0.450	0.350	71.1	9.68	3.31	2.98	10.3	3.18	1.26
WT9	×	71.5	21.0	9.745	11.220	1.320	0.730	142	18.5	2.60	2.09	156	27.7	2.72
	×	65	19.1	9.625	11.160	1.200	0.670	127	16.7	2.58	2.02	139	24.9	2.70
	×	59.5	17.5	9.485	11.265	1.060	0.655	119	15.9	2.60	2.03	126	22.5	2.69
	×	53	15.6	9.365	11.200	0.940	0.590	104	14.1	2.59	1.97	110	19.7	2.66
	×	48.5	14.3	9.295	11.145	0.870	0.535	93.8	12.7	2.56	1.91	100	18.0	2.65
	×	43	12.7	9.195	11.090	0.770	0.480	82.4	11.2	2.55	1.86	87.6	15.8	2.63
	×	38	11.2	9.105	11.035	0.680	0.425	71.8	9.83	2.54	1.80	76.2	13.8	2.61
WT9	×	35.5	10.4	9.235	7.635	0.810	0.495	78.2	11.2	2.74	2.26	30.1	7.89	1.70
	×	32.5	9.55	9.175	7.590	0.750	0.450	70.7	10.1	2.72	2.20	27.4	7.22	1.69
	×	30	8.82	9.120	7.555	0.695	0.415	64.7	9.29	2.71	2.16	25.0	6.63	1.69
	×	27.5	8.10	9.055	7.530	0.630	0.390	59.5	8.63	2.71	2.16	22.5	5.97	1.67
	×	25	7.33	8.995	7.495	0.570	0.355	53.5	7.79	2.70	2.12	20.0	5.35	1.65
WT9	×	23	6.77	9.030	6.060	0.605	0.360	52.1	7.77	2.77	2.33	11.3	3.72	1.29
	×	20	5.88	8.950	6.015	0.525	0.315	44.8	6.73	2.76	2.29	9.55	3.17	1.27
	×	17.5	5.15	8.850	6.000	0.425	0.300	40.1	6.21	2.79	2.39	7.67	2.56	1.22

STRUCTURAL TEES CUT FROM W SHAPES: DIMENSIONS AND PROPERTIES FOR DESIGNING

Section Number	Weight Per Foot	Area of Section A (in.^{2})	Depth of Tee d (in.)	Flange		Stem Thickness t_w (in.)	Axis X-X				Axis Y-Y		
				Width b_f (in.)	Thickness t_f (in.)		I_x (in.^{4})	S_x (in.^{3})	r_x (in.)	$\bar{y}$ (in.)	I_y (in.^{4})	S_y (in.^{3})	r_y (in.)
WT8	× 50	14.7	8.485	10.425	0.985	0.585	76.8	11.4	2.28	1.76	93.1	17.9	2.51
	× 44.5	13.1	8.375	10.365	0.875	0.525	67.2	10.1	2.27	1.70	81.3	15.7	2.49
	× 38.5	11.3	8.260	10.295	0.760	0.455	56.9	8.59	2.24	1.63	69.2	13.4	2.47
	× 33.5	9.8	8.165	10.235	0.665	0.395	48.6	7.36	2.22	1.56	59.5	11.6	2.46
WT8	× 28.5	8.38	8.215	7.120	0.715	0.430	48.7	7.77	2.41	1.94	21.6	6.06	1.60
	× 25	7.37	8.130	7.070	0.630	0.380	42.3	6.78	2.40	1.89	18.6	5.26	1.59
	× 22.5	6.63	8.065	7.035	0.565	0.345	37.8	6.10	2.39	1.86	16.4	4.67	1.57
	× 20	5.89	8.005	6.995	0.505	0.305	33.1	5.35	2.37	1.81	14.4	4.12	1.57
	× 18	5.28	7.930	6.985	0.430	0.295	30.6	5.05	2.41	1.88	12.2	3.50	1.52
WT8	× 15.5	4.56	7.940	5.525	0.440	0.275	27.4	4.64	2.45	2.02	6.20	2.24	1.17
	× 13	3.84	7.845	5.500	0.345	0.250	23.5	4.09	2.47	2.09	4.80	1.74	1.12
WT7	× 66	19.4	7.330	14.725	1.030	0.645	57.8	9.57	1.73	1.29	274	37.2	3.76
	× 60	17.7	7.240	14.670	0.940	0.590	51.7	8.61	1.71	1.24	247	33.7	3.74
	× 54.5	16.0	7.160	14.605	0.860	0.525	45.3	7.56	1.68	1.17	223	30.6	3.73
	× 49.5	14.6	7.080	14.565	0.780	0.485	40.9	6.88	1.67	1.14	201	27.6	3.71
	× 45	13.2	7.010	14.520	0.710	0.440	36.4	6.16	1.66	1.09	181	25.0	3.70
WT7	× 41	12.0	7.155	10.130	0.855	0.510	41.2	7.14	1.85	1.39	74.2	14.6	2.48
	× 37	10.9	7.085	10.070	0.785	0.450	36.0	6.25	1.82	1.32	66.9	13.3	2.48
	× 34	9.9	7.020	10.035	0.720	0.415	32.6	5.69	1.81	1.29	60.7	12.1	2.46
	× 30.5	8.9	6.945	9.995	0.645	0.375	28.9	5.07	1.80	1.25	53.7	10.7	2.45
WT7	× 26.5	7.81	6.960	8.060	0.660	0.370	27.6	4.94	1.88	1.38	28.8	7.16	1.92
	× 24	7.07	6.895	8.030	0.595	0.340	24.9	4.48	1.87	1.35	25.7	6.40	1.91
	× 21.5	6.31	6.830	7.995	0.530	0.305	21.9	3.98	1.86	1.31	22.6	5.65	1.89
WT7	× 19	5.58	7.050	6.770	0.515	0.310	23.3	4.22	2.04	1.54	13.3	3.94	1.55
	× 17	5.00	6.990	6.745	0.455	0.285	20.9	3.83	2.04	1.53	11.7	3.45	1.53
	× 15	4.42	6.920	6.730	0.385	0.270	19.0	3.55	2.07	1.58	9.79	2.91	1.49
WT7	× 13	3.85	6.995	5.025	0.420	0.255	17.3	3.31	2.12	1.72	4.45	1.77	1.08
	× 11	3.25	6.870	5.000	0.335	0.230	14.8	2.91	2.14	1.76	3.50	1.40	1.04
WT6	× 168	49.4	8.410	13.385	2.955	1.775	190	31.2	1.96	2.31	593	88.6	3.47
	× 152.5	44.8	8.160	13.235	2.705	1.625	162	27.0	1.90	2.16	525	79.3	3.42
	× 139.5	41.0	7.925	13.140	2.470	1.530	141	24.1	1.86	2.05	469	71.3	3.38
	× 126	37.0	7.705	13.005	2.250	1.395	121	20.9	1.81	1.92	414	63.6	3.34
	× 115	33.9	7.525	12.895	2.070	1.285	106	18.5	1.77	1.82	371	57.5	3.31
	× 105	30.9	7.355	12.790	1.900	1.180	92.1	16.4	1.73	1.72	332	51.9	3.28
	× 95	27.9	7.190	12.670	1.735	1.060	79.0	14.2	1.68	1.62	295	46.5	3.25
	× 85	25.0	7.015	12.570	1.560	0.960	67.8	12.3	1.65	1.52	259	41.2	3.22
	× 76	22.4	6.855	12.480	1.400	0.870	58.5	10.8	1.62	1.43	227	36.4	3.19

STRUCTURAL TEES CUT FROM W SHAPES: DIMENSIONS AND PROPERTIES FOR DESIGNING

Section Number	Weight Per Foot	Area of Section A (in.^{2})	Depth of Tee d (in.)	Flange Width b_f (in.)	Flange Thickness t_f (in.)	Stem Thickness t_w (in.)	Axis X-X I_x (in.^{4})	Axis X-X S_x (in.^{3})	Axis X-X r_x (in.)	Axis X-X $\bar{y}$ (in.)	Axis Y-Y I_y (in.^{4})	Axis Y-Y S_y (in.^{3})	Axis Y-Y r_y (in.)
WT6	× 68	20.0	6.705	12.400	1.250	0.790	50.6	9.46	1.59	1.35	199	32.1	3.16
	× 60	17.6	6.560	12.320	1.105	0.710	43.4	8.22	1.57	1.28	172	28.0	3.13
	× 53	15.6	6.445	12.220	0.990	0.610	36.3	6.91	1.53	1.19	151	24.7	3.11
	× 48	14.1	6.355	12.160	0.900	0.550	32.0	6.12	1.51	1.13	135	22.2	3.09
	× 43.5	12.8	6.265	12.125	0.810	0.515	28.9	5.60	1.50	1.10	120	19.9	3.07
	× 39.5	11.6	6.190	12.080	0.735	0.470	25.8	5.03	1.49	1.06	108	17.9	3.05
	× 36	10.6	6.125	12.040	0.670	0.430	23.2	4.54	1.48	1.02	97.5	16.2	3.04
	× 32.5	9.5	6.060	12.000	0.605	0.390	20.6	4.06	1.47	0.985	87.2	14.5	3.02
WT6	× 29	8.52	6.095	10.010	0.640	0.360	19.1	3.76	1.50	1.03	53.5	10.7	2.51
	× 26.5	7.78	6.030	9.995	0.575	0.345	17.7	3.54	1.51	1.02	47.9	9.58	2.48
WT6	× 25	7.34	6.095	8.080	0.640	0.370	18.7	3.79	1.60	1.17	28.2	6.97	1.96
	× 22.5	6.61	6.030	8.045	0.575	0.335	16.6	3.39	1.58	1.13	25.0	6.21	1.94
	× 20	5.89	5.970	8.005	0.515	0.295	14.4	2.95	1.57	1.08	22.0	5.51	1.93
WT6	× 17.5	5.17	6.250	6.560	0.520	0.300	16.0	3.23	1.76	1.30	12.2	3.73	1.54
	× 15	4.40	6.170	6.520	0.440	0.260	13.5	2.75	1.75	1.27	10.2	3.12	1.52
	× 13	3.82	6.110	6.490	0.380	0.230	11.7	2.40	1.75	1.25	8.66	2.67	1.51
WT6	× 11	3.24	6.155	4.030	0.425	0.260	11.7	2.59	1.90	1.63	2.33	1.16	0.847
	× 9.5	2.79	6.080	4.005	0.350	0.235	10.1	2.28	1.90	1.65	1.88	0.939	0.822
	× 8	2.36	5.995	3.990	0.265	0.220	8.70	2.04	1.92	1.74	1.41	0.706	0.773
	× 7	2.08	5.995	3.970	0.225	0.200	7.67	1.83	1.92	1.76	1.18	0.594	0.753
WT5	× 56	16.5	5.680	10.415	1.250	0.755	28.6	6.40	1.32	1.21	118	22.6	2.68
	× 50	14.7	5.550	10.340	1.120	0.680	24.5	5.56	1.29	1.13	103	20.0	2.65
	× 44	12.9	5.420	10.265	0.990	0.605	20.8	4.77	1.27	1.06	89.3	17.4	2.63
	× 38.5	11.3	5.300	10.190	0.870	0.530	17.4	4.04	1.24	0.990	76.8	15.1	2.60
	× 34	9.99	5.200	10.130	0.770	0.470	14.9	3.49	1.22	0.932	66.8	13.2	2.59
	× 30	8.82	5.110	10.080	0.680	0.420	12.9	3.04	1.21	0.884	58.1	11.5	2.57
	× 27	7.91	5.045	10.030	0.615	0.370	11.1	2.64	1.19	0.836	51.7	10.3	2.56
	× 24.5	7.21	4.990	10.000	0.560	0.340	10.0	2.39	1.18	0.807	46.7	9.34	2.54
WT5	× 22.5	6.63	5.050	8.020	0.620	0.350	10.2	2.47	1.24	0.907	26.7	6.65	2.01
	× 19.5	5.73	4.960	7.985	0.530	0.315	8.84	2.16	1.24	0.876	22.5	5.64	1.98
	× 16.5	4.85	4.865	7.960	0.435	0.290	7.71	1.93	1.26	0.869	18.3	4.60	1.94
WT5	× 15	4.42	5.235	5.810	0.510	0.300	9.28	2.24	1.45	1.10	8.35	2.87	1.37
	× 13	3.81	5.165	5.770	0.440	0.260	7.86	1.91	1.44	1.06	7.05	2.44	1.36
	× 11	3.24	5.085	5.750	0.360	0.240	6.88	1.72	1.46	1.07	5.71	1.99	1.33
WT5	× 9.5	2.81	5.120	4.020	0.395	0.250	6.68	1.74	1.54	1.28	2.15	1.07	0.874
	× 8.5	2.50	5.055	4.010	0.330	0.240	6.06	1.62	1.56	1.32	1.78	0.888	0.844
	× 7.5	2.21	4.995	4.000	0.270	0.230	5.45	1.50	1.57	1.37	1.45	0.723	0.810
	× 6	1.77	4.935	3.960	0.210	0.190	4.35	1.22	1.57	1.36	1.09	0.551	0.785

ANGLES, EQUAL LEGS

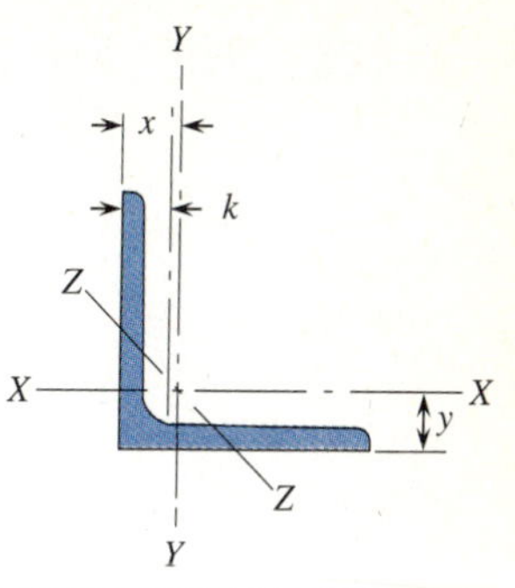

THEORETICAL DIMENSIONS AND PROPERTIES FOR DESIGNING

Section Number and Size (in.)	Thickness (in.)	Weight per Foot (lb)	Area of Section (in.²)	k (in.)	Axis X-X and Axis Y-Y $I_{x,y}$ (in.⁴)	$S_{x,y}$ (in.³)	$r_{x,y}$ (in.)	x or y (in.)	Axis Z-Z r_z (in.)
L8 × 8 ×	1⅛	56.9	16.7	1¾	98.0	17.5	2.42	2.41	1.56
	1	51.0	15.0	1⅝	89.0	15.8	2.44	2.37	1.56
	⅞	45.0	13.2	1½	79.6	14.0	2.45	2.32	1.57
	¾	38.9	11.4	1⅜	69.7	12.2	2.47	2.28	1.58
	⅝	32.7	9.61	1¼	59.4	10.3	2.49	2.23	1.58
	9/16	29.6	8.68	1 3/16	54.1	9.34	2.50	2.21	1.59
	½	26.4	7.75	1⅛	48.6	8.36	2.50	2.19	1.59
L6 × 6 ×	1	37.4	11.0	1½	35.5	8.57	1.80	1.86	1.17
	⅞	33.1	9.73	1⅜	31.9	7.63	1.81	1.82	1.17
	¾	28.7	8.44	1¼	28.2	6.66	1.83	1.78	1.17
	⅝	24.2	7.11	1⅛	24.2	5.66	1.84	1.73	1.18
	9/16	21.9	6.43	1 1/16	22.1	5.14	1.85	1.71	1.18
	½	19.6	5.75	1	19.9	4.61	1.86	1.68	1.18
	7/16	17.2	5.06	15/16	17.7	4.08	1.87	1.66	1.19
	⅜	14.9	4.36	⅞	15.4	3.53	1.88	1.64	1.19
	5/16	12.4	3.65	13/16	13.0	2.97	1.89	1.62	1.20
L5 × 5 ×	⅞	27.2	7.98	1⅜	17.8	5.17	1.49	1.57	0.973
	¾	23.6	6.94	1¼	15.7	4.53	1.51	1.52	0.975
	⅝	20.0	5.86	1⅛	13.6	3.86	1.52	1.48	0.978
	½	16.2	4.75	1	11.3	3.16	1.54	1.43	0.983
	7/16	14.3	4.18	15/16	10.0	2.79	1.55	1.41	0.986
	⅜	12.3	3.61	⅞	8.74	2.42	1.56	1.39	0.990
	5/16	10.3	3.03	13/16	7.42	2.04	1.57	1.37	0.994

ANGLES, EQUAL LEGS

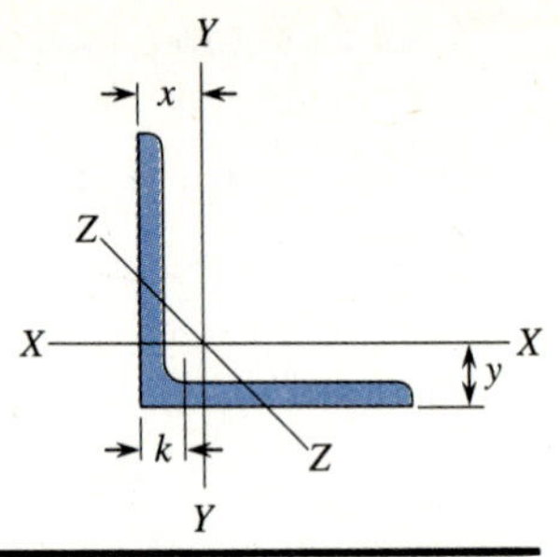

PROPERTIES FOR DESIGNING

Section Number and Size (in.)	Thickness (in.)	Weight per Foot (lb)	Area of Section (in.²)	k (in.)	Axis X-X and Axis Y-Y $I_{x,y}$ (in.⁴)	$S_{x,y}$ (in.³)	$r_{x,y}$ (in.)	x or y (in.)	Axis Z-Z r_z (in.)
L3½ × 3½ ×	½	11.1	3.25	⅞	3.64	1.49	1.06	1.06	.683
	7/16	9.8	2.87	13/16	3.26	1.32	1.07	1.04	.684
	⅜	8.5	2.48	¾	2.87	1.15	1.07	1.01	.687
	5/16	7.2	2.09	11/16	2.45	.976	1.08	.990	.690
	¼	5.8	1.69	⅝	2.01	.794	1.09	.968	.694
L3 × 3 ×	½	9.4	2.75	13/16	2.22	1.07	.898	.932	.584
	7/16	8.3	2.43	¾	1.99	.954	.905	.910	.585
	⅜	7.2	2.11	11/16	1.76	.833	.913	.888	.587
	5/16	6.1	1.78	⅝	1.51	.707	.922	.869	.589
	¼	4.9	1.44	9/16	1.24	.577	.930	.842	.592
	3/16	3.71	1.09	½	.962	.441	.939	.820	.596
L2½ × 2½ ×	½	7.7	2.25	13/16	1.23	.724	.739	.806	.487
	⅜	5.9	1.73	11/16	.984	.566	.753	.762	.487
	5/16	5.0	1.46	⅝	.849	.482	.761	.740	.489
	¼	4.1	1.19	9/16	.703	.394	.769	.717	.491
	3/16	3.07	0.902	½	.547	.303	.778	.694	.495
L2 × 2 ×	⅜	4.7	1.36	⅞	.479	.351	.594	.636	.389
	5/16	3.92	1.15	13/16	.416	.300	.601	.614	.390
	¼	3.19	.938	¾	.348	.247	.609	.592	.391
	3/16	2.44	.715	11/16	.272	.190	.617	.569	.394
	⅛	1.65	.484	⅝	.190	.131	.626	.546	.398
L1¾ × 1¾ ×	¼	2.77	.813	11/16	.227	.186	.529	.529	.341
	3/16	2.12	.621	⅝	.179	.144	.537	.506	.343
	⅛	1.44	.422	9/16	.126	.099	.546	.484	.347
L1½ × 1½ ×	¼	2.34	.688	⅝	.139	.134	.449	.466	.292
	3/16	1.80	.527	9/16	.110	.104	.457	.444	.293
L1¼ × 1¼ ×	¼	1.92	.563	⅝	.077	.091	.369	.403	.243
	3/16	1.48	.434	9/16	.061	.071	.377	.381	.244
L1 × 1 ×	⅛	.80	.234	½	.022	.031	.304	.296	.196

ANGLES, UNEQUAL LEGS

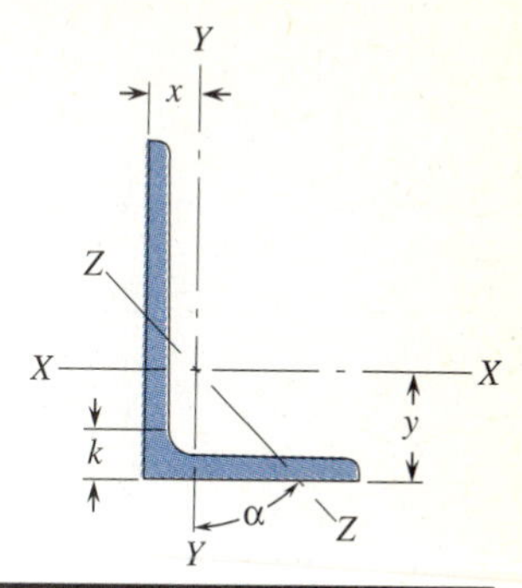

THEORETICAL DIMENSIONS AND PROPERTIES FOR DESIGNING

Section Number and Size (in.)	Thickness (in.)	Weight per Foot (lb)	Area of Section (in.²)	k (in.)	Axis X-X: I_x (in.⁴)	Axis X-X: S_x (in.³)	Axis X-X: r_x (in.)	Axis X-X: y (in.)	Axis Y-Y: I_y (in.⁴)	Axis Y-Y: S_y (in.³)	Axis Y-Y: r_y (in.)	Axis Y-Y: x (in.)	Axis Z-Z: r_z (in.)	Axis Z-Z: Tan α
L8 × 6 ×	1	44.2	13.0	1½	80.8	15.1	2.49	2.65	38.8	8.92	1.73	1.65	1.28	0.543
	⅞	39.1	11.5	1⅜	72.3	13.4	2.51	2.61	34.9	7.94	1.74	1.61	1.28	0.547
	¾	33.8	9.94	1¼	63.4	11.7	2.53	2.56	30.7	6.92	1.76	1.56	1.29	0.551
	3/16	25.7	7.56	1 1/16	49.3	8.95	2.55	2.50	24.0	5.34	1.78	1.50	1.30	0.556
	½	23.0	6.75	1	44.3	8.02	2.56	2.47	21.7	4.79	1.79	1.47	1.30	0.558
	7/16	20.2	5.93	15/16	39.2	7.07	2.57	2.45	19.3	4.23	1.80	1.45	1.31	0.640
L8 × 4 ×	1	37.4	11.0	1½	69.6	14.1	2.52	3.05	11.6	3.94	1.03	1.05	0.846	0.247
	¾	28.7	8.44	1¼	54.9	10.9	2.55	2.95	9.36	3.07	1.05	0.953	0.852	0.258
	3/16	21.9	6.43	1 1/16	42.8	8.35	2.58	2.88	7.43	2.38	1.07	0.882	0.861	0.265
	½	19.6	5.75	1	38.5	7.49	2.59	2.86	6.74	2.15	1.08	0.859	0.865	0.267
L7 × 4 ×	¾	26.2	7.69	1¼	37.8	8.42	2.22	2.51	9.05	3.03	1.09	1.01	0.860	0.324
	⅝	22.1	6.48	1⅛	32.4	7.14	2.24	2.46	7.84	2.58	1.10	0.963	0.865	0.329
	½	17.9	5.25	1	26.7	5.81	2.25	2.42	6.53	2.12	1.11	0.917	0.872	0.335
	⅜	13.6	3.98	⅞	20.6	4.44	2.27	2.37	5.10	1.63	1.13	0.870	0.880	0.340
L6 × 4 ×	⅞	27.2	7.98	1⅜	27.7	7.15	1.86	2.12	9.75	3.39	1.11	1.12	0.857	0.421
	¼	23.6	6.94	1¼	24.5	6.25	1.88	2.08	8.68	2.97	1.12	1.08	0.860	0.428
	⅝	20.0	5.86	1⅛	21.1	5.31	1.90	2.03	7.52	2.54	1.13	1.03	0.864	0.435
	9/16	18.1	5.31	1 1/16	19.3	4.83	1.90	2.01	6.91	2.31	1.14	1.01	0.866	0.438
	½	16.2	4.75	1	17.4	4.33	1.91	1.99	6.27	2.08	1.15	0.987	0.870	0.440
	7/16	14.3	4.18	15/16	15.5	3.83	1.92	1.96	5.60	1.85	1.16	0.964	0.873	0.443
	⅜	12.3	3.61	⅞	13.5	3.32	1.93	1.94	4.90	1.60	1.17	0.941	0.877	0.446
	5/16	10.3	3.03	13/16	11.4	2.79	1.94	1.92	4.18	1.35	1.17	0.918	0.882	0.448
L6 × 3½ ×	½	15.3	4.50	1	16.6	4.24	1.92	2.08	4.25	1.59	0.972	0.833	0.759	0.344
	⅜	11.7	3.42	⅞	12.9	3.24	1.94	2.04	3.34	1.23	0.988	0.787	0.767	0.350
	5/16	9.8	2.87	13/16	10.9	2.73	1.95	2.01	2.85	1.04	0.996	0.763	0.772	0.352
L5 × 3 ×	⅝	15.7	4.61	1	11.4	3.55	1.57	1.80	3.06	1.39	0.815	0.796	0.644	0.349
	½	12.8	3.75	1	9.45	2.91	1.59	1.75	2.58	1.15	0.829	0.750	0.648	0.357
	7/16	11.3	3.31	15/16	8.43	2.58	1.60	1.73	2.32	1.02	0.837	0.727	0.651	0.361
	⅜	9.8	2.86	⅞	7.37	2.24	1.61	1.70	2.04	0.888	0.845	0.704	0.654	0.364
	5/16	8.2	2.40	13/16	6.26	1.89	1.61	1.68	1.75	0.753	0.853	0.681	0.658	0.368
	¼	6.6	1.94	¾	5.11	1.53	1.62	1.66	1.44	0.614	0.861	0.657	0.663	0.371

PIPE

DIMENSIONS AND PROPERTIES

Dimensions				Weight per Ft. Lbs. Plain Ends	Properties				Schedule No.
Nominal Diameter in.	Outside Diameter in.	Inside Diameter in.	Wall Thickness in.		*A* in.^{2}	*I* in.^{4}	*S* in.^{3}	*r* in.	
Standard Weight									
½	.840	.622	.109	.85	.250	.017	.041	.261	40
¾	1.050	.824	.113	1.13	.333	.037	.071	.334	40
1	1.315	1.049	.133	1.68	.494	.087	.133	.421	40
1¼	1.660	1.380	.140	2.27	.669	.195	.235	.540	40
1½	1.900	1.610	.145	2.72	.799	.310	.326	.623	40
2	2.375	2.067	.154	3.65	1.07	.666	.561	.787	40
2½	2.875	2.469	.203	5.79	1.70	1.53	1.06	.947	40
3	3.500	3.068	.216	7.58	2.23	3.02	1.72	1.16	40
3½	4.000	3.548	.226	9.11	2.68	4.79	2.39	1.34	40
4	4.500	4.026	.237	10.79	3.17	7.23	3.21	1.51	40
5	5.563	5.047	.258	14.62	4.30	15.2	5.45	1.88	40
6	6.625	6.065	.280	18.97	5.58	28.1	8.50	2.25	40
8	8.625	7.981	.322	28.55	8.40	72.5	16.8	2.94	40
10	10.750	10.020	.365	40.48	11.9	161	29.9	3.67	40
12	12.750	12.000	.375	49.56	14.6	279	43.8	4.38	—
Extra Strong									
½	.840	.546	.147	1.09	.320	.020	.048	.250	80
¾	1.050	.742	.154	1.47	.433	.045	.085	.321	80
1	1.315	.957	.179	2.17	.639	.106	.161	.407	80
1¼	1.660	1.278	.191	3.00	.881	.242	.291	.524	80
1½	1.900	1.500	.200	3.63	1.07	.391	.412	.605	80
2	2.375	1.939	.218	5.02	1.48	.868	.731	.766	80
2½	2.875	2.323	.276	7.66	2.25	1.92	1.34	.924	80
3	3.500	2.900	.300	10.25	3.02	3.89	2.23	1.14	80
3½	4.000	3.364	.318	12.50	3.68	6.28	3.14	1.31	80
4	4.500	3.826	.337	14.98	4.41	9.61	4.27	1.48	80
5	5.563	4.813	.375	20.78	6.11	20.7	7.43	1.84	80
6	6.625	5.761	.432	28.57	8.40	40.5	12.2	2.19	80
8	8.625	7.625	.500	43.39	12.8	106	24.5	2.88	80
10	10.750	9.750	.500	54.74	16.1	212	39.4	3.63	80
12	12.750	11.750	.500	65.42	19.2	362	56.7	4.33	—
Double-Extra Strong									
2	2.375	1.503	.436	9.03	2.66	1.31	1.10	.703	—
2½	2.875	1.771	.552	13.69	4.03	2.87	2.00	.844	—
3	3.500	2.300	.600	18.58	5.47	5.99	3.42	1.05	—
4	4.500	3.152	.674	27.54	8.10	15.3	6.79	1.37	—
5	5.563	4.063	.750	38.55	11.3	33.6	12.1	1.72	—
6	6.625	4.897	.864	53.16	15.6	66.3	20.0	2.06	—
8	8.625	6.875	.875	72.42	21.3	162	37.6	2.76	—

The listed sections are available in conformance with ASTM Specification A53 Grade B or A501. Other sections are made to these specifications. Consult with pipe manufacturers or distributors for availability. Printed with permission AISC.

ALUMINUM ASSOCIATION STANDARD CHANNELS

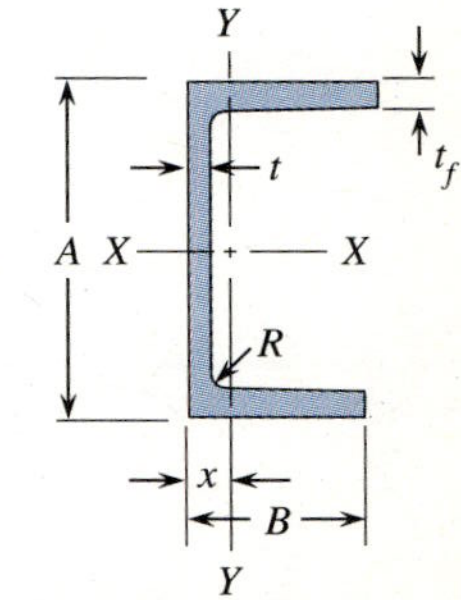

DIMENSIONS, AREAS, WEIGHTS AND SECTION PROPERTIES④

Size		Area①	Weight②	Flange Thick-ness	Web Thick-ness	Fillet Radius	Section Properties③						
							Axis X-X			Axis Y-Y			
Depth A	Width B			t_f	t	R	I	S	r	I	S	r	x
in.	in.	in.^{2}	lb/ft	in.	in.	in.	in.^{4}	in.^{3}	in.	in.^{4}	in.^{3}	in.	in.
2.00	1.00	0.491	0.557	0.13	0.13	0.10	0.288	0.288	0.766	0.045	0.064	0.303	0.298
2.00	1.25	0.911	1.071	0.26	0.17	0.15	0.546	0.546	0.774	0.139	0.178	0.391	0.471
3.00	1.50	0.965	1.135	0.20	0.13	0.25	1.41	0.94	1.21	0.22	0.22	0.47	0.49
3.00	1.75	1.358	1.597	0.26	0.17	0.25	1.97	1.31	1.20	0.42	0.37	0.55	0.62
4.00	2.00	1.478	1.738	0.23	0.15	0.25	3.91	1.95	1.63	0.60	0.45	0.64	0.65
4.00	2.25	1.982	2.331	0.29	0.19	0.25	5.21	2.60	1.62	1.02	0.69	0.72	0.78
5.00	2.25	1.881	2.212	0.26	0.15	0.30	7.88	3.15	2.05	0.98	0.64	0.72	0.73
5.00	2.75	2.627	3.089	0.32	0.19	0.30	11.14	4.45	2.06	2.05	1.14	0.88	0.95
6.00	2.50	2.410	2.834	0.29	0.17	0.30	14.35	4.78	2.44	1.53	0.90	0.80	0.79
6.00	3.25	3.427	4.030	0.35	0.21	0.30	21.04	7.01	2.48	3.76	1.76	1.05	1.12
7.00	2.75	2.725	3.205	0.29	0.17	0.30	22.09	6.31	2.85	2.10	1.10	0.88	0.84
7.00	3.50	4.009	4.715	0.38	0.21	0.30	33.79	9.65	2.90	5.13	2.23	1.13	1.20
8.00	3.00	3.526	4.147	0.35	0.19	0.30	37.40	9.35	3.26	3.25	1.57	0.96	0.93
8.00	3.75	4.923	5.789	0.41	0.25	0.35	52.69	13.17	3.27	7.13	2.82	1.20	1.22
9.00	3.25	4.237	4.983	0.35	0.23	0.35	54.41	12.09	3.58	4.40	1.89	1.02	0.93
9.00	4.00	5.927	6.970	0.44	0.29	0.35	78.31	17.40	3.63	9.61	3.49	1.27	1.25
10.00	3.50	5.218	6.136	0.41	0.25	0.35	83.22	16.64	3.99	6.33	2.56	1.10	1.02
10.00	4.25	7.109	8.360	0.50	0.31	0.40	116.15	23.23	4.04	13.02	4.47	1.35	1.34
12.00	4.00	7.036	8.274	0.47	0.29	0.40	159.76	26.63	4.77	11.03	3.86	1.25	1.14
12.00	5.00	10.053	11.822	0.62	0.35	0.45	239.69	39.95	4.88	25.74	7.60	1.60	1.61

①Areas listed are based on nominal dimensions.
②Weights per foot area are based on nominal dimensions and a density of 0.098 pound per cubic inch which is the density of alloy 6061.
③I = moment of inertia; S = section modulus; r = radius of gyration.
④Users are encouraged to ascertain current availability of particular structural shapes through inquiries to their suppliers.
Printed with permission of the Aluminum Association from 1988 Ed., Aluminum Standards and Data

ALUMINUM ASSOCIATION STANDARD I-BEAMS

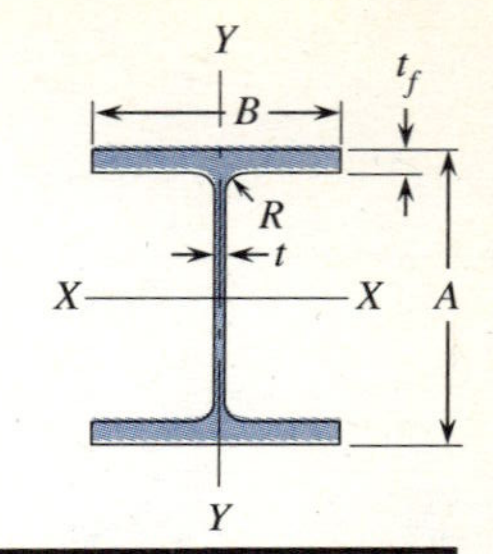

DIMENSIONS, AREAS, WEIGHTS AND SECTION PROPERTIES④

Size		Area①	Weight②	Flange Thick-ness	Web Thick-ness	Fillet Radius	Section Properties③					
							Axis X-X			Axis-Y-Y		
Depth A	Width B			t_f	t	R	I	S	r	I	S	r
in.	in.	in.^{2}	lb/ft	in.	in.	in.	in.^{4}	in.^{3}	in.	in.^{4}	in.^{3}	in.
3.00	2.50	1.392	1.637	0.20	0.13	0.25	2.24	1.49	1.27	0.52	0.42	0.61
3.00	2.50	1.726	2.030	0.26	0.15	0.25	2.71	1.81	1.25	0.68	0.54	0.63
4.00	3.00	1.965	2.311	0.23	0.15	0.25	5.62	2.81	1.69	1.04	0.69	0.73
4.00	3.00	2.375	2.793	0.29	0.17	0.25	6.71	3.36	1.68	1.31	0.87	0.74
5.00	3.50	3.146	3.700	0.32	0.19	0.30	13.94	5.58	2.11	2.29	1.31	0.85
6.00	4.00	3.427	4.030	0.29	0.19	0.30	21.99	7.33	2.53	3.10	1.55	0.95
6.00	4.00	3.990	4.692	0.35	0.21	0.30	25.50	8.50	2.53	3.74	1.87	0.97
7.00	4.50	4.932	5.800	0.38	0.23	0.30	42.89	12.25	2.95	5.78	2.57	1.08
8.00	5.00	5.256	6.181	0.35	0.23	0.30	59.69	14.92	3.37	7.30	2.92	1.18
8.00	5.00	5.972	7.023	0.41	0.25	0.30	67.78	16.94	3.37	8.55	3.42	1.20
9.00	5.50	7.110	8.361	0.44	0.27	0.30	102.02	22.67	3.79	12.22	4.44	1.31
10.00	6.00	7.352	8.646	0.41	0.25	0.40	132.09	26.42	4.24	14.78	4.93	1.42
10.00	6.00	8.747	10.286	0.50	0.29	0.40	155.79	31.16	4.22	18.03	6.01	1.44
12.00	7.00	9.925	11.672	0.47	0.29	0.40	255.57	42.60	5.07	26.90	7.69	1.65
12.00	7.00	12.153	14.292	0.62	0.31	0.40	317.33	52.89	5.11	35.48	10.14	1.71

①Areas listed are based on nominal dimensions.

②Weights per foot area are based on nominal dimensions and a density of 0.098 pound per cubic inch which is the density of alloy 6061.

③I = moment of inertia; S = section modulus; r = radius of gyration.

④Users are encouraged to ascertain current availability of particular structural shapes through inquiries to their suppliers.

PROPERTIES OF STRUCTURAL LUMBER

SECTIONAL PROPERTIES OF STANDARD DRESSED (S4S) SIZES

Nominal size b(inches)d	Standard dressed size (S4S) b(inches)d	Area of Section A	Moment of inertia I	Section modulus S	Approximate weight in pounds per linear foot of piece when weight of wood per cubic foot equals:					
					25 lb.	30 lb.	35 lb.	40 lb.	45 lb.	50 lb.
1 × 3	3/4 × 2-1/2	1.875	0.977	0.781	0.326	0.391	0.456	0.521	0.586	0.651
1 × 4	3/4 × 3-1/2	2.625	2.680	1.531	0.456	0.547	0.638	0.729	0.820	0.911
1 × 6	3/4 × 5-1/2	4.125	10.398	3.781	0.716	0.859	1.003	1.146	1.289	1.432
1 × 8	3/4 × 7-1/4	5.438	23.817	6.570	0.944	1.133	1.322	1.510	1.699	1.888
1 × 10	3/4 × 9-1/4	6.938	49.466	10.695	1.204	1.445	1.686	1.927	2.168	2.409
1 × 12	3/4 × 11-1/4	8.438	88.989	15.820	1.465	1.758	2.051	2.344	2.637	2.930
2 × 3	1-1/2 × 2-1/2	3.750	1.953	1.563	0.651	0.781	0.911	1.042	1.172	1.302
2 × 4	1-1/2 × 3-1/2	5.250	5.359	3.063	0.911	1.094	1.276	1.458	1.641	1.823
2 × 5	1-1/2 × 4-1/2	6.750	11.391	5.063	1.172	1.406	1.641	1.875	2.109	2.344
2 × 6	1-1/2 × 5-1/2	8.250	20.797	7.563	1.432	1.719	2.005	2.292	2.578	2.865
2 × 8	1-1/2 × 7-1/4	10.875	47.635	13.141	1.888	2.266	2.643	3.021	3.398	3.776
2 × 10	1-1/2 × 9-1/4	13.875	98.932	21.391	2.409	2.891	3.372	3.854	4.336	4.818
2 × 12	1-1/2 × 11-1/4	16.875	177.979	31.641	2.930	3.516	4.102	4.688	5.273	5.859
2 × 14	1-1/2 × 13-1/4	19.875	290.775	43.891	3.451	4.141	4.831	5.521	6.211	6.901
3 × 1	2-1/2 × 3/4	1.875	0.088	0.234	0.326	0.391	0.456	0.521	0.586	0.651
3 × 2	2-1/2 × 1-1/2	3.750	0.703	0.938	0.651	0.781	0.911	1.042	1.172	1.302
3 × 4	2-1/2 × 3-1/2	8.750	8.932	5.104	1.519	1.823	2.127	2.431	2.734	3.038
3 × 5	2-1/2 × 4-1/2	11.250	18.984	8.438	1.953	2.344	2.734	3.125	3.516	3.906
3 × 6	2-1/2 × 5-1/2	13.750	34.661	12.604	2.387	2.865	3.342	3.819	4.297	4.774
3 × 8	2-1/2 × 7-1/4	18.125	79.391	21.901	3.147	3.776	4.405	5.035	5.664	6.293
3 × 10	2-1/2 × 9-1/4	23.125	164.886	35.651	4.015	4.818	5.621	6.424	7.227	8.030
3 × 12	2-1/2 × 11-1/4	28.125	296.631	52.734	4.883	5.859	6.836	7.813	8.789	9.766
3 × 14	2-1/2 × 13-1/4	33.125	484.625	73.151	5.751	6.901	8.051	9.201	10.352	11.502
3 × 16	2-1/2 × 15-1/4	38.125	738.870	96.901	6.619	7.943	9.266	10.590	11.914	13.238

PROPERTIES OF STRUCTURAL LUMBER: SECTIONAL PROPERTIES OF STANDARD DRESSED (S4S) SIZES (CONT'D)

Nominal size b (inches) d	Standard dressed size (S4S) b (inches) d	Area of Section A	Moment of inertia I	Section modulus S	Approximate weight in pounds per linear foot of piece when weight of wood per cubic foot equals:					
					25 lb.	30 lb.	35 lb.	40 lb.	45 lb.	50 lb.
4 × 1	3-1/2 × 3/4	2.625	0.123	0.328	0.456	0.547	0.638	0.729	0.820	0.911
4 × 2	3-1/2 × 1-1/2	5.250	0.984	1.313	0.911	1.094	1.276	1.458	1.641	1.823
4 × 3	3-1/2 × 2-1/2	8.750	4.557	3.646	1.519	1.823	2.127	2.431	2.734	3.038
4 × 4	3-1/2 × 3-1/2	12.250	12.505	7.146	2.127	2.552	2.977	3.403	3.828	4.253
4 × 5	3-1/2 × 4-1/2	15.750	26.578	11.813	2.734	3.281	3.828	4.375	4.922	5.469
4 × 6	3-1/2 × 5-1/2	19.250	48.526	17.646	3.342	4.010	4.679	5.347	6.016	6.684
4 × 8	3-1/2 × 7-1/4	25.375	111.148	30.661	4.405	5.286	6.168	7.049	7.930	8.811
4 × 10	3-1/2 × 9-1/4	32.375	230.840	49.911	5.621	6.745	7.869	8.933	10.117	11.241
4 × 12	3-1/2 × 11-1/4	39.375	415.283	73.828	6.836	8.203	9.570	10.938	12.305	13.672
4 × 14	3-1/2 × 13-1/4	46.375	678.475	102.411	8.047	9.657	11.266	12.877	14.485	16.094
4 × 16	3-1/2 × 15-1/4	53.375	1034.418	135.66	9.267	11.121	12.975	14.828	16.682	18.536
5 × 2	4-1/2 × 1-1/2	6.750	1.266	1.688	1.172	1.406	1.641	1.875	2.109	2.344
5 × 3	4-1/2 × 2-1/2	11.250	5.859	4.688	1.953	2.344	2.734	3.125	3.516	3.906
5 × 4	4-1/2 × 3-1/2	15.750	16.078	9.188	2.734	3.281	3.828	4.375	4.922	5.469
5 × 5	4-1/2 × 4-1/2	20.250	34.172	15.188	3.516	4.219	4.922	5.675	6.328	7.031
6 × 1	5-1/2 × 3/4	4.125	0.193	0.516	0.716	0.859	1.003	1.146	1.289	1.432
6 × 2	5-1/2 × 1-1/2	8.250	1.547	2.063	1.432	1.719	2.005	2.292	2.578	2.865
6 × 3	5-1/2 × 2-1/2	13.750	7.161	5.729	2.387	2.865	3.342	3.819	4.297	4.774
6 × 4	5-1/2 × 3-1/2	19.250	19.651	11.229	3.342	4.010	4.679	5.347	6.016	6.684
6 × 6	5-1/2 × 5-1/2	30.250	76.255	27.729	5.252	6.302	7.352	8.403	9.453	10.503
6 × 8	5-1/2 × 7-1/2	41.250	193.359	51.563	7.161	8.594	10.026	11.458	12.891	14.323
6 × 10	5-1/2 × 9-1/2	52.250	392.963	82.729	9.071	10.885	12.700	14.514	16.328	18.142
6 × 12	5-1/2 × 11-1/2	63.250	697.068	121.229	10.981	13.177	15.373	17.569	19.766	21.962
6 × 14	5-1/2 × 13-1/2	74.250	1127.672	167.063	12.891	15.469	18.047	20.625	23.203	25.781
6 × 16	5-1/2 × 15-1/2	85.250	1706.776	220.229	14.800	17.760	20.720	23.681	26.641	29.601
6 × 18	5-1/2 × 17-1/2	96.250	2456.380	280.729	16.710	20.052	23.394	26.736	30.078	33.420
6 × 20	5-1/2 × 19-1/2	107.250	3398.484	348.563	18.620	22.344	26.068	29.792	33.516	37.240
6 × 22	5-1/2 × 21-1/2	118.250	4555.086	423.729	20.530	24.635	28.741	32.847	36.953	41.059
6 × 24	5-1/2 × 23-1/2	129.250	5948.191	506.229	22.439	26.927	31.415	35.903	40.391	44.878

PROPERTIES OF STRUCTURAL LUMBER: SECTIONAL PROPERTIES OF STANDARD DRESSED (S4S) SIZES (CONT'D)

Nominal size b(inches)d	Standard dressed size (S4S) b(inches)d	Area of Section A	Moment of inertia I	Section modulus S	Approximate weight in pounds per linear foot of piece when weight of wood per cubic foot equals:					
					25 lb.	30 lb.	35 lb.	40 lb.	45 lb.	50 lb.
8 × 1	7-1/4 × 3/4	5.438	0.255	0.680	0.944	1.133	1.322	1.510	1.699	1.888
8 × 2	7-1/4 × 1-1/2	10.875	2.039	2.719	1.888	2.266	2.643	3.021	3.398	3.776
8 × 3	7-1/4 × 2-1/2	18.125	9.440	7.552	3.147	3.776	4.405	5.035	5.664	6.293
8 × 4	7-1/4 × 3-1/2	25.375	25.904	14.803	4.405	5.286	6.168	7.049	7.930	8.811
8 × 6	7-1/2 × 5-1/2	41.250	103.984	37.813	7.161	8.594	10.026	11.458	12.891	14.323
8 × 8	7-1/2 × 7-1/2	56.250	263.672	70.313	9.766	11.719	13.672	15.625	17.578	19.531
8 × 10	7-1/2 × 9-1/2	71.250	535.859	112.813	12.370	14.844	17.318	19.792	22.266	24.740
8 × 12	7-1/2 × 11-1/2	86.250	950.547	165.313	14.974	17.969	20.964	23.958	26.953	29.948
8 × 14	7-1/2 × 13-1/2	101.250	1537.734	227.813	17.578	21.094	24.609	28.125	31.641	35.156
8 × 16	7-1/2 × 15-1/2	116.250	2327.422	300.313	20.182	24.219	28.255	32.292	36.328	40.365
8 × 18	7-1/2 × 17-1/2	131.250	3349.609	382.813	22.786	27.344	31.901	36.458	41.016	45.573
8 × 20	7-1/2 × 19-1/2	146.250	4634.297	475.313	25.391	30.469	35.547	40.625	45.703	50.781
8 × 22	7-1/2 × 21-1/2	161.250	6211.484	577.813	27.995	33.594	39.193	44.792	50.391	55.990
8 × 24	7-1/2 × 23-1/2	176.250	8111.172	690.313	30.599	36.719	42.839	48.958	55.078	61.198
10 × 1	9-1/4 × 3/4	6.938	0.325	0.867	1.204	1.445	1.686	1.927	2.168	2.409
10 × 2	9-1/4 × 1-1/2	13.875	2.602	3.469	2.409	2.891	3.372	3.854	4.336	4.818
10 × 3	9-1/4 × 2-1/2	23.125	12.044	9.635	4.015	4.818	5.621	6.424	7.227	8.030
10 × 4	9-1/4 × 3-1/2	32.375	33.049	18.885	5.621	6.745	7.869	8.993	10.117	11.241
10 × 6	9-1/2 × 5-1/2	52.250	131.714	47.896	9.071	10.885	12.700	14.514	16.328	18.142
10 × 8	9-1/2 × 7-1/2	71.250	333.984	89.063	12.370	14.844	17.318	19.792	22.266	24.740
10 × 10	9-1/2 × 9-1/2	90.250	678.755	142.896	15.668	18.802	21.936	25.069	28.203	31.337
10 × 12	9-1/2 × 11-1/2	109.250	1204.026	209.396	18.967	22.760	26.554	30.347	34.141	37.934
10 × 14	9-1/2 × 13-1/2	128.250	1947.797	288.563	22.266	26.719	31.172	35.625	40.078	44.531
10 × 16	9-1/2 × 15-1/2	147.250	2948.068	380.396	25.564	30.677	35.790	40.903	46.016	51.128
10 × 18	9-1/2 × 17-1/2	166.250	4242.836	484.896	28.863	34.635	40.408	46.181	51.953	57.726
10 × 20	9-1/2 × 19-1/2	185.250	5870.109	602.063	32.161	38.594	45.026	51.458	57.891	64.323
10 × 22	9-1/2 × 21-1/2	204.250	7867.879	731.896	35.460	42.552	49.644	56.736	63.828	70.920
10 × 24	9-1/2 × 23-1/2	223.250	10274.148	874.396	38.759	46.510	54.262	62.014	69.766	77.517

PROPERTIES OF STRUCTURAL LUMBER: SECTIONAL PROPERTIES OF STANDARD DRESSED (S4S) SIZES (CONT'D)

Nominal size b(inches)d	Standard dressed size (S4S) b(inches)d	Area of Section A	Moment of inertia I	Section modulus S	Approximate weight in pounds per linear foot of piece when weight of wood per cubic foot equals:					
					25 lb.	30 lb.	35 lb.	40 lb.	45 lb.	50 lb.
12 × 1	11-1/4 × 3/4	8.438	0.396	1.055	1.465	1.758	2.051	2.344	2.637	2.930
12 × 2	11-1/4 × 1-1/2	16.875	3.164	4.219	2.930	3.516	4.102	4.688	5.273	5.859
12 × 3	11-1/4 × 2-1/2	28.125	14.648	11.719	4.883	5.859	6.836	7.813	8.789	9.766
12 × 4	11-1/4 × 3-1/2	39.375	40.195	22.969	6.836	8.203	9.570	10.938	12.305	13.672
12 × 6	11-1/2 × 5-1/2	63.250	159.443	57.979	10.981	13.177	15.373	17.569	19.766	21.962
12 × 8	11-1/2 × 7-1/2	86.250	404.297	107.813	14.974	17.969	20.964	23.958	26.953	29.948
12 × 10	11-1/2 × 9-1/2	109.250	821.651	172.979	18.967	22.760	26.554	30.347	34.141	37.934
12 × 12	11-1/2 × 11-1/2	132.250	1457.505	253.479	22.960	27.552	32.144	36.736	41.328	45.920
12 × 14	11-1/2 × 13-1/2	155.250	2357.859	349.313	26.953	32.344	37.734	43.125	48.516	53.906
12 × 16	11-1/2 × 15-1/2	178.250	3568.713	460.479	30.946	37.135	43.325	49.514	55.703	61.892
12 × 18	11-1/2 × 17-1/2	201.250	5136.066	586.979	34.939	41.927	48.915	55.903	62.891	69.878
12 × 20	11-1/2 × 19-1/2	224.250	7105.922	728.813	38.932	46.719	54.505	62.292	70.078	77.865
12 × 22	11-1/2 × 21-1/2	247.250	9524.273	885.979	42.925	51.510	60.095	68.681	77.266	85.851
12 × 24	11-1/2 × 23-1/2	270.250	12437.129	1058.479	46.918	56.302	65.686	75.069	84.453	93.837
14 × 2	13-1/4 × 1-1/2	19.875	3.727	4.969	3.451	4.141	4.831	5.521	6.211	6.901
14 × 3	13-1/4 × 2-1/2	33.125	17.253	13.802	5.751	6.901	8.051	9.201	10.352	11.502
14 × 4	13-1/4 × 3-1/2	46.375	47.34	27.052	8.047	9.657	11.266	12.877	14.485	16.094
14 × 6	13-1/2 × 5-1/2	74.250	187.172	68.063	12.891	15.469	18.047	20.625	23.203	25.781
14 × 8	13-1/2 × 7-1/2	101.250	474.609	126.563	17.578	21.094	24.609	28.125	31.641	35.156
14 × 10	13-1/2 × 9-1/2	128.250	964.547	203.063	22.266	26.719	31.172	35.625	40.078	44.531
14 × 12	13-1/2 × 11-1/2	155.250	1710.984	297.563	26.953	32.344	37.734	43.125	48.516	53.906
14 × 14	13-1/2 × 13-1/2	182.250	2767.922	410.063	31.641	37.969	44.297	50.625	56.953	63.281
14 × 16	13-1/2 × 15-1/2	209.250	4189.359	540.563	36.328	43.594	50.859	58.125	65.391	72.656
14 × 18	13-1/2 × 17-1/2	236.250	6029.297	689.063	41.016	49.219	57.422	65.625	73.828	82.031
14 × 20	13-1/2 × 19-1/2	263.250	8341.734	855.563	45.703	54.844	63.984	73.125	82.266	91.406
14 × 22	13-1/2 × 21-1/2	290.250	11180.672	1040.063	50.391	60.469	70.547	80.625	90.703	100.781
14 × 24	13-1/2 × 23-1/2	317.250	14600.109	1242.563	55.078	66.094	77.109	88.125	99.141	110.156

PROPERTIES OF STRUCTURAL LUMBER: SECTIONAL PROPERTIES OF STANDARD DRESSED (S4S) SIZES (CONT'D)

Nominal size b(inches)d	Standard dressed size (S4S) b(inches)d	Area of Section A	Moment of inertia I	Section modulus S	Approximate weight in pounds per linear foot of piece when weight of wood per cubic foot equals:					
					25 lb.	30 lb.	35 lb.	40 lb.	45 lb.	50 lb.
16 × 3	15-1/4 × 2-1/2	38.125	19.857	15.885	6.619	7.944	9.267	10.592	11.915	13.240
16 × 4	15-1/4 × 3-1/2	53.375	54.487	31.135	9.267	11.121	12.975	14.828	16.682	18.536
16 × 6	15-1/2 × 5-1/2	85.250	214.901	78.146	14.800	17.760	20.720	23.681	26.641	29.601
16 × 8	15-1/2 × 7-1/2	116.250	544.922	145.313	20.182	24.219	28.255	32.292	36.328	40.365
16 × 10	15-1/2 × 9-1/2	147.250	1107.443	233.146	25.564	30.677	35.790	40.903	46.016	51.128
16 × 12	15-1/2 × 11-1/2	178.250	1964.463	341.646	30.946	37.135	43.325	49.514	55.703	61.892
16 × 14	15-1/2 × 13-1/2	209.250	3177.984	470.813	36.328	43.594	50.859	58.125	65.391	72.656
16 × 16	15-1/2 × 15-1/2	240.250	4810.004	620.646	41.710	50.052	58.394	66.736	75.078	83.420
16 × 18	15-1/2 × 17-1/2	271.250	6922.523	791.146	47.092	56.510	65.929	75.347	84.766	94.184
16 × 20	15-1/2 × 19-1/2	302.250	9577.547	982.313	52.474	62.969	73.464	83.958	94.453	104.948
16 × 22	15-1/2 × 21-1/2	333.250	12837.066	1194.146	57.856	69.427	80.998	92.569	104.141	115.712
16 × 24	15-1/2 × 23-1/2	364.250	16763.086	1426.646	63.238	75.885	88.533	101.181	113.828	126.476
18 × 6	17-1/2 × 5-1/2	96.250	242.630	88.229	16.710	20.052	23.394	26.736	30.078	33.420
18 × 8	17-1/2 × 7-1/2	131.250	615.234	164.063	22.786	27.344	31.901	36.458	41.016	45.573
18 × 10	17-1/2 × 9-1/2	166.250	1250.338	263.229	28.863	34.635	40.408	46.181	51.953	57.726
18 × 12	17-1/2 × 11-1/2	201.250	2217.943	385.729	34.939	41.927	48.915	55.903	62.891	69.878
18 × 14	17-1/2 × 13-1/2	236.250	3588.047	531.563	41.016	49.219	57.422	65.625	73.828	82.031
18 × 16	17-1/2 × 15-1/2	271.250	5430.648	700.729	47.092	56.510	65.929	75.347	84.766	94.184
18 × 18	17-1/2 × 17-1/2	306.250	7815.754	893.229	53.168	63.802	74.436	85.069	95.703	106.337
18 × 20	17-1/2 × 19-1/2	341.250	10813.359	1109.063	59.245	71.094	82.943	94.792	106.641	118.490
18 × 22	17-1/2 × 21-1/2	376.250	14493.461	1348.229	65.321	78.385	91.450	104.514	117.578	130.642
18 × 24	17-1/2 × 23-1/2	411.250	18926.066	1610.729	71.398	85.677	99.957	114.236	128.516	142.795
20 × 6	19-1/2 × 5-1/2	107.250	270.359	98.313	18.620	22.344	26.068	29.792	33.516	37.240
20 × 8	19-1/2 × 7-1/2	146.250	685.547	182.813	25.391	30.469	35.547	40.625	45.703	50.781
20 × 10	19-1/2 × 9-1/2	185.250	1393.234	293.313	32.161	38.594	45.026	51.458	57.891	64.323
20 × 12	19-1/2 × 11-1/2	224.250	2471.422	429.813	38.932	46.719	54.505	62.292	70.078	77.865
20 × 14	19-1/2 × 13-1/2	263.250	3998.109	592.313	45.703	54.844	63.984	73.125	82.266	91.406
20 × 16	19-1/2 × 15-1/2	302.250	6051.297	780.813	52.474	62.969	73.464	83.958	94.453	104.948
20 × 18	19-1/2 × 17-1/2	341.250	8708.984	995.313	59.245	71.094	82.943	94.792	106.641	118.490
20 × 20	19-1/2 × 19-1/2	380.250	12049.172	1235.813	66.016	79.219	92.422	105.625	118.828	132.031
20 × 22	19-1/2 × 21-1/2	419.250	16149.859	1502.313	72.786	87.344	101.901	116.458	131.016	145.573
20 × 24	19-1/2 × 23-1/2	458.250	21089.047	1794.813	79.557	95.469	111.380	127.292	243.203	159.115

PROPERTIES OF STRUCTURAL LUMBER: SECTIONAL PROPERTIES OF STANDARD DRESSED (S4S) SIZES (CONT'D)

Nominal size b (inches) d	Standard dressed size (S4S) b (inches) d	Area of Section A	Moment of inertia I	Section modulus S	Approximate weight in pounds per linear foot of piece when weight of wood per cubic foot equals:					
					25 lb.	30 lb.	35 lb.	40 lb.	45 lb.	50 lb.
22 × 6	21-1/2 × 5-1/2	118.250	298.088	108.396	20.530	24.635	28.741	32.847	36.953	41.059
22 × 8	21-1/2 × 7-1/2	161.250	755.859	201.563	27.995	33.594	39.193	44.792	50.391	55.990
22 × 10	21-1/2 × 9-1/2	204.250	1536.130	323.396	35.460	42.552	49.644	56.736	63.828	70.920
22 × 12	21-1/2 × 11-1/2	247.250	2724.901	473.896	42.925	51.510	60.095	68.681	77.266	85.851
22 × 14	21-1/2 × 13-1/2	290.250	4408.172	653.063	50.391	60.469	70.547	80.625	90.703	100.781
22 × 16	21-1/2 × 15-1/2	333.250	6671.941	860.896	57.856	69.427	80.998	92.569	104.141	115.712
22 × 18	21-1/2 × 17-1/2	376.250	9602.211	1097.396	65.321	78.385	91.450	104.514	117.578	130.642
22 × 20	21-1/2 × 19-1/2	419.250	13284.984	1362.563	72.786	87.344	101.901	116.458	131.016	145.573
22 × 22	21-1/2 × 21-1/2	462.250	17806.254	1656.396	80.252	96.302	112.352	128.403	144.453	160.503
22 × 24	21-1/2 × 23-1/2	505.250	23252.023	1978.896	87.717	105.260	122.804	140.347	157.891	175.434
24 × 6	23-1/2 × 5-1/2	129.250	325.818	118.479	22.439	26.927	31.415	35.903	40.391	44.878
24 × 8	23-1/2 × 7-1/2	176.250	826.172	220.313	30.599	36.719	42.839	48.958	55.078	61.198
24 × 10	23-1/2 × 9-1/2	223.250	1679.026	353.479	38.759	46.510	54.262	62.014	69.766	77.517
24 × 12	23-1/2 × 11-1/2	270.250	2978.380	517.979	46.918	56.302	65.686	75.069	84.453	93.837
24 × 14	23-1/2 × 13-1/2	317.250	4818.234	713.813	55.078	66.094	77.109	88.125	99.141	110.156
24 × 16	23-1/2 × 15-1/2	364.250	7292.586	940.979	63.238	75.885	88.533	101.181	113.828	126.476
24 × 18	23-1/2 × 17-1/2	411.250	10495.441	1199.479	71.398	85.677	99.957	114.236	128.516	142.795
24 × 20	23-1/2 × 19-1/2	458.250	14520.797	1489.313	79.557	95.469	111.380	127.292	143.203	159.115
24 × 22	23-1/2 × 21-1/2	505.250	19462.648	1810.479	87.717	105.260	122.804	140.347	157.891	175.434
24 × 24	23-1/2 × 23-1/2	552.250	25415.004	2162.979	95.877	115.052	134.227	153.403	172.578	191.753

Printed with permission of National Forest Products Association, Washington, D.C. from National Design Specification for Wood Construction

Beam Diagrams and Formulas

CASE 1 SIMPLY SUPPORTED BEAM—CONCENTRATED FORCE AT CENTER

$$R_L = R_R = V = \frac{P}{2}$$

$$M_{max} = \frac{PL}{4} \quad \text{at} \quad x = \frac{L}{2}$$

$$\theta_L = -\theta_R = \frac{-PL^2}{16EI}$$

$$y = \frac{-Px}{48EI}(3L^2 - 4x^2) \qquad 0 < x < \frac{L}{2}$$

$$y_{max} \text{ (at centerline)} = \frac{-PL^3}{48EI}$$

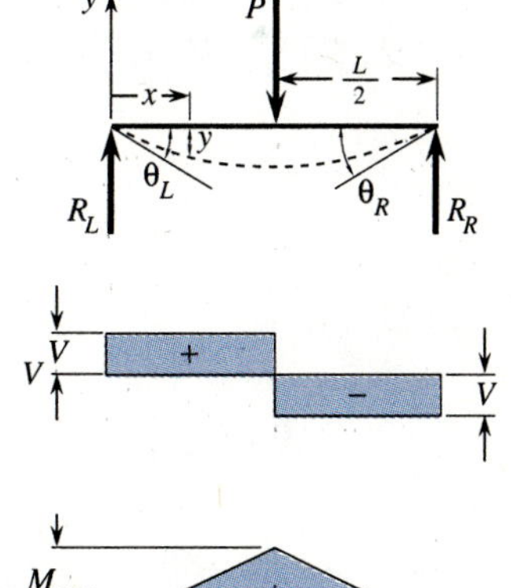

NOTE: In all equations:

1. positive deflections, y, are upward, while negative ones are downward.
2. positive rotations, θ or y', are counterclockwise, while negative ones are clockwise.

CASE 2 SIMPLY SUPPORTED BEAM—CONCENTRATED FORCE AT ANY POINT

$$R_L = V_L = \frac{Pb}{L}$$

$$R_R = V_R = \frac{Pa}{L}$$

$$M_{\max} = \frac{Pab}{L} \text{ (at point of application of load)}$$

$$\theta_L = \frac{-Pb(L^2 - b^2)}{6LEI}$$

$$\theta_R = \frac{Pab(2L - b)}{6LEI}$$

$$y = \frac{-Pbx}{6LEI}(L^2 - x^2 - b^2) \qquad 0 < x < a$$

$$y = \frac{-Pb}{6LEI}\left[\frac{L}{b}(x - a)^3 + (L^2 - b^2)x - x^3\right]$$

$$a < x < L$$

$$y_{\max} = \frac{-Pab(a + 2b)\sqrt{3a(a + 2b)}}{27LEI}$$

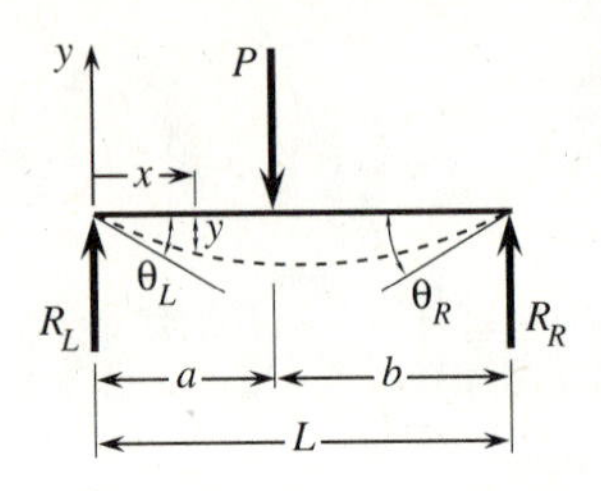

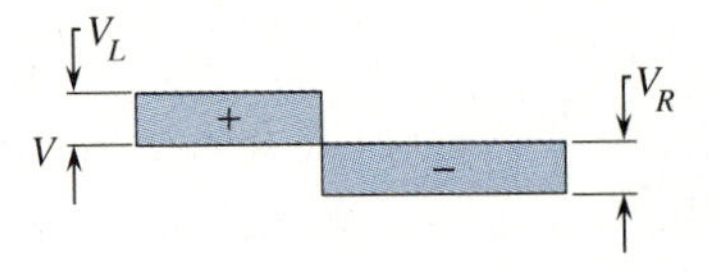

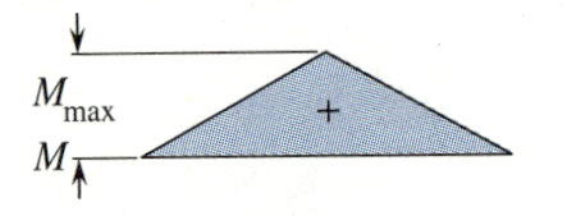

CASE 3 SIMPLY SUPPORTED BEAM—TWO EQUAL CONCENTRATED FORCES SYMMETRICALLY PLACED

$$R_L = R_R = V = P$$

$$M_{\max} = Pa$$

$$\theta_L = -\theta_R = \frac{-Pa}{2EI}(L - a)$$

$$y = \frac{-Px}{6EI}(3La - 3a^2 - x^2) \qquad 0 < x < a$$

$$y_{\max} = \frac{-Pa}{24EI}(3L^2 - 4a^2) \quad \text{at centerline}$$

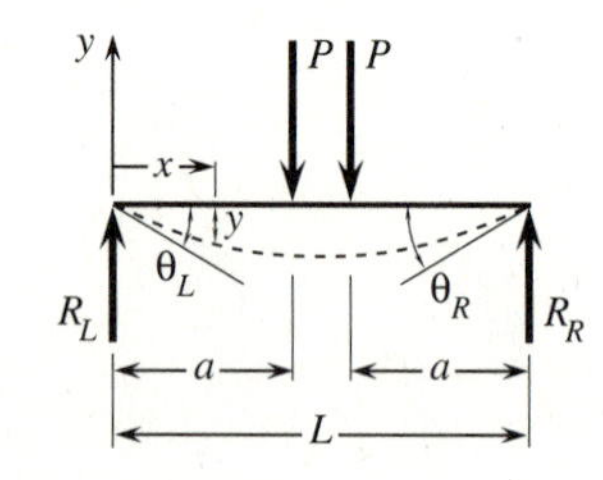

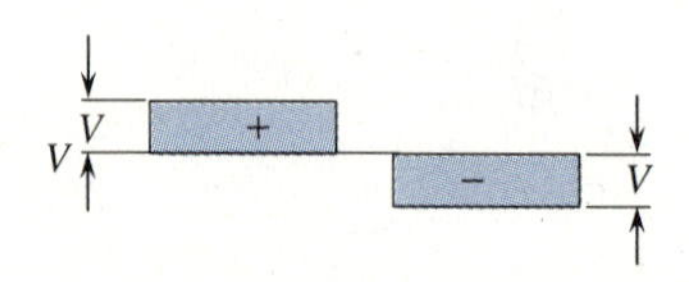

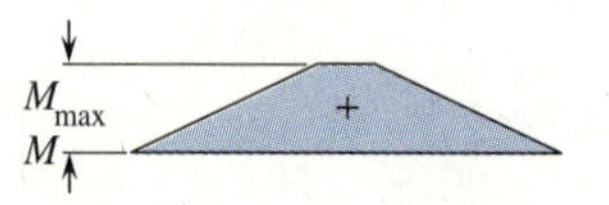

CASE 4 SIMPLY SUPPORTED BEAM—UNIFORMLY DISTRIBUTED LOAD

$$R_L = R_R = V = \frac{wL}{2}$$

$$M_{\max} = \frac{1}{8}wL^2 \quad \text{at centerline}$$

$$\theta_L = -\theta_R = \frac{-wL^3}{24EI}$$

$$y = \frac{-wx}{24EI}(L^3 - 2Lx^2 + x^3)$$

$$y_{\max} = \frac{-5wL^4}{384EI} \quad \text{at centerline}$$

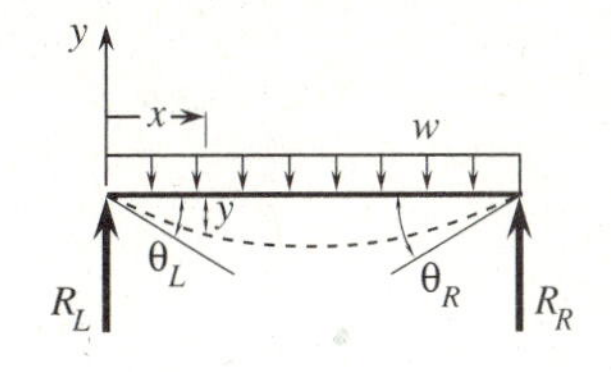

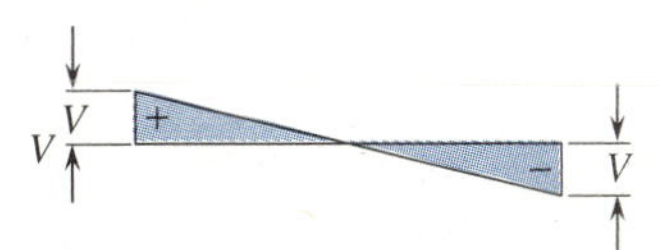

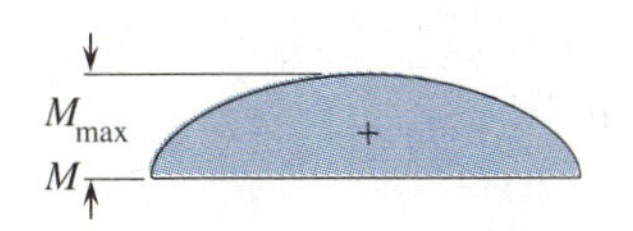

CASE 5 SIMPLY SUPPORTED BEAM—UNIFORMLY DISTRIBUTED LOAD OVER ONE-HALF BEAM

$$R_L = \frac{3wL}{8} = V_L$$

$$R_R = \frac{wL}{8} = V_R$$

$$M_{\max} = \frac{9wL^2}{128} \quad \text{at} \quad x = \frac{3L}{8}$$

$$y = \frac{-wx}{384EI}(9L^3 - 24Lx^2 + 16x^3) \qquad 0 \le x \le \frac{L}{2}$$

$$y' = \frac{-w}{384EI}(9L^3 - 72Lx^2 + 64x^3) \qquad 0 \le x \le \frac{L}{2}$$

$$y = \frac{-wL}{384EI}(8x^3 - 24Lx^2 + 17L^2x - L^3)$$

$$\frac{L}{2} \le x \le L$$

$$y' = \frac{wL}{384EI}(24x^2 - 48Lx + 17L^2)$$

$$\frac{L}{2} \le x \le L$$

$$y_c = \frac{-5wL^4}{768EI} \qquad \theta_L = \frac{-3wL^3}{128EI} \qquad \theta_R = \frac{7wL^3}{384EI}$$

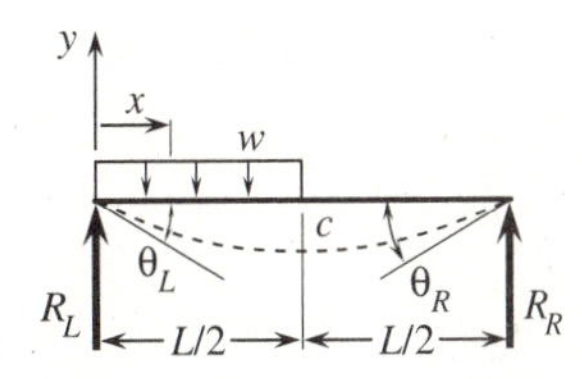

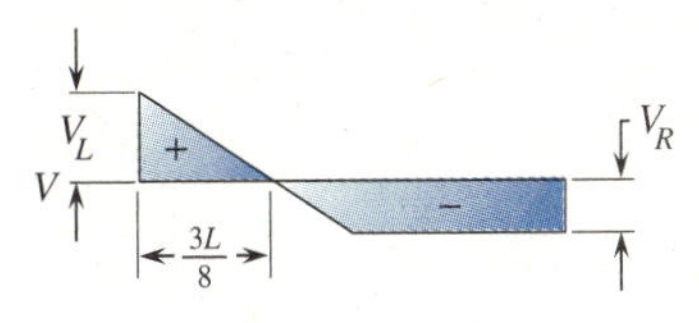

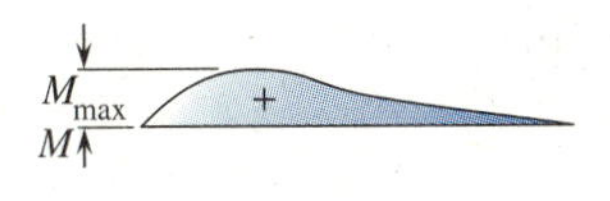

CASE 6 SIMPLY SUPPORTED BEAM—UNIFORMLY DISTRIBUTED LOAD OVER PORTION OF BEAM

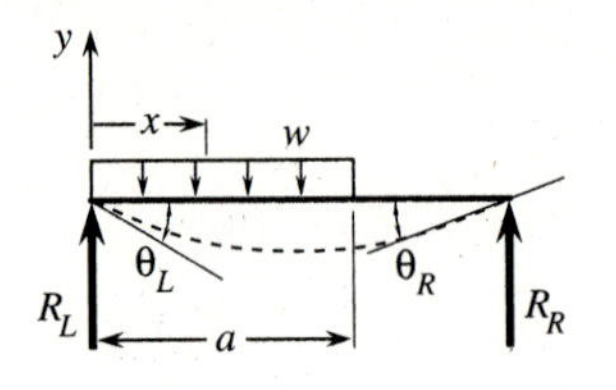

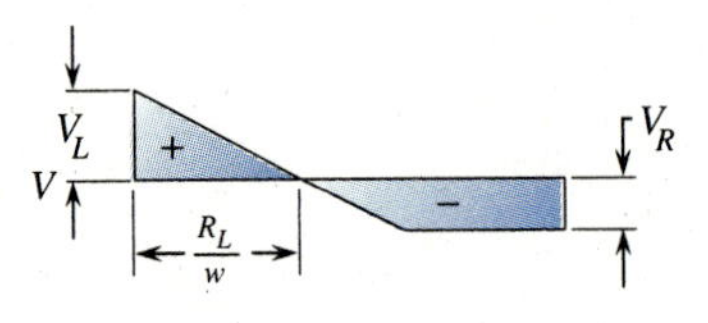

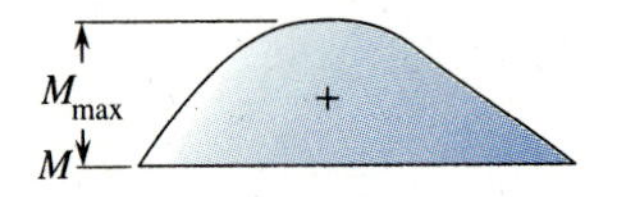

$$R_L = \frac{wa}{2L}(2L - a) = V_L$$

$$R_R = \frac{wa^2}{2L} = V_R$$

$$M_{max} = \frac{R_L^2}{2w} \quad \text{at} \quad x = \frac{R_1}{w}$$

$$y = \frac{-wx}{24LEI}(a^4 - 4a^3L + 4a^2L^2 + 2a^2x^2 - 4aLx^2 + Lx^3) \qquad 0 \le x \le a$$

$$y' = \frac{-w}{24LEI}(a^4 - 4a^3L + 4a^2L^2 + 6a^2x^2 - 12aLx^2 + 4Lx^3) \qquad 0 \le x \le a$$

$$y = \frac{-wa^2}{24LEI}(-a^2L + 4L^2x + a^2x - 6Lx^2 + 2x^3) \qquad a \le x \le L$$

$$y' = \frac{wa^2}{24LEI}(4L^2 + a^2 - 12Lx + 6x^2) \qquad a \le x \le L$$

$$\theta_L = \frac{-wa^2}{24LEI}(2L - a)^2 \qquad \theta_R = \frac{wa^2}{24LEI}(2L^2 - a^2)$$

CASE 7 SIMPLY SUPPORTED BEAM—LINEARLY VARYING DISTRIBUTED LOAD

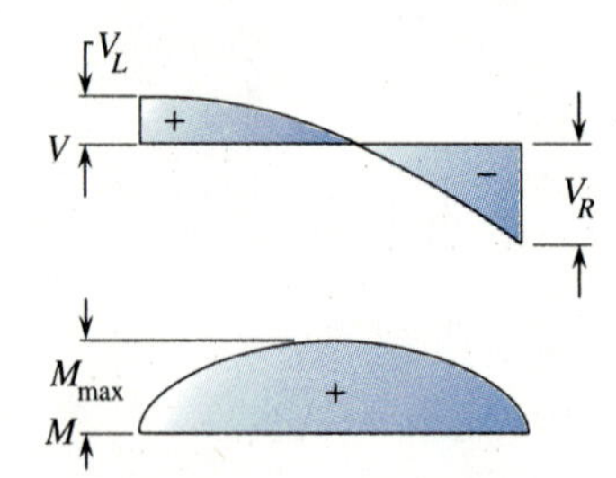

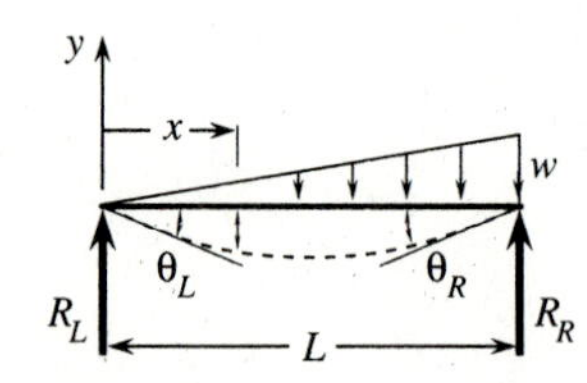

$$R_L = V_L = \frac{wL}{6}$$

$$R_R = V_R = \frac{wL}{3}$$

$$M_{max} = 0.0642wL^2 \quad \text{at} \quad x = 0.577L$$

$$\theta_L = \frac{-7wL^3}{360EI}$$

$$\theta_R = \frac{wL^3}{45EI}$$

$$y = \frac{-wx}{360LEI}(7L^4 - 10L^2x^2 + 3x^4)$$

$$y_{max} = -0.00652\frac{wL^4}{EI} \quad \text{at} \quad x = 0.519L$$

CASE 8 SIMPLY SUPPORTED BEAM—TWO LINEARLY VARYING DISTRIBUTED LOADS

$$R_L = R_R = V = \frac{wL}{2}$$

$$M_{\max} = \frac{wL^2}{6} \quad \text{at} \quad x = \frac{L}{2}$$

$$y = \frac{-w_0 x}{960LEI}(5L^2 - 4x^2)^2 \qquad 0 \le x \le \frac{L}{2}$$

$$y' = \frac{w_0}{192LEI}(5L^2 - 4x^2)(L^2 - 4x^2)$$

$$0 \le x \le \frac{L}{2}$$

$$y_c = y_{\max} = \frac{-w_0 L^4}{120EI} \qquad \theta_L = -\theta_R = \frac{-5w_0 L^3}{192EI}$$

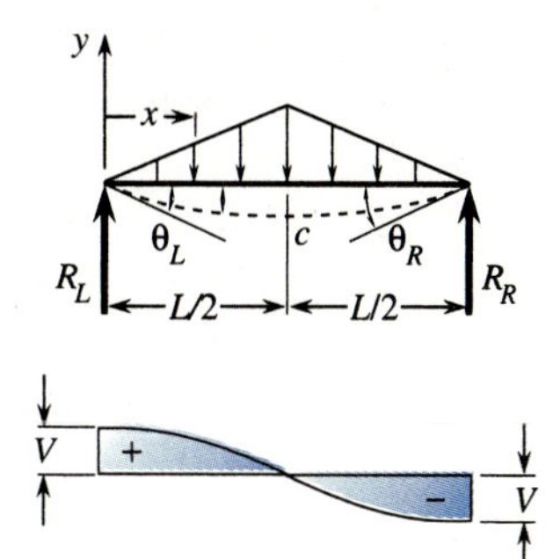

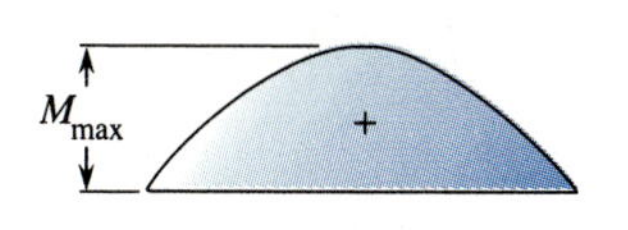

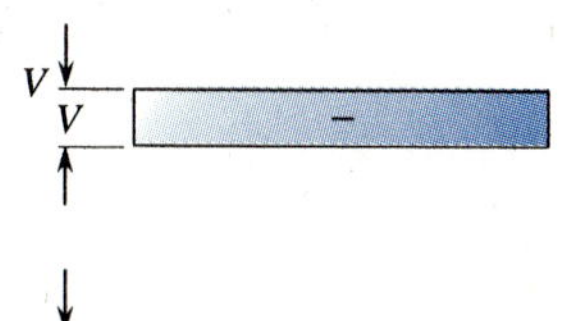

CASE 9 SIMPLY SUPPORTED BEAM—MOMENT AT ONE END

$$R_L = R_R = V = \frac{M}{L}$$

$$M_{\max} = M$$

$$\theta_L = \frac{ML}{3EI}$$

$$\theta_R = \frac{ML}{6EI}$$

$$y = \frac{-Mx}{6LEI}(2L^2 - 3Lx + x^2)$$

$$y_{\max} = \frac{-0.064ML^2}{EI} \quad \text{at} \quad x = 0.423L$$

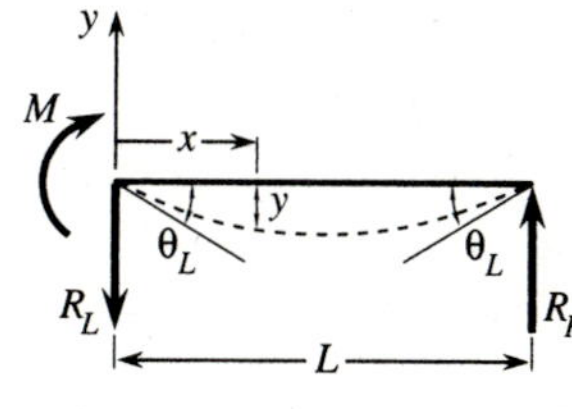

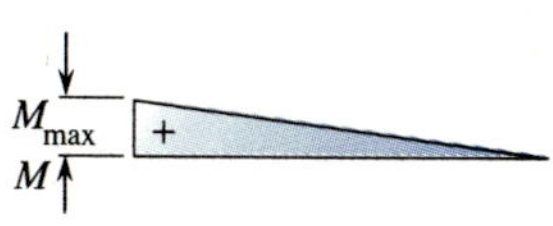

CASE 10 SIMPLY SUPPORTED BEAM—MOMENT AT MIDDLE

$$R_L = \frac{M}{L} = R_R = V$$

$$M_{\max} = \frac{M}{2}$$

$$y = \frac{-Mx}{24LEI}(L^2 - 4x^2) \qquad 0 \le x \le \frac{L}{2}$$

$$\theta = \frac{-M}{24LEI}(L^2 - 12x^2) \qquad 0 \le x \le \frac{L}{2}$$

$$y_c = 0 \qquad \theta_L = \frac{-ML}{24EI} \qquad \theta_R = \frac{ML}{24EI}$$

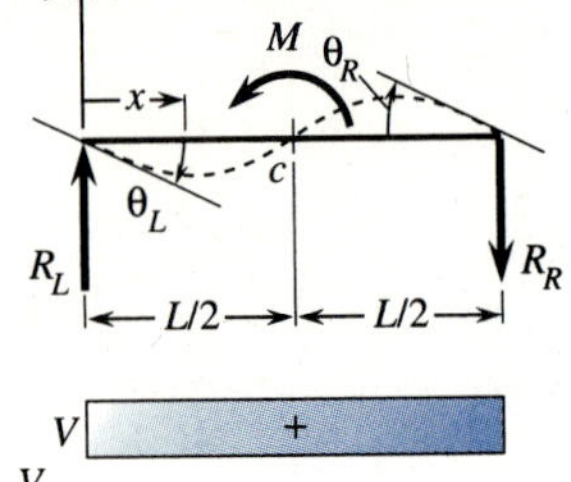

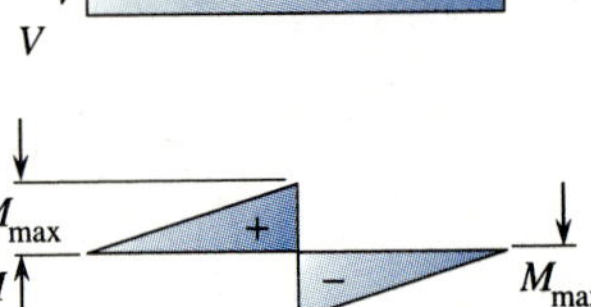

CASE 11 SIMPLY SUPPORTED BEAM—MOMENT AT ANY LOCATION

$$R_L = \frac{M}{L} = R_R = V$$

$$M_{\max} = \frac{Ma}{L} \quad \text{at } x = a \quad \text{for } a > b$$

$$y = \frac{Mx}{6LEI}(6aL - 3a^2 - 2L^2 - x^2) \qquad 0 \le x \le a$$

$$\theta = \frac{M}{6LEI}(6aL - 3a^2 - 2L^2 - 3x^2) \qquad 0 \le x \le a$$

$$\text{At } x = a\text{:} \quad y = \frac{-Mab}{3LEI}(2a - L)$$

$$\text{At } x = a\text{:} \quad \theta = \frac{M}{3LEI}(3aL - 3a^2 - L^2)$$

$$\theta_L = \frac{-M}{6LEI}(6aL - 3a^2 - 2L^2)$$

$$\theta_R = \frac{-M}{6LEI}(3a^2 - L^2)$$

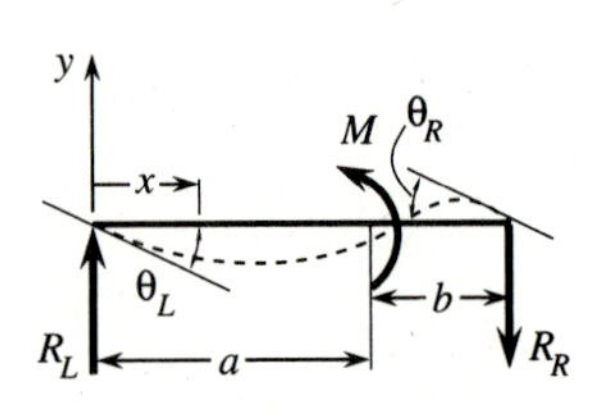

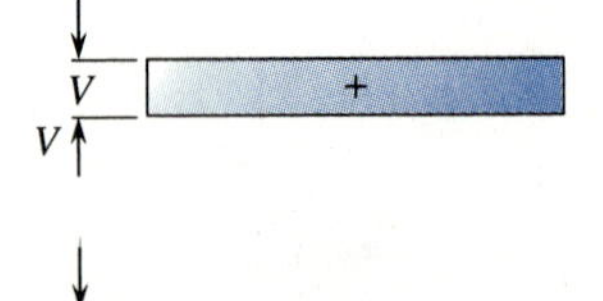

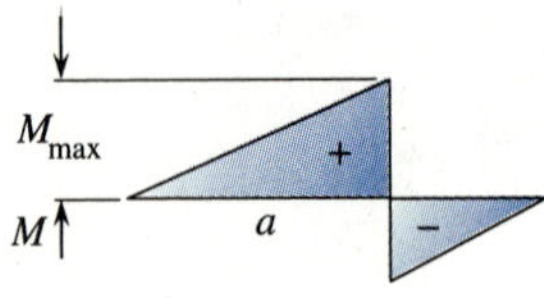

CASE 12 CANTILEVERED BEAM—CONCENTRATED FORCE AT THE FREE END

$$R_L = V = P$$

$$M_{\max} = PL \quad \text{at} \quad x = 0$$

$$\theta_{\text{end}} = \frac{-PL^2}{2EI}$$

$$y = \frac{-Px^2}{6EI}(3L - x)$$

$$y_{\max} = \frac{-PL^3}{3EI} \quad \text{at free end}$$

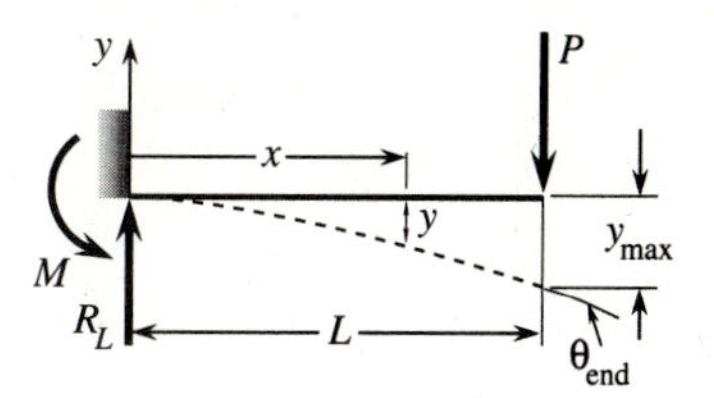

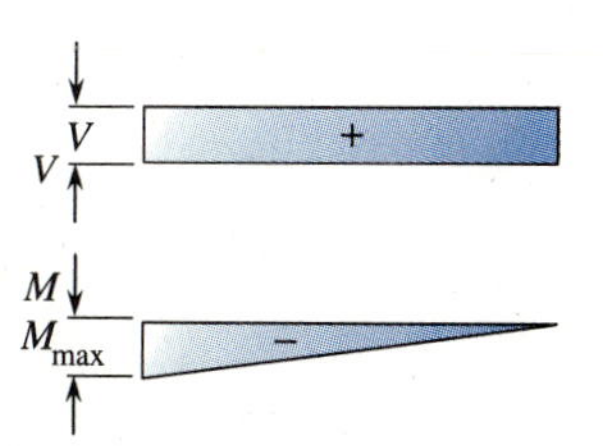

CASE 13 CANTILEVERED BEAM—CONCENTRATED FORCE AT ANY LOCATION

$$R_L = V = P$$

$$M_{\max} = Pa \quad \text{at} \quad x = 0$$

$$\theta_{\text{end}} = \frac{-Pa^2}{2EI}$$

$$y = \frac{-Px^2}{6EI}(3a - x) \qquad 0 < x < a$$

$$y = \frac{-Pa^2}{6EI}(3x - a) \qquad a < x < L$$

$$y_{\max} = \frac{-Pa^2}{6EI}(3L - a) \quad \text{at free end}$$

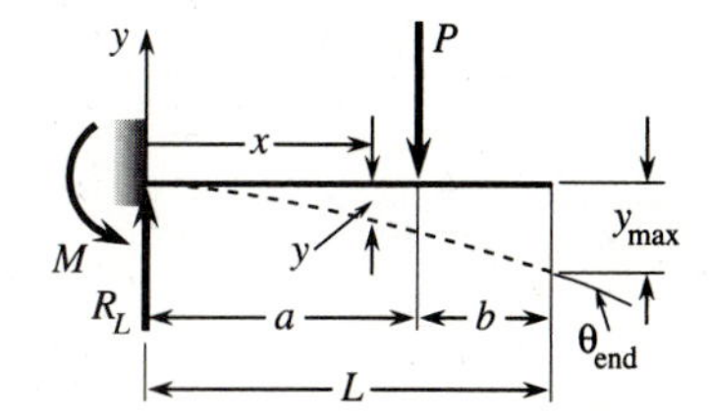

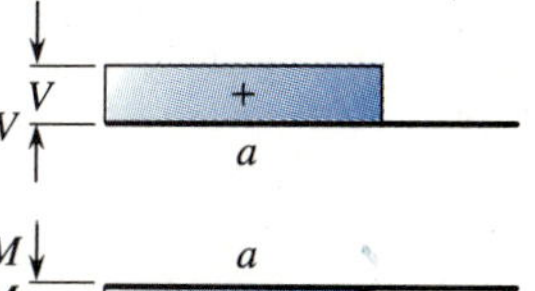

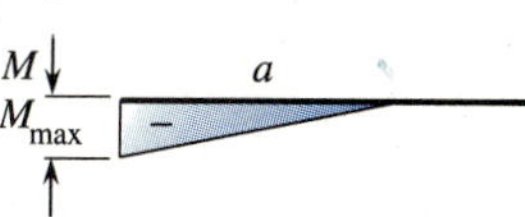

CASE 14 CANTILEVERED BEAM—MOMENT AT FREE END

$$R_L = 0 = V$$

$$M_{\max} = M$$

$$y = \frac{Mx^2}{2EI} \qquad \theta = \frac{Mx}{EI}$$

$$y_{\max} = \frac{ML^2}{2EI} \qquad \theta_{\text{end}} = \frac{ML}{EI}$$

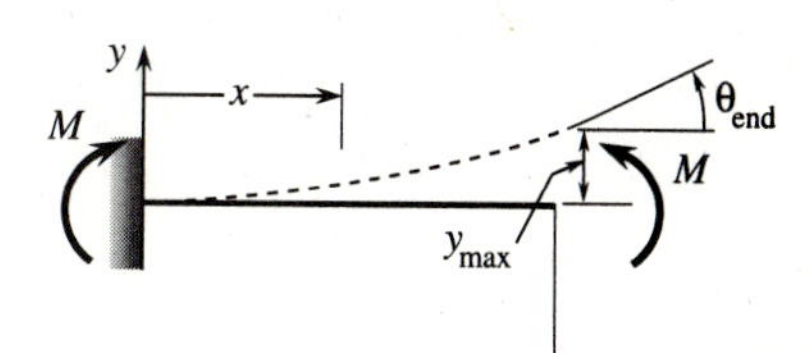

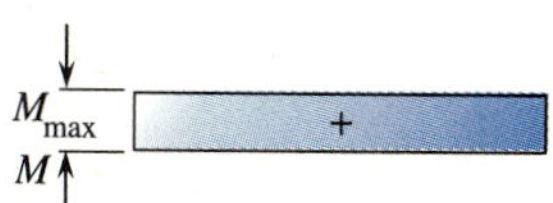

CASE 15 CANTILEVERED BEAM—MOMENT AT ANY LOCATION

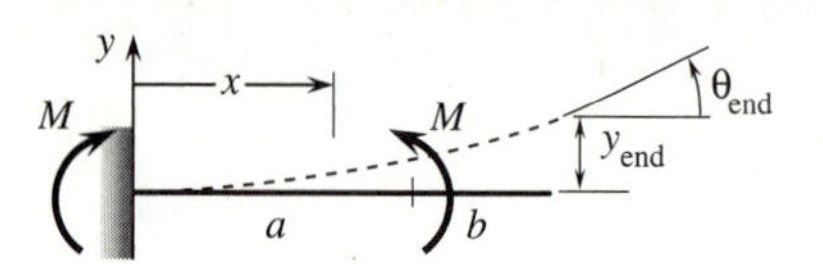

$R_L = 0 = V$

$M_{max} = M$

$y = \frac{Mx^2}{2EI} \qquad \theta = \frac{Mx}{EI} \qquad 0 \leq x \leq a$

$y = \frac{Ma}{2EI}(2x - a) \qquad \theta = \frac{Ma}{EI} \qquad a \leq x \leq L$

At $x = a$: $\quad y = \frac{Ma^2}{2EI} \qquad \theta = \frac{Ma}{EI}$

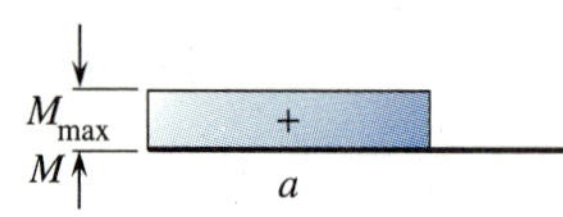

$y_{end} = \frac{Ma}{2EI}(2L - a) \qquad \theta_{end} = \frac{Ma}{EI}$

CASE 16 CANTILEVERED BEAM—UNIFORMLY DISTRIBUTED LOAD

$R_L = V = wL$

$M_{max} = \frac{wL^2}{2}$

$\theta_{end} = \frac{-wL^3}{6EI}$

$y = \frac{-wx^2}{24EI}(6L^2 - 4Lx + x^2)$

$y_{max} = \frac{-wL^4}{8EI}$

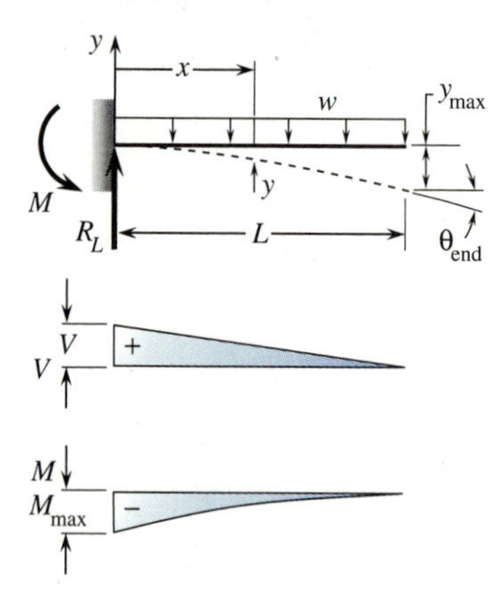

CASE 17 CANTILEVERED BEAM—UNIFORMLY DISTRIBUTED LOAD OVER PORTION OF BEAM

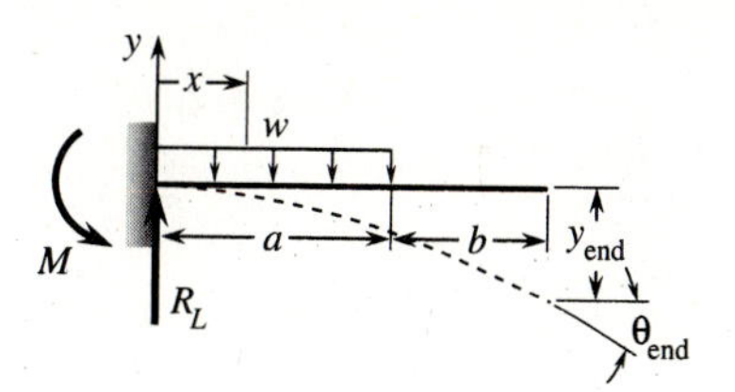

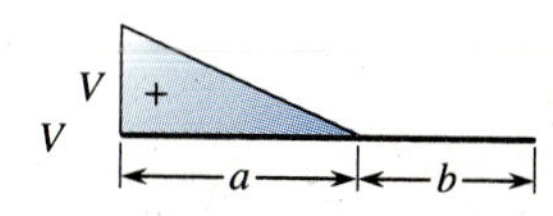

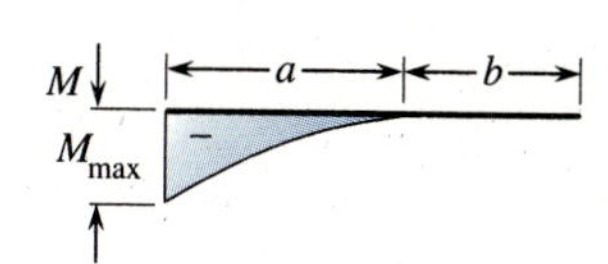

$R_L = V = wa$

$M_{max} = \dfrac{wa^2}{2}$

$y = \dfrac{wx^2}{24EI}(6a^2 - 4ax + x^2) \qquad 0 \le x \le a$

$\theta = \dfrac{wx}{6EI}(3a^2 - 3ax + x^2) \qquad 0 \le x \le a$

$y = \dfrac{-wa^3}{24EI}(4x - a) \qquad \theta = \dfrac{-wa^3}{6EI}$

$a \le x \le L$

At $x = a$: $\quad y = \dfrac{-wa^4}{8EI} \qquad \theta = \dfrac{-wa^3}{6EI}$

$y_{end} = \dfrac{-wa^3}{24EI}(4L - a) \qquad \theta_{end} = \dfrac{-wa^3}{6EI}$

CASE 18 CANTILEVERED BEAM—UNIFORMLY DISTRIBUTED LOAD OVER PORTION OF BEAM

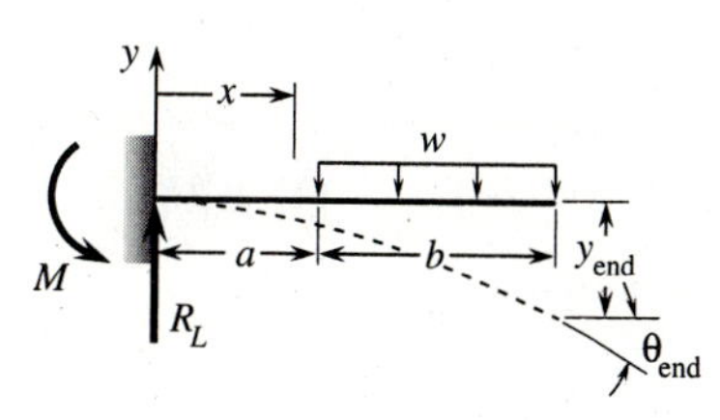

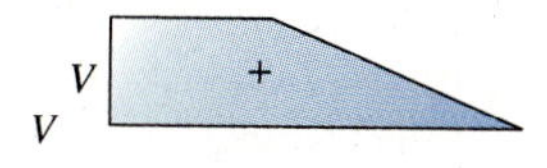

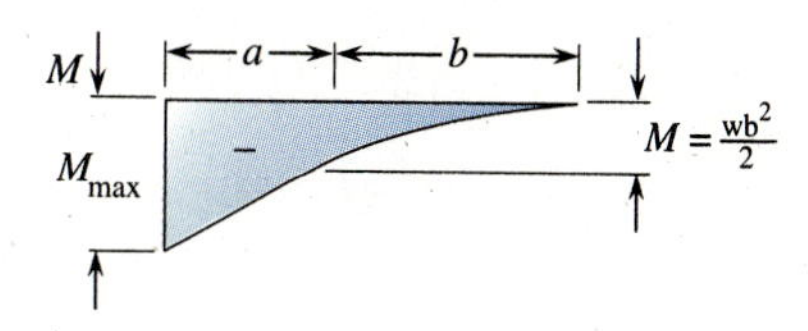

$R_L = V = wb$

$M_{max} = wb\left(a + \dfrac{b}{2}\right)$

$y = \dfrac{-wbx^2}{12EI}(3L + 3a - 2x) \qquad 0 \le x \le a$

$\theta = \dfrac{-wbx}{2EI}(L + a - x) \qquad 0 \le x \le a$

$y = \dfrac{-w}{24EI}(x^4 - 4Lx^3 + 6L^2x^2 - 4a^3x + a^4)$

$a \le x \le L$

$\theta = \dfrac{-w}{6EI}(x^3 - 3Lx^2 + 3L^2x - a^3)$

$a \le x \le L$

At $x = a$: $\quad y = \dfrac{-wa^2b}{12EI}(3L + a)$

$\theta = \dfrac{-wabL}{2EI}$

$y_{end} = \dfrac{-w}{24EI}(3L^4 - 4a^3L + a^4)$

$\theta_{end} = \dfrac{-w}{6EI}(L^2 - a^3)$

CASE 19 CANTILEVERED BEAM—LINEARLY VARYING DISTRIBUTED LOAD

$$R_L = V = \frac{w_0 L}{2}$$

$$M_{\max} = \frac{w_0 L^2}{6}$$

$$y = \frac{-w_0 x^2}{120LEI}(10L^3 - 10L^2 x + 5Lx^2 - x^3)$$

$$\theta = \frac{-w_0 x}{24LEI}(4L^3 - 6L^2 x + 4Lx^2 - x^3)$$

$$y_{\text{end}} = \frac{-w_0 L^4}{30EI} \qquad \theta_{\text{end}} = \frac{-w_0 L^3}{24EI}$$

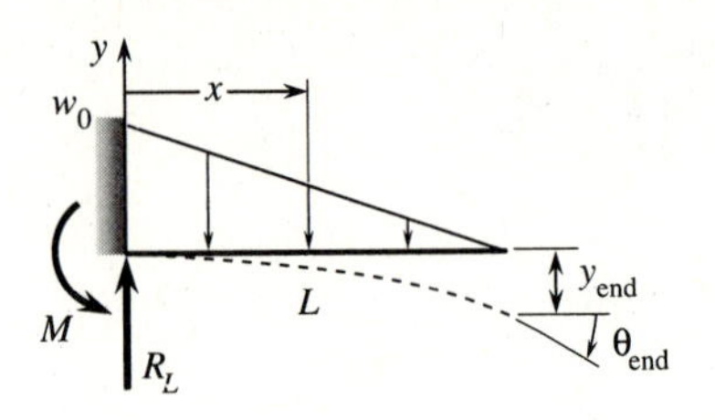

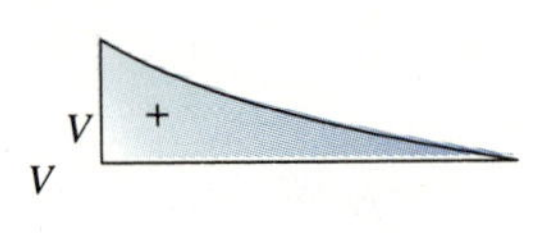

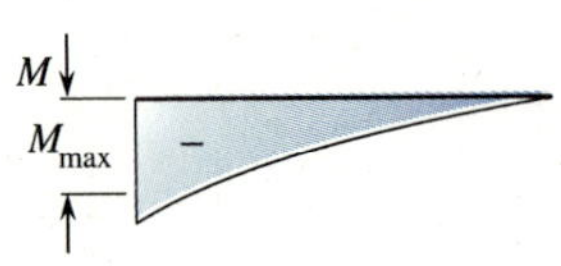

CASE 20 CANTILEVERED BEAM—LINEARLY VARYING DISTRIBUTED LOAD

$$R_L = V = \frac{w_0 L}{2}$$

$$M_{\max} = \frac{w_0 L^2}{3}$$

$$y = \frac{-w_0 x^2}{120LEI}(20L^3 - 10L^2 x + x^3)$$

$$\theta = \frac{-w_0 x}{24LEI}(8L^3 - 6L^2 x + x^3)$$

$$y_{\text{end}} = \frac{-11 w_0 L^4}{120EI} \qquad \theta_{\text{end}} = \frac{-w_0 L^3}{8EI}$$

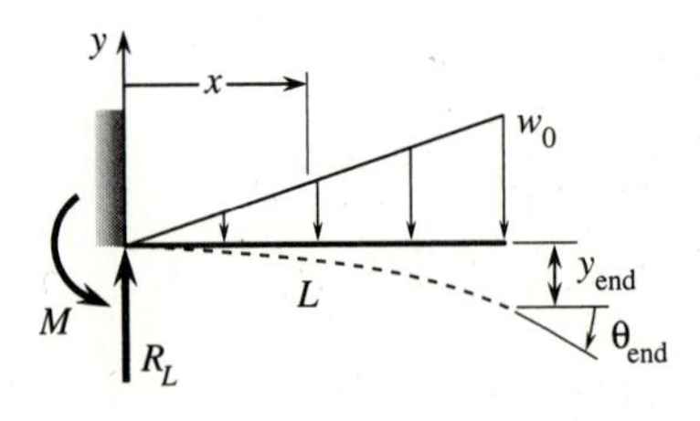

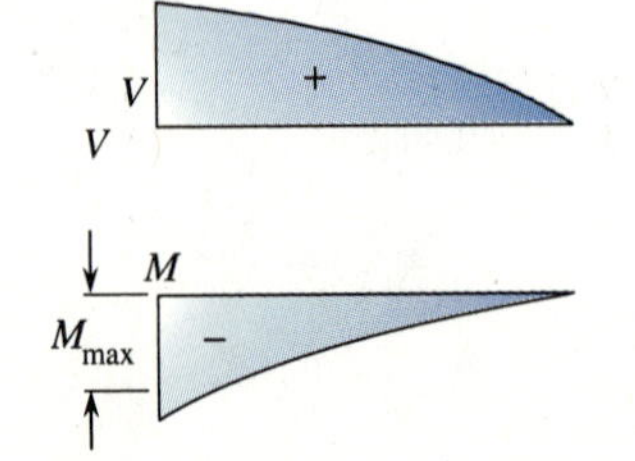

CASE 21 OVERHANGING BEAM—CONCENTRATED LOAD AT END OF OVERHANG

$$R_1 = V_1 = \frac{Pa}{L}$$

$$R_2 = V_1 + V_2 = \frac{P}{L}(L + a)$$

$$V_2 = P$$

$$M_{max} \text{ (at } R_2) = Pa$$

$$M_x \text{ (between supports)} = \frac{Pax}{L}$$

$$M_{x_1} \text{ (for overhang)} = P(a - x_1)$$

$$y_{max} \left(\text{between supports at } x = \frac{L}{\sqrt{3}}\right) =$$

$$\frac{PaL^2}{9\sqrt{3}EI} = 0.06415\frac{PaL^2}{EI}$$

$$y_{max} = \text{ (for overhang at } x_1 = a) = \frac{-Pa^2}{3EI}(L + a)$$

$$y \text{ (between supports)} = \frac{Pax}{6EIL}(L^2 - x^2)$$

$$y \text{ (for overhang)} = \frac{-Px_1}{6EI}(2aL + 3ax_1 - x_1^2)$$

CASE 22 OVERHANGING BEAM—UNIFORMLY DISTRIBUTED LOAD BETWEEN SUPPORTS

$$R = V = \frac{wL}{2}$$

$$V_x = w\left(\frac{L}{2} - x\right)$$

$$M_{max} \text{ (at center)} = \frac{wL^2}{8}$$

$$M_x = \frac{wx}{2}(L - x)$$

$$y_{max} \text{ (at center)} = \frac{-5wL^4}{384EI}$$

$$y \text{ (between supports)} = \frac{-wx}{24EI}(L^3 - 2Lx^2 + x^3)$$

$$y \text{ (overhang)} = \frac{wL^3x_1}{24EI}$$

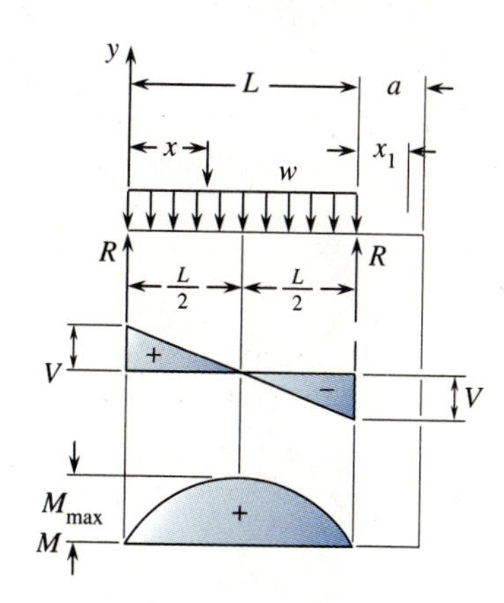

CASE 23 OVERHANGING BEAM—CONCENTRATED LOAD AT ANY POINT BETWEEN SUPPORTS

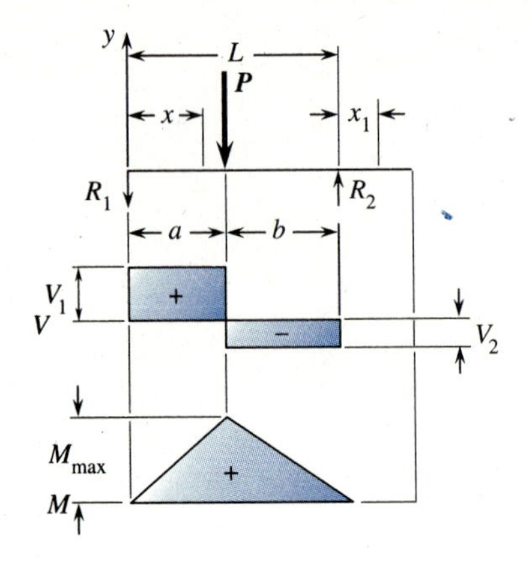

$R_1 = V_1$ (max. when $a < b$) $= \dfrac{Pb}{L}$

$R_2 = V_2$ (max. when $a > b$) $= \dfrac{Pa}{L}$

M_{max} (at point of load) $= \dfrac{Pab}{L}$

M_x (when $x < a$) $= \dfrac{Pbx}{L}$

$$y_{max}\left(\text{at } x = \sqrt{\frac{a(a+2b)}{3}} \text{ when } a > b\right) = \frac{Pab(a+2b)\sqrt{3a(a+2b)}}{27EIL}$$

y_a (at point of load) $= \dfrac{-Pa^2b^2}{3EIL}$

y_x (when $x < a$) $= \dfrac{-Pbx}{6EIL}(L^2 - b^2 - x^2)$

y (when $x > a$) $= \dfrac{-Pa(L-x)}{6EIL}(2Lx - x^2 - a^2)$

y (overhang) $= \dfrac{Pabx_1}{6EIL}(L + a)$

Numbering Systems for Some Metals

Standard numbering systems exist to identify various metals and their properties. In this appendix, we briefly describe the most common numbering systems used for the steels, cast irons, and aluminums. The Society of Automotive Engineers (SAE) first established a numbering system for steels [1]. Later the American Iron and Steel Institute (AISI) adopted a similar system. In 1975 the SAE developed a Unified Numbering System for Metals and Alloys (UNS). The American Society for Testing and Materials (ASTM) also established a numbering system for cast irons and steels [2]. However a single designation system for universal use has still not been fully accepted.

The SAE uses a four-digit identification and designation system for all steels produced and used in the United States. The steels are divided into two general categories: carbon steels and alloy steels.

Carbon Steels

There are four classifications for carbon steels. They are identified by the first two digits of the four digit identification. These are:

10xx which stands for nonsulfurized carbon steel. This series is the basic structural low-carbon or mild steel.

11xx which stands for resulfurized carbon steel. These are the free machining steels. They are inherently brittle and are of little practical use.

12xx which stands for resulfurized and rephosphorized carbon steel. They are also of little practical use.

15xx which stands for nonsulfurized, high-manganese carbon steel. This is the basic carbon steel used for low cost forgings.

The last two digits of the designator, represented by the xx, indicate the approximate midpoint amount of carbon content. For example SAE 1010 steel contains from 0.07% to 0.13% carbon, while SAE 1020 contains approximately 0.17% to 0.23% carbon.

Alloy Steels

Alloy steels are also specified by a four digit number. The last two numbers indicate the major alloying element or elements. The groups are:

13xx Manganese—1.75%
40xx Molybdenum—0.20% or 0.25%
41xx Chromium—0.50%, 0.80% or 0.95% plus Molybdenum 0.25%
43xx Nickel—1.83% plus Chromium 0.50% or 0.80% plus Molybdenum 0.25%
44xx Molybdenum—0.53%
46xx Nickel—0.85% or 1.83% plus Molybdenum 0.20% or 0.25%
61xx Chromium—0.60% or 0.95% plus Vanadium 0.13% or 0.15%
86xx Nickel—0.55% plus Chromium 0.50% plus Molybdenum 0.20%
87xx Nickel—0.55% plus Chromium 0.50% plus Molybdenum 0.25%
88xx Nickel—0.55% plus Chromium 0.50% plus Molybdenum 0.35%
92xx Silicon—2.00%

Stainless Steels

Stainless steels are ferrous alloys which contain 10.5% or more chromium plus other alloying elements. They are normally highly corrosive resistant and retain their mechanical properties at high temperatures. The chromium imparts the corrosive resistance to the stainless steels by combining with oxygen in the atmosphere to form a thin and transparent protective coating of chromium oxide on the surface of the metal. There are four basic groups of stainless steels:

1. The austenitic stainless steels are of the 300 series, such as 302 and 321. They are basically chromium-nickel alloys, can be hardened by cold working, have excellent corrosion resistance, and are quite formable. They are available in sheet and tube form. The 321 type is excellent for exhaust systems of vehicles.
2. The ferritic stainless steels of the 400 series are chromium alloys. They can be only moderately hardened by cold working and exhibit good ductility and corrosion resistance.
3. The martensitic stainless steels are chromium based alloys with reasonable ductility and good corrosion resistance.
4. The precipitation hardening stainless steels are chromium-nickel alloys which can be hardened to very high strengths.

Cast Irons

The ASTM numbering system for cast iron is in common use. The system is guided by the approximate tensile strength of the cast iron. For instance, ASTM A-47 malleable cast iron has a tensile strength of approximately 50 ksi. (See Appendix B or the back inside cover of the text.)

Aluminum Alloys

Similar to the steel industry, the aluminum industry established a four-digit system to designate and identify the numerous wrought aluminum alloys. In the U.S. System, the first digit identifies the major alloying element. These alloys are:

1xxx which stands for 99% pure aluminum. For instance, aluminum alloy 1100 is 99% pure aluminum with some specified amount of impurities. The second digit indicates modification in impurity limits while the last two digits designate the absolute minimum aluminum content allowed in 1/100's of 1% beyond the 99%.

In series 2xxx through 7xxx the major alloying element is identified by the first digit as follows:

2xxx Copper is the major alloying element
3xxx Manganese is the major alloying element
4xxx Silicon is the major alloying element
5xxx Magnesium is the major alloying element
6xxx Magnesium and silicon are the major alloying elements
7xxx Zinc is the major alloying element

Each of these alloys produces different mechanical properties in the aluminum, such as strength, ductility, and weldability. The 2, 6, and 7 series alloys are heat treatable while the 3, 4, and 5 series are not heat treatable. For more on properties of aluminum alloys consult reference [3].

The UNS uses a letter prefix to designate the kind of material. Standard designations have been set for some materials. For example, G is used for the carbon and alloy steels, A is used for aluminum alloys, C for copper-based alloys, and S for stainless or corrosion resistant steels.

For steels, the first two numbers following the prefix indicate the composition, excluding the carbon content. Some of the different compositions are listed as follows:

G10 Plain carbon
G11 Free-cutting carbon steel with more sulfur or phosphorus
G13 Manganese
G23 Nickel
G25 Nickel
G31 Nickel-chromium
G33 Nickel-chromium
G40 Molybdenum
G41 Chromium-molybdenum
G43 Nickel-chromium-molybdenum
G46 Nickel-molybdenum
G48 Nickel-molybdenum
G50 Chromium

The second pair of numbers indicates the approximate amount of carbon content. For instance, G10400 is a plain carbon steel with a carbon content of 0.37 to 0.44 percent as indicated by the pair of numbers 40 in the designation. The fifth or last digit following the number is used for special situations.

The UNS designation for stainless steels and aluminum alloys can be found in references [1] and [4] along with cross reference numbers for other material specifications.

REFERENCES

1. SAE Handbook, latest edition, Society of Automotive Engineers, Inc., 400 Commonwealth Drive, Warrendale, PA 15096.
2. ASTM Annual Book of Standards, American Society for Testing and Materials, 1916 Race Street, Philadelphia, PA 19103.
3. Aluminum Standards and Data, 9th edition, The Aluminum Association, Inc., 900 19th Street, N.W., Washington, D.C. 20006, 1988.
4. ASM Metals Reference Book, latest edition, American Society for Metals, Metals Park, Ohio 44073.

Index

A

AASHTO (American Association of State Highway and Transportation Officials), 94, 105, 279, 284, 586, 591
Absolute maximum shear stress, 505–13, 551
Absolute system of units, 2
Acceleration of gravity, 4
ACI (American Concrete Institute), 93–94, 104
AISC (American Institute of Steel Construction), 92, 94, 104, 132, 141, 153, 273, 286, 326, 346, 417, 584–85, 590–91
AITC (American Institute of Timber Construction), 94, 104, 273, 326
Alclad, 587
Allowable load, 92
Allowable stress, 91–92
  in pressure vessels, 455
  in welds, 141
Aluminum alloys, 48
Aluminum Association, 94, 104, 273, 326, 587, 591, 764
Aluminum Association:
  standard I-beams, 744
  standard channels, 743
Aluminum columns, design of, 587–89
Aluminum, design specification for, 104, 326, 587, 591, 764
American standard beam, 273, 728–29
American standard channels, 273, 730–32
American standard shapes (also see S-shapes), 728–29
Angle beams, 308–11, 414
Angle of twist, 166
  in circular shaft, 166–68
    in elastic range, 171, 210
    in plastic range, 194
  for narrow rectangular cross section, 197, 210
  for rectangular bar, 196, 210
  for thin-walled closed tube, 202–206
Angle shapes, 273
  properties of, 740–42
Anisotropic material, 50
Annealing, 58
ANSI (American National Standards Institute), 94, 104
ANSYS, 515, 551
AREA (American Railway Engineers Association), 94, 105
Area-moment method (see Moment-area method)
Area-moment of inertia, 246
Areas of common shapes, 715–18, inside back cover
Askeland, D.R., 104
ASM (American Society of Metals), 56, 105, 764
ASME Boiler and Pressure Vessel Code, 94, 105, 458
ASTM (American Society for Testing Materials), 44–45, 104, 451, 574, 764
Atmospheric pressure, inside back cover
Auger, 211, 517
Automobile hood mechanism, 154
Automobile lift, 595
AWS (American Welding Society), 141, 153
Axial loading
  centric, 31
    elastic deformations under, 61–70
    of columns, 561–89
    plastic deformations under, 100–101
    strain due to, 42
    stress due to, 30–37
  eccentric (see Combined loading)
Axle, 440, 483

B

Bar with rectangular cross section
  torsion of, 195–97
BBC (Basic Building Code), 94
Beach marks, 693–94
Beam shear stress formula, 259
Beams
  boundary conditions, 352–53, 416
  cantilevered (see Cantilevered beams)

Beams *(continued)*
composites, 286–92
continuity conditions, 352–53, 416
deflection of (also see Deflection of beams), 348
design of, 273–79
long, 351, 613
made of several materials (see Composite beams)
moving loads on, 279–86
normal stresses in, 244–56
shear stresses in, 256–73
slope of (also see Slope of beams), 349
strain in, 243–44
with three supports, 410
Beam-column, 24
Beam diagrams, 751–62
Bearing stress, 126, 146–53
Bending
deformations in, 348
in plane of symmetry, 241
of members made of several materials (see Composite beams)
stiffness, 350, 385
stress in (see Normal stress, in bending)
unsymmetric, 296–304
Bending moment
sign convention for, 224, 351
Bending moment diagram, 223, 226–40
Bending shear stress (also see Shear stress in beam), 261
Bending stiffness, 350, 385
Bent pipe, 13
Biaxial stress (also see Plane stress), 513, 519
Bleich, F., 578, 591
Block compression allowable, 579
Blodgett, O.W., 153, 458
Bolts, stress in, 127–32, 133–38
Boundary conditions
for statically determinate beams, 223, 352–53, 416
Box beam, 3, 465
Bracing, lateral, 568
Bredt, Rudolph, 203
Breaking strength (see Fracture strength)
Bridge pier, 222, 562
Brinell hardness number, 56
Brittle material, 39, 53, 175
fracture theory for, 518
Buckling load, 565
critical, 563
Buckling modes, 565, 567–68
Buckled column in load frame, 562
Buckled shape, 569
Building codes, 94, 104–5
Building frame, 3, 347
Bulk modulus, 550–51
Bumper stop, 687
Butt weld, 451, 456

C

Cantilevered beams, 12, 221, 234–37, 270–71, 348, 354, 357, 366, 374, 376, 414, 644
deflection analysis of, 354–59
stepped cantilever, 380
with spring support, 657–58
Cap channel, 302, 368
Case hardening, 60
Cast iron frame, 432–33
Castigliano, Alberto, 635
Castigliano's theorem, 635–54
analysis of statically indeterminate structures by, 654–62
deflection analysis
of beams and frames, 641–51, 676
of curved frames, 647–49
of torsion bar, 651–54
of trusses, 638–41, 676
Catwalk, 68
Centric load or centric loading (see Axial loading, centric)
Centroid
of a cross section, 250, 715–18, inside back cover
of an area, 245, 250
Centroidal axis
of an area, 715–18, inside back cover
principal, 245
Centroids of common shapes, tables of, 715–18
Changes in temperature, 83–91
Channel section, 274
shear center of, 318–21
shear stresses in, 318–21
Channel shapes, 730–32
Circular hole, stress distribution near, 96–98
Circular shafts
angle of twist in, 166–68
elastic torsion formula for, 170, 210
made of an elastoplastic material, 191–92
of variable cross section, 168
plastic deformations in, 191–94
statically indeterminate, 183–88
strains in, 167–68
stress-concentration factors for, 207
stresses in, 168–76
Clamp, 25, 460
Clevis, 135, 158
CMAA (Crane Manufacturers Association of America), 94, 105, 150, 153, 346, 417
Coefficient of thermal expansion, 83
of selected materials, 719–21
Cold working, 57
Columns, 561
design of under a centric load, 584
of aluminum alloys, 587
of structural steel, 584
of timber, 589
effective length of, 568
equivalent lengths, 573
Euler's formula for, 563–65, 573, 579
fixed-fixed, 569–70
fixed-free, 568–69
fixed-pinned, 570–72
pinned end, 563–68
inelastic, 575
intermediate, 580
long, 578
short, 580

Combined loadings, stresses under, 426–43
Combined normal stresses, 427–34, 457
Combined shear stresses, 434–43, 457
Common shapes, tables of areas and centroids of, 715–18
Compatibility, 70
Components of stress, 472–73
Composite beams, 286–92
Composite shafts, 186
Compression, modulus of, 550
Computer problems, 122–23, 163–64, 219–20, 344–45, 424–25, 469–70, 559–60, 598–99, 691–92, 714
Concrete, design specifications for, 104
Connecting rod, 10, 32
Connections
  bolted, 127–32
  scarf, 40
  welded, 140–43
Conservation of energy, 618, 676
Constitutive equations, 514–43
Continuity conditions, 352–53
Continuous beam, 405, 410
Conversion factors, 6, inside front cover
Coulomb-Mohr theory, 520
Coupling, flange, 158
Crane beam, 108, 347, 463
Crane beams, design specifications for, (also see CMAA) 153, 346, 417
Crane boom, 460
Crane hook, 539
Crane, hydraulic, 28–29, 136–37, 156, 158, 161
Crane truss, 562
Crankshaft, 465
Critical load, 565, 568–70, 573, 575
Critical column stress, 573, 575, 577, 591
C shapes, properties of, 730–32
Curvature
  of a beam, 415
  of neutral axis, 349–50
Curved frame, 647, 685
Cycles of loading, 695
Cycles to failure, 697–98
Cyclic loading, 695
Cylindrical pressure vessels, 444–49
  allowable stresses in, 455–56
  allowable thickness of, 454
  hoop stress in, 447–48, 452–53
  axial stress in, 447, 453

D

Deflection (see Deformation)
Deflection curve, 348, 375
Deflection of beams
  by Castigliano's theorem, 641–47
  by discontinuity functions, 400–404
  by energy method, 628–34
  by integration method, 352–64
  by moment-area method, 371–85
  by superposition, 365–71
  in unsymmetric bending members, 412–15
  in stepped shaft, 632–33
  maximum of common beams, 416
  table of, 751–62
Deflection of curved frame
  due to bending moment, 647–49
  due to normal force, 647–49
  due to shear force, 647–49
Deformations
  due to shear, 145–46, 153, 647–49
  due to temperature change, 83, 85–91, 104
  in bending, 346–404, 628–47
  in circular shafts (see Angle of twist)
  of bar under own weight, 66–67
  under axial loading, 61–70, 103
Density of oil, 453
Density of selected materials, table of, 719–21
Design
  for impact loads, 675
  of beams, 273–79
  of columns: under a centric load, 584–90
  of prismatic beams, 273–79
  of transmission shafts, 189–90
Design load, 92
Design specifications
  for aluminum alloy columns, 94, 587–89
  for boiler and pressure vessels, 94, 458
  for crane runway beams, 94, 153, 346, 417
  for highway bridges, 94, 279, 284, 586, 591
  for reinforced concrete structures, 94, 104
  for steel structures, 92, 94, 104, 153, 273, 286, 326, 346, 417, 584–85, 590–91
  for timber structures, 94, 104, 273, 326, 589
Deviation, tangential, 374
Dilatation, 548, 551
Dirac delta function, 385
Direct shear, 124
Discontinuity functions, 385, 393
  for beam deflections, 400–404
  for bending moment, 391, 393–400
  for shear force, 391, 393–400
Distortion energy, 515–16, 616, 676
Distortion strain energy density, 515–16, 616, 676
Distortion energy envelope, 516
Distributed load, 16, 18
  sign convention for, 351
Double integration, 352
Double shear, 128–29
Drive pin, 124–25
Ductile material, 39, 51, 175
  yield theory for, 513
Ductility, 51
Dynamic loadings (see Loadings, impact)
Dynamic test, 45

E

Eccentric load or eccentric loading (see Axial loading, eccentric)

Effective length of column, 568, 573
for various end conditions, 573
Effective slenderness ratio (see also Equivalent slenderness ratio), 591
Elastic curve (also see Deflection curve), 348
Elastic flexure formulas (see Flexure formula)
Elastic limit, 46
Elastic recovery, 47
Elastic section modulus (see Section modulus)
Elastic strain (see strain, elastic)
Elastic strain energy, 606–8, 610–11, 613
Elastic torsion formulas (see Torsion formula)
Elasticity, modulus of, 47
of selected materials, 719–21
Elastomeric pad, 144
Elastoplastic material, 191
torsion of shaft made of, 191–92
Electrode used in welding, 141
Elongation percent, 51
End conditions for columns, 568–73
Endurance limit, 695–706, 710
of aluminum, 696
of cast iron and cast steels, 696
of copper, 697
of magnesium alloys, 697
of steels, 696
of titanium and its alloys, 697
factors affecting, 699
load factor, 700, 710
size effect factor, 700, 710
stress concentration factor, 703–5, 711
surface finish factor, 699, 710
temperature factor, 700
Energy-absorbing capacity, 53–54, 605
Energy density (see Strain energy density)
Energy loads (see also impact loads), 600
Energy methods, 600–675
analysis of statically indeterminate structures, 656–62
deflections by, 628–34
Engesser, F., 575, 578, 591
Engineering stress, 46
Equations of statics or equilibrium, 21–22
Equivalent column length, 573, 591
for fixed-fixed, 573
for free-fixed, 573
for pinned-fixed, 573
for pinned-pinned, 573
Euler load, 565
Euler, Leonhard, 565
Euler-Engesser formula, 565, 591
Euler's columns, summary of, 573
Euler's formula, 565, 591
for other end conditions, 568–75
limitations on, 579
External forces, defined, 6
Extrusion, 58–59

F

Factor of safety, 91–96, 104, 134
for steel columns, 585
Failure envelope, 514, 516, 519–520
Failure theories, 513–520
Coulomb-Mohr theory, 520
Maximum normal stress theory, 518–19
Maximum octahedral shear stress theory, 515
Maximum shear stress theory, 513–14
Tresca yield theory, 513–14
von Mises' theory, 515
von Mises'-Hencky's theory, 515
Failure, modes of, 93
Fatigue
factors affecting, 699–710
load factor, 700, 710
size factor, 700, 710
surface finish factor, 699, 710
temperature factor, 700
Fatigue failure, 693, 710
Fatigue fracture surface, 694
Fatigue, high cycle, 697, 710
Fatigue loading, 693
Fatigue specimen, 695
Fatigue strength, 694–95
Fatigue stress concentration factor, 704
Fatigue test, 695
Fatigue zone, 693
Fillet, stress distribution near
in circular shaft, 294
in flat bar: in bending, 293–96
under axial loading, 99
Fillet weld, 153
Finite element method, 97
First moment
of area, 259
Fixed-fixed beam, 405, 408–10
Flange coupling, 158
Flange of W-beam, stresses in, 254–55
Flat bars, stress-concentration factors for
in bending, 292–96
under axial loading, 99
due to circular holes, 98
due to fillets, 98
due to grooves, 98
Flexibility method, 76–91
Flexural rigidity, 350
Flexural stiffness, 350
Flexural stress, 244, 246, 324
Flexure formula, 244, 246, 324
Fluctuating loading, 706
Forging, 58
Form factor in shear, 613
for rectangular cross sections, 613
for solid circular cross section, 613
for thin tube cross section, 613
for thin-walled webbed sections, 613
Fracture strength, 49
Fracture theory for brittle materials, 518–21
Frames, 647–50
Free-body diagram, 9
Fuchs, H.O., 711
Fully plastic shear stress, 193

G

Gauge length, 45–46
Gears, 214
General steps in problem solving, 21

Generalized Hooke's Law (also see Hooke's law), 548
Gere, J.M., 591
Gigapascal, see inside front cover
Goodier, J.N., 210, 261, 326
Goodman diagram (modified), 706–11
Graham J.A., 711
Groove weld, 153
Gyration, radius of (see Radius, of gyration)

H

Hacksaw, 459
Hardness number:
  Brinell, 56–57
  Rockwell, 56–57
Heat treatment, types of, 58–60
Helical spring, 466
Heaviside function, 385
Highway bridges, design specifications for, 279, 284, 591
Hoist, 159, 596
Homogeneous material, 49
Hooke, Robert, 48, 541
Hooke's law
  for uniaxial loading, 48, 103, 541
  for general state of stress, 548
  for multiaxial loading, 541–43
  for plane stress, 541–43
  for shear stress and strain, 145, 153, 169, 209, 534, 541
Hoop stress, 445
Horizontal shear stress, 261
Horsepower, 188
Hot rolling, 58
Hot working, 58
HS20-44 loading, 279–80
Hydrostatic pressure, 449, 550
Hydrostatic stress state, 550, 616

I

Impact factor, 665, 677
Impact forces, 662
Impact loading
  deflection in
    bars (rods), 666–69, 677
    beams, 671–75
    springs, 664–66, 677
    torsion bars, 669–71, 677
  stress in
    bars, 663, 676
    beams, 673
    torsion bars, 667, 669
  design for, 675
Impulsive loading (see Impact loading)
Inboard motor, 2
Incompressible material, 550
Indeterminate (see entries beginning with term: Statically indeterminate)
Inelastic column behavior, 575–84
Inertia moment of (see Moment of inertia)
Influence coefficients, 635
Instantaneous zone, 693
Integration method for deflection of beams, 352, 415
Internal forces, 6, 8, 10, 14, 20–21
Internal hinge, 353
Internal moment, 8, 12–13, 14, 16, 20–21
Internal reactions, 21
Internal shear force, 12–13, 14, 16, 20–21
Invariant property, 4
Isotropic material, 49

J

Jib crane, 1, 15, 461
Johnson, J.B., 579
Johnson's formula, 579–80, 583–84
Joint efficiency factor, 454–56
Joints
  circumferential, 455
  longitudinal, 455
Joule (unit), see inside front cover
Juvinall, R.C., 711

K

Keyway, 207–8
Kilopascal, 30
Kilowatt, 188
Kinematics of bending, 242–44
Kinetic energy, 662
Kip, defined, 5, 30

L

Lally-column, 72
Landing gear, 2, 28–29, 36–37
Lateral strain, 49–50
Lift ripper, 462
Limit:
  elastic, 46
  endurance, 695–96, 698–99, 701
  proportional, 46
Load:
  critical, 565, 568–70, 573, 575
  distributed, 16, 18
Load-deformation diagram, 601–3
Load frame, 26–27
Loading(s)
  axial (see Axial loading)
  centric (see Axial loading, centric)
  combined, 426
  cyclic, 695
  dynamic (see impact, below)
  eccentric (see Axial loading, eccentric)
  energy (see impact, below)
  fatigue, 693, 695
  fluctuating, 706
  HS20-44, 279–80
  impact, 93, 662–75
  repeated, 93, 693
  reverse, 701
  static, 93
  sustained, 93
  transverse
Loading cycles, 695
Log hoist, 159
Long beams, 351, 613
Long columns, 578
Longitudinal stress in cylindrical pressure vessels, 447

M

Macauley, W., 390
Macauley's brackets (see Discontinuity functions)

Macauley functions, 390
Manufacturing of seamless tubing, 59
Margin of safety, 92, 104
Material
brittle, 39, 53
fracture theory for a, 518–20
ductile, 39, 51
yield theory for, 513–15
elastoplastic (see Elastoplastic material)
Materials, table of typical properties of, 719–21
Maximum bending moments, common beams, 324
Maximum bending moments, table of, 751–62
Maximum bending moment under moving load, 280–86
Maximum deflection of beam, 416
by integration, 354–64
by moment-area method, 376–84
by discontinuity functions, 400–404
table of, 751–62
Maximum-distortion-energy theory, 515, 551
for general state of stress, 617
for plane stress, 618
Maximum in-plane shear strain, 526, 530, 532–34
Maximum in-plane shear stress, 485
Maximum normal stress theory, 518–19, 551
Maximum shear force, common beams, 324
Maximum shear force, table of, 751–62
Maximum shear force under moving load, 280, 284–86
Maximum shear strain, 526, 530, 532–34
Maximum shear stress, 484–92, 551
Maximum shear stress theory, 513–14, 551
Maxwell, James C., 637
Maxwell's law of reciprocity (reciprocal theorem), 637
Mechanics of materials (defined), 1
Megapascal, 30
Method of neutral equilibrium, 563–64
Method of virtual forces, 621
Method of virtual work, 621
Miscellaneous channels, 731–32
Miscellaneous shapes (also see M shapes), 728
Mitchell, L.D., 699, 711
Mode shapes, 569
Modulus
bulk, 550–51
of compression, 550
of elasticity, 47, 55, 103, 561
for selected materials, 719–21
of resilience, 53, 103, 605, 606, 675
of rigidity, 145, 153, 169
for selected materials, 719–21
of toughness, 54, 103, 605, 675
section (see Section modulus)
shear, 145, 153, 169
Young's (also see Modulus of elasticity), 47
Mohr, O., 492
Mohr's circle, 492–505
for tension test, 515
for plane strain, 531–35
for plane stress, 492–505
for pure shear, 515
for stresses: in cylindrical pressure vessels, 501–5, 510–12
sign convention, 493–96
Moment
bending (see Bending moment)
Moment-area method for deflection of beams, 371–85
Moment-area theorems, 372–74, 417
First moment-area theorem, 372
Second moment-area theorem, 373
Moment of inertia:
of area (table of), 715–18, back inside cover
of composite area, 251
polar, 169, 209
rectangular (see Moment of Inertia of area above)
Moments of inertia of common shapes, 715–18, inside back cover
Moore, R.R., 695
Moving loads, 279–86, 325
M shapes (also see Miscellaneous shapes), 728
Multiaxial loading:
generalized Hooke's law for, 541–43
Multiaxial stress, 541–43

N

National Forest Products Association, 104, 591
NDS (National Design Specification for Wood Structures), 94, 104–5, 267, 589
Necking, 51
Neutral axis
for symmetric bending, 242
for unsymmetric bending, 303
Neutral surface
of prismatic beam, 243
radius of curvature of, 243–44
Newton, defined, 4
Noncircular members
torsion of elliptical, 194–95, 210
torsion of solid rectangular, 195–96, 210
torsion of thin-walled hollow, 198, 200–206, 210
Nonlinear stress distribution, 190–94
Normal strain
in bending, 243–44
under axial loading, 43, 103, 535
Normal stress
due to centric axial loading, 30, 102
due to combined loading, 427, 429, 457
due to eccentric axial loading (see Combined loading)
due to transverse loading, 429
due to torsion, 174–76, 210
in bending, 241–56
maximum, in beams, 246
Notch sensitivity, 704–5
Numbering system for some metals, 763–66

O

Overload protection device, 160
Overhanging beam, 368, 382, 394, 396, 399, 403

P

Parallel-axis theorem (also called Transfer axis theorem), 250–51
Pascal, 22, 30
Paz, M., 677
Percent elongation, 51
Percent reduction in area, 51
Perfectly plastic, 48
Peterson, R.E., 97, 105, 210, 293, 326
Physical effects of forces, 9
  compression, 9
  shear, 9
  tension, 9
  torsion (twisting), 9
Physical properties of materials, see inside back cover
Pin-ended columns, 563–68
Pins, stress in, 133–34, 137–38
Plane strain, 521–28
  comparison to plane stress, 526–27
  example of, 522
  Mohr's circle for, 531–35
  transformation of, 523–28
Plane stress, 471, 513, 522
  comparison to plane strain, 526–27
  transformation of, 472–78
Plastic deformations
  made of an elastoplastic material for circular shafts, 191–92
Plastic strain, 47
Plastic torque
  in circular shaft, 193–94, 210
Pliers, 137–38, 154
Poisson effect, 542
Poisson's ratio, 49–51, 103
  limits on, 550
Polar moment of inertia
  equivalent, 203
  of area, 169
  of cross section of shaft, 169, 209
  for solid circular cross section, 209
  for hollow circular cross section, 209
Power, 188–90, 210
Power transmission, 188–90
Prefixes, see inside front cover
Press frame, 105, 472
Pressure, hydrostatic, 550
Pressure vessels, examples of, 3, 445, 448
Pressure vessels, stresses in thin-walled, 444–57
  Code design of, 454–57
Principal angle, 479, 551
Principal axes of strain, 532
Principal centroidal axes, 245
Principal moments of inertia, 304–11, 325
Principal planes of stress, 479, 495, 498–99
Principal strains, 526
Principal stresses, 478–84, 495, 497, 499–500, 503, 550
Principal of superposition, 365
Principal of work-energy, 618–21, 676
Prismatic beams, design of, 273–79
Progressive fracture (also see Fatigue failure), 693
Properties
  of aluminum standard channels, 743
  of aluminum standard I-beams, 744
  of pipe (standard), 742
  of plane areas, 715–18
  of rolled-steel shapes, 722–42
  of selected engineering materials, 719–21
  of structural lumber, 366, 746–49
Propeller of ship, 2
Proportional limit, 46
Propped cantilever, 405–7, 654–55
Punching operation, 125, 132

Q

Quenching, 58

R

Radius
  of curvature of neutral surface for prismatic beam, 415
  of gyration, 565, 573
  of cross section of column, 565, 573
Rate of loading, 60
Reciprocity, law of, 637
Reduction in area, percent, 51
Redundant reaction
  for a beam, 405, 408, 410, 654–58
  under axial loading, 660–61
Reinforced-concrete beams, 287
Relations
  among E, ν and G, 145
  between load and shear, 351, 385, 400
  between shear and bending moment, 351, 385, 400
Repeated loadings, 693
Resilience, modulus of, 53, 103, 605–06, 675
Reverse loading, 701
Right-hand rule, 7–8
Rigidity
  flexural (see Bending or Flexural stiffness)
  modulus of, 145
  for selected materials, 719–21
Rockwell hardness number, 56
Rolled steel sections, 274
Rolled-steel shapes, properties of, 722–42
Round bars, fillet, 99
Rosette, strain (also see Strain rosette), 535
Rotating beam fatigue test, 695

S

SAE Handbook, 764
Sachs, N.W., 210
Safety, factor of, 92, 104
  for steel columns, 585
Saint-Venant, Barre de, 97
Saint-Venant's Principle, 97, 293
SBCC (Standard Building Code), 94, 104

Second moment of an area (see
Area moment of inertia)
Section modulus, 247, 274, 325
Semicircular curved frame, 647, 685
Shafts
circular (see Circular shafts)
design of (see design of shafts)
noncircular (see Noncircular
members)
stepped, 632–33
transmission (see Transmission
shafts)
Shanley, F.R., 578, 591
Shear
deformation, 144–45
double, 128–29, 153
horizontal, 261
single, 128, 153
stress, 127–53
Shear center
defined, 316, 325
of angle shape, 323
of channel shape, 318–21
of semi-circular cross section,
321–22
of T shape, 323
of wide flange, 323
of Z shape, 323
Shear failure of bolt, 127
Shear force diagrams, 226–40
Shear flow, 200–202, 259, 314
in thin-walled hollow shafts,
200–205
under transverse loading, 259
in thin-walled members, 200–
205, 312–324
Shear force
diagrams, 223, 226–41
in bolts, 268
sign convention for, 224, 351
Shear strain-angle of twist, 209
Shear strain
due to direct shear, 143–45
in circular shafts, 167–68, 209
Shear stress
direct, 127–38
due to axial loading, 37–42
due to combined loading, 434–44
due to transverse loading, 256–73
of beam, 256–73
of circular cross section, 260–
61
of channel shape, 316
of rectangular cross section,
260–61
of S- and W-beams, 260–61,
314
of thin-walled member, 442–44
in torsion
of circular shafts, 168
of elliptical cross section,
194–95
of rectangular bar, 195–97
of thin-walled hollow shaft,
200–205
maximum, 484, 551
maximum in beam, 260–61
maximum in-plane, 484–92
Shear stress and strain, Hooke's
law for, 145, 153, 169, 209
Shear stress formula, 259
Shear stress-strain diagram, 170
Shear modulus, 145
Shigley, J.E., 699, 711
Shrink fitting, 208
Shot peening, 699–700
S-N diagram, 694, 697, 710
Sign convention:
for bending moment, 224, 351
for Mohr's circle, 493–96
in strain, 530–31
in stress, 492–93
for shear strain, 531
for shear stress, 224
Simply supported beams, 359,
361, 367, 369, 377, 401,
643, 672
deflection analysis of, 359–64
Single shear, 127
Singularity functions (also see
Discontinuity functions), 385,
390, 562
SI system, 2
SI units, table of, (see inside front
cover)
Shaft, elliptical, 198
Shaft, machine, 472
Slenderness ratio, 566, 579
effective, 579
transition, 580
Slip planes, 514
Slope of beams:
by Castigliano's theorem, 642,
644–45
by integration method, 349, 355,
358
by moment-area method, 372,
376, 379–80, 383–84
defined, 349
sign convention for, 351
Slug, 4–5
Socket wrench, 555
Specific weight of selected materials,
719–21
Spherical pressure vessels, 449–53
Spreader bar, 592
S shapes, properties of, 728–29
Stability, 561
conditions of, 562
of columns, 562
neutral, 562, 564
Stabilizer bar, 462
Stairway, 461
Standard test specimen
tension, 44
shear, 154
Standpipe, 556
Statically determinate beams,
boundary conditions for,
352–53, 416
Statically determinate, defined, 70
Statically indeterminate beams,
404–12
analysis of: by Castigliano's
theorem, 654–60
by integration, 404–10
by superposition, 410–12
Statically indeterminate problems,
70–83, 404–12, 656–62
Statically indeterminate shafts,
183–88
Static test, 45
Statics, 7, 21
Steel beams
design of, 277–78
properties of, 722–42
Steel columns, design of, 584–86
Steel, design specifications for, 92,
104, 153, 273, 326, 346, 417,
584–85, 590–91

Steel, design specifications for *(continued)*
- low alloy, high strength, 46
- mild, 46
- tool, 46

Step ladder, 160
Steps in problem solving, general, 21
Stiffness (see Bending stiffness)
Strain
- compressive, 43
- due to temperature change, 84, 104
- elastic, 47
- lateral, 49
- measurement, 46
- nominal, 46
- normal (also see Normal strain), 43–44, 103, 243–44
- offset, 48
- permanent, 46–47
- plastic, 47–48
- principal, 526
- principal axes of, 532
- residual, 46
- shear (see Shear strain)
- tensile, 43
- transformation of, 523
- true, 46

Strain energy
- distortional, 617, 676
- due to shear stresses, 608–9
- due to torsion, 610–11, 675
- due to transverse shear stress, 613–15, 676
- for general state of stress, 615
- for normal stresses, 604
  - due to axial loading, 604, 607, 675
  - due to bending, 611, 675
- in direct shear, 609, 675
- in stepped bar, 607–8
- in uniform bar, 607
- volumetric, 617, 676

Strain-energy density
- for general state of stress, 615
- for normal stresses, 604
- for shear stresses, 608
- for uniaxial stress, 675

Strain-hardening, 47, 49
Strain rosette, 535
- equiangular or delta, 537, 539
- rectangular or 45°, 536–37

Strength
- breaking (see fracture below)
- fracture, 49
- ultimate, 49
  - of selected materials, 719–21
- yield, 48
  - of selected materials, 719–21

Stress
- allowable, 91–92
- amplitude, 706–7
- axial, 445, 519
- bearing, 146–53
- biaxial (see Plane stress)
- components of, 472–73
- compressive, 31
- critical, 573, 575, 577, 591
- definition of, 30, 102
- due to axial loading (see under Normal stress)
- due to combined loading, 426–44
- due to temperature change, 85–91, 104
- due to transverse loading (see Shear stress due to transverse loading)
- engineering, 46
- fatigue, 697
- flow of, 100
- fluctuating, 706
- flexural, 244, 246, 324
- fracture, 49
- general state of, 472–73
- hoop, 445
- in bending (see Normal stress, in bending)
- in bolts, 127–32
- in pins, 137–38
- in thin-walled pressure vessels, 444–57
- in torsion (see Shear stress in torsion)
- in weld, 138–43, 503–5
- longitudinal (see Axial stress above)
- mean, 706–7
- multiaxial, 541–43
- normal (see Normal stress)
- on inclined plane under axial loading, 37, 103
- plane, 472–77
- principal (also see Principal stresses), 478–84, 495, 497, 499–500, 503, 550
- principal planes of, 479
- shear (see Shear stress)
- sign convention for, 472
- tensile, 31
- transformation of, 475
- triaxial, 473
- true, 46
- ultimate, 49
- uniaxial (see axial)
- uniform distribution of, 31–32
- yield (also see Yield stress), 48

Stress concentration factor, 98–99, 104, 207–10
Stress-concentration factors
- for circular shafts, 206
- for flat bars
  - in bending, 292–96
  - under axial loading, 98
- for circular bars
  - in bending, 294
  - under axial loading, 99
- in fatigue, 703–6

Stress concentrations
- in circular shafts, 206
- for flat bars
  - in bending, 293–94
  - under axial loading, 96–102
- for round bars
  - under axial loading, 99

Stress-strain diagram, 44–61
- for aluminum alloys, 48, 576
- for cast iron, 52, 518
- for concrete, 52
- for fiberglass, 52
- for low alloy, high strength steel, 47, 101
- for reinforced plastics, 52
- for stainless steel, 61
- for structural steels, 47, 101
- for tool steel, 47
- properties affecting:
  - rate of loading, 60
  - temperature, 61
  - type of heat treatment, 59

Stress-strain equations
plane, 541, 551
three-dimensional, 548, 551
Stress trajectory, 96
Structural compatibility, 70
Structural lumber
properties of, 746–49
Structural Plastics Design Manual, 94, 105
Structural steel, 46
Structural tees, 733–39
Successive integration (also see Integration method), 352
Superposition method
for deflection of beams, 365–71
for determination of stresses, 427–44, 453–54
of normal stresses, 427–34
of shear stresses, 434–44
Summary of important definitions and equations, 22, 102–4, 153, 209–10, 324–26, 415–17, 457, 550–51, 591, 675–77, 710–11
Superposition principle of, 365
Swanson, J.A., 551

T

Tangential deviation, 374
Tangent modulus formula, 575
Tank
compressed-air, 501
cylindrical, 511
scuba diver's, 452, 468–69
oil storage, 453
pressure, 545
propane, 468
water, 562, 586
Temperature changes, 83–91
Temperature effects, 83–91
Temperature of materials, 60–61
Tempering, 60
Tensile test, 44
Testing machine
tensile, 45
torsion, 170
Thermal expansion, coefficient of
of selected materials, 719–21
Thermal strain, 84
Thermally compatible, 83
Thin-walled cross sections, 311
Thin-walled hollow shafts, 200–206
Thin-walled pressure vessels, 444, 457
cylindrical, 457
spherical, 457
Thin-walled tube, 442–44
Thin-walled vessel, defined, 445
Three-dimensional
state of stress, 473, 615
Throat area, 140
Timber, 589
Timber columns, design of, 589–90
Timber, design specifications for, 104, 273, 326, 589, 591
Timber, properties of (also see Structural lumber, properties of), 745–49
Timoshenko, S.P., 210, 261, 326, 591
Toggle press, 162, 594
Tool steel, 46
Torsion bar, 543
Torsion bar spring, 651–52
Torsion bar suspension, 182, 610–11
Torque
maximum elastic, 170
normal stress due to, 174–76
plastic, 192
Torque-shear stress relationship, 168
Torque tube, 204
Torque-twist formula, 177
Torsion
of circular shafts, 165–94
of narrow rectangular sections, 196–97, 210
of noncircular members, 194–200
of elliptical cross section, 194–95, 210
of rectangular bars, 195–96, 210
of thin-walled closed tubes, 198, 200–206, 210
Torsion experiment, 170
Torsion formula, 170–174, 210
Torsion testing machine, 170
Toughness, modulus of, 54, 103, 605, 675
Tower, 562
Transcendental equations, 572
Transformation
of plane strain, 523–28
of plane stress, 472–78
Transformation equations for moments of inertia, 306
Transformed section
of beam made of several materials, 288–92
Transmission shafts, 11, 188–90
design of, 189–90
Transverse cross section of beam
curvature of, 350
Tresca's yield theory, 513
Triaxial stress, 473
Tributary area, 277
True strain, 46
True stress, 46
Truss, 27–29, 34–36, 107, 113, 562, 619
Twist, angle of (see Angle of twist)
Twisting of thin-walled members under transverse loading (see Shear center)
Two-force members, 28–29, 36, 131, 152

U

Ultimate strength
of selected materials, 719–21
Ultimate stress, 47, 49, 55
U.S. customary units, see inside front cover
Unit doublet function, 386
Unit impulse function, 386
Unit ramp function, 388
Unit step function, 387
Unsymmetric bending, 296–304, 412
Unsymmetric loading
deflection of beam under, 412
of thin-walled members, 309–10
USCS system, 2

V

Variable cross section
  bars of, 63
  shafts of, 178
Vehicle jack, 594
Vibration isolation support, 161
Virtual work method, 621, 676
  due to temperature change, 624, 676
  for beams and frames, 628–35, 676
  for trusses, 623–28, 676
Vise, 27, 598
Volumetric strain energy density, 617, 676
von Karman, T., 578, 591
von Mises' theory, 515

W

Warehouse floor, 277
Water tank, 562, 586
Watt (unit), see inside front cover
Web of S- or W-beam, stresses in, 260–261, 312–14, 316
edge method, 478
Weight, 4
Weld,
  butt, 138–40, 451–56, 555
  fillet, 138
    uses of, 139
  groove, (see butt above) 555
  inspection, 456
  joints, 138
  legs, 139
  normal stress in groove weld, 140, 153
  shear stress in fillet weld, 140–143, 153, 271–73, 505
  types of, 138–39
Wide-flange beam, 273, 722–27
Wide-flange shapes (also see W shapes), 722–27
Woodruff Keyway, 207
Work
  of a moment, 603, 675
  of a force, 601, 675
Work-energy method for deflection under a single load, 616–21
  for a bar, 618–19
  for a beam, 620
  for a truss, 619–20
Working load, 92
Working stress, 92
W shape, defined, 273
W shapes, properties of, 722–27

Y

Yield theory for ductile materials, 513–18
Yield or yielding, 48, 100
Yield stress point, 48
Yield strength, 48
  of selected materials, 719–21
Yield stress (see Yield strength)
Young's modulus, 47

Z

Z section, 323

Answers to Selected Problems

CHAPTER 1

1.1 (a) inch, in. (b) inch^2, in.^{2} (c) pound, lb (d) slug (e) pound/inch^3, lb/in.^{3} (f) inch^4, in.^{4} (g) pound/inch^2, lb/in.^{2} (psi) (h) foot · pound/second, ft · lb/s or horsepower, hp (i) degrees Fahrenheit, °F (j) pound · inch, lb · in. or pound · foot, lb · ft
1.3 19,600 N
1.5 9800 N (each front wheel)
14 700 N (each rear wheel)
1.7 4400 lb
1.8 2200 lb (front wheel), 3300 lb (rear wheel)
1.9 619 kg/m^3
1.10 90.7 kg, 890 N
1.11 (a) 55.32 MN (b) 46.8 km (c) 1.0 kg
1.13 1 Pa = 1.45×10^{-4} lb/in.^{2}
101 kPa
1.15 1290 lb
1.17 1167 kg
1.18 10.34×10^6 Pa (10.34 MPa)
1.19 149 MPa
1.21 1290 mm^2
1.22 2.29 mm
1.23 (a) 2560 in.^{3} (b) 1.49 ft^3 (c) 41.9×10^6 mm^3 (d) 41.9×10^{-3} m^3
1.25 $P_A = 3$ kN; $P_B = -5$ kN; $P_C = 2$ kN
1.27 $T_C = 50$ lb · ft (ccw); $T_D = 65$ lb · ft (cw); $T_E = 10$ lb · ft (ccw) (all answers looking in from end B)
1.29 $T_A = 350$ N · m (ccw); $T_B = 200$ N · m (ccw)
$T_C = 400$ N · m (ccw); $T_D = 0$
(All answers looking in from fixed end)
1.31 $P_a = 520$ lb (C), $V_a = 700$ lb, $M_a = -3900$ lb · ft
$P_b = 520$ lb (C), $V_b = 300$ lb, $M_b = -900$ lb · ft
1.33 $V_a = 400$ lb, $M_a = 6000$ lb · ft
$V_b = -800$ lb, $M_b = 4800$ lb · ft
1.34 $V_a = -3$ kN; $M_a = 1.5$ kN · m; $V_b = -6$ kN, $M_b = -12$ kN · m
1.35 $V_c = 158$ lb, $M_c = 632$ lb · ft
1.36 $P_A = 95$ lb, $V_A = 0$, $M_A = 285$ lb · in.
1.37 $P_D = 1600$ N (C), $V_D = 0$, $M_D = 2400$ N · m
1.39 $P_A = 24$ N (C), $V_{xA} = 0$, $V_{yA} = 0$
$M_{xA} = 0$, $M_{yA} = 0.84$ N · m, $M_{zA} = 0$
1.41 $P_A = 11.4$ kN (C), $V_A = 1.29$ kN,
$M_A = -10.8$ kN · m, $P_B = 8.50$ kN (C),
$V_B = -7.67$ kN, $M_B = -3.48$ kN · m
1.43 $P_C = 0$, $V_{xC} = 20$ lb, $V_{zC} = 148$ lb
$M_{xC} = 326$ lb · ft, $M_{yC} = -144$ lb · ft,
$M_{zC} = -40$ lb · ft
1.45 $P_a = 0$, $V_a = 9.71$ kN, $M_a = 9.71$ kN · m
$P_b = 25.4$ kN (C), $V_b = 0$, $M_b = 0$
1.47 $P_a = 31.2$ kN (C), $P_b = 39.3$ kN (T)
1.48 $P_a = 17.9$ kips (T), $P_b = 16$ kips (C)
1.49 $P_a = 981$ N (C), $V_a = 0$, $M_a = -4900$ N · m
1.50 $P_a = 0$, $V_a = 600$ lb, $M_a = 3600$ lb · in.;
$P_b = 1680$ lb (C)
1.51 $P_a = 3000$ lb (T), $V_a = 1000$ lb, $M_a = 8000$ lb · ft

CHAPTER 2

2.1 81.5 MPa
2.3 133 MPa
2.4 400 psi
2.5 18.1 ksi
2.6 952 psi
2.7 62.2 MPa
2.9 49.1 MPa
2.10 1020 psi
2.11 816 psi
2.13 25 700 lb
2.15 7.14 kN
2.16 25 ksi (C) in AB
2.17 $\sigma_{DE} = 2000$ psi (T), $\sigma_{DG} = 4240$ psi (T),
$\sigma_{HG} = 5000$ psi (C)
2.18 $\sigma = 6000$ psi, $\tau = 3460$ psi
2.19 3560 lb

2.21 $\tau = 54.1$ psi, $\sigma = 31.3$ psi
2.23 $\sigma = 82.9$ MPa, $\theta = 23.2°$
2.25 (a) $\sigma = 2400$ psi, (b) $\tau = 1200$ psi
2.27 (a) 30 000 psi, (b) 15 psi
2.28 (a) 0.29, (b) 58.2 GPa
2.29 (a) 0.20, (b) 0.03996 m, (c) 500 MPa, (d) 250 MPa, (e) 100 GPa
2.30 6.85×10^{-4}, 2.28×10^{-4}
2.31 25.4×10^6 psi, 0.20
2.32 (a) 24.4×10^3 ksi, (b) 0.24, (c) 76 psi
2.33 6.85×10^{-4} in./in.
2.35 115 kN
2.36 112 kN
2.37 93.5 kips
2.38 102 ksi, 83.5 ksi
2.39 (a) 29×10^6 psi, (b) 65 ksi, (c) 62 ksi, (d) 30 ksi
2.40 (a) 49.6%, (b) 22.5%
2.41 (a) 15 psi, (b) 19 ksi
2.44 (a) 26.0 MPa, (b) 38.0 MPa
2.46 6.0%, 7.0%, brittle; 23.5%, 36.3%, ductile; 16%, 74.2% ductile
2.47 28.4×10^6 psi
2.48 Need 11.4 mm (Use 12 mm)
2.49 (a) 1.30 mm, (b) 0.292 mm
2.51 22.037 in.
2.53 0.182 mm
2.55 2 in. × 2 in. × 0.25 in. angle, 0.106 in.
2.57 0.00142 in., 0.000171 in.
2.58 AB: −0.0554 in., BC: 0.0586 in.
2.59 1.77 mm
2.60 1.25 mm, elongates
2.61 0.0299 in., elongates
2.62 1.45 kN
2.63 0.0707 in., 0.0602 in.
2.65 1319 ft
2.66 7.22 mm
2.67 1.6 mm
2.68 138 MPa (steel), 13.9 MPa (concrete)
2.69 272 kips
2.71 1/2
2.73 1740 kN
2.74 $\sigma_{AB} = 22.2$ MPa (T), $\sigma_{BC} = 26.6$ MPa (C)
2.75 33.3 kips (to left)
2.77 9.33 kips
2.79 1 ksi (T), 2 ksi (T)
2.80 $\sigma_{BC} = 5.55$ MPa (T)
2.81 6.94 ksi (C) tube, 10.42 ksi (T) bolt
2.82 2475 psi (T) in steel, 2475 psi (C) in aluminum
2.83 0.0558 ft
2.85 189 MPa (C)
2.87 (a) 0.351 mm, (b) 234 MPa
2.89 114°F
2.91 11 700 psi
2.94 0.0241 ft
2.96 5.6 mm, 5.15 mm
2.97 0.78 in.
2.99 (a) 16 250 lb (C), (b) 16 250 psi (C), (c) −0.0052 in.
2.102 5850 psi (C), 2925 psi
2.104 9210 psi, 4605 psi
2.105 11 060 psi
2.107 0.278 in.^{2}
2.109 $A_{AB} = 183$ mm^2
2.111 8000 psi
2.113 49.0 kN
2.115 68.25 MPa
2.117 0.75 in.
2.119 5.38 ksi
2.121 529 kN
2.123 0.18 in. (use 0.25 in. in practice)

CHAPTER 3

3.1 191 MPa
3.2 125 psi
3.3 14.3 MPa
3.5 159 MPa
3.6 15.9 MPa
3.7 66 kN
3.8 4530 psi
3.9 5.6 MPa
3.11 8890 psi
3.13 51.4 kN
3.14 6.66 in.
3.15 190 psi, No
3.17 283 kN
3.18 3.56 MPa
3.19 78.6 mm
3.21 323 MPa
3.23 0.422 in. (use 0.50 in. in practice)
3.25 35.4 MPa
3.27 119 MPa
3.28 30 kip · in.
3.29 18.9 MPa
3.30 10.1 in. (use 10.25 in. in practice)
3.32 at B 3.62 ksi, at C 2.75 ksi
3.33 1.88 MPa
3.34 0.26 in.
3.36 342 MPa
3.37 $G = Ph/(2A\delta)$
3.38 0.0191 in.
3.39 $h = 33.3$ mm, $L = 200$ mm
3.40 5 in.
3.41 160 psi
3.42 2.08 in. (use 2.125 in. in practice)
3.43 (a) 13.2 MPa, (b) 23.1 MPa, (c) 20.8 MPa
3.45 need t = 0.267 in.
3.46 197 MPa

3.47 110 MPa
3.48 (a) 120 psi, (b) need 28.3 in. × 28.3 in. (use 30 in. × 30 in.)
4.69 7.70 ksi
4.71 5.33 kN · m
4.73 12 mm

CHAPTER 4

4.1 38.2 ksi
4.3 30.7 kip · in.
4.5 (a) 2.44 MPa, (b) 3.26 MPa
4.7 67.9 psi in section AB
4.8 22.6 MPa
4.9 3.06 ksi
4.10 76.7 N · m
4.11 745 psi
4.13 1.62 in.
4.14 5.14 in.
4.15 58.8 mm (use 60 mm)
4.16 1.96 kips
4.17 73.6 lb · in.
4.19 0.344 kip · in.
4.21 90 mm
4.22 15.7 N · m
4.23 37 mm
4.24 1390 lb · ft
4.25 1.26×10^{-2} rad cw looking in from free end
4.27 2.99×10^{-3} rad cw looking in from free end
4.29 0.936 in. (use 1 in.)
4.31 1.81 kip · in.
4.34 0.00896 rad cw looking in from C to A, 14.2 MPa
4.35 $TL_BL_AN_A/(L_BN_BG_AJ_A + L_AN_AG_BJ_B)$
4.36 (a) 1.37×10^{-3} rad cw (looking in from free end), (b) 525 psi, (c) 1049 psi
4.37 88.8 lb · in.
4.39 95 MPa, 39.5 MPa
4.41 9170 psi
4.43 3.5 in.
4.45 (a) 2535 psi, (b) 0.0071 rad
4.46 35.6 ksi
4.47 15.2 MPa
4.48 12.2 mm
4.49 (a) 0.0487 rad, (b) 101 MPa
4.50 4960 lb · ft
4.51 (a) 6.37 ksi, (b) 20 ksi
4.52 (a) 386 lb · in., (b) 4650 psi
4.53 (a) 2165 lb · in., (b) 20 ksi
4.54 175 kip · in.
4.55 3.26 MPa
4.56 61.6 MPa
4.57 15.7 kip · in.
4.59 (a) 900 psi, (b) 0.0012 rad/in.
4.61 (a) 23.8 ksi, (b) 22.4 ksi
4.62 (a) 208 lb/in., (b) 1670 psi, (c) 4.63×10^{-4} rad/in.
4.64 (a) 265 lb/in., (b) 2120 psi, (c) 3.19×10^{-4} rad/in.
4.67 (a) 25.5 kip · in., (b) 36.6 kip · in.

CHAPTER 5

5.1 $V_{max} = -wL$, $M_{max} = -wL^2/2$
5.3 $V_{max} = -10$ kips, $M_{max} = -30$ kip · ft
5.4 $V_{max} = 4.17$ kN, $M_{max} = 12.5$ kN · m
5.5 $V_{max} = -5$ kips, $M_{max} = 30$ kip · ft
5.7 $V_{max} = -20$ kN, $M_{max} = -60$ kN · m
5.8 $V_{max} = 4400$ lb, $M_{max} = 66$ kip · ft
5.9 $V_{max} = 4$ kN, $M_{max} = 5$ kN · m
5.11 $V_{max} = P$, $M_{max} = PL/3$
5.13 $V_{max} = 21.5$ kN, $M_{max} = 42$ kN · m
5.15 $V = 0$, $M_{max} = 50$ kN · m
5.16 $V_{max} = -1875$ lb, $M_{max} = -2250$ lb · ft
5.17 $V_{max} = -123$ lb, $M_{max} = -960$ lb · in.
5.19 $V_{max} = 4590$ lb, $M_{max} = -15$ kip · ft
5.21 $V_{max} = 10$ kips, $M_{max} = 80$ kip · ft
5.22 1770 psi
5.23 110 MPa, 215 MPa
5.24 4780 psi (T), 11,900 psi (C)
5.25 19.8 ksi
5.26 313 psi
5.27 313 psi
5.29 688 psi
5.31 23.4 psi
5.33 556 kPa
5.35 278 kPa
5.37 2.33 MPa
5.39 2.78 kPa
5.41 15 ksi
5.43 1790 psi
5.44 25.5 ksi
5.45 960 psi
5.46 19.2 MPa
5.47 (a) 2250 psi, (b) 1125 psi
5.48 260 psi
5.49 $b = \dfrac{d}{\sqrt{3}}$, $h = \sqrt{\dfrac{2}{3}}\,d$
5.51 1080 psi
5.53 (a) 43.6 MPa, (b) 15.4 MPa
5.54 19.6 ksi, 2.52 ksi
5.55 (a) 12.1 ksi, (b) 0, (c) 10.1 ksi
5.57 26 psi
5.59 22.9 psi
5.61 9.76 psi
5.63 34.75 kPa
5.65 33.3 kPa
5.69 0
5.71 312.5 psi
5.73 485 psi, 121 psi
5.74 (a) 3810 psi, (b) −1900 psi, (c) 383 psi, (d) 424 psi

5.75 16.0 ksi, 5050 psi
5.77 329 kips
5.78 (a) 600 psi, (b) 48.1 psi, (c) 41.25 psi
5.79 195 lb
5.80 6.85 in.
5.81 9760 N
5.83 use 1.50 in. diameter × 0.145 in. thick standard pipe
5.84 W 12 × 19
5.85 56.8 mm (use 60 mm diameter)
5.86 4 in. × 8 in. nominal size cross beams, 8 in. × 24 in. nominal size main beams
5.87 ST 3 × 8.625 lb
5.88 W 27 × 94 for load position shown in Fig. P5.88
5.89 W 18 × 97
5.90 Use 6 in. diameter × 0.280 in. thick standard pipe
5.91 (a) 6.53 in., (b) 6.89 in., (c) 4.34 in. × 8.68 in. (d) W 18 × 35
5.93 6 × 18 timber (based on shear)
5.95 Need 211 mm × 422 mm (use 225 mm × 450 mm)
5.97 1930 kN · m
5.99 20.3 kip · ft, 2850 lb
5.101 23.6 kip · ft under 800 lb load farthest to right
5.103 15.63 kip · ft under 4 kip load
5.105 156 psi in wood, 4.27 ksi in steel, 1320 psi wood alone
5.107 8250 lb
5.109 584 psi in wood, 12.6 ksi in steel
5.111 764 kip · in.
5.113 1550 psi in wood, 10.4 ksi in steel
5.115 (a) 6650 lb · in., (b) 7824 lb · in.
5.117 28.8 ksi
5.119 35.6 ksi
5.121 33.2 lb
5.123 178 N · m
5.125 5.25 MPa
5.127 1170 psi (C)
5.128 506 MPa (T) at A, 746 MPa (C) at B
5.129 28.8 MPa (C) at A, 159 MPa (C) at B
5.130 2.6 MPa (T) at A, 66.2 MPa (C) at B
5.133 24.25 ksi
5.135 2.62 kips
5.137 224 in.^{4}, 33.64 in.^{4}
5.138 133 in.^{4}, 12.3 in.^{4}
5.139 2.91×10^{-6} m^4, 0.390×10^{-6} m^4
5.141 515 in.^{4}, 44.3 in.^{4}
5.143 $e = b(2h + 3b)/(2h + 6b)$
5.144 $e = 1.125$ in.
5.145 $e = 2r$
5.147 $e = 1.125$ in.

CHAPTER 6

6.1 $y = \frac{M_0}{2EI}x^2$, $\theta = \frac{M_0}{EI}x$
$y(10\text{ ft}) = 0.144$ in., $\theta\,(10\text{ ft}) = 0.0024$ rad

6.2 $y_{end} = -\frac{w_0L^4}{30EI} = -208$ mm,
$\theta_{end} = \frac{w_0L^3}{24EI} = -0.065$ rad

6.4 $y_c = -\frac{5PL^3}{48EI}$

6.5 $y_{max} = -0.00652\frac{w_0L^4}{EI} = -0.0389$ in. at $x = 0.52L$

6.6 $y_{max} = -0.0737$ in. at $x = 0.423L$

6.7 $y = -\frac{Pda}{3EIL}(-3aL + a^2 + L^2)$

6.9 $y_{max} = -\frac{7wL^4}{384EI}$

6.11 (b) $y_{max} = -\frac{19wd^4}{8EI}$

6.15 $y_1 = \frac{1}{EI}(-x^3 + 40x^2 - 400x)$

6.16 $y = \frac{1}{EI}\left(-\frac{11}{3}x^3 + 50x^2 - 158.8x\right)$

6.17 $y = -\frac{7PL^3}{16EI}$

6.18 $y_{max} = -0.737$ mm

6.19 $y = -\frac{L^3}{6EI}\left(\frac{w_0L}{5} + \frac{5P}{8}\right)$

6.21 $y = -1.29$ in.
6.23 $y = -19.03$ mm
6.25 $y = -0.511$ mm
6.26 $y = -4.77$ in.
6.27 $y = -19.13$ mm
6.28 $y_{max} = -2.02$ in.
6.29 $y_{max} = -1.17$ in.

6.31 $\theta_B = \frac{M_0L}{EI}$, $y_B = \frac{M_0L^2}{2EI}$

6.33 $y = \frac{49wL^4}{3840EI}$

6.35 $y = -\frac{M_LL^2}{16EI}$

6.37 $y_c = -\frac{w_0L^4}{128EI} = -0.015$ in.

6.39 $w(x) = -5\langle x\rangle^{-1} + 2\langle x - 3\rangle^{-1} + \langle x\rangle^0 + 10.5\langle x\rangle^{-2}$
6.41 $w(x) = -60\langle x\rangle^{-1} + 20\langle x\rangle^0 - 20\langle x - 3\rangle^0 + 90\langle x\rangle^{-2}$
6.43 $w(x) = -7.25\langle x\rangle^{-1} + 4\langle x\rangle^0 - 4\langle x - 2\rangle^0 + 5\langle x - 3\rangle^{-1} - 5.75\langle x - 4\rangle^{-1}$
6.45 $w(x) = -7.25\langle x\rangle^{-1} + 8\langle x - 0.5\rangle^{-1} + 5\langle x - 1\rangle^0 - 5\langle x - 2\rangle^0 - 5.75\langle x - 2\rangle^{-1}$

6.47 $y = \frac{1}{EI}\left[\frac{5}{6}\langle x\rangle^3 - \frac{\langle x - 3\rangle^3}{3} - \frac{\langle x\rangle^4}{24} - 5.25\langle x\rangle^2\right]$
$y_{end} = 33.7$ mm (down)

6.49 $y = \frac{1}{EI}\left[10\langle x\rangle^3 - 45\langle x\rangle^2 - \frac{5}{6}\langle x\rangle^4 + \frac{5}{6}\langle x - 3\rangle^4\right]$
$y_{end} = 390$ mm (down)

6.51 $$y = \frac{1}{EI}\left\{\frac{7.25}{6}\langle x\rangle^3 - \frac{\langle x\rangle^4}{6} + \frac{\langle x-2\rangle^4}{6} - \frac{5}{6}\langle x-3\rangle^3 + \frac{5.75}{6}\langle x-4\rangle^3 + \left[-\frac{7.25}{6}L^2 + \frac{L^3}{6} - \frac{(L-2)^4}{6L} + \frac{5}{6}\frac{(L-3)^3}{L}\right]x\right\}$$

$y_c = 5.61$ mm (down)

6.55 $y_c = -5.27$ mm (down), $y_d = 5.63$ mm (up)

6.57 $y\left(\frac{L}{2}\right) = -\frac{wL^4}{768EI}$, $R_B = \frac{3}{20}w_0L$, $M_B = -\frac{w_0L^2}{30}$

6.59 $y = \frac{M_0x}{EI}\left[-\frac{x^2}{4L} + \frac{x}{2} - \frac{L}{4}\right]$

6.61 $R_A = \frac{7}{20}w_0L$, $M_A = \frac{w_0L^2}{20}$

6.63 $R_A = \frac{3Pd}{2L}$ (down)

6.65 $R_B = 5.0$ kN, $y_C = -0.300$ mm (down)

CHAPTER 7

7.1 13.71 ksi
7.2 9.87 ksi
7.3 9175 psi
7.4 56.4 MPa
7.5 6400 psi (T) at A, 4835 psi (C) at B
7.7 178 MPa (C)
7.9 169 MPa (C)
7.10 497 N
7.11 9177 psi (T), 6453 psi (C)
7.13 5.71 kN
7.15 194 kN
7.17 27.4 ksi (C)
7.18 3010 psi (C)
7.19 115 kPa (T)
7.21 6530 psi (C)
7.23 13.75 ksi (T)
7.24 11.5 ksi
7.25 7400 psi (T)
7.27 57.8 MPa (C)
7.29 5500 psi (T)
7.31 1.01 MPa (C)
7.33 15 ksi > 13.5 ksi not safe
7.35 24.9 MPa
7.37 184 psi
7.38 $\tau_E = 6652$ psi, $\tau_F = 7600$ psi, $\tau_H = 7130$ psi
7.39 6826 psi at A, 6086 psi at B, 2180 psi at C
7.41 36.5 MPa
7.43 7215 psi
7.45 97.6 MPa
7.47 26.7 MPa
7.48 417 psi
7.49 73.6 MPa
7.51 50 MPa
7.52 382 psi
7.53 552 kPa
7.55 260 psi
7.57 1-5/8 in. thick
7.59 3/16 in. thick
7.61 326 psi

CHAPTER 8

8.1 −4.93 ksi, −4 ksi
8.2 1.33 ksi, −4.23 ksi
8.3 10.4 MPa, −46.0 MPa
8.5 4.7 ksi, 6.84 ksi
8.7 −4.93 ksi, −4 ksi
8.9 10.4 MPa, −46.0 MPa
8.11 4.7 ksi, 6.84 ksi
8.13 12.8 ksi, −29.5°; 1.17 ksi, 60.5°; 5.83 ksi, −74.5°
8.14 6.41 ksi, 25.1°; −1.41 ksi, 115.1°; 3.91 ksi, −19.9°
8.15 67.6 MPa, 75.7°; −57.6 MPa, −14.3°; 62.6 MPa, 30.7°
8.17 4.14 ksi, −22.5°; −24.14 ksi, 67.5°; 14.14 ksi, −67.5°
8.19 50 MPa, 45°; −50 MPa, −45°; 50 MPa, 0°
8.21 (a) −3.72 MPa (b) −0.771 MPa
8.23 16.9 ksi, −2.6 ksi
8.25 1560 lb
8.27 −32.4 MPa, −42.5 MPa
8.29 −4.93 ksi, −4 ksi
8.31 10.4 MPa, −46.0 MPa
8.33 4.7 ksi, 6.84 ksi
8.35 same as 8.13
8.37 same as 8.15
8.39 same as 8.17
8.41 same as 8.19
8.43 $\sigma_x = \sigma_1$, $\sigma_2 = 0$, $\tau_{max} = \sigma_x/2$
8.45 5 ksi, 8.66 ksi (ccw)
8.46 89.7 MPa, 58.0 MPa (ccw)
8.47 −4.0 ksi, 1.73 ksi (ccw)
8.49 −10.4 ksi, 6 ksi (ccw)
8.51 9.96 ksi, −5.72 ksi, 7.84 ksi
8.53 38.4 ksi, 19.2 ksi, 19.2 ksi
8.54 75 MPa, −2.6 MPa, 38.8 MPa
8.55 $\sigma_{weld} = 6360$ psi, $\tau_{weld} = 2430$ psi
8.57 (a) 13.45 ksi, −0.45 ksi, 6.95 ksi
8.59 15.75 ksi
8.61 108.2 MPa
8.63 9.4 ksi
8.65 14.7 MPa, 0, 7.36 MPa
8.67 (a) (1) 13 ksi < 36 ksi, no, (2) 11.8 ksi < 36 ksi, no
(b) (1) 10 ksi < 36 ksi, no, (2) 8.89 ksi < 36 ksi, no
(c) (1) 16 ksi < 36 ksi, no, (2) 13.9 ksi < 36 ksi, no
(d) (1) 18 ksi < 36 ksi, no, (2) 17.3 ksi < 36 ksi, no

8.69 8 ksi
8.71 16.9 kip · in.
8.73 213 MPa > 95 MPa, yes
8.75 (a) 0.306 < 1, no; (b) 0.226 < 1, no; (c) 40 MPa < 60 MPa, no, (d) 1.13 > 1, yes
8.76 (a) 10 ksi, (b) 8 ksi
8.77 500μ, 0
8.79 974μ, −245μ
8.81 −93.3μ, −296μ
8.83 500μ, 0
8.85 974μ, 245μ
8.87 −93.3μ, −296μ
8.89 500μ, 500μ, 0
8.91 1005μ, 495μ, 510μ
8.93 256μ, 156μ, 412μ
8.95 790μ, −140μ, 930μ
8.97 731μ, 269μ, 462μ
8.99 (a) 18.2 ksi, 12.4 ksi; (b) 20.4 ksi, 10.2 ksi; (c) 213 psi
8.101 35.05 ksi, 13.5 ksi
8.103 700μ
8.104 24.17×10^6 psi, 3.31μ
8.105 7250 psi, 24.17×10^6 psi

CHAPTER 9

9.1 12.65 kips
9.2 890 N
9.3 1816 N
9.5 480 lb < 502 lb safe
9.6 12.8 kips
9.7 91 kips
9.9 3.98 lb
9.11 108 kips
9.12 (a) 173 (b) 200
9.13 13.8 kips for square column, 8.14 kips for circular column
9.15 178 kips
9.17 18.4 kN
9.19 need 0.213 in (use 0.25 in.)
9.21 Circular is best, I = 87.5 in.^{4}
9.22 26.85 kips > 6 kips, pipe is safe
9.23 105°F
9.28 250 kips
9.29 (a) 60.0 kN, (b) 41.2 kN, (c) 16.5 kN
9.30 (a) 25.4 ksi, 19.0 kips, (b) 7.57 ksi, 5.68 kips
9.31 158 in.
9.33 153 kN (based on member AB)
9.35 13.2 kips (based on long diagonal member)
9.36 576 kips
9.37 1764 kips
9.39 405 kips
9.41 237 kips
9.43 W 8 × 35
9.45 100 mm × 100 mm
9.47 100 mm
9.49 35 mm
9.51 2 in.
9.53 70 mm
9.54 3/16 in.
9.55 8 × 8
9.57 12 × 12
9.59 Yes

CHAPTER 10

10.1 315 in. · lb
10.3 1.315 kN · m
10.5 $1.49P^2L/(AE)$
10.7 $3.8P^2L/(AE)$
10.9 $437/(AE)$
10.10 84.8 in. · lb
10.11 12.57 in. · lb
10.13 0.30 N · m, 252 N
10.15 4.23 in. · kips
10.17 5.16 kN · m
10.19 94.9 N · m
10.21 23.2 in. · lb
10.23 0.430 in. · kips (using web area only)
10.25 0.422 N · m
10.27 1.27 N · m
10.29 2.61 in. · lb
10.31 0.00121 in. (down)
10.33 0.0314 in. (down)
10.34 0.914 mm (down)
10.37 $PL^3/(24EI)$ (down)
10.38 $PL^3/(48EI)$ (down)
10.39 $PL^3/(3EI)$ (down)
10.41 $M(L^2 - 3La + 3a^2)/(3EIL)$ (clockwise)
10.42 $ML/(3EI)$ (clockwise)
10.43 $2ML/(3EI)$ (clockwise)
10.45 $3.672/(AE)$ (down)
10.47 $761/(AE)$ (down)
10.49 $432/(AE)$ (down)
10.51 $1.835PL/(AE)$ (down)
10.53 0.823 mm (down)
10.55 0.351 mm (up)
10.57 0.054 in. (up)
10.59 0.0366 in. (to the right)
10.60 0.0312 in. (to the right)
10.61 0.0270 in. (to the right)
10.63 $Pb^2(L + b)/(3EI)$ (down)
10.65 $41wL^4/(384EI)$ (down)
10.67 $7PL^3/(54EI)$ (down)
10.69 $5wL^4/(384EI)$ (down)
10.71 $PL^3/(48EI)$ (down)
10.73 0.1125 mm (up)
10.75 0.102 in. (up)

10.77 0.278 in. (down)
10.79 $5.828/(AE)$ (down)
10.81 $731.5/(AE)$ (down)
10.83 $3.83PL/(AE)$ (down)
10.86 $wL^4/(8EI)$ (down)
10.87 $Pb^2(L + b)/(3EI)$ (down)
10.89 $41wL^4/(384EI)$ (down)
10.91 $7PL^3/(54EI)$ (down)
10.93 $5wL^4/(384EI)$ (down)
10.95 $PL^3/(48EI)$ (down)
10.97 $PL^3/(6EI) = 0.1125$ mm (up)
10.99 0.102 in. (up)
10.101 0.278 in. (down)
10.103 (a) $PR^2\pi/(EI)$ (cw) (b) $3\pi PR^3/(2EI)$ (down)
10.104 (a) $Pb(b/2 + a)/(EI)$ (counterclockwise), (b) $Pba^2/(2EI)$ (up)
10.105 0.122 rad ccw looking in from C to B
10.107 0.0127 rad cw looking in from C to A
10.109 $R_A = P\left(1 - \frac{3a^2}{L^2} + \frac{2a^3}{L^3}\right)$, $M_A = P\left(\frac{a^3}{L^2} - 2\frac{a^2}{L} + a\right)$
10.111 $A_y = wL/18$, $B_y = 5wL/12$, $C_y = 19wL/36$
10.113 $A_y = C_y = 5P/16$, $B_y = 11P/8$
10.115 $F_{BC} = 3.75$ kips (T), $F_{BD} = 1.77$ kips (T), $F_{BE} = 1.25$ kips (C)
10.117 7.07 in., 14.14 kips
10.119 0.134 m, 26.8 kN
10.121 26.7 lb
10.123 0.104 m
10.124 6.59 lb
10.125 13.2 lbs
10.126 25.8 N
10.127 51.6 N
10.129 (a) 2520 lb, (b) 0.185 in., (c) 13.6 ksi
10.131 (a) 1.11 mm, (b) 21.5 kN · m (c) 23.2 MPa
10.133 2.05 in., 3560 psi; 0.518 in. and 900 psi (static load)
10.135 (a) 81 MPa, (b) 0.041 rad
10.137 (a) 5.87 kip · in., (b) 0.0479 rad
10.139 (a) 44.9 MPa, (b) 0.0545 rad

CHAPTER 11

11.1 98 800 cycles
11.3 14 400 cycles
11.5 9280 cycles
11.7 14.2 ksi (expensive steel), 13.4 ksi (less expensive steel)
11.8 24.4 ksi
11.9 15.3 ksi
11.10 86.9 ksi
11.11 235 MPa
11.13 17.1 ksi, shaft is safe
11.15 2.49
11.17 3.92
11.18 1.85
11.19 3.00
11.21 3.00
11.23 part is not safe against fatigue failure

PROPERTIES OF PLANE AREAS

Rectangle

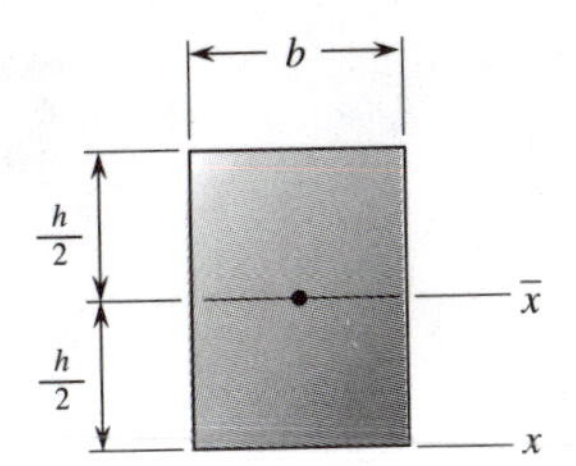

$$I_{\bar{x}} = \frac{bh^3}{12}$$

$$I_x = \frac{bh^3}{3}$$

Triangle

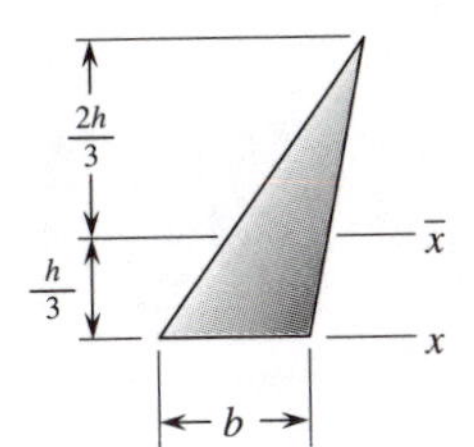

$$A = \frac{1}{2}bh$$

$$I_{\bar{x}} = \frac{bh^3}{36}$$

$$I_x = \frac{bh^3}{12}$$

Circle

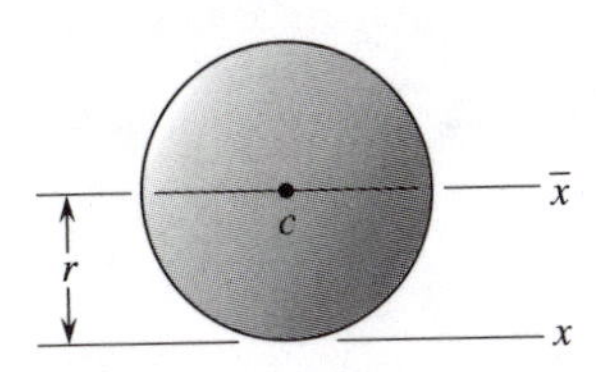

$$I_{\bar{x}} = \frac{\pi r^4}{4}$$

$$J_c = \frac{\pi r^4}{2}$$

$$I_x = \frac{5\pi r^4}{4}$$

Thin ring

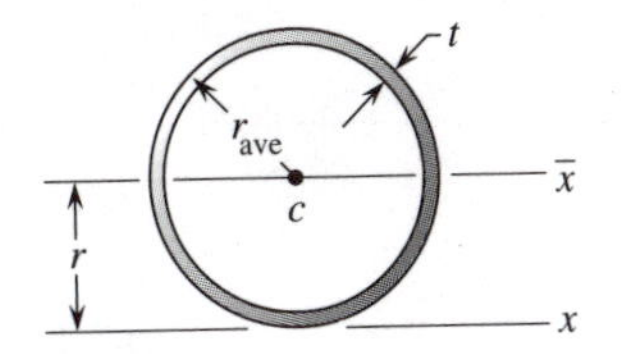

$$A = 2\pi r_{ave} t$$

$$J_c = 2\pi r_{ave}^3 t$$

$$I_{\bar{x}} = \pi r_{ave}^3 t$$

Semicircle

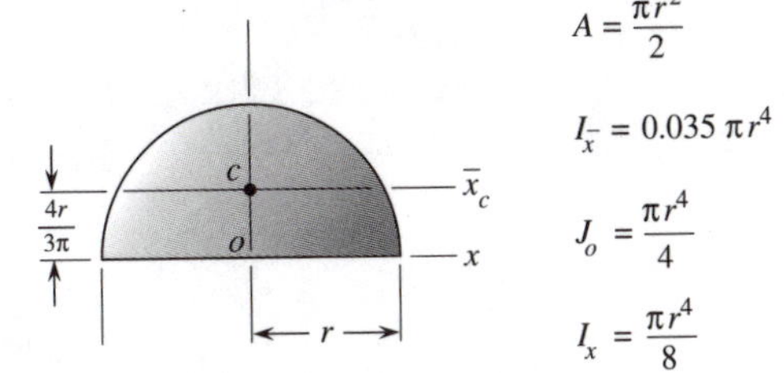

$$A = \frac{\pi r^2}{2}$$

$$I_{\bar{x}} = 0.035\,\pi r^4$$

$$J_o = \frac{\pi r^4}{4}$$

$$I_x = \frac{\pi r^4}{8}$$

Quarter ellipse

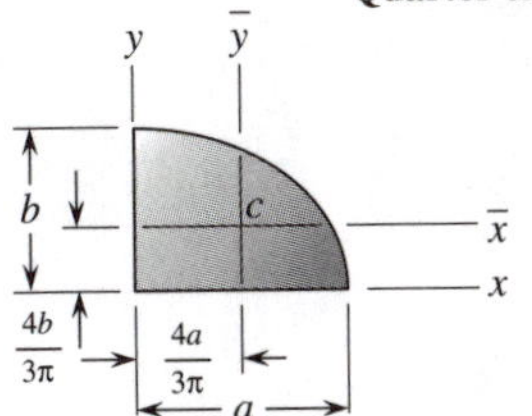

$$A = \frac{\pi ab}{4}$$

$$I_{\bar{x}} = 0.0175\pi ab^3$$

$$I_x = \frac{\pi ab^3}{16}$$

$$I_y = \frac{\pi a^3 b}{16}$$

Quadrant of parabola

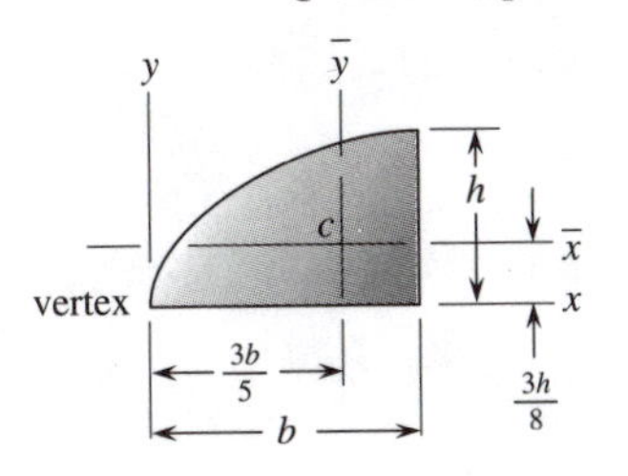

$$A = \frac{2}{3}bh$$

$$I_{\bar{x}} = 0.04\,bh^3$$

$$I_x = \frac{2bh^3}{15}$$

$$I_y = \frac{2hb^3}{7}$$

Parabolic Spandrel

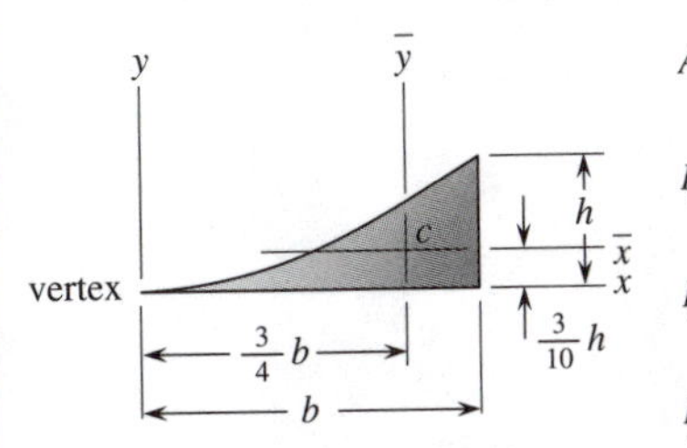

$$A = \frac{bh}{3}$$

$$I_{\bar{x}} = 0.0176\,bh^3$$

$$I_x = \frac{bh^3}{21}$$

$$I_y = \frac{hb^3}{5}$$

*Also see Appendix A for more complete properties

PHYSICAL PROPERTIES IN SI AND USCS UNITS

Property	SI	USCS
Water (fresh)		
specific weight	9.81 kN/m^3	62.4 lb/ft^3
mass density	1000 kg/m^3	1.94 slugs/ft^3
Aluminum		
specific weight	26.6 kN/m^3	169 lb/ft^3
mass density	2710 kg/m^3	5.26 slugs/ft^3
Steel		
specific weight	77.0 kN/m^3	490 lb/ft^3
mass density	7850 kg/m^3	15.2 slugs/ft^3
Reinforced concrete		
specific weight	23.6 kN/m^3	150 lb/ft^3
mass density	2400 kg/m^3	4.66 slugs/ft^3
Acceleration of gravity (on the earth's surface)		
Recommended value	9.81 m/s^2	32.2 ft/s^2
Atmospheric pressure (at sea level)		
Recommended value	101 kPa	14.7 psi

TYPICAL PROPERTIES OF SELECTED ENGINEERING MATERIALS

Material	Ultimate Strength σ_u		0.2% Yield Strength σ_Y		Modulus of Elasticity E		Shear Modulus G	Coefficient of Thermal Expansion, α		Density, ρ	
	ksi	MPa	ksi	MPa	(10^6 psi	GPa)	(10^6 psi)	10^{-6}/°F	10^{-6}/°C	lb/in.^{3}	kg/m^3
Aluminum											
Alloy 1100-H14 (99% Al)	14	110(T)	14	95	10.1	70	3.7	13.1	23.6	0.098	2710
Alloy 2024-T3 (sheet and plate)	70	480(T)	50	340	10.6	73	4.0	12.6	22.7	0.100	2763
Alloy 6061-T6 (extruded)	42	260(T)	37	255	10.0	69	3.7	13.1	23.6	0.098	2710
Alloy 7075-T6 (sheet and plate)	80	550(T)	70	480	10.4	72	3.9	12.9	23.2	0.101	2795
Yellow brass (65% Cu, 35% Zn)											
Cold-rolled	78	540(T)	63	435	15	105	5.6	11.3	20.0	0.306	8470
Annealed	48	330(T)	15	105	15	105	5.6	11.3	20.0	0.306	8470
Phosphor bronze											
Cold-rolled (510)	81	560(T)	75	520	15.9	110	5.9	9.9	17.8	0.320	8860
Spring-tempered (524)	122	840(T)	—	—	16	110	5.9	10.2	18.4	0.317	8780
Cast iron											
Gray, 4.5%C, ASTM A-48	25	170(T)	—	—	10	70	4.1	6.7	12.1	0.260	7200
	95	650(C)									
Malleable, ASTM A-47	50	340(T)	33	230	24	165	9.3	6.7	12.1	0.264	7300
	90	620(C)	—	—							